FUNDAMENTAL NEUROPATHOLOGY FOR PATHOLOGISTS AND TOXICOLOGISTS

FUNDAMENTAL NEUROPATHOLOGY FOR PATHOLOGISTS AND TOXICOLOGISTS

Principles and Techniques

Edited by

BRAD BOLON
GEMpath, Inc.
Longmont, Colorado

MARK T. BUTT
Tox Path Specialists, LLC
Hagerstown, Maryland

WILEY

A JOHN WILEY & SONS, INC., PUBLICATION

Published by John Wiley & Sons, Inc., Hoboken, New Jersey.
Published simultaneously in Canada.

For general information on our other products and services or for technical support, please contact our Customer Care Department within the United States at (800) 762-2974, outside the United States at (317) 572-3993 or fax (317) 572-4002.

Wiley also publishes its books in a variety of electronic formats. Some content that appears in print may not be available in electronic formats. For more information about Wiley products, visit our web site at www.wiley.com.

Library of Congress Cataloging-in-Publication Data:

Fundamental neuropathology for pathologists and toxicologists: principles and techniques / [edited] by Brad Bolon, Mark T. Butt.
    p. ; cm.
  Includes bibliographical references.
  ISBN 978-0-470-22733-6 (cloth)
  1. Neurotoxicology 2. Nervous system–Diseases . I. Bolon, Brad. II. Butt, Mark T.
  [DNLM: 1. Neurotoxicity Syndromes–diagnosis. 2. Neurotoxicity Syndromes–physiopathology
3. Models, Animal. 4. Nervous System–drug effects. 5. Neurotoxins–adverse effects. WL
140]
RC347.5F86 2011
616.8071–dc22

                                                                               2010033335

Printed in Singapore

oBook ISBN: 9780470939956
ePDF ISBN: 9780470939949
ePub ISBN: 9781118002230

10 9 8 7 6 5 4 3 2 1

This book is dedicated to all those who have endured the joys and miseries of neuropathology:

- *the mentors who struggled mightily to instill some neuropathological knowledge in us*
- *the colleagues who have suffered (and still suffer) our lifelong passion for neuropathology*
- *the many contributors that lent their time and expertise to this book*
- *the institutions that continue to invest research dollars to treat and cure the many devastating diseases of the nervous system*
- *the professionals who have long awaited a comprehensive toxicologic neuropathology reference*
- *our families, who have experienced neuropathology "up close and personal" during our months of isolation and distraction while we composed, edited, and formatted the many chapters, tables, figures, and legends that eventually resulted in this tome*

We hope that the effort fulfills all your hopes and fuels your dreams.
Also, we hope that the second edition is at least several years off.

CONTENTS

CONTRIBUTORS

Ana Alcaraz, DVM, PhD, DACVP, College of Veterinary Medicine, Western University of Health Sciences, Pomona, California

Douglas C. Anthony, MD, PhD, FCAP, University of Missouri School of Medicine, Columbia, Missouri

Brad Bolon, DVM, MS, PhD, DACVP, DABT, FIATP, GEMpath, Inc., Longmont, Colorado

Rogely Waite Boyce, DVM, PhD, WIL Research Laboratories, LLC, Ashland, Ohio

Mark T. Butt, DVM, DACVP, Tox Path Specialists, LLC, Hagerstown, Maryland

Paul W. Czoty, PhD, Department of Physiology and Pharmacology, Center for the Neurobiology of Addiction Treatments, Wake Forest University School of Medicine, Winston-Salem, North Carolina

David C. Dorman, DVM, PhD, DABT, DABVT, College of Veterinary Medicine, North Carolina State University, Raleigh, North Carolina

Karl-Anton Dorph-Petersen, MD, PhD, Centre for Psychiatric Research, Aarhus University Hospital, Risskov, Risskov, Denmark

Craig Fletcher, DVM, PhD, DACLAM, Division of Laboratory Animal Medicine, Department of Pathology and Laboratory Medicine, University of North Carolina School of Medicine, Chapel Hill, North Carolina

Andrew Forge, BSc, MSc, PhD, Centre for Auditory Research, UCL Ear Institute, London, United Kingdom

Kathy Gabrielson, DVM, PhD, DACVP, Departments of Molecular and Comparative Pathobiology and Environmental Health Sciences, School of Medicine and Bloomberg School of Public Health, Johns Hopkins University, Baltimore, Maryland

Robert H. Garman, DVM, DACVP, Consultants in Veterinary Pathology, Inc., Murrysville, Pennsylvania

Mary Beth Genter, PhD, DABT, Department of Environmental Health, University of Cincinnati, Cincinnati, Ohio

Tracy Gluckman, MS, DVM, DACLAM, Northwestern University, Chicago, Illinois

Doyle G. Graham, MD, PhD, Duke–NUS Graduate Medical School, Singapore

Hans Jørgen G. Gundersen, PhD, Stereology and Electron Microscopy Research Laboratory and MIND Center, Aarhus University, Aarhus, Denmark

D. Greg Hall, DVM, PhD, DACVP, Lilly Research Laboratories, Indianapolis, Indiana

Elizabeth Head, MA, PhD, Sanders–Brown Center on Aging, Department of Molecular and Biomedical Pharmacology, University of Kentucky, Lexington, Kentucky

John W. Hermanson, BS, MS, PhD, Department of Biomedical Sciences, College of Veterinary Medicine, Cornell University, Ithaca, New York

Monty J. Hyten, HT (ASCP), Covance Laboratories, Inc., Greenfield, Indiana

Karl F. Jensen, PhD, Neurotoxicology Division, National Health and Environmental Effects Research Laboratory, Office of Research and Development, U.S. Environmental Protection Agency, Research Triangle Park, North Carolina

William H. Jordan, DVM, PhD, DACVP, Vet Path Services, Inc., Mason, Ohio

Bernard S. Jortner, VMD, DACVP, Laboratory for Neurotoxicity Studies, Virginia–Maryland Regional College of Veterinary Medicine, Virginia Tech, Blacksburg, Virginia

Mary Jeanne Kallman, PhD, Investigative Toxicology, Covance Laboratories, Inc., Greenfield, Indiana

Wolfgang Kaufmann, Dr. med. vet., FTA Path, DECVP, Non-clinical Development–Global Toxicology, Merck Serono Research & Development, Merck KGaA, Darmstadt, Germany

Georg J. Krinke, MVDr, DECVP, Pathology Evaluations, Frenkendorf, Switzerland

Stephanie A. Lahousse, PhD, Cellular and Molecular Pathology Branch, National Toxicology Program, National Institute of Environmental Health Sciences, Research Triangle Park, North Carolina

Peter B. Little, DVM, MS, PhD, DACVP, Pathology Associates, Charles River Laboratories, Durham, North Carolina; Professor Emeritus, Department of Pathobiology, Ontario Veterinary College, University of Guelph, Guelph, Ontario, Canada

Lise Lyck, MSc, PhD, Human Health and Safety, DHI, Hørsholm, Denmark

Thomas J. Montine, MD, PhD, Department of Pathology, University of Washington, Seattle, Washington

Virginia C. Moser, PhD, DABT, Toxicity Assessment Division, National Health and Environmental Effects Research Laboratory, Office of Research and Development, U.S. Environmental Protection Agency, Research Triangle Park, North Carolina

M. Paul Murphy, MA, PhD, Sanders–Brown Center on Aging, Department of Molecular and Cellular Biochemistry, University of Kentucky, Lexington, Kentucky

Michael A. Nader, PhD, Department of Physiology and Pharmacology, Wake Forest University School of Medicine, Winston-Salem, North Carolina

Dennis O'Brien, DVM, PhD, DACVIM (Neurology), College of Veterinary Medicine, University of Missouri, Columbia, Missouri

Anna Oevermann, Dr. med. vet., DECVP, Neurocenter, Department of Clinical Research and Veterinary Public Health, Vetsuisse Faculty, University of Bern, Bern, Switzerland

Arun R. Pandiri, BVSc & AH, MS, PhD, DACVP, Cellular and Molecular Pathology Branch, National Toxicology Program, National Institute of Environmental Health Sciences, Research Triangle Park, North Carolina

Tucker A. Patterson, PhD, Division of Neurotoxicology, National Center for Toxicological Research, U.S. Food and Drug Administration, Jefferson, Arkansas

Merle G. Paule, PhD, Division of Neurotoxicology, National Center for Toxicological Research, U.S. Food and Drug Administration, Jefferson, Arkansas

Kathleen C. Raffaele, MPH, PhD, National Center for Environmental Assessment, Office of Research and Development, U.S. Environmental Protection Agency, Washington, DC

Meg Ramos, DVM, PhD, DACVP, Drug Safety Evaluation, Allergan, Inc., Irvine, California

Christopher M. Reilly, DVM, DACVP, Department of Pathology, Microbiology, and Immunology, School of Veterinary Medicine, University of California–Davis, Davis, California

Hope Salvo, MS, RAC, SAIC-Frederick, Inc., Frederick, Maryland

Sumit Sarkar, PhD, Division of Neurotoxicology, National Center for Toxicological Research, U.S. Food and Drug Administration, Jefferson, Arkansas

Larry Schmued, PhD, Division of Neurotoxicology, National Center for Toxicological Research, U.S. Food and Drug Administartion, Jefferson, Arkansas

Robert C. Sills, DVM, PhD, DACVP, Cellular and Molecular Pathology Branch, National Toxicology Program, National Institute of Environmental Health Sciences, Research Triangle Park, North Carolina

Michael H. Stoffel, Prof. Dr. med. vet. habil, Department of Clinical Research and Veterinary Public Health, Vetsuisse Faculty, University of Bern, Bern, Switzerland

Robert C. Switzer III, PhD, NeuroScience Associates, Knoxville, Tennessee

Ruth Taylor, BSc, MSc, PhD, Centre for Auditory Research, UCL Ear Institute, London, United Kingdom

Beth A. Valentine, DVM, PhD, DACVP, Department of Biomedical Sciences, College of Veterinary Medicine, Oregon State University, Corvallis, Oregon

William Valentine, PhD, DVM, DABT, DABVT, Department of Pathology, Vanderbilt University Medical Center, Nashville, Tennessee

Marc Vandevelde, Prof. Dr. med. vet., DECVN, Department of Clinical Veterinary Medicine, Vetsuisse Faculty, University of Bern, Bern, Switzerland

Karen M. Vernau, DVM, MSc, DACVIM (Neurology), Department of Surgical and Radiological Sciences, School of Veterinary Medicine, University of California–Davis, Davis, California

William Vernau, BVSc, BVMS, DVSc, PhD, DACVP, Department of Pathology, Microbiology and Immunology, School of Veterinary Medicine, University of California–Davis, Davis, California

Tony L. Yaksh, PhD, Department of Anesthesiology, University of California–San Diego, La Jolla, California

Jamie K. Young, DVM, PhD, DACVP, Department of Pathology, Covance Laboratories, Inc., Greenfield, Indiana

FOREWORD

Although many textbooks include an explanatory preface provided by the senior authors or editors, the practice of including a foreword seems to be on the decline. Thus I was especially pleased to be invited by Drs. Bolon and Butt to pen a few words for their new offspring. It has been said before that a good book needs no justification, whereas a bad book cannot be justified. I have no doubt that this new volume will gain Brad Bolon and Mark Butt many new friends. In a world where the concept of one medicine holds sway, perhaps in toxicological pathology as much as in any medical discipline do physicians, veterinarians, and Ph.D. scientists work side by side. It is no coincidence, then, that Brad and Mark have gathered around them a large and formidable stable of experts from all three professional backgrounds to provide us with a remarkably comprehensive and diverse text that embraces all creatures great and small. Whether in search of knowledge pertaining to neural development, neuroimaging, clinicopathologic correlations, neurobehavioral assays, teratology, species idiosyncrasies, toxicological pathology of the eye, ear, and olfactory systems, regulatory issues, stereology in neurotoxicology, and much, much more—it's all here.

Drs. Bolon and Butt, mid-1980s graduates of veterinary colleges in North America, are seasoned veterinary pathologists who have followed somewhat different career paths but clearly saw the need for a contemporary reference text of toxicological pathology. In recent decades, progressively more trained veterinary pathologists have found employment and made careers in toxicological pathology, especially as it intermeshes with the pharmaceutical industry. This book fills an important need, for it is clear that the nuances of the central and peripheral nervous systems overwhelm many capable anatomical pathologists who will be perfectly comfortable with other tissues and their expressions of disease. In contrast to some organs, the central nervous system (CNS) can be subtle and to many is just a vast wasteland in which the discrimination of normal from lesion and lesion from artifact is a constant dilemma. Indeed, as one colleague put it, the CNS is the only organ that autolyses before death! The first professor of veterinary pathology whom I encountered would examine a case up to the level of the dura mater, at which point the brakes were firmly applied and any evaluation of brain or spinal cord was delegated to "the specialist." *Fundamental Neuropathology for Pathologists and Toxicologist* will help to make all pathologists at home with this organ system.

Drs. Bolon and Butt bring to this new volume a broad perspective, honed by many years engaged in industrial pathology in which they have focused on neuropathology and the potential for chemicals to perturb the nervous system, including experimental investigations and studies in several species, including genetically modified laboratory animals. Guided by the premise "first do no harm," Brad, Mark, and their colleagues have devoted their careers to the proposal that in vivo and in vitro studies of potential new therapeutic compounds can best identify the safest and (what will prove to be) the most efficacious next generation of medicines while often throwing light on disease mechanisms and pathways along the way. Although in recent decades there have been notable successes in human health, there remains a massive need for pharmaceutical products with which to intercede in cardiovascular disease, neoplastic disorders (now potentially with compounds tailored to individual tumors), and the burgeoning area of dementing diseases as life expectancy in some areas of the world stretches beyond 80 years of age. I am delighted to see the birth of this new textbook, which is certain to secure a home on the shelves of many professionals, including neurologists, toxicologists, pathologists, laboratory investigators, and scientists in our regulatory agencies.

BRIAN A. SUMMERS, BVSc, PhD, MRCVS, FRCPath

Professor of Comparative Neuropathology
Royal Veterinary College, University of London
United Kingdom

PREFACE

The goal of this work was to provide, as much as possible, an essentially complete reference on the design and interpretation of studies involving toxicological neuropathology. It is a book not only for pathologists, but also for toxicologists and other scientists involved in the investigation of neurotoxicity. A series of journal articles to achieve this goal was just too fragmented an approach. A textbook was needed, and here it is.

Although there are numerous descriptions and illustrations of tissue changes, this book is not an atlas of lesions. There are other complete references, in print and online, with a plethora of images and diagnostic terms. Although the chapters contain many literature references, this book relates primarily the knowledge of many veterans of the neuropathology discipline with the hope that others can learn from our experience and mistakes. (Note: We did not specifically mention all of our mistakes.)

As with most projects that are either different or of a greater magnitude than someone is used to, the collection, assembly, and editing of this book took longer than predicted. However, the end result achieved both our vision and our goal.

Many years ago, while serving as chairman for a pathology working group for the National Toxicology Program, I (M.T.B.) pulled out my copy of *Pathology of the Fischer Rat* (Boorman GA, Eustis SL, Elwell MR, Montgomery CA, MacKenzie WF, eds., Academic Press, Inc., San Diego, CA, 1990). Several of the contributing authors were in the group.

Each was pleased to see that the binding of my book had long ago broken down because of the number of times I had grabbed it off the shelf for a consult. It is our hope that this book will receive just as much use long before the second edition makes its way into print.

ACKNOWLEDGMENTS

We would like to acknowledge Jake Butt and James Reickel, whose lengthy (and frequently late-night) formatting sessions provided so much assistance toward the finished product. We also thank our co-workers, who had to manage their schedules around our need to write, edit, format, and compile all the material that formed this volume.

We thank Amanda Amanullah and Jonathan T. Rose (Wiley) for their patience and support in serving as the publisher's editors of this volume, and Colin Moore (Longmont, Colorado; www.colinmoore.us) for preparing the graphic art in Chapter 4. Although not mentioned by name, we thank the many others who helped with editing, formatting, and illustrations.

Finally, and most important, we gratefully and humbly acknowledge the many contributors who took time from their already too full schedules to provide chapters of this book.

Mark T. Butt
Brad Bolon

Color versions of selected figures can be found online at
ftp://ftp.wiley.com/public/sci_tech_med/fundamental_neuropathology

INTRODUCTION

The 90-day study is over, the pathology data have been collected, and there is a problem: a microscopic change in the brain. The incidence is slightly higher in the treated groups than in the controls, or perhaps it is not present in the control groups at all. Still, the lesion does not seem like much of an issue. There were no in-life changes to suggest a problem. Brain weights were within normal limits. Even the cerebrospinal fluid was normal. No instances of seizures or abnormal ambulation were noted during cage-side observations. But was enough information gathered to make an accurate assessment? Does the lesion, which seems inconsequential to the animals, represent something that will affect the more complex lives of humans in a different and profound way? Is it possible that earlier time points need to be looked at to truly characterize the test article as "safe"? Could additional stains or the collection of morphometric data help? Did this study validate previous studies, and if not, are the differences real or due simply to variable interpretation or dissimilar terms—which, in actuality, mean the same thing? If clinical trials are able to begin, is there any way of tracking, assessing, or following this potential change in patients? These questions are what this book is about.

Diseases of the nervous system have a major impact on individuals and society. For many, perhaps for most, of these diseases we are not much closer to an effective treatment or cure than we were 10 years ago. Research moves slowly, but we need to make sure that it moves progressively and as quickly as possible. Lives are literally in the balance. We now know that Huntington's disease is a trinucleotide repeat disorder. Based on the number of repeats, we can even predict the probable progression of the disease. But we have very little to offer in terms of treatment. Using whatever tools necessary, we need to change that.

There is one nervous system. It consists of the brain and the spinal cord and the motor nerves. It is also comprised of sensory and autonomic ganglia, intraepidermal nerve fibers, and neurons that act more like endocrine glands than like traditional bipolar cell bodies. The brain is attached to the eye, to the ears, and to the pelvic ganglia, which all interact. The brain is a group of physically, chemically, and metabolically diverse groups of neurons that may look similar and may be located in generally the same place (inside the skull), but the differences end there. These groups of cells require examination. You cannot think of the brain as a liver or a kidney: one section of a brain (or even three sections) does not allow for adequate examination of the brain. The nervous system is a complex system that requires more than a lifetime to understand. This complexity is one of the central themes of this book.

This volume has been assembled to provide neuropathologists and neurotoxicologists, as well as toxicological pathologists and general toxicologists with an interest in the field, with a single resource that provides the introductory and advanced information needed to develop proficiency in the design, analysis, and interpretation of toxicologic neuropathology experiments.

Part 1 provides information on fundamental neurobiology. Since an understanding of the anatomy of the nervous system is paramount to assessment, Chapters 2 to 4 focus on neuroanatomy. Neuroanatomy is daunting, but once the anatomy is known, you can often predict where a lesion may be, based on the presenting clinical signs. That is the thrust of Chapter 5. Because the morphological evaluation of tissues is only one step toward understanding nervous system function, Chapters 6 and 7 focus on behavioral systems and cognitive assessments, and Chapter 8 deals with the effects of aging on brain structure and function. These are areas in which the neuropathologist and neurotoxicologist require general familiarity as well as specific knowledge on a study-by-study basis. Finally, Chapter 9 provides a framework for understanding the issues relevant to the design of a neurotoxicity study that will capture important morphological endpoints.

Part 2 deals primarily with methodology: how, when, and why. Chapters on practical methods of neurohistology, specialized (rapidly becoming commonplace) markers for neurotoxicity, processing and evaluation of peripheral nerves and muscle, and cerebrospinal fluid are all included, as is an essential chapter on stereology. You won't find everything you need in Part 2 (that's why second editions were invented), but it's a very good start.

Part 3 continues with more methodology, providing chapters on evaluation of the adult nervous system, the developing nervous system, the peripheral nervous system, and the ophthalmic otic and olfactory systems. The techniques described in Part 3 provide an excellent foundation for an evaluation of the nervous system.

Part 4 includes those chapters considered important to a thorough understanding of the issues facing neuropathologists and, by extension, study directors and investigators involved in running neurotoxicity studies. Direct delivery is an increasingly common means of getting various test articles, especially proteins, antibodies, lipophobic drugs, and stem cells to the central and peripheral components of the nervous system. Of those methods, spinal delivery is one of the most common, so a chapter on that delivery system and the drug safety implications is included. Other direct delivery methods, including intracerebroventricular catheters, direct parenchymal infusions, deep brain stimulation devices, and stem cell implants, were not included, due to space limitations. But through the context of intrathecal delivery, Chapter 27 does include many of the issues, real and potential, that are encountered when delivering drugs directly to the central nervous system.

The regulatory aspects of toxicologic neuropathology comprise two of the Part 4 chapters: one describes what is important to include in regulatory submissions (from the viewpoint of regulatory officials), the other provides useful suggestions for navigating the various regulatory guidelines pertinent to the conduct and interpretation of neurotoxicity studies involving pathology. Additionally in Part 4 are chapters on neuropathology in veterinary and medical practice, a chapter on diagnostic neuropathology (primarily of spontaneous diseases), and a guide to training personnel, primarily technical staff, who will be involved in the conduct of studies with neuropathological endpoints. Finally, there is a chapter on the neuropathology report. Since everyone has his or her own style and notions of what comprises a great report, this chapter is likely to be controversial. But even if only useful for stimulating debate, a chapter on reporting was needed.

Finally, Part 4 ends with a chapter on the future of neuropathology. It will be particularly interesting to dust off this first edition in 20 years and see how close the authors were in predicting changes in the dynamic field of toxicological neuropathology.

The morphological examination of the nervous system is the task of the neuropathologist. Coordinating that pathologist with all the other contributing scientists is the task of the researcher and/or study director. That is not an easy job. For all those involved in neurotoxicology in general and toxicologic neuropathology in particular, this book seeks to better inform you, to make your career even more interesting, and to provide you with an increased opportunity for success in gathering and interpreting the data necessary to make good decisions.

We hope that this book will be a boon to both experts and novices in toxicological neuropathology, and that this information will assist all those engaged in protecting the health of humans and animals from the devastating damage that can follow genetic, degenerative, physical, and toxic injury to the nervous system.

PART 1

FUNDAMENTALS OF NEUROBIOLOGY

1

FUNDAMENTAL NEUROPATHOLOGY FOR PATHOLOGISTS AND TOXICOLOGISTS: AN INTRODUCTION

BRAD BOLON

GEMpath, Inc., Longmont, Colorado

DOYLE G. GRAHAM

Duke–NUS Graduate Medical School, Singapore

THE IMPORTANCE OF NEUROTOXICOLOGICAL RESEARCH

Neurotoxicology is the study of the undesirable consequences that develop in the central nervous system (CNS) or peripheral nervous system (PNS) or both after an organism is exposed to a neurotoxic agent during development or adulthood. Such agents may be exogenous materials such as chemicals contaminating the external habitat (e.g., agrochemicals, pesticides, solvents) or introduced purposely into the internal environment (i.e., drugs); metals; or peptides/proteins (e.g., microbial toxins, biopharmaceuticals). Alternatively, neurotoxic agents may be produced endogenously (e.g., ammonia, unconjugated bilirubin) during the course of certain diseases. Thus, the nervous system is likely to experience constant exposure to a range of neurotoxic agents, although in many instances the level of exposure will be insignificant.

The potential scope of toxicant-induced neuropathology is immense. Each year in the United States, industries manufacture about 85,000 chemicals and register another 2000 to 3000 new compounds.[1] Approximately 3 to 5% of chemicals (between 2500 and 5000 entities) are estimated to be neurotoxic to some degree.[2] This estimate has serious implications for human, animal, and environmental health, because up to two-thirds of high-production-volume chemicals (those made yearly in quantities exceeding 1 million pounds) have never been tested sufficiently for neurotoxic potential.[3] The recognition that neurological dysfunction is a major occupational hazard for adults[4,5] and a common congenital occurrence in children[6] has engendered a wide-ranging global effort to identify and eliminate possible sources of neural damage—principally, sources of neurotoxicant exposure.

Neurotoxicity can present as aberrations in neural structure (i.e., toxicological neuropathology) or function (including altered behavior, biochemistry, cognition, or impulse conduction), or both.[7–11] All structural changes and any persistent functional deficits associated with xenobiotic exposure are judged to be neurotoxic because such effects cannot be countered by the meager regenerative capabilities of the CNS.[12] Reversible functional deficits linked to a recognized neurotoxicological mechanism (e.g., outright neurodegeneration or exaggerated neuropharmacological activity) or that might jeopardize occupational health (for adults) or scholastic performance (especially for children) are also considered to be neurotoxic manifestations. The current "best practice" in conducting risk assessments for potential neurotoxicants is to integrate all available structural and functional evidence in reaching a verdict.[9,13,14] Nevertheless, the permanence of toxicant-induced structural changes in the CNS typically leads regulators to place more emphasis on morphological data rather than on behavioral or biochemical alterations to determine reference doses for

Fundamental Neuropathology for Pathologists and Toxicologists: Principles and Techniques, First Edition. Edited by Brad Bolon and Mark T. Butt.
© 2011 John Wiley & Sons, Inc. Published 2011 by John Wiley & Sons, Inc.

managing neurotoxic risk.[15] Therefore, a comprehensive toxicological neuropathology evaluation is and will remain a critical element of the risk assessment process for novel xenobiotics.[16]

The catastrophic outcome of neurotoxic damage to affected persons, and the strain placed on the resources (money, time) of their immediate caretakers and the societal entities that must often fund chronic health care, has led to the expanded use of neurotoxicity endpoints as major criteria for assessing the risks posed by exposure to xenobiotics.[17] This approach is a direct result of two factors. First and foremost, an unfortunate aspect of human history from ancient times through the twentieth century is that the neurotoxic effects of many agents [e.g., ethanol, n-hexane, lead, mercury, polychlorinated biphenyls (PCBs)] have been identified first in humans.[18] Second, exposure to potential neurotoxicants remains a common feature of human existence. Slightly less than a third of all high-volume industrial chemicals can elicit neurotoxic syndromes in the workplace.[19] Similarly, many drugs [antiepileptics (e.g., valproic acid), antineoplastics (e.g., vincristine)] can induce neurotoxic sequelae as an undesirable side effect.[18,20,21] Thus, a primary goal of current neurotoxicological research is to prospectively recognize the neurotoxic potential of novel compounds in laboratory animals rather than to discover it retrospectively after epidemics of neurotoxicity in humans.

THE EVOLUTION OF TOXICOLOGICAL NEUROPATHOLOGY

People have exhibited an interest in fundamental neuroscience for millennia (Table 1).[22,23] Initial neurobiology investigations concentrated on gross anatomical characterization of the CNS and its PNS projections as well as the clinical detection and treatment of diseases affecting the nervous system. Neurohistological evaluations were first undertaken in a piecemeal sense early in the eighteenth century, and more systematic assessments of discrete neural regions were begun in the 1840s. These early studies were organized as descriptive studies of the normal nervous system anatomy. The first neuropathology reports examined neuroanatomical alterations resulting from physical disruption (e.g., Wallerian degeneration in transected axons, first described in 1850) rather than toxicant-mediated neural damage. This emphasis reflected the close alliance between neuropathology and clinical neurology in the European (mainly German) medical schools in which neuropathological research was formalized in the modern era.

Human interest in toxicology also dates from antiquity.[24] The impact of widespread neurotoxicity on the advance of civilization became clear with the rise of industrialization in medieval and Renaissance Europe, when chronic exposure to lead and mercury represented a substantial occupational hazard to members of many professions (alchemists, goldsmiths, hatters, and millworkers, to name a few). Toxicological inquiry progressed in fits and starts during the nineteenth and early twentieth centuries before ultimately evolving into the hypothesis-driven applied science that exists today. The intermittent progress in toxicology stemmed from its expansive approach to experimentation; the field grew from a synthesis of most other basic biological and chemical disciplines, and at its inception the numbers of people with the time and money to excel in such diverse intellectual arenas were few.

Thus, toxicological neuropathology represents the modern-era intersection of three scientific fields: basic neurobiology, applied toxicology, and pathology. The rise of toxicological neuropathology was delayed until the first decades of the twentieth century because it required the advent of both technical advances in neuroanatomical handling and processing techniques (Table 1) and the availability of well-trained scientists versed in the fundamental concepts of all three disciplines. These prerequisites were clearly attained by 1906, as indicated by the publication in that year of a detailed neuropathological description of presenile neurodegeneration by Alois Alzheimer as well as the presentation of the Nobel Prize in Physiology or Medicine to Camillo Golgi and Santiago Ramón y Cajal for their studies of nervous system cytoarchitecture. The subsequent founders of toxicological neuropathology built on these accomplishments by developing significant expertise in morphological pathology, dedicating decades of research to defining the experimental conventions and procedures used in modern toxicological neuropathology investigations, and familiarizing ever greater numbers of colleagues with these conventions and procedures (via their numerous publications and many graduate students).

In this regard, two scientists in particular served as major role models for the growth of toxicological neuropathology during the mid-twentieth century, ultimately influencing several generations of modern toxicological neuropathologists (including the careers of the two authors). One person was John B. Cavanagh, a British physician and professor who devised many of the routine morphological approaches used to evaluate toxicants (especially metals, pesticides, and solvents) associated with occupational neurotoxicity.[25–30] The other founder was Adalbert Koestner, a German veterinarian and professor at several U.S. universities whose interests ranged from the morphology and mechanisms of mutagen-induced neural neoplasms to the potential utility and safety of food additives and novel neurotherapeutics.[31–34] Modern investigations in toxicological neuropathology have since evolved to incorporate many other innovative neuropathology endpoints in addition to the traditional morphological techniques (Table 2). Nevertheless, the methods pioneered by these two

TABLE 1   Selected Historical Landmarks in the Evolution of Toxicological Neuropathology

Date	Event
ca. 1700 B.C.E.	First written record about the nervous system
ca. 1000 B.C.E.	First written treatise describing surgical treatments for some neurological disorders (Al-Zahrawi, also known as Abulcasis or Albucasis)
ca. 500 B.C.E.	First descriptions of nervous system dissection (cranial and sensory nerves) (Alcmaion of Crotona)
ca. 80	First description linking lead exposure to neurological disease (Dioscorides)
1549	Publication of *De Cerebri Morbis*, an early book devoted to neurological disease (Jason Pratensis)
1660–1700	First publications dedicated to neuroanatomy: *Cerebri Anatome* (Thomas Willis, 1664), *Neurographia Universalis* (Raymond Vieussens, 1684) and *The Anatomy of the Brain* (Humphrey Ridley, 1695)
1684	First record of a special preservation technique for neural tissue (boiling oil as a hardening agent, by Raymond Vieussens)
1717	First description of the nerve fiber in cross section (Anton van Leeuwenhoek)
1760	Initial demonstration that cerebellar damage affects motor coordination (Arne-Charles Lorry)
1766	Earliest scientific description of the cerebrospinal fluid (Albrecht von Haller)
1810–1825	First functional–structural correlates for many CNS regions are defined
1836	Neuron nucleus and nucleolus first differentiated by microscopy (Gabriel Gustav Valentin) Myelinated and unmyelinated axons are discerned (Robert Remak)
1837	Cerebellar neurons and their processes first investigated (Jan Purkinje)
1838	Myelin-forming cells in the peripheral nervous system described (Theodor Schwann)
1842	Spinal cord anatomy first studied in serial sections (Benedikt Stilling)
1844	First illustration provided of the six cerebrocortical layers (Robert Remak)
1850	Initial experimental investigation of axonal degeneration (Augustus Waller)
1859	The term *neuroglia* is coined (Rudolph Virchow)
1861	Functional localization in the cerebral cortex is described (Paul Broca)
1865	Axons and dendrites are first differentiated (Otto Friedrich Karl Deiters)
1873	First work on the silver nitrate method to enhance neuronal contrast (Camillo Golgi)
1878	Regular interruptions in the peripheral nerve myelin are first appreciated (Louis-Antoine Ranvier)
1884	Granular endoplasmic reticulum is discriminated in neurons (Franz Nissl)
1889	Nerve cells are proposed to be independent functional elements (Santiago Ramón y Cajal)
1891	The lumbar puncture (spinal tap) is developed (Heinrich Quinke) *Journal of Comparative Neurology* is founded
1897	Formaldehyde is employed as a brain fixative (Ferdinand Blum)
1906	First description of Alzheimer's disease (Alois Alzheimer) Nobel Prize in Physiology or Medicine awarded to Camillo Golgi and Santiago Ramón y Cajal for their work on neural cytoarchitecture
1921	Microglia described (Pío del Río-Hortega)
1929	Correlation between nerve fiber size and function is identified (Joseph Erlanger and Herbert Spencer Gasser)
1949	National Institute of Mental Health (NIMH) is launched at the U.S. National Institutes of Health (NIH)
1950	National Institute of Neurological Disorders and Stroke (NINDS) is established at the NIH
1959	Methylmercury from industrial effluent identified as the cause of a neurotoxicity epidemic in humans and feral cats living in villages lining Minamata Bay in Japan
1961	International Brain Research Organization (IBRO) is formed as an independent, nongovernmental organization
1964	Methylnitrosourea (MNU) identified as a relatively selective model neurocarcinogen in rats
1968	Neurotoxic potential of polychlorinated biphenyls (PCBs) is first recognized in Japan among people who have ingested rice oil that was contaminated during manufacturing
1969	Society for Neuroscience (SfN) is founded in the United States
1973	*Fetal alcohol syndrome* (FAS) is coined as the term for a distinct pattern of craniofacial (including brain), limb, and cardiovascular defects in children born to alcoholic mothers
1982	1-Methyl-4-phenyl-1,2,3,6-tetrahydropyridine (MPTP) shown to be the etiology of Parkinsonism in young illicit drug users who used an improperly synthesized bootleg version of an opioid analgesic
1990	"Decade of the Brain" is declared in the United States by presidential proclamation

Source: Adapted in part from Chudler.[22]

TABLE 2   Morphological Techniques Used in the Modern Practice of Toxicological Neuropathology

Test	Type of Neuropathology Data
Gross evaluation	Identifies overt lesions within neural tissues (via subjective gross examination of surface and internal features)
	May provide a crude quantitative estimate of large-scale cell loss (via organ weights or linear or areal morphometric measurements)
Light microscopy	Identifies region-specific vulnerability and susceptible cell populations [routine stains such as H&E, Fluoro-Jade (for neuronal degeneration), and anti-GFAP (for reactive astrocytes in affected regions)]
	Characterizes the nature, location, and quantity of macromolecules, and provides insights into neurotoxic mechanisms [special histochemical, immunohistochemical, and molecular methods, especially if employed in conjunction with such specialized microscopy methods as laser capture microdissection (LCM)]
Electron microscopy	Identifies subcellular targets of neurotoxicity and provides an indication of the metabolic state of the nervous system (transmission electron microscopy)
	Addresses the subcellular distribution of xenobiotics (specialized autoradiographic and immunoelectron microscopy techniques, elemental composition analysis)
Noninvasive imaging	Allows in vivo assessment of neuroanatomic integrity [computed tomography (CT), magnetic resonance imaging (MRI) and microscopy (MRM), ultrasound (US)]
	Permits in vivo investigation of region-specific neurochemistry and function, offering insights into mechanisms of neurotoxicity [optical imaging, positron emission tomography (PET), single photon-emission computed tomography (SPECT)]

men and others are—and will remain—the foundation of toxicologic neuropathology investigations in the foreseeable future.

REQUIREMENTS FOR PROFICIENCY IN TOXICOLOGICAL NEUROPATHOLOGY

More than for any of the other subdisciplines of toxicology or pathology, competent practitioners of toxicological neuropathology must have the appropriate educational and work-related experiences to succeed. Advanced theoretical and practical training in neurobiology and experience in neuropathology will appreciably enhance the pathologist's ability to recognize abnormalities in neural tissues.[35] Proficiency as a toxicological neuropathologist requires comprehension at multiple levels of biological organization (e.g., whole animal, cellular, biochemical, and molecular) and the ability to integrate this information with basic medical tenets to formulate differential diagnoses as well as to identify and characterize etiologies and mechanisms of neural disease. Acceptable "entry-level" proficiency in toxicological neuropathology requires that a person have expertise in (1) comparative and correlative aspects of normal neuroanatomy and neurophysiology, (2) causes and mechanisms of major background and neurotoxicant-induced diseases of humans and common laboratory animal species, and (3) principal techniques used for evaluating neuropathological changes (e.g., gross dissection, light and electron microscopy, immunocytochemistry, advanced in situ molecular methods, and morphometry). Therefore, the most direct means of acquiring sufficient expertise in toxicological neuropathology is to complete a clinical degree in either medicine or veterinary medicine and then pursue postgraduate training in either

diagnostic pathology (e.g., a residency) or toxicological pathology (such as an advanced research degree or a clinical fellowship) in a program that specializes in nervous system investigations. Fundamental research in the field can also be done by Ph.D. biologists with in-depth training in a relevant discipline (e.g., comparative pathology, neurotoxicology) as long as the focus emphasizes an integrative strategy for nervous system assessment (i.e., investigating questions at the whole animal, organ, cellular, and biochemical/molecular levels, as necessary) rather than a reductionist approach (e.g., limited to cellular or molecular studies).

In current practice, however, general toxicological pathologists and toxicologists must often undertake their own instruction in toxicological neuropathology. Such exposure is usually acquired via self-study or through mentored on-the-job experience, and may be gained in several fashions. The most straightforward way is to study standard references in the field, and in allied biological disciplines (see Appendixes 2, 3, 4, and 5). Indeed, people engaged in toxicological neuropathology on a regular basis will require ready access to many of these references, particularly to neuroanatomy atlases (Appendix 2), in order to undertake meaningful analyses of neurotoxicant-induced lesions. Two other paths are to find Web sites (Appendix 4) or to read classical literature reports related to specific research questions. In the authors' experience, however, the two latter routes are suitable only if one has sufficient prior familiarity with the field for efficient and effective sifting of many possible citations to find those that are most useful. Thus, we recommend that generalists tasked with learning toxicological neuropathology spend the effort, money, and time to understand the relationship between various neural structures and functions (a correlative approach), and to do so across species (a comparative approach).[36]

FUNDAMENTAL PRINCIPLES OF TOXICOLOGICAL NEUROPATHOLOGY

The complexity of the nervous system is a key factor in its vulnerability to toxicant insult.[37] Moreover, these same structural and functional intricacies render even the simplest assessments quite challenging.[13,38] Success in research in toxicological neuropathology thus requires strict adherence to a few fundamental principles. In the remainder of the chapter we list these basic concepts and suggest some practical steps to implement them in toxicological neuropathology research. These principles and practices are described in much greater detail in later chapters.

Principle 1: Learn the lingo. As with many technical fields, neurotoxicological research has developed a jargon that is typically the unique domain of experts in the field. It goes without saying that a solid knowledge of this nomenclature is a mandatory prerequisite to achieving proficiency as either a toxicological neuropathologist or a neurotoxicologist.

A topic that has caused some confusion is the difference between the naming conventions for neural structures in humans (and nonhuman primates) and other animals. The misunderstanding arises from the dissimilar body orientations of these species. Primates are bipedal, with a nervous system arranged along a vertical (upright) axis, whereas other laboratory animals commonly employed in toxicological neuropathology research are quadrupeds having a horizontal nervous system axis. These divergent body carriages dictate different naming conventions for neural structures in primates and other vertebrates (Table 3). To avoid confusion, publications and reports that describe toxicological neuropathology findings should invoke the correct nomenclature for the species being investigated. A compromise that can be applied when naming neural structures in animals is to include the medical (*nomina anatomica*) term for the structure in parentheses behind the veterinary (*nomina anatomica*

TABLE 3 Species-Specific Directional Nomenclature for Designating Neural Structures

Direction	Biped (Humans, Nonhuman Primates)	Quadruped (Carnivores, Lagomorphs, Rodents)
Up	Superior	Dorsal
Down	Inferior	Ventral
Front	Cranial (outside the skull)	Cranial (outside the skull)
	Anterior (inside the skull)	Rostral (inside the skull)
Back	Posterior	Caudal

veterinaria) term. For example, the *superior cervical ganglion* in humans should be designated in animals as either the *cranial cervical ganglion* (the recognized term) or the *cranial cervical ganglion* (*superior cervical ganglion*). Descriptive anatomical terms should be used rather than eponyms (e.g., *mesencephalic aqueduct* in preference to *aqueduct of Sylvius*) when identifying neural structures to promote clarity in communication of neuropathology findings across all species.

Principle 2: Responses are restricted. Neuropathological lesions resulting from neurotoxicant exposure have been implicated in acute[21,39] and delayed[18,40–43] neurodegeneration, neuronal heterotopia,[44] and neural neoplasia.[32,45,46] The same lesion generally is elicited by many structurally different neurotoxicants, because these agents often act via a common molecular mechanism (e.g., peripheral axonopathy as a consequence of cytoskeletal cross-linking following exposure to *n*-hexane or carbon disulfide[47]). Therefore, the pathologist who is able reliably to discern a few basic lesions (Table 4) in neural tissue is reasonably well equipped to participate in toxicological neuropathology assessment.

TABLE 4 Fundamental Structural Alterations in Neural Tissues from Toxicological Neuropathology Studies

Cell Type	Lesion Type	Preferred Method of Neuropathology Analysis[a]
Neuron	Cell death	Light microscopy of specially stained sections (Fluoro-Jade, silver impregnation)
	Cell loss	Light microscopy of specially processed sections [IHC for cell type–specific markers (e.g., enzymes or neurotransmitters)]
		Morphometric measurements of specific regions on tissue sections
		Stereological counts of specific cell populations
	Cell displacement (ectopia)	Light microscopy of routinely stained sections (H&E) or sections processed to reveal cell type–specific markers
	Abnormal neurite conformation	Light microscopy of specially stained sections (Fluoro-Jade, silver impregnation)
	Altered axonal size	Light microscopy of specially stained sections (IHC for cell type–specific cytoskeletal markers, silver stains)
Glia	Numerical changes	Light microscopy of specially processed sections (IHC for cell type–specific markers)
	Myelin amount/integrity	Light microscopy of specially stained sections (Luxol fast blue, IHC for cell type–specific markers)

[a] H&E, hematoxylin and eosin; IHC, immunohistochemistry.

Considerable care must be taken when investigating chronic neural diseases, as the damage to the principal target cell population may elicit secondary changes in other parts of the affected cells (e.g., central chromatolysis of the neuron cell body after transection of its axon) and/or in nearby groups of healthy cells (e.g., Schwann cells, which proliferate as a normal response to degeneration and dissolution of their associated axon).[48,49] The extent of the secondary repair processes may substantially exceed the reaction by the primary target cells, especially if the long-standing neural disease has already obliterated the target cells. The complete absence of a defined cell population may be obvious [e.g., selective loss of neurons in specific CA (cornu ammonis domains of the hippocampus)], but more often it is quite subtle and may easily be missed if more complex structures are evaluated by subjective estimates of cell number rather than objective quantification (e.g., reduction in neuronal numbers within the layers of the cerebral cortex). Special immunohistochemical procedures to detect markers specific for reactive astrocytes or activated microglia (Appendix 1) are often needed to detect neuronal lesions reliably, as expression of these glial markers is typically elevated in regions where neuronal degeneration has occurred.

Principle 3: Some sectors are selectively sensitive. Certain neural structures are much more susceptible to injury induced by many etiological agents, including neurotoxicants. Perhaps the most important attribute of toxicological neuropathologists and neurotoxicologists is their knowledge of the basic lesion patterns that can develop following neurotoxicant exposure.

"Hot spots" for neurotoxic damage can arise from many different factors.[37] One mechanism of enhanced regional susceptibility is the intricacy of the neural circuitry in a given structure. The more complex interconnections and dense synaptic beds that are characteristic of the cerebral cortex, hippocampus, and cerebellum render these regions quite sensitive to neurotoxic insult, particularly in periods of rapid cell proliferation during development.[50,51] Another factor leading to differential vulnerability of various neuron populations is the markedly high metabolic rate of the brain. This organ consumes disproportionate shares of the total cardiac output and blood-borne oxygen supply (approximately 15% and 20%, respectively) even though the brain mass represents only about 2% of the total body mass.[52] This tremendous metabolic rate makes the brain as a whole especially vulnerable to neurotoxicants that disrupt intracellular energy production.[53] That said, zonal variations in basal metabolic rate among neuronal populations predispose certain brain regions [especially gray matter (nuclei) of the thalamus, mammillary bodies, periaqueductal and periventricular brain stem, and cerebellar vermis] to toxicant-induced injury, above and beyond the level of vulnerability for the bulk of the brain.[25,26] An important ancillary consideration is that dependence

on a high rate of oxidative metabolism rate is a property also shared by the heart. Toxicants that injure the brain often injure the heart, and vice versa, so that in many instances (e.g., cyanide toxicity) it is difficult to distinguish a primary neurotoxic event from neural damage that results from primary cardiac toxicity. A third factor contributing to augmented regional vulnerability is the existence of cell type–specific neurochemical machinery. A biochemical example of such compartmentalization is the selective sensitivity of dopaminergic neurons to toxicants such as 6-hydroxydopamine[54] and 1-methyl-4-phenyl-1,2,3,6-tetrahydropyridine (MPTP),[55] both of which result in selective degeneration of dopaminergic neurons. A related biochemical route leading to higher susceptibility results from variations in chemical composition among cells; the extensive lipid content of neural cell membranes, especially myelin, provides an abundant target for the oxidizing actions of certain neurotoxicants.[56] Enhanced regional vulnerability may also reflect disparities in local blood flow and repair mechanisms. The efficiency of most biochemical, metabolic, and reparative processes decreases with age, which can further magnify zonal differences in neural tissue sensitivity to neurotoxicants.

Principle 4: What gets wrecked depends on when it gets whacked. As in other organs, toxicant-induced damage in the nervous system occurs only if the agent reaches a target cell population at a time when those cells are vulnerable. Such critical periods of sensitivity to toxicant exposure have been well documented in the developing nervous system following exposure to many different chemicals. For example, grossly evident malformations of the neuraxis happen in mouse embryos only if exposure occurs during neurulation,[57–59] which is the stage at which the cranial neuropore closes to form the brain primordium. As the brain continues to evolve during late gestation, each nucleus has one or two other critical periods for neuron production; a brief toxicant exposure during region-specific neurogenesis can thus decimate a structure engaged in its peak effort at neuronal production while causing minimal or no disruption in nearby quiescent regions.[60–62] Critical periods for some neuronal populations and processes extend well after birth,[61,63] including neuronal and glial expansion and migration, axonogenesis, synaptogenesis, and myelin formation over the first several years of postnatal life in human infants.[64–66]

Principle 5: When assessing acute lesions in neurons, "red and dead" is the real deal. In our experience, the majority of neurotoxicant-induced lesions in neurons are the outcome of primary degeneration. The main evidence for such a process is often the presence of dead and dying neurons, usually in clusters or dispersed throughout a given brain region. These degenerating cells exhibit a characteristic constellation of changes dominated by cytoplasmic hypereosinophilia in conjunction with either pyknosis (condensation and shrinkage) or karyorrhexis (fragmentation) of the nucleus. Such disintegrating neurons are typically termed *acidophilic neurons, eosinophilic neurons,* or *"red dead" neurons* (Chapter 13, Figure 2C and D).

This change must be distinguished from *dark neuron* artifact (Chapter 13, Figure 2A and B). Dark neurons indisputably embody the most common CNS artifact encountered by neuropathologists. Unfortunately, dark neuron artifacts have often mistakenly been judged by inexperienced pathologists, toxicologists, and neuroscientists to be evidence of neurodegeneration, and has been reported as such in the neuroscience and neurotoxicology literature.[67] Such reports have misidentified artifacts as neurotoxic injury, with subsequent unnecessary regulatory and public health alarm. Dark neuron artifact is usually observed with larger cells, such as the pyramidal neurons in the cerebral cortex and motor neurons in the spinal cord, and is characterized by darkly stained cytoplasm (especially intense in the apical dendrite) and nucleoplasm and shrunken cell bodies. If dark neurons represent the only visible alteration in sections of neural tissue, it is generally safe to interpret the change as artifactual. The main exception to this rule is for studies designed specifically to detect hyperacute neuron damage, as the earliest evidence of incipient neurodegeneration is transient cytoplasmic basophilia (Chapter 13), but even here later time points can be used to verify the nonartifactual nature of the alteration. Any practicing neuropathologist knows that dark neurons are readily produced by even mild trauma to the tissue before fixation is achieved, possibly as a consequence of localized ischemia, hypoglycemia, and excitatory neurotoxicity.[68] The simplest prospective way to avoid misinterpretation of dark neurons is to ensure that the neural tissues are fixed properly before they are handled and processed (Chapter 10). A post hoc means of distinguishing genuine lesions from dark neuron artifact is to process a serial section of each sample to reveal reactive astrocytes or activated microglia using immunohistochemical markers (Appendix 1), either or both of which may collect in areas where true neurodegeneration has transpired.

Principle 6: Make "special" stains part of your routine. The workhorse stain for screening most organs for toxicant-induced lesions is hematoxylin and eosin (H&E). This stain works well in the brain but is not suitable for detecting the entire spectrum of neurotoxic lesions that is ordinarily induced in neural tissue. The most readily recognized lesion in H&E-stained neural sections are neoplasms, but such sizable masses are an infrequent consequence of toxicant exposure. The more common toxicant-induced neural lesions—especially neuronal degeneration, myelin disruption, and glial hypertrophy/hyperplasia—may be recognized on H&E-stained sections, but the low contrast between the affected cells and the adjacent neuropil makes such evaluations relatively laborious and prone to false-negative errors.

The cure for this difficulty is to expand the menu of routine procedures that are used to screen neural tissues for neurotoxic lesions to include certain special stains. For most hypothesis-driven animal studies, the H&E-stained section should automatically be accompanied by serial sections processed to reveal degenerating neurons (e.g., Fluoro-Jade; Chapter 11) and reactive astrocytes [e.g., antiglial fibrillary acidic protein (GFAP); Chapters 10 and 21]. Inclusion of these two additional methods as a matter of course when conducting prospective neurotoxicity studies rather than waiting to request them based on the outcome of the examination using the H&E-stained section will substantially shorten the length of the analytical phase, because these two "special" procedures greatly simplify the neuropathologist's efforts to identify lesions, especially subtle ones. The choice regarding whether or not to include these additional stains in a diagnostic neuropathology setting can be left to the discretion of the pathologist.

Principle 7: Seeing is believing, but don't believe everything you see. Neuropathologists and neurotoxicologists have been educated to possess built-in biases to detect toxicant-induced alterations in cells. However, not all structural changes observed in neural tissues after exposure to potential neurotoxicants during a carefully controlled neurotoxicity study are the result of exposure to that agent.

We have seen several spurious causes of neurological dysfunction in toxicant-treated individuals which had nothing to do with the test agent. One example is spinal cord trauma and paralysis in incompletely restrained rabbits, which can kick so hard when handled that they fracture their vertebral column. A second instance is the incidental occurrence at necropsy of widespread neuronal necrosis in the cerebral cortex, hippocampus, and thalamus of some transgenic mice generated on the FVB genetic background. This spontaneous lesion has been attributed to intermittent seizure activity[69] rather than toxicant exposure, as the identical finding is evident in untreated control animals that have the same genetic background; the high susceptibility of FVB mice to chemically induced seizures indicates that great care will be required to confirm that neurodegenerative changes of this nature are truly related to toxicant exposure rather than to background neural overactivity. Finally, we have observed rodents treated with a known neurotoxicant to develop disorientation and ataxia as a sequel to acute bacterial meningitis. The point of these anecdotes is that the toxicological neuropathologist cannot set aside fundamental diagnostic skills when analyzing neural tissues from toxicant-exposed individuals.

Principle 8: Don't limit yourself to the pathology perspective. Although toxicant-induced neuroanatomical changes are often emphasized by regulators in managing neurotoxic risk,[15] reliance solely on neuropathological changes to identify neurotoxicants can be misleading. Some well-known neurotoxicants induce profound functional changes in the absence of recognizable structural alterations.[9,10] Some classic instances include chlorinated hydrocarbons (e.g., dieldrin), pyrethroids, and strychnine, all of which incite excessive synaptic excitation but no neuropathology, as well as barbiturates, lithium, and organic solvents (e.g., xylene), which cause neuronal depression in the absence of neuromorphological changes. On the other hand, clinical observation of functional deficits can signal the presence of subtle structural lesions. An example is the ability of early reductions in hindlimb grip strength and later progression to paralysis to indicate the presence of a distal axonopathy.

Furthermore, neurological signs in a toxicant-exposed individual are not necessarily evidence of direct neurotoxicity. Anorexia and associated weight loss in rodents are associated with many behavioral changes, including increased motor activity and escape behaviors, decreased hindlimb grip strength, and cognitive learning deficits.[70–72] In like manner, chemically induced injury to some extraneural organs (especially the kidney and liver) can lead to the induction of secondary neurological dysfunction via the accretion of unprocessed neurotoxic waste products. The classic example of this scenario is hepatic encephalopathy, in which severe liver damage permits ammonia accumulation in the blood and brain and ultimately disrupts many CNS metabolic pathways (especially in astrocytes) and glutamatergic excitatory neurotransmission.[73,74] Similarly, renal failure leads to uremic encephalopathy following increased circulating levels of many amino acids and protein metabolites. These examples again underscore the importance of integrating all available structural and functional evidence in reaching a conclusion regarding the risk posed by a potential neurotoxicant.[9,13,14,16]

Principle 9: Carry on with care. In practice, screening studies for neurotoxicity generally administer high doses of test agent to a small-animal species (typically, rats) over relatively short periods of time, assuming that the data gained in the exercise can be extrapolated from high doses to low and from animals to humans. This approach has worked reasonably well but is obviously not perfect, as a number of neurotoxicants have been detected first by epidemic intoxications in humans.[18]

Efforts at extrapolation among species are complicated by the large divergence in responsiveness following neurotoxicant exposure. For example, MPTP depletes nigrostriatal dopaminergic cells in humans and nonhuman primates, eliminates nigrostriatal synaptic terminals in mice, but has a minimal impact on comparable structures in the rat.[39] Species differences in MPTP neurotoxicity reflect variations in the rate and sites at which it is converted to its toxic metabolite, MPP$^+$ (1-methyl-4-phenylpyridine), by monoamine oxidase type B (MAO-B); in rats, the enzyme is localized in brain microvessels to exclude the highly polar MPP$^+$ from the neuropil, whereas in primates it is

concentrated in astrocytes and acts as a bioactivator of MPTP.[39] Dopaminergic neurons containing a higher quantity of neuromelanin (NM) also may be more susceptible to MPP$^+$ neurotoxicity, as the NM is thought to serve as a depot for extended release of the toxic metabolite within the target cells.[75,76] Similarly, differential responses between various rodent strains can affect the outcome of neurotoxicity screening studies.[77–79] For example, following amphetamine exposure, both Long–Evans and Sprague–Dawley rats have been shown to develop a comparable dose-dependent reduction of nigrostriatal dopaminergic terminals as well as damage to pyramidal cells and the somatosensory cortex, but only Long–Evans rats exhibit dense axonal degeneration and occasional degenerating cells in the frontal motor cortex.[80] Such differences in neurotoxicity between species and strains probably stem from many factors, including divergent pharmacokinetic profiles, dissimilar genetic backgrounds, unique neurophysiological processes (e.g., binding or uptake sites for toxicants, neural connectivity, naturally occurring neuroprotective agents, enhanced repair processes), or such environmental elements as husbandry and degree of stress. Discrimination among these factors requires a thorough understanding of each agent's mechanism of neurotoxic action. Interspecies extrapolation is especially difficult if xenobiotic-induced neural responses differ among animals of different strains, genders, and ages.[9,81–86]

Principle 10: Garbage in, garbage out. The prime directive for toxicologic neuropathology investigations is to have a preset plan, and to follow it. Professionals in this field follow a standard study design (e.g., Chapter 21; see also the article by Bolon et al.[38] for potential design considerations) that they use for conducting all routine screening studies, which they can then adapt for any specialized follow-up studies that may be required. Standardization of experimental methodology is essential so that the neuropathologist can become familiar with normal neuroanatomical features and neurobiological variations within and among various species and strains, and the technical staff can develop confidence in their prosection and processing skills via repeated practice.

Adequate assessment of neural tissues requires distinctive harvesting and processing techniques[87–89] to preserve structural detail and to avoid confusing artifacts.[90] Precise regional dissection,[91,92] special stains,[89,93–96] intricate morphological measurements,[97–101] and/or noninvasive imaging methods[102–109] may prove useful if biochemical and molecular (i.e., functional) information is to be evaluated in the context of its neuroanatomical localization. A major advantage of noninvasive imaging is that it allows researchers to conduct time-course experiments and more targeted neuroanatomical investigations with fewer animals, as the ability to screen for neuroanatomical changes in vivo helps in selecting subjects that actually have lesions.

Relative to other organs and body systems, the elements of the normal CNS are anatomically diverse, exhibiting major structural changes at both macroscopic and microscopic levels over very short distances and in all three dimensions. Thus, it may be difficult to discern the complete pattern of neural damage induced by a toxicant when the neuropathology screen is performed in two dimensions with only a few brain sections per animal. The complexity is magnified if the toxicant-induced lesions are very subtle[110,111] and the sections available for analysis are not taken at comparable levels for all individuals. The latter issue may be ameliorated by careful attention to detail during sampling[112] and the use of precision sectioning methods to assess multiple animals simultaneously.[13]

CONCLUDING REMARKS

Exposure to potential neurotoxicants in the home, the workplace, or the community is a fact of modern human life. Neurotoxicity epidemics have been induced in humans during recent decades by chemicals,[18] drugs,[39] and metals.[18] Thus, neurotoxicological research to identify and characterize the risk from existing and new compounds of unknown neurotoxic potential is a pressing need.

Toxicological neuropathology is a major consideration for neurotoxicity testing because structural effects are typically permanent in the CNS and at best slowly repaired in the PNS. Researchers who evaluate toxicological neuropathology endpoints must be thoroughly educated in multiple aspects of basic and applied neurobiology so that they can readily identify lesions and then correlate the neuroanatomical changes with biochemical and functional endpoints to provide an integrated, mechanistically based risk assessment. In most settings, toxicological neuropathologists and their neurotoxicologist colleagues are members of interdisciplinary teams rather than solo practitioners. This volume and the current chapter are designed to help neuropathologists and neurotoxicologists achieve a common baseline understanding of major principles and practices in toxicological neuropathology.

REFERENCES

1. Goldman LR. Linking research and policy to ensure children's environmental health. *Environ Health Perspect.* 1998;106 (Suppl 3):857–862.

2. Claudio L. An analysis of the U. S. Environmental Protection Agency neurotoxicity testing guidelines. *Regul Toxicol Pharmacol.* 1992;16:202–212.

3. Environmental Defense, Fund. *Toxic Ignorance: The Continuing Absence of Basic Health Testing for Top-Selling Chemicals in the United States.* Washington, DC: Environment Defense Fund; 1997.

4. Connelly JM, Malkin MG. Environmental risk factors for brain tumors. *Curr Neurol Neurosci Rep.* 2007;7:208–214.

5. Gobba F. Occupational exposure to chemicals and sensory organs: a neglected research field. *Neurotoxicology.* 2003; 24:675–691.

6. Bearer CF. Developmental neurotoxicity: illustration of principles. *Pediatr Clin N Am.* 2001;48:1199–1213.

7. Bouldin TW, Cavanaugh JB. Organophosphorous neuropathy: I. A teased-fiber study of the spatio-temporal spread of axonal degeneration. *Am J Pathol.* 1979;94:241–252.

8. Broxup B, et al. Correlation between behavioral and pathological changes in the evaluation of neurotoxicity. *Toxicol Appl Pharmacol.* 1989;101:510–520.

9. Dorman DC. An integrative approach to neurotoxicology. *Toxicol Pathol.* 2000;28:37–42.

10. Ray DE. Function in neurotoxicity: index of effect and also determinant of vulnerability. *Clin Exp Pharmacol Physiol.* 1997;24:857–860.

11. Spencer PS, Schaumburg HH. Classification of neurotoxic disease: a morphologic approach. In: Spencer PS, Schaumburg HH, eds. *Experimental and Clinical Neurotoxicology.* Baltimore: Williams & Wilkins; 1980:92–99.

12. Sette WF, MacPhail RC. Qualitative and quantitative issues in assessment of neurotoxic effects. In: Tilson HA, Mitchell CL, eds. *Neurotoxicology.* New York: Raven Press; 1992: 345–361.

13. Fix AS, et al. Integrated evaluation of central nervous system lesions: stains for neurons, astrocytes and microglia reveal the spatial and temporal features of MK-801-induced neuronal necrosis in the rat cerebral cortex. *Toxicol Pathol.* 1996;24:291–304.

14. Sette WF. Complexity of neurotoxicological assessment. *Neurotoxicol Teratol.* 1987;9:411–416.

15. Crofton KM, Jarabek AM. Neurotoxicity as the basis for chronic dose–response estimates. *Toxicologist.* 1995;15:34.

16. Eisenbrandt DL, et al. Evaluation of the neurotoxic potential of chemicals in animals. *Food Chem Toxicol.* 1994;32: 655–669.

17. McMaster SB. Developmental toxicity, reproductive toxicity, and neurotoxicity as regulatory endpoints. *Toxicol Lett.* 1993;68:225–230.

18. Costa LG, et al. Developmental neuropathology of environmental agents. *Annu Rev Pharmacol Toxicol.* 2004;44:87–110.

19. Campbell IC, Abdulla EM. Issues in in vitro neurotoxicity testing. *In Vitro Toxicol.* 1995;8:177–186.

20. Bittigau P, et al. Antiepileptic drugs and apoptotic neurodegeneration in the developing brain. *Proc Natl Acad Sci USA.* 2002;99:15089–15094.

21. Weiss B. Chemobrain: a translational challenge for neurotoxicology. *Neurotoxicology.* 2008;29:891–898.

22. Chudler EH. Milestones in Neuroscience Research. Available at: http://faculty.washington.edu/chudler/hist.html. Accessed Sept 11, 2009.

23. Henry JM. Neurons and Nobel Prizes: a centennial history of neuropathology. *Neurosurgery.* 1998;42:143–155 (discussion, 155–156).

24. Gallo MA. History and scope of toxicology. In: Klaassen CD, ed. *Casarett and Doull's Toxicology: The Basic Science of Poisons*. New York: McGraw-Hill; 1996:3–11.

25. Cavanagh JB. Lesion localisation: implications for the study of functional effects and mechanisms of action. *Toxicology*. 1988;49:131–136.

26. Cavanagh JB. Selective vulnerability in acute energy deprivation syndromes. *Neuropathol Appl Neurobiol*. 1993;19: 461–470.

27. Cavanagh JB, et al. Selective damage to the cerebellar vermis in chronic alcoholism: a contribution from neurotoxicology to an old problem of selective vulnerability. *Neuropathol Appl Neurobiol*. 1997;23:355–363.

28. Hawkes CH, et al. Motoneuron disease: a disorder secondary to solvent exposure? *Lancet*. 1989;1:73–76.

29. Tomiwa K, et al. The effects of cisplatin on rat spinal ganglia: a study by light and electron microscopy and by morphometry. *Acta Neuropathol*. 1986;69:295–308.

30. Wadsworth PF, et al. The topography, structure and incidence of mineralized bodies in the basal ganglia of the brain of cynomolgus monkeys (*Macaca fascicularis*). *Lab Anim*. 1995;29:276–281.

31. Butchko HH, et al. Aspartame: review of safety. *Regul Toxicol Pharmacol*. 2002;35:S1–S93.

32. Koestner A. The brain-tumour issue in long-term toxicity studies in rats. *Food Chem Toxicol*. 1986;24:139–143.

33. Koestner A. Characterization of *N*-nitrosourea-induced tumors of the nervous system: their prospective value for studies of neurocarcinogenesis and brain tumor therapy. *Toxicol Pathol*. 1990;18:186–192.

34. Yaeger MJ, et al. The use of nerve growth factor as a reverse transforming agent for the treatment of neurogenic tumors: in vivo results. *Acta Neuropathol*. 1992;83:624–629.

35. Katelaris A, et al. Brains at necropsy: to fix or not to fix? *J Clin Pathol*. 1994;47:718–720.

36. Bolon B. Comparative and correlative neuroanatomy for the toxicologic pathologist. *Toxicol Pathol*. 2000;28:6–27.

37. Bolon B, et al. Current pathology techniques symposium review: advances and issues in neuropathology. *Toxicol Pathol*. 2008;36:871–889.

38. Bolon B, et al. A "best practices" approach to neuropathologic assessment in developmental neurotoxicity testing—for today. *Toxicol Pathol*. 2006;34:296–313.

39. Kopin IJ. MPTP: an industrial chemical and contaminant of illicit narcotics stimulates a new era in research on Parkinson's disease. *Environ Health Perspect*. 1987;75:45–51.

40. Ekino S, et al. Minamata disease revisited: an update on the acute and chronic manifestations of methyl mercury poisoning. *J Neurol Sci*. 2007;262:131–144.

41. Jones DC, Miller GW. The effects of environmental neurotoxicants on the dopaminergic system: a possible role in drug addiction. *Biochem Pharmacol*. 2008;76:569–581.

42. Migliore L, Coppedè F. Environmental-induced oxidative stress in neurodegenerative disorders and aging. *Mutat Res*. 2009;674:73–84.

43. Rice D, Barone S Jr. Critical periods of vulnerability for the developing nervous system: evidence from humans and animal models. *Environ Health Perspect*. 2000;108(Suppl 3):511–533.

44. Battaglia G, et al. Neurogenesis in cerebral heterotopia induced in rats by prenatal methylazoxymethanol treatment. *Cereb Cortex*. 2003;13:736–748.

45. Bassil KL, et al. Cancer health effects of pesticides: systematic review. *Can Fam Physician*. 2007;53:1704–1711.

46. Rice JM, Wilbourn JD. Tumors of the nervous system in carcinogenic hazard identification. *Toxicol Pathol*. 2000;28:202–214.

47. Graham DG, et al. Pathogenetic studies of hexane and carbon disulfide neurotoxicity. *Crit Rev Toxicol*. 1995;25:91–112.

48. Adams JH, Duchen LW. *Greenfield's Neuropathology*, 5th ed. Oxford, UK: Oxford University Press; 1992.

49. Summers BA, et al. *Veterinary Neuropathology*. St. Louis, MO: Mosby; 1995.

50. Barone SF, et al. Vulnerable processes of the nervous system development: a review of markers and methods. *Neurotoxicology*. 2000;21:15–36.

51. Rodier PM, et al. Morphologic effects of interference with cell proliferation in the early fetal period. *Neurobehav Toxicol*. 1979;1:129–135.

52. Heiss WD. Cerebral blood flow: physiology, pathophysiology and pharmacological effects. *Adv Otorhinolaryngol*. 1981; 27:26–39.

53. Nicklas WJ, et al. Mitochondrial mechanisms of neurotoxicity. *Ann NY Acad Sci*. 1992;648:28–36.

54. Kostrzewa RM, Brus R. Destruction of catecholamine-containing neurons by 6-hydroxydopa, an endogenous amine oxidase cofactor. *Amino Acids*. 1998;14:175–179.

55. Tomac A, et al. Protection and repair of the nigrostriatal dopaminergic system by GDNF in vivo. *Nature*. 1995;373:335–339.

56. Juurlink BH, Paterson PG. Review of oxidative stress in brain and spinal cord injury: suggestions for pharmacological and nutritional management strategies. *J Spinal Cord Med*. 1988;21:309–334.

57. Bolon B, et al. Phase-specific developmental toxicity in mice following maternal methanol inhalation. *Fundam Appl Toxicol*. 1993;21:508–516.

58. Kotch LE, Sulik KK. Experimental fetal alcohol syndrome: proposed pathogenic basis for a variety of associated facial and brain anomalies. *Am J Med Genet*. 1992; 44:168–176.

59. Nau H, et al. Valproic acid-induced neural tube defects in mouse and human: aspects of chirality, alternative drug development, pharmacokinetics and possible mechanisms. *Pharmacol Toxicol*. 1991;69:310–321.

60. Altman J, Bayer SA. *Atlas of Prenatal Rat Brain Development*. Boca Raton, FL: CRC Press; 1995.

61. Rodier PM. Chronology of neuron development: animal studies and their clinical implications. *Dev Med Child Neurol*. 1980;22:525–545.

62. Rodier PM. Vulnerable periods and processes during central nervous system development. *Environ Health Perspect.* 1994;102(Suppl 2):121–124.

63. Dobbing J, Sands J. Comparative aspects of the brain growth spurt. *Early Hum Dev.* 1979;3:79–83.

64. Herschkowitz N, et al. Neurobiological bases of behavioral development in the first year. *Neuropediatrics.* 1997;28:296–306.

65. Herschkowitz N, et al. Neurobiological bases of behavioral development in the second year. *Neuropediatrics.* 1999;30:221–230.

66. Qiao D, et al. Developmental neurotoxicity of chlorpyrifos: What is the vulnerable period? *Environ Health Perspect.* 2002;110:1097–1103.

67. Jortner BS. The return of the dark neuron: a histological artifact complicating contemporary neurotoxicologic evaluation. *Neurotoxicology.* 2006;27:628–634 [see comment in *Neurotoxicology.* 2006;27:1126].

68. Kherani ZS, Auer RN. Pharmacologic analysis of the mechanism of dark neuron production in cerebral cortex. *Acta Neuropathol.* 2008;116:447–452.

69. Goelz MF, et al. Neuropathologic findings associated with seizures in FVB mice. *Lab Anim Sci.* 1998;48:34–37.

70. Albee RR, et al. Neurobehavioral effects of dietary restriction in rats. *Neurotoxicol Teratol.* 1987;9:203–211.

71. Dobbing J, Sands J. Vulnerability of developing brain: IX. The effect of nutritional growth retardation on the timing of the brain growth-spurt. *Biol Neonate.* 1971;19:363–378.

72. Gerber GJ, O'Shaughnessy D. Comparison of the behavioral effects of neurotoxic and systemically toxic agents: How discriminatory are behavioral tests of neurotoxicity? *Neurobehav Toxicol.* 1986;8:703–710.

73. Butterworth RF. Portal-systemic encephalopathy: a disorder of neuron-astrocytic metabolic trafficking. *Dev Neurosci.* 1993;15:313–319.

74. Felipo V, et al. Neurotoxicity of ammonia and glutamate: molecular mechanisms and prevention. *Neurotoxicology.* 1998;19:675–681.

75. Herrero MT, et al. Does neuromelanin contribute to the vulnerability of catecholaminergic neurons in monkeys intoxicated with MPTP? *Neuroscience.* 1993;56:499–511.

76. Zecca L, et al. The neuromelanin of human substantia nigra: structure, synthesis and molecular behaviour. *J Neural Transm.* 2003;Suppl:145–155.

77. Fink JS, Reis DJ. Genetic variations in midbrain dopamine cell number: parallel with differences in responses to dopaminergic agonists and in naturalistic behaviors mediated by central dopaminergic systems. *Brain Res.* 1981;222:335–349.

78. Mohajeri MH, et al. The impact of genetic background on neurodegeneration and behavior in seizured mice. *Genes Brain Behav.* 2004;3:228–239.

79. Stöhr T, et al. Rat strain differences in open-field behavior and the locomotor stimulating and rewarding effects of amphetamine. *Pharmacol Biochem Behav.* 1998;59:813–818.

80. Ryan LJ, et al. Histological and ultrastructural evidence that D-amphetamine causes degeneration in neostriatum and frontal cortex of rats. *Brain Res.* 1990;518:67–77.

81. Bondy SC. Especial considerations for neurotoxicological research. *Crit Rev Toxicol.* 1985;14:381–402.

82. Claassen V. *Neglected Factors in Pharmacology and Neuroscience Research.* New York: Elsevier; 1994.

83. Dyer RS. Cross species extrapolation and hazard identification in neurotoxicology. *Neurobehav Toxicol Teratol.* 1984;6:409–411.

84. Kacew S, Festing MFW. Role of rat strain in the differential sensitivity to pharmaceutical agents and naturally occurring substances. *J Toxicol Environ Health.* 1996;47:1–30.

85. Kacew S, et al. Strain as a determinant factor in the differential responsiveness of rats to chemicals. *Toxicol Pathol.* 1995;23:701–715.

86. Tipton KF, Singer TP. Advances in our understanding of the mechanisms of the neurotoxicity of MPTP and related compounds. *J Neurochem.* 1993;61:1191–1206.

87. Krinke GJ, et al. Optimal conduct of the neuropathology evaluation of organophosphorus induced delayed neuropathy in hens. *Exp Toxicol Pathol.* 1997;49:451–458.

88. Krinke GJ, et al. Teased-fiber technique for peripheral myelinated nerves: methodology and interpretation. *Toxicol Pathol.* 2000;28:113–121.

89. Switzer RC III. Application of silver degeneration stains to neurotoxicity testing. *Toxicol Pathol.* 2000;28:70–83.

90. Garman RH. Artifacts in routinely immersion-fixed nervous tissue. *Toxicol Pathol.* 1990;18:149–153.

91. Palkovits M, Brownstein MJ. *Maps and Guide to Microdissection of the Rat Brain.* New York: Elsevier; 1988.

92. Toga AW. Brain-mapping neurotoxicity and neuropathology. *Ann NY Acad Sci.* 1997;820:1–13.

93. Downing AE. Neuropathological histotechnology. In: Prophet EB, et al., eds. *Laboratory Methods in Histotechnology.* Washington, DC: American Registry of Pathology; 1992:81–104.

94. Morgan KT, et al. Vascular leakage produced in the brains of mice by *Clostridium welchii* type D toxin. *J Comp Pathol.* 1975;85:461–466.

95. O'Callaghan J, Sriram K. Glial fibrillary acidic protein and related glial proteins as biomarkers of neurotoxicity. *Expert Opin Drug Saf.* 2005;4:433–442.

96. O'Callaghan JP. Quantitative features of reactive gliosis following toxicant-induced damage to the CNS. In: Johannessen JN, ed. *Markers of Neuronal Injury and Degeneration.* New York: New York Academy of Sciences; 1993:195–210.

97. Broxup BR, et al. Quantitative techniques in neuropathology. *Toxicol Pathol.* 1990;18:105–114.

98. Coggeshall RE, Lekan HA. Methods for determining numbers of cells and synapses: a case for more uniform standards of review. *J Comp Neurol.* 1996;364:6–15.

99. Diemer NH. Quantitative morphological studies of neuropathological changes: Part 1. *Crit Rev Toxicol.* 1982;10:215–263.

100. Hyman BT, et al. Stereology: a practical primer for neuropathology. *J Neuropathol Exp Neurol.* 1998;57:305–310.

101. Murray JM. Neuropathology in depth: the role of confocal microscopy. *J Neuropathol Exp Neurol.* 1992;51:475–487.

102. Heckl S, et al. Molecular imaging: bridging the gap between neuroradiology and neurohistology. *Histol Histopathol.* 2004;19:651–668.

103. Hesselbarth D, et al. High resolution MRI and MRS: a feasibility study for the investigation of focal cerebral ischemia in mice. *NMR Biomed.* 1998;11:423–429.

104. Kornguth S, et al. Near-microscopic magnetic resonance imaging of the brains of phenylalanine hydroxylase–deficient mice, normal littermates, and of normal BALB/c mice at 9.4 tesla. *Neuroimage.* 1994;1:220–229.

105. Sharif NA, Eglen RM. Quantitative autoradiography: a tool to visualize and quantify receptors, enzymes, transporters, and second messenger systems. In: Sharif NA, ed. *Molecular Imaging in Neuroscience: A Practical Approach.* Oxford, UK: IRL Press; 1993: 71–138.

106. Sills RC, et al. Contribution of magnetic resonance microscopy in the 12-week neurotoxicity evaluation of carbonyl sulfide in Fischer 344 rats. *Toxicol Pathol.* 2004;32:501–510.

107. Suehiro M, et al. A PET radiotracer for studying serotonin uptake sites: carbon-11-McN-5652Z. *J Nucl Med.* 1993;34:120–127.

108. Willerman L, et al. Hemisphere size asymmetry predicts relative verbal and nonverbal intelligence differently in the sexes: an MRI study of structure–function relations. *Intelligence.* 1992;16:315–328.

109. Woodroofe MN, et al. In situ hybridization in neuropathology. *Neuropathol Appl Neurobiol.* 1994;20:562–572.

110. Balaban CD. Central neurotoxic effects of intraperitoneally administered 3-acetylpyridine, harmaline and niacinamide in Sprague–Dawley and Long–Evans rats: a critical review of central 3-acetylpyridine neurotoxicity. *Brain Res.* 1985;356:21–42.

111. Fix AS, et al. Quantitative analysis of factors influencing neuronal necrosis induced by MK-801 in the rat posterior cingulate/retrosplenial cortex. *Brain Res.* 1995;696:194–204.

112. Broxup B. Neuropathology as a screen for neurotoxicity assessment. *J Am Coll Toxicol.* 1991;10:689–695.

2

FUNCTIONAL NEUROANATOMY

MICHAEL H. STOFFEL
Department of Clinical Research and Veterinary Public Health, Vetsuisse Faculty, University of Bern, Bern, Switzerland

ANNA OEVERMANN
Neurocenter, Department of Clinical Research and Veterinary Public Health, Vetsuisse Faculty, University of Bern, Bern, Switzerland

MARC VANDEVELDE
Department of Clinical Veterinary Medicine, Vetsuisse Faculty, University of Bern, Bern, Switzerland

INTRODUCTION

The function of the nervous system is to enable the organism to respond to internal and external stimuli in order to survive. Stimuli are perceived by specific receptors that may be organized into sense organs (e.g., ear, eye, olfactory mucosa) or dispersed as individual elements. Data collected are analyzed and integrated by the central nervous system, which then initiates an appropriate reaction. This response in turn relies on definite effector organs (e.g., skeletal muscle, smooth muscle). Receptors and effector organs are connected to the central nervous system by peripheral nerves. The basis for data collection by receptors is their property of irritability. The precondition for transport of information and data processing, in turn, is conductivity of electrical impulses by neuronal processes.

Stimuli may arise both from the environment and from within the individual. Receptors for external stimuli include mechanoreceptors, thermoreceptors, and nociceptors. To monitor the status of the organism itself, appropriate receptors collect data on physical conditions (e.g., distension of viscera, body temperature, blood pressure) but also on chemical parameters (e.g., pH, osmolarity, oxygen saturation). Awareness of stimuli is always limited, as a considerable part of the information collected is processed unconsciously. Notwithstanding, all stimuli are integrated and contribute to initiating appropriate responses. Effector organs include skeletal muscles, cardiac muscle, smooth muscles, and glands. Thus, the nervous system regulates all vital functions, including homeostasis (internal steady state) and behavior (interaction with the environment). The elements of the nervous system maintaining homeostasis are defined as the visceral nervous system, while those involved in interplay with the environment are designated as the somatic nervous system.

In summary, the functional unit of the nervous system is the stimulus–response apparatus. It consists of five elements that are outlined here: (1) Receptors and (2) afferent neurons constitute the input limb. (3) The central nervous system ensures adequate processing of the information provided. The output is mediated through (4) efferent neurons and (5) effector organs. Afferent and efferent elements belong to the peripheral nervous system (PNS) as opposed to the brain and spinal cord, which together form the central nervous system (CNS) or neuraxis. More specifically, the efferent limb of the stimulus–response apparatus is also called the *lower motor neuron* as opposed to the higher-ranking (or superordinate) neurons in the central nervous system, which are collectively named the *upper motor neurons* for the sake of simplicity.

Fundamental Neuropathology for Pathologists and Toxicologists: Principles and Techniques, First Edition. Edited by Brad Bolon and Mark T. Butt.
© 2011 John Wiley & Sons, Inc. Published 2011 by John Wiley & Sons, Inc.

MORPHOGENESIS AND DESCRIPTIVE ANATOMY OF THE CENTRAL NERVOUS SYSTEM

In view of a functional understanding, however, two aspects of morphogenesis must be linked to descriptive neuroanatomy if neural organization is to be fully understood. These features are the shaping of the CNS and the emergence of the final distribution of the central gray matter. Concomitantly, the most important aspects of brain and spinal cord evolution are reviewed here.

Although the central nervous system should always be perceived as a single integrated whole, it is customary to subdivide it into anatomically distinct parts for the sake of facilitating communication. The spinal cord extends within the vertebral canal and is continuous with the brain stem. The caudal part of the brain, the rhombencephalon, consists of the myelencephalon (medulla oblongata) and metencephalon (comprised of the cerebellum dorsally and the pons ventrally). The cerebellum overlies the other parts of the rhombencephalon. The midbrain, or mesencephalon, is a compact section of the brain stem connecting the rhombencephalon and prosencephalon. The rhombencephalon and mesencephalon share many similarities with the spinal cord, and they give off the cranial nerves III to XII. The rostral end of the CNS is formed by the prosencephalon, which is composed of the diencephalon ventrally [associated with the optic nerve (cranial nerve II)] and the telencephalon [the origin of the olfactory nerve (cranial nerve I)].

FORMATION OF THE CENTRAL NERVOUS SYSTEM

Dorsal to the notochord, a thickening of the ectoderm occurs very early in embryonic development (i.e., prior to the closure of the amnion). The lateral edges of this neural plate fold up and bend inward. Fusion of the neural folds in the dorsal midline starts in the prospective neck region and proceeds both rostrally and caudally, thus yielding the neural tube. Failure of the neuropores (the open ends of the neural tube) to close results in malformations such as anencephaly or exencephaly of the brain, or spina bifida of the spinal cord.

The germinal layer for the production of neurons is adjacent to the central lumen. In most regions of the neural tube, neurons will remain next to the ventricular system. In the cerebellum and the telencephalon, however, substantial emigration to the outer surface of the neural tube will occur and generate stratified cortex. Unstratified gray matter in the vicinity of the ventricular system yields the basal ganglia of the prosencephalon, the nuclei of the brain stem and cerebellum, as well as the columns of the spinal cord (Figure 1).

The edges of the surface ectoderm seal to form a continuous epithelial sheet that covers the primordium of the CNS. Simultaneously, cells from the transitional area between the neuroectoderm and the surface ectoderm emigrate into the mesenchyme to generate two dorsolateral cords, the neural crests. The neural crests extend along the neural tube up to the prospective mesencephalon. They disintegrate regionally into segmental cell clusters from which the dorsal root ganglia (DRG) will develop. Other cells migrate away from the neural crests to give rise to the peripheral ganglia of the autonomic nervous system, the adrenal cortex, other paraganglia, and also to the glial cells (Schwann cells) of the PNS, to melanocytes, and to the leptomeninges. Furthermore, cells from the clusters flanking the brain stem will develop into the mesenchyme of the head.

Differential mitotic activity at the wider rostral end of the neural tube results in the formation of local expansions even before closure of the rostral neuropore. Three primary vesicles can be differentiated: the prosencephalon, the mesencephalon, and the rhombencephalon. The prosencephalon may later be subdivided further into telencephalon and diencephalon, while the rhombencephalon develops into metencephalon (including the cerebellum and pons) and into the myelencephalon. Many terms, such as *spinal bulb, medulla oblongata*, or simply *medulla*, are often used as synonyms for the myelencephalon. In parallel with the formation of the brain vesicles, dorsal and ventral curvatures develop. From caudal to rostral, these are the cervical flexure, the pontine flexure, and the mesencephalic flexure. The ventrally oriented cervical flexure at the junction with the spinal cord and the ventral bend of the mesencephalic flexure result in a U-shape of the developing brain. The dorsal pontine flexure later confers an S-shape on the organizing brain.

In its earliest stage, the neural tube is an unpaired structure. Later, however, two paired outgrowths supervene at the level of the prosencephalon. These are the optic vesicles, budding from the future diencephalons, and the telencephalic vesicles, which will develop bilaterally into the cerebral hemispheres. The telencephalic vesicles expand rostrally and in an initial caudodorsal direction, later being deflected ventrolaterally and rostrally again. Thereby, the hemispheres will overlap the unpaired structures of the basal brain dorsolaterally.

VENTRICULAR SYSTEM

As a result of its organogenesis, the central nervous system surrounds a continuous cavity, the ventricular system (Figure 2). These interconnected channels serve as reservoirs for the production and passage of cerebrospinal fluid (CSF). The ventricular system within the spinal cord is called the *central canal*. In mammals it narrows substantially and often becomes occluded with age. The central canal widens rostrally to become the fourth ventricle within the rhombencephalon. The CSF can escape the fourth ventricle into the subarachnoid space (specifically, the cisterna

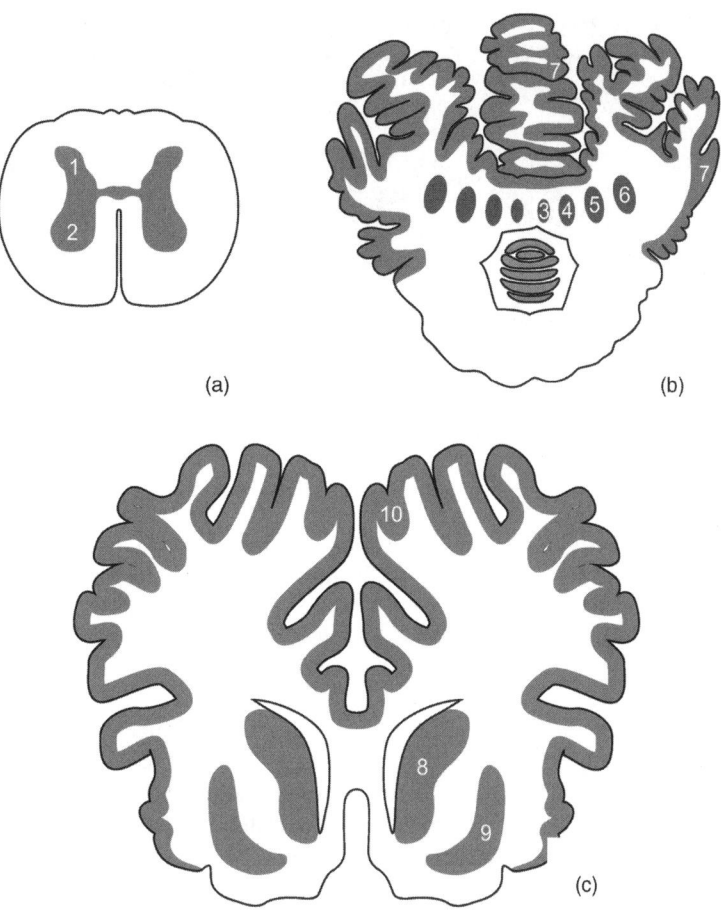

FIGURE 1   Distribution of gray matter in major divisions of the mammalian (dog) CNS. In the spinal cord (a), gray matter is adjacent to the central lumen. In cerebellum (b) and telencephalon (c), unorganized gray matter in the vicinity of the ventricular system is present as cerebellar nuclei and basal ganglia, respectively. In addition, emigration of proneurons from the periventricular region produces the stratified cortices of the cerebellum and the telencephalon. 1, 2, Gray matter of spinal cord (1, dorsal column; 2, ventral column); 3–6, cerebellar nuclei (3, nucleus fastigii; 4, nucleus interpositus medialis; 5, nucleus interpositus lateralis; 6, nucleus dentatus); 7, cerebellar cortex; 8, 9, subcortical gray matter of telencephalon (Basal ganglia) (8, nucleus caudatus; 9, putamen); 10, cerebral cortex. (Courtesy of Enke Verlag in MVS Medizinverlage Stuttgart GmbH & Co. KG.)

cerebellomedullaris s. magna) via the lateral apertures (foramina of Luschka) and, in primates, the median aperture (foramen of Magendie). The mesencephalic aqueduct (or aqueduct of Sylvius) of the midbrain is a narrow canal connecting the fourth ventricle to the third ventricle. This part of the ventricular system develops as a midsagittal cleft within the diencephalon. As left and right thalami expand, they come into contact with each other at the thalamic adhesion. As a result, the third ventricle is converted into an annular space extending into the suprapineal recess, the optic chiasm, and the infundibulum. In addition, the third ventricle is connected to the lateral ventricles within each cerebral hemisphere through the interventricular foramina (of Monro) underneath the fornix. The lateral ventricles inevitably adopt the curved shape being imposed by the

extension of the cerebral hemispheres. From their central part in the parietal lobe (pars centralis), the lateral ventricles extend into the olfactory bulbs (cornu rostrale) and into the temporal lobes (cornu temporale). All four ventricles are endowed with a choroid plexus where cerebrospinal fluid is formed as an ultrafiltrate of the blood. The choroid plexuses of the lateral ventricles extend caudally from the interventricular foramina, the choroid plexus of the third ventricle is found in the suprapineal recess, and of the fourth ventricle is located beneath the caudal medullary velum.

Although basically similar to that in mammals, the ventricular system of birds displays some distinctive features. The lateral ventricles are large; they extend along the medial and caudal border of the cerebral hemispheres and are located just underneath the brain surface, as the pallium is very thin in

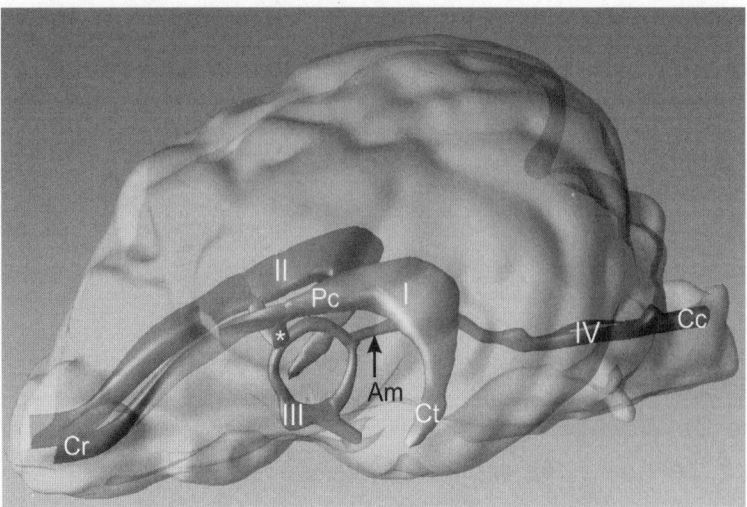

FIGURE 2   Mammalian ventricular system (modeled after the dog). I, II, Lateral ventricles (with Pc, pars centralis; Cr, cornu rostrale; Ct: cornu temporale); III, third ventricle (circling the interthalamic adhesion); Am, aqueductus mesencephali; IV, fourth ventricle; Cc, canalis centralis of the spinal cord; *, foramen interventriculare. (Courtesy of Enke Verlag in MVS Medizinverlage Stuttgart GmbH & Co. KG.) (*See insert for color representation of the figure.*)

birds. The choroid plexuses of the lateral ventricles are minute and limited to the region adjacent to the interventricular foramen. The third ventricle is a triangular cleft due to the absence of an interthalamic adhesion. The mesencephalic aqueduct is much more spacious then in mammals and extends into the tectum (ventriculus s. recessus tecti mesencephali). The central canal in the spinal cord is narrow but stays open over its full length.

SPINAL CORD

After the formation of the neural tube, its caudal part does not undergo further fundamental changes. The spinal cord thus remains a tube with some variations in diameter and thickness. In domestic species, the inner cavity persists as the central canal. In humans and primates, however, the central canal is often obliterated. In mammals, the morphogenesis of the spinal cord is concluded well before development of the vertebral column comes to an end. Therefore, the originally established relative position between spinal cord segments and corresponding vertebrae is not maintained [i.e., allometric growth (in which the relative proportions of structures change as development proceeds)].

The segmental organization of the spinal cord is betrayed by the spinal nerves, which emanate bilaterally at regular intervals and course to the corresponding intervertebral foramina. In principle, the spinal cord segments correspond to the somites, so spinal cord segments and vertebrae are homologous. As a rule, however, spinal nerves and their corresponding spinal cord segments bear the designation of the vertebra cranial to the intervertebral foramen by which

the nerve leaves the vertebral canal. This does not apply to the neck because the first cervical nerve passes through the lateral vertebral foramen at the cranial extremity of the atlas. The intervertebral foramen between atlas and axis is traversed by the second cervical nerve and the intervertebral foramen at the cervicothoracic junction by an eighth cervical nerve. Furthermore, the number of coccygeal spinal cord segments stays behind the number of caudal vertebrae.

Due to the allometric growth of vertebral column and spinal cord, not all spinal nerves originate opposite the intervertebral foramina intended for them (Figure 3). In all species, the spinal cord segment C_8 faces the intervertebral foramen between the last cervical and the first thoracic vertebra. The border between the last thoracic and the first lumbar spinal cord segments faces the cranial end of the first lumbar vertebra. Location of the caudal end of the spinal cord within the vertebral canal is subject to interspecies variation. In the dog, spinal cord segments S_{1-3} are located within the fifth lumbar vertebra and the tapering end of the medullary cone is to be found within L_6. In rats, the lumbar intumescence (L_{4-5}) is usually situated inside vertebrae L_{1-2} in young adults but inside vertebrae L_{4-5} in weanlings. In all mammalian species, however, the roots of the sacral and coccygeal spinal nerves run longitudinally through the vertebral canal to reach the intervertebral foramina they are meant for. This phenomenon is referred to by the descriptive term *cauda equina*. The cauda equina is absent in birds, as the spinal cord extends from one end of the vertebral column to the other.

Although basically cylindrical, the transverse diameter of the spinal cord is somewhat larger than its dorsoventral axis of symmetry. In addition, thickenings occur wherever limbs need to be supplied in addition to the trunk. Thus, a cervical

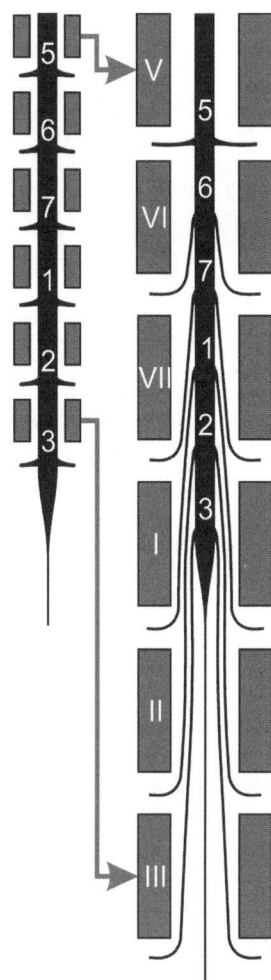

FIGURE 3 Development of the cauda equina. Early in gestation (small image on left), the spinal cord segments are aligned with the corresponding vertebrae. After birth (large image on right), allometric growth of the spinal cord and vertebral column results in progressive cranial displacement of spinal cord segments relative to the corresponding vertebrae. 5–7, Lumbar spinal cord segments 5 through 7; 1–3, sacral spinal cord segments 1 through 3; V–VII, lumbar vertebrae V through VII; I–III, sacral vertebrae I through III. (*See insert for color representation of the figure.*)

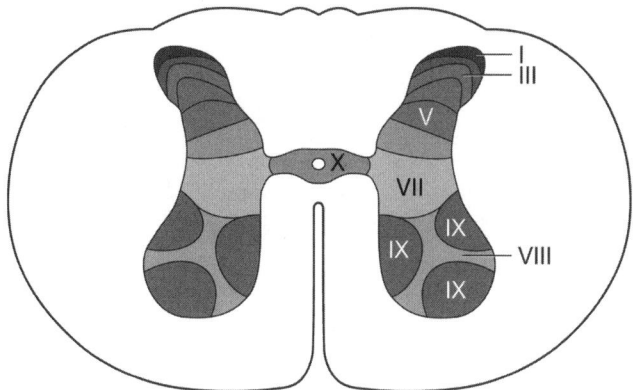

FIGURE 4 Laminar arrangement of spinal cord gray matter, according to Rexed. Laminae I to VI receive sensory input from pseudounipolar neurons, whereas visceral somatic efferents originate from layers VII to IX. Layer X contributes fibers to the spinoreticular and spinothalamic tracts. (Courtesy of Enke Verlag in MVS Medizinverlage Stuttgart GmbH & Co. KG.) (*See insert for color representation of the figure.*)

horn. The dorsal column (anterior column in humans) contains the general somatic afferent (GSA) neurons, while the ventral column (posterior column in humans) is formed by somatic efferent (SE) neurons (Figure 5a). Dorsal and ventral columns extend over the entire length of the spinal cord. A thin bridge of gray matter next to the central canal connects the dorsal and ventral columns. It extends over the full length of the spinal cord and receives visceral afferents. At the level of vertebrae T_1 to L_4, as well as in sacral segments, this bridge extends into the adjacent white matter and forms the lateral column (intermediate column in humans), which contains sympathetic neurons in the thoracolumbar part and parasympathetic neurons in the sacral segments. A shallow groove in the lateral wall of the central canal, the sulcus limitans, is a convenient landmark separating dorsal afferent columns from ventral efferent gray matter (Figure 5). A more sophisticated approach to parsing the spinal cord gray matter is based on the cytoarchitecture of neurons, and leads to a subdivision of the gray matter into 10 laminae according to Rexed (Figure 4). The laminae I to IX constitute the columns, and lamina X corresponds to the central intermediate gray matter surrounding the central canal. Laminae I to VI receive sensory input from pseudounipolar neurons, whereas the lamina VII encompasses the lateral intermediate zone. It extends into the ventral column at the level of the cervical and lumbar intumescences and includes visceral neurons. The ventral column harbors laminae VIII and layer IX, which is not a continuous sheet but, rather, comprises various aggregations of lower motor neurons in the ventral column. Laminae VII to IX contain the cell bodies of lower motor neurons from which efferent neurites originate, while lamina X contributes fibers to the spinoreticular and spinothalamic tracts.

The white matter (the outer spinal cord layer) is comprised of nerve fibers and is subdivided into three funiculi on each

intumescence occurs at the level of spinal cord segments C_6 to T_2, which generally reside within vertebrae C_5 to T_1. The ventral branches of the corresponding spinal nerves form the brachial plexus. Correspondingly, the major part of the lumbosacral plexus originates from the lumbar intumescence, which includes the thickened spinal cord segments L_3 to S_3.

Gray matter, which is characterized by the presence of neuronal bodies, surrounds the central canal and exhibits a resemblance to a butterfly on cross sections (Figure 4). Taking into account its longitudinal extension throughout the spinal cord, gray matter is best described in three dimensions with the term *column*, the two-dimensional cross section of which corresponds to the conventional designation

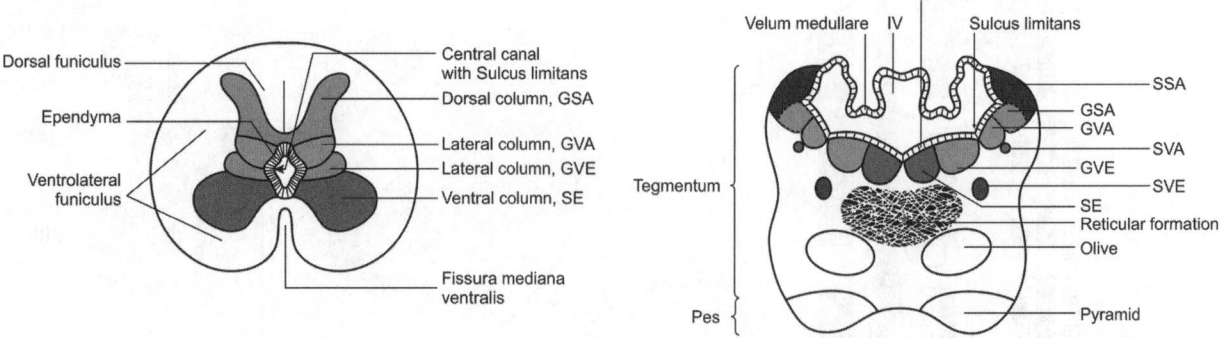

FIGURE 5 Functional organization of gray matter in the mammalian spinal cord (a) and rhomben-cephalon (b). As a consequence of the flattening of the rhombencephalon, dorsoventral alignment of modalities as seen in the spinal cord is modified into a lateromedial sequence. Nuclei dealing with the special modalities that are unique to the head are added at the lateral aspect (SSA) and deep to the general modalities in the tegmentum (SVA, SVE). GSA, General somatic afferent; GVA, general visceral afferent; GVE, general visceral efferent; SE, somatic efferent; SSA, special somatic afferent; SVA, special visceral afferent; SVE, special visceral efferent; IV, fourth ventricle. (Courtesy of Enke Verlag in MVS Medizinverlage Stuttgart GmbH & Co. KG.) (*See insert for color representation of the figure.*)

side (Figure 5). The dorsal funiculi extend from the dorsal midline to the origin of the dorsal spinal nerve roots. The lateral funiculi are confined by the dorsal roots and the indistinct ventral spinal nerve roots, whereas the ventral funiculi occupy the remaining area extending from the ventral roots to the ventral fissure. Because of their diffuse definition by the ventral roots, lateral and ventral funiculi are often considered together as the ventrolateral funiculi. The unmyelinated white matter adjacent to the gray substance connects different segments of the spinal cord. The more peripheral parts of the white matter are composed of myelinated ascending and descending tracts linking the spinal cord with the brain.

In birds, the spinal cord basically extends all along the vertebral column. Therefore, the origins of the spinal nerves face the corresponding intervertebral foramina, so no cauda equina is formed. In chickens, the diameter of the spinal cord is about 5 and 7 mm in the cervical and lumbosacral intu-mescences, respectively, and 3 to 4 mm outside these swel-lings. The number of spinal cord segments depends on the number of vertebrae. However, the distinction of the different segments of the vertebral column is complicated by the fact that extended ankyloses between vertebrae are physiological ("normal") features in birds. The synsacrum may include vertebrae from the thoracic to the caudal region, and the spinal nerves are the only indication of the boundaries between adjacent vertebrae. Therefore, only the cervical and thoracic segments may easily be identified; accordingly, spinal nerves are specified as cervical, thoracic, synsacral, and caudal. Depending on the species, the number of cervical vertebrae varies between 11 and 24 (e.g., 12 in pigeons and budgerigars, 14 in chickens and ducks, 17 in geese, and 24 in swans). As in mammals, the first cervical nerve exits the

vertebral canal between the occiput and the atlas. Consid-ering thoracic vertebrae to be those that are connected to the ribs, the number of thoracic segments is 7 in chickens and pigeons, and 9 in geese and ducks. Organization of gray and white matter, as seen on cross sections, is basically similar to the situation in mammals. A specific feature of the lumbar spinal cord in birds, however, is the gelatinous body: a clear mass harboring large amounts of glycogen-rich glial cells and unmyelinated nerve fibers. The gelatinous body is dorsal to the central canal and largest at the level of segments S_{3-6} (synsacral 4 to 7), thereby pushing apart the dorsal columns of gray matter in the region of the lumbo-sacral intumescence. From a dorsal perspective, the gelati-nous body stands out as the fossa rhomboidea spinalis. Its significance is still obscure, but the presence of argentaffin cells suggests that it may serve a neuroendocrine function.

BRAIN

Rhombencephalon and Mesencephalon

The rhombencephalon (hindbrain) consists of the metenceph-alon (cerebellum and pons) and myelencephalon (medulla oblongata). The roof of the rhombencephalon is attenuated and forms the medullary vela (velum medullare rostrale and velum medullare caudale) of the fourth ventricle (Figure 5b). Whereas the metencephalic velum rostral to the cerebellum contains minute amounts of white matter, its myenlencephalic equivalent consists of nothing more than the unistratified ependymal epithelium. This lamina epithelialis rhombence-phali is invaginated into the fourth ventricle, where it forms a choroid plexus. In combination with the weakening of the roof plate, the dorsal pontine flexure will cause the rhombencephalic vesicle to flatten and the side walls to splay outward. From a dorsal perspective, the floor of the lumen adopts a diamond shape and is adequately referred to as the rhomboidal fossa. The core part of the rhombencephalon ventral to the fourth ventricle thickens and is called the *tegmentum* (Figure 5b). It harbors all the nuclei of cranial nerves V to XII, the rhombencephalic part of the reticular formation, and the ascending and descending fiber tracts.

Functional organization of gray matter in the rhombencephalon is similar to the arrangement for the spinal cord. The sulcus limitans serves as the boundary between afferent and efferent nuclei up to the mesencephalon (Figure 5b). However, a number of important structural modifications in the brainstem relative to the spinal cord must be taken into account. First, sagittal alignment of dorsal, lateral, and ventral columns for the general modalities as seen in spinal cord is modified into a lateromedial sequence as a consequence of the flattening of the rhombencephalon (Figure 5b). Second, as there is no place left next to the floor of the ventricle, nuclei dealing with the special modalities that are unique to the head must be accommodated elsewhere (Figure 5b). Special somatic afferent (SSA) nuclei (cochlear and vestibular nuclei) are appended dorsolaterally at the junction between the metencephalon and the myelencephalon. The gustatory nucleus for the special visceral afferents (SVAs) (nucleus of solitary tract) is located deep to the general afferent modalities and is found ventrolateral to the sulcus limitans. Similarly, special visceral efferent (SVE) nuclei for pharyngeal arch muscles are located ventromedial to the sulcus limitans (Figure 5b). Third, the continuity of most columns of the spinal cord is abandoned in favor of distinct and separate nuclei. Thus, every single modality of each cranial nerve is associated with a distinct nucleus of gray matter. Ventral to the tegmentum, descending fiber tracts shape the pons of the metencephalon and the pyramid of the myelencephalon. Regardless of the brainstem region, the white matter ventral to the tegmentum is generically called the *pes*.

In contrast to the rhombencephalon, the mesencephalon maintains its tube shape. Its roof develops into a bulky plate with four colliculi, collectively termed the *corpora* (or *lamina*) *quadrigemina* or the *tectum*. The rostral colliculi mediate subconscious eye movements under the control of the visual centers in the occipital cortex, while the caudal colliculi control is a part of the pathway connecting the cochlea to the auditory centers in the temporal cortex. The mesencephalic core, or tegmentum, contains the nuclei of cranial nerves III and IV in an arrangement similar to that observed for the cranial nerve nuclei within the rhombencephalon. It also encloses the red nucleus, which plays a prominent role in motor activity. The most ventral part of the mesencephalon, called the *cerebral peduncles*, consists of descending fiber tracts that connect the cerebral cortex to the hindbrain and spinal cord.

In birds, the pons is continuous with the medulla oblongata. Therefore, the ventral aspect of the rhombencephalon does not allow a distinction to be made between myelencephalon and metencephalon. The medulla oblongata is connected to the cerebellum through the caudal cerebellar peduncle. The most prominent part of the avian mesencephalon is the tectum. This dorsal region is connected directly to the optic tract, and its extension reflects the prominent role that vision plays in avian life. The tectum in birds may thus be likened to the rostral colliculi of mammals, except that unlike its mammalian counterpart it is arranged in six layers that serve to integrate visual stimuli in advance of conscious visual perception within the telencephalon. The dorsal part of the lateral mesencephalic nucleus is the equivalent of the mammalian caudal colliculi but is completely concealed by the optical part of the tectum. Distribution of gray matter in the ventrally located tegmentum is basically similar to the situation in mammals (nuclei of the cranial nerves III and IV, nucleus of the mesencephalic tract of the trigeminal nerve, reticular formation, red nucleus). The mesencephalic aqueduct in birds is spacious and extends ventrolaterally as the ventriculus tecti mesencephali.

Cerebellum

The cerebellum develops from the rostral lips of the rhomboidal fossa (i.e., from the metencephalon). Their expansion yields a superstructure above the fourth ventricle and the medullary vela. The cerebellum is connected to the brain stem by three peduncles on either side. The rostral peduncle is linked to the mesencephalon, the middle one to the pons of the metencephalon, and the caudal one to the myelencephalon. In mammals, the cerebellum consists of two lateral hemispheres and the unpaired median vermis (Figure 6). From a functional point of view, it is more productive to distinguish three transverse slices, each encompassing corresponding areas of the hemispheres and the vermis. The caudoventral flocculonodular lobe is the phylogenetically oldest part of the cerebellum. It is separated from the caudal lobe by the uvulonodular fissure. The primary fissure is another transverse landmark separating the rostral lobe from the caudal lobe (Figure 6).

Histogenesis of the cerebellar gray matter exhibits a number of unique features. During early development, a subpopulation of the progenitor cells from the periventricular germinal layer migrates to the outer surface of the neural tube while other neuronal stem cells remain in their original location. This results in the formation of two distinct germinal layers, the internal and the external. The internal germinal layer yields the Purkinje cells and the neurons of the deep cerebellar nuclei. From medial to lateral, the cerebellar nuclei in each half of the cerebellar medulla are the fastigial nucleus, the nuclei interpositi (medial and lateral), and the dentate nucleus (Figure 1b). The granular

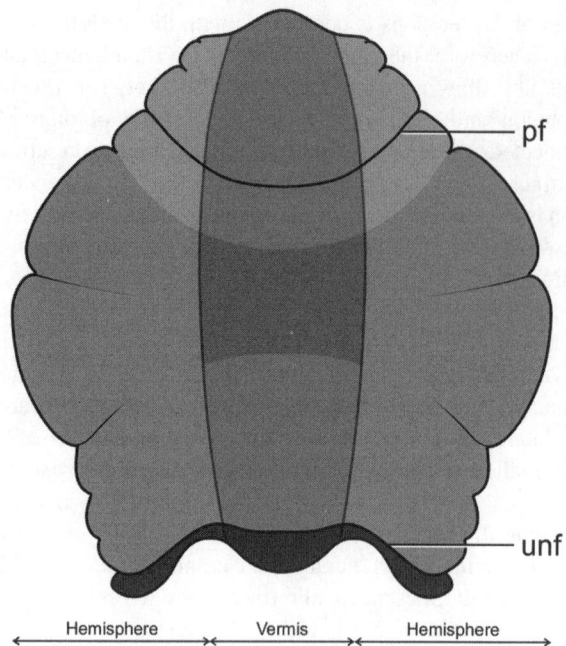

Hemisphere   Vermis   Hemisphere

FIGURE 6 Morphological and functional subdivisions of the cerebellum. The vestibulocerebellum consists of the flocculonodular lobe (dark blue) and deals with input from the vestibular apparatus. The spinocerebellum (green) receives proprioceptive (and exteroceptive) information. The cerebrocerebellum [s. pontocerebellum (brown)] is involved in processing collateral information on corticofugal motor activity. pf, Primary fissure; unf, uvulonodular fissure. (Courtesy of Enke Verlag in MVS Medizinverlage Stuttgart GmbH & Co. KG.) (*See insert for color representation of the figure.*)

cell layer underneath the Purkinje cells develops from the transient external germinal layer. Migration of proneurons from the external germinal layer along specialized glial cells (Bergmann cells) results in its depletion as the granular cell layer is populated. Precursors from the external germinal layer also provide neurons in the outer molecular layer, including the stellate cells and basket cells. Thus, the strata of the three-layered cerebellar cortex have two different origins: the outer (molecular layer) and inner (granular layer) portions evolve from the embryonic external germinal layer, whereas the intermediate stratum of Purkinje cells arises from the internal germinal layer. Formation of the cerebellar gray matter is completed only after birth.

The cerebellum in birds extends beyond the rhombencephalon. As the pons and its corresponding nuclei are poorly developed, the cerebellum is linked to the brainstem on either side by two peduncles only. The caudal peduncle provides the connection with the myelencephalon, while the rostral peduncle is related to the mesencephalon. The few pontocerebellar fibers present in birds are found within the above-mentioned peduncles. Furthermore, both medullary vela contain fibers allowing the exchange of information with the mesencephalon and the medullary bulb. The avian equivalent of the vermis is named the *corpus cerebelli*. Hemispheres are lacking, but the auricula cerebelli are lateral appendages of the corpus cerebelli. Deep fissures affect not only the cortex but extend into the medullary region. These grooves delineate a variable number of foliae according to the species. Stratification of the cerebellar cortex is typical and three cerebellar nuclei on each side are also found in the medullary zone.

Prosencephalon

The unpaired median part of the prosencephalon consists of the diencephalon and of a minor part of the telencephalon, notably the lamina terminalis and the rostral roof of the third ventricle. Thus, most of the telencephalon is formed by the paired cerebral hemispheres. Each hemisphere consists of a cerebral stem and the pallium. Both contain white and gray matter (Figures 1c and 7).

The basal nuclei are the gray matter of the cerebral stem. Initially compact, this gray matter is split into distinct nuclei as nerve fibers connecting the cerebrocortical layers and various subcortical centers (e.g., thalamus) grow through this region (Figures 1c and 7a). The developing white matter in this area is described as the *internal capsule*. The gray matter medial to the internal capsule is known as the *caudate nucleus*. It has a comma shape, bulges into the floor of the lateral ventricle rostrally with its head, and follows the curved extension of the hemisphere caudally with its body and tail. The caudate nucleus is craniolateral to the choroid plexus of the corresponding lateral ventricle. Nuclei located lateral to the internal capsule include the telencephalic putamen and amygdala as well as the globus pallidus, which belongs to the diencephalon (Figure 7). Because the caudate nucleus and putamen remain interconnected and constitute a single functional unit, they are often termed the *corpus striatum*.

The cortex is the gray substance of the pallium. Based on phylogenetic considerations, the pallium may be further subdivided into the paleopallium (evolutionarily oldest), archipallium, and neopallium (evolutionarily youngest). In fish, the paleopallium comprises most of the cerebrum. Amphibians develop an archipallium and paleopallium, whereas phylogenetically more recent groups develop all three regions (although the neopallium is relatively primitive in reptiles and birds). The primary function of the paleopallium and archipallium is generally thought to be integration of olfactory signals and limbic functions, while the neopallium is engaged in higher-order processing. In lissencephalic species (those without surface gyri and sulci on the brain) such as birds, rodents, or lagomorphs, the brain surface remains smooth. In contrast, carnivores, ungulates, and primates display extensive folding of their neocortex. Even in these gyrencephalic species, however, the paleopallium remains devoid of gyri and sulci. During development, the archipallium in mammals extends as a band along the medial border of either brain vesicle, dorsal to the choroid plexus of the lateral ventricle (Figure 7). Along with growth of the cerebral hemispheres, the

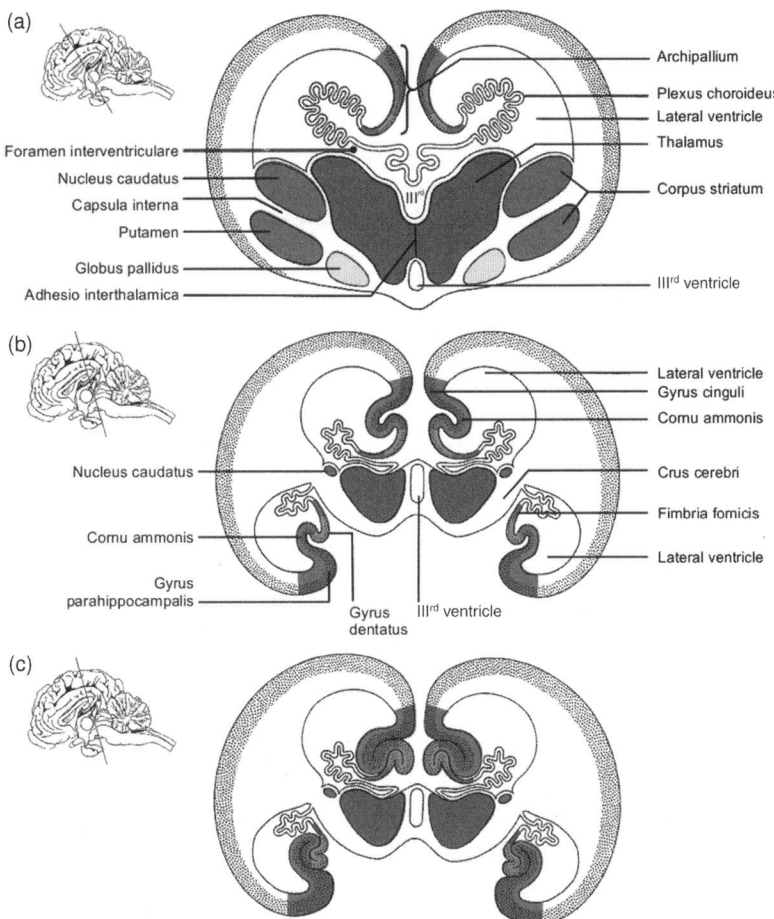

FIGURE 7 Embryonic development in the mammalian prosencephalon. Section at the level of the interventricular foramen (a), and sections caudal to the interthalamic adhesion at an early (b) and at a later stage (c) of development. (a) The basal nuclei include the nucleus caudatus, the putamen, and the globus pallidus. They develop from an originally compact mass of gray matter which is later split into separate nuclei by the entrance of axons that comprise the internal capsule. (b, c) The archipallium extends as a band along the medial border of either brain vesicle, dorsal to the choroid plexus of the lateral ventricle (a). During further development (b, c), the archipallium is invaginated into the lateral ventricle, thus forming the cornu ammonis (hippocampus proper). (Courtesy of Enke Verlag in MVS Medizinverlage Stuttgart GmbH & Co. KG.) (*See insert for color representation of the figure.*)

archipallium extends caudodorsally and then bends ventrally and rostrally again, thereby assuming its characteristic comma shape. During further development, the archipallium is then invaginated into the lateral ventricle, thus forming the hippocampus proper (Figure 7). As a result, the choroid plexus on the floor of the lateral ventricle is flanked by the caudate nucleus rostrolaterally and by the hippocampus caudomedially. The hippocampus is part of the limbic system. The paleopallium is the cortex area of the piriform lobe. Upon histological examination, the cortices of the paleopallium and of the archipallium are three-layered, whereas the cortex of the neopallium is six-layered.

Similar to the rhombencephalon, the dorsomedial wall of the prosencephalon is attenuated to the point of a mere unistratified epithelium. This lamina epithelialis prosencepha-

li is invaginated into the lateral ventricles on the medial aspect of the hemispheres caudal to the interventricular foramina and extends dorsally into the roof of the third ventricle (Figure 7). From these invaginations, the choroid plexus of the lateral ventricles and of the third ventricle will develop.

The core of the diencephalon is the thalamus. In mammals, it expands within the lateral walls of the third ventricle to such an extent that left and right thalami meet in the midline. The contact area is exempt of any exchange of fibers and is called the *massa intermedia thalami* (adhaesio interthalamica). This site obliterates the center of the third ventricle and transforms it into an annular space. All other parts of the diencephalon are designated with reference to the position of the thalamus. *Metathalamus* is an umbrella term for the medial and lateral geniculate bodies. The epithalamus includes the pineal gland, the habenulae, and their commissure. The hypothalamus contains a number of primary control nuclei for the visceral nervous system. A look at the ventral surface of the diencephalon reveals the optic chiasma (the site at which optic nerve fibers partially decussate in mammals), the mamillary body, and the tuber cinereum, which connects the hypothalamus to the hypophysis. Taking into account the extension of the ventricular system as well as the outer contours of the various brain regions, the origin of cross sections of the brain is easily identified.

Although it may also be subdivided into a diencephalic and a telencephalic part, compared with mammals the prosencephalon of birds exhibits profound structural differences, as the growth of the thalami is less extensive and no contact between left and right sides is established in the midline. The third ventricle thus persists as a vertical cleft. The epithalamus with its pineal gland and the hypothalamus may be differentiated in birds. The optic chiasm constitutes a complete decussation of the optic fibers. As in mammals, the telencephalon of birds consists of an unpaired stem and paired hemispheres. The pallium is lissencephalic, as only a shallow longitudinal depression, the vallecula telencephali, may be discerned. The olfactory bulb in birds is very small commensurate with the importance they put on smell. On the ventrolateral aspect, the fovea limbica betrays the proximity of the orbits to the brain. The proportions between the components of the cerebral hemispheres are fundamentally different from those seen in mammals. Thus, the cerebral stem containing the basal nuclei is of considerable size, whereas the pallium remains very thin (about 1 mm in chickens), so that the extensive lateral ventricle is displaced to a peripheral location. Also, stratification of the avian cortex is difficult to discern; most of the cortex pertains to the hippocampus, which basically fills the medial wall of the hemispheres. The corpus callosum is missing in birds, but the hemispheres are interconnected through the commissura rostralis and the commissura palli dorsal to the optic chiasm. As opposed to the pallium, the cerebral stem in birds is highly developed and performs a number of functions that are assigned, instead, to cortical regions in mammals. Thus, despite similar designations, the basal nuclei in birds should not be considered as the functional equivalents of the corpus striatum in mammals. The avian basal nuclei are subdivided into the archistriatum, the paleostriatum, the neostriatum, and the hyperstriatum. Although it is covered by the thin pallium, the hyerstriatum produces the sagittal eminence medial to the vallecula telencephali.

SOMATOTOPIC ORGANIZATION IN THE CNS

The nervous system is arranged so that specific structures of the CNS act via particular PNS pathways to serve preset regions of the body. Neurons with identical function are grouped together to form spinal cord columns, brainstem nuclei, or cortex areas. Axons conveying functionally similar information to or from a given body region are combined into tracts or fascicles. In like manner, topographical relationships in the periphery are mirrored at the brain surface. Receptors in neighboring skin regions will project to neighboring spots of the cerebral cortex. The same somatotopic arrangement applies to efferent connections reaching effector organs, and this principle is maintained within CNS and PNS tracts as well.

During embryonic development, short connections are established first. Fibers linking adjacent segments in the spinal cord are found adjacent to the gray matter, and longer fibers are added successively later as outer layers. Thus, the longer a connection, the more peripheral its location will be. As a consequence, external compression of the spinal cord will affect the connections to and from the hind limbs first, as these nerve fibers are the longest. The only exception to this rule is the relative position of the gracile and the cuneate fascicles, the latter of which is linked to the forelimb but located lateral to the gracile fasciculus (which serves mainly the hindlimb).

VASCULAR SUPPLY

Arterial blood supply to the spinal cord is segmental. Spinal branches enter the vertebral canal through the intervertebral foramina or (where present) the lateral vertebral foramina. Spinal branches arise from lumbar, dorsal intercostal, thoracic vertebral or supreme intercostal, and vertebral arteries. They divide into dorsal and ventral branches, which give rise to an anastomosing network of small arteries surrounding the spinal cord. The ventral rami feed the ventral spinal artery, from which numerous branches arise and enter the nervous tissue through the ventral fissure. Three concentric zones may be distinguished on cross sections through the spinal cord with respect to blood supply (Figure 8). The innermost area comprises the gray matter adjacent to the central canal. The intermediate region encompasses most of the gray matter (less the tips of the dorsal horns) and the adjacent white matter, including the fasciculi proprii. The outer ring of white matter underlying the pia mater constitutes the peripheral zone. The inner zone is fed by the branches arising from the ventral spinal artery only. Similarly, blood supply to the outer zone depends fully on the superficial network. As for the intermediate zone, both central and superficial systems contribute arterial vessels to its supply (Figure 8).

Arterial blood supply to the brain relies on two distinct routes, the internal carotids and the vertebral arteries (Figure 9). All these vessels eventually reach the circulus arteriosus cerebri (circle of Willis) at the base of the diencephalon. The vertebral arteries join the ventral spinal artery within the atlas to form the basilar artery. This unpaired

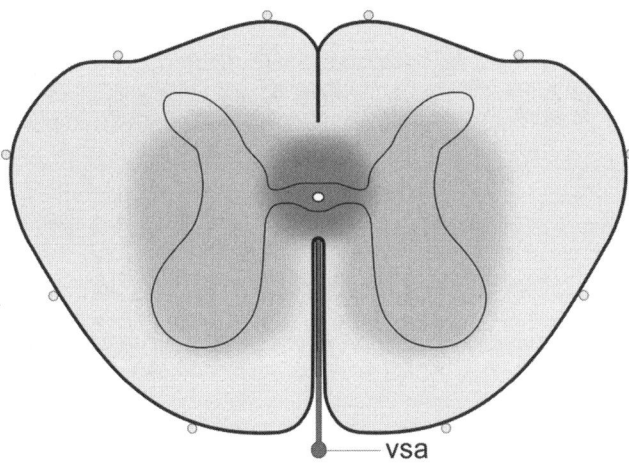

FIGURE 8 Blood supply to the spinal cord. The innermost zone (dark orange) is fed by branches arising from the ventral spinal artery (vsa), which enters the gray matter from the ventral fissure. The outer zone (yellow) depends on blood supplied from vessels penetrating the nervous tissue from the outer surface. The intermediate zone (pale orange) is supplied by both routes. (Courtesy of Enke Verlag in MVS Medizinverlage Stuttgart GmbH & Co. KG.) (*See insert for color representation of the figure.*)

vessel runs rostrally and gives off the caudal cerebellar and the labyrinthine arteries. The basilar artery splits at the level of the mesencephalon. Its branches, the caudal communicating arteries, emit the rostral cerebellar and the caudal cerebral arteries. The circulus arteriosus cerebri is completed by the rostral cerebral arteries, which are connected through the rostral communicating artery. On each side of the optic chiasm, the internal carotid arteries are connected to the circulus arteriosus cerebri. The middle cerebral arteries arise from the rostral half of the cerebral arterial circle next to the carotid arteries. Despite these anastomoses to form an interconnected blood supply, angiograms of vertebral and carotid arteries, respectively, reveal that areas supplied by these vessels are distinct. Thus, the rostral and middle portions of the cerebral hemispheres are provided with blood from the internal carotids, while the caudal cerebral hemispheres, the cerebellum, and the brainstem are fed with blood from the vertebral supply.

FUNCTIONAL NEUROANATOMY

Modalities

The physicochemical process of impulse transmission in neurons is always the same regardless of the information being conveyed. However, as long as a chain of neurons can be tracked unequivocally to either a receptor or an effector organ, the nature of the information being transmitted may be inferred. The various types of information are called *modalities*. General afferent and efferent modalities are mediated

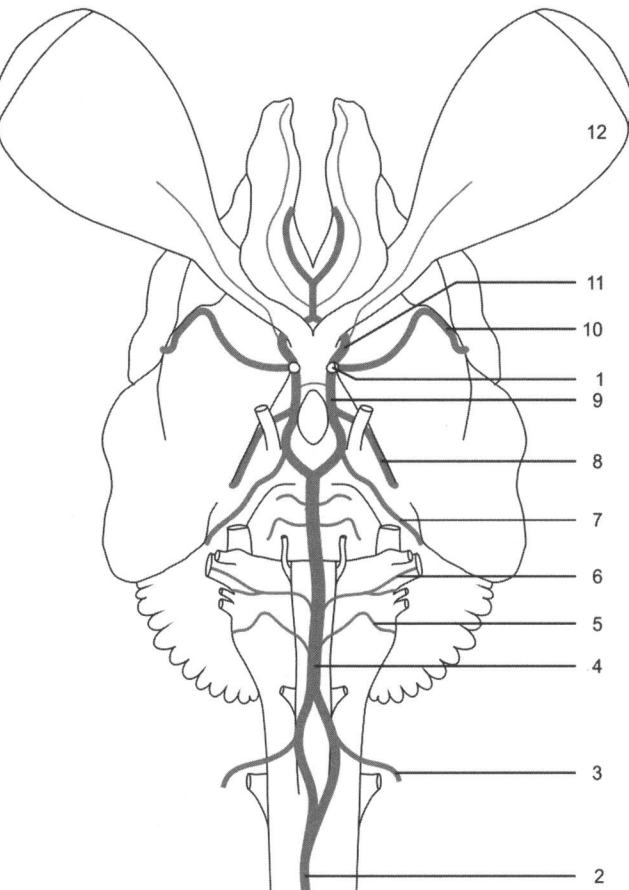

FIGURE 9 Blood supply to the mammalian brain. 1, internal carotid artery; 2, ventral spinal artery; 3, vertebral artery; 4, basilar artery; 5, caudal cerebellar artery; 6, labyrinthine artery; 7, rostral cerebellar artery; 8, caudal cerebral artery; 9, caudal communicating artery; 10, middle cerebral artery; 11, rostral cerebral artery; 12, left eye. (Courtesy of Enke Verlag in MVS Medizinverlage Stuttgart GmbH & Co. KG.) (*See insert for color representation of the figure.*)

through all the spinal nerves as well as through selected cranial nerves. Receptors and effector organs specific to the head are qualified as being "special." Consequently, special afferent and efferent modalities are conveyed by corresponding cranial nerves exclusively. A survey of the association between the various modalities and the gray matter in the brain and spinal cord is presented in Table 1.

General Modalities Information arising from receptors within the skin and the locomotor apparatus is described as general somatic afferent (GSA). According to their location, the receptors for general somatic afferents are classified into two groups. The skin harbors receptors that are sensitive to touch, pressure, vibration, temperature, and noxious stimuli and thus are configured to detect changes in the external environment (i.e., exteroception). Proprioception, in contrast, represents a survey of certain aspects of the internal

TABLE 1 Modalities Present in the Gray Matter in the Brain and Spinal Cord

	Special Somatic Afferent (SSA)	General Somatic Afferent (GSA)	Special Visceral Afferent (SVA)	General Visceral Afferent (GVA)	General Visceral Efferent (GVE)	Special Visceral Efferent (SVE)	Somatic Efferent (SE)
Information conveyed	Vision, hearing	Touch, temperature, pain	Olfaction, taste	Visceral sensory information	Secretion, heart rhythm, smooth muscle contraction	Branchiomotor activity	Somatomotor activity
Telencephalon	Area optica Area acustica	Area sensoria	Area olfactoria Amygdala Area sensoria	Systema limbicum	Systema limbicum	Cortex cerebri Area motoria	Cortex cerebri Area motoria Nucleus caudatus Putamen Capsula interna
Diencephalon	Corpus geniculatum laterale Corpus geniculatum mediale	Thalamus (nucleus ventralis caudalis thalami)	Thalamus (nucleus ventralis caudalis thalami)	Hypothalamus	Hypothalamus Hypophysis	Capsula interna	Capsula interna Globus pallidus Nucleus subthalamicus
Mesencephalon	Colliculus rostralis Colliculus caudalis	Lemniscus medialis Ascending reticular activating system (ARAS)		Formatio reticularis	Nucleus accessorius nervi III Formatio reticularis Tractus reticulospinalis		Nucleus ruber Substantia nigra
Metencephalon	Nuclei cochleares Nuclei vestibulares	Lemniscus medialis Ascending reticular activating system (ARAS)		Formatio reticularis	Nucleus parasympathicus nervi VII Formatio reticularis Tractus reticulospinalis	Nucleus motorius nervi V Nucleus motorius nervi VII	Nuclei pontis
Myelencephalon	Formatio reticularis	Nuclei gracilis et cuneatus Fibrae arcuatae internae Ascending reticular activating system (ARAS)	Nucleus tractus solitarii Tractus solitarius	Nucleus tractus solitarii	Nucleus parasympathicus nervi IX Nucleus parasympathicus nervi X Formatio reticularis Tractus reticulospinalis	Nucleus ambiguus nervi IX et X Nucleus motorius radicis cranialis nervi XI (caudal end of nucleus ambiguus)	Pyramides

Spinal cord		Fasciculi gracilis et cuneatus Tractus spinothalmicus Tractus spinocervicalis et cervicothalamicus Tractus spinoreticularis		Spinothalamic tract	Tractus reticulospinalis Cornu laterale T_1–L_4 Cornu laterale $S_{1/2}$–$S_{3/4}$	Tractus corticospinalis ventralis et lateralis
Afferent fibers/efferent fibers	Nervus opticus II Nervus vestibulocochlearis VIII	Nervi spinales Nervus trigeminus V Nervus facialis VII Nervus vagus X	Nervus olfactorius I Nervus trigeminus V Nervus facialis VII Nervus glossopharyngeus IX	Nervi sympathici et parasympathici Nervi spinales Nervus trigeminus V Nervus glossopharyngeus IX Nervus vagus X	Nervi spinales Nervi splanchnici Nervus vagus	Nervus facialis VII Nervus glossopharyngeus IX Nervus vagus X Nervus oculomotorius III Nervus trochlearis IV Nervus abducens VI Nervus accessorius XI Nervus hypoglossus XII
Receptors/effectors	Rod and cone cells of retina Hair cells of spiral organ	Mechanoreceptors Temperature and pain receptors	Olfactory neuroepithelium Taste buds	Visceral mechanoreceptors, chemoreceptors, and nociceptors	Glands Heart muscle Smooth muscle cells	Facial muscles Muscles of mastication Larynge muscle Muscle stapedius Muscle tensor tympani

environment, employing receptors such as muscle spindles, Golgi tendon organs, and joint capsule receptors in the locomotor system. Consequently, proprioception is concerned in large part with processing detailed data on muscle length and tension as well as on joint angulation; such information is mainly processed unconsciously within the cerebellum. However, proprioception also provides data used for conscious sense of movement and of the position of the limbs, which is integrated in the cerebral cortex.

General visceral afferents (GVAs) arise from mechanoreceptors, chemoreceptors, and nociceptors present in hollow organs as well as in blood vessels. Afferent fibers course centrally over autonomic nerves, spinal nerves, and cranial nerves.

The locomotor system is the only existing somatic effector organ. It therefore implements the general somatic efferents. Because there are no somatic effector organs specific to the head, the category "special" is not applicable to somatic efferents. Thus, with respect to somatic efferents (SEs), the adjective "general" is usually omitted.

General visceral efferents (GVEs) provide connections with heart and smooth muscle as well as exocrine and endocrine glands all over the body. The neuronal pathways of general visceral efferents are known as the *sympathetic* and *parasympathetic systems*. As the hypothalamus is an important integrating center for the visceral nervous system, neuroendocrine regulation also may justifiably be considered as an efferent limb of the visceral nervous system.

Special Modalities Modalities unique to the head are called *special*. Consequently, both afferent and efferent special modalities rely on cranial nerves. Special somatic afferents (SSAs) involve circumscribed organs located below the body surface and yet responsive to external stimuli. These include the exteroceptive input from the eye (optic nerve II) and ear (vestibulocochlear nerve VIII) as well as the proprioceptive input from the vestibular apparatus (vestibulocochlear nerve VIII). Similar to the general proprioceptive input, the cerebellum is concerned with unconscious processing of vestibular input. Particular chemoreceptors of the head provide the special visceral afferents (SVAs) of smell (olfactory nerve I) and taste (facial nerve VII, glossopharyngeal nerve IX, vagus nerve X).

The head is endowed not only with unique receptors but also with particular visceral effector organs. These are the striated muscles derived from the pharyngeal arches. Although histologically indistinguishable from skeletal muscles, their particular embryological origin and associated innervation patterns provide a rational basis for considering them as a separate entity. Special visceral efferents (SVEs) course through the pharyngeal nerves. These include the trigeminal nerve V, the facial nerve VII, the glossopharyngeal nerve IX, the vagus nerve X, and the internal branch of the accessory nerve XI.

CONSCIOUS AFFERENTS

Exteroception: Touch, Temperature, and Pain (GSA)

Exteroception is mediated by distinct receptors for pressure, touch, vibration, temperature, and pain. Information is conveyed to the spinal cord through pseudounipolar neurons in the spinal nerves and to the brain through cranial nerves V, VII, and X. Several major ascending pathways are involved in conveying conscious perception of these modalities. The lemniscal system is concerned with accurate perception of touch, pressure, vibration, and pinprick pain with a high degree of spatial discrimination (discriminative touch). A more global and coarse, arousing, and potentially irritating awareness of touch, pressure, temperature, itching, and pain is imparted by the spinothalamic, spinocervicothalamic, and spinoreticulothalamic systems.

Connections of the lemniscal system are unambiguous (Figure 10). The central processes of primary afferent neurons reach the spinal cord through the dorsal roots. After entering the white mater, they send collaterals to adjacent segments via the dorsolateral tract and emit collaterals that ascend within the dorsal fasciculi. Fibers belonging to coccygeal, sacral, lumbar, and thoracic spinal cord segments form the ipsilateral gracile fascicle, which is adjacent to the median septum. Fibers associated with cranial thoracic and cervical segments constitute the cuneate fascicle, which is located between the gracile fascicle and the dorsal column of the spinal gray matter. Information is relayed to second-order neurons located in the like-named gracile and cuneate nuclei of the myelencephalon. Corresponding axons decussate immediately after leaving the nuclei and run to the contralateral thalamus (nucleus ventralis caudalis) in a tract known as the *medial lemniscus*. Perikarya of pseudounipolar neurons in cranial nerves are located in the trigeminal (V), geniculate (VII), and proximal vagal (X) ganglia. Central processes of these primary afferent neurons continue to the pontine sensory nucleus of the trigeminal nerve, where they synapse with second-order neurons. Their axons decussate and travel along with the medial lemniscus. In the contralateral caudoventral thalamic nucleus, GSA information is transmitted to third-order neurons, which project to the ipsilateral somatosensory cortex. Thus, the pontine sensory nucleus of the trigeminal nerve is functionally homologous to the cuneate and gracile nuclei.

Transmission of coarse touch, temperature, itch, and deep pain is much more involved and subject to some interspecies variation (Figure 11). Primary sensory neurons in spinal nerves synapse with projection neurons in the gray matter of the corresponding and adjacent spinal cord segments (Rexed layers III and IV). These second-order neurons decussate immediately and form the contralateral spinothalamic tract in the lateral funiculus of the spinal cord. After another synapse in the caudoventral thalamic nucleus,

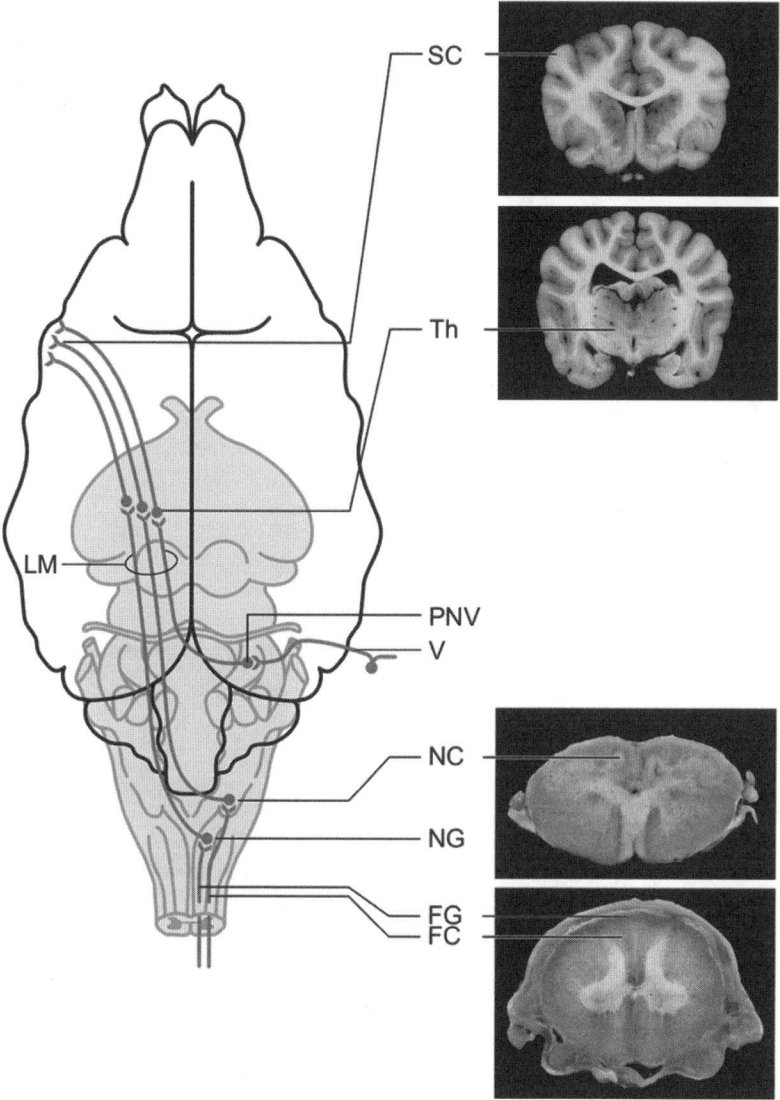

FIGURE 10 CNS pathways mediating accurate exteroception. SC, somatosensory cortex; Th, thalamus; LM, lemniscus medialis; PNV, pontine sensory nucleus of trigeminal nerve V; V, trigeminal nerve V; NC, nucleus cuneatus; NG, nucleus gracilis; FG, fasciculus gracilis; FC, fasciculus cuneatus.

third-order neurons again project to the somatosensory cortex of the corresponding hemisphere. As for the spinocervicothalamic tract, second-order neurons (spinocervical tract) ascend ipsilaterally to the lateral cervical nucleus in the first two cervical segments. Third-order neurons decussate immediately and follow the medial lemniscus to the thalamus (cervicothalamic tract). The primary sensory neurons conveying coarse exteroception through cranial nerves V, VII, and X project to the second-order neurons in the spinal sensory nucleus of the trigeminal nerve, which again relay data to the thalamus. As opposed to the pathways for accurate exteroception, somatotopic organization is largely lost along the spinothalamic connections.

The sensation of pain is a cerebral interpretation of afferent information originating from various receptors. Stimuli evoking a pain signal are indicative of potential or real damage to tissue. The spinothalamic tract in the lateral funiculus is considered to be the major conscious pain pathway in mammals. In dogs and cats, however, the spinocervicothalamic pathway in the lateral funiculus is regarded as being at least as important. The spinoreticulothalamic tract is a bilateral connection that is activated by collaterals from all sensory systems and reaches the thalamus through a polysynaptic pathway. This system is also known as the *ascending reticular activating system* (ARAS), as it is involved in regulating the degree of alertness in general.

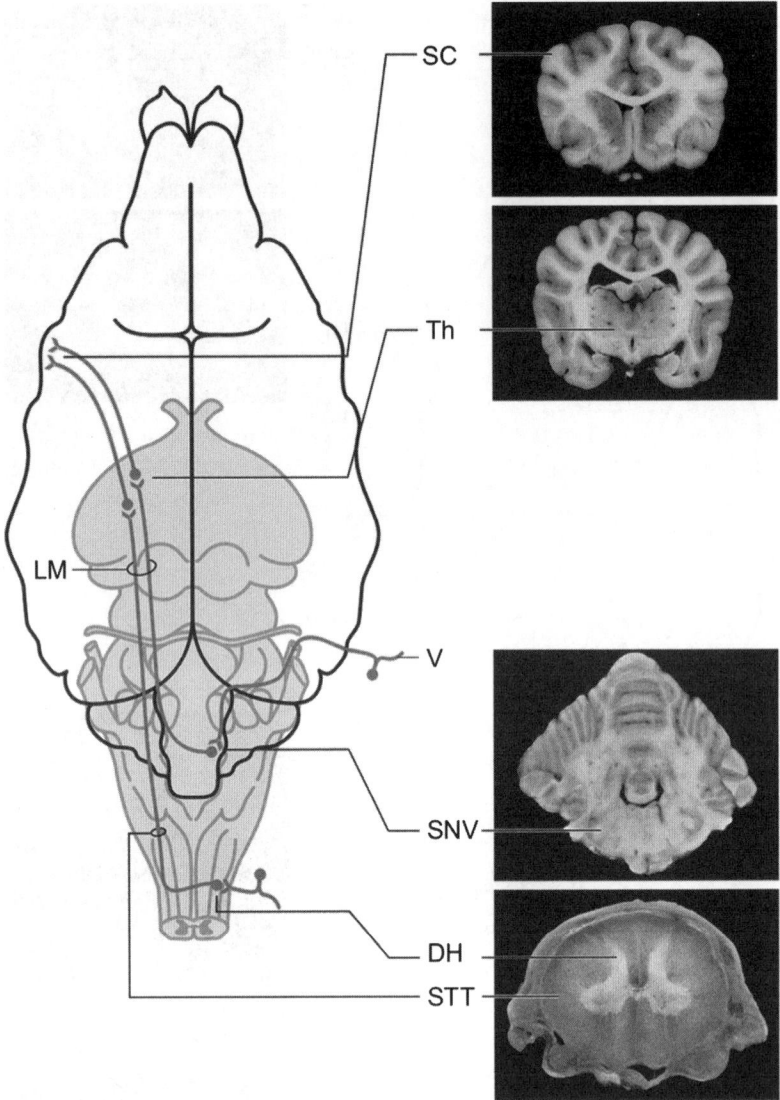

FIGURE 11 CNS pathways mediating coarse exteroception. SC, primary somatosensory cortex; Th, thalamus; LM, lemniscus medialis; V, trigeminal nerve V; SNV, spinal sensory nucleus of trigeminal nerve V; DH, dorsal horn; STT spinothalamic tract.

In summary, all the pathways for conscious exteroception decussate and thus end contralaterally to the receptors.

Proprioception (GSA)

Proprioception refers to all afferents relating to posture and motion (i.e., all information reporting on the relative positions and movements of different body parts). Receptors for proprioception include muscle spindles, Golgi tendon organs and joint capsule receptors as well as the vestibular system. Most proprioceptive data are processed by the cerebellum and thus are acted on at the subconscious level.

A minor proportion of the proprioceptive input, however, is dealt with at the conscious level. Corresponding afferents arise from receptors in the locomotor system as well as from

the vestibular apparatus. Conduction of conscious perception of posture and movement is intimately linked to the exteroceptive pathways. Conscious proprioception originating from the cervical region and the forelimb is transmitted through the cuneate fascicle, with fibers ending in the ventral part of the medial cuneate nucleus (nucleus X). Neurites of second-order neurons decussate in the myelencephalon and follow the medial lemniscus farther to reach the thalamus, from where information is projected to the somatosensory cortex. Similarly, conscious proprioception from the caudal body is transmitted by collaterals in the gracile fascicle and the dorsal spinocerebellar tract (spinomedullary tract, located in the lateral funiculus). They synapse in the rostral part of the gracile nucleus (nucleus Z), after which information follows the lemniscal pathway described above.

In summary, the conscious proprioceptive pathway ends contralaterally in the somotosensory cortex. As for the conscious perception of vestibular input, it is conveyed through the vestibular nuclei and the thalamus, which in turn projects bilaterally to the vestibular cortex. Regions involved in conscious perception of vestibular input include parietal, parietoinsular, and temporal areas of the cerebral cortex. These regions are considered to serve multiple sensory modalities.

Smell (SVA)

The olfactory nerve is composed of all neurites arising from the central processes of the bipolar neurons located in the olfactory neuroepithelium. Olfactory neuron axons pass through the cribriform plate and synapse with mitral cells and brush cells in the olfactory bulb. These second-order neurons send their neurites through the olfactory tract, which splits into medial and lateral divisions. The size of the tract varies with the importance of smell to the animal in question; thus, rodents and dogs have large olfactory tracts, while their counterparts in chickens and primates (including humans) are relatively small. The lateral olfactory tract conveys the information to the ipsilateral piriform lobe for conscious perception of smell. Thus, olfaction is linked directly to the primary olfactory cortex and is not relayed through the thalamus. The olfactory system is also linked to the hypothalamus, the reticular formation, and the parasympathetic nuclei of cranial nerves VII, IX, and X. Furthermore, olfaction is closely associated with the limbic system. The medial olfactory tract joins the septal area (area subcallosa and nuclei septi) and some fibers of the lateral olfactory tract enter the amygdala. From there, axons project to nuclei of the limbic system. Olfactory bulbs and piriform cortices are coupled through the rostral commissure.

Taste (SVA)

Gustatory input is conveyed through pseudounipolar neurons whose cell bodies are located in the geniculate ganglion of facial nerve VII (rostral two-thirds of the tongue) and the distal ganglia of the glosspharyngeal nerve IX and vagus nerve X (caudal third of the tongue and pharynx). Central processes of these pseudounipolar neurons project to the rostral (gustatory) part of the nucleus of the solitary tract. The gustatory pathway is relayed in the caudomedial nucleus of the thalamus and eventually reaches the somesthetic cortex. Besides sensory cells in taste buds, mechanoreceptors in the oral mucosa also contribute to taste perception.

Vision (SSA)

The extent of fiber decussation in the optic chiasm depends on the species and correlates with positioning of the eyes and the resulting capability for binocular vision. The more frontal the eyes are positioned, the fewer axons will cross. Approximately 50% of fibers cross in primates and 75% in the dog, while 100% decussate in nonmammalian vertebrates. Axons that cross arise from the nasal segment of the retina and thus convey information originating from the lateral part of the visual field. Consequently, every optic tract contains axons from the medial retina of the contralateral eye and axons from the lateral retina of the ipsilateral eye (Figure 12).

Conscious visual perception requires approximately four-fifths of the fibers (Figure 12), the remaining ones being involved in reflex control. The optic tracts begin at the optic chiasm and surround the thalamus. Fibers mediating conscious perception end in the lateral geniculate body, which is the primary processing center for visual information from the retina. Perikarya in this specialized thalamic nucleus display a lamination consistent with their retinotopic organization. The neurons in the geniculate nucleus project through the caudal part of the internal capsule to the primary optic cortex in the occipital lobe (marginal, ectomarginal, occipital, and splenial gyri). The visual cortex, in turn, is connected to the motor cortex, the cerebellum, the tectum (rostral colliculi), the nuclei of the motor nerves that control eye movements (III, IV, VI), and the tegmentum. These associations provide the basis for behavioral responses to visual input.

Subconscious visual pathways largely bypass the geniculate nucleus and terminate in the rostral colliculus (eye movements), the pretectal area (pupillary reflexes and accommodation), the hypothalamus (circadian rhythm), the vestibular nuclei (optokinetic nystagmus), and the reticular formation (arousal effect). Fibers running in the medial part of the optic tract project directly to the pretectal area. From there, bilateral connections to both parasympathetic nuclei of the oculomotor nerve III as well as to the contralateral pretectal area provide the basis for direct and consensual pupillary reactions. For eye accommodation, however, input to the preoptic area is relayed through the lateral geniculate body to the visual cortex. Neurons in the preoptic area then project to the nuclei of the oculomotor nerve III (i.e., the parasympathetic nucleus controlling the sphincter muscle of the pupil and the ciliary muscle) and to the motor nucleus for the medial rectus muscle.

Hearing (SSA)

Bipolar neurons in the spiral ganglion of the inner ear convey acoustic information to the cochlear nuclei through the cochlear division of the vestibulocochlear nerve VIII. Second-order neurons project to the ipsilateral trapezoid body and dorsal olivary nucleus as well as bilaterally to both caudal colliculi through the lateral lemniscus. Axons from neurons in the brain stem nuclei also ascend to the tectum and, in part, to the medial geniculate nucleus. From the

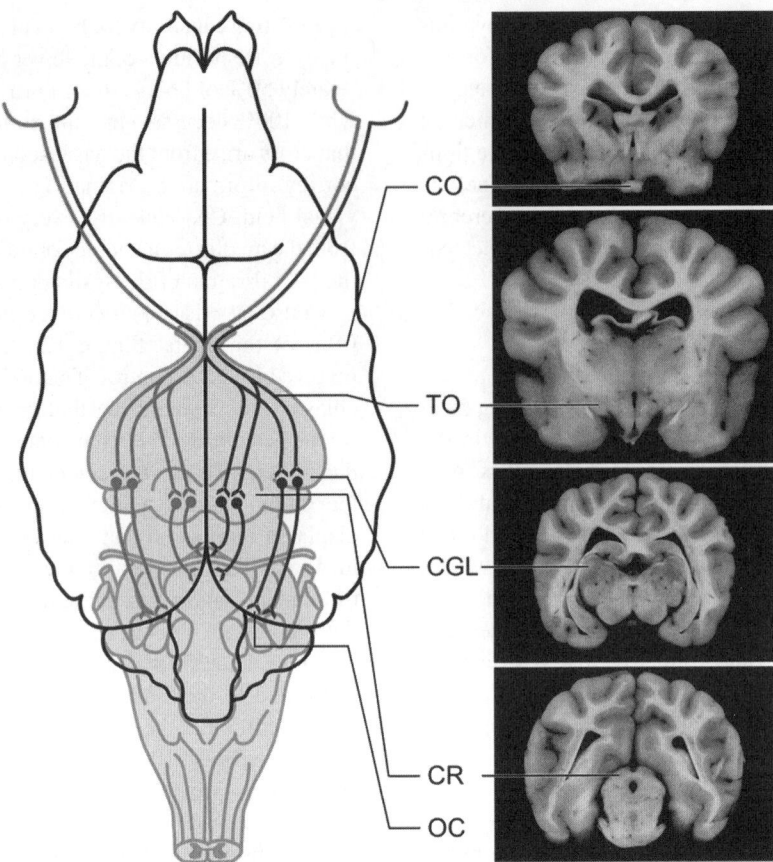

FIGURE 12   CNS pathways mediating vision. CO, Chiasma opticum; TO, tractus opticus; CGL, corpus geniculatum laterale; CR, colliculus rostralis; OC, pimary optic cortex.

caudal colliculi, information is forwarded to the medial geniculate nucleus, from whence it is carried to the primary auditory cortex in the temporal lobe (sylvian and ectosylvian gyri) through the acoustic radiation (Figure 13). As is the case for the other sensory pathways, a tonotopic organization is observed. High-pitched sounds are mapped rostrally, while low frequencies are distributed caudally. Several commissural tracts interconnect the olivary nuclei, the lateral lemnisci, and the caudal colliculi. Notwithstanding its bilateral projection, auditory input is represented predominantly in the contralateral cortex. Besides conscious perception, a reflex pathway originates from the dorsal olivary nucleus. It is relayed in the motor nucleus of the facial nerve VII and mediates a reflexive contraction of the stapedius muscle in response to an excessive acoustic input.

MOTOR SYSTEM (SE AND SVE)

The motor system is involved with voluntary and involuntary control of movement. Somatomotor responses to stimuli are mediated through striated musculature. It is common practice to subdivide the somatic efferent pathway into two hierar-

chical sections. The motor neurons connecting the central nervous system to the striated muscles to be innervated are called *lower motor neurons*. Neurites of lower motor neurons reach the motor unit without interruption and travel in myelinated peripheral nerves. Thus, the lower motor neuron includes the alpha motor neurons (skeletomotor neurons), innervating extrafusal muscle fibers, as well as the gamma motor neurons (fusimotor neurons), which adjust contraction of intrafusal muscle fibers. The cell bodies of lower motor neurons are located in the ventral gray column of the spinal cord gray matter (including those in the first five or six cervical segments from which the spinal root for the external branch of spinal accessory nerve XI arises) as well as in the motor nuclei of cranial nerves III, IV, VI and XII and pharyngeal nerves V, VII, IX, and X. The term *upper motor neuron* lumps together all neurons of the CNS that control the activity of the lower motor neurons. To become manifest, the activity of upper motor neurons is dependent on the compliance of the lower motor neurons; functional integrity of upper motor neurons may not be examined on its own but only in combination with the lower motor neurons. In contrast, extensor and flexor reflexes provide a means to assess the lower motor neuron in a fairly detached way.

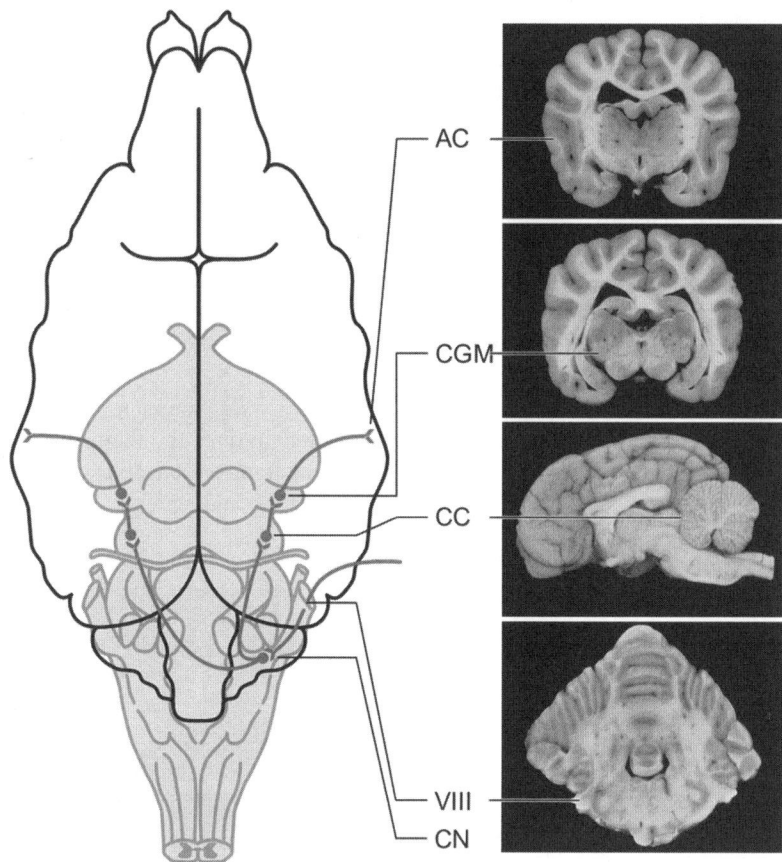

FIGURE 13 CNS pathways mediating hearing. AC, Primary auditory cortex; CGM, medial geniculate body; CC, caudal colliculus; VIII, vestibulocochlear nerve (cranial nerve VIII); CN cochlear nuclei.

Lower Motor Neuron

Although usually directed by upper motor neurons, lower motor neurons also act as the efferent limb of unconscious reflex arcs. Data processing in the central nervous system is minimal for reflex activities, so the afferent and efferent limbs of reflex arcs are closely linked in space. Pseudounipolar sensory neurons may synapse directly with lower motor neurons, as is the case for the extensor reflex ("monosynaptic reflex"). However, most reflexes (e.g., flexor or eye reflexes) rely on sophisticated neuronal pathways involving several interneurons. The neurons in reflex arcs typically project to interneurons that will carry knowledge of the in-progress reflex action toward the brain.

Extensor Reflex

Extensor reflexes present the simplest layout of a stimulus–response apparatus. Muscle spindles and Golgi tendon organs provide the proprioceptive receptors that collect data on muscle length and tension, respectively. Myelinated axons in peripheral nerves convey afferent information to the spinal cord, including segments at and immediately adjacent to the nerve's entry portal. After entering the spinal cord through the dorsal roots, central processes of pseudounipolar neurons may synapse directly with lower motor neurons (i.e., called a monosynaptic reflex) to yield the most rapid response. This is the only instance in which lower motor neurons are fed directly with sensory input. Most often, however, a number of interneurons will be intercalated between sensory and motor neurons and thus provide for some degree of higher-level integration and control. This reflex aims at keeping both muscle length and tension constant (i.e., it is myostatic). Reaction is fast, stereotypical, and resistant to fatigue. It is primarily involved in the control of muscles working against the effects of gravity, hence the name *extensor reflex*. Consequently, this reflex arc plays a pivotal role during the stance phase of locomotion.

Flexor Reflex

Flexor reflexes represent an answer to noxious stimuli. They originate from nociceptors and typically result in a flexion of the entire limb. Exteroceptive information is conveyed to the spinal cord through small myelinated and nonmyelinated axons in peripheral nerves. Upon entering the dorsolateral funiculus, central processes bifurcate to distribute the afferent information to several segments. Interneurons then extend through the spinal cord and also cross to the contralateral side (i.e., a polysynaptic reflex arc). With respect to the stimulated limb, this results in excitatory input to lower motor neurons controlling flexor muscles and inhibitory input to their extensor antagonists. On principle, withdrawal of the stimulated limb is associated with extension of the contralateral limb, although this effect is suppressed in adults by upper motor neurons unless spinal tracts are damaged. Thus, flexor reflex circuitry provides the basis for the swing phase of locomotion. Furthermore, activity of interneurons also produces upholding of limb flexion beyond the duration of the stimulus. As compared with extensor reflexes, withdrawal reflexes are much more complicated and rely on complex interconnections, ensuring adequate integration and modulation by higher neural centers. As a result, the flexor response is comparatively slow, variable, and subject to fatigue.

Upper Motor Neuron

Pyramidal System   The term *pyramidal system* refers to a connection between the neocortex and the lower motor neurons (Figure 14). It provides voluntary control of muscles, specifically by mediating contractions of extrafusal fibers via alpha motor neurons, maintaining muscle spindle sensitivity via gamma motor neurons, and inhibiting spinal reflexes to prevent their interference with the cerebral control of movement and posture. It is the basis for performing skilled movements. The connection starts from the

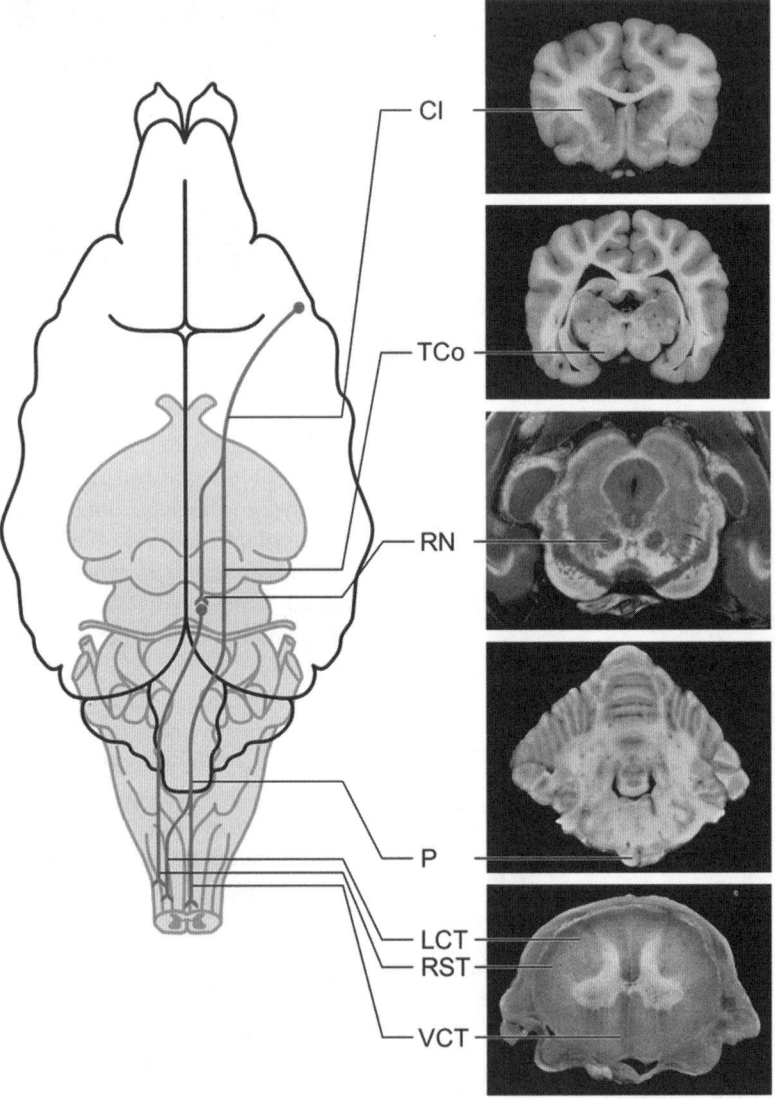

FIGURE 14   Pyramidal system. CI, Capsula interna; Tco, tractus corticospinalis; RN, red nucleus; P, pyramid; LCT, lateral corticospinal tract; RST, rubrospinal tract; VCT, ventral corticospinal tract.

primary motor cortex in the frontal lobe and includes the corticospinal tract as well as the corticonuclear tract (corticobulbar tract).

In descriptive terms, the corticospinal tract may be further subdivided. In the prosencephalon, it appears as an important part of the internal capsule and in the mesencephalon as the crura cerebri. After passing the pons, it may be sighted as the pyramid at the ventral margin of the myelencephalon. Fibers decussate either immediately caudal to the pyramid or close to the target spinal cord segment. Neurites remaining on the ipsilateral side emerge next to the ventral fissure as the ventral corticospinal tract (up to the thoracic region). On the other hand, fibers crossing within the myelencephalon, form the lateral corticospinal tract, which runs lateral to the dorsal column of the gray matter. About half the fibers end in the cervical segments, while the remaining ones descend to synapse in the thoracic and lumbosacral gray matter. Collaterals relayed via the pontine nuclei provide the cerebrocerebellar cortex with information on pyramidal motor activity. The corticonuclear tract serves a comparable function within the brain by connecting the motor cortex to the motor nuclei of contralateral cranial nerves III, IV, VI, and XII. Both the corticospinal and the corticonuclear tracts show a somatotopic organization. Their axons synapse with lower motor neurons directly or, as a rule, indirectly through one or a very few local interneurons.

Collateral projections of the corticospinal tract also link the motor cortex to the red nucleus in the mesencephalon. Rubrospinal fibers decussate, display somatotopic organization, and project upon lower motor neurons in a manner very similar to that of the corticospinal tract itself. This corticorubrospinal pathway plays a pivotal role in voluntary movement in animals. Although often considered to be a part of the extrapyramidal system, the involvement of this tract in directed motor activity and its basic similarity to the corticospinal tract make a good case for including the pathway in the pyramidal system. As motor activity in birds is far less influenced by the brain, a direct connection from the thin neocortex to lower motor neurons is not present.

Extrapyramidal System   The extrapyramidal system is involved in the control of nonreflexive but stereotyped movement patterns. Therefore, it governs the repertoire of constitutive motion sequences. Although involuntary, this type of motor activity is perceived consciously.

The extrapyramidal system includes several polysynaptic pathways eventually affecting the activity of the lower motor neurons (Figure 15). Neurons involved are located in a

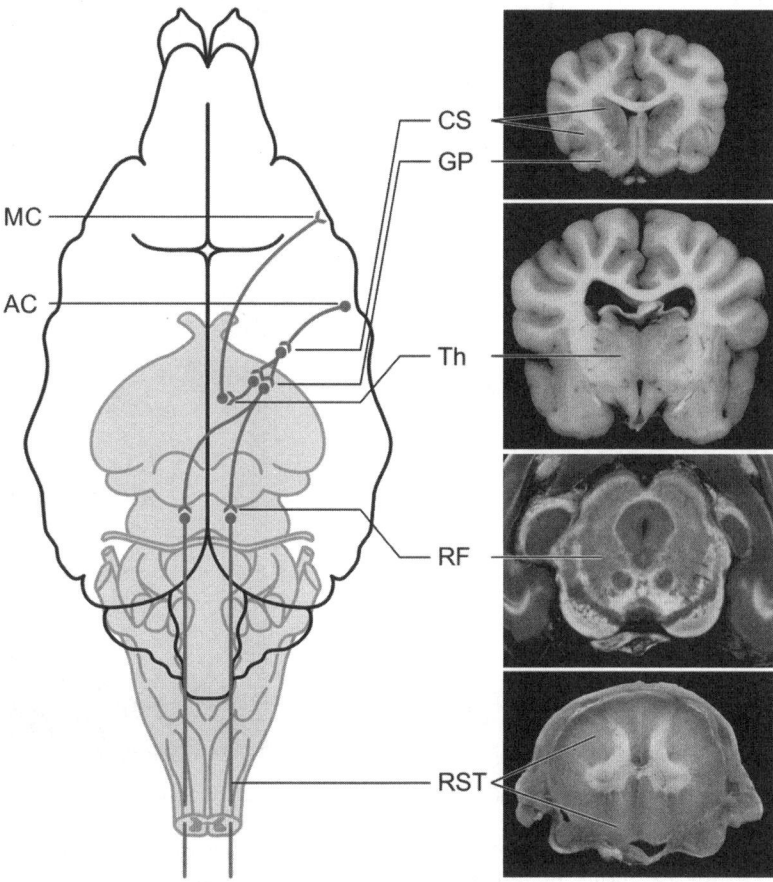

FIGURE 15   Extrapyramidal system. CS, Corpus striatum; GP, globus pallidus; MC, primary motor cortex; AC association cortex; Th, thalamus; RF, reticular formation; RST, reticulospinal tracts.

number of brainstem nuclei. Information transfer through the extrapyramidal system may originate from many cortex areas, especially the association cortex in the frontal and parietal lobes. From there, information is conveyed to the corpus striatum (caudate nucleus and putamen), which exerts an inhibiting effect on the pallidum (globus pallidus). Lesions of the corpus striatum result in hyperkinetic motor disorders. From the pallidum, the pathway then divides into two branches. The ansa lenticularis forwards the information to the thalamus (nucleus subthalamicus and several ventral nuclei). Interestingly, the thalamus projects back to the primary motor cortex and thus the extrapyramidal system establishes a feedback connection to the pyramidal pathway. The second branch arising from the pallidum ends bilaterally in the reticular formation, from where the reticulospinal tracts descend to the lower motor neurons. The extrapyramidal pathways are also linked to the cerebrocerebellar cortex via the olive nuclei.

At every relay, extrapyramidal motor pathways may be modulated by several feedback loops. Thus, a connection exists between the GABA-ergic neurons in the striatum and the dopaminergic neurons in the substantia nigra of the mesencephalon. Similar reciprocal connections link the pallidum to the subthalamic nucleus.

Cerebellar System

The cerebellum integrates a broad range of sensory input in order to regulate balance, coordinate and control motor activity, and adjust movements (see Figures 16 and 17). From

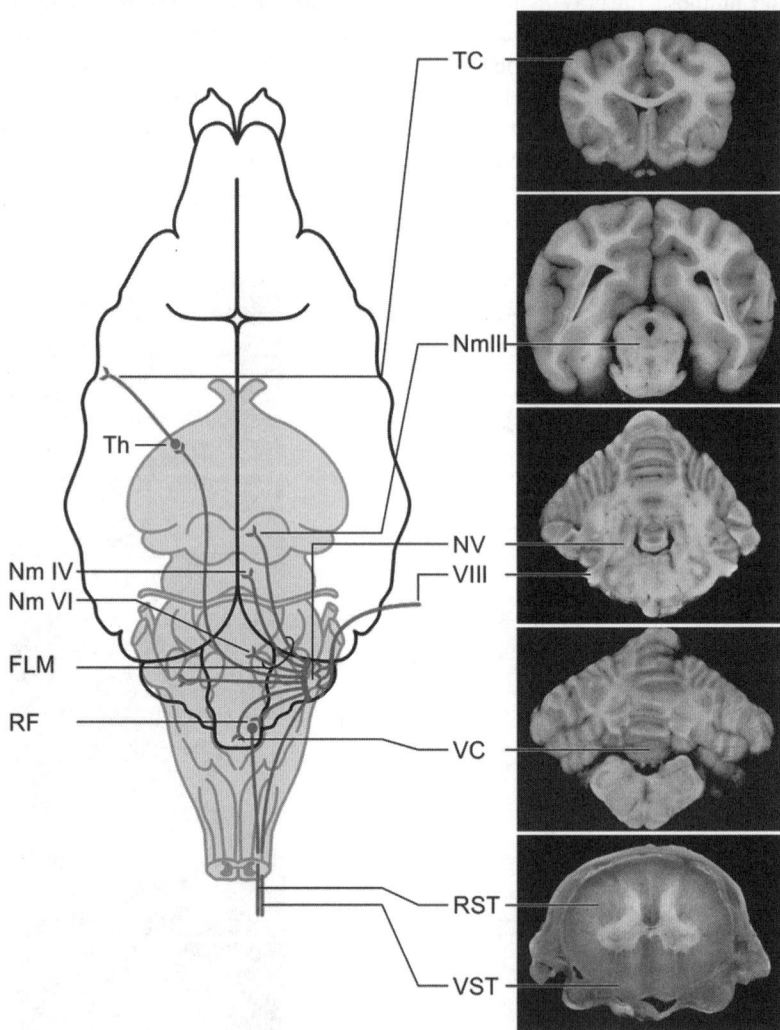

FIGURE 16 Vestibular pathways. Th, Thalamus; Nm IV, nucleus motorius nervi trochlearis IV; Nm VI, nucleus motorius nervi abducentis VI; FLM, fasciculus longitudinalis medialis; TC, temporal cortex; Nm III, nucleus motorius nervi oculomotori III; NV, nuclei vestibulares; VIII, vestibulocochlear nervi VIII; RF, reticular formation; VC, vestibulocerebellar cortex; RST, reticulospinal tract; VST, vestibulospinal tract.

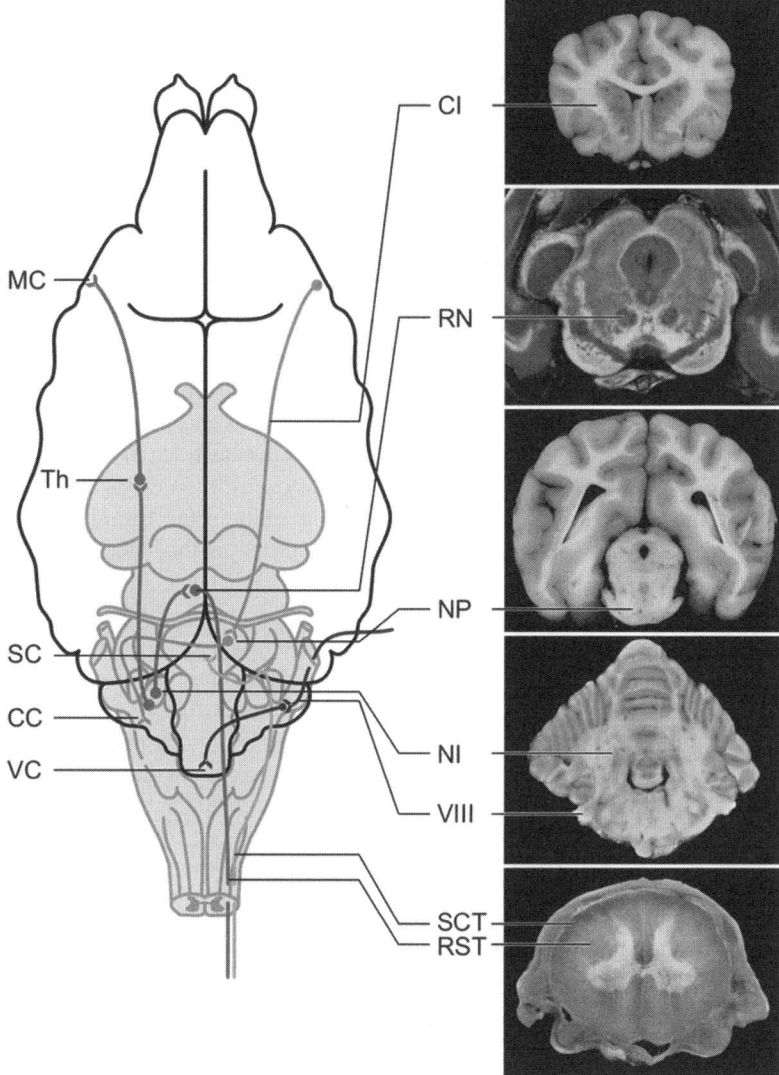

FIGURE 17   Cerebellar pathways. CI, Capsula interna; MC, primary motor cortex; RN, red nucleus; Th, thalamus; NP, nuclei pontis; SC, spinocerebellar cortex; NI, nuclei interpositi; VIII, vestibulocochlear nerve VIII; CC, cerebrospinal cortex; VC, vestibulospinal cortex; SCT, spinocerebellar tract; RST, rubrospinal tract.

a functional perspective, the cerebellum is best subdivided into three parts based on the main input channels. Thus, the vestibulocerebellum deals with vestibular data, the spinocerebellum is involved in processing proprioceptive (and exteroceptive) information, and the cerebrocerebellum (s. pontocerebellum) receives input from the forebrain.

AFFERENTS

Vestibular Pathways (SSA) and Vestibulocerebellum

In conjunction with the cerebellum, the vestibular system (Figure 16) is in charge of maintaining equilibrium and of harmonizing movements of the head, trunk, and limbs. It also

ensures coordinated eye movements. The vestibular apparatus has two main components: the maculae staticae, which collect data about linear acceleration (gravity and movement), and the cristae ampullares, which gather data on angular acceleration. Hair cells in these two structures signal to bipolar neurons. The corresponding perikarya are located within the vestibular ganglion and the axons form the vestibular division of the vestibulocochlear nerve VIII, which enters the brain stem caudal to the pons at the cerebellomedullary angle. Most axons synapse in the vestibular nuclei, but a few run through the caudal peduncle and synapse in the fastigial nucleus (direct vestibulocerebellar tract).

The four vestibular nuclei on either side (rostral, medial, lateral, and caudal) exhibit a somatotopic organization. They extend throughout the rhombencephalon. Besides vestibular input, vestibular nuclei also receive optic and proprioceptive afferents. These collateral connections provide the basis for the pivotal role the vestibular nuclei play in maintaining equilibrium.

Output from the vestibular nuclei is manifold:

1. Second-order neurons send axons (fasciculus uncinatus) to the vestibulocerebellum (lobus flocculonodularis) and, in parallel, to the nucleus fastigii.

2. The ipsilateral vestibulospinal tract connects the lateral vestibular nuclei to the ventral gray column of the spinal cord. This tract stimulates lower motor neurons to extensor muscles and inhibits those innervating flexor muscles; some interneurons also decussate to inhibit contralateral lower motor neurons to extensor muscles. This results in ipsilateral facilitation and contralateral inhibition of extensor reflexes and allows motor activity of the limbs to be coordinated with head movements.

3. Ascending fibers in the medial longitudinal fascicle reach the motor nuclei of cranial nerves III, IV, and VI and modulate the activity of the external muscles of the eye. This allows coordinated adjustment of the eye and head movements (nystagmus).

4. Some fibers end in the reticular formation, from where output is forwarded to lower motor neurons through the reticulospinal tract. Vestibuloreticular connections include fibers to the emetic center.

5. Input to vestibular nuclei on one side is conveyed to contralateral nuclei through commissural fibers. A modest proportion of vestibular information also reaches the conscious level.

Subconscious Exteroception, Proprioception, and Spinocerebellum

Exteroception refers to the perception of external sensations, including touch, temperature, pressure, vibration, and itch sensitivity, whereas *proprioception* relates to the relative position and movement of different body parts (especially the limbs). The rule that information is processed on conscious and subconscious levels simultaneously also applies to these senses. Both exteroceptive and proprioceptive data are conveyed to the thalamus and to the somatosensory cortex, thus reaching conscious perception. The subconscious components of proprioceptive and exteroceptive afferents are processed in the cerebellum (Figure 17). Thus, exteroceptive input is not only dealt with consciously in the forebrain but also without conscious involvement in the cerebellum. This is particularly obvious in withdrawal reflexes but also applies to spinocerebellar pathways. The concomitance of exteroceptive and proprioceptive components should be kept in mind whenever considering spinocerebellar tracts.

Just like exteroception, proprioception encompasses precise and more global aspects. Precise proprioception originates primarily from muscle spindles. A more global proprioceptive input arises from Golgi tendon organs and joint capsule receptors and provides information on the overall position of whole limbs. Information from proprioceptors is forwarded by IA and IB fibers of pseudounipolar neurons. Several tracts are involved in conveying proprioceptive and exteroceptive data to the cerebellum for subconscious processing.

Interestingly, precise proprioception originating from the thoracic limb and from the neck ascends as the lateral part of the cuneate tract. It is formed by central processes of primary sensory neurons corresponding to segments C_1 to T_8 and ends in the lateral cuneate nucleus. Neurites from the neurons in the lateral cuneate nucleus project to the granule cell layer of the spinocerebellum through the superficial arcuate fibers which travel through the caudal cerebellar peduncle. This spinocuneocerebellar tract remains ipsilateral. Precise proprioception from the pelvic limb and trunk follows a different pattern: Central processes of pseudounipolar neurons synapse in initial and adjacent segments at the medial basis of the dorsal column (lamina VII). This is the thoracic nucleus (Clarke's nucleus), which extends from C_8 to L_4. Central processes of spinal nerves caudal to L_4 ascend in the gracile fascicle to reach the thoracic nucleus of the cranial lumbar segments. Axons from those second-order neurons that shape the thoracic nucleus form the dorsal spinocerebellar tract, which ascends ipsilaterally in the dorsolateral part of the lateral fascicle. It continues on the surface of the myelencephalon and contributes to the formation of the caudal cerebellar peduncle. Besides these spinocerebellar connections, central processes of pseudounipolar sensory neurons conveying input from muscle spindles also synapse with lower motor neurons in the ventral column to provide the circuit for the extensor reflexes.

Global proprioceptive information enters the spinal cord via central processes of spinal nerves, all of which synapse at the lateral base of the dorsal column. Neurites from second-order neurons in segments caudal to the cervical intumescence decussate and ascend contralaterally as the ventral spinocerebellar tract along the ventrolateral margin of the lateral funiculus. The ventral spinocerebellar tract passes through the rhombencephalon before turning back within the mesencephalon and joining the rostral cerebellar peduncle. Its mossy fibers cross the median plane again. The ventral spinocerebellar pathway thus decussates twice and ends ipsilaterally as well. Corresponding pseudounipolar neurons conveying global proprioception from the thoracic limb also synapse in initial and adjacent segments. However, neurites of second-order neurons form the cranial spinocerebellar tract, which ascends ipsilaterally medial and ventral to the ventral spinocerebellar tract. In summary, unconscious proprioception is processed ipsilaterally in the cerebellum.

Perikarya of first-order neurons conveying proprioceptive input from the head muscles via cranial nerves are located in the nucleus of the mesencephalic tract of the trigeminal nerve V. Thus, perikarya of the primary afferent neurons conveying proprioception from the muscles of mastication, the facial and extraocular muscles, are localized to the brain stem. This is an exception to the rule that cell bodies of peripheral afferent nerve cells are to be found in peripheral ganglia. Information is passed on to the pontine sensory nucleus of the trigeminal nerve and through second-order neurons to corresponding nuclei of cranial nerves and to the cerebellum.

Collateral Information from the Motor System (UMN) and Cerebrocerebellum

The third category of input to the cerebellum arises as collateral information from the pyramidal and extrapyramidal systems (Figure 17). Such signals are relayed primarily via the nuclei pontis and in the olive nuclei. Thus, collaterals of the corticospinal tract synapse in the pontine nuclei. Corresponding axons constitute the middle cerebellar peduncle (mossy fibers). With respect to the extrapyramidal system, axons from the pallidum, the red nucleus, and the reticular formation forward information on motor activity to the olive nucleus (tractus tegmenti centralis). Neurons in the olives send their neurites as climbing fibers through the caudal cerebellar peduncle to the dentate nucleus and to the cerebellar cortex of the lateral parts of the caudal lobe. Climbing fibers decussate and synapse directly with the dendrites of the Purkinje cells in the molecular layer (see below).

EFFERENTS

The cerebellum lacks any direct means to shape the activity of the lower motor neurons (Figure 17). Instead, its major function is to compare initiated muscle activities with intended movements, taking into account ongoing action and posture status, and then provide corrective feedback to the neurons that initiate motor activity. The cerebellum therefore receives proprioceptive information from both the vestibular system and the locomotor apparatus as well as complete information regarding the progress of intended motor activity. These data are processed in different brain areas. The vestibulocerebellum dealing with input from the vestibular system includes the most caudal part of the vermis, the nodulus, and adjacent parts of the hemispheres, the flocculi

(flocculonodular lobe). The entire rostral lobe, the most rostral part of the caudal lobe adjacent to the primary fissure, as well as the caudomedial segment of the caudal lobe (intermediate part) receive proprioceptive (and some exteroceptive) afferents through spinocerebellar pathways and are classified as the *spinocerebellum*. The remaining parts of the caudal lobe receive collaterals from the pyramidal and extrapyramidal pathways and therefore are categorized as the cerebrocerebellum or pontocerebellum (Figure 6). The cerebellum thus processes vestibular input, spinal, and cranial nerve proprioceptive and tactile information as well as input from the cerebral cortex regarding intended movements.

In the cerebellum, gray matter is present both as cortex and as relay nuclei near the confluence of the three peduncles. These deep nuclei are located bilaterally and include the nuclei fastigii for the vestibulocerebellum, the nuclei interpositi for the spinocerebellum, and the nuclei dentati for the cerebrocerebellum (Figure 1). Incoming information is always transmitted in parallel to the respective nuclei and to corresponding cortex regions. Whereas mossy fibers are relayed in the granular cell layer of the cerebellar cortex, climbing fibers synapse directly with the dendrites of the Purkinje cells. As a consequence, excitatory input to the nuclei is modulated by the inhibitory effect of the Purkinje cells in the cerebellar cortex. Eventually, all the efferents from the cerebellum arise exclusively from the cerebellar nuclei. Output from the cerebrocerebellum is conveyed to the red nucleus, from where it descends through the rubrospinal tract. Furthermore, information is forwarded to the thalamus (dentothalamic tract to the ventral rostral and ventral lateral thalamic nuclei) and is then projected to the motor cortex (Figure 17). Finally, efferents from the cerebellum also reach the reticular formation and the rostral colliculi.

VISCERAL NERVOUS SYSTEM

Unlike the somatic nervous system, the visceral nervous system is equipped with direct feedback loops that operate automatically in peripheral organs without CNS involvement. Thus, distension of an isolated piece of intestine will elicit a contraction despite the absence of any connection to the spinal cord or brain. Three levels of integration may be distinguished in the visceral nervous system. The peripheral level is fully incorporated within the wall of the gastrointestinal tract. The spinotegmental level includes the spinal cord and the reticular formation in the tegmentum of the brain stem. Its effects are mediated through the sympathetic and parasympathetic pathways. The highest echelon is the prosencephalic level of integration, which embraces the hypothalamus, hypophysis, and the limbic system. It modulates visceral function through descending pathways but mainly via the endocrine system.

Visceral Afferent Pathways

Any stimulus–response feedback loop necessarily relies on information collected. Because only a minor part of visceral afferent signaling is perceived consciously, these pathways are commonly neglected in the study of neuroanatomy. However, receptors in the viscera and blood vessels are of vital importance with respect to maintaining homeostasis and health, and in preventing or reducing disease. Visceral afferents are mediated by mechanoreceptors, chemoreceptors, and nociceptors. They respond to pressure, stretch, oxygen or carbon dioxide content, pH, and osmolarity. Nociception in this system does not detect direct tissue damage (e.g., cutting or burning) but, rather, excessive distension, tissue ischemia, or inflammation. Although morphologically indistinguishable, visceral receptors are specific. However, some mechanoreceptors and chemoreceptors will also produce visceral pain sensation when stimulation exceeds a certain threshold level.

General visceral afferents (GVAs) are included in sympathetic and parasympathetic pathways as well as in all somatic nerves. They enter the central nervous system through the dorsal roots of the spinal nerves and via cranial nerves V, IX, and X. Perikarya of these pseudounipolar viscerosensory neurons are located within the corresponding cranial nerve (sensory) ganglia and within dorsal root ganglia for the spinal nerves. As a rule, visceral pain sensation is conveyed through afferent axons in sympathetic nerves and the white communicating branches, whereas fibers mediating afferents for functional regulation of the viscera follow parasympathetic nerves. Within the central nervous system, visceral pain signals are carried to the contralateral thalamus through the spinothalamic tract. Most afferents involved in the regulation of homeostasis travel through the glossopharyngeal IX and vagus X nerves. Their cell bodies are located within the corresponding ganglia, and the central processes end in the nucleus of the solitary tract, which is the most important sensory nucleus of the visceral nervous system. Parasympathetic afferents of the pelvic organs are conveyed to sacral spinal cord segments through pelvic nerves. They are involved in local reflex arcs but are also linked to the spinothalamic tract.

Visceral Efferent Pathways

Visceral output includes the general visceral efferents (GVEs), which are arranged in sympathetic and parasympathetic pathways, the special visceral efferents (SVEs) (innervating the striated muscles derived from the pharyngeal arches), and the endocrine system. Whereas the special visceral efferents are mediated through a single neuron, general visceral efferents include two motor neurons in succession, the myelinated preganglionic and the thinly myelinated postganglionic neurons. Except for the noradrenergic postganglionic neurons of the sympathetic pathways, all neurons of the general visceral pathways are cholinergic. Many postganglionic fibers form mixed sympathetic and parasympathetic perivascular plexuses.

INTEGRATION

Peripheral Level of Integration

Integration at the peripheral level consists of the nonmyelinated fibers and ganglia of the myenteric and submucosal plexuses of the gastrointestinal tract. This enteric nervous system contains as many neurons as the spinal cord and provides a powerful local feedback loop.

Spinotegmental Level of Integration

The reticular formation contains numerous integrating components, such as the cardiovascular, respiratory, emetic, and micturition centers. The mesencephalic aqueduct of the midbrain is surrounded by a layer of gray matter, the periaqueductal gray, which is involved in controlling defensive and reproductive behaviors. Input is gathered from the cerebral cortex, limbic system, and spinal cord.

Control of the circulatory system is twofold. The cardiovascular excitatory center occupies the rostral and ventrolateral part of the myelencephalic reticular formation. It has an excitatory effect on sympathetic lower motor neurons in the lateral column of the thoracolumber spinal cord segments. It is complemented by the cardiovascular inhibitory (or depressor) center, which is localized to the ventrolateral part of the nucleus ambiguous. Efferent fibers follow the vagus nerve (cranial nerve X).

Disseminated neurons in the rhombencephalic reticular formation constitute the respiratory center. Input is collected from pulmonary stretch receptors, from cardiovascular centers, and from higher brain regions such as the limbic system. However, blood oxygen and carbon dioxide concentrations in the respiratory center of the brain stem play a key role in modulating the activity of the respiratory neurons. They project to the lower motor neurons of the phrenic nerve in cervical segments C_5 to C_7 and to the segments from which the innervation of thoracic and abdominal muscles engaged in breathing arises.

The rhombencephalon also accommodates the emetic center. Chemoreceptors of the oral cavity, the stomach, and/or the small intestine incite a sensory input through neurons following the vagus nerve X. They terminate in the nucleus of the solitary tract. Alternatively, circulating toxins may diffuse to neurons of the area postrema (chemoreceptive trigger zone) due to the absence of the blood–brain barrier in this area. From either relay site, signals are transmitted to the emetic center in the lateral reticular formation ventral to the nucleus of the solitary tract. The emetic center in turn connects to the dorsal vagal nucleus (which can function to reverse bowel peristalsis) and to the spinal cord (which can initiate contractions in somatic muscles required for oral expulsion of digestive tract contents).

Distension of the urinary bladder is signaled to the sacral spinal cord via pelvic nerves. Sensory input is translated into inhibition of the sphincter muscle and stimulation of the detrusor muscle in the bladder wall. This expulsion reflex arc is, however, controlled by the micturition center in the rostral metencephalon. It receives input from the urinary bladder itself but also from higher centers, such as the cortex, hypothalamus, and reticular formation. Neurons in the dorsolateral region promote sphincter contraction and urinary bladder relaxation. The opposite effect is produced by cells in a more medial position in the metencephalon. Micturition involves the contraction of bladder smooth muscle cells and relaxation of the detrusor muscle. Contraction of bladder smooth muscle is mediated by parasympathetic motor neurons originating from the sacral segments. Relaxation of the detrusor muscle, in turn, results from sympathetic inhibition conveyed by motor neurons from the last thoracic and first lumbar spinal cord segments. Some sympathetic neurons will also contribute to urine storage by inhibiting signal transmission in parasympathetic ganglia. Occlusion of the bladder is twofold, by both smooth and striated muscle. Contraction of the internal urethral sphincter is promoted by sympathetic neurons. Somatic lower motor neurons in the pudendal nerve innervate the external urethral sphincter.

Prosencephalic Level of Integration

The hypothalamus plays a critical role in the control of homeostasis, eating, drinking, and reproduction. It is widely connected with many areas of the central nervous system, such as the limbic system and the reticular formation and, most directly, with the hypophysis. As hypothalamic nuclei are poorly delimited, it is helpful to adopt a functional perspective and to distinguish between hypophyseal and nonhypophyseal nuclei.

Hypophyseal gray matter may be further subdivided into magnocellular and parvocellular nuclei. Supraoptic and paraventricular nuclei are magnocellular. They produce vasopressin and oxytocin, which are released into the blood after axoplasmic transport to the neurohypophysis. Vasopressin is a key player in regulating circulating blood volume, pressure, and fluid balance. Oxytocin exhibits an endocrine effect on smooth muscle cells of the female genital tract and mammary gland and also serves as a neurotransmitter in a pathway connecting the hypothalamus to autonomic centers in the brain stem. Parvocellular nuclei

(dorsomedial, ventromedial, and infundibular nuclei) produce liberins and statins to control the endocrine activity of the adenohypophysis.

Nonhypophyseal nuclei are involved in regulating heat balance, feeding, drinking, and reproduction. Regulation of body temperature is controlled by the rostromedial hypothalamus. Responses to changes in body temperature include peripheral vasodilation or vasoconstriction, sweating, piloerection, shivering, and adaption of metabolic activity. Thus, they are mediated through visceral, somatic, and behavioral reactions. Together with the limbic system, the hypothalamus affects feed uptake and feeding behavior. Mechanisms involved include modulation of insulin release and fat metabolism via connections to the dorsal vagal nucleus and sympathetic motor neurons. With respect to reproduction, the hypothalamus is involved in the control of hormone secretion and sexual response. Sexual behavior is regulated by areas of gray matter in the medial preoptic region.

LIMBIC SYSTEM

The limbic system constitutes an extended interface between the somatic and visceral parts of the nervous system. It plays a pivotal role in emotional expression and behaviors such as escape, feeding, and reproduction, and it is concerned with memory functions. Besides olfaction, the limbic system receives optic, auditory, exteroceptive, and viscerosensory inputs from various cortical areas, and it delivers its output to the hypothalamus as well as to both visceral and somatic lower motor neurons. The processing of information in this system remains subconscious to a considerable extent. The limbic system includes a broad variety of structures (Figure 18). As it is widely ramified, it is difficult to delimit the limbic system concisely.

The cortex region pertaining to the limbic system is located at the medial surface of the hemispheres, where it forms a ring at the junction between telencephalon and diencephalon (Figure 7). The term *limbus* (Greek for "edge" or "fringe," e.g., of a garment) refers to this location at the very edge of the brain vesicles. Two concentric arches may be distinguished. The geniculate gyrus, the indusium griseum, the dentate gyrus, the hippocampus proper, and the fornices constitute the inner part, which is surrounded on its convex side by the area subcallosa, the cingulate gyrus, and the gyrus parahippocampalis. The hippocampus proper is unique in that it projects into the lateral ventricle.

Besides these cortex regions, a number of nuclei connected with the cortical structures are also included in the limbic system. These subcortical parts comprise the amygdalae, the septal nuclei, and the habenular nuclei in the telencephalon; several thalamic nuclei and also the preoptic

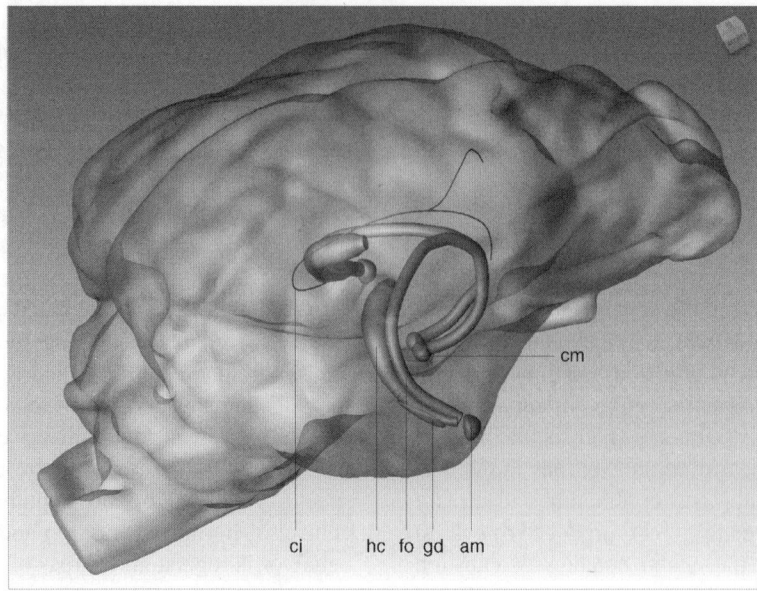

FIGURE 18   Selected structures of the mammalian limbic system (modeled on the dog). am, Amygdala; ci, cingulum; gd, gyrus dentatus; hc, hippocampus proper (cornu ammonis); fo, fornix; cm, corpus mamillare.

and mamillary components of the diencephalon; and the interpeduncular nucleus and reticular formation of the mesencephalon.

Sensory input to the limbic system arises directly from the olfactory system (amygdala and piriform lobe) and indirectly for all other modalities from neocortical association areas. Output from the hippocampus is conveyed through the fornices, which end in the mamillary bodies. These relay nuclei are linked to thalamic nuclei, which then project to the cingulum and to the reticular formation. Information is subsequently forwarded via descending tracts to visceral and somatic lower motor neurons. Furthermore, the hypothalamus obviously modulates endocrine functions.

RECOMMENDED READING

Alstermark B, Isa T. Premotoneuronal and direct corticomotoneuronal control in the cat and macaque monkey. *Adv Exp Med Biol.* 2002;508:281–297.

Amayasu H, et al. Cornu Ammonis of the dog: a rudimentary CA2 field is only present in a small part of the dorsal division, and is absent in the ventral division of the cornu ammonis. *J Hirnforsch.* 1999;39(3):355–367.

Apps R. Columnar organisation of the inferior olive projection to the posterior lobe of the rat cerebellum. *J Comp Neurol.* 1990;302 (2):236–254.

Badonic T, et al. Changes in the dog corticospinal tract after ligation of the middle cerebral artery. *Folia Morphol (Praha).* 1986;34 (2):119–123.

Barone R. Anatomie comparée des mammifères domestiques. In: *Neurologie I: Système nerveux central.* Paris: Vigot; 2004.

Bosco G, Poppele RE. Modulation of dorsal spinocerebellar responses to limb movement: II. Effect of sensory input. *J Neurophysiol.* 2003;90(5):3372–3383.

Boyle R. Morphology of lumbar-projecting lateral vestibulospinal neurons in the brainstem and cervical spinal cord in the squirrel monkey. *Arch Ital Biol.* 2000;138(2):107–122.

Cervero F. Sensory innervation of the viscera: peripheral basis of visceral pain. *Physiol Rev.* 1994;74(1):95–138.

Clarac F, et al. The maturation of locomotor networks. *Prog Brain Res.* 2004;143:57–66.

Craig AD Jr., Spinal and medullary input to the lateral cervical nucleus. *J Comp Neurol.* 1978;181(4):729–743.

Edge AL, et al. Lateral cerebellum: functional localization within crus I and correspondence to cortical zones. *Eur J Neurosci.* 2003;18(6):1468–1485.

Edgley SA, Grant GM. Inputs to spinocerebellar tract neurones located in Stilling's nucleus in the sacral segments of the rat spinal cord. *J Comp Neurol.* 1991;305(1):130–138.

Evans HE. *Anatomy of the Dog,* 3rd ed. Philadelphia: W.B. Saunders; 1993.

Gorska T, et al. Motor effects of cortical stimulation after chronic lesion of medullary pyramid in the dog. *Acta Neurobiol Exp (Wars).* 1980;40(5):861–879.

Hukuda S, et al. The dog corticospinal tract: anatomic and neurologic correlations. *Surg Forum*. 1968;19:421–423.

Kirkwood PA, et al. Interspecies comparisons for the C3–C4 propriospinal system: unresolved issues. *Adv Exp Med Biol*. 2002;508:299–308.

Kowalska DM. Cognitive functions of the temporal lobe in the dog: a review. *Prog Neuropsychopharmacol Biol Psychiatry*. 2000;24(5):855–880.

Kruger L, et al. *Photographic Atlas of the Rat Brain*. Cambridge, UK: Cambridge University Press; 1995.

Leigh EJ, et al. Clinical anatomy of the canine brain using magnetic resonance imaging. *Vet Radiol Ultrasound*. 2008 Mar–Apr;49(2):113–121.

Lemon RN, et al. Direct and indirect pathways for corticospinal control of upper limb motoneurons in the primate. *Prog Brain Res*. 2004;143:263–279.

Manocha SL, et al. *A Stereotaxic Atlas of the Brain of the Cebus Monkey (Cebus apella)*. Oxford, UK: Oxford University Press; 1968.

Nishizawa O, et al. Pontine micturition center in the dog. *J Urol*. 1988;140(4):872–874.

Paxinos G, Watson C. *The Rat Brain in Stereotactic Coordinates*, 3rd ed. San Diego, CA: Academic Press; 1997.

Prechtl JC, Powley TL. B-afferents: a fundamental division of the nervous system mediating homeostasis? *Behav Brain Sci*. 1994;13:289–331.

Puelles L, et al. *The Chick Brain in Stereotaxic Coordinates: An Atlas Featuring Neuromeric Subdivisions and Mammalian Homologies*. San Diego, CA: Academic Press; 2007.

Singer M. *The Brain of the Dog in Section*. Philadelphia: W.B. Saunders; 1962.

Spooner L. Cerebral angiography in the dog. *J Small Anim Pract*. 1961;2:243–252.

Stevens RT, et al. The location of spinothalamic axons within spinal cord white matter in cat and squirrel monkey. *Somatosens Mot Res*. 1991;8(2):97–102.

Thomas CE, Combs CM. Spinal cord segments: B. Gross structure in the adult monkey. *Am J Anat*. 2005;116:205–216.

Van Tienhoven A, Juhasz LP. The chicken telencephalon, diencephalon and mesencephalon in stereotaxic coordinates. *J Comp Neurol*. 1962;118:185–197.

Verburgh CA, et al. Spinocerebellar neurons and propriospinal neurons in the cervical spinal cord: a fluorescent double-labeling study in the rat and the cat. *Exp Brain Res*. 1989;75(1):73–82.

3

ATLAS OF COMPARATIVE NEUROANATOMY

Anna Oevermann

Neurocenter, Department of Clinical Research and Veterinary Public Health, Vetsuisse Faculty, University of Bern, Bern, Switzerland

Marc Vandevelde

Department of Clinical Veterinary Medicine, Vetsuisse Faculty, University of Bern, Bern, Switzerland

Michael H. Stoffel

Department of Clinical Research and Veterinary Public Health, Vetsuisse Faculty, University of Bern, Bern, Switzerland

HOW TO USE THIS ATLAS

The anatomy of the nervous system is highly complex in all three dimensions, which makes pathological examination of macroscopic and microscopic specimens much more challenging than is the case for other organs. The scope of this photographic atlas is to give a comparative overview of the major morphological landmarks and functional divisions that are macroscopically identifiable in the central nervous systems of the chicken, rat, dog, and primate. The majority of pictures present formalin-fixed brain sections where the visible features resemble those with which the reader will be confronted during a routine neuropathological examination. This comparison will help the reader to examine the brain in a standardized way, to readily identify morphological abnormalities, and to correlate observed lesions with the clinical history of neurological alterations.

The cerebral hemispheres, cerebellum, brainstem, olfactory lobes, pituitary, optic chiasm, pons, and pyramids are easily identified surface structures and serve as general landmarks for locating major internal features (Figures 1 to 3). Placement of the transverse cuts is shown for the three mammalian species in Figure 4 (primate, dog, and rat, respectively). The small numbers at the base of each level indicate the figure pertinent to that cut.

In transverse sections (Figures 5 to 28), the ventricular system serves as a useful reference for anatomical orientation. In the mammalian brain, the lateral ventricles are confined to the cerebral hemispheres, where they are bordered by the cerebral cortex, hippocampus, and at their ventral extensions, the basal nuclei. The circular third ventricle appears in transverse sections as two cross-sectioned cavities in the midline of the thalamus. The small mesencephalic aqueduct that connects the third and fourth ventricles is surrounded by the midbrain (mesencephalon). Finally, the fourth ventricle is enclosed ventrally by the brain stem and dorsally by the vela medullaria and the cerebellum.

The transverse brain sections (Figures 5 to 28) in this atlas are arranged from rostral to caudal. All sections were made perpendicular to the base of the brain, in concordance with the common convention used in magnetic resonance tomography. Analogous brain sections of mammalian species are shown side by side, to emphasize neuroanatomical similarities and differences among these species. For that reason, cut intervals may appear somewhat irregular. For orientation, schematic drawings showing the plane of section with reference to the ventricular system are given. In addition to standard gross preparations, selected sections of the canine midbrain, brain stem, and cerebellum are stained with Mulligan stain to define the positions of cranial nerve nuclei (Figures 17 to 20). In these sections, gray

---

Fundamental Neuropathology for Pathologists and Toxicologists: Principles and Techniques, First Edition. Edited by Brad Bolon and Mark T. Butt.

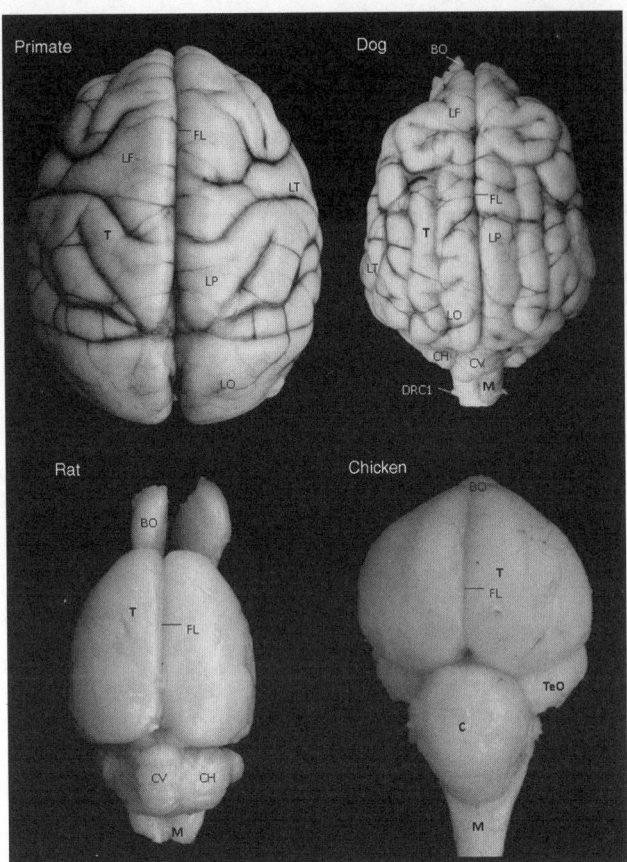

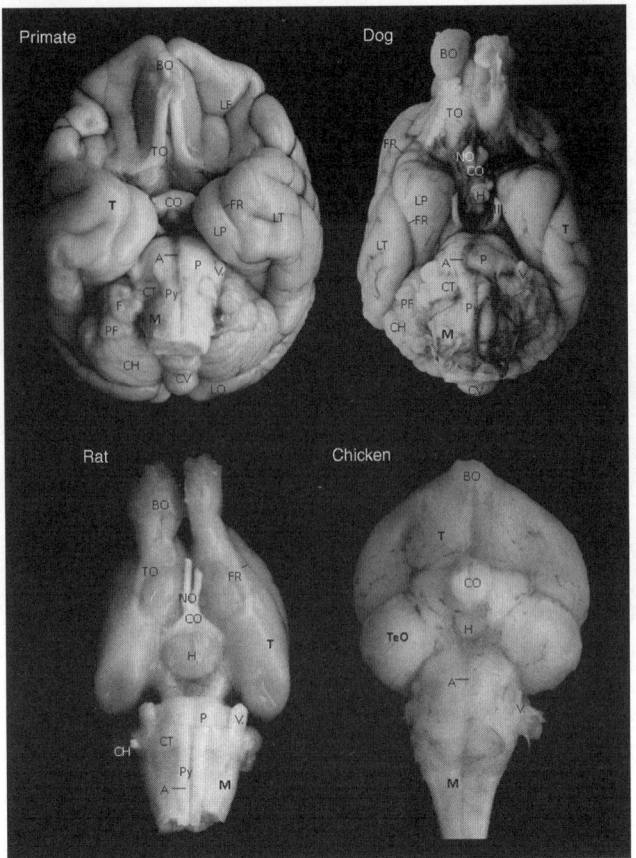

FIGURE 1   Dorsal view. BO, bulbus olfactorius; C, cerebellum; CH, cerebellar hemisphere; CV, cerebellar vermis; DRC1, dorsal root of the spinal nerve of C1; FL, fissura longitudinalis cerebri; LF, lobus frontalis; LO, lobus occipitalis; LP, lobus parietalis; LT, lobus temporalis; M, myelencephalon; T, telencephalon; TeO, tectum mesencephali (tectum opticum).

FIGURE 2   Ventral view. A, arteria basilaris; BO, bulbus olfactorius; CH, cerebellar hemisphere; CO, chiasma opticum; CT, corpus trapezoideum; CV, cerebellar vermis; F, Flocculus; FR, fissura rhinalis; H, stalk of the hypophysis; LF, lobus frontalis; LO, lobus occipitalis; LP, lobus piriformis; LT, lobus temporalis; M, myelencephalon; NO, nervus opticus; P, pons; PF, paraflocculus; Py, pyramis; T, telencephalon; TeO, tectum mesencephali (tectum opticum); TO, tractus olfactorius; III. nervus oculomotorius; V. nervus trigeminus.

matter is stained deep blue, whereas white matter remains unstained.

The transverse sections of the chicken brain are presented separately (Figures 22 to 28) due to their greater divergence from the mammalian pattern of regional brain organization. The levels of those transverse cuts are shown in Figure 21.

SPECIMEN DERIVATION AND PREPARATION

Brains were acquired from a vervet monkey (*Chlorocebus pygerythrus*), beagle dog, Sprague–Dawley rat, and a white leghorn chicken. All brains are from young adult animals. For the Mulligan stain, brain was fixed by immersion in formalin and then sectioned into slices (approximately

0.5 cm thick). Slices were rinsed for 4 h in cold running tap water (CRTW) and then incubated in Mulligan solution [4 mL of phenol, 0.125 mL of hydrochloric acid, and 0.5 g of copper sulfate ($CuSO_4$) in 100 mL of distilled water] at 60 to 65°C for 2 min. The slices were then rinsed in CRTW for 1 min and incubated in 1% ferric chloride ($FeCl_3$) for 3 min (until the gray matter darkened). Slices were rinsed again in CRTW for 5 min and incubated in 1% potassium ferrocyanide ($K_4[Fe(CN)_6]$) for 1 to 3 min (until the gray matter turned blue). Finally, slices were rinsed in CRTW for 1 min. Excess uptake of stain by the gray matter may be reversed gradually by immersion in 10% ammonia at room temperature (RT), and brown discoloration of white matter can be removed by incubating the sections in 2% hydrochloric acid over night at RT. Stained slices may be stored either in formalin or 70% ethanol.

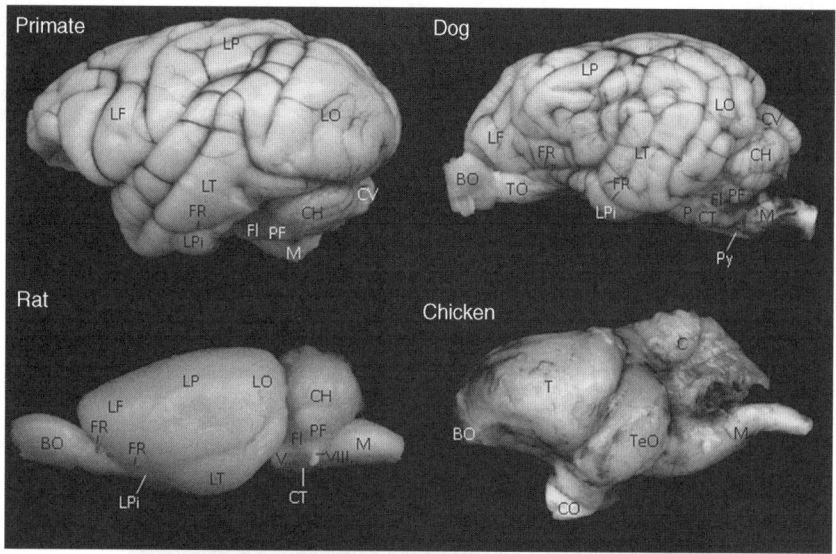

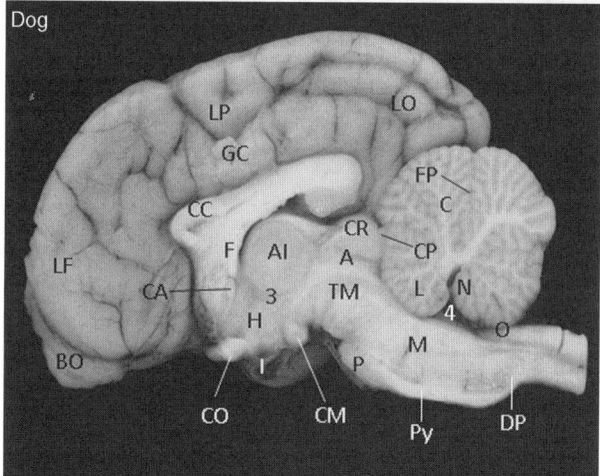

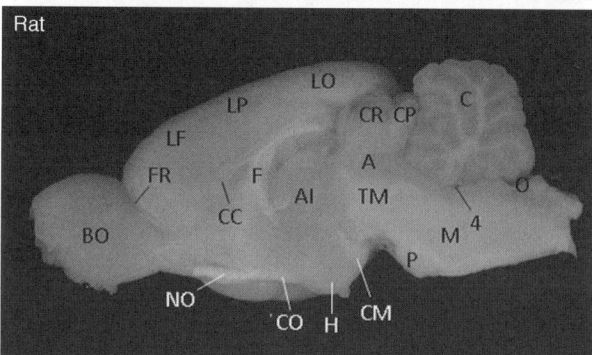

FIGURE 3 Lateral and medial view. A, aquaeductus mesencephali; AI, adhesio interthalamica; BO, bulbus olfactorius; C, cerebellum; CA, commissura anterior; CC, corpus callosum; CH, cerebellar hemisphere; CM, corpus mamillare; CO, chiasma opticum; CP, colliculus caudalis; CR, colliculus rostralis; CT, corpus trapezoideum; CV, cerebellar vermis; DP, decussatio pyramidum; F, fornix; Fl, flocculus; FP, fissura prima cerebelli; FR, fissura rhinalis; GC, gyrus cinguli; H, hypothalamus; I, infundibulum hypophisis; L, lingula cerebelli; LF, lobus frontalis; LO, lobus occipitalis; LP, lobus parietalis; LPi, lobus piriformis; LT, lobus temporalis; M, myelencephalon; N, nodulus cerebelli; NO, nervus opticus; O, obex; P, pons; PF, paraflocculus; Py, pyramis; T, telencephalon; TeO, tectum mesencephali (tectum opticum); TM, tegmentum mesencephali; TO, tractus olfactorius; 3, third ventricle; 4, fourth ventricle; V. nervus trigeminus; VIII. nervus vestibulocochlearis.

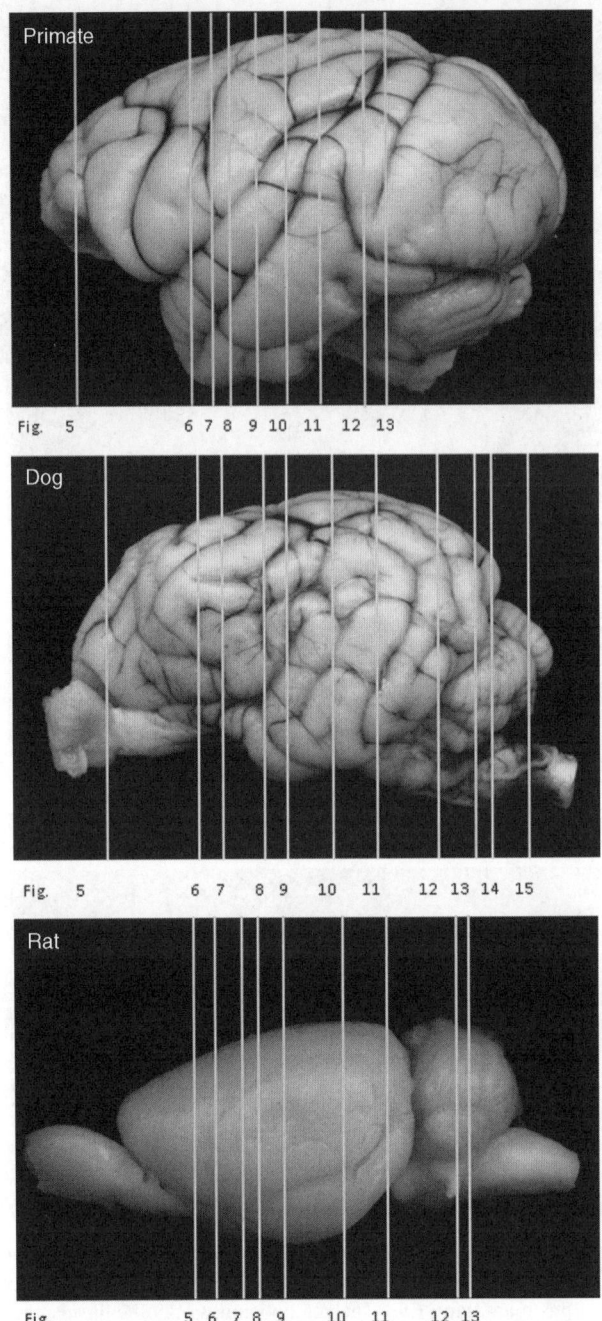

FIGURE 4   Planes of sections in the primate, dog, and rat.

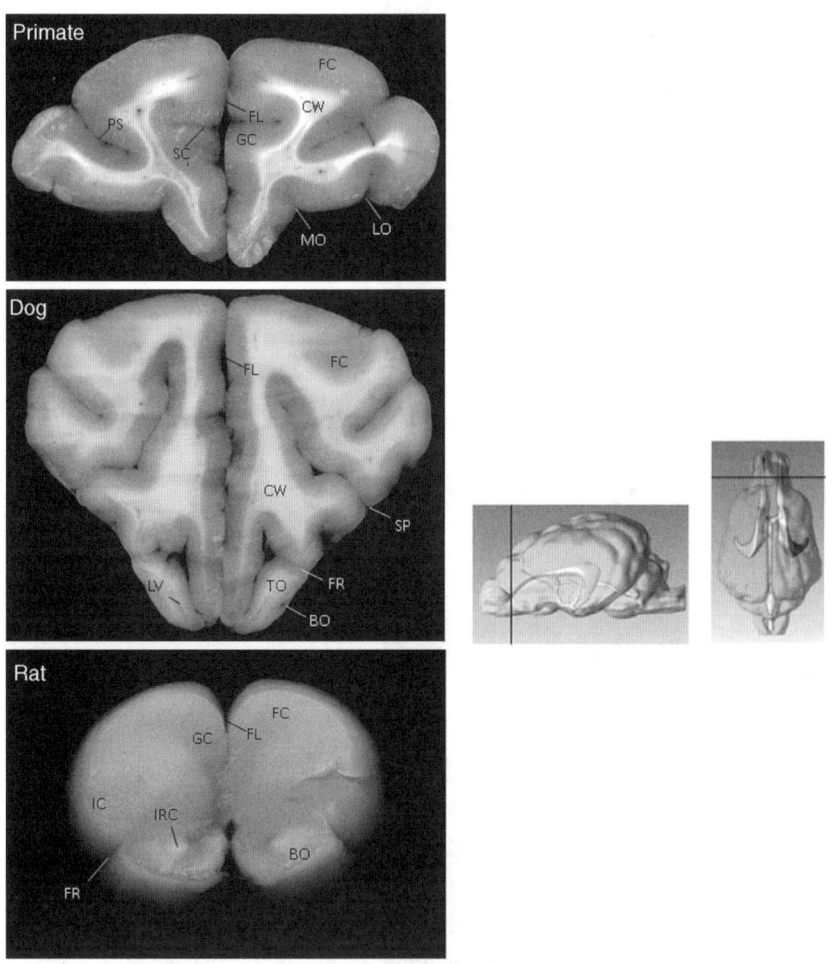

FIGURE 5   Telencephalon, olfactory area. BO, bulbus olfactorius; CW, cerebral white matter; FC, frontal cortex; FL, fissura longitudinalis cerebri; FR, fissura rhinalis; GC, gyrus cinguli; cingulate cortex (rat); IC, insular cortex; IRC, commissura rostralis (intrabulbar); LO, lateral orbital sulcus; LV, lateral ventricle (recessus olfactorius); MO, medial orbital sulcus; PS, principal sulcus; SC, sulcus cinguli; SP, sulcus presylvius; TO, tractus olfactorius.

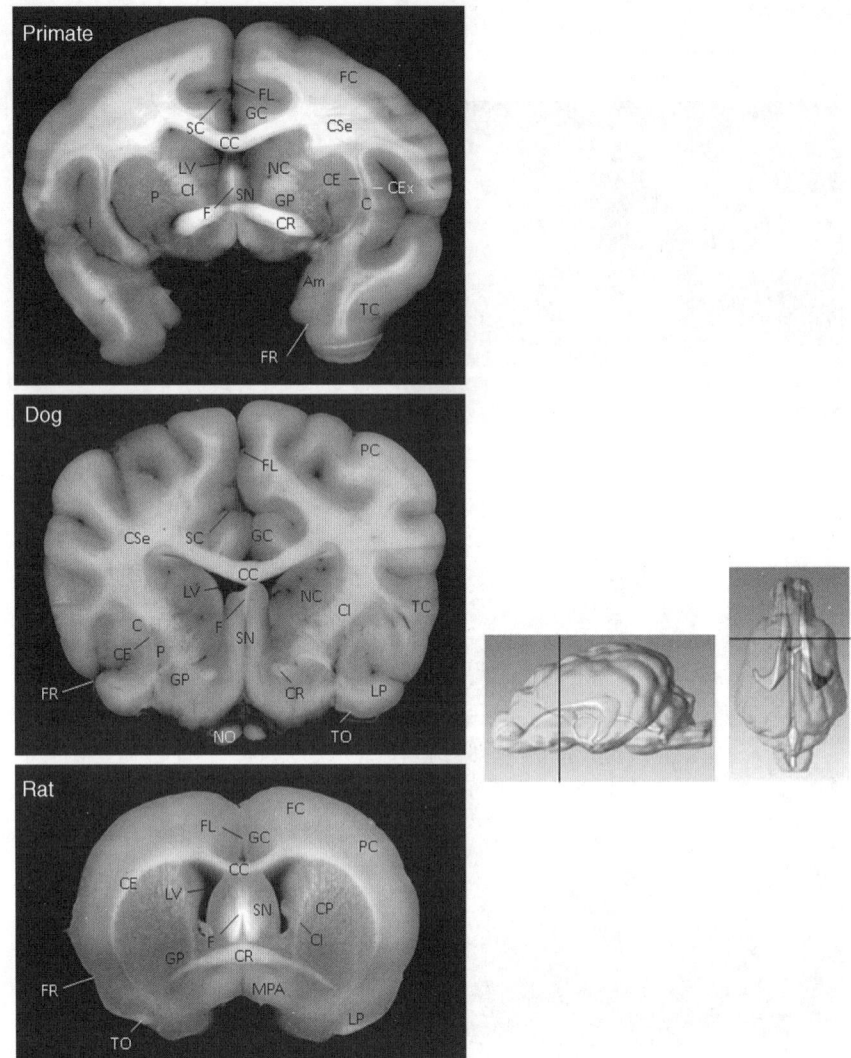

FIGURE 6   Telencephalon, basal nuclei. Am, amygdala; C, claustrum; CC, corpus callosum; CE, capsula externa; CEx, capsula extrema; CI, capsula interna; CP, caudate putamen; CR, commissura rostralis; CSe, centrum semiovale; F, fornix; FC, frontal cortex; FL, fissura longitudinalis cerebri; FR, fissura rhinalis; GC, gyrus cinguli; cingulate cortex (rat); GP, globus pallidus; I, insula; LP, lobus piriformis; LV, lateral ventricle; MPA, medial preoptic area; NC, nucleus caudatus; NO, nervus opticus; P, putamen; PC, parietal cortex; SC, sulcus cinguli; SN, septal nuclei; TC, temporal cortex; TO, tractus olfactorius.

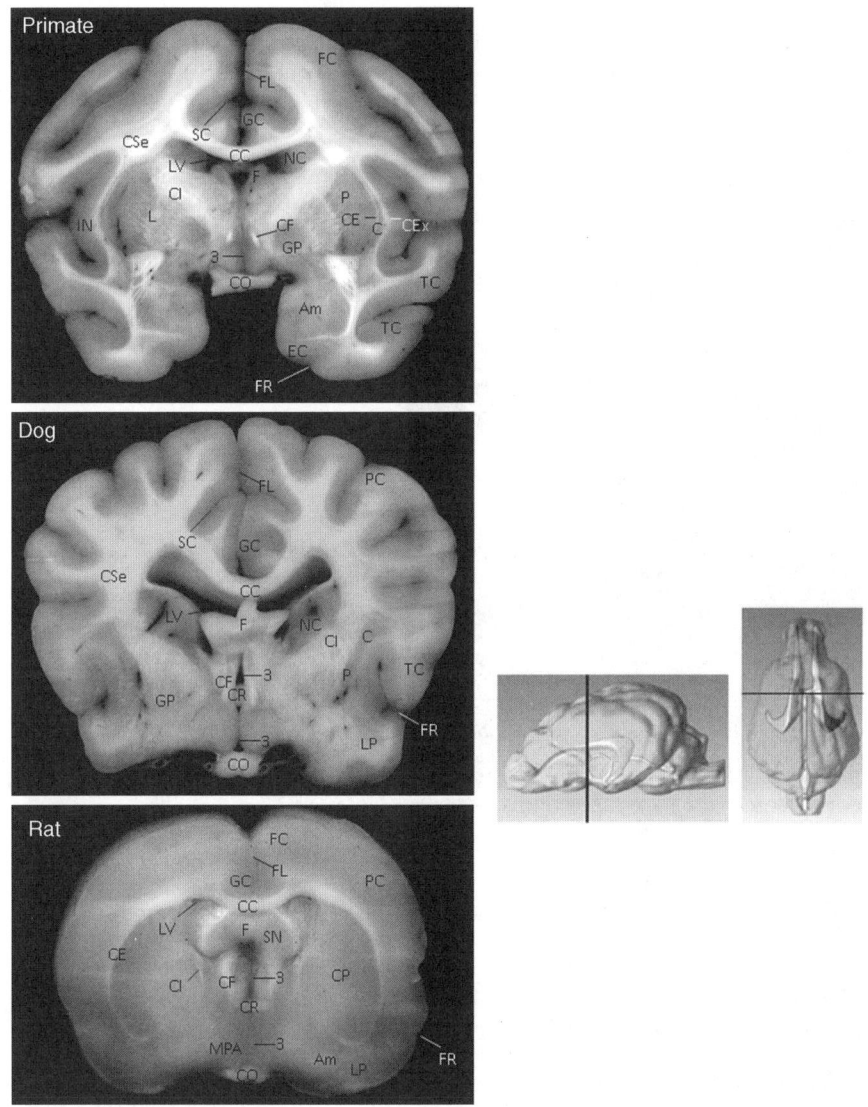

FIGURE 7 Telencephalon, transition basal nuclei/thalamus. Am, amygdala; C, claustrum; CC, corpus callosum; CE, capsula externa; CEx, capsula extrema; CF, columna fornicis; CI, capsula interna; CO, chiasma opticum; CP, caudate putamen; CR, commissura rostralis; CSe, centrum semiovale; EC, entorhinal cortex; F, fornix; FC, frontal cortex; FL, fissura longitudinalis cerebri; FR, fissura rhinalis; GC, gyrus cinguli; cingulate cortex (rat); GP, globus pallidus; IN, insula; L, lateral medullary lamina; LP, lobus piriformis; LV, lateral ventricle; MPA, medial preoptic area; NC, nucleus caudatus; P, putamen; PC, parietal cortex; SC, sulcus cinguli; SN, septal nuclei; TC, temporal cortex; 3, third ventricle.

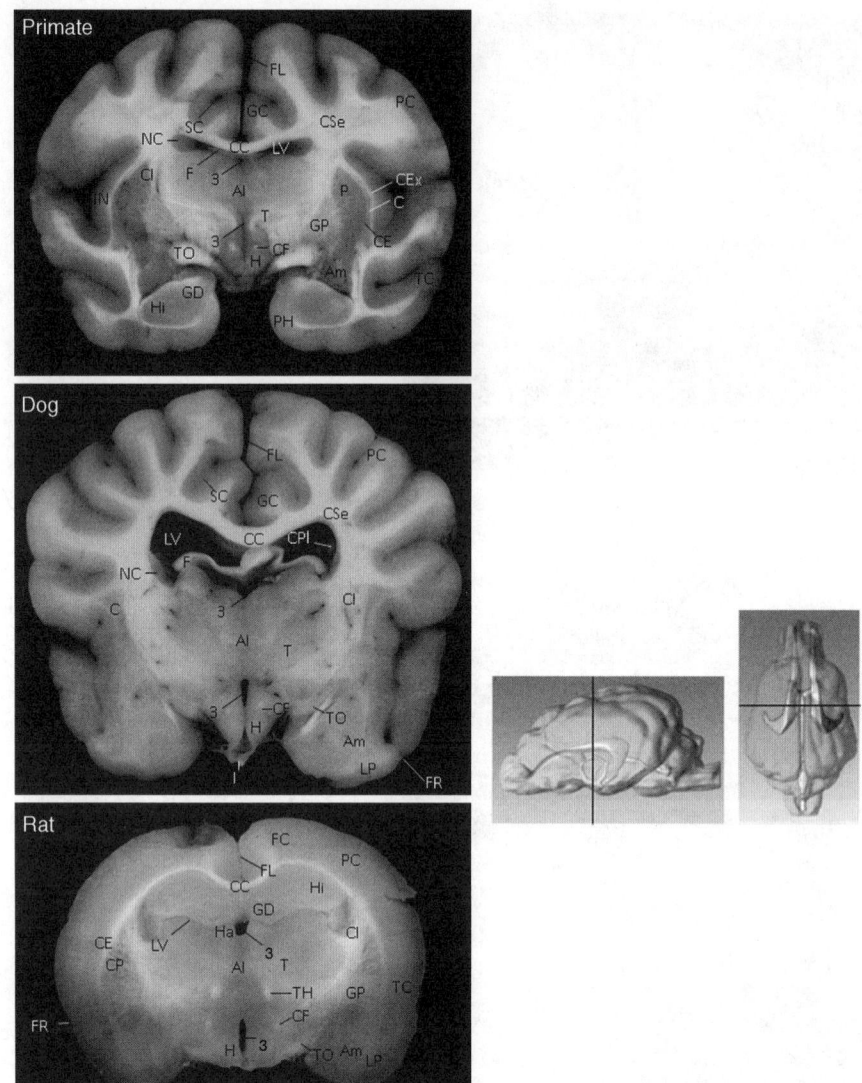

FIGURE 8   Telencephalon, thalamus. Am, amygdala; AI, adhesio interthalamica; C, claustrum; CC, corpus callosum; CE, capsula externa; CEx, capsula extrema; CF, columna fornicis; CI, capsula interna; CP, caudate putamen; CPl, choroid plexus; CSe, centrum semiovale; F, fornix; FC, frontal cortex; FL, fissura longitudinalis cerebri; FR, fissura rhinalis; GC, gyrus cinguli; GD, gyrus dentatus; GP, globus pallidus; H, hypothalamus; Ha, habenula; Hi, hippocampus; I, infundibulum hypophysis; IN, insula; LP, lobus piriformis; LV, lateral ventricle; NC, nucleus caudatus; P, putamen; PC, parietal cortex; PH, parahippocampal gyrus; SC, sulcus cinguli; T, thalamus; TC, temporal cortex; TH, tractus habenulointerpeduncularis; TO, tractus opticus; 3, third ventricle.

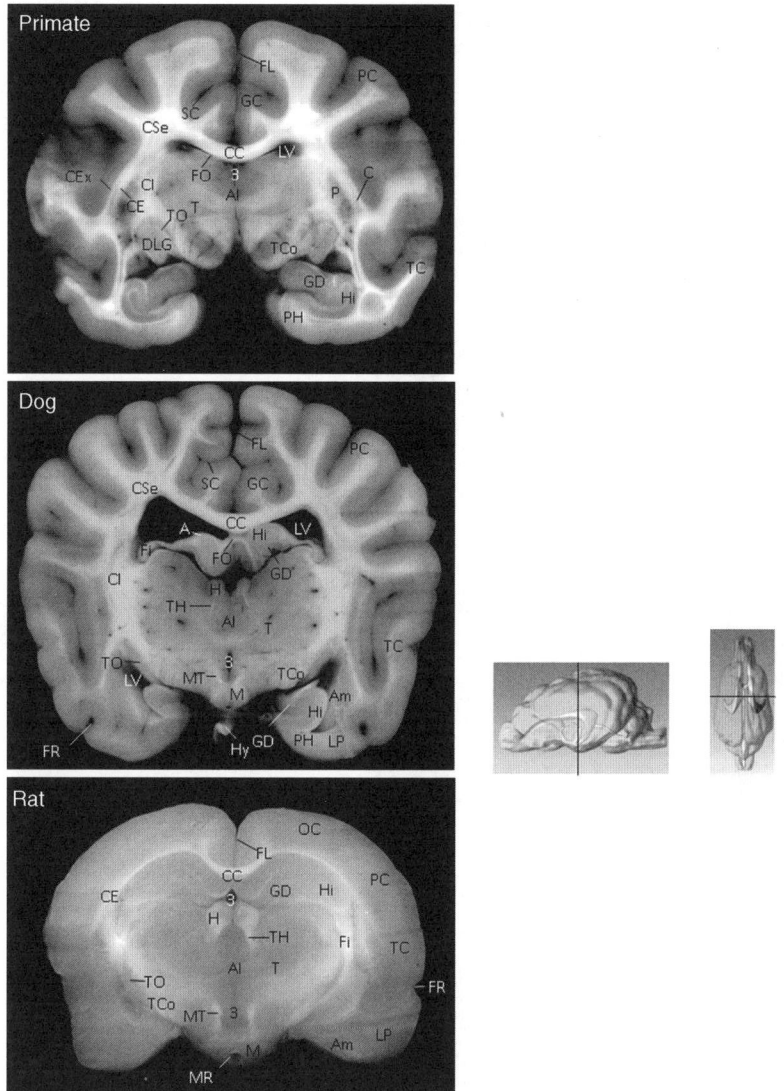

FIGURE 9 Telencephalon, thalamus. A, alveus; AI, adhesio interthalamica; Am, amygdala; C, claustrum; CC, corpus callosum; CE, capsula externa; CEx, capsula extrema; CI, capsula interna; CSe, centrum semiovale; DLG, dorsal nucleus of lateral geniculate body; Fi, fimbria fornicis; FL, fissura longitudinalis cerebri; FO, fornix; FR, fissura rhinalis; GC, gyrus cinguli; GD, gyrus dentatus; H, habenula; Hi, hippocampus; Hy, hypophysis; LP, lobus piriformis; LV, lateral ventricle; M, corpus mammillare; MR, mammillary recess of the third ventricle; MT, mammillothalamic tract; OC, occipital cortex; P, putamen; PC, parietal cortex; PH, parahippocampal gyrus; SC, sulcus cinguli; T, thalamus; TC, temporal cortex; TCo, tractus corticospinalis; TH, tractus habenulointerpeduncularis; TO, tractus opticus; 3, third ventricle.

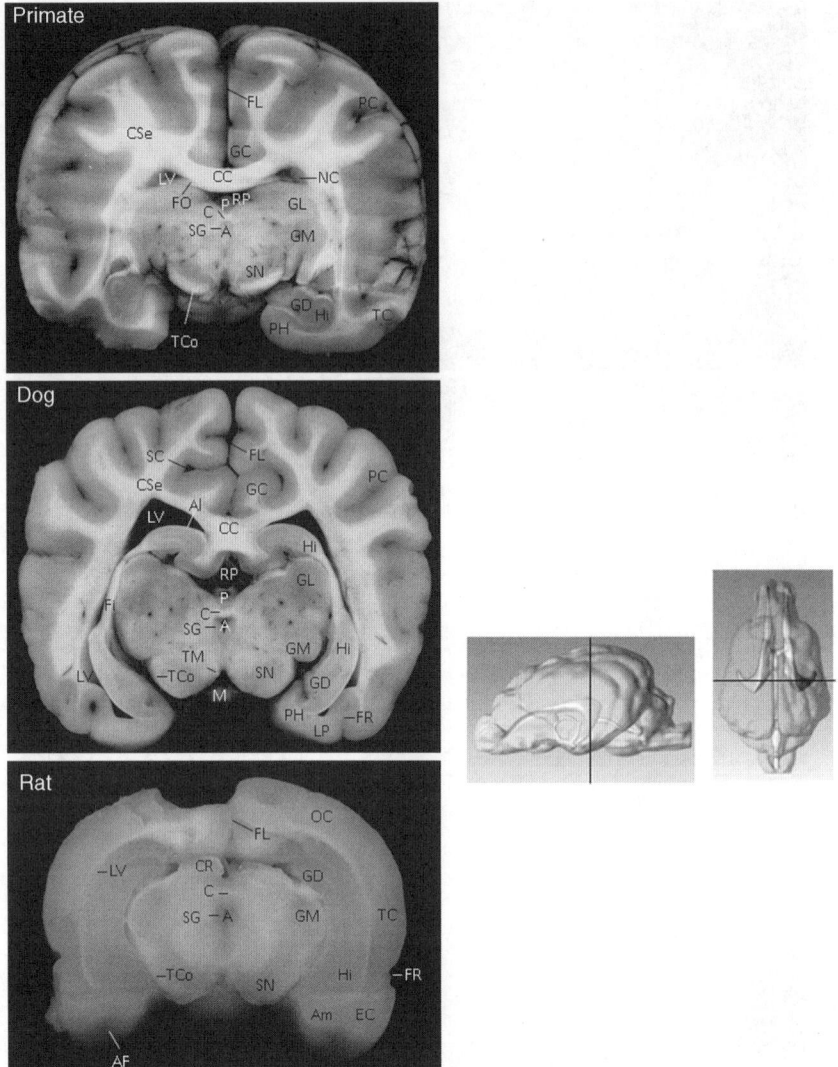

FIGURE 10   Telencephalon, transition caudal thalamus/midbrain. A, aquaeduct; AF, amygdaloid fissure; Al, alveus; Am, amygdala; C, commissura caudalis; CC, corpus callosum; CR, colliculus rostralis; CSe, centrum semiovale; EC, entorhinal cortex; Fi, fimbria fornicis; FL, fissura long- itudinalis cerebri; FO, fornix; FR, fissura rhinalis; GC, gyrus cinguli; GD, gyrus dentatus; GL, corpus geniculatum laterale; GM, corpus geniculatum mediale; Hi, hippocampus; LP, lobus piriformis; LV, lateral ventricle; M, corpus mammillare; NC, nucleus caudatus; OC, occipital cortex; P, corpus pinealis; PC, parietal cortex; PH, gyrus parahippocampalis; RP, recessus suprapinealis; SC, sulcus cinguli; SG, substantia grisea centralis; SN, substantia nigra; TC, temporal cortex; TCo, tractus corticospinalis; TM, tractus mammillothalamicus.

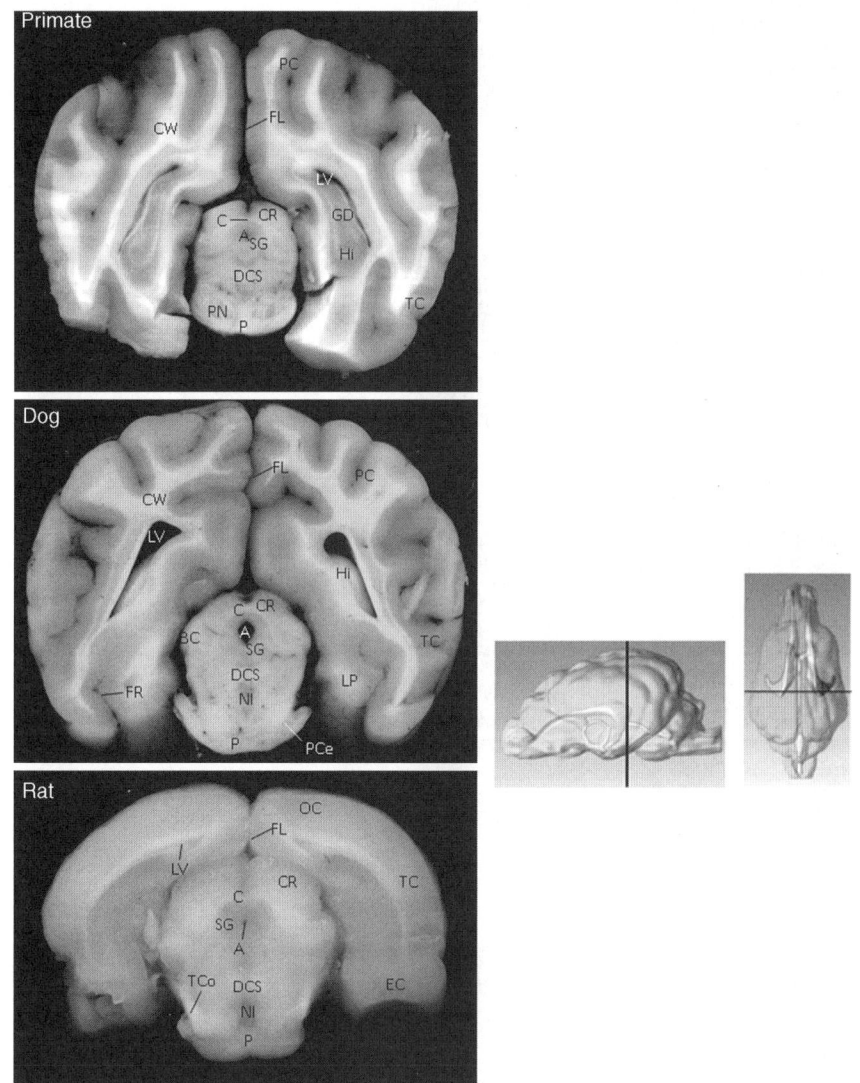

FIGURE 11   Telencephalon, mesencephalon. A, aquaeductus mesencephali; BC, brachium colliculi caudalis; C, commissura caudalis; CR, colliculus rostralis; CW, cerebral white matter; DCS, decussatio pedunculorum cerebellarium superiorum; EC, entorhinal cortex; FL, fissura longitudinalis cerebri; FR, fissura rhinalis; GD, gyrus dentatus; Hi, hippocampus; LP, lobus piriformis; LV, lateral ventricle; NI, nucleus interpeduncularis; OC, occipital cortex; P, pons; PC, parietal cortex; PCe, pedunculus cerebellaris medius; PN, nuclei pontis; SG, substantia grisea centralis; TC, temporal cortex; TCo, tractus corticospinalis (crus cerebri).

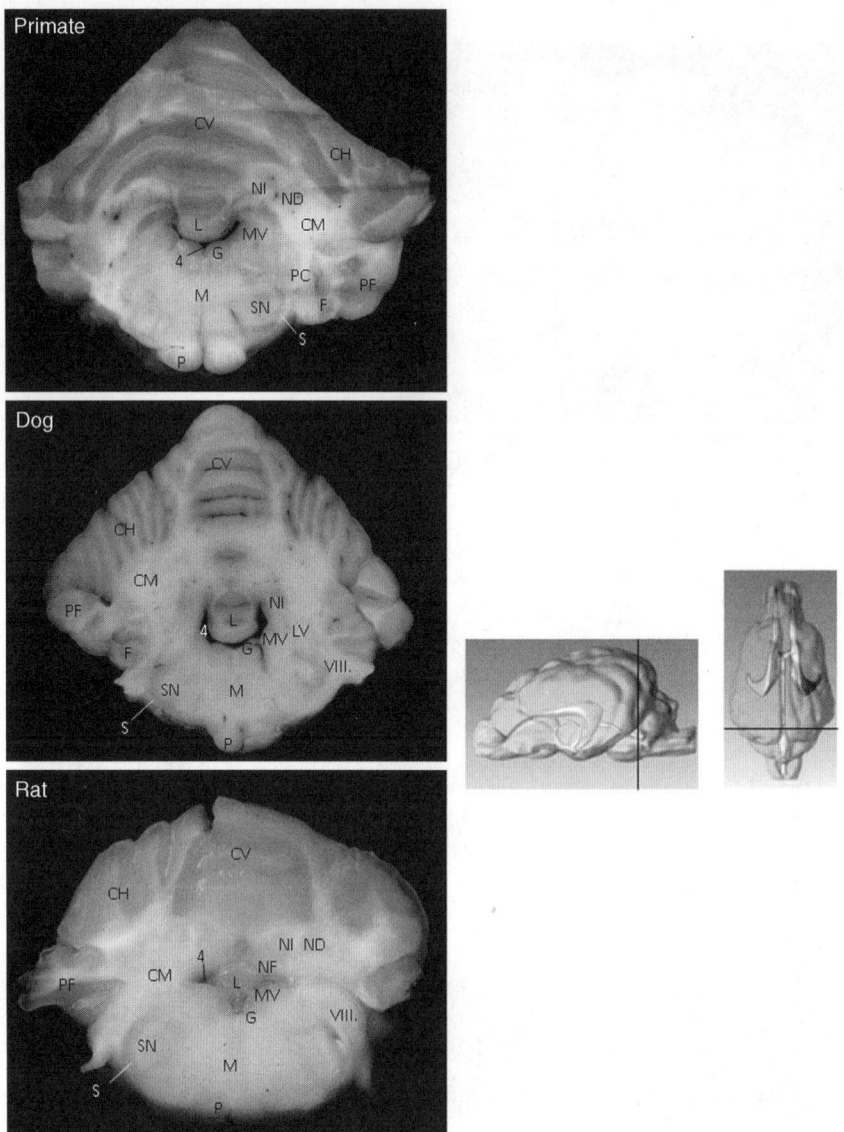

FIGURE 12   Myelencephalon, cerebellum. CH, cerebellar hemispheres; CM, corpus medullare cerebelli; CV, cerebellar vermis; F, flocculus; G, genu nervi facialis; L, lingula; LV, nucleus vestibularis lateralis; M, myelencephalon; MV, nucleus vestibularis medialis; ND, nucleus dentatus; NF, nucleus fastigii; NI, nucleus interpositus; P, pyramid; PC, pedunculus cerebelli caudalis; PF, paraflocculus; S, spinal tract of the trigeminal nerve; SN, spinal tract nucleus of the trigeminal nerve; 4, fourth ventricle; VIII., nervus vestibulocochlearis.

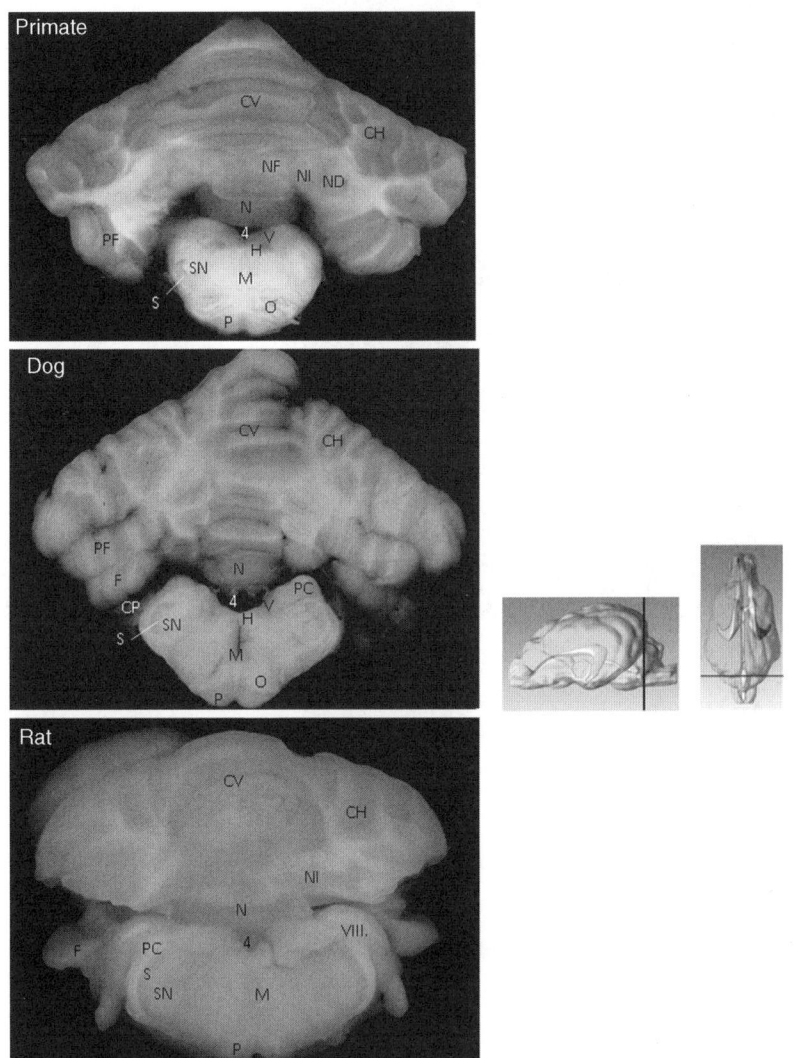

FIGURE 13   Myelencephalon, cerebellum. CH, cerebellar hemispheres; CP, choroid plexus of the fourth ventricle; CV, cerebellar vermis; F, flocculus; H, nucleus motorius nervi hypoglossi; M, myelencephalon; N, nodulus; ND, nucleus dentatus; NF, nucleus fastigii; NI, nucleus interpositus; O, olives; P, pyramid; PC, pedunculus cerebellaris caudalis; PF, paraflocculus; S, spinal tract of the trigeminal nerve; SN, spinal tract nucleus of the trigeminal nerve; V, nucleus parasympathicus n. vagi; VIII. nervus vestibulocochlearis; 4, fourth ventricle.

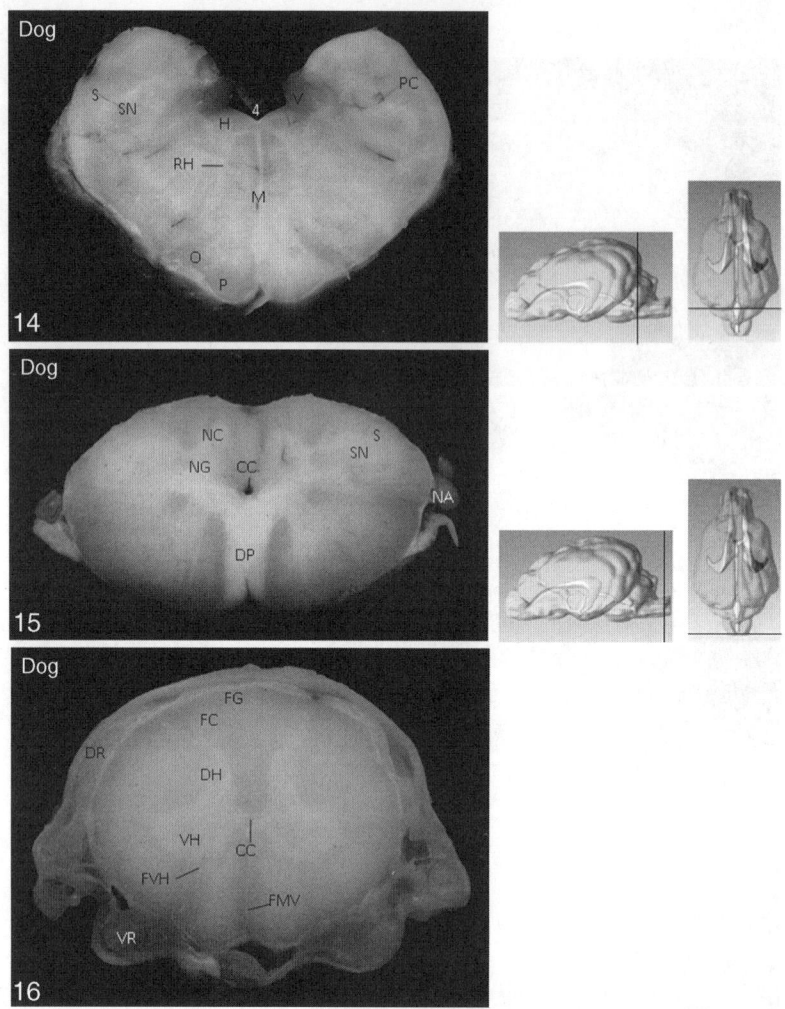

FIGURE 14   Myelencephalon. H, nucleus motorius nervi hypoglossi; M, myelencephalon; O, nucleus olivaris; P, pyramid; PC, pedunculus cerebelli caudalis; RH, radix n. hypoglossi; S, spinal tract of the trigeminal nerve; SN, spinal tract nucleus of the trigeminal nerve; V, nucleus parasympathicus n. vagi; 4, fourth ventricle.

FIGURE 15   Spinal cord, C1. CC, central canal; DP, decussatio pyramidum; NA, nervus accessorius; NC, nucleus cuneatus; NG, nucleus gracilis; S, spinal tract of the trigeminal nerve; SN, spinal tract nucleus of the trigeminal nerve.

FIGURE 16   Spinal cord, cervical intumescence. CC, central canal; DH, dorsal horn of the central grey matter; DR, dorsal nerve root; FC, fasciculus cuneatus; FG, fasciculus gracilis; FMV, fissura mediana ventralis; FVH, fibers of the ventral horn motor neurons; VH, ventral horn of the central grey matter; VR, ventral nerve root.

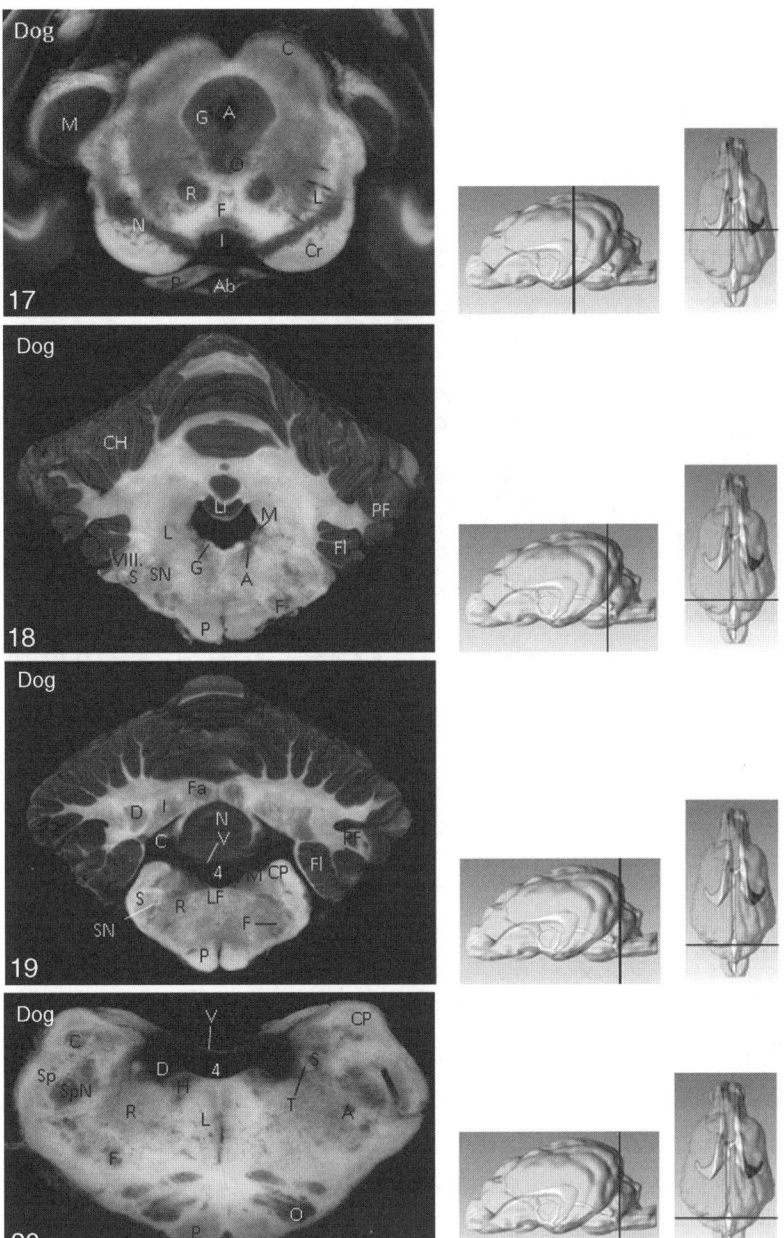

FIGURE 17 Dog, mesencephalon. A, aquaeductus mesencephali; Ab, arteria basilaris; C, colliculus rostralis; Cr, crus cerebri/tractus corticospinalis; F, fibrae tractus rubrospinalis; G, substantia grisea centralis; I, nucleus interpeduncularis; L, lemniscus medialis; M, corpus geniculatum mediale; N, substantia nigra; O, nucleus nervi oculomotorii; P, pons; R, nucleus ruber.

FIGURE 18 Dog, myelencephalon and cerebellum. A, nucleus motorius nervi abducentis; CH, cerebellar hemisphere; F, nucleus motorius nervi facialis; Fl, flocculus; G, genu nervi facialis; L, nucleus vestibularis lateralis; Li, lingula; M, nucleus vestibularis medialis; P, pyramis; PF, paraflocculus; S, spinal tract of trigeminal nerve; SN, spinal tract nucleus of trigeminal nerve; VIII, nervus vestibulocochlearis.

Figures 17 to 20 were stained using the Mulligan procedure (Mulligan, 1931).

FIGURE 19 Dog, myelencephalon and cerebellum. C, choroid plexus; CP, caudal cerebellar peduncle; D, nucleus dentatus; F, nucleus nervi facialis; Fa, nucleus fastigii; Fl, flocculus; I, nucleus interpositus; LF, fasciculus longitudinalis medialis; M, nucleus vestibularis medialis; N, nodulus; P, pyramis; R, formatio reticularis; S, spinal tract of trigeminal nerve; SN, spinal tract nucleus of trigeminal nerve; V, velum medullare; 4, fourth ventricle.

FIGURE 20 Dog, myelencephalon. A, nucleus ambiguus; C, nucleus cuneatus lateralis; CP, caudal cerebellar peduncle; D, nucleus parasympathicus of vagus nerve (DMNV); F, nucleus motorius nervi facialis; H, nucleus motorius nervi hypoglossi; L, fasciculus longitudinalis medialis; O, nucleus olivaris; P, pyramis; R, formatio reticularis; S, nucleus tractus solitarii; Sp, spinal tract of trigeminal nerve; SpN, spinal tract nucleus of trigeminal nerve; T, tractus solitarii; V, velum medullare; 4, fourth ventricle.

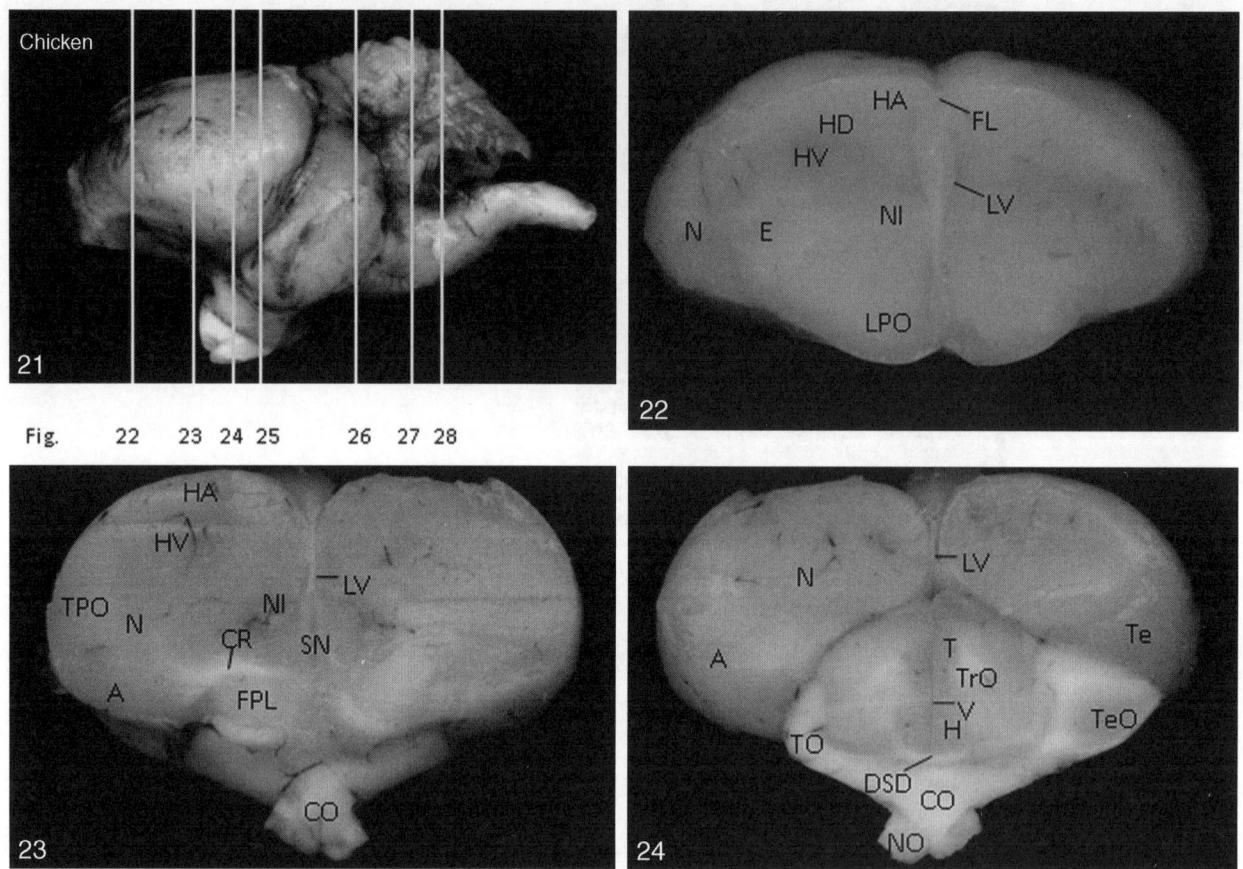

FIGURE 21 Planes of sections in the chicken.

FIGURE 22 Telencephalon. E, ectostriatum; FL, fissura longitudinalis cerebri; HA, hyperstriatum accessorium; HD, hyperstriatum dorsale; HV, hyperstriatum ventrale; LPO, lobus paraolfactorius; LV, lateral ventricle; N, neostriatum; NI, neostriatum intermedium.

FIGURE 23 Telencephalon. A, archistriatum; CO, chiasma opticum; CR, commissura rostralis; FPL, fasciculus prosencephali lateralis; HA, hyperstriatum accessorium; HV, hyperstriatum ventrale; LV, lateral ventricle; N, neostriatum; NI, neostriatum intermedium; SN, septal nuclei; TPO, area temporo-parieto-occipitalis.

FIGURE 24 Telencephalon, tectum mesencephali (tectum opticum). A, archistriatum; CO, chiasma opticum; DSD, decussatio supraoptica dorsalis; H, hypothalamus; LV, lateral ventricle; N, neostriatum; NO, nervus opticus; T, thalamus; Te, telencephalon; TeO, tectum mesencephali (opticum); TO, tractus opticus; TrO, tractus occipitomesencephalicus; V, ventricle.

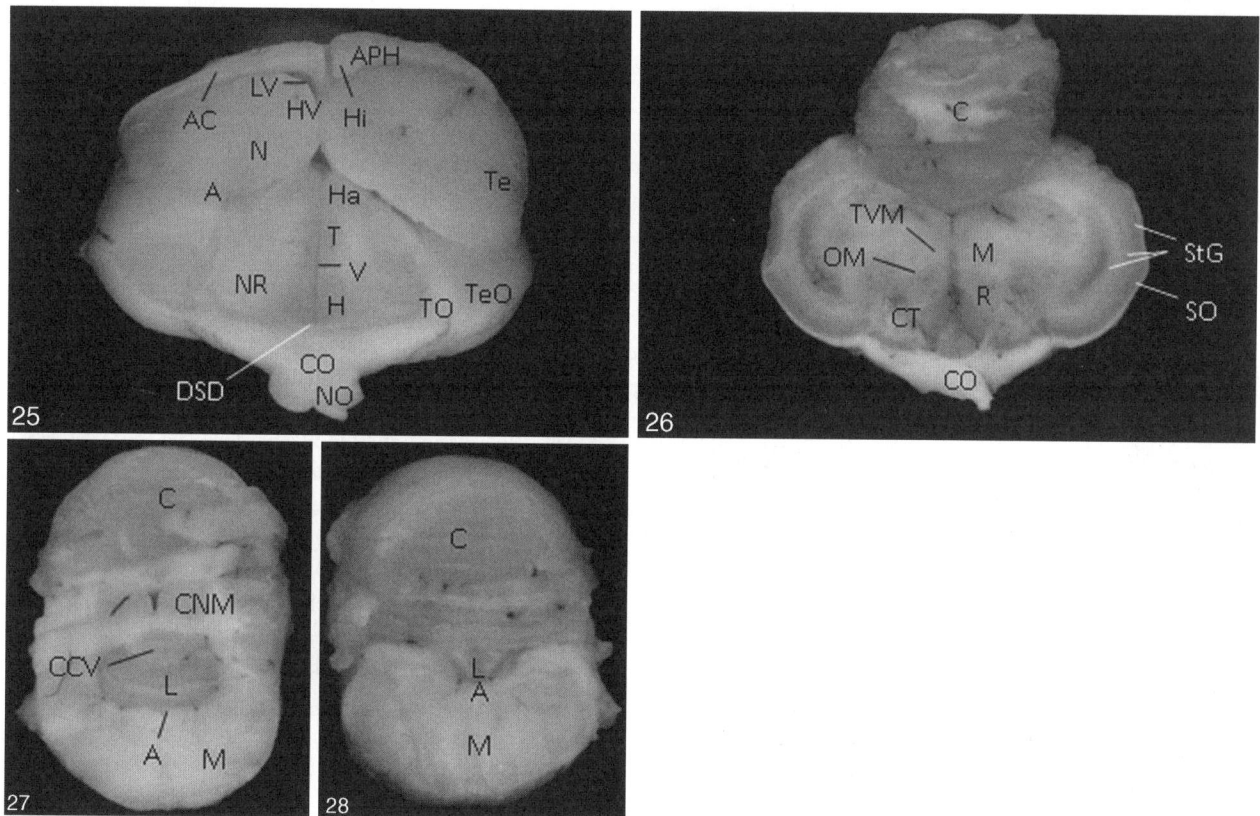

FIGURE 25   Telencephalon, tectum mesencephali (tectum opticum). A, archistriatum; AC, area corticoidea dorsolateralis; APH, area parahippocampalis; CO, chiasma opticum; DSD, decussatio supraoptica dorsalis; H, hypothalamus; Ha, habenula; Hi, hippocampus; HV, hyperstriatum ventrale; LV, lateral ventricle; N, neostriatum; NO, nervus opticus; NR, nucleus rotundus; T, thalamus; Te, telencephalon; TeO, tectum mesencephali (opticum); TO, tractus opticus; V, ventricle.

FIGURE 26   Tectum mesencephali (tectum opticum), cerebellum. C, cerebellum; CO, chiasma opticum; CT, commissura tectalis; M, mesencephalon; OM, tractus occipitomesencephalicus; R, nucleus ruber; StG, strata grisea; SO, striatum opticum; TVM, tractus vestibulomesencephalicus.

FIGURE 27   Cerebellum, myelencephalon. A, aquaeductus cerebri; C, cerebellum; CCV, commissura cerebellaris ventralis; CNM, cerebellar nucleus medialis; L, lingula; M, myelencephalon.

FIGURE 28   Cerebellum, myelencephalon. A, aquaeductus cerebri; C, cerebellum; L, lingula; M, myelencephalon.

RECOMMENDED READING

Barnard JW, Roberts JO, Brown JG. A simple macroscopic staining and mounting procedure for wet sections from cadaver brains. *Anat Rec.* 1949;105(1):11–17.

Davis R, Huffman RD. *A Stereotactic Atlas of the Brain of the Baboon (Papio papio).* University of Texas Press, Austin; 1968.

DeArmond SJ, Fusco MM, Dewey MM. *Structure of the Human Brain. A Photographic Atlas.* Oxford University Press, Oxford; 1974.

Dua-Sharma S, Sharma S, Jacobs HL. *The Canine Brain in Stereotactic Coordinates.* MIT Press, Cambridge, Massachusetts; 1970.

Kim RKS, Liu CN, Moffitt RL. *A Stereotaxic Atlas of the Dog's Brain.* Charles C Thomas, Springfield, Illinois; 1960.

Kruger L, Saporta S, Swanson LW. *Photographic Atlas of the Rat Brain.* Cambridge University Press, Cambridge; 1995.

Leigh EJ, Mackillop E, Robertson ID, Hudson LC. Clinical anatomy of the canine brain using magnetic resonance imaging. *Vet Radiol Ultrasound.* 2008;49 (2):113–21.

LeMasurier HE. Simple method of staining macroscopic brain sections. *Arch Neurol Psychiatry.* 1935;34:1065–1067.

Manocha SL, Shantha TR, Bourne GH. *A Stereotatic Atlas of the Brain of the Cebus Monkey (Cebus apella).* Oxford University Press, Oxford; 1968.

Mulligan JH. A method of staining the brain for macroscopic study. *J Anat.* 1931;65:468–472.

Paxinos G, Watson C. *The Rat Brain in Stereotactic Coordinates.* 3rd edition, Academic Press, San Diego, California; 1997.

Puelles L, Martinez-de-la-Torre M, Paxinos G, Watson C, Martinez S. *The Chick Brain in Stereotaxic Coordinates: An Atlas Featuring Neuromeric Subdivisions and Mammalian Homologies.* Academic Press, San Diego, California; 2007.

Shryock EH. An application of le Masurier's method of staining macroscopic brain sections: a photographic aid for the teaching laboratory. *Anat Rec.* 2005;76(3):291–293.

Singer M. *The Brain of the Dog in Section.* W.B. Saunders Company, Philadelphia, Pennsylvania; 1962.

Snider RS, Lee JC. *A Stereotactic Atlas of the Monkey Brain (Macaca mulatta).* University of Chicago Press, Chicago, Illinois; 1961.

Stephan H, Baron G, Schwerdtfeger WK. *The Brain of the Common Marmoset (Callithrix jacchus): A Stereotactic Atlas.* Springer Verlag, Berlin; 1980.

Van Tienhoven A, Juhasz LP. The chicken telencephalon, diencephalon and mesencephalon in stereotactic coordinates. *J Comp Neurol.* 1962;118:185–197.

INTERNET SOURCES

Atlas of the primate brain.http://braininfo.rprc.washington.edu/PrimateBrainMaps/atlas/Mapcorindex.html

Primate MRI brain atlas.http://brainmeta.com/mri_primate/154.html

LONI Rat brain atlas.http://www.loni.ucla.edu/Atlases/Atlas_Detail.jsp?atlas_id=1/Rat.html

Avian brain org. A resource for brain researchers.http://www.avianbrain.org/index.html

Canine brain transections.http://vanat.cvm.umn.edu/brainsect/

Brain biodiversity bank. Michigan State University.https://www.msu.edu/user/brains/

Brain maps org.http://brain-maps.org/index.php?p=timeline

Comparative brain collections.http://brainmuseum.org/

Human neuroanatomy and neuropathology.http://www.neuropat.dote.hu/

4

PRINCIPLES OF COMPARATIVE AND CORRELATIVE NEURODEVELOPMENT AND THEIR IMPLICATIONS FOR DEVELOPMENTAL NEUROTOXICITY

BRAD BOLON

GEMpath, Inc., Longmont, Colorado

INTRODUCTION

Distinct functions have been associated with discrete neural regions for over a century.[1,2] No animal species provides a direct correlation to the anatomical and functional complexities that are characteristic of human neurological development and function. Nevertheless, the basic structural blueprint for nervous system development is comparable across all mammalian species, including humans,[3-10] and similar neural structures may be discerned in avian species.[9,11] In like manner, major neural functions—motion, sensation, association ("higher thought"), and controlling some aspects of homeostasis—are remarkably similar across species.[8,12-18] Thus, all principal neural regions and general functions in humans are also present in laboratory animals.[19]

The nervous system is doubtless the most anatomically and functionally intricate organ in the body, and the processes involved in its development are equally complex. The neural plate, the progenitor of the central nervous system, is the first recognizable organ precursor to develop in the vertebrate embryo.[20] This "simple" flat disk will eventually evolve to form the complex, multidimensional, unique domains of the nervous system, all of which have their own distinctive structures and tasks. Both the anatomical features and the roles they serve can change markedly over very short distances in any or all three physical dimensions, and they change significantly over time as well. The problems faced by neuroscientists are to comprehend how this tremendous diversity in form is generated and how each distinct structure

of the nervous system is integrated into the vast neural complex to produce a system capable of smoothly and reliably controlling the simultaneous input from and activities by billions of integrated neurons.

The developing brain is vulnerable to neurotoxicant exposure for many reasons, among them the high degree of cell specialization, the slow pace of maturation, the immaturity of protective mechanisms, and a limited capacity for regeneration.[21-23] In this chapter I review the chief events that take place during development of the vertebrate nervous system to provide an easily mastered set of fundamental principles for understanding the complex anatomy of developing and mature neural organs. Emphasis is placed on defining a rational basis for designing toxicological neuropathology assessments in developing mammals.

PRINCIPLES OF ANATOMICAL ORGANIZATION IN THE DEVELOPING NERVOUS SYSTEM

Organogenesis (the process whereby anatomical structures evolve from primordia into structurally and functionally mature organs) as applied to the nervous system is termed *neurogenesis*. As in other developing organs, the mainstays of tissue reorganization during neurogenesis are such basic processes as cell proliferation, cell migration, cell differentiation, and cell death. Region-specific gross and/or cytoarchitectural alterations occur when one or more of these basic processes is disrupted, and such structural anomalies

Fundamental Neuropathology for Pathologists and Toxicologists: Principles and Techniques, First Edition. Edited by Brad Bolon and Mark T. Butt.
© 2011 John Wiley & Sons, Inc. Published 2011 by John Wiley & Sons, Inc.

have been linked to neural dysfunction (behavioral, electrophysiological, neurochemical, and/or other abnormalities). Structural aberrations are driven by both intrinsic ["nature" (neural-specific gene expression patterns, cell-to-cell interactions, hormonal status, neurotrophic factors, etc.)] and extrinsic ["nurture" (extraneural variables such as behavioral experience, nutrition, trauma, etc.)] factors. Thus, the primary issues in neurobiology today are (1) to identify the relative contributions of nature and nurture in directing the development of specific neural structures and their functions and (2) to negate or minimize the impact of preventable insults, such as neurotoxicants, to the developing nervous system.

A thorough understanding of the spatial and temporal events that occur during normal neural development greatly simplifies the tasks of pathologists and toxicologists engaged in neurotoxicity research. The main advantage of such a neuroembryological approach to study design and interpretation is that the presence of a given functional deficit can be used to predict the affected neural structure, and vice versa. A corollary benefit is that such information can be employed to forecast what neural structures and functions might be affected by exposure to a potential neurotoxicant during a given developmental stage.

EARLY SPECIFICATION OF THE NERVOUS SYSTEM

The basic sequence of neurodevelopmental events is largely conserved in vertebrates.[11,24–37] The exact timing and ultimate anatomical appearance of different neural structures varies with such factors as the species, the gestational length,

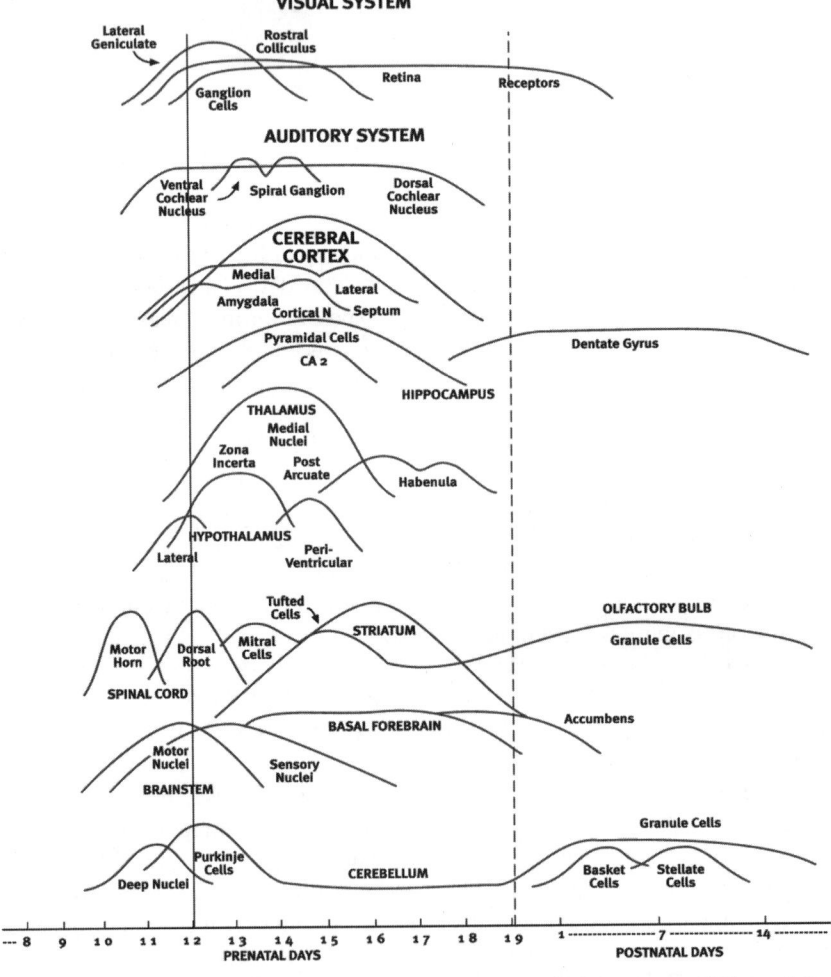

FIGURE 1 Sequence of neuron production in various domains of the mouse CNS (as defined by autoradiographic data). Curves represent the burst of neuronogenesis for a particular region and/or cell type. The peak of each curve in relative terms may be referenced to developmental age (*x*-axis); the area under the curves is not indicative of the number of cells generated. The solid vertical line denotes the last embryonic day (E12) on which gross malformations may be induced by disrupting neuron proliferation, while the dashed vertical line defines the day of birth. The position of the regions and cell groups on the *y*-axis is arbitrary. (Adapted from Rodier,[119] by permission of the publisher.)

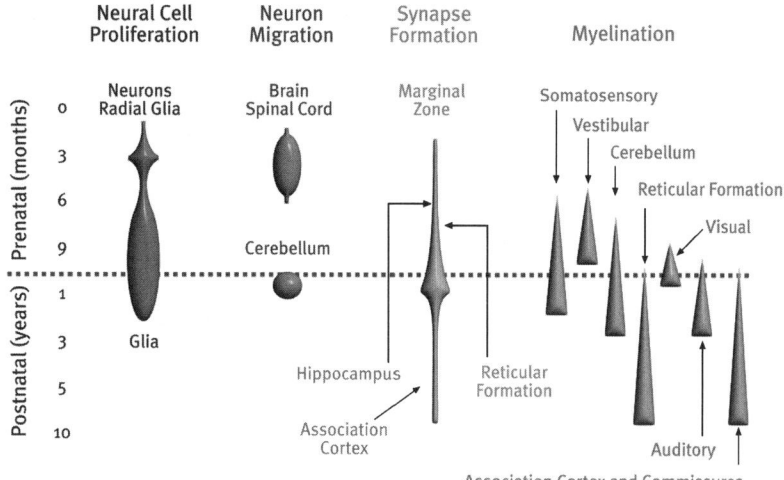

FIGURE 2 Timing and extent of major neurodevelopmental processes in the human CNS. The width of the graphic symbols reflects the relative intensity of the events. (Adapted from Herschkowitz et al.,[44] by permission of Georg Thieme Verlag.)

and the degree of neural development at birth (see Figures 1 and 2). A comparative chronology of major neurodevelopmental events is given in Table 1.

Neurulation [the initial process of forming the central nervous system (CNS)] is first evident in the formation of the neural plate, a shoe-shaped plate of thickened dorsal ectoderm. The cephalic and caudal ends of the embryo have already been decided, and different domains along the longitudinal neuraxis are preparing to proceed through the neurulation process at their region-specific rates. Soon thereafter the edges of the neural plate begin to elevate, thus forming the neural folds. With time the neural folds continue to rise, ultimately approaching each other and fusing along the midline to form the neural tube. Fusion typically begins in the caudal cervical or cranial thoracic region and proceeds in both the cephalic and caudal directions. Fusion is delayed transiently at both the cranial (anterior) and caudal (posterior) neuropores, thus maintaining a temporary communication between the lumen of the neural tube and the amniotic cavity. The cranial neuropore closes at a slightly earlier stage than the caudal neuropore, in keeping with the slightly faster development of cephalic structures relative to caudal organs and tissues.

The brain and spinal cord are each derived from approximately 50% of the neural tube. The organization of these two regions during their subsequent development diverges once the neural tube has formed. The spinal cord assumes a segmental pattern in which each region serves a predefined and limited portion of the body. In contrast, the brain follows a suprasegmental template in which repeating segments are lacking and the various brain domains can control multiple targets throughout the body. The brain stem and midbrain represent an intermediate pattern, part segmental (e.g., in the columnar arrangement of the motor and sensory

nuclei of the cranial nerves) and part suprasegmental. The molecular basis for such segmentation is the restricted expression patterns of many molecules, the best example of which is the Hox gene family, which designates the locations of hindbrain rhombomeres.[38]

Soon after the cranial neuropore closes, the cephalic region of the neural tube dilates in three contiguous regions to form the three primary brain vesicles. These bulges are the prosencephalon (forebrain primordium), mesencephalon (midbrain primordium), and rhombencephalon (hindbrain primordium). Two prominent flexures of the neural tube arise at the same time: the cephalic flexure in the midbrain area and the cervical flexure at the junction of the hindbrain and spinal cord. With additional time, the three primary brain vesicles are reorganized to produce the secondary brain vesicles, which represent the initial zonal specification that directs subsequent brain compartmentalization; the prominence of the suprasegmental brain division observed in some species (e.g., primates) is initiated at this time and is often designated *encephalization*. The prosencephalic vesicle is partitioned into two distinct domains: the dorsally located telencephalon or primordial cerebral hemispheres, which arise as projecting hollow bulges with constricted necks, and the ventrally placed diencephalon, from which the optic vesicles arise. The mesencephalon, which is the most primitive brain vesicle, is not partitioned at this time except for the genesis of a deep furrow (the rhombencephalic isthmus) which separates it from the rhombencephalon. Concomitantly, the rhombencephalic vesicle is divided into the cranially located metencephalon, from which the cerebellum and pons will form, and the caudally positioned myelencephalon or primitive medulla oblongata. The boundary between the metencephalon and myelencephalon is marked by the pontine flexure.

TABLE 1   Timing of Major Neurodevelopmental Events in Species Used for Developmental Neurotoxicological Research[a]

Event	Developmental Age (Time After Conception)						
	Chicken	Mouse	Rat	Guinea Pig	Rabbit	Rhesus Macaque	Human
Length of gestation/*in ovo* incubation	21	19	22	67	32	167	267
Implantation occurs	None	4.5–5.0	5.5–6.0	6.0–6.5	7.0–7.5	9	6–7.5
Primitive streak forms	7–19 h	6.5–7.0	8.5–9.0	12.0–13.0	7.25	15–17	13.5–17
Neural plate develops	20–24 h	7.0	9.0–9.5	13.5	8.0	20	18–19
Neural folds evident	22–26 h	7.5	9.0–9.5	14.0–14.5	7.75–8.25	20–21	19–21
First somite visible	24 h	7.75–8.0	9.5–10.0	14.5	7.75–8.25	20–21	19–21
Neural tube closure begins	26–29 h	8.0	10.25–10.75	14.0–15.0	8.5–9.0	21–23	22–24
Primary (3) brain vesicles present	33–38 h	8.0	10.5–12.0	15.0–15.5	8.25–9.5	25–30	25–27
Otic primordium observed	29–40 h	8.9–9.0	10.5–10.75	15.0–15.5	8.25–9.0	23–25	27–29
Neuropore closure: cranial (anterior)	48–54 h	9.0–9.1	10.5–10.75	15.2–15.5	9.0–9.5	24–26	24–26
Optic primordium seen	29–33 h	8.0–8.5	10.5–11.0	15–16	8.5–9.0	21–23	24–26
Neuropore closure: caudal (posterior)	38–44 h	9.0–9.5	10.5–11.5	15.3–15.5	9.5–10.5	25–31	25–28
Rathke's pouch (adenohypophysis) forms	50–52 h	8.5–9.0	10.5	15.5	9.5–11.0	28–32	28–34
Secondary (5) brain vesicles evident	50–56 h	10.0	11.5–12.0	17.0–17.5	11.0	29–33	30–35
Neurohypophysis evagination begins	3.0	11.5	11.5–11.75	18.5	12.0	30–34	30–42
Ventricular zone forms	—	9.25–10.0	11.5–12.0	—	—	—	35–42
Cerebral hemispheres bulge	3.0	10.0	12.0–12.2	17.5	11.0	29–33	30–35
Subventricular zone develops	—	10.5	12.5	—	—	—	35–42
Pontine flexure evident	—	10.0	13.5	—	10.0	30–33	35–38
Olfactory bulbs observed	—	11.0	13.5	23.0	14.0	38	37
Cochlea and vestibular apparatus seen	—	11.0–11.5	12.0–13.0	—	—	30–34	35–42
Olfactory neuroepithelium forms	4–6	12.0	12.5	—	—	—	35–40
Choroid plexus (lateral/ fourth ventricles)	—	10.5–13.0	13.5	—	—	36–42	48–55
Cerebellum (primordium) initiated	4.5	12.0	14.0	19.0	15.0	30–36	37
Purkinje cells (cerebellum) formed	—	10.75–13.5	13–15	—	—	—	35–50
Retinal differentiation begins	4.0–4.25	11.5–12	—	—	—	—	42–44
Cerebral arteries sprout	—	10.5–11.5	—	—	—	—	42–49
Optic nerve present	4.0	13.0	14.0	21.5	15.0	39	46–48
Fourth ventricle develops	—	9.0–9.5	—	—	—	—	35–42
Pineal gland (primordium) forms	4.0–4.5	11.5	14.0–14.5	—	—	32–38	33–48
Internal capsule observed	—	—	—	—	—	—	50–55

TABLE 1 (*Continued*)

Event	Developmental Age (Time After Conception)						
	Chicken	Mouse	Rat	Guinea Pig	Rabbit	Rhesus Macaque	Human
Meninges evolve	—	12.5–14.5	—	—	—	—	50–55
Spinal nerve roots present	—	—	—	—	—	—	50–55
Optic chiasm expands	—	13.5–14.5	—	—	—	—	60–70
External granular cell layer (cerebellum)	—	13–17	—	—	—	—	100–120
Myelination begins (caudal CNS)	—	22	26	—	—	—	100
Peak growth, diencephalon	—	—	—	—	—	—	260
Peak growth, telencephalon	—	—	—	—	—	—	360
Peak growth, rhombencephalon	—	—	—	—	—	—	450
Neuronogenesis in specific domains							
Birth	21	19	22	67	32	167	267
Brain growth spurt (peak)	—	26	29	50	45	120	280
Amygdala	—	11–15	12–22	—	—	—	30–130
Cerebellum	—	11–35	12–40	—	—	—	30–270
Granular cells produced	—	20–35	26–40	—	—	—	170–270
Purkinje cells produced	8–12	11–13	14–15	—	—	—	40–50
Cerebral cortex (neocortex)	—	11–17	14–22				40–130
Gyri form	—	N.A.	N.A.	—	—	—	100
Lamination occurs	—	14.5–19	16–21	—	—	—	140–170
Corpus striatum	—	12.5–20	14–25	—	—	—	50–170
Entorhinal cortex	—	—	14–17	—	—	—	40–55
Hippocampus (CA 1–3)	—	12.5–16	15–20				45–100
Hippocampus (dentate gyrus)	—	17.5–35	19–40	—	—	—	70–270
Hypothalamus	—	11–14	12–19	—	—	—	30–85
Limbic cortex	—	—	14–22	—	—	—	40–130
Medulla	—	—	11–16	—	—	—	24–52
Mesencephalic tectum	—	10–16	12–25	—	—	—	30–55
Mesencephalic tegmentum	—	—	12–17	—	—	—	30–170
Pallidum	—	—	12–16	—	—	—	30–52
Piriform cortex	—	—	13–18	—	—	—	35–70
Pons	—	—	11–16	—	—	—	24–52
Spinal cord	—	9.5–14	11–16	—	—	—	24–52
Subiculum	—	—	15–19	—	—	—	45–85
Thalamus	—	11–16	13–18	—	—	—	40–70
References	251,252	33,133,197,251, 253–255	19,50,130, 251,253, 256	251–253	251–253	251–253	19,50,130, 251,253, 257

[a] Times are in hours (h) or days (no unit given). Day of conception is designated gestational day 0. CNS, central nervous system, N.A., not applicable.

The cavities within the secondary brain vesicles, as well as the central canal of the spinal cord, form the ventricular system. The lateral ventricles develop inside the cerebral hemispheres of the telencephalon and communicate via the interventricular foramina (of Monro) with the third ventricle inside the diencephalon. The third ventricle is connected in turn through the narrow mesencephalic aqueduct (of Sylvius) with the fourth ventricle, a cistern bounded dorsally by a thin membranous roof, as well as laterally and ventrally by the metencephalon and myelencephalon. The fourth ventricle drains into the subarachnoid space via the lateral foramina (of Luschka) and (in nonhuman primates and humans) the median aperture (foramen of Magendie) in the membranous roof. The fourth ventricle maintains its connection to the central canal of the spinal cord. The choroid plexus first forms when the secondary brain vesicles are evolving, appearing first in the fourth ventricle and subsequently in the lateral ventricles.[39]

The spinal cord is the descendant of the caudal neural tube and retains a patent central canal throughout its length. The layering of the spinal cord is equivalent throughout the early developmental stages. However, regional specialization does develop over time as various spinal cord domains become affiliated with distinct segmental functions [e.g., the cervical and lumbar intumescences, or enlargements, which are associated with the junction of the peripheral nervous system (PNS) trunks from the forelimbs and hindlimbs, respectively].

Throughout the CNS, neurons arise from the neuroepithelium, a pseudostratified layer of small, dark, rapidly proliferating stem cells located adjacent to the lumen of the nearest cavity of the ventricular system. Soon after neural tube closure, neuroepithelial stem cells begin asymmetric divisions to yield two daughter cells: a neuroblast (i.e., a committed neuron progenitor) and a new stem cell. Neuroblasts have large round nuclei and pale nucleoli and come to reside in the mantle layer located just lateral to the neuroepithelium. This mantle layer will ultimately form the gray matter. In the spinal cord, the nerve fibers arising from neuroblasts in the mantle layer pass outward to reach the marginal layer, which eventually forms the white matter tracts. In the brain, neurons are formed in successive waves, with neurons from a given population usually arising from the same wave.

With continued production of neuroblasts, the mantle layer of the neural tube develops distinct bilateral thickenings throughout most of its length. The basal plates are prominent expansions (especially in the spinal cord) that contain the nascent motor neurons, while the alar plates are distinct but generally less-pronounced bulges that harbor sensory neurons. The basal and alar plates are demarcated by a shallow longitudinal groove, the sulcus limitans. In the thoracic region of the spinal cord, an additional column of autonomic neurons forms an intermediate (or lateral) horn at the junction of the alar and basal plates. The regional organization of the alar and basal plates in the brain stem (metencephalon and myelencephalon) resembles that of the spinal cord (Chapter 2,

Figure 5),[30,34] although the incomplete dorsal closure of this region has displaced the alar plates away from the midline to instead rest lateral to the basal plates. Derivatives of the alar and basal plates are intermingled in some portions of the pons and medulla. The diencephalon is thought to harbor only alar plates, as this region's primary functional specialization is nonmotor tasks. The presence of distinct alar and basal plates cannot be discerned in the telencephalon.

CORRELATIVE NEURODEVELOPMENT

Neural structures, particularly in the CNS, consist of populations of functionally similar neurons arranged in stereotypical ("somatotopic") patterns. These arrangements are indicated both by obvious external or internal anatomical landmarks (e.g., surface gyri delineating the lobes of the cerebral cortex and cerebellum),[4,5,7] cytoarchitectural features [e.g., regional variations in the morphology of cerebrocortical granular cells (small intracortical interneurons) and pyramidal cells (large projection neurons)[6,40]], and also by differential expression of different molecules (e.g., cytochrome oxidase,[41] 5'-nucleotidase,[42] and zebrins[41,43] in the cerebellum). In general, these morphological zones correspond to evolutionary and functional partitions between various brain regions.

Basic structural domains in the nervous system are often highly conserved among species, although the degree of similarity may diverge substantially, depending on evolutionary imperatives that dictate each species' lifestyle and habits. The degree of development within the sensory centers in the cerebral cortex provides the most conspicuous example of this principle. For example, approximately half the primate cortex is dedicated to vision (the primary sensory modality in these species). In contrast, in rodents, the same proportion or more of the cortex is directed toward the auditory, olfactory, and tactile [for vibrissae (whiskers)] senses, while the small visual domain provides sight but not fine spatial resolution or color perception.[19] The proportional sizes of the corresponding neural tracts are scaled to match the relative sizes of the sensory cortical domains. For example, the olfactory mucosa and olfactory tract in macrosmatic (olfactory-oriented) rodents and dogs are much larger than those in microsmatic primates. This divergence reflects the greater importance of olfaction in the two former species and the reliance on vision in the latter.

Regardless of the species, cortical functions develop in a regimented fashion.[16,17,44,45] The somatic motor region in vertebrates is always located rostral to the somatosensory cortex. Within both these zones, body regions are projected in a somatotopic fashion, with areas representing the head localized in ventrolateral positions and those for the lower pelvic limb residing in dorsomedial domains just off the midline. Contiguous cortical regions begin to function in a distinct sequence. For example, maturation of the entire somesthetic cortex occurs before that of the auditory cortex, which in turn develops more precociously than does the visual cortex. This temporal pattern follows the order in which an altricial neonate (i.e., one that is not developmentally mature at birth) first begins to use sensory pathways to explore the world.

Rolling acquisition of full function in various neural domains closely follows regional waves of myelination. In the rat, for example, myelin first appears in the cerebrum within the large white matter tracts and commissures, spreads to other portions of the ventral forebrain (such as the thalamus and corpus striatum), and reaches the cerebral cortex much later in development.[46] Fiber systems serving similar tasks are myelinated at the same time. As far as is known, this process is approximately comparable in mammals of relevance toxicological testing.

In general, neural function is acquired in more cranial divisions at a slightly earlier stage of development than are the comparable roles of caudal divisions.[6,47] For example, activation of thoracic limb domains occurs both morphologically and electrophysiologically at an earlier time than does that of the pelvic limb. The rostral portion of the corpus callosum precedes that of the caudal part.[7] The corticospinal fibers of the pyramidal tracts develop caudally in a similar sequence. In all regions, the fields serving the upper extremity develop first, followed by those for the lower extremity, digits, and finally, the face.[6,47]

The immature brain not only lacks some structures for extended periods during development (e.g., a fully formed cerebellum, genesis of which is a late prenatal or early postnatal event, depending on the species), but also has some transient neural features that are evident early during neurogenesis but which regress at some point after the nervous system has been formed.[48] Prominent cytoarchitectural examples of such temporary structures that are evident in conventionally stained brain sections include the thick layer of neuroepithelium (germinal matrix) located beneath the ependymal lining of the lateral ventricles, the subplate zone of "waiting" neurons between the cerebral cortex and white matter in human neonates,[49,50] and the external granular cell layer on the surface of the cerebellar folia (reviewed by Rodier[16]). The numerous extra neurons and dense synaptic beds that develop in multiple brain regions during early postnatal life before regressing during adolescence also fit this description. In general, these transient structures serve either as sources of precursor cells for newly forming neural regions or as committed but incompletely differentiated neurons to help repair sites with damage.

COMPARATIVE NEURODEVELOPMENT

In general, the anatomical features, functions, and endogenous chemical and molecular composition of a given neural region are comparable among species, and within a given species and strain they are almost identical among normal (control or "wild type") individuals. Thus, stereotaxic coordinates reliably localize the positions of CNS structures within a species, if not their precise margins, and for individuals of like size such coordinates may be extrapolated between strains with considerable accuracy.[51] However, minor discrepancies in the locations and extents have been demonstrated in various brain regions (even in age-, gender-, and weight-matched individuals). Such subtle distinctions have been observed in mice,[52–54] rats,[51] and primates,[55–57] including the brains of monozygotic human twins.[58]

Among species, structural differences may exist in the absence of functional variations, while dramatic functional divergence may occur in the presence of apparent structural homology.[59] Many neurobiological differences have been reported that could affect the choice of one species over another for neurotoxicity bioassays (reviewed in several books and articles[60–64]). Examples include disparities in neurochemistry, behavior, and xenobiotic responsiveness that exist between rats of different strains, genders, and ages,[65–68] as well as those noted between rats and mice.[69,70] Subtle differences also exist between various primate species.[71–73]

Structural Divergence Among Mammalian Species

Neural organization among adults of the different mammalian species utilized in neurotoxicological research follows one of three general models based on the prominence of the cerebrocortical subdivisions: the archicortex [the most primitive phylogenetic derivative, consisting of the olfactory bulbs, olfactory tracts, and olfactory cortex (the piriform cortex in mammals)], paleocortex (the next oldest region), and the neocortex.[74] The first is the *rodent* model (guinea pig, mouse, rat, rabbits, etc.), which is characterized by a lissencephalic (smooth) cerebral surface and large archicortex. The second pattern is the *carnivore* (e.g., cat and dog), in which the cerebrum exhibits prominent gyri and sulci but abuts the cerebellum without extending over it. The final model is the *primate* (monkeys, humans), where the cerebrum features many gyri and sulci and is expanded to such a degree that it covers the cerebellum. The primate category can be further subdivided into modestly fissured platyrrhine (e.g., aotus, cebus, marmoset, squirrel monkey), intermediate catarrhine (e.g., baboon, cynomolgus, macaque), and markedly fissured catharine (chimpanzee, human) groups.[71,73] For practical purposes, however, the various primate subdivisions may be considered as a group because the least partitioned (platyrrhine) conformation still possesses a markedly greater degree of cerebrocortical complexity relative to the carnivore-type brain.

Despite the similarity of various brain domains in adults, the pace at which the adult neural arrangement is attained during neurodevelopment varies considerably among mammalian species. With the exception of guinea pigs and sheep (i.e., "precocial" species that are relatively mature and mobile at birth), common laboratory animals as well as humans are altricial (i.e., born with developmentally immature nervous systems). The brain growth spurt (BGS), the period of most rapid volumetric expansion and neurogenesis, occurs at different times in different species (Table 1). The BGS in humans peaks in the days just before and after birth, while the rodent BGS occurs about postnatal day (PND) 10.[75] This relationship is the rationale for equating the three trimesters of a full-term human gestation to the first 30 postconception days in rodents; in other words, in mice and rats the first 10 postnatal days are approximately comparable to the third trimester. The BGS tapers off slowly over the first three to four weeks of life in rodents and the first two years in human infants.

The spinal cord is modified during development to supply the appropriate number of fibers needed to innervate the body area. Thus, the quantity of white matter (assessed in cross section) increases more quickly than gray matter in larger species.[3] This pattern is more prominent in the dorsal (posterior) funiculus, the main position for the tracts transmitting most tactile and proprioceptive impulses, and is especially true in carnivores and primates relative to rodents. In addition, the dorsal funiculus as a percentage of total white matter content is much bigger in carnivores and primates (almost double that of rodents),[4] probably because it serves to relay fine sensation from the digits in these animals.

The development of the spinal motor tracts exhibits distinct differences in mammalian species used in neurotoxicity research. The two principal pathways are the corticospinal tracts (which provide cortical control of motor functions) and the rubrospinal tract (which manages subcortical direction of motor activities). The size of the two corticospinal tracts is increased progressively in phylogenetically advanced species, forming about 10% of the total white matter in carnivores, 20% in nonhuman primates, and 30% in humans.[3] The larger segment of this pathway travels in the lateral funiculus as the lateral corticospinal tract, while the functionally equivalent fibers in the rat reside in the dorsal funiculus as the dorsal corticospinal tract.[76] The corticospinal axons that form the lateral tract decussate in the pyramids of the pons, which are most apparent in primates. Essentially all axons cross in the rodent and dog, while only 85% decussate in primates. The rubrospinal tract is much smaller in primates than in rodents and carnivores. The two direct routes from the spinal cord to the thalamus, the medial lemniscus (which carries sensory information from the contralateral nuclei gracilis and cuneatus) and the spinal or lateral lemniscus (representing the conjoined sensory fibers of the spinothalamic and spinotectal tracts), are enhanced in carnivores and primates because their thalamic and cortical targets are larger.

The structure of the cerebellum is highly correlated with an animal's type of limb movement, center of gravity, and posture.[10] The sizes and shapes of different cerebellar divisions vary in mammals of toxicological significance. Examples include the greater lateral extension of the cranial lobe in primates relative to rodents and carnivores, and the larger size of the lingula in animals with long tails. The parafloccular lobule is relatively large in rodents, carnivores, and monkeys, but is reduced to insignificance in humans. The lobules of the middle lobe are unvaryingly small in rodents and carnivores, but their lateral portions are much larger in primates, due to an increase in the length and number of the cerebellar folia. The profile of the vermis on the caudal surface of the cerebellum is generally straight in rodents and primates, but takes a twisting course in carnivores. The overriding principle of cerebellar development is that mammals with well-developed limbs capable of extensive independent movement (especially primates) have expanded lateral cerebellar hemispheres. The corticopontocerebellar pathway that connects the cerebellum to higher brain centers is most highly developed in primates. The cerebellar nuclei of rodents and carnivores are not as discrete as those found in primates.

The subcortical centers exhibit both structural and functional differences among mammals. The corpus striatum (basal ganglia) is unified in rodents but divided by the internal capsule into distinct caudate and lentiform (i.e., putamen and globus pallidus in combination) nuclei in large mammals. The subcortical centers of primates cannot function independently to the degree that the same domains do in other animals. Therefore, extensive lesions in the motor cortex may have a negligible impact in lower animals but result in permanent deficits in primates. The vulnerability of large masses of gray matter (cerebral cortex, basal ganglia) in humans increases near the end of gestation and for several months after birth,[77] due to the intense synaptogenesis in these areas. It is interesting that leg functions often recover more quickly than those of the arms in human patients with unilateral lesions of the motor cortex. This disparity is thought to reflect the greater control of pelvic limb tasks by local spinal cord reflexes.

The prominence of various cerebral cortex structures varies with the species. For example, the volume multiplies as a cubic function, while the brain surface area (excluding the formation of folds) increases as the square.[3] Similarly, the quantity of white matter increases as a cubic function while that of the gray matter increases as the square.[3] Thus, bigger animals have a higher white matter/gray matter ratio than that of smaller ones. Primates have large brains for their body size,[78] due mainly to extensive cerebrocortical expansion. It is interesting that one cerebral hemisphere (usually the right) is larger and weighs more than the other in most species (e.g., cat,[79] dog,[80,81] human,[82,83] mouse,[79] rabbit,[79] rat[79]), regardless of "hand" preference. Such size discrepancies may or may not correspond to the known hemispheric preference of certain functions (e.g., left-sided location of the speech center in humans), as the sizes of the major functional domains (e.g., motor and sensory centers) that are conserved across species are bilaterally symmetrical.[81] Such structural asymmetry develops before birth, and functional asymmetry becomes evident soon after birth.[45]

Basic cerebrocortical topography among mammalian species is equivalent, although some pronounced structural distinctions exist. The most obvious neuroanatomical difference is the variable presence of sulci on the brain surface. The gyri covering the cerebrocortical surface give the primate brain a surface area 400 times that of the mouse.[84] Where present, the pattern of gyri and sulci provide a species-specific signature[10] that does not change with age,[84] so the nomenclature of all but major gyri and sulci are not applied readily across species, especially when comparing the human pattern to those found in carnivores. For a given species, the pattern of partitioning also varies between the two hemispheres.[4,81] At the simplest level, gyri and sulci are absent in rodents (lissencephaly) but numerous in larger mammals. However, fetal carnivores and primates are lissencephalic until relatively late in gestation.[30,31,34,85,86] The increased folding of the primate brain surface stems almost entirely from brain enlargement rather than from the addition of new connections and functions.[87]

Brain regions delineated by macroscopic structural landmarks differ in their prominence across species. In the rat, the caudal margins of the cerebral hemispheres do not hide the pineal gland and caudal colliculus, while in larger animals the close apposition (carnivores) or overgrowth (primates) of the cerebral cortex and cerebellum hides the midbrain. The olfactory brain (archicortex) of macrosmatic (olfactory-dependent) rodents and carnivores is much larger than the corresponding structures in vision-dependent (microsmatic) primates.[2] The relative sizes of the frontal cortex surface in the dog and human are similar. However, the occipital pole of the dog extends much further rostrally than does the corresponding region of the human brain. This results in the parietal lobe being reduced in size in the dog.[10] The temporal lobe of the dog also extends farther dorsally than does that of humans, further reducing the size of the parietal lobe in the former.

Cytoarchitectural features vary among neural cell types (especially neuronal populations) in each brain region. For example, neurons with divergent morphological characteristics are thought to serve different functions: small, dark, round cells (e.g., cerebrocortical granular cells, which are much more numerous in primates[84]) are thought to serve as regional interneurons for associative tracts, whereas large, pale, polygonal neurons with large nuclei (e.g., pyramidal cells) are considered to be projection neurons.[6,40] In many instances equivalent positions within the cerebral cortex and basal ganglia of rats and primates have morphologically dissimilar neurons,[88] and such cell type–specific character-

istics appear to reflect structural specialization to perform a defined function. The mediodorsal thalamic nucleus projects to different locations in the frontal cortices of rats[89] and monkeys,[90] so location-matched cortical fields of the two species differ in their cytoarchitecture. In like manner, functionally equivalent projections of the visual field occur in different cortical regions of the cat[91] and monkey.[92] Altering the projection of functional brain connections may simply shift the cytoarchitectonic features to other brain domains, as has been shown for the visual field projections in different species of monkeys.[92,93] Thus, functional areas of the brain are much less bound to gross anatomical landmarks than to specific connectional pathways (architectonic units). This reorganization may occur on an individual basis[57,94] or be characteristic of all members of a given species. Accordingly, care must be taken that the apparent anatomical differences are true and not just based on inadequate or biased sampling.[95] Neural cells with apparently unique cellular features have been described, and their presence can vary among species. For example, while all mammals have conventional stellate astrocytes, monkeys but not rats or dogs contain astrocytes with very long processes that bridge multiple cortical laminae in certain regions of the cerebral cortex.[96,97] Such unique elements probably reflect distinct functional specialization in glial populations[98–100] similar to that expressed by different neuron populations.[6]

The location of the major functional domains in the cerebral cortex are equivalent among all mammalian species, but their extent varies widely.[4] The most telling interspecies difference in cortical function is the ratio of paleocortex (which is engaged largely in tasks mediated by the limbic system) to that of neocortex (which participates in associative and cognitive activities). The cerebral cortex of rodents and rabbits consists almost entirely of paleocortex, so these species possess few associative and cognitive abilities. The carnivore cortex dedicates 80% of its surface area to the primary sensory and motor areas and olfactory cortex, while these same functions in the primate encompass less than 20% of the area. The relationship is reversed for the associative areas of the neocortex in these two species. Accordingly, cortical interneurons play a much larger role in the primate brain than in the rodent brain.[88] The association fields related to the primary somatic domains are prominent in primates and much less so in carnivores.[2] The visual cortex is located more laterally in rodents than in carnivores and primates, where the visual field has been shifted over the tip of the occipital pole by the caudal expansion of other brain regions.[2] The primary motor and sensory areas of the rat and cat are rostral to those of the dog, resulting in a substantially reduced frontal "lobe" in the two former species. Taken together, these principles indicate that rodent data should be used with care (if at all) in studies designed to explore the impact of neurotoxicants on higher cognitive functions.[60,101]

Structural Divergence Among Strains

Within a species, strain differences in regional neuroanatomy are relatively minimal. Known examples include the small brain size of purpose-bred albino rats relative to wild Norway rats[61] and the modest variance in positioning[51] of specific structures in the cerebral cortex. In contrast, numerous disparities among the mouse[102–104] and rat[67,68] strains have been documented in behavior (especially motor and stereotyped actions), neurochemistry (hormone production, regional neurotransmitter levels, etc.), and responsiveness to xenobiotics. The ordering of strains with respect to behavior, quantity of a neurochemical, or susceptibility or resistance to a xenobiotic varies with the endpoint being measured as well as among the strains being tested.[62] Differences in neurological function between strains may not be associated with neuronal changes, although glial cell numbers may be increased in some instances in which functional deficits become progressively worse with age.[105] Similarly, apparently incidental structural changes such as mineral deposition, neuronal inclusions,[106] or protein deposits[107] may develop in some strains over time.

Avian Neuroanatomy

The chicken is a common model used in testing organophosphate chemicals for delayed neuropathy,[108,109] although in general neither fertilized eggs nor chicks are employed in developmental neurotoxicity research. Nevertheless, the basic features of neurogenesis in chick embryos are consistent with the process in mammalian embryos. Accordingly, a brief review of avian neuroanatomy is warranted, even though birds exhibit several distinctive neuroanatomical characteristics that differ markedly from the mammalian pattern. Each spinal cord segment in mammals is found rostral (superior) to the like-numbered vertebra because the spinal cord does not extend into the caudal lumbar and sacral vertebrae (which instead contain the cauda equina). In contrast, the avian spinal cord occupies the entire length of the vertebral canal. The cervical and lumbosacral portions of the spinal cord are twofold longer in birds than in mammals, while the thoracic cord is reduced by half. The prominent sinus rhomboidalis is unique to the lumbosacral region of the bird, where it opens dorsally along the midline and is filled with a clear jelly (the glycogen body). The dorsal funiculus of birds (the cuneate and gracilis tracts for sensory fibers traveling to the cerebral cortex) is much smaller than that of mammals, presumably because widely spaced feathers do not permit as much fine skin sensation as does densely haired skin.[3] In like manner, the medial lemniscus, which represents the rostral extension of the dorsal funiculus into the brain, is smaller in birds. Modulation of spinal motor neurons via higher brain centers depends more on cerebellar than on cortical input in birds, so the cerebellospinal tract

(located just medial to the spinocerebellar tract in the lateral funiculus) is much more prominent in avians to compensate for the absence of the corticospinal tract. The bird cerebellum has a well-developed middle portion, equivalent to the mammalian vermis, which regulates symmetrical limb movements.[10] Unlike mammals, the most laminated portion of the avian brain is the optic tectum (dorsal midbrain), which processes visual signals. The bird cerebral cortex is poorly laminated and thin in comparison to the mammalian brain, while the basal ganglia are massive (reflecting the high degree of automated reflex motor activity that is managed at subcortical levels). In the chicken, the basilar artery is derived from only one caudal communicating artery.[7] The side from which the communicating vessel arises appears to be sex-linked. Birds have only two meninges, whereas mammals have three.[110]

Structural Divergence Between Genders

A variety of gender-related differences in behavior and drug responsiveness may be appreciated in neural structures and functions, including humans.[111,112] Anatomical variations include the existence of sexually dimorphic nuclei and white matter tracts as well as distinctions in the number and location of sulci and the neuronal density in various brain regions.[83,112–115] For example, aged female C57BL/6 mice produce significantly more astrocytes and microglia (by approximately 25 to 40%) in the hippocampus relative to age-matched males.[116] Biochemical disparities, such as regional differences in the quantities and circadian cycling of hormones[117] and neurotransmitters[62,118] have been recorded in rodents[62] and primates.[115] The explanation for these differences is presumed to be the influence of the individual's complement of gender-specific hormones.[116] These gender-based differences first arise during gestation. Human males do not produce as many extra neurons during neurodevelopment as do age-matched females, but the males retain more neurons throughout adulthood.[112] These numerical differences in neuron number have been postulated to have clinical significance, as young males are predisposed to learning deficits and mental retardation (due to reduced neural connectivity during development), while elderly females experience a higher incidence of dementia (via reduced functional reserve over time).

Structural Divergence During Neurodevelopment

The most rapid changes in neural anatomy and physiology occur in utero and during the first few months (or years, in long-lived species) of postnatal life.[44,45,119] When the periods in which specific neural changes occur are related to the lifespan of the animal, a pattern emerges in mammalian species which suggests that age-related changes after maturation result from an extension of the neural and endocrine mechanisms that specify earlier development.[120] A plethora of progressive anatomical changes related to increasing developmental age have been documented, including the gradual increase in gyri,[30,31,34,85,86] regional differences in the rate of CNS myelination,[46] variable neuronal densities in different brain regions,[115] and the reduction in the number of glial cell nests in the subependymal layer of the paleocortex.[121] Functional variations among age groups may be considerable. For example, the blood–brain barrier in developing[61] and older[122,123] individuals is more permeable than that of mature but not senescent adults. Levels of neurotransmitters,[62] their receptors,[118] and their synthetic pathways[124] increase with age in the brains of rodents and primates. Age-related differences in neurotoxicity of xenobiotics may be considerable, due to differences in the efficiency of local and systemic detoxifying enzymes.[62]

PRINCIPLES OF VERTEBRATE NEURODEVELOPMENT

The entire process of neural development represents a cascade of overlapping events. Many of the major events are stereotypical across vertebrate species, and thus serve as examples of fundamental principles of neurodevelopment. A firm understanding of such principles is imperative when designing and interpreting developmental neurotoxicity studies.

Neurodevelopmental events are subject to two chronological scales, one reflecting the absolute time that has passed since conception and the other reflecting an event's timing relative to earlier, concurrent, and later events. Within a species, the absolute time frame with respect to conception [typically defined as "embryonic day" (E) or "gestational day" (GD)] or birth [generally termed "postnatal day" (abbreviated P or PND)] is a convenient means of describing when a given developmental process is taking place; in humans, the absolute time frame for gestation is usually founded on the date of the last menstrual cycle. However, a more convenient mechanism for comparisons across species, breeds or strains, and differing husbandry conditions is to use staging systems based on readily observable morphological criteria. Indeed, the progressive alteration of macroscopic CNS structures during neurogenesis may serve as major landmarks for assigning development age to embryos and fetuses.

An obvious but often underappreciated concept is that the sequence of time during development is one-dimensional. An early event can affect one or more processes that occur either concurrently or subsequently, but later events cannot affect earlier ones. For example, the number of neurons produced (an early event) coupled with the subsequent removal of improperly integrated neurons (an intermediate event) work sequentially to specify the number of neurons residing in various regions of the nervous system (a late outcome). Another general principle is that the ultimate outcome of a toxicant exposure may depend on the interaction between two or more events anywhere along the cascade, although typically the result reflects damage that can be associated directly with the timing of the exposure. A consequence of these concepts is that the production of any given population of neural cells is controlled by the summation of preceding events that affect their proliferation, migration, differentiation, and/or survival. Another corollary principle is that the earlier during development that an event takes place, the greater its potential to influence later events and the final conformation and function of the nervous system for both good (i.e., in stipulating the proper development of the nervous system) and ill (e.g., the induction of congenital anatomic defects and/or functional deficiencies if the event is disrupted). Indeed, gross malformations of the brain and spinal cord typically require that the insult occur during neurulation,[125,126] the earliest stage of organizing the nervous system.

Another broad time-related tenet is that the nervous system and the various domains within it are differentially susceptible to adverse stimuli, including toxicants, based on their diverse critical periods.[23,127] Neurons are produced continually in the brain during development, but production in specific domains waxes and wanes depending on the structure being built.[119] Neuron production in all vertebrate species is highest prior to birth,[128] often occurring in advance of the BGS, but neuronogenesis continues after birth in all mammalian species used in toxicological neuropathology research (especially highly altricial ones such as rodents) as well as in humans.[44,45,129] Some CNS regions experience an extended phase or multiple periods of elevated neuron generation.[119,130] In general, times of region-specific neuron production are critical periods of vulnerability to many neurotoxicants.[127] Glial cells (especially astrocytes and oligodendrocytes) are generally fabricated in higher numbers near and after birth (i.e., in association with the brain growth spurt)[129,131] and can be produced well into senescence. Therefore, neurotoxicant exposures that affect glial cells typically occur late in gestation or after birth during the major phase of glial production.[131,132]

The CNS is still immature at birth by most functional and some structural measures (e.g., degree of cerebellar formation), although the extent of maturation varies substantially across species. Events that happen before birth in some precocious species are postnatal developments in other species (Table 1). For example, in rodents, neurons of the cerebral cortex are produced during the second half of gestation,[133] whereas in primates (including humans) the same neuronal populations are in place by the end of the second trimester.[130,133] Therefore, general conclusions regarding neurodevelopmental processes are best made using relative timing of events, and even then they must be compared rigorously to experimental findings in the species in which the events of interest are being evaluated.

Neural structure can be affected by postnatal experiences. For example, an enriched environment results in an increased number and density of dendritic spines projecting from rat neocortical[134] or hippocampal[135] neurons. This effect can be produced in rodents by stimulation during the preweaning period[136] (i.e., equivalent to the third trimester and first postnatal years in humans) and is mediated in part by the activity of regional astrocytes.[135] Furthermore, the increased neuronal connectivity is sufficient to preserve function substantially following neurotoxicant exposure.[134] The proposed mechanism for this change is that behavioral enrichment leads to an increase in the retention of synapses and/or a reduction in programmed cell death of underutilized neurons—in short, to reduced pruning of connectivity during adolescence and adulthood. Accordingly, to avoid the induction of structural (and functional) artifacts in developmental neurotoxicity research, considerable care will be required (especially in designing studies in which quantitative morphometry will be a neuropathology endpoint) to ensure that all treatment groups experience comparable environmental conditions.

MECHANISMS OF NEURODEVELOPMENTAL VULNERABILITY

The CNS is the most vulnerable of all body systems to developmental injury, including toxicity.[137] The susceptibility of emergent neural tissues to toxic damage is dependent on many different mechanisms. Toxicants may affect one or many events. The severity and extent of the resulting neurodevelopmental dysfunction will depend on the dose of toxicant received, the timing of exposure,[19,119,127] and the number of cellular and molecular processes that are disrupted.[20,21] In general, developmental neurotoxicants act by upsetting the production, positioning, and connectivity of one or more neural cell populations.[22,138] These processes are unique to the immature nervous system in many respects. The impact of developmental neurotoxicants is aggravated by the large number of potential neural targets, the long periods required to generate the mature nervous system, and the imperfect protection afforded by immature shielding systems. In the remainder of this section are outline the fundamental mechanisms that help delineate the basic structure of the nervous system.

Cell Proliferation

Production of neural cells is a paramount feature of neurodevelopment. The cerebral cortex of the adult human contains approximately 10^{10} neurons and 10^{11} glial cells, all of which originate from a one-celled zygote. To achieve these numbers, the developing nervous system produces almost double these numbers.[139] The bulk of neurons that provide the foundation for the adult nervous system are produced prior to birth,[44,119,130] while (depending on the species) many to most adult-type glia are produced just before or after birth.[129,131] That said, populations of comparable brain regions among species may occur at different times relative to birth; in the hippocampus, rodent neonates have only 15% of the total cell complement at this site, whereas newborn primates have 80%.[140] As a general rule, large neurons begin forming before small ones do.[50] Radial glia (for scaffolding to guide neural organization) and microglia (to scavenge apoptotic debris) develop in parallel to neurons in most brain structures.[21] Depending on the species, other classes of glia form predominantly late in gestation (astrocytes before oligodendrocytes) or even after birth (especially rodent oligodendrocytes).

The earliest neural cells are generated from precursors residing in the neuroepithelium (or ventricular zone), a thick layer of uniformly small, dark, rapidly proliferating stem cells usually located adjacent to the lumen of the nearest cavity of the ventricular system. Soon after neural tube closure, neuroepithelial stem cells commence asymmetric divisions to generate two daughter cells: a neuroblast (committed neuron progenitor) and a new stem cell. Neuroblasts have large, round nuclei and pale nucleoli and come to reside in the mantle layer (located just lateral to the neuroepithelium), which ultimately forms the CNS gray matter. In the brain, the neurons are formed in successive waves, with cells that arise first settling in the deep gray matter layers while younger cells migrate through the deep regions to populate progressively more superficial gray matter zones.[141] The ventricular zone is the sole proliferative domain to appear in some parts of the developing CNS, such as the CA (cornu ammonis) divisions of the hippocampus.[142,143] In such areas, the ventricular zone produces both neurons and glia throughout neurodevelopment.

However, in other regions neural cell production shifts over time from the neuroepithelium to a supplemental area such as the subventricular zone (SVZ, a layer that is contiguous but superficial to the ventricular zone). These secondary domains remain active into adulthood in rats[144] and primates[145] to supply stem cells for replacement interneurons[146] and glia.[147] The SVZ may be distinguished from the ventricular zone using cytoarchitectural features (SVZ cells are pale and of variable size) and cell density (SVZ cells are less tightly packed). The SVZ has been proposed to be a more recent phylogenetic adaptation than the ventricular zone because older brain structures (e.g., hippocampus, hypothalamus) lack this domain entirely, whereas newer brain areas (e.g., neocortex, other diencephalic subdivisions) have a large SVZ.[148] Where both proliferative domains are present, the ventricular zone is thought to produce mainly neurons, while the SVZ contributes chiefly glia.[149] However, the SVZ does produce some neurons in the adult brain.[150] The presence of these different cell proliferation options beginning just as the first neurons are being produced[142] indicates that functional differentiation of neural cell types and major CNS subdivisions is well under way at a very early stage of development (approximately the second month of gestation in humans).

The cell cycle dynamics for both the ventricular zone and the SVZ have been explored in the developing CNS. The cell cycle in the mouse neocortical ventricular zone progressively lengthens from about 8.1 h to 18.4 h during the six-day period from E11–E12 to E17–E18 when neurons are being produced.[133,151] The length of the cell cycle in the mouse SVZ is similar.[149] The cell cycle is longer in primates,[152] probably as a consequence of the greater period during which neurons can be generated (about four gestational months in the human brain).[153] Interestingly, the number of cell cycles required to produce all the neurons in the human neocortex is only threefold higher than that needed to populate the mouse neocortex.[153]

Alterations in the number of cells produced during development is a primary anatomical finding in many instances of developmental neurotoxicity.[154] It is a particularly critical injury, as decreases in cell number tend to affect all of the subsequent neurodevelopmental events that are required to organize the normal neural circuitry successfully in the brain region affected.[21] Agents that have been shown to alter neuronogenesis typically disrupt DNA synthesis (e.g., antineoplastics, carcinogens), inhibit mitosis (e.g., colchicine, methyl mercury), or selectively kill cells (e.g., ionizing radiation, retinoic acid, valproic acid).[21,137] These toxicants tend to reduce organ volume rather than affect cell density, as developing neurons are preprogrammed to maintain their spacing. Deficiencies in tissue volume are induced by lower doses of toxicant than are alterations in cell density or frank structural lesions.[154] The reductions in cell number are generally permanent. Furthermore, they are typically not accompanied by the reactive gliosis response that is characteristic of degenerative and necrotizing lesions in the adult brain,[21,50,155,156] probably because the late-forming glia have yet to become available at the time of exposure. The capacity for reactive gliosis is acquired over time, beginning with the third trimester in humans[50] and after birth in rodents.[50] Accordingly, the most useful means of detecting a change in cell number is to take a quantitative measurement (e.g., total brain weight, regional brain weight, morphometrical assessment of brain/regional thickness or area or volume) rather than to rely on qualitative histopathological evaluation.[155,156]

Cell Migration

Basic neural circuitry has to be laid down during neurodevelopment, and in doing so postmitotic neural cells journey widely to reach the far-flung sites where they will reside during adulthood.[157] Two mechanisms have been demonstrated to account for this orderly relocation. The first is that newly produced cells move a short distance away from their site of origin before being passively shifted farther outward as additional cells are produced beneath them. Thus, the neurons that form first are located farthest from the proliferative zone and usually rest nearest the outer surface of the neural tube. This scheme is typical of poorly laminated or unlayered structures such as the hippocampal dentate gyrus, diencephalic divisions (e.g., thalamus, hypothalamus), many regions of the brainstem, and the spinal cord. The second mechanism is that differentiating cells migrate actively over extended distances. The neurons that arise early in development form the layers located nearest the proliferating zones, while later neurons move through the layers formed first to populate more superficial layers. This scheme is generally found in well-laminated CNS structures such as the cerebral cortex and several subcortical areas. Migration may occur in either a radial (outward) or tangential (sideways) direction, depending on the specific cell population.[158] Multiple modes and directions of migration may be active in a single neural domain.[159] In general, migratory cells in the telencephalon can migrate in both radial and tangential dimensions, but they are restricted to differentiating along one of a narrow set of lineages.[160] In contrast, migratory cells in other regions tend to travel strictly in the outward direction but have more options for the final lineage they assume once they arrive.[160]

Neuron migration occurs in phases. First, newly generated daughter cells commit to neuronal differentiation. These postmitotic neurons then move into position near radial glia (or another support cell) and establish an intracellular axis of polarity that prepares them to travel away from the proliferating zone.[161] These specialized glia are derived from the earliest neuroepithelial divisions and will remain in predefined positions until migration ends, their processes spanning from the ventricular zone to the brain surface[151]; these glia provide the migrating neurons with both spatial and temporal cues.[100,160] Next, the new neurons migrate away from the proliferating zone along the outward-radiating processes of the glial cells. The moving cells can alter their shape, rate, and mode of movement as needed in response to local environmental cues.[159] Multiple neurons in succession can travel along the same radial glial cell. Once the neurons attain their final destinations, they cease migrating and detach from their guiding glial cells, complete their differentiation program, and begin establishing connections with their predetermined target cells.

The neural crest (NC) is a special neural tube derivative that arises from the site where the neural folds fuse. Cells migrating from this tissue establish, among other structures, the peripheral connections of the parasympathetic and sympathetic divisions of the autonomic nervous system (ANS) as well as multiple cranial nerves.[162,163] The NC cells begin their exodus shortly after neural tube closure is completed. They migrate differentially, depending on their site of origin, with those from the incompletely segmented midbrain moving as a continuous sheet, while those leaving the specific hindbrain and trunk segments travel in groups within their assigned segment. The trunk NC cells move in two distinct planes, one group moving ventrally and inward to innervate viscera and the other cohort heading in a dorsal direction to generate melanocytes in the skin.

Disruption of neural cell migration results in neuronal heterotopia (or ectopia, i.e., displacement of normal cells to abnormal locations). Specific mechanisms of migration that might be affected by developmental neurotoxicants include disordered cell-to-cell interactions (e.g., damaged cell adhesion molecules), aberrant guide cell (glia) function, and death of the migrating cells. Neurons that settle into an improper site are subject to two potential fates. One is to attempt to establish their normal afferent and efferent connections despite their dislocation, and the other is to seek connections with whatever nearby cells are available. The outcome of this choice dictates the functional capabilities of the brain structures that harbor large numbers of ectopic neurons. Misplaced cells can still connect to their appropriate targets over long distances—often using greatly altered trajectories—even if the target also occupies a markedly abnormal position.[164] The function of such distorted circuits may be normal. Defective neuronal migration in the cerebral cortex has been linked to multiple neurological conditions in humans, including behavioral disorders (e.g., attention-deficit hyperactivity disorder, autism, dyslexia, epilepsy, schizophrenia, obsessive-compulsive disorder, Tourette's syndrome) and mental retardation (Down syndrome).[165–167]

Changes in cell orientation or position (ectopia) are a less commonly recognized consequence of developmental neurotoxicity.[154] This lack probably reflects the tendency of large aggregates of misplaced cells to disperse over time rather than the resistance of migration to toxicant-induced interruption. Agents that have been shown to induce neuronal ectopia include ionizing radiation (which preferentially targets migrating cells) as well as ethanol, lead, methylmercury, and toluene.[21,137,168] Many of these agents also destroy neurons, so the mechanism may eliminate trophic signals needed to guide the migration of late-forming cells.[137] The induction of ectopia generally requires a high-dose exposure since migratory deficits occur only if neurodevelopment has been massively disturbed; less extensive insults often allow surviving cells to reach (or at least approach) their normal position but may prevent them from assuming a normal orientation or conformation. The most serviceable screen for heterotopias is a qualitative histopathological analysis using conventional histological stains [e.g., hematoxylin and eosin (H&E), cresyl violet] by a neuropathologist who possesses a good mental picture of normal region- and developmental-stage specific neuroanatomy.[156] If necessary,

the examination may be performed on sections labeled using a special procedure to assess whether or not a cell type-- specific marker of interest is evident in an abnormal location.

Cell Differentiation

Neurons and glia that have committed to a particular fate begin the process of differentiation while migrating.[21] Acting in response to both internal and external cues, committed cells engage in a series of coordinated molecular alterations leading to the cytoarchitectural remodeling necessary to assume their preprogrammed function. Specific activities include alterations in conformation (shape and size), gene expression (especially the complement of neurotransmitters, receptors, and signaling cascades), and polarity. The differentiating cells also extend neurites to their target cells and begin producing an assortment of molecules to signal adjacent and/or distant partners (both neural and nonneural derivatives). The principal function served by such neuron-derived signaling molecules in the CNS and PNS is communication (e.g., neurotransmitters and their receptors, as well as the enzymes that make and remove them). In contrast, glial-derived molecules support many functions, including communication (enzymes and transport proteins that remove neurotransmitters, myelin sheath proteins that insulate neuron processes) and defense (proteins required to mount and maintain an immunologic response, scavenge debris, or bolster barrier mechanisms). The pace of cell differentiation for a given neural cell population varies substantially across time and space. For example, myelination in the cerebral cortex occurs in ordered waves, spreading in sequence from the primary centers to the secondary regions to the association areas for an extended period after birth.

Differentiation of a given neural cell population typically requires the genesis of many transient connections as branching processes extend from their cell of origin to their target cell. These temporary links often connect to cells that are not normally contacted in the mature CNS. Such temporary associations serve two purposes. Convergent connections result from the innervations of a target neuron by several neurons or neuronal populations, only one of which maintains its circuit with the target in the adult. In contrast, divergent connections form when a single neuron or neuronal population innervates more cells or a greater expanse of the immature CNS than it does in the adult CNS. Both classes of transient projections may arise from a single neuronal population. Axonal extension in both convergent and divergent associations is specified by the gradient of morphogens (i.e., molecules that guide the generation of neural structures) established by the target. In the end, the neurites that follow the optimal path to the target will survive, whereas the extra collaterals (typically those that receive less exposure to target-derived growth factors due to a suboptimal approach to the target) will be removed. This pruning process is a major mechanism for shaping the form of the fully differentiated nervous system as the individual moves from infancy through childhood to adolescence.[44,45]

Disruption of cell differentiation is a common occurrence following exposure to developmental neurotoxicants.[21] Many agents have been shown to affect differentiation, although it is often difficult to separate primary from secondary effects. Two classes of toxicants that appear specifically to affect differentiation are organochlorine pesticides and polychlorinated biphenyls (PCBs).[21] A standard means of evaluating region-specific neural cell differentiation is qualitative histopathologic examination by an experienced neuropathologist to ensure that cell morphology at given sites is within normal limits. Complete exploration of a cell's state of differentiation may require special procedures (e.g., immunohistochemistry or in situ hybridization for cell type–specific markers). Such direct exploration is often directed to evaluating one or a few cell types (e.g., those expressing a specific gene or protein) rather than all cells indiscriminately, and thus is typically reserved for research projects with well-defined mechanistic questions.[156]

Production of Neurites and Synapses

Recently committed neural cells have only a few processes that exhibit minimal branching.[20] To assume their full function, these elements usually must extend one (oligodendrocytes, some neurons) or more (most neurons, glial cells) major processes to interact with a specific target site. The number, shape, and arborization of these connections are defined by the cell's specific genetic heritage as modified by chemical and physical stimuli within their immediate surroundings.

The largest numbers of intercellular connections within the developing nervous system are made by differentiating neurons. In general, neurons with short axons develop later than do neurons with long processes. Dendritic extension is maximal from the late fetal period (i.e., after neuron production is nearly completed) through early infancy, which corresponds to approximately PND 10 in rats, approximately two months of age in nonhuman primates, and between 10 and 12 months of age in humans.[20,44] As differentiation progresses, the dendritic spines change their shapes from the conformation used for seeking a target to assume the contours needed to support synaptogenesis. The features of these "seeking" and "synaptogenic" profiles vary by region and among investigators (compare, e.g., Norton[169] and Purpura[170]).

Synapses are produced in great numbers in the developing human[48,171] and rhesus macaque[172,173] nervous system until an infant reaches approximately two years of age. This process is energy-intensive, as indicated by the progressive increase in the rate of brain glucose metabolism (reaching twice the adult level) during this period.[174] In general, synapse formation during this period exceeds by 40 to 50% the number needed for effective function in the mature nervous system[44]; similar rates of overproduction have been demonstrated in other species. The hyperconnectivity supported by this exuberant synaptogenesis is required to ensure that extending neurites reach their appropriate targets by the optimal route. Waves of synapses are formed in most neural regions, with production peaks in a given region often separated by days[20] and different neural domains exhibiting different rates of synaptic production.[175] Synapses that form initially are immature; sufficient time is required for a new neuron to produce processes of satisfactory length (about a week) and with enough synapses (another week) to permit successful integration of the new cell into the developing neural circuitry (which requires another one or two weeks). The dense synaptic beds reorganize their connections over time (three to four weeks in rodents, through the adolescent years in primates) to attain the final form and function of the mature nervous system.[21,176,177]

The characteristics of particular synapse types are directed by the complement of factors and receptors that they and their neighbors generate. Synaptic production occurs in a specific sequence [e.g., those utilizing the inhibitory neurotransmitter γ-aminobutyric acid (GABA) appearing prior to those using the excitatory neurotransmitter glutamate[178]]. Inappropriate synapses may develop if neuron proliferation and/or migration are delayed.[21] The amplification in synapse number is paralleled by increases in neurotransmitter content and receptor density. This expansion starts at approximately the fifth month of gestation in humans,[44] although the exact time at which synaptogenesis is initiated varies with the brain region.[175] Existing receptors are capable of reconfiguring their structures to participate in activity-dependent synaptic plasticity.[48] This plasticity is a critical component of neurogenesis in the developing nervous system both before and after birth and continues in many regions of the adult nervous system.[59] Neuronal plasticity depends not only on inherent properties of the neuron but may also depend on active astrocyte participation.[179]

Neurological dysfunction arising from dendritic and/or synaptic abnormalities may be associated with subtle microscopic lesions.[170] The usual state is reminiscent of the situation in early development when new neurites are seeking targets but have not begun to form many synapses. Exposure to various neurotoxicants can affect the synaptic connectivity, neurotransmitter activity, and receptor number within the developing nervous system.[21,137] Examples include antimitotics, cholinesterase inhibitors, endocrine disruptors, ethanol, hormonal imbalances (e.g., hyper- and hypothyroidism, which yield either increased but abnormal or reduced synaptic numbers, respectively[180]), ionizing radiation, heavy metals (e.g., lead, methyl mercury, triethyl tin), neuroactive drugs (e.g., diazepam, haloperidol), and nicotine (following exposure in utero or postnatally).[21,154] The degree of synaptogenesis in coronal tissue blocks or tissue sections may be inferred by qualitative or quantitative assessment of the thickness of various synapse-rich neural regions, such as the cerebral cortex (especially layer 1 in sections), hippocampus (dentate gyrus), and cerebellum (molecular layer).[21,181] If required for a specific research project, direct assessment of neurite and synapse conformation is best performed using special silver stains (see Garman et al.[156] and Switzer[182]).

Myelin Formation

Myelination is required if vertebrate neural circuits are to function effectively. The process peaks at different times in various species (e.g., the second postnatal week in rodents and during the last trimester in humans) but continues through adolescence in both rats[183] and humans.[184] The myelinating cells only begin to form once neurites are being formed, which makes them among the last neural cells to develop.[21] The process is progressive, leading to an approximately 20-fold increase in conduction velocity in two-year-old children with well-myelinated circuits.[45] The general sequence is comparable across mammalian species. Myelin is laid down first in a caudocranial order beginning in the spinal cord.[50] As the process reaches the brain, myelin develops initially in major functional areas (e.g., primary motor and sensory areas in the cerebral cortex, corpus callosum, optic chiasm) and subsequently in waves spreading outward to other portions of the nervous system. For example, in humans, myelination begins in certain spinal cord tracts by approximately four months of gestation, whereas the corticospinal and pyramidal tracts in the brain begin myelinating at birth.[50] In contrast, the associative circuitry of the cerebrum continues to acquire insulation into adulthood.[154,185] The human corpus callosum contains mature myelin by one year of age but continues to add myelin for another two decades.[45] In general, oligodendrocytes are produced only after most neurons and many astrocytes have been generated.

Aberrant myelination may result from abnormal neuronal production and neurite extension, as myelin is not needed if neuronal processes are not present. In addition, myelin production may be altered by malnutrition (especially copper deficiency) as well as a variety of toxic agents, including anticonvulsants (at high doses), cholesterol inhibitors, ethanol, or heavy metals (e.g., chronic exposure to lead or methyl mercury).[154] The degree of myelination may be estimated with reasonable success by evaluating the dimensions of myelinated tracts in routinely stained histopathological sections. However, more specific evaluation of myelin is best performed using a special procedure to enhance myelin, such as Luxol fast blue (LFB) or antimyelin basic protein (MBP).

Production of Neurotrophic Factors

Specification of the nervous system during development requires the correct region- and time-specific action of numerous morphogens (secreted or sometimes membrane-bound substances that direct the pattern of tissue development, especially the positions and orientations of specialized cell types). Examples of such signals include classical neurotransmitters[19,186,187] and neurotrophic factors.[21,188–190] These substances influence all aspects of neurogenesis, including neuronal and glial proliferation, migration, differentiation, and neurite and synapse formation. These signaling molecules are derived from neural cells (including neural crest derivatives) as well as adjacent mesoderm.[191] Sensitivity to regulation by neurotrophic factors continues beyond the period of neurogenesis. For example, the dura mater helps control developmental events in the cerebral cortex, hippocampus, and cerebellum via its production of stromal-derived factor 1, beginning about E15 in the rodent brain.[192] An interesting aspect of their neurodevelopmental functions is that the spectrum of roles undertaken by any given molecule is more broad than—and often different from—the number of tasks they serve in the adult.[21] For example, the classical adult inhibitory neurotransmitter GABA serves during neurodevelopment as a chemoattractant for cortical neurons[193] and an excitatory stimulus for new hypothalamic neurons.[194]

Aberrant production of one or more neurotrophic signals is sufficient to disrupt the timing and coordination of neurogenesis.[21,137] The effects of altering neurotrophic activity are often confined to one or a few regions, the sites depending on the stage of development during which the insult was received. Neurotoxicants that have been shown to alter the milieu of signaling molecules in the developing brain include ethanol and heavy metals (e.g., lead, methyl mercury).[21] Neuropathology evaluation of abnormal neurotrophic factor expression must be gathered using in situ molecular procedures (e.g., immunohistochemistry or in situ hybridization to detect a specific factor). Less specialized pathology techniques are capable of assessing only the secondary structural defects resulting from altered molecular expression.

Physiologic Cell Death

It is estimated that the developing human nervous systems produces almost twice as many neurons as are needed in the adult nervous system by the time a child is two years old.[171] The continuation of normal neurodevelopment thus requires pruning of underutilized or unused elements until the appropriate organization of the nervous system has been attained. The pattern and rate of cell death is both region- and time-specific,[175,195,196] a fact that probably reflects the different timing of peak neuronal generation among various domains in the nervous system.[119,197]

Unnecessary neural cells and transient connections are eliminated via two processes. The first is selective cell death such that neurons that fail to establish suitable projections to their normal target fields will follow a preset cell death program. This mechanism, carried out by apoptosis, is widespread throughout the developing vertebrate CNS and PNS.[195,196,198,199] Approximately 50% of postmitotic neurons (and their axons) have been demonstrated to die naturally during embryonic and fetal development in both avian and mammalian species, including humans;[195] indeed, 70% of recently generated cerebrocortical neurons are apoptotic at the peak of cortical neuronogenesis in mice.[200] However, the majority of dying cells actually represent stem cells in the proliferating zones rather than committed neurons.[200,201] The extent of apoptosis is controlled by a number of factors,[202] both positive [e.g., hormone balance (growth hormone, thyroid hormone), neurotrophic factors, local mitogenic effects] and negative [e.g., local inhibitory effects of undifferentiated precursor cells (via secreted molecules) or matrix-specific proteins]. Glial numbers are pruned similarly to match the need specified by the neurons retained.[202]

The second mechanism for removing transient connections is by simple retraction of the improperly placed collateral branches arising from an optimally positioned neuron. This trims the exuberant terminal arborizations that arose when cell processes were first sent forth while retaining the best positioned connection(s) and the cell itself. This process is particularly active late in development, especially after birth.[203] The refinement required to achieve the appropriate point-to-point circuits necessary for normal function of the adult nervous system indicates that synaptic pruning takes place gradually over an extended period—generally, the entire decade of late childhood and adolescence in humans.[171,175]

Disruption of apoptosis during nervous system development results in widespread and usually severe anatomical and functional deficits in the immature nervous system.[204–206] Certain neurotoxicants, including ethanol, methanol, methyl mercury, and some pesticides, are capable of altering the rate of region-specific neural cell death, typically by accelerating the degree of apoptosis.[21] The presence of apoptosis may be detected by routine histopathological evaluation for fragmenting nuclei, but the rapid rate at which apoptotic debris is removed does not permit a reliable estimate of its true extent. When assessment of programmed cell death is an endpoint of interest, the most dependable means of analysis is by using a special procedure [e.g., commercial terminal deoxynucleotidyl transferase dUTP nick end labeling (TUNEL) kits] to selectively label dying cells for enumeration.

Immature Protective Mechanisms

The developing nervous system is vulnerable to many neurotoxicants because some neural defense systems are only partially activated or inactive in utero and soon after birth, which become fully operational in adulthood. The primary protective measures available to the developing nervous system include anatomical obstacles such as the blood–brain barrier (BBB) and various metabolic pathways.

Relative to the adult, the structural barriers that protect the developing nervous system are incompletely established.[139] Brain regions that possess relatively little BBB in the adult [e.g., circumventricular organs, dorsal root ganglia (DRG)] also lack a BBB during development.[207] In most other regions, the BBB begins to assemble in utero at about the time the secondary brain vesicles begin to form[39,208] but does not become fully functional until well after birth (approximately six months of age in humans).[23] The embryonic BBB develops transient structures and functions that regress after birth. Examples include the intracellular route for transporting albumin into the cerebrospinal fluid (CSF), thereby providing very high (relative to adult) levels of CSF protein for local cell support during neurogenesis, and the fabrication of unique intercellular junctions (plate, strap, and wafer types) at some barrier surfaces.[39] Some toxicants (e.g., pesticides[209]) can alter the permeability of the immature BBB. Such effects may persist for an extended period. Once an agent gains entry through the BBB, the immature neural tissue is more permeable to small lipid-permeable molecules than is the adult nervous system. The relatively greater water content of the developing nervous system[156] also permits greater concentration of water-soluble molecules. Taken together, the developing nervous system in utero is apt to experience more exposure to systemic toxicants than will the better protected nervous system of the mother.

Metabolic detoxification takes place in the developing brain, but to a lesser degree than in the adult nervous system and nonneural sites within the fetus and mother. Many cytochromes P450 isoforms (involved in phase I modification of xenobiotics) are present in the choroid plexus, ependyma, and endothelial cells throughout the nervous system, while specific isoforms are expressed selectively in particular neuronal or glial populations.[210] In like manner, glutathione (GSH) and glutathione-S-transferases (GST, involved in phase II processing) are common throughout the developing nervous system, particularly during embryogenesis.[210] For example, all rodent CNS regions (and also the DRG) at E13 exhibit uniform GSH expression in neurons and glia. The degree of GSH expression starts to regress in some CNS regions [hippocampus, ventral (anterior) gray matter of the spinal cord] by E17, eventually receding in neurons and glia by P5 to P10 but remaining in the neuropil. It seems reasonable that the marked differences in metabolic capacity among various neural cell types may be responsible for some portion of enhanced regional susceptibility to certain developmental neurotoxicants.

Vascular Development

Neurodevelopment can be altered indirectly by primary abnormalities of the developing vascular system. The two main processes that participate in the formation of embryonic and fetal vessels are vasculogenesis (production of vessels from in situ differentiation of angioblasts) and angiogenesis (sprouting of new capillaries from preexisting vessels).[211] Many soluble molecules, including various growth factors[212] and elements of the Notch signaling pathway,[212] participate in the vascularization of the embryonic brain. They are elaborated by neighboring neural cells as well as adjacent mesodermal domains.

Vascular defects can seriously affect neurodevelopment, especially later in gestation and after birth when dense neural tissues cannot be nourished adequately by diffusion alone. Fragile capillaries may be prone to rupture, which can cause hemorrhage into the ventricular zone and or SVZ that will destroy the ability to produce new neural cells and may result in posthemorrhagic hydrocephalus.[213] Alternatively, vascular insufficiency in the developing nervous system may precipitate ischemia, which will culminate in periventricular leukomalacia (due to the heightened vulnerability of myelinating oligodendrocytes to hypoxia and/or excitotoxicity)[214] or stroke. Some toxicants (e.g., methyl mercury) have been shown to affect vascular integrity in the developing nervous system.[154]

DEVELOPMENTAL NEUROTOXICITY: A LIFELONG MENACE

Children are at greater risk for developmental neurotoxicity. Young beings eat more food, drink more fluids (water and formulated drinks), breath more air, explore more by tasting, and spend more time closer to the ground than do adults.[215] In addition to a greater propensity for exposure, the developing nervous system of children is more vulnerable to many toxicants than is the adult nervous system.[216] Developmental neurotoxicants may also damage the adult nervous system. However, these agents often induce developmental damage at lower doses than those required to affect the adult system, and some developmental neurotoxicants do not damage the adult nervous system.[16,22]

The immature nervous system has a different set of basic tissue reactions relative to neural tissue responses in the adult.[50] Early damage usually causes gross anomalies (neural tube defects, microcephaly, polymicrogyria); the earlier the damage (assuming that it is not lethal), the more profound the effects.[20,139] Later insults produce a variety of histopathological lesions that vary with the location and timing. Neural cell degeneration characterized by chromatolysis, cell swelling, and nuclear pyknosis is more common than frank necrosis. Liquefaction and dissolution of dead tissue is rapid,

typically resolving within two weeks of the initial insult. Such lesions are resolved by microglia and macrophages, which differentiate well before (typically by the fifth month of gestation in humans) other glial lineages. Gliosis is minimal or absent until astrocytes are produced in quantity (typically, one to two weeks after birth in rodents, and during the last trimester in humans). When they occur, glial reactions in the immature nervous system typically have less pronounced astrocytic hypertrophy and variable (and often low) levels of glial fibrillary acid protein (GFAP) expression.

Dozens of agents induce neurotoxicity in animals and people. In general, the largest number of developmental neurotoxicants are pesticides, organic solvents, and related compounds, followed closely by metals.[217] Many pharmaceuticals and illicit drugs also elicit developmental neurotoxicity,[218,219] as do certain biological toxins, such as fungal[220,221] and plant[34,222–224] derivatives, as well as some endogenous molecules (e.g., unconjugated bilirubin).[225] Developmental neurotoxicity is a particularly likely outcome for neuroactive agents (e.g., hormones, psychomodulatory drugs), small biomolecules (e.g., amino acids, peptides), and compounds known to cause neurotoxicity in adults, as well as structurally related materials.[226] Similar anatomic changes can be precipitated by different developmental neurotoxicants, while some agents (e.g., ethanol, heavy metals) can affect multiple processes that contribute to the production of neural lesions.[20,21] In general, structural lesions in the nervous system will induce one or more functional deficits as well. Although less common, early exposure to neurotoxicants that cause functional alterations can lead to structural changes later in life (e.g., growth hormone deficiency, which prevents the escalation of body and brain growth, which normally begins at about two years of age in humans).[197]

The developing nervous system exhibits two major periods of vulnerability to xenobiotics. Early exposure (especially at high levels) produces frank malformations by affecting the ability of the neural plate, neural folds, and neural tube to fully set the general shape of the neuraxis. The most common gross lesion induced by neuroteratogens (i.e., agents that cause visible malformations) is the neural tube closure defect (NTD), which is readily observed without magnification during autopsy/necropsy. The usual locations for such anomalies are at the cranial (anterior) and caudal (posterior) neuropores, which are the last portions of the neural tube to fuse.[50] Persistent patency of the cranial neuropore results in anencephaly [absence of the rostral brain, more typical of mammals with long gestations (e.g., primates)], exencephaly [exposure of the brain, most common in animals with short gestations (rodents)], or encephalocele (bulging of the brain through a midline skull defect); the presumed difference in incidence in the two former lesions results from the length of time the delicate neural tissue spends in contact with the corrosive amniotic fluid. The same fusion defect at the caudal neuropore leads to spina bifida (exposure of the spinal cord) or myelocele (bulging of the spinal cord through a midline vertebral defect). Rachischisis results from failed fusion of the caudal neural tube (i.e., spinal cord precursor), while craniorachischisis is the extreme case where neural tube closure failed completely. Common causative mechanisms proposed for NTD are increased cell death or altered molecular expression at the neural crest (thereby thwarting neural tube fusion) or abnormal cytoskeletal function in cells of the neural folds (thus preventing their elevation). Heavily polluted communities may experience NTD rates up to eightfold higher than the typical incidence (about 1 in 1000 live human births), presumably due to excessive exposure to environmental neurotoxicants.[227,228] Prenatal exposure to known human neuroteratogens generally induces a range of functional and structural defects in the animal nervous system that resembles the lesions produced in human embryos.[229,230]

The second major period of neurodevelopmental susceptibility occurs later in gestation. The timing of this window varies by the species, but typical boundaries are the latter half of gestation and early postnatal period in rodents, and the last two trimesters in human fetuses. Late exposure (particularly at lower levels) will commonly induce microscopic damage and/or functional deficits but rarely elicit overt malformations. This outcome occurs because the general shape of the nervous system has already been established during neurulation; accordingly, the only processes that remain to be affected are those components of neurogenesis (described in the section above) that dictate the growth and maturation of the established neuraxis. Such late-stage neurotoxicity may be initiated by altering any aspect—or several—of neurogenesis. Common lesions that result from developmental neurotoxicity during this period include microcephaly (reduced brain size, due to depleted cell number), neuronal heterotopia (via altered cell migration), region-specific cell depletion (caused by insufficient production or excessive apoptosis), shrinkage of synapse-rich domains (due to reduced formation of neurites or synapses), and myelination defects (the consequence of inadequate production or excessive loss of oligodendrocytes). Structural evaluation to detect such changes typically involves a detailed qualitative histopathological examination of multiple brain blocks as well as morphometric assessment of selected features. Particular attention should be paid to regions where cell production, migration, differentiation, and synaptogenesis are concentrated (e.g., cerebral cortex, hippocampus, cerebellum),[21,181,231] sites where major neurotransmitters are stored in high concentrations (e.g., the basal ganglia[113]), and major white matter tracts (e.g., corpus callosum, internal capsule, corticospinal and pyramidal tracts).

The immature nervous system is quite vulnerable to excitotoxic insults, probably even more so than the adult nervous system. Multiple factors contribute to this enhanced susceptibility. Most neurons as well as many oligodendrocytes and astrocytes possess functional glutamate receptors.[232] In many regions, 60% to 70% of the synapses utilize glutamate as their primary neurotransmitter.[233] The density of glutamate receptors is often higher in such regions (e.g., cerebral cortex,[234,235] thalamus,[232] basal ganglia,[236,237] hippocampus[236,237]) in the developing brain. The newly formed glutamate receptors on immature neural cells are more easily stimulated by glutamate.[238,239] This amplified receptivity helps maintain the developing nervous system in a more active condition than the corresponding adult nervous system, a state that is required to achieve optimal connectivity given the rapid pace of neurogenesis.[48,240,241] However, the heightened activity is associated with a greater turnover of Ca^{2+}, a major mediator of excitotoxicity.[242] Causes of developmental excitotoxicity include hypoxia/ischemia, hypoglycemia, kernicterus, and trauma.[232] Neurotoxicants that can interact with glutamate receptors should have no difficulty acting in the developing nervous system.

Neurodevelopment as a process merges seamlessly over time into neural maturity and eventual senescence. A current hypothesis is that many neurodegenerative diseases that occur late in life may represent the unmasking of previously hidden neurodevelopmental damage.[19,243] Such "silent neurotoxicity" conditions have been postulated to result from either (1) the inability to initiate a late-manifesting function due to earlier structural damage or (2) persistent or progressive biochemical or morphological injury, the result of which is a reduced capacity to compensate for other exogenous factors, such as concurrent disease, continued toxicant exposure, stress, or even normal aging. Developmental toxicants that have been linked to delayed neurotoxicity in rodents include ionizing radiation,[244] methyl mercury,[245] and triethyl tin.[246] Methyl mercury has also been implicated in delayed neurotoxicity in human and nonhuman primates.[247] Interestingly, developmental exposure to some agents (e.g., pesticides) can cause changes that not only persist into adulthood but that also appear to potentiate the adult nervous system's susceptibility to additional, low-dose exposure to the same or other neurotoxicants.[248] Further work will be required to completely define the role of developmental neurotoxicity in aged-related neurological diseases, particularly the implications of the expanding number of high-production-volume chemicals (greater than 1 million pounds per year) that are used throughout the world each year (approximately 2900 pounds in the United States[249]), for which sufficient developmental neurotoxicity data often have not been acquired.[250]

REFERENCES

1. Campbell AW. *Histological Studies on the Location of Cerebral Function*. New York: Cambridge University Press; 1905.

2. Papez JW. *Comparative Neurology*. New York: Hafner; 1929.

3. Ariëns Kappers CU, et al. *The Comparative Anatomy of the Nervous System of Vertebrates*. New York: Hafner; 1960.

4. Bolon B. Comparative and correlative neuroanatomy for the toxicologic pathologist. *Toxicol Pathol*. 2000;28:6–27.

5. Butler AB, Hodos W. *Comparative Vertebrate Neuroanatomy: Evolution and Adaptation*, 2nd ed. Hoboken, NJ: Wiley-Liss; 2005.

6. Crosby EC, et al. *Correlative Anatomy of the Nervous System*. New York: Macmillan; 1962.

7. Crosby EC, Schnitzlein HN. *Comparative Correlative Neuroanatomy of the Vertebrate Telencephalon*. New York: Macmillan; 1982.

8. de Lahunta A, Glass E. *Veterinary Neuroanatomy and Clinical Neurology*, 3rd ed. Philadelphia: W.B. Saunders; 2009.

9. Gilbert SF. *Developmental Biology*, 8th ed. Sunderland, MA: Sinauer Associates; 2006.

10. Jenkins TW. *Functional Mammalian Neuroanatomy*. Philadelphia: Lea & Febiger; 1972.

11. Schoenwolf GC. *Atlas of Descriptive Embryology*, 7th ed. San Francisco: Pearson Benjamin Cummings; 2008.

12. FitzGerald MJT, et al. *Clinical Neuroanatomy and Neuroscience*, 5th ed. Philadelphia: W.B. Saunders; 2007.

13. Lorenz MD, Kornegay JN. *Handbook of Veterinary Neurology*, 4th ed. Philadelphia: W.B. Saunders; 2004.

14. Moser VC. The functional observational battery in adult and developing rats. *Neurotoxicology*. 2000;21:989–996.

15. Oliver JE, et al. *Veterinary Neurology*. Philadelphia: W.B. Saunders; 1987.

16. Rodier PM. Comparative postnatal neurologic development. In: Needleman HL, Bellinger D, eds. *Prenatal Exposure to Toxicants: Developmental Consequences*. Baltimore: Johns Hopkins University Press; 1994:3–23.

17. Wood SL, et al. Species comparison of postnatal CNS development: functional measures. *Birth Defects Res B Dev Reprod Toxical*. 2003;68:391–407.

18. Young PA, et al. *Basic Clinical Neuroscience*, 2nd ed. Baltimore: Lippincott Williams & Wilkins; 2008.

19. Rice D, Barone S Jr. Critical periods of vulnerability for the developing nervous system: evidence from humans and animal models. *Environ Health Perspect*. 2000;108(Suppl 3): 511–533.

20. Suzuki K. Special vulnerabilities of the developing nervous system. In: Spencer PS, Schaumburg HH, eds. *Experimental and Clinical Neurotoxicology*. Baltimore: Williams & Wilkins; 1980:48–61.

21. Barone SF, et al. Vulnerable processes of the nervous system development: a review of markers and methods. *Neurotoxicology*. 2000;21:15–36.

22. Claudio L, et al. Testing methods for developmental neurotoxicity of environmental chemicals. *Toxicol Appl Pharmacol*. 2000;164:1–14.

23. Rodier PM. Developing brain as a target of toxicity. *Environ Health Perspect*. 1995;103(Suppl 6):73–76.

24. Bayer SA, Altman J. The spinal cord from gestational week 4 to the 4th postnatal month. In: Bayer SA, Altman J, eds. *Atlas of Human Nervous System Development*. Vol 1. Boca Raton, FL: CRC Press; 2002.

25. Bayer SA, Altman J. The human brain during the third trimester. In: Bayer SA, Altman J, eds. *Atlas of Human Nervous System Development*. Vol 2. Boca Raton, FL: CRC Press; 2004.

26. Bayer SA, Altman J. The human brain during the second trimester. In: Bayer SA, Altman J, eds. *Atlas of Human Nervous System Development*, Vol 3. Boca Raton, FL: CRC Press; 2005.

27. Bayer SA, Altman J. The human brain during the late first trimester. In: Bayer SA, Altman J, eds. *Atlas of Human Nervous System Development*, Vol 4. Boca Raton, FL: CRC Press, 2006.

28. Bayer SA, Altman J. The human brain during the early first trimester. In: Bayer SA, Altman J, eds. *Atlas of Human Nervous System Development*, Vol 5. Boca Raton, FL: CRC Press; 2008.

29. Cochard LR. *Netter's Atlas of Human Embryology*. Teterboro, NJ: Icon Learning Systems; 2002.

30. de Lahunta A, Glass E. Development of the nervous system: malformation. In: de Lahunta A, Glass E, eds. *Veterinary Neuroanatomy and Clinical Neurology*. Philadelphia: W. B. Saunders; 2009:23–53.

31. Drews U. *Color Atlas of Embryology*. New York: Thieme Medical Publishers; 1995.

32. Kaufman MH. *The Atlas of Mouse Development*, 2nd ed. San Diego, CA: Academic Press; 1992.

33. Kaufman MH, Bard JBL. *The Anatomical Basis of Mouse Development*. San Diego, CA: Academic Press; 1999.

34. McGeady TA, et al. *Veterinary Embryology*. Oxford, UK: Blackwell Publishing; 2006.

35. O'Rahilly R, Müller F. *The Embryonic Human Brain: An Atlas of Developmental Stages*. New York: Wiley-Liss; 1994.

36. Paxinos G, et al. *Atlas of the Developing Rat Nervous System*, 2nd ed. San Diego, CA: Academic Press; 1994.

37. Steding G. *The Anatomy of the Human Embryo: A Scanning Electron-Microscopic Atlas*. Basel, Switzerland: S. Karger; 2009.

38. Narita Y, Rijli FM. Hox genes in neural patterning and circuit formation in the mouse hindbrain. *Curr Top Dev Biol*. 2009;88:139–167.

39. Saunders NR, et al. Barrier mechanisms in the brain: II. immature brain. *Clin Exp Pharmacol Physiol*. 1999; 26:85–91.

40. Brodmann K. Beitrage zur histologischen Lokalisation der Grosshirnrinde. Dritte Mitteilung: Die Rindenfelder der niederen Affen. *J Psychol Neurol Lpz*. 1905;4:177–226.

41. Leclerc N, et al. The compartmentalization of the monkey and rat cerebellar cortex: zebrin I and cytochrome oxidase. *Brain Res.* 1990;506:70–78.

42. Marani E. Topographic histochemistry of the cerebellum. *Prog Histol Cytochem.* 1986;16:1–169.

43. Hawkes R, Leclerc N. Purkinje cell axonal distributions reflect the chemical compartmentalization of the rat cerebellar cortex. *Brain Res.* 1989;476:279–290.

44. Herschkowitz N, et al. Neurobiological bases of behavioral development in the first year. *Neuropediatrics.* 1997;28: 296–306.

45. Herschkowitz N, et al. Neurobiological bases of behavioral development in the second year. *Neuropediatrics.* 1999;30: 221–230.

46. Zeman W, Innes JRM. *Craigie's Neuroanatomy of the Rat.* New York: Academic Press; 1963.

47. Conel JL. *The Cortex of the Newborn: The Postnatal Development of the Human Cerebral Cortex*, Vol I. Cambridge, MA: Harvard University Press; 1939.

48. Johnston MV. Neurotransmitters and vulnerability of the developing brain. *Brain Dev.* 1995;17:301–306.

49. Kostović I, et al. Structural basis of the developmental plasticity in the human cerebral cortex: the role of the transient subplate zone. *Metab Brain Dis.* 1989;4:17–23.

50. Friede RL. *Developmental Neuropathology*, 2nd ed. New York: Springer-Verlag; 1989.

51. Zilles K. *The Cortex of the Rat: A Stereotaxic Atlas.* New York: Springer-Verlag; 1985.

52. Hayes NL, Nowakowski RS. Dynamics of cell proliferation in the adult dentate gyrus of two inbred strains of mice. *Brain Res Dev Brain Res.* 2002;134:77–85.

53. Mohajeri MH, et al. The impact of genetic background on neurodegeneration and behavior in seizured mice. *Genes Brain Behav.* 2004;3:228–239.

54. Wu WJ, et al. Aminoglycoside ototoxicity in adult CBA, C57BL and BALB mice and the Sprague–Dawley rat. *Hear Res.* 2001;158:165–178.

55. Gergen JA, MacLean PD. *A Stereotaxic Atlas of the Squirrel Monkey's Brain (Saimiri sciureus).* Public Health Service Publication 933. Bethesda, MD: National Institutes of Health; 1962.

56. Toga AW, Mazzioatta JC. *Brain Mapping: The Systems.* San Diego, CA: Academic Press; 2000.

57. Witelson SF, et al. The exceptional brain of Albert Einstein. *Lancet.* 1999;353:2149–2153.

58. Steinmetz H, et al. Brain (A) symmetry in monozygotic twins. *Cereb Cortex.* 1995;5:296–300.

59. Finch CE. Neurons, glia, and plasticity in normal brain aging. *Adv Gerontol.* 2002;10:35–39.

60. Berger B, et al. Dopaminergic innervation of the cerebral cortex: unexpected differences between rodents and primates. *Trends Neurosci.* 1991;14:21–27.

61. Calabrese EJ. *Principles of Animal Extrapolation.* Chelsea, MI: Lewis; 1991.

62. Claassen V. *Neglected Factors in Pharmacology and Neuroscience Research.* New York: Elsevier; 1994.

63. Frank DW. Physiological data of laboratory animals. In: Melby ECJ, Altman NH, eds. *Handbook of Laboratory Animal Science.* Boca Raton, FL: CRC Press; 1976:23–64.

64. Krinke GJ. Neuropathologic screening in rodent and other species. *J Am Coll Toxicol.* 1989;8:141–146.

65. Fix AS, et al. Quantitative analysis of factors influencing neuronal necrosis induced by MK-801 in the rat posterior cingulate/retrosplenial cortex. *Brain Res.* 1995;696: 194–204.

66. Iwaski H, et al. Strain differences in vulnerability of hippocampal neurons to transient cerebral ischaemia in the rat. *Int J Exp Pathol.* 1995;76:171–178.

67. Kacew S, et al. Strain as a determinant factor in the differential responsiveness of rats to chemicals. *Toxicol Pathol.* 1995;23:701–715.

68. Kacew S, Festing MFW. Role of rat strain in the differential sensitivity to pharmaceutical agents and naturally occurring substances. *J Toxicol Environ Health.* 1996;47:1–30.

69. McNamara RK, et al. Distinctions between hippocampus of mouse and rat: protein F1/GAP-43 gene expression, promoter activity, and spatial memory. *Mol Brain Res.* 1996;40: 177–187.

70. Setola V, Roth BL. Why mice are neither miniature humans nor small rats: a cautionary tale involving 5-hydroxytryptamine-6 serotonin receptor species variants. *Mol Pharmacol.* 2003;64:1277–1278.

71. Manocha SL, et al. *A Stereotaxic Atlas of the Brain of the Cebus Monkey (Cebus apella).* Oxford, UK: Oxford University Press; 1968.

72. Markowitsch HJ, et al. Cortical and subcortical afferent connections of the primate's temporal pole: a study of rhesus monkeys, squirrel monkeys, and marmosets. *J Comp Neurol.* 1985;242:425–458.

73. Riche D, et al. *Atlas Stéréotaxique du Cerveau de Babourin (Papio papio).* Paris: Centre National de la Recherche Scientifique; 1968.

74. MacLean PD. On the origin and progressive evolution of the triune brain. In: Armstrong E, Falk D, eds. *Primate Brain Evolution: Methods and Concepts.* New York: Plenum Press; 1982:291–316.

75. Dobbing J, Sands J. Comparative aspects of the brain growth spurt. *Early Hum Dev.* 1979;3:79–83.

76. Paxinos G. *The Rat Nervous System.* 2nd ed. San Diego, CA Academic Press; 1995.

77. Barkovich AJ. MR and CT evaluation of profound neonatal and infantile asphyxia. *Am J Neuroradiol.* 1992;13:959–972. [discussion, 973–975; comment in *Am J Neuroradiol.* 2000;921:979–981].

78. Radinsky L. Primate brain evolution. *Sci Am.* 1975;63: 656–663.

79. Kolb B, et al. Asymmetry in the cerebral hemispheres of the rat, mouse, rabbit, and cat: the right hemisphere is larger. *Exp Neurol.* 1982;78:348–359.

80. Tan Ü, Çaliskan S. Allometry and asymmetry in the dog brain: the right hemisphere is heavier regardless of paw preference. *Int J Neurosci.* 1987;35:189–194.

81. Tan Ü, Çaliskan S. Asymmetries in the cerebral dimensions and fissures of the dog. *Int J Neurosci.* 1987;32:943–952.

82. Murphy GMJ. Volumetric asymmetry in the human striate cortex. *Exp Neurol.* 1985;88:288–302.

83. Willerman L, et al. Hemisphere size asymmetry predicts relative verbal and nonverbal intelligence differently in the sexes: an MRI study of structure–function relations. *Intelligence.* 1992;16:315–328.

84. Haug H. Brain sizes, surfaces, and neuronal sizes of the cortex cerebri: a stereological investigation of man and his variability and a comparison with some mammals (primates, whales, marsupials, insectivores, and one elephant). *Am J Anat.* 1987;180:126–142.

85. Netter FH. Nervous system: I. Anatomy and physiology. In: *CIBA Collection of Medical Illustrations.* Vol 1. East Hanover, NJ: Novartis Medical Education; 1986.

86. Patten BM. *Foundations of Embryology*, 2nd ed. New York: McGraw-Hill; 1964.

87. Elias H, Schwartz D. Cerebrocortical surface areas, volumes, lengths of gyri, and their interdependence in mammals, including man. *Z. Säugetierkd.* 1971;36:147–163.

88. Graveland GA, Difiglia M. The frequency and distribution of medium-sized neurons with indented nuclei in the primate and rodent neostriatum. *Brain Res.* 1985;327:307–311.

89. Leonard CM. The prefrontal cortex of the rat: I. Cortical projections of the mediodorsal nucleus. *Brain Res.* 1969;12:321–343.

90. Galaburda AM, Pandya DN. Role of architectonics and connections in the study of primate brain evolution. In: Armstrong E, Falk D, eds. *Primate Brain Evolution: Methods and Concepts.* New York: Plenum Press; 1982:203–216.

91. Clare MH, Bishop GH. Potential wave mechanisms in the cat cortex. *Electroencephalog Clin Neurophysiol.* 1956;8: 583–602.

92. Seltzer B, Pandya DN. Afferent cortical connections and architectonics of the superior temporal sulcus and surrounding cortex of the rhesus monkey. *Brain Res.* 1978;149:1–24.

93. Allman JM, Kaas JH. A representation of the visual field in the caudal third of the middle temporal gyrus of the owl monkey (*Aotus trivirgatus*). *Brain Res.* 1971;31:85–105.

94. Bentivoglio M. Cortical structure and mental skills: Oskar Vogt and the legacy of Lenin's brain. *Brain Res Bull.* 1998;47:291–296.

95. Anderson B, Harvey T. Alterations in cortical thickness and neuronal density in the frontal cortex of Albert Einstein. *Neurosci Lett.* 1996;210:161–164.

96. Colombo JA. Interlaminar astroglial processes in the cerebral cortex of adult monkeys but not of adult rats. *Acta Anat.* 1996;155:57–62.

97. Colombo JA, et al. "Rodent-like" and "primate-like" types of astroglial architecture in the adult cerebral cortex of mammals: a comparative study. *Anat Embryol.* 2000;201:111–120.

98. Hartfuss E, et al. Characterization of CNS precursor subtypes and radial glia. *Dev Biol.* 2001;229:15–30.

99. Noctor SC, et al. Interference with the development of early generated neocortex results in disruption of radial glia and abnormal formation of neocortical layers. *Cereb Cortex.* 1999;9:121–136.

100. Rakic P. Specification of cerebral cortical areas. *Science.* 1988;241:170–176.

101. Meyer J. Behavioral assessment in developmental neurotoxicology. In: Slikker WJ, Chang LW, eds. *Handbook of Developmental Neurotoxicity.* San Diego, CA: Academic Press; 1998:403–426.

102. Crawley JN, et al. Behavioral phenotypes of inbred mouse strains: implications and recommendations for molecular studies. *Psychopharmacology.* 1997;132:107–124.

103. Crawley JN. *What's Wrong with My Mouse? Behavioral Phenotyping of Transgenic and Knockout Mice.* New York: Wiley; 2000.

104. Paylor R, et al. Developmental differences in place-learning performance between C57BL/6 and DBA/2 mice parallel the ontogeny of hippocampal protein kinase C. *Behav Neurosci.* 1996;110:1415–1425.

105. Jucker M, et al. Structural brain aging in inbred mice: potential for genetic linkage. *Exp Gerontol.* 2000;35:1383–1388.

106. Yanai T, et al. Eosinophilic neuronal inclusions in the thalamus of ageing B6C3F1 mice. *J Comp Pathol.* 1995; 113:287–290.

107. Jucker M, et al. Age-related deposition of glia-associated fibrillar material in brains of C57BL/6 mice. *Neuroscience.* 1994;60:875–889.

108. Jortner BS. Selected aspects of the anatomy and response to injury of the chicken (*Gallus domesticus*) nervous system. *Neurotoxicology.* 1982;3:299–310.

109. Krinke GJ, et al. Optimal conduct of the neuropathology evaluation of organophosphorus induced delayed neuropathy in hens. *Exp Toxicol Pathol.* 1997;49:451–458.

110. Wake MH, ed. *Hyman's Comparative Vertebrate Anatomy*, 3rd ed. Chicago: University of Chicago Press; 1979.

111. Cahill L. His brain, her brain. *Sci Am.* 2005;292:40–47.

112. de Courten-Myers GM. The human cerebral cortex: gender differences in structure and function. *J Neuropathol Exp Neurol.* 1999;58:217–226.

113. Paxinos G. *The Rat Nervous System*, 3rd ed. San Diego, CA: Academic Press; 2004.

114. Steinmetz H, et al. Corpus callosum and brain volume in women and men. *Neuroreport.* 1995;6:1002–1004.

115. Zhou JN, Swaab DF. Activation and degeneration during aging: a morphometric study of the human hypothalamus. *Microsc Res Tech.* 1999;44:36–48.

116. Mouton PR, et al. Age and gender effects on microglia and astrocyte numbers in brains of mice. *Brain Res.* 2002; 956:30–35.

117. Wise PM. Changing neuroendocrine function during aging: impact on diurnal and pulsatile rhythms. *Exp Gerontol.* 1994;29:13–19.

118. Nowak G, et al. Age dependent day/night variations of α_1- and β-adrenoceptors in the rat cerebral cortex. *Physiol Behav.* 1986;38:53–55.

119. Rodier PM. Chronology of neuron development: animal studies and their clinical implications. *Dev Med Child Neurol.* 1980;22:525–545.

120. Finch CE. The regulation of physiological changes during mammalian aging. *Quart Rev Biol.* 1976;5:49–83.

121. Reihlen A, et al. Age-dependent changes in the glial cell nests of the canine rhinencephalic allocortex. *Anat Rec.* 1994;238:415–423.

122. Gerhart DZ, et al. Localization of glucose transporter GLUT 3 in brain: comparison of rodent and dog using species-specific carboxyl-terminal antisera. *Neuroscience.* 1995;66:237–246.

123. Mooradian AD. Effect of aging on the blood–brain barrier. *Neurobiol Aging.* 1988;9:31–39.

124. Amenta F, et al. Neuroanatomy of aging brain: influence of treatment with L-deprenyl. *Ann NY Acad Sci.* 1994;717:33–44.

125. Bolon B, et al. Methanol-induced neural tube defects in mice: pathogenesis during neurulation. *Teratology.* 1994;49:497–517.

126. Hovland DNJ, et al. Differential sensitivity of the SWV and C57BL/6 mouse strains to the teratogenic action of single administrations of cadmium given throughout the period of anterior neuropore closure. *Teratology.* 1999;60:13–21.

127. Rodier PM. Vulnerable periods and processes during central nervous system development. *Environ Health Perspect.* 1994;102(Suppl 2):121–124.

128. Dobbing J. The later development of the brain and its vulnerability. In: Davis JA, Dobbing J, eds. *Scientific Foundations of Pediatrics.* Philadelphia: W.B. Saunders; 1974:565–577.

129. Bayer SA, Altman J. *Neocortical Development.* New York: Raven Press; 1991.

130. Bayer SA, et al. Timetables of neurogenesis in the human brain based on experimentally determined patterns in the rat. *Neurotoxicology.* 1993;14:83–144.

131. Aschner M, Kimelberg HK. *The Role of Glia in Neurotoxicity.* Boca Raton, FL: CRC Press; 1996.

132. Aschner M, et al. Glial cells in neurotoxicity development. *Annu Rev Pharmacol Toxicol.* 1999;39:151–173.

133. Takahashi T, et al. The cell cycle of the pseudostratified ventricular epithelium of the embryonic murine cerebral wall. *J Neurosci.* 1995;15:6046–6057.

134. Mandolesi L, et al. Environmental enrichment provides a cognitive reserve to be spent in the case of brain lesion. *J Alzheimer's Dis.* 2008;15:11–28.

135. Viola GG, et al. Morphological changes in hippocampal astrocytes induced by environmental enrichment in mice. *Brain Res.* 2009;1274:47–54.

136. Pascual R, Figueroa H. Effects of preweaning sensorimotor stimulation on behavioral and neuronal development in motor and visual cortex of the rat. *Biol Neonate.* 1996;69:399–404.

137. Rodier PM. Environmental causes of central nervous system maldevelopment. *Pediatrics.* 2004;113(Suppl 4):1076–1083.

138. Monk CS, et al. Prenatal neurobiological development: molecular mechanisms and anatomical change. *Dev Neuropsychol.* 2001;19:211–236.

139. Costa LG, et al. Developmental neuropathology of environmental agents. *Annu Rev Pharmacol Toxicol.* 2004;44:87–110.

140. Diamond A. Rate of maturation of the hippocampus and the developmental progression of children's performance on the delayed non-matching to sample and visual paired comparison tasks. *Ann NY Acad Sci.* 1990;608:394–426. (discussion: 426–433).

141. Angevin JBJ, Sidman RL. Autoradiographic study of the cell migration during histogenesis of cerebral cortex in the mouse. *Nature.* 1961;192:766–768.

142. Nowakowski RS, Rakic P. The site of origin and route and rate of migration of neurons to the hippocampal region of the rhesus monkey. *J Comp Neurol.* 1981;196:129–154.

143. Reznikov KY. Cell proliferation and cytogenesis in the mouse hippocampus. *Adv Anat Embryol Cell Biol.* 1991;122:1–74.

144. Bolon B, et al. Region-specific DNA synthesis in brains of F344 rats following a six-day bromodeoxyuridine infusion. *Cell Prolif.* 1996;29:505–511.

145. McDermott KW, Lantos PL. Cell proliferation in the subependymal layer of the postnatal marmoset, *Callithrix jacchus. Dev Brain Res.* 1990;57:269–277.

146. Temple S, Alvarez-Buylla A. Stem cells in the adult mammalian nervous system. *Curr Opin Neurobiol.* 1999;9:135–141.

147. Earle KL, Mitrofanis J. Development of glia and blood vessels in the internal capsule of rats. *J Neurocytol.* 1998;27:127–139.

148. Nowakowski RS, et al. Population dynamics during cell proliferation and neuronogenesis in the developing murine neocortex. *Results Probl Cell Differ.* 2002;39:1–25.

149. Takahashi T, et al. Early ontogeny of the secondary proliferative population of the embryonic murine cerebral wall. *J Neurosci.* 1995;15:6058–6068.

150. García-Verdugo JM, et al. Architecture and cell types of the adult subventricular zone: in search of the stem cells. *J Neurobiol.* 1998;36:234–248.

151. Miller FD, Gauthier AS. Timing is everything: making neurons versus glia in the developing cortex. *Neuron.* 2007;54:357–369.

152. Kornack DR, Rakic P. Changes in cell-cycle kinetics during the development and evolution of primate neocortex. *Proc Natl Acad Sci USA.* 1998;95:1242–1246.

153. Caviness VSJ, et al. Numbers, time and neocortical neuronogenesis: a general developmental and evolutionary model. *Trends Neurosci.* 1995;18:379–383.

154. Rodier PM. Developmental neurotoxicology. *Toxicol Pathol.* 1990;18(1 Pt 2):89–95.

155. de Groot DMG, et al. Regulatory developmental neurotoxicity testing: a model study focussing [sic] on conventional neuropathology endpoints and other perspectives. *Environ Toxicol Pathol.* 2005;19:745–755.

156. Garman RH, et al. Methods to identify and characterize developmental neurotoxicity for human health risk assessment: II. Neuropathology. *Environ Health Perspect.* 2001;109 (Suppl 1):93–100.

157. Sidman RL, Rakic P. Neuronal migration, with special reference to developing human brain: a review. *Brain Res.* 1973;62:1–35.

158. Huang Z. Molecular regulation of neuronal migration during neocortical development. *Mol Cell Neurosci.* 2009;42: 11–22.

159. Komuro H, Rakic P. Distinct modes of neuronal migration in different domains of developing cerebellar cortex. *J Neurosci.* 1998;18:1478–1490.

160. Götz M. Getting there and being there in the cerebral cortex. *Experientia.* 1995;51:301–316.

161. Hatten ME. Central nervous system neuronal migration. *Annu Rev Neurosci.* 1999;22:511–539.

162. Bronner-Fraser M. Neural crest cell formation and migration in the developing embryo. *FASEB J.* 1994;8:699–706.

163. Farlie PG, et al. The neural crest: basic biology and clinical relationships in the craniofacial and enteric nervous systems. *Birth Defects Res C Embryo Today.* 2004;72:173–189.

164. Stanfield BB, et al. The organization of certain afferents to the hippocampus and dentate gyrus in normal and reeler mice. *J Comp Neurol.* 1979;185:461–483.

165. Kuwagata M, et al. Observation of fetal brain in a rat valproate-induced autism model: a developmental neurotoxicity study. *Int J Dev Neurosci.* 2009;27:399–405.

166. Korkmaz B, et al. Migration abnormality in the left cingulate gyrus presenting with autistic disorder. *J Child Neurol.* 2006;21:600–604.

167. Peterson BS. Neuroimaging in child and adolescent neuropsychiatric disorders. *J Am Acad Child Adolesc Psychiatry.* 1995;34:1560–1576.

168. Reuhl KR, Lowndes HE. Factors influencing morphological expression of neurotoxicity. In: Tilson HA, Mitchell C, eds. *Neurotoxicology.* New York: Raven Press; 1992:67–81.

169. Norton S. Is behavior or morphology a more sensitive indicator of central nervous system toxicity? *Environ Health Perspect.* 1978;26:21–27.

170. Purpura DP. Dendritic spine "dysgenesis" and mental retardation. *Science.* 1974;186:1126–1128.

171. Huttenlocher PR, de Courten C. The development of synapses in striate cortex of man. *Hum Neurobiol.* 1978;6:1–9.

172. Bourgeois JP, Rakic P. Changes of synaptic density in the primary visual cortex of the macaque monkey from fetal to adult stage. *J Neurosci.* 1993;13:2801–2820.

173. Zecevic N, et al. Changes in synaptic density in motor cortex of rhesus monkey during fetal and postnatal life. *Brain Res Dev Brain Res.* 1989;50:11–32.

174. Chugani HT, et al. Positron emission tomography study of human brain functional development. *Ann Neurol.* 1987;22: 487–497.

175. Huttenlocher PR, Dabholkar AS. Regional differences in synaptogenesis in human cerebral cortex. *J Comp Neurol.* 1997;387:167–178.

176. Johnson FR, Armstrong-James MA. Morphology of superficial postnatal cerebral cortex with special reference to synapses. *Z Zellforsch Mikrosk Anat.* 1970;110:540–558.

177. Uylings HB, van Eden CG. Qualitative and quantitative comparison of the prefrontal cortex in rat and in primates, including humans. *Prog Brain Res.* 1990;85:31–62.

178. Espósito MS, et al. Neuronal differentiation in the adult hippocampus recapitulates embryonic development. *J Neurosci.* 2005;25:10074–10086.

179. Vernadakis A. Glia–neuron intercommunications and synaptic plasticity. *Prog Neurobiol.* 1996;49:185–214.

180. Nicholson JL, Altman J. The effects of early hypo- and hyperthyroidism on the development of rat cerebellar cortex: I. Cell proliferation and differentiation. *Brain Res.* 1972;44:13–23.

181. Rodier PM, Gramann WJ. Morphologic effects of interference with cell proliferation in the early fetal period. *Neurobehav Toxicol.* 1979;1:129–135.

182. Switzer RC III. Application of silver degeneration stains to neurotoxicity testing. *Toxicol Pathol.* 2000;28:70–83.

183. Wiggins RC. Myelin development and nutritional insufficiency. *Brain Res Rev.* 1982;4:151–175.

184. Paus T, et al. Structural maturation of neural pathways in children and adolescents: in vivo study. *Science.* 1999;283: 1908–1911.

185. Kaufmann W. Developmental neurotoxicity. In: Krinke GJ, ed. *The Laboratory Rat.* San Diego, CA: Academic Press; 2000:227–250.

186. Buznikov GA, et al. Changes in the physiological roles of neurotransmitters during individual development. *Neurosci Behav Physiol.* 1999;29:11–21.

187. Buznikov GA, et al. Serotonin and serotonin-like substances as regulators of early embryogenesis and morphogenesis. *Cell Tissue Res.* 2001;305:177–186.

188. Ernsberger U. The role of GDNF family ligand signalling in the differentiation of sympathetic and dorsal root ganglion neurons. *Cell Tissue Res.* 2008;333:353–371.

189. Brockington A, et al. Vascular endothelial growth factor and the nervous system. *Neuropathol Appl Neurobiol.* 2004;30: 427–446.

190. Ernfors P. Local and target-derived actions of neurotrophins during peripheral nervous system development. *Cell Mol Life Sci.* 2001;58:1036–1044.

191. Keynes R, Lumsden A. Segmentation and the origin of regional diversity in the vertebrate central nervous system. *Neuron.* 1990;2:1–9.

192. Gagan JR, et al. Cellular dynamics and tissue interactions of the dura mater during head development. *Birth Defects Res C Embryo Today.* 2007;81:297–307.

193. Behar TN, et al. GABA stimulates chemotaxis and chemokinesis of embryonic cortical neurons via calcium-dependent mechanisms. *J Neurosci.* 1996;16:1808–1818.

194. Gao XB, van den Pol AN. GABA, not glutamate, a primary transmitter driving action potentials in developing hypothalamic neurons. *J Neurophysiol.* 2001;85:425–434.

195. Lo AC, et al. Apoptosis in the nervous system: morphological features, methods, pathology, and prevention. *Arch Histol Cytol.* 1995;58:139–149.

196. Narayanan V. Apoptosis in development and disease of the nervous system: 1. Naturally occurring cell death in the developing nervous system. *Pediatr Neurol.* 1997;16:9–13.

197. Rodier PM. Structural–functional relationships in experimentally induced brain damage. *Prog Brain Res.* 1988;73: 335–348.

198. Coleman PD, Flood DG. Neuron numbers and dendritic extent in normal aging and Alzheimer's disease. *Neurobiol Aging.* 1987;8:521–545.

199. Buss RR, et al. Adaptive roles of programmed cell death during nervous system development. *Annu Rev Neurosci.* 2006;29:1–35.

200. Blaschke AJ, et al. Widespread programmed cell death in proliferative and postmitotic regions of the fetal cerebral cortex. *Development.* 1996;122:1165–1174.

201. Blaschke AJ, et al. Programmed cell death is a universal feature of embryonic and postnatal neuroproliferative regions throughout the central nervous system. *J Comp Neurol.* 1998;396:39–50.

202. Williams RW, Herrup K. The control of neuron number. *Annu Rev Neurosci.* 1988;11:423–453.

203. Webb SJ, et al. Mechanisms of postnatal neurobiological development: implications for human development. *Dev Neuropsychol.* 2001;19:147–171.

204. Kuida K, et al. Decreased apoptosis in the brain and premature lethality in CPP32-deficient mice. *Nature.* 1996;384: 368–372.

205. Tenkova T, et al. Ethanol-induced apoptosis in the developing visual system during synaptogenesis. *Invest Ophthalmol Vis Sci.* 2003;44:2809–2817.

206. Yoshida H, et al. Apaf1 is required for mitochondrial pathways of apoptosis and brain development. *Cell.* 1998;94:739–750.

207. Jacobs JM. Vascular permeability and neurotoxicity. In: Mitchell CL, ed. *Nervous System Toxicology.* New York: Raven Press; 1982:285–298.

208. Strazielle N, Ghersi-Egea JF. Choroid plexus in the central nervous system: biology and physiopathology. *J Neuropathol Exp Neurol.* 2000;59:561–574.

209. Gupta A, et al. Functional impairment of blood–brain barrier following pesticide exposure during early development in rats. *Hum Exp Toxicol.* 1999;18:174–179.

210. Lowndes HE, et al. Substrates for neural metabolism of xenobiotics in adult and developing brain. *Neurotoxicology.* 1994;15:61–73.

211. Risau W. Embryonic angiogenesis factors. *Pharmacol Ther.* 1991;51:371–376.

212. Lasky JL, Wu H. Notch signaling, brain development, and human disease. *Pediatr Res.* 2005;57:104R–109R.

213. Volpe JJ. Intraventricular hemorrhage in the premature infant—current concepts: I. *Ann Neurol.* 1989;25:3–11.

214. Oka A, et al. Vulnerability of oligodendroglia to glutamate: pharmacology, mechanisms, and prevention. *J Neurosci.* 1993;13:1441–1453.

215. National Research Council. *Pesticides in the Diets of Infants and Children.* Washington, DC: National Academy Press; 1993.

216. Tilson HA. Developmental neurotoxicology of endocrine disruptors and pesticides: identification of information gaps and research needs. *Environ Health Perspect.* 1998;106(Suppl 3):807–811.

217. Andersen HR, et al. Toxicologic evidence of developmental neurotoxicity of environmental chemicals. *Toxicology.* 2000;144:121–127.

218. Vorhees CV. Developmental neurotoxicity induced by therapeutic and illicit drugs. *Environ Health Perspect.* 1994;102 (Suppl 2):145–153.

219. Costa LG, et al. Structural effects and neurofunctional sequelae of developmental exposure to psychotherapeutic drugs: experimental and clinical aspects. *Pharmacol Rev.* 2004;56:103–147.

220. Chen A, et al. A comparison of behavioural and histological outcomes of periventricular injection of ibotenic acid in neonatal rats at postnatal days 5 and 7. *Brain Res.* 2008; 1201:187–195.

221. Miki T, et al. Regional difference in the neurotoxicity of ochratoxin A on the developing cerebral cortex in mice. *Brain Res Dev Brain Res.* 1994;82:259–264.

222. Golden JA. Towards a greater understanding of the pathogenesis of holoprosencephaly. *Brain Dev.* 1999;21:513–521.

223. Eteng MU, et al. Recent advances in caffeine and theobromine toxicities: a review. *Plant Foods Hum Nutr.* 1997;51: 231–243.

224. Furukawa S, et al. Indole-3-acetic acid induces microencephaly in mouse fetuses. *Exp Toxicol Pathol.* 2007;59:43–52.

225. Brites D, et al. Biological risks for neurological abnormalities associated with hyperbilirubinemia. *J Perinatol.* 2009;29: S8–S13.

226. Levine TE, Butcher RE (1990). Workshop on the qualitative and quantitative comparability of human and animal developmental neurotoxicity, Work Group IV report: triggers for developmental neurotoxicity testing. *Neurotoxicol Teratol.* 1990;12:281–284.

227. Campbell LR, et al. Neural tube defects: a review of human and animal studies on the etiology of neural tube defects. *Teratology.* 1986;34:171–187.

228. Anonymous. Poisoning the border: many American-owned factories in Mexico are fouling the environment, and their workers aren't prospering. *US News World Rep.* 1991:32–36.

229. Stanton ME, Spear LP. Workshop on the qualitative and quantitative comparability of human and animal developmental neurotoxicity, Work Group I report: comparability of measures of developmental neurotoxicity in humans and laboratory animals. *Neurotoxicol Teratol.* 1990;12: 261–267.

230. Scharden JL. Animal/human concordance. In: Slikker WJ, Chang LW, eds. *Handbook of Developmental Neurotoxicology.* San Diego, CA: Academic Press; 1998:687–708.

231. Bolon B, et al. A "best practices" approach to neuropathologic assessment in developmental neurotoxicity testing—for today. *Toxicol Pathol.* 2006;34:296–313.

232. Johnston MV. Excitotoxicity in perinatal brain injury. *Brain Pathol.* 2005;15:234–240.

233. Fonnum F. Glutamate: a neurotransmitter in mammalian brain. *J Neurochem*. 1984;42:1–11.

234. Blue ME, Johnston MV. The ontogeny of glutamate receptors in rat barrel field cortex. *Dev Brain Res*. 1995;84:11–25.

235. Tremblay E, et al. Transient increased density of NMDA binding sites in the developing rat hippocampus. *Brain Res*. 1988;461:393–396.

236. McDonald JW, Johnston MV. Physiological and pathophysiological roles of excitatory amino acids during central nervous system. *Brain Res Brain Res Rev*. 1990;15:41–70.

237. Trescher WH, et al. Quinolinate-induced injury is enhanced in developing rat brain. *Brain Res Dev Brain Res*. 1994;83:224–232.

238. Burgard EC, Hablitz JJ. Developmental changes in NMDA and non-NMDA receptor-mediated synaptic potentials in rat neocortex. *J Neurophysiol*. 1993;69:230–240.

239. Sheng M, et al. Changing subunit composition of heteromeric NMDA receptors during development of rat cortex. *Nature*. 1994;368:144–147.

240. Chen C-K, et al. Perinatal hypoxic–ischemic brain injury enhances quisqualic acid-stimulated phosphoinositide turnover. *J Neurochem*. 1988;51:353–359.

241. Johnston MV, et al. Hypoxic and ischemic central nervous system disorders in infants and children. *Adv Pediatr*. 1995;42:1–45.

242. Burnashev N, et al. Control by asparagine residues of calcium permeability and magnesium blockade in the NMDA receptor. *Science*. 1992;257:1415–1419.

243. Reuhl KR. Delayed expression of neurotoxicity: the problem of silent damage. *Neurotoxicology*. 1991;12:341–346.

244. Wallace RB, et al. Behavioral effects of neonatal irradiation of the cerebellum: 3. Qualitative observations in aged rats. *Dev Psychobiol*. 1972;5:35–41.

245. Spyker JM. Behavioral teratology and toxicology. In: Weiss B, Laties VG, eds. *Behavioral Toxicology*. New York: Plenum Press; 1975:311–344.

246. Barone S Jr, et al. Neurotoxic effects of neonatal triethyltin (TET) exposure are exacerbated with aging. *Neurobiol Aging*. 1995;16:723–735.

247. Rice DC. Evidence for delayed neurotoxicity produced by methylmercury. *Neurotoxicology*. 1996;17:583–596.

248. Eriksson P, Talts U. Neonatal exposure to neurotoxic pesticides increases adult susceptibility: a review of current findings. *Neurotoxicology*. 2000;21:37–47.

249. Goldman LR, Koduru S. Chemicals in the environment and developmental toxicity to children: a public health and policy perspective. *Environ Health Perspect*. 2000;108(Suppl 3):443–448.

250. Environmental Defense Fund. *Toxic Ignorance: The Continuing Absence of Basic Health Testing for Top-Selling Chemicals in the United States*. Washington, DC: Environmental Defense Fund; 1997.

251. DeSesso JM. Comparative embryology. In: Hood RD, ed. *Developmental and Reproductive Toxicology: A Practical Approach*. Boca Raton, FL: CRC Press; 2006:147–197.

252. Butler H, Juurlink BHJ. *An Atlas for Staging Mammalian and Chick Embryos*. Boca Raton, FL: CRC Press; 1987.

253. MacKenzie KM, Hoar RM. Developmental toxicology. In: Derelanko MJ, Hollinger MA, eds. *CRC Handbook of Toxicology*. Boca Raton, FL: CRC Press; 1995:403–450.

254. Jacque C, et al. Myelin basic protein deposition in the optic and sciatic nerves of dysmyelinating mutants quaking, jimpy, trembler, mld, and shiverer during development. *J Neurochem*. 1983;41:1335–1340.

255. McLone DG, Bondareff W. Developmental morphology of the subarachnoid space and contiguous structures in the mouse. *Am J Anat*. 1975;142:273–293.

256. Rozeik C, Von Keyserlingk D. The sequence of myelination in the brainstem of the rat monitored by myelin basic protein immunohistochemistry. *Brain Res*. 1987;432:183–190.

257. Nishimura H. *Atlas of Prenatal Histology*. Tokyo: Igaku-Shoin; 1983.

5

LOCALIZING NEUROPATHOLOGICAL LESIONS USING NEUROLOGICAL FINDINGS

BRAD BOLON

GEMpath, Inc., Longmont, Colorado

DENNIS O'BRIEN

College of Veterinary Medicine, University of Missouri, Columbia, Missouri

INTRODUCTION

Specialists in various neuroscience disciplines typically approach neurobiological investigations using the prism of their own expertise. For example, neuropathologists generally emphasize the etiology and mechanisms of neural diseases, and the most effective biochemical (for clinical pathology) and/or morphological (anatomical pathology) means for discerning lesions in neural tissues. In many cases, people who employ such a defined focus fail to appreciate the full potential for clinical findings to predict the specific structural changes that are induced by exposure to neurotoxicants. This deficiency is unfortunate, as neuroscientists with some skill in localizing structural lesions using functional data will substantially enhance their ability to find and describe neural lesions, and to prioritize their importance with respect to identifying and ranking potential hazards to human and animal health.

The ability to effectively localize the neural regions that harbor significant toxicant-induced lesions requires a solid understanding of fundamental neuroanatomical and neurophysiological concepts. The correlation between neural structure and function has been reviewed in depth elsewhere in this volume and in many other references.[1–14] Numerous texts provide diagnostic algorithms and case studies for conducting clinical neurology investigations.[1,3,8,10,14] In this current chapter we presume that for a given diagnostic or experimental neurobiology problem the clinical neurology data will be acquired using a conventional investigational plan (Table 1) by an experienced neurologist and then delivered to the neuropathologist and neurotoxicologist. Accordingly, we offer a brief correlative review of significant clinical neurology findings to provide neuropathologists and neurotoxicologists with a ready reference for aligning abnormal neural function with lesions in specific neuroanatomical regions. The neurological "syndrome" defined by the neurologist will assist the neuropathologist and/or neurotoxicologist to design their portions of the study (Table 1) to obtain the toxicological neuropathology data necessary to support a definitive diagnosis (in the clinical setting), mechanistic understanding (in the experimental setting), or best possible risk assessment (in the regulatory setting).

LESION LOCALIZATION USING THE SYNDROME CONCEPT OF CLINICAL NEUROLOGY

Damage to discrete neuroanatomical structures within the central nervous system (CNS) or peripheral nervous system (PNS) results in predictable patterns of neurological signs. In turn, these specific collections of neurological abnormalities, termed *syndromes*,[15] can be used to define the location of neural lesions and provides the basis for formulating a list of differential diagnoses, diagnostic plans, and treatment regimens. Cumulative experience in medical and veterinary medical practice as well as neuropathology research (for all

---

Fundamental Neuropathology for Pathologists and Toxicologists: Principles and Techniques, First Edition. Edited by Brad Bolon and Mark T. Butt.
© 2011 John Wiley & Sons, Inc. Published 2011 by John Wiley & Sons, Inc.

TABLE 1   Sequential Diagnostic Plan for Neurological Assessment During Toxicological Neuropathology Investigations

Preliminary Assessment for Neurological Dysfunction and Neuropathology

Collect minimum diagnostic database for initial analysis.
   Detailed history
   Physical examination
   Neurological evaluation (e.g., observation, postural reactions, and reflexes)
   Clinical pathology analysis
      Complete blood count (CBC), including automated differential count
      Chemistry profile for analytes that may be altered in diseases that can induce neurological symptoms
         Alanine aminotransferase (ALT) activity
         Albumin level
         Bilirubin level
         Blood urea nitrogen (BUN) level
         Calcium level
         Electrolytes: chloride (Cl^-), potassium (K^+), sodium (Na^+)
         Glucose level (fasting)
         Total protein level
Identify existence of neurological deficit(s).
Tentative localization of neuraxial level(s) responsible for neurological deficits.
Estimate extent of lesions within each level affected.

Diagnostic Neuropathology Practice

Neurotoxicant exposure is possible but not certain; main goals are to preserve life, and ideally, reverse neurological damage.

Compile a list of differential diagnoses.
Assemble an antemortem investigative plan (clinical testing)—employ as indicated.
   Biochemical assays [serum ammonia levels, neuropathy target esterase (NTE) activity, tissue levels of putative toxicant, serum titer/
      polymerase chain reaction (PCR) for infectious diseases, etc.]
   Functional tests [electroencephalography (EEG), electromyography (EMG), electroretinography (ERG), nerve conduction testing,
      repetitive nerve stimulation, etc.]
   Imaging with intrathecal contrast (e.g., myelogram) or without [e.g., computed tomography (CT), magnetic resonance imaging (MRI)]
   Neuropathology analysis (clinical pathology), if indicated
      Cerebrospinal fluid analysis
         Cytological examination
         Differential cell count
Confirm the diagnosis of neurotoxicity.
   Neuropathology analysis (anatomic pathology), if indicated
      Antemortem: neural biopsy
      Postmortem: autopsy/necropsy with complete neuropathology evaluation
   Neurotoxicity screen to confirm tissue levels of putative toxicant(s)

Experimental Neuropathology Setting

Neurotoxicity has been induced intentionally; primary goals are to define the spectrum of neurological changes (facilitate risk assessment),
   determine mechanisms of toxicant action, and/or to test new diagnostic methods and/or treatments.

Assemble a postmortem investigative plan (pathology analyses $\gg$ clinical testing).
   Antemortem tests
      Neurophysiological analysis by way of relevant functional testing [e.g., behavioral evaluation, cognitive assessment, EEG, EMG, ERG,
         functional observational battery (FOB), nerve conduction tests, etc.] to define the full spectrum of neurological deficits induced by the
         toxicant
      Neuroradiography to correlate anatomical and functional abnormalities in situ [e.g., positron-emission tomography (PET)] and/or to
         follow lesion evolution over time [e.g., CT, MRI, ultrasonography (US)]
   Postmortem assessments
      Neurochemical analysis to differentiate altered molecular and neurochemical pathways that cause primary toxicant-induced neural
         injury or secondary damage at other neural/nonneural sites

TABLE 1 (*Continued*)

Evaluation of the impact of toxicant dose on expression of various molecular markers (e.g., gene expression, protein production) in normal and toxicant-damaged neural tissues

Identification of potential biomarkers of toxicant-induced neural damage

Neuropathology analysis to characterize the entire spectrum of structural lesions

Documentation of the site(s) of toxicant-induced neural damage

Definition of the relationship between neurotoxicant dose and lesion severity

In situ correlation of molecular and neurochemical alterations with anatomic at lesions (e.g., altered immunohistochemical patterns for neural-specific proteins)

---

disease etiologies, including toxicants) has provided sufficient data to identify:

1. Neural regions that are particularly sensitive to toxic insult (e.g., cerebral cortex, hippocampus, cerebellum)[16]
2. Patterns of clinical neurological deficits indicative of lesions in particular neural regions (Table 2)
3. Key clinical signs that foretell the presence of structural lesions to a particular neural region (Table 3)
4. Neurotoxicants with predilections for specific neural domains.

By this rationale, unique syndromes defined by one or several key clinical signs have been identified for multiple neural domains.

The syndrome classification scheme functions in advance of neuropathology testing to predict likely etiologies for specific constellations of neurological signs. Neuropathology endpoints serve to provide additional diagnostic information (clinical pathology analysis of cerebrospinal fluid, and sometimes neural biopsy), to confirm the nature of neural diseases (surgical biopsies and/or postmortem gross and microscopic evaluations), to define the complete catalog of affected neural regions (postmortem gross and microscopic evaluations, including molecular techniques for neural cell type–specific molecules), and to understand the mechanisms by which neurotoxicants induce neural damage (postmortem gross and microscopic evaluations, including molecular techniques for neural cell type–specific molecules, and sometimes clinical pathology analysis of cerebrospinal fluid and tissue preparations).

Neurologists know, and neuropathologists and neurotoxicologists must remember, that a given neurological syndrome may be present even if only a portion of its characteristic clinical sign spectrum is observed. Accordingly, the information given below emphasizes the key signs that are sufficient for the diagnosis of a particular syndrome (Tables 2 and 4).

Correlation of Clinical Signs and Lesion Localization in the Brain

In the brain, distinctive neurological syndromes have been defined for the cerebrum, diencephalon, mesencephalon,

brain stem, cerebellum, and vestibular system. The correlation between clinical signs and affected brain structures is given in Tables 2 and 3, and the use of clinical sign groupings to localize brain lesions is shown in Table 4.

Cerebral syndrome is a common consequence of damage to the cerebral cortex or, occasionally, the subcortical centers that feed information to the cerebrocortical layers or regulate consciousness. The cerebral syndrome is typically characterized by an altered mental state—apathy, aggression, depression, disorientation, or hyperexcitability (including seizures)—in conjunction with such gait and/or postural abnormalities as circling (usually toward the damaged side), continual pacing, and head pressing. Even if the gait is unaffected under normal conditions, postural reactions in response to diagnostic tests (e.g., hemi-walking, hopping) may be depressed on the side opposite the cerebral damage. Focal lesions may produce visual deficits (depressed menace response, blindness) in the eye contralateral to the lesion without affecting the pupillary light reflex. The cerebral syndrome is most commonly induced by any of numerous nontoxic neural insults, including cranial trauma, hydrocephalus, inflammation (bacterial or viral etiology), ischemia, neoplasia, nutritional deficiency [e.g., polioencephalomalacia of ruminants, a consequence of insufficient thiamine (vitamin B_1)], and vascular accidents. The cerebral syndrome is also a regular complication when dehydration results in ionic imbalances in the neuropil (salt poisoning in swine), nitrogen-rich metabolic by-products cannot be excreted (e.g., in hepatic encephalopathy or uremic encephalopathy) and following exposure to exogenous toxicants such as heavy metals (e.g., lead, mercury), certain mycotoxins (e.g., fumonisin B_1), and some solvents (e.g., hexachlorophene).[17,18]

Diencephalic syndrome is relatively uncommon. Clinical features include altered behavior and mental status (e.g., aggression, coma, disorientation, hyperexcitability) in the absence of major gait or postural abnormalities. However, the defining traits of lesions in this region are abnormalities of autonomic functions [e.g., disrupted thermoregulation, uncontrolled appetite (either anorexia or hyperphagia/obesity)] or endocrine systems (often, diabetes insipidus or hyperadrenocorticism) that are centered in the hypothalamus. Lesions in this region may impinge on the optic chiasm, where they produce bilateral visual deficits that are accompanied by

TABLE 2 Association Between Lesion Localization and Major Neurological Signs[a]

Localization of Neuroanatomical Lesions	Clinical Signs
Cerebral cortex	Altered mental status
	Aggression
	Behavioral abnormalities [e.g., circling (toward the damaged side), continual pacing, head pressing, vocalization]
	Depression (but often also a result of anemia, fever, or metabolic disease)
	Disorientation (but often also a result of fever or metabolic disease)
	Hyperexcitability
	Aberrant movement
	Lameness (UMN) (but often also orthopedic)
	Paresis/paralysis (UMN): one side (typically, contralateral) or all limbs
	Abnormal posture or proprioception (position)
	Head
	Limbs (UMN)
	Weakness (UMN) (but also metabolic or muscular disease)
	Altered perception
Temporal region (contralateral)	Hearing deficit (no vestibular signs)
Occipital region (contralateral)	Visual deficit (pupillary reflexes normal)
	Seizures
Basal ganglia	**Tremors** (but common also in drug reactions, fatigue, fear, hypothermia, and primary muscular disease)
Diencephalon	Altered mental status
Hypothalamus	**Aggression**
	Coma
	Disorientation (but often also a result of fever or metabolic disease)
	Hyperexcitability
Thalamus	Altered perception
	Taste deficit
	Touch abnormality (generalized)
Medial geniculate nucleus	Hearing deficit (no vestibular signs)
Lateral geniculate nucleus	Visual deficit (pupillary reflexes normal)
Hypothalamus	**Autonomic dysfunction**: deranged appetite (anorexia or hyperphagia), disrupted thermoregulation
Hypothalamus	**Endocrine dysfunction**: diabetes insipidus, hyperadrenocorticism
Thalamus	**Seizures**
Limbic system	Altered mental status
	Behavioral abnormalities
Midbrain	Altered mental status
	Coma
	Abnormal position of eye
CN III	**Strabismus** (deviation), ventrolateral
	Abnormal posture or proprioception (position)
	Limbs: opisthotonos (rigid extension of all limbs)
	Altered perception
Cerebellum	Aberrant movement
	Ataxia
	Dysmetria
	Tremors (common also in drug reactions, fatigue, fear, hypothermia, and primary muscular disease)
Brainstem	Altered mental status
Reticular formation	**Coma or stupor**
	Aberrant movement
	Lameness (UMN) (but often also orthopedic)
	Paresis/paralysis (UMN)
	Weakness (UMN) (but also metabolic or muscular disease)

TABLE 2   (*Continued*)

Localization of Neuroanatomical Lesions	Clinical Signs

Nucleus solitarius	Altered perception
	  Taste deficit (caudal one–third of tongue)
	Abnormal posture or proprioception (position)
	  Head
	  Limbs (UMN)
	Syncope (but usually cardiovascular or metabolic disease)
	Urinary incontinence (but also urinary tract disease)

Vestibular system	Aberrant movement
	  Ataxia
	  Nystagmus (involuntary rhythmic movements of the eyes)—more common in central (brain) than peripheral (labyrinth) injury
	Altered perception
	  ± Hearing deficit (with vestibular signs)
	Abnormal posture or proprioception (position)
	  Head tilt (toward side with lesion)

Spinal cord	Aberrant movement
	  Lameness (LMN) (but often also orthopedic)
	  Paresis/paralysis (LMN)
	  Weakness (LMN) (but also metabolic or muscular disease)
	Altered perception
	  Touch abnormality (localized: usual pattern is normal sensation above a lesion, reduced sensation below it, ± a local increase in sensation over the lesion)
	Abnormal posture or proprioception (position)
	  Trunk (LMN) (but also vertebral column injury)
	  Limbs (LMN)
	Urinary incontinence (but also urinary tract disease)

Meninges	Altered perception Touch abnormality (generalized)
Cranial nerves (nuclei and/or trunks)	
CN I and/or olfactory neuroepithelium	Altered perception: olfactory deficit (anosmia) (but often nasal passage damage)
CN II and/or retina and/or optic chiasm	Altered perception: **visual deficit** (with abnormal pupillary reflexes)

CN III	Aberrant movement of extraocular muscles
	  Paresis/paralysis [especially ptosis (drooping of upper eyelid)]
	Abnormal position of eye
	  Strabismus (deviation), ventrolateral

CN IV	Aberrant movement of extraocular muscles Paresis/paralysis (leading to abnormal lateral rotation of eye, typically of modest degree)

CN V	Altered perception
	  Altered response to direct stimulation of nasal mucosa (maxillary)
	  Altered palpebral reflexes
	    Medial canthus stimulation (ophthalmic branch)
	    Lateral canthus stimulation (maxillary branch)
	  Touch abnormality (localized over jaw) (mandibular branch)
	Aberrant movement: **paresis/paralysis** of masticatory muscles (leading to permanent drooping of the mandible)
	  Atrophy (masseter and temporal muscles)

CN VI	Aberrant movement of extraocular muscles
	  Paresis/paralysis (especially inability to gaze laterally)
	Abnormal position of eye
	  Strabismus (deviation), medial

CN VII	Altered perception
	  Taste deficit (rostral two–third of tongue)
	Aberrant movement of facial muscles
	  Paresis/paralysis (leading to drooping of ears, eyes, and/or lips)

CN VIII (cochlear division) and/or cochlea and/or spiral ganglion	Altered perception: **hearing deficit with no vestibular signs**

TABLE 2   (*Continued*)

Localization of Neuroanatomical Lesions	Clinical Signs
CN VIII (vestibular division)	Aberrant movement
	Ataxia
	Nystagmus (involuntary rhythmic movements of the eyes)
	Altered perception
	± Hearing deficit (with vestibular signs)
	Abnormal posture or proprioception (position)
	Head tilt (toward side with lesion)
CN IX	Altered perception: taste deficit (caudal one–third of tongue)
CN X	Altered perception: taste deficit (caudal one–third of tongue)
CN XI	Atrophy (neck and shoulder muscles), rarely evident
CN XII	Atrophy (muscles of the tongue)
Peripheral somatic nerves	Aberrant movement
	Lameness (but often also orthopedic)
	Paresis/paralysis with depressed or absent spinal reflexes
	Weakness (but also metabolic or muscular disease)
	Altered perception: **touch abnormality** (localized: reduced or absent sensation ± altered sensation [paraesthesia]) in the dermatone innervated by the affected nerve
Autonomic nerves (nuclei and/or trunks)	
Parasympathetic	
CN III	**Pupillary dilation** (fixed)
CN VII	Loss of tear production
CN X	Aberrant movement: **paresis/paralysis** of laryngeal muscles
Sympathetic	
Vagosympathetic trunk and CN V	**Horner's syndrome**
	Enophthalmos (sunken eye)
	Miosis (small pupil)
	Ptosis (drooping of upper eyelid)
	Prolapse of the third eyelid
Pelvic/pudendal nerves	Altered anal sphincter tone (debated)
	Urinary incontinence (but also urinary tract disease)

[a] Key signs for each neural region are given in boldface type. CN, cranial nerve; LMN, lower motor neuron; UMN, upper motor neuron.

dilated pupils and a weak or absent pupillary response to light. Hearing deficits (ipsilateral) may be present if a lesion affects the medial geniculate nucleus (hearing relay center in the thalamus), while visual dysfunction (contralateral) will be evident if the lateral geniculate nucleus (thalamic visual relay center) is damaged. The usual cause of diencephalic syndrome is a space-occupying mass such as an abscess, granuloma, or neoplasm. Thus, this pattern of neurological abnormalities is an uncommon outcome of neurotoxicant exposure.

Mesencephalon (midbrain) syndrome is also infrequent except with tentorial herniation secondary to increased intracranial pressure. Affected persons may have altered mental status (coma, depression). Widespread lesions may produce opisthotonos (rigid extension of all four limbs, caused by interruption of descending motor control circuitry from the cerebrum and midbrain), while focal damage may elicit hemiparesis (weakness) of the contralateral limbs. Multiple signs usually indicate involvement of cranial nerve (CN) III, including widely dilated pupils that will not constrict in response to light, ptosis (drooping of the upper eyelids), and ventrolateral strabismus (deviation of the eye due to paralysis of some extraocular muscles); the location and extent of the midbrain lesion will dictate whether the signs are ipsilateral or bilateral. Causes of mesencephalic lesions include cranial trauma, inflammation (viral-induced), nutritional (thiamine) deficiency, and certain toxicants (e.g., carboplatin,[19] carbon monoxide,[20] carbonyl sulfide,[21] 1,3-dintrobenzene,[22,23] ethanol,[24] thiophene,[25] some venoms[26]).

Brainstem syndrome is a relatively frequent collection of neurological deficits resulting from damage to the pons and/ or medulla. Regular features include altered mental status (coma, depression, or stupor), abnormal respiration (shallow and rapid, with intermittent periods of apnea), and aberrant movement (usually tetraparesis or tetraplegia) with normal reflexes. Dysfunction of multiple cranial nerves is common, including signs related to deficits in:

- *CN V.* reduced palpebral reflexes, lessened facial sensation, and jaw paralysis

TABLE 3   Association Between Major Neurological Signs and Lesion Localization[a]

Clinical Presentation	Lesion Localization
Always or Often Originating in the Nervous System	

Abnormal posture or proprioception (position)

  Eye

Pupillary dilation	CN III
Strabismus (deviation): lateral rotation	CN IV
Strabismus: medial	CN VI
Strabismus: ventrolateral	CN III
Head	Cerebral cortex (especially head pressing)
	Brain stem or cerebellum
Tilt (toward side with lesion)	Vestibular apparatus or CN VIII
Limbs	Cerebral cortex (UMN)
	Brain stem (UMN)
	Spinal cord (UMN or LMN)
Lips (drooping)	CN VII
Trunk	Spinal cord (LMN) (or vertebral column injury)
Altered mental status	
Behavioral abnormalities	Cerebrum
	Limbic system
	Diencephalon (hypothalamus)
Coma or stupor	Cerebrum
	Diencephalon
	Brain stem (reticular formation)
Aberrant movement	
Altered response to palpebral stimulation	CN VII
Ataxia	Cerebellum
	Vestibular system
Circling (toward side with lesion)	Cerebral cortex
	Vestibular system
Dysmetria	Cerebellum
Intention tremor	Cerebellum
Nystagmus (involuntary rhythmic movements of the eyes)	Vestibular system
Paresis/paralysis	Cerebral cortex (UMN)
	Brain stem (UMN)
	CN V
Laryngeal muscles	CN X
	Spinal cord (LMN)
	Peripheral somatic nerves (localized)
Ptosis [drooping of the upper eyelid (Horner's syndrome)]	Sympathetic system: CN III
Altered perception	
Ataxia (proprioceptive)	Brain stem
	Spinal cord
Altered response to palpebral stimulation	CN V
Hearing deficit	
No vestibular signs	Cerebral cortex: temporal region
	Midbrain: medial geniculate nucleus (ipsilateral)
	Inner ear: cochlea, spiral ganglion, CN VIII (cochlear division)
Vestibular signs	Vestibular apparatus (labyrinth)
	CN VIII (vestibular division)
Taste deficit	Thalamus
Rostral two–third of tongue	Nucleus solitarius
	Brain stem
	CN VII
Caudal one–third of tongue	CN IX
	CN X

(continued)

TABLE 3 (*Continued*)

Clinical Presentation	Lesion Localization
Touch abnormality	
Generalized	Thalamus
	Meninges
	CN V
Localized (usual pattern is normal sensation above a lesion, reduced sensation below ± locally increased sensation over the lesion)	
	Spinal cord
	Peripheral somatic nerves
Visual deficit	
Pupillary reflexes normal	Occipital cortex (contralateral)
	Thalamus: lateral geniculate nucleus (contralateral)
Pupillary reflexes abnormal	Retina
	Optic nerve (CN II)
	Optic chiasm
Autonomic dysfunction	
Deranged appetite (anorexia, hyperphagia)	Hypothalamus
Disrupted thermoregulation	Hypothalamus
Seizures	Cerebrum
Possibly Originating in Nervous System	
Depression (but often anemia, fever, or metabolic disease)	Cerebral cortex
Endocrine dysfunction (diabetes insipidus, hyperadrenocorticism) (but must be distinguished by similar presentations arising from adrenal or pancreatic injury)	Hypothalamus
Lameness (but typically, orthopedic disease)	Cerebral cortex (UMN)
	Brain stem (UMN)
	Spinal cord (LMN)
	Peripheral somatic nerves (localized)
Olfactory deficit (anosmia) (but often nasal passage disease)	Olfactory neuroepithelium
	CN I
Syncope (usually cardiovascular or metabolic disease)	Brain stem
Tremors (common also in drug reactions, fatigue, fear, hypothermia, and primary muscular disease)	Basal ganglia
	Cerebellum
Urinary incontinence (also urinary tract disease)	Brain stem
	Spinal cord
	Sympathetic system: pelvic/pudendal nerves
Weakness (also metabolic or muscular disease)	Cerebral cortex (UMN)
	Brain stem (UMN)
	Spinal cord (LMN)
	Peripheral somatic nerves (localized)

ᵃ CN, cranial nerve; LMN, lower motor neuron; UMN, upper motor neuron.

- *CN VI.* medial strabismus
- *CN VII.* drooping ears, paralysis of the eyelids and lips
- *CN VIII.* head tilt (toward the damaged side), nystagmus (involuntary rhythmic movements of the eyes), and rolling
- *CNs IX and X.* laryngeal and pharyngeal paralysis (leading to dysphonia, dysphagia, and a diminished gag reflex)
- *CN XII.* tongue paralysis

Brain stem syndrome generally results from cranial trauma, inflammation (viral-induced), or the presence of a space-occupying mass. Xenobiotics may induce specific brain stem lesions following local (intrathecal)[27,28] or systemic[29,30] delivery of small-molecule drugs, and as a delayed consequence of organophosphate exposure.[17] However, in many cases systemic exposure to potential neurotoxicants appears to affect the brain stem only in conjunction with effects to other neural regions.[22,26,31]

TABLE 4 Typical Signs for Localizing Lesions Affecting the Brain[a]

Site of Lesion	Movement	Posture	Postural Reactions	Mental Status	Cranial Nerve (CN) Involvement
Cerebral cortex	Gait normal to slight hemiparesis (contralateral)	Normal	Normal or abnormal (contralateral)	**Altered behavior, depression, and/or seizures**	Normal usually or central blindness
Diencephalon (thalamus and hypothalamus)	Gait normal to hemiparesis (contralateral) or tetraparesis	Normal	Abnormal (contralateral)	**Altered behavior,** depression, and/or **autonomic and endocrine deficits**	**CN II**
Brainstem (midbrain, pons, medulla oblongata)	Hemiparesis (contralateral) or tetraparesis, ataxia	Normal or circling or falling	Abnormal (ipsilateral or contralateral)	**Coma, stupor**	**CN III to XII**[b]
Cerebellum	**Ataxia, dysmetria, tremors, but no paresis**	Normal usually	Normal or abnormal (due to dysmetria)	**Normal**	Normal usually
Vestibular, central (medulla oblongata)	Hemiparesis (usually ipsilateral), ataxia	**Head tilt, circling, falling** (all toward affected side)	**Abnormal** (ipsilateral)	**Depression** (global lesions) or normal (for focal lesions)	**CN VIII (nystagmus is very common)**, sometimes CN V and CN VII
Vestibular, peripheral (labyrinth)[c]	Normal to ataxia	**Head tilt, circling, falling** (all toward affected side)	**Normal usually**	Normal	CN VIII (nystagmus occurs occasionally), sometimes CN VII

Source: Adapted from Lorenz et al.[8]

[a] Key signs for each neural region are given in boldface type.

[b] Deficits most easily recognized for CN III, V, VII, VIII, IX, and X.

[c] Structure adjacent to but outside brain (included here for comparison to central vestibular disease).

Vestibular syndrome is another readily distinguished pattern of neurological deficits. This constellation of signs may originate centrally (medulla oblongata) or peripherally [vestibular labyrinth or CN VIII (vestibular division)], with the latter variant being more common. Prominent features include circling, head tilting, and/or rolling (all toward the side with the lesion) as well as nystagmus (with the quick phase presenting as a horizontal rotation away from the damaged side). Multiple signs attributable to cranial nerve dysfunction may be evident. Both central and peripheral disease may be associated with facial paralysis (CN VII), as both these nerves pass through the middle ear. Horner's syndrome—enophthalmos (recession of the eye into the orbit), miosis (constriction of the pupil), ptosis (drooping eyelid), and elevation of the third eyelid—can occur with central lesions but more commonly reflects damage to the sympathetic nerves as they course through the middle ear. Central vestibular disease may also induce deficiencies in CN V (decreased facial sensation, lessened palpebral reflexes, jaw weakness), CN VI (medial strabismus), and CNs IX and X (altered laryngeal and pharyngeal function). Common causes of vestibular syndrome include inflammation [particularly bacterial-induced otitis media (i.e., peripheral)] and neoplasms. Toxicants that produce vestibular syndrome usually do so by causing hair cell degeneration in the sensory epithelia of the vestibular labyrinth.[32–34]

Cerebellar syndrome is a common neurological pattern characterized by an easily identifiable collection of clinical signs. A key feature is abnormal motion [e.g., exaggerated limb movements, especially when starting to move (intention tremors, especially of the head), in conjunction with exaggerated movements (hypermetria)]. Fine side-to-side eye oscillations and a reduced or absent menace response may occur in some cases, but vision is normal. Another major characteristic is altered posture, including such deficits as a truncal ataxia (swaying of the torso when walking) and a broad-based stance when at rest. Signs are generally bilateral if cerebellar lesions are widespread and ipsilateral (including menace deficits) if damage is confined to a single focus. With damage to the floccular or nodular lobes, vestibular signs will be present but the signs will be the opposite of that of a peripheral vestibular lesion (paradoxical vestibular syndrome). Cerebellar syndrome is typically the outcome of defective development (e.g., abiotrophy, hypoplasia), inflammation (viral-induced), or a space-occupying mass, but is associated with exposure to certain toxicants (e.g., bromethalin,[35] hexachlorophene,[36] methyl mercury,[17,37] trimethyl tin[38]) and toxins (e.g., plants of the genus *Solanum*[17]).

Correlation of Clinical Signs and Lesion Localization in the Spinal Cord

The clinical syndromes localized to the spinal cord exhibit different patterns of neurological signs for five major domains:

upper cervical (C_1 to C_5), cervicothoracic (C_6 to T_2), thoracolumbar (T_3 to L_3), lumbosacral (L_4 to S_2), and caudal (Cd_1 to Cd_x). The relationships between spinal cord lesions and clinical signs are given in Tables 2 and 3. The use of neurological signs to localize spinal cord lesions is shown in Table 5, while localizing lesions to the brain or spinal cord using patterns of clinical deficits is given in Table 6.

The *upper cervical syndrome* (C_1 to C_5) is indicated by a pattern of neurological signs in which the most important aspect is abnormal movement. Widespread lesions cause weakness or paralysis of all four limbs, while focal lesions are associated with ipsilateral weakness or paralysis. Spinal reflexes are often increased because the connection to the upper motor neurons in the cerebrum are disrupted. If the lesion is cranial to the origin of the phrenic nerve, death from respiratory paralysis will ensue. Upper cervical syndrome is typically caused by cervical trauma, developmental malformations of the associated vertebrae, inflammation (bacterial, protozoal, or viral etiologies), or the presence of a space-occupying mass. Certain toxic agents (e.g., some organoarsenic and organophosphorus compounds, a few solvents, cycad toxins) are known for inducing changes in this region.[17,39–42] The character of the toxicant-mediated lesions reflects damage to the peripheral elements in the ascending sensory tracts in the dorsal (posterior) columns of C_1 to C_5 as well as neuronal changes in the cuneate and gracile nuclei (located in the caudal medulla oblongata) to which the columns project.

The *cervicothoracic syndrome* (C_6 to T_2) is noted chiefly for depressed or absent reflexes and muscle tone with weakness and muscle atrophy in the thoracic limbs. In contrast, in the pelvic limbs reflexes and tone are normal or increased, and muscle mass is unaffected. Another common sign is Horner's syndrome (enophthalmos, miosis, ptosis, and prolapse of the third eyelid). Diffuse lesions cause bilateral deficits, while focal lesions cause ipsilateral effects. Cervicothoracic syndrome is usually the outcome of spinal cord damage secondary to vertebral injury (e.g., malformations, ruptured intervertebral disks), or brachial plexus destruction caused by neoplasia or trauma (avulsion, catheter- or injection-associated laceration). Local[43] or systemic[44] delivery of a few small therapeutic molecules (especially local anesthetics) has been shown to cause this neurological pattern by affecting the cervical spinal cord and/or brachial plexus. Selenium and 6-aminonicotinamide [6-AN, a model compound that produces nicotinamide (the amide form of niacin, a B-complex vitamin) deficiency] also target this spinal cord region in young pigs.[17]

The *thoracolumbar syndrome* (T_3 to L_3) is a common and readily identified clinical pattern featuring profound pelvic limb dysfunction in conjunction with normal thoracic limb capabilities. Pelvic abnormalities include increased reflexes and tone (especially affecting extensor muscles), decreased postural responses, and spastic weakness or paralysis. Cutaneous sensation is depressed in dermatomes located

TABLE 5   Signs for Localizing Lesions Affecting the Spinal Cord

Site of Lesion	Upper Motor Neuron (Long Tract Disease)	Lower Motor Neuron (Segmental Disease)
Coccygeal/Caudal	—	Tail: flaccidity
Sacral		
Sacral (S_1–S_3)	Tail: flaccidity	
Pelvic plexus	—	Altered autonomic tone: anal sphincter, urinary bladder
Lumbar		
Lumbar intumescence (L_4–S_2)	Tail: flaccidity	
Lumbar plexus	Altered autonomic tone: anal sphincter, urinary bladder	Altered autonomic tone: anal sphincter, urinary bladder
	Normal thoracic limb function	Pelvic limb dysfunction: altered proprioception, decreased reflexes; paresis or paralysis
Thoracolumbar (T_3–L_3)	Altered autonomic tone: urinary bladder	Altered mass and/or tone of associated spinal muscles
	Normal thoracic limb function	
	Pelvic limb dysfunction: altered proprioception, increased reflexes; paresis or paralysis	
Cervicothoracic		
Cervical intumescence (C_6–T_2)	Altered autonomic tone: urinary bladder only	Thoracic limb dysfunction: altered proprioception, decreased reflexes; paresis or paralysis
	Pelvic limb dysfunction: altered proprioception, increased reflexes; paresis or paralysis	
Brachial plexus		Thoracic limb dysfunction (ipsilateral): altered proprioception, decreased reflexes; paresis or paralysis
Upper cervical (C_1–C_5 and also medulla oblongata)	Altered autonomic tone: urinary bladder only	
	Dysfunction of all four limbs: altered proprioception, increased reflexes; paresis or paralysis	

caudal to the level of the spinal cord lesion with severe lesions. Urinary continence is common and can be associated with brief incomplete attempts at urination ("spastic bladder"). Principal factors that induce thoracolumbar syndrome are degeneration (myelomalacia), inflammation (viral-induced), metastatic neoplasia, and vertebral damage (discospondylitis, fractures, ruptured disks). Toxicants do not specifically damage this spinal cord domain, although damage to upper motor

TABLE 6   Differentiating Signs of Motor Neuron Disease at Different Neuraxial Levels

Function	Upper Motor Neuron (Long Tract Disease)	Lower Motor Neuron (Segmental Disease)
Motor		
Movement	Paresis to paralysis: loss of voluntary movements	Paresis to paralysis: flaccidity, loss of power
Reflexes	Normal to increased	Decreased to absent
Tone	Normal to increased	Decreased
Muscle maintenance	Atrophy: late and mild (disuse)	Atrophy: early and severe (neurogenic), with late contracture
Electromyography	No changes	Abnormal potentials
Sensory		
Perception	Decreased (superficial and deep pain) caudal to lesion	Anesthesia of innervated region; ± altered (paresthesia) or increased (hyperesthesia) sensation prior to loss
Proprioception	Decreased to absent	Decreased to absent

neuron (UMN) and lower motor neuron (LMN) centers will be reflected in axonal lesions within the thoracolumbar region.

The *lumbosacral syndrome* (L_4 to S_2) is associated with a range of clinical signs related to the pelvic limbs, regional sphincters, and tail. Key features include flaccid weakness or paralysis of the pelvic limbs and tail, with reduced or absent pelvic limb reflexes, sensation, postural reactions, and tone, leading to muscle atrophy. In contrast, thoracic limb function is normal. The anal sphincter may be paralyzed and dilated, resulting in fecal incontinence, while paralysis of the urinary bladder causes urine retention with eventual passive overflow (urinary incontinence). Common causes of lumbosacral syndrome include regional disk disease, inflammation (of the nerve roots or vertebrae), neoplasia, congenital anomalies, or trauma (fractures, luxations), as well as local (intrathecal) delivery of anesthetic agents.[45,46] In young pigs, this spinal cord domain is also susceptible to 6-AN and selenium toxicities.[17]

Caudal (coccygeal) syndrome is generally limited to animals with tails and is defined by flaccid paralysis or pain in this appendage. Causes include abnormal axonal regeneration (e.g., "amputation neuroma"), inflammation (of the cauda equina nerve roots or tail tissues), neoplasia (e.g., chordomas), and trauma (fractures, vascular injections). Local delivery of xenobiotics into the soft tissues rather than tail vasculature may incite regional nerve injury.

Correlation of Clinical Signs and Lesion Localization in Peripheral Nervous Systems

The neurological deficits that develop in the PNS typically induce clinical signs related to only one PNS domain (autonomic parasympathetic, autonomic sympathetic, somatic motor and/or somatic sensory), and often affect only a portion of the domain affected. Certain PNS syndromes are actually the consequence of lesions in peripheral target organs rather than the PNS proper. The association between key clinical signs and PNS lesion localization are shown in Tables 2 and 3.

Neuropathic syndrome caused by damage to one or more peripheral (and occasionally, cranial) nerves is a common occurrence in the clinical setting. Key features of this pattern are typically observed in the head or appendicular (trunk) muscles and include reduced or absent reflexes and tone, flaccid weakness or paralysis, and eventually, neurogenic atrophy. With neuromuscular junction diseases, the weakness may be better characterized as fatigability with normal muscle strength after rest. Sensory loss in the affected dermatome(s) is variable. Autonomic effects may be variable (e.g., botulism) or the primary symptom (e.g., anticholinergic drugs). Etiologies of neuropathic syndrome include trauma or neoplasia affecting one or a few closely associated nerve trunks and toxic blockade of the neuromuscular junctions (e.g., botulism, tick paralysis) affecting most or all nerves.

Myopathic syndrome is an uncommon neurological presentation reflecting damage to skeletal muscle rather than the nerves that innervate them. Hallmarks include generalized weakness, abnormal movements (stiff gait, worsened by exercise), and altered muscle mass (increased or decreased, depending on the etiology). Muscle changes tend to be bilateral but are often asymmetric. The functions of reflex circuits and sensory pathways are usually normal, although myalgia (with or without cramping) may occur. Myopathic syndrome usually results from congenital dystrophies, inflammation (autoimmune, idiopathic), pharmaceutical toxicity,[47,48] or prolonged inactivity. The primary factors used to classify toxicant-induced myopathies are histopathologic lesions—atrophy (e.g., steroids),[49–51] inflammation (certain immunosuppressants, thiols),[47,49] necrosis [e.g., lipid-lowering drugs (fibrates, statins)],[47,48] ragged red fibers (characteristic of mitochondrial injury),[47,48] and vacuolation (anti-malarials)[47]—or electrolyte imbalances (e.g., hypokalemia).[47] Toxic myopathies are usually reversible if exposure is discontinued.

Correlation of Clinical Signs with Multifocal Lesions

In some cases, careful consideration of the clinical sign spectrum reveals a complex set of neurological abnormalities that cannot be consigned to a single clinical syndrome. This mixture is characteristic of lesions in multiple neural regions, leading to the simultaneous occurrence of two (or more) clinical syndromes. In almost all cases, *overlapping syndromes* can be explained by a single etiologic agent, such as disseminated infection, metabolic disorders (e.g., storage diseases associated with a congenital or toxicant-induced enzyme deficiency), metastatic neoplasia, or trauma (e.g., explosive intervertebral disk prolapse leading to widespread myelomalacia). Neurotoxic agents that cause neuronal storage diseases include the active principles of certain plant species (*Astragalus* sp., *Swainsona* sp., *Trachyandra* sp.) and the unidentified environmental toxicant responsible for Gomen disease.[17] Strychnine and tetanus toxin also induce overlapping clinical syndromes due to their diffuse interaction with synapses. Exposure of young pigs to either 6-AN or selenium is associated with bilaterally symmetrical cavitation of the ventral (anterior) gray columns in the cervical and lumbar intumescences, thus resulting in concurrent onset of cervicothoracic and lumbosacral syndromes.[17]

Clinical Syndromes Not Associated with Structural Lesions

Certain patterns of neurological dysfunction combine to reliably produce distinct clinical syndromes in the absence of corresponding morphologic lesions. These disorders tend to occur at intermittent intervals. Between episodes, the person affected is neurologically normal and responsive. These conditions are termed *paroxysmal syndromes* due to the sporadic and temporary nature of the clinical deficits.

Paroxysmal syndromes of neural origin often present with seizures, suggesting that the cerebrum is the probable location of the neural damage. The seizures typically begin and end suddenly; if the affected individual survives, they may recur at irregular intervals without warning. Episodes are sometimes preceded by a brief pre-ictal period of apprehension and restlessness and/or followed by a post-ictal phase (minutes to 24 h or more) characterized by disorientation, depression, or prolonged sleep. Other neural-based clinical signs include narcolepsy [cataplexy (sudden flaccid paralysis, in dogs) or episodic sleep (humans)], syncope (sudden loss of consciousness due to temporary disturbances in cerebral circulation), and tremors (characterized by abnormal motion). Paroxysmal syndromes centered in the nervous system usually have no evident origin (i.e., are idiopathic); develop secondary to cardiovascular (especially for syncope), metabolic (e.g., hypoglycemia), or nutritional (e.g., thiamine deficiency) disorders; or are induced by exposure to certain neurotoxicants (e.g., heavy metals, some organophosphorus compounds, certain pesticides).[15] Some neurotoxic agents induce a single, paroxysmal episode of variable duration. Death may be so abrupt that structural changes do not have time to develop (e.g., chlorinated hydrocarbon insecticides), or the condition may develop more gradually in response to biochemical imbalances outside the nervous system [e.g., ethylene glycol toxicity, in which acidosis, uremia, and perhaps hypocalcemic tetany arise when calcium oxalate crystals precipitate inside renal tubules and systemic microvessels (including those of the brain)].[17] Altered neurotransmission of other functional abnormalities are likely explanations for other conditions in which structural lesions are lacking, such as perennial rye grass staggers and phalaris staggers (caused by natural toxins) and ivermectin toxicity in collie dogs [the result of blocking the inhibitory neurotransmitter γ-aminobutyric acid (GABA)].[17]

Some paroxysmal syndromes affect skeletal muscle rather than the nervous system. These conditions are associated with episodic weakness and excessive fatigue that are induced or exaggerated by exercise, with regression of the neurological signs when the person is at rest. Common causes of muscle-centered paroxysmal syndromes include cardiovascular conditions (e.g., arrthymias, conduction blockade, congestive heart failure), immune-mediated neuromuscular junction diseases (e.g., myasthenia gravis), metabolic imbalances (acidosis, hypoglycemia, hyper- or hypokalemia), and exposure to many pharmaceutical agents[52] and natural toxins.[53]

PRACTICAL APPLICATION OF NEUROLOGICAL SYNDROMES BY NEUROPATHOLOGISTS AND NEUROTOXICOLOGISTS

As noted above, the neurological examination is generally performed by an experienced clinical scientist, after which the results are given to the neuropathology and neurotoxicology representatives of the research team. Nevertheless, pathologists and toxicologists should become familiar with (1) the investigational strategy by which a given set of clinical signs may be used most efficiently and effectively to localize neural lesions (Table 1), and their roles in the research team undertaking the investigation; (2) the most easily distinguished patterns of neurological dysfunction; and (3) the more common groupings of clinical deficiencies that arise following exposure to neurotoxicants. The unique clinical features associated with almost all of the brain-based syndromes (Tables 2 to 4) allow lesions to be localized to most intracranial CNS domains with some confidence. However, only the cerebral, cerebellar, brain stem, and vestibular syndromes are commonly observed in the clinical setting. Of these, neurotoxicants are the inciting agent mainly for lesions localized to the cerebrum (systemic exposure), cerebellum (systemic delivery), and brain stem (intrathecal injection). In the spinal cord (Tables 2, 3 and 5), the most readily identified neurological patterns occur in the cervicothoracic, thoracolumbar, and lumbosacral syndromes, although the thoracolumbar pattern is the most common. Toxic lesions in these regions may be primary for locally delivered agents (affecting primarily the cervicothoracic and lumbosacral areas) but more commonly are secondary to primary damage originating in distant neurons and/or axons. Neuropathic and myopathic syndromes [representing damage to the PNS or skeletal muscle (appendicular or axial), respectively] are easy to recognize. The neuropathic syndrome occurs more commonly, but the myopathic syndrome is known to be caused by a greater number of toxicants.

In the final evaluation, the ability to correlate clinical deficits to specific structural lesions is an important aspect of any toxicological neuropathology study. The responsibility for the final integration of the clinical data with the pathology data set will probably vary with the setting, falling to the neurologist in medical or veterinary medical practice but often residing with the neuropathologist and/or neurotoxicologist in organizations engaged in product development and risk assessment. A basic understanding of these principles will facilitate communication among biomedical professionals with different areas of expertise and improve the value of literature and regulatory reports generated from toxicological neuropathology studies.

REFERENCES

1. de Lahunta A, Glass E. *Veterinary Neuroanatomy and Clinical Neurology*, 3rd ed. Philadelphia: W.B. Saunders; 2009.

2. FitzGerald MJT. *Neuroanatomy: Basic and Clinical*, 3rd ed. Philadelphia: W.B. Saunders; 1996.

3. FitzGerald MJT, et al. *Clinical Neuroanatomy and Neuroscience*, 5th ed. Philadelphia: W.B. Saunders; 2007.

4. Getty R. *Sisson and Grossman's The Anatomy of the Domestic Animals*, 5th ed. Philadelphia: W.B. Saunders; 1975.

5. Greenstein B, Greenstein A. *Color Atlas of Neuroscience: Neuroanatomy and Neurophysiology*. New York: Theime; 1999.

6. Guyton AC, Hall JE. *Textbook of Medical Physiology*, 11th ed. Philadelphia: W.B. Saunders; 2005.

7. Kandel ER, et al. *Principles of Neural Science*, 4th ed. New York: McGraw-Hill; 2000.

8. Lorenz MD, et al. *Handbook of Veterinary Neurology*, 5th ed. Philadelphia: W.B. Saunders; 2010.

9. Nestler EJ, et al. *Molecular Neuropharmacology: A Foundation for Clinical Neuroscience*, 2nd ed. New York: McGraw-Hill; 2008.

10. Oliver JE, et al. *Veterinary Neurology*. Philadelphia: W.B. Saunders; 1987.

11. Paxinos G. *The Rat Nervous System*, 3rd ed. San Diego, CA: Academic Press; 2004.

12. Paxinos G, Mai JK. *The Human Nervous System*, 2nd ed. San Diego, CA: Academic Press; 2004.

13. Williams PL, Warwick R. *Gray's Anatomy*, 36th ed. Philadelphia: W.B. Saunders; 1980.

14. Young PA, et al. *Basic Clinical Neuroscience*, 2nd ed. Baltimore: Lippincott Williams & Wilkins; 2008.

15. Braund KG. *Clinical Syndromes in Veterinary Neurology*. Baltimore: Williams & Wilkins; 1986.

16. Bolon B, et al. A "best practices" approach to neuropathologic assessment in developmental neurotoxicity testing—for today. *Toxicol Pathol.* 2006;34:296–313.

17. Summers BA, et al. *Veterinary Neuropathology*. St. Louis, MO: Mosby; 1995.

18. Spencer PS, et al. *Experimental and Clinical Neurotoxicology*, 2nd ed. New York: Oxford University Press; 2000.

19. Husain K, et al. Carboplatin-induced oxidative injury in rat inferior colliculus. *Int J Toxicol.* 2003;22:335–342.

20. Webber DS, et al. Mild carbon monoxide exposure impairs the developing auditory system of the rat. *J Neurosci Res.* 2003;74:655–665.

21. Morgan DL, et al. Neurotoxicity of carbonyl sulfide in F344 rats following inhalation exposure for up to 12 weeks. *Toxicol Appl Pharmacol.* 2004;200:131–145.

22. Philbert MA, et al. 1,3-Dinitrobenzene-induced encephalopathy in rats. *Neuropathol Appl Neurobiol.* 1987;13:371–389.

23. Romero I, et al. Vascular factors in the neurotoxic damage caused by 1,3-dinitrobenzene in the rat. *Neuropathol Appl Neurobiol.* 1991;17:495–508.

24. Tenkova T, et al. Ethanol-induced apoptosis in the developing visual system during synaptogenesis. *Invest Ophthalmol Vis Sci.* 2003;44:2809–2817.

25. Mori F, et al. Thiophene, a sulfur-containing heterocyclic hydrocarbon, causes widespread neuronal degeneration in rats. *Neuropathology.* 2000;20:283–288.

26. Tan CK, Gopalakrishnakone P. Experimental clinicopathological study of the habu (*Trimeresurus flavoviridis*) venom: I. A light microscopic study of terminal degeneration in some brainstem nuclei of the cat following intracisternal injection of the venom. *Jpn J Exp Med.* 1985;55:137–142.

27. Watanabe I, et al. Neurotoxicity of intrathecal gentamicin: a case report and experimental study. *Ann Neurol.* 1978;4:564–572.

28. Watterson J, et al. Fatal brain stem necrosis after standard posterior fossa radiation and aggressive chemotherapy for metastatic medulloblastoma. *Cancer.* 1993;71:4111–4117.

29. Brewer TG, et al. Fatal neurotoxicity of arteether and artemether. *Am J Trop Med Hyg.* 1994;51:251–259.

30. Petras JM, et al. Arteether-induced brain injury in *Macaca mulatta*: I. The precerebellar nuclei: the lateral reticular nuclei, paramedian reticular nuclei, and perihypoglossal nuclei. *Anat Embryol (Berl).* 2000;201:383–397.

31. Höglinger GU, et al. Chronic systemic complex I inhibition induces a hypokinetic multisystem degeneration in rats. *J Neurochem.* 2003;84:491–502.

32. Cunningham LL. The adult mouse utricle as an in vitro preparation for studies of ototoxic-drug-induced sensory hair cell death. *Brain Res.* 2006;1091:277–281.

33. Kanda T, Igarashi M. Ultrastructural changes in vestibular sensory end organs after viomycin sulfate intoxication. *Acta Otolaryngol.* 1969;68:474–488.

34. Wright CG, Schaefer SD. Inner ear histopathology in patients treated with cis-platinum. *Laryngoscope.* 1982;92:1408–1413.

35. Dorman DC, et al. Neuropathologic findings of bromethalin toxicosis in the cat. *Vet Pathol.* 1992;29:139–144.

36. Kimbrough RD, Gaines TB. Hexachlorophene effects on the rat brain: study of high doses by light and electron microscopy. *Arch Environ Health.* 1971;23:114–118.

37. Philbert MA, et al. Mechanisms of injury in the central nervous system. *Toxicol Pathol.* 2000;28:43–53.

38. Reuhl KR, et al. Developmental effects of trimethyltin intoxication in the neonatal mouse: I. Light microscopic studies. *Neurotoxicology.* 1983;4:19–28.

39. Itoh H, et al. Studies on the delayed neurotoxicity of organophosphorus compounds (III). *J Toxicol Sci.* 1985;10:67–82.

40. Tanaka DJ, et al. Selective axonal and terminal degeneration in the chicken brainstem and cerebellum following exposure to bis (1-methylethyl)phosphorofluoridate (DFP). *Brain Res.* 1990;519:200–208.

41. Lehning EJ, et al. Triphenyl phosphite and diisopropylphosphorofluoridate produce separate and distinct axonal degeneration patterns in the central nervous system of the rat. *Fundam Appl Toxicol.* 1996;29:110–118.

42. Ichihara G, et al. 1-Bromopropane, an alternative to ozone layer depleting solvents, is dose-dependently neurotoxic to rats in long-term inhalation exposure. *Toxicol Sci.* 2000;55: 116–123.

43. Selander D. Neurotoxicity of local anesthetics: animal data. *Reg Anesth.* 1993;18:461–468.

44. Al Masri O, et al. Recovery of tacrolimus-associated brachial neuritis after conversion to everolimus in a pediatric renal transplant recipient: case report and review of the literature. *Pediatr Transplant.* 2008;12:914–917.

45. Pollock JE. Neurotoxicity of intrathecal local anaesthetics and transient neurological symptoms. *Best Pract Res Clin Anaesthesiol.* 2003;17:471–484.

46. Takenami T, et al. Neurotoxicity of intrathecally administered bupivacaine involves the posterior roots/posterior white matter and is milder than lidocaine in rats. *Reg Anesth Pain Med.* 2005;30:464–472.

47. Coquet M, et al. Drug-induced and toxic myopathies [in French]. *Rev Prat.* 2001;51:278–283.

48. Kuncl RW. Agents and mechanisms of toxic myopathy. *Curr Opin Neurol.* 2009;22:506–515.

49. Courtney AE, et al. Acute polymyositis following renal transplantation. *Am J Transplant.* 2004;4:1204–1207.

50. Amaya-Villar R, et al. Steroid-induced myopathy in patients intubated due to exacerbation of chronic obstructive pulmonary disease. *Intensive Care Med.* 2005;31:157–161.

51. Kanda F, et al. Steroid myopathy: pathogenesis and effects of growth hormone and insulin-like growth factor-I administration. *Horm Res.* 2001;56:24–28.

52. Kaeser HE. Drug-induced myasthenic syndromes. *Acta Neurol Scand Suppl.* 1984;100:39–47.

53. Senanayake N, Román GC. Disorders of neuromuscular transmission due to natural environmental toxins. *J Neurol Sci.* 1992;107:1–13.

6

BEHAVIORAL MODEL SYSTEMS FOR EVALUATING NEUROPATHOLOGY

VIRGINIA C. MOSER

Toxicity Assessment Division, National Health and Environmental Effects Research Laboratory, Office of Research and Development, U.S. Environmental Protection Agency, Research Triangle Park, North Carolina

Disclaimer: The information in this document has been funded wholly by the U.S. Environmental Protection Agency (EPA). It has been reviewed by the National Health and Environmental Effects Research Laboratory and approved for publication. Approval does not signify that the contents necessarily reflect the views of the EPA, nor does mention of trade names or commercial products constitute endorsement or recommendation for use.

INTRODUCTION

Neurobehavioral as well as pathological evaluations of the nervous system are essential components of toxicity testing and basic research. The approaches are complementary and provide confidence that chemical effects on the nervous system are detected and characterized.[1] There is no a priori indication of whether behavioral or pathological evaluations will be the most appropriate and sensitive for neurotoxic assessment of a given chemical.[2] While neuropathological assessments provide insight as to long-term cellular changes in neurons, behavioral methods evaluate the functional consequences of disruption of neuronal communications. In some cases the underlying causes of certain behavioral alterations may be understood, but many do not have known direct or one-to-one associations with specific brain pathologies. Indeed, given the intricacy and complexity of the central and peripheral nervous systems, such simple relationships should not be expected. In this chapter, common procedures for behavioral testing are discussed and examples given of chemical-specific neurobehavioral–pathological correlations, in order to interpret and integrate neuropathological and behavioral outcomes.

The use of behavioral assays has a rich history over many decades in psychological and pharmacological research, whereas their use in toxicology assessments gained popularity only after recommendations by several expert panels[3–6] and scientists[7–10] in the late 1970s and early 1980s. Behavior represents the integrated sum of activities mediated by the nervous system, which cannot be assessed using only neurochemical, histological, or physiological techniques.[11] This convergence of influences on behavior often makes it a sensitive marker of nervous system dysfunction. Measurements of complex behaviors, such as motor activity, tap multiple neuronal functions that contribute to sensitivity of the approach, but suffer lack of specificity. In contrast, tests of simple reflex behaviors (e.g., simple sensory responses) may be more specific but only be altered by a few neurotoxic agents. Thus, a battery of tests that includes both types of behaviors is often suggested for screening or hazard identification purposes. Follow-up studies to more closely characterize toxicity or delineate mechanisms may employ more focused tests.[12–14]

The strength of functional assessments has been exploited by many investigators and regulatory agencies, and they are now used routinely in the assessment of the neurological effects of chemicals in regulatory and safety pharmacology testing.[15–17] Specific methods include functional observational batteries (FOBs), Irwin screens, tests of motor activity, and expanded clinical observations.[18,19] These tests have an

Fundamental Neuropathology for Pathologists and Toxicologists: Principles and Techniques, First Edition. Edited by Brad Bolon and Mark T. Butt.
© 2011 John Wiley & Sons, Inc. Published 2011 by John Wiley & Sons, Inc.

advantage over biochemical and pathological endpoints in that they permit evaluation of a single animal over longitudinal studies to determine onset, progression, duration, and reversibility of a neurotoxic injury. Cross-laboratory comparisons of a defined FOB protocol with a limited numbers of compounds[20,21] suggest that these methods can identify neurotoxic compounds reliably.

Screening for motor activity and observational batteries in laboratory rodents are described. It is important to note, however, that all species have specific behavioral repertoires, and observational tests may be developed and modified to focus on these species-specific behaviors. Using this approach, behavioral screening batteries have also been described for nonrodent species, including dogs,[22,23] guinea pigs,[24] and nonhuman primates.[25,26]

COMMON BEHAVIORAL TESTS

Motor Activity

Locomotor activity may be considered a simple behavior—an animal moves from one place to another—but it actually requires coordinated assimilation of motor, sensory, and integrative functions. In addition, there are several aspects of activity that are often measured, including exploration, habituation, horizontally vs. vertically directed activity, and spatial patterning. Most motor activity tests use automated equipment providing objective data, which is often considered an improvement over subjective evaluations. A search through instrumentation catalogs reveals any number of sizes and shapes of activity chambers. In general, detection systems are either based on photocells (simple counts of photobeam breaks), infrared capacitance, movement transducers (jiggle cages), or imaging systems. A complicating factor with activity measures is that the shape and size of the arena may influence motor behaviors. The reliability, reproducibility, and sensitivity of motor activity evaluations have been demonstrated in several cross-laboratory collaborations.[20,27] Activity levels and habituation characteristics follow an established ontogeny during development, supporting the use of activity in developmental studies. While activity tests may be considered a stand-alone assay, current U.S. Environmental Protection Agency (EPA) Test Guidelines[17] recommend that additional functional evaluations be taken in the same animals that undergo activity measurements.

Being an apical reflection of central and peripheral motor function as well as general well-being, no specific brain region or neural substrate can be implicated when activity levels are altered. However, certain patterns of change may be suggestive; for example, increases in motor activity are mostly observed with pharmacological actions in the dopaminergic or cholinergic systems (e.g., cocaine, amphetamine, scopolamine). Habituation during a test session (decrease

over time in response to a novel stimulus) has been considered a crude indicator of learning, and ontogeny mirrors development of the cholinergic system.[28] Thus, changes in habituation in cases where overall activity levels are not greatly altered could imply cognitive effects. Since rodents typically spend most of their exploration near the outer portion of an arena, changes in this spatial pattern could indicate disinhibitory (entering into the center more often) or anxiogenic (thigmotaxis, closely staying against the outside region) responses. However, these explanations are generalities and should not be considered exclusive.

Functional Observational Batteries

Observational batteries provide a systematic and close evaluation of an animal's behavior and function, and there can simply be no substitute for looking closely at a test subject on some sort of open field (as opposed to remaining in the home cage). Such tests allow for detection of unanticipated effects and thus are critical for screening in the absence of anticipated outcomes. In addition, compiling the effects in terms of functional domains that are altered allows better characterization of the effects. An FOB is most often used in toxicity testing laboratories in first-tier tests, whereas the testing battery popularized by Irwin[30] is used in safety pharmacology and evaluation of nervous system side effects. In reality, these approaches are very similar, and in fact the FOB development[29] was based on the published Irwin screen.[30,31] There are also simplified examinations termed *expanded clinical observations*.[1,32] The tests have been likened to a neurological examination, which allows a more informed extrapolation across species. Numerous protocols have been published.[29,33–38]

Generally speaking, it is FOB-type evaluations that will be used in toxicity studies that include neuropathologic evaluations. The FOB, being a multifaceted and standardized battery of tests, is sensitive in detecting neurological and behavioral effects of chemicals.[18,19,29,39] Observations take place in the animal's home cage and in an open-field arena, followed by reflex testing and handling manipulations. Many of the behaviors are scored subjectively using defined rating scales. Care must to be taken to assure that the observer is "blind" to the treatment, to prevent bias in scoring. Other endpoints are quantified, such as the landing foot splay[40] and fore- and hindlimb grip strength.[41] The neurological domains that are evaluated include neuromuscular function, sensory reflexes, autonomic changes, excitability states, and general activity levels. Some common endpoints are listed in Table 1.

Neuromotor function is described with observations and measurements of gait, posture, and grip strength. Gait evaluations play a particularly important role in behavioral examinations. Because maintaining steady gait requires a range of neurological functions, this evaluation provides a

TABLE 1   Behavioral Evaluations Commonly Included in Observational Batteries for Neurotoxicity Testing

Neuromotor
   Gait (e.g., ataxia, altered limb placement)
   Posture (e.g., hunched, tiptoe)
   Forelimb and hindlimb grip strength
   Landing foot splay (also proprioceptive)
Sensorimotor
   Auditory response (e.g., click, automated startle)
   Pain response (e.g., tail or toe pinch)
   Somatosensory response
   Olfactory response
   Visual response (e.g., placing)
   Postural reactions
   Proprioceptive positioning
Reflexes
   Righting
   Palpebral closure
   Pinna
   Pupillary
   Extensor thrust
   Grasp
Activity and excitability
   Open-field activity and arousal
   Rearing
   Reactivity to handling or removal from cage
Autonomic
   Salivation
   Lacrimation
   Respiration rate
Convulsions, tremors, myoclonus, fasciculations

sensitive detector of chemical effects. Gait requires the integrity of long motor axons, and is therefore sensitive to axonopathies, but it also requires central integration and is therefore sensitive to alterations in motor cortex, nigrostriatum, and other motor pathways. As was described for motor activity, there is no direct congruence between gait changes and a particular neural substrate. A quantitative measure of grip strength in rodents using a strain gauge[41] is analogous to a neurologist's evaluation of a patient's grip force, and has been used to measure muscle strength in animal models of arthritis, for example.[42] Grip strength is influenced by several factors, including central nervous system depression,[43] spinal pathology,[44] peripheral neuropathy,[45] neuromuscular junction dysfunction,[46] as well as nonspecific factors.[47] Vestibular function is evaluated by scoring or timing the righting reflex, or the ability of the animal to right itself from a supine position. Other indicators of vestibular involvement include ataxia (staggering gait) and inability to rear.

Sensory evaluations are difficult to conduct in rodents, especially when using tests that require a motor output to indicate the perception of the sensory input. Treatments that decrease activity or have various other motor effects may confound the sensorimotor test results. Typically, the sensory examinations measure the function of small fibers carrying sensations of pain (or sometimes temperature), as well as large fibers for proprioception. Nociception (awareness of pain) may be tested subjectively by observing the response to a pinch of the tail or toe, or by using an objective instrument such as a tail flick monitor or hot plate. The landing foot splay is considered a measure of proprioception as well as neuromuscular function, since it requires an intact sense of body positioning for the animal to land with its limbs correctly placed. The distance between the paws upon landing is the dependent measure.[40] Of the senses, only auditory function is typically measured well in rodents, by either observing the response to a sudden sound or using an automated acoustic response apparatus. Since rodents have low visual acuity, tests such as placing or response to an approaching object are not particularly useful. However, rodents are highly olfactory oriented, and tests of odor could be incorporated into sensory evaluations, although few research laboratories are currently including odor-based tests. Observational batteries may also focus on specific reflex responses or cranial nerve function, such as flexor reflex or palpebral reflex.[1] These tests are focused more on neuroanatomical pathways. Since these tests may add time to the overall study conduct, some researchers choose to perform these procedures only on animals that have already shown other signs of toxicity.[36,37]

MODELS OF NEUROTOXICITY: BEHAVIORS AND UNDERLYING NEUROPATHOLOGY

Neuronopathies

Certain features that make neurons more vulnerable to injury include their high metabolic rate, the demands of maintaining physical features such as very long axons, and chemical processes that maintain the excitatory properties of the cell membrane. A number of toxicants attack specific neurons, or particular groups of neurons, resulting in their injury and/or cell death. Neurotoxicants that are selective in their action and that affect only a subpopulation of neurons may lead to particular or characteristic patterns of functional changes. A few examples are listed below.

3-AcetylPyridine   A single dose of 3-acetylpyridine (3-AP) produces irreversible lesions in the climbing fibers of the cerebellum by destroying the inferior olivary nuclei.[48,49] It has been used as a model for cerebellar ataxia.[50,51]

We conducted a study using a single dose of 3-AP to adult Long–Evans rats, and evaluated them using the FOB at 4 h, then again at 1, 3, 10, and 23 days. Signs of neuromuscular toxicity were noted at all doses (40, 50, 60 mg/kg), with ataxia, altered equilibrium (righting reflex), and decreased

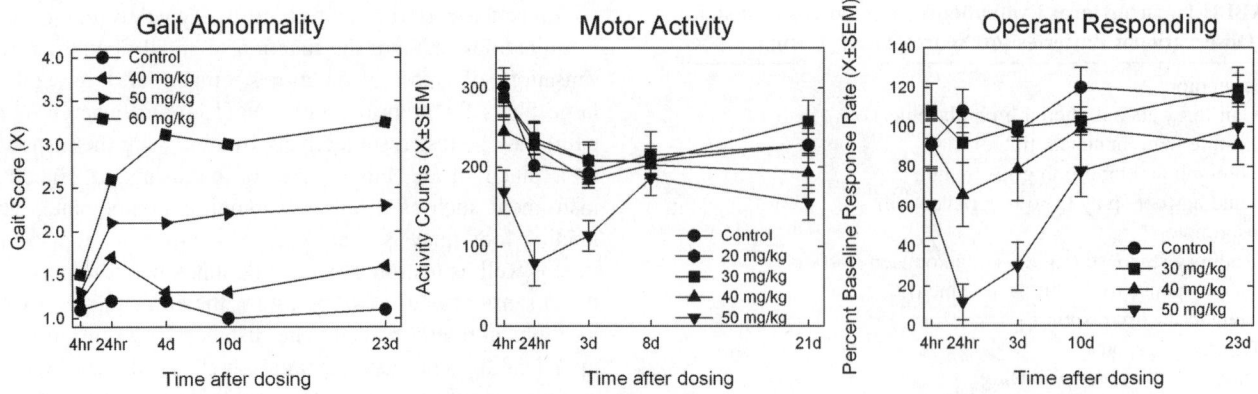

FIGURE 1   Time course of 3-acetylpyridine neurotoxicity as assessed by gait score (ranked 1 to 5; mean score shown), motor activity (counts over a 1-h session, mean ± SEM), and operant responding (response rate as a percent of each rat's control value on a fixed interval schedule of reinforcement, mean ± SEM).

grip strength. In a separate group of rats tested in motor activity chambers, the 50-mg/kg dose group (highest dose tested), but not 40 mg/kg, decreased motor activity (males only) and disrupted operant responding (males and females). These data suggest that the cerebellar signs are better detected with observations of the animals than in automated activity devices.

Our data with 3-AP also provide an excellent example of behavioral compensation. Effects on motor activity and operant performance showed recovery when tested 10 days after dosing, whereas the neuromotor signs measured in the FOB were observed throughout testing. These comparisons are shown in Figure 1. The brain lesions produced by 3-AP are irreversible.[48] If only operant performance or motor activity measures had been used, it is likely that the persistent motor and vestibular effects would not have been observed, leading to erroneous conclusions regarding the reversibility of the effects as well as the sensitivity of the behavioral measures in relation to pathology.

Trimethyl Tin   Trimethyl tin (TMT) is a prototypic neurotoxicant that produces clear pathological damage in the hippocampus, pyriform cortex, amygdala, and neocortex.[52,53] The corresponding behavioral profile is indicative of limbic dysfunction, including hyperactivity, impaired learning, and performance.[54–56]

In a study of rats administered a single intravenous dose of TMT (4, 6, or 8 mg/kg),[57] the profile of FOB changes mirrored the behavioral effects reported in earlier studies. Specifically, motor activity, rearing, arousal, and reactivity were all increased for up to 42 days. Interestingly, the magnitude and time course of these effects were greater in F344 rats compared to Long–Evans rats. The relatively subtle and selective sensory dysfunction reported by others[58,59] was not observed, possibly because the appropriate stimuli (somatosensory, high-frequency tone) were not employed.

Gait and righting reflex were also altered but showed significant compensation after 21 days and were not different from control behaviors at 42 days. Thus, functional recovery was similar to that observed with 3-AP described earlier. Other neuromuscular endpoints (hindlimb grip strength, landing foot splay) as well as tremors showed similarly transient changes. These latter effects had not been reported in the literature, and since this particular study did not include corresponding pathological evaluations, it is impossible to state whether they were due to central lesions of the central nervous system (CNS) or reflected acute generalized toxicity.

IDPN   β,β′-Iminodipropionitrile (IDPN) produces a "waltzing syndrome" in rats and other mammals that includes excitement, circling, head weaving, and overalertness.[60–62] IDPN produces a neurofilamentous proximal axonopathy as well as vestibular hair cell degeneration, and the cause of the classic IDPN behavioral syndrome appears due to the latter.[63–65]

Short-term repeated IDPN dosing is sufficient to produce perhaps permanent behavioral and pathological changes. For example, 3 days of IDPN 100, 200, or 400 mg/kg produced marked neurobehavioral changes, including the IDPN dyskinetic syndrome (circling, retropulsion, vertical and lateral head movements), equilibrium changes, hyperactivity, increased excitability, and neuromuscular weakness.[66] For many of these measurements, quantitative differences between male and female rats were noted, and effects persisted throughout three months of testing. These behavioral reports have been fairly consistent among studies in which similar evaluations were conducted.[60,67–69] In our study we also detected sensory deficits confined to the auditory (click) and visual (approach) stimuli. These auditory changes have been reported previously,[70] and in our study the apparent visual effects correlated with changes in visual evoked potentials in the same animals.[66]

Peripheral Axonopathies

A number of neurotoxic chemicals produce disorders known as *axonopathies*, in which the axon and surrounding myelin sheath degenerate.[71] The central long axons, such as ascending sensory axons in the posterior columns or descending motor axons, along with peripheral long sensory and motor axons, are most vulnerable in these situations. As these axons degenerate, the result is most often the clinical condition of peripheral neuropathy. In humans, sensations and motor strength are first impaired in the longest and most distal axonal processes: the feet and hands. With time and continued injury, the deficit progresses to involve more proximal areas of the body and the long axons of the spinal cord. In laboratory animals, the corresponding syndrome is a pattern of behavioral changes typical of "length-related gradient of nerve pathology."[72] Well-studied neurotoxicants that fall into this class are carbon disulfide, acrylamide, *n*-hexane, and 2,5-hexanedione (the active metabolite of *n*-hexane).

Carbon Disulfide   The clinical effects of exposure to carbon disulfide, which are very similar to those of hexane exposure, illustrate the development of sensory and motor symptoms occurring initially in a "stocking-and-glove" distribution.[73] Laboratory animals show similar signs of neuropathy with accompanying motor dysfunction.[74]

We conducted a series of studies to characterize the time course and dose–response of biochemical, behavioral, electrophysiological, and neuropathological effects of inhaled carbon disulfide.[75–82] Rats were tested with the FOB at the end of 2, 4, 8, and 13 weeks of exposure. Immediately following each exposure time point, nerve conduction velocity measures were taken, and the rats were then euthanized and perfused for microscopic evaluation of central and peripheral nerves. Neuromuscular deficits were more pronounced in the hindlimbs (hindlimb grip strength, altered hindlimb placement) and showed a clear progression with longer exposures. Lesser effects were evident as tremors, ataxia, and changes in reactivity and a crude visual test. Mild gait changes were evident after only two weeks of exposure, and became more severe with longer exposures. These rats had axonal swellings in the spinal cord and muscular branch of the posterior tibial nerve at eight and 13 weeks, with progression to degeneration and regeneration of tibial nerve axons at 13 weeks. In this same time frame, electrophysiological recordings documented decreased nerve conduction velocity. Similar effects were reported in mice exposed to carbon disulfide.[83] Thus, the preferential sensitivity of the hindlimbs correlated well with pathological involvement, but occurred earlier in the time course.

Acrylamide   Many positive-control studies have utilized repeated dosing with acrylamide to produce a characteristic neuromotor syndrome[69,84–86] or to exemplify detailed gait analyses.[51,87] Such studies report altered gait, decreased hindlimb grip strength, increased landing foot splay, altered righting reflex, and decreased rearing. The latter three tests have a significant sensory component as well as measuring neuromotor function.

We dosed rats for 5 days/week for 13 weeks (1, 4, and 12 mg/kg per day i.p.) and tested them with the FOB at monthly intervals. The effects listed above were noted in the same rats that under neuropathologic examination showed a severe distal axonopathy. Peripherally, degeneration was evident in the sciatic nerve at the level of the sciatic notch and at all more distal points, along with axonal degeneration in the long tracts of the spinal cord.[45] Thus, as with carbon disulfide, the behavioral profile of neuromotor changes correlated well with peripheral neuropathies.

A comparison of acrylamide and carbon disulfide illustrates reliable and specific differences and similarities. In addition, 2,5-hexanedione produces FOB effects that mirror those of carbon disulfide.[86] While gait changes (ataxic, uncoordinated hindlimb placement) and decreased grip strengths (greater in hindlimbs) are obvious with these chemicals, the effects on landing foot splay (increased only with acrylamide) serve to differentiate their neuromotor syndromes.[72] A comparison of dose–response data for acrylamide[45] and carbon disulfide[75] is shown in Figure 2. Neuropathological evaluations have documented greater and earlier involvement of sensory neurons with acrylamide, and this correlates nicely with the differential foot splay effects.[61,88]

INTERPRETATION OF NEUROBEHAVIORAL DATA

Changes in behavior have sometimes not been considered "adverse," especially when direct clinical correlations are not clear. Given the similarities of neuronal function across species, any indication of altered function should be considered seriously. Comparisons of human symptomatology and toxicity signs in laboratory animals produced by specific chemicals have shown good concordance in many cases.[89] For interpreting neurotoxicity data, the U.S. Environmental Protection Agency[90,91] has stated that:

- Any change in the structure or function of the central and/or peripheral nervous system should be considered an adverse effect.
- Adverse effects may appear as an increase or decrease in a measured behavior, or an acceleration or delay in the appearance of the behavior.
- Effects that are transient or occur only at specific times during development should be considered adverse.
- Behavioral changes may occur before, along with, or in the absence of morphological and/or pathological changes.

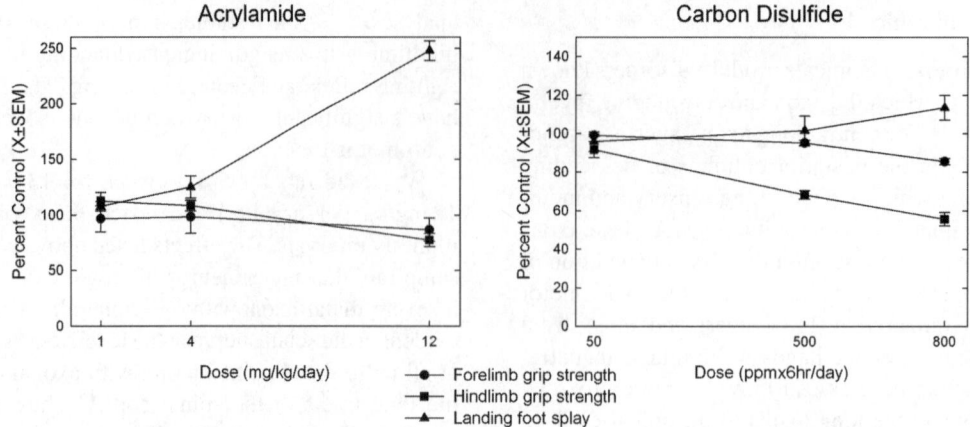

FIGURE 2 Dose– or concentration–response data for acrylamide and carbon disulfide (90-day exposures for both chemicals) on measures of forelimb and hindlimb grip strength and landing foot splay. Data expressed as percent of control values (mean ± SEM). Note differences in scale.

- Differences in effects may occur as a function of gender, age, strain or species, prior exposures and so on.
- Any behavioral changes should be considered neurotoxicity unless there are data supporting a nonneuronal mechanism.

Evaluating the pattern of changes and understanding the advantages and limitations of behavioral tests are important considerations for interpreting behavioral changes. For example, with observational or simple sensory tests, it is difficult to detect specific effects, such as threshold shifts or frequency-dependent changes.[92] Sensory tests also depend on both the ability to detect the stimulus (sensory input) and the ability to respond (motor output). Therefore, indications of specific sensory effects could present as decreased response to specific stimuli with normal responses to all other stimuli, along with normal gait and activity levels. Alterations in motor endpoints may show patterns or profiles of changes that imply underlying neural substrates. Although generalized nervous system depression also presents as neuromotor changes, most measures are affected equally and no specificity is apparent. It is important to recognize that at high doses of many chemicals, generalized effects in several or many behaviors may be evident along with overt toxicity.

Behavior may be altered by any number of nervous system alterations, of which neuropathological changes represent only one. There are several reasons why comparisons between behavioral and pathological outcomes may not provide correlative data (listed in Table 2). One example is the transient nature of some forms of toxicity. Many behavioral changes are due to acute changes in neuronal function by altering neurotransmission or cellular excitability; these changes typically do not produce alterations visible using routine neuropathology, and recovery may be evident within hours to days. Thus, behavior is often more sensitive to acute pharmacological effects of agents. Persistent effects, or those

with prolonged exposures, may be more likely to be accompanied by pathology. Behavior may be altered earlier during exposure than pathological damage, in which case it may appear more sensitive due simply to the timing of the assessments. As another example, behavioral changes may be observed but pathological sampling is not adequate to visualize specific or focal lesions. Similarly, diffuse pathology (e.g., hydrocephalus) may not be detected in histological samples, but may produce generalized neurobehavioral changes.

Neuropathology may be documented without corresponding behavioral changes, but this does not mean that the behavioral tests were not useful or sensitive. As mentioned above, timing of assessments is an important factor. There is considerable functional reserve in the behavioral system, due to redundancy, compensation, residual capacity, and plasticity of neuronal function, which allows for recovery despite potentially permanent pathology. Another reason for recovery of behavior is tolerance. Tolerance may involve changes

TABLE 2 Potential Factors That Could Produce Noncorrelations Between Behavior and Pathology Outcomes

Behavioral changes without neuropathology
    Transient or acute actions
    Changes in neuronal function but not structure
    Timing of measurement
    Inadequate sampling of nervous tissue
    Dose of test substance
    Insufficient duration of exposure
    Diffuse pathology
Neuropathological changes without behavior
    Functional reserve and/or compensation
    Timing of measurement
    Inappropriate choice of behavioral test(s)
    Tolerance or adaptation
    Species of test subjects

in kinetic parameters (e.g., increased liver detoxification) or altered functional responses (e.g., downregulation of neuro-transmitter receptors). The possibility also exists that the behavioral changes being measured are inadequate or inappropriate for the pathological change. Species-specific behavioral responses have been documented for some agents, with pathology being more consistent for extrapolation. For example, some organophosphorus pesticides produce a delayed neuropathy that is characterized by axonopathy in both hens and rodents, yet the rodents do not show the ataxia and gait changes that are easily observed in hens.[93,94]

SUMMARY

Common procedures for characterizing neurobehavioral toxicity include testing with the FOB and/or an automated measure of motor activity, although a wide range and variety of other tests can and have been used as well. The data show considerable sensitivity, reliability, reproducibility, and validity for the purposes of neurotoxicity research, yet interpretation of the data can be a challenge. Even if a behavior may appear simple, one must avoid the temptation to assume that behavior has a simple relationship to an underlying physiological or neuronal substrate. Some profiles of neurological changes may reveal or give clues as to the neuronal substrate or underlying pathology, but many may not. Direct relations between behavior and neuropathology should not be assumed or expected, and where correlational agreement does exist, it does not prove cause and effect. However, functional and morphological/pathological assessments provide valuable information about an agent's effect on the nervous system.

REFERENCES

1. Ross JF. ECOs, FOBs, and UFOs: making sense of observational data. *Toxicol Pathol*. 2000;28:132–136.

2. Norton S. Is behavior or morphology a more sensitive indicator of central nervous system toxicity? *Environ Health Perspect*. 1978;26:21–27.

3. National Academy of Sciences. *Principles for Evaluating Chemicals in the Environment*. Washington, DC: National Academy of Sciences Press; 1975.

4. National Academy of Sciences. *Toxicity Testing: Strategies to Determine Needs and Priorities*. Washington, DC: National Academy of Sciences Press; 1984.

5. National Research Council. *Principles and Procedures for Evaluating the Toxicity of Household Substances*. Washington, DC: National Academy of Sciences Press; 1977.

6. World Health Organization. *Principles and Methods for the Assesment of Neurotoxicity Associated with Exposure to Chemicals*. Geneva, Switzerland: WHO; 1986.

7. Brimblecombe RW. Behavioral tests in acute and chronic toxicity studies. *Pharmacol Ther B*. 1979;5:413–415.

8. Fowler JS, et al. The rat toxicity screen. *Pharmacol Ther B*. 1979;5:461–466.

9. Mitchell CL, Tilson HA. Behavioral toxicology in risk assessment: problems and research needs. *Crit Rev Toxicol*. 1982;10:265–274.

10. Schaeppi U, Hess R. What can specific behavioural testing procedures contribute to the assessment of neurotoxicity in laboratory animals? *Agents Actions*. 1984;14:131–138.

11. Mello NK. Behavioral toxicology: a developing discipline. *Fed Proc*. 1975;34:1832–1834.

12. Kulig BM. Comprehensive neurotoxicity assessment. *Environ Health Perspect*. 1996;104 (Suppl 2):317–322.

13. Kulig B, et al. Animal behavioral methods in neurotoxicity assessment: SGOMSEC joint report. *Environ Health Perspect*. 1996;104 (Suppl 2):193–204.

14. Tilson HA, Cabe PA. Strategy for the assessment of neurobehavioral consequences of environmental factors. *Environ Health Perspect*. 1978;26:287–299.

15. International Conference on Harmonisation. *Safety Pharmacology Studies for Human Pharmaceuticals*. Guideline CPMP/ICH/539/00-ICH S7A. Geneva: International Conference on Harmonisation of Technical Requirements for Registration of Pharmaceuticals for Human Use; 2000. Available at: http://www.ich.org/LOB/media/MEDIA504.pdf. Accessed Mar 26, 2010.

16. Organisation for Economic Co-operation and Development. *Neurotoxicity Study in Rodents*. Guideline 424. Paris: OECD; 1997.

17. US Environmental Protection Agency. *Health Effects Test Guidelines: Neurotoxicity Screening Battery*. OPPTS 870.6200. EPA 712-C-98-238. Washington, DC: US EPA; 1998.

18. Moser VC. Observational batteries in neurotoxicity testing. *J Am Coll Toxicol*. 2000;19:407–411.

19. Tilson HA, Moser VC. Comparison of screening approaches. *Neurotoxicology*. 1992;13:1–13.

20. Moser VC, et al. The IPCS Collaborative Study on Neurobehavioral Screening Methods: V. Results of chemical testing. Steering Group. *Neurotoxicology*. 1997;18:969–1055.

21. Moser VC, et al. The IPCS Collaborative Study on Neurobehavioral Screening Methods: II. Protocol design and testing procedures. *Neurotoxicology*. 1997;18:929–938.

22. Gad SC, Gad SE. A functional observational battery for use in canine toxicity studies: development and validation. *Int J Toxicol*. 2003;22:415–422.

23. Schaeppi U, Fitzgerald R. Practical procedure of testing for neurotoxicity. *J Am Coll Toxicol*. 1989;8:29–34.

24. Hulet SW, et al. The dose–response effects of repeated subacute sarin exposure on guinea pigs. *Pharmacol Biochem Behav*. 2002;72:835–845.

25. O'Keefe RT, Lifshitz K. Nonhuman primates in neurotoxicity screening and neurobehavioral toxicity studies. *J Am Coll Toxicol*. 1989;8:127–140.

26. Gauvin DV, Baird TJ. A functional observational battery in non-human primates for regulatory-required neurobehavioral assessments. *J Pharmacol Toxicol Methods.* 2008;58:88–93.

27. Crofton KM, et al. Interlaboratory comparison of motor activity experiments: implications for neurotoxicological assessments. *Neurotoxicol Teratol.* 1991;13:599–609.

28. Grant M. Cholinergic influences on habituation of exploratory activity in mice. *J Comp Physiol Psychol.* 1974;86:853–857.

29. Moser VC, et al. Comparison of chlordimeform and carbaryl using a functional observational battery. *Fundam Appl Toxicol.* 1988;11:189–206.

30. Irwin S. Comprehensive observational assessment: Ia. A systematic, quantitative procedure for assessing the behavioral and physiologic state of the mouse. *Psychopharmacologia.* 1968;13:222–257.

31. Irwin S. Drug screening and evaluative procedures. *Science.* 1962;136:123–128.

32. Ross JF, et al. Expanded clinical observations in toxicity studies: historical perspectives and contemporary issues. *Regul Toxicol Pharmacol.* 1998;28:17–26.

33. Haggerty GC. Development of tier I neurobehavioral testing capabilities for incorporation into pivotal rodent safety assessment studies. *J Am Coll Toxicol.* 1989;8:53–70.

34. Kulig BM. A neurofunctional test battery for evaluating the effects on long-term exposure to chemicals. *J Am Coll Toxicol.* 1989;8:71–83.

35. McDaniel KL, Moser VC. Utility of a neurobehavioral screening battery for differentiating the effects of two pyrethroids, permethrin and cypermethrin. *Neurotoxicol Teratol.* 1993;15:71–83.

36. O'Donoghue JL. Screening for neurotoxicity using a neurologically based examination and neuropathology. *J Am Coll Toxicol.* 1989;8:97–115.

37. O'Donoghue JL. Clinical neurologic indices of toxicity in animals. *Environ Health Perspect.* 1996;104 (Suppl 2): 323–330.

38. Tegeris JS, Balster RL. A comparison of the acute behavioral effects of alkylbenzenes using a functional observational battery in mice. *Fundam Appl Toxicol.* 1994;22:240–250.

39. Moser VC. Applications of a neurobehavioral screening battery. *J Am Coll Toxicol.* 1991;10:661–669.

40. Edwards PM, Parker VH. A simple, sensitive, and objective method for early assessment of acrylamide neuropathy in rats. *Toxicol Appl Pharmacol.* 1977;40:589–591.

41. Meyer OA, et al. A method for the routine assessment of fore- and hindlimb grip strength of rats and mice. *Neurobehav Toxicol.* 1979;1:233–236.

42. Williams JM, et al. Functional assessment of joint use in experimental inflammatory murine arthritis. *J Orthop Res.* 1993;11:172–180.

43. Nevins ME, et al. Quantitative grip strength assessment as a means of evaluating muscle relaxation in mice. *Psychopharmacology (Berl).* 1993;110:92–96.

44. Moser VC, et al. Neurotoxicity produced by dibromoacetic acid in drinking water of rats. *Toxicol Sci.* 2004;79:112–122.

45. Moser VC, et al. Comparison of subchronic neurotoxicity of 2-hydroxyethyl acrylate and acrylamide in rats. *Fundam Appl Toxicol.* 1992;18:343–352.

46. Crofton KM, et al. The effects of 2,4-dithiobiuret on sensory and motor function. *Fundam Appl Toxicol.* 1991;16:469–481.

47. Maurissen JP, et al. Factors affecting grip strength testing. *Neurotoxicol Teratol.* 2003;25:543–553.

48. Desclin JC, Escubi J. Effects of 3-acetylpyridine on the central nervous system of the rat, as demonstrated by silver methods. *Brain Res.* 1974;77:349–364.

49. Balaban CD. Central neurotoxic effects of intraperitoneally administered 3-acetylpyridine, harmaline and niacinamide in Sprague–Dawley and Long–Evans rats: a critical review of central 3-acetylpyridine neurotoxicity. *Brain Res.* 1985;356:21–42.

50. Butterworth RF, et al. Cerebellar ataxia produced by 3-acetyl pyridine in rat. *Can J Neurol Sci.* 1978;5:131–133.

51. Jolicoeur FB, et al. Measurement of ataxia and related neurological signs in the laboratory rat. *Can J Neurol Sci.* 1979;6:209–215.

52. Brown AW, et al. The behavioral and neuropathologic sequelae of intoxication by trimethyltin compounds in the rat. *Am J Pathol.* 1979;97:59–82.

53. Aldridge WN, et al. Brain damage due to trimethyltin compounds. *Lancet.* 1981;2:692–693.

54. Reiter LW, Ruppert PH. Behavioral toxicity of trialkyltin compounds: a review. *Neurotoxicology.* 1984;5:177–186.

55. McMillan DE, Wenger GR. Neurobehavioral toxicology of trialkyltins. *Pharmacol Rev.* 1985;37:365–379.

56. Perretta G, et al. Neuropathological and behavioral toxicology of trimethyltin exposure. *Ann Ist Super Sanita.* 1993;29:167–174.

57. Moser VC. Rat strain- and gender-related differences in neurobehavioral screening: acute trimethyltin neurotoxicity. *J Toxicol Environ Health.* 1996;47:567–586.

58. Howell WE, et al. Somatosensory dysfunction following acute trimethyltin exposure. *Neurobehav Toxicol Teratol.* 1982;4:197–201.

59. Crofton KM, et al. Trimethyltin effects on auditory function and cochlear morphology. *Toxicol Appl Pharmacol.* 1990;105:123–132.

60. Chou SM, Hartmann HA. Axonal lesions and waltzing syndrome after IDPN administration in rats: with a concept—"axostasis." *Acta Neuropathol.* 1964;3:428–450.

61. Cavanagh JB. Neurotoxicology of acrylamides, hexacarbons, IDPN, and carbon disulfide: summarizing remarks. *Neurotoxicology.* 1985;6:97–98.

62. Cadet JL. The iminodipropionitrile (IDPN)-induced dyskinetic syndrome: behavioral and biochemical pharmacology. *Neurosci Biobehav Rev.* 1989;13:39–45.

63. Chou SM, Hartmann HA. Electron microscopy of focal neuroaxonal lesions produced by β-β'-iminodipropionitrile (IDPN) in rats: I. The advanced lesions. *Acta Neuropathol.* 1965;4:590–603.

64. Llorens J, et al. The behavioral syndrome caused by 3,3'-iminodipropionitrile and related nitriles in the rat is

associated with degeneration of the vestibular sensory hair cells. *Toxicol Appl Pharmacol.* 1993;123:199–210.

65. Llorens J, Dememes D. Hair cell degeneration resulting from 3,3′-iminodipropionitrile toxicity in the rat vestibular epithelia. *Hear Res.* 1994;76:78–86.

66. Moser VC, Boyes WK. Prolonged neurobehavioral and visual effects of short-term exposure to 3,3′-iminodipropionitrile (IDPN) in rats. *Fundam Appl Toxicol.* 1993;21:277–290.

67. Ivens I. Neurotoxicity testing during long-term studies. *Neurotoxicol Teratol.* 1990;12:637–641.

68. Crofton KM, Knight T. Auditory deficits and motor dysfunction following iminodipropionitrile administration in the rat. *Neurotoxicol Teratol.* 1991;13:575–581.

69. Schulze GE, Boysen BG. A neurotoxicity screening battery for use in safety evaluation: effects of acrylamide and 3′,3′-iminodipropionitrile. *Fundam Appl Toxicol.* 1991;16:602–615.

70. Crofton KM, et al. Characterization of the effects of *N*-hydroxy-IDPN on the auditory, vestibular, and olfactory systems in rats. *Neurotoxicol Teratol.* 1996;18:297–303.

71. Moser VC, et al. Toxic responses of the nervous system. In: Klaassen CD, ed. *Casarett and Doull's Toxicology: The Basic Science of Poisons,* 7th ed. New York: McGraw-Hill; 2008: 631–664.

72. Sterman AB. The pathology of toxic axonal neuropathy: a clinical–experimental link. *Neurobehav Toxicol Teratol.* 1984;6:463–466.

73. Spencer PS, Schaumburg HH. Experimental models of primary axonal disease induced by toxic chemicals. In: Dyck PJ, et al., eds. *Peripheral Neuropathy.* Philadelphia: W.B. Saunders; 1984:636–649.

74. Wood RW. Neurobehavioral toxicity of carbon disulfide. *Neurobehav Toxicol Teratol.* 1981;3:397–405.

75. Moser VC, et al. Carbon disulfide neurotoxicity in rats: VII. Behavioral evaluations using a functional observational battery. *Neurotoxicology.* 1998;19:147–157.

76. Harry GJ, et al. Carbon disulfide neurotoxicity in rats: VIII. Summary. *Neurotoxicology.* 1998;19:159–161.

77. Herr DW, et al. Carbon disulfide neurotoxicity in rats: VI. Electrophysiological examination of caudal tail nerve compound action potentials and nerve conduction velocity. *Neurotoxicology.* 1998;19:129–146.

78. Sills RC, et al. Carbon disulfide neurotoxicity in rats: V. Morphology of axonal swelling in the muscular branch of the posterior tibial nerve and spinal cord. *Neurotoxicology.* 1998;19:117–127.

79. Toews AD, et al. Carbon disulfide neurotoxicity in rats: IV. Increased mRNA expression of low-affinity nerve growth factor receptor—a sensitive and early indicator of PNS damage. *Neurotoxicology.* 1998;19:109–116.

80. Valentine WM, et al. Covalent modification of hemoglobin by carbon disulfide: III. A potential biomarker of effect. *Neurotoxicology.* 1998;19:99–107.

81. Sills RC, et al. Carbon disulfide neurotoxicity in rats: I. Introduction and study design. *Neurotoxicology.* 1998; 19:83–87.

82. Moorman MP, et al. Carbon disulfide neurotoxicity in rats: II. Toxicokinetics. *Neurotoxicology.* 1998;19:89–97.

83. Sills RC, et al. Characterization of carbon disulfide neurotoxicity in C57BL6 mice: behavioral, morphologic, and molecular effects. *Toxicol Pathol.* 2000;28:142–148.

84. Broxup B, et al. Correlation between behavioral and pathological changes in the evaluation of neurotoxicity. *Toxicol Appl Pharmacol.* 1989;101:510–520.

85. Moser VC, et al. The IPCS Collaborative Study on Neurobehavioral Screening Methods: III. Results of proficiency studies. *Neurotoxicology.* 1997;18:939–946.

86. Shell L, et al. Neurotoxicity of acrylamide and 2,5-hexanedione in rats evaluated using a functional observational battery and pathological examination. *Neurotoxicol Teratol.* 1992; 14:273–283.

87. Youssef AF, Santi BW. Simple neurobehavioral functional observational battery and objective gait analysis validation by the use of acrylamide and methanol with a built-in recovery period. *Environ Res.* 1997;73:52–62.

88. Sterman AB, Sheppard RC. A correlative neurobehavioral-morphological model of acrylamide neuropathy. *Neurobehav Toxicol Teratol.* 1983;5:151–159.

89. Moser VC. Approaches for assessing the validity of a functional observational battery. *Neurotoxicol Teratol.* 1990;12:483–488.

90. US Environmental Protection Agency. *Guidelines for Neurotoxicity Risk Assessment.* 630/R-95/001F. Washington, DC: US EPA; 1998.

91. Boyes WK, et al. EPA's neurotoxicity risk assessment guidelines. *Fundam Appl Toxicol.* 1997;40:175–184.

92. Fukumura M, et al. Effects of the neurotoxin 3,3′-iminodipropionitrile on acoustic startle and locomotor activity in rats: a comparison of functional observational and automated startle assessment methods. *Neurotoxicol Teratol.* 1998;20:203–211.

93. Dyer KR, et al. Comparative dose-response studies of organophosphorus ester-induced delayed neuropathy in rats and hens administered mipafox. *Neurotoxicology.* 1992;13:745–755.

94. Padilla S, Veronesi B. Biochemical and morphological validation of a rodent model of organophosphorus-induced delayed neuropathy. *Toxicol Ind Health.* 1988;4:361–371.

7

COGNITIVE ASSESSMENTS IN NONHUMAN PRIMATES

TUCKER A. PATTERSON AND MERLE G. PAULE

Division of Neurotoxicology, National Center for Toxicological Research, U.S. Food and Drug Administration, Jefferson, Arkansas

Disclaimer: This document has been reviewed in accordance with U.S. Food and Drug Administration (FDA) policy and approved for publication. Approval does not signify that the contents necessarily reflect the position or opinions of the FDA, nor does mention of trade names or commercial products constitute endorsement or recommendation for use. The findings and conclusions in this report are those of the authors and do not necessarily represent the views of the FDA.

INTRODUCTION

The ability to assess cognitive function in nonhuman primates is incredibly valuable not only because it provides an opportunity to learn about the biological substrates that subserve critical brain function, but also because it provides researchers with invaluable metrics of nervous system integrity. These metrics can then serve as biomarkers of health and act as sensitive indicators of the effects of chemicals that affect the nervous system. Before a discussion on this topic can proceed, however, some definitions about specific aspects of cognitive function are warranted. What follows are relatively cursory descriptions of the topics to be addressed in this chapter; more detailed descriptions can be found elsewhere.[1–3]

Cognition

Cognition is a term for "process of thought": *cognoscere*, "to know"; *intellegere*, "to understand." The mental processes subserving cognitive function are thought to include atten-

tion, comprehension, inference, decision making, memory, learning, and planning. Intelligence is the variable mental capacity that underlies individual differences in reasoning, problem-solving, abstract thinking, comprehension, and learning. It is a broad term used to describe a property of the brain that encompasses many of these related abilities. As with most of these terms, there are several ways to define intelligence. A simple definition is the ability to apply knowledge in order to perform better in a given environment.

A *modal model of the mind* is a depiction of the brain as a set of memory storage compartments and control processes for manipulating and moving information. It has long served as a standard framework for thinking about the human brain. The mind can be viewed as the entire set of a person's sensations, perceptions, memories, thoughts, dreams, motives, emotional feelings, and other subjective experiences.[1]

Attention

Attention is a process or set of processes by which the brain chooses from among the various stimuli that strike the senses at any given moment, allowing only some of those stimuli to proceed to higher stages of information processing. In a modal model of the mind, attention is the process that controls the flow of information from the sensory store into working memory. More broadly, attention is the focusing of mental activity along a specific track, whether that track consists purely of inner memories and knowledge or is based on external stimuli.[1] Simply put, attention is the cognitive process of selectively concentrating or focusing on one aspect of the environment while ignoring others.

Fundamental Neuropathology for Pathologists and Toxicologists: Principles and Techniques, First Edition. Edited by Brad Bolon and Mark T. Butt.
© 2011 John Wiley & Sons, Inc. Published 2011 by John Wiley & Sons, Inc.

Learning

Learning is a process or set of processes through which sensory experience at one time can affect an organism's behavior at a future time.[1] The ability to learn is possessed by humans, animals, and even some machines. Learning is the acquisition of new knowledge, behaviors, skills, preferences, or understanding and involves the synthesis of different types of information. There are multiple types of learning, but here the focus will be on associative learning as opposed to simple, nonassociative learning. Nonassociative learning is learning through habituation and sensitization, whereas associative learning is the process by which learning is accomplished through association with a separate preoccurring element. Associate learning includes classical (Pavlovian) conditioning and operant conditioning. Operant conditioning has been used widely in the assessment of cognition in nonhuman primates and will be the primary behavioral paradigm discussed here. B. F. Skinner used the term *operant conditioning* to describe the effects of the consequences of a particular behavior on the future occurrence of that behavior. Operant conditioning is a method of learning that occurs as a consequence of reward and punishment: an association is made between a behavior and the consequences of that behavior. There are four key components or types of operant conditioning: positive reinforcement, negative reinforcement, punishment, and extinction. A reinforcer is any event that strengthens or weakens the behavior it follows. A positive reinforcer (i.e., food to a hungry animal) increases the likelihood that the behavior that produced it (i.e., a lever press) will happen again; a negative reinforcer (electric shock) decreases the likelihood that the behavior that produced it (a lever press) will happen again. Removal of a negative reinforcer can serve as a positive reinforcer. Thus, animals (and people) will learn to press a lever to terminate or avoid shock. The juxtaposition of negative reinforcement following a behavior is also called *punishment*. Under extinction paradigms, no reinforcement is provided after the occurrence of behaviors that had previously been reinforced. Thus, the contingencies under which reinforcements occur can profoundly influence the behavior that precedes them. Only positive reinforcement is used in the examples provided in this chapter.

Memory

Memory refers to all the information in an organism's brain and the brain's capacity to store and retrieve that information.[1] In a modal model of the mind, the mind is portrayed as containing three types of memory stores: sensory memory, working (short-term) memory, and long-term memory. Each type of memory is characterized by the role it plays in the overall function of the brain, the amount of information it can hold at any given instance (capacity), and the length of time it can hold an item of information (duration). In addition, a set of control processes exists that include attention, rehearsal, encoding, and retrieval, and these govern the processing of information within the memory stores and the movement of information from one store to another.

Sensory memory lasts only for up to 2 s, has unlimited capacity, and is characterized by being outside of conscious control. A sensory-memory store is believed to exist for each sensory system, although only the auditory (echoic memory) and visual (iconic memory) systems have been examined extensively. The function of the store is presumably to hold on to sensory information for a long enough period of time for it to be analyzed by unconscious mental processes and for a decision to be made on whether or not to bring that information into the working-memory store.[1]

The selective process of attention determines what information moves from sensory memory to working memory. *Working memory* is the store where conscious mental work takes place on information brought in from sensory memory and long-term memory. An alternative name for this compartment is *short-term memory*, due to the relatively fleeting nature of information in this store. Information quickly fades and is lost within seconds if it is no longer attended to or thought about actively.[1] Working memory is generally considered to have limited capacity. The frontal cortex, parietal cortex, anterior cingulated, and parts of the basal ganglia appear crucial in the function of working memory. The prefrontal cortex appears to be especially involved in working memory and has been found to be active in a variety of tasks that require executive functions.[4] This has led to the argument that the role of the prefrontal cortex in working memory is in controlling attention, selecting strategies, and manipulating information, but not in the maintenance of that information. The maintenance function is attributed to more posterior areas of the brain, including the parietal cortex.[5,6] Activity in the parietal cortex has also been interpreted as reflecting executive functions (see below), because the same area is also activated in other tasks requiring executive attention but no memory.[7] Working memory is thought to be a temporary potentiation of neural connections that can become long-term memory through the process of rehearsal and meaningful association. Once an item has passed from sensory memory to working memory, it may or may not be encoded into long-term memory. Although the biological mechanisms of long-term memory have not been elucidated fully, the process of long-term potentiation, which involves change in the structure and excitability of neurons, has been proposed as the mechanism by which information in working memory is converted into long-term memory.

Long-term memory is the stored representation of all that a person knows and has unlimited capacity.[1] It differs both structurally and functionally from working memory. Long-term memory is typically divided into declarative memory and procedural memory. Declarative memory refers to all the memories that are consciously available, such as facts and dates, and is encoded by the hippocampus, entorhinal cortex, and perirhinal cortex. However, this information is probably consolidated in the temporal cortex. *Procedural memory* refers to the memory of how to use objects or perform particular movements of the body, such as tying one's shoes or riding a bike. This information is probably encoded and stored by the cerebellum and the striatum. Long-term memory and working memory are sharply differentiated, with long-term memory being passive (a "storehouse" of information) and working memory being active (a place where information is processed). Long-term memory can last for a lifetime, whereas in working memory, information can disappear within seconds if not rehearsed or otherwise kept activated.

Executive Functions

The term *executive functions* refers to those cognitive abilities that enable a subject to engage successfully in independent, goal-directed, problem-solving behavior. They are high-level abilities that influence more basic functions, such as attention, memory, and motor skills. Executive functions facilitate the connection of past experiences with present actions. They are used when planning, organizing, strategizing, paying attention to, and remembering details, and are functions performed by the prefrontal lobes of the cerebral cortex in conjunction with the limbic system. A few of the aspects of executive function include orchestrating working memory resources, directing the storage of information into long-term memory, directing retrieval of information from long-term memory, monitoring and regulating the speed of information processing, directing and sustaining attention while screening out interfering stimuli, and interrupting and returning to an ongoing activity. Executive functions have a tremendous impact on the capacity to learn new information, perform learned actions, and adapt to new environments and challenges.

MODELING SPECIFIC FUNCTIONS OR BEHAVIORS

More than 20 years ago, during preparations for the conduction of a large-scale monkey study in which some of the primary endpoints of interest were aspects of complex brain function, a battery of operant behavioral tasks was developed at the National Center for Toxicological Research (NCTR), with each task designed to engender responding indicative of a different brain function. This battery of cognitive function tasks has since become known as the NCTR Operant Test Battery (OTB).[8–12] The NCTR OTB contains several food-reinforced tasks, in which correct performance is thought to depend on relatively specific and important brain functions. It has been shown that the tasks in the OTB are differentially sensitive to the acute effects of drugs from different pharmacological classes and that performance of each task is not highly correlated with performance of the other tasks. Thus, each task is measuring different aspects of function.

The term *operant* indicates that subjects have to operate or manipulate something in their environment (in this case, response levers and press-plates; see Figure 1) in order to obtain reinforcers (for the procedures described in this chapter, monkey behavior is positively reinforced with banana-flavored food pellets).

Operant tasks require some degree of training and therefore differ from untrained, spontaneous, or ethological behaviors that can also serve as metrics of brain function. The beauty and strength of operant behaviors, however, stems from the fact that they can be made extremely specific, allowing for the isolation of brain functions deemed especially important or noteworthy, and they can be generated at the will of the experimenter because their elicitation comes under strong stimulus control after appropriate training (i.e., the presence or absence of cue lights or tones can be used to indicate the availability of reinforcement and prompt subject participation). This quality allows for precise scheduling of experimental events providing maximal efficiency of data collection. Operant tasks also lend themselves very nicely to automation; the NCTR OTB, for example, is completely computer-driven. This aspect of task administration not only frees up study personnel for other tasks, but also eliminates the potential confounding interactions that can occur between experimenters and subjects under circumstances where direct observations and/or other types of human–animal interactions are part of the assessment. This occurs, for example, when using the traditional Wisconsin General Test Apparatus (WGTA) in testing the cognitive abilities of nonhuman primates.[13,14] In addition, the automated nature of task administration and data collection eliminates subjective interpretation of responses and easily provides for repeated testing under identical conditions. The specific brain functions thought to be modeled by performance in the NCTR OTB (motivation, learning, color and position discrimination, and short-term memory) and the tasks utilized to assess them are described below.

Aspects of Motivation Are Assessed by Monitoring Performance in a Progressive Ratio Task

In the interpretation of the effects of an experimental manipulation on a given behavioral measure, it is important to have some indication of the level of motivation of the experimental subjects to perform the task of interest. If, for example, a drug treatment causes a decrease in overall response rate, an increase in reaction time, and so on, it may be that the subject is simply less motivated to perform that

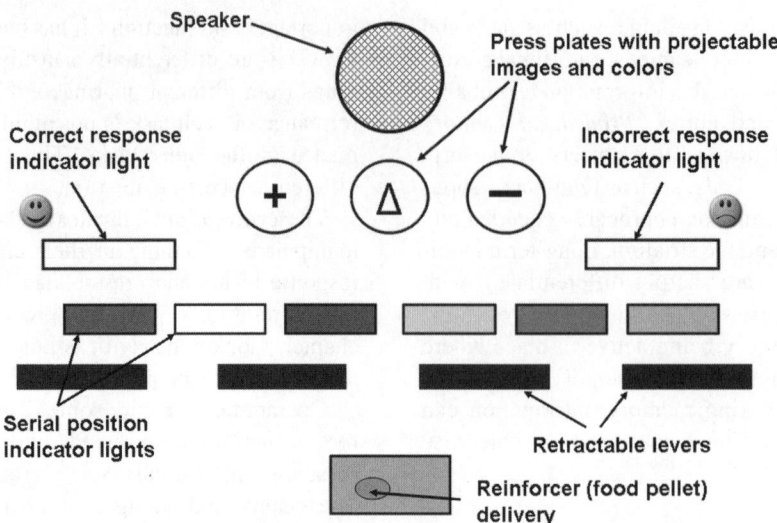

FIGURE 1 Schematic of the operant test panel. Press-plates are used for the conditioned position responding (CPR) and delayed-matching-to-sample (DMTS) tasks; retractable response levers are used for the progressive ratio (PR), incremental repeated acquisition (IRA), and temporal response differentiation (TRD) tasks. The correct response, incorrect response, and serial position indicator lights are used in the IRA task. Masking white noise is piped into the testing chamber through the speaker at the top of the panel.

task, not that the drug has otherwise affected specific "cognitive" processes or psychomotor speed associated with it. The NCTR OTB has incorporated a progressive ratio (PR) task specifically for the purpose of obtaining an assessment of behavior thought to be highly associated with a subject's motivation to "work" for the banana-flavored food pellet reinforcers. Progressive ratio tasks have been used in a variety of experimental settings to study processes associated with reinforcer efficacy and the interaction of deprivation states on PR performance.[15,16] Obviously, if subjects are not motivated to perform for the reinforcer, the behavior of interest will be difficult to elicit. Performance will not be maintained if the reinforcer becomes ineffective, as can be the case when subjects working for food are fully satiated or are sick. To maintain a reproducible level of motivation to work for food, access to food is often restricted such that subjects earn a portion of their daily rations during the behavioral test session and are then provided the remainder of their daily ration immediately following the test session. In this way, 22 to 23 h of food restriction can be attained easily and routinely while allowing subjects to maintain healthy body weights or rates of growth.

For the progressive ratio task, responses are made on a single response lever. In the NCTR monkey OTB it is the rightmost of four retractable levers that are aligned horizontally across a response panel (the other three levers remain retracted during this task). Here, the number of lever presses required for reinforcer delivery (the response/reinforcer ratio) increases with each reinforcer. Initially, some small number of lever presses (e.g., two) is required for the first

food pellet. The next reinforcer requires four lever presses, the next six, and so on, up to the completion of a predetermined ratio, the attainment of a specific number of reinforcers, or some maximum session length. The task can be defined in terms of both the initial ratio requirement (in this case, two) and the ratio increment (also two). The task just described is a PR 2 + 2 task. As used in the NCTR OTB, the initial ratio and ratio increments are kept equal for a given subject (e.g., PR 3 + 3 or PR 5 + 5), but the ratios can be adjusted for individual subjects such that they generally earn a specific number of reinforcers during any given session. Measures obtained include percent task completed (PTC, number of reinforcers obtained/maximum number obtainable × 100), response rate (RR, lever presses per second), reinforcers earned, breakpoint (the size of the last ratio completed during the session), and post-reinforcement pause (time from reinforcer delivery to reinitiation of lever pressing). As part of the NCTR OTB, this task lasts only 10 min, to allow for the administration of the other OTB tasks within a relatively short period of time. It has been determined that PR behavior obtained in 10-min sessions correlates highly and significantly with PR behavior generated in much longer sessions of 30, 45, and 60 min (unpublished observations).

Learning Is Modeled by Performance of an Incremental Repeated Acquisition Task

An organism's ability to survive is related directly to its ability to learn; thus, learning is a critical brain function. The learning task used in the NCTR OTB is referred to as the *incremental repeated acquisition* (IRA) *task*. The IRA task is a modification of more traditional repeated acquisition procedures[17] and is very similar to the children's game Simon. As the name implies, subjects performing this task must repeatedly acquire knowledge in an incremental fashion. Here, four horizontally aligned response levers serve as the manipulanda, and a row of six colored lights above the levers indicate how many more correct lever presses are needed to obtain the next reinforcer. Correct and incorrect responses are also indicated by the illumination of white stimulus lights located slightly above and to the left and right of the response levers, respectively. Subjects are required to learn a new sequence of lever presses (generated randomly) every test session. The IRA session starts with presentation of a one-lever response "sequence"; a response to the correct one of the four levers results in reinforcer delivery. After this sequence has been mastered (e.g., after 20 reinforcers have been earned), the sequence length (and presumably task difficulty) is incremented to a two-lever response sequence. The subject must now learn which of the four levers is the "new" lever and remember to follow a response to it with a response to the correct lever from the previously learned one-lever sequence. Once this two-lever sequence has been performed correctly 20 times, the response requirement is again incremented to a three-lever sequence, and so on, up to a six-lever sequence or until the session times out.

Data for the IRA task include PTC, overall (collapsed across all lever sequence lengths) accuracy (ACC) and RR, and ACC and RR for each lever sequence length. Additionally, the number of acquisition (searching for the new lever) and recall (remembering the lever already learned)-type errors are obtained for lever sequences of all lengths, as are learning curves (errors vs. number of correct sequences completed). As part of the NCTR OTB, this task is presented for 35 min per session.

Color and Position Discrimination Is Modeled by Performance of a Conditioned Position Responding Task

It is often of great interest to know how well subjects can discriminate between objects and make decisions based on those differences. Processes underlying these types of abilities are thought to reside in frontal-cortical brain areas.[18,19] In the NCTR OTB, a conditioned position responding (CPR) task has been adapted to provide information as to how subjects discriminate colored stimuli. In this task, three press-plates that can be illuminated from behind are used as manipulanda. They are aligned horizontally on the response panel above the four retractable levers that are used in the IRA and other tasks. Initially, a red, yellow, blue, or green color is presented on the center press-plate. Subjects acknowledge the presence of this stimulus by pressing the plate (making an observing response to it), after which it is

immediately extinguished and the two side keys are illuminated white. If the center press-plate had been red or yellow, a left-choice response is reinforced; if it had been blue or green, a right-choice response is reinforced. Incorrect choices result in a 10-s timeout (timeouts from positive reinforcement can take on attributes of negative reinforcements) and presentation of the next trial (colors presented randomly). Task solution is relatively simple for most subjects, as evidenced by rapid responding and greater than 90% average choice accuracies. PTC, observing response latency (ORL), choice response latency (CRL), and choice accuracy (ACC) are the primary endpoints for this task. Choice response latencies can be further categorized as either correct or incorrect (CCRL and ICRL, respectively). As part of the NCTR OTB, this task is presented for no more than 5 min per test session, yet typically generates over 100 choice trials.

Aspects of Short-Term Memory are Modeled by Performance in a Delayed Matching-to-Sample Task

Memory is clearly another very important brain function. Within the confines of even a relatively short test session, it is quite possible to assess processes associated with short-term memory using a delayed matching-to-sample (DMTS) task (for a review of DMTS procedures, see Paule et al.,[20] and for a discussion of experimental approaches to the study of short-term memory, see Rodriguez and Paule[21]). For this task, each trial begins with the illumination of the center of the three horizontally aligned press-plates described previously for the CPR task. Instead of using colors for this task, however, the press-plates are illuminated with one of seven white-on-black geometrical shapes. Subjects make observing responses to the shape (the sample stimulus) by pushing the illuminated plate, after which it is extinguished immediately. Following an interval (recall delay) that varies randomly (e.g., from 2 to 32 s), all three plates are illuminated, each with a different stimulus shape but only one of which matches the sample stimulus. A choice response to the plate illuminated with the matching stimulus results in reinforcer delivery. Incorrect choices are followed by a 10-s timeout followed by initiation of a new trial.

Data from this task include PTC, ACC, RR, ORL, and CRL, as well as others associated with specific aspects of task performance (e.g., attention, encoding, forgetting). Accuracy of matching after no delays or very short delays (i.e., with little or no opportunity to forget or be distracted) is thought to represent a measure of an organism's ability to attend to and encode the information to be remembered. Typically, response accuracy at very short delays is quite high. As the delay intervals increase, response accuracy decreases, with the slope of the decrease in accuracy over time representing a decay or forgetting function. Increases in the negative slope of the decay line without changes in the origin (position on the *y*-axis) are thought to represent increases in the rate of forgetting in the absence of any difficulty in attention or encoding. Changes in the origin with no change in slope would indicate changes in attention or encoding but no change in the rate of forgetting. Changes in both the slope and the origin indicate changes in both attention or encoding and recall. As part of the NCTR OTB, this task is presented for 30 min per session.

Timing Behavior Is Assessed by the Temporal Response Differentiation Task

The ability to time events is advantageous because it provides for the anticipation and prediction of events in time and space. Many types of learning are sensitive to the timing of events, suggesting that this capability provides the tools needed to make the correct response at the most optimal time (see Paule et al.[22] for a review of experimental aspects of timing ability). The task used in the NCTR OTB for the assessment of timing behavior is referred to as a temporal response differentiation (TRD) task. In this task, the subject is required to press and hold a response lever in a depressed position for a minimum of 10 s but not more than 14 s. Thus, to obtain a reinforcer, the subject must target a 4-s window (release the lever between 10 and 14 s). Releasing the lever too early or too late has no programmed consequence and the subject may begin the next trial at any time.

Data obtained from the TRD task include PTC, RR, ACC, and a variety of measures associated with the distribution of lever-hold durations. Typically, in well-trained subjects, the largest proportion of response durations (lever holds) occurs within the 4-s reinforced window. These response durations are typically characterized by Gaussian distributions. It is thought that the distribution averages represent the timing accuracy of the subject and that the spread or standard deviation of the distribution represents the precision of timing. Additionally, it has been suggested that the peak height (e.g., number of responses in the average timing bin) is associated with aspects of motivation. Alterations in the characteristics of the response duration distribution are thought to provide insights into the mechanisms of timing. Shifts in the mean, for example, are thought to indicate changes in the speed of an "internal clock": leftward shifts indicate speeding up of the clock (8 s feels like 10 s), whereas rightward shifts indicate slowing of the clock.[23] As part of the NCTR OTB, this task is presented for 20 min per session.

EXPERIMENTAL MANIPULATIONS: CONSEQUENCES OF DRUGS, TOXICANTS, AND LESIONS

When employing a battery approach to data collection, it is important that each component of the battery provides information not attainable from the other components of the battery. One way of determining whether the behaviors

contained in the NCTR OTB are independent of each other involves determining whether experimental manipulations can preferentially affect the components of the battery. Toward this end, a series of experiments were conducted in monkeys in which the effects of psychoactive compounds on OTB performance were determined (summarized by Paule[8]). In these studies,[24–36] a variety of doses of each compound were given and the relative sensitivities of each OTB task to these treatments were determined. In short, the behavioral profiles obtained were specific for each particular compound tested. These profiles demonstrated that certain drugs affect different behaviors preferentially, and suggest that the OTB behaviors are subserved by different neurological substrates since they can be manipulated independently of one another. In these studies, the order of task sensitivity indicates which behavioral task was affected at doses that did not affect the other tasks. For example, the TRD task (timing behavior) was disrupted significantly by doses of Δ^9-tetrahydrocannabinol (THC) that had no effect on the other OTB behaviors. Learning (IRA), short-term memory (DMTS), and color and position discrimination (CPR) were all affected at the same dose of THC, but at a higher dose than that required to affect timing. Motivation (PR) was unaffected by doses of THC that significantly affected all other tasks. Thus, the profile of THC effects on OTB responding was TRD > IRA = DMTS = CPR > PR. For cocaine, TRD and PR were affected at the lowest dose tested, whereas a higher dose was required to affect IRA, a still higher dose was required to affect DMTS, and a still higher dose was required to affect CPR. Thus, the profile of cocaine effects on OTB responding was TRD = PR > IRA > DMTS > CPR. In the case of atropine (a cholinergic antagonist), the profile of drug effect was IRA > CPR > DMTS = TRD = PR. This observation is interesting because learning (IRA) behavior is thought to involve brain cholinergic systems, atropine is a cholinergic antagonist, and learning behavior is the most sensitive of the OTB tasks to atropine. These types of observations serve to validate the use of OTB tasks.

Thus, depending on the dose and drug administered, a particular OTB behavior can be significantly disrupted or totally unaffected, showing that it is possible to affect specific behaviors differentially. Interestingly, even compounds with similar mechanisms of action (e.g., d-amphetamine, cocaine, and methylphenidate) can show differential behavioral effects, suggesting that the behaviors monitored using the OTB are sensitive to apparently subtle differences in drug action even among drugs in the same pharmacological classes.

RELEVANCE TO HUMANS

Since the primary reason for using animal models in research is to obtain information that is relevant to humans, it is important to address the issues of relevance and validity as they pertain not only to the model being used but also the endpoints being collected. Animal models have been used extensively based on their face validity: They are more like humans than not like humans. In reality, remarkably few animal models have been validated thoroughly in terms of their relevance to the human condition. This is one potential reason that many drugs, developed and tested in animal models, fail during clinical trials in humans. For a variety of obvious and not-so-obvious reasons that will not be detailed here (see Paule[37]), the monkey is the laboratory animal that most often best mimics humans. Presuming then that the monkey is probably the best animal model, we seek to explore details of cognitive function that we hope will be relevant to humans. Thus, the NCTR OTB was developed in the belief that the tasks of which it is comprised would generate behaviors dependent on specific brain functions that are comparable to, if not homologous with, those same functions in humans. In addition to utilizing the psychoactive drugs mentioned above to generate OTB behavioral profiles, when possible, they were also used to help validate the monkey model. For those drugs for which comparable human data exist, it was found that the monkey model was very predictive of drug effects in humans.[8] For example, when monkeys were given THC, the active ingredient in marijuana smoke, it was observed that monkeys overestimated the passage of time: they indicated that 8-s durations were perceived as 10 s. These types of observations are exactly like those observed in humans. Marijuana smoke itself disrupted short-term memory in the monkey model, and this is a well-known effect in humans. Chronic marijuana smoking in the monkey model resulted in an amotivational syndrome, an oft-reported finding in human teenagers.

Other examples (see Paule[8]) serve to further validate the monkey model. The issue concerning endpoint relevance was addressed by having human subjects perform the same OTB tasks as performed by our monkey subjects. Initially, children 4 to 12 years old were tested using money (nickels) instead of banana-flavored food pellets as reinforcers.[38] Perhaps not so surprising was the remarkable similarity in task performance across species.[39] Thus, the same tasks are useful and applicable across species. In addition, the observation that the performance of most OTB tasks correlates highly and significantly with measures of intelligence in children is perhaps most important of all.[40] In fact, using creative computational techniques it is possible to generate algorithms that can estimate children's IQs based on OTB performance and then, by analogy, estimate the IQs of nonhuman primates.[41]

Other than the NCTR OTB, there are only a few computerized tasks that have been developed to assess cognitive processes in nonhuman primates, especially those in which cross-species comparisons can be made utilizing the same or a similar system. With the Cambridge Neuropyschological Test Automated Battery (CANTAB), computerized neuropsychological tasks can be presented on a touch-sensitive computer screen. This system has been used to assess cognitive processes in both humans and nonhuman primates, primarily rhesus monkeys and marmosets.[42–44] The CANTAB includes tests for memory [(delayed nonmatching to sample (DNMS), self-ordered spatial search (SOSS), reaction time (RT), motivation (PR), and fine motor coordination (bimanual)]. When using this battery to test nonhuman primates, subjects are either left in a transport cage or transferred to a cage outside their home cage. The test cage is modified to allow the animal easily to reach out of the cage to access a computer monitor fitted with a touch-sensitive screen. The animals are trained to reach out of the cage to touch the location on the screen at which visual stimuli are presented to obtain a food reward. Findings in rhesus monkeys have demonstrated that the CANTAB is useful for distinguishing the effects of neuropharmacological manipulation on various aspects of cognitive performance.[45]

A recently developed software and training procedure has been designed to teach infant monkeys to interact with a touch-screen computer.[46,47] With this procedure, animals are tested in a cage similar to their home cage. The test end of the cage has a template with 11 square openings that are aligned with 11 response areas (nine test areas and two task control areas) of the touch screen. The template helps to restrict the sweeping motor responses to potentially active areas on the screen. The main goal of training with this system is to have the animal touch a stimulus to receive a reward. The results of these studies demonstrate the feasibility of using computers to assess cognitive and perceptual abilities early in development, but also reveal some apparent limitations of computer-based testing with infant monkeys.[46,47] Compared to performance in the WGTA, infants tested in a computerized touch-screen environment appeared to have difficulty in a task that required them to form response strategies. However, infants performed comparably on simple discriminations, reversal learning, and delayed nonmatch to sample rule learning.[47]

In summary, the use of the NCTR OTB and other computerized tasks for cognitive assessments in nonhuman primates produces information both relevant to and predictive of important aspects of brain function in humans. The results produced from these assessments demonstrate that a nonhuman primate can serve as an invaluable surrogate for the study of human brain function and dysfunction. By examining the effects of potentially neuroactive or neurotoxic agents on cognitive function in nonhuman primates, our ability to predict the adverse effects of such agents on related brain functions in humans is greatly enhanced.

REFERENCES

1. Gray P. *Psychology*. New York: Worth Publishers; 2007.

2. Cowan N. *Attention and Memory: An Integrated Framework*. New York: Oxford University Press; 1995.

3. Leahey TH, Harris RJ. *Learning and Cognition*, 5th ed. Upper Saddle River, NJ: Prentice Hall; 2000.

4. Kane MJ, Engle RW. The role of prefrontal cortex in working-memory capacity, executive attention, and general fluid intelligence: an individual-differences perspective. *Psychon Bull Rev.* 2002;9:637–671.

5. Curtis CE, D'Esposito M. Persistent activity in the prefrontal cortex during working memory. *Trends Cogn Sci.* 2003;7: 415–423.

6. Postle BR. Working memory as an emergent property of the mind and brain. *Neuroscience.* 2006;139:23–28.

7. Collette F, et al. Exploration of the neural substrates of executive functioning by functional neuroimaging. *Neuroscience.* 2006;139:209–221.

8. Paule MG. Validation of a behavioral test battery for monkeys. In: Buccafusco JJ, ed. *Methods of Behavioral Analysis in Neuroscience.* Boca Raton, FL: CRC Press; 2001: 281–294.

9. Schulze GE, et al. Acute effects of delta-9-tetrahydrocannabinol in rhesus monkeys as measured by performance in a battery of complex operant tests. *J Pharmacol Exp Ther.* 1988; 245:178–186.

10. Schulze GE, et al. Effects of marijuana smoke on complex operant behavior in rhesus monkeys. *Life Sci.* 1989; 45:465–475.

11. Paule MG, et al. Complex brain function in monkeys as a baseline for studying the effects of exogenous compounds. *Neurotoxicology.* 1988;9:463–470.

12. Paule MG, et al. Chronic marijuana smoke exposure in the rhesus monkey: II. Effects on progressive ratio and conditioned position responding. *J Pharmacol Exp Ther.* 1992;260: 210–222.

13. Harder JA, et al. Learning impairments induced by glutamate blockade using dizocilpine (MK-801) in monkeys. *Br J Pharmacol.* 1998;125:1013–1018.

14. Machado CJ, Bachevalier J. The effects of selective amygdala, orbital frontal cortex or hippocampal formation lesions on reward assessment in nonhuman primates. *Eur J Neurosci.* 2007;25:2885–2904.

15. Hodos W. Progressive ratio as a measure of reward strength. *Science.* 1961;134:943–944.

16. Hodos W, Kalman G. Effects of increment size and reinforcer volume on progressive ratio performance. *J Exp Anal Behav.* 1963;6:387–392.

17. Cohn J, Paule MG. Repeated acquisition: the analysis of behavior in transition. *Neurosci Biobehav Rev.* 1995; 19:397–406.

18. Goldman PS, et al. Selective sparing of function following prefrontal lobectomy in infant monkeys. *Exp Neurol.* 1970;29:221–226.

19. Kojima S, et al. Operant behavioral analysis of memory loss in monkeys with prefrontal lesions. *Brain Res.* 1982;248:51–59.

20. Paule MG, et al. Symposium overview: the use of delayed matching-to-sample procedures in studies of short-term memory in animals and humans. *Neurotoxicol Teratol.* 1998; 20:493–502.

21. Rodriguez J, Paule MG. Learning and memory: delayed response tasks in monkeys. In: Buccafusco JJ, ed. *Methods of Behavioral Analysis in Neuroscience*, 2nd ed. Boca Raton, FL: CRC Press; 2008: 247–265.

22. Paule MG, et al. The use of timing behaviors in animals and humans to detect drug and/or toxicant effects. *Neurotoxicol Teratol.* 1999;21:491–502.

23. Meck WH. Neuropharmacology of timing and time perception. *Brain Res Cogn Brain Res.* 1996;3:227–242.

24. Paule MG, et al. Cocaine (COC) effects on several "cognitive" functions in monkeys. *Pharmacologist.* 1992;34:137.

25. Morris P, et al. Acute behavioral effects of methylphenidate on operant behavior in the rhesus monkey. *Soc Neurosci Abstr* 1995;21:1465.

26. Schulze GE, Paule MG. Acute effects of *d*-amphetamine in a monkey operant behavioral test battery. *Pharmacol Biochem Behav.* 1990;35:759–765.

27. Ferguson SA, Paule MG. Acute effects of chlorpromazine in a monkey operant behavioral test battery. *Pharmacol Biochem Behav.* 1992;42:333–341.

28. Frederick DL, Ali et al. Acute effects of dexfenfluramine (D-FEN) and methylenedioxymethamphetamine (MDMA) before and after short-course, high-dose treatment. *Ann NY Acad Sci.* 1998;844:183–190.

29. Frederick DL, et al. Acute effects of methylenedioxymethamphetamine (MDMA) on several complex brain functions in monkeys. *Pharmacol Biochem Behav.* 1995;51:301–307.

30. Frederick DL, Gillam MP, Lensing S, Paule MG. Acute effects of LSD on rhesus monkey operant test battery performance. *Pharmacol Biochem Behav.* 1997;57:633–641.

31. Schulze GE, et al. Effects of atropine on operant test battery performance in rhesus monkeys. *Life Sci.* 1992;51:487–497.

32. Frederick DL, et al. Acute effects of physostigmine on complex operant behavior in rhesus monkeys. *Pharmacol Biochem Behav.* 1995;50:641–648.

33. Schulze GE, Paule MG. Effects of morphine sulfate on operant behavior in rhesus monkeys. *Pharmacol Biochem Behav.* 1991;38:77–83.

34. Morris P, et al. Acute effects of naloxone on operant behaviors in the rhesus monkey. *FASEB J.* 1995;9:A101.

35. Frederick DL, et al. Acute behavioral effects of phencyclidine on rhesus monkey performance in an operant test battery. *Pharmacol Biochem Behav.* 1995;52:789–797.

36. Buffalo EA, et al. Acute behavioral effects of MK-801 in rhesus monkeys: assessment using an operant test battery. *Pharmacol Biochem Behav.* 1994;48:935–940.

37. Paule MG. The nonhuman primate as a translational model for pesticide research. In: Krieger R, ed. *Hayes' Handbook of Pesticide Toxicology.* St. Louis, MO: Elsevier; 2009.

38. Paule MG, et al. Quantitation of complex brain function in children: preliminary evaluation using a nonhuman primate behavioral test battery. *Neurotoxicology.* 1988;9:367–378.

39. Paule MG, et al. Monkey versus human performance in the NCTR operant test battery. *Neurotoxicol Teratol.* 1990;12: 503–507.

40. Paule MG, et al. Operant test battery performance in children: correlation with IQ. *Neurotoxicol Teratol.* 1999;21:223–230.

41. Hashemi RR, et al. IQ estimation of monkeys based on human data using rough sets. In: Proceedings of the International Workshop on Rough Sets and Soft Computing. 1994;400–407.

42. Weed MR, et al. Performance norms for a rhesus monkey neuropsychological testing battery: acquisition and long-term performance. *Brain Res Cogn Brain Res*. 1999;8:185–201.

43. Katner SN, et al. Effects of nicotine and mecamylamine on cognition in rhesus monkeys. *Psychopharmacology (Berl)*. 2004;175:225–240.

44. Spinelli S, et al. Enhancing effects of nicotine and impairing effects of scopolamine on distinct aspects of performance in computerized attention and working memory tasks in marmoset monkeys. *Neuropharmacology*. 2006;51:238–250.

45. Taffe MA, et al. Scopolamine alters rhesus monkey performance on a novel neuropsychological test battery. *Brain Res Cogn Brain Res*. 1999;8:203–212.

46. Mandell DJ, Sackett GP. A computer touch screen system and training procedure for use with primate infants: results from pigtail monkeys (*Macaca nemestrina*). *Dev Psychobiol*. 2008;50:160–170.

47. Mandell DJ, Sackett GP. Comparability of developmental cognitive assessments between standard and computer testing methods. *Dev Psychobiol*. 2009;51:1–13.

8

IMPACT OF AGING ON BRAIN STRUCTURE AND FUNCTION IN RODENTS AND CANINES

M. Paul Murphy and Elizabeth Head

Sanders–Brown Center on Aging, Department of Molecular and Biomedical Pharmacology, University of Kentucky, Lexington, Kentucky

INTRODUCTION

A major area of public health concern is age-related diseases of the nervous system. One of the more challenging of these diseases in humans is Alzheimer's disease (AD), which is the most common cause of dementia in the elderly. As the population of the developed world ages, the number of elderly developing AD is rising rapidly. Patients show a progressive loss of memory and worsening of symptoms until end-stage dementia is reached. Currently, there are no treatments to slow or halt the progression of AD. The aged human brain is also afflicted with a number of other age-related neuropathology conditions, as described in Table 1. Lewy bodies, intracellular inclusions that contain neurofilament proteins, ubiquitin, and α-synuclein are found in dementia with Lewy bodies and in Parkinson's disease.[1,2] In hippocampal sclerosis, a selective loss of neurons is also associated with gliosis and the accumulation of trans-acting responsive element (TAR)-DNA binding protein of 43 kDa (TDP-43).[3] Nonspecific, age-associated changes include the presence of corpora amylacea, which are spherical inclusions that predominantly accumulate within glial processes and stain with hematoxylin and cresyl violet.[4]

The most common approach to understanding AD in humans and to developing preventive or disease-modifying treatments is to use animal model systems. Rodents are the most commonly used model, including both normal rats and mice and genetically modified (transgenic) mice. Other models include aging dogs and nonhuman primates. As a consequence, we are learning a great deal about animal

aging, which is beneficial to the study of both AD and normal brain aging. Interesting overlaps in the aging process occur across species, but there are also important differences, highlighting the advantages to using several models and combining information.[5]

In this chapter we describe aging and AD in humans to provide an outline and rationale for the specific aspects of aging that we have focused on in both rodents and canines. We then describe rodent and canine aging, including discussions of both behavioral/cognitive phenotypes and possible neurobiological substrates. When possible, we make direct comparisons across species and summarize commonalities and differences.

HUMAN AGING AND ALZHEIMER'S DISEASE

Aging in humans may be accompanied by selective cognitive impairment, although this is usually quite mild (e.g., age-associated memory impairment).[6] However, not all types of learning and memory are vulnerable. For example, spatial memory declines with age in humans,[7] whereas other types of memory remain intact, including long-term memories and language ability.[8] In addition, not all aged persons show deficits in cognition, with some learning and remembering as well as their younger counterparts,[9] whereas others develop severe cognitive decline (dementia). Neurobiologically, the aged human brain shows increasing cortical atrophy (loss of tissue), as seen by magnetic resonance imaging (MRI).[10,11] Interestingly, neuron loss may not account for atrophy

Fundamental Neuropathology for Pathologists and Toxicologists: Principles and Techniques, First Edition. Edited by Brad Bolon and Mark T. Butt.
© 2011 John Wiley & Sons, Inc. Published 2011 by John Wiley & Sons, Inc.

TABLE 1   Pathological Lesions in Human Brain Aging

Category	Neurological Diseases of Aging	Pathognomonical Features	Regions Affected
Neurodegenerative diseases	Alzheimer's disease	Plaques and tangles	Limbic system, neocortex, brainstem
	Dementia with Lewy bodies	Lewy bodies	Limbic system, neocortex
	Parkinson's disease	Lewy bodies	Substantia nigra
	Hippocampal sclerosis	Neuron loss, gliosis, TDP-43	Hippocampus
	Prion diseases	Neuronal vacuolation, cell loss	Neocortex, thalamus, cerebellum
Neurovascular changes	Large ischemic infarcts	—	Neocortex and subcortical structures
	Lacunar infarcts	—	Neocortex and subcortical structures
	Microinfarcts	—	Neocortex and subcortical structures
	Hemorrhagic infarcts	—	Neocortex and subcortical structures
	Arteriolosclerosis	—	Neocortex and subcortical structures
	Cerebral amyloid angiopathy	Blood vessel amyloid	Neocortex (favors occipital cortex), limbic system
Other including structural changes	Brain atrophy	—	Neocortex, limbic system
	Normal pressure hydrocephalus	—	Lateral ventricles
	Contusions and subdural hemorrhages	—	Trauma-dependent
	Neoplasms	—	Primary: entire neuraxis Metastatic (nonneural): entire brain
Nonspecific changes	Corpora amylacea	—	Neocortex and subcortical structures
	Involution of pineal gland	—	Pineal gland
	Calcification of choroid plexus	—	Choroid plexus
	Basal ganglia calcifications	—	Basal ganglia

observed by MRI as initially thought (see Peters et al.[12] for a thorough review), which leads to the question of what the atrophy observed by MRI represents. Selective neuron loss appears to occur in the hippocampus, a region critically involved with memory.[13,14] Conversely, the mature human brain can generate new neurons, particularly in the hippocampus,[15] but it is not known if this ability declines with age. Cortical neurogenesis appears to be minimal and does not occur postnatally.[16]

Tissue or neuron loss with age may be due to the progressive accumulation of neuropathology or damage both within and around vulnerable neurons. Progressive oxidative damage accumulates in the human brain, resulting in damage to proteins, lipids, DNA, and RNA.[17] Downstream consequences of these oxidative modifications include reduced protein synthesis,[18] altered proteasome function,[19] and impaired protein and enzyme function.[20,21] Further, selective oxidative modifications to key proteins may lead to neuronal dysfunction through abnormalities in pathways associated with energy metabolism, excitotoxicity, proteasomal dysfunction, lipid metabolism, synaptic dysfunction, and pH buffering.[22] Additionally, normal aged human brain frequently contains extracellular plaque deposits of the naturally occurring β-amyloid (Aβ) protein and intracellular neurofibrillary tangles, which are defining features of AD because they are found in much higher numbers in AD brain relative to unaffected aged brain.

The most common form of AD is sporadic, has no clear etiology, and occurs late in life. Possible environmental contributions to AD development or vulnerability have included events such as exposure to pesticides or to aluminum or other metals,[23,24] although a direct causal link between these exposures and disease has not been clearly established.[25] In addition, dietary and lifestyle factors can be significant contributors that are both beneficial and detrimental to the aging process.[26–29] Much less common are cases of inherited, autosomal dominant familial AD (FAD), which usually occur at a younger age. Other than age of onset, the clinical course and neuropathology of sporadic AD and FAD are essentially identical. For this reason, the study of the genes mutated in FAD has been indispensable in elucidating the molecular basis of the disease process[30,31] and designing suitable animal models. Although simpler organisms with very short lifespans [e.g., fruit fly (*Drosophila melanogaster*) and nematode (*Caenorhabditis elegans*)] have made substantial contributions to our understanding of the molecular basis of disease, only rodents offer advantages of both mammalian physiology and amenability to genetic manipulation.

There are two hallmark features of AD within the human brain. First, extracellular deposits of Aβ occur in plaques[32] (Figure 1A and C), and frequently, Aβ also accumulates around blood vessels [cerebral amyloid angiopathy (CAA)]. Second, neurofibrillary tangles (NFT) composed

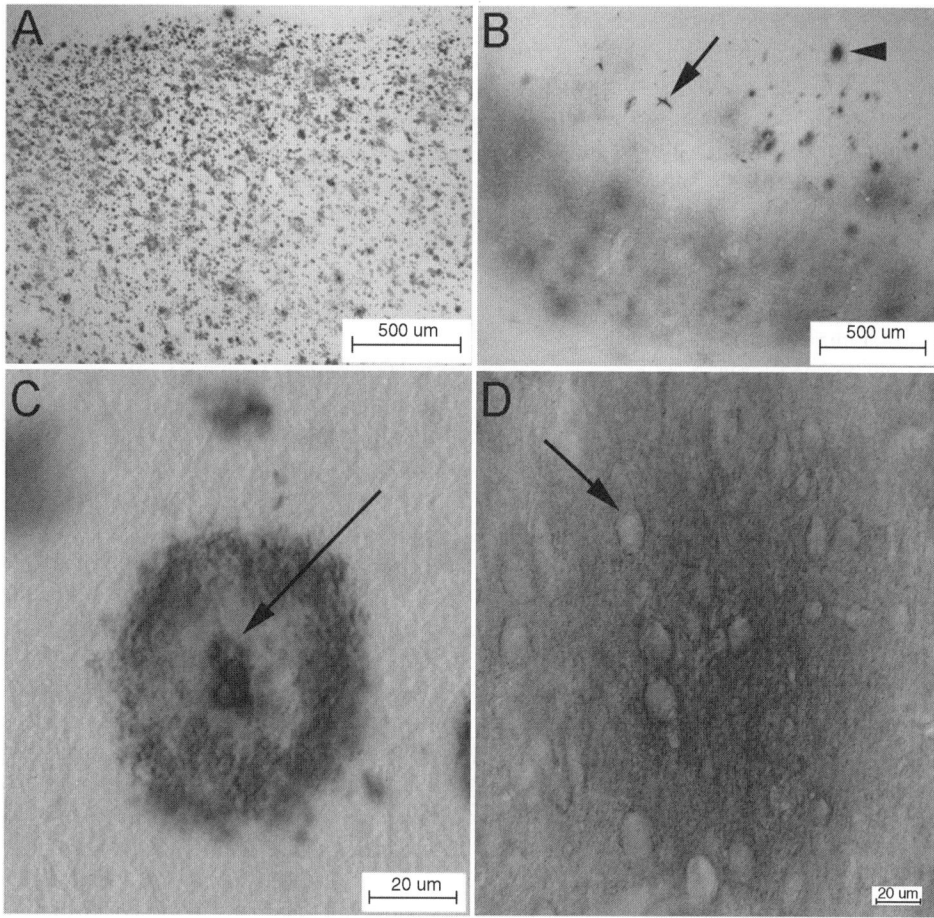

FIGURE 1   Amyloid-beta (Aβ) neuropathology in the frontal cortex of a human Alzheimer's disease (AD) patient and an aged beagle dog. (A) Extensive Aβ pathology is observed in the AD brain, identified on the basis of immunohistochemistry using antibodies against Aβ1–16. (B) Significant Aβ is also observed in the aged beagle brain but as a diffuse cloud in deep cortical layers with smaller, more condensed deposits in the superficial layer (arrowhead). Note also the presence of cerebral amyloid angiopathy (CAA) in gray matter (arrow). (C) Higher magnification of a single Aβ deposit in a plaque of the AD brain with a dense core of Aβ (arrow). (D) In contrast, canine Aβ plaques are diffuse and often contain intact neurons (arrow).

of hyperphosphorylated τ protein also can be observed within cerebrocortical neurons, a lesion also seen in a variety of other disorders.[33,34] Plaques and tangles develop in characteristic patterns within the affected brain.[35] Plaques are thought to accumulate first in the frontal cortex and forebrain and then move posterior to affect the limbic system and other regions of neocortex.[36] Tangles, in contrast, are observed first in the transentorhinal cortex and subsequently move into the hippocampus and then, later in the disease, into the neocortex.[37] The question of which of these lesions precedes the other has been a subject of debate within the AD research community, essentially since their first description in the early years of the twentieth century. Here again, animal models of the disease have been instrumental in resolving this question. For example, transgenic mice that develop age-related amyloid deposits in the brain have been used to

demonstrate that Aβ accumulation both precedes and accelerates the development of tangle pathology.[38,39]

ANIMAL MODELS OF HUMAN AGING AND AD

The study of AD as an age-related disorder raises two issues that are relevant to our discussion of model systems and the study of pathology in the aging brain. The utility of rodent models of both age-related disease and normal aging processes is that they can be modified to study the role of different genes and molecular pathways of interest in the context of mammalian biology and physiology (Table 2). Rodents are also among the shortest-lived mammals, as their brief lifespan (less than three years) and small size permit lifelong studies in a laboratory setting. However, these features do not make rodents the perfect model of brain aging and disease for human neurological disorders. With a few exceptions, the introduction of a single mutant gene or single mutation into a rodent model does not fully recapitulate the human disease from which it was identified. For these studies, therefore, we need to turn to other mammalian models. Although nonhuman primates are evolutionarily closer to humans, ethical and practical (commonly cost) considerations usually make these models prohibitive for most investigators. For the study of brain aging and neurological illness, a better suited intermediate between rodents and humans is the aging canine (Table 2).

Behavioral Aging in Rodents

The prime utility of rats and mice are their short lifespans and amenity to genetic manipulation. These two features make rodents ideal tools for dissection of the molecular pathways involved in the development of age-associated brain pathology. At the genetic level, rodents are similar to humans, with humans, mice, and rats sharing greater than 90% sequence identity in homologous genes.[40–42] This degree of similarity indicates a high likelihood that many physiological processes in humans and rodents are nearly identical. Although functional similarity extends to *Drosophila* and *C. elegans*,[43] rodents share far more physiological and anatomical parallels with humans in all body systems, including the nervous system.

Individual performance variability is a very consistent and largely task-independent change in rodent behavior with age.[44] In addition, age-related cognitive decline in rodents has been well documented.[45] For example, aged rats show deficits in remembering the location of a hidden platform in a water maze task, a measure of spatial memory.[45] Aging rodents display a steady decline in a variety of tests of sensorimotor ability,[46–49] independent of changes in cognitive function.[48] Spontaneous activity, such as locomotion and rearing, also decrease with age in both mice and rats.[50–53] Although the two are highly correlated, spontaneous activity is distinct from exploratory activity, which has a strong cognitive component.[54] It is not clear how much of the age-related decline in both sensorimotor function and activity are related to the development of pathology in the peripheral nervous system and skeletal musculature as opposed to dysfunction in the brain alone. Rather, it is likely that accumulating lesions in the central and peripheral nervous systems as well as the musculoskeletal system contribute.

Other noncognitive behaviors are known to change markedly with age. For example, when rodents are placed in a novel environment, they explore their surroundings and show a variety of related behaviors, such as rearing. These are thought to represent a form of displacement activity for lowering high arousal in response to mild stress.[55] In comparison to young control animals, older rats also show a large increase in grooming behavior under these circumstances.[56,57] As with age-related cognitive decline,[44] perhaps the most striking occurrence with noncognitive behaviors is an increase in the amount of individual variability.[48] Since some people show patterns of behavior that are very similar to those of younger animals, this is frequently referred to as "successful" vs. "pathologic" aging.

TABLE 2   Pathological Lesions in Animal Models of Aging

Lesion	Rat/Mouse	Representative Reference[a]	Dog	Representative Reference[a]
Total brain atrophy	None	60	12 years	124
Frontal cortex atrophy	None	60	8 years	124
Hippocampal atrophy	None	191	12 years	124
Neuron loss	None	191	13 years[a]	192
Neurogenesis decrease	>12 months	193	13 years[a]	133
β-Amyloid accumulation	None	5	8 years	194
Cerebral amyloid angiopathy	None	195	10 years	196
Oxidative damage increase	Linear increase across lifespan	197	Linear increase across lifespan	137
DNA damage increase	Linear increase across lifespan	197	>6 years	125
Substantia nigra neuromelanin accumulation	No change	105	>1 year	105
Lafora-like inclusion bodies	None	4	>8 years	198
Lipofuscin accumulation	Linear increase across lifespan	199	Linear increase across lifespan	164
Corpora amylacea accumulation	>5 months	119, 200	>13 years	119

[a] May occur at an earlier age, but younger animals not included in the study.

Brain Aging in Rodents

Presumably, age-related decline in cognitive function in rodents is related to an underlying neurobiological lesion. However, the aging rodent brain is remarkable in that, in the absence of disease and despite behavioral changes, there is strikingly little neuropathology. Some rodents (especially rats) continue to grow through most of their lifespan. The degree of continued growth is not consistent across all rat strains. The net result of this is that the brain gets proportionally larger in some strains (e.g., F344) but not others (e.g., Sprague–Dawley).[58–60] However, there are subtle neuroanatomic changes in the aged rodent brain that can be linked to behavioral decline. Aged rodents have brain lesions that are similar to those observed in humans (and dogs) (Table 2), but the accumulation of AD-like pathology does not occur spontaneously in rodents. In contrast to humans, little if any loss of actual neurons or synapses occurs in the rodent brain[44,61] despite the progressive increase in more subtle structural changes, including small declines in hippocampal volume.[62] Thus, dynamic replacement through continued neurogenesis may offset and mask an ongoing process of cell loss in multiple domains within the aged rodent brain. Although neurogenesis may decrease with aging, this effect does not appear to account for the cognitive decline that is commonly seen in older animals.[63–65]

There have been many documented changes at the molecular level in the aged rodent brain, although these can often be quite subtle. For example, decreased axonal transport can be observed in aging rats by magnetic resonance imaging (MRI).[66] A wide range of damage to both genomic[67] and mitochondrial[68] DNA accumulates with age in rodents. However, it is not always clear whether DNA alterations are a direct contributing cause to brain aging or another consequence of the aging process. The accumulation of damage may be related to a gradual decline in ability to maintain and repair DNA as the animal ages.[69] It is worth noting that a significant proportion of the DNA damage in the rodent brain appears to be oxidative in nature. Additionally, genetic modifications that markedly accelerate age-related phenotypes affect genes that are involved in both mediating oxidative stress and maintaining the integrity of DNA.[70–72] A consistent observation across many laboratories and using many different techniques is that rodent brain aging is accompanied by the progressive accumulation of oxidative damage to proteins, lipids, and DNA/RNA.[20,73]

Despite minimal accumulation of neuropathology, significant neuronal loss, or other overt changes, functional deficits are clearly a feature of the aged rodent brain. One of the advantages to studying rats and mice is that they are excellent model systems for in vivo electrophysiological studies, which are rarely possible in humans and challenging to accomplish in dogs. For example, direct measurement of neuron activity or changes in response to electrical stimula-tion is one of the most clear-cut tests of neuronal function. Electrophysiological studies clearly demonstrate that the hippocampal neurons of aged rats do not function as well as those of younger rodents. Consistent deficits are observed particularly with long-term potentiation, an electrophysiological correlate of memory.[45] In turn, these functional deficits may be related to changes in gene expression and modifications to signal transduction pathways.[74]

Rodents as Genetic Models of Aging

Relative ease of genetic manipulation is the strongest advantage to rodents over other models of mammalian aging. Insights from transgenic rodent models have been indispensible for our understanding of the aging brain in mammals.[75] However, there are several caveats with studying genetically altered mice (and occasionally, rats) as models of the human aging process in the central nervous system. Studies of the impact of a genetic acceleration on the aging process in the brain are appealing from an economic and logistical standpoint and potentially saves enormous amounts of time and resources. However, genetic manipulation necessarily requires the introduction of some combination of mutating, overexpressing, or removing (knocking out) a normal gene. The hidden problem with this approach is that alterations are being introduced into proteins and molecular pathways for which, frequently, functions have not been defined. The consequences of these mutations on normal physiology are unknown, or at best are only vaguely appreciated. The problem is compounded when a mutant protein is overexpressed to obtain the desired phenotype, given that this is usually accomplished through the use of an ectopic promoter designed to drive expression at relatively high levels. Hence, it is often the case that a protein of poorly understood function is being overexpressed at physiologically irrelevant levels in areas where it may not normally be found. In addition, the introduction of foreign DNA into the rodent genome causes disruption of the surrounding chromatin and may contribute to the phenotype observed.[76]

Consideration of background strain is also important, given that different strains of rats and mice have wide differences in health characteristics, lifespan, and behavior.[77] Inbred rodent strains have a reasonable amount of genetic uniformity, leading to better reproducibility between and within labs. However, this genetic homogeneity can sometimes lead to erroneous and misleading conclusions about the underlying causes of a particular neural phenotype. A related issue is that of strain-specific pathologies, which are often overlooked by many investigators. These can be relatively well known (e.g., many strains of albino rats and mice are nearly blind) or somewhat less so (e.g., aging C57BL/6 mice are highly susceptible to ulcerative dermatitis[78]; FVB mice may develop seizures secondary to spontaneous neural degeneration[79]), but often lead to mistakes in data interpretation. Hybrid strains are better for avoiding some of these problems, but the introduction of hybrid vigor into an existing genetically modified strain often leads to diminished expression of the phenotype.[80]

All of these issues tie into a central problem with the study of accelerated aging in genetically modified rodent models: it is not clear how much of the phenotype observed is due to dysfunction of normal physiological processes and how much is a bona fide acceleration of a much broader general aging process. For example, the various forms of the rare genetic condition associated with accelerated aging in humans, progeria, can be modeled in mice to various degrees.[81] However, progeria is associated with accelerated segmental aging (apparent aging of some physiological systems and not others) rather than an increase in organism-wide aging. A similar problem exists with the study of various strains of senescence-accelerated mice (SAMP8, etc.) developed more than 30 years ago.[82] In a broad sense, in rodent models of aging based on genetic and strain variables, it is very difficult to determine if changes in lifespan are due to actual changes in the rate of aging or are caused by a critical alteration in a physiological pathway that leads to changes in the health status of the animal.

Despite these caveats, the insights derived from genetically modified mice have been critically important in moving the fields of aging and age-related disease forward. This is especially true for establishing the order of events in age-related disease, a feat that is not possible from observing end-stage neuropathology in the human brain. Further, genetically modified mice are indispensible as low-cost organisms in which preclinical interventions can be tested before they are introduced to higher mammals or proceed to human clinical trials. An alternative approach is to turn to higher mammalian models of aging that naturally develop some aspects of the human aging phenotype. Many animals develop brain changes with age that are similar to humans, including cats, goats, sheep, nonhuman primates, and others.[83,84] Our laboratory has been studying dog aging for more than 15 years, as dogs show consistent cognitive changes with age as well as the progressive development of brain pathology that can be strikingly similar to human aging.

Cognitive Aging in Dogs

Studying aging in dogs provides unique research opportunities and benefits that can translate not only to human clinics but also to the companion animal clinic. There can be significant variation in longevity in dogs depending on their breed, their total body weight, and their environment. For example, larger breeds of dogs typically have shorter lifespans than do smaller breeds.[85–87] Cognitive aging in dogs has several key features, including the observations that not all types of learning and memory ability are equally affected by aging and that significant individual variability exists in the extent of neurological decline over time (i.e., not all old dogs become cognitively impaired). These cognitive features of canine brain aging are also very consistent with rodent and human aging.

We have developed laboratory tasks that use reward as a motivation to detect cognitive deterioration with age in dogs. Our testing procedures typically involve showing animals stimulus objects (e.g., wooden blocks differing in size) that they can push aside to reveal a food reward. To prevent animals from selecting the correct object based on smell, we also put a food reward in the incorrect object that cannot be obtained by the animal. Using these tests, we find that aged dogs show deficits in complex learning tasks, including size concept learning (being able to select objects correctly based on their size),[88,89] oddity discrimination learning (selecting one of three objects that is different from two other identical objects),[90,91] size discrimination learning (selecting the larger or smaller of two objects),[92,93] and spatial learning (selecting objects on their left or right based on which side a reward had been obtained previously).[94] Tasks sensitive to prefrontal cortex function, including reversal learning (dogs must learn to stop responding to a previously rewarded object) and visuospatial working memory (learning to select a position they recognize as having been seen before) also deteriorate with age.[92,93,95] Further, egocentric spatial learning and reversal, which measures the ability of animals to select a correct object based on their own body orientation, is also age-sensitive.[94] We have optimized landmark discrimination, a measure of spatial attention originally described in nonhuman primates, for use in dogs and have shown that this cognitive domain is also vulnerable to aging.[96,97] Interestingly, on simple learning tasks and procedural learning measures, aged dogs perform as well as do younger animals,[98] suggesting that a subset of cognitive function remains intact with age and that sensory deficits do not contribute significantly to increased error scores. However, as with rodents, aged dogs show decreases in spontaneous locomotion and exploratory behavior,[99–103] suggesting changes in sensorimotor function.

To test declining memory in aging dogs, we have used an object recognition task based on a procedure similar to that described in nonhuman primates.[104] This task measures the ability of dogs to select which of two objects they have not seen before (i.e., to remember having seen one object previously and subsequently to choose the novel object).[98] These age-dependent cognitive deficits are not linked to obvious sensory deficits or locomotor impairment.[99] The neurobiological substrates for locomotor or sensory decline in dogs may be related to increased neuromelanin[105] and ubiquitin-positive degenerate neurites[106] in the substantia nigra, increased gliosis in subcortical nuclei,[107] and decreased numbers of olfactory cells[108] and inner ear cells.[109] Perhaps the most useful age-sensitive task we have used is a spatial memory task, in which dogs are required to remember the location of an object and then respond to a different location during the test trial. Earlier results suggested that the task was age-sensitive and that old dogs overall would receive higher error scores.[110] However, we have now identified the time course of progressive cognitive decline and have found that deterioration in spatial ability occurs early in the aging process, between 6 and 7 years of age in beagles.[95] Thus, cognitive decline in aged dogs depends on the type of function assessed and involves memory, similar to other animal models and aging humans.

As with human cognitive aging, we observe increased individual variability in error scores in dogs beginning in middle age,[111] with the most variability among old animals. Using spatial learning and memory tasks, we are able to identify three groups of old dogs, categorized as (1) successful agers, (2) impaired dogs that perform more poorly than young dogs but can learn and remember information with sufficient training, and (3) severely impaired dogs that fail to learn the task or show significant memory deficits.[83] This clustering of aged dogs on the basis of cognitive ability is consistent with cognitive aging in rats, nonhuman primates[112–116] and humans.[9]

Neurobiological Changes Underlying Cognitive Aging in Dogs

Several morphologic features of aging in the dog brain are similar to those observed in normal aging in both rodents and in humans and in early/mild AD[83,117–120] and correlate with cognitive decline in descriptive studies (Table 2). In vivo noninvasive imaging of the canine brain shows significant cortical atrophy, neuron loss within the hippocampus, reduced neurogenesis, and progressive oxidative damage. In contrast to the rodent brain, the aged dog brain does accumulate Aβ in plaques within the neuropil and also within cerebral vessels in patterns that resemble those observed in aged human brain.

Cortical Atrophy and Neuron Loss or Neurogenesis

In vivo brain imaging studies show that cortical atrophy[121] and ventricular widening[121–123] occur with age in dogs. More recent MRI studies suggest differential vulnerabilities of specific brain areas to aging. For example, in aging dogs the prefrontal cortex loses tissue volume at an earlier age than does the hippocampus.[124] Furthermore, more extensive prefrontal cortical volume loss is associated with poorer cognition.[124] However, the hippocampus shows progressive atrophy, reaching significantly lower volumes in dogs older than 11 years of age than in younger adult dogs. The extent of cortical atrophy is associated significantly with cognition; animals with more extensive atrophy perform more poorly on tests of learning and memory,[125] a performance similar to that of elderly humans with dementia.[126,127] Atrophy, par-

ticularly in the hippocampus, may result as a consequence of neuron loss, as reported in normal human brain aging,[13,128] although more extensive neuronal loss occurs in AD.[129,130] Furthermore, the pattern of neuron loss is distinguishable between normal human aging (where neurons are generally lost in the hilus of the hippocampal dentate gyrus in the hippocampus[13,14]) and AD (where neuron loss is concentrated in hippocampal area CA1[14,129–131] and in the subiculum[130]). When neurons were counted using unbiased stereological methods within individual subfields of the hippocampus of young (age range 3.4 to 4.5 years) and old (age range 13.0 to 15.0 years) dogs, the hilus of the dentate gryus showed a significant loss (∼30%) in the aged brain compared to young dogs.[132] Although the sample size was relatively small ($n = 5$ young, $n = 5$ old) and there was individual variability in numbers of hilar neurons in aged animals, only one aged dog (14.2 years) had hilar neuron counts within the range of those of the young dogs. In preliminary studies this variability related to cognitive function; dogs with higher numbers of hippocampal neurons performed a size discrimination task with fewer errors.[132] A much larger study is required to determine if old dogs with larger neuron losses represent severely impaired animals (much like AD) and if neuron losses first appear when aged dogs are functionally impaired relative to the performance of unimpaired aged dogs. Reduced neurogenesis in the mature brain may contribute to age-associated cognitive decline, as there could be slower replacement of dying neurons. In counts of new neurons in the hippocampus of aged beagles, a 90 to 95% decline in neurogenesis was measured with age in dogs.[133] Further, the degree of neurogenesis was correlated with cognitive function; animals with fewer new neurons had higher error scores in measures of learning and memory and poorer learning ability.[134] Similar reductions in neurogenesis in aged dogs have been reported in other laboratories.[135,136]

Oxidative Damage

In dog brain, the accumulation of carbonyl groups increases with age.[137,138] This is a measure of oxidative damage to proteins and is associated with reduced endogenous antioxidant enzyme activity or protein levels, including reduced levels of glutamine synthetase and superoxide dismutase (SOD).[137,139–141] In several studies, a relation between age and increased oxidative damage has been inferred by measuring the end products of lipid peroxidation (oxidative damage to lipids), including 4-hydroxynonenal (4HNE),[125,141–143] lipofuscin,[125] lipofuscin-like pigments (LFP),[142–143] or malondialdehyde.[137] Additionally, evidence of increased oxidative damage to DNA or RNA (8OHdG) in aged dog brain has been reported.[125,144] Oxidative damage may also be associated with behavioral decline in dogs. Increased oxidative end products (lipofuscin-like pigment and protein carbonyls) in aged companion dog brain are correlated with a greater severity of behavioral changes due to cognitive dysfunction.[125,143,145] Similarly, in our own studies of aging beagles, higher protein oxidative damage (3-nitrotyrosine) and lower endogenous antioxidant capacity (SOD and glutathione-*S*-transferase activities) are associated with poorer prefrontal-dependent and spatial learning.[140] Thus, aged dogs and rodents develop oxidative damage that is consistent with that observed in humans with age-related neurological dysfunction.

Aβ Pathology

Neuron loss and cortical atrophy in vulnerable brain regions of the aged dog may be due to the accumulation of abnormal proteins. Indeed, the observation of Aβ-containing lesions with identical amino acid sequence to that of humans[146,147] first stimulated interested in the use of the dog to model human aging and disease (Figure 1B and D).[118] Our work and the work of others demonstrate that specific brain regions show differential accumulation of Aβ in the aging dog brain, paralleling some reports in the aged human brain.[36,147–154] When cortical regions are sampled for Aβ deposition, each region shows a different age of Aβ onset.[152] Aβ deposition occurs earliest in the prefrontal cortex of the dog and later in temporal and occipital cortex, which is similar to previous reports in human brain aging.[36] It is critical to note that initial Aβ deposits occur in a three- to four-year time window in dogs, starting between 8 and 9 years of age, suggesting that longitudinal studies for evaluating interventions to slow or halt Aβ are feasible. The extent of Aβ plaque deposition in the dog brain is linked to the severity of cognitive deficits.[92,125,155]

Age and cognitive status can predict Aβ accumulation in discrete brain structures. For example, dogs with prefrontal cortex-dependent reversal learning deficits show significantly higher amounts of Aβ in this brain region.[92,156] On the other hand, dogs who did poorly in a size- discrimination learning task show large amounts of Aβ deposition in the entorhinal cortex.[92] As in laboratory beagles, the extent of Aβ plaques varies as a function of age in companion dogs.[125,143,157] Further, the extent of Aβ plaques correlates with behavior changes that persist even if age is removed as a covariate.[125,155] Aβ plaques in the parietal lobe of aged companion animals correlate with behavioral changes related to appetite, drinking, incontinence, day and night rhythm, social behavior (interaction with owners and other dogs; personality), orientation, perception, and memory.[125] Thus, similar to humans, cognitive decline in aged dogs correlates with the age-related accumulation of Aβ.[92,158–165]

A common type of pathology observed in both normal human brain aging and particularly in AD is the presence of CAA, which is characterized by the accumulation of Aβ in the walls of cerebral vessels.[166–168] CAA does not occur in the aged rat brain but was first observed in the dog brain[169] as early as 1956 and was subsequently confirmed (Figure 1B).[151] Vascular and perivascular abnormalities and CAA pathology are frequently found in aged dogs.[148,149,170–177] CAA may compromise the blood–brain barrier, impair vascular function,[178] and cause microhemorrhages.[175,176,179] The distribution of CAA in dog brain is similar to that in humans, with particular vulnerability in the occipital cortex.[167] The extent of CAA in aged dog brains can correlate with clinical signs of cognitive dysfunction in companion dogs.[155] Thus, aged dogs develop cerebrovascular aging abnormalities that are consistent with those reported in humans.

ENVIRONMENTAL NEUROTOXICANTS AS POTENTIAL CONTRIBUTORS TO NEURODEGENERATIVE DISEASE

As mentioned above, genetic causes of age-associated neurodegenerative diseases are relatively rare. However, environmental stimuli or events such as acute or long-term exposure to natural toxins, heavy metals, or chemicals may be important for causing or enhancing the vulnerability of the aging brain to degenerative disease. For example, acute onset Parkinson's disease (PD) developed in a drug abuser who had been injected meperidine containing the derivative 1-methyl-4-phenyl-1,2,3,6-tetrahydropyridine (MPTP).[180] At autopsy, a severe loss of dopaminergic neurons in the substantia nigra was reported. Additional reports of acute PD following inadvertent MPTP exposure[181] has led to the use of MPTP as a tool for eliciting PD in nonhuman primate models.[182] Similarly, rotenone, a chemical used in insecticides and pesticides, is also a potent neurotoxicant that can elicit PD in rodents.[183] Many other chemicals, including paraquat and maneb, have also been linked to PD in humans.[184] With the exception of MPTP, the typical clinical onset of chemically induced PD appears to follow exposures that have occurred over the lifespan. That said, PD late in life may result from either progressive accumulation of neural damage over time or from the long-lasting effects after acute exposures, or even from both scenarios acting additively.

Similarly, in AD, several environmental stimuli, such as exposure to heavy metals or even factors including dietary habits, may lead to altered (enhanced or reduced) risk of developing the disease. Aluminum has long been linked to AD, beginning with early reports of increased metal concen-trations within AD brains, particularly in regions vulnerable to neurofibrillary tangles.[185] Since then, multiple epidemiologic studies have been performed analyzing the presence and amount of aluminum in drinking water and the risk of developing AD, with variably positive and negative associations.[24] The current consensus is that there is no causal link between aluminum accumulation and the risk of developing AD. However, heavy metals may exacerbate the development of AD pathology, as zinc, copper, aluminum, and iron are all found within Aβ plaques in the brain[186] and may enhance the formation of plaques or other assembly states of Aβ (such as oligomers) that are known to be neurotoxic.[187] Based on the latter mechanistic association between Aβ and metals, clinical trials have been proposed to explore the utility of metal chelators as a possible treatment for AD.[188]

In short, many environmental influences may affect the risk of developing an age-related neurodegenerative disease. These factors include not only toxicants (such as pollution or pesticides) but also other medical conditions (obesity, malnutrition, infection).[189,190] At this time, however, the extent of the relationship between neurotoxicants, environmental parameters, and the nervous system's vulnerability to disease has not been clearly established for neurodegenerative conditions of humans.

SUMMARY

Cognitive and behavioral aging in humans, dogs, and rodents share two common features: (1) increasing individual variability with age and (2) selective vulnerability of specific neural functions with advancing age. Only a subset of humans, rats, mice, and dogs show impairments in cognition with age, leaving a subset that retain neural function as well as younger cohorts. In addition, not all types of learning and memory are equally vulnerable to the aging process in all of these species. For example, certain types of memory (e.g., spatial ability) are consistently age-sensitive across species. In terms of brain pathology, the rodent shows the least overlap with human brain aging, but is particularly well suited for electrophysiology studies to evaluate neuronal function and offer the unique opportunity to use genetic modifications to test the role of specific proteins or molecular pathways in the aging process. In contrast, aged dogs have similar brain aging features as humans and naturally develop Aβ both in plaques within the neuropil and in the cerebral vasculature. Thus, each animal model of human brain aging has unique characteristics and is well suited for addressing specific questions related to the study of brain aging.

Acknowledgments

The authors appreciate helpful suggestions during the preparation of this chapter from Amy Dowling (University of Kentucky) and assistance from Peter Nelson (University of Kentucky) for Table 1.

REFERENCES

1. Baba M, et al. Aggregation of alpha-synuclein in Lewy bodies of sporadic Parkinson's disease and dementia with Lewy bodies. *Am J Pathol.* 1998;152:879–884.
2. McKeith I, et al. Dementia with Lewy bodies. *Lancet Neurol.* 2004;3:19–28.
3. Zarow C, et al. Understanding hippocampal sclerosis in the elderly: epidemiology, characterization, and diagnostic issues. *Curr Neurol Neurosci Rep.* 2008;8:363–370.
4. Cavanagh JB. Corpora-amylacea and the family of polyglucosan diseases. *Brain Res Rev.* 1999;29:265–295.
5. Woodruff-Pak DS. Animal models of Alzheimer's disease: therapeutic implications. *J Alzheimer's Dis.* 2008;15:507–521.
6. Crook T, et al. Diagnosis and assessment of age-associated memory impairment. *Clin Neuropharmacol.* 1990;13:S81–S91.
7. Wilkniss SM, et al. Age-related differences in an ecologically based study of route learning. *Psychol Aging.* 1997;12:372–375.
8. Craik FI. Age differences in human memory. In: Birren J, Schaie K, eds. *Handbook of the Psychology of Ageing.* New York: Van Nostrand Reinhold; 1977:384–420.
9. Rowe JW, Kahn RL. Human aging: usual and successful. *Science.* 1987;237:143–149.
10. Raz N, et al. Neuroanatomical correlates of cognitive aging: evidence from structural magnetic resonance imaging. *Neuropsychology.* 1998;12:95–114.
11. Raz N, et al. Aging, sexual dimorphism, and hemispheric asymmetry of the cerebral cortex: replicability of regional differences in volume. *Neurobiol Aging.* 2004;25:377–396.
12. Peters A, et al. Feature article: Are neurons lost from the primate cerebral cortex during normal aging? *Cereb Cortex.* 1998;8:295–300.
13. West MJ. Regionally specific loss of neurons in the aging human hippocampus. *Neurobiol Aging.* 1993;14:287–293.
14. West MJ, et al. Differences in the pattern of hippocampal neuronal loss in normal ageing and Alzheimer's disease. *Lancet.* 1994;344:769–772.
15. Eriksson PS, et al. Neurogenesis in the adult human hippocampus. *Nat Med.* 1998;4:1313–1317.
16. Bhardwaj RD, et al. Neocortical neurogenesis in humans is restricted to development. *Proc Natl Acad Sci USA.* 2006;103:12564–12568.
17. Poon HF, et al. Free radicals and brain aging. *Clin Geriatr Med.* 2004;20:329–359.
18. Ding Q, et al. Oxidative damage, protein synthesis, and protein degradation in Alzheimer's disease. *Curr Alzheimer Res.* 2007;4:73–79.
19. Ding Q, Keller JN. Proteosomes and proteosome inhibition in the central nervous system. *Free Radic Biol Med.* 2001;31:574–584.
20. Stadtman ER. Protein oxidation and aging. *Science.* 1992;257:1220–1224.

21. Stadtman ER, Berlett BS. Reactive oxygen-mediated protein oxidation in aging and disease. *Chem Res Toxicol.* 1997; 10:485–494.

22. Butterfield DA, Sultana R. Redox proteomics identification of oxidatively modified brain proteins in Alzheimer's disease and mild cognitive impairment: insights into the progression of this dementing disorder. *J Alzheimer's Dis.* 2007;12:61–72.

23. Gauthier E, et al. Environmental pesticide exposure as a risk factor for Alzheimer's disease: a case–control study. *Environ Res.* 2001;86:37–45.

24. Shcherbatykh I, Carpenter DO. The role of metals in the etiology of Alzheimer's disease. *J Alzheimer's Dis.* 2007; 11:191–205.

25. Wu J, Basha MR, Zawia NH. The environment, epigenetics and amyloidogenesis. *J Mol Neurosci.* 2008;34:1–7.

26. Wilson RS, et al. Participation in cognitively stimulating activities and risk of incident Alzheimer disease. *JAMA.* 2002;287:742–748.

27. Barberger-Gateau P, et al. Dietary patterns and risk of dementia: the three-city cohort study. *Neurology.* 2007;69: 1921–1930.

28. Engelhart MJ, et al. Dietary intake of antioxidants and risk of Alzheimer disease. *JAMA.* 2002;287:3223–3229.

29. Warsama Jama J, et al. Dietary antioxidants and cognitive function in a population-based sample of older persons. *Am J Epidemiol.* 1996;144:275–280.

30. Hardy J. Amyloid, the presenilins and Alzheimer's disease. *Trends Neurosci.* 1997;20:154–159.

31. Selkoe DJ. Alzheimer's disease: genes, proteins, and therapy. *Physiol Rev.* 2001;81:741–766.

32. McGeer PL, et al. Pathological proteins in senile plaques. *Tohoku J Exp. Med.* 1994;174:269–277.

33. Lee VM, et al. Neurodegenerative tauopathies. *Annu Rev Neurosci.* 2001;24:1121–1159.

34. Maurage CA, et al. Similar brain tau pathology in DM2/PROMM and DM1/Steinert disease. *Neurology.* 2005; 65:1636–1638.

35. Braak H, Braak E. Neuropathological stageing of Alzheimer-related changes. *Acta Neuropathol.* 1991;82:239–259.

36. Thal DR, et al. Phases of Abeta-deposition in the human brain and its relevance for the development of AD. *Neurology.* 2002;58:1791–1800.

37. Braak H, Braak E. Staging of Alzheimer's disease–related neurofibrillary changes. *Neurobiol Aging.* 1995;16:271–284.

38. Gotz J, et al. Formation of neurofibrillary tangles in P301l tau transgenic mice induced by Abeta 42 fibrils. *Science.* 2001;293:1491–1495.

39. Lewis J, et al. Enhanced neurofibrillary degeneration in transgenic mice expressing mutant tau and APP. *Science.* 2001;293:1487–1491.

40. Gibbs RA, et al. Genome sequence of the brown Norway rat yields insights into mammalian evolution. *Nature.* 2004; 428:493–521.

41. Lander ES, et al. Initial sequencing and analysis of the human genome. *Nature.* 2001;409:860–921.

42. Waterston RH, et al. Initial sequencing and comparative analysis of the mouse genome. *Nature.* 2002;420:520–562.

43. Kim SK. Common aging pathways in worms, flies, mice and humans. *J Exp Biol.* 2007;210:1607–1612.

44. Bizon JL, Nicolle MM. Rat models of age-related cognitive decline. In: Conn PM, ed. *Handbook of Models for Human Aging.* San Diego, CA: Academic Press; 2006:379–391.

45. Erickson CA, Barnes CA. The neurobiology of memory changes in normal aging. *Exp Gerontol.* 2003;38:61–69.

46. Dean RL, et al. Age-related differences in behavior across the life span of the C57BL/6J mouse. *Exp Aging Res.* 1981;7: 427–451.

47. Janicke B, et al. Motor performance achievements in rats of different ages. *Exp Gerontol.* 1983;18:393–407.

48. Gage FH, et al. Age-related impairments in spatial memory are independent of those in sensorimotor skills. *Neurobiol Aging.* 1989;10:347–352.

49. Murphy MP, et al. A simple and rapid test of sensorimotor function in the aged rat. *Neurobiol Learn Mem.* 1995;64: 181–186.

50. Gage FH, et al. Spatial learning and motor deficits in aged rats. *Neurobiol Aging.* 1984;5:43–48.

51. Goodrick CL. Free exploration and adaptation within an open field as a function of trials and between-trial-interval for mature-young, mature-old, and senescent Wistar rats. *J Gerontol.* 1971;26:58–62.

52. Sprott RL, Eleftheriou BE. Open-field behavior in aging inbred mice. *Gerontologia.* 1970;20:155–162.

53. Willig F, et al. Short-term memory, exploration and locomotor activity in aged rats. *Neurobiol Aging.* 1987;8:393–402.

54. Lalonde R. The neurobiological basis of spontaneous alternation. *Neurosci Biobehav Rev.* 2002;26:91–104.

55. Jolles J, et al. ACTH-induced excessive grooming in the rat: the influence of environmental and motivational factors. *Horm Behav.* 1979;12:60–72.

56. Kametani H, et al. Increased novelty-induced grooming in aged rats: a preliminary observation. *Behav Neural Biol.* 1984;42:73–80.

57. Continella G, et al. Quantitative alteration of grooming behavior in aged male rats. *Physiol Behav.* 1985;35:839–841.

58. Roberts J, Goldberg PB. Some aspects of the central nervous system of the rat during aging. *Exp Aging Res.* 1976; 2: 531–542.

59. Eurich RE, Lindner J. Body weights, absolute and relative organ weights by maturation and ageing (with sexual differences), and their importance as measures of reference for metabolic investigations. *Z Gerontol.* 1984;17:60–68.

60. DeKosky ST, et al. Strain differences and laminar localization of structural neurochemical changes in aging rat. *Neurobiol Aging.* 1985;6:277–286.

61. Scheff SW, et al. Strain comparison of synaptic density in hippocampal CA1 of aged rats. *Neurobiol Aging.* 1985;6: 29–34.

62. Driscoll I, et al. The aging hippocampus: a multi-level analysis in the rat. *Neuroscience.* 2006;139:1173–1185.

63. Bizon JL, Gallagher M. Production of new cells in the rat dentate gyrus over the lifespan: relation to cognitive decline. *Eur J Neurosci.* 2003;18:215–219.

64. Bizon JL, Gallagher M. More is less: neurogenesis and age-related cognitive decline in Long-Evans rats. *Sci Aging Knowledge Environ.* 2005:re2.

65. Bizon JL, et al. Neurogenesis in a rat model of age-related cognitive decline. *Aging Cell.* 2004;3:227–234.

66. Cross DJ, et al. Age-related decrease in axonal transport measured by MR imaging in vivo. *Neuroimage.* 2008; 39:915–926.

67. Schumacher B, et al. Age to survive: DNA damage and aging. *Trends Genet.* 2008;24:77–85.

68. Trifunovic A. Mitochondrial DNA and ageing. *Biochim Biophys Acta.* 2006;1757:611–617.

69. Ruzankina Y, et al. Replicative stress, stem cells and aging. *Mech Ageing Dev.* 2008;129:460–466.

70. Oberdoerffer P, et al. SIRT1 redistribution on chromatin promotes genomic stability but alters gene expression during aging. *Cell.* 2008;135:907–918.

71. de Magalhaes JP, Faragher RG. Cell divisions and mammalian aging: integrative biology insights from genes that regulate longevity. *Bioessays.* 2008;30:567–578.

72. Kuro-o M. Klotho as a regulator of oxidative stress and senescence. *Biol Chem.* 2008;389:233–241.

73. Beckman KB, Ames BN. The free radical theory of aging matures. *Physiol Rev.* 1998;78:547–581.

74. Gallagher M, et al. Effects of aging on the hippocampal formation in a naturally occuring animal model of mild cognitive impairment. *Exp Gerontol.* 2003;38:71–77.

75. Bartke A. New findings in gene knockout, mutant and transgenic mice. *Exp Gerontol.* 2008;43:11–14.

76. Matthaei KI. Genetically manipulated mice: a powerful tool with unsuspected caveats. *J Physiol.* 2007;582:481–488.

77. Ingram DK, Jucker M. Developing mouse models of aging: a consideration of strain differences in age-related behavioral and neural parameters. *Neurobiol Aging.* 1999; 20:137–145.

78. Andrews AG, et al. Immune complex vasculitis with secondary ulcerative dermatitis in aged C57BL/6NNia mice. *Vet Pathol.* 1994;31:293–300.

79. Goelz MF, et al. Neuropathologic findings associated with seizures in FVB mice. *Lab Anim Sci.* 1998;48:34–37.

80. Taft RA, et al. Know thy mouse. *Trends Genet.* 2006;22: 649–653.

81. Osorio FG, et al. Accelerated ageing: from mechanism to therapy through animal models. *Transgen Res.* 2009; 18:7–15.

82. Nomura Y, Takeda T. *The Senescence Accelerated Mouse (SAM).* Amsterdam: Elsevier Health Sciences; 2004: 448.

83. Head E, et al. Neurobiological models of aging in the dog and other vertebrate species. In: Hof P, Mobbs C, eds. *Functional Neurobiology of Aging.* San Diego, CA: Academic Press; 2001:457–468.

84. Peters A, et al. Neurobiological bases of age-related cognitive decline in the rhesus monkey. *J Neuropathol Exp Neurol.* 1996;55:861–874.

85. Greer KA, et al. Statistical analysis regarding the effects of height and weight on life span of the domestic dog. *Res Vet Sci.* 2007;82:208–214.

86. Galis F, et al. Do large dogs die young? *J Exp Zoolog B.* 2007;308:119–126.

87. Patronek GJ, et al. Comparative longevity of pet dogs and humans: implications for gerontology research. *J Gerontol A.* 1997;52:B171–B178.

88. Tapp PD, et al. Concept abstraction in the aging dog: development of a protocol using successive discrimination and size concept tasks. *Behav Brain Res.* 2004;153:199–210.

89. Tapp D, et al. An antioxidant enriched diet improves concept learning in aged dogs. Abstract 836.13. Presented at 2003 Society for Neuroscience; Nov. 12, 2003, New Orleans, LA.

90. Milgram NW, et al. Dietary enrichment counteracts age-associated cognitive dysfunction in canines. *Neurobiol Aging.* 2002;23:737–745.

91. Cotman CW, et al. Brain aging in the canine: a diet enriched in antioxidants reduces cognitive dysfunction. *Neurobiol Aging.* 2002;23:809–818.

92. Head E, et al. Visual-discrimination learning ability and beta-amyloid accumulation in the dog. *Neurobiol Aging.* 1998; 19: 415–425.

93. Tapp PD, et al. Size and reversal learning in the beagle dog as a measure of executive function and inhibitory control in aging. *Learn Mem.* 2003;10:64–73.

94. Christie LA, et al. A comparison of egocentric and allocentric age-dependent spatial learning in the beagle dog. *Prog Neuropsychopharmacol Biol Psychiatry.* 2005;29:361–369.

95. Studzinski CM, et al. Visuospatial function in the beagle dog: an early marker of cognitive decline in a model of human aging and dementia. *Neurobiol Learn Mem.* 2006;86: 197–204.

96. Milgram NW, et al. Landmark discrimination learning in the dog: effects of age, an antioxidant fortified diet, and cognitive strategy. *Neurosci Biobehav Rev.* 2002;26:679–695.

97. Milgram NW, et al. Landmark discrimination learning in the dog. *Learn Mem.* 1999;6:54–61.

98. Milgram NW, et al. Cognitive functions and aging in the dog: acquisition of nonspatial visual tasks. *Behav Neurosci.* 1994;108:57–68.

99. Head E, et al. Open field activity and human interaction as a function of age and breed in dogs. *Physiol Behav.* 1997;62: 963–971.

100. Siwak C, et al. A comparison of open field and home cage behavior in young and aged dogs. International Behavioral Neuroscience Society Meeting, Nancy, France: 1999.

101. Siwak CT, et al. Effect of age and level of cognitive function on spontaneous and exploratory behaviors in the beagle dog. *Learn Mem.* 2001;8:317–325.

102. Siwak CT, et al. Age-dependent decline in locomotor activity in dogs is environment specific. *Physiol Behav.* 2002;75: 65–70.

103. Siwak CT, et al. Locomotor activity rhythms in dogs vary with age and cognitive status. *Behav Neurosci.* 2003;117:813–824.

104. Mishkin M, Delacour J. An analysis of short-term visual memory in the monkey. *J Exp. Psychol Animal Behav Proc.* 1975;1:326–334.

105. DeMattei M, et al. Neuromelanic pigment in substantia nigra neurons of rats and dogs. *Neurosci Lett.* 1986;72:37–42.

106. Uchida K, et al. Age-related histological changes in the canine substantia nigra. *J Vet Med Sci.* 2003;65:179–185.

107. Shimada A, et al. An immunohistochemical and ultrastructural study on age-related astrocytic gliosis in the central nervous system of dogs. *J Vet Med Sci.* 1992;54:29–36.

108. Hirai T, et al. Age-related changes in the olfactory system of dogs. *Neuropathol Appl Neurobiol.* 1996;22:531–539.

109. Le T, Keithley EM. Effects of antioxidants on the aging inner ear. *Hear Res.* 2007;226:194–202.

110. Head E, et al. Spatial learning and memory as a function of age in the dog. *Behav Neurosci.* 1995;109:851–888.

111. Adams B, et al. The canine as a model of human cognitive aging: recent developments. *Prog Neuropsychopharmacol Biol Psychiatry.* 2002;24:675–692.

112. Baxter MG, Gallagher M. Neurobiological substrates of behavioral decline: models and data analytic strategies for individual differences in aging. *Neurobiol Aging.* 1996; 17:491–495.

113. Markowska AL, et al. Individual differences in aging: behavioral and neurobiological correlates. *Neurobiol Aging.* 1989; 10:31–43.

114. Rapp PR, Amaral DG. Recognition memory deficits in a subpopulation of aged monkeys resemble the effects of medial temporal lobe damage. *Neurobiol Aging.* 1991;12:481–486.

115. Rapp PR. Neuropsychological analysis of learning and memory in aged nonhuman primates. *Neurobiol Aging.* 1993; 14:627–629.

116. Rapp PR, et al. New directions for studying cognitive decline in old monkeys. *Semin Neurosci.* 1994;6:369–377.

117. Cummings BJ, et al. The canine as an animal model of human aging and dementia. *Neurobiol Aging.* 1996;17:259–268.

118. Wisniewski HM, et al. Aged dogs: an animal model to study beta-protein amyloidogenesis. In: Maurer PR, Beckman H, eds. *Alzheimer's Disease Epidemiology, Neuropathology, Neurochemistry and Clinics.* New York: Springer-Verlag; 1990:151–167.

119. Borras D, et al. Age-related changes in the brain of the dog. *Vet Pathol.* 1999;36:202–211.

120. Tapp PD, Siwak C. The canine model of human brain aging: cognition, behavior and neuropathology. In: Conn PM, ed. *Handbook of Models of Human Aging.* New York: Elsevier; 2006:415–434.

121. Su M-Y, et al. MR Imaging of anatomic and vascular characteristics in a canine model of human aging. *Neurobiol Aging.* 1998;19:479–485.

122. Kimotsuki T, et al. Changes of magnetic resonance imaging on the brain in beagle dogs with aging. *J Vet Med Sci.* 2005; 67:961–967.

123. Gonzalez-Soriano J, et al. Age-related changes in the ventricular system of the dog brain. *Ann Anat.* 2001;183:283–291.

124. Tapp PD, et al. Frontal lobe volume, function, and beta-amyloid pathology in a canine model of aging. *J Neurosci.* 2004;24:8205–8213.

125. Rofina JE, et al. Cognitive disturbances in old dogs suffering from the canine counterpart of Alzheimer's disease. *Brain Res.* 2006;1069:216–226.

126. Du AT, et al. White matter lesions are associated with cortical atrophy more than entorhinal and hippocampal atrophy. *Neurobiol Aging.* 2005;26:553–559.

127. Ezekiel F, et al. Comparisons between global and focal brain atrophy rates in normal aging and Alzheimer disease: boundary shift integral versus tracing of the entorhinal cortex and hippocampus. *Alzheimer Dis Assoc Disord.* 2004;18: 196–201.

128. Simic G, et al. Volume and number of neurons of the human hippocampal formation in normal aging and Alzheimer's disease. *J Comp Neurol.* 1997;379:482–494.

129. West MJ, et al. The CA1 region of the human hippocampus is a hot spot in Alzheimer's disease. *Ann NY Acad Sci.* 2000;908:255–259.

130. Bobinski M, et al. Relationships between regional neuronal loss and neurofibrillary changes in the hippocampal formation and duration and severity of Alzheimer disease. *J Neuropathol Exp Neurol.* 1997;56:414–420.

131. Price JL, et al. Neuron number in the entorhinal cortex and CA1 in preclinical Alzheimer disease. *Arch Neurol.* 2001;58:1395–402.

132. Siwak-Tapp CT, et al. Region specific neuron loss in the aged canine hippocampus is reduced by enrichment. *Neurobiol Aging.* 2008;29:521–528.

133. Siwak-Tapp CT, et al. Neurogenesis decreases with age in the canine hippocampus and correlates with cognitive function. *Neurobiol Learn Mem.* 2007;88:249–259.

134. Hwang IK, et al. Differences in doublecortin immunoreactivity and protein levels in the hippocampal dentate gyrus between adult and aged dogs. *Neurochem Res.* 2007;32: 1604–1609.

135. Pekcec A, et al. Effect of aging on neurogenesis in the canine brain. *Aging Cell.* 2008;7:368–374.

136. Head E, et al. Oxidative damage increases with age in a canine model of human brain aging. *J Neurochem.* 2002;82:375–381.

137. Skoumalova A, et al. The role of free radicals in canine counterpart of senile dementia of the Alzheimer type. *Exp Gerontol.* 2003;38:711–719.

138. Kiatipattanasakul W, et al. Immunohistochemical detection of anti-oxidative stress enzymes in the dog brain. *Neuropathology.* 1997;17:307–312.

139. Opii WO, et al. Proteomic identification of brain proteins in the canine model of human aging following a long-term treatment with antioxidants and a program of behavioral enrichment: relevance to Alzheimer's disease. *Neurobiol Aging.* 2008;29:51–70.

140. Hwang IK, et al. Differences in lipid peroxidation and Cu, Zn-superoxide dismutase in the hippocampal CA1 region

between adult and aged dogs. *J Vet Med Sci.* 2008; 70: 273–277.

141. Papaioannou N, et al. Immunohistochemical investigation of the brain of aged dogs: I. Detection of neurofibrillary tangles and of 4-hydroxynonenal protein, an oxidative damage product, in senile plaques. *Amyloid: J Protein Fold Disord.* 2001;8:11–21.

142. Rofina JE, et al. Histochemical accumulation of oxidative damage products is associated with Alzheimer-like pathology in the canine. *Amyloid.* 2004;11:90–100.

143. Cotman CW, Head E. The canine (dog) model of human aging and disease: dietary, environmental and immunotherapy approaches. *J Alzheimer's Dis.* 2008;15:685–707.

144. Skoumalova A, et al. The role of free radicals in canine counterpart of senile dementia of the Alzheimer type. *Exp Gerontol.* 2003;38:711–719.

145. Johnstone EM, et al. Conservation of the sequence of the Alzheimer's disease amyloid peptide in dog, polar bear and five other mammals by cross-species polymerase chain reaction analysis. *Brain Res Mol Brain Res.* 1991; 10: 299–305.

146. Selkoe DJ, et al. Conservation of brain amyloid proteins in aged mammals and humans with Alzheimer's disease. *Science.* 1987;235:873–877.

147. Giaccone G, et al. Cerebral preamyloid deposits and congophilic angiopathy in aged dogs. *Neurosci Lett.* 1990;114: 178–183.

148. Ishihara T, et al. Immunohistochemical and immunoelectron microscopial characterization of cerebrovascular and senile plaque amyloid in aged dogs' brains. *Brain Res.* 1991;548: 196–205.

149. Wisniewski HM, et al. Senile plaques and cerebral amyloidosis in aged dogs. *Lab Invest.* 1970;23:287–296.

150. Wisniewski HM, et al. Aged dogs: an animal model to study beta-protein amyloidogenesis. In: Maurer PR, Beckman H, eds. *Alzheimer's Disease Epidemiology, Neuropathology, Neurochemistry and Clinics.* New York: Springer-Verlag; 1990:151–167.

151. Head E, et al. Region-specific age at onset of beta-amyloid in dogs. *Neurobiol Aging.* 2000;21:89–96.

152. Braak H, Braak E. Neuropathological stageing of Alzheimer-related changes. *Acta Neuropathol.* 1991;82:239–259.

153. Braak H, et al. Staging of Alzheimer-related cortical destruction. *Rev Clin Neurosci.* 1993;33:403–408.

154. Colle MA, et al. Vascular and parenchymal Aβ deposition in the aging dog: correlation with behavior. *Neurobiol Aging.* 2000;21:695–704.

155. Cummings BJ, et al. Beta-amyloid accumulation correlates with cognitive dysfunction in the aged canine. *Neurobiol Learn Mem.* 1996;66:11–23.

156. Rofina J, et al. Canine counterpart of senile dementia of the Alzheimer type: amyloid plaques near capillaries but lack of spatial relationship with activated microglia and macrophages. *Amyloid.* 2003;10:86–96.

157. Alafuzoff L, et al. Histopathological criteria for progressive dementia disorders: clinical-pathological correlation and classification by multivariate analysis. *Acta Neuropathol (Berl).* 1987;74:209–225.

158. Delaere P, et al. Large amounts of neocortical beta A4 deposits without neuritic plaques nor tangles in a psychometrically assessed, non-demented person. *Neurosci Lett.* 1990;116: 87–93.

159. Dayan AD. Quantitative histological studies on the aged human brain: I. Senile plaques and neurofibrillary tangles in "normal" patients. *Acta Neuropathol (Berl).* 1970;16: 85–94.

160. Dickson DW, et al. Correlations of synaptic and pathological markers with cognition of the elderly. *Neurobiol Aging.* 1995;16:285–304.

161. Langui D, et al. Alzheimer's changes in non-demented and demented patients: a statistical approach to their relationships. *Acta Neuropathol.* 1995;89:57–62.

162. Tomlinson BE, et al. Observations on the brains of non-demented old people. *J Neurol Sci.* 1968;7:331–356.

163. Wisniewski HM. The aging brain. In: Andrews EJ, Ward BC, Altman NH, eds. *Spontaneous Animal Models of Human Disease.* New York: Academic Press; 1979:148–152.

164. Cummings BJ, Cotman CW. Image analysis of beta-amyloid "load" in Alzheimer's disease and relation to dementia severity. *Lancet.* 1995;346:1524–1528.

165. Attems J. Sporadic cerebral amyloid angiopathy: pathology, clinical implications, and possible pathomechanisms. *Acta Neuropathol.* 2005;110:345–359.

166. Attems J, et al. Alzheimer's disease pathology influences severity and topographical distribution of cerebral amyloid angiopathy. *Acta Neuropathol.* 2005;110:222–231.

167. Herzig MC, et al. Mechanism of cerebral beta-amyloid angiopathy: murine and cellular models. *Brain Pathol.* 2006;16: 40–54.

168. Braunmuhl A. Kongophile Angiopathie und senile Plaques bei greisen Hunden. *Arch Psychiatr Nervenkr.* 1956;194: 395–414.

169. Shimada A, et al. Topographic relationship between senile plaques and cerebrovascular amyloidosis in the brain of aged dogs. *J Vet Med Sci.* 1992;54:137–44.

170. Uchida K, et al. Immunohistochemical studies on canine cerebral amyloid angiopathy and senile plaques. *J Vet Med Sci.* 1992;54:659–667.

171. Uchida K, et al. Immunohistochemical analysis of constituents of senile plaques and cerebro-vascular amyloid in aged dogs. *J Vet Med Sci.* 1992;54:1023–1029.

172. Uchida K, et al. Double-labeling immunohistochemical studies on canine senile plaques and cerebral amyloid angiopathy. *J Vet Med Sci.* 1993;55:637–642.

173. Uchida K, et al. Immunohistochemical study of constituents other than beta-protein in canine senile plaques and cerebral amyloid angiopathy. *Acta Neuropathol.* 1997;93:277–284.

174. Uchida K, et al. Amyloid angiopathy with cerebral hemorrhage and senile plaque in aged dogs. *Nippon Juigaku Zasshi.* 1990;52:605–611.

175. Uchida K, et al. Pathological studies on cerebral amyloid angiopathy, senile plaques and amyloid deposition in visceral organs in aged dogs. *J Vet Med Sci.* 1991;53: 1037–1042.

176. Yoshino T, et al. A retrospective study of canine senile plaques and cerebral amyloid angiopathy. *Vet Pathol.* 1996;33: 230–234.

177. Prior R, et al. Loss of vessel wall viability in cerebral amyloid angiopathy. *Neuro report.* 1996;7:562.

178. Deane R, Zlokovic BV. Role of the blood–brain barrier in the pathogenesis of Alzheimer's disease. *Curr Alzheimer Res.* 2007;4:191–197.

179. Davis GC, et al. Chronic Parkinsonism secondary to intravenous injection of meperidine analogues. *Psychiatry Res.* 1979;1:249–254.

180. Langston JW, Ballard P. Parkinsonism induced by 1-methyl-4-phenyl-1,2,3,6-tetrahydropyridine (MPTP): implications for treatment and the pathogenesis of Parkinson's disease. *Can J Neurol Sci.* 1984;11:160–165.

181. Lane E, Dunnett S. Animal models of Parkinson's disease and L-dopa induced dyskinesia: How close are we to the clinic? *Psychopharmacol (Berl).* 2008;199:303–312.

182. Schmidt WJ, Alam M. Controversies on new animal models of Parkinson's disease pro and con: the rotenone model of Parkinson's disease (PD). *J Neural Transm Suppl.* 2006; 273–276.

183. Landrigan PJ, et al. Early environmental origins of neurodegenerative disease in later life. *Environ Health Perspect.* 2005;113:1230–1233.

184. Crapper DR, et al. Brain aluminum distribution in Alzheimer's disease and experimental neurofibrillary degeneration. *Science.* 1973;180:511–513.

185. Morante S. The role of metals in beta-amyloid peptide aggregation: x-ray spectroscopy and numerical simulations. *Curr Alzheimer Res.* 2008;5:508–524.

186. Drago D, et al. Role of metal ions in the Abeta oligomerization in Alzheimer's disease and in other neurological disorders. *Curr Alzheimer Res.* 2008;5:500–507.

187. Bush AI. Drug development based on the metals hypothesis of Alzheimer's disease. *J Alzheimer's Dis.* 2008;15:223–240.

188. Miller DB, O'Callaghan JP. Do early-life insults contribute to the late-life development of Parkinson and Alzheimer diseases? *Metabolism.* 2008;57(Suppl 2):S44–S49.

189. Dosunmu R, et al. Environmental and dietary risk factors in Alzheimer's disease. *Expert Rev Neurother.* 2007;7: 887–900.

190. Rosenzweig ES, Barnes CA. Impact of aging on hippocampal function: plasticity, network dynamics, and cognition. *Prog Neurobiol.* 2003;69:143–179.

191. Siwak-Tapp CT, et al. Region specific neuron loss in the aged canine hippocampus is reduced by enrichment. *Neurobiol Aging.* 2008;29:39–50.

192. Kuhn HG, et al. Neurogenesis in the dentate gyrus of the adult rat: age-related decrease of neuronal progenitor proliferation. *J Neurosci.* 1996;16:2027–2033.

193. Cummings BJ, et al. Beta-amyloid accumulation in aged canine brain: a model of plaque formation in Alzheimer's disease. *Neurobiol Aging.* 1993;14:547–560.

194. Van Vickle GD, et al. Tg-SwDI transgenic mice exhibit novel alterations in AbetaPP processing, Abeta degradation, and resilient amyloid angiopathy. *Am J Pathol.* 2008;173: 483–493.

195. Head E, et al. A two-year study with fibrillar beta-amyloid (Abeta) immunization in aged canines: effects on cognitive function and brain Abeta. *J Neurosci.* 2008;28: 3555–3566.

196. Butterfield DA, et al. Elevated oxidative stress in models of normal brain aging and Alzheimer's disease. *Life Sci.* 1999;65:1883–1892.

197. Suzuki Y, et al. Lafora-like inclusion bodies in the CNS of aged dogs. *Acta Neuropathol. (Berl).* 1978;44:217–222.

198. Moore WA, Ivy GO. Implications of increased intercellular variability of lipofuscin content with age in dentate gyrus granule cells in the mouse. *Gerontology.* 1995;41 (Suppl 2): 187–199.

199. Cavanagh JB, Jones HB. Glycogenosomes in the aging rat brain: their occurrence in the visual pathways. *Acta Neuropathol.* 2000;99:496–502.

200. Schipper HM. Experimental induction of corpora amylacea in adult rat brain. *Microsc Res Tech.* 1998;43:43–48.

9

FUNDAMENTALS OF NEUROTOXICITY DETECTION

ROBERT C. SWITZER III

NeuroScience Associates, Knoxville, Tennessee

INTRODUCTION

The goal of any neurotoxicity assessment is simply to determine the safety risk of the test article. Accomplishing this successfully relies on being able to answer four key questions: what, when, where, and how. *What* refers to precisely what endpoints are valid to determine neurotoxicity in a given study. *When* is the timing of meaningful observations for the endpoint chosen. *Where*, in pathological terms, defines the scope of the locations of the central nervous system (CNS) in which to search for the endpoint. *How* is the method: a histological stain, behavioral panel, and so on, that will very specifically provide the evaluation of neurotoxicity for the endpoint defined in the what if used in the proper when and where. Although this seems straightforward, all too often studies rely on only one or two of these four components and ignore some or all of the others. The most common misapplication is when the method such as a stain is chosen as the primary tool for a study without regard for what it can detect, and where and when to look. The best tool can be rendered essentially useless if used in an improper manner or context.

In proper sequence, what, where, and why are addressed in this chapter, with how addressed in detail in subsequent chapters by additional contributors.

IDENTIFICATION OF THE ENDPOINT FOR NEUROTOXICITY

Leveraging proper principles of study design in architecting a neurotoxicity study is the foundation to a successful study.

Within the neuroscience and regulatory community, natural biases exist regarding favored methods for providing the answer to a neurotoxicity evaluation. Specialists in any given realm of neuroscience have a natural tendency to favor methods and practices within their area as the "holy grail" for assessing neurotoxicity. In reality, most of this bias is misplaced; it is more accurate to consider that each unique approach or tool is conditional and specialized to reveal a particular endpoint. No single method or endpoint is capable of revealing all expressions of neurotoxicity, but utilizing a proper combination of endpoints has the ability to assess neurotoxicity to meet the needs of a given study.

One common topic of debate and confusion in a neurotoxicity assessment approach has been in choosing between pathological or behavioral endpoints as the method to use in a safety study. The correct answer is—both; they should be considered complementary approaches. There are certain expressions of neurotoxicity that are most easily appreciated in the course of behavioral testing. Other effects do not lead to observable differences in behavior, but may have observable temporary or permanent changes in the pathology of the brain. Finally, there are overlapping expressions where a behavioral change and its corresponding pathological change can be detected, so either method may suffice. Because this overlap is not complete, the proper use of each method should be employed during safety studies (Figure 1).

The specifics and approaches to behavioral testing are best addressed by experts in that area of neuroscience. This discussion focuses on the design for pathological expressions of neurotoxicity.

There are an ever-increasing number of neuropathological tools for assessing neurotoxicity in the form of various

Fundamental Neuropathology for Pathologists and Toxicologists: Principles and Techniques, First Edition. Edited by Brad Bolon and Mark T. Butt.
© 2011 John Wiley & Sons, Inc. Published 2011 by John Wiley & Sons, Inc.

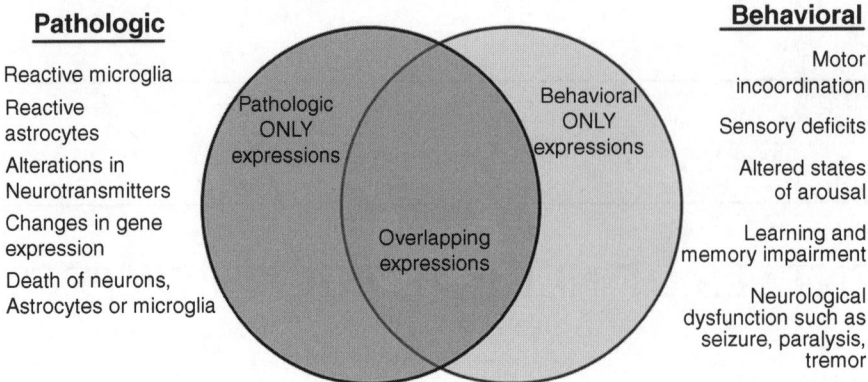

FIGURE 1 Graphic representation of individual expressions of pathological and behavioral components of a complete neurotoxicity evaluation is shown. Although overlap exists, for some endpoints there are no structural correlates to explain behavior and no behavior changes related to structural abnormalities. Both areas must be assessed independently, then considered together in a truly comprehensive evaluation.

stains and immunohistochemistry. It is tempting to choose a method as the first design element in a neurotoxicity study, but specific method selection should be one of the final elements in crafting a study design. Before a method can be selected, the scope of what will be evaluated must be defined.

SPECTRUM OF PATHOLOGICAL ENDPOINT DETECTION

Unfortunately, there is not a single endpoint that is both valid and comprehensive to meet the needs of all pathological safety evaluations. Rather, there is a spectrum of safety considerations in the pathology of the brain.[1–3]

At one end of the spectrum, there are no detectable changes to the CNS (safe), and at the other end of the spectrum are permanent changes to the CNS (unsafe). In between these two extremes, for example, are detectable changes in chemistry, neurotransmitter responses, and perturbations leading to an inflammatory response. In Figure 2, the farther a change falls to the right in the spectrum of changes, the more likely the change may represent a true safety concern. At the left end of the spectrum (i.e., chemistry changes), there are many potential pathological endpoints from which to choose, and some

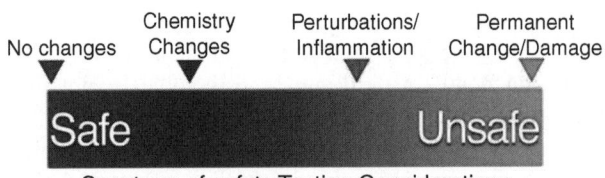

FIGURE 2 Spectrum of potential safety changes: as alterations progress to the right, the more likely they are to be of a safety concern.

have greater safety implications than others. Moving to the right (unsafe) side of the spectrum, the number of detection methods decreases and endpoints are more specific and definitive as a safety concern.

While each safety study has different approaches, all study designs should include the evaluation of a selected point on this spectrum as well as all assessments to the right of that point. For example, an evaluation of "safe" could not be concluded if a specific chemistry change were evaluated (with a favorable result) without also evaluating for perturbations and permanent damage. The minimum scope for a safety study would focus solely on the most conclusive endpoint of permanent damage at the far right of the spectrum. Studies with a lower threshold for neurotoxicity expressions can also include evaluations progressing toward the left of the spectrum as long as those evaluations are cumulative and include everything to the right.[1–3]

Evaluation of Chemical Changes or Other Changes from Normal

In its most literal form, neurotoxicity could be defined as any departure from normal in the CNS. However, all departures from normal or changes in CNS are not negative. In fact, many compounds deliberately cause changes to the CNS as therapy. Judgment is required as to what changes constitute a negative impact and should be evaluated in the scope of a safety study. Most often, compounds that rely on a specific mechanism of action are evaluated with respect to their impact on the brain relative to that mechanism to make sure that intended effects are not also accompanied by unintended effects. There are an incredible number of endpoints that could be evaluated to include a departure from normal. A common approach is to evaluate endpoints specific to the known mechanism of a compound to specifically quantify their impact on the brain.

Evaluation of Perturbations or Inflammation

The detection of perturbations in the brain is often employed as a key tool in safety assessments. The presence of a perturbation signal is an undesirable finding in safety testing and represents a degree of harm that has been caused to the brain. In some approaches, perturbations are used as an indicator to search for signs of permanent damage. In the context of some studies, the detection of a perturbation in itself is enough to warrant a conclusive safety concern. A perturbation represents a stress to the health of the brain, but a perturbation signal alone does not indicate the infliction of permanent damage to the brain.

Evaluation of Permanent Damage

Many alterations to the brain can be negative and potentially long-lasting, but not all can be considered permanent. Some injuries can be treated or will resolve after time if the source of the injury is removed (i.e., a harmful compound ceases to be administered), whereas other injuries lead to permanent implications. The most definitive and easy way to interpret evidence of permanent damage is the destruction of neurons. Even though the brain has amazing functional compensatory power, the loss of irreplaceable neurons creates a permanent deficit. In the case of a truly recoverable injury, neurons are not destroyed, the brain actually returns to "health," and there is no permanent implication.

Potential neurotoxins may result in many detectable endpoints related to chemistry changes, perturbations, and others prior to reaching the point of neuronal destruction. The sequence of events and pathways followed by any given compound or insult vary.

These and other endpoints may occur in a variety of sequences and combinations preceding cell death. In some cases, these perturbations resolve and are considered recoverable. For other compounds and insults, a "point of no return" is reached after which the neurons begin a pathway leading to cell death as the common final endpoint of permanent neurological damage.

Following permanent damage, the brain is also often resilient and able to compensate functionally for the permanent injury. Whether or not compensation occurs, any permanent damage is of significant consequence. Functional compensation may mask the significance of damage, but the damage persists as a permanent deficit. Once damage has occurred, the brain becomes less capable of compensating for future insults. Depending on the scope defined for a safety study, it may be crucial to differentiate between a permanent injury and one that results in a reversible perturbation that may be considered "acceptable" in certain circumstances. For cell death itself, the important keys to remember are:

- There is no recovery from cell death.
- Cell death is the universal single profile of unrecoverable events in the brain.
- Pathological detection of cell death is definitive for neurotoxicity.

Even within the specific endpoint of cell death, there are important considerations of scope in determining what to look for. The scope of the neuron itself is not defined consistently in safety studies. Due primarily to the staining technologies available in the past, the traditional definition of a neuron to be assessed in a neurotoxicity study included just the cell body itself. With the availability of a new class of tools (degeneration stains) and armed with better understanding of the neuronal system, a more contemporary scope of assessment of the neuron is to consider all of the neuronal elements: cell body, dendrites, axons, and axon terminals.

This evolution in assessing neurotoxicity is due to the fact that

1. the technology to utilize this approach was not available previously but is now, and.
2. the destruction of any of the neuronal elements is as devastating as destruction of the cell body itself.

The dendrites, axon, and axons terminals are the communication infrastructure of the neuron. Destruction of any of these elements renders the cell isolated, and a neuron that has been isolated sometimes dies as a result. The dendrites, axon, and axons terminals are occasionally capable of regrowing, but they may not establish the same connectivity, so a permanent deficit has been created by their destruction. Clearly, disconnecting a neuron cell body from the CNS network renders it as useless as if it were dead.

Occasionally, just one element will be destroyed, but more often all four are destroyed, providing an increased opportunity for detecting damaged attributes of the neuron's large surface area and distinct morphology. The destruction of these elements most often occurs in sequential time lines, so it is not always possible to witness all four elements disintegrating for a given neuron. Rather, the elements often appear one or two at a time, and observation of any one of these elements is a strong indicator that complete neuronal destruction has or will occur. Illustrations of the appearance of disintegrating cellular elements are provided in Chapter 12.

STUDY DESIGN PRINCIPLES: LOCATION, LOCATION, LOCATION

It is commonly accepted that different parts of the body have unique vulnerability to various forms of toxicity. The heart, liver, kidney, brain, and so on, are each considered separately during toxicity assessments. Although some organs may be affected in a homogeneous fashion, organs like the heart and brain have unique anatomical elements that are specifically and uniquely vulnerable to toxic agents.

There are major categories of structure such as arteries, valves, chambers in the heart and cortex, hippocampus and cerebellum in the brain, but each of these must be divided further into more specific elements for consideration of vulnerability to a toxin. For example, the hippocampus of the brain includes subpopulations such as CA1, CA2, CA3, ventral dentate gyrus, and dorsal dentate gyrus. Due to the known specificity of toxins, each of these elements must be considered during a safety assessment, as any of them is potentially at risk.

Using the analogy of the different structures of the heart, we have introduced the fact that the brain also has unique elements and populations. Armed with that basic understanding the next step is to appreciate the targets of neurotoxins in their specificity to major or minor populations in the brain. Most often, neurotoxins target specific subpopulations rather than an entire major population in the brain.

The descriptive text that follows cites Figures 3 to 16 which have been reprinted with permission from the very fine brain atlas compiled/authored by George Paxinos and Charles Watson (*The Rat Brain in Stereotaxic Coordinates*, 6th. ed., Academic Press/Elsevier Science, London, 2008). These figures are used with the publisher's permission and my gratitude.

Major recognizable regions highlighted in the sagittal section of the brain (Figure 3) include cerebral cortex, hippocampus, thalamus, hypothalamus, cerebellum, and the brain stem. Not all of the major regions can be represented in a single level of course, but the major divisions, are frequently referred to in the brain.

Figure 4 displays the same level as Figure 3, but Figure 4 includes the identification of all of the subpopulations (vs. just the major divisions in Figure 3). The total number of subpopulations is far greater than the number of major divisions, and only a limited number are visible at any given level of the brain. Each division of the brain has different cell types, connectivity, and functionality. In the past several decades, our understanding of the brain has been increasing at a seemingly exponential rate, but we still do not fully understand the comprehensive functions of each population, the interactions between all of the populations, or The symptoms or functional impact of damage to any specific population.

Brain cells in different populations of the brain exhibit unique vulnerabilities to neurotoxic compounds. We are not yet capable of predicting with reliability which subpopulations will be affected by a neurotoxin, nor are we always able to predict or even assess the functional significance of the impairment caused by damage to each respective subpopulation. Although our understanding of the brain is perhaps not as complete as with other organs, we do not know of any regions of nonimportance.

The unique potential of vulnerability across subpopulations of the brain is a critical concept to understand when designing a safety study.[1-3] Simply put: One would never rely on the pathology of a kidney to qualify a compound as being safe for the heart, lungs, or brain, for example. The same

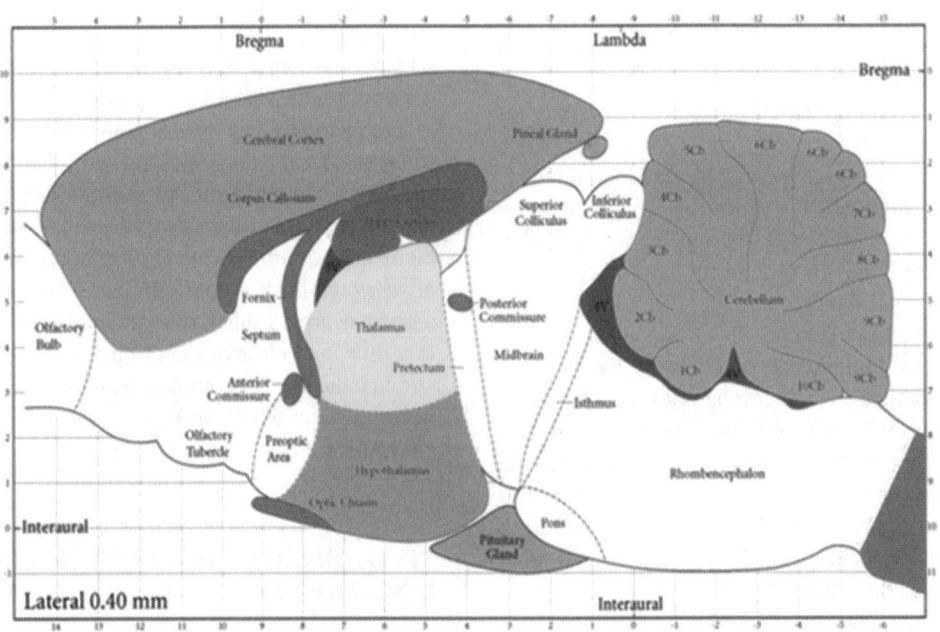

FIGURE 3 Major subdivisions of the brain shown in sagittal section: this single level (and there are many more levels) shows the complexity of brain organization. (*See insert for color representation of the figure.*)

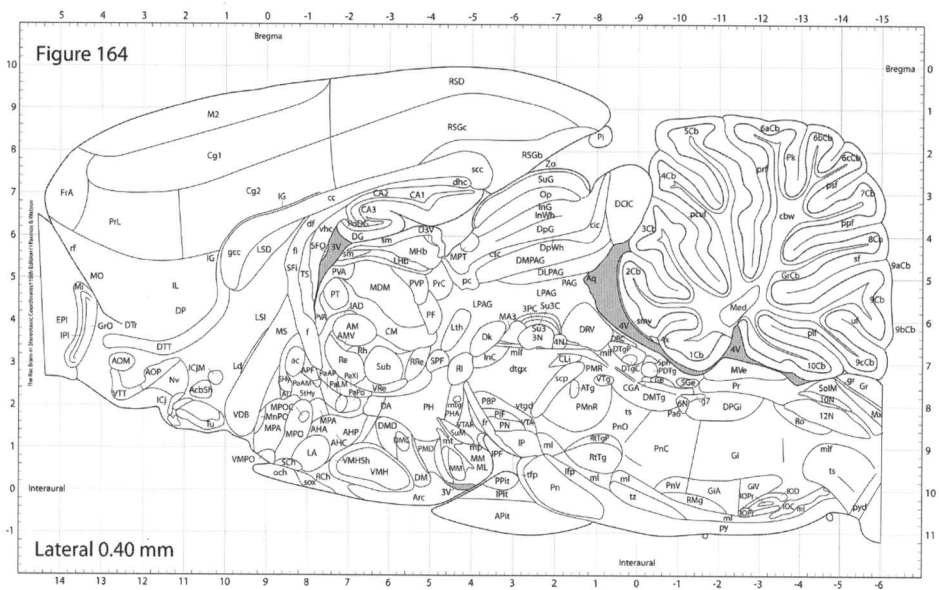

FIGURE 4 The various regions of the brain are shown in this sagittal section. Note the complexity and number of anatomically, chemically, and functionally distinct structures.

concept applies within the brain, as it is just as important to treat each subpopulation as if it were a unique organ for consideration. Each subpopulation of the brain must be assessed.

Examples of Neurotoxic Specificity

The location a particular toxin may affect in the brain is not always predictable. Figure 5 shows the neurotoxic profiles for nine well-known neurotoxins.[4–24] It is the particular area of the brain, not the relative size of the area, vulnerable to a neurotoxin that is most relevant in quantifying the importance or severity of the neurotoxic effect. But the potential relative small size of affected areas does have important implications in designing a safety assessment approach.

Brain sections for pathological examination can be cut and collected from any desired plane on a brain; however, coronal is most often used. But since coronal sections are taken (generally) in a rostral–caudal direction, sagittal illustrations (primarily) will be used to demonstrate what coronal levels would be necessary to detect a particular toxicity.

Sampling Approach

A safety assessment must have an approach for sampling the brain: What levels will be chosen for sampling? How many levels are required? In this section we discuss the principles from which an appropriate sampling strategy can be derived.

Figure 6 shows the areas affected by 3-nitropropionic acid (3-NPA).[45] Also shown are 11 coronal levels selected for sampling (as represented by the red lines in the sagittal view). It can be seen that eight of the 11 coronal levels chosen in this example would contain neuron populations that were affected by 3-NPA and detection of this effect would be quite likely. In this example, the damage is widespread in terms of location, and the chances of selecting a level(s) that contains an affected population are very good. Most neurotoxic compounds, however, affect smaller regions of the brain.

In the brain, size does not necessarily matter. Figure 7 shows the raphe nucleus (in red). $2'$-NH$_2$-MPTP is responsible for the destruction of cells in the raphe nuclei.[23] Nearly all serotonergic cell bodies in the brain lie in the raphe nuclei, and losing these cells can lead to profound long-term negative effects. Serotonin is a neurotransmitter involved in regulating normal functions as well as diseases (e.g., depression, anxiety, stress, sleep, vomiting).

While causing a major impact from a neurotoxicity standpoint, the area damaged by 2-NH$_2$-MPTP is relatively small and could easily be missed unless a coronal sample was assessed within the <2-mm anterior–posterior range in which the affected area occupies. The sampling strategy must be able to accommodate small structures as well as large ones.

Sampling at Major vs. Minor Divisions

Compounds do not necessarily affect an entire "major" population, and different compounds may affect different subpopulations of the same major population. Domoic acid, PCP, and alcohol[8,12–15,24] have all been show to destroy cells in the hippocampus. However, none destroys cells in all regions of the hippocampus, and in fact, each of the three neurotoxins affects unique, nonoverlapping areas of hippocampus.

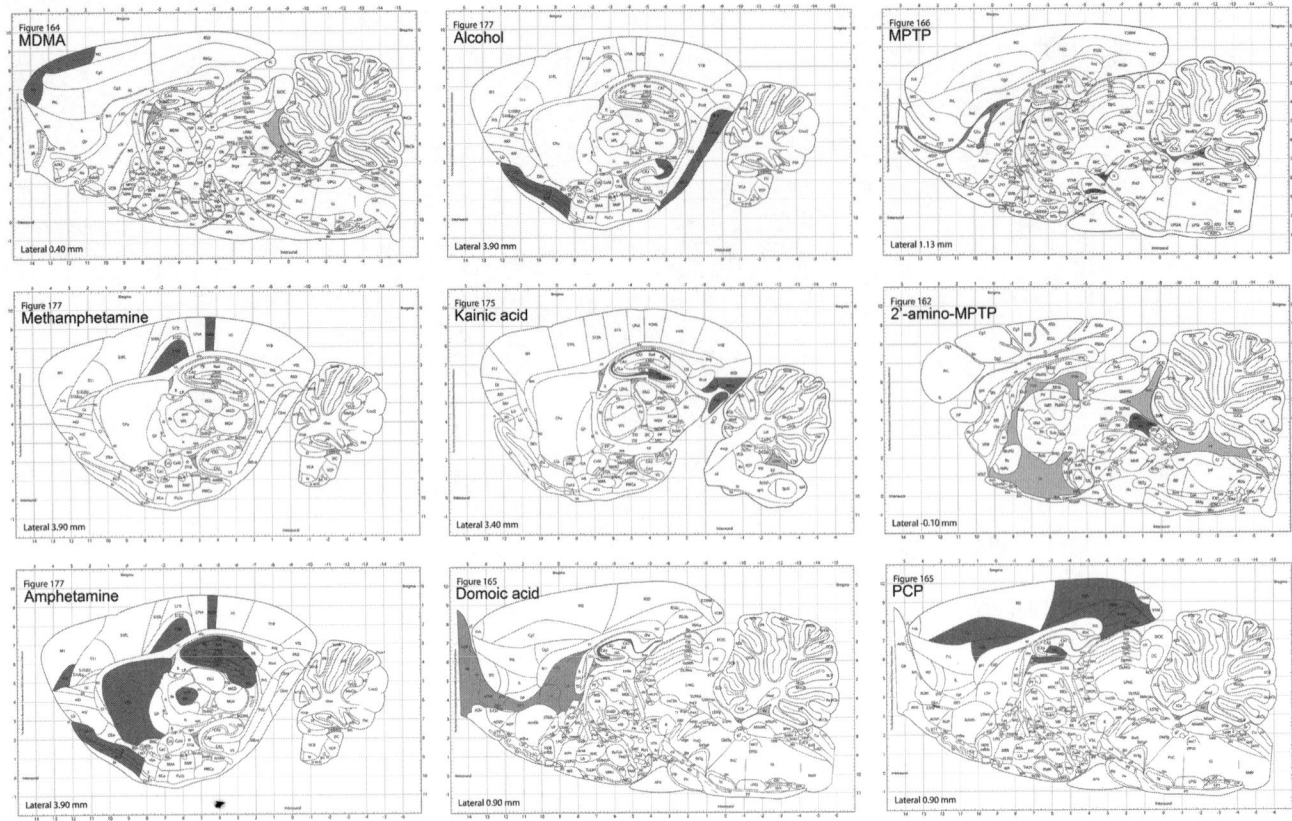

FIGURE 5   Areas of the brain affected by nine known neurotoxins; note that some of the areas are quite small and easily missed in all but the most thorough brain sectioning scheme. (*See insert for color representation of the figure.*)

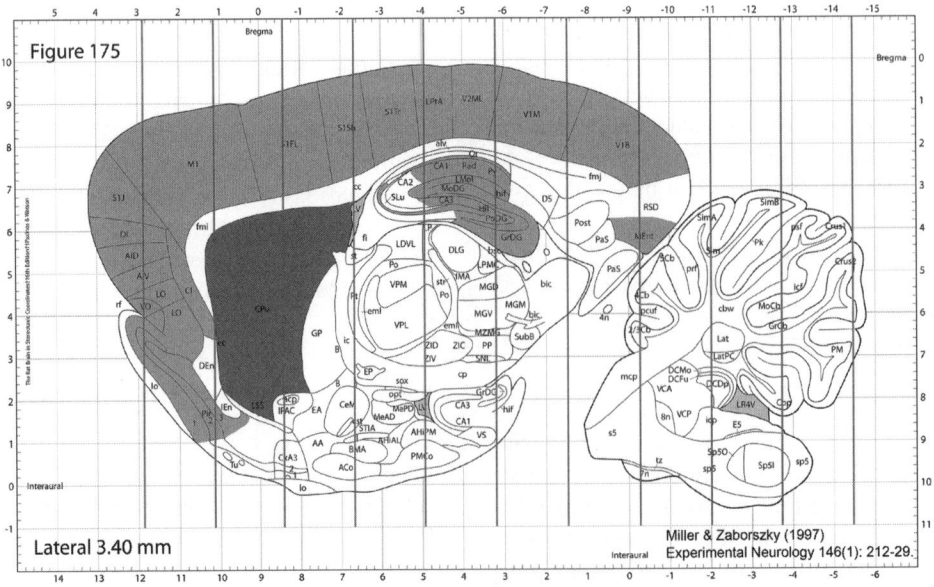

FIGURE 6   Area of the brain affected by the administration of 3NPA: the vertical lines represent the levels that would be sampled and evaluated in an 11-section scheme (sections approximately 1.8 mm apart). (*See insert for color representation of the figure.*)

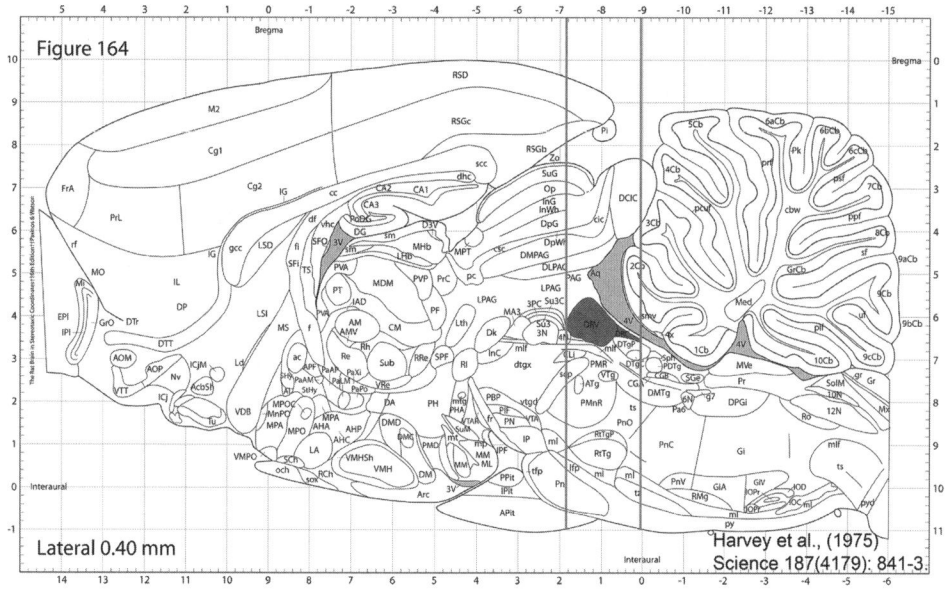

Raphe nuclei span less than 2mm A-P

FIGURE 7 The rahpe nucleus is highlighted. Most of the serotonin-producing cells are within the raphe nucleus. Neuronal death in this nucleus may have a profound effect on the animal. Note how easy it would be to miss this small area on anything but the most comprehensive sectioning schemes. (*See insert for color representation of the figure.*)

- Domoic acid destroys cells in the pyramidal layer (of hippocampus).
- PCP destroys cells in the dorsal dentate gyrus (of hippocampus).
- Alcohol destroys cells in the ventral dentate formation (of hippocampus).

The two coronal sections that correspond to the red lines in Figure 8 are shown in Figure 9. When a single level of hippocampus is chosen for sampling, the level on the left of Figure 9 is a typical (popular) level for selection. In this example, it is significant to note that the effects of alcohol can be appreciated only in a more ventral level containing hippocampus which appears at a more caudal level. The lesson of these examples is that major divisions of the brain are not homogeneous from a neurotoxicity standpoint and that the coronal levels containing each subpopulation must be included in the sampling scheme.

The Location of Damage in the Brain Is Unpredictable

One of the most common flaws in safety assessments is to look only in areas where damage is expected. Currently, there is no reliable way to model the complexity of the brain that can accurately predict where neurotoxicity is likely to occur. For example, the findings of two separate research groups studying the neurotoxicity of D-amphetamine varied.[4–11] In the Belcher study, the researchers anticipated, looked for,

and confirmed that D-amphetamine destroyed cells in parietal cortex and somatosensory barrel field cortex. Studies by Bowyer et al.[6–9] also assessed for neurotoxicity of D-amphetamine but did not always restrict the assessment to expected areas. The second study confirmed the findings of the first, but more important, also found cell death in frontal cortex, piriform cortex, hippocampus, caudate putamen, thalamus, and areas not shown in the figure, including tenia tecta, septum, and other thalamic nuclei.

A basis for restricting the location in the brain to assess can also be found when knowledge of the location of neurotoxicity from one compound in a class is extrapolated to others in the same class or with similar profiles. This is a flawed assumption. While these compounds may have similar neurotoxic (or nonneurotoxic) profiles, many exceptions exist and it is not appropriate to assume the safety, neurotoxicity, or the location of effects simply by extrapolation. An example can be seen below with MPTP (1-methyl-4-phenyl-1,2,3,6-tetrahydrophridine) and 2′-NH2-MPTP. Although these compounds may be considered similar and in fact both do cause neurotoxicity, the populations of the brain they affect are different and quite unique functionally (Figure 10).

MPTP damages the dopaminergic system whereas 2′-NH2-MPTP damages the serotonergic system.[23–42] Note that the coronal levels that would be used to sample either of these areas do not overlap the coronal levels of the other area. If a safety study were designed based on the neurotoxic profile of one of these to assess the other one, not only would

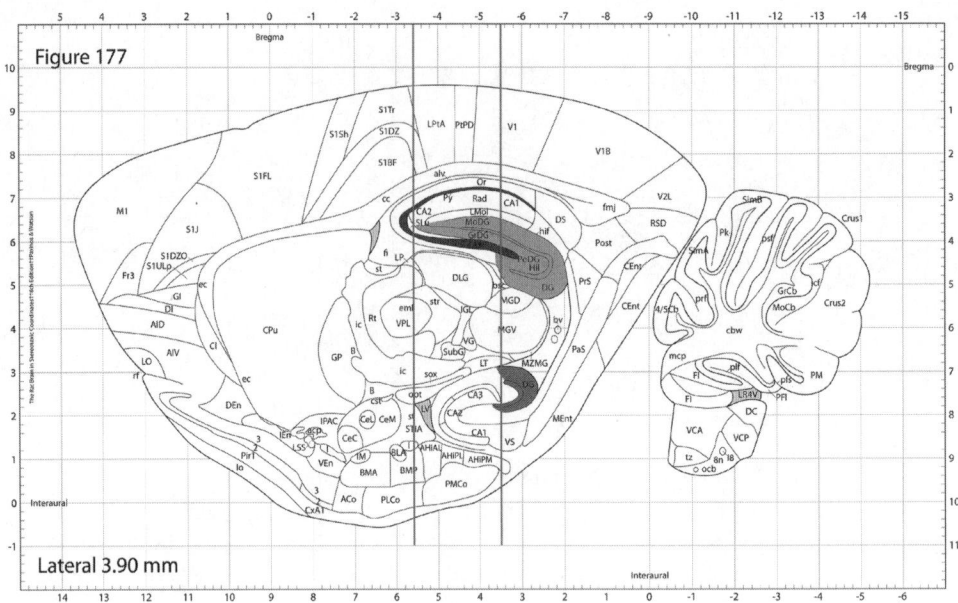

FIGURE 8 Areas of the hippocampus affected by domoic acid (pyramidal layer; thin, C-shaped region), PCP (dorsal dentate gyrus; thicker, reverse C-shaped area), and alcohol (lower area). (*See insert for color representation of the figure.*)

cell death not be found where expected, but the cell death that did occur would not be visible on the same level as expected and would not be detected.

Cell Death Can Only Be Witnessed in Locations That Are Assessed

One of the most studied neurotoxic compounds has been the NMDA (*N*-methyl-D-aspartic acid) receptor antagonist MK-801.[24–42] It is significant in its unique effectiveness as an NMDA receptor antagonist but was discovered to be neurotoxic, and its clinical potential for treating stroke was neutralized. The first indication of neurotoxicity discovered by John Olney was the observation of vacuoles in the posterior cingulate/retrosplenial cortex.

As MK-801 was studied further, it was discovered that permanent damage occurred from MK-801 in the form of cell death.[24–42] The location of cell death coincided with the location of the vacuoles, but more significantly, cell death was found in other regions of the brain distant from the vacuole sites, as shown in Figure 11.

Due to the neurotoxic profile of MK-801, the U.S. Food and Drug Administration routinely has requested safety studies to evaluate for the presence of the vacuoles when testing NMDA

In a commonly used rostral section of hippocampus, ventral structures are not present

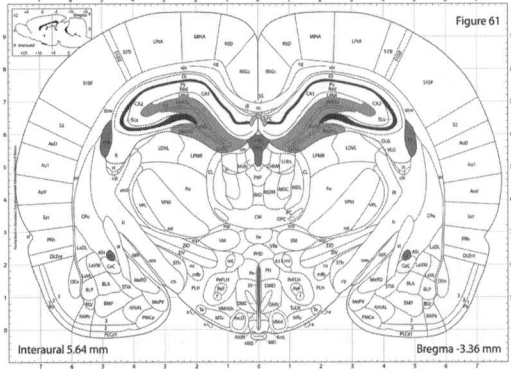

A more caudal section allows ventral structures to be seen

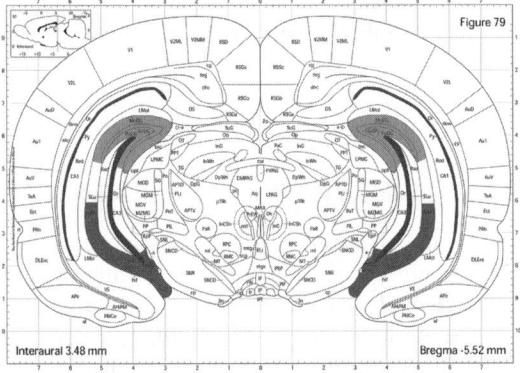

FIGURE 9 The two coronal sections indicated by the vertical lines in Figure 8. Note that the section of hippocampus on the left does not include the ventral portion of the hippocampus. Only the section on the right contains the area affected by alcohol. (*See insert for color representation of the figure.*)

MPTP:
Destroys cells in the VTA
and substantia nigra, pars compacta

2'-amino-MPTP:
Destroys cells in the dorsal raphe

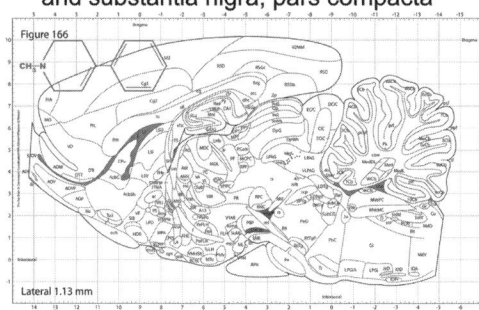
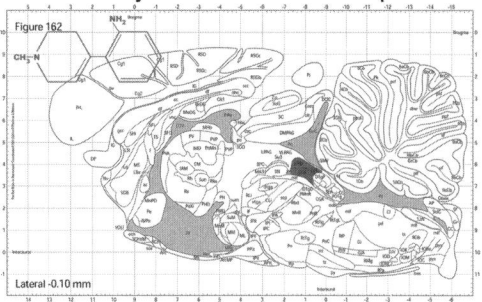

FIGURE 10   Neuronal areas primarily affected by MPTP (substantia nigra) and 2′-NH₂-MPTP (dorsal raphe nucleus). Note that although the structure of the compounds is similar, the selective areas affected are quite different. Note also that although each may affect an important population of neurons adversely, the overall size of the areas affected is small. (*See insert for color representation of the figure.*)

Vacuole location: retrosplenial cortex

Cell death locations:

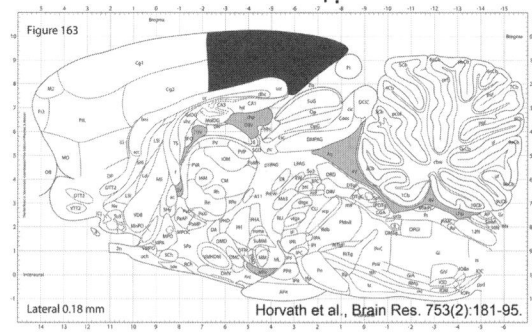
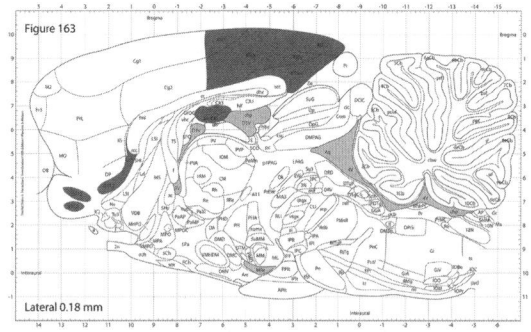

MK-801 destroys cells in:

*Retrosplenial cortex
*Tenia tecta
*Dentate gyrus
*Pyriform cortex
*Amygdala
*Entorhinal cortex
*Ventral CA1 and CA3 of hippocampus

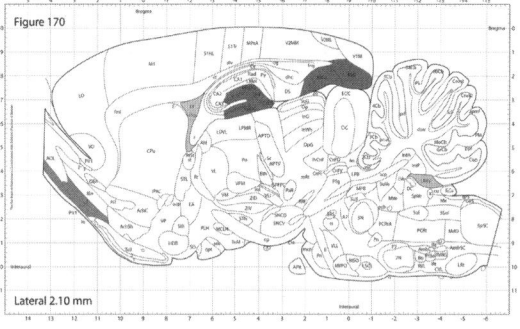
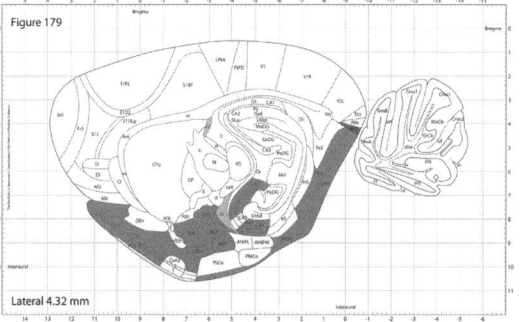

FIGURE 11   MK 801 (NMDA receptor antagonist) administration: The area affected (retrosplenial cortex) by neuronal vacuolation (the Olney effect) in the sagittal diagram on the left as compared to the areas where cell death may be observed (three sagittal sections on the right). Full knowledge of a compounds full capacity to produce neurotoxicity is assessed only after a very thorough sectioning and evaluation of the brain. (*See insert for color representation of the figure.*)

antagonists. As more was learned about the correlation of vacuoles with subsequent death of cells, markers of cell death have become the basis for these safety studies. While originally restricted to the specific areas affected by MK-801, the trend in these safety studies is to assess beyond the anticipated areas and now includes a more appropriate comprehensive safety assessment of the regions of the brain.

Study Design Principles for Location

In summary:

- The brain is heterogeneous. Each of the 600 + populations has unique functions.
- Neurotoxins often affect just one or perhaps several distinct and possibly distant regions.
- Affected regions can be very small but functionally significant.
- The location of effects is unpredictable:
  o Based on other pathological and behavioral indicators
  o Between compounds that share similar structures (same class)

The key principles to apply in a successful study design with respect to sampling the brain are now understood and an approach can be considered. In summary, all populations of the brain must be evaluated for each compound, but how can that be done effectively? A consistent, systematic approach is the most practical option and can be achieved by evaluating full cross sections of the brain (levels) at regular intervals from one end of the brain to the other. Defining the interval

spacing between the samples is the final variable in a sampling approach.

Using the sagittal view of rat brain once again (Figure 12), a single cross section represented by the vertical red line yields a given coronal level. Any single level crosses a relatively small percentage of brain cell populations. The populations in the brain differ dramatically between levels that are separated by very short intervals. In Figure 13 the populations present at these four levels in the rat brain are discussed. The interval between each level is 1 mm.

Figure 13 displays the coronal levels represented by levels 1 to 4 in Figure 12. The shaded areas in each coronal level represent populations that exist in that level of the rat brain that did not exist in the prior level. The differences in populations present in one but not the next are quite dramatic.

Although 1 mm between levels (across a 21-mm-long rat brain) seems very close together, the illustration above indicates that there is much variance between even 1-mm levels in rat brains. Defining a sampling approach for pathological safety assessments is really a trade-off exercise. To sample and analyze every level (each section cut) throughout the brain would be completely thorough, but is impractical and seemingly unnecessary for most studies. Sampling levels at too great an interval can leave gaps and there would be populations that were missed entirely. A compromise must be achieved that delivers reasonable safety assurance without imposing an excessive burden on both neurohistologists and neuropathologists. As discussed previously and as displayed in Figure 14, 1-mm sampling through a rat brain can be shown to leave broad gaps between samples. Sampling every 1 mm yields about 20 to 23 levels or sections per brain.

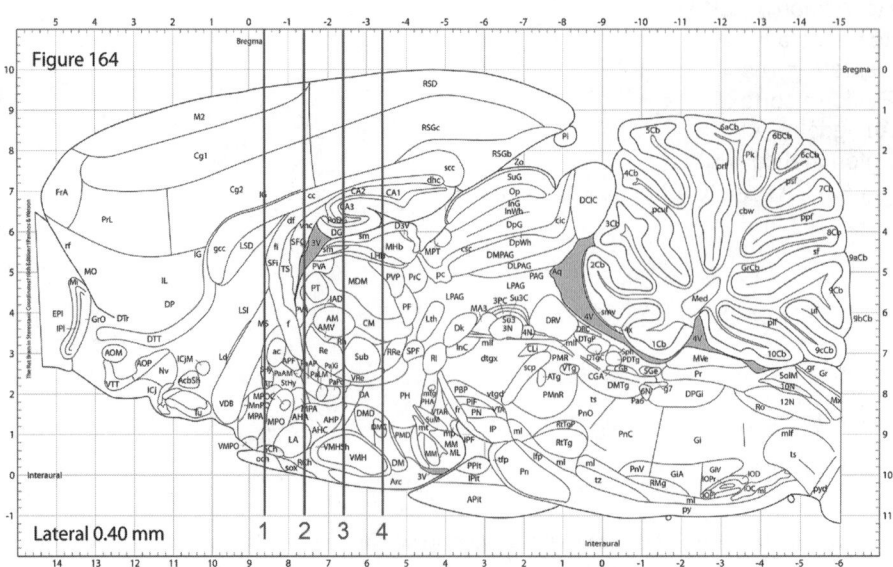

FIGURE 12 Sagittal section with four levels indicated with vertical lines; the distance between these lines is 1 mm. (*See insert for color representation of the figure.*)

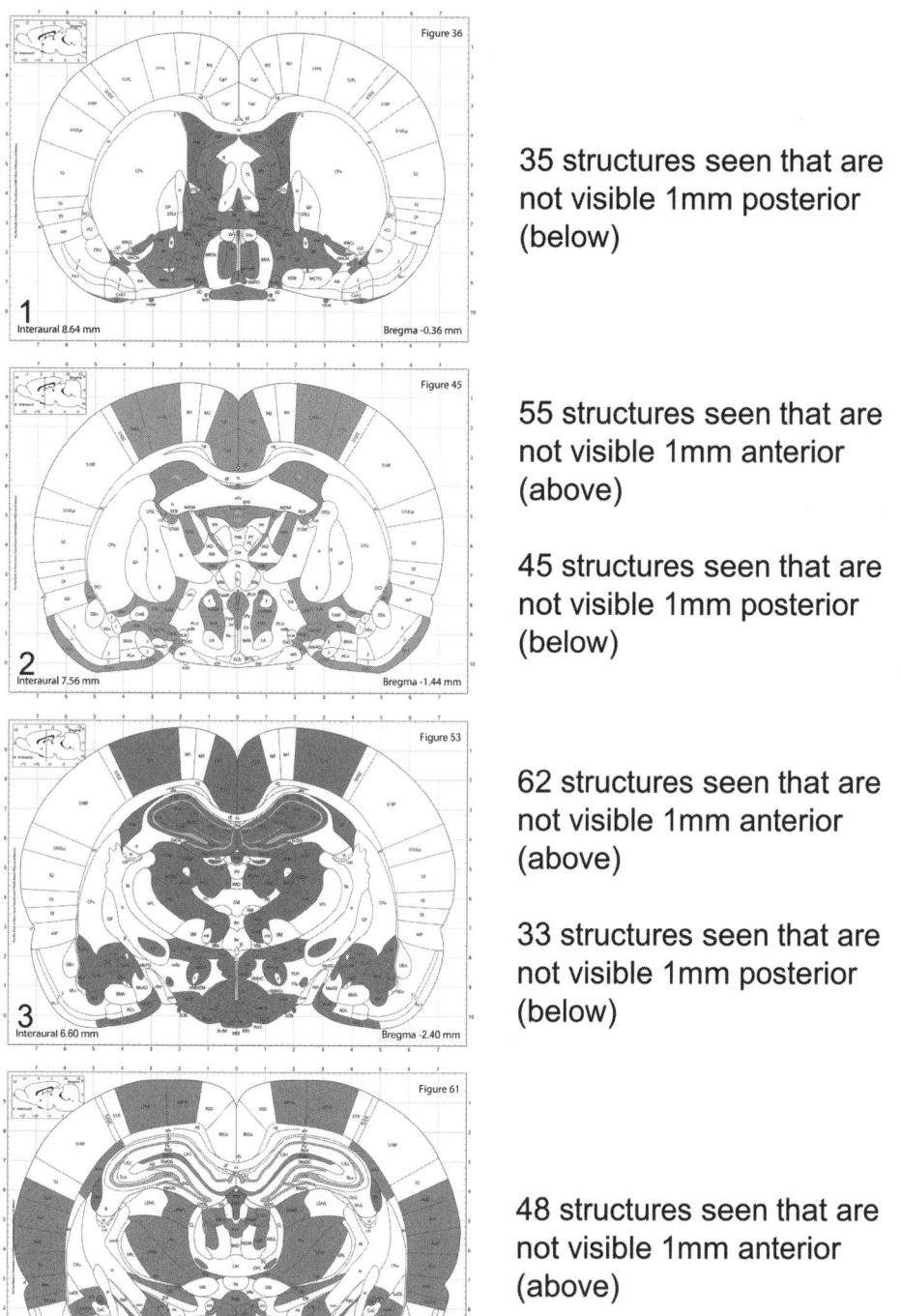

35 structures seen that are not visible 1mm posterior (below)

55 structures seen that are not visible 1mm anterior (above)

45 structures seen that are not visible 1mm posterior (below)

62 structures seen that are not visible 1mm anterior (above)

33 structures seen that are not visible 1mm posterior (below)

48 structures seen that are not visible 1mm anterior (above)

FIGURE 13 The four coronal levels that correspond to the four vertical lines from Figure 12. The shaded areas are those that were not visualized in the prior section, demonstrating the marked difference even 1 mm may make regarding the presence or absence of brain structures. (*See insert for color representation of the figure.*)

FIGURES 14 to 16 Sagittal diagrams showing sectioning intervals of 1 mm (Figure 14), 0.5 mm (Figure 15), and 0.32 mm (Figure 16). Sectioning at 0.32 mm provides approximately 60 coronal sections of the brain and is satisfactory in rats to provide adequate representation of the major neuronal areas. (*See insert for color representation of the figures.*)

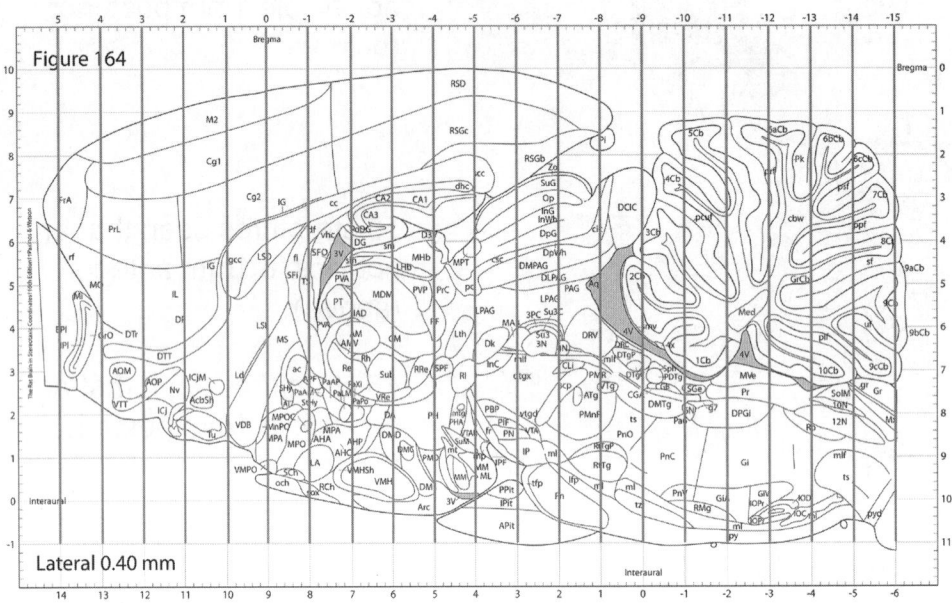

FIGURE 14

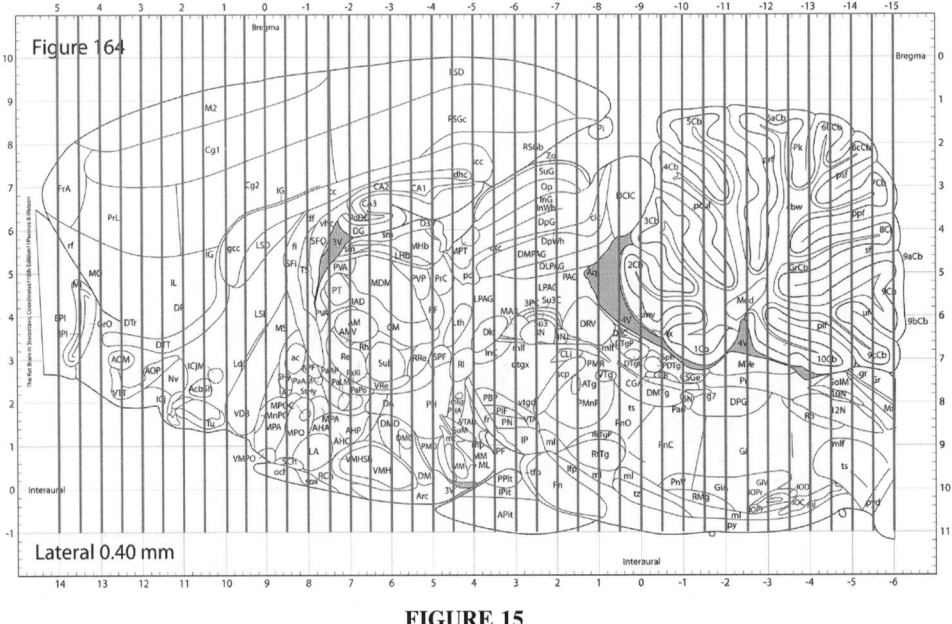

FIGURE 15

Doubling the sample rate to every 0.5 mm between levels greatly improves the opportunity to sample all populations, but gaps can still occur. Sampling every 0.5 mm yields about 40 to 46 levels or sections per rat brain, as shown in Figure 15.

Doubling the sampling rate once again yields an interval of 0.25 mm between levels. This sampling rate is very thorough, with most populations sampled multiple times. This sampling rate is the one reflected in the original rat brain

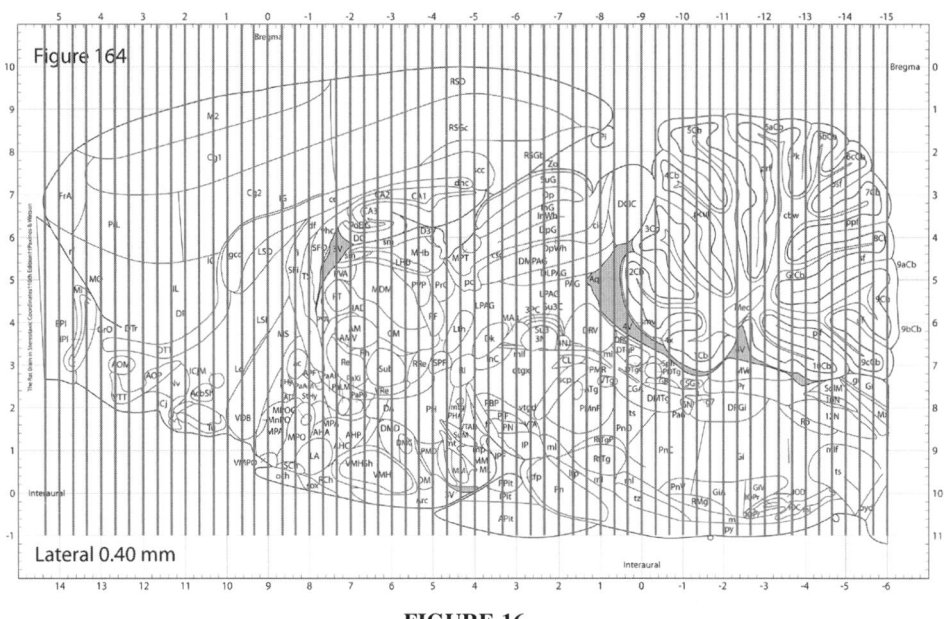

FIGURE 16

atlas by Paxinos and Watson and yields about 80 to 90 levels or sections per rat brain.

In practice, 0.25-mm sampling between levels is probably more than should be required. The most commonly used interval (in the author's laboratory) for R&D purposes for rat brain is 0.32 mm (Figure 16). This approach ensures adequate representation through most regions of the brain and yields about 60 to 65 sections in a rat brain.

In the distant past, the prospect of performing the neurohistology on 60 to 65 levels of each rat brain in a safety study may have been cost and time prohibitive and considered an obstacle to feasibility. Service companies have begun to penetrate the neurohistology arena over the past 20 years, and it is now possible to process hundreds of brains in a matter of weeks using mass processing techniques.

The rat brain has been discussed, but certainly other animals are used for safety testing and they need to be considered also. Conveniently, a rule of thumb has emerged by which any species brain can be sampled, providing comparable levels of representation. The number of levels should be the same regardless of species to achieve comparable representation as shown in Figure 17.

Location of Other Markers for Neurotoxicity

This section has focused on specific examples related to the location of cell death due to known neurotoxins. Cell death is the most extreme indication of neurotoxicity, but it may not be the only relevant endpoint for a given study. The mechanics and locations of perturbations, inflammation, or chemical changes may coincide with the locations for cell death or may be remote. The principles for sampling remain the same regardless of the endpoint:

A thorough sampling scheme, as detailed in the section above, prevents the exclusion of findings that occur in unexpected locations.

TIME COURSE FOR OBSERVATIONS OF NEUROTOXICITY

The second crucial factor in assessing neurotoxicity is determining when the test animals should be sacrificed following exposure to a test article. The observables of chemical changes, perturbations, and cell death do not necessarily occur at the same time. For most neurotoxins, at the moment of insult there is no pathology to assess. The brain has not yet had the opportunity to react to the insult and develop observable pathology. After a period of time that varies by neurotoxin and by pathological endpoint, the pathology becomes visible. Some time later, the pathology disappears as the brain clears away debris and removes evidence of the insult. For each pathological endpoint, there is a temporal window of opportunity during which it is possible to observe that endpoint. This makes observation a challenging exercise; however, trends in the timing of the onset and the duration of each pathological endpoint make it possible to design an efficient and effective study design that accounts for these variations.

The timing of specific perturbation expressions is discussed later in this section. To demonstrate the importance of timing, specific examples of the timing of observable cell ddeath are the emphasis of discussion. To simplify the example further, the discussion will focus on the detectable disintegration of the cell nucleus as our endpoint, as could be observed using a hematoxylin and eosin (H&E) stain.

Species	Brain Length (mm)	Sampling Interval (in mm)		
		Using 40 samples	Using 60 samples	Using 80 samples
Mouse	12	0.30	0.20	0.15
Rat	21	0.53	0.35	0.26
Monkey	65	1.63	1.08	0.81
Dog	75	1.88	1.25	0.94

FIGURE 17 The approximate brain length of common laboratory species is in the left column. The sectioning interval necessary to achieve approximately 60 overall sections is provided. For example, in an average-size dog, sectioning every 1.25 mm will produce around 60 total sections, which will include most of the major neuronal groups in the brain. (*See insert for color representation of the figure.*)

The study design specifics necessary to observe the event of cell death due to acute, subchronic, or chronic dosing differ. However, once a cell begins the dying or disintegration process, the time line of events is very similar across all neuronal cells. Depending on the insult or neurotoxin, cells may survive for a period of time in a perturbed state before beginning the disintegration cycle, whereas other insults may result in an immediate degenerative response. Once the point of no return has passed, the neuronal cell body begins an observable pattern of disintegration that is histologically detectable and visible to the pathologist. Figure 18 depicts the appearance of a cell visualized with an H&E stain as it progresses through this time line:

- *Day 1.* The cell appears normal and healthy
- *Day 2.* The nucleus of the disintegrating cell body becomes visible.
- *Day 3.* The nucleus begins to fragment.

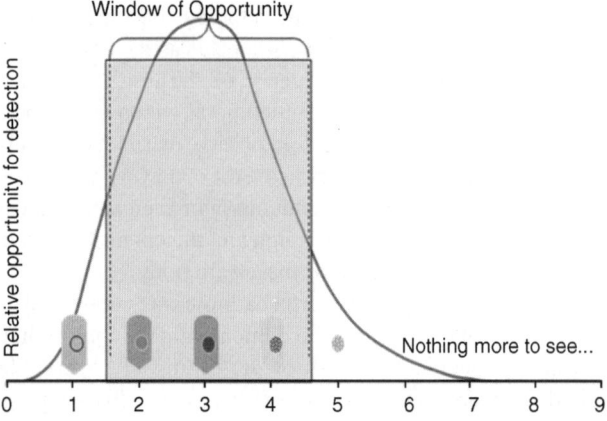

FIGURE 18 Typical time line of cell death that occurs with a variety of neurotoxicants. The cartoons represent cell body changes. Note that the ability to detect changes in the cell body vanishes around day 7. (*See insert for color representation of the figure.*)

- *Day 4.* The cell body debris begins to be removed.
- *Day 5.* The final debris is removed.
- *Day 6.* The remaining debris of the nucleus has been cleared away. Nearby cells begin to shift physically to consume the space left by the vacated cell, and the pathology appears as if no damage had occurred.

During this time line, the cell death event is detectable by various methods. Early and late in the cycle there is minimal pathology available for detection, and the event could easily be missed. In the middle of the cycle, however, the evidence of cell disintegration is more conspicuous and detection is more reliable. In general, there is about a 3-day window of opportunity, during which detection is likely to occur using the H&E stain. The time line described can begin immediately upon insult or exposure to a neurotoxin, or there may be a delay of several days before the observable disintegrative activity begins.

Acute Cell Death

Cell death due to an acute reaction to a neurotoxin is the simplest to detect and the most common type to be expressed, but may be the endpoint missed most often during neurotoxicity studies. Cell death due to an acute exposure or insult follows specific temporal patterns, making detection relatively simple. In acute cell death examples, all cells that are vulnerable to a neurotoxin tend to begin dying at about the same time.[4–45] Observable pathology is detectable within 1 to 5 days of exposure to a neurotoxin and persists for 2 to 3 days. By 5 to 10 days following exposure, there is no remaining pathology left to observe which indicates that cell death occurred. Since no evidence persists beyond day 10, only the time frame of days 1–10 (following exposure) is required for assessment of cell death. This makes detection simpler because there are many data points (cells) reacting in a similar manner at a point in time (the chosen time of sacrifice). Detection can be simple in that the assessment window is limited to days 1 to 10.

Although it is possible to experience cell death due to subchronic and chronic dosing, in the author's experience and based on published reports, all (publicly) documented pharmaceutical class compounds that cause neuronal cell death have done so following acute dosing (i.e., the necrosis was detectable in a narrow time frame following the initiation of dosing). The peak of neuronal cell death detection was consistently within 2 to 4 days following exposure to each neurotoxin (Figure 19).[4–45]

The evidence of cell death is transient, but the effects of cell death are permanent. In the CNS it is important to recognize that pathological examinations reveal a snapshot in time rather than a cumulative picture of past events. Dying

Days post-administration

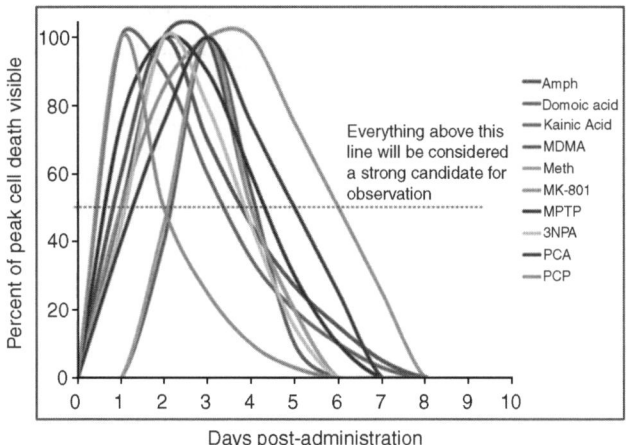

FIGURE 19 Percentage of peak cell death visible against days postadministration. Although the lines are difficult to differentiate, the graph clearly shows that most cell death, with a variety of neurotoxicants, takes place between 1 and 4 days following initial administration of a compound.

neurons have a short window (several days) of opportunity for detection, and after that the window closes; the neurons destroyed are no longer visible. Unless the lesion is huge, there is no scarring and the cells are not replaced. Most compounds known to cause cell death do so following the first dose, and therefore the time line for observation is a 2 to 3 day window sometime during the first week.[46]

There is not a single time point at which all acute neurotoxins will have an observable effect; rather, there is a range of when to look. There are factors that can skew the observability curve slightly earlier or later in the time line. Each compound can elicit different pathways leading to cell death and therefore can have a unique timing profile. Higher doses of a compound can sometimes accelerate the pathway events leading to cell death. Species, strain, gender, and age would all be expected to have a potential impact on the observability curve. It is important therefore to leverage the predictability that statistical evidence affords while still accommodating the potential for variation. Cell death timing can be relatively predictable, but not necessarily precise with different variables. The full potential temporal range of cell death must be assessed.

False-negative results for neurotoxicity can easily be concluded if brain sections are not observed at the correct time. Observing sections on or after day 7 following the initiation of dosing may well miss the main event. The lack of observable cell death at a specific point in time is not definitive. Rather, such a finding should be qualified as "not evident at that point in time." An accurate conclusion that no cell death occurred is appropriate only when all applicable time ranges have been assessed. Once vulnerable cells die,

subsequent administration of a compound may not induce further cell death.

It has been established that most documented cases of cell death are due to an acute response from first or early exposures. With this response, most or all of the cells vulnerable to a particular neurotoxin die and future exposure usually does not lead to future cell death. In a case study of alcohol,[12] three groups of rats were exposed to a "binge" of heavy alcohol exposure for 5 days. The first group of rats was evaluated 72 h following this exposure on day 8. The second and third groups were allowed a 7-day recovery period, and then repeated the binge. Group 2 was evaluated 72 h after this second binge cycle. The third group repeated the recovery one more week, followed by a third binge period of 5 days, and was then evaluated 72 h later.

Cell death was found to have occurred in group 1 and was evident 3 days after a 5-day binge of heavy alcohol exposure. In this case cell death did not occur immediately upon first exposure but, rather, as a response to multiple or persistent exposure over the 5 days. Three days later, consistent with an acute cell death model, the vulnerable cells began to die in a synchronous pattern. In groups 2 and 3, however, there was no cell death evident following their second and third binges, respectively. This suggests that repeated exposure to a neurotoxin does not lead to continued cell death over time; rather, all vulnerable cells can be destroyed during early exposure.

As a consequence to this lack of perpetual cell death, cell death as a result of an acute response is the most often missed event in neurotoxicity evaluations. Specifically, study designs that evaluate tissues 14, 28, or 60 days (as examples) after the initiation of dosing may well fail to detect the cell death that occurred shortly after the initiation of dosing but was no longer evident once the animals were necropsied. As the alcohol example illustrates, looking for cell death at any point in time later than when the event occurs (which is usually in the first 10 days of exposure) will not result in the detection of cell death. This does not imply that 14-, 28-, 60- day, and so on, studies are invalid; rather, that in addition to evaluating those time points, earlier time points must also always be evaluated.

Different toxicity observations require unique timing intervals for appropriate assessments and measurements. For many measures, evaluations for a time period are only valid if they are conducted during the time period rather than one time at the end. Pharmacokinetic analysis, for example, requires sampling over time to tell a complete story. Functional tests, cage-side observations, functional observation batteries (FOBs), and so on, are likewise conducted throughout the duration of a study. The observation of neuropathic endpoints entails sacrificing subsets of animals and evaluating their brains at periodic intervals during the course of the study in order to assess completely the time course of cell death.

Figure 19 shows that most of the profiles of cell death overlap in their ability to be observed sometime during days 2 to 4. However, there is no single point in time during which all of the compounds would have a high probability of detection. Two sacrifice times may be necessary to capture both the early and late cell death cycles. Assessing a group of animals at about 48 h and about 96 h creates the highest probability of witnessing acute cell death. This approach achieves a high probability of success. The only potential gap would be a late occurrence from an acute response that could occur as late as 10 days from exposure. Adding a third time point of assessment at 7 to 8 days would virtually eliminate the potential of missing a cell death event due to acute exposure.

Considerations for Subchronic and Chronic Exposure

The temporal attributes of cell death are more varied in the subchronic and chronic time frames (vs. acute); however, many of the same principles can be adapted. As described earlier, cell death due to acute exposure occurs fairly rapidly, and the cascade of events leading to cell death occurs at a fairly consistent time line for all cells, leading to a seemingly synchronous cell death event of all vulnerable cells.

Longer exposures to a neurotoxin may trigger multiple factors or may affect cells at different rates. The result is that vulnerable cells have the potential to die in a wider variability of time following insult or exposure. The longer it takes for cells to die, the more variance is possible from one cell to another in the timing of cell death.

As a result of this variation, detection of events during the subchronic and chronic time frames is much more challenging. Although the same volume of cells in a population may be affected (as compared to an acute response), fewer are observable at any point in time. The effect is that observations of cell death may appear to be minimal in quantity, may not be noticed at all, or may be attributed to natural attrition of cells and not be recognized as a concern. Technological advances in staining technology can help with this.

Despite the challenges, periodic intervals of sacrifice during the course of administration are still the most effective method of achieving a reasonably adequate observation. Temporal sampling in subchronic and chronic studies is a trade-off between thorough and practical. To be completely thorough, as with acute studies, animals could be evaluated every 2 to 3 days. Although thorough, this approach is impractical; however, evaluating animals only at the end of the study risks missing an event that probably occurred at some point during the study. The very fact that cell death is not likely to be synchronous in these cases allows for a reasonable and effective compromise. The interval between evaluations can increase later in the study just as the likely variance of cell death timing among a population increases.

The most important consideration in designing a subchronic or chronic study is not to ignore the potential for acute effects. Always perform an acute evaluation, regardless of the anticipated time line of events or the method of administration and expected pattern of use. Following the initial two acute evaluations at 2 and 4 days, additional evaluations should occur at 7 to 10 days, 16 to 20 days, 25 to 30 days, and then monthly from 30 to 90 days and quarterly for studies lasting longer than 90 days.

The most common pitfalls of neurotoxicity study designs with respect to timing are:

- Sacrificing and assessing animals only at the end of a study. This almost ensures a negative evaluation for neurotoxicity.
- Spreading the sacrifice points too far apart. For the acute range, this means no more than 48 to 72 h between time points.

Timing of Other Markers for Neurotoxicity

This section has focused on specific examples related to the timing of cell death (neuronal) due to known neurotoxins. There are timing considerations for all markers of toxicity since few are persistent and many are observable at unique points in time. Each endpoint has the potential for unique temporal profiles of expression, and each endpoint marker must be used in accordance with the known characteristics of the profile for that expression. Next to cell death markers, the two most common markers of neurotoxicity are to look for hypertrophy of microglia or astrocytes, indicating a perturbation in the brain. Glial reactions vary considerably depending on the nature of the neurotoxin. Astrocyte perturbation may be detected as early as 36 h following exposure, peaking at about 72 h and persisting for several weeks. Microglial perturbation may also be detected very early, but often begins at about 4 days, peaks at 5 to 7 days, and may persist for several weeks.

For any given marker of neurotoxicity, the typical timing of the expression of that marker must be understood and accounted for in the study design. A single time point is rarely adequate in these assessments. Instead, a series of two or more time points are used to sample during the most likely times of expression.

REFERENCES

1. Switzer RC III. Neurotoxicity Study Design. 2007. http://www.neuroscienceassociates.com/Presentations/neurotox_study_design.zip.

2. Franssen CL, Switzer RC III. Neurotoxicity Study Design: The Essential Element of Sacrifice Times. Poster presentation. 37th Annual Meeting of the Society for Neuroscience. San Diego, CA: NeuroScience Associates; 2007. Available at: http://

www.neuroscienceassociates.com/Documents/Publications/2007_sfn_neurotox_timing_poster.pdf

3. Switzer RC III, Miller D. Neurotoxicity Study Design: The Essential Element of Location Assessment. Annual Meeting Society of Toxicology, Seattle, WA, 2008. Society of Toxicology Annual Proceedings, p. 378. Available at: http://www.neuroscienceassociates.com/ntoxdesign.html.

4. Belcher AM, O'Dell SJ, Marshall JF. Impaired object recognition memory following methamphetamine, but not *p*-chloroamphetamine- or *d*-amphetamine–induced neurotoxicity. *Neuropsychopharmacology*. 2005;30(11):2026–2034.

5. Bowyer JF. Neuronal degeneration in the limbic system of weanling rats exposed to saline, hyperthermia or *d*-amphetamine. *Brain Res*. 2000;885(2):166–171.

6. Bowyer JF, et al. Neuronal degeneration in rat forebrain resulting from amphetamine-induced convulsions is dependent on seizure severity and age. *Brain Res*. 1998;809(1):77–90.

7. Bowyer JF, Delongchamp RR, Jakab RL. Glutamate N-methyl-D-aspartate and dopamine receptors have contrasting effects on the limbic versus the somatosensory cortex with respect to amphetamine-induced neurodegeneration. *Brain Res*. 1998;1030(2):234–246.

8. Carlson J, et al. Selective neurotoxic effects of nicotine on axons in fasciculus retroflexus further support evidence that this is a weak link in brain across multiple drugs of abuse. *Neuropharmacology*. 2000;39(13):2792–2798.

9. Ellison G. Neural degeneration following chronic stimulant abuse reveals a weak link in brain, fasciculus retroflexus, implying the loss of forebrain control circuitry. *Eur. Pharmacol*. 2002;12:287–297.

10. Jakab RL, Bowyer JF. Parvalbumin neuron circuits and microglia in three dopamine-poor cortical regions remain sensitive to amphetamine exposure in the absence of hyperthermia, seizure and stroke. *Brain Res*. 2002;958(1):52–69.

11. Jakab RL, Bowyer JF. The injured neuron/phagocytic microglia ration "R" reveals the progression and sequence of neurodegeneration. *Toxicol. Sci*. 2004;78(S-1).

12. Crews FT, et al. Binge ethanol consumption causes differential brain damage in young adolescent rats compared with adult rats. *Alcoholism Clin. Exp. Res*. 2000;24(11):1712–1723.

13. Han JY, et al. Ethanol induces cell death by activating caspase-3 in the rat cerebral cortex. *Molecules Cells*. 2005;20(2):189–195.

14. Ikegami Y, et al. Increased TUNEL positive cells in human alcoholic brains. *Neurosci Lett*. 2003;349:201–205.

15. Colman JR, et al. Mapping and reconstruction of domoic acid–induced neurodegeneration in the mouse brain. *Neurotoxicol Teratol*. 2005;27:753–767.

16. Benkovic SA, O'Callaghan JP, Miller DB. Regional neuropathology following kainic acid intoxication in adult and aged C57BL/6J mice. *Brain Res*. 2006;1070:215–231.

17. Benkovic SA, O'Callaghan JP, Miller DB. Sensitive indicators of injury reveal hippocampal damage in C57BL/6J mice treated with kainic acid in the absence of tonic–clonic seizures. *Brain Res*. 2004;1024(1–2):59–76.

18. Schmued LC, Bowyer JF. Methamphetamine exposure can produce neuronal degeneration in mouse hippocampal remnants. *Brain Res*. 1997;759(1):135–140.

19. Jensen KF, et al. Mapping toxicant-induced nervous system damage with a cupric silver stain: a quantitative analysis of neural degeneration induced by 3,4-methylenedioxymethamphetamine. In: L Erinoff ed. *Assessing Neurotoxicity of Drugs of Abuse*. Rockville, MD: U.S. Department of Health and Human Services; 1993;133–149.

20. Johnson EA, et al. *d*-MDMA during vitamin E deficiency: effects on dopaminergic neurotoxicity and hepatotoxicity. *Brain Res*. 2002;933:150–163.

21. Johnson EA, O'Callaghan JP, Miller DB. Chronic treatment with supraphysiological levels of corticosterone enhances D-MDMA–induced dopaminergic neurotoxicity in the C57BL/6J female mouse. *Brain Res*. 2002;933:130–138.

22. O'Shea E, et al. The relationship between the degree of neurodegeneration of rat brain 5-HT nerve terminals and the dose and frequency of administration of MDMA ("ecstasy"). *Neuropharmacology*. 1998;37:919–926.

23. Luellen BA, et al. Neuronal and astroglial responses to the serotonin and norepinephrine neurotoxin: 1-methyl-4-(2-aminophenyl)-1,2,3,6-tetrahydropyridine. *J Pharmacol Exp Ther*. 2003;307(3):923–931.

24. Ellison G. The N-methyl-D-aspartate antagonists phencyclidine, ketamine and dizocilpine as both behavioral and anatomical models of the dementias. *Brain Res Rev*. 1995;20(2):250–267.

25. Fix AS, et al. Integrated evaluation of central nervous system lesions: stains for neurons, astrocytes, and microglia reveal the spatial and temporal features of MK-801–induced neuronal necrosis in the rat cerebral cortex. *Toxicol Pathol*. 1996;24(3):291–304.

26. Fix AS, et al. Neuronal vacuolization and necrosis induced by the noncompetitve N-methyl-D-aspartate (NMDA) antagonist MK(+)801 (dizolcilpine maleate): a light and electron microscope evaluation of the rat retrosplenial cortex. *Exp Neurol*. 1993;123(2):204–215.

27. Fix AS, et al. Pathomorphologic effects of N-methyl-D-aspartate antagonists in the rat posterior ingulate/retrosplenial cerebral cortex: a review. *Drug Dev Res*. 1994;32(3):147–152.

28. Fix AS, et al. Quantitative analysis of factors influencing neuronal necrosis induced by MK-801 in the rat posterior cingulate/retrosplenial cortex. *Brain Res*. 1995;696:194–204.

29. Olney JW. Excitotoxicity, apoptosis and neuropsychiatric disorders. *Curr Opin Pharmacol*. 2003;3(1):101–109.

30. Olney JW, et al. Do pediatric drugs cause developing neurons to commit suicide? *Trends Pharmacol Sci*. 2004;25(3):135–139.

31. Olney JW, et al. Environmental agents that have the potential to trigger massive apoptotic neurodegeneration in the developing brain. *Environ Health Perspect*. 2000;108 (Suppl 3):383–388.

32. Olney JW, et al. MK-801 powerfully protects against N-methyl-aspartate neurotoxicity. *Eur J Pharmacol*. 1987;141:357–361.

33. Olney JW, et al. MK-801 prevents hypobaric–ischemic neuronal degeneration in infant rat brain. *Neurosci.* 1989; 9(5):1701–1704.

34. Olney JW, Labruyere J, Price MT. Pathological changes induced in cerebrocortical neurons by phencyclidine and related drugs. *Science.* 1989;244(4910):1360–1362.

35. Wozniak DF, et al. Disseminated corticolimbic neuronal degeneration induced in rat brain by MK801. *Neurobiol of Dis.* 1998;5(5):305–322.

36. Horvath ZC, Czopf J, Buzsaki G. MK-801–induced neuronal damage in rats. *Brain Res.* 1997;753(2):181–195.

37. Creeley CE, et al. Donezepil markedly potentiates memantine neurotoxicity in the adult rat brain. *Neurobiol Aging.* 2008;29 (2):153–167.

38. Creeley C, et al. Low doses of memantine disrupt memory in adult rats. *J Neurosci.* 2006;26(15):3923–3932.

39. Jevtovic-Todorovic V, et al. Early exposure to common anesthetic agents causes widespread neurodegeneration in the developing rat brain and persistent learning deficits. *J Neurosci.* 2003;23(3):876–882.

40. Jevtovic-Todorovic V, et al. Prolonged exposure to inhalational anesthetic nitrous oxide kills neurons in adult rat brain. *Neuroscience.* 2003;122(3):609–616.

41. Jevtovic-Todorovic V, Benshoff N, Olney JW. Ketamine potentiates cerebrocortical damage induced by the common anaesthetic agent nitrous oxide in adult rats. *Br J Pharmacol.* 2000;130(7):1692–1698.

42. Maas J.W., et al. Calcium-stimulated adenylyl cyclases modulate ethanol-induced neurodegeneration in the neonatal brain. *J. Neurosci.* 2005;25(9):2376–2385.

43. Wilson MA, Molliver ME. Microglial response to degeneration of serotonergic axon terminals. *Glia.* 1994;11:18–34.

44. Harvey JA, McMaster SE, Yunger LM. *p*-Chloroamphetamine: selective neurotoxic action in brain. *Science.* 1975;187 (4179):841–843.

45. Miller PJ, Zaborsky L. 3-Nitropropionic acid neurotoxicity: visualization by silver staining and implications for use as an animal model of Huntingtons disease. *Exp Neurol.* 1997;1146:212–229.

46. Allen HL, et al. Phencyclidine, dizocilpine, and cerebrocortical neurons. *Science.* 1990;247(4939):221.

PART 2

TOXICOLOGIC NEUROPATHOLOGY: METHODOLOGY

10

PRACTICAL NEUROPATHOLOGY OF THE RAT AND OTHER SPECIES

WILLIAM H. JORDAN
Vet Path Services, Inc., Mason, Ohio

D. GREG HALL
Lilly Research Laboratories, Indianapolis, Indiana

MONTY J. HYTEN
Covance Laboratories, Inc., Greenfield, Indiana

JAMIE K. YOUNG
Department of Pathology, Covance Laboratories, Inc., Greenfield, Indiana

Source Acknowledgments: The work represented herein was conducted at and supported by Lilly Research Laboratories. Much of this information was originally published in a special issue of the *Journal of Histotechnology*.[1] Portions of this chapter and illustrations within are reproduced with permission of the *Journal of Histotechnology*.[1]

INTRODUCTION

This chapter focuses on details for a technical approach to efficient preparation of specimens for neuropathological evaluations from relatively large numbers of animals. It also touches on some specialized techniques, a practical approach for consistent brain trimming based on recognition of landmarks on the brain surface, and a review of basic neuroanatomy to facilitate more precise localization of brain structures and lesions using a brain atlas. Although the presentation is primarily in the context of a research and regulatory setting, some of the general principles can also be pertinent to diagnostic settings. The discussion is directed primarily to investigations using the rat, with occasional reference to other species, including dogs and nonhuman primates.

Rats are used extensively as models of human disease, including many diseases of the nervous system, as well as for safety testing of pharmaceutical compounds. The nervous system is composed of closely interrelated, yet anatomically and functionally heterogeneous structures which tend to be more sensitive to histological artifacts than are most other tissues. Although specific design details for neuropathological evaluations will be driven by the goals of the specific study, many features of tissue collection, processing, and examination will be common to all neuropathological evaluations.

Optimal neuropathological evaluations require detailed clinical signs and gross postmortem observations, careful and timely collection and processing of tissues, and a working knowledge of both neuroanatomy and neuropathology. Dorman et al.[2] provide a review of toxicological neuropathology, and Fix and Garman[3] address specific histological techniques and artifacts. The structural and functional organizational features of the rat brain, including the special senses, are reviewed extensively by multiple authors in a

Fundamental Neuropathology for Pathologists and Toxicologists: Principles and Techniques, First Edition. Edited by Brad Bolon and Mark T. Butt.
© 2011 John Wiley & Sons, Inc. Published 2011 by John Wiley & Sons, Inc.

book edited by Paxinos.[4] Solleveld and Boorman[5] and Mitsumori and Boorman[6] present basic techniques and many of the common changes of the central (CNS) and peripheral (PNS) nervous systems, respectively, of the Fischer 344 rat. Brain atlases are also readily available for the mouse[7] and nonhuman primate, but the most comprehensive dog brain atlas (Singer; see Table 1) has been out of print for many years. Table 1 lists selected print and online brain atlas resources for the dog and nonhuman primate.

Although special needs may require more detailed efforts, cost-effective high-quality histological evaluations can occur if animals are exsanguinated quickly, brains removed with minimal handling to maintain structural integrity and avoid dark neuron artifacts, immersion fixed quickly and thoroughly, and trimmed and processed to survey multiple areas consistently in histological sections. Most evaluations can be based on hematoxylin and eosin (H&E)–stained sections; however, availability of an epifluorescent microscope is useful for identification of neurodegeneration in H&E-stained sections, and special stains may be required to answer specific questions. Quality evaluations also require that both the technical staff and investigator have at least a working knowledge of a few easily identified neuroanatomical sites based on detailed atlases such as those referenced in this chapter.

SPECIMEN PREPARATION: SPECIAL CONSIDERATIONS

While the level of effort and attention to detail for a specific study will be driven by the goals of that study, minimizing common collection-derived artifacts will increase the detection sensitivity and quality of answers obtained for all studies. Common collection artifacts include incomplete tissue fixation, "dark neurons," instrument-induced tissue distortions, and the presence of erythrocytes remaining in tissues. The use of controls is very important. Controls should be matched for age, strain, and study or batch. At each stage, from necropsy through slide preparation and evaluation, terminal procedures must occur either randomly or consistently across treatment groups to avoid artifacts due to differences between and within operators (both technical and professional staff).

Inadequate preservation of tissues will often result in tearing and folding artifacts due to difficulties in microtomy. Slow penetration of fixative will often result in perinuclear vacuoles in thalamic and other nuclei. High concentrations of alcohol may also lead to artifactual vacuolation, especially of white matter.[8] While routine processing in 70% alcohol does not cause vacuolation, storage for as few as 5 days in 70% alcohol can lead to artifactual vacuolation.[9] Storage at higher

---

TABLE 1   Selected Print and Online Brain Atlas Resources for Dogs and Nonhuman Primates

Dogs

Print:[a]

  Singer M. *The Brain of the Dog in Section.* Philadelphia: W.B. Saunders; 1962 (out of print).

  Lim RKS, et al. *A Stereotaxic Atlas of the Dog's Brain.* Springfield, IL: Charles C Thomas; 1960 (out of print).

  Dua-Sharma S, et al. *The Canine Brain in Stereotaxic Coordinates; Full Sections in Frontal, Sagittal, and Horizontal Planes.* Cambridge, MA: MIT Press; 1970 (occasional used copies available).

Online:

  http://brainmaps.org/index.php?p=speciesdata&species=canis-lupus (a coronal series of Nissl-stained sections)

  http://vanat.cvm.umn.edu/brainsect/ (selected images of Nissl-stained coronal sections from the series published at http://www.brainmaps.org)[b]

  http://www.brainmuseum.org/sections/index.html[c]

Nonhuman Primates

Print:[d]

  Paxinos G, et al. *The Rhesus Monkey Brain in Stereotaxic Coordinates.* 2nd ed. San Diego, CA: Academic Press; 2009.

  Snider RS, Lee JC. *A Stereotaxic Atlas of the Monkey Brain (Macaca mulatta).* Chicago: University of Chicago Press; 1961.

  Saleem KS, Logothetis NK. *A Combined MRI and Histology Atlas of the Rhesus Monkey Brain in Stereotaxic Coordinates.* San Diego, CA: Academic Press; 2007.

  Szaba J, Cowan WM. A stereotaxic atlas of the brain of the cynomolgus monkey (*Macaca fascicularis*). *J Comp Neurol.* 1984;222:265–300.

Online:

  http://brainmaps.org/index.php?p=speciesdata&species=macaca-mulatta (includes coronal, sagittal, and horizontal series of Nissl-stained sections)[e]

  http://www.brainmuseum.org/sections/index.html[c]

---

[a] See also http://www.kopfinstruments.com/Atlas/Dog.htm.

[b] Transverse sections are related to their appropriate position on images of an unsectioned dog brain photographed from the lateral aspect and the sagittal midline.

[c] Series of Nissl-stained coronal sections of Basenji dog brain without labels.

[d] See also http://www.kopfinstruments.com/Atlas/Macaque.htm.

[e] Coronal sections with various antibody labels of cynomolgus monkey (*Macaca fasciculata*) also available.

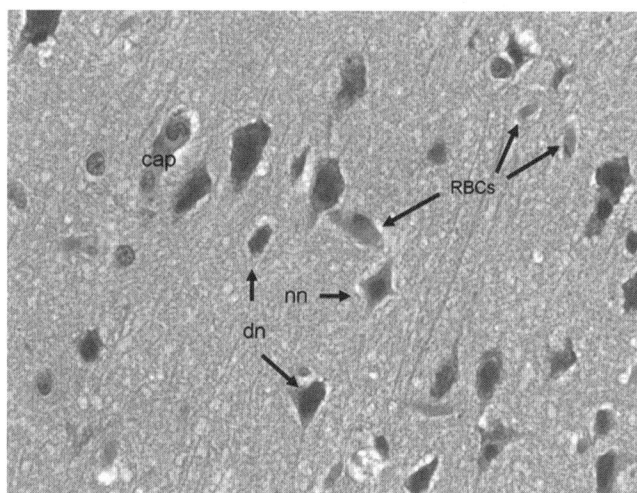

FIGURE 1 A single degenerating neuron (nn) is surrounded by many artifactually created dark neurons (dn). Several capillaries (cap), often containing erythrocytes (RBCs) are also present, structures whose shape and size can resemble degenerate neurons. (*See insert for color representation of the figure.*)

concentrations for longer than 24 h increases the probability of artifactual vacuolation.

Some of the most common artifacts observed histologically in rat brains are dark neurons (Figure 1), which are sometimes mistaken for degenerating or necrotic neurons.[10] Dark neurons have amphophilic cytoplasm (with H&E stain, cytoplasm is slightly pink to deep purple in color), well defined, often angular cytoplasmic boundaries, and slightly shrunken to normal-size nuclei, the centers of which often match the color of the altered cytoplasm. Dark neurons are produced by direct contact or pressure at any time between the death of the animal and the time the brain is completely fixed, including procedures involving handling of the brain, such as weighing or tissue collection list reconciliation. The effect of pressure on the surface of the brain can often be traced from the surface through the hippocampus and into the deeply underlying tissues. Even the pressure caused by human fingers in picking up a rat brain prior to complete fixation can cause moderate numbers of dark neurons. Fresh rodent brains may be picked up with very little damage by gently cradling with a slightly opened pair of blunt-tipped curved scissors rather than grasping with fingers or forceps.

Punctures, tears, and forceps teeth indentations are commonly induced iatrogenic (artifactual) changes that must be differentiated from disease or compound induced lesions in rats. Considerable time may be devoted by the pathologist in deciding whether focal hemorrhages in the brain are iatrogenic or a result of treatment. At the very least, pressure from the extraction instruments may cause dark neurons or destroy areas critical to the evaluation of the brain.

Dark neurons are useful for determining the source of forceps-induced artifacts in brain sections so that good

handling procedures can be reinforced. Artifacts induced by instruments at necropsy are almost always associated with abundant dark neurons and often with release of erythrocytes into the tissues without evidence for erythrophagocytosis. Instrument-induced artifacts caused by handling partially fixed brains generally produce dark neurons with minimal numbers of free erythrocytes and often instrument-shaped indentations, whereas those induced after fixation do not cause dark neurons but tend to leave indentations matching the shape of the instrument surface.

The importance of exsanguination for routine tissue collection is often overlooked. Erythrocytes in histological preparations, when viewed at low magnifications, have tinctorial qualities similar to those of degenerating neurons and activated astrocytes. While abnormal neurons and astrocytes can readily be distinguished from erythrocytes, higher microscopic magnifications are required resulting in decreased efficiency due to increased pathologist's time spent in the evaluation. Even more important, there is an increased chance of missing degenerating neurons or increased numbers of astrocytes hidden among the blood-filled vascular spaces. Because erythrocytes often autofluoresce, their presence becomes even more challenging when evaluating fluorescent-labeled stains. This is especially true with the Fluoro-Jade B (or C)-stained histological slides since the erythrocytes fluoresce at about the same intensity and whose size can be similar to that of degenerating neurons. The challenges associated with autofluorescence of erythrocytes while evaluating any of the Fluoro-Jade stains can be offset by using a narrower band filter, but the decrease in light increases the challenge of maintaining neuroanatomical perspective. Thus, rats should be exsanguinated when euthanized for histopathological evaluation. Tissue quality is best preserved when exsanguination occurs immediately after euthanasia, whether induced by carbon dioxide asphyxiation, overdose of injectable anesthetic, or by another commonly used euthanasia method. For rapid exsanguination, the abdomen of the anesthetized or freshly euthanized rat can be opened and the abdominal vena cava and descending aorta severed. If a portion of the brain is to be processed for chemical analysis, flushing of the brain with buffered saline to remove the blood will also produce erythrocyte-free histologic sections as will perfusion fixation discussed later, but both perfusion procedures are labor intensive.

COLLECTION AND PRESERVATION

Collection and preservation of nervous system tissues requires greater care than for most other tissues. Fixation by immersion in 10% neutral buffered formalin (NBF) is most efficient for many routine studies. Tissue preservation and minimization of artifacts can be improved by whole- or upper-body perfusion of the anesthetized rat.

Excellent perfusion fixation can be obtained with the same 10% NBF used for immersion fixation. Specific applications may be enhanced by perfusing with paraformaldehyde, glutaraldehyde,[11] or Bouin's fixative,[12] although the safety to technical staff becomes more of an issue, due to increased volatility of these fixatives and the resulting greater potential for exposure to toxic fumes. Contrary to recommendations by others, our experience suggests that skipping the preflush and infusing room temperature fixative directly into the heart of lightly anesthetized rodents produces optimal fixation with almost complete exsanguination. Pretreatment with vasodilators or heparin is not required. Seven minutes of intracardiac perfusion through a 21-gauge butterfly catheter with the perfusate 80 to 100 cm above the table is a good starting point. Elevation (i.e., pressure) of perfusate can be adjusted to optimize tissue blanching without causing fluids to flow from the nostrils (an indication that the pressure was too high).

If immersion fixed, the incidence of dark neurons will be reduced if the rat brains remain in fixative for at least 48 h before they are trimmed. If perfusion fixed, they should remain in the intact calvarium (immersed in the fixative) for at least 48 h prior to removal and trimming. Although it may seem logical to slightly open the cranial vault of a perfusion-fixed rat immediately after perfusion to allow the brain to come into better contact with the fixative, this will only result in increased numbers of dark neurons (Figure 2).

In recently euthanized rats, and even in rats that have been dead for a few hours, spinal cords may easily be collected by hydraulic extraction after the head has been removed. Removal of the first few cervical vertebrae by transecting the spinal column at an intervertebral disk with a scalpel to create

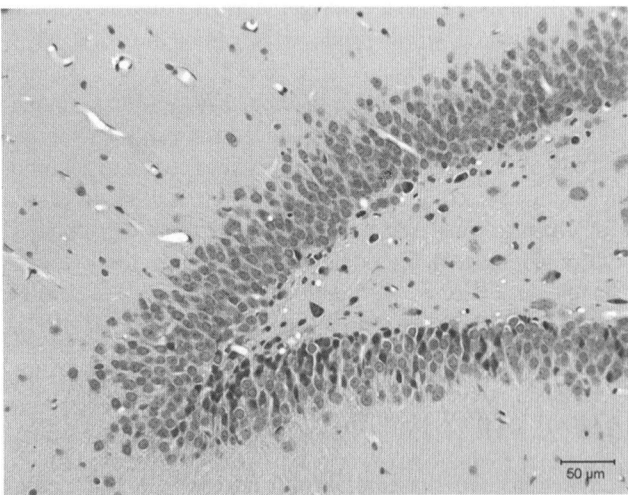

FIGURE 2 Dark neurons, a handling artifact, caused by removing the brain a short time (within 1 h) following systemic formalin perfusion. The artifact can be avoided by allowing sufficient fixation time prior to opening the calvarium. Rat dentate gyrus, H&E. (*See insert for color representation of the figure.*)

a clean exit route for the cord will help to assure success. A sharp push on the plunger of a 6- to 12-mL syringe filled with tap water and attached to a blunt 18- to 22-gauge needle placed into the vertebral canal transected at the midlumbar region will force the cord out through the cervical canal.[13] The hub of the needle should be placed firmly into the spinal column, and a few gentle pushes on the syringe plunger prior to the sharp push may be useful. Because the diameter of the spinal canal is greater in the cervical region, the cord should not be forced rostral to caudal. Do not extract with formalin because of the increased potential for a splash hazard. Although the process of hydraulic extraction appears as though it should produce an abundance of dark neurons or artifactual vacuolation, in practice, high-quality sections can be obtained in this manner. When hydraulic extraction occasionally fails, fix the column intact and dissect out the spinal cord, push out a short segment of fixed cord with a blunt probe, or decalcify and section the cord within the spinal column.

In an alternative method that allows microscopic evaluation of the cord without special collection or dissection, the spinal column with the cord in place is fixed and decalcified, then trimmed and processed for microscopic evaluation. This method provides the advantage of retaining the spinal nerves and ganglia within the specimen so that they are available for evaluation (depending on the trimming scheme used). However, focused evaluation of dorsal root ganglia may require careful dissection, usually following in situ fixation. The mild decalcifying properties of Davidson's[14] or Bouin's[12] fixatives also allow for sectioning of the spinal cord within the spinal column (or for the brain in the calvarium), especially from young rodents.

Sciatic and other peripheral nerves will hold their shape reasonably well if left attached to the underlying muscle during immersion fixation. Either or both (nerve and muscle) may be placed on a piece of paper and briefly air dried before fixation to help maintain anatomical features. Although tedious to perform, teased nerve preparations provide an excellent view of both axons and myelin. Preservation of one of a pair of nerves in glutaraldehyde is suggested if electron microscopy is contemplated.

TRIMMING AND PROCESSING

The approach to brain trimming is an important consideration that is influenced by the specific questions of the study: the desire to correlate the histological evaluation with imaging studies and the experience and preference of the pathologist. The planes and location of sectioning should be consistent between animals in the study and should facilitate the ability of the examiner to recognize histological landmarks and achieve the greatest benefit from reference brain atlases. Brains are typically trimmed in either coronal (transverse),

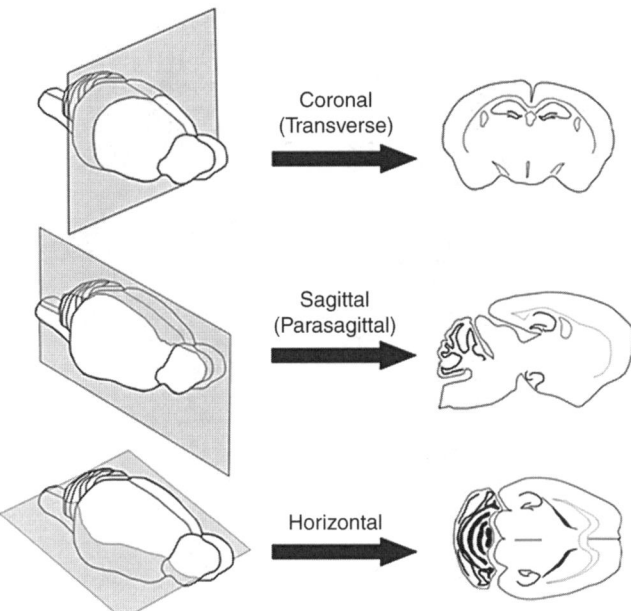

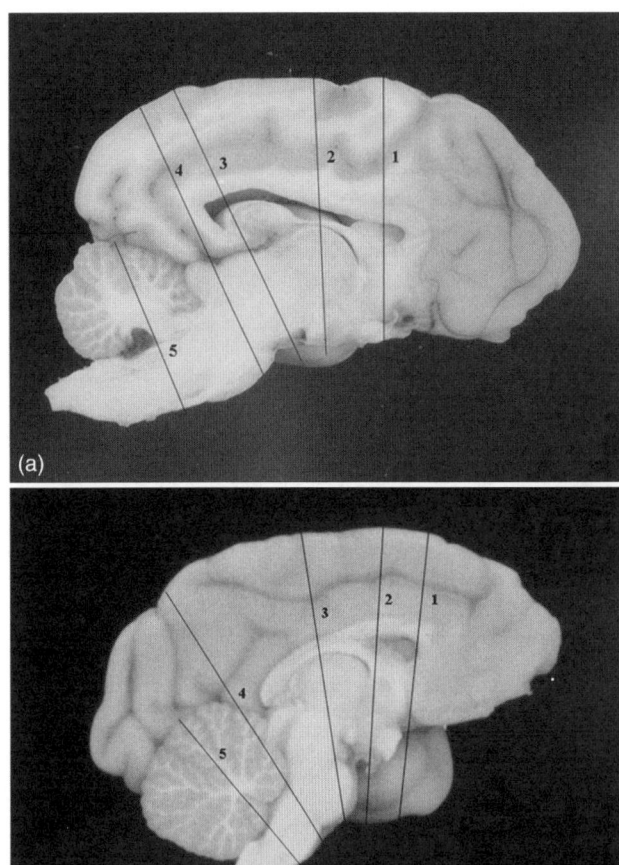

FIGURE 3 Coronal, sagittal, and horizontal sectioning planes applied to the rodent brain. (The olfactory bulbs are depicted on the illustration of the intact brain but not on the sagittal or horizontal slices.)

FIGURE 4 Sampling paradigms for dog (a) and nonhuman primate (b) designed to represent selected areas of interest in a minimum number of slices. Note that some of the planes of section intentionally deviate (tilt) from true coronal plane. Process only the ventral one-half of sections 1, 2, and 4 if examination of fewer cerebral cortical sections is desired. (*See insert for color representation of the figure.*)

sagittal, or horizontal planes (Figure 3), with coronal sectioning employed most commonly in toxicology studies.

Coronal planes are parallel to the plane that divides the body into front and back halves. Structures that appear in the coronal plane of section vary among species due to differences in body position and head carriage, with differences being particularly notable when rodents are compared to nonhuman primates. Since the coronal plane is in reference to the head rather than to the removed brain, histological coronal sections may not register exactly with coronal renderings collected by noninvasive imaging techniques such as MRI or PET. Additionally, brain sectioning paradigms that deviate intentionally from true coronal sectioning are sometimes chosen for species such as dogs (Figure 4a) and nonhuman primates (Figure 4b) in order to survey selected brain structures efficiently within a limited number of histological sections. This special sectioning must be done consistently between animals and with the knowledge that the resulting slides will differ from plates in reference atlases.

Many brain atlases present their series of coronal (transverse) sections in reference to a landmark known as Bregma, which is the location on the skull where the coronal and sagittal sutures meet (Figure 5). In the rat, Bregma 0.0 mm corresponds to a coronal section at the level of the optic chiasm. Sections going rostrally have progressively more positive Bregma designations, while those going caudally are progressively more negative.[15] The coronal series in most atlases represent the brain in parallel transverse planes of section, without tilting or skewing. The histotechnologist

must recognize that tilting or skewing the blade during brain trimming will result in sections that deviate from the intended coronal plane, and the slide evaluator must be aware that this practice may result in individual histological sections appearing on slides that contain structures separated by several plates in the atlas (Figure 6).

Consistent trimming of brains for histological sections is important to optimize the chances for lesion identification and to standardize the areas of the brain evaluated from multiple test animals. The three brain sections suggested by Solleveld and Boorman[5] are commonly used by many toxicological pathology laboratories and include (1) the section just anterior to the optic chiasm, (2) the caudal border of the mammillary body, and (3) 2 mm caudal to the pons as viewed on the ventral surface of the brain. Although these locations were probably selected to represent a practical

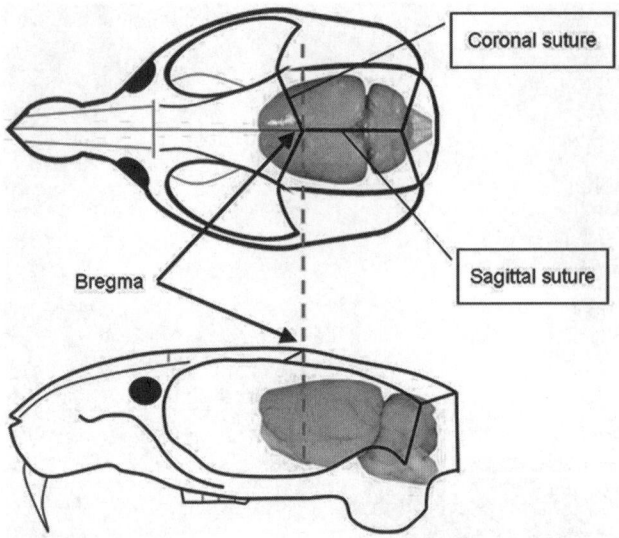

FIGURE 5   Bregma, the location on the skull where the coronal and sagittal sutures meet, provides a fixed reference point for the brain relative to the skull. (*See insert for color representation of the figure.*)

survey of the brain, the current authors have noted many variations among laboratories and even within studies based on this trimming approach. As implemented, the current three-section scheme sometimes misses the hippocampus and will include either midbrain or multiple thalamic and amygdaloid nuclei, but not both. We have found that trimming an additional cerebral/midbrain section is cost-effective since the sections still fit in one or two paraffin blocks and inclusion of the extra section helps to assure that important neuroanatomical structures are evaluated more

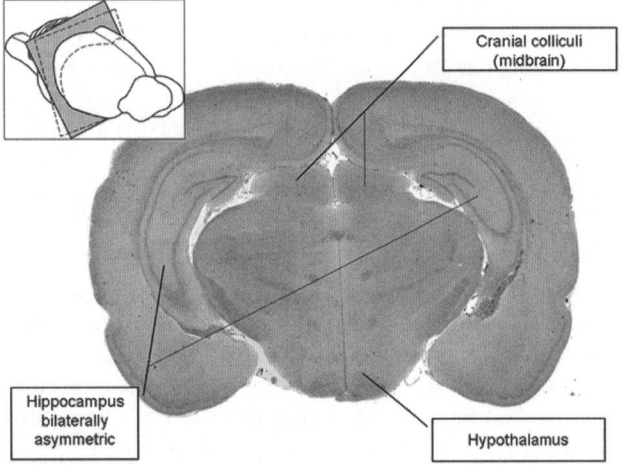

FIGURE 6   Section of brain that is both tilted and skewed. The histological section lacks bilateral symmetry of the hippocampus and displays both midbrain (cranial colliculus) and hypothalamic structures. The inset illustrates the intended plane of section and the tilted/skewed plane (hatched). (*See insert for color representation of the figure.*)

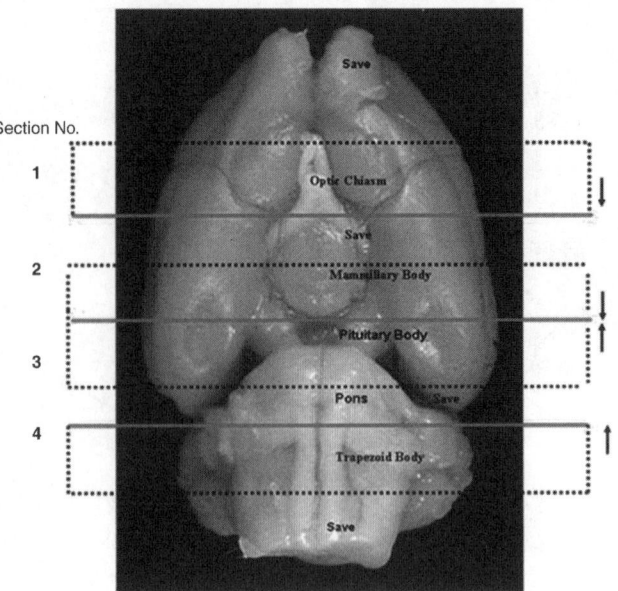

FIGURE 7   Diagram for a four-section rat brain trim. Solid red lines indicate positions for trim blade placement. Arrows indicate the surfaces to be microtomed. (*See insert for color representation of the figure.*)

consistently. See Figure 7 for a suggested four-section trim diagram for the rat. Note that the resulting histological sections would be similar to those in Figure 8 and that brain sections may be evaluated more efficiently if placed linearly in anatomical order on the microscopic slide rather than in the 2 × 2 pattern depicted in Figure 8.

If the focus of the study is on neuropathology and multiple areas of the brain are of interest, the use of a brain matrix (Figure 9, available from multiple vendors) assists consistent trimming. The matrix is used to guide slicing based on consistent ventral surface landmarks. This technique will help to obtain slices at precise locations and to minimize tilting and skewing of the plane of section, which can result in substantial deviation from the intended coronal plane. The use of tissue-marking dyes on the surface of each slice to be embedded away from the face of the block will help to assure that slices are not flipped in the embedding process. The use of four-well processing cassettes will facilitate placing the slices in consistent anatomical order on the slides. Consistent placement of the slices in a rostral-to-caudal or other logical neuroanatomical order will allow the researcher to focus on morphological changes without the distraction of relocating sections to maintain the desired sequence of evaluation. Two to four coronal slices of rat brain per block improve the efficiency over single slices per block. Limiting the trimming staff to one or a small group of skilled technicians for trimming the brains will further improve consistency.

Specific areas of brain to be examined should be based on the most likely neurochemical targets or other model features. However, if the purpose is to survey the brain more

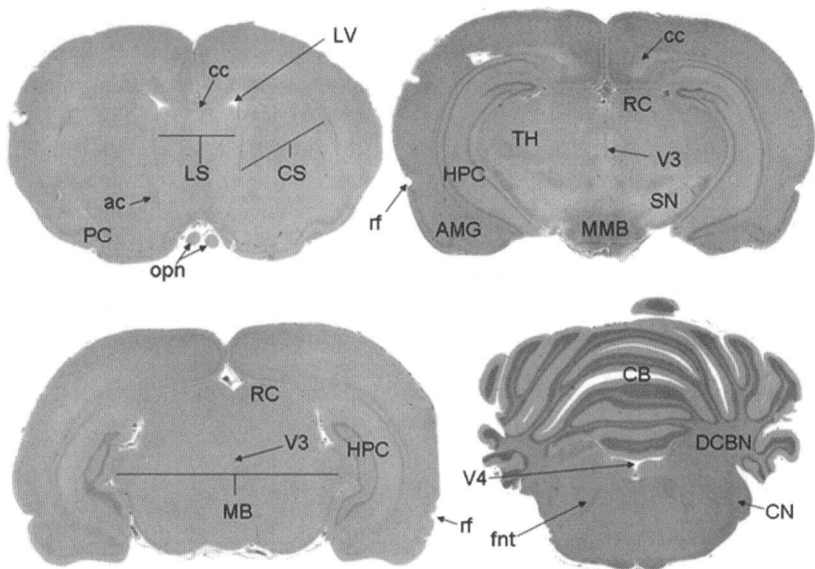

FIGURE 8   Location of important brain landmarks in histological sections in sections from a four-section trim. ac, Anterior commissure; AMG, amygdaloid region; CB, cerebellum; cc, corpus callosum; CN, cochlear nucleus; CS, striatum; DCBN, deep cerebellar nuclei; fnt, facial nerve tract; HPC, hippocampus; LS, septal nuclear area; LV, lateral ventricle; MB, midbrain; MMB, mammillary body; opn, optic nerve; PC, piriform cortex; RC, rostral colliculus; rf, rhinal fissure; SN, substantia nigra; TH, thalamus; V3, third ventricle; V4, fourth ventricle. (*See insert for color representation of the figure.*)

extensively, eight slices collected at 2- to 3-mm intervals provide a representative sampling. Brains should be placed in a brain matrix or other device to guide the razor blade, seating the specimens fully within the matrix to obtain consistent planes of sectioning between slices. To minimize variations, insertion of the first blade should be based on a centrally located ventral landmark such as the mammillary body. The trimmer then makes progressive rostral and caudal cuts from

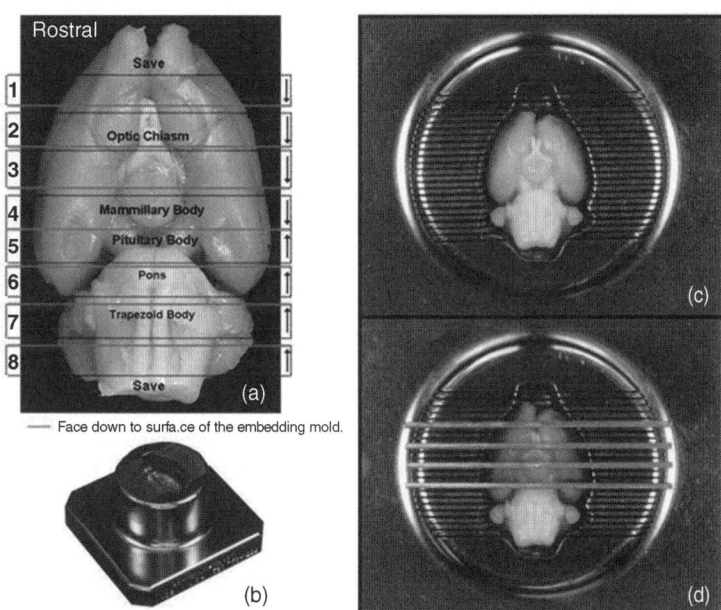

FIGURE 9   Detailed trimming of rat brains: (a) trim diagram to obtain eight histological sections; (b) brain matrix; (c) rat brain in matrix with ventral surface up; (d) with four of the five most rostral blades in place for an eight-section trim [blades are minimally farther apart than sectioning planes depicted in diagram (a) to obtain sufficient light to clearly illustrate brain features]. (*See insert for color representation of the figure.*)

the initial cut. To minimize shifting of the brain within the apparatus, all slices should remain in the matrix until all blades have been inserted. The blades and brain can then be removed individually or together while carefully assuring that the slices can be individually oriented into the processing cassettes while maintaining rostral-to-caudal sequence. Adjustments to the procedure described in the following paragraph may be made to focus on specific neuroanatomical features.

An eight-section (slice) brain trimming scheme (Figure 9a) that is practical for preclinical toxicological screening of neuroactive pharmaceutical compounds is as follows. Beginning with a coronal (transverse) cut at the caudal edge of the mammillary body, collect four coronal slices rostrally (slices 1 to 4) and four coronal slices caudally (slices 5 to 8), all at approximately 2- to 3-mm intervals. Use the optic chiasm, mammillary body, pons, and trapezoid body as landmarks for proper spacing of the slices. Embed the four rostral slices with the *caudal* surfaces down in one block (to become the block face) and the four caudal slices with the *rostral* surface down in another block. This orientation will result in opposing surfaces being microtomed at the level of the caudal thalamus or rostral midbrain. The most rostral and most caudal 2- to 4-mm portions of the brain will not be included. After orientation in the tissue cassette, mark the top cut surfaces of the brain slices with tissue ink to facilitate orientation during embedding. Embed the marked surface away from the surface to be microtomed.

The small size and high lipid content of neural tissues contribute to rapid dehydration; therefore, histology processing times are often shorter for tissues of both the central and peripheral nervous systems than for other tissues. A sample rat CNS processing schedule is provided in Table 2. Although some pathologists prefer thicker (e.g., 10-μm) sections, paraffin-embedded brains, like other tissues, may be microtomed at 4 to 6 μm and floated onto a water bath at 38 to 40°C. In high-production laboratories, cutting all tissues at the same thickness reduces mistakes and increases productivity. The authors also personally prefer 4- to 6-μm-thick sections. Clean tap or bottled spring water produces a more appropriate surface tension for wrinkle-free sections than does distilled water. Some tap water may contain contaminants (debris), which tend to interfere with special stains such as the Fluoro-Jade stains and are observed as randomly distributed foci of variable stain intensity.

If needed, fixed or snap frozen brains may be embedded in OCT (Sakura/Tissue-Tek, Torrence, CA). Allow the specimen to equilibrate with the cryostat temperature (−20°C) before cutting sections. Cryosections are easily cut at approximately −15 to −20°C and mounted on positively charged slides. Best sectioning results are achieved when frozen blocks are sectioned at 10 to 15 μm, but thin sections at 5 to 6 μm are ideal for immunohistochemistry. Sections air dried for approximately 1 h are ready for immediate use or can be stored in a −70°C freezer for up to one month.

SPECIAL STAINS AND TECHNIQUES

Most abnormalities of the CNS can be identified initially by an experienced pathologist with a combination of a good clinical history and evaluation of H&E-stained histological sections from paraffin-embedded blocks. Many additional stains can be used to characterize neuroanatomical changes. Commonly used relatively readily available histochemical and immunohistochemical stains to help screen for, or further characterize, specific common changes in the CNS are listed in Table 3.

TABLE 2 Rat CNS Processing Schedule

Station	Solution	Time in Solution	Temperature Setting (°C)
1	10% Formalin	6 h[a]	38
2	70% Dehydrant[b]	30 min	38
3	80% Dehydrant	30 min	38
4	80% Dehydrant	45 min	38
5	95% Dehydrant	30 min	38
6	95% Dehydrant	45 min	38
7	100% Dehydrant	30 min	38
8	100% Dehydrant	45 min	38
9	Xylene substitute	45 min	38
10	Xylene	45 min	38
11	Paraffin	30 min	60
12	Paraffin	45 min	60
13	Paraffin	45 min	60

[a] Or more for the processor to begin and complete at the appropriate time.
[b] The dehydrant is a Richard-Allan product, Kalamazoo, MI.

TABLE 3 Common Stains for Screening and Characterizing Changes in the CNS

Indication and Stain (Tissue)	Vendor or Reference
Screening	
H&E	20
Neurodegeneration	
Fluoro-Jade B (or C)	HistoChem, Jefferson, AK
Silver (axons/dendrites)	21
Luxol fast blue/cresyl violet (myelin and nuclei) or LFB/Holmes silver nitrate (myelin and axons)	22
Inflammation	
Glial fibrillary acidic protein	Dako, Carpinteria, CA
CD68(ED1) (activated microglia)	AbD Serotec, Raleigh, NC
Perls (hemosiderin)	23
Regeneration	
Timm's (hippocampal mossy fiber sprouting)	24,25
Synaptophysin	Dako, Carpinteria, CA

For detecting degenerating neurons, autofluorescence of damaged neurons and special stain-augmented fluorescence are useful tools. Autofluorescence of degenerating neurons (Figure 10) is apparent in histological sections stained with H&E, and some, but not all, other stains when viewed with a fluorescein isothiocyanate (FITC) blue excitation filter (450 to 490 nm). (The authors have noted variations in the autofluorescence intensity of both degenerative neurons and erythrocytes which appeared to be due to differences in the specific H&E stains used by different laboratories and/or differences in mounting media, but have not yet characterized these differences.) When the same FITC filter is used to examine the Fluoro-Jade B (FJB)–stained section,[10] yellow-green fluorescence of the degenerative/necrotic neurons is demarcated more clearly. The FJB stain is particularly useful for detection of low-grade degeneration of axons and

dendrites occurring without consistent changes in the soma (nerve cell body).

The importance of tissue quality is especially apparent in evaluating H&E- and FJB-stained sections with fluorescence microscopy. In addition to degenerating neurons, erythrocytes are also relatively intensely autofluorescent or fluorescent with most H&E and fluorescent stains, and dark neurons fluoresce at roughly one-half the intensity of degenerating neurons in both the H&E- and FJB-stained sections. Autofluorescent erythrocytes can generally be distinguished from degenerating neurons based on location within capillaries, uniform size, and smooth spherical shape. Note that the FJB stain[16] is better than the original Fluoro-Jade stain and that Fluoro-Jade C[17] has been developed to have less background staining and a longer stained section shelf life (although in the authors' experience, may not stain as

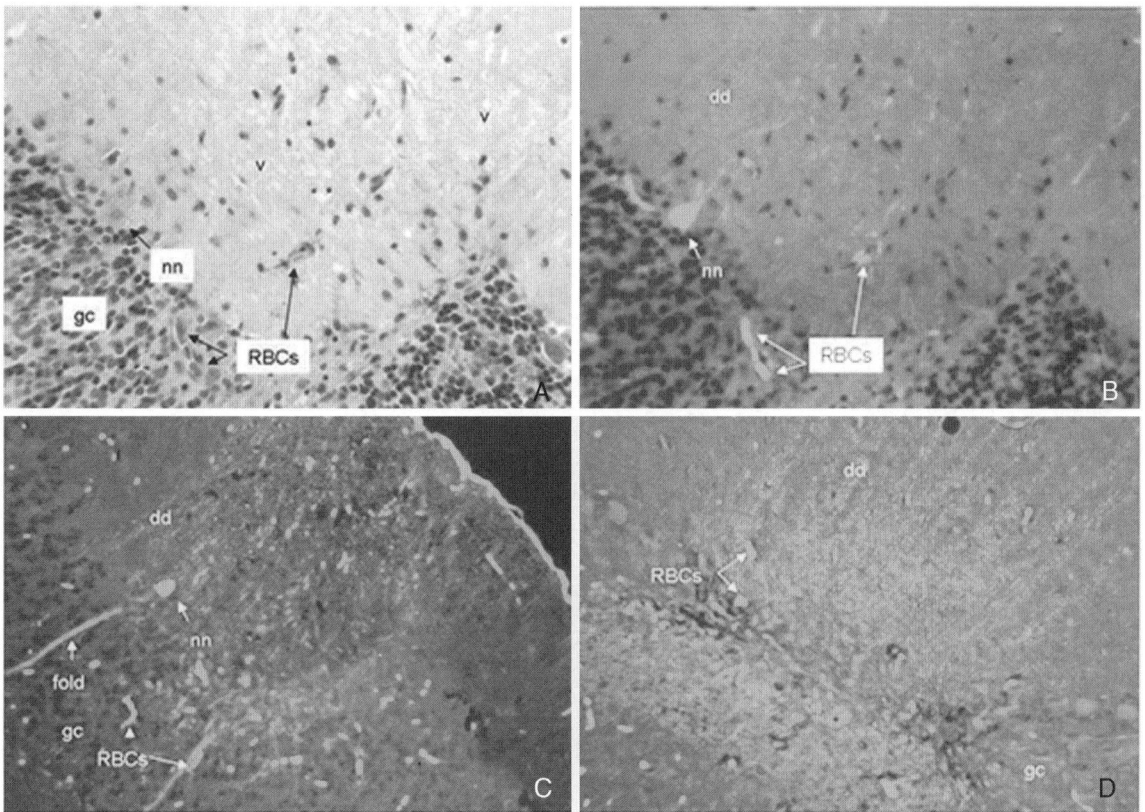

FIGURE 10 Histological sections of cerebellum from two rats with Purkinje neuronal degeneration and loss. (A) and (B) represent the same H&E-stained section viewed, respectively, with transmitted white light and with a FITC filter. In (A), note degenerating neuron (nn), paucity of Purkinje neurons overlying the granule cell (gc) layer, and vacuolation (v) of the overlying molecular layer. In (B), the degenerating neuron, overlying dendrites (dd), and a few erythrocytes (RBCs) are autofluorescent. (C) Serial section from the same region as in (A) and (B), but at slightly lower magnification. The autofluorescence of the degenerating neuron, overlying dendrites, and erythrocytes is more intense than background. Autofluorescence of two tissue folds is enhanced artifactually. (D) Fluoro-Jade B to illustrate enhancement of degenerating dendrites of injured Purkinje neurons (bodies of degenerating neurons not present in this section). (*See insert for color representation of the figure.*)

intensely as the B stain). Counterstaining with 4,6-diamidi-no-2-phenylindole dihydrochloride (DAPI) demonstrates nuclei when the violet (385 to 400 nm) excitation filter is used, and a fluorescent label for glial fibrillary acidic protein (GFAP) on astrocytes can also be added to the same sections to facilitate evaluation.

GFAP labels both resting and activated astrocytes. Therefore, consistent sectioning and good controls are critical for recognition of altered astrocytic size, distribution, and/or number. In contrast, increased expression of the macrophage marker CD68 (ED1) produces little background labeling since it is much more specific for activated microglia.

Hemorrhage often accompanies inflammation and degeneration; thus, the Perls stain for hemosiderin is useful in searching for small residual foci of chronic hemorrhage. The addition of diaminobenzidine will enhance identification of the ferric ferrocyanide product from the Perls stain.[18]

NEUROANATOMY

In addition to an understanding of both the expected time-courses for degeneration to develop and resolve, and the unique features of neuroinflammation, a working knowledge of neuroanatomy is important in finding and interpreting changes in the CNS. Important neuropathological changes may include neuronal loss or increases in cellularity, which can be appreciated only if the investigator recognizes the specific subanatomical site being evaluated and the range of normal features for that location. Even with a good brain atlas, the process of actually matching slides on the microscope with sites depicted in the atlas can be a daunting task. Histological sections frequently do not exactly match a specific plate in the atlas due to intended or unintended tilting and/or skewing of the plane of trimming; indeed, quadrants on the slide often span several plates. The gold standard for rat neuroanatomy appears to be a series of atlases by Paxinos and Watson. These atlases include the large-format fourth edition,[19] containing photomicrographs and labeled drawings in three planes, and the compact fifth edition, featuring coronal sections of the brain of a single rat at consistent 120-μm steps.[15]

A practical approach to neuroanatomy can begin with a relatively small number of readily recognizable structures, both at the gross level and in sections stained with H&E. The major subdivisions of the rodent brain in coronal section have unique shapes and relationships that provide readily recognizable landmarks (Figure 11). The

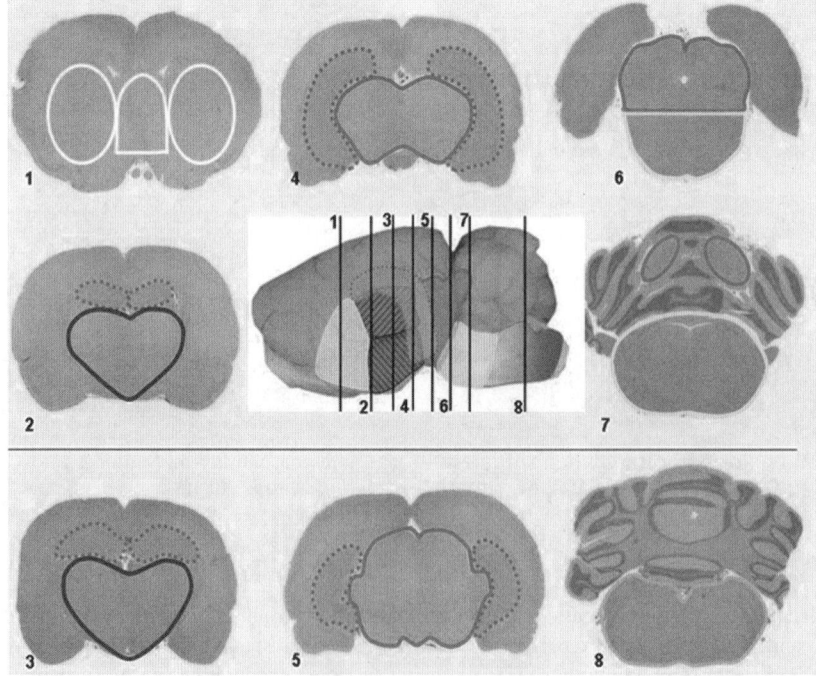

FIGURE 11   Moving rostrally to caudally, the major regions of the rat brain in coronal section have characteristic profiles. The central image depicts a lateral view of rat brain with an overlay of the septal-caudate (white), thalamus-hypothalamus (dark blue), hippocampus (orange), midbrain (green), pons (yellow), and medulla (light blue) regions. Coronal sectioning planes indicated on the central figure correspond to histological sections (1) through (8). Overlays on each of the sections, which match the color of the corresponding portion of the overlay on the central image, illustrate the readily recognizable changes in shape, contour, and position of these major brain regions in coronal section. (*See insert for color representation of the figure.*)

most rostral region of cerebrum is bilobed and bilaterally symmetric. Moving caudally, the bilobed caudate region is separated at the midline by the septum. Progressively more caudal regions include the thalamus and hypothalamus, which together are heart-shaped; the midbrain, which is a squat and lumpy oval; and the tall oval rostral pons. The thalamus/hypothalamus and midbrain are bordered dorsally or laterally by the hippocampus, one of the most easily recognizable structures in the brain. Bilateral differences in the shape and position of the hippocampus relative to the thalamus/hypothalamus and/or midbrain can be used to determine which side of a skewed section is more rostral. This can be very helpful for reconciling structures on a slide to features in an atlas. Slices through the most rostral regions of cerebellum are not attached to the underlying caudal pons, which has a flattened oval profile in this location. Slices including the rostral cerebellum are particularly distinct when lobes of caudal colliculi are present between cerebellar folia. Sections through the midregion of cerebellum are attached by the cerebellar peduncles to the underlying medulla. Slices through the most caudal portion of cerebellum are not attached to the medulla oblongata.

The shapes of a few of the key locating structures can be used to provide general locations of histological sections relative to figures in the atlas (Figure 8). Progressing from rostral to caudal, some of these landmarks include olfactory lobe, rhinal fissure, anterior commissure, corpus callosum, optic tract, lateral ventricle, hippocampus, third ventricle, mammillary bodies, pontine nuclei, fourth ventricle, rostral and caudal colliculi, cochlear nuclei, facial nerve, deep cerebellar nuclei, hypoglossal nuclei, and the central canal. Coronal sectioning provides the advantage of bilateral symmetry and minimal size of sections to fit on microscopic slides, but similar systematic approaches could be developed by the researcher interested in evaluations requiring sagittal or frontal (horizontal) planes of section.

In summary, although complex, the nervous system can be evaluated efficiently by investigators with a basic knowledge of neuroanatomy and neuropathology when sufficient effort is directed to obtaining good clinical observations and to careful and timely tissue collection and processing. Efforts to reduce dark neurons, erythrocytes, and other collection and processing artifacts can significantly improve both the efficiency and quality of neuropathological evaluations. Autofluorescence is a useful feature of degenerating neurons and special stains may be useful for addressing specific questions.

REFERENCES

1. Jordan W, et al. Practical rat neuropathology. *J Histotechnol.* 2007;30:115–120.

2. Dorman D, et al. Nervous system. In: Haschek WM, et al., eds. *Handbook of Toxicologic Pathology.* Vol 2. San Diego, CA: Academic Press; 2002:509–538.

3. Fix A, Garman R. Practical aspects of neuropathology: a technical guide for working with the nervous system. *Toxicol Pathol.* 2000;28:122–131.

4. Paxinos G. *The Rat Nervous System.* 3rd ed. San Diego, CA: Academic Press; 2004.

5. Solleveld HA, Boorman GA. Brain. In: Boorman GA, et al., eds. *Pathology of the Fischer Rat: Reference and Atlas.* San Diego, CA: Academic Press; 1990:155–177.

6. Mitsumori K, Boorman GA. Spinal cord and peripheral nerves. In: Boorman GA, et al., eds. *Pathology of the Fischer Rat: Reference and Atlas.* San Diego, CA: Academic Press; 1990:179–191.

7. Franklin KB, Paxinos G. *The Mouse Brain in Stereotaxic Coordinates,* 3rd ed. San Diego, CA: Academic Press; 2007.

8. Wells G, Wells M. Neuropil vacuolation in brain: a reproducible histological processing artifact. *J Comp Pathol.* 1989; 101:355–362.

9. Sterchi D, Jordan W. Test your knowledge. *J Histotechnol.* 2007;30:135.

10. Jortner B. The return of the dark neuron: a histological artifact complicating contemporary neurotoxicologic evaluation. *Neurotoxicology.* 2006;27:628–634.

11. Hayat M. *Aldehydes in Fixation for Electron Microscopy.* New York: Academic Press; 1981:66.

12. Bronson R. Pathologic characterization of neurological mutants. In: Ward JM, et al., eds. *Pathology of the Genetically Engineered Mice.* Ames, IA: Iowa State University Press; 2000: Chap 17.

13. Meikle AD, Martin AH. A rapid method for removal of the spinal cord. *Stain Technol.* 1981;6:235–237.

14. Woodland J. *National Wild Fish Health Survey: Laboratory Procedure Manual,* 3.1 ed. Pinetop, AZ: U.S. Fish and Wildlife Service; 2006.

15. Paxinos G, Watson C. *The Rat Brain in Stereotaxic Coordinates,* 5th ed. Burlington, MA: Elsevier Academic Press; 2005.

16. Schmued L, Hopkins K. Fluoro-Jade B: a high affinity fluorescent marker for the localization of neuronal degeneration. *Brain Res.* 2000;874:123–130.

17. Schmued L, et al. Fluoro-Jade C results in high resolution and contrast labeling of degenerating neurons. *Brain Res.* 2005;1035:24–31.

18. Racke M, et al. Exacerbation of cerebral amyloid angiopathy-associated microhemorrhage in amyloid precursor protein transgenic mice by immunotherapy is dependent on antibody recognition of deposited forms of amyloid β. *J Neurosci.* 2005;25:629–636.

19. Paxinos G, Watson C. *The Rat Brain in Stereotaxic Coordinates,* 4th ed. San Diego, CA: Academic Press; 1998.

20. Sheehan DC, Hrapchak BB. Nuclear and cytoplasmic stains. In: Hrapchak BB, ed. *Theory and Practice of Histotechnology,* 2nd ed. Columbus, OH: Batelle Press; 1980:137–158.

21. Luna LG. Methods for nerve cells and fibers. In: *Manual of Histologic Staining Methods of the Armed Forces Institute of Pathology*. New York: McGraw-Hill; 1968:189–216.

22. Carson F, et al. Nerve tissue. In: Hrapchak BB, ed. *Theory and Practice of Histotechnology*, 2nd ed. Columbus, OH: Batelle Press; 1980:137–158.

23. Luna LG. Methods for pigments and minerals. In: *Manual of Histologic Staining Methods of the Armed Forces Institute of Pathology*. New York: McGraw-Hill; 1968:174–188.

24. Sloviter R. A simplified Timm stain procedure compatible with formaldehyde fixation and routine paraffin embedding of rat brain. *Brain Res Bull*. 1982;8:771–774.

25. Babb T, et al. Synaptic reorganization by mossy fibers in human epileptic fascia dentate. *Neuroscience*. 1991;42:351–363.

11

FLUORO-JADE DYES: FLUOROCHROMES FOR THE HISTOCHEMICAL LOCALIZATION OF DEGENERATING NEURONS

SUMIT SARKAR AND LARRY SCHMUED

Division of Neurotoxicology, National Center for Toxicological Research, U.S. Food and Drug Administration, Jefferson, Arkansas

INTRODUCTION

Prior to the introduction of Fluoro-Jade for the detection of degenerating neurons, pathologists relied heavily on conventional histological techniques such as Nissl (cresyl violet, thionin, etc.), hematoxylin and eosin (H&E) staining (Figure 1A),[1] or suppressed silver methods (Figure 1B).[2] Each histochemical technique exhibits its own advantages and disadvantages. H&E and Nissl stains are generally simple and reproducible but lack the ability to specifically stain degenerating neurons.[3] Disadvantages of these techniques are that they stain all cells, and the presence of neuronal degeneration must be inferred from relatively subtle morphological changes such as vacuolation, shrinkage, or hyperchromatism.[4,5] Even these changes are not necessarily indicative of neuronal degeneration since they can also be due to either processing artifacts or nonlethal changes in cellular morphology. These techniques can therefore result in the production of both false-positive and false-negative results. Suppressed silver has the specific ability to detect degenerating neurons; however, the technique is labor intensive, time consuming, capricious and has potential to produce false positives. Furthermore, the suppressed silver stains cannot be performed reliably on paraffin-embedded tissues.

Another more recently developed marker of neuronal degeneration, caspase3 immunohistochemistry, is specific for the detection of apoptotic cells and therefore does not necessarily detect all degenerating neurons such as necrotic

neurons.[6] As all of the methods mentioned previously have limitations, we were prompted to develop a fluorochrome that is technically simple, sensitive, reproducible, and specific for the detection of all degenerating neurons. Three novel anionic fluorescein analogs, Fluoro-Jade (Figure 1C), Fluoro-Jade B (Figure 1D), and Fluoro-Jade C (Figure 1E), were subsequently developed to meet these requirements.

Numerous neuropathological conditions have been demonstrated by using Fluoro-Jade since its discovery in 1997.[3] Fluoro-Jade has been used extensively to demonstrate neuronal degeneration upon exposure to various insults, including kainic acid (Figure 1C), 3-nitropropoinic acid, ibogaine (Figure 1H), and ennucleation of one eye (Figure 2A).[3] This histochemical stain has also been used to localize neuronal degeneration following administration of methamphetamine (Figure 2E),[3] ketamine,[7] D-fenfluramine,[8] and 1-methyl-4-phenyl-1,2,3,6-tetrahydropyridine (MPTP).[9,10] Subsequent developments include the synthesis of the next-generation anionic fluorescein derivatives, Fluoro-Jade B[10] and Fluoro-Jade C.[11] We have used Fluoro-Jade B to characterize neuronal degeneration associated with exposure to aurothioglucose[12] (Figure 2B), MDMA[13] (Figure 11.2C and D), and kainic acid at various time points.[14] Fluoro-Jade C[11] has been used to localize degenerating neurons following exposure to methamphetamine,[15] kainic acid (Figure 1E and G),[11] and 3-NPA (Figure 1F).[11] These three fluorochromes have the ability to stain all degenerating neurons following neurotoxic

Fundamental Neuropathology for Pathologists and Toxicologists: Principles and Techniques, First Edition. Edited by Brad Bolon and Mark T. Butt.
© 2011 John Wiley & Sons, Inc. Published 2011 by John Wiley & Sons, Inc.

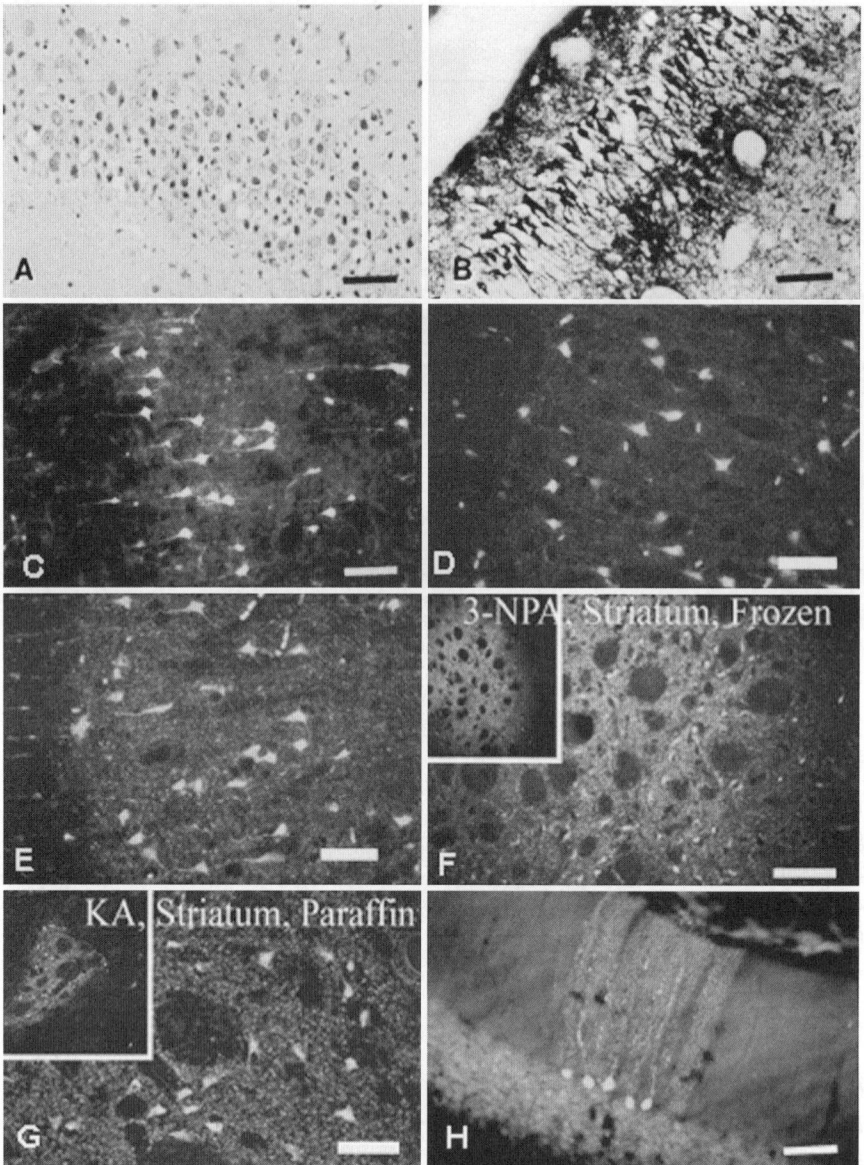

FIGURE 1   (A) Hematoxylin and eosin were used to stain this paraffin-embedded section of the hippocampus of a rat 24 h after exposure to kainic acid. Note the variety of cellular morphologies and dye affinities, making interpretation difficult; bar = 200 μm. (B) The de Olmos suppressed cupric silver technique was used to label degenerating pyramidal cells in frozen sections taken from the hippocampus contralateral to that seen in (A). Degenerating neurons appear black, while normal neurons are unstained; bar = 200 μm. (C) Typical Fluoro-Jade labeling of parietal cortex following kainic acid injection; bar = 200 μm. (D) Typical Fluoro-Jade B labeling of cingulate cortex following kainic acid injection (i.p.); bar = 200 μm. (E) Typical Fluoro-Jade C labeling cingulate cortex following kainic acid injection (i.p.); bar = 200 μm. (F) Fluoro-Jade C labeling in the central striatum of a frozen section following 3-NPA injection (s.c.); bar = 400 μm. (G) Fluoro-Jade C labeling in the central striatum of a paraffin-embedded section following KA injection (i.p.); bar = 200 μm. (H) Patches of Purkinje cells and their dendrites in the medial cerebellum demonstrate ibogaine-induced degeneration as revealed by Fluoro-Jade labeling; bar = 400 μm. (*See insert for color representation of the figure.*)

insults. Therefore, the patterns of neuronal degeneration observed following exposure to either kainic acid or methamphetamine, for example, are quite similar for all of the Fluoro-Jade dyes. However, qualitative differences are observed in the staining characteristics of these fluorochromes. Specifically, Fluoro-Jade C displays a greater signal-to-noise

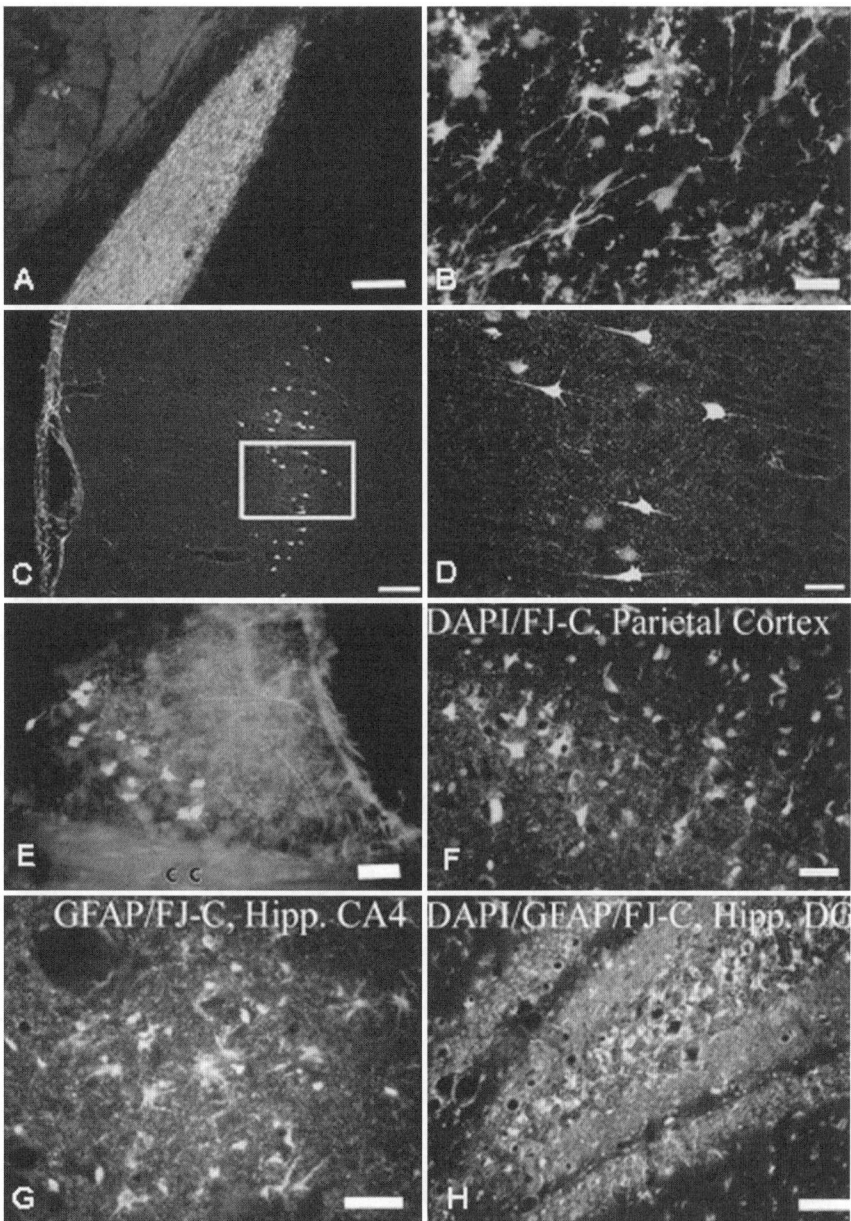

FIGURE 2 (A) Fluoro-Jade labeling of the optic tract following ennucleation the contralateral of eye; bar $= 400 \,\mu m$. (B) Double labeling of rostral arcuate nucleus of the rat brain showing GFAP (red) and Fluoro-Jade B (green) following aurothioglucose injection (i.p.); bar $= 40 \,\mu m$. (C) Low magnification of Fluoro-Jade B positive neurons in the parietal cortex of rat exposed to MDMA; bar $= 400 \,\mu m$. (D) High magnification of the boxed area as shown in (C) reveals Fluoro-Jade B-positive neuronal cell bodies and axon terminals; bar $= 40 \,\mu m$. (E) Tenia tecta of mouse brain exposed to methamphetamine showing Fluoro-Jade-labeled neuronal cell bodies at left and terminals at right. CC, corpus callosum; bar $= 40 \,\mu m$. (F) Double fluorescent labeling with Fluoro-Jade and DAPI result in the respective localization within the parietal cortex of degenerating neurons (green) and viable cell nuclei (blue) following kainic acid injection (i.p); bar $= 40 \,\mu m$. (G) High magnification of the central nucleus of the amygdala is visualized by using sequential blue and green illumination to reveal red GFAP labeled activated astrocytes in association with green Fluoro-Jade C-positive degenerating cells and terminals; bar $= 40 \,\mu m$. (H) A triple exposure of the hippocampal dentate gyrus illuminated with green, blue, and UV light reveals Fluoro-Jade-positive degenerating polymorphic neurons in the hialar region, DAPI staining of nuclei, which is especially conspicuous among the intact granular cells and GFAP-positive activated astrocytes; mag $= 100 \,\mu m$. (*See insert for color representation of the figure.*)

ratio as well as higher resolution than Fluoro-Jade B,[11] which in turn displays higher contrast and resolution than the original Fluoro-Jade.[10] Therefore, we generally prefer Fluoro-Jade C when localizing not only nerve cell bodies but dendrites, axons, and terminals as well, although high-quality staining can also be achieved with Fluoro-Jade B.

CHEMICAL IDENTITY

Structurally, all three Fluoro-Jade dyes are related. The parent structure of these fluorochromes is Fluoro-Jade,[16] which is a mixture of 5-carboxyfluorescein disodium salts and 6-carboxyfluorescein disodium salts. Fluoro-Jade B[16] is a mixture of three related fluorochromes: (1) trisodium 5-(6-hydroxy-3-oxo-3*H*-xanthen-9yl)benzene, 1,2,4-tricarboxylic acid, (2) disodium 2-(6-hydroxy-3-oxo-3*H*-xanthen-9yl)-5-(2,4-dihydroxybenzol)terepthalic acid, and (3) disodium 2,5-bis(6-hydroxy-3-oxo-3*H*-xanthen-9yl)terepthalic acid. The exact chemical structure of Fluoro-Jade C is under investigation, but preliminary observations suggest that it is the sulfate ester of one or more of the above-mentioned components of Fluoro-Jade B. Figure 3 represents the respective excitation and emission profiles for each of the Fluoro-Jade dyes. Excitation and emission peaks are given in the figure legend.

STAINING METHODS

For staining with Fluoro-Jade dyes, the following procedures should be followed to achieve optimal results. Briefly, the tissue should be fixed in neutral phosphate-buffered 10% formalin or 4% paraformaldehye. Although immersion fixation can be adequate, we typically first perfuse the animal intracardially with fixative (1 mL/g body weight, with descending aorta clamped) and then remove the brain. It is then postfixed and cryoprotected in the same fixative solution plus 20% sucrose for at least overnight. The sucrose is omitted from the postfix if the tissue is to be paraffin embedded. Nonembedded tissue can be cut with a freeze sliding microtome, vibratome, or cryostat at approximately 25 μm thickness. Prior to staining, sections are mounted from distilled water onto gelatin-coated slides. Gelatin-coated slides can be prepared by immersion of the clean slides in a 60°C solution of 1% pig-skin gelatin (Sigma, 300 Bloom) and then dried overnight in a paraffin-free convection oven at the same temperature. The tissue sections are mounted from distilled water and air dried for at least 30 min on a slide warmer at 50°C.

FROZEN SECTIONS

Slides with frozen sections should first be immersed in a basic alcohol solution. Traditionally, this involves transferring

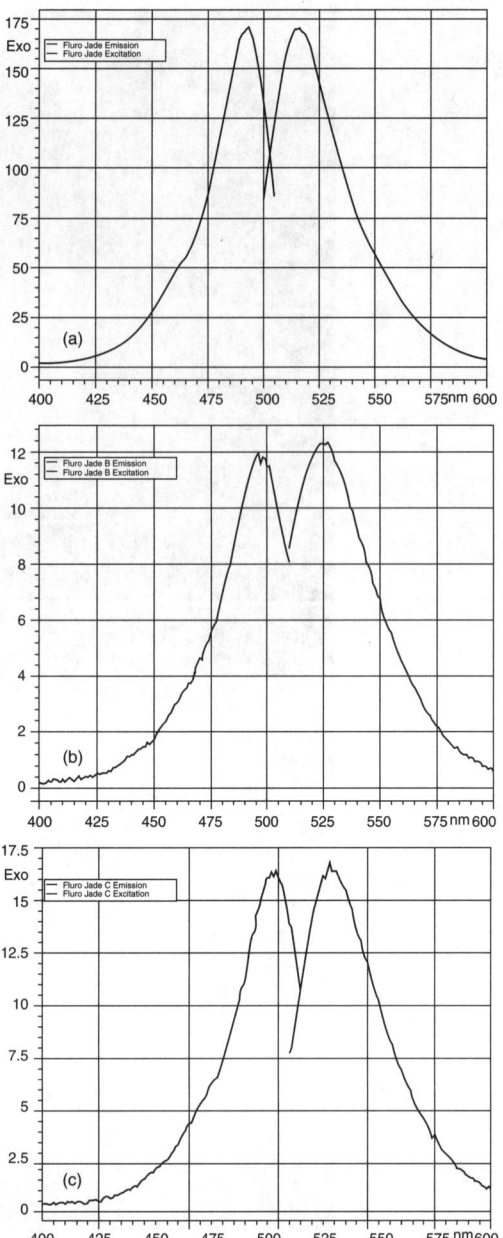

FIGURE 3 Spectrofluorometric profiles of (a) Fluoro-Jade (excitation peak, 492 nm; emission peak, 516 nm), (b) Fluoro-Jade B (excitation peak, 493 nm; emission peak, 526 nm), and (c) Fluoro-Jade C (excitation peak, 502 nm; emission peak, 529 nm).

the slides to a solution consisting of 20 mL of 1% sodium hydroxide in 80 mL of ethanol for 5 min. Because this solution is stable for a few hours only, we have recently substituted this for another more stable basic alcohol solution consisting of 0.03 M tripotassium phosphate in 70% ethanol (unpublished personal observations), which is stable for several months. The tissue should then be rinsed in 70% ethanol for 2 min, rinsed in distilled water for 2 min, and then incubated in 0.06% potassium permanganate solution for 10 min. After a 2-min distilled water rinse, slides are then immersed for 10 or 15 min in Fluoro-Jade, Fluoro-Jade B, or Fluoro-Jade C (source: Histo-Chem, Inc., Jefferson, AR) solution prepared in a 0.1% acetic acid vehicle. The respective recommended concentrations and staining times for each dye are as follows: Fluoro-Jade: 0.001% for 15 min; Fluoro-Jade B: 0.0004% for 10 min; and Fluoro-Jade C: 0.0001% for 10 min. Initially, a 0.01% stock solution is prepared by adding 10 mg of the dye to 100 mL of distilled water. The stock solution can be kept for at least a month in the refrigerator but should not be used if it turns cloudy. The working solution is made by diluting the stock solution to the respective dye concentration indicated above in a 0.1% acetic acid vehicle. Thus, to make up 100 mL of Fluoro-Jade C staining solution, 1 mL of stock solution is added to 99 mL of an acetic acid vehicle. The working solution should be made fresh prior to use and used within 8 h of preparation. The slides should be rinsed three times for 2 min per change in distilled water and then air dried on a slide warmer at 50°C for at least 5 min. The air-dried slides are cleared in xylene for 1 min and cover-slipped with DPX (Fluka or Sigma) nonpolar mounting media. Polar cover-slipping media that contain water, alcohol, or glycerol should not be used.

PARAFFIN SECTIONS

For paraffin-embedded tissue, the sections are first deparaffinized through two 10-min changes in xylene and then rehydrated through a graduated series of alcohol solutions, omitting the basic alcohol solution. Since the xylene used for deparaffinizing the tissue serves the same purpose as the basic alcohol. After one rinse in distilled water, the sections are transferred to the potassium permanganate solution, followed by the staining steps described above for frozen sections.

STAINING VARIANTS

Although we use Fluoro-Jade C routinely as just described, variations can be incorporated in the following steps:

1. The basic alcohol solution serves to delipidize the tissue. As mentioned previously, we typically replace the ETOH/NaOH (80 mL ethanol + 20 mL of 1% NaOH) solution with either an ETOH/K_3PO_4 (70 mL ethanol + 30 mL of 0.1 M K_3PO_4) solution for frozen sections, or two changes of xylene for paraffin-processed tissue. These steps can be omitted, although this may result in a higher background staining of myelinated tracts.

2. The $KMnO_4$ solution serves to improve the contrast of the stain by increasing the signal-to-background ratio. The $KMnO_4$ treatment may, however, inhibit the labeling with other histological or immunocytochemical probes. In these cases it may be necessary to reduce or even omit exposure to this solution. Omission will, however, result in higher Fluoro-Jade background staining, which may partially be compensated for by reducing the dye concentration to one-half to one-tenth of its usual concentration.

3. All of the Fluoro-Jade staining solutions are used in a 0.1% acetic acid vehicle. The stain will not work at a neutral or basic pH. The earlier dyes (e.g., Fluoro-Jade) tended to employ longer staining times (15 min) and higher dye concentrations (0.001%), while the later variants requires shorter staining times (10 min) and lower concentrations (0.0001%). We find the latter concentration to be about ideal for Fluoro-Jade C, although it can be doubled if the staining intensity is weak, or may be cut in half if the background staining is high.

4. Dehydration of the stained and rinsed slides is typically accomplished by placing the wet slides on a slide warmer at 50°C for at least 5 min. Attempts to dehydrate sections through graduated series of ethanol resulted in stain degradation. Far better results (unpublished observations) were obtained using butanol dehydration, which was judged to produce a stain of almost as high quality as that produced by air dehydration. This was accomplished by transferring the slides from the staining solution to the following solutions for 1 min per solution: one change of distilled water, followed by one change of equal parts water/ethanol/butanol, followed by two changes of butanol, followed by two changes of xylene.

5. Slides should always be cover-slipped with a nonpolar mounting medium (such as DPX) and not an aqueous or glycerol-based mounting medium.

COMPARISON OF FLUORO-JADE, FLUORO-JADE B, AND FLUORO-JADE C

New methods continue to evolve for visualizing degenerating neurons. The original Fluoro-Jade dye was introduced initially[3] to localize the degenerating neurons following diverse insults such as kainic acid, domoic acid, ibogaine, 3-NPA, MPTP, and physical trauma. Subsequently, we have shown that this dye can also be used to visualize the degenerating neurons following methamphetamine and *d*-fenfluramine exposure.

Fluoro-Jade B, an analog of Fluoro-Jade, was subsequently developed.[10] Like Fluoro-Jade, Fluoro-Jade B has an affinity for the entire degenerating neuron, including cell body, dendrites, axon, and axon terminals. When used to stain kainic acid–induced neurodegeneration, higher resolution and contrast was obtained with Fluoro-Jade B than with Fluoro-Jade. Apparently, these two dyes share an affinity for the same biomolecule but differ in the strength of their affinity.

Most recently, we have developed Fluoro-Jade C,[11] which in our experience exhibits the highest affinity for degenerating neurons and results in the highest resolution and contrast of all the Fluoro-Jade dyes. It can be used at a very low concentration and requires less staining time than is required by Fluoro-Jade. However, some investigators still prefer Fluoro-Jade B, possibly reflecting different imaging systems in different labs.

MULTIPLE LABELING TECHNIQUES

Fluoro-Jade and Fluorescent Nissl Stains

The Fluoro-Jade dyes can be used readily for multiple labeling purposes. The combination with a fluorescent Nissl stain allows visualization of both degenerating and viable cells.[3,11] The Fluoro-Jade dyes can be combined with ethidium bromide, resulting in good color contrast whereby degenerating neurons appear green and viable cells appear red. Ethidium bromide counterstaining, however, is not compatible with the use of $KMnO_4$, which must be omitted, resulting in higher Fluoro-Jade background staining. DAPI counterstain, however, is compatible with the $KMnO_4$ pretreatment procedure, although the treatment alters the staining, so only the nuclear (DNA) rather than Nissl body (DNA + RNA) staining will be seen. With this combination, degenerating cells appear green whereas viable nuclei are blue. The counterstains are accomplished simply by incorporating 0.0001% ethidium bromide or DAPI directly into the Fluoro-Jade working solution. This is accomplished by making a 0.01% stock solution of the counterstain, which is then diluted at 1 : 100 with the Fluoro-Jade staining solution. The counterstains contrast well with green fluorescent Fluoro-Jade labeled degenerating neurons (Figures 11.2H and 2F).

Fluoro-Jade and Immunofluorescence

The Fluoro-Jade dyes can also be combined with immunofluorescence. For example, we have demonstrated that kanic acid and aurothioglucose-activated astrocytes can be localized by using an antibody directed against glial fibrillary acidic protein (GFAP) (Figures 2B and 2G). A secondary antibody conjugated with rhodamine will allow localization of red labeled astrocytes adjacent to green Fluoro-Jade-labeled degenerating neurons[3,10] (Figure 2H). Avoidance of potassium permanganate pretreatment allowed extensive GFAP labeling of astrocytes but increased background staining with Fluoro-Jade. Conversely, pretreatment of sec-

tions with potassium permanganate yields high-contrast Fluoro-Jade labeling but diminished GFAP labeling.[3,11] Therefore, the time in the potassium permanganate step may have to be reduced, depending on the stability of the epitope being studied. To partially compensate for the resulting increased background fluorescence, the Fluoro-Jade C solution is typically used at one-half to one-tenth the standard dye concentration.

COMPARISON OF FLUORO-JADE DYES WITH OTHER HISTOLOGICAL MARKERS: ADVANTAGES AND DISADVANTAGES

Hematoxylin and Eosin

H&E staining has been used extensively for decades because of its simplicity, reliability, and permanence. However, the main disadvantage of this technique in regard to localizing degenerating neurons is that it stains all brain structures rather than only degenerating neurons. Inferences are based mostly on morphological criteria, such as shrinkage of neurons, vacuolation, and hyperchromatism, all of which must be distinguished from handling and processing artifacts.[4,5,17] This technique can give not only false-positive results, but also false negatives, as it is possible to miss degenerating neurons because all cells stain and there is often a subtle morphological difference between normal and degenerating neurons. H&E also stains the neuropil a rather uniform pink, which can hinder visualization of degenerating neurons.

Suppressed Silver Stains

The main advantage of suppressed silver stains is that ideally, they stain only degenerating neurons. This stain is permanent and yields high-contrast labeling of degenerating neurons. The main drawbacks of suppressed silver methods are that they are capricious and can produce false-positive results.[17,18] Also, suppressed silver techniques are very labor intensive, time consuming, and often employ toxic reagents, including arsenic-based buffers for tissue perfusion and collection. Specific fixation techniques and the inability of suppressed silver to be used satisfactorily on paraffin-embedded tissues often prevents the possibility of retrospective evaluations. Also, it is less suitable for double labeling studies because of the harsh chemicals employed and by virtue of being a brightfield technique.

Fluoro-Jade Dyes

The main advantages of Fluoro-Jade stains are their simplicity, sensitivity, reliability, and specificity for localizing degenerating neurons in the brain. While many fluorescent stains fade (sometimes rapidly), we have not observed archival fading in sections stained with the Fluoro-Jade dyes. Although prolonged exposure at high magnification under an ultraviolet light can result in detectable fading, this was minimal, as no difference was observed when comparing photomicrographs of stained sections before or after 1 h of constant epifluorescent illumination.[11] The dyes can also be used to either screen various drugs for neurotoxicity or for demonstrating neuroprotection. Neuronal degeneration is detected regardless of the underlying mechanism of toxicity.[19] These tracers will also detect neurons undergoing natural apoptosis as well as trauma or ischemia-induced neuronal degeneration. These dyes also benefit toxicological pathologists, who utilize this technique to verify the results obtained with more traditional, procedures.

SPECIALIZED APPLICATIONS

In several collaborative studies, we have investigated the possible use of Fluoro-Jade staining techniques in a wide variety of applications, summarized below.

Detection of Developmental Apoptosis When sections from normal brains of six-day-old rat pups were labeled with Fluoro-Jade dyes, we observed sparse scattered small, round undifferentiated Fluoro-Jade-positive cells, mainly in the thalamus and cortex.

Detection of β-Amyloid Plaques When human temporal lobe autopsy tissue from Alzheimer's disease patients was subjected to Fluoro-Jade labeling, many plaque-like Fluoro-Jade-positive structures were found within the hippocampal complex, in addition to rarely observed Fluoro-Jade-positive neurons. These structures are β-amyloid positive and were not observed in age-matched controls.[10]

Astrocytic Labeling Following Chronic Prion Infection or Neurotoxic Insult When rat brain tissue sections were labeled with Fluoro-Jade dyes after two months of intracranial injection of scrapie prion, numerous Fluoro-Jade-positive astrocytic cells were observed. Prion infections, like a number of neurotoxicants, can result in the relatively faint labeling of hypertrophied astrocytes adjacent to labeled degenerating neurons. The greatest numbers were observed in the gray matter adjacent to the third ventricle.[20]

Detection of Neuronal Degeneration in Tissue Culture When PC-12 cells were exposed to MPTP in a conventional tissue-culture medium, they exhibited conspicuous Fluoro-Jade labeling, although no labeled cells were detected in the vehicle-treated controls.[10]

Fluoro-Jade in Primary Neuronal Cell Cultures Fluoro-Jade and its derivative can be used for in vitro neuronal cell culture. Recently, Schmuck and Kahl[21] have demonstrated that Fluoro-Jade can be used as a sensitive and reliable indicator to detect compounds that induce neurodegenerative lesions.

Fluoro-Jade in Ophthalmological Research Chidlow et al.[22] have used Fluoro-Jade C to mark degenerating neurons in the rat retina and optic nerve.

DISCUSSION

The Fluoro-Jade dyes were designed specifically to stain only degenerating neurons (and their processes) in the brain following all types of neurotoxic insults or physical trauma. Among the three Fluoro-Jade dyes, Fluoro-Jade C has been reported to be the most sensitive fluorescent marker of neuronal degeneration, allowing for the localization not only of degenerating neuronal cell bodies but also dendrites, axons, and terminals. This suggests that Fluoro-Jade C has a higher affinity for the endogenous neurodegenerating molecule than its predecessor Fluoro-Jade B, which in turn exhibits better contrast and resolution than its predecessor, Fluoro-Jade.

Although the structures of the Fluoro-Jade dyes have been characterized to some extent,[15] minimal information is available regarding the chemical identity of the endogenous "neurodegeneration molecule" to which these dyes bind, but several plausible mechanisms have been proposed.[11] One of the more plausible potential modes of binding consistent with the present data is the notion that the dye is staining polyamines generated during the degeneration process. For example, when living tissue putrefies, amino acids with secondary amino groups are decarboxylated. Thus, lysine is converted to cadaverine, arginine is converted to spermidine or putrescine and histamine can be converted to histadine. These decarboxylated polyamines would be expected to exhibit a high affinity for the polyanionic Fluoro-Jade dyes. The unequivocal identification of the neurodegeneration markers bound by these dyes might be investigated further using standard two-dimensional gels or affinity chromatography methods.

The Fluoro-Jade dyes are quite stable, with no loss of signal even in one-year-old tissue sections that were kept at 4°C in phosphate buffer.[11] Although all the Fluoro-Jade dyes are compatible with fixed or unfixed tissue, formaldehyde-based fixation methods generally yield better morphological detail. These stains are compatible with a variety of sectioning procedures, including paraffin microtome, cryostat, freezing sliding microtome, and vibratome. Recently, we have observed that sections kept in freezing solutions (mixture of glycerol, ethylene glycol, and phosphate buffer) at −20°C did not adversely affect the Fluoro-Jade labeling in the brain (personal observations).

CONCLUSIONS

Fluoro-Jade dyes offer clear advantages over more tradi- tional histological methods for detecting neuronal degen- eration. They are as methodologically simple and reliable as H&E or Nissl stains, yet much more specific and reliable for localizing degenerating neurons. They are faster, simpler, and more reliable than the suppressed silver techniques, which are relatively specific for labeling degenerating neurons. The Fluoro-Jade dyes are also generally more applicable than immunologic markers such as caspase 3, which have been found to be specific for apoptic cells but will not detect neuronal death via alternative mechanisms such as necrosis. By virtue of their fluorescent nature, Fluoro-Jade dyes are ideally suited for multiple labeling studies in which one can localize degenerating neurons and another histochemical marker of a different color simulta- neously. Thus, one might combine Fluoro-Jade staining with a neuronal marker such as tyrosine hydroxylase (TH) to demonstrate, for example, that methamphetamine exposure resulted in a decrease of TH-labeled terminals in the striatum, accompanied by a proportional increase in Fluoro-Jade C-labeled terminals.[11,15] Alternatively, Fluoro-Jade labeling of degenerating neurons can be com- bined with immunohistochemical markers of other cell types, such as astrocytes, microglia, oligodendrocytes and vascular elements, to evaluate the effect of neuronal degen- eration on these adjacent cells.[11,12,14] In our lab, when the three Fluoro-Jade dyes were compared, the highest contrast and resolution was afforded by Fluoro-Jade C, followed by Fluoro-Jade B, followed by Fluoro-Jade. Fluoro-Jade B and C are both capable of labeling fine degenerating neuronal axons, distal dendrites, and terminals. The Fluoro-Jade dyes can detect neuronal degeneration following a wide variety of insults, including physical trauma, deafferentation, devel- opmental apoptosis, ischemia, and exposure to a wide variety of neurotoxins, including kainic acid, domoic acid, 3-NPA, methamphetamine, d-fenfluramine, MDMA, MPTP, ibogaine, ketamine, PCP, and aurothioglucose. Although this chapter was designed to be a fairly comprehensive review of past findings and applications of the Fluoro-Jade dyes, some novel information has been included with the hope that it might be of benefit to the researcher, such as the full emission/excitation profile of each dye, the develop- ment of a more stable basic ethanol solution, and an optional method for solvent dehydration.

Acknowledgments

We would like to thank David Heard for his technical assistance with histochemical processing and spectrofluoro- metric analysis. We also would like to acknowledge Elsevier publishers for allowing us to use the previously published photomicrographs. This work was supported in part by U.S. Food and Drug Administration protocol E7312.

REFERENCES

1. Lillie RD. *Histopathologic Technical and Practical Histochemistry*, 4th ed. New York: McGraw-Hill; 1976.
2. de Olmos JS, et al. Use of an amino-cupric-silver technique for the detection of early and semiacute neuronal degeneration caused by neurotoxicants, hypoxia, and physical trauma. *Neurotoxicol Teratol.* 1994;16:545–561.
3. Schmued LC, et al. Fluoro-Jade: a novel fluorochrome for the sensitive and reliable histochemical localization of neuronal degeneration. *Brain Res.* 1997;751:37–46.
4. Garcia JH, Kamijyo Y. Cerebral infarction: evolution of histopathological changes after occlusion of a middle cerebral artery in primates. *J Neuropathol Exp Neurol.* 1974;33:408–421.
5. Siesjo BK. Cell damage in the brain: a speculative synthesis. *J Cereb Blood Flow Metab.* 1981;1:155–185.
6. Nicholson DW. Caspase structure, proteolytic substrates, and function during apoptotic cell death. *Cell Death Differ.* 1999;6:1028–1042.
7. Scallet AC, et al. Developmental neurotoxicity of ketamine: morphometric confirmation, exposure parameters, and multiple fluorescent labeling of apoptotic neurons. *Toxicol Sci.* 2004;81:364–370.
8. Schmued L, et al. AT d-Fenfluramine produces neuronal degeneration in localized regions of the cortex, thalamus, and cerebellum of the rat. *Toxicol Sci.* 1999;48:100–106.
9. Freyaldenhoven TE, et al. Systemic administration of MPTP induces thalamic neuronal degeneration in mice. *Brain Res.* 1997;759:9–17.
10. Schmued LC, Hopkins KJ. Fluoro-Jade B: a high affinity fluorescent marker for the localization of neuronal degenera- tion. *Brain Res.* 2000;874:123–130.
11. Schmued LC, et al. Fluoro-Jade C results in ultra high resolu- tion and contrast labeling of degenerating neurons. *Brain Res.* 2005;1035:24–31.
12. Schmued LC. The progression of neuronal, myelin, astrocytic, and immunological changes in the rat brain following exposure to aurothioglucose. *Brain Res.* 2002;949:171–177.
13. Schmued LC. Demonstration and localization of neuronal degeneration in the rat forebrain following a single exposure to MDMA. *Brain Res.* 2003;974:127–133.
14. Hopkins KJ, et al. Temporal progression of kainic acid induced neuronal and myelin degeneration in the rat forebrain. *Brain Res.* 2000;864:69–80.
15. Bowyer JF, Schmued LC. Fluoro-Ruby labeling prior to an amphetamine neurotoxic insult shows a definitive massive loss of dopaminergic terminals and axons in the caudate-putamen. *Brain Res.* 2006;1075:236–239.
16. Xu L HT, et al. Isolation and characterization of Fluoro-Jade B, a selective histochemical stain for neuronal degeneration. *J Liq Chromatogr Rel Technol.* 2004;27:1627–1640.

17. Stensaas SS, et al. An experimental study of hyperchromic nerve cells in the cerebral cortex. *Exp Neurol.* 1972;36:472–487.

18. Cammermeyer J. The importance of avoiding the "dark" neurons in experimental neuropathology. *Acta Neuropathol.* 1961;1:245–270.

19. Schmued LC, Hopkins KJ. Fluoro-Jade: novel fluorochromes for detecting toxicant-induced neuronal degeneration. *Toxicol Pathol.* 2000;28:91–99.

20. Ye X, et al. Fluoro-Jade and silver methods: application to the neuropathology of scrapie, a transmissible spongiform encephalopathy. *Brain Res Brain Res Protoc.* 2001;8:104–112.

21. Schmuck G, Kahl R. The use of Fluoro-Jade in primary neuronal cell cultures. *Arch Toxicol.* 2009;83:397–403.

22. Chidlow G, et al. Evaluation of Fluoro-Jade C as a marker of degenerating neurons in the rat retina and optic nerve. *Exp Eye Res.* 2009;88:426–437.

12

HISTOLOGICAL MARKERS OF NEUROTOXICITY (NONFLUORESCENT)

ROBERT C. SWITZER III

NeuroScience Associates, Knoxville, Tennessee

MARK T. BUTT

Tox Path Specialists, LLC, Hagerstown, Maryland

INTRODUCTION

Once the threshold for safety testing for a study has been determined (see Chapter 9 for a detailed explanation), the scope of the evaluation to be performed should include methods for the detection of general and specific neurotoxicity. Those methods should be sufficient to determine those dosing levels likely to produce adverse changes, those dosing levels that might be considered safe, and those dosing levels where no morphologic changes are detectable. These methods should almost always include the means of detecting specific markers of neurotoxicity.

Some tissue-staining methods are used in a general sense. The most common example is the hematoxylin and eosin (H&E) stain. But common general staining methods have many limitations, especially when applied to central nervous system tissues. More specific staining methods, designed to detect a specific marker, are often necessary. So although a general stain may be the most useful for the overall evaluation of a tissue, the ideal ancillary or additional stains may be those that have very high specificity (highlight a single marker), high sensitivity (produce little nonspecific and/or background staining), and can be evaluated quickly (because the marker is readily visible on the stained section). Once the specific endpoints have been determined for a study, the selection of stains that can best augment the general morphological evaluation can proceed. A general classification of some of the most common neurohistology stains and

their corresponding endpoints is depicted in Table 1. The stains listed are those with which the authors' have personal experience. In the table, the notations E (for "excellent"), P (for "possible"), and a blank cell (for not generally possible or useful) represent the experience of the authors, but the notations are well supported in the literature.

MARKERS OF CHEMICAL CHANGE

In the spectrum of safety testing considerations for neurotoxicity, the most basic change to be considered is generally any change in chemistry in the brain. Quite a long list could be derived that meets this classification, but perhaps the most useful tools in this category are the neurotransmitter markers. These markers are typically assessed using biochemical assays or with immunohistochemistry (IHC). The evaluation consists of comparing the normal expression of each neurotransmitter against what is to be considered a normal expression (e.g., typically, animals treated with a specific test article are compared to untreated control animals). The timing of this evaluation with respect to compound administration should be spread over a period sufficient to determine any persistent or transient alterations. The determination of toxicity or adversity of treatment may then be determined based on the intended actions of the compound and any potentially unexpected changes in neurotransmitter activity. In this respect, the determination as to whether or not

Fundamental Neuropathology for Pathologists and Toxicologists: Principles and Techniques, First Edition. Edited by Brad Bolon and Mark T. Butt.
© 2011 John Wiley & Sons, Inc. Published 2011 by John Wiley & Sons, Inc.

TABLE 1 Common Markers of Neurotoxicity[a]

Stain	Chemistry Changes				Pertubations/ Inflammation		Neuron Cell Body	Cell Death		
	Dopaminergic	Cholinergic	Serotonergic	Noradrenergic	Astrocytic	Microglial		Axons	Axon Terminals	Dendrites
Tyrosine hydroxylase IHC[1–3]	E									
ChAT IHC[4]		E								
5-HT serotonin IHC[5]			E							
Norepinephrin IHC[6]				E						
GFAP IHC[7]					E					
IBA1 IHC[8]						E				
CD68[9]						E				
Cresyl violet							P			
Thionine							P			
H&E					P	P	E			
NeuN[10]							P			
TUNEL[11]							E			
Fluoro-Jade[12,13]					P[b]		E	E	E	E
Cupric silver[14]							E	E	E	E

[a]IHC, immunohistochemistry; ChAT, choline acetyltransferase; 5HT, 5-hydroxytryptamine; GFAP, glial fibrillary acidic protein; Iba-1, ion binding; TUNEL, terminal deoxynucleotidye transferase DUTP nick end labeling. E, excellent; P, possible; blank, not generally possible or useful for the specific endpoint.

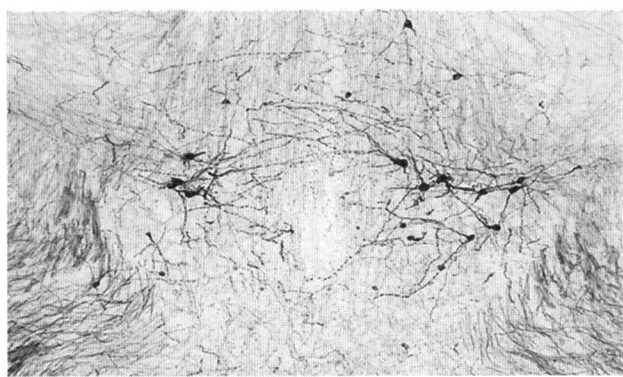

FIGURE 1 Dopaminergic neurons in the brain stem of rat revealed by IHC with an antibody against tyrosine hydroxylase. (*See insert for color representation of the figure.*)

an alteration of neurotransmitter activity is adverse would be considered similarly to other effects in a study.[15] Figures 1 to 4 show examples of stains that detect a particular neurohistochemical subpopulation of neurons or axons. When interpreting the level of neuronal staining for any neurohistochemical stain, but particularly for immunohistochemical stains, it is important to realize that apparent decreases may require careful quantification (using stereological techniques; see Chapter 15), as decreases may be due to changes in expression (i.e., changes in phenotype) and not necessarily to cell loss.

PERTURBATIONS AND INFLAMMATION

The primary neurohistological indicators of perturbations in the brain, often associated with a diagnosis of inflammation, are astrocytes and microglia. Each can be visualized with their respective immunohistochemistry (IHC)-specific antibodies. The most commonly used antibodies reveal both

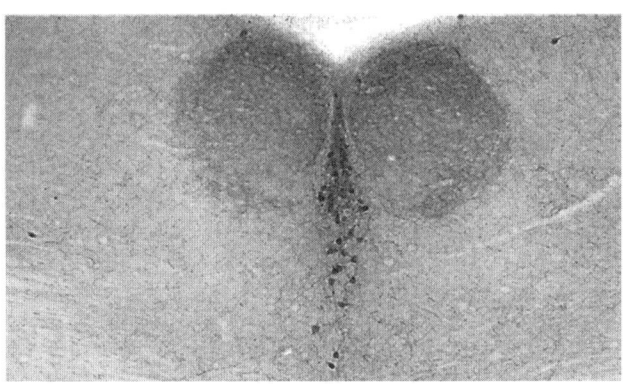

FIGURE 3 Serotonergic neurons in the dorsal raphe nucleus (rat) revealed by IHC with an antibody against serotonin (5HT). (*See insert for color representation of the figure.*)

normal and hypertrophic microglia or astrocytes. The morphology determines whether or not there is perturbation.

The primary method for detection of perturbed astrocytes is through GFAP (glial fibrillary acidic protein) immunohistochemistry which reveals the cytoskeletal protein unique to astrocytes.[7,16] Under normal circumstances, the astrocytes perform a variety of functions, including contributing to the blood–brain barrier, movement of nutrients and waste within the brain, and glutamate transport.[17] Under perturbation or damage, they respond with a change in state, as they are presumably being called upon to perform additional tasks.[17] These perturbed astrocytes are more intensely stained due to an increased expression of GFAP; their processes become thicker, "gnarly," and more highly branched. Figures 5 and 6 show the difference in GFAP staining between normal and reactive astrocytes.

Changes in the attributes of GFAP staining is most noticeable beginning 36 to 48 h following an insult, peaking at about 72 h, and persisting in potentially diminishing fashion for weeks to months. Due to some areas with endogenous staining, such as the hippocampus and white

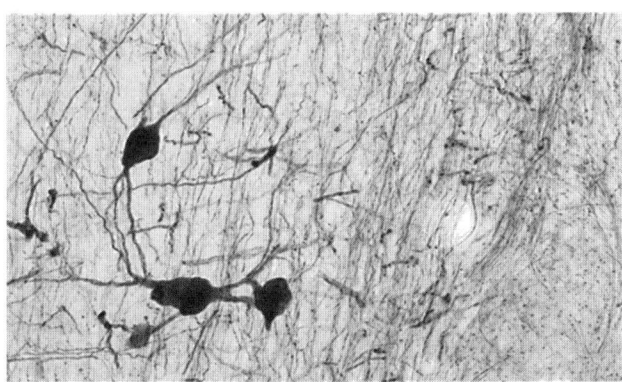

FIGURE 2 Cholinergic neurons in the ventral limb of the diagonal band of Brocca (rat) revealed by IHC with an antibody against choline acetyltransferase (ChAT). (*See insert for color representation of the figure.*)

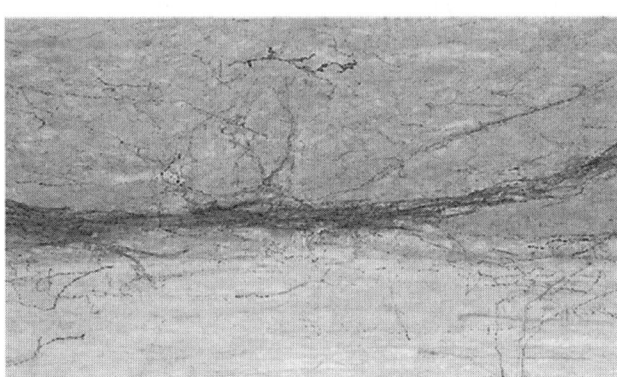

FIGURE 4 Serotonergic axons in a horizontal section of the spinal cord of a rat revealed by IHC with an antibody against serotonin (5HT). (*See insert for color representation of the figure.*)

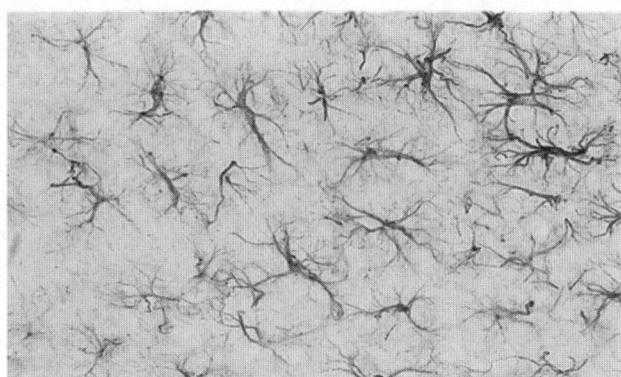

FIGURE 5 Normal morphology of astrocytes revealed by IHC with an antibody against glial fibrillary acidic protein (GFAP), a cytoskeletal protein mostly unique to astrocytes. This antibody cross reacts with many species. (*See insert for color representation of the figure.*)

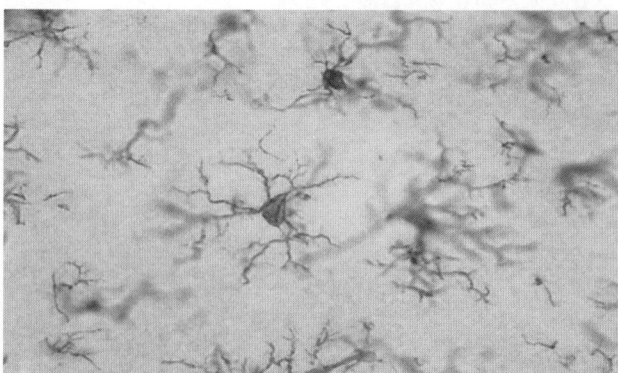

FIGURE 7 Normal morphology of microglia revealed by IHC with an antibody against IBA1. This antibody cross reacts with many species. (*See insert for color representation of the figure.*)

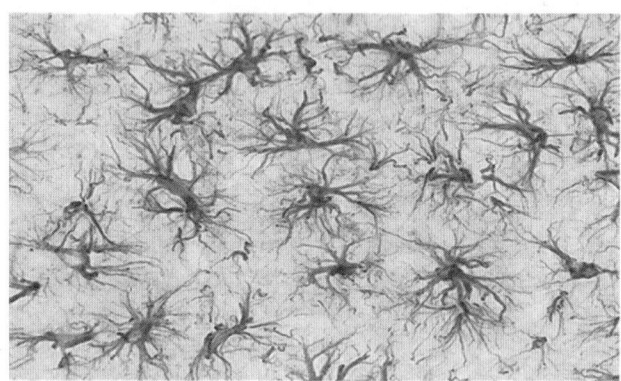

FIGURE 6 These reactive astrocytes display more numerous and thickened processes and enhanced density of staining. (*See insert for color representation of the figure.*)

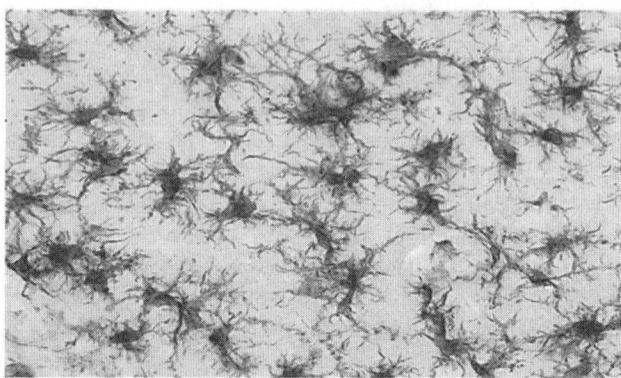

FIGURE 8 Hypertrophic morphology of reactive microglia revealed by IHC with an antibody against IBA1. The morphologic changes range from appearing near normal, to progressive loss of processes, to enlarged cell body, to finally an amoebiform phenotype. (*See insert for color representation of the figure.*)

matter, the ability to detect subtle changes in expression can be difficult. The presence of reactive (hypertrophied) astrocytes signals the occurrence of perturbations, but there are also those circumstances for which the astrocytes themselves are damaged or destroyed.

Depending on the timing of the investigation, a change in astrocyte staining may indicate that central nervous system damage has occurred but provides no specific information regarding what the initial insult may have been, such as prior death of neurons.

The primary method for detection of perturbed microglia, the brain's resident macrophages, is through IBA1 IHC, which shows all microglia, or through CD68 IHC, which reveals microglia that are of a more specific state of reactivity.[8,9] When microglia become perturbed, they may be capable of emitting both beneficial and harmful substances into the brain and thus may exacerbate or ameliorate neurotoxicity.[18] Whereas in the short term their presence

may have beneficial results, prolonged activation of microglia (i.e., due to infections, prion disease, etc.) may coincide with detrimental endpoints of neuronal or astrocytic death.[19] Whether causative or indicative of the secondary damage, revealing activated microglia is a helpful tool in identifying consequences of perturbations that may have resulted in permanent deleterious effects.

Reactive microglia responding to perturbations display a hypertrophy shown by IHC with IBA1 antibody. The cell body becomes enlarged, processes become fewer until there may be none (amoebiform state), and there is a tendency to cluster. Figures 7 and 8 illustrate unperturbed and reactive microglia. IBA1 IHC reveals the most activity beginning about 4 to 5 days after an insult or exposure, peaking at 5 to 7 days, and persisting for two weeks or more.

CD68 (clone FA-11 for mice and ED-1 for rats) reveals a subset of the microglia visible with IBA1 and at the very least exhibits a bias toward the reactive microglia, making it a very tempting tool to use as a replacement for IBA1. It is unclear, however, whether this subset of microglia is capable of showing all reactive microglia or a subset. While providing a better signal-to-noise ratio by seemingly revealing only the activated microglia, CD68 may also risk missing hypertrophied microglia that could have been visualized with IBA1. It is advisable to stain with IBA1 first to get a more comprehensive view of the microglia population, then follow with CD68 to probe for the activated microglia as defined by this antibody. As with GFAP, alterations in IBA1 staining provides a clue that permanent damage (cell death) may have occurred and should also be probed.

MARKERS OF PERMANENT CHANGE OR DAMAGE

The most important routine evaluation to be performed in a safety study is to evaluate for permanent damage, which is expressed as cell death. The two basic approaches in evaluating cell death are by omission or through witnessing disintegration. Evaluation by omission consists of using a cellular or Nissl stain and determining whether or not cells are missing. This method has the obvious flaw of being highly influenced by the magnitude of cell loss (subtle and even prominent cell loss can easily be missed) and the degree of background cellularity. When considering neurons in brain sections, detecting a prior loss of neurons may be considered analogous to looking at a starry sky and being asked to determine whether stars are missing. There are times when evaluation by omission is the only option, but in most neurotoxicity studies, the option of evaluation by witnessing cell death is a valid and more favorable and affirmative alternative. Witnessing cell death is observing the pathology of a cell going through the disintegrative process, but this must be done prior to the debris of the dead cell being cleared from the brain or spinal cord.

There are multiple pathological alternatives to witnessing cell death. They can be grouped into categories according to the types of pathology they reveal. The two basic categories are cell body–specific stains and disintegrative degeneration stains, which reveal the degeneration of all of the elements of a neuron. Cell body–specific stains include cresyl violet, thionine, H&E, NeuN IHC, TUNEL (terminal deoxynucleotidyl transferase dUTP nick end labeling), and others. It is quite valid to reach a conclusion of neurotoxicity when using such staining tools. However, detecting subtle cell death may be challenging with these stains. This is due to the uniformity of staining across sections, the sometimes large quantity of tissue to be examined, the need to use higher-power objectives (higher-power and thus smaller fields than may be necessary when viewing specific disintegration stains) to detect or verify cell death, and confusion of the characteristics of cell death with artifactual cell changes that may

occur from suboptimal handling, fixation, and processing.[20] (See Chapter 13 for an updated and thorough treatise of artifacts in brain sections.)

Most important, if the safety scope includes all of the neuronal elements (which it should), these general, nonspecific markers do not evaluate the entire scope of neurotoxicity. Specifically, these general methods (H&E staining is the most common example) reveal only cell body disintegration and are not capable of revealing a comprehensive picture of degeneration or disintegration, including the axons, axon (synaptic) terminals, and dendrites. Fulfilling this requirement is the disintegrative degeneration class of stains, which include the Fluoro-Jade stains and various cupric silver methods (collectively known earlier as derivatives of the Nauta reduced silver methods).

There have been no published comprehensive comparisons of Fluoro-Jade vs. cupric silver methods to determine if their scope matches identically or whether they have precisely the same sensitivity. Anecdotally, there have been subjective comparisons made by researchers who favor one method or the other. A final determination of which to use is most likely to be a matter of preference or familiarity by those performing the staining, what other endpoints are required in the study (as the cupric silver stain requires the use of free floating freeze-cut sections), equipment availability (Fluoro-Jade requires the use of a fluorescent microscope), and the experience of the pathologist who will evaluate the staining. Both staining methods have the advantage of staining degenerating elements without ambiguous background staining of normal healthy cells.

So although the Fluoro-Jade and cupric silver methods are relatively comparable, there are significant differences in visualization. Fluoro-Jade is a fluorescent marker, requiring the use of a fluorescent microscope with the appropriate filter in place.[12,13,21] The cupric silver methods provide optically dense profiles (black staining) very easily visible with bright-field (traditional light) microscopy. Both methods are amenable to digital analysis and both provide a high signal-to-noise ratio, although the Fluoro-Jade method does have the disadvantage of some structures being autofluorescent (red blood cells), which can complicate evaluation and analysis. (Note: Red blood cells also stain with the amino cupric silver stain, but the distraction is generally less because of the counter stain and the use of light microscopy as opposed to fluorescence.) Both stains are capable of revealing the full scope of disintegration, although in the experience of the author (M.T.B.), disintegrating neuronal processes (axons, dendrites, terminals) are usually more readily and more easily visualized using the cupric silver staining methods. The decision of which to use most likely comes down to the decision of which method of visualization is preferred by the pathologist. The Fluoro-Jade methods are discussed in Chapter 11, in the remainder of this section we focus on the cupric silver methods.

The cupric silver methods for disintegration are one of the many applications using silver as a marker in pathology. It is not the silver itself that is significant; rather, it is how it is directed to stain. Much like the use of DAB or fluorescent markers used in IHC, silver particles are designed to be the visual marker after being directed to a specific point of attachment. Silver is used on pathology stains to reveal Alzheimer's plaques,[22] neurofilaments, and other structures. Of interest for neurotoxicity are the disintegrative degeneration applications of the cupric silver method.[14,16,23] Within silver degeneration methods, there are two classes: de Olmos's CuAg and the amino CuAg. The protocols are quite different to conduct, but the end result is that they both stain the same scope of degenerating elements. The specific element (cell body, axon, dendrite, axon/synaptic terminal) being stained can be identified through morphology. Positive staining of neuronal elements is conclusive evidence of cell death and permanent damage.[14] The exact moiety being stained in these disintegrating neuronal processes is not precisely known. It is believed to be (primarily) fragments of disintegrated proteins that have an increased affinity for silver particles as a neuron dies.[14] In a healthy cell, the globular shape of proteins block these binding sites. But the proteolysis associated with cell death (or leading to cell death) dismantles the proteins and exposes the sites for silver binding.

Visually, the CuAg method yields an amber background, while the Amino CuAg method does not have any background staining, so it appears white or blank (Figure 9). Due to the absent background in the amino CuAg method, it most often receives a counterstain with neutral red to reveal cell bodies and landmarks as a reference for any positive degeneration staining that also occurs. The pale background of the amino CuAg is more desirable from a signal-to-noise ratio perspective and it has replaced the application of the previous generation of CuAg in most applications. The exception to this is for neonate or fragile tissue. The protocol for CuAg[23] is physically less demanding on the tissue than the amino CuAg method, and for very frail tissue the CuAg may yield better results. In the author's (M.T.E) laboratory, amino CuAg is used for 95% + of disintegrative degeneration staining and specific CuAg only for very young animals.

Staining for all of the neuronal elements also provides a larger "footprint" from which to recognize degeneration. A comparison of adjacent slides stained with H&E vs. the amino CuAg method (Figure 10) illustrates the ease of interpretation and the increased spatial footprint of the positive staining.

The timing for cell death has already been discussed (Chapter 9) as (typically) having a span of several days with a peak at 3 to 4 days. That statement applies primarily to the cell body; other neuronal elements each have its own specific time lines. The windows of opportunity to view each element vary slightly but overlap considerably. However, depending on the timing of an evaluation, it is possible to witness just one or two of the elements. Compared to the possibility of witnessing just the cell body disintegration, the temporal

CuAg Method   Amino CuAg Method

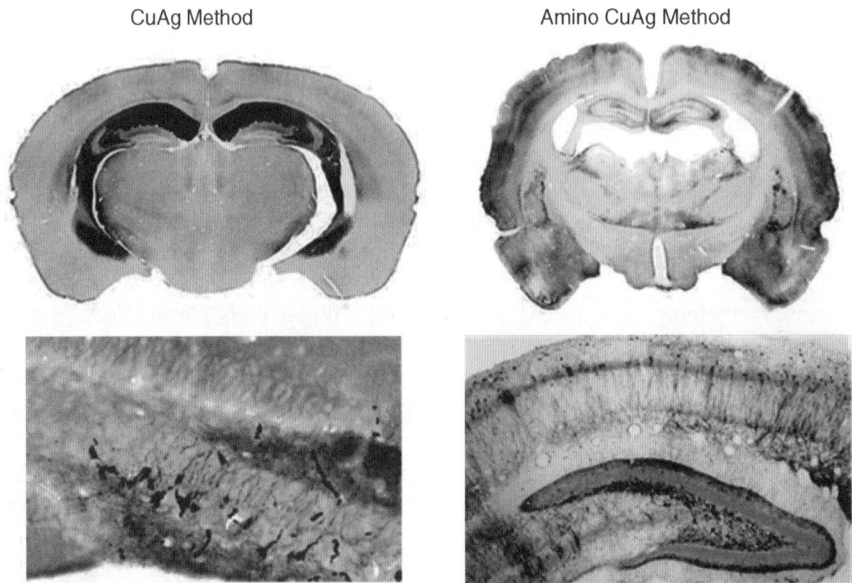

FIGURE 9 Comparison of the background staining of the cupric silver (CuAg) method on the left and the amino cupric silver (amino CuAg) method on the right. The amino CuAg section is counterstained with neutral red. In each stain, the black areas demonstrate areas of neuronal degeneration. In general, detail and staining specificity is improved with the amino CuAg technique. (*See insert for color representation of the figure.*)

Control Cases   Affected Cases   Higher Mag.

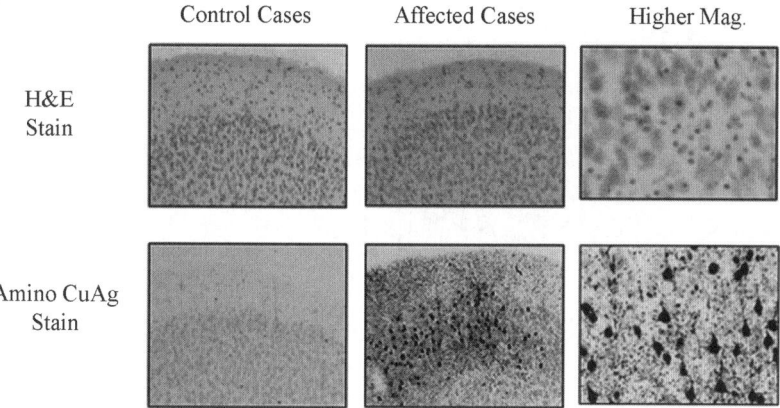

FIGURE 10 Comparison between what is visible using traditional H&E staining as compared to the much greater visualization of neuronal changes possible using the amino CuAg stain. The panels on the left are from controls; there is no evidence of degeneration or disintegration in either stain. In the middle panels, at low power, there are no readily apparent morphological changes to the cells with the H&E stain but abundant disintegrating neurons are easily observed with the amino CuAg stain. The numerous black dots at the top of the section are disintegrating synaptic terminals, a change not visible at any power with H&E. The panels to the right are a closer view of the changes in the middle panels. At this power, the florid neuronal disintegration or necrosis is evident with the H&E stain, but the affected cells are still not easily visualized. With the amino CuAg stain, the black (disintegrating or necrotic) neurons are very readily apparent. (*See insert for color representation of the figure.*)

window of opportunity to be able to witness any of the neuronal elements is much wider (Figure 11).

Interpretation of Cupric Silver Degeneration Staining

The cupric silver degeneration staining methods are generally a dichotomous, all-or-nothing stain for revealing degenerating neurons. However, other features are stained, and it is important to differentiate between endogenous neurodegeneration and treatment or test article–related neu-

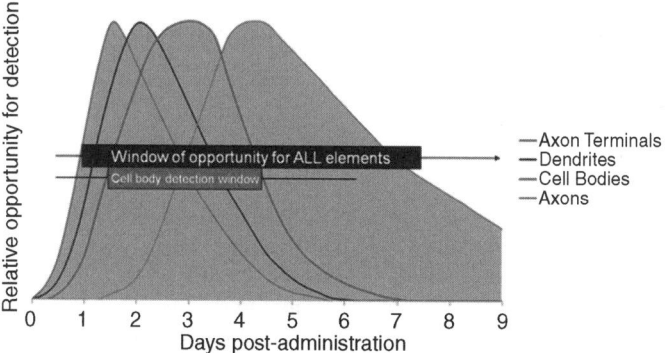

FIGURE 11 Typical time course of disintegrating elements. Note that by day 7, the disintegration of synaptic terminals, dendrites, and cell bodies are no longer apparent. Disintegrating elements in axons are frequently visible for several more days and even weeks. This allows for the detection of neuronal death even after the cell bodies are gone from view in a section. (*See insert for color representation of the figure.*)

rodegeneration. It is also important to differentiate between actual degeneration and any potential staining artifact that can occur.

EVALUATION OF TRUE DEGENERATION OR DISINTEGRATION

The four neuronal elements that can be revealed as they undergo disintegration can be seen in Figure 12. These are recognizable as degeneration both by the fact that they are stained and by the morphology of each element. Disintegrating cell bodies are typically visualized as prominent angular black structures connected to more elongated, but usually equally black, axons and dendrites. In the authors' experience, if there is a cluster of affected cells, the black area caused by the combined staining of the cell body and the associated axons and dendrites is frequently visible on the section even without the aid of magnification (i.e., the affected area is visible prior to placing the slide under the microscope). Examples of macroscopically visible cell body disintegration include the substantia nigra of MPTP-treated mice and primates and zones III and IV in the cerebral cortex of rats treated with some NMDA receptor antagonists (personal observation of M.T.B.). Although the macroscopic observation of stained regions does not eliminate the need for thorough microscopic examination, it does illustrate the tremendous ability of the disintegrative silver stains to differentiate disintegrating elements of the neuron (and sometimes astrocytes; see below) from the surrounding neuropil.

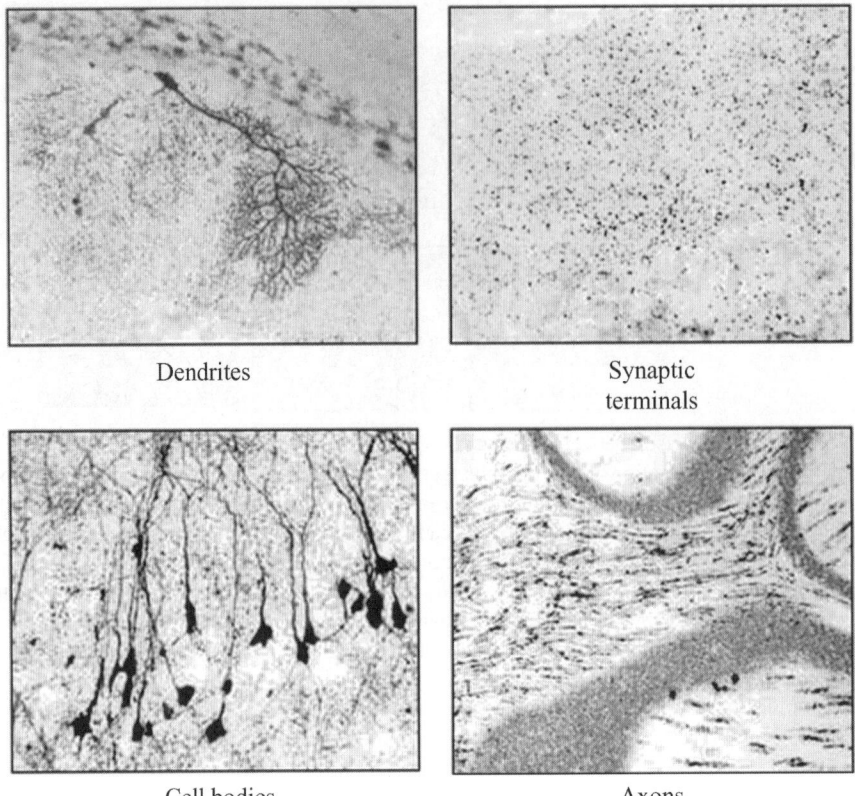

Dendrites

Synaptic terminals

Cell bodies

Axons

FIGURE 12   General appearance of the disintegration of the four cell elements that can be visualized using the CuAg staining methods. (*See insert for color representation of the figure.*)

Disintegrating axons and dendrites appear as variably thick black lines that may or may not be attached visibly to the cell body (depending on the location of the section). As indicated in Figure 11, detection of remaining, persistent disintegrating portions of axons may be possible for many days after disintegrating terminals, cell bodies, and dendrites are no longer visible. In the authors' experience, the ability to detect disintegrating axons may persist for several weeks to even months.[14]

Disintegrating axon and synaptic terminals are generally visible as clusters of black dots. When these clusters are large enough, they may be macroscopically visible on the stained section. Examples of some macroscopically visible synaptic terminal disintegration include the striatum of MPTP-treated mice (certain strains) and zone 1 of the retrosplenial cortex of rats treated with some NMDA receptor antagonists.[16]

EVALUATION OF BACKGROUND STAINING

Some background staining of brain structures in animals may occur; that is, not all staining by the silver disintegrative stains is necessarily indicative of a test article effect. While recognition of the background staining patterns in the various species (and for the various ages of animals) logically becomes easier with increasing experience with this staining method, careful comparison to controls for any given study generally provides all the necessary information to differentiate background from actual test article effects. To assist the examining pathologist, several examples of the most common background staining noted by the authors are provided here.

There is fine granular staining of the glomeruli in the olfactory bulb and the accessory olfactory formation (AOF), and this is real degeneration of synaptic terminals. The sensory cells in the epithelium corresponding to both structures are constantly being turned over. Hence, their synaptic terminals in the glomeruli show up. It is typically at a low level but nearly always present. Some circumstances lead to very dense degeneration in one or more glomeruli—typically, this is random and fairly rare. Something that damages the epithelium in general will elicit profound degeneration in the glomeruli. When this happens, the AOF glomeruli will show a denser pattern of terminal degeneration.

The trapezoid body in the brain stem and medulla nearly always has large axons in a state of degeneration. This actually serves as a nice internal control to show that the stain is working properly, at least for demonstrating axons. The reason for this degeneration in this area is unknown. In

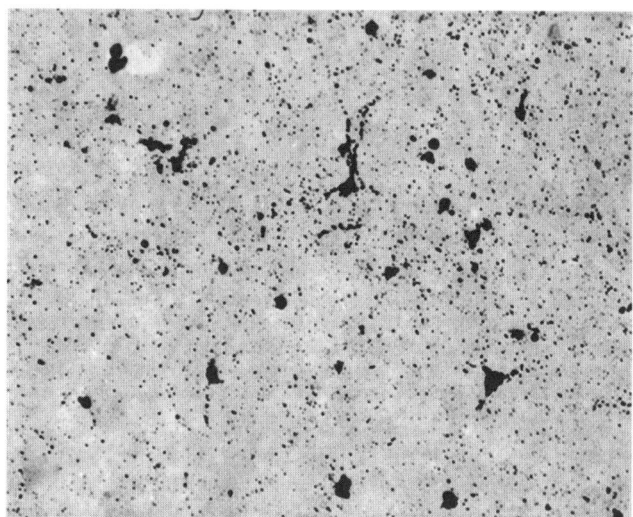

FIGURE 13 The CuAg stain reveals abundant apoptotic neurons in a neonate. (*See insert for color representation of the figure.*)

the lateral curve of the hippocampus (inside the lateral sharp curvature in the CA3 zone), somewhat coarse terminals may be seen. These stained terminals are common and should not be considered an effect of a test article unless other evidence suggests associated neuronal death. In the mammillary bodies there are synaptic terminal-like profiles similar to those seen in the glomeruli of the olfactory bulb.

Large neurons in the reticular formation and red nucleus will often be stained. With Nissl staining methods, neurons in the same locations can be darkly stained and appear dystrophic. The staining effect is believed to be due to physical distortion near the time of death. These neurons do not have the characteristics of being in a state of disintegration. [Note: In the author's (R.C.S.) experience, this artifactual staining can be ameliorated by leaving the brain of a perfusion-fixed animal in the skull for 18 to 24 h before removing it.]

In neonate brains up until about 14 days of age, abundant degenerative staining can be found in different locations, depending on the age. These are neurons undergoing programmed cell death (apoptosis) (Figure 13). Also abundant in neonate brains are degenerative profiles that appear to be disintegrating astrocytes. The same profiles may be found in adult brains proximal to areas with damage.

There are stained features in a few areas of the brain that are not degeneration or disintegration. These appear as somewhat diffuse staining with the amino cupric stain that de Olmos et al. have described.[24] The affected areas include the globus pallidus, the lateral part of the bed nucleus of the stria terminalis (just behind the anterior commissure), and the central nucleus of the amygdala. In the lateral hypothalamus there is a population of neurons whose cytoplasm has a fine, dense granular appearance that stains with these silver procedures.

Neuromelanin in the cell body of the substantia nigra and locus ceruleus of primates stains quite intensely and should be disregarded, as it is a normal condition. In the author's (M.T.B.) experience, when there is disintegration of neurons in the substantia nigra (as can occur, for example, with the administration of MPTP to some strains of mice and to primates), the affected cells stain intensely with the animo cupric stain and are easily recognized as dead or disintegrating neurons.

In *Macaque mulattas* there is a granular staining of neurons in layer 5 of the motor cortex and intralaminar nuclei of the thalamus. This staining does not occur in other macaque subspecies. This is a normal feature and is to be disregarded as being test article related.

ARTIFACTUAL STAINING

Artifactual staining is possible with virtually any method, and the cupric silver degeneration methods are no exception. Artifactual staining is best excluded through either morphology or by consistency of staining. One staining artifact is caused by the presence of red blood cells. Perfusion can minimize or eliminate red blood cells from the brain, but if present, they do stain, and although it is a distraction at low magnification, the features can be excluded on a morphological basis. Heavy staining of this type of artifact can obscure "true" degeneration staining in some cases, so it is imperative to properly perfuse tissue destined to be processed with a disintegrative degeneration stain.

Other types of artifactual staining may exist due to a contaminant or variation in staining conditions on a particular section(s). Suspicion of artifact due to these causes can be confirmed or refuted by comparing nearby sections that contain the same structure. Consistency across multiple slides suggests a true positive signal vs. a staining artifact.

REFERENCES

1. Luellen BA, et al. Neuronal and astroglial responses to the serotonin and norepinephrine neurotoxin: 1-methyl-4-(2′-aminophenyl)-1,2,3,6-tetrahydropyridine. *J Pharmacology Exp Ther*. 2003;307(3):923–931.

2. Jackson-Lewis V, Jakowec M, Burke RE, Prezedborski S. Time course and morphology of dopaminergic neuronal death caused by the neurotoxin 1-methys-4-phenyl-1,2,3,6-tetrahydropyridine. *Neurodegeneration*. 1995;4:257–269.

3. Scott SA, Diaz NM, Ahmad SO. Stereologic analysis of cell number and size during postnatal development in the rat substantia nigra. *Neurosci Lett*. 2007;419:34–37.

4. Capsoni S, Ugolini G, Comparini A, Ruberti F, Berardi N, Cattaneo A. Alzheimer-like neurodegeneration in aged antinerve growth factor transgenic mice. *Proc Nat Acad Sci USA*. 2000;97(12):6826–6831.

5. Shen WZ, Luo ZB, Zheng ER, Yew DT. Immunohistochemical studies on the development of 5-HT (serotonin) neurons in the nuclei of the reticular formations of human fetuses. *Pediatr Neurosurg.* 1989;15(6):291–295.

6. Werhofstad AA, Steinbusch HW, Penke B, Varga J, Joosten HW. Use of antibodies to norepinephrine and epinephrine in immunohistochemistry. *Adv Biochem Psychopharmacol.* 1980;25:185–193.

7. O'Callaghan J, Sriram K. Glial fibrillary acidic protein and related glial proteins as biomarkers of neurotoxicity. *Expert Opin Drug Saf.* 2005;4:433–442.

8. Daisuke I, Tanaka K, Suzuki S, Dembo T, Fukuuchi Y. Enhanced expression of Iba1, ionized calium-binding adapter molecule 1, after transient focal cerebral ischemia in rat brain. *Stroke.* 2001;32:1208–1215.

9. Lemstra AW, Groen in't Woud J, Hoozemans J, et al. *J Neuroinflamm.* 2007;4:4–12.

10. Wolf HK, Buslei R, Schmidt-Kastner R, et al. NeuN: a useful neuronal marker for diagnostic histopathology. *J Histochem Cytochem.* 1996;44(10):1167–1171.

11. Deng X, Wang Y, Chou J, Cadet J. Methamphetamine causes widespread apoptosis in the mouse brain: evidence from using an improved TUNEL histochemical method. *Mol Brain Res.* 2001;93(1):64–69.

12. Schmued LC, Stowers CC, Scallet AC. Fluoro-Jade results in ultra high resolution and contrast labeling of degenerating neurons. *J Brain Res.* 2005;1035:24–31.

13. Damjanac M, Rioux BA, Barrier L, et al. Fluoro-Jade B stianins as useful tool to identify activated microglia and astrocytes in a mouse transgenic model of Alzheimer's disease. *Brain Res.* 2007;11(2891):40–49.

14. Switzer R. Application of silver degeneration stains for neurotoxicity testing. *Toxicol Pathol.* 2000;28:70–83.

15. Lewis RW, Billington R, Debryune E, et al. Recognition of adverse and nonadverse effects in toxicity studies. *Toxicol Pathol.* 2002;30(1):66–74.

16. Fix A, Ross J, Stitzel S, Switzer R. Integrated evaluation of central nervous system lesions: stains for neurons, astrocytes, and microglia reveal the spatial and temporal features of MK-801–induced neuronal necrosis in the rat cerebral cortex. *Toxicol Pathol.* 1996;24:291–304.

17. Morenberg MD. The reactive astrocyte. In: Aschner M, Costa LG, eds. *The Role of Glia in Neurotoxicity.* Boca Raton, FL: CRC Press; 2005:73–92.

18. Streit WJ. The role of microglia in neurotoxicity. In: Aschner M, Costa LG, eds. *The Role of Glia in Neurotoxicity.* Boca Raton, FL: CRC Press; 2005:29–40.

19. Giese A, Brown DR, Groschup MH, Feldmann C, Haist I, Kretzschmar HA. Role of microglia in neuronal cell death in prion disease. *Brain Pathol.* 2006;8:(3):449–457.

20. Garman R. Artifacts in routinely immersion fixed nervous tissue. *Toxicol Pathol.* 1990;18:149–153.

21. Schmued LC, Hopkins KJ. Fluoro-Jade B: a high affinity fluorescent marker for the localization of neuronal degeneration. *Brain Res.* 2000;874:123–130.

22. Tsamis K, Mytilinaios D, Njau SN, et al. The combination of silver techniques for studying the pathology of Alzheimer's disease. *Int J Neurosci.* 2008;118(2):257–266.

23. de Olmos JS, Ingram WR. Silver methods for the impregnation of degenerating axonal and terminal degeneration. *Brain Res.* 1972;33:523–529.

24. de Olmos JD, Beltramino CA, de Olmos-de Lorenzo S. Use of an amino-cupric-silver technique for the detection of early and semiacute neuronal degeneration caused by neurotoxicants, hypoxia, and physical trauma. *Neurotoxicol Teratol.* 1994;16:545–561.

13

COMMON HISTOLOGICAL ARTIFACTS IN NERVOUS SYSTEM TISSUES

ROBERT H. GARMAN

Consultants in Veterinary Pathology, Inc., Murrysville, Pennsylvania

INTRODUCTION

Recognition of histological artifacts in sections of the nervous system is critical in avoiding false positives and false negatives in both basic research and safety evaluation studies. False positives are common and typically result from misinterpretation of an artifact as representing a degenerative event. False negatives, by their very nature, are more difficult to substantiate but undoubtedly occur. The reader may ask why so much emphasis is placed on recognition of histological artifacts in the nervous system: Are artifacts more common in the nervous system than in other tissues? Are such artifacts easy to confuse with neurodegenerative processes? Or is it merely that many pathologists, toxicologists, and neuroscientists are inexperienced in the microscopic evaluation of sections of the nervous system? All three of these situations apply.

Because the tissues of the central nervous system (CNS) are highly susceptible to handling-induced artifacts, perfusion fixation is the method of choice for critical evaluation of the brain and spinal cord by light microscopy. However, since perfusion fixation is usually not employed in routine histomorphological studies in which CNS lesions are not suspected initially, it is important to be aware of the types of artifactual change that are encountered in immersion-fixed CNS tissues. Furthermore, histological artifacts cannot be completely eliminated by perfusion fixation, and perfusion fixation followed by rapid handling of CNS tissues may even enhance the development of certain artifacts. Histological artifacts commonly encountered at the light microscopic level are highlighted in this chapter.

RETRACTION SPACES AROUND NEURONS, VESSELS, AND GLIAL CELLS

Several types of artifact are commonly encountered within sections from brains fixed by immersion in aldehyde fixatives (Figure 1). Clearing, or retraction, of the cytoplasm of oligodendrocytes results in the classic "fried egg" appearance (Figure 1A) that is readily recognized by most pathologists and is occasionally helpful in the identification of oligodendrocytes (e.g., as in oligodendrogliomas). Separation of neuronal layers from the adjacent neuropil is seen most frequently where small granule neurons are in abundance. Two of the most common brain regions for this pattern of separation are along the blades of the dentate gyrus bordering on the hilus of the hippocampus (Figure 1B) and within the Purkinje neuron layer of the cerebellar cortex (Figure 1C). On first glance, these clefts appear merely to represent a separation of cell layers; however, if this artifact is examined critically, it will be seen that the clefts usually comprise numerous distinct vacuoles. This pattern suggests that the vacuoles may represent swollen astrocytic cell processes. A similar pattern of neuropil retraction with vacuole formation may be seen around small to medium-sized blood vessels (Figure 1D). Once again, it is likely that at least some of

Fundamental Neuropathology for Pathologists and Toxicologists: Principles and Techniques, First Edition. Edited by Brad Bolon and Mark T. Butt.
© 2011 John Wiley & Sons, Inc. Published 2011 by John Wiley & Sons, Inc.

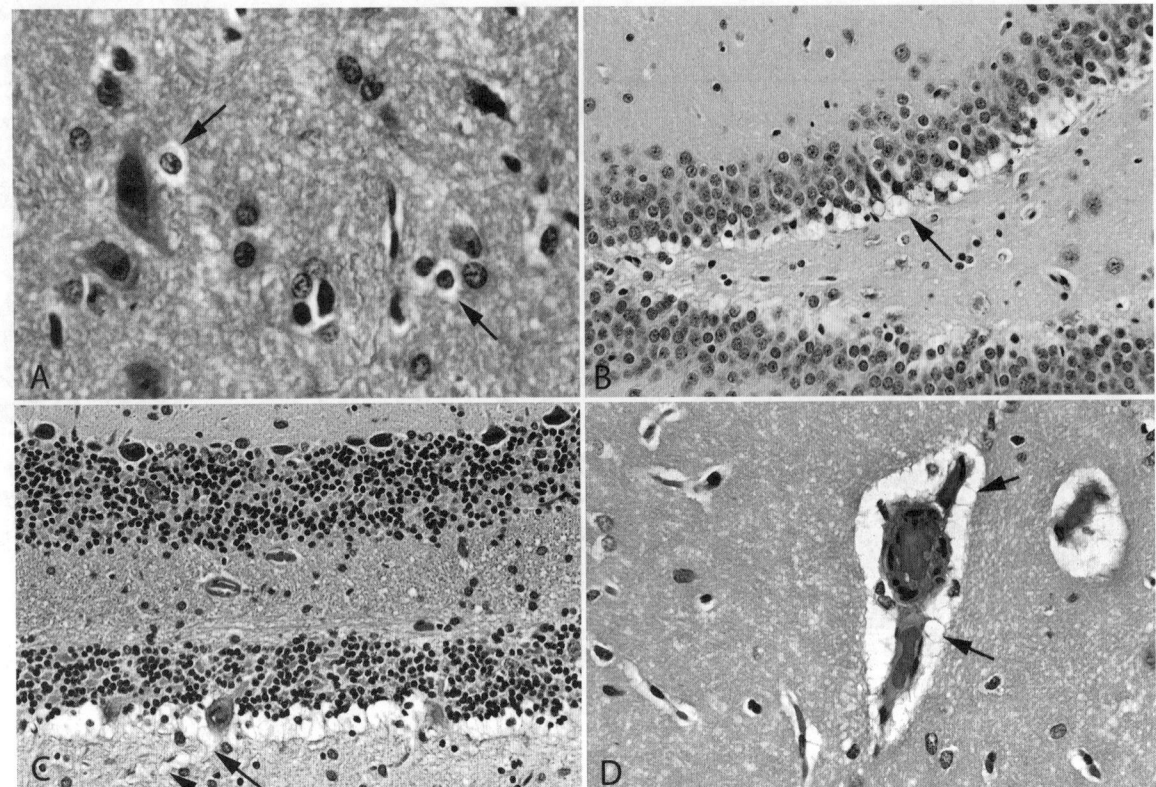

FIGURE 1   Several artifacts commonly observed in immersion-fixed brains: (A) pale rings around the nuclei of oligodendrocytes, resulting in a classic "fried egg" appearance (arrows); (B) separation of the blades of the hippocampal dentate gyrus from the adjacent hilar neuropil; (C) similar separation of the Purkinje neuron layer from the molecular layer of the cerebellum; (D) separation of vessels from the surrounding neuropil. Careful examination of the zones of separation in panels (B), (C), and (D) indicates a pattern of vacuolation (arrows), suggesting the possibility that a component of this separation may represent swelling of astrocytic processes.

this perivascular retraction may represent swelling of astrocytic cell processes.

DARK (BASOPHILIC) NEURONS

Dark neurons undoubtedly represent the most important of the various CNS artifacts. These neurons have frequently been misinterpreted by inexperienced pathologists, neuroscientists, and toxicologists as degenerating neurons. Examples of such misinterpretation are common in the published toxicology and neuroscience literature; some of these erroneous reports have been cited in other publications.[1-3]

Dark neurons, also referred to as *dark spiky neurons* or *basophilic neurons* even though typically amphophilic in their staining character, are commonly encountered in sections of CNS tissue. Dark neurons are shrunken and typically have irregular contours as well as increased staining of their processes (particularly the apical dendrites). Both the nuclei and cytoplasm of these neurons stain darkly, often to the point where the darkened nucleus blends

imperceptibly with the darkened cytoplasm (Figure 2A). Dark neuron alteration is most commonly seen in larger neurons such as pyramidal neurons and Purkinje neurons, but is also common within the granule neurons of the dentate gyrus. Typically, the dark neuron alteration involves clusters of neurons, or at least numerous neurons, scattered about within selected brain regions (Figure 2B). Dark neuron alteration involving small neurons with little cytoplasm (such as the granular neurons of the cerebellum) may be manifested solely by nuclear condensation with little cytoplasmic staining.

In hemotoxylin and eosin (H&E)–stained sections, the intensely basophilic cytoplasm of the dark neuron is in sharp contrast to the bright acidophilic cytoplasm of degenerating neurons (typically termed *acidophilic neurons*, *eosinophilic neurons*, *red dead neurons*, and *acute eosinophilic neuron degeneration*), so these two processes will typically not be confused (Figure 2C and D). (Use of the term *eosinophilic neuron* naturally refers to the appearance of these neurons within H&E-stained sections, but the cytoplasm of dying neurons will be stained by other acidophilic dyes in addition to eosin.) However, ischemic neurons in peracute

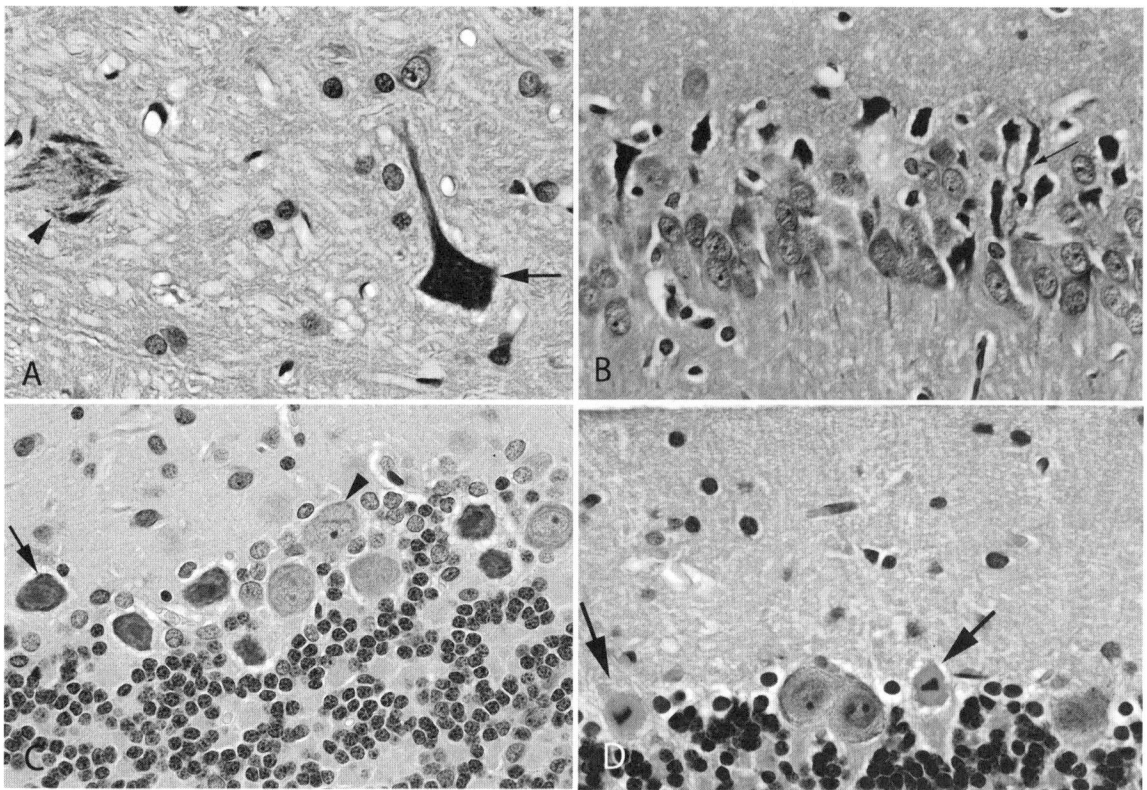

FIGURE 2   Distinguishing dark neurons from eosinophilic (red dead) neurons. (A) In dark neuron formation, the neuron shrinks, and its major dendritic process often becomes visible (full arrow). Contrast the dark neuron on the right side with the normal neuron (with prominent Nissl substance) on the left (arrowhead). (B) Dark neurons tend to cluster, and the dendritic processes are often irregularly contoured or corkscrew-shaped (arrow). (C, D) The appearance of normal Purkinje neurons (arrowhead in C), dark Purkinje neurons (arrow in C), and those undergoing acidophilic degeneration (arrows in D) is strikingly different. Necrotic Purkinje neurons are characterized by pyknotic nuclei and bright eosinophilic (rather than dark) cytoplasm in H&E-stained sections (arrows in D).

stages of degeneration may be indistinguishable from basophilic neurons.

In a series of six manuscripts published between 1960 and 1978, Cammermeyer drew attention to dark neurons as significant artifacts in the CNS.[4–9] Dark neurons are readily produced by handling either fresh CNS tissues or tissues that have been perfused with fixative but not allowed sufficient time for completion of the fixation process.[10] Most dark neuron artifacts can be eliminated by perfusion fixation as long as the removal of the tissues is delayed for between 4 and 24 h (depending on the fixative) after the perfusion has been performed.[4]

In human brain tissue, dark neurons are most frequently seen in surgical biopsies and are less common in autopsy material.[2] However, in animal toxicity studies, dark neurons are common in immersion-fixed brains. This is probably due to the fact that freshly harvested brains from animals in toxicity studies are immersed in fixative relatively promptly (typically between 30 and 60 min after euthanasia) so that neurons (particularly within small brains such as those of mice or rats) are exposed to fixatives while some of their

structural proteins and enzymatic machinery are still intact. In fact, one hypothesis for the generation of dark neurons is that the fixation process may stimulate the intracellular actin that is present within both the nucleus and cytoplasm.[2]

Dark neurons have been associated with trauma to the head or brain, ischemia (including postischemic reperfusion injury), hypoglycemia, epilepsy, electrical stimulation, excitatory amino acids, and spreading depression.[2,11–21] Pretreatment (prior to fixation) of rat cortical biopsies with glutamate receptor blocking agents has been shown to minimize dark neuron transformation.[2] The results of that study suggest that a possible final common pathway to dark neuron formation may be depolarization, and that the depolarization may, in turn, be due to the release of glutamate or to some other excitatory stimulus. Another hypothesis is that the dark neuron alteration is caused by structural alteration of cytoplasmic proteins since dark neurons can be eliminated by treatment of the fixed tissues with proteases.[21]

A variety of postmortem tissue perturbations are capable of generating dark neurons. These include traumatic postmortem handling, perfusion with cold fixative, immersion

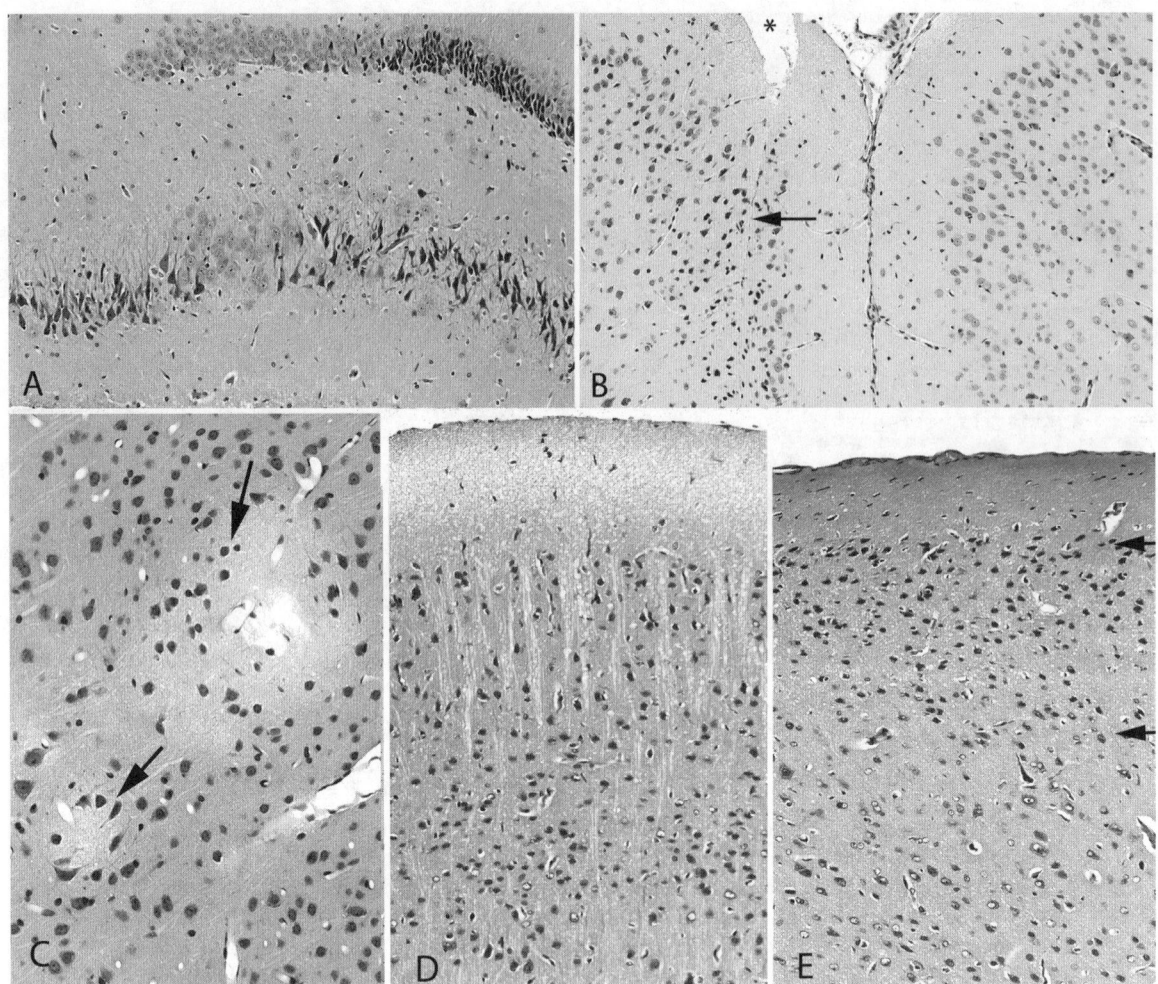

FIGURE 3 Dark neurons are associated with a variety of postmortem brain perturbations. (A) Clusters of dark neurons can be seen within the granule neuron layer of the dentate gyrus (top neuronal layer) and in the underlying CA$_3$ (left side of bottom layer) and CA$_4$ (right side of bottom layer) sectors of the hippocampal pyramidal neuron layer, suggesting that the dark neurons were produced by a compressive force generated across these layers (from the overlying cortical surface). Corresponding zones of dark neurons were also present in the overlying CA$_1$ sector and cortex in this brain. (B) Pressure on the surface of the frontal cortex on the left side created a surface defect (*) as well as a "fault line" through the cortex (arrow). Note how dark neurons cluster around this fault line but are absent within the contralateral intact cortex. (C) Perfusion with ice-cold fixative solution resulted in pallor of the brain neuropil around numerous blood vessels. Dark neurons are abundant in the regions of pallor, suggesting that these may have developed as a result of thermal shock. (D) Immersion of this rat brain in physiologic saline for 30 min induced extensive vacuolation of cortical layer I (top) and the production of numerous dark neurons throughout the full cortical thickness. (E) Drying of freshly harvested rat brain for approximately 1 h at room temperature prior to immersion in fixative resulted in hypereosinophilic (i.e., dessicatation) artifact in cortical layer I of this H&E-stained section and abundant dark neurons in layers II to IV (between the arrows) but not deeper within the cortex.

of the brain in saline, and relatively short-term drying of the fresh brain (Figure 3). In addition to premature removal of perfusion-fixed brains from the cranial vaults, other factors contributing to an increased frequency of dark neurons in perfusion-fixed tissues include the maturational state of the neurons (with neurons from younger animals being less susceptible to this change[22]), a delayed or slow saline perfusion, and prolonged perfusion with Bouin's fixative.[9]

Figure 4 contrasts the appearance of dark neurons and degenerating neurons in sections stained with cresyl violet or with the amino cupric silver stain (one of the most frequently employed silver degeneration stains). The cresyl violet stain is frequently used by neuroscientists for performing neuronal

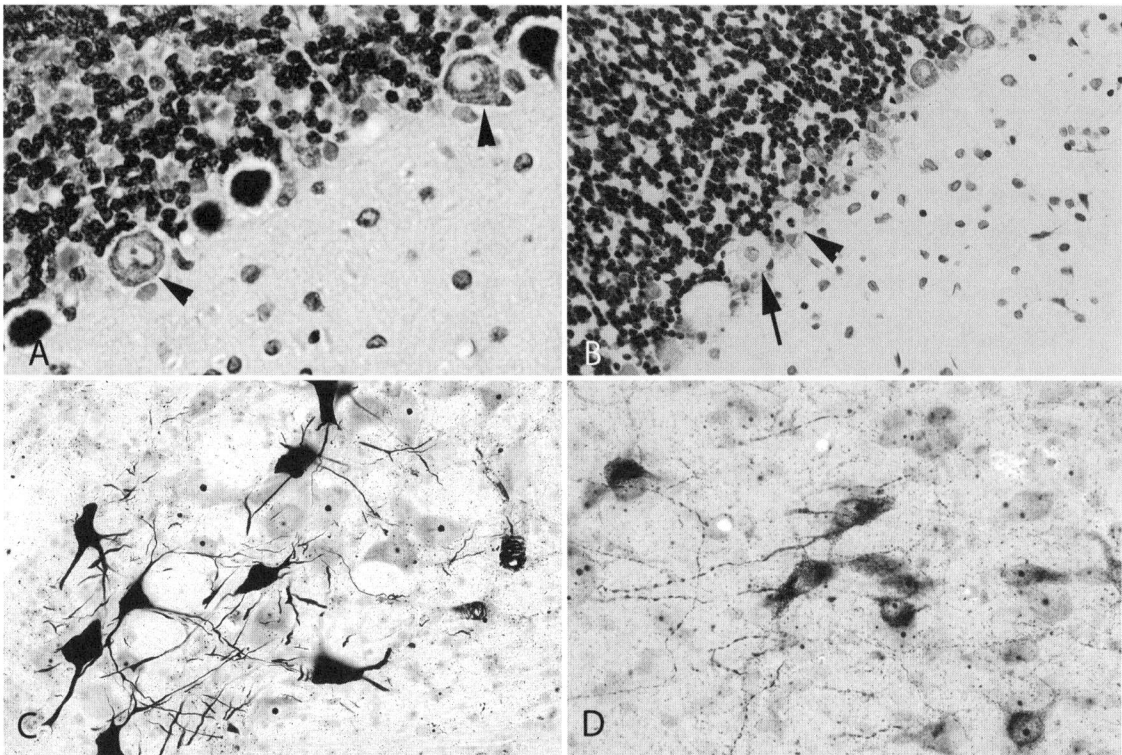

FIGURE 4   Appearance of dark neurons *vs.* degenerating neurons with special neuronal stains: cresyl violet and cupric silver stains. (A) In cresyl violet–stained sections, dark neurons (in this case, Purkinje cells) are dark blue in color with little contrast between the nucleus and cytoplasm relative to the appearance of normal-appearing Purkinje neurons (arrowheads). (B) In contrast, the cytoplasm of degenerating neurons do not stain with cresyl violet since this technique primarily stains Nissl substance (which is rapidly dispersed during cell degeneration). The full arrow points to a Purkinje neuron in an early stage of degeneration (characterized by a normal-appearing nucleus but essentially unstained cytoplasm), while the arrowhead points to a Purkinje neuron in a slightly later stage of degeneration (with a pyknotic nucleus and unstained cytoplasm). Darkly stained neurons in cresyl violet–stained sections should never by interpreted as degenerative. (C) Homogeneously stained (i.e., argyrophilic) artifactual dark neurons in amino cupric silver–stained sections have even contours and nonfragmented processes (red nucleus from a rat brain). (D) Degenerating neurons in an amino cupric silver–stained brain section have granular staining patterns, and the neuronal processes are both beaded in appearance and have an affinity for silver impregnation at considerable distances from the neuronal somas.

cell counts, judging regional brain cellularity, or delineating various brain nuclei. It is surprising that dark neurons in cresyl violet–stained sections are frequently misinterpreted as degenerative because cresyl violet stains Nissl substance (i.e., rough endoplasmic reticulum), and Nissl substance is rapidly dispersed during neuron degeneration (Figure 4B). In this author's experience, dark neurons have not been particularly common in sections stained with amino cupric silver if the brains have been optimally fixed and handled.[23] When present within amino cupric silver–stained sections, the dark argyrophilic neurons are usually restricted to small numbers of large neurons such as those within the red nucleus of the midbrain (Figure 4C). However, extensive dark neuron formation has been described using a variety of

silver staining methodologies.[13,16,21,24–30] Argyrophilic dark neurons are characterized by an even homogeneous staining pattern (Figure 4C), whereas degenerating neurons typically have a more granular pattern of argyrophilia, which extends along the neuronal processes for greater distances from the neuronal cell bodies (Figure 4D). Depending on the stage of neuronal degeneration, the argyrophilic neuronal processes of dying neurons will often also appear fragmented (Figure 4D).

If dark neurons represent the only cellular alteration that is present within the sections evaluated, it is generally safe to assume that the change in these neurons is artifactual in nature unless the study was specifically designed to detect peracute neuronal injury. While dark neuron transformation

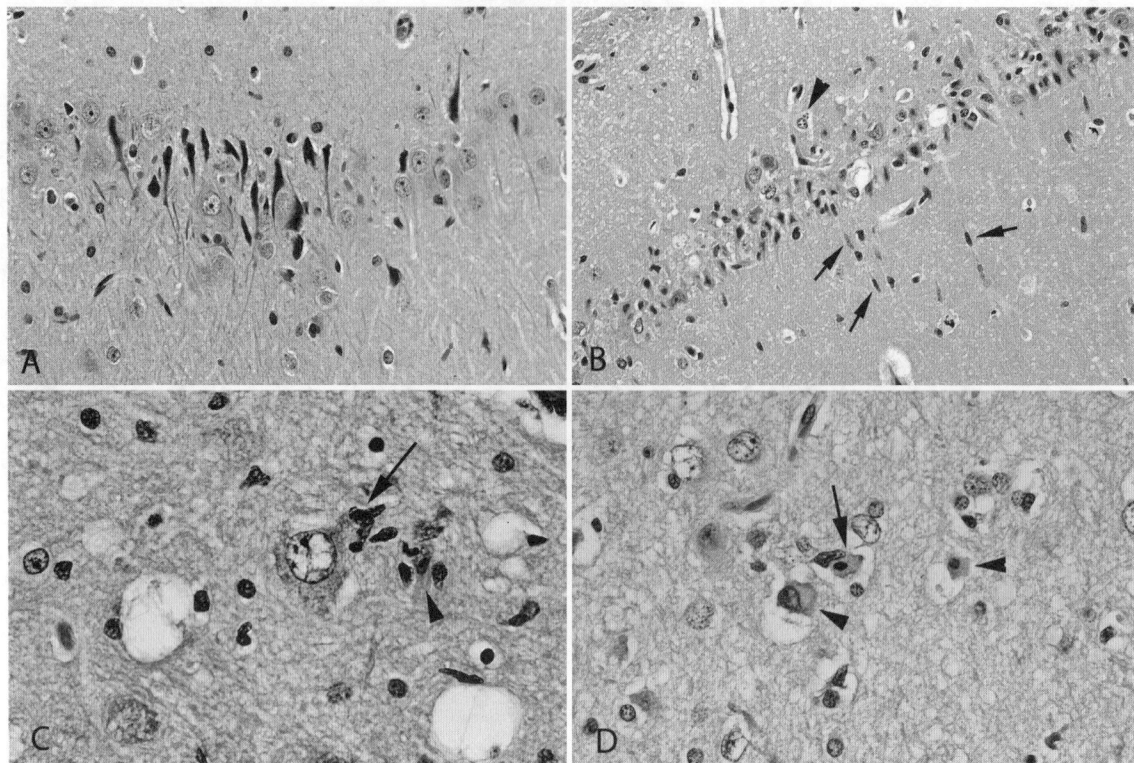

FIGURE 5   Neuronal degeneration is characterized by a heterogeneous histological pattern. (A) Typical dark neuron artifact within the hippocampal pyramidal neuron layer of a rat. The process of dark neuron formation is homogeneous, with all of the affected neurons being in the same "stages of darkness" and with no secondary cellular reactions being present. (B) Hippocampal pyramidal layer of a rat that had a history of seizure activity. Large numbers of degenerating "acidophilic" ("eosinophilic") pyramidal neurons are present in H&E-stained sections. However, there is also a secondary glial cell reaction comprised of both microglia (arrows) and reactive astrocytes (arrowhead). The reactive astrocytes would be better demonstrated with a stain for glial fibrillary acidic protein (GFAP), and the microglia by stains such as ionized calcium-binding adapter molecule 1 (IBA1) or CD68 (or others). Depending on the cause of the neuronal degeneration, microglia may arrive promptly after the neuronal insult such as in (C), where a cluster of microglial cells is associated with a neuron which has a swollen nucleus but whose cytoplasm is not yet eosinophilic (arrow). Other microglia are associated with a more typically-appearing eosinophilic neuron with a pyknotic nucleus (C, arrowhead). An additional finding in this field is neuropil vacuolation, with a microglial cell being associated with the vacuole at the lower right. [While neuropil vacuoles may represent artifact in some situations (see Figure 1), the association of activated microglials cells with vacuoles and degenerating neurons indicates that the vacuoles are lesions in this particular field.] (D) The progression of eosinophilic neuron change is demonstrated by three neurons in the center of the photomicrograph. At the left (arrowhead), a degenerating neuron (with eosinophilic cytoplasm in an H&E-stained section) still has a prominent nucleus, whereas the dead neuron on the right (arrowhead) has a pyknotic nucleus. The central neuron (full arrow) has an associated microglial cell. This spectrum of change represents the hallmark of a genuine degenerative process.

may represent the earliest stage of neuron degeneration, it will generally last only a few hours prior to the development of the cytoplasmic acidophilia that represents the hallmark of neuron degeneration. In studies in which neuron degeneration was generated either by electric shock or by mechanical injury, dark neurons disappeared (and cytoplasmic acidophilia developed) within 4 h.[13,21] Neuron degeneration of one to several days in duration is characterized by heterogeneous

cellular patterns (Figure 5). Dead neurons will develop pyknotic and later karyorrhectic nuclei, which typically disappear prior to removal of the acidophilic cytoplasm remnants of the dead neurons. Neuron degeneration will also provoke reactions in microglial cells (i.e., increased numbers and cellular enlargement) as well as astrocytosis, characterized by enlarged reactive astrocytes. Some aspects of these glial cell reactions will be readily seen within

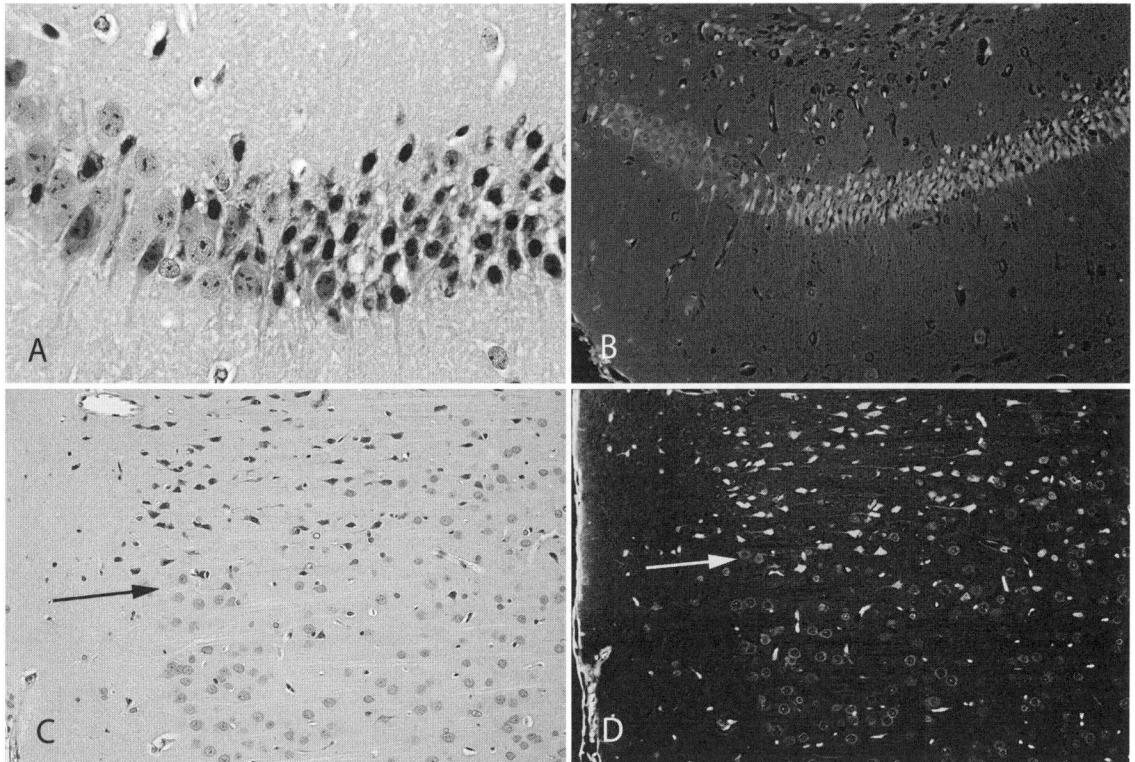

FIGURE 6   Special stains assist in differentiating dark neuron alteration from peracute neuronal degeneration. Peracute neuron degeneration (up to several hours post-insult) may resemble dark neuron "artifact." (A) H&E-stained section of the hippocampus from a mouse that was euthanatized immediately after a 2-h grand mal seizure. Although the dark staining of many pyramidal neurons in this field is primarily nuclear, little cytoplasmic eosinophilia was evident in this H&E-stained section. However, a Fluoro-Jade B-stained section of the same hippocampus (panel (B) at a lower magnification) showed bright staining of many pyramidal neurons within the CA_1 sector. (C) H&E-stained section of rat temporal cortex characterized by dark neuron artifact (above the boundary denoted by the arrow) caused by pressure on the cortex during brain removal. An adjacent section stained with Fluoro-Jade B shows the dark neurons to have slightly brighter cytoplasmic staining but only because the cytoplasm of these cells was contracted or condensed. What cannot be appreciated in this black-and-white image is that the cytoplasm of these dark neurons stained pale green rather than the bright yellow color of the degenerating neurons in panel (B). Note that the small, more intensely-stained cells in panel (D) represent intravascular erythrocytes (which are autofluorescent). A silver degeneration stain would also be helpful for differentiating peracute neuron degeneration from dark neuron formation.

H&E-stained sections but may be enhanced by selective staining for the reactive cells [e.g., with immunostains such as glial fibrillary acidic protein (GFAP) for astrocytes, or CD68 or ionized calcium-binding adapter molecule 1 (IBA1) for microglia, to name a few]. Because peracute neuronal degeneration of only a few hours' duration may simulate the appearance of dark neuron artifact, stains that are more specific for neuron degeneration are quite useful (Figure 6).[31] For paraffin-embedded tissues, this will generally be one of the Fluoro-Jade stains[32–35] since silver degeneration stains are generally not compatible with paraffin embedding. (A silver degeneration stain that utilizes a physical developer and that may be used on paraffin-embedded

tissue has been described, but the author has had no experience with this stain.[21])

ARTIFACTS INVOLVING MYELIN, AXONS, AND SENSORY GANGLION NEURONS

Vacuoles of uncertain cellular location are often encountered within the CNS at the light microscopic level. It is well recognized by electron microscopists that exposure of CNS tissues to alcohol results in significant degrees of myelin artifact (primarily myelin sheath splitting) and is the reason that commercial preparations of buffered formalin (which

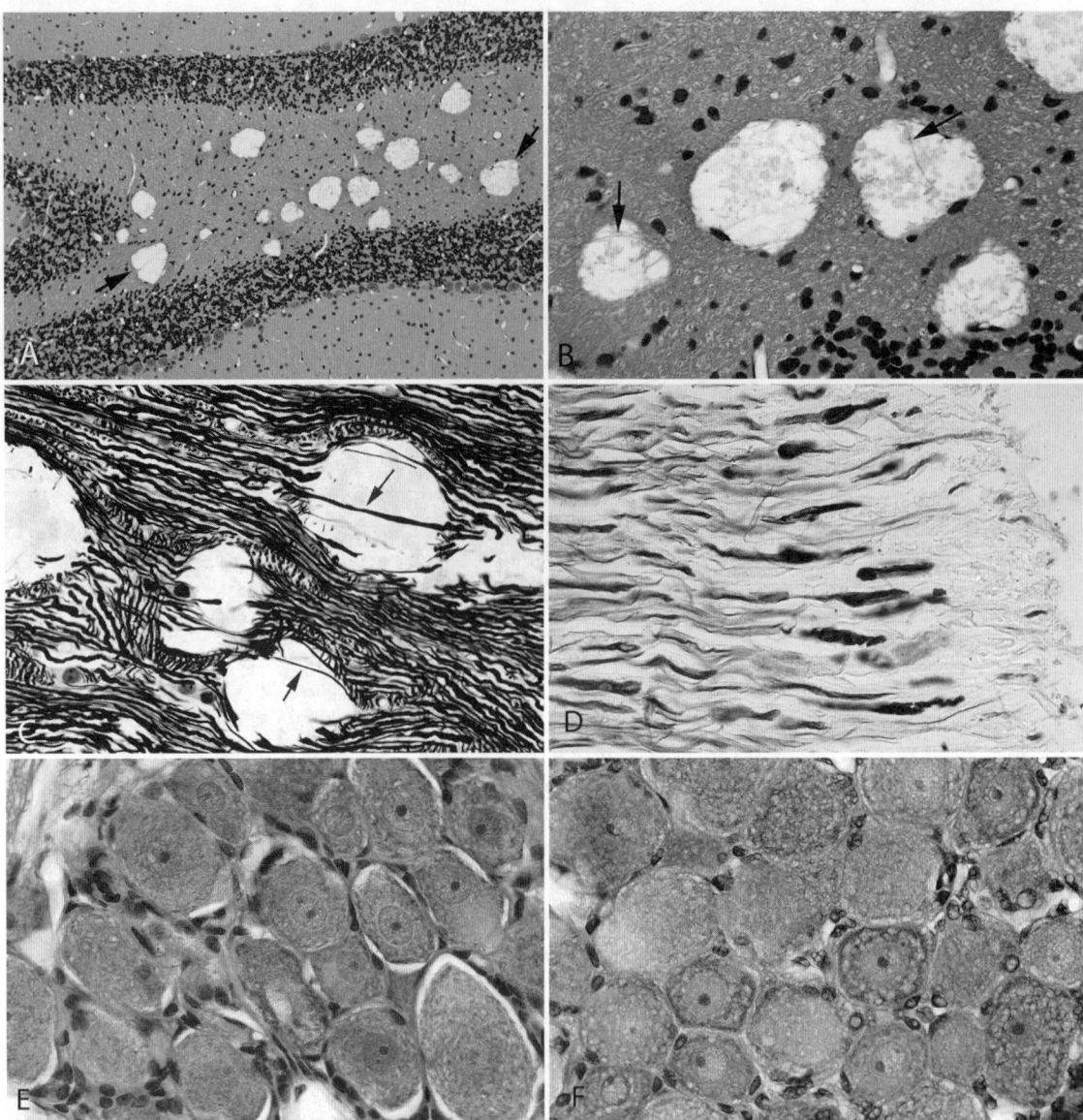

FIGURE 7   Artifacts involving myelin, axons, and sensory ganglion neurons. Artifactual vacuoles are most common in regions of white matter. In perfusion-fixed rodent brains, these are most frequently present in the deep cerebellar white matter and brain stem, probably resulting from pressure placed on the brain surface during removal of the calvaria. These vacuoles (representing Buscaino bodies) are generally thought to be the result of handling-induced alteration in poorly fixed myelin, although some patterns of myelin vacuolation may also be induced by excessive exposure to alcohol during tissue fixation and processing. (A) Numerous vacuoles are present in the deep cerebellar white matter between the two arrows. (B) Higher magnification view of (A). Many vacuoles contain amorphous material which differs in appearance from the myelin debris seen in instances of myelin degeneration. Such vacuoles are often partially birefringent when viewed with polarized light. Note that intact axons of normal dimensions (arrows) traverse the vacuoles. (C) Vacuolar artifact in myelin within a Bielschowsky's silver–stained section of a rat spinal cord. The presence of normal-appearing axons (arrows) and the lack of any cellular response contribute to the interpretation that this alteration represents artifact. (D) Postmortem crush artifact at the site of nerve transection in a Bielschowsky's-stained longitudinal section of peripheral nerve is characterized by numerous enlarged, irregularly contoured axons near the cut end of nerve (at right). Such axons are common near the severed end of nerves, particularly if these nerves have been cut with scissors. (E) H&E-stained section of a normal trigeminal (gasserian) ganglion from a perfusion-fixed rat showing large ganglion neurons with distinct nuclei and nucleoli and abundant cytoplasm; the pericellular retraction spaces are an artifact. (F) Extensive microvacuolation of neuronal cytoplasm in an H&E-stained section of a rat dorsal root ganglion represents swollen organelles such as mitochondria and is an artifactual consequence of suboptimal (delayed) fixation.

contain alcohol as a stabilizing agent) are avoided as fixatives for ultrastructural preparations. However, it is less well known that holding trimmed CNS tissues in alcohol may also produce vacuolation of the white matter.[36] A common example of this artifact is extensive vacuolation of the deep cerebellar white matter and cerebellar peduncles in brain specimens that have been held in an alcohol bath on a tissue processor over a weekend.

A very common artifact that is seen most frequently in brains handled immediately after perfusion fixation is the Buscaino body, also known as a mucocyte or metachromatic body.[36–38] (The term *mucocyte* is misleading and will not be used further, because the artifact being described is not related to accumulation of mucin.) Buscaino bodies are rounded, oval, or lobulated bodies that are often described as being arranged in grapelike clusters. These range in size from approximately 100 μm in diameter to considerably larger (Figure 7). They are most frequently found within the white matter of the brain and spinal cord, although smaller similar-appearing vacuolar artifacts may be seen in the optic nerve as well. Buscaino bodies develop as a result of postmortem traumatization of the nervous system[6] and are thought to be caused by solubilization and subsequent precipitation (by fixation) of some myelin component.[39] Buscaino bodies are typically pale but may be slightly basophilic or gray in color, as well as metachromatic or periodic acid-Schiff (PAS)-positive.[38] They sometimes (although not always) are partially birefringent.[40] In this author's experience, one of the most common brain regions in which Buscaino bodies are encountered in rodents is the cerebellar white matter (Figure 7A and B). When present at this site, these bodies are typically also present in the heavily myelinated tracts within the underlying brain stem. In these two hindbrain locations, Buscaino bodies probably result from pressure placed on the cerebellum when the calvaria is pried off the posterior brain surface using a rongeur or other instrument inserted through the foramen magnum (a dissection practice that should be discouraged). Although Buscaino bodies may appear empty in tissues processed to paraffin, they often contain poorly stained flocculent densities (Figure 7B). However, there is no associated distortion of the axons that traverse these bodies (Figure 7C).

Myelin vacuolation comparable to the Buscaino body artifact present within the CNS is also frequently encountered in the peripheral nervous system and has a similar genesis (i.e., trauma-induced myelin alteration that manifests as a vacuole after tissue processing). Although not illustrated in this chapter, these peripheral nerve artifacts will appear similar to the Buscaino bodies that develop in the spinal cord (Figure 7C). These artifacts are distinguished from nerve fiber (axonal) degeneration by the absence of "digestion chambers" (cellular debris and macrophages within the vacuoles), which are a common feature within degenerating nerves. Although Buscaino bodies are not associated with

axonal distortion, irregular thickening of axons may be seen where nervous system tissues have been crushed during removal. This is most frequently encountered at the cut ends of peripheral nerves, particularly if these have been grasped with forceps or severed with scissors (Figure 7D). The ends of nerves severed during dissection should be discarded during trimming.

Even when perfusion fixation is employed for optimal preservation of the histomorphological features of nervous system tissues, the sensory ganglia (particularly the dorsal root ganglia) may not receive adequate exposure to the fixative. Normal ganglionic neurons are large round-to-oval cells with abundant cytoplasm, a central nucleus, and a single prominent nucleolus (Figure 7E). Delayed fixation of these sensory ganglia results in microvacuolation of the cytoplasm (Figure 7F). These small cytoplasmic vacuoles have been demonstrated ultrastructurally to represent swollen mitochondria.[41]

MISCELLANEOUS ARTIFACTS

Where saws (either oscillating electric or manual) are used to remove brains from the larger animal species, contact between the saw blade and the brain surface produces sharp lines of neuropil displacement/compression accompanied by deposition of bone fragments, mineralized debris, hair, or other structures within the brain.[36] Clefts in the neuropil (e.g., Figure 3B) are easy to produce in fresh or freshly fixed brains by even light pressure. If the spinal cord is bent or kinked prior to fixation, displacement of the neuropil may simulate the appearance of intrinsic tumors or developmental anomalies. Intussusception of the spinal cord may also be seen, resulting in a prominent pattern of distortion referred to as "toothpaste artifact."[38] In addition to the artifacts discussed above, many other nonpathologic alterations may be induced in nervous tissue as a result of suboptimal dissection, processing, embedding, sectioning, mounting, and staining procedures.[6,10,36,42]

Due to space and figure limitations, only selected artifacts are illustrated in this chapter but most of those not illustrated will be recognized readily by pathologists experienced in the microscopic evaluation of nervous system tissues. Inexperienced pathologists will continue to be challenged by histological alterations of uncertain genesis, some of which will undoubtedly be artifactual in nature. Careful attention to the proper fixation, handling, and processing of nervous system tissues is imperative, as is the utilization of supplementary staining procedures when indicated. Furthermore, interpretation of neuronal alterations should make sense within the context of the temporal aspects of the study design. For example, whereas dark neurons may represent a peracute stage of neuron degeneration, the dark neuron alteration seen in neurons that go on to die is transitional and of brief duration.

Pathologists performing safety evaluations on sections of the nervous tissue must have an appreciation for both the sequence of cytological alterations that are seen in neurodegenerative processes as well as the types of cellular reaction that accompany such degeneration. Furthermore, reviewers of journal articles and study reports should be suspicious of photomicrographic representations of neuronal degeneration characterized by monomorphic populations of dark neurons unless the authors can provide sufficient evidence that these represent a peracute effect of the treatment.

Acknowledgments

The author is grateful to Rosalyn Garman for the majority of the histological preparations illustrated in this chapter (except for the cupric silver–stained sections, which were prepared by NeuroScience Associates in Knoxville, TN).

REFERENCES

1. Jortner BS. The return of the dark neuron: a histological artifact complicating contemporary neurotoxicologic evaluation. *Neurotoxicology.* 2006;27:628–634 [see comment in *Neurotoxicology* 2006;627:1126].

2. Kherani ZS, Auer RN. Pharmacologic analysis of the mechanism of dark neuron production in cerebral cortex. *Acta Neuropathol.* 2008;116:447–452.

3. Jortner BS. Neuropathologic assessment in acute neurotoxic states: the "dark" neuron. *J Med Chem Biol Radiol Def.* 2005;3. Available at: http://www.jmedcbr.org/Issue_0301/Jortner_0405.html. Accessed Mar 24, 2010

4. Cammermeyer J. A critique of neuronal hyperchromatosis. *J Neuropathol.* 1960;19:141.

5. Cammermeyer J. The importance of avoiding "dark" neurons in experimental neuropathology. *Acta Neuropathol.* 1961;245–270.

6. Cammermeyer J. Nonspecific changes of the central nervous system in normal and experimental material. In: Bourne GH, ed. *The Structure and Function of the Nervous System,* Vol 6. New York: Academic Press; 1972:131–251.

7. Cammermeyer J. Schemic neuronal disease: of Spielmeyer: a reevaluation. *Arch Neurol.* 1973;23:391–393.

8. Cammermeyer J. Histochemical phospholipid reaction in ischemic neurons as an indication of exposure to postmortem trauma. *Exp Neurol.* 1975;49:252–271.

9. Cammermeyer J. Is the solitary dark neuron a manifestation of postmortem trauma to the brain inadequately fixed by perfusion? *Histochemistry.* 1978;56:97–115.

10. Garman RH. Artifacts in routinely immersion fixed nervous tissue. *Toxicol Pathol.* 1990;18:149–153.

11. Auer RN, et al. The temporal evolution of hypoglycemic brain damage: I. Light- and electron-microscopic findings in the rat cerebral cortex. *Acta Neuropathol.* 1985;7:13–24.

12. Auer RN, Sutherland GR. General pathology of the central nervous system. In: Love S, et al., eds. *Greenfield's Neuropathology,* Vol 1. London: Edward Arnold; 2008: 63–119.

13. Csordás A, et al. Recovery versus death of "dark" (compacted) neurons in non-impaired parenchymal environment: light and electron microscopic observations. *Acta Neuropathol.* 2003;106:37–49.

14. Czurko A, Nishino H. "Collapsed" (argyrophilic, dark) neurons in rat model of transient focal cerebral ischemia. *Neurosci Lett.* 1993;162:71–74.

15. Dietrich WD, et al. Intraventricular infusion of *N*-methyl-D-aspartate: 2. Acute neuronal consequences. *Acta Neuropathol.* 1992;84:630–637.

16. Gallyas F, et al. Four modified silver methods for thick sections of formaldehyde-fixed mammalian central nervous tissue: "dark" neurons, perikarya of all neurons, microglial cells and capillaries. *J Neurosci Methods.* 1993;50:159–164.

17. Jenkins LW, et al. The role of postischemic recirculation in the development of ischemic neuronal injury following complete cerebral ischemia. *Acta Neuropathol.* 1981;55:205–220.

18. Kövesdii E, et al. The fate of "dark" neurons produced by transient focal cerebral ischemia in a non-necrotic and non-excitotoxic environment: neurobiological aspects. *Brain Res.* 2007;1147:272–283.

19. Ooigawa H, et al. The fate of Nissl-stained dark neurons following traumatic brain injury in rats: difference between neocortex and hippocampus regarding survival rate. *Acta Neuropathol.* 2006;112:471–481.

20. Söderfeldt B, et al. Bicuculline-induced epileptic brain injury: transient and persistent cell changes in rat cerebral cortex in the early recovery period. *Acta Neuropathol.* 1983;62:87–95.

21. Zsombok A, et al. Basophilia, acidophilia and argyrophilia of "dark" (compacted) neurons during their formation, recovery or death in an otherwise undamaged environment. *J Neurosci Methods.* 2005;142:145–152.

22. Ebels EJ. Dark neurons, a significant artifact: the influence of the maturational state of neurons on the occurrence of the phenomenon. *Acta Neuropathol.* 1975;33:271–273.

23. Switzer RCI. Application of silver degeneration stains to neurotoxicity testing. *Toxicol Pathol.* 2000;28:70–83.

24. Gallyas F, et al. Golgi-like demonstration of "dark" neurons with an argyrophil III method for experimental neuropathology. *Acta Neuropathol.* 1990;79:620–628.

25. Gallyas F, et al. An immediate light microscopic response of neuronal somata, dendrites and axons to contusing concussive head injury in the rat. *Acta Neurophathol.* 1992;83:394–401.

26. Gallyas F, et al. Light microscopic response of neuronal somata, dendrites and axons to post-mortem concussive head injury. *Acta Neurophathol.* 1992;83:499–503.

27. Gallyas F, et al. Formation of "dark" (argyrophilic) neurons of various origin proceeds with a common mechanism of biophysical nature (a novel hypothesis). *Acta Neuropathol.* 1992;83:504–509.

28. Gallyas F, Zoltay G. An immediate light microscopic response of neuronal somata, dendrites and axons to non-contusing concussive head injury in the rat. *Acta Neuropathol.* 1992;83:386–393.

29. Gallyas F, et al. An immediate morphopathologic response of neurons to electroshock; a reliable model for producing "dark" neurons in experimental neuropathology. *Neurobiology.* 1993;1:133–146.

30. Gallyas F, et al. Gel-to-gel phase transition may occur in mammalian cells: mechanism of formation of "dark" (compacted) neurons. *Biol Cell.* 2004;96:313–324.

31. Poirier JL, et al. Differential progression of dark neuron and Fluoro-Jade labeling in the rat hippocampus following pilocarpine-induced status epilepticus. *Neuroscience.* 2000;97:59–68.

32. Schmued LC, et al. Fluoro-Jade: a novel fluorochrome for the sensitive and reliable histochemical localization of neuronal degeneration. *Brain Res.* 1997;751:7–46.

33. Schmued LC, Hopkins KJ. Fluoro-Jade B: a high affinity fluorescent marker for the localization of neuronal degeneration. *Brain Res.* 2000;874:123–130.

34. Schmued LC, Hopkins KJ. Fluoro-Jade: novel fluorochromes for detecting toxicant-induced neuronal degeneration. *Toxicol Pathol.* 2000;28:91–99.

35. Schmued LC, et al. Fluoro-Jade C results in ultra high resolution and contrast labeling of degenerating neurons. *Brain Res.* 2005;1035:24–31.

36. McInnes E. Artefacts in histopathology. *Comp Clin Pathol.* 2005;13:100–108.

37. Adornato B, Lampert P. Status spongiosus of nervous tissue, electron microscopic studies. *Acta Neuropathol.* 1971;19:271–289.

38. Vinters HV, Kleinschmidt-DeMasters BK. General pathology of the central nervous system. In: Love S, et al., eds. *Greenfield's Neuropathology*, Vol 1. London: Edward Arnold; 2008:1–62.

39. Ibrahim MZW, Levine S. Effect of cyanide intoxication on the metachromatic material found in the central nervous system. *J Neurol Neurosurg Psychiatry.* 1967;30:545–555.

40. Fix AS, Garman RH. Practical aspects of neuropathology: a technical guide for working with the nervous system. *Toxicol Pathol.* 2000;28:122–131.

41. Li XG, Zochodne DW. Microvacuolar neuronopathy: a postmortem artifact of sensory neurons. *J Neurocytol.* 2003;32:393–398.

42. Thompson SW, Luna LG. *An Atlas of Artifacts Encountered in the Preparation of Microscopic Tissue Sections.* Springfield, IL: Charles C Thomas; 1978.

14

HIGH-DEFINITION MICROSCOPIC ANALYSIS OF THE NERVOUS SYSTEM

BERNARD S. JORTNER

Laboratory for Neurotoxicity Studies, Virginia–Maryland Regional College of Veterinary Medicine, Virginia Tech, Blacksburg, Virginia

Contemporary neuropathology relies on microscopic study of tissue for diagnostic and research purposes. In addition to the skill of the pathologist, the effectiveness of this approach is related to the methods and equipment available for these procedures. The use of sections cut from formalin-fixed and paraffin-embedded tissue has long been, and remains, a mainstay of such evaluation. In addition to use in routine stains and a wide variety of special stains for elements of the nervous system, such sections are also suitable for immunohistochemistry. The rise of electron microscopy in the mid-twentieth century demonstrated a need for more rigid and durable embedding media, which was largely met by epoxy resins such as epon. It quickly became apparent that semithin sections (1 μm thick, also called "thick" sections) provided preparations that allowed greater definition and resolution when examined by light microscopy. Transmission electron microscopy (all references to electron microscopy will be to the transmission form of that procedure) of tissues enhanced the resolution and greatly expanded the magnification of these samples. The use of such resin-embedded nervous system tissue for light and electron microscopic neuropathologic study is the subject of this chapter, focusing on several issues, specifically specimen preparation and examples of the utility of these procedures in neurotoxicology.

Historically, tissue fixatives evolved from alcohol or chromic acid (used for much of the nineteenth century) to formalin.[18] The latter, with its cross-linking of proteins, was introduced in the 1890s and soon became the standard tissue fixative.[7] Tissue sections for microscopic study needed to

be sufficiently thin so that individual layers of cells could be observed in sufficient detail. These were initially cut by hand using a razor blade, providing sections 50 to 100 μm thick, which often obscured cellular detail. Improving techniques of sectioning led to the 15-μm-thick sections used by Nissl and Brodmann for their late nineteenth and early twentieth century observations of cerebral cortical structure. Development of more refined microtomes and paraffin embedding media have led to the 5 to 7-μm-thick sections that are used routinely in modern neuropathology.[7]

Contemporary neuropathology is largely based on the use of formalin-fixed tissue, which is embedded in paraffin, sectioned, stained with hematoxylin and eosin (H&E), and examined by light microscopy. This is often supplemented by use of a variety of special stains or immunohistochemical procedures, which highlight elements of the nervous system, such as neuronal cell bodies, their axons, myelin astrocytes, and other glia and microglia. This approach has been productive and has contributed prominently to the current state of knowledge of the nervous system.

Two morphological approaches allow higher definition for more detailed views of nervous (and other) tissues that are useful in toxicological studies. These are the use of semithin (1-μm-thick) sections for light microscopy and transmission electron microscopy.[4] For both of these procedures, transcardial whole-body perfusion fixation is optimal. Methods for this have been described.[1,2,4] Excellent fixation is the basis for such sophisticated pathological assessments of the nervous system. Thus, for experimental animals (such as the rat), perfusion fixation is optimal. The basic idea in this

Fundamental Neuropathology for Pathologists and Toxicologists: Principles and Techniques, First Edition. Edited by Brad Bolon and Mark T. Butt.

procedure is to allow fixation to occur with minimal autolytic or handling artifact, and thus the fixative perfuses through the blood circulatory system with the animal under deep anesthesia. Although there are several approaches to whole-body perfusion,[1] common basic steps include anesthesia, cannulation of the cardiac left ventricle/aorta, washout of the blood, and administration of the fixative.

Fixatives for perfusion may vary depending on the desired use of tissue. As an example, the five-carbon aliphatic dialdehyde glutaraldehyde enhances the quality of samples for transmission electron microscopy, since it provides superior cross-linking of proteins.[1,5] This is important in the nervous system, since it has high lipid content, making fixation more difficult, although high glutaraldehyde concentrations can interfere with immunohistochemistry. The concentration of the glutaraldehyde fixative for perfusion can vary, but 2.5 to 5.0% seems optimal.[1] The monoaldehyde paraformaldehyde is another commonly used fixative, employed in concentrations of 1 to 4%.[1,5] This is the same molecule as formaldehyde, but comes in powdered form. Glutaraldehyde and paraformaldehyde combinations may also be used for electron microscopy, using mixtures varying from 1% of each to 4% paraformaldehyde and 3% glutaraldehyde. If immunohistochemistry and electronic microscopy are the objectives of the study, paraformaldehyde with a small amount of glutaraldehyde (such as 4% paraformaldehyde, 0.2% glutaraldehyde) or paraformaldehyde alone may be used. All of the above are also suitable for light microscopy. In addition to fixation concentration, the osmolarity of the fixative and the buffer in which it is diluted also affect microscopic quality. Osmolarity of fixative solutions in the range of 400 to 700 milliosmols is considered optimal for nervous system tissues.[1,5] Hayat considers fixation for electron microscopy in considerable detail.[5]

Following successful perfusion fixation, the tissue can be left in situ at refrigerator temperature for a number of hours before dissection.[1] This avoids handling artifacts such as the "dark" neuron.[10] If perfusion is suboptimal, quick tissue removal and immersion fixation should be undertaken. Peripheral nerve that is immersion fixed can also be used for these higher-definition preparations.[9,13] Subsequent to any of the fixation methods above, tissues are postfixed in osmium tetroxide and embedded in epoxy resin such as epon,[4] as would be done for transmission electron microscopy. For routine electron microscopy it is common to make small cubes of tissue, about 1 mm in maximum diameter, which allows the osmium and epoxy resin to infiltrate the tissue. For the nervous system, an alternative approach may be used; since neuropathology requires some anatomical orientation for microscopic interpretation, the use of tiny cubes of tissue is often inadequate. However, the perfusion-fixed brain or spinal cord can be dissected to provide a block face of up to 0.5 cm, which can be successfully embedded in

epoxy resin if one dimension of the sample is 1 mm or less. This does require care (and sharp blades) in the dissection. The processing time is increased slightly to allow a longer time for reagent and resin penetration.[4] Sections that are 1 μm thick from such blocks can be cut on a standard ultramicrotome, using glass knives, and are stained with toluidine blue alone or in combination with safranin.[4] The more rigid embedding medium (relative to paraffin) allows for thinner sections, which provide elegant microscopic preparations. The greater light-microscopic resolution provided by such sections give dramatic views of the nervous system, allowing one to better distinguish its components (Figure 1). In addition, the size of the block face allows better neuroanatomical inferences to be made.

The use of 1-μm-thick sections is a standard component of light-microscopic pathological examination of the peripheral nervous system, in both clinical and experimental settings.[13] It is combined with evaluation of sections of frozen and paraffin-embedded tissue, employing stains such as hematoxylin and eosin, special stains such as luxol fast blue, Glees and Marsland's, and Holmes' procedures. Although of value in highlighting components of the nerve, they do not provide the greater resolution inherent in semithin sections.

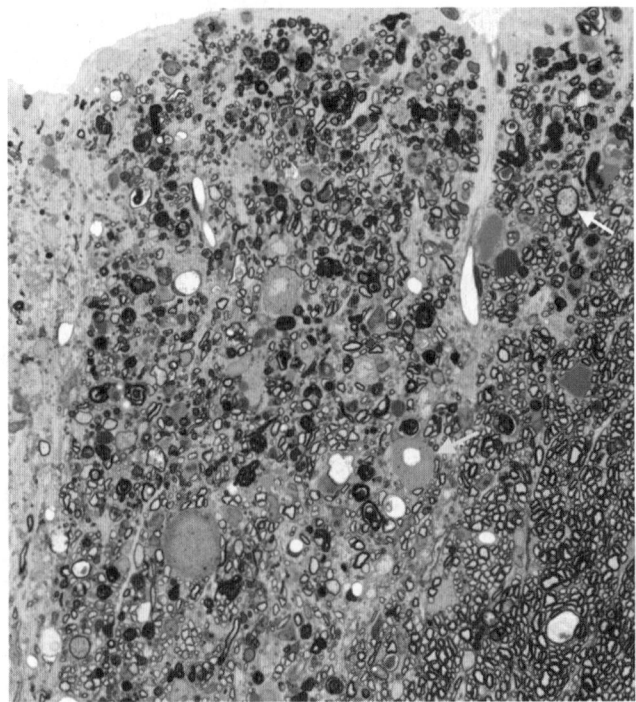

FIGURE 1   Cross section of gracile nucleus from a rat administered multiple doses of tri-*o*-tolyl phosphate and chlorpyrifos. Lesions, such as dark-stained myelin-rich degenerating fibers, swollen myelinated axons, and large, pale-stained dystrophic axons, are well-demonstrated (arrows). A 1-μm-thick section stained with toluidine blue and safranin. (From Jortner et al.[12])

The latter allows the observer to distinguish more cellular detail in peripheral nerve fibers and their supporting glia than can be determined with the preparations noted above. For example, axon size and myelin sheath thickness, features critical for determination of the *g* ratio (axon diameter/fiber diameter), are best determined by the semithin preparations. Degradative and reparative processes such as axonal degeneration and regeneration, demyelination, and remyelination are readily observed. The differential staining of myelin and collagen permits observation of the extent of myelinated fiber preservation in chronic neuropathies. The technique is suitable to evaluate peripheral ganglia as well. While employed routinely for light microscopy of the peripheral nerve, 1-μm-thick sections have also been used to assess changes in the central nervous system. Jortner et al.[12] have examined axonal degeneration and dystrophic axon formation in terminal levels of the gracile fasciculus in rats exposed to neurotoxic organophosphates over long periods of time (Figure 1).

Transmission electron microscopy of appropriately fixed and prepared tissue sections provides levels of resolution and magnification not achievable by other methods. As commercial electron microscopes became available in the 1960s and 1970s, this instrument became the ultimate research and diagnostic tool in anatomical pathology.[17] This enabled investigators to address issues in dispute regarding the cytology of the central nervous system. As an example, electron microscopy proved the validity of the neuron doctrine that had eluded earlier investigators by demonstrating discontinuity between neurons.[19] A key component of this was the determination of synaptic ultrastructure, demonstrating the anatomical basis of intercellular communication between neurons.

As expected, electron microscopy has contributed to contemporary understanding of many toxicant-related neuropathologies. A lesion that is often noted but seldom explained is axonal atrophy. This is an important event in hexacarbon neuropathy,[15] where some mechanistic evidence for this lesion has been obtained. Studies have shown that the active γ-diketone metabolite of toxic hexacarbons reacts with neurofilaments triplet proteins to form pyrrole adducts, diminishing the axonal levels of these cytoskeletal proteins.[16] Since neurofilament content, especially in larger fibers, is related to axon caliber,[3] such toxicant-induced loss of these cytoskeletal elements leads to axon atrophy, a feature that is well demonstrated by electron microscopy (Figure 2).

In contrast to atrophy, axonal dystrophy is a toxicant-associated lesion characterized by marked swelling of terminal axons. Axonal dystrophy is a dramatic form of nerve fiber degeneration, manifest by marked enlargement of affected neurites associated with aggregations of intraaxonal debris.[8] These lesions are seen in terminal regions of long central nervous system myelinated fibers in conditions such

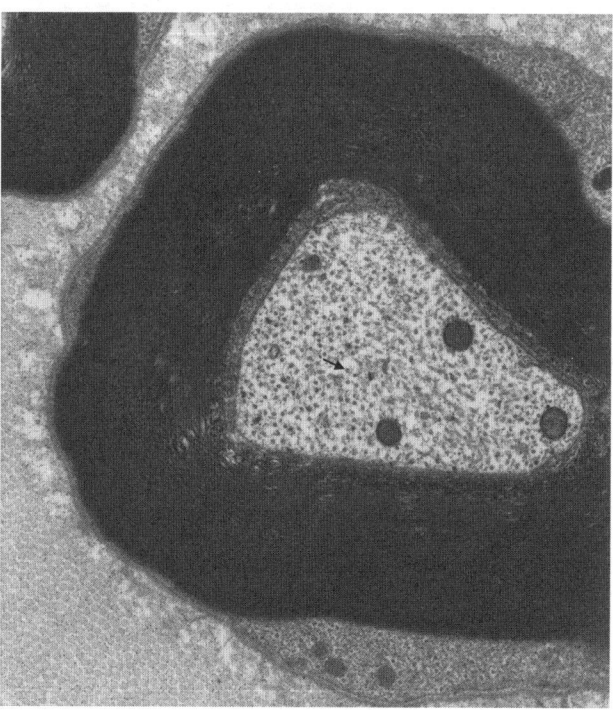

FIGURE 2 Electron micrograph of a cross section of a tibial nerve atrophic myelinated axon with diminished neurofilaments and closely approximated microtubules (arrow). The size of the axon is small relative to the thickness of the myelin sheath. Rat with 98 days of gavage exposure to 175 mg/kg 2,5-hexanedione.

as aging, vitamin E deficiency, diabetes mellitus, several genetic disorders, and some intoxications. Electron microscopy has helped delineate the nature of the process by defining the contents of affected neurites. These lesions have been generated in rats given multiple doses of tri-*o*-tolyl phosphate over a 63-day period.[12] In these animals, dystrophic fibers were prominent in the terminal regions of the gracile fasciculus and in the gracile nucleus. Ultrastructural features included the presence of abnormal mitochondria, tubulovesicular elements, and granular to amorphous matrix (Figure 3).

Ultrastructural pathology has played a major role in clarifying the fine structure of myelin and its formation, as well as in demonstrating the nature and mechanisms of toxicant-induced myelinopathy and its repair.[6,19] A classical white-matter lesion induced by agents such as triethyltin, hexachlorophene, and bromethalin is intramyelinic edema, a myelinopathy that was identified by using electron microscopy to discern its nature. At the light-microscopic level these lesions appear as microvacuoles in affected white matter. Electron microscopy revealed the nature of the basic lesion of intramyelinic edema to be splits in the myelin lamellae at the intraperiod line.[20] The intraperiod line of myelin is the fused outer layers of oligodendroglial or Schwann cell plasma membrane. This

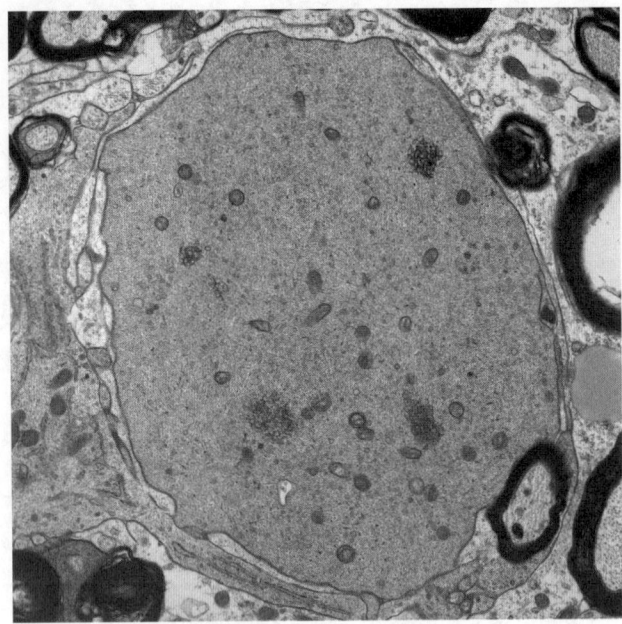

FIGURE 3 Electron micrograph of a cross section of a dystrophic terminal level axon in the gracile nucleus. The swollen axon contains aggregates of tubulovesicular elements, abnormal mitochondria, and granular debris. The dystrophic axon has lost its myelin sheath. Rat exposed to multiple doses of tri-*o*-tolyl phosphate. (From Jortner et al.[12])

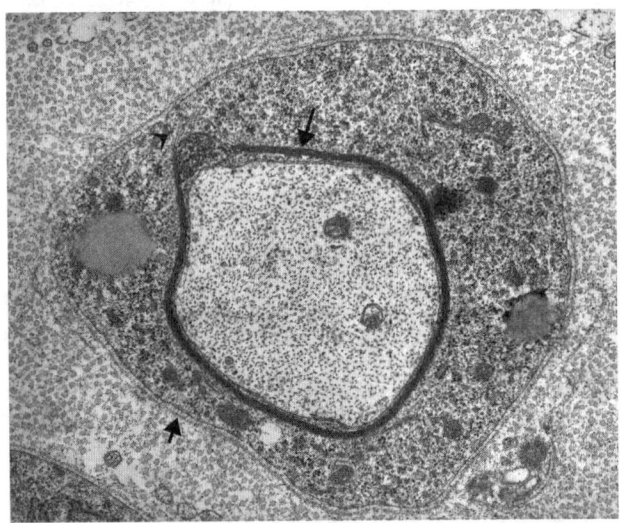

FIGURE 4 Electron micrograph of a cross section of a remyelinating peripheral nerve fiber internode, showing the thin newly formed myelin sheath (long arrow), the mesaxon (arrowhead), and the basal lamina of the Schwann cell (short arrow). Chicken following treated dietary riboflavin deficiency.

alteration is thought to reflect a direct detergent effect of at least one of the causative agents, triethyltin.[14] In segmental demyelination the axon is preserved, allowing repair by remyelination, whereby myelin-forming cells engage the

denuded axonal segment, and replacement myelin sheaths are formed. The latter is demonstrated in Figure 4 to demonstrate the value of electron microscopy in assessing myelin lesions.

In summary, in this chapter we examined an approach to histopathology of the nervous system based on the use of 1-μm-thick sections of epoxy resin-embedded tissue, which provide greater definition at the light-microscopic level. This can be supplemented by electron microscopy. Such procedures provide valuable information that may be used in conjunction with contemporary molecular techniques. Indeed, there can be a synergistic combination of such approaches, as in immunoelectron microscopy.

REFERENCES

1. Fix AS, Garman RH. Practical aspects of neuropathology: a technical guide for working with the nervous system. *Toxicol Pathol*. 2000;28:122–131.

2. Friedrich VL, Mugnaini E. Electron microscopy: preparation of neural tissues for electron microscopy. In: Heimer L, Robards MJ,eds. *NeuroAnatomical Tract-Tracing Methods*. New York: Plenum Press; 1981: 345–375.

3. Griffin JW, Hoeke A. The control of axon caliber. In: Dyck PJ, Thomas PK, eds. *Peripheral Neuropathy*. Philadelphia: W.B. Saunders; 2005: 433–446.

4. Hancock SK, et al. Morphological measurement of neurotoxic injury in the peripheral nervous system: preparation of material for light and electron microscopic study. In: Costa L, et al., eds. *Current Protocols in Toxicology*. Hoboken, NJ: Wiley; 2004: 12.12.1–12.12.19.

5. Hayat MA. Principles and techniques of electron microscopy. In: *Biological Applications*, 4th ed. Cambridge UK: Cambridge University Press; 2000.

6. Hirano A. The role of electron microscopy in neuropathology: a personal historical note. *Acta Neuropathol*. 2005;109:115–123.

7. James H. Neurons and Nobel prizes: a centennial history of neuropathology. *Neurosurgery*. 1998;42:143–156.

8. Jellinger K. Neuronal dystrophy: its natural history and related diseases. *Prog Neuropathol*. 1973;2:129–180.

9. Jortner BS. Mechanisms of toxic injury in the peripheral nervous system: neuropathologic considerations. *Toxicol Pathol*. 2000;28:54–69.

10. Jortner BS. The return of the dark neuron. A histological artifact complicating contemporary neurotoxicologic evaluation. *Neurotoxicology*. 2006;27:628–634.

11. Jortner BS, et al. Peripheral neuropathy of dietary riboflavin deficiency in chickens. *J Neuropathol Exp Neurol*. 1987;46: 544–555.

12. Jortner BS, et al. Neuropathological studies of rats following multiple exposure to tri-*ortho*-tolyl phosphate, chlorpyrifos and stress. *Toxicol Pathol*. 2005;33:378–385.

13. King RHM. *Atlas of Peripheral Neuropathology*. London: Hodder Arnold; 1999.

14. Krinke GJ. Triethyltin. In: Spencer PS, et al., eds. *Experimental and Clinical Neurotoxicology*, 2nd ed. New York: Oxford University Press; 2000: 1206–1208.

15. Lehning EJ, et al. γ-Diketone peripheral neuropathy: I. Quantitative morphometric analyses of axonal atrophy and swelling. *Toxicol Appl Pharmacol*. 2000;193: 29–46.

16. LoPachin RM, DeCaprio AP. γ-Diketone neuropathy: axon atrophy and the role of cytoskeletal protein adduction. *Toxicol Appl Pharmacol*. 2004;199:20–34.

17. Mrak RE. The big eye in the 21st century: the role of electron microscopy in modern diagnostic neuropathology. *J Neuropath Exp Neurol*. 2002;61:1027–1039.

18. Papez JW. Bernhard Aloys von Gudden. In: Haymaker W, ed. *The Founders of Neurology*. Springfield, IL: Charles C Thomas; 1953: 44–48.

19. Peters A, et al. *The Fine Structure of the Nervous System. Neurons and Their Supporting Glia*, 3rd ed. New York: Oxford University Press; 1991.

20. Watanabe I. Organotins (triethyltin). In: Spencer PS, Schaumburg HH, eds. *Experimental and Clinical Neurotoxicology*. New York: Oxford University Press; 1980: 545–557.

15

STEREOLOGICAL SOLUTIONS FOR COMMON QUANTITATIVE ENDPOINTS IN NEUROTOXICOLOGY

ROGELY WAITE BOYCE

WIL Research Laboratories, LLC, Ashland, Ohio

KARL-ANTON DORPH-PETERSEN

Centre for Psychiatric Research, Aarhus University Hospital, Risskov, Risskov, Denmark; Centre for Stochastic Geometry and Advanced Bioimaging, Aarhus University, Aarhus, Denmark; and Department of Psychiatry, University of Pittsburgh, Pittsburgh, Pennsylvania

LISE LYCK

DHI, Hørsholm, Denmark

HANS JØRGEN G. GUNDERSEN

Stereology and Electron Microscopy Research Laboratory and MIND Center, Aarhus University, Aarhus, Denmark

INTRODUCTION

Neurotoxicology is essentially the study of (1) what effect a given test article has on nervous tissues, and (2) via risk assessment, what a safe or no-observed-effect level (NOEL) is for a given test article. The definition of test article–related effects may critically depend on functional testing, biochemical assays, molecular biology tools, and the assessment of tissue sections by a skilled neuropathologist. However, none of these techniques may have the sensitivity or precision required to establish safe levels of treatment in a given setting. Because of the very limited regenerative capacity of nervous tissues, test article–related effects can frequently be adverse and irreversible where small tissue structural changes can have significant consequences. Methods that provide precise, reproducible, quantitative morphological estimates of changes in tissue structure (i.e., that can detect differences with high sensitivity in global tissue volumes, cell numbers, or nerve fiber number and size) may be required in the risk assessment process.

Over the past 20 years, there has been an explosion of new three-dimensional, sensitive, quantitative statistical methods, broadly known as *design-based stereology*, that have revolutionized approaches to obtaining quantitative information about tissue morphology in the central and peripheral nervous systems. These methods differ markedly from traditional model-based methods by combining statistical sampling with geometric probes to generate valid three-dimensional structural data without making *any* assumptions about the size, shape, distribution, or orientation of the tissue structure analyzed.

The application of design-based stereology has historically been technically difficult, labor-intensive, and simply not routinely applicable in most laboratories, particularly in a regulatory toxicology setting. Fortunately, recent advances in stereological theory, software, hardware, and histological techniques have increased the efficiency of data collection and provided for automation of many of the labor-intensive elements of conducting stereological analyses. These advances substantially reduce time and resource requirements to generate quantitative, stereological estimates of three-dimensional structures from sections and offer some practical solutions for incorporation of neurosterelogical endpoints in a toxicology setting.

Fundamental Neuropathology for Pathologists and Toxicologists: Principles and Techniques, First Edition. Edited by Brad Bolon and Mark T. Butt.
© 2011 John Wiley & Sons, Inc. Published 2011 by John Wiley & Sons, Inc.

The intent of this chapter is to provide the neurotoxicologist and neuropathologist with an introduction to design-based stereology and the methodologies required to estimate global tissue volume, total cell numbers, and specifically for the peripheral nervous system, myelinated nerve fiber numbers and axonal size distributions. Hopefully, the reader will gain an appreciation of how common pitfalls in the generation of quantitative structural data can be overcome and how issues such as accuracy and precision of the data can be addressed.

WHAT IS STEREOLOGY?

Morphological studies of organs typically rely on histological tissue sections. Unfortunately, when a very complex three-dimensional (3-D) structure such as the brain is sectioned with a microtome into thin histological sections, there is a loss of 3-D structural information, as illustrated conceptually in Figure 1. Sectioning a complex 3-D structure with a microtome corresponds to introducing a series of two-dimensional (2-D) planes in the structure that effectively reduces 3-D structures to 2-D profiles in the section plane. The size, shape, and frequency of these profiles depend on a variety of parameters, including the position and orientation of the sectioning plane. Sectioning produces widely different 2-D profiles from structurally similar 3-D structures and very similar 2-D profiles from dissimilar 3-D structures. The neuropathologist can study these sections and conceptualize, in a qualitative manner, 3-D relationships. However, the generation of robust quantitative data relevant to the 3-D organ is nontrivial. One classical method is serial reconstruction: for some parameters an elegant, but very labor-intensive technique not practical for routine use. For a general and robust approach, stereology is needed.

Stereology is a mathematical science that combines statistical sampling principles with geometric probes providing efficient and robust tools for estimation of 3-D geometric quantities such as volume, surface area, length, and number of objects contained within an entire organ from measurements made on sections. An important aspect of stereology is that accurate (i.e., unbiased) and precise estimates can be obtained with remarkably few measurements made on a small number of sections (e.g., between 100 and 200 "counts" made on 6 to 12 sections). The statistical sampling efficiently reduces the amount of tissue to be analyzed without reducing the precision of the estimated value below a preset level, and the geometric probes ensure unbiasedness and three-dimensionality of the measurements. N.B.: The sections and measurements *must* meet special requirements to ensure unbiasedness—i.e., that the final estimates, on average, do not deviate from the true values to be assessed. A conceptual overview of these special requirements can best be conveyed by the analogous science of opinion polling.

Stereological methods are similar to methods used in statistical population studies such as opinion polling.[1] The success and accuracy of an opinion poll depends on obtaining a representative sample of the population that ensures all citizens in the population an equal probability of being sampled (uniform random sampling) as well as proper questionnaires to avoid biasing or influencing the response. The precision of the opinion poll can be improved by sampling more citizens and can reliably be predicted due to statistical principles used in sampling. The same is true for stereology, which can conceptually be considered a 3-D opinion poll in an organ.

Similar to a pollster polling a small sample of citizens from an entire population to estimate the possibility of a republican or democratic victory, the neuroscientist may seek to estimate changes in neuron number by counting neurons in a small sample of a particular brain gray matter area (e.g., the substantia nigra compacta) to characterize the severity of a particular condition (e.g., a Parkinson's disease–like condition). The probable success of either process is dependent on the accuracy and precision of the estimate. Accuracy of the

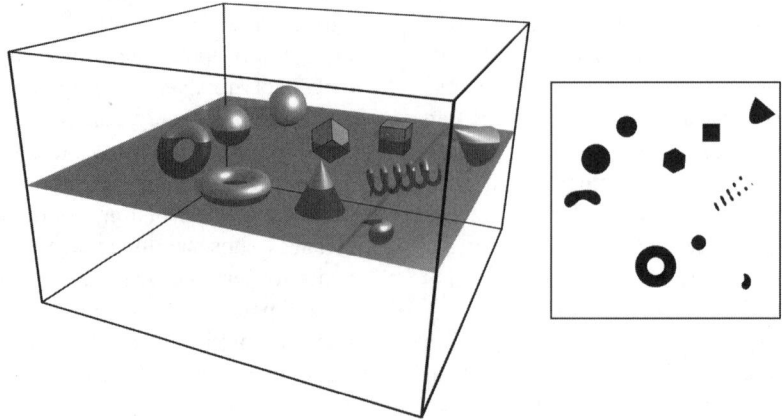

FIGURE 1   Sectioning a complex three-dimensional structure into two-dimensional sections results in loss of three-dimensional information.

poll is ensured by adhering strictly to sampling protocols that are proven mathematically to provide results that, on average, equal the true or real population value. For most biological questions, accuracy cannot easily be judged empirically because the real number (i.e., the true population value) is rarely known. Fortunately, the statistical principles used in both polling and stereology can guarantee accuracy without knowledge of the true population value. The guarantee of accuracy is critically dependent on (1) selecting a sample from the entire population that is representative (uniform random), and (2) asking questions of that sample that will yield truthful answers . The reproducibility of the estimate is determined by the precision of the estimate, commonly referred to as the sampling error.

What is a representative sample? The pollster might like to poll polite well-dressed professionals in a single "typical city," and the bench pathologist may have a strong preference for working with "optimum sections" through the center of a brain region, but either approach is very unlikely to give an accurate estimate. A representative sample of the population is one that is prospectively designed to give each and every citizen, or each and every neuron, a known (often an equal) opportunity to be polled or counted. Hence, the entire population or region of interest must be available for sampling. For the poll results to accurately estimate the opinion of the entire population in an unbiased manner, or the stereological assessment from the neuroscientist to estimate, without bias, the total number of neurons in the substantia nigra in human brains, proper application of random sampling is required. This means that the population or region of interest will be sampled with known probabilities (often sampled uniformly random). *Any deviation from this sampling principle will result in systematic error or bias that directly abrogates the guarantee that the results are accurate.* This translates in political polling to "all people are created equal" and in neuronal counting in tissue sections to all neurons must have an equal chance of being counted, regardless of size, shape, distribution, or orientation.

The second requirement to guarantee that the results of an election opinion poll will be accurate is that the citizen sampled must be questioned or "probed" in such a way as not to influence or bias the response. If questioned in a manner that favors a certain response, there will be systematic error or bias in the results. The pollster might ask: "Will you be voting for the Democrat or the Republican?" yielding an answer likely to be influenced by the party affiliation of the person questioned. In contrast, asking: "Will you be voting for Mr. X or Ms. Y?" will possibly be biased by both name recognition or gender considerations. Similarly, the bench pathologist might count cell profiles in 2-D sections, unaware that cell size directly affects the probability of a cell being counted in a section, or measure cell profile diameters as an expression of cell sizes, under the assumption that cells are round or ovoid. Used correctly, the geometrical

probes of stereology help the scientist to "question" the tissue structures sampled without making assumptions.

GENERATING A REPRESENTATIVE SAMPLE IN STEREOLOGY: SYSTEMATIC UNIFORM RANDOM SAMPLING

To obtain a uniform random sample from an organ, the first requirement is that the entire organ must be available for sampling. Stereologists generally sample the organ using *systematic uniform random sampling* (SURS). In this form of uniform random sampling, the first sampling position in an organ is chosen uniformly randomly within a predetermined sampling interval with subsequent positions chosen systematically at predetermined constant intervals. This sampling method ensures that all structures in all positions in the organ are given equal probability of being sampled and is more efficient than simple random sampling.[2] Efficiency in this context means the sampling method yields an estimate of a given precision while doing minimal (i.e., optimal in terms of efficiency) work, meaning that the fewest sections and fewest measurements required for accuracy are needed.

Figure 2 illustrates the principle of SURS in a hypothetical organ compared with simple uniform random sampling. With simple uniform random sampling, the position of each of the cuts in the organ is uniform randomly along the length of the organ (i.e., all positions have an equal probability of being sampled; synonymous with random sampling as known to most biologists). Although the sampling is uniformly random, it will take many sections to reach a satisfactory level of precision. With SURS, the organ is sliced at regular uniform intervals. The uniform random element of the sampling comes with positioning the first cut uniformly random within the first interval (e.g., by using a random number table to select the position). All successive cuts are then made systematically at constant predetermined intervals. Intuitively, this form of sampling will almost always reduce overall variance and produce an estimate of a given precision with fewer sections.

Further subsampling using SURS can continue to reduce overall sample size. Organ slices can be cut into bars, and bars, can be cut into cubes. Sampled cubes are embedded and cut, yielding a set of sections for analysis where microscopic fields are ultimately selected using SURS for probing for the quantity of the geometric structure of interest.

Strictly speaking, unbiased estimation requires that the organ is sampled using procedures which ensure that all structures of interest are given a known or equal probability of being sampled, independent of size, shape, orientation, and spatial distribution. This independence means that no assumptions are necessary regarding the structures in the organ and is fundamental to design-based stereology. Most stereological

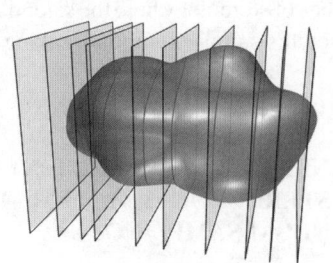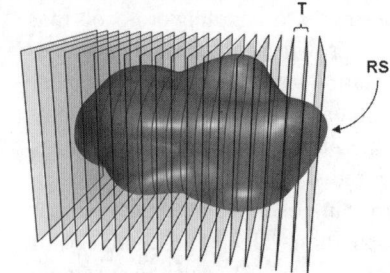

Random Sampling   Systematic Uniform Random Sampling

FIGURE 2   The schematic on the left depicts uniform random sampling of an organ. Positions of sampling cuts are positioned uniformly random along the longitudinal axis of the organ. The schematic to the right depicts systematic uniform random sampling of an organ. The organ is cut at a uniform constant interval (T) with the position of the first cut positioned uniformly random (RS) in the interval 0 to T, selected from a random number table or from randomly positioning of the organ in an agar block prior to sampling.

methods are based on uniform random sampling where all structures are sampled with the same probability. However, recently, very efficient stereological estimators based on nonuniform sampling have been developed.[3–5] To keep things simple, in this chapter we focus on classical uniform random sampling, except for a brief description of the proportionator method (discussed later in the chapter).

QUESTIONNAIRE IN STEREOLOGY: GEOMETRIC PROBES

In stereology, the "questionnaire" for estimating 3-D structural content consists of geometric probes that "sense" the structural feature (Table 1). Once the organ has been sampled uniformly random, the geometric structure of interest is probed in sections with a randomized geometric probe that interacts with the structure. The number of interactions between the geometric probe and the structural feature of interest is proportional to the quantity of that structure in 3-D. For this mathematical relationship to be valid and for the results to be accurate (i.e., estimate the true value), the entire sampling and probing must give all objects or all positions in the organ equal opportunity or known probability of being sampled for analysis.

For the analysis of different structural geometric features to be unbiased, the appropriate probe must be applied. Table 1 shows the relationship of geometric probes with the associated geometric structure of interest. To estimate volume, a 3-D geometric feature, the section is probed with a zero-dimensional probe, a point. The number of points interacting with the volume appearing as an area in tissue section is proportional to volume in 3-D. To estimate surface area, a 2-D geometric feature, the surface appearing as a boundary in a tissue section is probed with 1-D lines. The number of intersections of the lines with the boundary is proportional to the surface area in 3-D. To estimate the length of a structure, a

1-D geometric feature, the number of profiles created in tissue section by a 2-D plane (the sectioning plane of the microtome knife) is counted. The number of profiles created by the plane is proportional to length in 3-D. For estimation of cell number, a zero-dimensional geometric feature, the tissue must be probed with a 3-D probe called a *disector*.[7] This concept is elaborated upon in our discussion of approaches to estimating cell number.

A general theme should be evident in reviewing Table 1: The dimension of the geometric structure of interest plus the dimension of the geometric probe used to sense its quantity is equal to 3, the dimension of ordinary space. To estimate the quantity of a 3-D structure, all three dimensions must be accounted for by the structure and the probe (i.e., interactions of probe and structure happen only if the sum of the dimensions of the probe and the structure is at least 3). If the sum equals 3, the interactions are simple countable events (see Figure 3 in West[6]). By design, the position and (if needed) orientation of the relevant probe is randomized relative to the structure, ensuring the needed randomness in probe–structure interaction. This forms the basis of performing unbiased "questioning" in a 3-D opinion poll conducted in an organ using design-based stereology.

Also noted in Table 1 are special section orientation requirements for surface area and length estimation. Because both of these geometric features can have preferred orientation in tissue, sections must be produced with random orientation to give the structures equal probability of being sampled to ensure that the "questioning" or probing of the sections will not be biased. These randomly oriented sections are known as *isotropic uniform random sections*[8,9] and *vertical sections*,[10] respectively. Because surface area and length estimates are not discussed here, the interested reader is referred to a list of review articles and books at the end of the chapter that detail approaches to estimating these structures. Methods for analysis of volume and number of objects are described later in the chapter.

TABLE 1   Relationship of Geometric Structures and Probes[a]

Geometric Structural Feature and Dimension				
	Volume (3)	Surface Area (2)	Length (1)	Number (0)
Probe and Dimension	Point (0)	Line (1)	Plane (2)	Volume (3)
Interaction of Structure and Probe and Sum of Dimensions	(3)	(3)	(3)	(3)
Section Orientation Requirements	None	Isotropic Uniform Random or Vertical	Isotropic Uniform Random	None

Source: Modified from West.[6]

[a] Volume, surface area, length, and number are sensed by points, lines, planes, and volumes, respectively. The number of interactions of the probe with associated geometric structural feature is proportional to the quantity of the geometric structure in 3-D if the sum of dimensions of the geometric structural feature and probe is 3.

WHY STEREOLOGY? ACCURACY, PRECISION, VARIANCE, AND EFFICIENCY

The value of stereology is that it provides estimates of the true value of volume, surface area, length, or number in the entire organ or any (well-defined) region of interest. These unbiased methods will give an accurate numerical estimate; the average of these estimates will become more precise, or closer to the true value of the population, with replication (i.e., repeating sampling and measurements in the same animal) as well as by increasing sample size (i.e., increased sampling and increased number of counts per animal). In contrast, replicating or increasing the sampling of a biased method increases the precision of the average result, but it cannot make the average result accurate. Inaccurate estimates only become more precisely inaccurate with replication. Applying stereological principles guarantees that the estimate will be accurate and that when decisions need to be based on

quantitative data, they can be made with confidence. Figure 3 illustrates the statistical concepts of accuracy, unbiasedness, and precision.

Another important feature is that the error associated with the sampling and counting can be assessed for the most commonly used stereological estimators. Thus, the sources of variance in a study can be analyzed. This information is used to determine where best to focus efforts to reduce the variance and increase the precision of the estimate. A key formula in stereology that allows the sources of variance in a data set to be identified is

$$OCV^2 = BCV^2 + \overline{CE}^2 \qquad (1)$$

where OCV^2, the squared observed coefficient of variation of a data set from a group of animals is equal to the sum of BCV^2, the squared biological coefficient of variation, and $\overline{CE}^2$, the mean squared coefficient of error of the

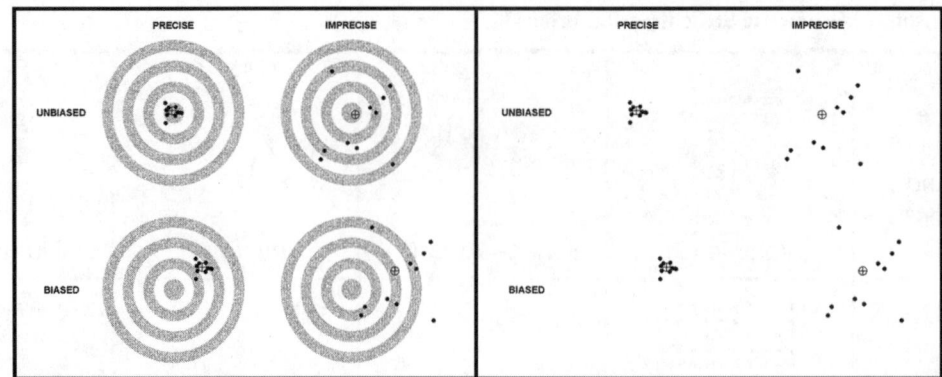

FIGURE 3   Illustration of the statistical concepts of accuracy, unbiasedness, and precision. The targets in the top row of the left panel illustrate the statistical consequences of application of unbiased methods that guarantee the mean values, indicated by the crosshatch, on average, estimate the true population value. The true population value is located within the bull's-eye of the target. The data may have low variance and be very precise, as indicated in the left target, or may have higher variation, as indicated in the right target. Regardless, unbiased methods guarantee that the mean value will estimate the true population value. The statistical consequences of biased methods are depicted in the lower row of targets in the left panel. Data may appear either very precise (left target) or highly variable (right target); however, the mean value for biased estimates regardless of the variance, on average, will never provide an accurate estimate of the true population value (i.e., the mean value never hits the bull's-eye). The right panel is a duplication of the left panel that depicts real life where the population mean is unknown and there is no bull's-eye to inform the researcher if data are biased. Unbiased methods are required to guarantee the estimate, on average, predicted or estimated the true population value. (Modified from Figure 2 in Gundersen.[11])

stereological procedures across all animals or individuals in a group. Basically, the relationship conveys that the overall observed variance in a data set is the sum of the inherent biological variation and the variance contributed by the stereological methods (i.e., sampling and counting).

The observed coefficient of variation OCV is calculated from the standard deviation (SD) and mean of a data set:

$$OCV = \frac{SD}{mean} \qquad (2)$$

For simple independent variables, CE is calculated as

$$CE = \frac{SEM}{mean} = \frac{CV}{\sqrt{n}} \qquad (3)$$

This formula conveys that the variance due to sampling and counting can be reduced proportional to $1/\sqrt{n}$, where n is the number of measurements or counts. However, systematic uniform random sampling used widely in stereology is by definition not independent. Therefore, for stereological estimators, CE may be proportional to $1/n^2$ or even better. This underscores the striking efficiency of many stereological estimators where an estimate of certain precision can be achieved with relatively few measurements. The actual formulas to calculate the CE for many estimators are substantially more complex than those for simple independent variables, and the reader is referred to Gundersen et al.[2,12] and Cruz-Orive and Geiser[13] for more details. By inserting data generated during sampling and counting procedures into

these formulas, it is possible to obtain robust approximations to the actual CE for each estimated value, allowing the researcher to report stereological data along with the CE, similar to the reporting of biochemical endpoints with the confidence interval for the measurement.

Information about the variation introduced by the sampling and counting procedures is used to determine where best to focus efforts to reduce variance and increase the precision of the data estimated for individual animals in the experimental group. This also allows for statistical planning of experiments and to appropriately power a definitive study on which to base critical decisions. To work efficiently, the intensity of sampling or how much work is done should be targeted such that the stereological sampling is just not inflating the observed variance (no more, no less). As a rule of thumb, this is typically achieved when the CE values are $\leq \frac{1}{2} OCV$. Thus, if the OCV is about 20%, a CE of about 10% will typically be of sufficient precision and in most cases it would then be adequate to count approximately 200 events (counts of probe interactions with geometric structures). Under these conditions, it would not be efficient to do additional sampling and counting because it will do little to reduce the overall variance, as the greatest contribution to the OCV then comes from the biological variation, i.e., the variation between animals. If a suitable CE has been achieved, the only way to increase the precision of the group mean [i.e., decrease the variance of the group mean (SEM)] is to increase the number of animals. This is the mantra of design-based stereology: "Do more less well."[14]

It is always advisable to conduct a pilot study prior to a main study to optimize sampling procedures and assess the sources of variance, in order to perform a proper statistical planning of the definitive experiment. If nothing is known about the biological variation of the structure of interest, it has been suggested that five animals per group and five blocks per animal be used as a starting point.[15]

WHEN STEREOLOGY?

In toxicological pathology, quantitative evaluation of structural changes in a tissue may be performed with various levels of sensitivity. The first level of evaluation, the level most often used to satisfy scientific and regulatory requirements, is the qualitative/semiquantitative histopathological evaluation by an experienced pathologist. However, if additional quantitative endpoints are needed, measurements may be performed on the routine sections collected for histopathology. This second level of quantification can be most useful to corroborate impressions or interpretations of the pathologist and may be sufficiently sensitive to detect a signal if the magnitude of the change is large. However, the latter endpoints do not convey information about quantities of 3-D structure at the organ level, and the results may be difficult to interpret and potentially biased, even in such a way as to lead to erroneous conclusions.

The third level of quantification would include data generated using design-based stereological approaches implemented when valid quantitative data are needed to drive critical decisions in drug development or risk assessment. These are cases where the accuracy, and hence veracity, of the quantitative data must be guaranteed. Only unbiased stereological methods provide this guarantee. This is especially important in neurotoxicology because of limited regenerative capabilities of nervous tissue. Minor test article–related morphologic changes in the nervous system can be adverse, irreversible, and unmonitorable. Stereological methods have the necessary sensitivity to detect subtle changes and provide unbiased estimates with a known precision on which critical decisions about a test article can be based. Biased 2-D measurements such as counting cell profiles on "optimal" sections collected at prescribed locations do not provide estimates with that guarantee.

STEREOLOGICAL ESTIMATORS AS SOLUTIONS TO COMMON QUANTITATIVE ENDPOINTS IN NEUROTOXICOLOGY

Cavalieri's Estimator of Volume

The size or volume of a structure such as a brain nucleus or peripheral ganglion is frequently an informative parameter to monitor test article-related effects. Many neuroanatomical structures are contained in larger structures defined by fine, anatomical features necessitating methods for volume estimation at the microscopic level. *Cavalieri's principle* is a simple and robust estimator of volume regardless of how irregularly shaped the object of interest is or if it is contained within a larger structure.[2,16] The estimation principle involves exhaustive slicing or sectioning of the object of interest by a series of parallel section planes separated by a known fixed distance, T, with the position of the first section plane chosen at random in the interval 0 to T. This step ensures an unbiased approach by generating a systematic, uniform random sample of slabs or histologic sections. The area of the object of interest is determined on each slab or section and the sum of the areas multiplied by T is an unbiased estimate of the total volume. Areas can be estimated with the aid of an image analysis system or by using a point probe. A grid of points where each point represents a known area, a, is superimposed on the section with random orientation and the number of points overlaying the region of interest is counted. The volume (V) of the region of interest is then estimated as the product of T and either the sum of the areas (ΣA) or the sum of points (ΣP) across all slices multiplied by the area associated with each point (a):

$$\begin{aligned} V &:= T \cdot \sum A \\ &:= T \cdot a \cdot \sum P \end{aligned} \quad (4)$$

Here and elsewhere $:=$ indicates that the entity at the left is estimated by the expression at the right.

In practice, the Cavalieri estimator of volume can be applied to slabs prepared from whole brains or to histological sections collected at known fixed distances, T, through the complete tissue region of interest. The brain can be embedded in agar and cut into parallel slabs of uniform thickness, T, with the position of the first cut uniform randomly in the interval 0 to T. Every subsequent cut is separated by distance T. Alternatively, small brains or complete regions of interest can be embedded and sectioned with a microtome collecting histological sections at fixed intervals (e.g., every tenth section). In the latter case the thickness, T, will be the microtome block advance (which determines section thickness) multiplied by the sampling period (e.g., $80\,\mu m \times 10$).[17] If the sampling is done by histological sectioning, the starting point may be selected by choosing a random number between 0 and the sampling period. As noted in Table 1, there are no special section orientation requirements for estimating volume; therefore, any standard section orientation can be used. Figure 4 schematically illustrates the Cavalieri principle to estimate the total volume of a region of the mouse neocortex using histological sections.

An important requirement for estimating the volume of a structure is that the neuroanatomic limits of the object of interest are definable and identifiable in their complete 3-D

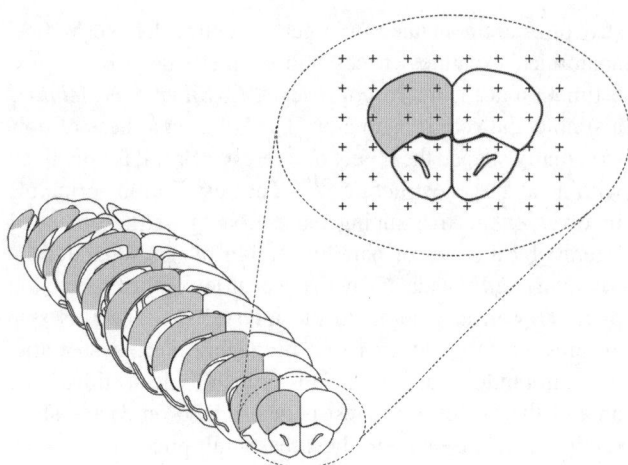

FIGURE 4   Application of Cavalieri's principle to determine the total volume of a defined region of a brain (shaded area of neocortex). A mouse brain is sectioned into a series of coronal sections in which the distance between the sections is uniform (T = section thickness × sampling period). The position of the first section is randomized through random selection of a number between 1 and the sampling period. The area of the neuroanatomically definable region of interest (gray shaded regions) is estimated in each section, as illustrated for the first section using a point grid where each point represents a known area. The total volume is estimated by summing the number of points across all sections, multiplying by the area associated with each point, then multiplying by T. (Modified from Lyck et al.[17])

extent. For analysis of the cross-sectional areas in the series of slabs, one of the cut surfaces should be selected consistently (i.e., the posterior cut surface); this means that all slices are analyzed by application of the point grid on the posterior cut surface of the slice (hence, the last slice in the organ will not contain a posterior cut surface). In this case, the point grid can be printed on clear acetate film and placed over the tissue slices, a method suitable for volume estimation in large structures.[18,19] If histological sections are used, sections normally are thin enough compared with the intersectional distance to represent planes through the organ, and the distinction between cut surfaces is not necessary. With histological sections, the point grid is typically superimposed onto the live microscopic image of the section using commercially available software. In general, a precise estimate can be achieved by counting approximately 200 total points collected over 6 to 12 sections, depending on the complexity of the shape of the structure.[2]

Progress has been made toward automating the analysis process using automated image capture combined with image analysis and segmentation for measurement of cross-sectional areas of the structure of interest in a series of histological sections. If the sampled set of sections meets the requirements of SURS, the challenging step in this approach is creation of robust segmentation protocols for reliable identification of the structure of interest and measurement of the area in each section.

When estimating volume in tissue sections, it is important to prevent shrinkage or swelling of the tissue, and this should be carefully avoided during section preparation. Tissue processing and embedding can cause profound tissue shrinkage that will bias volume estimates. This is particularly true for paraffin processing, which can cause 40 to 50% reduction in tissue volume.[20–22] It is a dangerous assumption that shrinkage effects will cancel out if the tissues from control and test article–treated groups are handled identically. Test articles cause known and unknown cellular effects that may affect the way a tissue responds to processing and embedding. The pitfalls of assuming that treatment and controls will respond similarly to these steps have been elegantly demonstrated by Mendis-Handagama and Ewing.[23] Shrinkage concerns can be minimized if plastic, frozen, or vibratome sections are used and processed carefully[24] or if the Cavalieri principle is implemented using unprocessed slabs. Finally, the reader should notice that by using the principle of Archimedes, very fast and precise assessment of total organ volume may be obtained and can be used to keep track of processing-induced shrinkage (see Dorph-Petersen et al.[25] for a discussion of processing-induced shrinkage).

Estimation of Total Cell Number

The Problem of Counting Cell Profiles and the Solution: The Disector   A test article effect on cell numbers is one of the most critical quantitative endpoints in neurotoxicology. Routine histopathologic evaluations are very insensitive with respect to the estimation of changes in cell numbers in the nervous system. Early stereological methodologies for cell quantification, while valid and sensitive, were simply too cumbersome and challenging to practically implement in a regulatory toxicology setting. Consequently, two alternative endpoints based on counting cell profiles in routine sections have been used to quantify changes in cell number: (1) profile density (e.g., number of 2-D cell profiles per unit area), and (2) labeling index (e.g., the number of histochemically labeled cell profiles per total number of cell profiles) . These methods can certainly detect a signal if very large changes in cell profile counts have occurred, such as a three- or four-fold increase in Ki-67 labeling index would be detected as a proliferation signal. However, these methods are highly biased and insensitive, potentially leading to incorrect conclusions.

The challenge of counting cell profiles in sections containing a population of small and large cells is illustrated in Figure 5. When serial sections are cut as indicated by the parallel horizontal planes, the collective (arbitrarily arranged) profiles produced in the tissue sections are shown. Larger cells make more profiles and are overrepresented because a section plane samples particles or cells propor-

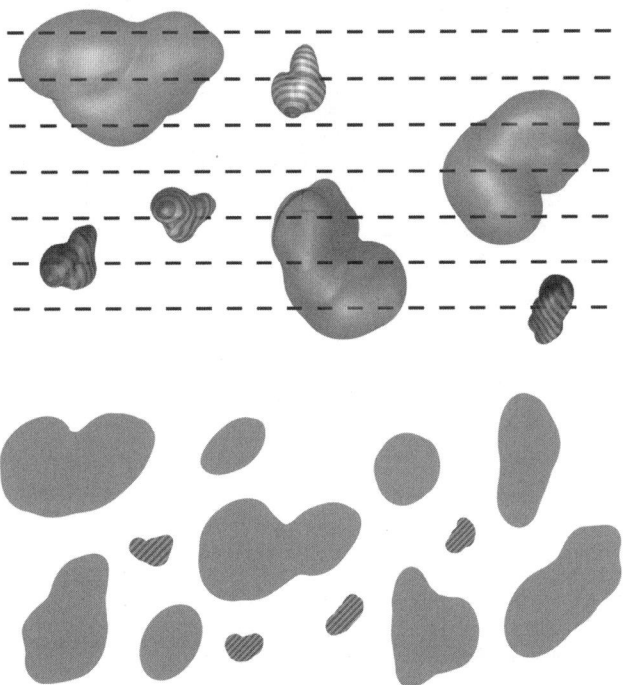

FIGURE 5 Sectioning cells with two-dimensional sectioning planes samples cells proportional to their height or size, not their number. A tissue containing a population of three large and four small cells is sectioned exhaustively by uniform sectioning planes by the microtome indicated by the parallel hatched lines. The total number of profiles of large (nine) and small (four) cells counted on all the sectioning planes through the tissue is illustrated at the bottom of the figure. Clearly, the numbers of profiles of the respective cell types do not in any simple way reflect the actual number of these cells or their relative proportion.

tional to their height. Hence, a section plane does not probe or sense cells proportional to their number but rather to their height perpendicular to the section plane. By analogy with election polling, the questioning or probing is biased toward the large cells. This bias will not average out by taking more sections and counting more profiles; every section will overrepresent larger cells.

As discussed previously, the sum of the dimensions of a geometric feature and the probe that interacts with the feature proportional to its 3-D quantity must be at least 3. Because cell number is zero-dimensional, a 3-D or volume probe is required to interact with cells proportional to their number. It is not possible to probe a single thin histologic section with a 3-D probe. The problem of how to count cells using thin histological sections was solved in 1984 with the invention of the disector, an unbiased 3-D sampling probe.[7] Figure 6 illustrates the disector counting principle. The disector consists of (at least) two parallel planes, either physical (two thin tissue sections separated by a known distance), or optical (a stack of focal planes scanning a known distance in a thick section). These are known as *physical* or *optical disectors*, respectively. The distance

between the physical sections (i.e., from section face to section face) or the height of the stack of focal planes is the disector height. The area of the section planes and the tissue between them, the disector height, constitutes a volume of tissue. Cells are counted only if they are present in one section, the sampling or counting section, and not in the other, the look-up section. This rule ensures cells are only counted once. The last component of the counting principle is sampling the sections with an unbiased counting frame.[26] Cells are only counted in the tissue area defined by the frame if the cells are contained in the frame or touch the hatched or "accepted" lines. Cells that touch the solid or "forbidden" lines or their extensions are excluded (Figure 7).

Using unbiased methods to estimate changes in total cell number in contrast to biased 2-D profile counting can be of critical importance when determining the potential neurotoxic effects of a test article. Estimates of neuron number obtained using 2-D profile counts with assumptions about cell size and shape have been compared to estimates obtained using unbiased methods. These comparisons indicate that 2-D methods can be biased by as much as 40%.[27,28] Notably, changes in the volume of the analyzed structure can mask the true changes in cell numbers occurring during pathological[29,30] and nonpathological processes[17,31,32] in the brain. The sensitivity of unbiased methods to estimate test article-related changes in cell number is underscored by the work of de Groot et al. In neonatal rats exposed in utero to methyl mercury, unbiased stereological evaluation revealed an approximate 11% reduction in cerebellar granular neurons, an effect that occurred but was not detected by conventional neuropathology microscopic evaluations or behavioral endpoints applied as specified in guidelines for neurotoxicity testing (D. de Groot, personal communication). The potential bias in 2-D cell profile counts is not acceptable when minor reductions in neuron number may represent an adverse, irreversible, and unmonitorable test article–related effect.

Density Estimates and Two-Step Estimation of Total Cell Number

When cell number is estimated using either physical or optical disectors in SURS microscopic fields using an unbiased counting frame, the intermediate result may be expressed as mean cell density (N_V) or the average number of cells per unit volume of tissue. Total cell number in an organ or object can be estimated in a two-step process by determining the mean cell density, then multiplying cell density by the reference volume (V_{ref}), the volume of the organ:

$$N := N_V \cdot V_{ref} \qquad (5)$$

Most frequently, reference volume is determined using Cavalieri's principle. Although this two-step process is widely applicable,[29,31,33,34] it can be affected or biased

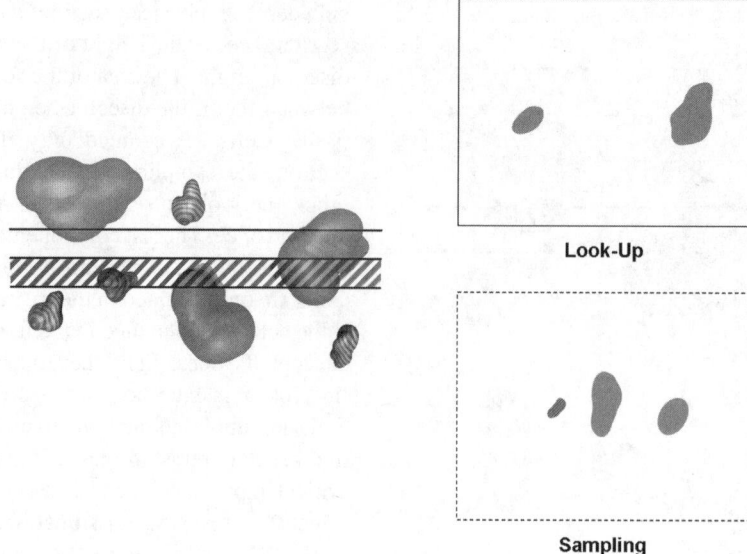

Look-Up

Sampling

FIGURE 6   A disector samples cells proportional to their number. The figure to the left depicts two serial sections collected from a tissue (seen on edge). The hatched section is the sampling section and the section immediately above is the look-up section. The profiles of particles present in the respective sections are illustrated on the right. Particles present in the sampling section are counted only if they are not present in the look-up section. This ensures that particles or cells are sampled proportional to their number and not their size and that they are only counted once. In this illustration, two particles are counted in the sampling section that are not present in the look-up section. The area sampled for counting in the section planes and the distance separating them, in this case the thickness of the hatched section, constitute a known volume.

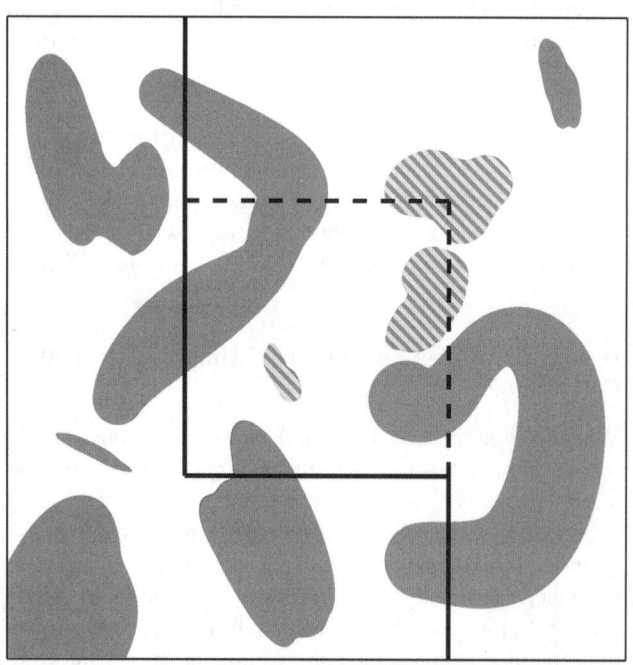

FIGURE 7   The unbiased counting frame ensures that only objects that belong to the area of section sampled by the frame are counted. Only objects that are contained in the frame, including those that touch the dashed, accepted lines, are counted (hatched objects). Those objects that touch the solid, exclusion line or their extensions are excluded (solid objects). The latter objects belong to adjacent areas that would be counted if the frame was translated.

significantly by tissue shrinkage. Therefore, great care must be taken to control tissue shrinkage between the time of volume estimation and the subsequent density assessment, especially when using paraffin sections.[24] In addition, density estimates alone can be misleading because of potential changes in the total volume of the organ or object, known as *the reference trap*.[35] Density estimates may be similar between treatment groups, but total cell number in an object such as a brain nucleus can be changed drastically because there has been a change in the total volume of the nucleus. From a practical point of view, this can be cumbersome and is largely why this method has been replaced by the fractionator, a direct estimator of cell number that obviates the issues of tissue shrinkage and the need to estimate total organ or reference volume.

THE FRACTIONATOR PRINCIPLE

Estimation of total cell number has been greatly simplified by a robust sampling principle called the *fractionator*, which is insensitive to shrinkage and eliminates the need to measure the reference or organ volume.[36] The basic principle of the fractionator is that a known fraction of an organ is sampled, in one or more sampling steps, and in the final sample the total number of cells is counted using either physical or optical directors; the estimators are known, respectively, as the *physical* or *optical fractionator*. The total number of cells in the organ is then estimated by multiplying the number of cells counted in the final sample, $\sum Q^-$, by the inverse of the sampling fraction (i.e., the inverse sampling probability for each sampling step).

The key feature of the fractionator is that all types of shrinkage are allowed to occur before the final sample is taken and the cell number estimated. Because the cell number is zero-dimensional, no matter how the tissue shrinks or deforms during processing, the total number of cells present is unaffected and a known fraction is sampled and counted. This freedom from tissue shrinkage concerns makes paraffin sections suitable for this estimator.

A very efficient strategy for the initial sampling steps is the *smooth fractionator*. At each sampling step, a large number of tissue blocks or strips are created which are then arranged in a smooth pattern, frequently based on size, before systematic sampling. The reader is referred to Gundersen[37] for further details and examples.

Physical Disector and the Physical Fractionator

Physical disectors for cell counting typically consist of paired serial 2- or 3-μm-thick sections. These constitute the counting and look-up sections, and the distance between section faces determines the disector height (Figure 8). The optimal height of the disector depends on the height of the objects to be counted and is generally targeted at one-fourth to one-third of the object height. In order of preference, the nucleolus, the nucleus, or the cell top can be used as the unique counting feature.

Thin paired serial sections are most easily prepared from plastic- and paraffin-embedded tissue. However, tissue shrinkage with paraffin is a consideration if physical disectors are used with the two-step estimation method.[24] A specific requirement for unbiased cell counts using physical disectors is that section thickness is constant and uniform. Plastic-embedded sections must be cut essentially dry because wetting the block surface causes block swelling and variable section thickness. Similarly, paraffin sections must be prepared at a constant temperature because intermittent icing of the block face causes variable thermal shrinkage and variable section thickness. Paraffin sections can be prepared either chilled in a cryotome or at room temperature. Hydrating the paraffin block face is permissible because it does not influence section thickness (HJG Gundersen, unpublished observation).

Physical disectors have several advantages. The use of thin sections obviates concerns of "lost caps," *z*-axis deformation, and stain penetration that can be encountered with optical disectors (see the section "Optical Disectors and Optical Fractionators" for details). The histological procedures used for physical disectors are in line with standard procedures in a regulatory pathology laboratory, and staining protocols for thin sections are amenable to automation.

On the other hand, if dimensional stability is needed in combination with immunohistochemical stains (e.g., when estimating cell size), frozen or vibratome sections are to be used, typically as thick sections using optical disectors.[24] Application of a physical disector in frozen or vibratome sections is difficult because it may be technically challenging to cut sufficiently thin cryo- and vibratome sections.

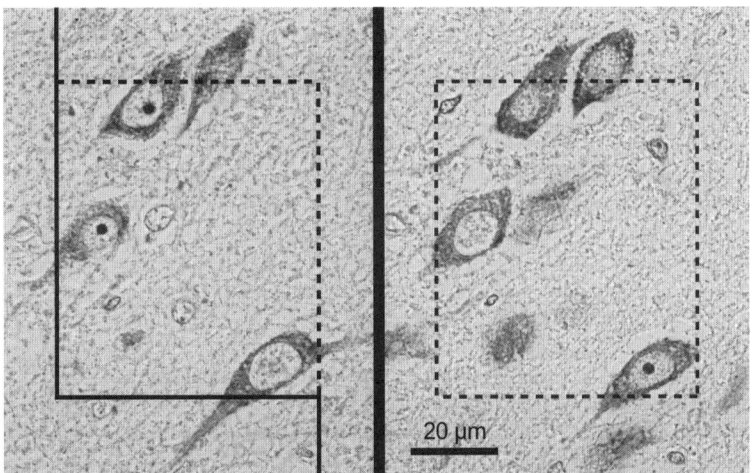

FIGURE 8 Physical disector consisting of a pair of two consecutive 2.5-μm sections from the hilus region of a human hippocampus. The width of the vertical black line between the two sections corresponds to the disector height of 2.5 μm. Two neurons are counted in the disector as their nucleoli are sampled by the unbiased counting frame in the left sampling section, *and* these nucleoli are not present in the right look-up section.

Simple Physical Fractionator with Exhaustive Sectioning of the Organ

A simple implementation of the physical fractionator is illustrated in Figure 9. This very straightforward fractionator design is possible when the entire organ can be processed and embedded in toto [e.g., for analysis of small brains and brain regions (or other small organs)]. The entire organ or region is processed, embedded, and subjected to shrinkage before sampling, then serially sectioned to exhaustion. Beginning at a uniform random section among the first $1/ssf$ sections, where ssf is the section sampling fraction (every one hundredth section, for example), paired disector sections are collected at uniform intervals until the entire region of interest has been sectioned. Note that adjacent sections can be collected for routine histopathology evaluations, special stains, or other special purposes. In the set of disector sections, a known fraction of the area of the sections, the areal sampling fraction (asf), is sampled by an unbiased counting frame, and the cell number is counted in physical disectors ($\Sigma\, Q^-$). The total cell number (N) in the organ is then estimated as

$$N := \frac{1}{ssf} \cdot \frac{1}{asf} \cdot \sum Q^- \tag{6}$$

As an example, if 3-μm disector pairs are collected every 300 μm, ssf is 3/300 or 1/100 with a random start within the interval 0 to 99. If asf is 2% or 1/50 and a total of 200 cells are counted, then

$$\begin{aligned} N &:= 100 \cdot 50 \cdot 200 \\ &= 1,000,000 \end{aligned} \tag{7}$$

The physical fractionator has many applications, but its use for the estimation of cell number has been limited due to the requirement for registration of matched high-magnification microscopic fields in the paired sections. Until recently, registration was performed manually or by coarse registration with some software packages. Hence, the registration was labor intensive and time consuming. However, recent software advances now provide for automated sampling, capture, and registration of matching high-magnification microscopic fields, making implementation of the "automated" physical fractionator for cell counts practical and efficient (Auto-Disector™ and NewCast™ software, VisioPharm, Hørsholm, Denmark).

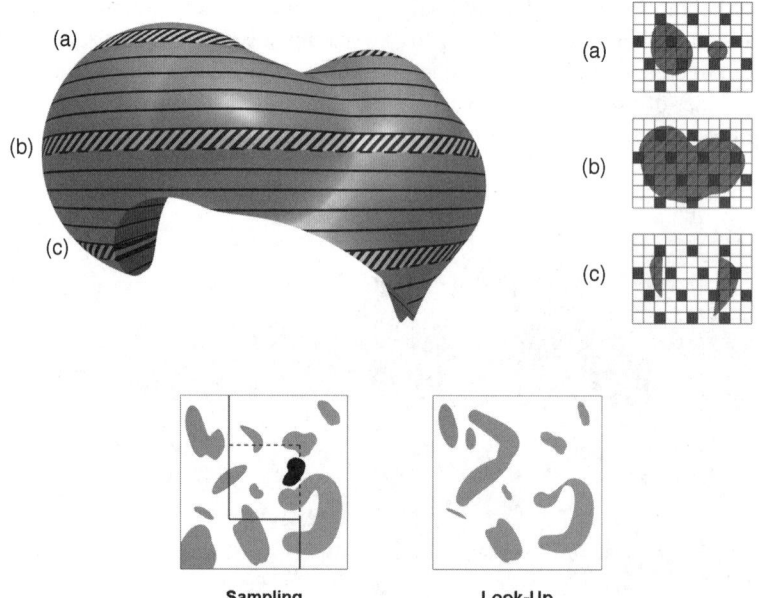

Sampling   **Look-Up**

FIGURE 9 With the physical fractionator design, an organ or anatomically defined region of an organ may be embedded in toto. The organ is sectioned exhaustively and a known fraction of sections (ssf) is collected for quantification. In this example, every sixth section was collected with the position of the first section selected randomly in the interval of 1 to 6 [hatched sections (a), (b), and (c)]. In this case, a random number of 2 was chosen and every subsequent sixth was collected. A known fraction of the section areas (asf) of sections (a), (b), and (c) is sampled by a sampling frame indicated by the solid squares in the overlaying tessellation. Within the matched disector microscopic fields sampled by the counting frame, cells are counted using the disector counting principle. In the sampling and look-up microscopic fields illustrated, only one object profile is only present in the sampling field and hence counted.

Physical Fractionator with Subsampling

When organs or objects of interest are too large to embed and section in toto (e.g., when studying brains from larger animals such as dog and monkey), additional levels of subsampling of tissue blocks are needed. The tissue sample collected after one or more subsampling steps can be processed and embedded, then sectioned to exhaustion, collecting a known fraction of sections for analysis. To avoid the need for exhaustive sectioning to collect the final fractionator sample, a modification of this classical fractionator design has been developed (H. J. G. Gundersen, unpublished) and is an approach to consider for studies of brains from larger animals (or any other large organ). Tissue pieces (usually, bars) constituting the penultimate fractionator sample are processed through paraffin and embedded in a closely apposed arrangement in a long rectangular embedding mold. At room temperature, the paraffin block is marked at predefined constant intervals (T) with a random start 0 to T. Using a small handsaw, the block is cut at intervals of width T (Figure 10). The block segments are then carefully mounted onto preformed paraffin blocks ensuring that the

front face of the block, from which sections for physical disectors (the final sample of the fractionator) will be collected, is not subjected to any deformation or temperature change. This is a critical step because no thermal contraction or expansion of the block face is allowed. Otherwise, the fraction of the block of (mean) thickness T represented by the physical disector collected from the block, will be biased and inaccurate. From each of the mounted block segments, a constant number of sections (k) are first removed in order to produce a complete tissue section. Then a pair of thin serial sections, constituting the counting and look-up sections for a physical disector, are prepared from each block at room temperature where section thickness (i.e., disector height) represents a known fraction of T. In this case, the micrometer setting of the sectioning device is not sufficiently accurate to use as a measure of section thickness. It is the microtome block advance (MA), synonymous with block advance, that actually determines what is removed (i.e., sampled) from the block, and this must be determined for the microtome or sectioning device.[24] Depending on the microtome, there can be marked discrepancies between MA

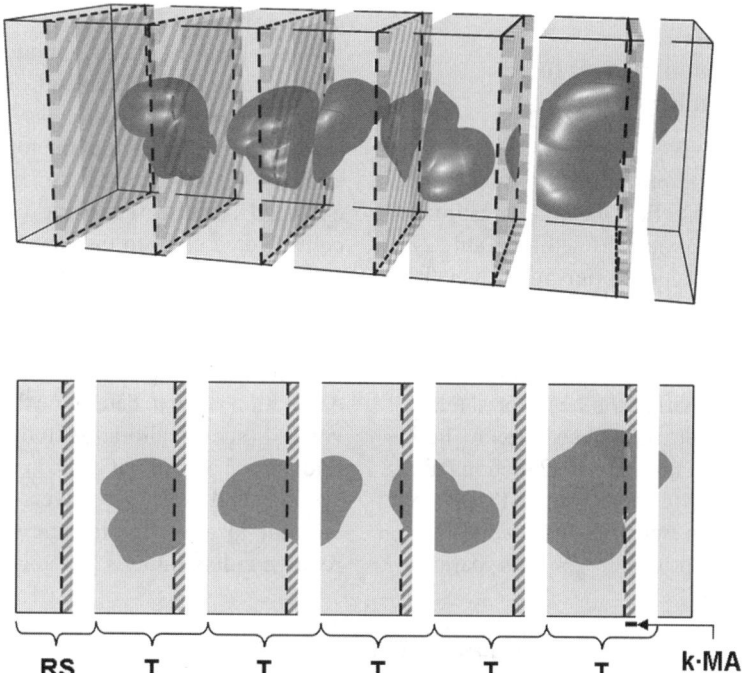

FIGURE 10 Principles of a novel fractionator method with subsampling that avoids the requirement for exhaustive sectioning of the final sample, developed by Gundersen (unpublished). After one or more sampling steps, final tissue pieces are processed through paraffin and embedded in a long rectangular embedding mold. The paraffin block is marked at predefined constant intervals (T) with a random start (RS) within the interval 0 to T at room temperature. The block is cut into segments of width T using a handsaw. Segments are mounted onto preformed paraffin blocks and a constant number of sections (k) are first removed in order to produce a complete tissue section. Then a pair of thin serial sections that constitutes the sampling and look-up sections for a physical disector are prepared from each block at room temperature. Section thickness is determined by the microtome block advance (MA), which represents a known fraction of T.

and micrometer settings. Cell counts (Q^-) are performed in physical disectors in a known fraction of section area (*asf*) using an unbiased counting frame. As noted above, software may automate the capture and registration of matched high-magnification microscopic fields. The total cell number N is estimated by

$$N := \frac{1}{bsf} \cdot \frac{1}{ssf} \cdot \frac{1}{asf} \cdot \sum Q^-$$
$$= \frac{1}{bsf} \cdot \frac{T}{MA} \cdot \frac{1}{asf} \cdot \sum Q^- \qquad (8)$$

where BSF indicates the block sampling fraction due to the initial subsampling of tissue blocks.

The foregoing technique is a novel solution for the generation of systematic uniformly random paraffin sections without exhaustive sectioning of the region of interest solving the problem of tissue shrinkage due to the paraffin embedding. However, methods for generating systematic, uniformly random, thick plastic, frozen, or vibratome sections without exhaustive sectioning have commonly been used for years.[38–40]

Benefits of Physical Fractionator Designs for Estimation of Cell Number in Regulatory Toxicology Studies

There are several features of the physical fractionator that makes it a highly efficient and practical method for estimating cell number in a regulatory toxicology setting (Table 2). The ability to use thin paraffin sections offers many benefits that enhance its utility in this setting. Several sets of routine paraffin sections can be collected for histopathology, using all kinds of stains and antibodies, from the same animal in the simple fractionator design, obviating the need for satellite groups. Time and resources can be reduced due to both section staining requirements (at most about 12 section pairs are required) and software and hardware advances that either automate or accelerate sampling steps. Manual staining can be avoided because staining protocols for thin paraffin

sections can be standardized and performed by automated stainers.[41] Software advances allow for the unattended capture, registration, and electronic archiving of matched high-magnification physical disector microscopic fields, allowing sampling to occur without supervision and for cell counting to be performed off-line at leisure. Further, the thin histological sections are suitable for whole-slide imaging technology, allowing for rapid *in silico* sampling and enhanced electronic documentation of sampling sites.

OPTICAL DISECTOR AND OPTICAL FRACTIONATOR

Optical disectors are created in thick sections by scanning the section with thin optical focal planes using a high-numerical-aperture oil (or water) immersion objective.[36] A 100× oil immersion lens with a numerical aperture of 1.4 has a focal plane of thickness of about 0.5 μm. As the focal plane is moved a known distance through the section, cells are counted as they come into focus (Figure 11). A unique structural feature of the cell, such as the nucleolus or the large, crisp equatorial profile of the nucleus, is used as the counting feature.

Thick paraffin, plastic, vibratome, or frozen sections may be used for optical disectors. Counts must be confined to the central region of the section as illustrated in Figure 12. The shaded regions located at the top and bottom of the section, known as *guard zones*, are excluded. These regions are excluded because sectioning can extract cells or parts of cells located close to the sectioning plane. These extracted cells are referred to as *lost caps*.[36] The width of the guard zones may vary with tissue and embedding media and can be optimized to avoid problems due to lost caps.[42] Therefore, it is important as a part of the pilot work to perform a z-axis calibration study to determine the required sizes of the guard zones. In such a calibration study, cells are sampled in the full thickness (i.e., without guard zones) in a representative set of sections and their respective z-positions recorded together with the local section thickness. A plot of the absolute and relative z-distributions of the cells will indicate the size of

TABLE 2 Comparative Features of the Optical and Physical Fractionators

	Optical Fractionator	Physical Fractionator
Processing and staining requirements	Section thickness ≥ 25 μm Verification of stain penetration in thick sections Float staining of thick sections	Section thickness ≤ 3 μm Standardized staining protocols
Time lines and Resources	Manual staining required Counting in real time or in stored z-stacks	Automated staining Counting live or off-line on stored disector image pairs Amenable to virtual slide imaging and *in silico* sampling
Documentation	Stored aqueous-based sections may progressively shrink (i.e., may be less suitable as raw data)	Stored thin paraffin sections constitute raw data

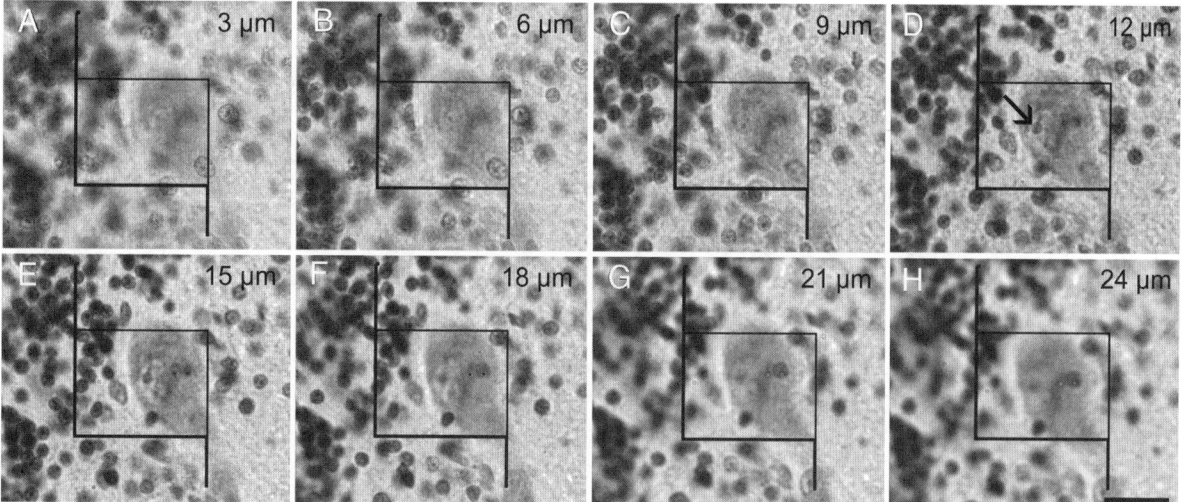

FIGURE 11 Counting of Purkinje neurons in human cerebellum in an optical disector. From the top of the thick paraffin section the focal plane is moved down through the section, creating a stack of optical planes, here illustrated by high magnification micrographs at 3, 6, 9, 12, 15, 18, 21, and 24 μm from the top plane of the section. The nucleolus of a Purkinje neuron comes into focus at 12 μm below the section top plane [indicated by an arrow in (D)] and is counted. Note smaller granular neurons coming into focus in one panel, but being out of focus in the neighboring panels. Scale bar = 20 μm.

the zones affected by lost caps as well as verify whether full z-axis penetration of the used stain has been obtained (see Figure 4 in the article by Dorph-Petersen et al.[43]).

Section thickness is another important consideration. Section thickness at the time of analysis should typically be a minimum of 25 μm to allow for adequate guard zones and a sufficient disector height ($\geq 10\,\mu$m). A section thickness of 20 μm may be adequate for plastic sections. With frozen or vibratome sections, shrinkage in the z-axis can be substantial following preparation (see, e.g., the article by Lyck et al.[17]), although steps may be taken to reduce the z-axis shrinkage (see the article by Konopaske et al.[44] and references therein). Therefore, to ensure adequate thickness at the time of analysis, frozen and vibratome sections should be cut at 40 to 80 μm (depending on the tissue and the staining protocol). Z-axis shrinkage of frozen and vibratome sections has also been demonstrated to be nonuniform,[24] as illustrated

in Figure 13. Local cell counts are influenced by the degree of local deformation. Nevertheless, unbiased cell number estimation remains straightforward using estimators based on the number-weighted mean section thickness. However, the method depends on frequent measurements of the local section thickness.[24]

There are several important technical considerations when using optical disectors. A fundamental requirement for any stereological analysis is that objects can be counted only if they can be identified unambiguously. Therefore, penetration of the stain through the complete section thickness is a key requirement for unbiased counting. Because immunophenotyping is frequently required to identify specific cell types in neural tissue, antibody and reagent penetration in thick sections is critical. Staining protocols for thick sections typically require that sections be stained free floating with extended incubation times and specialized

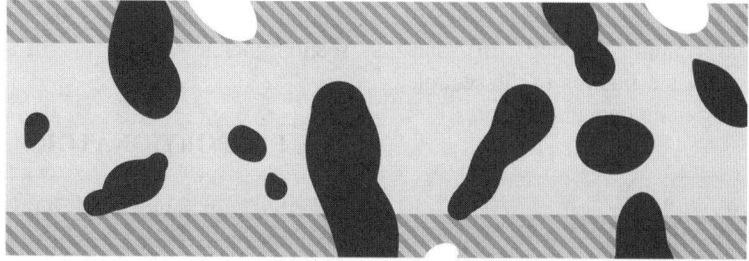

FIGURE 12 Thick section seen edge on. A defined region of the upper and lower boundaries of a thick section close to the sectioning plane should not be included in the optical disector. These regions, the guard zones, are excluded because sectioning can remove cells or portion of cells. If guard zones were included, the cell number would be underestimated. The size of the zones depends upon the tissue and the preparation. (Modified from Figure 2.3 in Gundersen.[36])

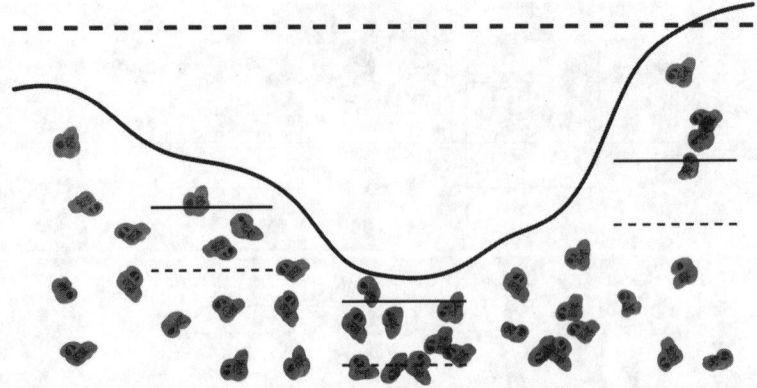

FIGURE 13   Thick section seen edge on. Tissue section shrinkage in the z-axis can bias cell counts in an optical disector of fixed height. Section deformation, a local shrinkage in the z-axis, increases the local density of cell number sampled by the optical disector. Local section thickness must be measured concurrently with cell number to obtain an unbiased estimate of cell number (cf. Eqs. 9 through 11). (Modified from Figure 1 in Dorph-Petersen et al.[24])

methods for epitope retrieval, mounting, and cover-slipping,[45,46] all of which is done manually.

Optical disectors and thick sections have the advantage of being truly 3-D, making it possible to observe cells in their full 3-D extent. This allows for easy and robust cell classification based on morphologic criteria. Also, delineation of subtle regions of interest may be easier in thick sections than in thin sections.

Sampling designs described for the physical fractionator can all be performed with thick sections and optical disectors. In the past 15 years, the optical fractionator has been the gold standard for the estimation of cell numbers in experimental neurosciences that solved the problem of using physical disectors for cell counts: namely, the registration of matching high-magnification microscopic fields in paired sections.[24,47] Although the method has gained great popularity in basic research due to its simplicity and efficiency, some of the aspects of optical disectors mentioned previously may limit its utility in a regulatory toxicology setting.

In the optical fractionator, the total cell number in the organ (N) is estimated by the product of the number of cells counted in the final sample, $\sum Q^-$, and the inverse of the sampling fraction for each sampling step, which includes section sampling fraction (*ssf*), areal sampling fraction (*asf*), and height sampling fraction (*hsf*), or the fraction of the section thickness used for cell counting in an optical disector:

$$N := \frac{1}{ssf} \cdot \frac{1}{asf} \cdot \frac{1}{hsf} \cdot \sum Q^- \tag{9}$$

$$hsf := \frac{h}{\overline{t_{Q^-}}} \tag{10}$$

$$\overline{t_{Q^-}} = \frac{\sum (t_i \cdot q_i^-)}{\sum q_i^-} \tag{11}$$

where h is the disector height, $\overline{t_{Q^-}}$ is the number-weighted mean section thickness, and t_i is the (final) local section thickness in the ith counting frame with a corresponding disector count of q_i^-. Notice that $\overline{t_{Q^-}}$ may be estimated from thickness measures in a systematic random subsample of all counting frames (e.g., every fourth). The estimator above is also correct in the face of nonuniform shrinkage in the z-axis that may be observed in almost all thick sections.[24] Technical considerations discussed for optical disectors regarding staining requirements, guard zones, and so on, all applies to the optical fractionator. A wide range of tissue preparations can be used with optical fractionators and would be the preferred estimator if frozen or vibratome sections are required.[39,40,44] However, frozen, vibratome, or plastic sections in a simple optical fractionator design may not be ideal in a regulatory toxicology setting where routine paraffin sections may be required for concomitant histopathologic evaluation. To avoid or minimize the potential for progressive z-axis shrinkage of thick sections, particularly when using vibratome or frozen sections, meticulous attention to cover-slipping and sealing is required. If progressive z-axis collapse is detected, temporal coordination of section preparation and analysis may be required to ensure adequate section thickness at the time of analysis.

THE PROPORTIONATOR

A further gain in efficiency is possible using a novel form of sampling called the *Proportionator*.[4,5] This method uses non-uniform random sampling guided by image analysis-detected features of interest in the stained sections. The image analysis feature defined could be, for example, immunohistochemical staining intensity. Simplified, all possible sampling fields are assigned a value for the image feature and subsampled based on a mathematical relationship with the assigned value. Compared with SURS, this unbiased method can increase the intrasection sampling efficiency up to 15-fold. The Proportionator sampling can be applied to both thin and thick sections—and for all estimators of structural quantities,[4,5] but the method holds particular promise for cell number analysis. So far, the sampling method is implemented in one commercially available software package (Proportionator™, Visiopharm, Hørsholm, Denmark).

ESTIMATION OF TOTAL MYELINATED NERVE FIBER NUMBER AND ABSOLUTE SIZE DISTRIBUTION

Peripheral neuropathy with axonal atrophy and loss may be a consequence of test article treatment. Slight changes in axon number and size in peripheral nerves may go undetected by qualitative morphological evaluation even in the face of functional changes. Sensitive quantitative methods are required to detect these subtle changes. Stereological techniques can provide accurate and precise estimates of both axon number and size. When performed with a computer-driven microscope interfaced with stereological software, these data can be generated in a time-effective way.

Excellent comprehensive stereological approaches to quantification of peripheral nerve have been developed by Larsen.[48] Because axons are single continuous structures in peripheral nerves, the axon number can be estimated in 2-D, and counting in disectors is not required. Therefore, axon number can be estimated in standard crosssections of peripheral nerve. This is a special case where 2-D profile numbers provide an accurate estimate of 3-D total number. Changes in axon size can be assessed by changes in the absolute size distribution of axon diameter or cross-sectional areas.

For quantitative evaluation of myelinated fibers, standard segments of the nerve should be removed following perfusion and immersion fixation, embedded in hard plastic such as epon, sectioned at 1 μm, and stained with toluidine blue. Measurements are performed with a 100× oil objective. Because cross-sectional areal measurements can be affected by shrinkage and deformation, plastic embedding is required to minimize tissue shrinkage. When nerve segments are prepared, cuts should be made as close to the ideal perpendicular plane relative to the longitudinal axis as possible. This will avoid oblique sections of axon profiles that will affect areal measurements.

The most simple and direct estimator of total number of myelinated nerve fibers is the fractionator. Myelinated fibers are counted (Q) in a known areal fraction of the total nerve cross-sectional area (*asf*, cf. Figure 9) using an unbiased counting frame. The total number (N) is estimated as

$$N := \frac{1}{asf} \cdot Q \qquad (12)$$

The size of the unbiased counting frame should be adjusted by the software to sample approximately five axons per frame in homogeneous nerves. In heterogeneous nerves, the counting frame size is adjusted to sample a few axon profiles per frame, reflecting that more sampling sites are needed in the heterogeneous case to capture the variation in the nerve (while keeping the *asf* constant). The *asf* should typically be optimized to yield total counts of approximately 100 to 200 to obtain a CE of about 7 to 10%.[48]

A very simple and efficient estimate of axonal size distribution may be obtained as the diameter distribution. This is due to the fact that for ideal cylinders, the shortest diameter perpendicular to the longest aspect of the profile is the 3-D diameter, irrespective of the cutting angle. For each of the profiles sampled (as described above), the diameter is measured and a set of such data is the diameter distribution. A more elaborate, but also more accurate estimator of axon size is the distribution of axonal cross-sectional areas. Estimates of individual myelinated axon cross-sectional area are easily obtained using a four-way 2-D nucleator (Figure 14).[48,49] Axon profiles are sampled using the fractionator and unbiased counting frame to guarantee that all axons are given equal probability of being sampled (as described above). Axon profiles sampled by the counting frame are used to estimate axon cross-sectional area. From the approximate central point of an axon profile, four lines through the point are generated by the computer and software in systematic random directions to ensure that all directions are given equal probability. The intersections of the lines with the axon boundary are marked and the computer records the lengths of the lines (l) from the central point to the intersection with the boundary. The axon area (a) is estimated from the mean of the squared lengths as

$$a := \pi \cdot \overline{l^2} \qquad (13)$$

A symmetrical distribution of absolute size distribution of axonal areas is best achieved by plotting the axonal areas on a logarithmic abscissa and the absolute axon number on the ordinate. In addition to providing the geometrical mean axonal cross-sectional area, this approach can reveal changes in axon number in a specific subpopulation of axons that can be difficult to detect by shifts in relative frequency plots.[48]

Additional endpoints for evaluation of myelinated fibers as well as the ultrastructural approaches required for quantification of unmyelinated fibers are reviewed by Larsen.[48]

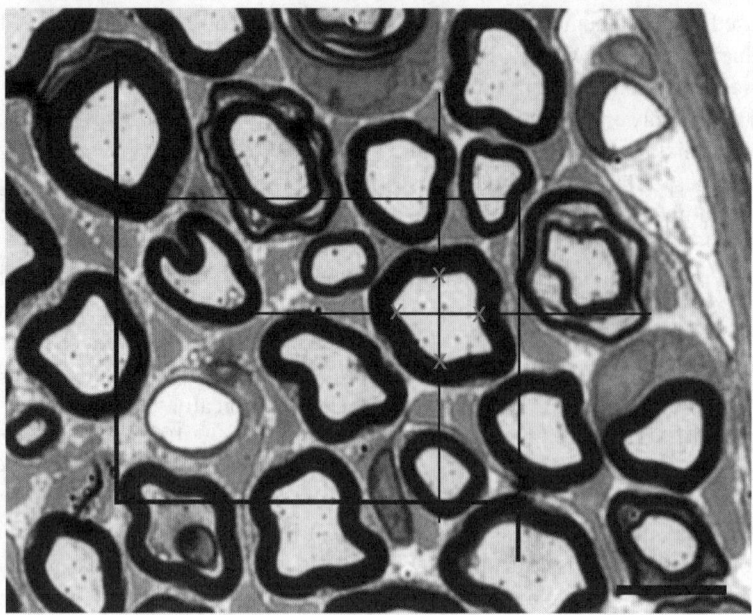

FIGURE 14 The number of myelinated axons and their size distribution in a peripheral nerve is measured using an unbiased counting frame for sampling of axon profiles. For estimation of axonal size distribution, the four-way planar nucleator is employed on all sampled profiles. Intersections between the nucleator test lines and the inner border of the myelin sheath are marked by gray X's. Note that the four-way two-dimensional nucleator must have random rotation. Scale bar = 10 μm.

CONCLUSIONS

The estimators and methods discussed in this chapter represent a major paradigm shift in the application of stereological approaches in a regulatory toxicology setting. With the integration of precise and efficient stereological estimators with automated slide scanning, automated section sampling, computer-assisted measurements, automated capture and registration of physical disectors, and novel image analysis-based sampling, it is now feasible to include stereological endpoints in regulatory toxicology studies. In regard to estimation of cell number using the physical fractionator, a significant recent advance is the development of software that provides for unattended, automated capture and registration of matched high-magnification sampling sites. Because thin paraffin sections can be used, sections for stereological analyses and routine histopathology can be generated from the same animal, reducing animal use. All current technologies that automate and accelerate slide staining and sampling, such as automated immunostainers and *in silico* sampling of virtual slide images, can be used with this slide format to maximize sampling efficiency and provide for enhance study-specific electronic documentation.

Acknowledgments

The authors would like to thank Jytte Overgaard Larsen for the stained section of peripheral nerve used in Figure 14,

John Boyce for his helpful editorial comments, and J. P. Long for his expert assistance in the preparation of illustrations. The first author (R.W.B.) would like to acknowledge and thank Patrick Wier and GlaxoSmithKline for supporting the development of the AutoDisector™ software.

RECOMMENDED READING

Cruz-Orive LM, Weibel ER. Recent stereological methods for cell biology: a brief survey. *Am J Physiol.* 1990;258:L148–L156.

Dorph-Petersen K-A, Lewis DA: Stereological approaches to identifying neuropathology in psychosis. *Biological Psychiatry* 2011;69:113–126.

Evans SM, et al. *Quantitative Methods in Neuroscience: A Neuroanatomical Approach.* Oxford, UK: Oxford University Press; 2004.

Gundersen HJG, et al. Some new, simple and efficient stereological methods and their use in pathological research and diagnosis. *APMIS,* 1988;96:379–394.

Gundersen HJG, et al. The new stereological tools: Disector, fractionator, nucleator and point sampled intercepts and their use in pathological research and diagnosis. *APMIS,* 1988;96:857–881.

Howard CV, Reed MG. *Unbiased Stereology, Three-dimensional Measurements in Microscopy.* 2nd ed. Garland Science/Bios Scientific Publishers, Abingdon, UK; 2005.

Schmitz C, Hof PR. Design-based stereology in neuroscience. *Neuroscience.* 2005;130:813–831.

Baddeley A, Vedel Jensen EB. *Stereology for Statisticians*. Boca Raton, FL: Chapman & Hall/CRC Press; 2005. (This text is recommended for the statistically inclined reader.)

REFERENCES

1. Baddeley AJ. Stereology and survey sampling theory. *Bull Inter Statis Inst, Proc 49th Session, Florence*. 1993;50. Book 2:435–449.

2. Gundersen HJG, Jensen EB. The efficiency of systematic sampling in stereology and its prediction. *J Microsc*. 1987;147:229–263.

3. Dorph-Petersen K-A, et al. Non-uniform systematic sampling in stereology. *J Microsc*. 2000;200:148–157.

4. Gardi JE, et al. Automatic sampling for unbiased and efficient stereological estimation using the proportionator in biological studies. *J Microsc*. 2008;230:108–120.

5. Gardi JE, et al. The proportionator: Unbiased stereological estimation using biased automatic image analysis and non-uniform probability proportional to size sampling. *Comput Biol Med*. 2008;38:313–328.

6. West MJ. New stereological methods for counting neurons. *Neurobiol Aging*. 1993;14:275–285.

7. Sterio DC. The unbiased estimation of number and sizes of arbitrary particles using the disector. *J Microsc*. 1984;134:127–136.

8. Mattfeldt T, et al. Stereological investigations of anisotropic structures with the orientator. *Acta Stereol*. 1989;8:671–676.

9. Nyengaard JR, Gundersen HJG. The isector: a simple and direct method for generating isotropic, uniform random sections from small specimens. *J Microsc*. 1992;165:427–431.

10. Baddeley AJ, et al. Estimation of surface area from vertical sections. *J Microsc*. 1986;142:259–276.

11. Gundersen HJG. Stereology: The fast lane between neuroanatomy and brain function–or still only a tightrope? *Acta Neurol Scand Suppl*. 1992;137:8–13.

12. Gundersen HJG, et al. The efficiency of systematic sampling in stereology—reconsidered. *J Microsc*. 1999;193:199–211.

13. Cruz-Orive LM, Geiser M. Estimation of particle number by stereology: an update. *J Aerosol Med*. 2004;17:197–212.

14. Gundersen HJG, Østerby R. Optimizing sampling efficiency of stereological studies in biology: or "Do more less well!" *J Microsc*. 1981;121:65–73.

15. Cruz-Orive LM, Weibel ERW. Recent stereological methods for cell biology: a brief survey. *Am J Physiol*. 1990;258:L148–L156.

16. Cavalieri B. *Geometria Indivisibilibus Continuorum. Nova Quadam Ratione Promota*. Bononiæ: Typis Clementis Feronij; 1635. Reprinted as *Geometria degli Indivisibili*. Torino, Italy: Unione Tipografico-Editrice Torinese, 1966.

17. Lyck L, et al. Unbiased cell quantification reveals a continued increase in the number of neocortical neurons during early postnatal development in mice. *Eur J Neurosci*. 2007;26: 1749–1764.

18. Regeur L, Pakkenberg B. Optimizing sampling designs for volume measurements of components of human brain using a stereological method. *J Microsc*. 1989;155:113–121.

19. Jelsing J, et al. Assessment of in vivo MR imaging compared to physical sections in vitro—a quantitative study of brain volumes using stereology. *Neuroimage*. 2005;26:57–65.

20. Haug H, et al. The significance of morphometric procedures in the investigation of age changes in cytoarchitectonic structures of human brain. *J Hirnforsch*. 1984;25:353–374.

21. Iwadare T, et al. Dimensional changes of tissues in the course of processing. *J Microsc*. 1984;136:323–327.

22. Miller PL, Meyer TW. Effects of tissue preparation on glomerular volume and capillary structure in the rat. *Lab Invest*. 1990;63:862–866.

23. Mendis-Handagama SMLC, Ewing LL. Sources of error in the estimation of Leydig cell numbers in control and atrophied mammalian testes. *J Microsc*. 1990;159:73–82.

24. Dorph-Petersen K-A, et al. Tissue shrinkage and unbiased stereological estimation of particle number and size. *J Microsc*. 2001;204:232–246.

25. Dorph-Petersen K-A, et al. The influence of chronic exposure to antipsychotic medications on brain size before and after tissue fixation: A comparison of haloperidol and olanzapine in macaque monkeys. *J Neuropsychopharmacol*. 2005;30: 1649–1661.

26. Gundersen HJG. Notes on the estimation of the numerical density of arbitrary profiles: the edge effect. *J Microsc*. 1977;111:219–223.

27. Pakkenberg B, et al. The absolute number of nerve cells in substantia nigra in normal subjects and in patients with Parkinson's disease estimated with an unbiased stereological method. *J Neurol Neurosurg Psychiatry*. 1991;54: 30–33.

28. Coggeshall RE. A consideration of neural counting methods. *Trends Neurosci*. 1992;15:9–13.

29. Reguer L, et al. No global neocortical nerve cell loss in brains from patients with senile dementia of Alzheimer's type. *Neurobiol Aging*. 1994;15:347–352.

30. West MJ, et al. Differences in the pattern of hippocampal neuronal loss in normal ageing and Alzheimer's disease. *Lancet*. 1994;344:769–772.

31. Pakkenberg B, Gundersen HJG. Neocortical neuron number in humans: effect of sex and age. *J Comp Neurol*. 1997;384: 312–320.

32. Jelsing J, et al. The postnatal development of neocortical neurons and glial cells in the Göttingen minipig and the domestic pig brain. *J Exp Biol*. 2006;209:1454–1462.

33. Pelvig DP, et al. Neocortical glial cell numbers in Alzheimer's disease: a stereological study. *Dement Geriatr Cogn Disord*. 2003;16:212–219.

34. Pelvig DP, et al. Neocortical glial cell numbers in human brains. *Neurobiol Aging*. 2008;29:1754–1762.

35. Brændgaard H, Gundersen HJG. The impact of recent stereological advances on quantitative studies of the nervous system. *J Neurosci Methods*. 1986;18:39–78.

36. Gundersen HJG. Stereology of arbitrary particles: A review of unbiased number and size estimators and the presentation of some new ones, in memory of William R. Thompson. *J Microsc.* 1986;143:3–45.

37. Gundersen HJG. The smooth fractionator. *J Microsc.* 2002;207:191–210.

38. West MJ, Gundersen HJG. Unbiased stereological estimation of the number of neurons in the human hippocampus. *J Comp Neurol.* 1990;296:1–22.

39. Dorph-Petersen K-A. Stereological estimation using vertical sections in a complex tissue. *J Microsc.* 1999;195: 79–86.

40. Dorph-Petersen K-A, et al. Primary visual cortex volume and total neuron number are reduced in schizophrenia. *J Comp Neurol.* 2007;501:290–301.

41. Lyck L, et al. Immunohistochemical markers for quantitative studies of neurons and glia in human neocortex. *J Histochem Cytochem.* 2008;56:201–221.

42. Andersen BB, Gundersen HJG. Pronounced loss of cell nuclei and anisotropic deformation of thick sections. *J Microsc.* 1999;196:69–73.

43. Dorph-Petersen K-A, et al. Volume and neuron number of the lateral geniculate nucleus in schizophrenia and mood disorders. *Acta Neuropathol.* 2009;117:369–384.

44. Konopaske GT, et al. Effect of chronic antipsychotic exposure on astrocyte and oligodendrocyte numbers in macaqueiatry monkeys. *Biol Psychiatry.* 2008;63:759–765.

45. Müller GJ, et al. Stereological cell counts of GABAergic neurons in rat dentate hilus following transient cerebral ischemia. *Exp Brain Res.* 2001;141:380–388.

46. Lyck L, et al. Immunohistochemical visualization of neurons and specific glial cells for stereological application in the porcine neocortex. *J Neurosci Methods.* 2006;152: 229–242.

47. West MJ, et al. Unbiased stereological estimation of the total number of neurons in the subdivisions of the rat hippocampus using the optical fractionator. *Anat Rec.* 1991;231:482–497.

48. Larsen JO. Stereology of nerve cross sections. *J Neurosci Methods.* 1998;85:107–118.

49. Gundersen HJG. The nucleator. *J Microsc.* 1988;151:3–21.

16

ANATOMY AND PROCESSING OF PERIPHERAL NERVE TISSUES

WILLIAM VALENTINE

Department of Pathology, Vanderbilt University Medical Center, Nashville, Tennessee

BRAD BOLON

GEMpath, Inc., Longmont, Colorado

INTRODUCTION

The peripheral nervous system (PNS) includes the cranial nerves, the spinal nerves with their roots and ganglia, the somatic nerves, and the autonomic nerves and their ganglia (Figure 1A). Specific details of PNS distribution and physiology have been described in general neuroscience,[1] medical,[2–4] and veterinary medical[5,6] texts, and are not reiterated here.

Peripherally projecting neurons include some of the largest and longest cells in mammals with their soma located either in the spinal cord gray matter or cranial nerve nuclei, or in the sensory (dorsal root) or autonomic ganglia. Somatic motor neurons are located within the ventral horns of the spinal cord, and their efferent axons travel caudally and then exit through the ventral spinal root. The dorsal spinal roots contain sensory efferent fibers from the dorsal root ganglia (DRG). Both dorsal and ventral spinal roots pass through a region where there is a transition from central nervous system myelin to peripheral nerve myelin (Figure 1B). Distal to the DRG, the afferent sensory and efferent motor fibers join to form a segmental nerve that leaves the spinal canal through an intervertebral foramen to become a mixed peripheral nerve.

ANATOMY OF THE PNS

Within the DRG, the somas of sensory neurons and satellite cells (glia), as well as myelinated axons, can readily be identified by light microscopy (Figure 1C and D). The sensory neurons in the DRG are large spherical cells that display a wide range of diameters within a given ganglion. They are distinguished by their size, centrally located nuclei, and single prominent nucleoli. In the cytoplasm of these neurons are mitochondria, microtubules, neurofilaments, and Nissl substance (composed of rough endoplasmic reticulum). The DRG cells are classified as pseudounipolar, due to their lack of dendrites and the presence of a single axon that bifurcates into a centrally directed efferent branch and a peripherally directed branch that functions as a dendrite to carry afferent sensory signals. These neuronal appendages are apparent within dorsal root ganglia as myelinated axons with their associated Schwann cells. Satellite cells are neuroglial cells within the ganglia that encircle the sensory neuron cell bodies.

Peripheral nerves are extended tubular structures that contain bundles of individual nerve fibers or axons. These axons serve as conduits for action potentials traveling to and from the central nervous system (CNS). In addition to axons, peripheral nerves contain a complement of elements—Schwann cells, fibroblasts, mast cells, and macrophages—that perform many diverse structural, trophic, and immunological functions. Collectively, these components serve effectively to promote rapid intercellular communication over extended distances, maintain structural integrity of the delicate nerve trunk, and facilitate regeneration following injury.

Fundamental Neuropathology for Pathologists and Toxicologists: Principles and Techniques, First Edition. Edited by Brad Bolon and Mark T. Butt.
© 2011 John Wiley & Sons, Inc. Published 2011 by John Wiley & Sons, Inc.

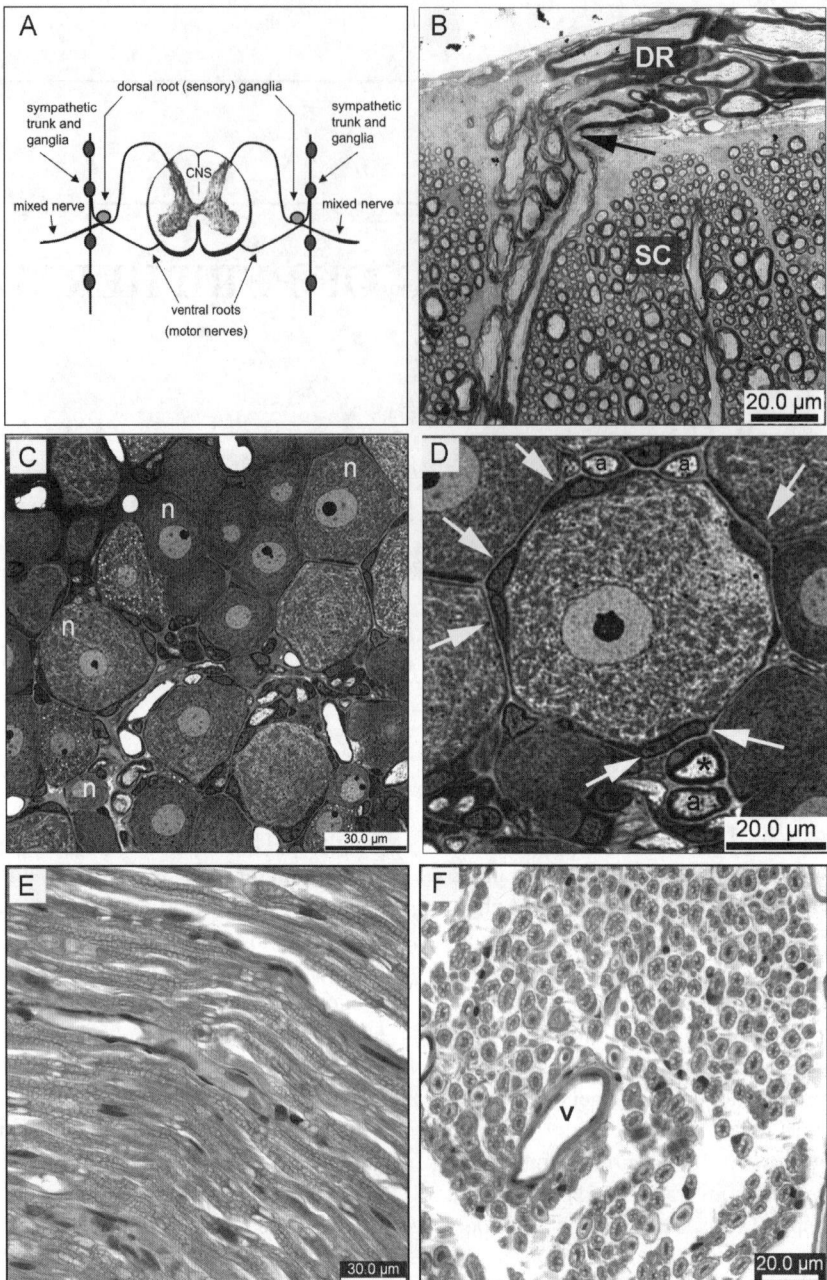

FIGURE 1 (A) Ventral motor and dorsal sensory spinal nerve roots emerging from the spinal cord. These nerve roots eventually join to form a mixed (motor and sensory) somatic nerve. Neuron cell bodies of peripheral nerves are either located in the CNS within the ventral (anterior) motor horns of the spinal cord or in peripheral sensory dorsal root ganglia (DRG) or autonomic ganglia. (B) When spinal roots (DR = dorsal spinal root) exit the spinal cord (SC), they pass through a transition zone (arrow) where the myelin changes from CNS myelin produced by oligodendroglia to PNS myelin made by Schwann cells; the myelin from both cell types is morphologically similar. (C) Low-power image of a plastic-embedded DRG stained with toluidine blue, illustrating the wide range of neuron (n) diameters. (D) Higher magnification of a plastic-embedded DRG illustrating characteristic features such as neurons with large central nuclei, prominent nucleoli, and abundant Nissl substance (rough endoplasmic reticulum) within the cytoplasm, as well as multiple satellite (glial) cells (arrows) and myelinated axons (a). One axon (∗) is accompanied by its associated Schwann cell (nucleus at the 10 o'clock position relative to the nerve fiber). (E) Longitudinal section of a formalin-fixed, paraffin-embedded rat sciatic nerve stained with hematoxylin and eosin. Note the typical "bubbled" appearance of the myelin sheaths in this type of preparation. (F) Same block and processing as (E), but with the specimen in cross section with a large blood vessel (v) present. Note the large artifactual spaces between adjacent axons.

In cross section, the major components of a peripheral nerve that can be identified by light microscopy include the connective tissue layers (comprising the epineurium, perineurium, and endoneurium), myelinated axons, and (with sufficient magnification) unmyelinated axons (Figures 2A to C). Nucleated cells inside peripheral nerves include Schwann cells (both myelinating and nonmyelinating), macrophages, fibroblasts, mast cells, and blood vessel constituents (chiefly capillary endothelium). The Schwann cells are associated with axons, and the remaining cells are scattered throughout the connective tissue layers.

The epineurium is the outermost connective tissue matrix that confines all other components of the nerve. The epineurium has two layers, the epifascicular and interfascicular layers, and is thicker on large nerves and near joints. Mast cells, adipocytes, blood and lymph vessels, and fibroblasts are scattered within the epineurium. Collagens I and III, along with elastin fibers, provide tensile strength for this structure.

The perineurium groups bundles of axons into distinct fascicles within the nerve trunk. It consists of concentric cylindrical sheaths of perineurial cells separated by collagen fibrils; the thickness varies in proportion to the fascicle size. The perineurial cells are both metabolically active and connected by tight junctions, thus constituting a contiguous diffusion barrier around the exterior of each fascicle. Thus, the perineurium functions both actively and passively to maintain a constant microenvironment within the endoneurial space of the enclosed fascicle.

The endoneurium consists of the connective tissue matrix within each fascicle. It supports the individual fibers (i.e., axons and their associated Schwann cell covering). Approximately half of the endoneurial compartment is occupied by nerve fibers and about one-third by fluid and extracellular collagen fibrils. The remainder of the endoneurial space is occupied by several populations of nonneural nucleated cells and small blood vessels (mostly capillaries). Similar to perineurial cells, the endothelial cells of the intrafasicular vessels are joined by tight junctions to generate another diffusion barrier. The specialized endothelium and the perineurial covering together comprise the "blood–nerve barrier." Collagen within the endoneurium is composed of collagens I and III and is organized into fibrils about 50 nm in diameter arranged longitudinally (Figure 2D to F).

Several populations of intrafascicular nucleated cells are associated with the endoneurium. Schwann cells are distinguished by their close apposition to axons and the presence of a basal lamina (Figure 2E). Mast cells are readily identified by their abundant dense metachromatic granules and pseudopodia (Figure 2F). Fibroblasts and macrophages can be challenging to differentiate in normal nerves. At higher electron microscopy magnifications, fibroblasts can be identified by the presence of short, dense, thin layers of uncertain origin on the cell surface (sometimes referred to as *plasmalemmal linear densities*).

Myelinated axons and unmyelinated axons in peripheral nerve trunks can readily be discriminated by light microscopy in properly prepared specimens. However, structural assessments of unmyelinated axons require electron microscopy due to their small diameter (0.3 to 3.0 μm) (Figure 2D). Unmyelinated axons are present both individually and in small groups of up to about 10 axons. In either case, the individual axons are encircled by a single layer of cytoplasm from a nonmyelinating Schwann cell. Groups of unmyelinated axons and their associated Schwann cell are called *Remak fibers*. Tight junctions connect the overlapping processes of the Schwann cells in a Remak fiber. Nonmyelinating Schwann cells possess an outer basal lamina identical to those of myelinating Schwann cells but otherwise exhibit a distinct phenotype characterized by the absence of myelin proteins and greater expression of certain proteins (e.g., S100) characteristic of undifferentiated Schwann cells.

The axons of myelinated and unmyelinated fibers are similar in composition and structure and differ primarily in their diameters. In fully developed nerves, unmyelinated axons exhibit a unimodal size distribution within 0.3 to 3.0 μm, while myelinated axons exhibit a bimodal size distribution within the range of 3 to 20 μm diameter. A three-layered 8-nm-thick axolemma bounded by a periaxonal space encloses the axoplasm and separates the axon from the adaxonal Schwann cell. Present within the axoplasm are organelles and components of the cytoskeleton (Figure 3A). Mitochondria 0.1 to 0.3 μm in diameter and smooth endoplasmic reticulum are the most readily recognized organelles, while rough endoplasmic reticulum and Golgi apparatus are not present. Neurofilaments (NFs) are the intermediate filament of the axon and are unique to this cell type. Neurofilaments are 10 nm in diameter and are formed by the noncovalent association of three protein subunits that differ in the length of their carboxyl terminal tails (Figure 3B). These subunits are designated light (NFL), medium (NFM), and heavy (NFH) based on their apparent approximate molecular weights of 70, 145 to 160, and 200 to 220 kDa, respectively; they are very argyrophilic and typically represent the major structures stained by certain silver-based methods. Microtubules, analogous to those in other cells, are 20-nm-diameter hollow cylinders formed by the polymerization of tubulin. These tubules serve as tracts for the transport of mitochondria, vesicles, and cytoskeletal components along the axon by kinesin and dynein motors. Microfilaments (5 to 7 nm thick) formed by chains of actin are associated with the axolemma. They are often poorly fixed by standard methods and thus difficult to identify. Whereas the density of neurofilaments is about 150 to 200 μm^{-2} and does not vary with axon diameter, the density of microtubules ranges from 10 to 40 μm^{-2} and is

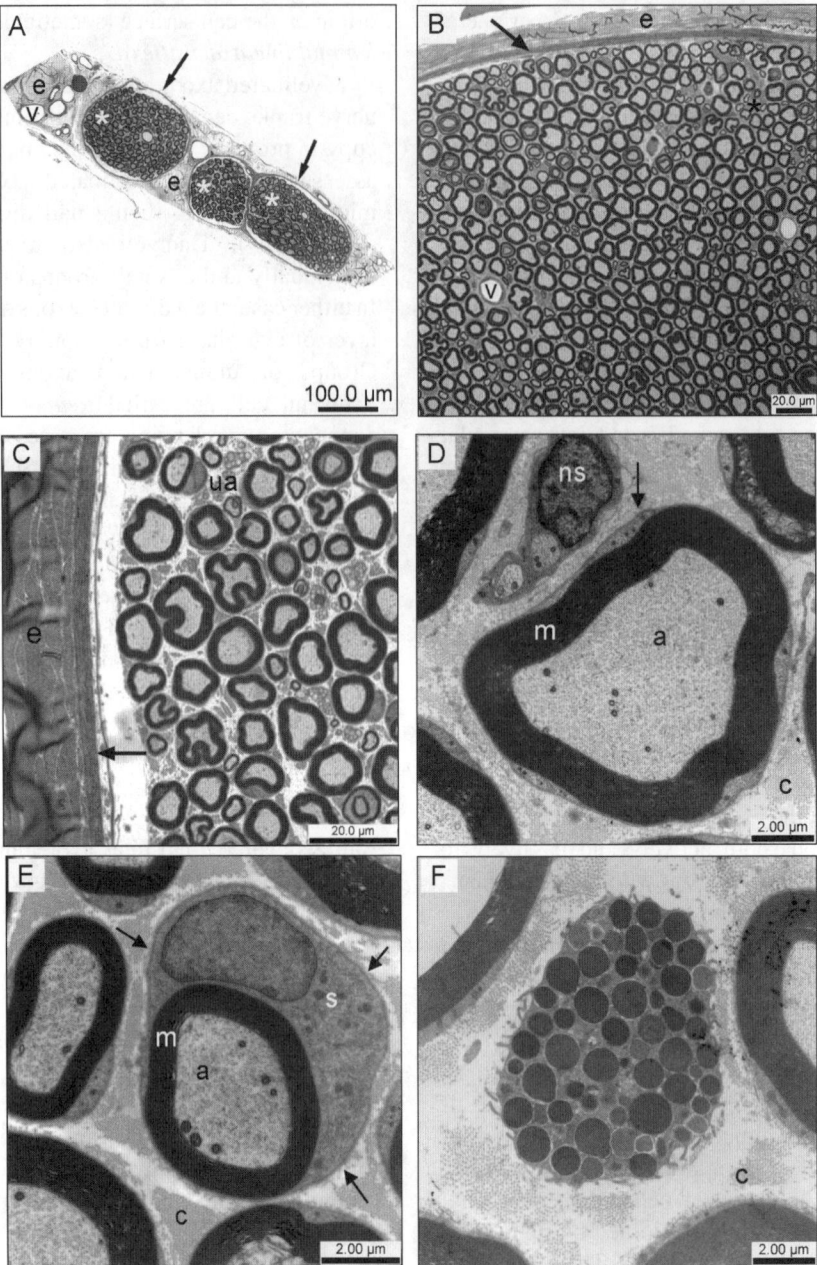

FIGURE 2 (A) Plastic-embedded, toluidine blue–stained, 1-µm-thick cross sections of the muscular branch of the caudal (posterior) tibial nerve obtained from an adult rat. The section shows three fascicles (∗), each surrounded by a dense fibrous perineurium (arrows) and encompassed in a looser epineurial matrix (e). Vessels (v) and fat can be identified within the epineurium. (B) A portion of a fascicle at intermediate magnification showing the epineurium (e), perineurium (arrow), endoneurial space, intrafascicular vessels (v), and numerous myelinated axons. (C) Higher magnification of a fascicle in which unmyelinated axons (ua) are intermingled with myelinated axons exhibiting a typical bimodal (large and small) distribution. Both the epineurium (e) and perineurium (arrow) are also present. (D) Electron micrograph of a peripheral nerve showing a large myelinated axon (a) surrounded by compact myelin (m) and the associated outer Schwann cell cytoplasm (arrow). A pair of unmyelinated axons surrounded by the plasma membrane of a nonmyelinating Schwann cell (ns) and collagen fibrils (c) are evident in the endoneurial space. (E) Electron micrograph showing a cross section of a large myelinated axon (a) obtained at the middle of an internode, as indicated by the adjacent Schwann cell soma (s) and nucleus. The axon is surrounded by compact myelin (m) and contains six dark, round mitochondria within the axoplasm. The Schwann cell is encircled by the basal lamina (arrows); collagen fibrils (c) are present in the endoneurial compartment. (F) Electron micrograph of a mast cell with characteristic dense granules on its surface. The cell is located within the endoneurial space along with collagen fibrils (c).

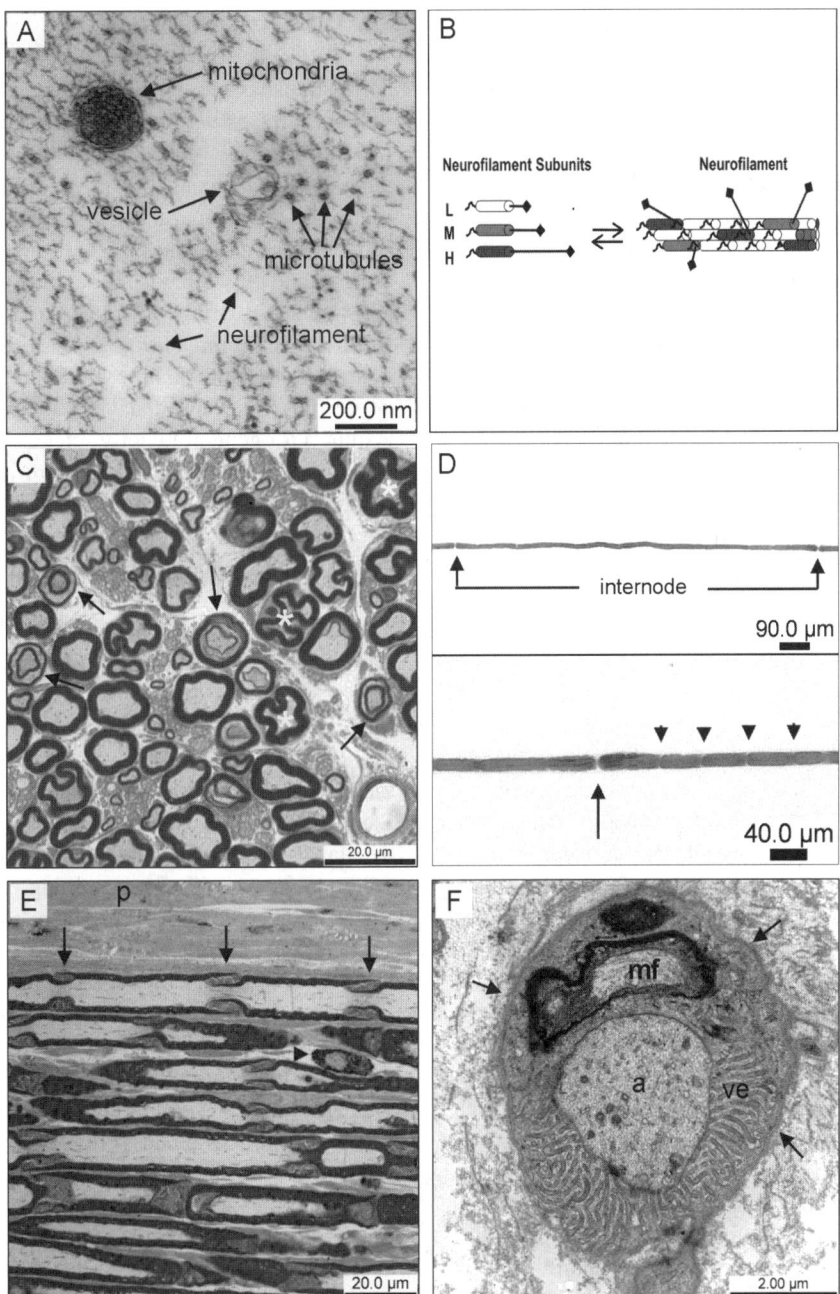

FIGURE 3 (A) Electron micrograph illustrating characteristic neuronal organelles present within the axoplasm. (B) Diagrammatic representation of the light (L), medium (M), and heavy (H) neurofilament subunits showing the variable length of their carboxy tail regions and their assembly into 10-nm-diameter neurofilaments (with their carboxy tails extended to the side). (C) Cross section of a plastic-embedded, toluidine blue–stained section of rat sciatic nerve illustrating Schmidt–Lanterman incisures (arrows) and fluting of myelinated axons (∗) as they approach paranodal regions. (D) Individual teased axons demonstrating nodes of Ranvier (arrows), Schmidt–Lanterman incisures (arrowheads), and a complete internode. (E) Longitudinal section of a plastic-embedded, toluidine blue–stained section of rat sciatic nerve showing the layered structure of the perineurium (p), Schmidt–Lanterman incisures (arrows) within the myelin, and a mast cell (arrowhead) in the endoneurial space. (F) Slightly oblique section through a myelinated axon showing paranodal (top) and nodal (bottom) regions. In the paranodal region, the myelin folds (mf) can be seen. The axon (a) is constricted and contains a high density of organelles and neurofilaments. In the nodal region, the villous extensions (ve) of the Schwann cell plasma membrane can be seen. The continuous basal lamina (arrows) extends over the node and encircles the axon and Schwann cell extensions.

inversely proportional to the diameter of the axon. In normal nerves, these axonal components are evenly distributed throughout the axoplasm, with the exception of microfilaments, which tend to be concentrated in the periphery near the axolemma.

Myelin in the PNS is formed when the extended processes of myelinating Schwann cells surround axons (Figure 2E). Extrusion of the Schwann cell cytoplasm within the processes produces compact myelin, a tightly spiraled, multilayered covering of the axon composed of lipid and protein. At high magnification in cross section, it appears as alternating light and dark bands. Intracellular membranes of the collapsed processes are compressed following extrusion of the cytoplasm to generate the dark-colored major dense line, and the juxtaposed adjacent exterior of the plasma membranes form the lighter-colored intraperiod lines. A single Schwann cell produces the myelin covering for a segment of one axon, termed an *internode* (Figure 3D). To myelinate an entire axon, multiple Schwann cells are aligned along its length, with the nucleus of each Schwann cell located at the middle of each internode (Figure 3E). The length of each internode is proportional to the diameter of the axon, suggesting a relatively greater metabolic demand on the Schwann cells associated with large-diameter axons. At multiple sites along the internode, the compact myelin is interrupted by regions of noncompact myelin, which are thought to provide a continuous cytoplasmic connection from the adaxonal (innermost)-to-abaxonal (outermost) surface for the passage of metabolites required to support the axon. These interruptions are termed *Schmidt–Lanterman incisures*, and their incidence is positively correlated with the diameter of the axon. In cross section, Schmidt–Lanterman incisures appear as concentric rings of compact myelin separated by an area of normal Schwann cell cytoplasm (Figure 3C). In longitudinal section, these incisures appear as diagonal separations of myelin running from the exterior to the interior surfaces of the myelin sheath (Figure 3D and E). On the exterior surface of myelin, a 20- to 30-nm-thick basal lamina forms a continuous tubular structure encircling the Schwann cells and axon along the entire length of the fiber.

The spaces between adjacent Schwann cells, termed *nodes of Ranvier*, are approximately 1 μm in length and feature constricted, nonmyelinated regions of axon. Compact myelin is absent at the nodes. However, Schwann cell villous cytoplasmic extensions extend into the region and (particularly for large axons) interdigitate with the processes from the adjacent Schwann cell, thus forming the bracelets of Nageotte (Figure 3F). The basal lamina is continuous from one Schwann cell to the next over the nodes. In cross section, paranodal and nodal regions present varying profiles, due to folds in the plasma membrane. Myelin becomes fluted as the node is approached, and the axon becomes constricted with an increased density of cytoplasmic contents (Figure 3C to F).

SAMPLE ACQUISITION

The extensive ramification of the PNS and the selective vulnerability of various PNS sites to toxic injury[7] require that samples be collected at multiple sites. The nerves and ganglia must be handled gently by one end to avoid tension-induced damage (particularly to myelin). The end used for sample transfer should be clearly identified since it will be crushed and thus will not be suitable for microscopic analysis.

In general, peripheral somatic nerves are routinely taken from the hindlimb during general toxicology and special neurotoxicity studies. In terminal animal studies, the sciatic nerve (proximal and distal portions) is the typical sample acquired for general toxicity screens, while the sciatic nerve sample is supplemented by additional specimens from the tibial (internal popliteal), common peroneal (external popliteal, fibular), sural, and plantar nerves (Figure 4) in specific neurotoxicity studies. These regions have inherently different susceptibilities.[8,9] In human clinical cases, suspected cases of neurotoxicity may be confirmed using a sural nerve biopsy, as this nerve can be sampled with minimal long-term impact on peripheral nerve function. For experimental studies in animals, some neuropathologists prefer to leave the nerve in the carcass so that the delicate PNS tissues are fixed without any possibility of distortion during removal, while others remove nerves at necropsy and then attach them to a flat card by stapling their ends prior to post-fixation. Both methods are suitable for screening experiments.[10] Unfixed nerve tissue will generally adhere to the card, due to the stickiness of the collagen in the epineurium, while perfusion-fixed nerves must be attached using pins or staples placed near the cut ends. In general, nerves should be embedded in both cross and longitudinal orientations.

Methods for sampling PNS ganglia vary somewhat with the experimenter's preference and the question being asked. Some pathologists process cranial nerve ganglia and DRG in situ and then evaluate the neural tissues after decalcification of the encasing bone (skull or vertebral column, respectively). The primary advantage of this approach is that the ganglia and their connections remain intact during tissue acquisition and processing. However, a major disadvantage is that prolonged decalcification may induce artifacts in ganglion cell morphology that might masquerade as toxicant-induced changes. A second disadvantage when sampling DRG is that few ganglia are available for examination, due to the difficulty in obtaining sections of vertebral column that contain more than a few DRG while still providing an acceptable orientation for the spinal cord. Other pathologists prefer to remove multiple DRG and either embed them in a single block (all spinal cord regions together) or embed several DRG from a given vertebral segment (cervical, thoracic, lumbar) in separate blocks. Regardless, a good practice is to assess ganglion cell morphology in at least two-step sections taken at least 200 μm apart (since the variable size of ganglia prevents them

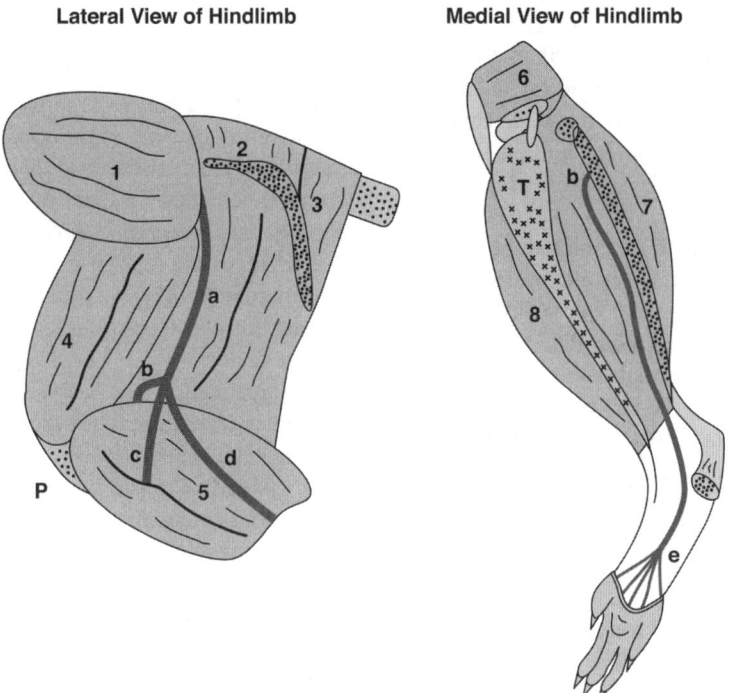

Lateral View of Hindlimb　　　**Medial View of Hindlimb**

FIGURE 4 Location of principal peripheral nerves to be evaluated for detailed toxicological neuropathology studies. *Muscles*: 1, gluteus medius; 2, biceps femoris; 3, semitendinosus; 4, quadriceps femoris; 5, gastrocnemius lateralis; 6, rectus femoris; 7, gastrocnemius medialis; 8, tibialis cranialis. *Nerves*: a, sciatic; b, tibial; c, common peroneal (fibular); d, lateral sural; e, plantar. *Bones*: P, patella; T, tibia. (Adapted from Popesko et al.[24] by G. Krinke for use in Bolon, et al.[22] Reprinted here with the permission of the authors; the original publisher, Saunders; and Dr. Krinke.) (*See insert for color representation of the figure.*)

from all being adequately sampled in a single section). A third option is to dissect and serially section a few specific ganglia (and ideally, their nerve roots); for example, L_4 and L_5 are often selected because they are the largest lumbar DRG and provide the greatest contribution to the rat sciatic nerve.[11–14] In fact, some pathologists only evaluate the lumbar DRG, based on the rationale that assessment of the cervical and thoracic DRG is meaningless because their peripheral nerve trunks are not routinely analyzed. Examination of one cervical (taken from C_4 to C_7) and one lumbar (from L_4 or L_5) DRG should satisfy the minimal requirements for DRG evaluation included in current regulatory guidelines.[15–18]

Elements of the autonomic nervous system (ANS) are seldom taken in a prospective fashion but, instead, are usually evaluated if they remain attached to or encompassed by other organs taken for histopathological examination. The ANS regions that are commonly available for assessment in this manner include the ganglia of the sympathetic chain (particularly the cranial cervical ganglion (superior cervical ganglion) in the neck), the celiacomesenteric ganglionic plexus (a sympathetic system center) at the base of the intestinal mesentery, and the intramural parasympathetic ganglia within the walls of many viscera (especially the intestine). These ganglia harbor the postganglionic (second)

neurons linking the ANS centers in the CNS with the periphery. When specific sampling of ANS nerves and ganglia is required, specimens are harvested and processed in a fashion similar to that of other PNS sites.

PROCESSING PNS SPECIMENS

Nerves and PNS ganglia require special handling procedures to ensure that morphological preservation is adequate for a meaningful examination. Conventional processing used for nonneural tissues—immersion fixation using neutral buffered 10% formalin (NBF), paraffin embedding, and staining with hematoxylin and eosin—is inadequate for PNS tissues as this sequence fails to stabilize the lipid component of myelin, thereby resulting in its extraction during processing and the induction of substantial artifacts (Figure 1E and F).

Fixation

As with other neural tissues, the PNS is best prepared by a multistep fixation protocol. The optimal first step is to undertake, antemortem, whole-body perfusion with a dilute aldehyde solution (NBF or, preferably, 2% to 3.5%

glutaraldehyde) to cross-link proteins. The nerves selected are then removed immediately and post-fixed for 24 to 48 h in fresh aldehyde fixative, followed by additional post-fixation for 1 to 3 h in 1% osmium tetroxide (in either 0.2 M cacodylate or phosphate-buffered saline, pH 7.4) to preserve lipids.[19–21] The aldehyde solution used for the antemortem and postmortem fixation steps need not be identical; for example, acceptable fixation can be attained by perfusion fixation with NBF followed by immersion post-fixation in 3.5% glutaraldehyde. Some investigators prefer immersion fixation for the initial PNS preservation step because the rate at which animals can be processed is greatly increased if perfusion is not employed. In this setting, researchers often choose to supplement the primary immersion fixative (typically, NBF) with glutaraldehyde (0.5% to 1.0%) to improve PNS preservation should electron microscopy be required (G. Krinke, personal communication). Perfusion fixation with 4% to 5% glutaraldehyde may be suitable for PNS preservation, but such specimens are usually too brittle to be used in routine paraffin embedding or teased-fiber preparations.

Embedding

The current guidance on PNS embedding offered by the various regulatory agencies is not identical. The U.S. Environmental Protection Agency (EPA) guidelines[16,17] require that PNS tissues be embedded in plastic, while the Organisation for Economic Co-operation and Development (OECD) guidelines[15,18] allow paraffin processing when the tissues have been properly fixed and carefully handled. The EPA insistence on hard plastic [e.g., epoxy resin (Epon)] is based on the historical requirement that a PNS ultrastructural examination be included as a component of the toxicological neuropathology evaluation in rodents. One-µm-thick hard plastic sections afford superior resolution of fine cellular detail, particularly for subtle changes induced at lower toxicant doses relative to that offered by conventional paraffin sections (5 to 8 µm thick). The OECD flexibility on PNS embedding stems from the recognition that special molecular stains to explore neurotoxic mechanisms are often desirable, but are possible only if tissues have been encased in soft plastic [e.g., glycol methacrylate (GMA)] or paraffin. Soft plastics provide sufficient PNS reinforcement for screening axonal and myelin morphology at light-microscopic magnifications, thus avoiding the high cost (in labor and money) of ultrastructural processing. A recent publication has reported that the resolution of PNS structures in soft plastic sections (usually cut at 2 µm) is comparable to that of paraffin sections[22] and has recommended that PNS tissues be embedded in paraffin in venues where guidelines permit this choice. Obviously, PNS tissues must continue to be embedded in plastic where required by existing regulations.

Staining

The architecture of myelinated axons is adequately evaluated in 1-µm-thick plastic sections stained with toluidine blue, although thicker (1.5 µm) sections are often cut to achieve improved contrast. Uranyl acetate can be used to further improve contrast for electron microscopy.

Teased-fiber preparations are useful to examine individual fibers over longer distances. Such preparations allow for the inspection of both axonal and myelin integrity, as well as evaluation of such cellular features as the internodal distance. These factors are often critical for discerning between states of demyelination and remyelination or axonal atrophy and axonal regeneration. The nerves are fixed with an appropriate aldehyde, isolated (ideally after fixation), and then separated into individual nerve bundles (fascicles) by removing the epineurial connective tissue sheath.[19–21,23] The bundles are then rinsed in phosphate-buffered sucrose and post-fixed in 1% osmium tetroxide. After removal of osmium and rinsing with phosphate-buffered sucrose, the bundles are dehydrated in graded alcohols and placed in cedar wood oil or glycerin for storage. Individual myelinated axons are teased apart in cedar wood oil or glycerin under a stereomicroscope, mounted on a slide, dried overnight between 63°C and 68°C to remove excess oil, and then cover-slipped with Permount.

REFERENCES

1. Kandel ER, et al. *Principles of Neural Science*, 4th ed. New York: McGraw-Hill; 2000.

2. Williams PL, Warwick R. *Gray's Anatomy*, 36th ed. Philadelphia: W.B. Saunders; 1980.

3. Paxinos G, Mai JK. *The Human Nervous System*, 2nd ed. San Diego, CA: Academic Press; 2004.

4. FitzGerald MJT, et al. *Clinical Neuroanatomy and Neuroscience*, 5th ed. Philadelphia: W.B. Saunders; 2007.

5. Evans HE. *Miller's Anatomy of the Dog*, 3rd ed. Philadelphia: W.B. Saunders; 1993.

6. Paxinos G. *The Rat Nervous System*, 3rd ed. San Diego, CA: Academic Press; 2004.

7. Anthony DC, et al. Toxic responses of the nervous system. In: Klaassen CD, ed. *Casarett & Doull's Toxicology: The Basic Science of Poisons*. New York: McGraw-Hill; 2002:535–564.

8. Krinke G, et al. Differential susceptibility of peripheral nerves of the hen to triorthocresyl phosphate and to trauma. *Agents Actions*. 1979;9:227–231.

9. Krinke GJ, et al. Detecting necrotic neurons with Fluoro-Jade stain. *Exp Toxicol Pathol*. 2001;53:365–372.

10. Kasukurthi R, et al. Transcardial perfusion versus immersion fixation for assessment of peripheral nerve regeneration. *J Neurosci Methods*. 2009;184:303–309.

11. Devor M, et al. Proliferation of primary sensory neurons in adult rat dorsal root ganglion and the kinetics of retrograde cell loss after sciatic nerve section. *Somatosens Res*. 1985;3:139–167.

12. Paul I, Devor M. Completeness and selectivity of ricin "suicide transport" lesions in rat dorsal root ganglia. *J Neurosci Methods*. 1987;22:103–111.

13. Schmalbruch H. The number of neurons in dorsal root ganglia L4–L6 of the rat. *Anat Rec*. 1987;219:315–322.

14. Aldskogius H, et al. Selective neuronal destruction by *Ricinus communis* agglutinin I and its use for the quantitative determination of sciatic nerve dorsal root ganglion cell numbers. *Brain Res*. 1988;461:215–220.

15. Organisation for Economic Co-operation and Development. *Neurotoxicity Study in Rodents*. Test Guideline 424. Paris: OECD; 1997.

16. US Environmental Protection Agency. *Health Effects Test Guidelines: Developmental Neurotoxicity Study*. OPPTS 870.6300. Washington, DC: US EPA; 1998.

17. US Environmental Protection Agency. *Health Effects Test Guidelines: Neurotoxicity Screening Battery*. OPPTS 870.6200. Washington, DC: US EPA; 1998.

18. Organisation for Economic Co-operation and Development. *OECD Guideline for the Testing of Chemicals: Developmental Neurotoxicity Study*. Guideline 426. Paris: OECD; 2007.

19. King R. *Atlas of Peripheral Nerve Pathology*. London: Hodder Arnold; 1999.

20. Graham DI, Lantos PL. *Greenfield's Neuropathology*, 7th ed. London: Hodder Arnold; 2002.

21. Dyck PJ, Thomas PK, eds. *Peripheral Neuropathy*. Philadelphia: W.B. Saunders; 2005.

22. Bolon B, et al. A "best practices" approach to neuropathologic assessment in developmental neurotoxicity testing—for today. *Toxicol Pathol*. 2006;34:296–313.

23. Krinke GJ, et al. Teased-fiber technique for peripheral myelinated nerves: methodology and interpretation. *Toxicol Pathol*. 2000;28:113–121.

24. Popesko P, et al. *Colour Atlas of Anatomy of Small Laboratory Animals*, Vol 2. London: W.B. Saunders; 2003.

17

PATHOLOGY METHODS IN NONCLINICAL NEUROTOXICITY STUDIES: EVALUATION OF MUSCLE

BETH A. VALENTINE
Department of Biomedical Sciences, College of Veterinary Medicine, Oregon State University, Corvallis, Oregon

JOHN W. HERMANSON
Department of Biomedical Sciences, College of Veterinary Medicine, Cornell University, Ithaca, New York

INTRODUCTION

Adequate evaluation of skeletal muscle forms an essential part of many types of studies. Skeletal muscle as a tissue has a vital role in maintaining respiration, locomotion, and posture. Skeletal muscle is also involved in the function and regulation of many endocrine hormones and of body metabolism. The response of skeletal muscle to insulin and the role of muscle in total body glucose metabolism in health and disease is an active field of research.[1] There is also a great deal of active research into mechanisms of, and treatments for, muscle fiber atrophy related to age, space travel, and disease.[2–5] Various pharmaceuticals can result in drug-induced myotoxicity, leading to research into mechanisms of muscle cell injury.[6–8] These are but a few examples to illustrate the importance of skeletal muscle in toxicological pathology in general. With respect to neurotoxicity studies, skeletal muscle is an important ancillary specimen based on its function as an effector end organ under direct neural control.

Skeletal muscle has unique features that affect sample selection, handling, and interpretation of findings. If samples are not properly handled, sectioned, and stained, the ability to obtain valid interpretation from sections will be severely impaired. Gross, microscopic, and ultrastructural features of normal skeletal muscle have been described elsewhere.[9–12]

Understanding normal physiological features of muscle is also vital to interpretation of muscle lesions but is not described in this chapter. Particularly important terms and anatomic and physiologic features are summarized in Table 1.

MUSCLE FIBER TYPES

Fiber-type properties were first elucidated during the 1970s[13] and soon became important in the evolution of our understanding of neuromuscular recruitment during posture, or locomotion, and of neuromuscular disease. Skeletal muscle can be designated as red (slow-twitch = oxidative = type I) and white (fast-twitch = glycolytic = type II) muscle. These visual correlates are really applicable only to some tasty avian species, and thus they are considered inappropriate scientific terms. Nevertheless, fiber-type properties have been useful as diagnostic aids for a number of neuromuscular disorders, and understanding such correlations has yielded insight into several diseases of the domestic mammals that may be applied as models of human disease.

What are the basic fiber types, and how are they assessed? At a basic level, myosin adenosine triphosphatase (ATPase) activity correlates with the speed at which muscle fibers shorten.[14] Analysis of myosin adenosine triphosphatase

Fundamental Neuropathology for Pathologists and Toxicologists: Principles and Techniques, First Edition. Edited by Brad Bolon and Mark T. Butt.

TABLE 1   Summary of Pertinent Skeletal Muscle Terminology, Anatomy, and Physiology

Terminology

- Myofiber = muscle fiber = skeletal muscle cell = myocyte.
- The term *fiber* is also appropriate and is used interchangeably with *myofiber* in this chapter.

Anatomical and Physiological Features

- Skeletal muscle cells are multinucleate cells with local nuclear domains allowing for development and maintenance of remarkably long cells. Pathological alterations can affect myofiber segments and not necessarily affect the entire cell.
- Myonuclei are peripherally located in fibers except during embryologic development, during early stages of regeneration (except for rodents, where internal nuclei are retained following regeneration), or following some types of chronic injury leading to chronic myopathic change. Any nucleus located one nuclear diameter or greater from the plasma membrane is considered an internal nucleus.
- Skeletal muscle cells have associated stem cells (satellite cells = resting myoblasts) that, with the proper environment (most important basal lamina integrity), allow for complete regeneration following necrosis of parts, or all, of the cell.
- Fiber diameter can vary depending on fiber type. Myofiber diameter is dependent on an intact associated peripheral nerve.
- Skeletal muscle cells have different fiber types that correspond to different physiological functions. The fiber-type composition of each muscle contributes to its function. Myofiber type is dependent on the electrical activity of the associated nerve. Fiber damage can be fiber-type selective, which can affect proper muscle sampling for diagnosis.
- Fiber type can change following denervation if reinnervation is by a nerve with different electrical activity. Fiber type can also be changed, but to a much lesser degree, by exercise conditioning and, occasionally, in association with chronic myopathic conditions.

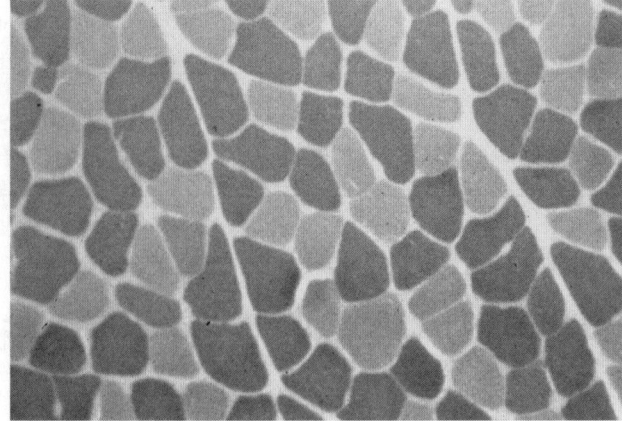

FIGURE 1 Fiber types are intermixed within skeletal muscle fascicles, thereby forming a mosaic pattern. In this alkaline-preincubated ATPase--reacted frozen section of canine muscle, type II fast-twitch fibers are dark and type I slow-twitch fibers are light.

Type II fibers can be subdivided into forms traditionally labeled IIA or IIB. The IIA fibers are fast-twitch, but also have relatively high aerobic potential [assayed with either a nicotinamide adenine dinucleotide hydrogenase–tetrazolium reductase (NADH-TR) or an succinate dehydrogenase (SDH) reaction for mitochondrial content; Figure 2] as well as high anaerobic potential. Thus, the IIA fibers tend to be fast, fatigue-resistant, and to have some capacity for

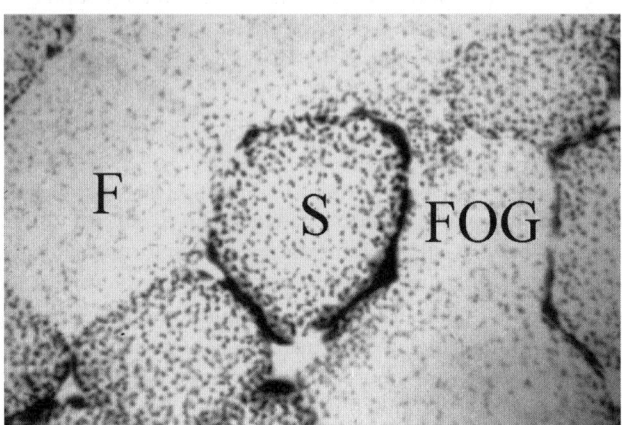

FIGURE 2 Fiber types can also be characterized by mitochondrial content. In this frozen section of cat muscle processed for a nicotinamide adenine dinucleotide hydrogenase (NADH) preparation, type I slow-twitch fibers (S) have the highest mitochondrial density, type IIB fast-twitch glycolytic fibers (F) have the fewest mitochondria, and type IIB fast oxidative-glycolytic fibers (FOG) have intermediate mitochondrial content. The difference in fiber type–specific diameter is readily apparent. This variation in mitochondrial content and fiber size is found in most species except the dog, where fibers are of relatively uniform diameter and all exhibit a high mitochondrial density.

(mATPase) is performed on frozen cryostat sections.[9,10,13] Serial sections of a muscle can be studied after preincubation at either alkaline (pH 9.4 to 10.4) or acidic (pH 4.2 to 4.6) levels in a suitable buffer. Subsequent reaction of the muscle sections with ATP at pH 9.4 yields a positive (darkly stained after a succession of steps, including ammonium sulfate as an end stage) result for muscle fibers with mATPase activity that remains stable after the preincubation and incubation stages, and yields a negative [unstained (pale)] result for fibers with mATPase that is labile after conclusion of the preincubation treatment (Figure 1). Type I slow-twitch fibers are generally stable after preincubation at lower pH levels (pH 4.3) but are alkaline labile at higher (pH 9.4 to 10.4) levels. Type II fibers are fast-twitch and are generally labile at higher acidic preincubation levels (e.g., pH 4.4 to 4.5) and are stable at alkaline preincubation levels (pH 9.4 to 10.4).

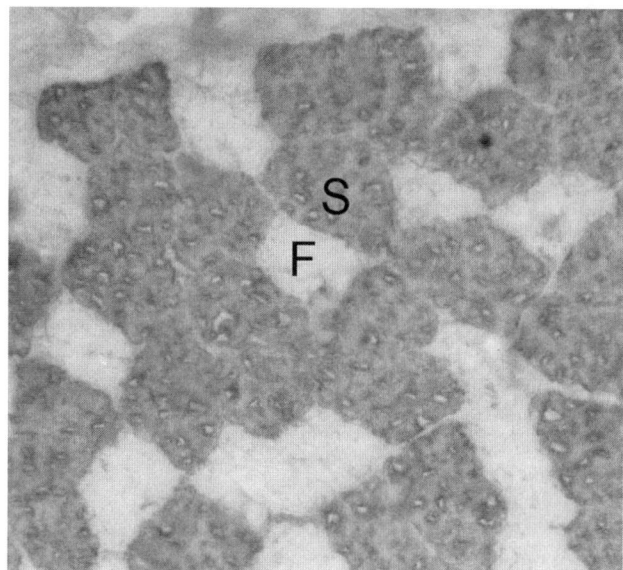

FIGURE 3

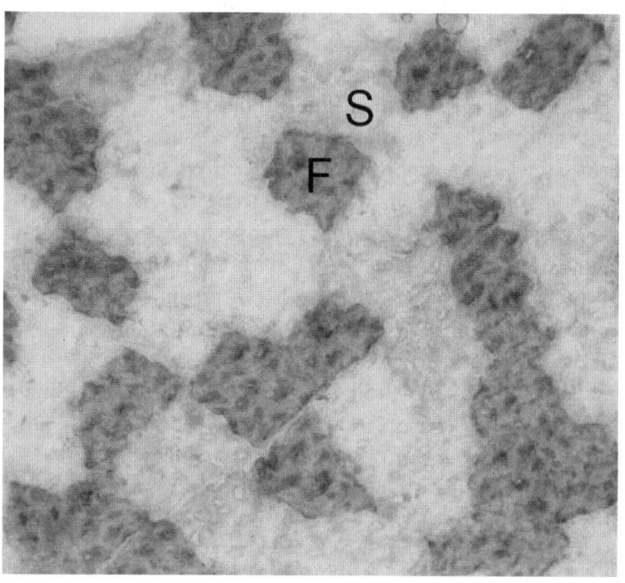

FIGURE 4

FIGURE 3 and FIGURE 4 Immunohistochemistry employing antibodies to demonstrate slow and fast myosin heavy chains can be performed, in this case on frozen muscle sections from a dog. Type IIA fast-twitch (F) and type I slow-twitch (S) fibers are identified within two serial sections.

short-term sprint activity (fast-oxidative-glycolytic fibers). In contrast, the IIB fibers are subject to fatigue (low activity demonstrated with the NADH-TR assay; Figure 2) and are relatively well suited for glycolytic or sprintlike performance. The IIB fibers are morphometrically the largest of the fiber types, having a larger cross-sectional area than most type I or IIA fibers, and thus are more capable of generating large forces.

Work in the last 20 years has demonstrated a correlation of the histochemical properties described above with reaction against specific antibodies to type I, IIA, and IIB myosin (Figures 3 and 4) as well as correlation with isoforms that can be isolated using electrophoretic protocols. Other fiber types have also been identified, including a IIX fiber that demonstrates fast-twitch aerobic force profiles and a type IIM or "superfast" fiber of uncertain function. The IIM fibers are found in specific muscles, such as the masticatory muscles of carnivores and in the extraocular muscles of many mammals, and have been associated with immune-based disease states in dogs and other carnivores that present with an inability to open the jaw.[11,12]

The usual flow of tissue collection during clinical and experimental studies often makes it necessary to study skeletal muscle samples subsequent to routine formalin fixation. Several myosin antibodies are not effective for muscle tissue after formalin fixation; however, selection of formalin-tolerant antibodies has been broadly applied. Antibodies to both type I slow and type II fast myosin are available which perform well for fiber typing of formalin-fixed muscle from rodents, cats, and nonhuman primates (Table 2).[15–18] There is limited information on the performance of these antibodies in other species. Only the anti-type I slow antibody appears to perform acceptably in formalin-fixed muscle sections from dogs. Preliminary studies by one author (B.A.V.) suggest that these antigens may be sensitive to postmortem change and delayed fixation, such as can occur in the center of muscle samples fixed in inappropriately large blocks.

SIGNIFICANCE OF SKELETAL MUSCLE FIBER TYPE IN TOXICOLOGICAL PATHOLOGY

Most skeletal muscles in all species have a mixed fiber type population. The fiber types are interspersed apparently at random, resulting in a mosaic pattern (Figures 1 to 4). Determination of fiber-type involvement (specific types affected, or all types involved) is very important for diagnostic purposes in disorders resulting in necrosis, atrophy, or hypertrophy. Detection of alterations in the normal mosaic pattern of fiber types is vital when studying disorders that result in denervation due to either peripheral neuropathy or motor neuronopathy (see the section "Atrophy" below). The most commonly encountered alteration in the fiber-type mosaic pattern is to find fiber groups all composed of the same fiber type, known as *fiber-type-grouping*. This lesion is virtually pathognomonic for skeletal muscle denervation followed by reinnervation.

TABLE 2   Examples of Antibodies for Detecting Myofiber Types by Immunohistochemistry

Antibody	Antigen Specificity	Species	Required Processing	Source
MY-32	Adult and neonatal fast myosin	Most vertebrates	Frozen section or formalin-fixed, paraffin-embedded	Sigma Chemical http://www.sigmaaldrich.com
Myosin heavy chain-Slow	Adult slow myosin	Mammals	Frozen section	Vector Laboratories http://www.vectorlabs.com
Fast	Adult fast myosin	Mammals	Frozen section	Vector Laboratories
Neonatal	Neonatal myosin	Mammals	Frozen section	Vector Laboratories
S58	Adult slow myosin	Birds, zebrafish, some mammals	Immunoblotting; frozen section	Developmental Studies Hybridoma Bank http://dshb.biology.uiowa.edu
NOQ7.5.4D	Adult slow myosin	Most mammals and birds	Formalin-fixed, paraffin-embedded	GeneTex http://www.genetex.com

SKELETAL MUSCLE SAMPLING

Skeletal muscles exhibit wide variation in function, often based on location [e.g., respiratory, masticatory, locomotory, extraocular, and support (trunk muscles)]. This variation is often reflected in differing fiber-type composition or specific myosin isoforms. Appropriate muscle sample selection will thus vary depending on the nature of the study.

Exactly which technique to select will depend chiefly on the nature of the anticipated, or suspected, lesion. Screening studies may involve only a small number of muscles (e. g., diaphragm and limb muscle). When testing xenobiotic compounds that are known or suspected to target skeletal muscle, more extensive and more selective testing should be employed.

Depending on the nature of the compound tested, selection for a type I predominant, a type II predominant (Table 3), or a mixed-type muscle (e.g., diaphragm) may be indicated.[19–23] For example, compounds affecting mitochondrial function or imposing oxidative stress will probably preferentially affect type I oxidative muscle fibers; compounds that might alter muscle glucose/glycogen metabolism, reduce thyroid function, or increase serum cortisol levels will be more likely to selectively affect type II glycolytic muscle fibers.

SKELETAL MUSCLE PROCESSING

Various approaches to preparing skeletal muscle samples are summarized in Table 4. Minimal additional effort is required to prepare appropriate sections that allow for accurate histopathological interpretation.

Whatever the type of study employed, three fundamental principles always apply.

Principle 1: Minimize processing artifacts. Skeletal muscle is a contractile tissue that is prone to severe artifactual change during processing, in particular the appearance of contraction band artifact. Ideal morphology can be achieved by employing frozen-section histopathology; this approach is time- and labor-intensive but will be necessary for many studies (e.g., mATPase histochemistry and immunohistochemistry with certain cell marker antibodies and most antibodies to dystrophin). Samples for frozen section do not need stabilization prior to processing (Figure 5). In contrast, fixed muscle tissue samples,

TABLE 3   Examples of Skeletal Muscles with a Predominance of a Single Myofiber Type

Species	Type I–Predominant Muscles	Type II–Predominant Muscles
Primate (human and nonhuman)	Soleus, intermediate vastus	Lateral vastus
Dog	Deep medial triceps, anconeus, superficial digital flexor	Extensor carpi radialis, lateral digital extensor (hindlimb)
Cat	Deep medial triceps, soleus	Cranial tibial, long head of the triceps, temporalis
Mouse	Soleus	Quadriceps; long digital extensor
Rat	Soleus	Cranial tibial; long digital extensor
Hamster	Soleus	Biceps brachii
Rabbit	Soleus	Psoas, long digital extensor
Guinea pig	Soleus, intermediate vastus	Lateral vastus
Horse	Medial triceps, intermediate vastus, sacrocaudalis dorsalis medialis	Semitendinosus, semimembranosus, lateral triceps, quadriceps

TABLE 4   Comparison of Processing Methods for Skeletal Muscle Samples

Procedure	Initial Processing	Advantages	Disadvantages
Fixed sections	Immersion in neutral buffered 10% formalin followed by paraffin embedding	Readily available, relatively short turnaround, inexpensive, can be performed in simply equipped laboratories, can prepare both longitudinal and transverse sections, multiple sections per slide possible, major classes of infiltrating leukocytes identifiable on routine sections	Prone to severe artifactual change; will not detect denervation and fiber type-specific disorders; cannot support some immunochistochemical methods that require frozen sections
Frozen sections	Snap-freezing in isopentane-cooled in liquid nitrogen	*The* gold standard for skeletal muscle pathology; minimal artifactual change, rapid turnaround, allows for multiple special procedures such as fiber typing, mitochondrial enzyme analysis, and use of many antibodies for routine immunohistochemical methods	Longitudinal sections not possible without artifact; technically difficult to prepare more than one section per slide, therefore increasing the likelihood of sampling error; identification of infiltrating leukocytes more difficult without cell markers; cryosectioning requires additional equipment, reagents, and skills; time- and labor-intensive, therefore expensive
Transmission electron microscopy	Immersion fixation in special liquids (e.g., 4% glutaraldehyde)	Minimal artifactual change, identifies subtle cytoarchitectural abnormalities; aids in determining nature of the injury in early stages of degenerative myopathy	Requires additional fixatives, reagents, equipment, and skills; longer processing time; time- and labor-intensive, therefore expensive
Biochemical analysis	Snap-freezing directly in liquid nitrogen	Protein analyses (e.g., enzyme level/activity, especially in mitochondria); glycogen analysis	Requires additional laboratory reagents, equipment, and skills
Molecular biology analysis	Snap-freezing directly in liquid nitrogen	Detect RNA, DNA, protein levels	Requires additional laboratory reagents, equipment, and skills

whether allocated for light microscopy or electron microscopy, require stabilization of the sample ends to minimize contraction during handling and immersion fixation (Figures 6 to 8). Stabilization of the sample to avoid artifact will also aid in achieving principle 2. (See the section "Fixed Tissues" below for more details.)

Principle 2: Achieve proper alignment of transverse and longitudinal sections. Samples must be aligned such that transverse and longitudinal sections can be obtained. Transverse sections allow evaluation of the entire cross section of muscles in small animals, demonstrate the largest number of individual fibers (thus enhancing overall sampling of the muscle being examined), and provide the best orientation for assessing the extent of a lesion within the entire muscle.

Precisely oriented transverse sections are also necessary for accurate determination of fiber size and shape, and appropriate examination of many cytoarchitectural changes (especially internal nuclei) and the fiber-type distribution. Evaluation of longitudinal sections aids in confirming the presence of segmental myofiber necrosis and regeneration as well as in determining the extent of damage along the length of the individual muscle fibers.

Principle 3: Avoid the vast pink wasteland. Few things are harder on a pathologist than evaluating the pink expanse that characterizes routine hematoxylin and eosin (H&E)–stained sections of fixed skeletal muscle (Figure 9). One of the advantages of frozen sections is the increased contrast between eosinophilia and basophilia obtained with H&E

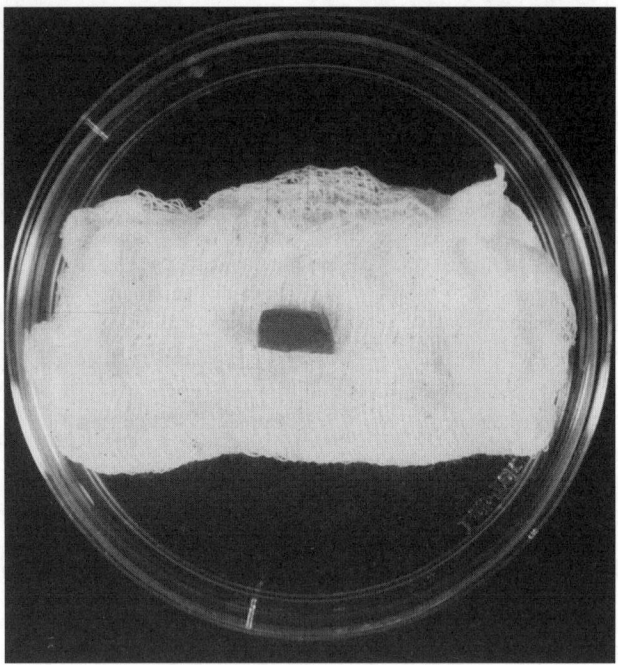

FIGURE 5 Muscle samples designated for frozen sectioning must be kept from drying out during processing. An isolated strip of skeletal muscle wrapped in a gauze sponge (barely moistened in saline, pH 7.4) and placed within a petri dish can be kept at 4°C (on ice or in a refrigerator) for up to 24 h before processing.

staining (Figure 10); another is the availability of additional histochemical procedures offering even greater contrast, especially modified Gomori's trichrome stain. Commonly employed procedures performed on formalin-fixed and on frozen sections of skeletal muscle are compared in Table 5. Details of such procedures can be found in standard textbooks.[9,10]

When evaluating skeletal muscle (and nervous tissue) in formalin-fixed, paraffin-embedded sections, at the very least the composition of the H&E stain should be adjusted to enhance basophilia of the stained sections. One author (B.A.V.) has found that replacing the last acid–alcohol rinse with a tap water rinse is very useful in this regard. An even greater degree of contrast can be achieved by utilizing polychrome stains, such as Masson's trichrome stain (Figure 11). This stain actually permits even better evaluation of intramuscular nerves than does H&E, due to the enhanced contrast between the red-stained myelin and the blue-dyed endoneurial collagen. The amylase-digested section of the periodic acid-Schiff (PAS) stain for glycogen can also be very useful for detecting myofiber margins and myopathic change (Figure 12).

Fixed Tissues

Minimizing artifact in fixed skeletal muscle can be accomplished by either high- or low-tech means.

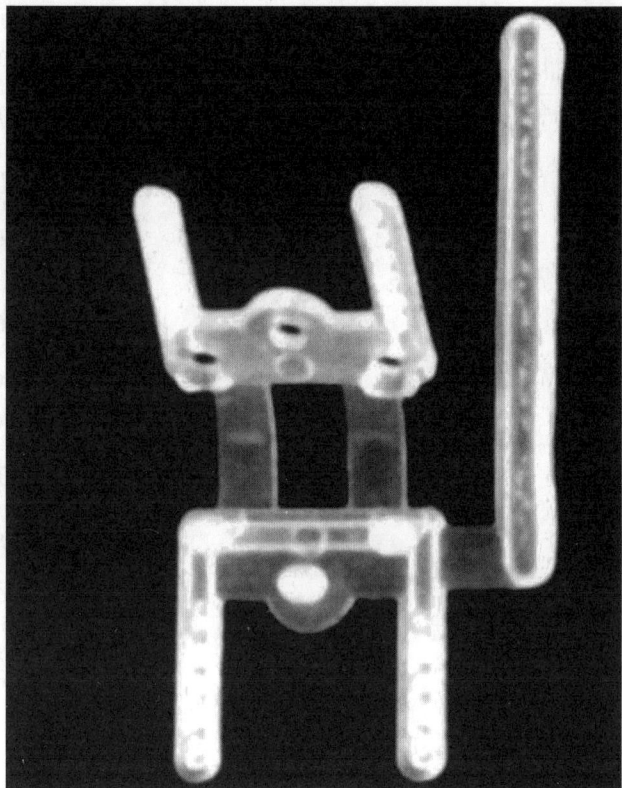

FIGURE 6 To avoid contraction band artifact when fixing muscle samples, specimens for routine light microscopy or ultrastructural examination require stabilization of their ends to maintain fiber length during fixation.

High-tech methods involve specially designed muscle clamps that can immobilize myofibers completely during the fixation step (Figure 6). A strip of muscle with fibers oriented longitudinally is obtained using incisions paralleling the length of the fibers. The arrangement of muscle fibers within the muscle will dictate how best to orient incisions that parallel the orientation of muscle fibers; the prosector must be aware of the contrasting fiber arrangements in parallel-fibered muscles, in which most fibers are aligned with the long axis of the muscle (e.g., biceps femoris, lateral vastus) and pinnate-fibered muscles, in which fibers are angled

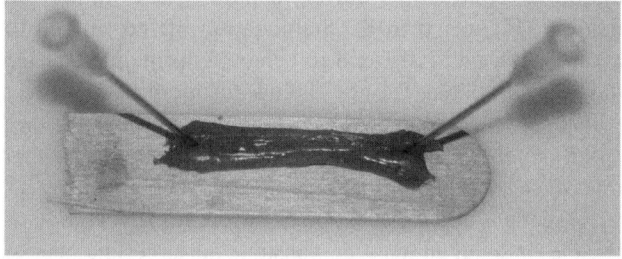

FIGURE 7 When muscle clamps are not available, muscle strips can be pinned onto a portion of wooden tongue depressor to maintain fiber length during fixation.

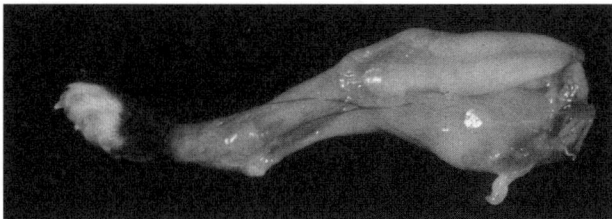

FIGURE 8 The simplest means of maintaining fiber distension when examining limb muscles of small animals is to fix the entire limb following removal of the skin. The bone serves as an internal stabilizing system.

acutely relative to the long axis of the muscle and may be packed into a herringbone arrangement (e.g., gastrocnemius or the equine superficial digital flexor). Following placement of the two initial longitudinal incisions, the strip is undermined using blunt dissection instruments, and the muscle clamp is applied to isolate the semidetached strip. The sample ends are not transected until after the strip has been clamped, thus maintaining the necessary tension on the stretched fibers to avoid contraction-related artifacts.

Low-tech methods involve a collection technique similar to that described above, with removal of a strip of muscle that is then pinned or sutured onto a portion of wooden tongue depressor before fixation (Figure 7). If multiple muscle samples are obtained, the sample site can be written on the plastic muscle clamp or on the piece of tongue depressor, allowing all samples to be fixed in the same container of fixative. A truly low-tech method, but one that is very effective for limb muscles of small animals such as rodents or neonatal dogs, cats, and rabbits, is to remove the skin and fix the entire limb, allowing the bone origin and insertion to stabilize the muscle during fixation (Figure 8).

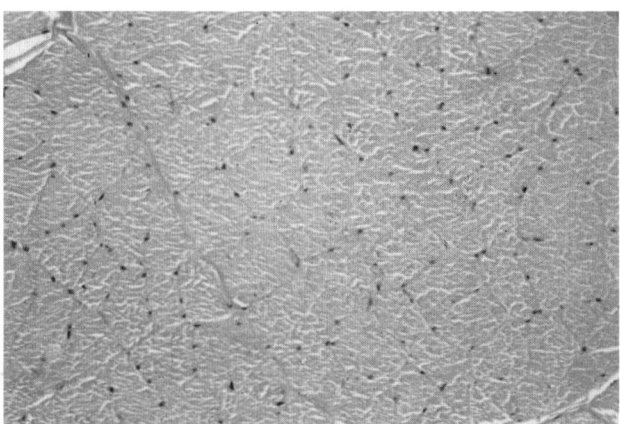

FIGURE 9 Routinely processed (formalin-fixed, paraffin-embedded) skeletal muscle stained with H&E tends to lack cellular detail. Cellular margins are partially to totally indistinct, thus making meaningful interpretation of such sections difficult.

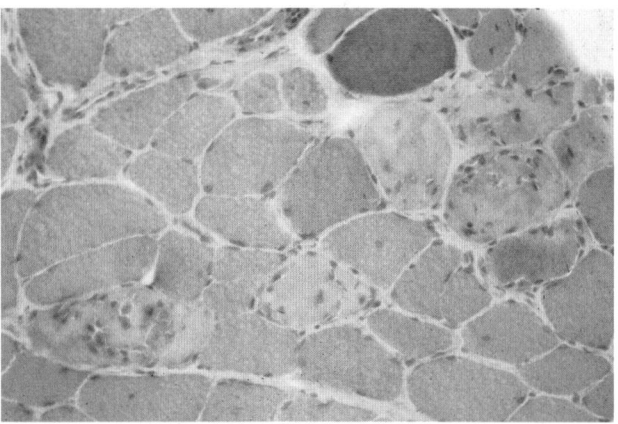

FIGURE 10 Frozen sections of muscle stained with H&E exhibit clarity and substantial morphologic detail, including myofiber margins. This section of muscle from a dog with dystrophin-deficiency muscular dystrophy contains many lesions, including excessive fiber size variation, presence of internal nuclei indicative of chronic myopathic change, large-diameter darkly stained ("large dark") fibers—thought to represent an early stage of necrosis—and pale-stained fibers undergoing coagulation necrosis and macrophage infiltration.

Regardless of the method employed, the preferred fixative for preserving skeletal muscle samples for routine analysis is commercial-grade neutral buffered 10% formalin. The ideal fixative for ultrastructural examination is 4% glutaraldehyde (pH 7.2) formulated using either 0.1 M phosphate or 0.1 M cacodylate as the buffer.[10]

Frozen Sections

Skeletal muscle may be the one tissue for which routine examination is best served by evaluating unfixed frozen sections. Nevertheless, frozen sections are prone to their own types of artifact (e.g., freeze artifact) and thus require special handling to minimize the risk of processing-induced damage. Equipment needed for freezing skeletal muscle samples for histochemistry are listed in Table 6.

Samples for frozen sections should be kept moist and at refrigerator temperature or on ice (~4°C) prior to freezing. Isolated samples should be wrapped loosely in gauze sponges that have *just barely* been dampened with saline, or in sponges without any saline at all; retention under conditions that are too dry is better than having them too wet. Wrapped samples are kept in a closed container (Figure 5) until processing. Appropriately handled samples can be kept at refrigerator temperature for up to 24 h before freezing. Be aware, though, that labile compounds such as glycogen and some enzymes (e.g., phosphofructokinase) will be depleted during storage. Therefore, if samples must be stored prior to freezing, appropriate control specimens must be handled in the same manner.

TABLE 5 Comparison of Common Processing and Light-Microscopic Procedures Performed on Skeletal Muscle Sections

Type of Processing	Procedure[a]	Use	Value
Routine processing	H&E	Routine stain for muscle histopathology evaluation in neurotoxicity studies	Barely adequate for skeletal muscle assessment. Internal nuclei will be visible, but indistinct myofiber margins often preclude meaningful morphometric measurements.
	Masson trichrome, Gomori trichrome	Provides more intense contrasting colors; muscle fibers and myelinated nerve fibers within intramuscular nerves are red, endoneurial collagen is blue (Masson) or green (Gomori)	Very valuable, especially for detection of denervation atrophy.
	Reticulin	Identifies fiber margins by staining reticulin mesh between myofibers	Useful, but trichrome or PAS with amylase digestion are just as good and provide more information.
	PTAH	Identifies cross striations in muscle fibers	Not very useful.
	PAS–glycogen procedure	Identifies glycogen and polyglucosan bodies in sarcoplasm	Used for detection of glycogen storage disorders and glycogen depletion; amylase-digested sections very useful for determining fiber size and the presence of internal nuclei. *Note:* Glycogen is highly prone to artifactual depletion, due to postmortem change or slow fixation.
	Alizarin red S, von Kossa	Identifies mineral	Detects myofiber mineralization in degenerative myopathies.
	Immunohistochemistry	Identifies epitopes of selected cytoskeletal and extracellular proteins of myofibers and infiltrating cells	Useful for evaluation of membrane and cytoarchitectural alterations and identification of infiltrating cells.

	Stain	Purpose	Comments
Frozen sections	H&E	Routine stain	Useful for evaluating overall architecture of fibers and presence of inflammation; has much more contrast than in routinely processed (fixed, paraffin-embedded) muscle sections.
	Modified Gomori's trichrome	Routine stain	Ideal for overall evaluation of muscle. Mitochondrial abnormalities can be detected, and collagen and nerves are readily identified.
	NADH, SDH	Stains to demonstrate mitochondrial enzymes	Detect changes in mitochondrial number and density; detect cytochrome oxidase deficiency.
	ATPase	Distinguishes between different myofiber types	Invaluable for detection of denervation and reinnervation; often utilized for fiber morphometry; detects fiber type selectivity in degenerative or atrophic disease.
	Acid phosphatase	Detects denervated fibers and macrophages	Useful, but not performed routinely by the authors.
	Nonspecific esterase	Detects macrophages and neuro-muscular junctions	Helps to distinguish primary inflammation (which typically has multiple cell types) from macrophage-rich infiltration as a secondary scavenging response following primary necrosis.
	Oil red O, sudan black	Detects lipid	Important for detection of lipid-storing myopathies, which can be induced by toxicants. Can be performed on frozen sections of routinely processed samples as well as on frozen sections. Intramyofiber lipid is best detected in nonfixed frozen sections.
	Alizarin red S, von Kossa	Identifies calcium (ARS) and mineral (both)	Alizarin red S can detect early calcium influx in degenerative myopathies.
	PAS	Same as for routine processing	Same as for routine processing.
	Immunohistochemistry	Same as for routine processing	Invaluable for many studies; some antigens and antibodies cannot be demonstrated except in frozen sections.

[a] ATPase, adenosine triphosphatase; H&E, hematoxylin and eosin; NADH, nicotinamide adenine dinucleotide hydrogenase; PAS, periodic acid-Schiff; PTAH, phosphotungstic acid hematoxylin; SDH, succinate dehydrogenase.

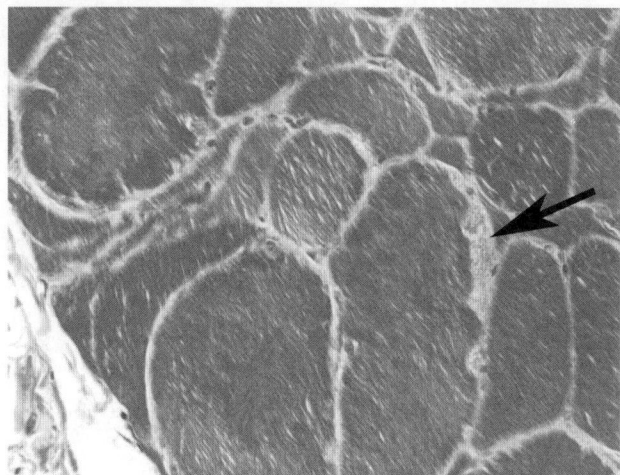

FIGURE 11 Routinely processed (formalin-fixed, paraffin-embedded) muscle sections stained with Masson's trichrome stain exhibit more contrast than is apparent in routinely processed H&E-stained sections (Figure 9). Fiber margins are more readily identified, and alterations in myofiber size and diameter can be readily detected. In this section of denervation atrophy from the sacrocaudalis dorsalis medialis muscle of a horse with motor neuron disease, small contiguous clusters of small myofibers affected by severe angular atrophy are admixed with hypertrophied myofibers, a constellation of changes characteristic of chronic denervation atrophy. There is also chronic myopathic change characterized by excessive fiber size variation and the presence of a pale subsarcolemmal zone (sarcoplasmic mass; arrow) within a hypertrophied fiber.

Skeletal muscle requires rapid and complete freezing to avoid severe freeze artifact.[9,10] Liquid nitrogen is too cold ($\sim -196°C$) to allow direct immersion of samples; freezing in this liquid alone results in gas "bubbles" at the specimen surface that actually slow freezing and promote artifact. One simple means of avoiding this problem is to cover samples in talcum powder prior to freezing. However, the most common procedure is to cool a container of a cryoprotective "quenching" solution such as 2-methylbutane (isopentane) in liquid nitrogen and then immerse the sample in the cooled quenching solution.

It is imperative that skeletal muscle samples destined for cryostat sectioning be positioned to allow cutting in the transverse orientation. Embedding samples in a drop of viscous medium [e.g., optimal cutting temperature (OCT)] or 5% gum tragacanth on blocks of cork allows for ready identification of the muscle sample (by writing on the cork) and storage of multiple samples in one container. Such thick embedding compounds provide a relatively "stiff" foundation in which to position the tissue during preparation and during immersion in cooled quenching solution. Mounting in such thick embedding compounds is also useful in helping the operator to align the muscle accurately for transverse sectioning.

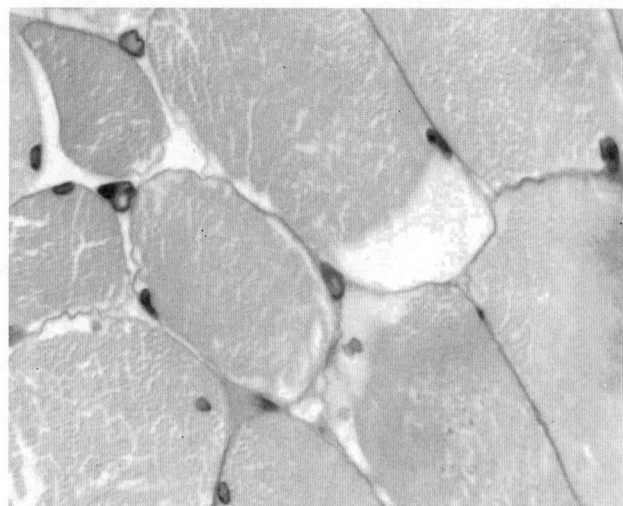

FIGURE 12 Amylase-digested periodic acid-Schiff (PAS)–stained sections of routinely processed (formalin-fixed, paraffin-embedded) muscle also allow for reliable evaluation of myofiber margins as well as fiber size and shape. This section exhibits excessive fiber size variation. Two fibers have internal nuclei, one (lower center region) within a subsarcolemmal clear zone that contained glycogen prior to processing. Section from the semimembranosus muscle of a horse with polysaccharide storage myopathy.

Cryosections are typically picked up on 22-mm^2 number 2 coverslips maintained at room temperature. Cut sections can be stored in a freezer ($-25°C$ or, ideally, $-80°C$) until staining, although it is recommended that the ATPase procedure for fiber typing be performed as soon as possible. Histochemical

TABLE 6 Equipment Needed for Freezing Skeletal Muscle Samples for Histochemistry

- Sealed container (e.g., vial, petri dish, plastic bags, Whirli-bags, etc.) for maintaining a moist environment for stored muscle samples until processed
- Sharp razor blade, EM quality
- Needle-nosed forceps, for handling muscle
- Dissecting microscope, for mounting muscle in transverse orientation
- Laboratory corks, sliced into disks (or other chuck for mounting specimens)
- Viscous tissue-embedding medium
- Quenching solution [e.g., isopentane (2-methylbutane)]
- Liquid nitrogen
- Shatterproof beaker
- Long forceps, for immersing samples in quenching solution
- Insulated gloves (to protect the hand holding the long forceps if multiple samples are to be frozen in one session)
- Scintillation vial or other container for sample storage at -70 to $-80°C$
- Ultralow freezer (-70 to $-80°C$)
- Glass coverslips, to pick up freshly cut cryosections

FIGURE 13 Frozen sections on coverslips can be handled easily using a coverslip carrier (left) and a Columbia staining jar (right).

reactions are performed in Columbia staining jars designed for coverslips (10 mL volume), which reduces the amount of reagents needed (Figure 13). For larger volumes of solutions, multiple coverslips are readily moved from container to container using a coverslip holder (Coors holder, part 8542E40, https://www.thomassci.com/catalog/product/4076; Thomas Scientific, Swedesboro, NJ) (Figure 13). Specific details of histochemical procedures for frozen-section muscle pathology may be found in published texts or by consulting a laboratory with experience in muscle histopathology.[9,10] Some stains available for frozen section histochemistry are also possible on formalin-fixed tissues (Table 5).

Histochemistry of frozen skeletal muscle sections involves a series of reactions on serial sections. Two coverslips with sections can be mounted simultaneously on one glass microscope slide (Figure 14). Alternatively, frozen sections can be mounted directly onto glass microscope slides for reaction in Coplin jars. These will require using more reagents (more fluid), but this may be mitigated by placing the muscle samples toward one end of the glass slide. The glass slide/Coplin jar method is useful when larger muscle samples are being processed.

OTHER PROCESSING TECHNIQUES

Depending on the nature of the study, other procedures can be employed to evaluate toxicant-induced pathology in skeletal muscle.

Biochemistry and Molecular Biology

Muscle samples are snap-frozen in liquid nitrogen. This is accomplished using small (1 mL) microcentrifuge tubes to hold the muscle sample. The tube and sample can be immersed directly in liquid nitrogen and held there for about 10 s to ensure rapid and complete freezing. Alternatively, samples can be frozen by direct immersion in liquid nitrogen and then stored in a suitable container, such as a scintillation vial. Frozen muscle samples can be stored at −80°C until the sample is ready for study.

Electron Microscopy

Clamped specimens are fixed in an appropriate fixative (i.e., 4% glutaraldehyde, pH 7.2) and processed for EM.

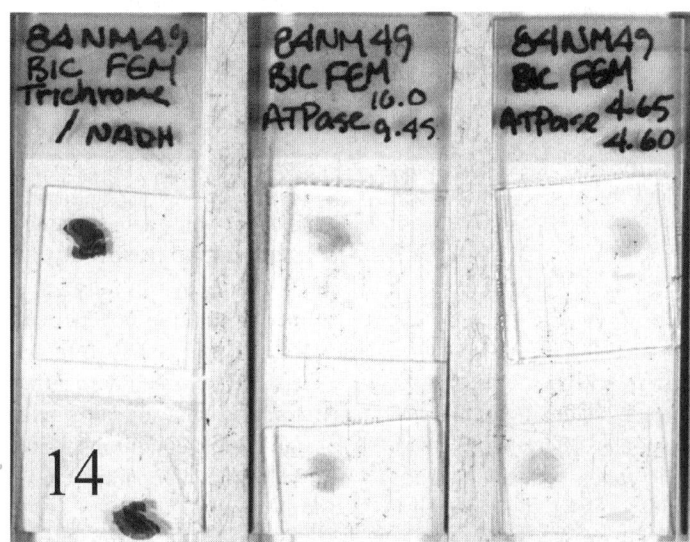

FIGURE 14 Serial frozen sections on coverslips can be mounted two to a slide for more efficient staining and evaluation.

Skeletal muscle is one organ in which transmission or scanning electron microscopy is still an extremely valuable tool. This is particularly true in disorders involving mitochondrial damage and cytoarchitectural alterations.

SUGGESTED APPROACHES TO SAMPLING AND PROCESSING

Studies with No Known Myotoxicity

Minimal sampling and routine processing of all organs and tissues can be employed. However, the investigator should still consider sampling multiple muscles (e.g., diaphragm, trunk, masticatory, proximal *and* distal limb) so that specimens taken include both type I- and type II–predominant muscles (Table 3). Samples should be stabilized during fixation (as described in the section "Fixed Tissues"). At least one transverse section and one longitudinal section from each muscle should be prepared from paraffin-embedded samples. Serial sections should be stained with H&E and amylase-digested PAS to evaluate the basic architecture of muscle fibers and intramuscular nerves. For large, general toxicity studies, conventional paraffin-embedded, H&E-stained sections are a suitable means to provide minimum screening for overt microscopic lesions in skeletal muscle.

Studies in Which Myotoxicity Is Suspected or Confirmed

Multiple samples from muscles representing both predominant fiber types should be obtained (see above and Table 3). In addition to routinely processed paraffin-embedded samples, unfixed specimens from adjacent muscle fields should be taken and divided, with one portion allocated for mounting in viscous medium (e.g., OCT) in preparation for cryosectioning and the other snap-frozen directly in liquid nitrogen for possible biochemistry or molecular biology assays. Acquisition, fixation, and storage of additional small specimens should be considered for transmission electron microscopy.

INTERPRETATION OF SKELETAL MUSCLE SECTIONS

It has often been said that skeletal muscle has a limited range of response to injury but, in fact, skeletal muscle is a remarkably plastic tissue that responds differently to a wide array of altered physiological conditions and insults. Detailed descriptions of potential skeletal muscle lesions are available in excellent standard references.[9–12] Those muscle lesions most typically encountered during toxicological pathology examinations are summarized in this section.

Skeletal muscle samples from toxicity studies that have been processed for light microscopy typically allow for identification of several prominent changes. For toxicity studies, detecting myofiber degeneration/necrosis and regeneration, fiber atrophy, mitochondrial dysfunction, and fiber type–specific disorders will be the most regularly seen and diagnostically important findings. Inflammation as a consequence of xenobiotic-induced myotoxicity is less likely to be encountered. Fiber type–selective lesions can be detected to some degree in routine toxicity studies, but only by examining samples of type I and type II predominant muscles (Table 2). Detection of mitochondrial changes will require frozen-section histochemistry (Figure 2), electron microscopy, or biochemical analysis.

Myofiber Degeneration or Necrosis and Regeneration

Unless muscle injury is global, such as that stemming from vascular occlusion or a regional crush injury, most degenerative disorders of skeletal muscle will involve segmental degeneration (i.e., visibly altered cell structure without outright cell death) or necrosis of myofibers rather than involvement of the entire fiber length. Characteristics of fiber degeneration include swelling and sarcoplasmic vacuolization, while features of fiber necrosis also encompass sarcoplasmic hypereosinophilia, loss of sarcoplasmic detail, and/or myofiber fragmentation, coiling, and formation of contraction bands.[9–12] Necrosis and fiber fragmentation are followed by macrophage infiltration to clear cellular debris (Figure 10). Necrosis can also recruit a small number of other inflammatory cell types (including neutrophils, lymphocytes, and eosinophils). Therefore, careful evaluation of the pattern and type of infiltrating cells is essential to distinguishing degenerative myopathy with secondary inflammation from primary necrotizing myositis.

Under most conditions, skeletal muscle fibers will regenerate affected segments and completely restore the integrity of the damaged fiber. Conditions that impair this restorative ability are those that cause damage to satellite cells, the basal lamina, or the vasculature. Regeneration of skeletal muscle recapitulates embryologic development. Activation of satellite cells, which act as resting myoblasts, leads to mitosis and fusion to form narrow myotubes with basophilic cytoplasm and chains of centrally located, euchromatic nuclei. With time, maturation of repaired fibers results in restoration of sarcoplasmic eosinophilia and, in most species, peripheral dispersal of the internal nuclei chains. In most species, internal nuclei are only retained in chronic myopathic or neuropathic conditions, and thus they constitute a feature of chronic myopathic change and not regeneration after toxicant-induced injury.

The skeletal muscle of rodents represents an exception to this pattern of changes, as internal nuclei are retained in normal rodent muscle following regeneration. This property of muscle provides a convenient measure of the number of myofibers that had previously undergone necrosis in these species.

Fiber Size and Size Variation

Evaluation for myofiber atrophy and myofiber hypertrophy is an essential feature in most studies of skeletal muscle, and must be performed in transverse sections. In small animals such as rodents, it is possible to remove and weigh entire muscles to evaluate organ weights as potential evidence of overall atrophy or hypertrophy.[24] Other common measures of muscle fiber size include fiber diameter and fiber area.[1,10,19–25]

For determination of individual fiber diameter, it is vital that fiber margins be visualized readily and reliably. Frozen sections prepared for ATPase evaluation are ideal for this purpose (Figure 1); one author (J.W.H.) employs this method routinely. In formalin-fixed, paraffin-embedded skeletal muscle tissue, the fiber margins are often difficult to discern in H&E-stained sections but are generally more readily detected using Masson's trichrome stain (Figure 11). One author (B.A.V.) has found that amylase-digested, PAS-stained sections from routinely processed tissue are even more useful than trichrome stain for detecting fiber diameter and revealing cytoarchitectural changes such as internal nuclei (Figure 12).

Morphometric analysis of fiber dimensions can readily be achieved using computer-based scans and software.[26] One of the authors (J.W.H.) commonly uses easily accessed shareware (NIH Image: http://rsb.info.nih.gov/nih-image/) to measure fiber-size profiles, working one myofiber at a time. This fiber-by-fiber approach is especially useful when the contrast between adjacent myofibers becomes poor and the operator's judgment is necessary to define fiber boundaries more accurately. Appropriate controls are essential when acquiring quantitative data. In particular, fiber size can vary markedly depending on multiple factors, including the muscle examined, species, age, nutritional status, and gender. Although variation is expected to be minimal between individual age-matched rodents, cats, and rabbits, there can be significant disparity in muscle fiber size when comparing different breeds of dogs.[22] Therefore, morphometric evaluations during prospective studies of potential myotoxicity require strict adherence to standard control procedures.

Atrophy

Atrophy of entire muscle fascicles, regardless of cause, tends to maintain a round to polygonal cross-sectional shape in affected myofibers. Detection of overall muscle atrophy often relies on comparison to control specimens. When individual or small groups of myofibers undergo atrophy, compression by adjacent unaffected fibers usually results in a shrunken, elongate fiber with angular margins,

a finding known as *angular atrophy* (Figure 11). Severe angular atrophy of myofibers is a hallmark of denervation, and both type I and type II fibers will be involved. Less severe angular atrophy often affects type II fibers selectively; it develops under a variety of circumstances, including disuse atrophy, endocrine disease, cachexia due to chronic disease, and starvation. When angular atrophy is detected, it will be important to carefully evaluate the associated peripheral nerves and motor neurons in the central nervous system, and to employ a technique to determine the fiber type of the affected myofibers.

Cytoarchitectural Changes

Internal nuclei in nonregenerative fibers can be considered a cytoarchitectural change (Figures 10 and 12). As described in the section "Myofiber Degeneration or Necrosis and Regeneration," internal nuclei are evidence of ongoing fiber regeneration in most species but are often retained in rodent muscle fibers well after regeneration is complete.

Sarcoplasmic masses are pale zones under the sarcolemma which represent loss or disarray of myofilaments (Figure 11). Such masses are best appreciated with transmission electron microscopy. Ring fibers are myofibers with peripheral sarcomeres oriented 90° from normal, which results in a "radiating spoke" appearance of the myofiber periphery in transverse sections. Whorled and coiled myofibers exhibit sarcoplasmic "swirling," often with internal nuclei, due to the presence of disordered sarcomeres. Split fibers are seen as multiple fibers with reduced diameters within a single basal lamina. They often occur under circumstances in which muscle load is persistently increased. All these findings are most often nonspecific chronic myopathic changes.

Inflammation

There are various infectious causes of myositis, but the most important forms of inflammatory myopathy encountered in experimental studies will be immune-mediated myositis. Specific lymphocytic subsets are often encountered in these cases, and immunohistochemistry for lymphocyte markers will be invaluable. Cases of florid myonecrosis with secondary inflammation (mainly numerous macrophages and fewer numbers of other white blood cells) may be difficult to distinguish from primary myositis. (Note: Macrophages within necrotic myofibers are not considered inflammatory cells.) Regardless of presentation, inflammatory conditions of skeletal muscle are rarely, if ever, the result of direct myotoxicity.

Glycogen

Detection of abnormal glycogen storage or depletion in myofiber sarcoplasm is important in studies of naturally occurring or induced metabolic myopathies. Biochemical analysis for glycogen content will be more accurate than cytochemical evaluation of tissue sections. Toxicants can affect glycogen stores in muscle fibers, but this change is rarely the sole or predominant finding.

CONCLUSIONS

Interpretation of skeletal muscle lesions that develop during the course of neurotoxicity studies is an ancillary but potentially critical endpoint supplementing the primary examination of neural tissues. Alterations in muscle fiber morphology may occur as a sequel to primary neural damage, and the pattern of muscle lesions can help define the distribution and nature of the nervous system injury. Unfortunately, evaluation of skeletal muscle by pathologists has often proven to be frustrating, unrewarding, and downright boring. The authors believe that this is due, in large part, to suboptimal handling and processing. The information in this chapter should improve and enhance the experience of pathologists engaged routinely in the evaluation of muscle samples.

REFERENCES

1. Aughsteen AA, et al. Quantitative morphometric study of the skeletal muscles of normal and streptozotocin-diabetic rats. *J Pancreas.* 2006;7:382–389.

2. Costelli P, et al. Ca^{2+}-dependent proteolysis in muscle wasting. *Int J Biochem Cell Biol.* 2005;37:2134–2146.

3. Glass DJ. Skeletal muscle hypertrophy and atrophy signaling pathways. *Int J Biochem Cell Biol.* 2005;37:1974–1984.

4. Jackman RW, Kandarian SC. The molecular basis of skeletal muscle atrophy. *Am J Physiol Cell Physiol.* 2004;287: C834–C843.

5. Riley DA, et al. Muscle sarcomere lesions and thrombosis after spaceflight and suspension unloading. *J Appl Physiol.* 1992;73 (Suppl):33S–43S.

6. Banwarth B. Drug-induced myopathies. *Expert Opin Drug Saf.* 2002;1:65–70.

7. Guis S, et al. Drug-induced and toxic myopathies. *Best Pract Res Clin Rheumatol.* 2003;17:877–907.

8. Walsh RJ, Amato AA. Toxic myopathies. *Neurol Clin.* 2005;23:397–428.

9. Cumming WJK, et al. *Color Atlas of Muscle Pathology.* London: Mosby-Wolfe; 1994: 3–90, 177–192.

10. Dubowitz V, Sewry C. *Muscle Biopsy: A Practical Approach,* 3rd ed. Philadelphia: W.B. Saunders; 2006: 3–219.

11. Valentine BA, McGavin MD. Skeletal muscle. In: McGavin MD, Zackary JF, eds. *Pathologic Basis of Veterinary Disease,* 4th ed. St. Louis, MO: Mosby; 2007: 973–1039.

12. Van Vleet JF, Valentine BA. Muscle and tendon. In: Maxie MG, ed. *Jubb, Kennedy, and Palmer's Pathology of Domestic Animals.* 5th ed. Edinburgh, UK: Elsevier Saunders; 2007;1: 185–280.

13. Brooke MH, Kaiser KK. Muscle fibre types: How many and what kind? *Arch Neurol (Chicago).* 1970;23:369–379.

14. Bárány M. ATPase activity of myosin correlated with speed of muscle shortening. *J Gen Physiol.* 1967;50 (Suppl):197–218.

15. Dodson A, et al. Monoclonal antibody that detects human type I muscle fibres in routinely fixed wax embedded sections. *J Clin Pathol.* 1987;40:172–174.

16. Havenith MG, et al. Muscle fiber typing in routinely processed skeletal muscle with monoclonal antibodies. *Histochemistry.* 1990;93:497–499.

17. Jay V, Becker LE. Fiber-type differentiation by myosin immunohistochemistry on paraffin-embedded skeletal muscle: a useful adjunct to fiber typing by the adenosine triphosphatase reaction. *Arch Pathol Lab Med.* 1994;118: 917–918.

18. Rojiani AM, Cho ES. Neuropathologic applications of immunohistochemical fiber typing in the non-neoplastic muscle biopsy. *Mod Pathol.* 1998;11:334–338.

19. Armstrong RB, et al. Distribution of fiber types in locomotory muscles of dogs. *Am J Anat.* 1982;163:87–98.

20. Braund KG, et al. Histochemical and morphometric study of fiber types in ten skeletal muscles of healthy young cats. *Am J Vet Res.* 1995;56:349–357.

21. Braund KG, et al. Observations on normal skeletal muscle of mature dogs: a cytochemical, histochemical, and morphometric study. *Vet Pathol.* 1982;19:577–595.

22. Kuzon Jr, WM, et al. A comparative histochemical and morphometric study of canine skeletal muscle. *Can J Vet Res.* 1989;53:125–132.

23. Ogato T. Morphological and cytochemical features of fiber types in vertebrate skeletal muscle. *CRC Crit Rev Anat Cell Biol.* 1988;1:229–275.

24. Molon-Noblot S, et al. The effects of ad libitum feeding and marked dietary restriction on spontaneous skeletal muscle pathology in Sprague–Dawley rats. *Toxicol Pathol.* 2005;33:600–608.

25. Panisello P, et al. Capillary supply, fibre types and fibre morphometry in rat tibialis anterior and diaphragm muscles after intermittent exposure to hypobaric hypoxia. *Eur J Appl Physiol.* 2008;103:203–213.

26. Kim Y-J, et al. Fully automated segmentation and morphometric analysis of muscle fiber images. *Cytometry A.* 2007; 71A:8–15.

18

IN VIVO IMAGING APPLICATIONS FOR THE NERVOUS SYSTEM IN ANIMAL MODELS

KATHY GABRIELSON

Departments of Molecular and Comparative Pathobiology and Environmental Health Sciences, School of Medicine and Bloomberg School of Public Health, Johns Hopkins University, Baltimore, Maryland

CRAIG FLETCHER

Division of Laboratory Animal Medicine, Department of Pathology and Laboratory Medicine, University of North Carolina School of Medicine, Chapel Hill, North Carolina

PAUL W. CZOTY

Department of Physiology and Pharmacology, Wake Forest University School of Medicine, Winston-Salem, North Carolina

MICHAEL A. NADER

Department of Physiology and Pharmacology and Department of Radiology, Wake Forest University School of Medicine, Winston-Salem, North Carolina

TRACY GLUCKMAN

Northwestern University, Chicago, Illinois

INTRODUCTION

Molecular imaging is now emerging as a valuable tool for elucidating basic biological processes and physiological pathways at the most fundamental level. The inherent value of imaging an antibody or a drug within the body allows for the ready detection of molecular distribution of these molecules within various organs and serves as an in vivo means of target validation. Molecular imaging is also able to detect disease biomarkers at earlier time points and can thus replace conventional survival curves, thereby serving as a refinement tool in biomedical research. This tool substantially reduces the costs of drug development and the number of animals needed for these studies. Moreover, animals can serve as their own control and can be imaged multiple times to follow disease progression or treatment efficacy.

The imaging modalities reviewed in this chapter include magnetic resonance imaging (MRI), positron-emission tomography (PET), single photon-emission computed tomography (SPECT), optical imaging, and ultrasound (US). Each of these modalities has the capability of assessing molecular expression, and all are currently used on a routine basis in human medicine with the exception of optical imaging. Table 1 compares and contrasts the important features of each modality as they apply to small-animal imaging. The breadth of imaging systems now available provides the opportunity to elucidate anatomic, functional, and molecular expression simultaneously within an animal model. The molecular imaging of diseases will allow for the increasingly rapid translation of disease pathogenesis and disease treatment from animal models to human therapeutic applications. In this chapter we focus on the fundamentals of each of these in vivo imaging methods as applied to rodent and nonhuman

Fundamental Neuropathology for Pathologists and Toxicologists: Principles and Techniques, First Edition. Edited by Brad Bolon and Mark T. Butt.
© 2011 John Wiley & Sons, Inc. Published 2011 by John Wiley & Sons, Inc.

TABLE 1   Comparison of Noninvasive Neuroimaging Modalities Used in Nonclinical Toxicological Neuropathology Testing[a]

	MRI	MR Microscopy	CT	PET/SPECT	Optical	US
Modality physics	Magnetic field	Magnetic field	X-ray	Radioactivity	Bioluminescence fluorescence	Sound waves
Noninvasive	✓	Ex vivo	✓	✓	✓	✓
Longitudinal	✓	—	✓	✓	✓	✓
Reduction of animals	✓	✓	✓	✓	✓	✓
Image time/animal[b]	30–60 min	30–60 min	Minutes	30–60 min[c]	Minutes	Minutes
Resolution	0.05 mm	0.04 mm	0.05 mm	1.5 mm	1.0 mm	0.03 mm
Sensitivity	μM–mM	—	—	nM	pM	?
Structure	✓	✓	✓	✓	✓	✓
Function	MR spectroscopy	—	✓	✓	✓	✓
Quantitative	✓	✓	✓	✓	✓	✓
Gene expression	✓	—	✓	✓	✓	✓
Translational to humans	✓	—	✓	✓	Not yet	✓
Therapeutic intervention	—	—	—	✓	—	✓
Cost	++++	+	+++	++++/++	+	+
Anesthesia required	✓	Occurs after necropsy	✓	✓	✓	Not always

[a] CT, computed tomography; MR, magnetic resonance; MRI, magnetic resonance imaging; PET, positron-emission tomography; SPECT, single photon-emission computed tomography; US, ultrasound.

[b] Time/image for rodents.

[c] PET image time up to 180 min for ^{18}F tracers in primates.

primate models of central nervous system (CNS) and peripheral nervous systme (PNS) diseases.

NUCLEAR IMAGING: PET AND SPECT IN NON-HUMAN PRIMATE STUDIES

In vitro techniques such as receptor autoradiography have been effective for characterizing brain changes after chronic drug administration in nonhuman primates (reviewed by Porrino et al.[1]). More recently, the development of noninvasive brain imaging techniques has enabled in vivo assessments of brain function in small laboratory animals. These techniques have enabled the systematic examination of variables that contribute to the etiology and progression of human diseases. Adapting brain imaging techniques to laboratory animals allows investigators to control a variety of potentially confounding variables related to, for example, subjects' environment and previous drug exposure.

In the effort to translate findings from laboratory animals to humans, nonhuman primates offer considerable advantages over other species because of their neurobiological similarity to humans and their longer lifespan, allowing for long-term longitudinal studies. In this section we provide a brief overview of PET in nonhuman primates with a focus on methodological issues pertaining to anesthesia. Although we discuss PET imaging studies that involve the dopamine (DA) system, with a special emphasis on DA D2 receptors, and focus on models of drug abuse, the research questions are relevant to all models of human disease. Information gained through use of techniques such as PET imaging will provide the foundation for future studies aimed at developing behavioral and pharmacological treatments for many human diseases.

BRAIN IMAGING IN ANIMAL SUBJECTS

As with any research technique, brain imaging in nonhuman primates has advantages and disadvantages. Like all animal models, PET imaging experiments provide for the control of many variables in order to better understand how one or a few affect brain function. However, this enhanced experimental control can also be a limitation: In studies on humans (as well as in disease states themselves), numerous factors can affect experimental or clinical outcomes; interactions between these factors are clearly relevant and need to be investigated. In most cases, however, such experimental control is an advantage, allowing scientists to perform experiments that are ethically or practically impossible in humans, such as comparing brain images of animals that have been exposed to drugs with those of drug-naive individuals, thereby eliminating a major confounding factor in clinical studies. The development of animal models and brain imaging methodologies continues to grow in sophistication such that future studies will minimize this limitation. An additional advantage is that using a relatively noninvasive imaging technique such as PET makes it possible to perform longitudinal studies by repeatedly imaging an animal to determine changes in a brain measure over time or in response to drug treatment.

As research subjects in brain imaging studies of addiction, nonhuman primates offer several advantages over rodents and other laboratory animal species. First, abundant evidence indicates that rodent and primate brains differ in the anatomy, physiology, and neurochemistry of brain neurotransmitter systems that mediate the abuse-related effects of drugs, including DA, norepinephrine, and opiate systems.[2-7] Moreover, compared to rodents, monkeys are more similar to humans in the pharmacokinetics and metabolism of several drugs, including opiates and serotonergics.[8] Imaging data have also demonstrated that the nonhuman primate brain differs substantially from the rodent brain in terms of cocaine-induced changes in brain metabolism.[9,10] This is apparent when using monkeys as subjects in longitudinal PET studies, first as drug-naive subjects, then over several years of cocaine exposure.[11] Finally, nonhuman primate colonies exhibit complex social behaviors (e.g., aggressive and affiliative interctions) that closely model human social interactions.[12] As such, monkeys are an ideal animal model for examining the influence of social variables on sensitivity to the abuse-related effects of drugs.[13]

PET IMAGING

The basic description below provides an introduction to PET imaging, which is perhaps the most frequently used modality with nonhuman primates. Studies that employ this technique require a large institutional investment; an investigator would not add PET imaging to a research program if the institution did not already have a PET camera and cyclotron (as well as physicists, radiologists, radiochemists, and other qualified personnel to support the multiuser PET facility). Readers can learn more about these techniques in recent textbooks.[14,15]

Methodological Considerations

PET derives its name from its function. After a radioactive PET ligand is injected into a subject, decay of the ligand produces positrons, which travel in space in a random fashion until they collide with an electron. The resulting annihilation causes the emission of gamma-ray photons that project outward at $180°$ (opposite directions) from the collision event. PET cameras have detectors arranged in a circle around the subject; the stimulation of two detectors (at $180°$) reveals in three dimensions the location of the annihilation and thus of the radioactive molecule or tracer, producing a two-dimensional image of a slice through the brain (e.g., tomography). The most frequently used radiotracers for PET studies contain ^{15}O (with a half-life of 2.04 min; e.g., ^{15}O water to assess blood flow), ^{11}C (with a half-life of 20 min), and ^{18}F (with a half-life of 110 min). The latter two isotopes are the most frequently used tracers for imaging specific brain substrates such as neurotransmitter receptors and

transporters. Radiolabeled fluorodeoxyglucose ($[^{18}F]FDG$) is also used to measure glucose metabolism in metabolically active organs such as brain.

Before analyzing PET data, the investigator must determine which areas of the brain to examine, using MRI or computed tomography (CT) images to identify these regions of interest (ROIs). The MRI and CT images provide structural data that impart anatomical relevance to the PET functional/molecular data. For example, after a PET study, the investigator superimposes the PET and MRI images and measures the tissue content of the radioactive molecule in the ROIs. Defining ROIs using MR images before examining the PET data enables an unbiased assessment of how an independent variable has affected PET measures in specific brain regions (Figure 1). For example, in studies using receptor ligands, analysis of the PET data follows a three-compartment model: radioactivity in the blood, in the extracellular space, and bound to the receptor. Rates of movement of the tracer between compartments are analyzed to generate a distribution volume (DV) in the ROI, which serves as a measure of the percent injected dose of radioactive substance in the ROI across time. To quantify and compare PET data after some manipulation, the DV for one ROI is studied in relation to a control region that contains relatively few receptors bound by that ligand. The ratio of these DVs constitutes the distribution volume ratio (DVR), which is the primary dependent variable in most PET imaging studies.

The DVR can be used in PET and SPECT to understand the biology of receptor–ligand interactions. In the past, such interactions were studied exclusively by in vitro receptor binding studies and autoradiography, but similar information can be acquired in vivo from nuclear imaging studies to measure the receptor binding potential (BP, a dependent variable that is defined as the DVR minus 1). These dependent variables, DVR and BP, are unitless numbers that represent the ratio of receptor density (B_{max}) to affinity (K_d) for a receptor. Thus, changes in DVR may result from changes in either the numerator (receptor numbers) or the denominator (affinity for the receptor). Importantly, this ability is considered an advantage of PET because it provides a measure of receptor binding in the presence of endogenous neurotransmitters that may compete with the radioligand. Depending on the interest of the investigator, the next research direction may be to determine the exact nature of the DVR change (i.e., increase or decrease in receptor number or changes in levels of the competing endogenous neurotransmitter).

The principles of PET imaging that lead to its utility in studies of brain function are also characteristic of a related imaging technique, SPECT. The primary difference between the techniques is that SPECT studies use different radioactive substances with longer decay times. In addition, because SPECT radiotracers emit single photons rather than pairs, the technique is less sensitive than PET, and the resulting images are less detailed. The primary advantage of SPECT

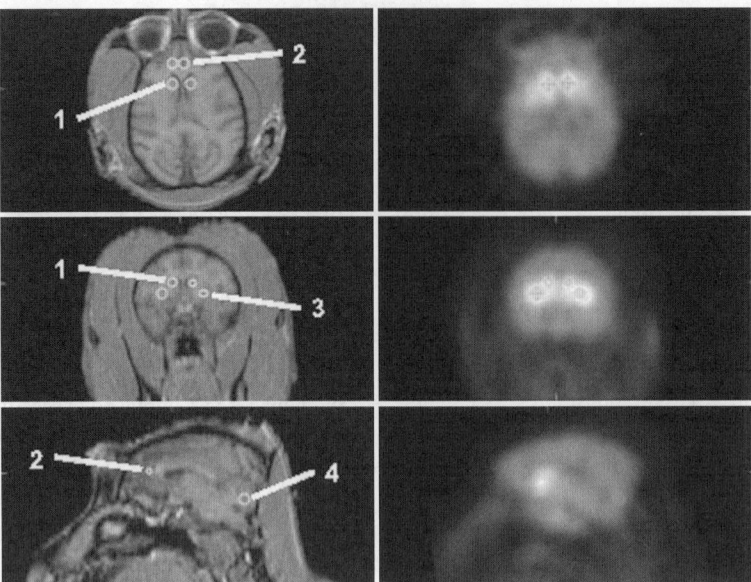

FIGURE 1   Combined presentation of functional and structural information in the macaque brain using overlapping images from two different neuroimaging modalities. *Left column*: T1-weighted magnetic resonance images (MRI) in the horizontal (top), coronal, and parasagittal (bottom) planes. Circles indicate regions of interest (ROIs) placed bilaterally in the caudate nucleus (labeled 1), anterior cingulate cortex (2), putamen (3), and cerebellum (4). *Right column*: Co-registered (overlapping) MRI and positron-emission tomographic (PET) images from the identical planes following administration of [18F] fluoroclebopride ([18F]FCP) to investigate dopamine D2 receptor availability. (From Porrino et al.[18])

versus PET is lower expense and a relatively greater ease of implementation.[17]

Imaging Conscious Versus Unconscious Subjects

A main issue in imaging studies is whether subjects should be anesthetized or awake during scanning. This independent variable naturally depends on the question under study. For many purposes, use of monkeys that are awake is optimal. However, most imaging procedures require subjects to be almost completely motionless. Adapting monkeys to this requirement requires several months of training. As discussed below, however, achieving this state is not impossible. In contrast, if anesthesia is used, a motionless state is readily attained, but the effects (if any) of the anesthetic agent on the dependent variable should be determined. Such questions are best addressed using a within-subjects design, comparing data in the presence and absence of the anesthetic agent. Examples of studies using awake and anesthetized monkeys are presented below.[18]

Only a few research groups have published results of PET imaging studies that were performed in conscious primates. Such studies require a restraint apparatus that permits trained monkeys to sit still in the PET scanner. The usual apparatus is a modification of a standard primate restraint chair,[19] in which a customized head holder mounted to the chair prevents head movements during the acquisition of PET data. Primates must be acclimated to the restraint apparatus over a

10- to 12-week period, so that the monkeys learn to remain motionless in the apparatus for up to 4 h per session, twice per week. Immobilization does not cause appreciable stress to the animal (a condition that itself could alter brain function and the resulting PET data), as monkeys trained to prolonged restraint do not exhibit significantly elevated cortisol concentrations.[19]

In the first study using this apparatus, the effects of acutely administered cocaine (1.0 mg/kg, i.v.) on cerebral blood flow in conscious monkeys were assessed using ^{15}O-labeled water. Significant increases in blood flow, induced by cocaine, were evident in the whole brain, striatal regions, and medial temporal and frontal cortical regions; these changes were reproducible from session to session. Additionally, replications of the experiment showed that pretreatment with alaproclate, a drug that increases extracellular levels of serotonin, blocked cocaine-induced increases in cerebral blood flow.[11] Interestingly, the effects of cocaine on cerebral blood flow in these previously cocaine-naive monkeys contrast with reports describing cocaine-induced decreases in cerebral blood flow in humans with an extensive history of cocaine use,[20–23] providing important information related to the consequences of chronic cocaine exposure. Such information highlights the strength of longitudinal designs.

Depending on the research question, it is not always necessary to use conscious monkeys in PET imaging studies. For example, our group has investigated environmental and pharmacological variables that influence DA receptor availability (specifically the D2-like receptor).[13,24] However, it is essential to consider carefully the proper anesthetic agent to use, and to ensure that effects on PET measures are due to independent variables and not to interaction between those variables and the anesthesia. The choice of agent depends on several factors, including the duration of the PET scan and the radiotracer used. For our studies, which were approximately 3 h long and employed radiotracers labeled with ^{18}F, we chose isoflurane because of its minimal effects on neurotransmitters that can influence DA levels[25] and because other studies have shown that isoflurane had no effect on the availability of the D2-selective radiotracer [^{11}C]raclopride.[27]

The interaction between the anesthetic agent and the radioligand depends on a compound's primary targets. For example, if the radioligand binds to opiate receptors, it is unlikely that an anesthetic that affects DA would produce significant effects on the PET measure. In contrast, that same anesthetic may affect measures of DA receptor availability. For example, the most frequently used anesthetic in veterinary medicine involving nonhuman primates is ketamine, which is the anesthetic of choice for immobilizing a monkey and transporting it to the PET center. However, ketamine can alter extracellular DA levels[28] and might therefore affect PET measures of DA receptor availability. In a PET study involving conscious rhesus monkeys,[29] a low dose of ketamine (5 mg/kg) administered 30 min before the D2 ligand [^{11}C]N-methylspiperone (NMSP) increased the binding of NMSP to D2 receptors. However, induction of anesthesia with ketamine followed by anesthesia with isoflurane (1.5 to 5%) does not significantly affect D2 receptor availability for the ligand [^{18}F]fluoroclebopride (FCP).[30]

Dopaminergic system imaging has been evaluated to understand the mechanism of certain environmental neurotoxicants. Acute manganese administration increases dopamine transporter (DAT) levels in the nonhuman primate striatum, a fact indicated by [^{11}C]WIN35,428 PET scans of baboons. Ancillary in vitro studies suggest that the in vivo change may be a compensatory response to $MnSO_4$ inhibitory action on DAT.[31]

MICROPET AND MICROSPECT ANIMAL MODEL APPLICATIONS

With the development of miniaturized equipment specialized for use in small laboratory animals, microSPECT and microPET studies are allowing for rapid understanding of a diversity of neurological diseases. The ability to use transgenic mice, reporter genes, and noninvasive neuroimaging[32] has greatly expanded the applications in this field. While nuclear imaging in nonhuman primates has centered on the understanding of brain chemistry, the biological processes that can be imaged with microPET and microSPECT are diverse, ranging from cell metabolism to cell death. For example, in vivo imaging to detect a radiolabeled glucose surrogate (FDG) by PET allows the evaluation of tumor growth and metabolism before and after treatment with an anticancer drug,[33] including glioma models (Figure 2).[34] Using a [^{125}I]-labeled monoclonal antibody (mAb) directed against hepatocyte growth factor/scatter factor (HGH/SF), microSPECT/CT images of human tumor xenografts orthotopically grafted into the right hemisphere of mouse brain demonstrated that HGH/SF[+] human U87 glioblastoma implants had a 42-fold higher uptake of antibody 2 days after antibody injection relative to HGH/SF SNB19 implants. This experiment demonstrates that [^{125}I]anti-HGH/SF mAb can be used to specifically target and image brain tumors that express a specific marker protein.

In a second example, microSPECT has been used to quantify the cell death dose–response. This method works because phosphatidylserine (normally hidden inside the cell membrane) flips out to the cell surface during the process of cell death, where it can serve as a binding target for annexin V. SPECT imaging of 99m-technetium-labeled annexin V can then be used to track and quantify cell death after antineoplastic treatments or to monitor a dose–response in unintended chemotherapy-induced cell death.[35]

Choosing MicroSPECT or MicroPET

Radionuclide selection, financial considerations, and available facilities all influence the decision on which method will be the most appropriate for rodent neuroimaging studies. A drawback to nuclear imaging is poor anatomical detail relevant to localizing the exact distribution of nuclear emissions. However, hybrid technologies such as SPECT-CT, PET-CT, or nuclear imaging combined with MR imaging are now available that incorporate multiple modalities to provide a composite image of higher resolution.[36,37] Nuclear images (PET or SPECT) are coaligned with structural images (MR or CT) to give sufficient anatomical detail to localize the nuclear imaging signal.

Spatial resolution of SPECT is also superior to PET, approaching 0.5 μm, compared to 1.0 μm with PET. In general, rodents require 30 to 60 min of anesthesia for the acquisition of the typical microPET or microSPECT scan, yet this depends on the $t_{1/2}$ value of the radiotracer and the signal-to-noise ratio, although nuclear imaging has been accomplished in conscious rodents.[38] Most radioactive ligands are injected intravenously in the tail vein, but intraperitoneal injections may be adequate to image the striatum in microSPECT.[39] It is important to scan at the peak uptake of the radiotracer. The temporal resolution of PET corresponds to the time chosen for PET scan acquisition. Characteristics of the radiotracer such as affinity and metabolism will determine the length of time the tracer is bound to the receptor. Many radioligands that have been developed for human medicine are also useful in microSPECT and

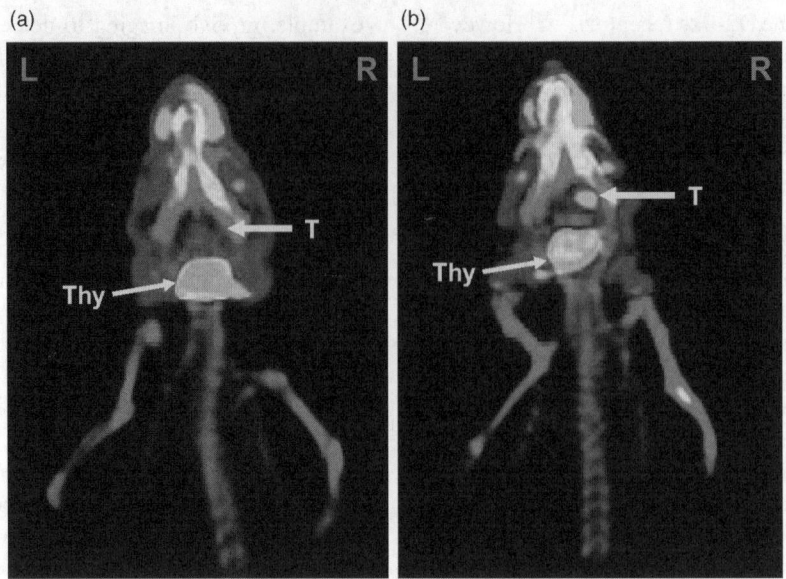

FIGURE 2 Single photon-emission tomography-computed tomography (SPECT-CT) imaging of hepatocyte growth factor/scatter factor (HGF/SF)-positive and HGF/SF-negative tumor xenografts in mouse brain using [^{125}I]anti-HGH/SF monoclonal antibody (mAb). One of two human glioblastoma cell lines, SNB19 (HGH/SF$^-$) and U87 (HGH/SF$^+$), was orthotopically grafted into the right (R) cerebral hemisphere. On day 21, mice were injected with 1 mCi of [^{125}I]anti-HGH/SF mAb intraveneously; quantitative SPECT-CT images were acquired 2 days after mAb injection. Antibody binding to specific tissues was calculated as the percent injected dose per gram (% ID/g); the value was 42-fold higher in the U87 tumor (0.5667 ± 0.113% ID/g; panel B) relative to the SNB19 tumor (0.0134 ± 0.0004% ID/g; panel A). L, left; R, right; T, tumor xenograft; Thy, thymus. [Courtesy of Catherine Foss (Department of Radiology) and Jonathan Laterra (Department of Neurosciences), Kennedy Krieger Institute, Johns Hopkins University.] (*See insert for color representation of the figure.*)

microPET in laboratory animals to study the activity of brain enzymes, receptors, or transporters.

Choosing a Radiotracer

In general, a candidate radiotracer for small-animal neuroimaging must have high specific activity when labeled. The ligand labeled should have as many binding sites as possible labeled with the isotope, leaving minimal free radionuclide in the injection. A too-low specific activity may compromise the specific binding of the radiotracer. The actual mass of labeled tracer is also very low, preventing a pharmacological effect on the animal (tracer principle). A radiopharmaceutical useful for imaging the brain must have relevant brain uptake either through active transporters or should have sufficient lipid solubility to pass the blood--brain barrier. Potential radiotracers should have (1) minimal labeled blood metabolites, (2) low blood penetration of metabolites, (3) high brain stability, (4) rapid washout of the metabolites in the brain, and (5) low metabolite binding in the brain. Currently, radiotracers are available for virtually all neurotransmitter receptors. In most cases, a probe that was developed for humans will also work in animals; however, there are a few exceptions.

The majority of animal and human ligand activation studies have been performed in the dopaminergic (DA) system; although ligands are available to most major neurotransmitter systems. In the DA system, many of these studies have used [^{11}C]raclopride, a compound that competes with endogenous dopamine because of its relatively low affinity to dopamine D2 receptor and can thus be used to monitor drug effects on the brain DA system. [^{11}C] Raclopride's binding to the D2 receptor is also decreased after amphetamine,[40–42] haloperidol,[43] nicotine,[26,44] alcohol,[45–47] and ketamine.[48] Thus, careful probe selection and study interpretation in receptor displacement studies are important.

Imaging Neuroinflammation in Rodents

In the diseased brain, activated microglia express binding sites for synthetic ligands designed to recognize 18-kDa translocator protein, TP-18, part of the peripheral benzodiazepine receptor (PBR) complex. The name is derived from the binding of specific benzodiazepines to this protein, which is distinct from the central benzodiazepine receptor-associated GABA$_A$- regulated channels. In the CNS, the PBR is found exclusively on nonneuronal cells, primarily activated microglia or cells of the monocytic lineage. More recently, the term *translocator protein* has been proposed to refer to the main molecular function of the 18-kDa protein: namely, binding and transport into the mitochondria of cholesterol, proteins, and porphyrin, with regulation of steroid synthesis being a primary area of interest. PD11195 is the prototype ligand used widely for the functional chararcterization of TP-18. Radiolabeled [^{11}C](R)-PK11195 has been used in PET imaging of the microglial activation that occurs with ethanol injury in the rat striatum[49] and in Parkinson's disease (PD) animal models.[50] New PBR radiotracers with higher specific binding to PBR are now under development.

Imaging Blood–Brain Barrier in Rodents

Recruitment of white blood cells in neuroinflammatory conditions such as multiple sclerosis, stroke, or viral diseases takes place at the intact but activated brain endothelium. Until recently, these changes could not be seen by in vivo imaging techniques because such methods rely on increased permeability of the blood–brain barrier (BBB) in the later stages of the diseases. Recent advances in imaging the BBB allow for presymptomatic in vivo imaging of brain diseases such as experimental autoimmune encephalomyelitis (EAE) using T2 MRI-visible glyco-nanoparticles (GNPs) composed of an iron-linked E-selectin ligand.[51] In these studies, the GNP detects CD62E (an adhesion molecule on activated endothelium) as a biomarker on the blood side of the BBB that is indicative of pathology on the brain side. Additionally, GNP does not induce breakdown of the BBB like that seen with conventional contrast agents such as gadolinium-diethylenetriaminepentaacetic acid (Gd-DPTA). Another important biomarker of the BBB is the transport molecule P-glycoprotein (also designated CD243), which is responsible for transport of multiple drugs across the BBB. Imaging of P-glycoprotein function at the BBB has been accomplished in macaques using PET.[52]

Imaging Alzheimer's Disease Mouse Models

Alzheimer's disease (AD) is increasingly common in aging populations worldwide, and diagnosis with certainty requires a postmortem neuropathology examination. Thus, the development of in vivo imaging probes for this disease is of great interest in basic and applied research for this condition. Mouse models of AD may be useful for probe development for human applications, but at present mouse-to-human translation is not always reliable. For example, the amyloid-binding compound thioflavin has been modified to produce a compound called Pittsburgh Compound B (PIB). This agent exhibits a two- to threefold higher retention in the association cortex of patients with AD than in healthy controls. Direct imaging of amyloid-beta (Aβ) plaques in transgenic mice has been performed with PIB[53] with equivocal results.

In multiple studies using the Tg2576 single transgenic or presenilin-1/amyloid precursor protein (PS1/APP) double transgenic mouse models of AD, the binding capacity of PIB was much less in mouse brain than in human brain. Ex vivo studies show that the B_{max} for binding of [^{3}H]PIB to PS1/APP mouse brain was more than 1000-fold lower than the B_{max} for binding in human AD brains. So far, no reliable model for preclinical in vivo evaluation of radiotracers for brain plaques has been established.[54–56]

PET imaging of AD mouse models is leading to the better understanding of this disease. Besides the direct imaging of Aβ plaques, microPET imaging can help characterize and phenotype animal models for AD by using radiotracers already used in the human clinical setting. Significant differences in [^{18}F]FDG uptake (glucose metabolism) were seen in APP23-transgenic mice 12 weeks after treatment with the locus coeruleus toxicant N-(2-chloroethyl)-N-ethylbromobenzylamine (dsp-4).[55] Additionally, acetylcholinesterase (AchE) activity alterations have been demonstrated in transgenic mice assessed using N-[^{11}C]methyl-4-piperidinylacetate ([^{11}C]MP4A-PET), an acetylcholine analog. After passing the BBB, [^{11}C]MP4A-PET is hydrolyzed by AchE and trapped within the specific brain regions in proportion to the regional level of AchE activity. By comparison, MRI has an advantage over PET in that it provides higher spatial resolution, and with it, the ability to detect individual amyloid plaques. Some molecular components of plaques such as iron render them detectable for MRI without the aid of contrast agents.[57]

In summary, investigators need to collect pilot data before undertaking a long-term PET or SPECT neuroimaging project. If anesthesia is a potential confounding factor, pilot experiments should compare the response of conscious animals with that of anesthetized animals; this requirement is most applicable to primates at present, although in the near future rodent systems will readily permit imaging the brain of conscious animals. If the differences are not statistically significant, it is probably not necessary to train the remaining experimental group(s) for conscious nuclear imaging studies unless the experiment calls for imaging in subjects who are performing a task. If anesthesia is to be used, it is important to consider any direct and indirect pharmacological effects of the agent on the primary target. Again, within-subject comparisons and a repeated-testing protocol are necessary design considerations to determine the best anesthetic to use in these studies. This basic information needs to be a priority in future neuroimaging studies.

MAGNETIC RESONANCE IMAGING

The sculpting of the nervous system from the neural plate to its final complex topology requires a concerted and coordinated expression of numerous genes. Failure of one of these early patterning events can result in the loss of large regions of the CNS or PNS. The anatomical heterogeniety of the CNS and PNS makes a complete evaluation by routine histopathology difficult in toxicological studies. A comprehensive MRI study of the brain or spinal cord offers a complementary approach to histopathology, which can provide a lower-resolution but more complete assessment of the nervous system. The MRI modality is especially well suited for neuroimaging because it can clearly differentiate gray and white matter[58] and can be used to quantitatively map brain regions, especially in the course of developmental neurobiology experiments and aging research. Monitoring tissue characteristics using a nondestructive imaging method such as MRI makes it feasible to conduct longitudinal studies on rodent models.[59]

Furthermore, MRI has many applications in studying rodent models of various neurological injuries and disorders and their structural consequences, including hydrocephalus, stroke, head injury, brain tumors, and spinal cord injuries.[60,61] MRI can be used for accurate determination of optimal times for interventional treatments, to validate functional endpoints, or in selecting the most appropriate sites for tissue harvest and pathology sampling, thus further reducing the number of animals required in research studies.

Physical and technical concepts

MRI relies on quantum mechanical properties of atomic nuclei when placed in a strong magnetic field. MRI provides much greater contrast between the different soft tissues of the body than CT does, making it especially useful in neurological (brain), musculoskeletal, cardiovascular, and oncological (cancer) imaging. An additional advantage is that MRI, unlike CT, uses no ionizing radiation.[62]

The basic MRI signal uses the properties of electromagnetic forces to align the nuclear magnetization of hydrogen (usually) nuclei or protons in the body water found in various tissues. Radio-frequency (RF) fields are then used to systematically alter the alignment of this magnetization, causing the hydrogen nuclei to produce a rotating magnetic field detectable by the scanner. This signal can be manipulated by additional magnetic fields to build enough magnetic information to construct a detailed image of the body. The major components of an MRI scanner consist of the following: (1) a strong static magnetic field that produces splitting of electronic energy levels in certain nuclei, (2) an RF coil transmitter and receiver that excite the sample and receive the back, signal emitted and (3) three orthogonal controllable magnetic field gradients within the field of view that provide the spatial information necessary to form an image. Magnetic field strength is one of the most important factors in determining image quality. Higher magnetic fields increase the signal-to-noise ratio, permitting higher spatial and temporal resolution or faster scanning. However, higher field strengths require more costly magnets with higher maintenance costs and have increased safety concerns.[62]

Typically, many institutions have imaging cores with dedicated MRI facilities for rodent imaging with magnetic field strengths ranging from 4.7 to 11 T.[63,64] The pulse of RF wavelengths provided by the RF coil excites the atomic nuclei corresponding to their two energy states (lower vs. higher). The excited nuclei in the tissue will emit RF radiation and lose energy over time while interacting with neighboring molecules. The variation in magnetic field gradient across the body (a field gradient) and the different spatial locations become associated with different frequencies. Usually, these field gradients are pulsed; it is the almost infinite variety of RF and gradient pulse sequences that gives MRI its versatility. The inherent isotropic three-dimensional nature of MRI allows detailed analysis of organs and retrospective studies through any arbitrary plane.[65] The recovery of longitudinal magnetization is called T_1 *relaxation*. The loss of phase coherence in the transverse plane is called T_2 *relaxation*. Different tissues vary widely in their T_1 and T_2 values, which show up with various intensities in the MRI image. Depending on the temporal sequence and duration of RF excitation of the sample and the time at which the emitted RF signal is sampled, either T_1 or T_2 contrast can be emphasized in the image to give MRI its tremendous soft tissue contrast.[66,67]

The most important determinant of the signal-to-noise ratio is the size of the RF coil (the ratio is linearly dependent on the RF coil diameter), which is used to detect the signal. Therefore, smaller-diameter coils (i.e., array coils) designed to image smaller fields are rapidly being developed for high-field animal systems and are becoming the standard for clinical machines in medical neuroimaging.[60,68] However, despite these major inherent advantages, low-signal-sensitivity problems continue to be a major limitation of some MRI studies. The use of contrast agents based on gadolinium (Gd) or iron oxide particles (causing decreased signal) influences the relaxation times of water in their immediate vicinity.[69–72] Gd-based agents have become extremely useful for imaging increases uses to evaluate permeability alterations caused by breakdown of the BBB and for measurements of cerebral perfusion.[73]

Earlier work used iron oxide nanoparticles to track migration, angiogenesis, apoptosis, and gene expression.[74–79] Continuing research on the development of iron-based contrast agents, including superparamagnetic nanoparticles as a negative contrast agent, has been useful to image tumor xenografts implanted orthotopically in the nude rat brain.[80] Both major groups of contrast agents have great clinical utility because they can leak into and highlight damaged tissue; however, they are restricted to extracellular spaces.[81] Targeted ("smart") contrast agents have been developed to focus particles to specific cell-surface receptors using conjugated antibodies and peptides.[82] Manganese (Mn^{2+}) is another paramagnetic MRI contrast agent that can be taken up through voltage-gated Ca^{2+} channels, transported along axons, and even across synapses. Injections of $MnCl_2$ at low concentrations into either the striatum or amygdala of mouse brains produced significant contrast enhancement along the known neuronal circuitry originating from these sites; indeed, Mn^{2+} selectively accumulated in the pathways upon activation. These combined properties allow for its use as an effective neuronal tract tracer suitable for MRI detection.[83]

Conventional (structural) MRI defines the borders between different tissues (e.g., gray and white matter in the CNS) based on divergence in both their water content and the physical properties of water–macromolecular interactions, and the content of magnetically interactive compounds such as iron. Functional MRI (fMRI) makes use of contrast mechanisms related to physiological changes in brain resulting from various stimuli. Tissue contrast is derived from the local changes in relative blood flow, blood volume, and oxygenation with brain activity [blood-oxygen-level-dependent (BOLD)] to provide a rapid, noninvasive approach to functionally assess neuronal activation.[84]

The application of fMRI in rodents has been limited due to the signal-attenuating effects of anesthesia. Most animal studies with MRI require that the animal be anesthetized to prevent motion artifacts. For fMRI studies, there is no perfect anesthesia protocol; however, chloral hydrate, α-chloralose, or propofol are commonly used in animals models because they interfere less with sensory-evoked potentials, including brain stem auditory potentials.[85]

Physiological monitoring is an important component of MRI experiments, due to the effect of anesthetic drugs on respiration and heart rate. Alterations in body movement resulting from changes in these two parameters can confound the contrast signal interpretation. There are now increasingly sophisticated anesthetic delivery systems that are able to provide physiological monitoring and supportive care for rodent species during imaging studies.[69,86–90] These monitoring devices typically employ fiber-optic technology (and therefore are MRI-compatible) and are able to record the respiratory rate, heart rate, body temperature, tidal volume, and electrocardiographic readings during image acquisition. Furthermore, spatial resolution can be significantly enhanced during MRI of rodents by cardiac gating and respiratory synchronization.[91,92] Complete control of respiration requires intubation or tracheotomy and mechanical ventilation of the experimental animals. Finally, cardiac (i.e., via ECG) and thoracic motion information is processed by a scan management system to control timing of the MR sequence recording.[93,94]

Magnetic resonance microscopy (MRM) is based on the same physical principles as MRI. However, to make higher-resolution images at the microscopic level, MRM must use much stronger magnetic gradients, approximately 50 to 100 times those of most clinical neuroimaging systems.

MRM resolution is usually less than $100\,\mu m$ in at least one spatial dimension.[95] Acquisition of a fixed mouse brain image with a resolution of $43\,\mu m$ would take roughly 30 min. Increasing the imaging time to 1 h will result in 21-μm resolution. MRM has been used to analyze both rat and mouse brain structures in vivo as well as in vitro.[60,68] A recently published paper used the MRI-sensitive $[^{19}F]$ amyloidophilic dye known as FSB, which crosses the BBB to specifically label in vivo amyloid β in plaques in the brains of APP-transgenic mice. The investigators compared the abilities of $[^{19}F]$- and $[^1H]$-FSB to label brain plaques in living mice relative to that of an anti-amyloid β antibody to label plaques in brain ex vivo. They found that ^{19}F-MRI was the most reliable indicator of amyloid plaque accumulation in this mouse model of AD.[96]

During development of both the CNS and the PNS, most cellular populations studied are found to undergo some cell death during the maturation process. The dysregulation of apoptosis can lead to the destruction of normal tissues in a variety of disorders, including autoimmune and neurodegenerative diseases (too much apoptosis) or the growth of tumors (too little apoptosis). Diffusion-weighted MRI (DWI) is an MRI modality that can image apoptosis (and putative edema within tissues) without the need for a contrast agent.[77,97,98] DWI may also be used to monitor changes in tissue volume and/or fluid content due to cell swelling, tumor shrinkage, necrosis, hydrocephalus, and other processes that can occur with therapy. Alternatively, MRI has been employed to detect apoptosis by using reagents in which annexin V (a molecule that interacts with the cell membrane constituent phosphatidylserine, which is externalized in cells undergoing apoptosis) is either conjugated to nanocrystals of iron oxides, coating the surface of Gd-containing liposomes, or incorporated into quantum dots with a paramagnetic lipidic coating[99–101]. Cell death that occurs with domoic acid[102] or carbonyl sulfide is easily identified with MR imaging.[68,103,104]

Neuroanatomical abnormalities can be mapped by MRI; Figure 3 contrasts the T1 scans at 21.5-μm resolution through the brain of a homozygous Reeler mutant mouse (Reln rl/rl) and an age-matched wild-type (WT) control. These scans emphasize the spectrum of lesions that develop in the Reln rl/rl mouse, including severe cerebellar atrophy, ventricular enlargement, and disorganization of cell layers in the hippocampus.[105]

OPTICAL IMAGING

Optical imaging is the only noninvasive neuroimaging method that is not presently being used in human medicine. Optical modalities currently use either bioluminescent or fluorescent signals for detection. These light-capturing techniques are safe alternatives to gamma-emitting systems and

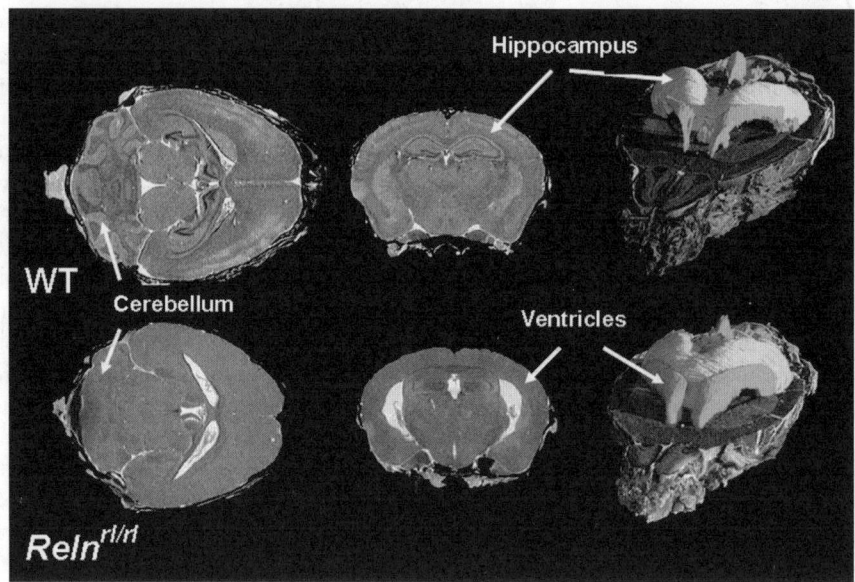

FIGURE 3 Magnetic resonance microscopy (MRM) distinguishes neuroanatomic differences between homozygous Reeler mutant mice (Reln rl/rl) and age-matched wild-type (WT) controls. MRM scans at 21.5-μm resolution demonstrate severe cerebellar atrophy, ventricular dilation, and disorganization of the hippocampal cell layers[105] in the Reln rl/rl mouse. The images on the right are a partial three-dimensional reconstruction of the brain to emphasize the altered shape and size of the hippocampus (yellow) and lateral ventricles (blue) in the mutant mice. [Courtesy of Andrea Badea (Department of Radiology), Duke University.] (*See insert for color representation of the figure.*)

have a limited impact on mammalian cells. Moreover, recent advances in bionanotechnologies, a discipline that fuses the biological and engineering fields, opens the door for examining the molecular aspects of infectious disease, immunology, cancer biology, and other diseases by evaluation of drug pharmacokinetics and physiology in the whole animal, yet at the nanoscale. In the context of preclinical drug studies, optical imaging is accelerating discovery and development by advances such as direct tracking of molecular targets, following drug distribution longitudinally within a single mammalian cell, and quantification of a drug effect on the cellular or subcellular target within a matter of minutes.[106] The relatively low cost and high speed of data acquisition makes optical technologies a valuable tool in high-throughput animal studies.

The basis of noninvasive bioluminescent imaging (BLI) is light detection emitted endogenously via a chemical reaction driven by the enzyme luciferase. This enzyme occurs naturally in several organisms; luciferases derived from nematodes, fireflies, and the sea pansy (*Renilla reniformis*) are the most commonly used genes in bioluminescent technology. Luciferase catalyzes the conversion of D-luciferin to oxyluciferin, resulting in emission of photons (light generation). Bioluminescent technology utilizes luciferase as a reporter construct, due to its ability to "report" its location by this emission of light. Cell transfection via nonpathogenic viral [e.g., adenovirus, adeno-associated virus (AAV), or cytomegalovirus (CMV)] or plasmid vector systems are the

most common approaches to expression of luciferase in mammalian cells. In most systems the cellular expression of luciferase is always active or "on," and light is emitted in the presence of the appropriate substrate. However, luciferase can undergo in vivo targeted expression by cleavage into its active form via the introduction of specific promoter elements during dynamic processes such as expression of the Cox-2-catalyzed inflammatory pathway[107] and apoptotic events.[108,109]

One example of BLI technology designed for toxicology drug screening is the use of light-producing transgenic animals [LPTAs; available from Caliper Life Sciences (formerly Xenogen, Inc.), Hopkinton, MA, http://www.caliperls.com/tech/optical-imaging]. These genetically engineered models have a luciferase reporter gene that is driven by specific enzymatic or genetic promoters. For example, several CYP3A (cytochrome P450 isoform 3A) human and mouse promoters linked to luciferase exist to study the mediation of drug–drug interactions. An Hmox1 (heme oxygenase-1) transgenic mouse has been developed to evaluate hypoxemia, iron-overload toxicities, and heavy metal and chemical toxicities. Several other transgenic systems have been generated to study the various isoforms of the stress-inducible gene that encodes the growth arrest and DNA-damage-inducible protein [Gadd; also known as CCAAT-enhancer-binding proteins (C/EBP)-homologous protein (Chop)]. Gene expression occurs following homeostatic disruption of the endoplasmic reticulum. Gadd153 expression is associated with a reduction in glutathione and upregulation of reactive oxygen species (ROS), the oxidative cellular events that occur during apoptosis.[110]

Bioluminescence analysis of Smad-dependent transforming growth factor β (TGF-β) signaling in live mice is possible using mice with a Smad-responsive luciferase reporter (SBE-luc, currently available through the Jackson Laboratory, Bar Harbor, ME).[111] Using SBE-luc mice, the brain has been shown to have the highest baseline activity of Smad proteins (a class of molecules that modulate the activity of TGF-β ligands) of any major organ in the mouse. In vivo excitatory stimulation of certain CNS neurons with kainic acid (KA) leads to an increase in the production of phosphorylated Smad2 (Smad2P) followed by its nuclear translocation in hippocampal CA3 neurons, which correlates with a significantly higher luciferase activity. Although this activation was most prominent 24 h after KA administration in neurons, Smad2P immunoreactivity gradually increased in astrocytes and microglial cells at 3 and 5 days as well, consistent with initiation of reactive gliosis in association with areas of neuron degeneration. Bioluminescence measured over the skull in living mice peaked at 12 to 72 h and was correlated with the extent of microglial activation (CD68 immunostaning) and astrocyte activation (GFAP staining) 3 days after injury. Treatment with the glutamate receptor antagonist MK-801 strongly reduced bioluminescence and neuropathology. These results show that Smad2 signaling is a sensitive marker of neuronal activation and CNS injury that can be used to monitor KA-induced neuronal degeneration.[112]

For all luciferase reporters, the detection of light occurs using a light-sensitive camera called a charge-coupled device (CCD). To reduce motion artifacts in localizing the source of the emitted light, general anesthesia is required to immobilize mice or rats for BLI studies. The sensitivity of BLI is dependent on the depth of the tissue, the presence of hair, the temperature of the targeted cells, and the presence of background phosphorescence or "noise." For example, chlorophylls from plant material in conventional rodent chows can cause significant background phosphorescence.[113] However, unlike fluorescent techniques, there is no inherent background light signal generated within living cells with bioluminescence, thus allowing for greater sensitivity during imaging.

Experimental autoimmune encephalomyelitis (EAE) is a widely used animal model to investigate basic mechanisms of multiple sclerosis. Typically, the disease in rodent models is scored by observing clinical signs of paralysis; therefore, in vivo imaging could allow for earlier, more humane selection of the most appropriate time point for euthanasia. Noninvasive BLI measurement of luciferase activity during EAE in transgenic mice expressing an injury-responsive luciferase reporter in astrocytes (GFAP-luc) showed that BLI readings in the brain and spinal cord correlated strongly with the severity of clinical disease (particularly at early time

points) and a number of pathological changes in the brain. These results highlight the potential use of BLI to monitor neuroinflammation for rapid and objective (human observer-- independent) drug screening and immunological studies in EAE and suggest that similar approaches could be applied to other animal models of autoimmune and inflammatory brain disorders.[114] Figure 4 demonstrates the use of optical imaging in this model.

A recombinant luciferase reporter has been developed that when expressed in mammalian cells, leads to attenuated levels of reporter activity in normal cells. In cells undergoing apoptosis, a caspase 3–specific cleavage of the recombinant product occurs, resulting in the restoration of luciferase activity that can be detected in living animals using BLI. The ability to image apoptosis noninvasively and dynamically over time provides an opportunity for high-throughput screening of pro- and anti-apoptotic compounds and for target validation in vivo in both cell lines and transgenic animals.[108]

Optical imaging using flurescence is another method of monitoring neurophysiological events in animal models. The excitation of an electron from its ground state also results in the formation of a photon. Fluorophores (flurescent molecules) absorb photons, and upon returning to their ground state, energy is released in the form of fluorescence emission. Similar to BLI, fluorescence detection requires a CCD camera to capture light emissions. A fluorophore may be a naturally occurring molecule or an engineered tag generated specifically to label cells or proteins through direct binding, gene expression, embryonic development, or viral transfection. The most commonly used fluorophores, GFP (green fluorescent protein) and RFP (red fluorescent protein), have specific, yet overlapping wavelengths. However, recent crossover from in vitro labeling uses has also introduced FITC (flurescein isothiocyanate)- and TRITC (tetramethyl flurescein isothiocyanate)-labeled probes for use with in vivo systems. Previously, weak or lost fluorescent signals precluded deep tissue imaging, due to failure of light penetration through tissue to the CCD. A solution for the weak signal is to use fluorophores with higher excitation and emission wavelengths (and thus more penetrating power) in the near-infrared region of 650 to 1000 nm, which allows for deeper tissue imaging.

One limitation of fluorophores is the reduced sensitivity of detection due to emission wavelength overlap. The use of semiconductor quantum dots (QDs) has enhanced tracking capabilities within deeper tissues. Quantum dots are synthetic nanocrystals (2 to 10 nm in diameter) with a variety of metalloid–crystalline complex cores [e.g., zinc–selenium (ZnSe), cadmium–selenium (CdSe), and zinc–sulfide (ZnS)] that can be attached to small molecules (nucleic acids, peptides, and proteins). The QD labels can emit a wide array of colored fluorescence with heightened sensitivity and reduced emission wavelength overlap. This allows multiple

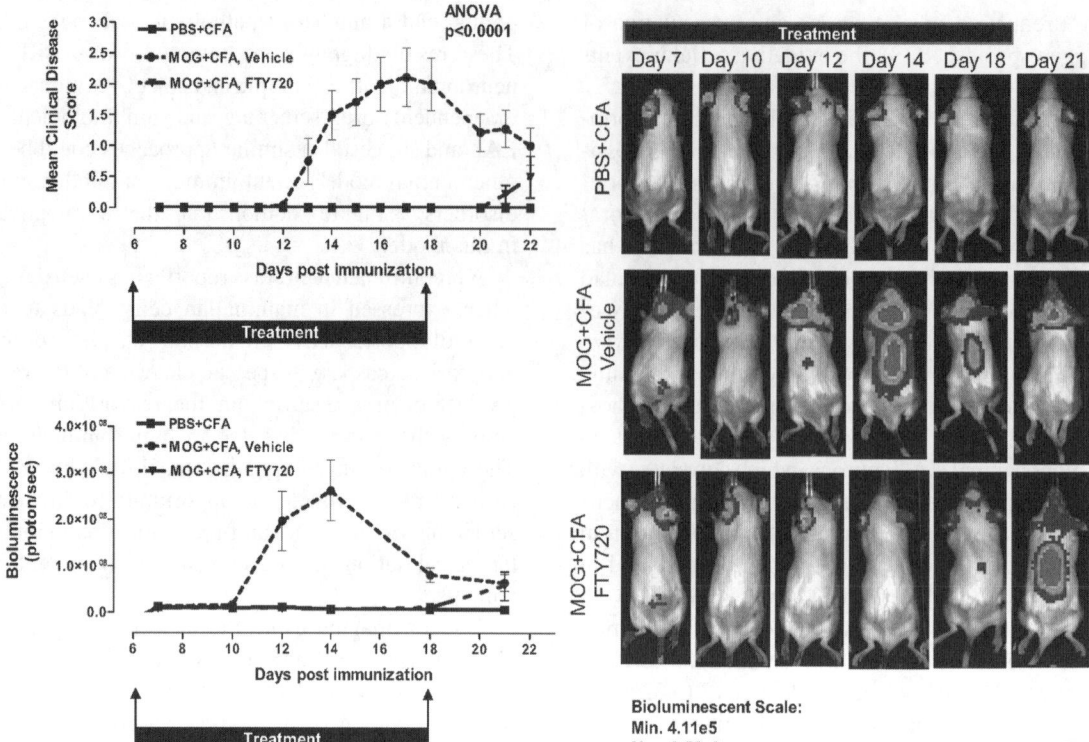

FIGURE 4 Whole-body bioluminescence imaging (BLI) of glial cell reactivity in the brain is correlated to the clinical progression of murine experimental autoimmune encephalitis (EAE). FVB/N mice [nonsusceptible strain due to major histocompatability complex (MHC) haplotype H^{2q}] engineered to express the GFAP-luc fusion transgene [linking the glial fibrillary acidic protein (GFAP) promoter to luciferase (luc)] were crossed with C57BL/6J-Tyrc-2J mice (susceptible strain via MHC haplotype H^{2b}). The FVB/N animals do not develop EAE in response to immunization with a myelin oligodendrocyte glycoprotein fragment (MOG35-55), while the transgenic F_1 offspring (H^{2q}/H^{2b} haplotype) readily develop the disease. EAE was induced by immunization with MOG35-55 peptide emulsified in complete Freund's adjuvant (CFA) followed by pertussis toxin 24 h later; the negative control group did not receive MOG35-55. One EAE group was gavaged daily with FTY720 (a reference immuno-modulating compound that reduces inflammation; Cayman Chemicals, Ann Arbor, MI) at 5 mg/kg per day; the negative control group was given phosphate-buffered saline (PBS) daily by oral gavage. The treatment (black bar) started on day 6 after immunization and continued for the duration of the study. The intensity of BLI signaling superimposed over the neuraxis (brain and/or spinal cord) of the mice is red (greatest) > yellow > green > blue (least). [Courtesy of Rukiye-Nazan E. Dogan, Stephen J. McAndrew, and Olesia Buiakova, Caliper Life Sciences, Discovery Alliances & Services Division (Xenogen Biosciences).] (*See insert for color representation of the figure.*)

molecular targets to be imaged simultaneously with high sensitivity. Several factors make QDs useful biologic labels, especially their hydrophilic and photostable properties. However, noncoated semiconductor nanocrystals have demonstrated toxicities and bioincompatibilities that preclude their current use in animal toxicology studies (reviewed by Aillon et al.[115]). Current discrepancies in the literature probably reflect a lack of toxicology-based studies, inconsistent QD dosages/exposure for prior in vivo work, and the wide variety of physiochemical properties that are unique to each type of QD.[116] Even though these inconsistencies exist, current work with quantum dots is geared toward reducing their toxicity and bioreactivity. One

approach involves encapsulating them in phospholipid micelles.[117] Once technological advances mitigate these difficulties, QD will become an increasingly popular means of performing noninvasive neuroimaging studies in animal models.

The ability to transduce neural cells with viral constructs coupled to fluorescent or bioluminescent markers has been used in vivo to track cell migration, changes in cell function, and protein expression. Such applications have immense potential in examining the progression and treatment of neurologic disorders. The in vivo and in vitro study of central neurons and intracellular neuronal components in the postnatal brain is best accomplished by the use of viral vector technology. The ability to transduce cells with viral vectors containing highly active promoters, coupled to fluorescent markers that target various cell types, allows for optical imaging in selected subpopulations of neurons. For in vivo imaging lentivirus (LV), adenovirus (AV), AAV, and CMV vectors are typically employed, since they lack the genes that encode for structural proteins and therefore prevent viral self-assembly. However, for rapid transgene expression (4 to 6 h), alphaviruses such as Sindbis virus (SV) and Semliki forest virus (SFV) are utilized for acute in vivo or brain slice imaging. Alphaviruses are currently the primary viral vector employed to express flurorescent proteins in nervous tissue.[118] However, these viruses are known to manipulate the host cell protein synthesis machinery, causing rapid cellular apoptosis. The limited time course of cellular survival restricts the use of this virus to acute in vivo and in vitro studies.

For example, the fate of human neural stem cells (hNSC) in an experimental murine glioma model has been explored using a variety of LV vectors coupled to fused marker proteins incorporating both flurorescent and bioluminescent labels.[119] Combining these two optical imaging modalities circumvented the limitations of bioluminescent sensitivity. These LV vectors are often used for such purposes because they offer stable integration of the transgene into the host genome regardless of the target cell's state of division. Using confocal and multiphoton microscopy, this optical imaging application demonstrated the homing properties of hNSC under the influence of gliomas. Multiphoton microscopy offers the best working distances of all the optical microscopy modalities; however, this is still limited to hundreds of micrometers. Signal loss at greater depth is partially related to light scatter and background.

Near-infared spectroscopy (NIRS) is a technology used previously as a noninvasive tool for in vivo monitoring of tissue oxygenation (pulse oximetry). Today, continuous-wavelength NIRS (CW-NIRS) is a low-cost imaging approach that provides a noninvasive, nonionizing modality for analyzing real-time vascular and metabolic functions of the CNS. Using wavelengths in the near-infrared range, NIRS measures the difference in light absorption between oxygenated hemoglobin, deoxygenated hemoglobin, water, and lipids as blood passes through the tissues. A sensitive CW-NIRS prototype using fiber-optic probes has been developed to investigate the brain physiology of rodents. Variations in tissue composition, vascularity, and metabolic activity of tissues can affect transmission of photons from the tissue source to the detector.[120] Nevertheless, CW-NIRS identifies this modality as a useful index for evaluating certain classes of neuroactive chemicals based on their brain penetration and pharmacokinetic profile.

ULTRASOUND

In general, ultrasound (US) is not well suited for neuroimaging, due to the impedance of the sound waves by the mineralized bone of the skull. Imaging the fetal brain in utero or the neonatal brain through the fontanelle might have potential application in primates. Application of low-intensity ($f = 0.69$ MHz), focused US administration with microbubbles (FUS-MB) has been used successfully to open the BBB to deliver radiotracers or drugs[121] and even therapeutic antibodies[122] into the brain. The delivery of tumor-specific antibodies conjugated to MR contrast agents can allow for molecular imaging of intracranial tumors. This same method has been applied with the MR-guided neuroimaging to target therapeutic agents directly into neurodegenerative lesions[123] in a mouse model of Alzheimer's disease. Antibodies were targeted to the hippocampus successfully using MRI guidance, but due to the size of the transducer (2×14 mm), some off-target delivery into the cortex and thalamus was observed.[123] Electron microscopy studies suggest that the microbubbles traverse the BBB through paraendothelial passages or by transendothelieal vescicular transport.[124,125] Sequelae of this treatment are small petechial hemorrhages in the neuropil; these structural findings have not been linked to postprocedure neurological dysfunction. Further understanding of the mechanisms and cellular response of FUS-MB therapy will be needed before translation of this technology to humans.

CONCLUSIONS

In summary, it is clear that noninvasive in vivo neuroimaging in animal models of human neurologic diseases will not only provide the basis for better understanding of genetic predispositions and environmental modulation of many human diseases but will also enhance the evaluation of novel treatment strategies. In this era of translational medicine, neuroimaging techniques involving animal subjects will provide unique insights into the diagnosis, treatment, and prognosis of human neural diseases.

Acknowledgments

P.C. and M.N. NIH/NIDA grants DA 10584, DA 17763, DA 14637, and DA 06634; K.G. RO1HL088649, Spore in Breast Cancer Developmental Research Award P50CA88843-04, NIEHS, PES003819, Department of Defense BC030727, AHA SDG 0335273N.

REFERENCES

1. Porrino LJ, et al. The expanding effects of cocaine: studies in a nonhuman primate model of cocaine self-administration. *Neurosci Biobehav Rev.* 2004;27(8):813–820.

2. Berger B, et al. Dopaminergic innervation of the cerebral cortex: unexpected differences between rodents and primates. *Trends Neurosci.* 1991;14(1):21–27.

3. Cragg SJ, et al. Dopamine release and uptake dynamics within nonhuman primate striatum in vitro. *J Neurosci.* 2000;20(21):8209–8217.

4. Haber SN, McFarland NR. The concept of the ventral striatum in nonhuman primates. *Ann NY Acad Sci.* 1999;877:33–48.

5. Joel D, Weiner I. The connections of the dopaminergic system with the striatum in rats and primates: an analysis with respect to the functional and compartmental organization of the striatum. *Neuroscience.* 2000;96(3):451–474.

6. Mansour A, et al. Anatomy of CNS opioid receptors. *Trends Neurosci.* 1988;11(7):308–314.

7. Smith HR, et al. Distribution of norepinephrine transporters in the non-human primate brain. *Neuroscience.* 2006;138(2):703–714.

8. Weerts EM, et al. The value of nonhuman primates in drug abuse research. *Exp Clin Psychopharmacol.* 2007;15(4):309–327.

9. Lyons D, et al. Cocaine alters cerebral metabolism within the ventral striatum and limbic cortex of monkeys. *J Neurosci.* 1996;16(3):1230–1238.

10. Porrino LJ, et al. Metabolic mapping of the effects of cocaine during the initial phases of self-administration in the nonhuman primate. *J Neurosci.* 2002;22(17):7687–7694.

11. Howell LL, et al. Cocaine-induced brain activation determined by positron emission tomography neuroimaging in conscious rhesus monkeys. *Psychopharmacology Berl.* 2002;159(2):154–160.

12. Kaplan JR, et al. Social status, environment, and atherosclerosis in cynomolgus monkeys. *Arteriosclerosis.* 1982;2(5):359–368.

13. Nader MA, Czoty PW. PET imaging of dopamine D2 receptors in monkey models of cocaine abuse: genetic predisposition versus environmental modulation. *Am J Psychiatry.* 2005;162(8):1473–1482.

14. Senda M, Herscovitch PKY. In: Senda KY, Herscovitch MP, eds. *Brain Imaging Using PET.* San Diego, CA: Academic Press; 2002.

15. Huettel SA, McCarthy G. In: Huettel SA, McCarthy G, eds. *Functional Magnetic Resonance Imaging.* Sunderland, MA: Sinauer Associates; 2003.

16. Klunk WE, et al. Imaging brain amyloid in Alzheimer's disease with Pittsburgh compound-B. *Ann Neurol.* 2004;55(3):306–319.

17. Kaufman MJ ed. *Brain Imaging in Substance Abuse: Research, Clinical and Forensic Applications.* Totowa, NJ: Humana Press; 2001.

18. Nader MA, Czoty PW. Brain imaging in nonhuman primates: insights into drug addiction. *Ilar J.* 2008;49(1):89–102.

19. Howell LL, et al. An apparatus and behavioral training protocol to conduct positron emission tomography (PET) neuroimaging in conscious rhesus monkeys. *J Neurosci Methods.* 2001;106(2):161–169.

20. Kaufman MJ, et al. Cocaine decreases relative cerebral blood volume in humans: a dynamic susceptibility contrast magnetic resonance imaging study. *Psychopharmacology (Berl).* 1998;138(1):76–81.

21. Kaufman MJ, et al. Cocaine-induced cerebral vasoconstriction detected in humans with magnetic resonance angiography. *JAMA.* 1998;279(5):376–380.

22. Pearlson GD, et al. Correlation of acute cocaine-induced changes in local cerebral blood flow with subjective effects. *Am J Psychiatry.* 1993;150(3):495–497.

23. Wallace EA, et al. Acute cocaine effects on absolute cerebral blood flow. *Psychopharmacology Berl.* 1996;128(1):17–20.

24. Nader MA, et al. Review: Positron emission tomography imaging studies of dopamine receptors in primate models of addiction. *Philos Trans R Soc Lond B* 2008;363(1507):3223–3232.

25. Lecharny JB, et al. Effects of thiopental, halothane and isoflurane on the calcium-dependent and -independent release of GABA from striatal synaptosomes in the rat. *Brain Res.* 1995;670(2):308–312.

26. Barrett SP, et al. The hedonic response to cigarette smoking is proportional to dopamine release in the human striatum as measured by positron emission tomography and [^{11}C]raclopride. *Synapse.* 2004;54(2):65–71.

27. Tsukada H, et al. Isoflurane anesthesia enhances the inhibitory effects of cocaine and GBR12909 on dopamine transporter: PET studies in combination with microdialysis in the monkey brain. *Brain Res.* 1999;849(1–2):85–96.

28. Lindefors N, et al. Differential effects of single and repeated ketamine administration on dopamine, serotonin and GABA transmission in rat medial prefrontal cortex. *Brain Res.* 1997;759(2):205–212.

29. Onoe H, et al. Ketamine increases the striatal *N*-[^{11}C]methylspiperone binding in vivo: positron emission tomography study using conscious rhesus monkey. *Brain Res.* 1994;663(2):191–198.

30. Nader MA, et al. PET imaging of dopamine D2 receptors with [^{18}F]fluoroclebopride in monkeys: effects of isoflurane- and ketamine-induced anesthesia. *Neuropsychopharmacology.* 1999;21(4):589–596.

31. Chen MK, et al. Acute manganese administration alters dopamine transporter levels in the non-human primate striatum. *Neurotoxicology.* 2006;27(2):229–236.

32. Massoud TF, et al. Noninvasive molecular neuroimaging using reporter genes: II. Experimental, current, and future applications. *Am J Neuroradiol.* 2008;29(3):409–418.

33. Lee JS, et al. FDG-PET for pharmacodynamic assessment of the fatty acid synthase inhibitor C75 in an experimental model of lung cancer. *Pharm Res.* 2007;24(6):1202–1207.

34. Lan Q, et al. Preclinical evaluation of SPECT imaging with ^{131}I-labeled monoclonal antibody SZ39 in nude mice bearing human glioma xenografts. *J Neuroimag.* 1996;6(2):131–135.

35. Gabrielson K, et al. Detection of dose response in chronic doxorubicin-mediated cell death with cardiac technetium 99m annexin V single-photon emission computed tomography. *Mol Imag.* 2008.

36. Riemann B, et al. Small animal PET in preclinical studies: opportunities and challenges. *Q J Nucl Med Mol Imag.* 2008;52(3):215–221.

37. Schober O, et al. Multimodality molecular imaging: from target description to clinical studies. *Eur J Nucl Med Mol Imag.* 2009.

38. Kornblum HI, et al. In vivo imaging of neuronal activation and plasticity in the rat brain by high resolution positron emission tomography (microPET). *Nat Biotechnol.* 2000;18(6): 655–660.

39. Meyer PT, et al. Comparison of intravenous and intraperitoneal [^{123}I]IBZM injection for dopamine D2 receptor imaging in mice. *Nucl Med Biol.* 2008;35(5):543–548.

40. Bertolino A, et al. The relationship between dorsolateral prefrontal neuronal *N*-acetylaspartate and evoked release of striatal dopamine in schizophrenia. *Neuropsychopharmacology.* 2000;22(2):125–132.

41. Cardenas L, et al. Oral D-amphetamine causes prolonged displacement of [^{11}C]raclopride as measured by PET. *Synapse.* 2004;51(1):27–31.

42. Leyton M, et al. Amphetamine-induced increases in extracellular dopamine, drug wanting, and novelty seeking: a PET/[^{11}C]raclopride study in healthy men. *Neuropsychopharmacology.* 2002;27(6):1027–1035.

43. Hirvonen J, et al. Measurement of striatal and thalamic dopamine D2 receptor binding with ^{11}C-raclopride. *Nucl Med Commun.* 2003;24(12):1207–1214.

44. Marenco S, et al. Nicotine-induced dopamine release in primates measured with [^{11}C]raclopride PET. *Neuropsychopharmacology.* 2004;29(2):259–268.

45. Boileau I, et al. Alcohol promotes dopamine release in the human nucleus accumbens. *Synapse.* 2003;49(4):226–231.

46. Constantinescu CC, et al. Estimation from PET data of transient changes in dopamine concentration induced by alcohol: support for a non-parametric signal estimation method. *Phys Med Biol.* 2008;53(5):1353–1367.

47. Yoder KK, et al. Heterogeneous effects of alcohol on dopamine release in the striatum: a PET study. *Alcohol Clin Exp Res.* 2007;31(6):965–973.

48. Breier A, et al. Effects of NMDA antagonism on striatal dopamine release in healthy subjects: application of a novel PET approach. *Synapse.* 1998;29(2):142–147.

49. Toyama H, et al. In vivo imaging of microglial activation using a peripheral benzodiazepine receptor ligand: [^{11}C]PK-11195 and animal PET following ethanol injury in rat striatum. *Ann Nucl Med.* 2008;22(5):417–424.

50. Cicchetti F, et al. Neuroinflammation of the nigrostriatal pathway during progressive 6-OHDA dopamine degeneration in rats monitored by immunohistochemistry and PET imaging. *Eur J Neurosci.* 2002;15(6):991–998.

51. van Kasteren SI, et al. Glyconanoparticles allow pre-symptomatic in vivo imaging of brain disease. *Proc Natl Acad Sci USA.* 2008.

52. Liow JS, et al. P-glycoprotein function at the blood–brain barrier imaged using ^{11}C-*N*-desmethyl-loperamide in monkeys. *J Nucl Med.* 2008.

53. Klunk WE, et al. Binding of the positron emission tomography tracer Pittsburgh compound-B reflects the amount of amyloid-beta in Alzheimer's disease brain but not in transgenic mouse brain. *J Neurosci.* 2005;25(46):10598–10606.

54. Henriksen G, et al. Development and evaluation of compounds for imaging of beta-amyloid plaque by means of positron emission tomography. *Eur J Nucl Med Mol Imag.* 2008; 35(Suppl. 1):S75–S81.

55. Winkeler A, et al. Imaging noradrenergic influence on amyloid pathology in mouse models of Alzheimer's disease. *Eur J Nucl Med Mol Imag.* 2008;35(Suppl 1): S107–S113.

56. Nordberg A. PET imaging of amyloid in Alzheimer's disease. *Lancet Neurol.* 2004;3(9):519–527.

57. Wengenack TM, et al. MR microimaging of amyloid plaques in Alzheimer's disease transgenic mice. *Eur J Nucl Med Mol Imag.* 2008;35(Suppl 1):S82–S88.

58. Natt O, et al. High-resolution 3D MRI of mouse brain reveals small cerebral structures in vivo. *J Neurosci Methods.* 2002; 120(2):203–209.

59. Johnson MH, et al. Neuroimaging of typical and atypical development: a perspective from multiple levels of analysis. *Dev Psychopathol.* 2002;14(3):521–536.

60. Benveniste H, Blackband SJ. Translational neuroscience and magnetic-resonance microscopy. *Lancet Neurol.* 2006;5(6): 536–544.

61. Matthews PM, et al. Applications of fMRI in translational medicine and clinical practice. *Nat Rev Neurosci.* 2006;7(9): 732–744.

62. Bock NA, et al. Multiple-mouse MRI. *Magn Reson Med.* 2003;49(1):158–167.

63. Catana C, et al. Simultaneous in vivo positron emission tomography and magnetic resonance imaging. *Proc Natl Acad Sci USA.* 2008;105(10):3705–3710.

64. Judenhofer MS, et al. PET/MR images acquired with a compact MR-compatible PET detector in a 7-T magnet. *Radiology.* 2007;244(3):807–814.

65. Johnson GA, et al. Histology by magnetic resonance microscopy. *Magn Reson Q.* 1993;9(1):1–30.

66. Aime S, et al. Insights into the use of paramagnetic Gd(III) complexes in MR-molecular imaging investigations. *J Magn Reson Imag.* 2002;16(4):394–406.

67. Terreno E, et al. Paramagnetic liposomes as innovative contrast agents for magnetic resonance (MR) molecular imaging applications. *Chem Biodivers.* 2008;5(10):1901–1912.

68. Maronpot RR, et al. Applications of magnetic resonance microscopy. *Toxicol Pathol.* 2004;32 (Suppl 2):42–48.

69. Bulte JW, Kraitchman DL. Iron oxide MR contrast agents for molecular and cellular imaging. *NMR Biomed.* 2004;17(7): 484–499.

70. Louie AY, et al. In vivo visualization of gene expression using magnetic resonance imaging. *Nat Biotechnol.* 2000;18(3): 321–325.

71. Modo M, et al. Cellular MR imaging. *Mol Imag.* 2005;4(3): 143–164.

72. Walczak P, et al. Instant MR labeling of stem cells using magnetoelectroporation. *Magn Reson Med.* 2005;54(4):769–774.

73. Natt O, et al. Use of phased array coils for a determination of absolute metabolite concentrations. *Magn Reson Med.* 2005;53(1):3–8.

74. Artemov D. Molecular magnetic resonance imaging with targeted contrast agents. *J Cell Biochem.* 2003;90(3):518–524.

75. Artemov D. et al. MR molecular imaging of the Her-2/neu receptor in breast cancer cells using targeted iron oxide nanoparticles. *Magn Reson Med.* 2003;49(3):403–408.

76. Bhujwalla ZM, et al. Reduction of vascular and permeable regions in solid tumors detected by macromolecular contrast magnetic resonance imaging after treatment with antiangiogenic agent TNP-470. *Clin Cancer Res.* 2003;9(1):355–362.

77. Blankenberg FG. In vivo detection of apoptosis. *J Nucl Med.* 2008;49(Suppl2):81S–95S.

78. Partlow KC, et al. ^{19}F magnetic resonance imaging for stem/progenitor cell tracking with multiple unique perfluorocarbon nanobeacons. *FASEB J.* 2007;21(8):1647–1654.

79. Verdijk P, et al. Sensitivity of magnetic resonance imaging of dendritic cells for in vivo tracking of cellular cancer vaccines. *Int J Cancer.* 2007;120(5):978–984.

80. Remsen LG, et al. MR of carcinoma-specific monoclonal antibody conjugated to monocrystalline iron oxide nanoparticles: the potential for noninvasive diagnosis. *Am J Neuroradiol.* 1996;17(3):411–418.

81. Pirko I, et al. Magnetic resonance imaging, microscopy, and spectroscopy of the central nervous system in experimental animals. *NeuroRx.* 2005;2(2):250–264.

82. Lee JH, et al. Artificially engineered magnetic nanoparticles for ultra-sensitive molecular imaging. *Nat Med.* 2007;13(1):95–99.

83. Silva AC, et al. Manganese-enhanced magnetic resonance imaging (MEMRI): methodological and practical considerations. *NMR Biomed.* 2004;17(8):532–543.

84. Buhmann C, et al. Pharmacologically modulated fMRI: cortical responsiveness to levodopa in drug-naive hemiparkinsonian patients. *Brain.* 2003;126(Pt 2):451–461.

85. van Zijl PC, et al. Quantitative assessment of blood flow, blood volume and blood oxygenation effects in functional magnetic resonance imaging. *Nat Med.* 1998;4(2):159–167.

86. Dazai J, et al. Multiple mouse biological loading and monitoring system for MRI. *Magn Reson Med.* 2004;52(4):709–715.

87. Driehuys B, et al. Small animal imaging with magnetic resonance microscopy. *Ilar J.* 2008;49(1):35–53.

88. Hedlund LW, et al. MR-compatible ventilator for small animals: computer-controlled ventilation for proton and noble gas imaging. *Magn Reson Imag.* 2000;18(6):753–759.

89. Hedlund LW, et al. A ventilator for magnetic resonance imaging. *Invest Radiol.* 1986;21(1):18–23.

90. Klaunberg BA, MJ Lizak. Considerations for setting up a small-animal imaging facility. *Lab Anim NY.* 2004;33(3):28–34.

91. Hockings PD, et al. Longitudinal magnetic resonance imaging quantitation of rat liver regeneration after partial hepatectomy. *Toxicol Pathol.* 2002;30(5):606–610.

92. Keilholz SD, et al. Comparison of first-pass Gd-DOTA and FAIRER MR perfusion imaging in a rabbit model of pulmonary embolism. *J Magn Reson Imag.* 2002;16(2):168–171.

93. Cassidy PJ, et al. Assessment of motion gating strategies for mouse magnetic resonance at high magnetic fields. *J Magn Reson Imag.* 2004;19(2):229–237.

94. Inderbitzin D, et al. Abdominal magnetic resonance imaging in small rodents using a clinical 1.5 T MR scanner. *Methods.* 2007;43(1):46–53.

95. Johnson GA, et al. High-throughput morphologic phenotyping of the mouse brain with magnetic resonance histology. *Neuroimage.* 2007;37(1):82–89.

96. Higuchi M, et al. ^{19}F and ^{1}H MRI detection of amyloid beta plaques in vivo. *Nat Neurosci.* 2005;8(4):527–533.

97. Charles-Edwards EM, deSouza NM. Diffusion-weighted magnetic resonance imaging and its application to cancer. *Cancer Imaging.* 2006;6:135–143.

98. Grohn OH, et al. Novel magnetic resonance imaging contrasts for monitoring response to gene therapy in rat glioma. *Cancer Res.* 2003;63(22):7571–7574.

99. Hiller KH, et al. Assessment of cardiovascular apoptosis in the isolated rat heart by magnetic resonance molecular imaging. *Mol Imag.* 2006;5(2):115–121.

100. Sosnovik DE, et al. Magnetic resonance imaging of cardiomyocyte apoptosis with a novel magneto-optical nanoparticle. *Magn Reson Med.* 2005;54(3):718–724.

101. van Tilborg GA, et al. Annexin A5-functionalized bimodal lipid-based contrast agents for the detection of apoptosis. *Bioconjug Chem.* 2006;17(3):741–749.

102. Lester DS, et al. Virtual neuropathology: three-dimensional visualization of lesions due to toxic insult. *Toxicol Pathol.* 2000;28(1):100–104.

103. Morgan DL, et al. Neurotoxicity of carbonyl sulfide in F344 rats following inhalation exposure for up to 12 weeks. *Toxicol Appl Pharmacol.* 2004;200(2):131–145.

104. Sills RC, et al. Contribution of magnetic resonance microscopy in the 12-week neurotoxicity evaluation of carbonyl sulfide in Fischer 344 rats. *Toxicol Pathol.* 2004;32(5):501–510.

105. Badea A, et al. Neuroanatomical phenotypes in the reeler mouse. *Neuroimage.* 2007;34(4):1363–1374.

106. Rudin M, Weissleder R. Molecular imaging in drug discovery and development. *Nat Rev Drug Discov.* 2003;2(2):123–131.

107. Ishikawa TO, et al. Imaging cyclooxygenase-2 (Cox-2) gene expression in living animals with a luciferase knock-in reporter gene. *Mol Imag Biol.* 2006;8(3):171–187.

108. Laxman B, et al. Noninvasive real-time imaging of apoptosis. *Proc Natl Acad Sci USA.* 2002;99(26):16551–16555.

109. Liu JJ, et al. Bioluminescent imaging of TRAIL-induced apoptosis through detection of caspase activation following cleavage of DEVD-aminoluciferin. *Cancer Biol Ther.* 2005;4(8):885–892.

110. McCullough KD, et al. Gadd153 sensitizes cells to endoplasmic reticulum stress by down-regulating Bcl2 and perturbing the cellular redox state. *Mol Cell Biol.* 2001;21(4):1249–1259.

111. Luo J, Wyss-Coray T. Bioluminescence analysis of smad-dependent tgf-Beta signaling in live mice. *Methods Mol Biol.* 2009;574:193–202.

112. Luo J, et al. Bioluminescence imaging of Smad signaling in living mice shows correlation with excitotoxic neuro-degeneration. *Proc Natl Acad Sci USA.* 2006;103(48): 18326–18331.

113. Zinn KR, et al. Noninvasive bioluminescence imaging in small animals. *Ilar J.* 2008;49(1):103–115.

114. Luo J, et al. Bioluminescence in vivo imaging of autoimmune encephalomyelitis predicts disease. *J Neuroinflamm.* 2008;5:6.

115. Aillon KL, et al. Effects of nanomaterial physicochemical properties on in vivo toxicity. *Adv Drug Deliv Rev.* 2009; 61(6):457–466.

116. Hardman R. A toxicologic review of quantum dots: toxicity depends on physicochemical and environmental factors. *Environ Health Perspect.* 2006;114(2):165–172.

117. Dubertret B, et al. In vivo imaging of quantum dots encapsulated in phospholipid micelles. *Science.* 2002;298(5599): 1759–1762.

118. Teschemacher AG, et al. Imaging living central neurones using viral gene transfer. *Adv Drug Deliv Rev.* 2005;57(1): 79–93.

119. Steffen D, Weinberg RA, The integrated genome of murine leukemia virus. *Cell.* 1978;15(3):1003–1010.

120. Crespi F. Near-infrared spectroscopy (NIRS): a non-invasive in vivo methodology for analysis of brain vascular and metabolic activities in real time in rodents. *Curr Vasc Pharmacol.* 2007;5(4):305–321.

121. Hynynen K, et al. Local and reversible blood–brain barrier disruption by noninvasive focused ultrasound at frequencies suitable for trans-skull sonications. *NeuroImage.* 2005;24(1): 12–20.

122. Kinoshita M, et al. Noninvasive localized delivery of herceptin to the mouse brain by MRI-guided focused ultrasound-induced blood–brain barrier disruption. *Proc Natl Acad Sci USA.* 2006;103(31):11719–11723.

123. Raymond SB, et al. Ultrasound enhanced delivery of molecular imaging and therapeutic agents in Alzheimer's disease mouse models. *PLoS One.* 2008;3(5):e2175.

124. Sheikov N, et al. Brain arterioles show more active vesicular transport of blood-borne tracer molecules than capillaries and venules after focused ultrasound-evoked opening of the blood–brain barrier. *Ultrasound Med Biol.* 2006;32(9): 1399–1409.

125. Sheikov N, et al. Cellular mechanisms of the blood–brain barrier opening induced by ultrasound in presence of microbubbles. *Ultrasound Med Biol.* 2004;30(7):979–989.

19

CEREBROSPINAL FLUID ANALYSIS IN TOXICOLOGICAL NEUROPATHOLOGY

WILLIAM VERNAU

Department of Pathology, Microbiology and Immunology, School of Veterinary Medicine, University of California–Davis, Davis, California

KAREN M. VERNAU

Department of Surgical and Radiological Sciences, School of Veterinary Medicine, University of California–Davis, Davis, California

BRAD BOLON

GEMpath, Inc., Longmont, Colorado

INTRODUCTION

Biomedical scientists engaged in neurotoxicity research are well acquainted with the basic properties of cerebrospinal fluid (CSF). In contrast, the importance of the interstitial fluid (ISF) within the brain and spinal cord is often unappreciated, even though ISF is the principal constituent in the microenvironment immediately surrounding neurons and glia within the CNS. Analysis of CSF as a general index of central nervous system (CNS) health is a tool with reasonable sensitivity but low specificity. The range of CSF abnormalities is relatively limited compared to the varieties of neurological diseases that exist (particularly if the analysis is restricted to total and differential cell counts and protein concentration determination). Furthermore, the type and degree of CSF alterations is affected by such factors as prior therapy[1] and the site of CSF collection relative to the location of the lesion within the CNS;[2] diseases centered in the meninges and pereventricular regions (the CSF compartment) generally produce greater abnormalities than do diseases in the deep neuropil (the ISF compartment). Finally, the CSF of people with neurological disease is not always abnormal.[3] Therefore, in most situations, the chief utility of CSF analysis is to rule out the presence of certain confounding disease processes. Examination of CSF is most useful when the results are correlated with the history, clinical findings, CNS imaging studies, and ancillary laboratory tests.

The rationale for sampling CSF as an index of CNS health is straightforward. The CSF within the ventricular system is continuous with the ISF that permeates the deep neuropil of the brain and spinal cord. Thus, a CSF sample provides an avenue for indirectly evaluating tissue responses within the CNS parenchyma as well as evaluating more directly diseases involving the meninges and ventricular system. In addition, acquisition of a CSF specimen is a much less invasive means of probing neuraxial lesions relative to a direct biopsy of CNS tissue. Data from CSF evaluation often offer the first definitive information about the nature of a neuropathological process.

FUNCTIONS OF CSF AND ISF IN THE CNS

The production and circulation of CSF and ISF within the CNS serves four primary purposes.[4–7] First, CSF acts as a cushion to physically support delicate neural structures. A 1500 g human brain buoyed by CSF weighs only about 50 g. The volume of the CSF cushion readily adapts to comply with volume changes in the other intracranial contents, particularly the rapid shifts in blood quantity arising from cyclic

Fundamental Neuropathology for Pathologists and Toxicologists: Principles and Techniques, First Edition. Edited by Brad Bolon and Mark T. Butt.
© 2011 John Wiley & Sons, Inc. Published 2011 by John Wiley & Sons, Inc.

waves of arterial and central venous pressure associated with movement (e.g., postural adjustments, respiration, and exertion). Acute or chronic pathological changes in intracranial contents can also be accommodated to some extent by changes in CSF volume.

Second, the dynamic flux that exists between blood plasma, CSF, and ISF maintains the chemical microenvironment in the neuropil within the relatively narrow limits required for optimal function of neural cells. The CNS is segregated from the "body" (i.e., the systemic circulation) by several semipermeable barriers. Selective carriers and ion pumps use energy to transport essential molecules across these barriers. The production, circulation, and absorption of CSF and ISF provide continual turnover of the molecules, proteins, and nonanchored cells in the neuropil and ventricular system, thereby preserving a relatively stable milieu despite continual changes in blood composition.

Third, CSF and ISF serve as an excretory conduit for removing metabolites, larger proteins, and cells from the neuropil. This function is accomplished by the bulk flow of ISF through the lipid-rich parenchyma into the ventricular system,[8] and is facilitated by the capacity of CSF to act as a "sink" for water-soluble molecules that have entered or been synthesized in the brain. Given their generally low concentrations in the CNS, such polarized solutes diffuse freely from the ISF into the CSF. Removal then occurs by bulk CSF resorption at the arachnoid villi or, in some cases, by fluid transfer into the capillaries of the choroid plexus.

Finally, CSF and ISF may serve as intracerebral carriers for neuroactive substances.[8] For example, hormone-releasing factors from the hypothalamus are discharged into the CSF within the third ventricle, from whence they can readily diffuse to their target neurons in the median eminence. This route also appears to be relevant for intracerebral transport of opiates. In like manner, ISF flow within the neuropil can distribute neurotoxic proteins widely throughout the CNS

(e.g., β-amyloid[8,9]), and can presumably serve the same function for other neurotoxic chemicals, drugs, and metals.

PHYSIOLOGY OF CSF AND ISF

Anatomy of Brain–Body Barriers

Fluid entry into the CNS is controlled by a series of structural obstacles (Table 1). During evolution, the barriers appear to have shifted from a glial-oriented to an endothelial-localized arrangement.[10] These physical barriers are regulated by the coordinated interaction of influences on both the blood side and the CNS side.[10,11]

The blood–ISF barrier within the neuropil is comprised of the blood–brain barrier (BBB) and corresponding blood–spinal cord barrier residing in the intraparenchymal capillaries in most regions of the mammalian[4–7] and avian[12,13] brain. These interfaces are characterized by special structural adaptations of endothelial cells: (1) the absence of fenestrae, (2) fewer pinocytotic pits and vesicles, (3) more mitochondria in individual cells, and (4) the presence of tight junctions between adjacent cells. The structural adaptations of endothelia are accompanied by enriched expression of numerous proteins that facilitate unidirectional flow of critical solutes (including transporters, receptors, and proteins involved in vesicle trafficking[14]); such molecular markers of barrier maturity appear simultaneously in both parenchymal and pial vessels.[15] The integrity of the blood–ISF barriers within the endothelia is enhanced by the close apposition of foot processes from perivascular astrocytes. These processes are much more robust surrounding parenchymal vessels than near pial vessels.[15] The endothelial and astrocytic cells provide reciprocal inductive signals to form and sustain these barriers.[16,17] This induction can be affected by neuropathologic conditions, including exposure to neurotoxic agents.[16]

TABLE 1   Composition of the Interfaces Between Cerebrospinal and Interstitial Fluids and the Nervous System

Interface	Cell Type	Barrier Characteristics
Blood–ISF ("blood–brain barrier")	Capillary endothelium	High expression of unidirectional transport proteins
		Intercellular tight junctions[a]
		Intracellular fenestrae are absent
		Reduced pinocytosis
	Astrocytes	Foot processes
Blood–CSF	Choroid plexus epithelium	Tight junction[a]
CSF–brain	Ependyma	Gap junction[a]
	Pia mater	Gap junction[a]
		Glial limitans (putative)
CSF–blood	Arachnoid cells	Tight junction[a]
		Valve (in villi)

Source: Adapted in part from Rosenberg.[7]

[a] Denotes the main barrier component at each site.

The blood–CSF barrier has two components, segregated to two distinct domains of the CNS. The first is located in the specialized epithelial cells lining the choroid plexus. These cuboidal cells are joined by intercellular tight junctions but contain an abundance of fenestrations and intracellular organelles to facilitate their dual barrier and secretory functions.[4–7] The choroid plexus epithelium also has the brush border (microvilli) and numerous basolateral infoldings typical of cells engaged in fluid transport. Similar epithelial adapations are present in the circumventricular organs (CVOs) that abut the ventricular system. The second part of the blood–CSF barrier resides in the arachnoid villi, which project into the lumina of the dural sinuses. As with choroid plexus epithelium, the arachnoid cells have organelles that contribute to their twin functions as a barrier (tight junctions) and transport site (giant intracellular vacuoles and pinocytotic vesicles). Endothelium-lined channels in the venous sinus may link directly with the subarachnoid space.[4,18] Arachnoid villi also penetrate the spinal veins near the spinal nerve roots.[6,18]

The CSF–brain barrier exists at the ependymal lining to inhibit the flow of CSF back into the neuropil. This hurdle consists of gap junctions between the ependymal cells lining the ventricles, and between the lining cells of the pia mater. The linings of the ventricles and the pia are relatively permeable and allow continuous circulation between the CSF and ISF.[4,6] However, the subpial glia limitans may restrict diffusional exchange between the CSF and ISF.[19] The contribution of the CVO ependyma to this function is insignificant, as the combined area of all seven CVOs on the ventricular surface represents less than 1% of the total cerebrovascular bed.[20]

Physiology of CSF Production, Circulation, and Absorption

Formation of CSF occurs at the choroid plexuses.[4,6] Fluid is filtered across the porous choroidal capillaries and then secreted by the overlying epithelium. Filtration utilizes both hydrostatic pressure and energy-dependent, ion-specific pumps, and occurs through both the intercellular spaces and by intracellular transport. Secretion of CSF is dependent mainly on the active unidirectional transport of sodium (Na^+), chloride (Cl^-), and bicarbonate (HCO_3^-) ions from the blood into the ventricles.[21] This osmotic gradient results from the coordinated activity of Na^+/K^+-ATPases, K^+ channels, and $Na^+/K^+/Cl^-$ co-transporters at the apical (ventricular) membrane and coupled co-transporters (e.g., for Na^+ and HCO_3^-, and for K^+ and Cl^-) at the basolateral (parenchymal) cell border,[21] and it is established despite a very high electrical resistance across the barriers (averaging about 2000 Ω).[22] Water is drawn along the osmotic gradient passively, passing through an unknown channel at the basolateral membrane and by aquaporin-1 at the apical membrane.[21] The function of the transporters may be con-

TABLE 2   Rate of CSF Formation in Various Species[a]

Species	Rate (µL/min)
Mouse	0.325
Rat	2.1–5.4
Guinea pig	3.5
Rabbit	10
Cat	20–22
Dog	47–66
Primate	
Nonhuman	29–41
Human	350–370

Source: Modified from Davson and Segal.[4]
[a] Estimated by ventriculocisternal perfusion.

trolled by autonomic nerve terminals in the choroid plexus.[5,23] A lesser quantity of CSF (reportedly as much as a third less[4,6]) arises by simple diffusion across the ependyma or pia mater. The relative capacities for CSF production at choroidal and extrachoroidal sites are not clear for either normal or pathological conditions.

The rate of CSF formation varies among species (Table 2) but is closely correlated to the weight of the choroid plexus.[24] Humans produce approximately 500 mL of CSF per day; therefore, the total CSF volume (between 120 and 150 mL) is completely recycled three to four times daily.[19] Production of CSF generally occurs at a constant rate regardless of wide variations in cerebral blood flow,[25] although the rate can be modified by certain agonists and xenobiotic agents (Table 3). Clinical utility of these agents is limited by either their time frame of action or their toxicity.[4,7] Modest elevations in intraventricular pressure probably do not affect CSF formation, but a chronic increase in intracranial pressure results in decreased CSF generation.[5]

Brain fluids flow continuously in bulk from their sites of formation to sites of absorption. The CSF formed in the lateral ventricles moves through the paired interventricular foramina (of Monro) into the third ventricle, then through the mesencephalic aqueduct (of Sylvius) to reach the fourth ventricle. The majority of CSF moves from the fourth ventricle into the subarachnoid space through the lateral apertures (foramina of Luschka) and the median aperture (foramen of Magendie); the median aperture is lacking in nonhuman species except for the anthropoid apes.[26,27] In general, CSF permeates the spinal cord by flowing from the spinal subarachnoid space into the spinal perivascular spaces, across the interstitial space, and then into the central canal.[28] However, a small amount may pass from the fourth ventricle directly into the central canal. Mechanisms for propelling the CSF through the ventricular system probably include (1) pressure from newly produced CSF, (2) ciliary action of the ventricular ependyma, (3) cyclic pulses from respiratory and vascular movements, and (4) the pressure differential that exists across the arachnoid villi.[6]

TABLE 3   Agents That Can Influence the Rate of CSF Formation[a]

Effect	Agent	Site of Action	Mechanism of Action
Increase	Cholera toxin	Gs-α protein	Constitutive Gs-α activation leads to chronic increases in Gs-α-mediated adenylate cyclase activity and cAMP production
	Phenylephrine[a]	Cholinergic pathways	α1-Adrenergic receptor agonist
Decrease	Acetazolamide, furosemide	Choroidal carbonic anhydrase	Enzyme inhibition removes intracellular HCO_3^- and H^+ ions necessary for ion transporters to maintain osmotic gradient
	Atrial natriuretic peptide	Guanylate cyclase (the major action of ANP receptors)	Elevated cGMP leads to abnormally enhanced intracellular kinase activity
	Diazepam analogs[b]	Choroidal benzodiazepine receptor	Inhibits $GABA_A$ receptor-mediated neuronal inhibition
	Dopamine D_1 receptor agonists	Choroidal dopamine receptor	Modulates acetylcholine levels in cholinergic pathways
	Hyperosmolarity (of CSF/ISF)	Choroidal capillaries	Enhanced fluid efflux into neuropil
	Norepinephrine[a]	Choroidal Na^+/K^+-ATPase	Modulation of osmotic gradient across choroid epithelium
	Omeprazole[b]	Choroid	Unknown (apparently not associated with inhibition of H^+/K^+-ATPase)
	Ouabain	Choroidal Na^+/K^+-ATPase	Modulation of osmotic gradient across choroid epithelium
	Serotonin receptor agonists	Choroidal serotonin receptor	Modulation of osmotic gradient across choroid epithelium
	Steroids[b]	Choroidal Na^+/K^+-ATPase	Modulation of osmotic gradient across choroid epithelium
	Vasopressin	Choroidal vasopressin (V_1) receptor	Modulation of water movement across capillary walls

Source: Modified from Fishman[5], unless specified otherwise.
[a] From Nilsson, et al.[23]
[b] From Davson and Segal.[4]

Absorption of CSF occurs by both unidirectional bulk flow of fluid across the arachnoid villi and specific uptake of individual constituents (i.e., ions, proteins, and xenobiotics) at the choroid plexus. Absorption is accomplished indiscriminately by vesicles (both micropinocytotic and giant) and possibly through intercellular clefts as well as for specific constituents by active transport; the mechanisms differ among species.[18] Small quantities of solutes may also be cleared from the CSF by diffusion into adjacent neural cells or intraparenchymal capillaries.[5,6] Particles ranging in size from colloidal gold (0.2 μm) to erythrocytes (7.5 μm) can be transported across normal arachnoid villi without difficulty, but in disease clumps of larger particles (e.g., protein aggregates, leukocytes) may block villar absorption and initiate hydrocephalus.[5,6] Bulk absorption is directly proportional to the CSF hydrostatic pressure[4] and occurs via arachnoid villi that project into the dural venous sinuses, dural lymphatic channels, and perineural sheaths of cranial (particularly the olfactory) and spinal nerves.[4,6,29,30] Solutes and fluid absorbed via these routes reach the regional (especially the deep cervical) lymph nodes within minutes.[29,31] Disorders of the peripheral lymphatic system have been proposed as a potential etiology for dysfunction of the ventricular system in the CNS.[32] Again, the importance of these absorption routes varies among species.[18] In humans, CSF resorption via the arachnoid villi into the blood is thought to substantially exceed lymphatic trafficking of CSF.[33]

Physiology of ISF Production, Circulation, and Absorption

Relative to CSF, the physiology of ISF within the CNS has been less well documented. Formation of ISF in the brain and spinal cord occurs at a rate approximately 1/100 as fast as the generation of CSF,[8] or at between 0.1 to 0.3 mL/min/g in the rat.[8,30,34] Its production is thought to occur chiefly by active secretion from endothelium lining the intraparenchymal capillaries,[8] although an alternative hypothesis is that ISF results from passive filtration across the capillary endothelium followed by active astrocytic transport of water away from the perivascular spaces.[6,7] Minor contributions to ISF production are thought to arise by recycling of the CSF into the neuropil [from the subarachnoid space into perivascular (mainly arterial) spaces on the ventral surface of the brain] and formation of water as a by-product of neural cell metabolism.[8] The composition of ISF is adjusted by transporters specialized for secretion into [e.g., glucose transporter 1 (GLUT-1), system L-amino acid transporter] or away from (e.g., P-glycoprotein) the CNS parenchyma.[17]

The ISF circulates freely by bulk flow within the CNS. In the rat, the velocity of movement toward the ventricles has been calculated as 10.5 mm/min.[35] The pressures of CSF and ISF in the normal (rat) brain are comparable (about 3.5 mmHg[36]), so no impediment exists for unrestricted ISF flow. Movement of ISF tends to follow major white matter tracts and perivascular spaces (especially of arterioles and arteries) in gray matter.[8,31,33,35] This flow pattern is consistent with the spread and clearance of malignant cells and microbes within the CNS and is responsible for the distribution of most xenobiotics within the neuropil.[8] Large molecules carried in the ISF exhibit comparable clearance half-lives [e.g., polyethylene glycol, 0.9 kDa, 14.4 h; albumin, 69 kDa, 12.2 h], suggesting that fluid movement is an efficient and unrestricted means of transporting materials within the neuropil.[34]

Absorption of ISF occurs chiefly by transfer into the CSF. The drainage site depends on the site of ISF formation; fluid is transferred into the ventricular system at a point relatively close to its origin.[30] The movement of ISF into the CSF is accelerated in neuropathological conditions (e.g., cerebral edema) in which the ISF hydrostatic pressure exceeds that of the ventricles, and is retarded in conditions (e.g., hydrocephalus) in which CSF pressure exceeds that of the ISF.[37]

COMPOSITION OF CSF DURING HEALTH

Cellular Constituents

Normally, CSF is relatively acellular. Erythrocytes in a CSF sample are usually iatrogenic, due to trauma associated with needle placement.[38–40] Nucleated cells (chiefly leukocytes) may be found in the CSF of clinically and histologically normal animals but should be rare.[1,41] The nucleated cell count may be lower for CSF samples acquired from the lumbar cistern versus the cerebellomedullary cistern.[41] In our experience with cerebellomedullary CSF samples of dogs, a nucleated cell count of 0 to 2 cells/μL is normal, a count of 3 nucleated cells/μL is possibly abnormal, and 4 or more nucleated cells/μL is definitely abnormal.

Differential counts of nucleated cells in CSF samples may be useful in diagnosing neural diseases.[1,39] Normal CSF consists of varying proportions of monocytes (macrophages) and small lymphocytes, although a very small number of neutrophils may be found in normal CSF samples.[5,42,43] The cell proportions are species- and age-dependent.[42,44] Other nucleated cells (i.e., activated lymphocytes, activated macrophages, plasma cells, and other granulocytes) are not seen in normal CSF,[5,42,45] and their presence is nonspecific evidence

of such inflammatory disorders as acute viral encephalitis or chronic bacterial infections or immune-mediated diseases (e.g., multiple sclerosis, Guillain–Barré syndrome). Reactive leukocytes may also be present with neoplastic involvement of the CNS.[39]

Nucleated cells other than leukocytes can be seen in both normal and abnormal CSF.[39,42,46,47] The usual types are cuboidal epithelium (from the choroid plexus) or ependymal cells (from the ventricular lining), which are indistinguishable by routine cytologic analysis,[39,42,46] or squamous cells (representing skin contamination or an underlying neuropathologic process such as an epidermoid cysts). These elements typically exist as single cells, small papillary fronds, or flat rafts. Traumatic collection of CSF samples by lumbar puncture occasionally results in contamination by chondrocytes from the intervertebral disk[48] or hematopoietic precursors from the vertebral bone marrow.[42,49] Another source for hematopoietic precursors in normal dogs is extramedullary hematopoiesis in the choroid plexus.[50] Finally, neurons and glial cells may be observed in the CSF of people[48] and animals,[51] especially if the collection procedure was demonstrably traumatic.

Principal Biochemical Constituents of Normal CSF

CSF may be considered a plasma ultrafiltrate in which certain constituents are concentrated by active secretion. This clear, colorless fluid contains low concentrations of certain ions and proteins. In health, CSF composition is relatively constant, although some modest fluctuations occur as the plasma composition changes.

In general, the CSF concentration of a protein is inversely related to its molecular weight. If the BBB is functioning normally, proteins with a molecular weight greater then 160 kDa (i.e., having a large hydrodynamic radius) are largely excluded.[52] Almost all the proteins normally present in CSF are derived from the serum. The exceptions are transthyretin (prealbumin) and transferrin,[23,53] which are also synthesized by the choroid plexus, and beta and gamma trace proteins, τ protein (β2-transferrin), glial fibrillary acidic protein (GFAP), and myelin basic protein (MBP), all of which are synthesized within the CNS. The choroid plexus also manufactures certain CSF-borne hormones and growth factors, including insulin-like growth factor II (IGF II), vasopressin (VP), and transforming growth factor β1 (TGF-β1).[53]

The major CSF protein is liver-derived albumin (MW, 69 kDa). When total CSF protein rises, the greatest contributor is an influx of albumin from the periphery, entering across leaky blood–brain and/or blood–CSF barriers. Multiple globulins are also present within the CSF. No correlation has been shown between changes in the concentrations of α- and β-globulins and neurological disease in animals or humans.[5,54] The most abundant γ-globulins in normal CSF are the immunoglobulins: IgG and, to a lesser extent, IgA and IgM. Immunoglobulin levels in CSF are increased in immune-mediated diseases (e.g., multiple sclerosis) and infectious diseases of the CNS due to augmented entry through increasingly porous blood–brain and blood–CSF barriers as well as intrathecal production by leukocytes that have come to reside in the brain (especially the choroid plexus) or the CSF.[5,42]

Numerous enzymes have been demonstrated in the CSF of animals and people. These proteins can reach the CSF from the blood or may be released from either normal or neoplastic, neural or nonneural cells within the CSF.[5,42] Enzyme levels in CSF are usually lower than those measured in the blood, even if their source in the CSF is not from the peripheral circulation. To date, CSF enzyme assays are not sufficiently sensitive or specific to warrant routine use in clinical practice.[5,42]

Because they are direct by-products of CNS neuronal activity, neurotransmitters [e.g., γ-aminobutyric acid (GABA), glutamate, aspartate, and dopamine] and their metabolites [e.g., 5-hydroxyindolacetic acid (5-HIAA), homovanillic acid (HVA), and dihydroxyphenylacetic acid (DHPA)] have been studied extensively in people[55] and domestic animals[56–58] for their potential use as markers of neurological and psychiatric disease. Their clinical utility as reliable biomarkers of neurologic dysfunction has yet to be established.[5,42]

Composition of CSF varies with age in animals (rats, rabbits, cats, dogs, pigs, and nonhuman primates) and humans. Neonatal CSF is usually xanthochromic (yellow), presumably a reflection of the higher levels of bilirubin and protein permitted to pass by the immature BBB (as bilirubin in solution is generally bound tightly to albumin). The leaky immature BBB also produces a higher concentration of glucose in neonatal CSF (approaching that of the plasma), in contrast to the adult situation where the need for transport by facilitated diffusion and relatively higher brain metabolic rate limit the normal CSF glucose level to a range between 60% to 80% of the blood glucose concentration.[53] With advancing age, the CSF/plasma concentration ratios (R_{CSF}) of Na^+, Mg^{2+}, and Cl^- generally increase, whereas the R_{CSF} of K^+, HCO_3^-, and urea, as well as total protein and individual proteins and amino acids, usually decrease. The divergence in CSF protein concentration compared to plasma protein levels is indicative of the maturing blood–brain and blood–CSF barriers. However, large interindividual variations exist at all ages, and for some entities the pattern is actually reversed. The amino acid taurine, for example, is higher in the CSF of adult rats than in neonates.

In addition to variations with age, the concentration of many CSF constituents varies along the length of the neuraxis. For example, the CSF concentrations of total protein, albumin, and globulin increase substantially from rostral to caudal in dogs,[43] rhesus macaques,[59] and humans.[4,5,60] The increased protein content in the caudal neuraxis may result

from greater permeability of the spinal blood–CSF barrier to albumin,[5] equilibration of CSF with plasma through the capillary walls,[60] and low flow rates of lumbar CSF.[4] In contrast, the highest CSF concentrations of glucose in humans[5,42] and neurotransmitter metabolites in dogs[61] occur rostrally. Gradients for these two components probably reflect their higher levels near their major brain sites for utilization (glucose) and production (neurotransmitters).

CONSIDERATIONS IN SAMPLING AND ANALYZING CSF

Collection

Descriptions of CSF collection techniques are not covered here because clinical procedures are not typically performed by pathologists and toxicologists. Detailed sampling methods are available for animal species used in neurotoxicology research, including mice,[62] rats,[63,64] rabbits,[65] dogs,[66,67] nonhuman primates,[68] and chickens.[69] Regardless of the species, several considerations should be kept in mind when preparing to harvest and analyze CSF.

The choice of collection site is influenced by the species, location of the neurological lesion, and anesthetic requirements. Cerebellomedullary puncture usually can be accomplished even in very large or obese animals. It should be done under general anesthesia. Lumbar puncture may be difficult in very large or obese animals. However, it can often be performed in sedated individuals under local anesthesia and is therefore the procedure of choice if general anesthesia is contraindicated. The collection site should be located as close to the suspected lesion as possible or else caudal to it. In animals with spinal cord disease, lumbar fluid is abnormal more often than cerebellomedullary fluid except when lesions exist in the cervical cord. With intracranial disease, CSF from both sites is usually abnormal, perhaps because both sites are caudal to the lesion.[2]

In general, CSF samples should be submitted in a plastic or silicon-coated glass tube to prevent attachment and activation of cells (especially monocytes) to the container walls[5] or degeneration of cells within the sample.[5,38,42,70,71] In practical terms the nature of the container matters little if samples are processed rapidly; the typical recommendation for both animals[2,39,72,73] and humans[42,71,72] is to initiate analysis within 30 min of collection. However, containers containing CSF samples may be held on ice or in a refrigerator for up to 48 h if the analysis will be delayed, although the differential cell counts will be altered in a time-dependent manner starting about 2 to 4 h after collection.[70,74] Less cellular degradation occurs in samples with a higher protein concentration (≥ 50 mg/dL). Extended delays in analysis may benefit from the addition of 20% fetal calf serum (FCS) as a stabilizer.[70] An option is to split the CSF sample into two aliquots, one of which will be treated by addition of 20%

FCS.[70,75] The unaltered aliquot will be used to measure the protein concentration, while the altered aliquot is used to assess the total nucleated cell count (taking into account dilutional effects), the differential cell count, and cell morphology.

Cell Analysis

The first stage of CSF analysis is to record physical characteristics such as color, clarity, and viscosity. Normal CSF in mature animals and humans is clear, colorless, and has the same viscosity as that of distilled water. In fact, a similar amount of distilled water in the same type of container may be used as a "normal control" sample for this purpose. Color and clarity are best judged by holding the containers against a white background, while viscosity is assessed by gentle shaking. If the CSF appears abnormal by these simple visual tests, the sample should be centrifuged and then reassessed.

The next step is to determine total and differential cell counts. These analyses are required routinely because very low total cell counts (i.e., within the normal range) may be associated with abnormalities in the differential cell count (e.g., larger percentage of neutrophils, or the presence of activated macrophages or plasma cells) and/or the cytoarchitectural features.[45] For human CSF, a laser-based cell counter is now used for these assessments[76,77] as the precision and accuracy are substantially superior to those of manual counting methods. While evaluated in animal CSF,[78] the high cost and unsuitability of the current software algorithms generally preclude an accurate differential count when using an automated cell counter. Therefore, enumeration of cell numbers and types in animal CSF requires a manual (hemocytometer) count and microscopic examination, typically a cytocentrifuge-prepared slide. The hemocytometer chamber is charged with undiluted CSF and, ideally, allowed to settle for 10 min in a humidified environment so that all the cells will be visible in the same focal plane. The cells in the nine largest squares on both sides of the chamber are counted (18 squares total) and the result multiplied by 0.55 to obtain the number of cells per microliter. Leukocytes are typically larger, have an irregular cell border, and appear granular due to the presence of a nucleus. In contrast, erythrocytes are smaller, smooth, and refractile, although their borders may become crenated if analysis is delayed.[1] Differentiating nucleated cells and erythrocytes in a hemocytometer chamber can be expedited by staining with new methylene blue prior to counting.[70] The latter technique can be used without significant dilutional effects.

Morphologic assessment of cells is the subsequent analytical step. Various methods to produce higher cell yields and thus facilitate this evaluation include simple centrifugation, sedimentation, membrane filtration, and cytocentrifugation. In general, cell damage during simple centrifugation precludes a meaningful cytological examination. Membrane filtration frequently yields cell recoveries of over 90%,[79] but the technique is too laborious and time consuming for high-throughput experiments, given the somewhat poor cell morphology and the inability to visualize completely the many cells that are trapped in the filter substrate. Cytocentrifugation[80] is the current method of choice for evaluating both human[5,42] and animal[1,45] CSF samples. This method is rapid and simple, and produces preparations with good cytomorphologic detail. The cell morphology may be enhanced by mixing the sample with an equal volume of protein-rich fluid (20% FCS) prior to centrifugation. The disadvantages of cytocentrifugation are the expense of the instrument and the relatively low cell yield (approximately 15% of membrane filtration[79]). Some types of sedimentation are reported to provide cell morphology comparable to that of cytocentrifugation but with a two- to threefold higher cell yield.[81,82] However, at least one study suggests that the yield is marginally higher for cytocentrifugation than for sedimentation.[83] Standard Romanowsky stains (e.g., Wright's and Wright–Giemsa) are recommended for processing CSF, as they provide good cell detail with readily recognizable cell features when used on air-dried preparations.[1]

Special staining procedures may enhance the information obtained from the cellular analysis. The most common supplemental procedure is immunocytochemistry.[42] This technique is particularly useful for immunophenotypic studies of cytocentrifuged CSF preparations as an aid to differentiating the origin of CNS neoplasms or defining leukocyte subpopulations; it is commonly used for human CSF but is employed less frequently, at present, for animal CSF. For example, in normal human CSF, approximately 85% of the lymphocytes are T-cells, most of which are T-helper cells[42]; T-cells also predominate in dog CSF, although they account for only 50 to 60% of cells.[84,85] Alterations in leukocyte percentages are correlated with different types of neurological disease.[42,86] Polymerase chain reaction (PCR) to detect specific antigens of infectious agents or neoplasms is another useful technique for CSF analysis, often providing more rapid diagnosis with superior sensitivity and specficity than that of standard techniques such as cell culture, serology, or cytomorphology.[87] The greatest limitation to special cytomorphological studies is the low volume and cellularity of the typical CSF specimen. Special procedures should be undertaken mainly to confirm preliminary diagnoses made using a combination of other diagnostic data (e.g., clinical information coupled with routine blood and CSF analyses).

The usual source of erythrocytes in CSF samples is traumatic puncture, particularly if the sample is taken from the lumbar region.[2,41,73] Leukocytes released by a traumatic puncture can affect both the total nucleated and the differential cell counts. Numerous correction factors have been used to correct CSF leukocyte counts for blood contamination, but the common finding that thousands of erythrocytes may be present in the absence of leukocytes suggests that these correction factors are not accurate and hence not useful.[40,43,66,88]

Biochemical Analysis

The first biochemical parameter to evaluate in CSF is the total protein concentration, which is a time-tested indicator of many neurological diseases. The concentration should be measured using a quantitative assay rather than by such semiquantitative methods as diagnostic dipsticks, which are prone to error.[89] The most accurate quantitative methods for CSF protein determination are microprotein dye-binding assays such as Coomassie brilliant blue, Ponceau S red, and pyrogallol red.[90,91] Even so, the total CSF protein values obtained vary noticeably with the methodology and among laboratories, necessitating the use of laboratory-specific reference intervals. After total protein has been defined, additional assessments of protein fractions may be warranted.[5] The most useful fraction to identify is albumin, as an increase in CSF albumin can only result from damage to the blood–brain and/or blood–CSF barriers, intrathecal hemorrhage, or a traumatic CSF tap. Because albumin leaks into the CSF in proportion to its concentration in serum, an increase in the ratio of CSF albumin to serum albumin can be used (in the absence of intrathecal hemorrhage) as an index of increased barrier permeability.[92,93] The utility of this indicator may be limited because of the large variability in CSF albumin among normal animals.[3,94,95] In humans, the albumin index is age-dependent, being highest in newborns and lowest in childhood before increasing again during adulthood.[87]

Other nonspecific analytes also may provide information regarding the type of neurological disease. Immunoglobulins (Ig) in CSF may be evaluated by either quantitative[5,42,92,93] or qualitative[5,42,87,94] methods, both of which are typically required as the total amount of CSF Ig may be within normal limits even though the qualitative distribution of the various Ig fractions may be abnormal.[43] An increased CSF glucose concentration usually reflects systemic hyperglycemia, while decreased CSF glucose concentration generally occurs with widespread meningeal involvement with an infectious (bacterial, fungal, or protozoal) or neoplastic (carcinomatosis) process. Analytical techniques for these parameters have been detailed elsewhere[5,42] and will not be repeated here, as alterations in CSF Ig and CSF glucose concentrations have not been reported in neurotoxicity syndromes.

Measurement of CNS-specific analytes in CSF has been undertaken commonly as an additional means for evaluating neurological diseases. Particular attention has been paid to major excitatory [e.g., glutamate (GLU) and inhibitory [e.g., γ-aminobutyric acid (GABA)] neurotransmitters, which have been implicated in the neural cell death that occurs in seizure disorders of both humans and animal models.[58,96,97] Altered serotonin [5-hydroxytryptamine (5-HT)] metabolism has been identified in the CSF of dogs with hepatic encephalopathy.[98] Proteins such as C-reactive protein, interferon, myelin basic protein (MBP), and S100 are increased in the CSF during many different CNS diseases, suggesting that the presence of these molecules in CSF may be useful general screens for neurologic dysfunction without pinpointing the exact nature of the disturbance.[5,42,99]

Ongoing research in defining the CSF proteome will probably result in the discovery of one or more novel biomarkers of neurological disease in the next few years.[100] Major inroads are currently being made in distinguishing among various neurodegenerative conditions, some of which (Alzheimer's disease, Parkinson's disease) have been linked in some cases to neurotoxic events. The CSF proteome attributed to a specific neural cell population can be altered by neurotoxicant exposure.[101] Accordingly, evaluation of the CSF proteome might eventually provide a valuable means of differentiating among potential target cell populations or among various divergent classes of neurotoxic agents.

GENERAL CHARACTERISTICS OF CSF IN NEUROLOGICAL DISEASE

Analysis of CSF can include an assessment of multiple parameters. In the following section we emphasize those features that have relevance to investigations of neurotoxic conditions. Other factors that might serve as useful diagnostic aids in other conditions unrelated to CNS toxicity (e.g., degeneration, inflammation, neoplasia) have been described elsewhere in detail[5,42,43] and are not treated at length here. Even though neurological signs may occur, the CSF associated with toxicity is usually normal.[102]

Cell Analysis

Normal CSF is clear, essentially colorless, and has the consistency of water. Under pathological conditions, these physical features may change. Hazy CSF usually results from an increased number of cells (about 200 white blood cells/μL or 400 red blood cell/μL will produce a visible change[43]), microorganisms, epidural fat, or the presence of myelographic contrast agent. Minor color variations in CSF include pink or orange (from oxyhemoglobin), yellow (from bilirubin), or brown (from methemoglobin), depending on the predominant pigment produced from proteins released from hemolyzed erythrocytes and the length of time that has passed since hemolysis.[43] Oxyhemoglobin may be detected about 2 h after erythrocytes enter the CSF, peaks about 36 h later, and disappears over the next 4 to 10 days. Bilirubin may be found about 10 h after erythrocytes enter the CSF, peaks at about 48 h, and may persist for two to four weeks. This molecule is the main pigment responsible for the abnormal color of CSF samples with a high protein content, as bilirubin binds tightly to the albumin that has crossed the disrupted blood–brain and/or blood–CSF barriers; the intensity of yellow CSF tint may also be enhanced during icterus. More intense xanthochromia (yellowing) of the CSF is also a characteristic of conditions that create CNS perivascular hemorrhages, such as moldy corn poisoning in horses.[103,104] Methemoglobin in CSF is a degradation product released from old sites of intrathecal hemorrhage.[5,42] Certain drugs may alter the color of CSF (and other body fluids); for example, rifampin can impart an orange-red color to CSF.[5] Increased viscosity of CSF usually results from a very high CSF protein content, particularly from fibrinogen. Indeed, if the CSF cell count is also higher, the CSF sample may clot. The presence of dispersed epidural fat or nucleus pulposus material may also raise CSF viscosity or result in globules floating within the fluid.[5,42]

An increase in CSF cellularity is termed *pleocytosis*. The degree of pleocytosis is dependent on several factors, including the nature of the neurological disease and the severity and location of the lesion with respect to the subarachnoid space or ventricular system.[39] A normal CSF analysis does not exclude the presence of CNS disease,[5,39,42] particularly where the lesions are deep within the neuropil and do not communicate with the ependymal surfaces or the leptomeninges (and hence the subarachnoid space). Abnormal CSF findings always indicate the presence of pathology. In general, increased CSF neutrophil numbers indicate either an acute meningitis of either bacterial or viral origin or an immune-mediated vasculitis,[5,42,43] while higher lymphocyte numbers are indicative of a subacute-to-chronic viral infection,[105] postantibiotic bacterial meningitis,[5,42] or primary inflammatory diseases such as granulomatous meningoencephalitis and necrotizing meningoencephalitis.[43] The presence of eosinophils in CSF is abnormal and evidence of underlying disease, but not a specific disease.[106] Eosinophils in CSF may occur in cases of parasitic disease but also at various stages of bacterial, fungal, and viral infections, among others.[1,107] Blood eosinophilia and CSF eosinophilic pleocytosis may not occur simultaneously, and if they do, their magnitudes often are not correlated.[106,107] Neoplastic cells may be evident in CSF, but the ability to diagnose CNS tumors via CSF cytology is quite variable. In humans, success rates range from 30% for primary CNS tumors to 20 to 60% for meningeal carcinomas to 70% for leukemia.[42] In animals, our experience suggests that the only tumors that can be detected readily via CSF analysis are lymphomas and leukemias.[43] Tumor cells may be observed in the CSF even when the CSF cell counts are within normal limits.[81]

With toxicities, alterations of CSF cell counts are less common than biochemical alterations. Mild toxicant-induced elevations of the CSF leukocyte count may accompany dysfunction of the blood–brain and/or blood–CSF barriers and/or may result from neural cell degeneration or necrosis. A prototypic example is lead poisoning[108,109] which selectively damages capillary endothelial cells[110] and cerebrocortical neurons.[111] If necrosis is severe, the white blood cell count can be increased markedly with a predominance of neutrophils, as with leukoencephalomalacia caused by moldy corn poisoning in horses.[104]

Biochemical Analysis

An increase in the total protein concentration is the single most useful screening test in CSF chemical composition for routine diagnostic purposes.[5] This change may result from (1) increased permeability of the blood–brain and/or blood–CSF barriers, allowing passage of serum proteins, (2) interruption of CSF flow and/or absorption, or (3) intrathecal production of globulins (typically in response to an infectious agent). In many CNS diseases, two or all three of these mechanisms are at work.[5,42] Increased CSF total protein concentration is a nonspecific finding, indicating the presence of a CNS disease (e.g., inflammation, neoplasia, trauma, compression, and occasionally, metabolic errors[54,94,95,112]) without differentiating a specific condition. Decreased CSF total protein occurs infrequently and is a consequence of more rapid removal when elevated intracranial pressure leads to greater bulk flow absorption of CSF. A potential confounder in diagnosing decreased protein content is the collection site, as CSF from the cerebellomedullary cistern has a lower protein concentration than lumbar CSF;[5,42] ideally, neuroscientists who regularly evaluate CSF should standardize their collection site so that they can develop a meaningful historical control database. In many CNS disease processes, the CSF cell count and total protein concentration have roughly parallel increases. In some disorders, however, the CSF cell count remains normal as the CSF total protein concentration rises greatly. This phenomenon, termed *albuminocytologic dissociation*, may be observed in neurodegenerative disorders, ischemia and infarction, immune-mediated disease (e.g., polyradiculoneuritis), neural compression, and neoplasia.[113] Mild elevations of the CSF protein concentration may occur during neurointoxication syndromes if the toxicant degrades the blood–brain and/or blood–CSF barriers, and/or induces neural degeneration or necrosis. The obvious examples are lead intoxication[108,109] and moldy corn poisoning in horses.[104]

Despite the utility of protein measurements, in neurotoxic conditions, biochemical alterations in the CSF may occur more commonly than alterations of either CSF cell counts or CSF protein concentration. Terminal neurological signs in cases of fatal lead poisoning in calves are accompanied by higher CSF concentrations of glucose, urea, creatinine, and creatine kinase,[109] presumably as a consequence of severe breakdown in the blood–brain and/or blood–CSF barriers and possibly release from injured cells within the CSF or neuropil. The adverse effects of neurotoxicants on the CNS can alter CSF concentrations of various neurotransmitters and their metabolites. Commonly measured molecules include GABA, GLU, and monoamines [e.g., dopamine (DA), norepinephrine (NE), and serotonin (5-HT)]. For example, increased levels of homovanillic acid (HVA, a DA metabolite) and 5-hydroxyindolacetic acid (5-HIAA, a 5-HT metabolite) have been measured in the CSF of Collie dogs that develop severe neurological deficits after experimental administration of invermectin.[114] Similarly, dogs with portosystemic shunts and clinical signs of hepatic encephalopathy have higher CSF levels of glutamine, tryptophan (a 5-HT precursor), 5-HIAA, and quinolinic acid (an excitotoxic L-tryptophan metabolite that is an agonist for N-methyl-D-aspartate receptors).[98] Goats that exhibit behavioral changes and seizures after experimental treatment with inorganic boron have elevated CSF levels of DA and 5-HT metabolites,[115] possibly reflecting excessive neuronal stimulation. Elevated CSF levels of some neurotransmitters might conceivably function as neurotoxicants in their own right, as suggested by a close correlation between elevated CSF concentrations of NE and the severity of clinical symptoms in combat veterans with post-traumatic stress syndrome.[116] Finally, certain neurotoxicants can displace chemically related normal constituents from peripheral compartments into the CSF. This capacity has been demonstrated in rats treated with manganese, which shifts iron from the serum into the CSF.[117]

RECOMMENDATIONS FOR CSF ANALYSIS IN NEUROTOXICITY EVALUATIONS

The toxicological neuropathology literature available today indicates that CSF analysis, while quite valuable in defining the nature and cause of some neurological conditions, probably will provide little information of relevance in standard neurotoxicity studies. This "deficiency" reflects the inability of CSF samples to reliably probe neurotoxic events deep in the neuropil, especially in those instances where small numbers of cells in a discrete region are affected. Therefore, we do not recommend routine CSF collection and analysis in conventional multianimal, prospective, regulatory-type neurotoxicity studies unless other endpoints (e.g., measurement of xenobiotics, their metabolites, or CNS-derived signaling molecules) are also to be assessed. Evaluation of CSF on an individual patient basis in the diagnostic setting may be warranted to rule out other neurological diseases, although most neurotoxic conditions will induce few or no changes in CSF biochemistry or cellularity.[67]

REFERENCES

1. Jamison EM, Lumsden JH. Cerebrospinal fluid analysis in the dog: methodology and interpretation. *Semin Vet Med Surg (Small Anim)*. 1988;3:122–132.

2. Thomson CE, et al. Analysis of cerebrospinal fluid from the cerebellomedullary and lumbar cisterns of dogs with focal neurologic disease: 145 cases (1985–1987). *J Am Vet Med Assoc. 1990;*196:1841–1844.

3. Tipold A. Diagnosis of inflammatory and infectious diseases of the central nervous system in dogs: a retrospective study. *J Vet Intern Med*. 1995;9:304–314.

4. Davson H, Segal MB. *Physiology of the CSF and Blood–Brain Barriers*. Boca Raton, FL: CRC Press; 1996.

5. Fishman RA. *Cerebrospinal Fluid in Diseases of the Nervous System.*, 2nd ed. Philadelphia: W. B. Saunders; 1992.

6. Milhorat TH. *Cerebrospinal Fluid and the Brain Edemas*. New York: Neuroscience Society of New York; 1987.

7. Rosenberg GA. *Brain Fluids and Metabolism*. New York: Oxford Unversity Press; 1990.

8. Abbott NJ. Evidence for bulk flow of brain interstitial fluid: significance for physiology and pathology. *Neurochem Int*. 2004;5:545–552.

9. Barten DM, et al. Dynamics of β-amyloid reductions in brain, cerebrospinal fluid, and plasma of β-amyloid precursor protein transgenic mice treated with a γ-secretase inhibitor. *J Pharmacol Exp Ther*. 2005;312:635–643.

10. Abbott NJ. Dynamics of CNS barriers: evolution, differentiation, and modulation. *Cell Mol Neurobiol*. 2005;25:5–23.

11. Abbott NJ. Inflammatory mediators and modulation of blood–brain barrier permeability. *Cell Mol Neurobiol*. 2000;20:131–147.

12. Bertossi M, et al. Effects of 6-aminonicotinamide gliotoxin on blood–brain barrier differentiation in the chick embryo cerebellum. *Anat Embryol (Berl)*. 2003;207:209–219.

13. Stewart PA, Wiley MJ. Structural and histochemical features of the avian blood–brain barrier. *J Comp Neurol*. 1981;202:157–167.

14. Enerson BE, Drewes LR. The rat blood–brain barrier transcriptome. *J Cereb Blood Flow Metab*. 2006;26:959–973.

15. Cassella JP, et al. Ontogeny of four blood–brain barrier markers: an immunocytochemical comparison of pial and cerebral cortical microvessels. *J Anat*. 1996;189:407–415.

16. Abbott NJ, et al. Astrocyte-endothelial interaction: physiology and pathology. *Neuropathol Appl Neurobio*. 1992;18:424–433.

17. Abbott NJ. Astrocyte–endothelial interactions and blood--brain barrier permeability. *J Anat*. 2002;200:629–638.

18. Bell WO. Cerebrospinal fluid reabsorption: a critical appraisal. 1990. *Pediatr Neurosurg*. 1995;23:42–53.

19. Abbott NJ. Modeling CSF/ISF flow and drug distribution: relevance to CNS drug delivery and neural repair, In: Allen JW, Yaksh TL, eds. *CNS Drug Delivery*. San Diego, CA: Society for Neuroscience; 2007: 8–15.

20. Drewes LR. Molecular architecture of the brain microvasculature: perspective on blood–brain transport. *J Mol Neurosci*. 2001;16:93–98 (discussion: 151–157).

21. Brown PD, et al. Molecular mechanisms of cerebrospinal fluid production. *Neuroscience*. 2004;129:957–970.

22. Crone C, Olesen SP. Electrical resistance of brain microvascular endothelium. *Brain Res*. 1982;241:49–55.

23. Nilsson C, et al. Neuroendocrine regulatory mechanisms in the choroid plexus–cerebrospinal fluid system. *Brain Res Rev*. 1992;17:109–138.

24. Cserr HF. Physiology of the choroid plexus. *Physiol Rev.* 1971;51:273–311.

25. Martins AN, et al. PCO$_2$ and rate of formation of cerebrospinal fluid in the monkey. *Am J Physiol.* 1976;231:127–131.

26. Fankhauser R. In: Innes JRM, Saunders LZ, eds. *Comparative Neuropathology.* New York: Academic Press; 1962: 21–54.

27. Fletcher TF. In: Evans HE, ed. *Miller's Anatomy of the Dog.* Philadelphia: W.B. Saunders; 1993: 800–828.

28. Stoodley MA, et al. Evidence for rapid fluid flow from the subarachnoid space into the spinal cord central canal in the rat. *Brain Res.* 1996;707:155–164.

29. Knopf PM, et al. Physiology and immunology of lymphatic drainage of interstitial and cerebrospinal fluid from the brain. *Neuropathol Appl Neurobiol.* 1995;21:175–180.

30. Szentistvanyi I, et al. Drainage of interstitial fluid from different regions of rat brain. *Am J Physiol Renal Physiol.* 1984;246:F835–F844.

31. Kida S, et al. Anatomical pathways for lymphatic drainage of the brain and their pathological significance. *Neuropathol Appl Neurobiol.* 1995;21:181–184.

32. Koh L, et al. Integration of the subarachnoid space and lymphatics: Is it time to embrace a new concept of cerebrospinal fluid absorption? *Cerebrospinal Fluid Res.* 2005;2: DOI: 10. 1186/1743-8454-1182-1186.

33. Weller RO, et al. Pathways of fluid drainage from the brain: morphological aspects and immunological significance in rat and man. *Brain Pathol.* 2008;2:277–284.

34. Cserr HF, et al. Efflux of radiolabeled polyethylene glycols and albumin from rat brain. *Am J Physiol.* 1981;240: F319–F328.

35. Rosenberg GA, et al. Bulk flow of brain interstitial fluid under normal and hyperosmolar conditions. *Am J Physiol Renal Physiol.* 1980;238:F42–F49.

36. Wiig H, Reed RK. Rat brain interstitial fluid pressure measured with micropipettes. *Am J Physiol Heart Circ Physiol.* 1983;244:H239–H246.

37. Reulen HJ, et al. Clearance of edema fluid into cerebrospinal fluid: a mechanism for resolution of vasogenic brain edema. *J Neurosurg.* 1978;48:754–764.

38. Chrisman CL. Cerebrospinal fluid analysis. *Vet Clin North Am Small Anim Pract.* 1992;22:781–810.

39. Cook JR Jr, DeNicola DB. Cerebrospinal fluid. *Vet Clin North Am Small Anim Pract.* 1988;18:475–499.

40. Wilson JW, Stevens JB. Effects of blood contamination on cerebrospinal fluid analysis. *J Am Vet Med Assoc.* 1977;171:256–258.

41. Bailey CS, Higgins RJ. Comparison of total white blood cell count and total protein content of lumbar and cisternal cerebrospinal fluid of healthy dogs. *Am J Vet Res.* 1985;46:1162–1165.

42. Kjeldsberg CR, Knight JA. *Body Fluids: Laboratory Examination of Amniotic, Cerebrospinal, Seminal, Serous & Synovial Fluids*, 3rd ed. Chicago: American Society of Clinical Pathologists; 1993.

43. Vernau W, et al. Cerebrospinal fluid. In: Kaneko JJ, et al., eds. *Clinical Biochemistry of Domestic Animals.* San Diego, CA: Academic Press; 2008: 769–819.

44. Meeks JC, et al. The maturation of canine cerebrospinal fluid. *J Vet Intern Med.* 1994;8:*177* [Abst 139].

45. Christopher MM, et al. Reassessment of cytologic values in canine cerebrospinal fluid by use of cytocentrifugation. *J Am Vet Med Assoc.* 1988;192:1726–1729.

46. Garma-Avina A. Cytology of the normal and abnormal choroid plexi in selected domestic mammals, wildlife species, and man. *J Vet Diagn Invest.* 2004;16:283–292.

47. Rand JS, et al. Reference intervals for feline cerebrospinal fluid: cell counts and cytologic features. *Am J Vet Res.* 1990;51:1044–1048.

48. Bigner SH, Jonston WW. The cytopathology of cerebrospinal fluid: I. Nonneoplastic conditions, lymphoma and leukemia. *Acta Cytol.* 1981;25:345–353.

49. Christopher MM. Bone marrow contamination of canine cerebrospinal fluid. *Vet Clin Pathol.* 1992;21:95–98.

50. Bienzle D, et al. Extramedullary hematopoiesis in the choroid plexus of five dogs. *Vet Pathol.* 1995;32:437–440.

51. Fallin CW, et al. Cytologic identification of neural tissue in the cerebrospinal fluid of two dogs. *Vet Clin Pathol.* 1996;25:127–129.

52. Felgenhauer K. Protein size and cerebrospinal fluid composition. *Klin Wochenschr.* 1974;52:1158–1164.

53. Redzic ZB, Segal MB. The structure of the choroid plexus and the physiology of the choroid plexus epithelium. *Adv Drug Deliv Rev.* 2004;56:1695–1716.

54. Sorjonen DC, et al. Cerebrospinal fluid protein electrophoresis: a clinical evaluation of a previously reported diagnostic technique. *Prog Vet Neurol.* 1991;2:261–268.

55. Davis BA. *Biogenic Monoamines and Their Metabolites in the Urine, Plasma, and Cerebrospinal Fluid of Normal, Psychiatric, and Neurological Subjects.* Boca Raton, FL: CRC Press; 1990.

56. Bardon T, Ruckebusch M. Changes in 5-HIAA and 5-HT levels in lumbar CSF following morphine administration to conscious dogs. *Neurosci Lett.* 1984;49:147–151.

57. Ellenberger C, et al. Inhibitory and excitatory neurotransmitters in the cerebrospinal fluid of epileptic dogs. *Am J Vet Res.* 2004;65:1108–1113.

58. Podell M, Hadjiconstantinou M. Cerebrospinal fluid gamma-aminobutyric acid and glutamate values in dogs with epilepsy. *Am J Vet Res.* 1997;58:451–456.

59. Smith MO, Lackner AA. Effects of sex, age, puncture site, and blood contamination on the clinical chemistry of cerebrospinal fluid in rhesus macaques (*Macaca mulatta*). *Am J Vet Res.* 1993;54:1845–1850.

60. Weisner B, Bernhardt W. Protein fractions of lumbar, cisternal, and ventricular cerebrospinal fluid: separate areas of reference. *J Neurol Sci.* 1978;37:205–214.

61. Vaughn DM, et al. A rostrocaudal gradient for neurotransmitter metabolites and a caudorostral gradient for protein in canine cerebrospinal fluid. *Am J Vet Res.* 1988;49:2134–2137.

62. Vogelweid CM, Kier AB. A technique for the collection of cerebrospinal fluid from mice. *Lab Anim Sci.* 1988;38:91–92.

63. Hudson LC, et al. Cerebrospinal fluid collection in rats: modification of a previous technique. *Lab Anim Sci.* 1994;44:358–361.

64. Sharma AK, et al. Development of a percutaneous cerebrospinal fluid collection technique in F-344 rats and evaluation of cell counts and total protein concentrations. *Toxicol Pathol.* 2006;34:393–395.

65. Vistelle R, et al. Rapid and simple cannulation technique for repeated sampling of cerebrospinal fluid in the conscious rabbit. *Lab Anim Sci.* 1994;44:362–364.

66. de Lahunta A, Glass E. Cerebrospinal fluid and hydrocephalus. In: de Lahunta A, Glass E, eds. *Veterinary Neuroanatomy and Clinical Neurology*, Philadelphia: W.B. Saunders; 2009: 54–76.

67. Lorenz MD, Kornegay JN. Confirming a diagnosis. In: Lorenz MD, Kornegay JN, eds. *Handbook of Veterinary Neurology.* Philadelphia: W.B. Saunders; 2004: 91–109.

68. Gilberto DB, et al. An alternative method of chronic cerebrospinal fluid collection via the cisterna magna in conscious rhesus monkeys. *Contemp Top Lab Anim Sci.* 2003; 42:53–59.

69. Skewes PA, et al. Avian cerebrospinal fluid: repeated collection and testing for a possible role in food intake regulation. *Poult Sci.* 1986;65:1172–1177.

70. Fry MM, et al. Effects of time, initial composition, and stabilizing agents on the results of canine cerebrospinal fluid analysis. *Vet Clin Pathol.* 2006;35:72–77.

71. Steele RW, et al. Leukocyte survival in cerebrospinal fluid. *J Clin Microbiol.* 1986;23:965–966.

72. Chrisman CL. Cerebrospinal fluid evaluation. In: Kirk RW, ed. *Current Veterinary Therapy VIII: Small Animal Practice.* Philadelphia: W.B. Saunders; 1983:676–681.

73. Oliver JE, Lorenz MD. *Handbook of Veterinary Neurology.* Philadelphia: W. B. Saunders; 1993.

74. Stokes HB, et al. An improved method for examination of cerebrospinal fluid cells. *Neurology.* 1975;25:901–906.

75. Bienzle D, et al. Analysis of cerebrospinal fluid from dogs and cats after 24 and 48 hours of storage. *J Am Vet Med Assoc.* 2000;216:1761–1764.

76. Aune MW, et al. Automated flow cytometric analysis of blood cells in cerebrospinal fluid: analytic performance. *Am J Clin Pathol.* 2004;121:690–700.

77. Mahieu S, et al. Evaluation of ADVIA 120 CSF assay (Bayer) vs. chamber counting of cerebrospinal fluid specimens. *Clin Lab Haematol.* 2004;26:195–199.

78. Ruotsalo K, et al. Evaluation of the Advia 120 CSF assay for analysis of canine cerebrospinal fluid. *Vet Clin Pathol.* 2005;34:282.

79. Barrett DL, King EB. Comparison of cellular recovery rates and morphologic detail obtained using membrane filter and cytocentrifuge techniques. *Acta Cytol.* 1976;20:174–180.

80. Hansen HH, et al. The cyto-centrifuge and cerebrospinal fluid cytology. *Acta Cytol.* 1974;18:259–262.

81. Grevel V, Machus B. Diagnosing brain tumors with a CSF sedimentation technique. *Vet Med Rep.* 1990;2:403–408.

82. Kolmel HW. A method for concentrating cerebrospinal fluid cells. *Acta Cytol.* 1977;21:154–157.

83. Ducos R, et al. Sedimentation versus cytocentrifugation in the cytologic study of craniospinal fluid. *Cancer.* 1979;3:1479–1482.

84. Duque C, et al. The immunophenotype of blood and cerebrospinal fluid mononuclear cells in dogs. *J Vet Intern Med.* 2002;16:714–719.

85. Tipold A, et al. Lymphocyte subsets and CD45RA positive T-cells in normal canine cerebrospinal fluid. *J Neuroimmunol.* 1998;82:90–95.

86. Tipold A, et al. Lymphocyte subset distribution in steroid responsive meningitis-arteriitis in comparison to different canine encephalitides. *Zentralbl Veterinarmed A.* 1999;46:75–85.

87. Deisenhammer F, et al. Guidelines on routine cerebrospinal fluid analysis: report from an EFNS task force. *Eur J Neurol.* 2006;13:913–922.

88. Novak RW. Lack of validity of standard corrections for white blood cell counts of blood-contaminated cerebrospinal fluid in infants. *Am J Clin Pathol.* 1984;82:95–97.

89. Jacobs RM, et al. Relationship of cerebrospinal fluid protein concentration determined by dye-binding and urinary dipstick methodologies. *Can Vet J.* 1990;31:587–588.

90. Marshall T, Williams KM. Protein determination in cerebrospinal fluid by protein dye-binding assay. *Br J Biomed Sci.* 2000;57:281–286.

91. Pesce MA, Strande CS. A new micromethod for determination of protein in cerebrospinal fluid and urine. *Clin Chem.* 1973;19:1265–1267.

92. Link H, Tibbling G. Principles of albumin and IgG analyses in neurological disorders: II. Relation of the concentration of the proteins in serum and cerebrospinal fluid. *Scand J Clin Lab Invest.* 1977;37:391–396.

93. Tibbling G, et al. Principles of albumin and IgG analyses in neurological disorders: I. Establishment of reference values. *Scand J Clin Lab Invest.* 1977;37:385–390.

94. Bichsel P, et al. Immunoelectrophoretic determination of albumin and IgG in serum and cerebrospinal fluid in dogs with neurological diseases. *Res Vet Sci.* 1984;37:101–107.

95. Krakowka S, et al. Quantitative determination of serum origin cerebrospinal fluid proteins in the dog. *Am J Vet Res.* 1981;42:1975–1977.

96. Griffith NC, et al. Interictal behavioral alterations and cerebrospinal fluid amino acid changes in a chronic seizure model of temporal lobe epilepsy. *Epilepsia.* 1991;32: 767–777.

97. Meldrum BS. Glutamate as a neurotransmitter in the brain: review of physiology and pathology. *J Nutr.* 2000;130: 1007S–1015S.

98. Holt DE, et al. Cerebrospinal fluid glutamine, tryptophan, and tryptophan metabolite concentrations in dogs with portosystemic shunts. *Am J Vet Res.* 2002;63:1167–1171.

99. Lowenthal A, et al. Cerebrospinal fluid proteins in neurology. *Int Rev Neurobiol.* 1984;25:95–138.

100. Romeo MJ, et al. CSF proteome: a protein repository for potential biomarker identification. *Exp Rev Proteom.* 2005;2:57–70.

101. Lafon-Cazal M, et al. Proteomic analysis of astrocytic secretion in the mouse: comparison with the cerebrospinal fluid proteome. *J Biol Chem.* 2003;278:24438–24448.

102. Feldman BF. Cerebrospinal fluid. In: Kaneko JJ, ed. *Clinical Biochemistry of Domestic Animals.* San Diego, CA: Academic Press; 1989: 835–865.

103. Masri MD, et al. Clinical, epidemiologic and pathologic evaluation of an outbreak of mycotoxic encephalomalacia in south Louisiana horses. *Am Assoc Eq Pract Proc.* 1987;33:367.

104. McCue PM. Equine leukoencephalomalacia. *Compend Cont Ed Pract Vet.* 1989;11:646–650.

105. Vandevelde M, Spano JS. Cerebrospinal fluid cytology in canine neurologic disease. *Am J Vet Res.* 1977;38: 1827–1832.

106. Bosch I, Oehmichen M. Eosinophilic granulocytes in cerebrospinal fluid: analysis of 94 cerebrospinal fluid specimens and review of the literature. *J Neurol.* 1978;219: 93–105.

107. Smith-Maxie LL, et al. Cerebrospinal fluid analysis and clinical outcome of eight dogs with eosinophilic meningoencephalomyelitis. *J Vet Intern Med.* 1989;3:167–174.

108. Mayhew IG. *Large Animal Neurology.* Philadelphia: Lea & Febiger; 1989.

109. Swarup D, Maiti SK. Changes in some biochemical constutuents in blood and cerebrospinal fluid of lead-intoxicated calves. *Ind J Anim Sci.* 1991;61:942–945.

110. Goldstein GW, et al. Isolated brain capillaries: a model for the study of lead encephalopathy. *Ann Neurol.* 1977;1:235–239.

111. Christian RG, Tryphonas L. Lead poisoning in cattle: brain lesions and hematologic changes. *Am J Vet Res.* 1970;32:203–216.

112. Sorjonen DC. Total protein, albumin quota, and electrophoretic patterns in cerebrospinal fluid of dogs with central nervous system disorders. *Am J Vet Res.* 1987;48:301–305.

113. Laterre DC. In: Vinken PJ, Bruyn GW, eds. *Handbook of Clinical Neurology. Part I. Tumors of the Spine and Spinal Cord.* New York: American Elsevier, 1996:125–138.

114. Vaughn DM, et al. Determination of homovanillic acid, 5-hydroxyindoleacetic acid and pressure in the cerebrospinal fluid of collie dogs following administration of ivermectin. *Vet Res Commun.* 1989;13:47–55.

115. Sisk DB, et al. Experimental acute inorganic boron toxicosis in the goat: effects on serum chemistry and CSF biogenic amines. *Vet Hum Toxicol.* 1990;32:205–211.

116. Geracioti TDJ, et al. CSF norepinephrine concentrations in posttraumatic stress disorder. *Am J Psychiatry.* 2001;158: 1227–1230.

117. Li GJ, et al. Molecular mechanism of distorted iron regulation in the blood–CSF barrier and regional blood–brain barrier following in vivo subchronic manganese exposure. *Neurotoxicology.* 2006;27:737–744.

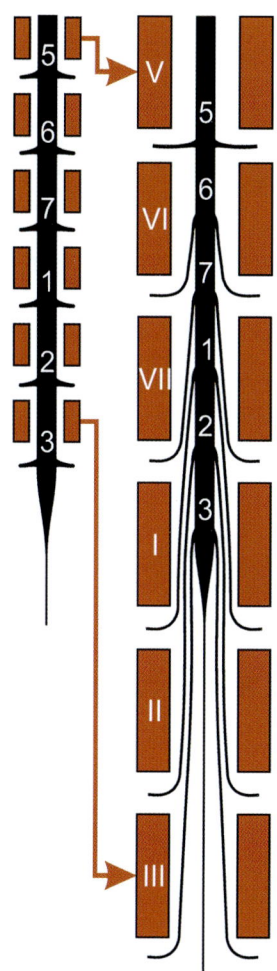

CHAPTER 2, FIGURE 3 Development of the cauda equina. Early in gestation (small image on left), the spinal cord segments are aligned with the corresponding vertebrae. After birth (large image on right), allometric growth of the spinal cord and vertebral column results in progressive cranial displacement of spinal cord segments relative to the corresponding vertebrae. 5–7, Lumbar spinal cord segments 5 through 7; 1–3, sacral spinal cord segments 1 through 3; V–VII, lumbar vertebrae V through VII; I–III, sacral vertebrae I through III.

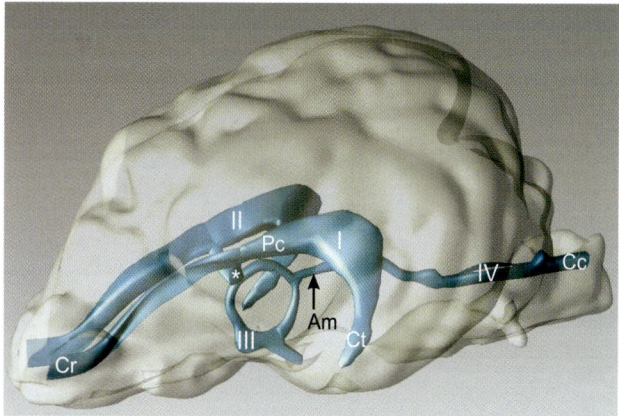

CHAPTER 2, FIGURE 2 Mammalian ventricular system (modeled after the dog). I, II, Lateral ventricles (with Pc, pars centralis; Cr, cornu rostrale; Ct: cornu temporale); III, third ventricle (circling the interthalamic adhesion); Am, aqueductus mesencephali; IV, fourth ventricle; Cc, canalis centralis of the spinal cord; ∗, foramen interventriculare.

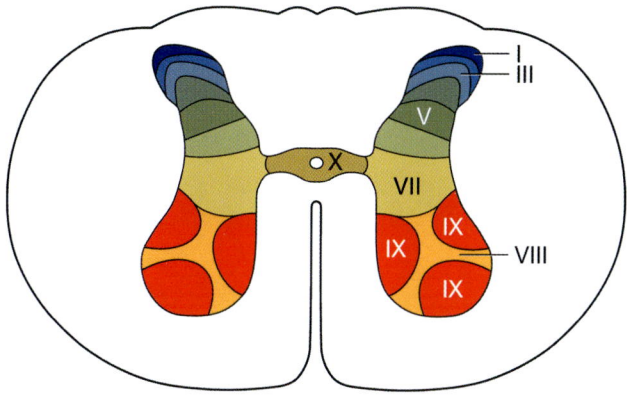

CHAPTER 2, FIGURE 4 Laminar arrangement of spinal cord gray matter, according to Rexed. Laminae I to VI receive sensory input from pseudounipolar neurons, whereas visceral somatic efferents originate from layers VII to IX. Layer X contributes fibers to the spinoreticular and spinothalamic tracts. (Courtesy of Enke Verlag in MVS Medizinverlage Stuttgart GmbH & Co. KG.).

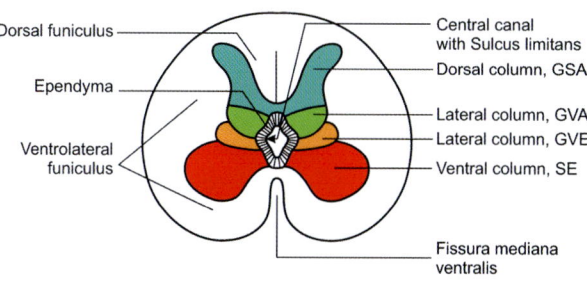

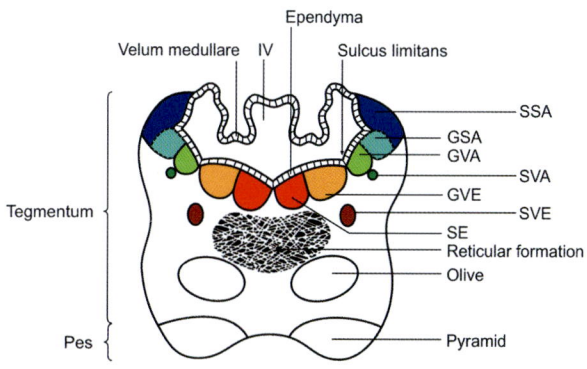

CHAPTER 2, FIGURE 5 Functional organization of gray matter in the mammalian spinal cord (a) and rhombencephalon (b). As a consequence of the flattening of the rhombencephalon, dorsoventral alignment of modalities as seen in the spinal cord is modified into a lateromedial sequence. Nuclei dealing with the special modalities that are unique to the head are added at the lateral aspect (SSA) and deep to the general modalities in the tegmentum (SVA, SVE). GSA, General somatic afferent; GVA, general visceral afferent; GVE, general visceral efferent; SE, somatic efferent; SSA, special somatic afferent; SVA, special visceral afferent; SVE, special visceral efferent; IV, fourth ventricle.

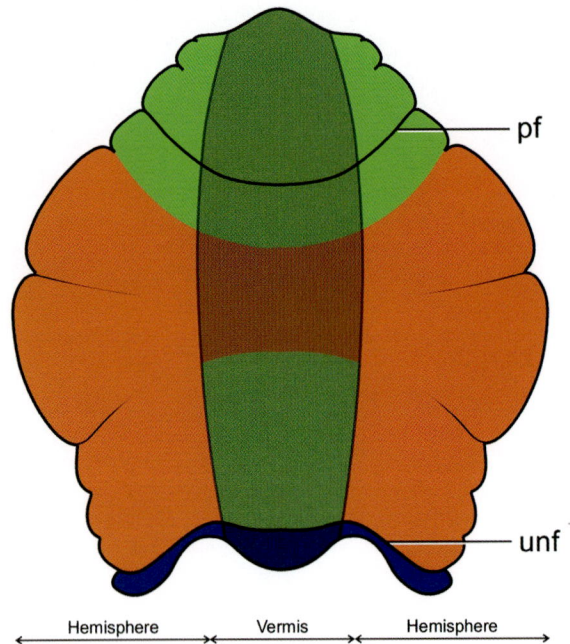

pf

unf

Hemisphere · Vermis · Hemisphere

CHAPTER 2, FIGURE 6 Morphological and functional subdivisions of the cerebellum. The vestibulocerebellum consists of the flocculonodular lobe (dark blue) and deals with input from the vestibular apparatus. The spinocerebellum (green) receives proprioceptive (and exteroceptive) information. The cerebrocerebellum [s. pontocerebellum (brown)] is involved in processing collateral information on corticofugal motor activity. pf, Primary fissure; unf, uvulonodular fissure. (Courtesy of Enke Verlag in MVS Medizinverlage Stuttgart GmbH & Co. KG.)

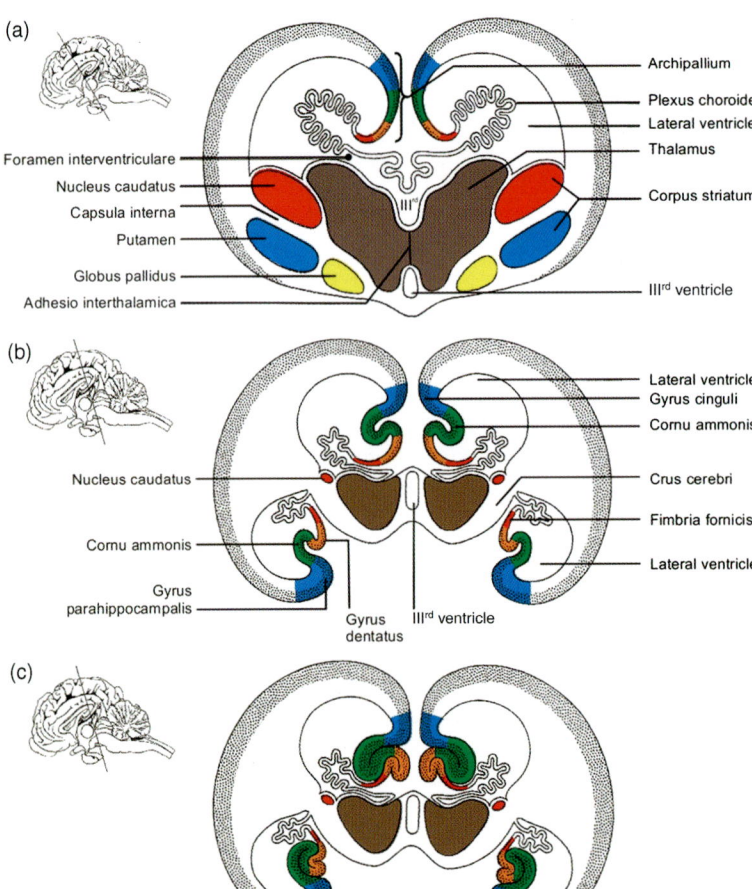

(a)

Archipallium
Plexus choroideus
Lateral ventricle
Thalamus

Foramen interventriculare
Nucleus caudatus
Capsula interna
Putamen
Globus pallidus
Adhesio interthalamica

Corpus striatum

IIIrd ventricle

(b)

Lateral ventricle
Gyrus cinguli
Cornu ammonis

Nucleus caudatus

Crus cerebri

Cornu ammonis

Fimbria fornicis

Gyrus parahippocampalis

Lateral ventricle

Gyrus dentatus · IIIrd ventricle

(c)

CHAPTER 2, FIGURE 7 Embryonic development in the mammalian prosencephalon. Section at the level of the interventricular foramen (a), and sections caudal to the interthalamic adhesion at an early (b) and at a later stage (c) of development. (a) The basal nuclei include the nucleus caudatus, the putamen, and the globus pallidus. They develop from an originally compact mass of gray matter which is later split into separate nuclei by the entrance of axons that comprise the internal capsule. (b, c) The archipallium extends as a band along the medial border of either brain vesicle, dorsal to the choroid plexus of the lateral ventricle. (a) During further development (b, c), the archipallium is invaginated into the lateral ventricle, thus forming the cornu ammonis (hippocampus proper). (Courtesy of Enke Verlag in MVS Medizinverlage Stuttgart GmbH & Co. KG.)

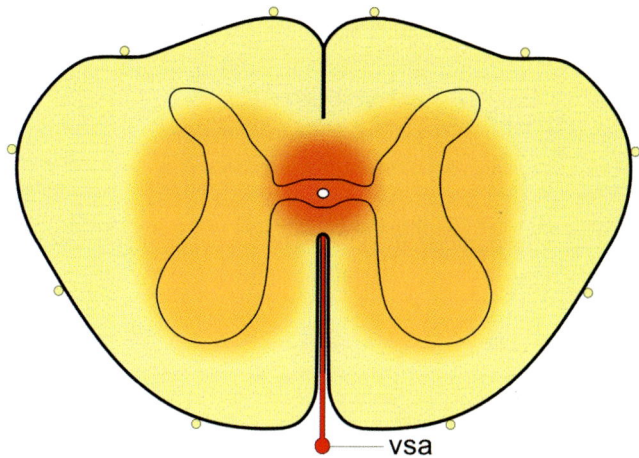

CHAPTER 2, FIGURE 8 Blood supply to the spinal cord. The innermost zone (dark orange) is fed by branches arising from the ventral spinal artery (vsa), which enters the gray matter from the ventral fissure. The outer zone (yellow) depends on blood supplied from vessels penetrating the nervous tissue from the outer surface. The intermediate zone (pale orange) is supplied by both routes. (Courtesy of Enke Verlag in MVS Medizinverlage Stuttgart GmbH & Co. KG.)

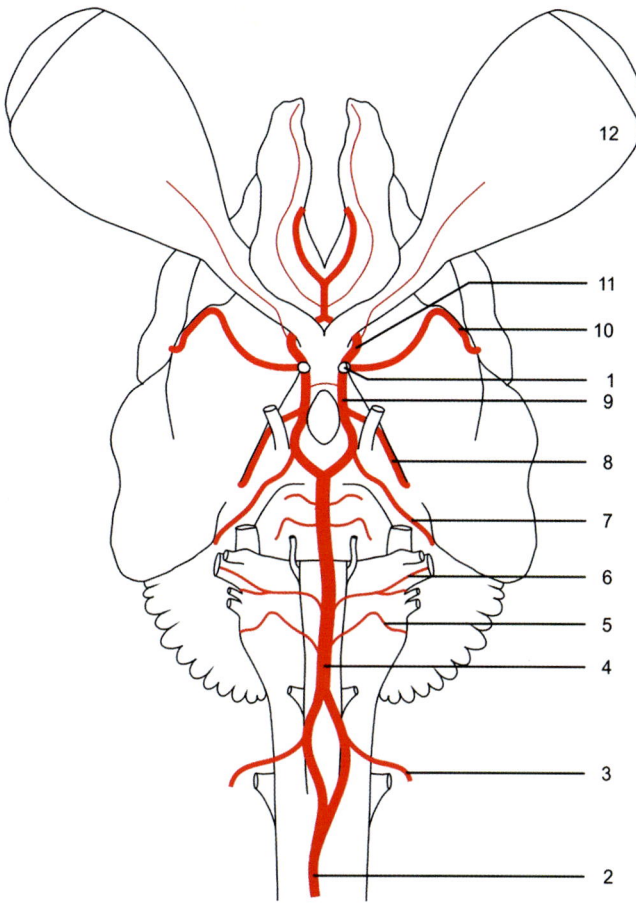

CHAPTER 2, FIGURE 9 Blood supply to the mammalian brain. 1, internal carotid artery; 2, ventral spinal artery; 3, vertebral artery; 4, basilar artery; 5, caudal cerebellar artery; 6, labyrinthine artery; 7, rostral cerebellar artery; 8, caudal cerebral artery; 9, caudal communicating artery; 10, middle cerebral artery; 11, rostral cerebral artery; 12, left eye. (Courtesy of Enke Verlag in MVS Medizinverlage Stuttgart GmbH & Co. KG.)

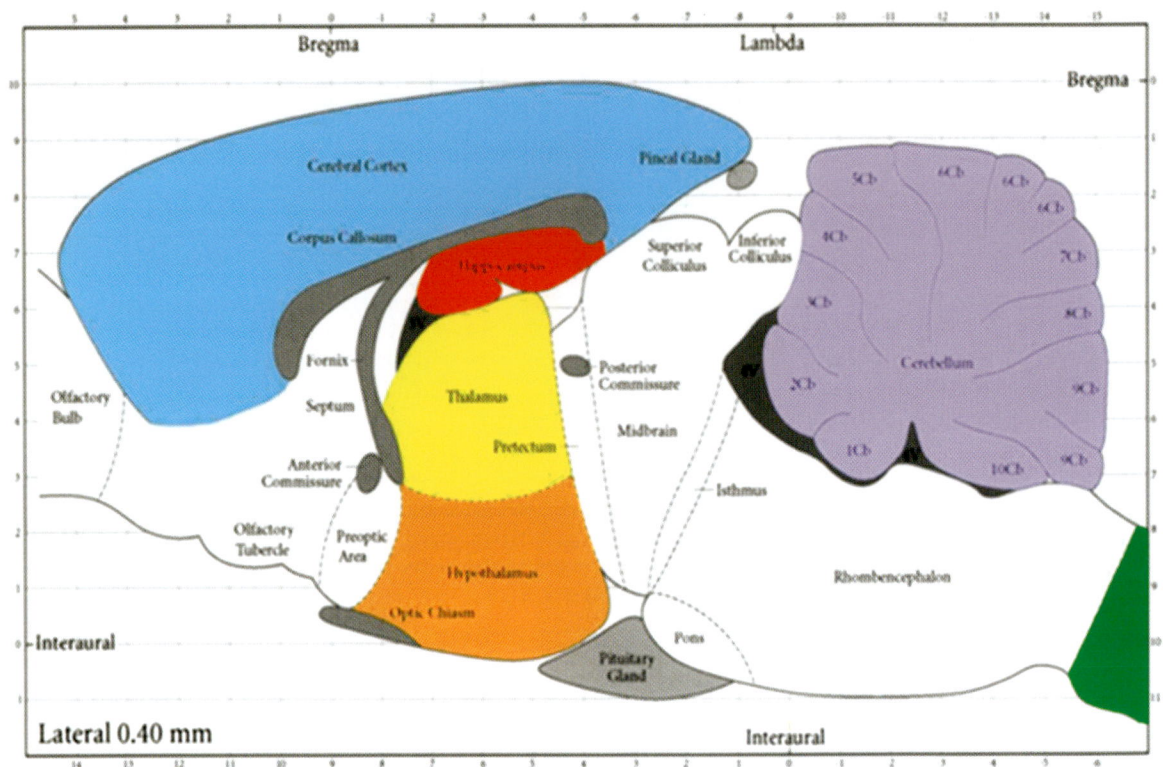

CHAPTER 9, FIGURE 3 Major subdivisions of the brain shown in sagittal section: this single level (and there are many more levels) shows the complexity of brain organization.

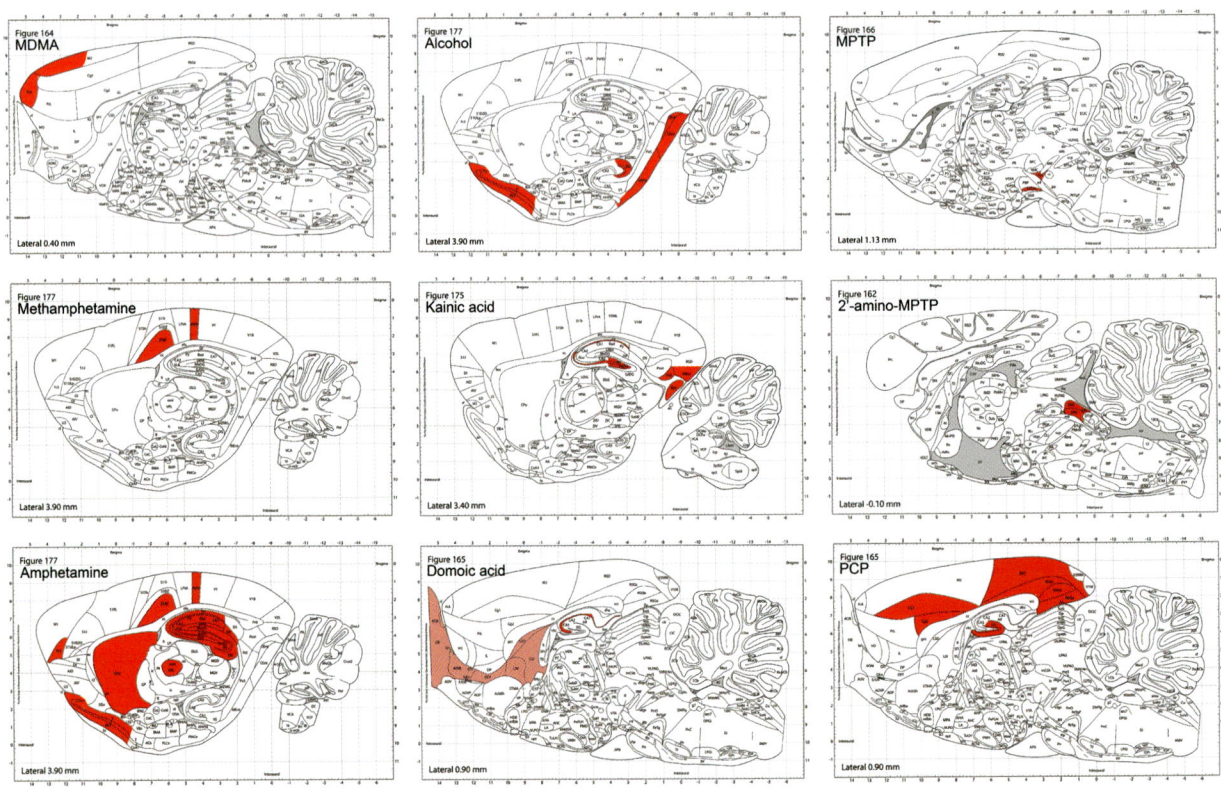

CHAPTER 9, FIGURE 5 Areas of the brain affected by nine known neurotoxins; note that some of the areas are quite small and easily missed in all but the most thorough brain sectioning scheme.

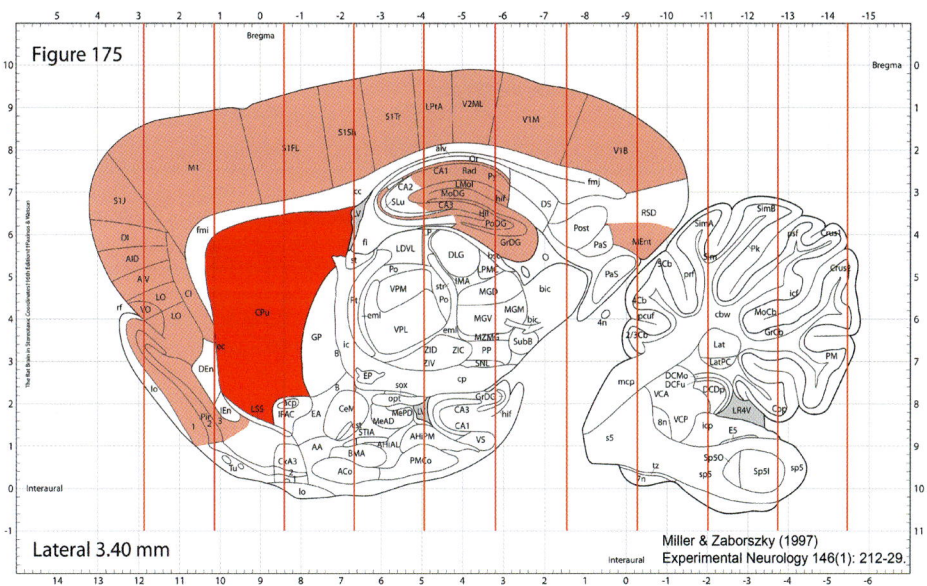

CHAPTER 9, FIGURE 6 Area of the brain affected by the administration of 3NPA: the vertical lines represent the levels that would be sampled and evaluated in an 11-section scheme (sections approximately 1.8 mm apart).

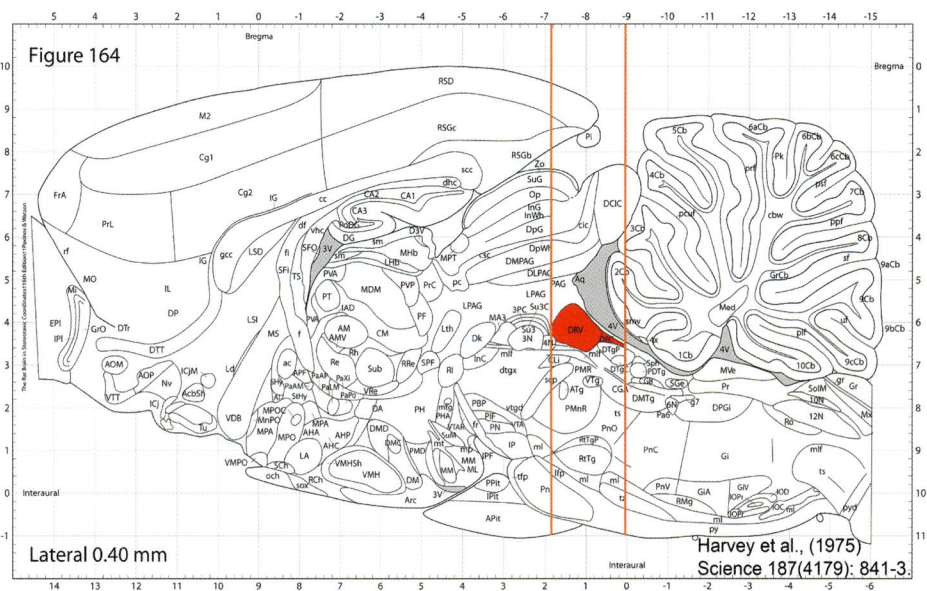

Raphe nuclei span less than 2mm A-P

CHAPTER 9, FIGURE 7 The rahpe nucleus is highlighted. Most of the serotonin-producing cells are within the raphe nucleus. Neuronal death in this nucleus may have a profound effect on the animal. Note how easy it would be to miss this small area on anything but the most comprehensive sectioning schemes.

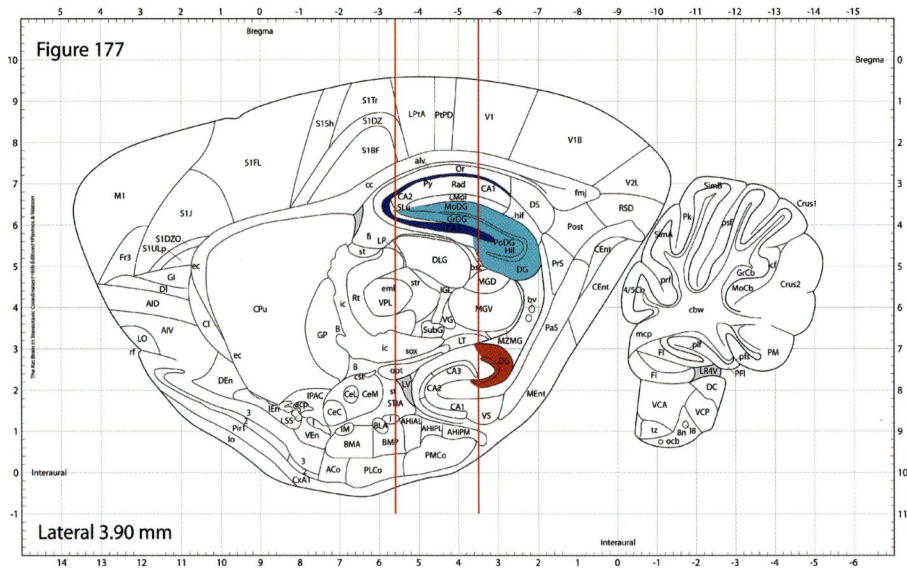

CHAPTER 9, FIGURE 8 Areas of the hippocampus affected by domoic acid (pyramidal layer; thin, C-shaped region), PCP (dorsal dentate gyrus; thicker, reverse C-shaped area), and alcohol (lower area).

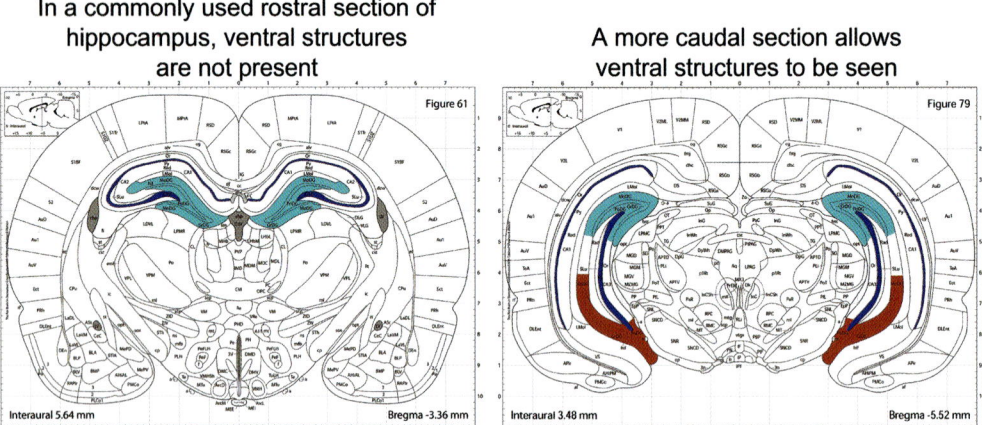

CHAPTER 9, FIGURE 9 The two coronal sections indicated by the vertical lines in Figure 8. Note that the section of hippocampus on the left does not include the ventral portion of the hippocampus. Only the section on the right contains the area affected by alcohol.

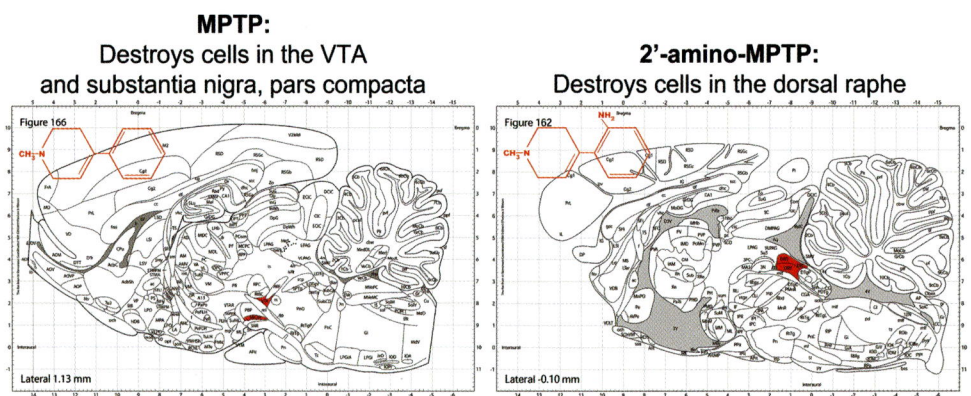

CHAPTER 9, FIGURE 10 Neuronal areas primarily affected by MPTP (substantia nigra) and 2′-NH₂-MPTP (dorsal raphe nucleus). Note that although the structure of the compounds is similar, the selective areas affected are quite different. Note also that although each may affect an important population of neurons adversely, the overall size of the areas affected is small.

Vacuole location: retrosplenial cortex

Cell death locations:

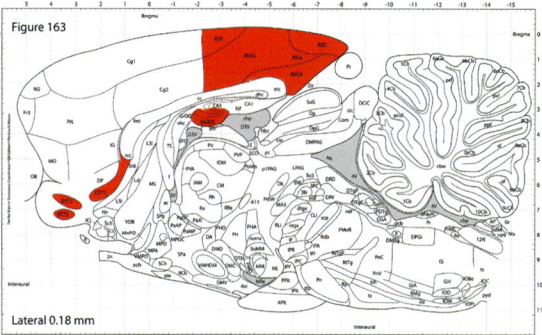

MK-801 destroys cells in:

*Retrosplenial cortex
*Tenia tecta
*Dentate gyrus
*Pyriform cortex
*Amygdala
*Entorhinal cortex
*Ventral CA1 and CA3 of hippocampus

CHAPTER 9, FIGURE 11 MK 801 (NMDA receptor antagonist) administration: The area affected (retrosplenial cortex) by neuronal vacuolation (the Olney effect) in the sagittal diagram on the left as compared to the areas where cell death may be observed (three sagittal sections on the right). Full knowledge of a compounds full capacity to produce neurotoxicity is assessed only after a very thorough sectioning and evaluation of the brain.

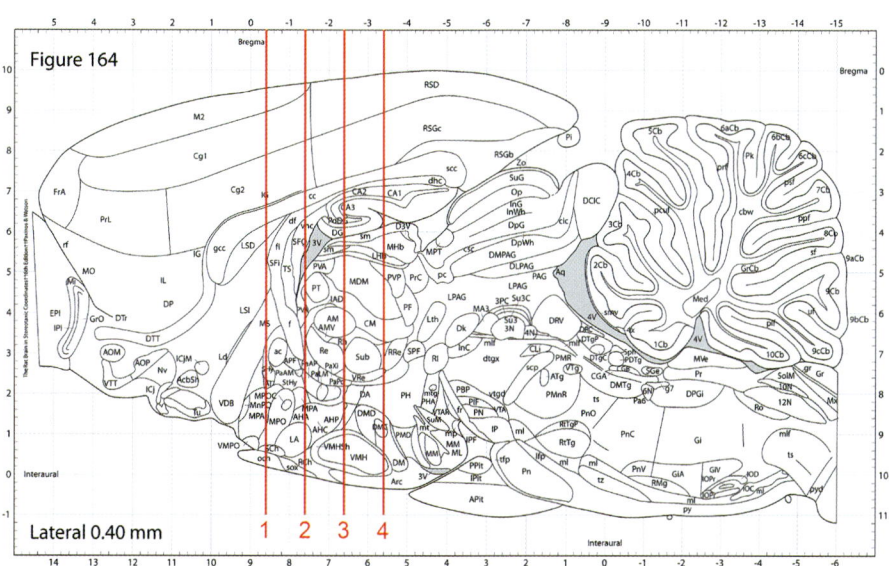

CHAPTER 9, FIGURE 12 Sagittal section with four levels indicated with vertical lines; the distance between these lines is 1 mm.

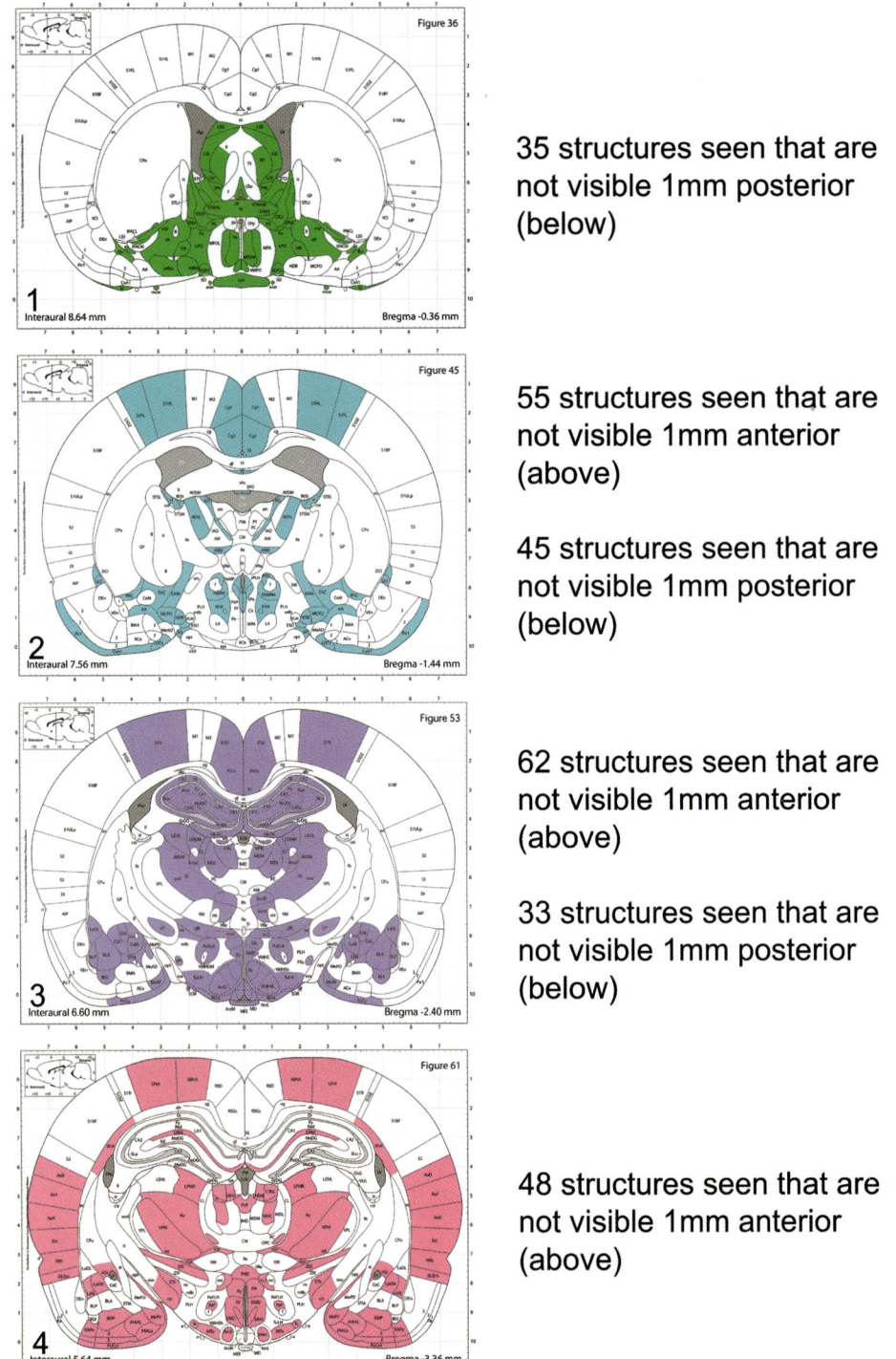

35 structures seen that are not visible 1mm posterior (below)

55 structures seen that are not visible 1mm anterior (above)

45 structures seen that are not visible 1mm posterior (below)

62 structures seen that are not visible 1mm anterior (above)

33 structures seen that are not visible 1mm posterior (below)

48 structures seen that are not visible 1mm anterior (above)

CHAPTER 9, FIGURE 13 The four coronal levels that correspond to the four vertical lines from Figure 12. The shaded areas are those that were not visualized in the prior section, demonstrating the marked difference even 1 mm may make regarding the presence or absence of brain structures.

CHAPTER 9, FIGURES 14 to 16 Sagittal diagrams showing sectioning intervals of 1 mm (Figure 14), 0.5 mm (Figure 15), and 0.32 mm (Figure 16). Sectioning at 0.32 mm provides approximately 60 coronal sections of the brain and is satisfactory in rats to provide adequate representation of the major neuronal areas.

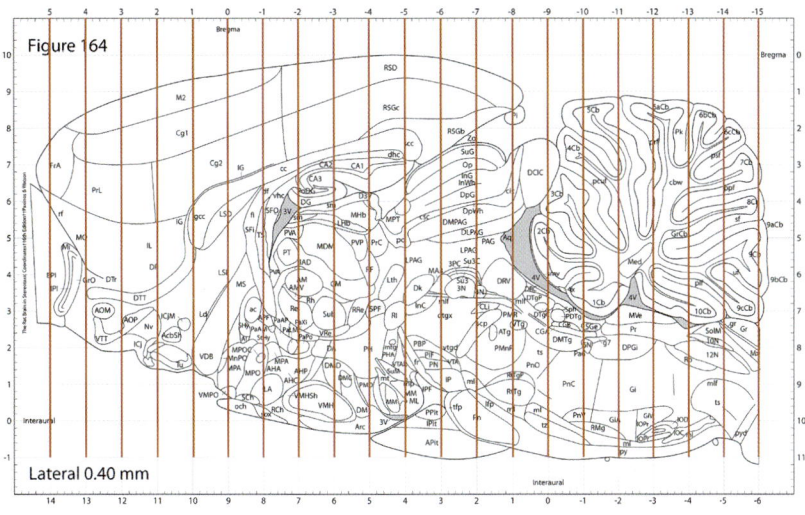

FIGURE 14

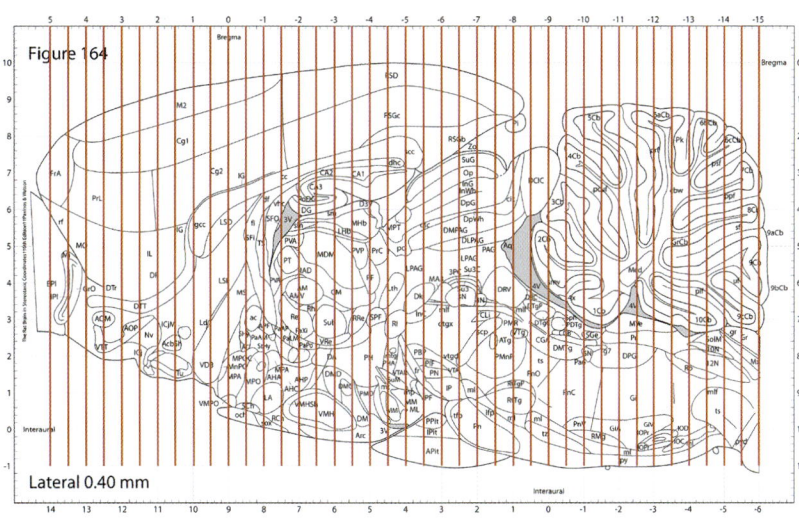

FIGURE 15

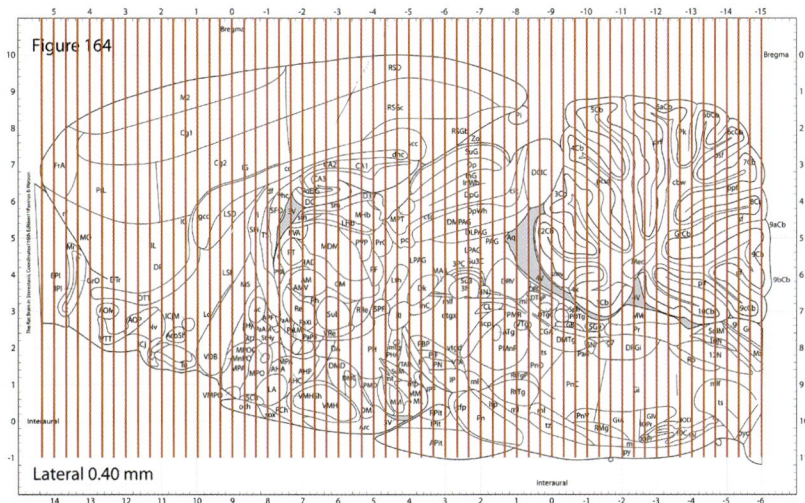

FIGURE 16

Species	Brain Length (mm)	Sampling Interval (in mm)		
		Using 40 samples	Using 60 samples	Using 80 samples
Mouse	12	0.30	0.20	0.15
Rat	21	0.53	0.35	0.26
Monkey	65	1.63	1.08	0.81
Dog	75	1.88	1.25	0.94

CHAPTER 9, FIGURE 17 The approximate brain length of common laboratory species is in the left column. The sectioning interval necessary to achieve approximately 60 overall sections is provided. For example, in an average-size dog, sectioning every 1.25mm will produce around 60 total sections, which will include most of the major neuronal groups in the brain.

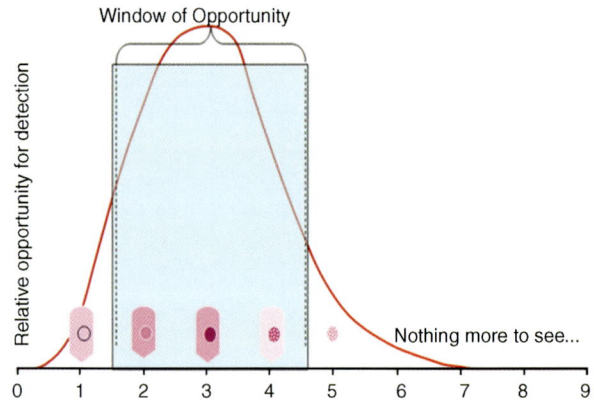

CHAPTER 9, FIGURE 18 Typical time line of cell death that occurs with a variety of neurotoxicants. The cartoons represent cell body changes. Note that the ability to detect changes in the cell body vanishes around day 7.

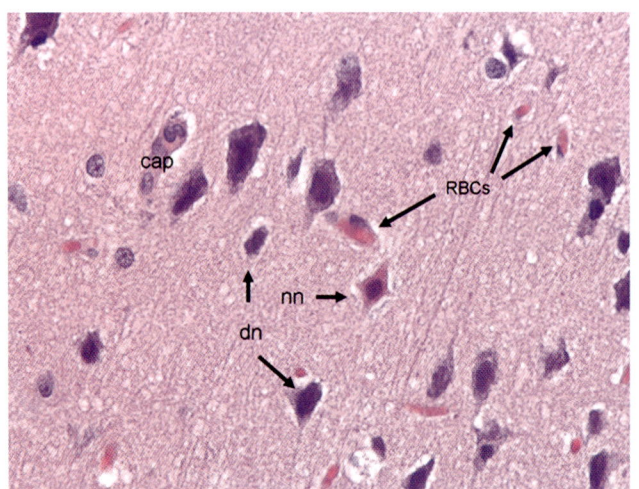

CHAPTER 10, FIGURE 1 A single degenerating neuron (nn) is surrounded by many artifactually created dark neurons (dn). Several capillaries (cap), often containing erythrocytes (RBCs) are also present, structures whose shape and size can resemble degenerate neurons.

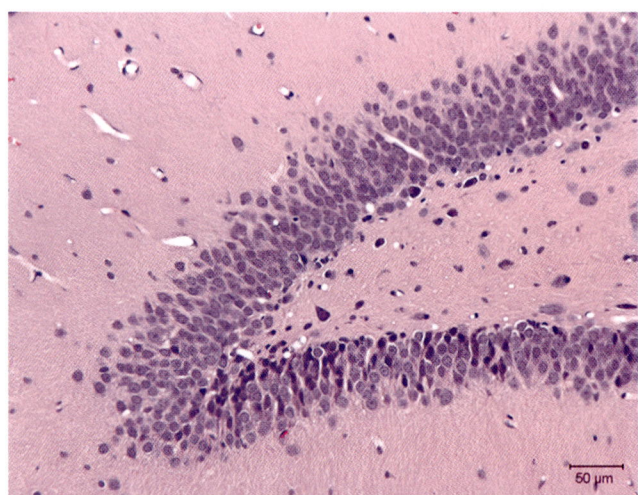

CHAPTER 10, FIGURE 2 Dark neurons, a handling artifact, caused by removing the brain a short time (within 1 h) following systemic formalin perfusion. The artifact can be avoided by allowing sufficient fixation time prior to opening the calvarium. Rat dentate gyrus, H&E.

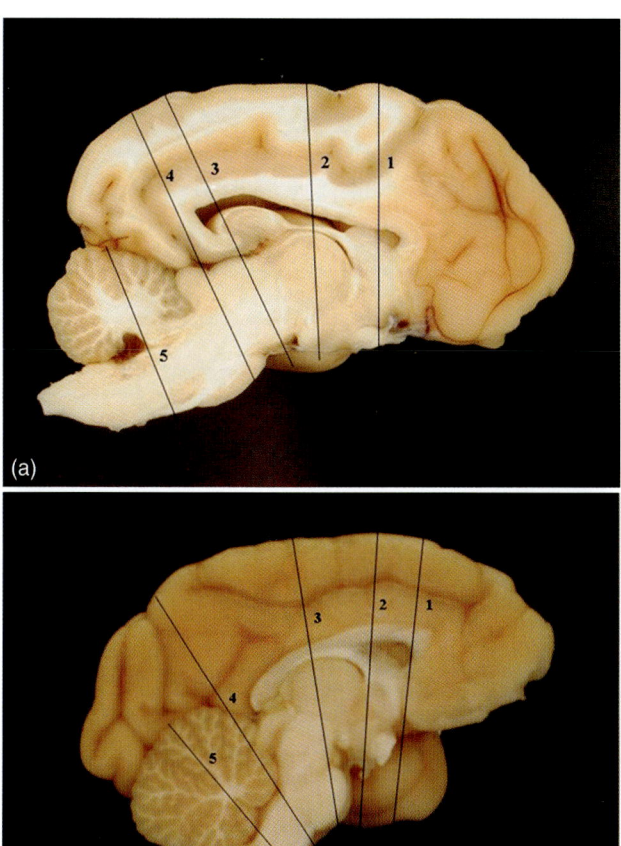

(a)

(b)

CHAPTER 10, FIGURE 4 Sampling paradigms for dog (a) and nonhuman primate (b) designed to represent selected areas of interest in a minimum number of slices. (*See text for full caption*).

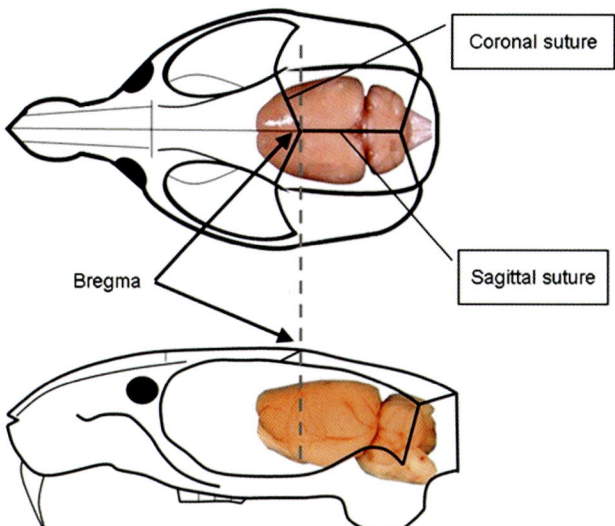

CHAPTER 10, FIGURE 5 Bregma, the location on the skull where the coronal and sagittal sutures meet, provides a fixed reference point for the brain relative to the skull.

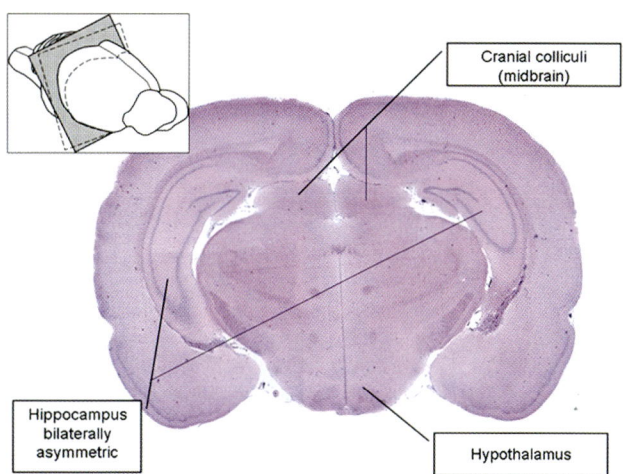

Cranial colliculi (midbrain)

Hippocampus bilaterally asymmetric

Hypothalamus

CHAPTER 10, FIGURE 6 Section of brain that is both tilted and skewed. The histological section lacks bilateral symmetry of the hippocampus and displays both midbrain (cranial colliculus) and hypothalamic structures. The inset illustrates the intended plane of section and the tilted/skewed plane (hatched).

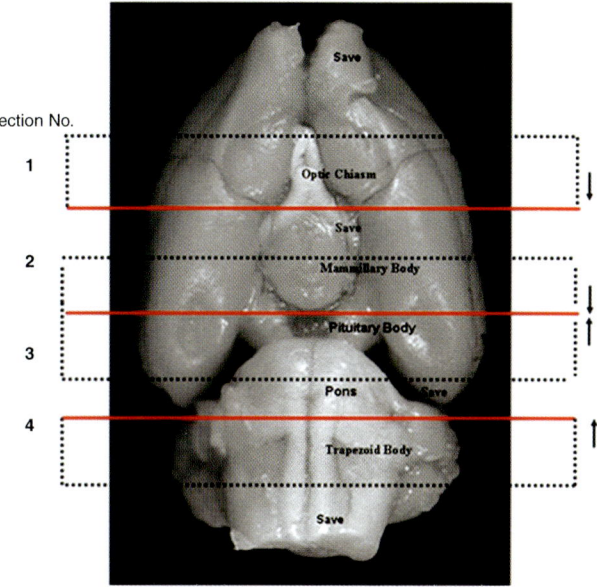

Section No.

Save

Optic Chiasm

Save

Mammillary Body

Pituitary Body

Pons

Save

Trapezoid Body

Save

CHAPTER 10, FIGURE 7 Diagram for a four-section rat brain trim. Solid red lines indicate positions for trim blade placement. Arrows indicate the surfaces to be microtomed.

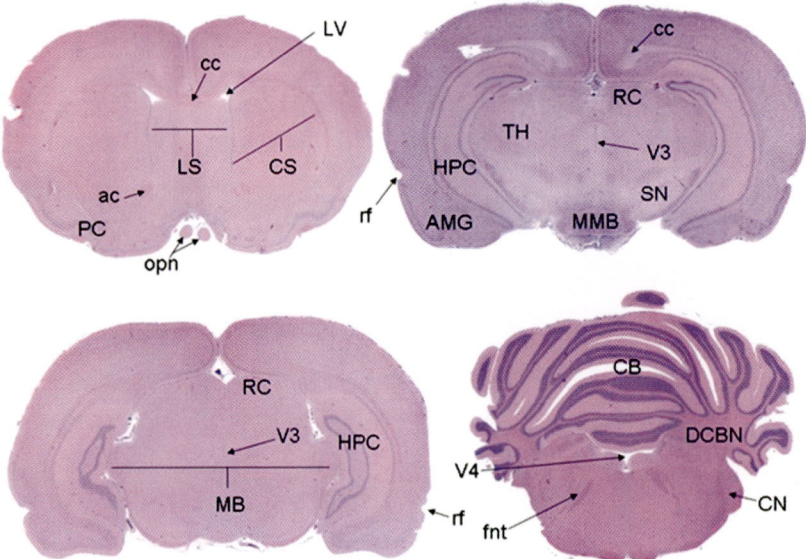

CHAPTER 10, FIGURE 8 Location of important brain landmarks in histological sections in sections from a four-section trim. ac, Anterior commissure; AMG, amygdaloid region; CB, cerebellum; cc, corpus callosum; CN, cochlear nucleus; CS, striatum; DCBN, deep cerebellar nuclei; fnt, facial nerve tract; HPC, hippocampus; LS, septal nuclear area; LV, lateral ventricle; MB, midbrain; MMB, mammillary body; opn, optic nerve; PC, piriform cortex; RC, rostral colliculus; rf, rhinal fissure; SN, substantia nigra; TH, thalamus; V3, third ventricle; V4, fourth ventricle.

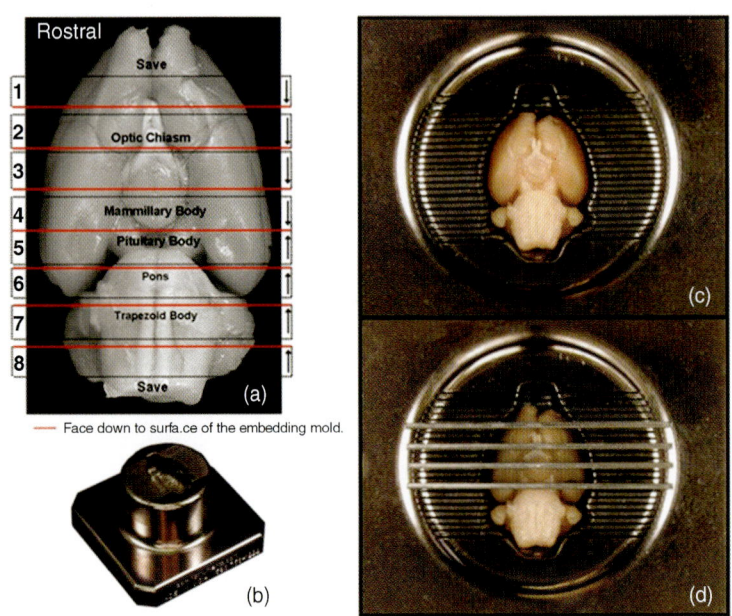

CHAPTER 10, FIGURE 9 Detailed trimming of rat brains: (a) trim diagram to obtain eight histological sections; (b) brain matrix; (c) rat brain in matrix with ventral surface up; (d) with four of the five most rostral blades in place for an eight-section trim [blades are minimally farther apart than sectioning planes depicted in diagram (a) to obtain sufficient light to clearly illustrate brain features].

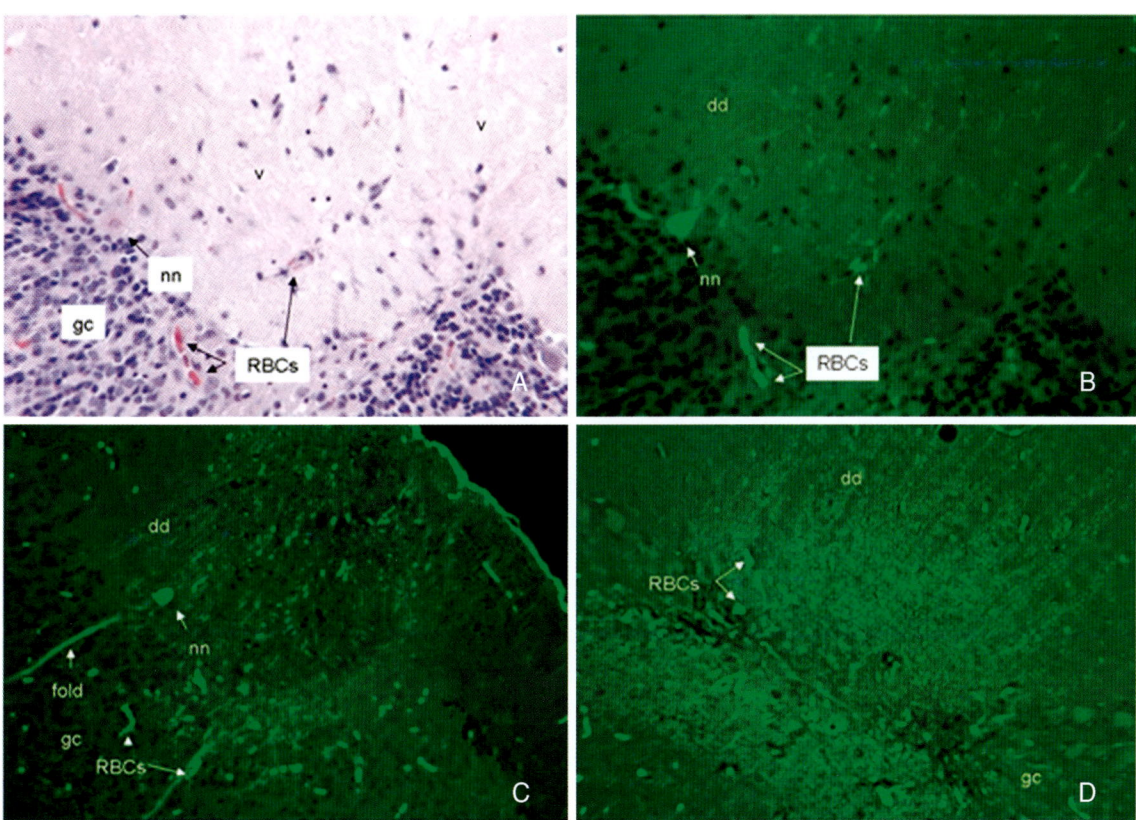

CHAPTER 10, FIGURE 10 Histological sections of cerebellum from two rats with Purkinje neuronal degeneration and loss. (A) and (B) represent the same H&E-stained section viewed, respectively, with transmitted white light and with a FITC filter. In (A), note degenerating neuron (nn), paucity of Purkinje neurons overlying the granule cell (gc) layer, and vacuolation (v) of the overlying molecular layer. In (B), the degenerating neuron, overlying dendrites (dd), and a few erythrocytes (RBCs) are autofluorescent. (C) Serial section from the same region as in (A) and (B), but at slightly lower magnification. The autofluorescence of the degenerating neuron, overlying dendrites, and erythrocytes is more intense than background. Autofluorescence of two tissue folds is enhanced artifactually. (D) Fluoro-Jade B to illustrate enhancement of degenerating dendrites of injured Purkinje neurons (bodies of degenerating neurons not present in this section).

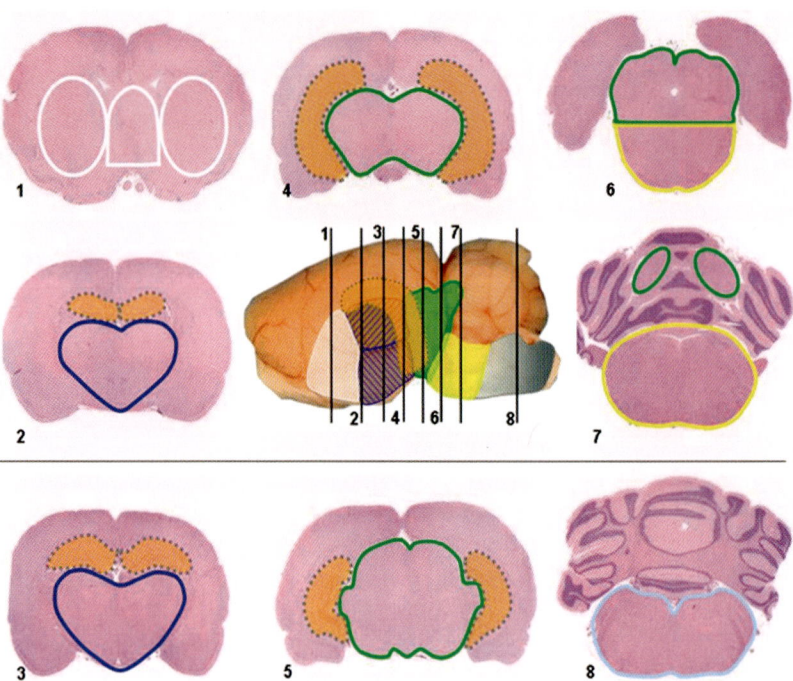

CHAPTER 10, FIGURE 11 Moving rostrally to caudally, the major regions of the rat brain in coronal section have characteristic profiles. The central image depicts a lateral view of rat brain with an overlay of the septal-caudate (white), thalamus-hypothalamus (dark blue), hippocampus (orange), midbrain (green), pons (yellow), and medulla (light blue) regions. Coronal sectioning planes indicated on the central figure correspond to histological sections (1) through (8). Overlays on each of the sections, which match the color of the corresponding portion of the overlay on the central image, illustrate the readily recognizable changes in shape, contour, and position of these major brain regions in coronal section.

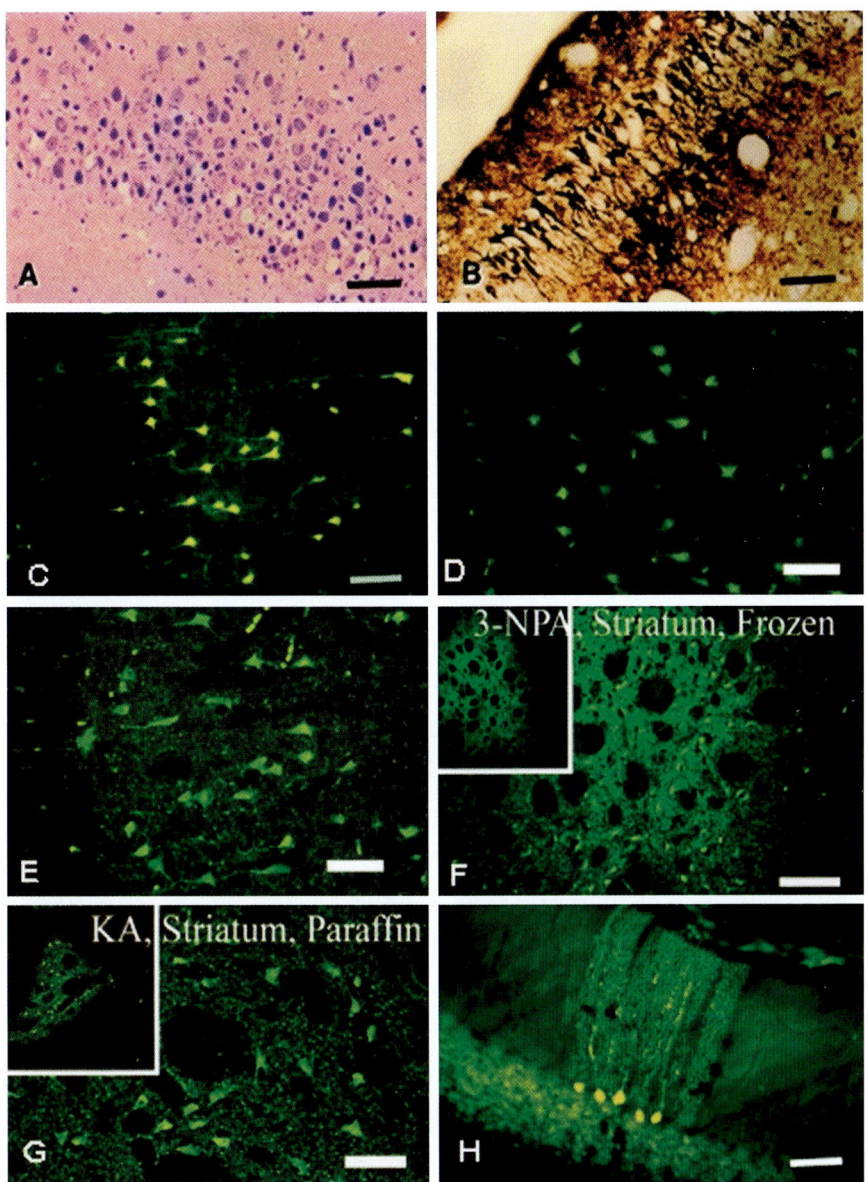

CHAPTER 11, FIGURE 1 (A) Hematoxylin and eosin were used to stain this paraffin-embedded section of the hippocampus of a rat 24 h after exposure to kainic acid. Note the variety of cellular morphologies and dye affinities, making interpretation difficult; bar = 200 μm. (B) The de Olmos suppressed cupric silver technique was used to label degenerating pyramidal cells in frozen sections taken from the hippocampus contralateral to that seen in (A). Degenerating neurons appear black, while normal neurons are unstained; bar = 200 μm. (C) Typical Fluoro-Jade labeling of parietal cortex following kainic acid injection; bar = 200 μm. (D) Typical Fluoro-Jade B labeling of cingulate cortex following kainic acid injection (i.p.); bar = 200 μm. (E) Typical Fluoro-Jade C labeling cingulate cortex following kainic acid injection (i.p.); bar = 200 μm. (F) Fluoro-Jade C labeling in the central striatum of a frozen section following 3-NPA injection (s.c.); bar = 400 μm. (G) Fluoro-Jade C labeling in the central striatum of a paraffin-embedded section following KA injection (i.p.); bar = 200 μm. (H) Patches of Purkinje cells and their dendrites in the medial cerebellum demonstrate ibogaine-induced degeneration as revealed by Fluoro-Jade labeling; bar = 400 μm.

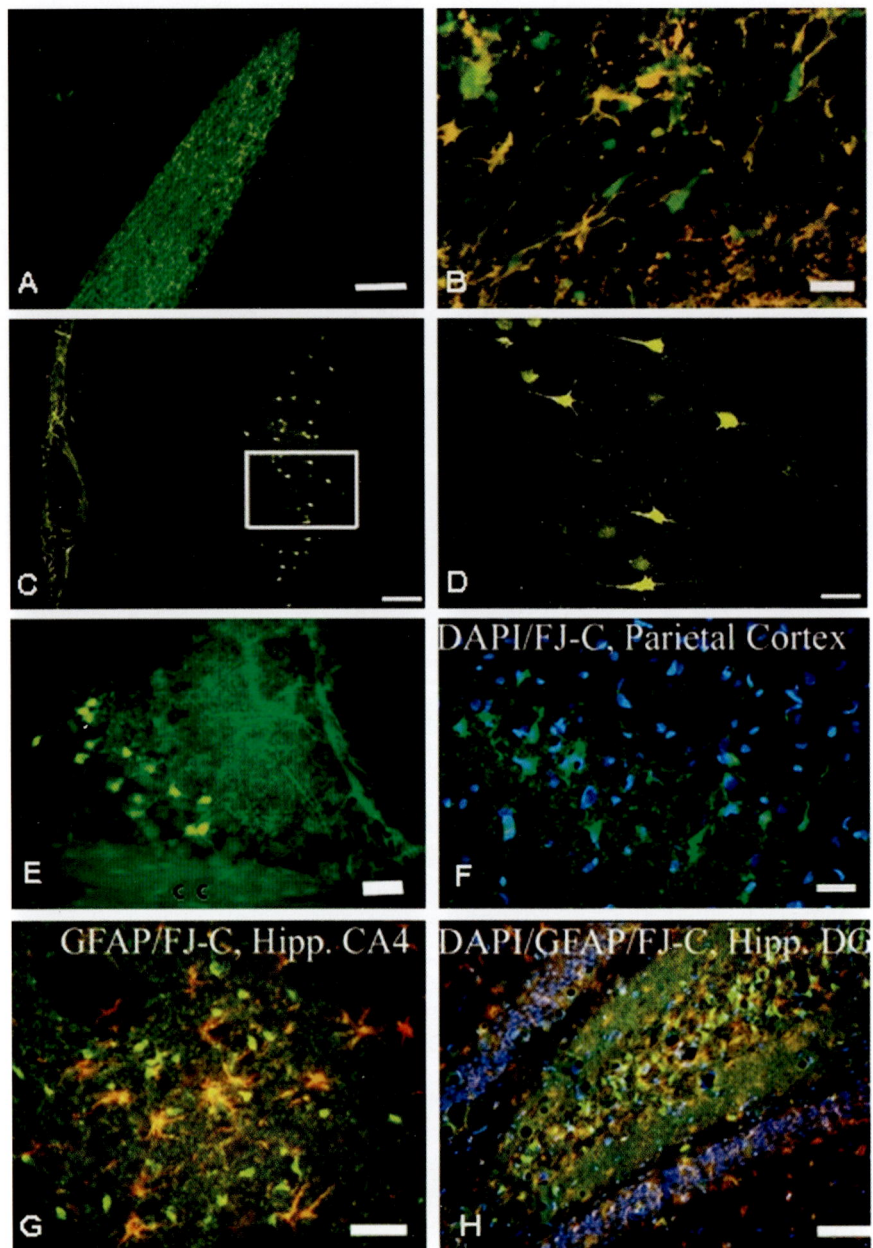

CHAPTER 11, FIGURE 2 (A) Fluoro-Jade labeling of the optic tract following ennucleation the contralateral of eye; bar = 400 μm. (B) Double labeling of rostral arcuate nucleus of the rat brain showing GFAP (red) and Fluoro-Jade B (green) following aurothioglucose injection (i.p.); bar = 40 μm. (C) Low magnification of Fluoro-Jade B positive neurons in the parietal cortex of rat exposed to MDMA; bar = 400 μm. (D) High magnification of the boxed area as shown in (C) reveals Fluoro-Jade B-positive neuronal cell bodies and axon terminals; bar = 40 μm. (E) Tenia tecta of mouse brain exposed to methamphetamine showing Fluoro-Jade-labeled neuronal cell bodies at left and terminals at right. CC, corpus callosum; bar = 40 μm. (F) Double fluorescent labeling with Fluoro-Jade and DAPI result in the respective localization within the parietal cortex of degenerating neurons (green) and viable cell nuclei (blue) following kainic acid injection (i.p); bar = 40 μm. (G) High magnification of the central nucleus of the amygdala is visualized by using sequential blue and green illumination to reveal red GFAP labeled activated astrocytes in association with green Fluoro-Jade C-positive degenerating cells and terminals; bar = 40 μm. (H) A triple exposure of the hippocampal dentate gyrus illuminated with green, blue, and UV light reveals Fluoro-Jade-positive degenerating polymorphic neurons in the hialar region, DAPI staining of nuclei, which is especially conspicuous among the intact granular cells and GFAP-positive activated astrocytes; mag = 100 μm.

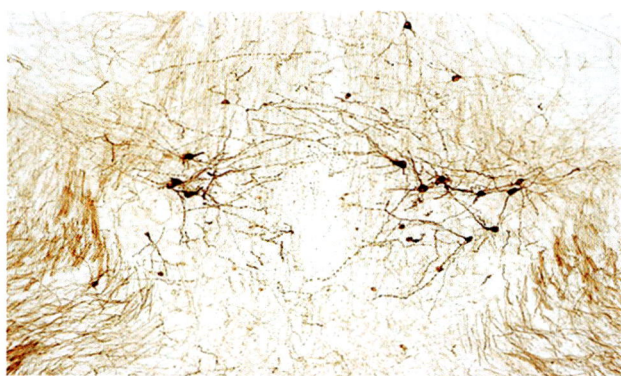

CHAPTER 12, FIGURE 1 Dopaminergic neurons in the brain stem of rat revealed by IHC with an antibody against tyrosine hydroxylase.

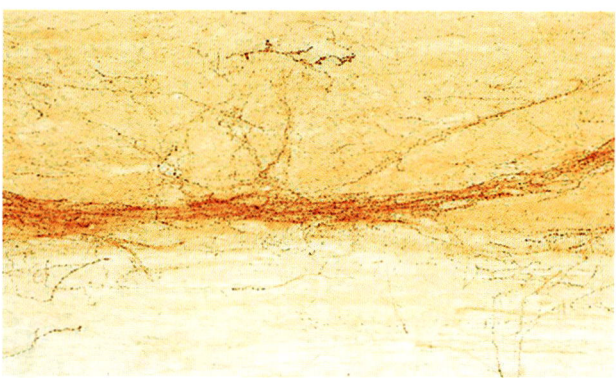

CHAPTER 12, FIGURE 4 Serotonergic axons in a horizontal section of the spinal cord of a rat revealed by IHC with an antibody against serotonin (5HT).

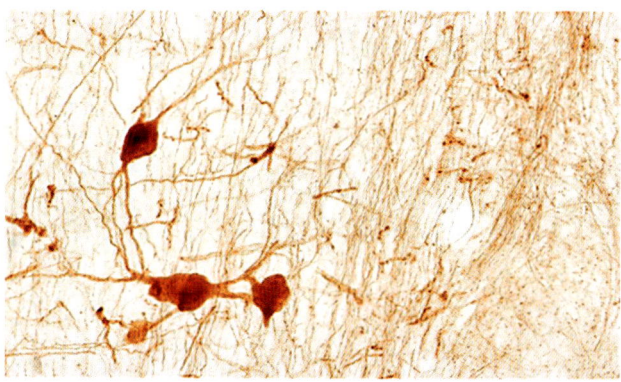

CHAPTER 12, FIGURE 2 Cholinergic neurons in the ventral limb of the diagonal band of Brocca (rat) revealed by IHC with an antibody against choline acetyltransferase (ChAT).

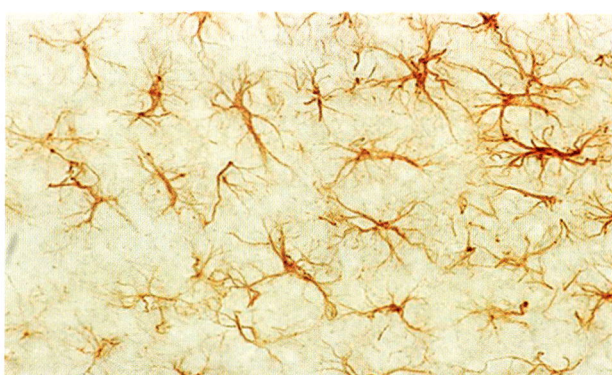

CHAPTER 12, FIGURE 5 Normal morphology of astrocytes revealed by IHC with an antibody against glial fibrillary acidic protein (GFAP), a cytoskeletal protein mostly unique to astrocytes. This antibody cross reacts with many species.

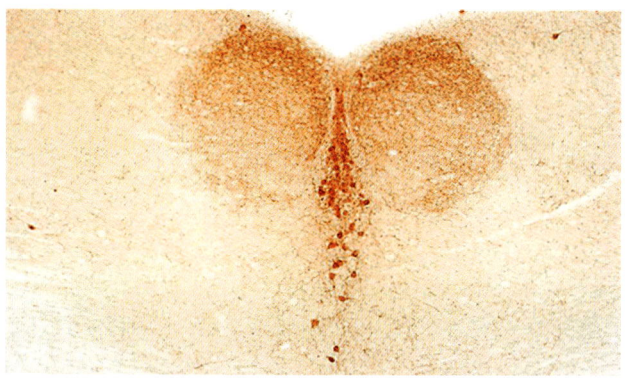

CHAPTER 12, FIGURE 3 Serotonergic neurons in the dorsal raphe nucleus (rat) revealed by IHC with an antibody against serotonin (5HT).

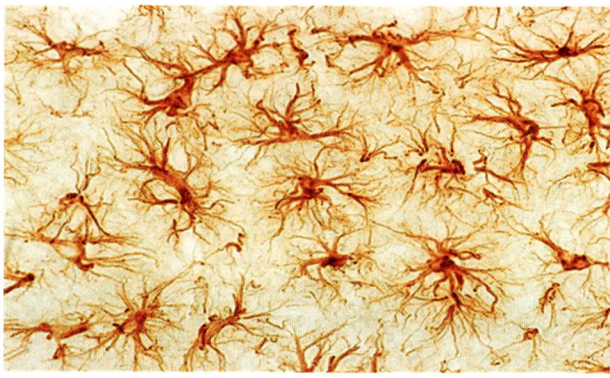

CHAPTER 12, FIGURE 6 These reactive astrocytes display more numerous and thickened processes and enhanced density of staining.

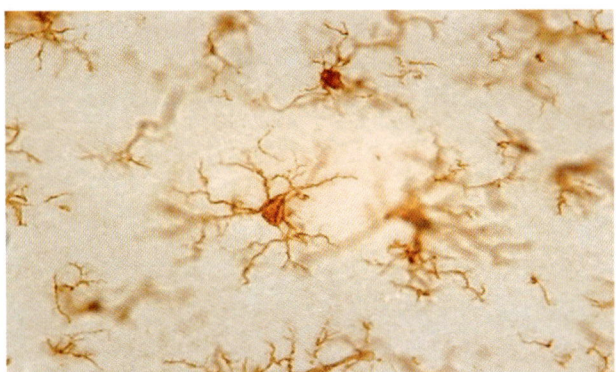

CHAPTER 12, FIGURE 7 Normal morphology of microglia revealed by IHC with an antibody against IBA1. This antibody cross reacts with many species.

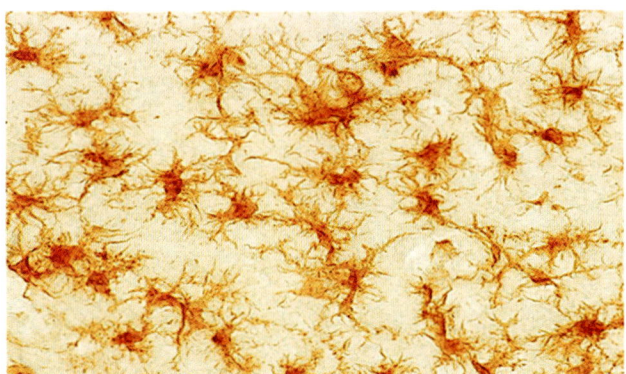

CHAPTER 12, FIGURE 8 Hypertrophic morphology of reactive microglia revealed by IHC with an antibody against IBA1. The morphologic changes range from appearing near normal, to progressive loss of processes, to enlarged cell body, to finally an amoebiform phenotype.

CuAg Method   Amino CuAg Method

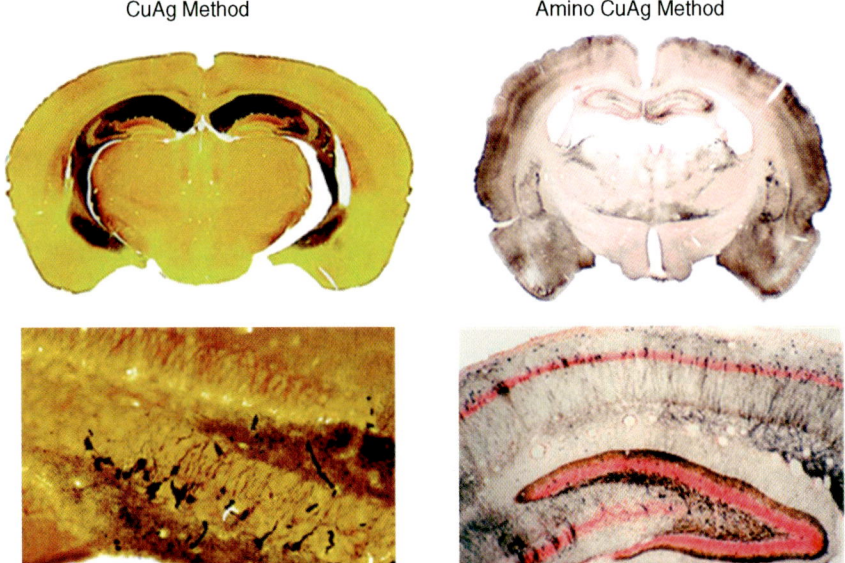

CHAPTER 12, FIGURE 9 Comparison of the background staining of the cupric silver (CuAg) method on the left and the amino cupric silver (amino CuAg) method on the right. The amino CuAg section is counterstained with neutral red. In each stain, the black areas demonstrate areas of neuronal degeneration. In general, detail and staining specificity is improved with the amino CuAg technique.

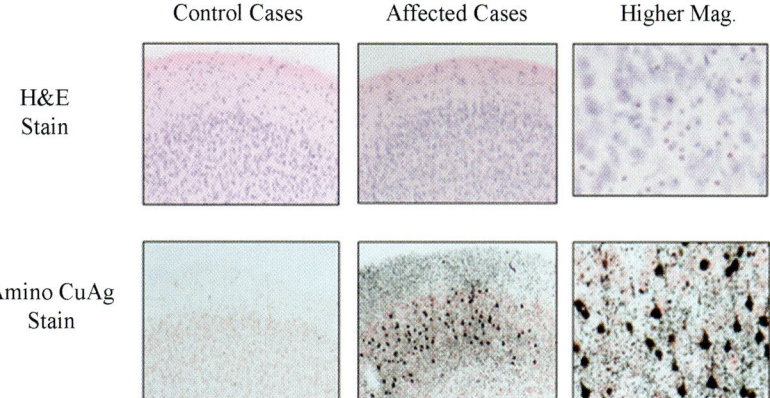

Control Cases   Affected Cases   Higher Mag.

H&E
Stain

Amino CuAg
Stain

CHAPTER 12, FIGURE 10 Comparison between what is visible using traditional H&E staining as compared to the much greater visualization of neuronal changes possible using the amino CuAg stain. The panels on the left are from controls; there is no evidence of degeneration or disintegration in either stain. In the middle panels, at low power, there are no readily apparent morphological changes to the cells with the H&E stain but abundant disintegrating neurons are easily observed with the amino CuAg stain. The numerous black dots at the top of the section are disintegrating synaptic terminals, a change not visible at any power with H&E. The panels to the right are a closer view of the changes in the middle panels. At this power, the florid neuronal disintegration or necrosis is evident with the H&E stain, but the affected cells are still not easily visualized. With the amino CuAg stain, the black (disintegrating or necrotic) neurons are very readily apparent.

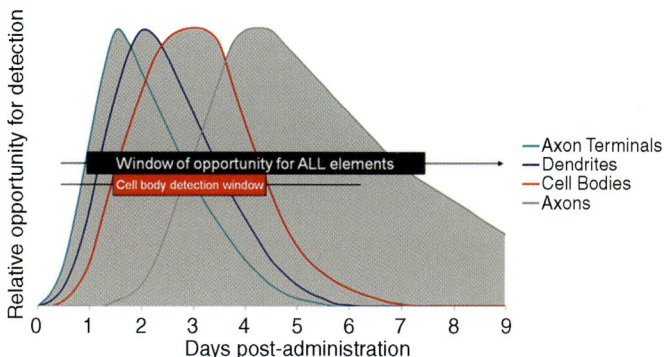

CHAPTER 12, FIGURE 11 Typical time course of disintegrating elements. Note that by day 7, the disintegration of synaptic terminals, dendrites, and cell bodies are no longer apparent. Disintegrating elements in axons are frequently visible for several more days and even weeks. This allows for the detection of neuronal death even after the cell bodies are gone from view in a section.

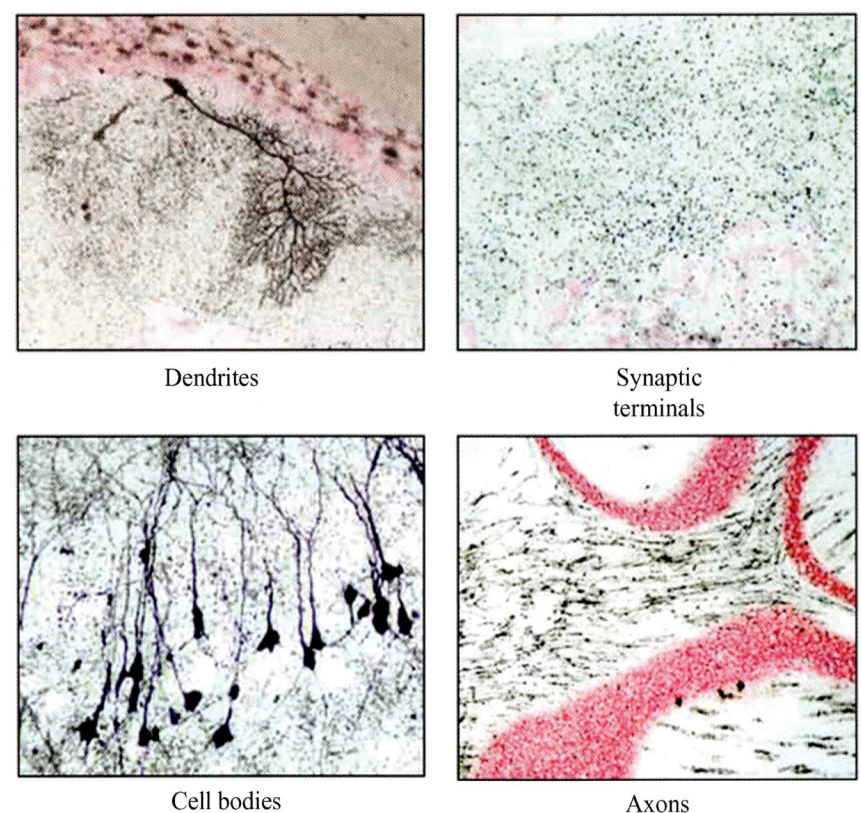

Dendrites

Synaptic
terminals

Cell bodies

Axons

CHAPTER 12, FIGURE 12 General appearance of the disintegration of the four cell elements that can be visualized using the CuAg staining methods.

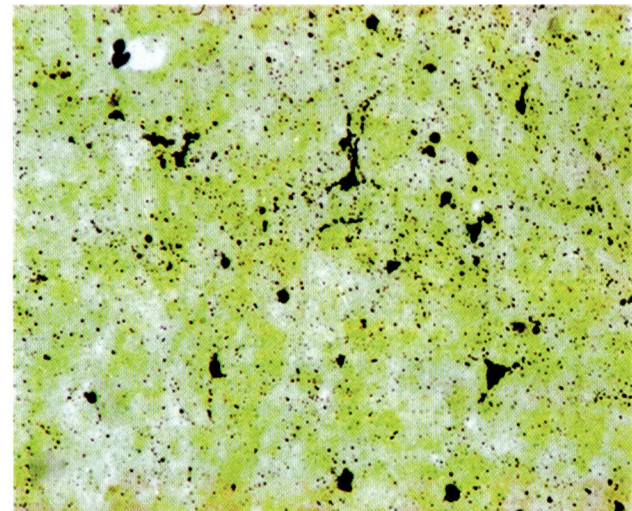

CHAPTER 12, FIGURE 13 The CuAg stain reveals abundant apoptotic neurons in a neonate.

Lateral View of Hindlimb

Medial View of Hindlimb

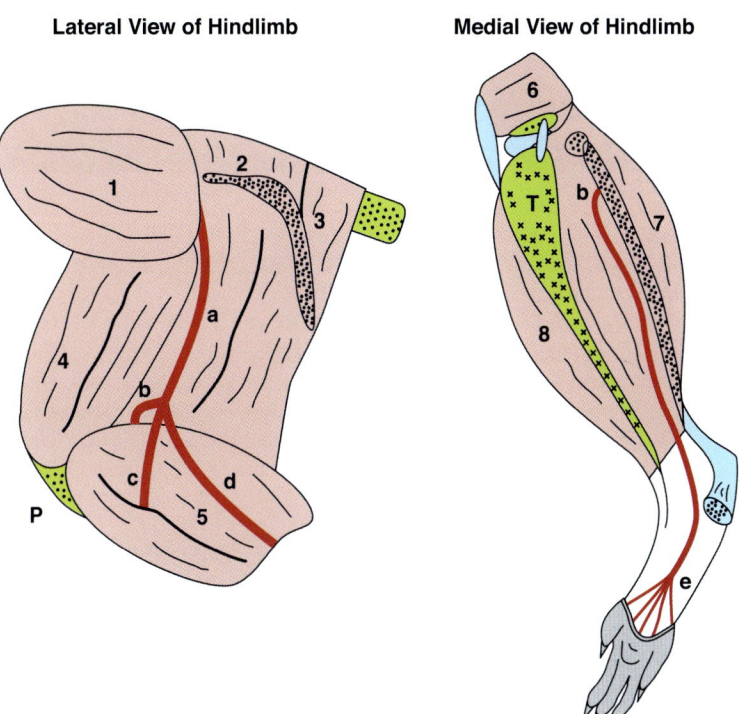

CHAPTER 16, FIGURE 4 Location of principal peripheral nerves to be evaluated for detailed toxicological neuropathology studies. Muscles: 1, gluteus medius; 2, biceps femoris; 3, semitendinosus; 4, quadriceps femoris; 5, gastrocnemius lateralis; 6, rectus femoris; 7, gastrocnemius medialis; 8, tibialis cranialis. Nerves: a, sciatic; b, tibial; c, common peroneal (fibular); d, lateral sural; e, plantar. Bones: P, patella; T, tibia. (Adapted from Popesko et al.[24] by G. Krinke for use in Bolon, et al.[22] Reprinted here with the permission of the authors; the original publisher, Saunders; and Dr. Krinke.)

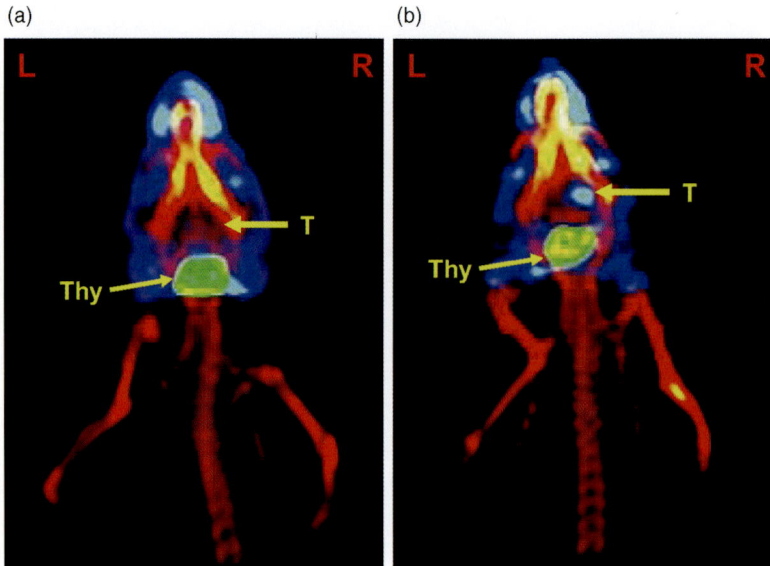

(a)   (b)

CHAPTER 18, FIGURE 2 Single photon-emission tomography-computed tomography (SPECT-CT) imaging of hepatocyte growth factor/scatter factor (HGF/SF)-positive and HGF/SF-negative tumor xenografts in mouse brain using [^{125}I]anti-HGH/SF monoclonal antibody (mAb). One of two human glioblastoma cell lines, SNB19 (HGH/SF$^-$) and U87 (HGH/SF$^+$), was orthotopically grafted into the right (R) cerebral hemisphere. On day 21, mice were injected with 1 mCi of [^{125}I]anti-HGH/SF mAb intraveneously; quantitative SPECT-CT images were acquired 2 days after mAb injection. Antibody binding to specific tissues was calculated as the percent injected dose per gram (% ID/g); the value was 42-fold higher in the U87 tumor (0.5667 ± 0.113% ID/g; panel B) relative to the SNB19 tumor (0.0134 ± 0.0004% ID/g; panel A). L, left; R, right; T, tumor xenograft; Thy, thymus. [Courtesy of Catherine Foss (Department of Radiology) and Jonathan Laterra (Department of Neurosciences), Kennedy Krieger Institute, Johns Hopkins University.]

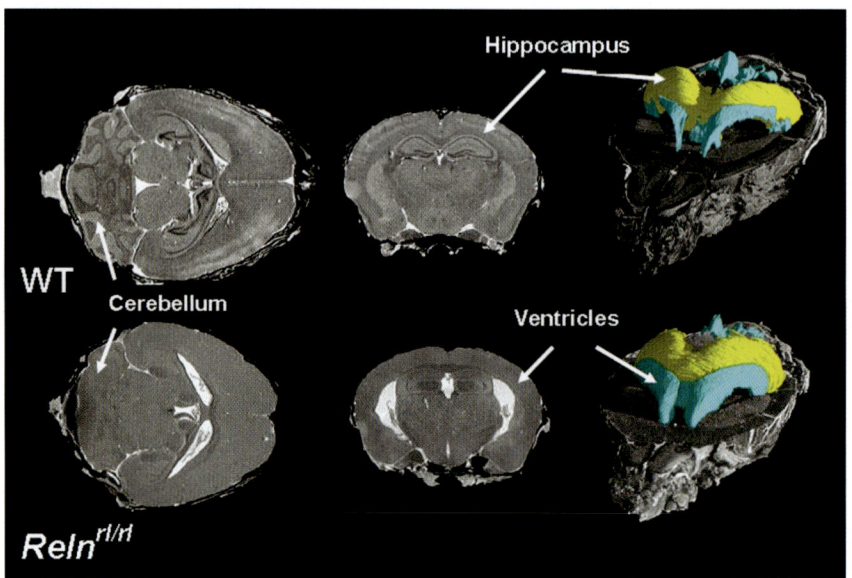

CHAPTER 18, FIGURE 3 Magnetic resonance microscopy (MRM) distinguishes neuroanatomic differences between homozygous Reeler mutant mice (Reln rl/rl) and age-matched wild-type (WT) controls. MRM scans at 21.5-μm resolution demonstrate severe cerebellar atrophy, ventricular dilation, and disorganization of the hippocampal cell layers[105] in the Reln rl/rl mouse. The images on the right are a partial three-dimensional reconstruction of the brain to emphasize the altered shape and size of the hippocampus (yellow) and lateral ventricles (blue) in the mutant mice. [Courtesy of Andrea Badea (Department of Radiology), Duke University.]

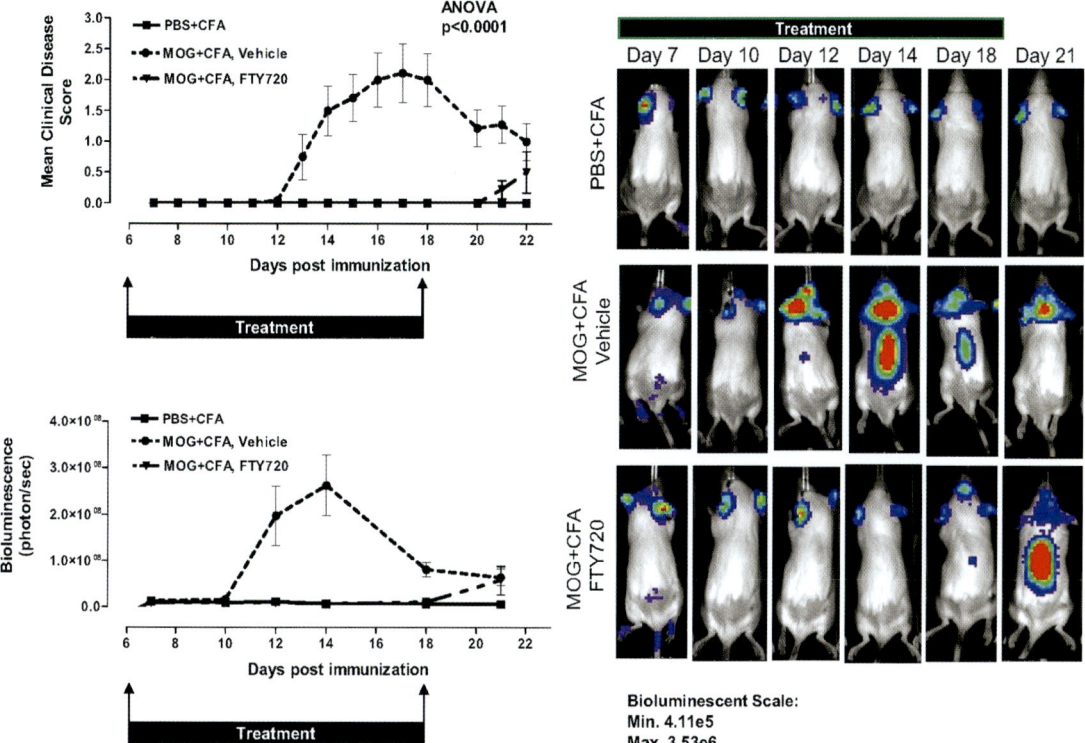

CHAPTER 18, FIGURE 4 Whole-body bioluminescence imaging (BLI) of glial cell reactivity in the brain is correlated to the clinical progression of murine experimental autoimmune encephalitis (EAE). FVB/N mice [nonsusceptible strain due to major histocompatability complex (MHC) haplotype H^{2q}] engineered to express the GFAP-luc fusion transgene [linking the glial fibrillary acidic protein (GFAP) promoter to luciferase (luc)] were crossed with C57BL/6J-Tyrc-2J mice (susceptible strain via MHC haplotype H^{2b}). The FVB/N animals do not develop EAE in response to immunization with a myelin oligodendrocyte glycoprotein fragment (MOG35-55), while the transgenic F_1 offspring (H^{2q}/H^{2b} haplotype) readily develop the disease. EAE was induced by immunization with MOG35-55 peptide emulsified in complete Freund's adjuvant (CFA) followed by pertussis toxin 24 h later; the negative control group did not receive MOG35-55. One EAE group was gavaged daily with FTY720 (a reference immuno-modulating compound that reduces inflammation; Cayman Chemicals, Ann Arbor, MI) at 5 mg/kg per day; the negative control group was given phosphate-buffered saline (PBS) daily by oral gavage. The treatment (black bar) started on day 6 after immunization and continued for the duration of the study. The intensity of BLI signaling superimposed over the neuraxis (brain and/or spinal cord) of the mice is red (greatest) > yellow > green > blue (least). [Courtesy of Rukiye-Nazan E. Dogan, Stephen J. McAndrew, and Olesia Buiakova, Caliper Life Sciences, Discovery Alliances & Services Division (Xenogen Biosciences).]

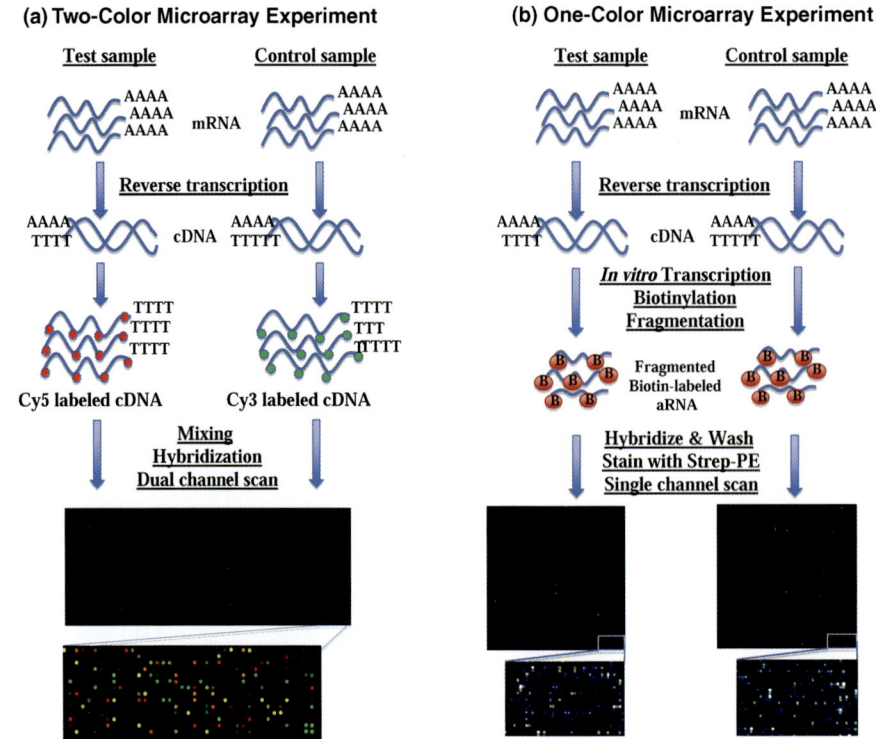

CHAPTER 20, FIGURE 2 Workflow in a typical one- and two-color microarray experiment. In the two-color experiment (a), the mRNA isolated from each of the test and control (or reference) samples is subjected to reverse transcription to generate cDNA targets. The single-stranded cDNA targets from each sample are labeled with fluorophores (Cy3 and Cy5) and mixed in equal proportions. Equal amounts of the labeled cDNA targets are hybridized with the cDNA microarray for a period of about 12 h to attain equilibrium. Ideally, at most probes on the cDNA microarray chip, equal amounts of the labeled cDNA targets should be available for hybridization. The hybridized microarray chip is scanned under a confocal laser microscope and a raw image that is pseudocolored is obtained for data analysis. When equal amounts hybridize, the color is yellow; if Cy5-labeled targets preferentially hybridize, the color is red and if Cy3-labeled targets preferentially hybridize, the color is green; and if none of the labeled targets hybridize, the color is black. The differential fluorescent signal ratios are calculated to derive the differential gene expression within the test and control (or reference) samples from a single microarray chip. Two-color experiments are also done with long oligomer nucleotide probes, such as in Agilent microarray chips. A one-color experiment (b) is performed in a similar manner except that the cDNA [or aRNA (antisense RNA)] from a single test sample is labeled with a single fluorophore and hybridized to a microarray [without a control (or reference) sample as in the two-color technique]. The hybridized microarray is scanned under a confocal laser microscope at an appropriate wavelength (single channel), and the fluorescent values of all the hybridized probes are calculated. This process is repeated with the external control (or reference) sample. The ratio of the fluorescent values of the test and control samples is calculated to derive the differential gene expression data. The Affymetrix Genechip uses a one-color method and includes labeling the aRNA with biotin. The labeled aRNA are fragmented and hybridized to a Genechip. The hybridized Genechip is washed and stained with Streptavidin-Phycoerythrin and subsequently scanned at a defined wavelength. (From Kevin Gerrish, with permission.)

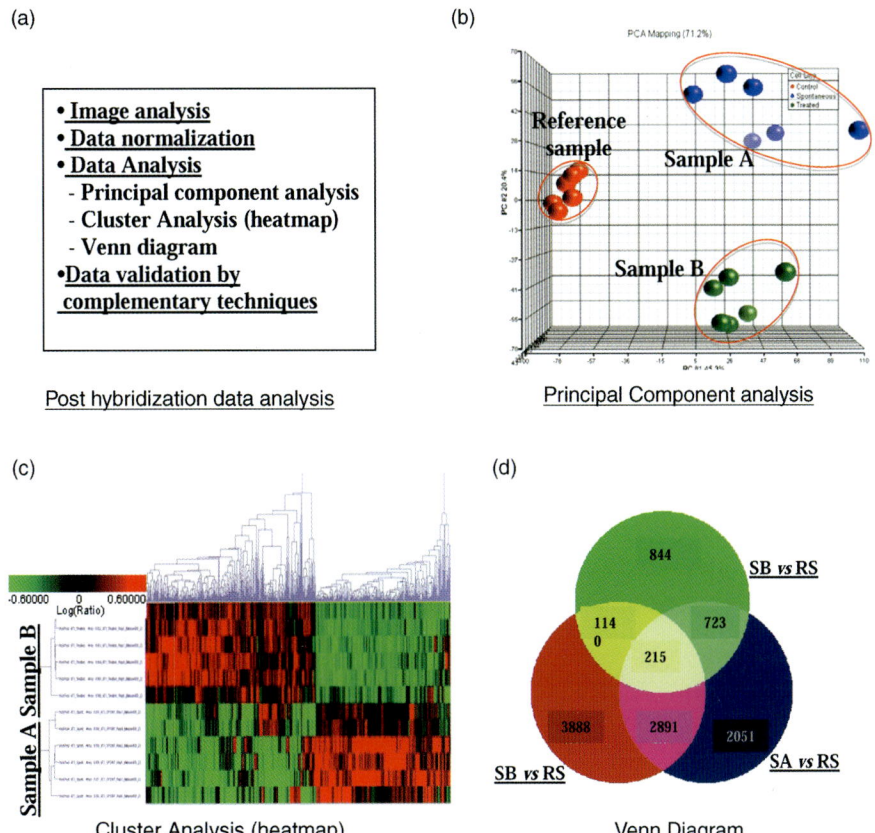

(a)

- • **Image analysis**
- • **Data normalization**
- • **Data Analysis**
  - - **Principal component analysis**
  - - **Cluster Analysis (heatmap)**
  - - **Venn diagram**
- •**Data validation by**
  complementary techniques

Post hybridization data analysis

(b)

Principal Component analysis

(c)

Cluster Analysis (heatmap)

(d)

Venn Diagram

CHAPTER 20, FIGURE 4 Data analysis post microarray experiment. After hybridization of the labeled targets with probes, the microarray chips are scanned with a confocal laser scanner to obtain fluorescent raw images. Subsequent analyses include image analysis, data normalization, and principle component analysis (b), hierarchical cluster analysis (heat maps) (c), and Venn diagrams (d) of selected genes. The principal component analysis helps in visualizing the prominent gene expression patterns and also in identifying the outlier samples. The hierarchical cluster analysis helps in visualizing segregation of the gene expression and comparison of the samples with the reference/ control samples. The Venn diagram helps in comparing the gene lists from different samples and aids in visualizing, at a glance, the degree of containment, intersections, and disjunction between gene lists from various samples. SA = sample A, SB = sample B, RS = reference sample. (From Mark Hoenerhoff and Pierre Bushell, with permission.)

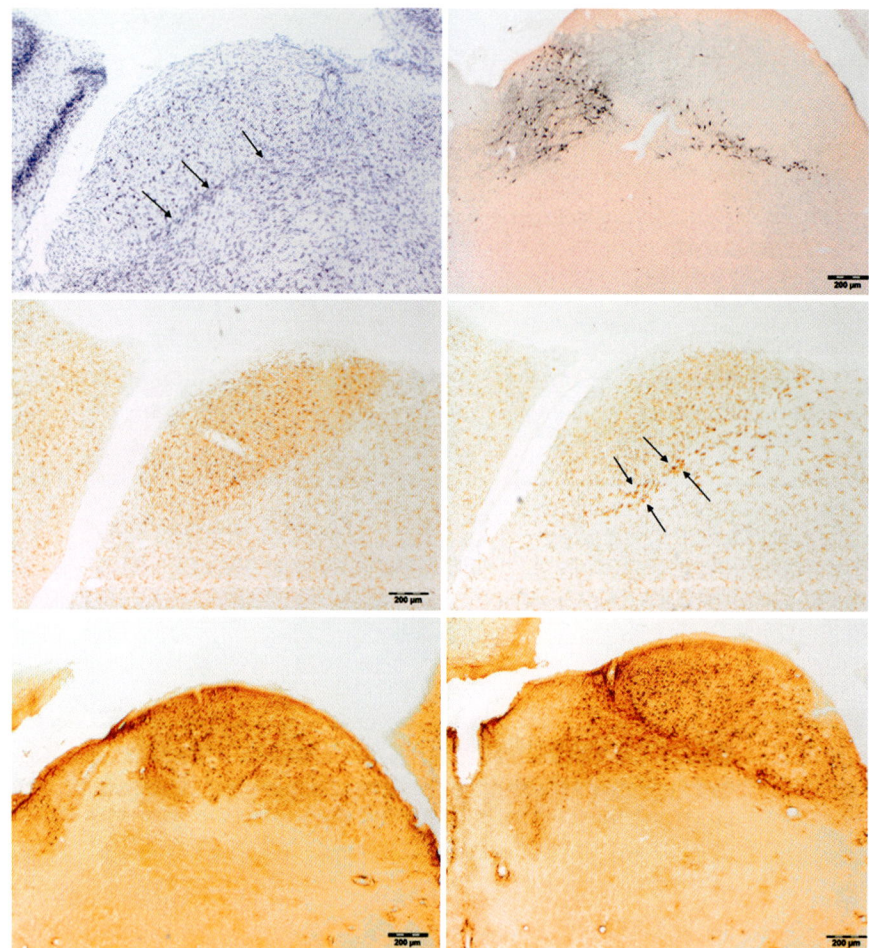

CHAPTER 21, FIGURE 4 Series of images (all taken with a 4× objective with the same camera, microscope, and software) from a mouse treated with MPTP (1-methyl-4-phenyl-1,2,3,6-tetrahydropyridine). MPTP can cause necrosis of dopaminergic neurons of the pars compacta region of the substantia nigra (and the adjacent, medial ventral tegmental region) when administered to some strains of mice and also subhuman primates and humans. These images show the usefulness of various special stains when applied at the correct time to the correct region of the brain. Both of the top two images show the region of the substantia nigra from the same animal. Neuronal disintegration is present in both, but the affected neurons are only visible, at this magnification, in the amino cupric silver stained section on the right where the affected neurons appear as black structures against a pale orange to pink background. The arrows in the left top image indicate the pars compacta region of the substantia nigra. The center images are IBA1-stained (for microglia) sections of the substantia nigra from two animals: a negative control on the left and a MPTP-treated animal on the right. The arrows on the right image show the increased IBA1 staining, indicating a microglial reaction in the pars compacta region. The bottom images are GFAP-stained (for astrocytes) sections of the substantia nigra from two animals: a negative control on the left and a MPTP-treated animal on the right. Note the increased GFAP staining in the right image, staining that forms a linear line in the pars compacta. There is also increased GFAP staining in the adjacent ventral tegmental area (to the left of the pars compacta) of the right lower image.

CHAPTER 22, FIGURES 1 to 3 Coronal sections of control Wistar rat brains showing the parietal cortex, hippocampus, and diencephalon. Myelination (blue stain) is absent at postnatal day (PND) 11 (Figure 1), slight at PND 21 (Figure 2), and clearly evident at PND 62 (Figure 3). Luxol fast blue stain, original magnification 10×.

CHAPTER 22, FIGURES 4 and 5 Coronal sections of control Wistar rat brains showing the anterior commissure. Myelination (blue stain) is absent at postnatal day (PND) 11 (Figure 4) but clearly evident at PND 62 (Figure 5). Luxol fast blue stain, original magnification 200×.

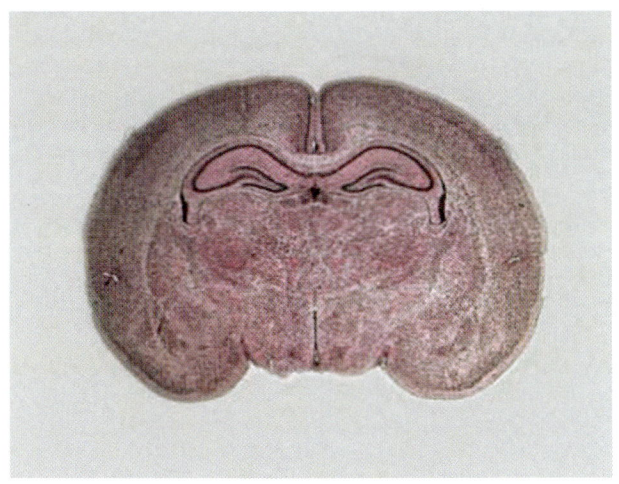

FIGURE 1

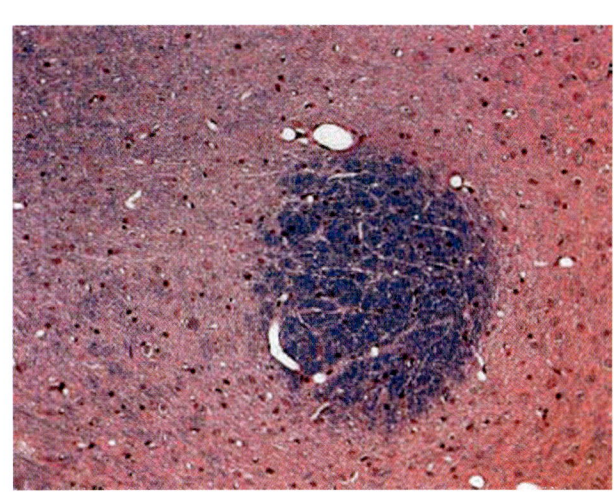

FIGURE 4

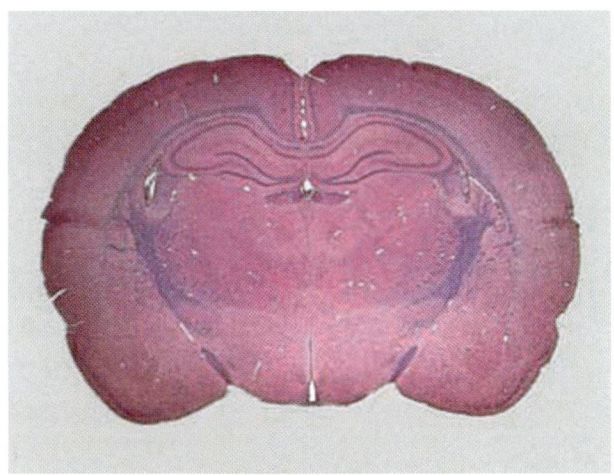

FIGURE 2

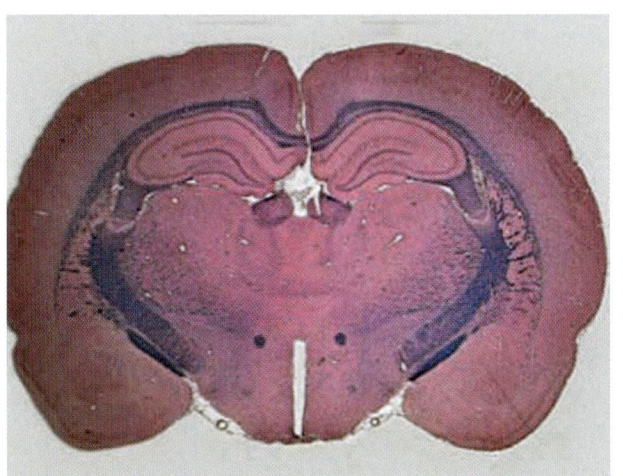

FIGURE 5

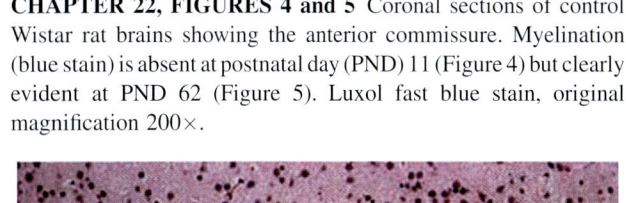

FIGURE 3

CHAPTER 22, FIGURES 7 to 9 Longitudinal (midsagittal) sections of control Wistar rat cerebellum. At postnatal day (PND) 11 (Figure 7), the cerebellum-specific external germinal matrix layer (arrows) is quite wide due to the high demand for neuronal production for this brain region. By PND 21 (Figure 8), only remnants of the external germinal matrix layer (arrows) are left, indicating that the end of neurogenesis stage is imminent. No external germinal matrix layer is found at PND 62 (Figure 9, arrows). White matter myelination (blue stain) of the white matter is lacking at PND 11, slight at PND 21, and prominent at PND 62. Luxol fast blue stain, original magnification 100×.

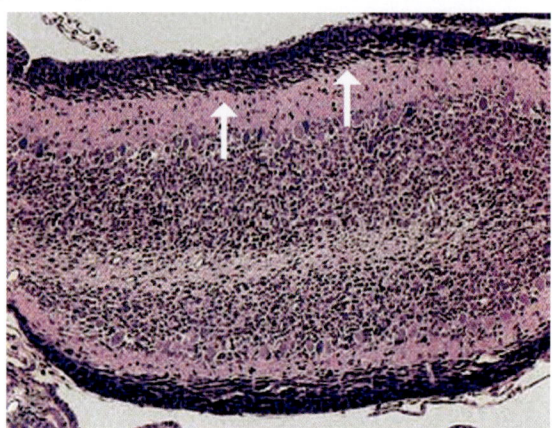

FIGURE 7

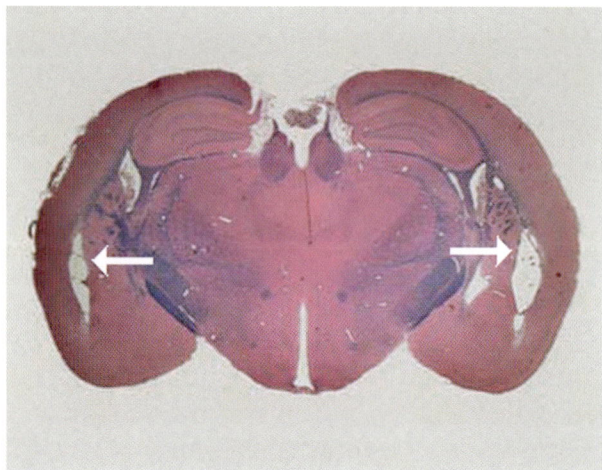

CHAPTER 22, FIGURE 6 Coronal section of a neurotoxicant-exposed Wistar rat brain at postnatal day (PND) 62 showing the parietal cortex, hippocampus, and diencephalon. Relative to an age-matched control animal (Figure 3), the size of the brain is reduced significantly, but the extent and intensity of myelination (blue stain) is not altered. Hypoplasia of the cerebral cortex results in hydrocephalus internus in the adult brains (arrows), the first stage of which is apparent in this section. The neurotoxicant, methylazoxymethanol, was given to the dam on gestational day 15 at 30 mg/kg body weight i.p.. Luxol fast blue stain, original magnification 10×.

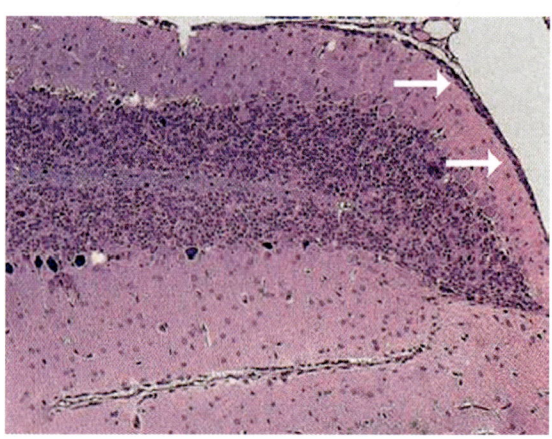

FIGURE 8

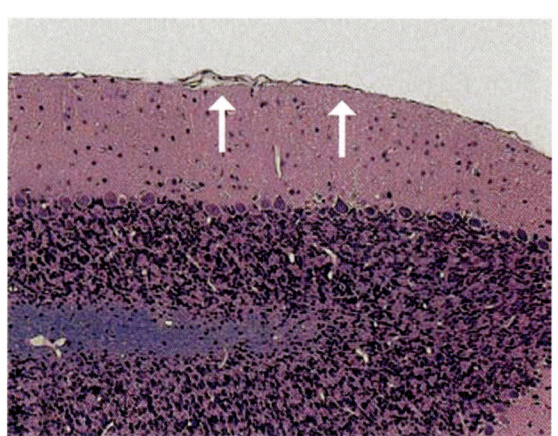

FIGURE 9

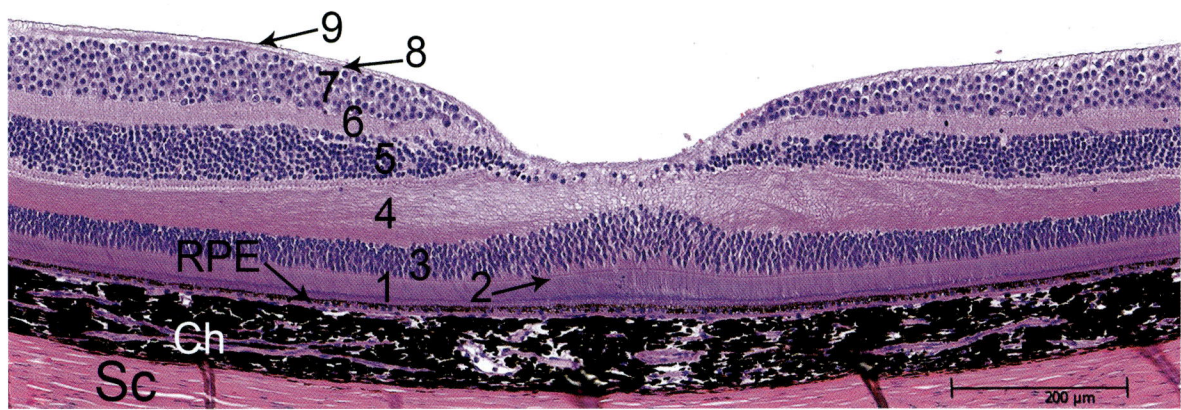

CHAPTER 24, FIGURE 1 Normal retina from an adult nonhuman primate near the fovea centralis (the central depressed region). The photosensitive retina has nine layers: 1, photoreceptors (rods and cones); 2, outer limiting membrane; 3, outer nuclear layer (photoreceptor cell bodies); 4, outer plexiform layer, 5, inner nuclear layer [amacrine, bipolar, and horizontal neurons as well as Müller (glial) cells]; 6, inner plexiform layer; 7, ganglion cell layer; 8, layer of optic nerve fibers; and 9, inner limiting membrane. The retinal pigmented epithelium (RPE) separates the photosensitive inner retina from the vascular choroid (Ch) and fibrous sclera (Sc). Paraffin-embedded section, H&E, 100×.

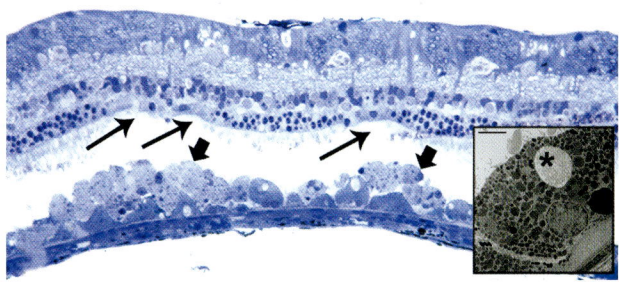

CHAPTER 24, FIGURE 5 Spontaneous degeneration of the retinal pigmented epithelium (RPE) with retinal detachment in a rabbit. Degenerated and detached RPE are present in the subretinal space (thick arrows), corresponding to regions of photoreceptor loss in the segments and outer nuclear layer (thin arrows). The remaining RPE are hypertrophied and lack cellular detail. Enlarged vacuoles/ inclusions (whose contents were lost during processing) are occasionally observed in the RPE. Plastic-embedded section, Richardson's stain, 200×. *Inset:* Transmission electron micrograph of a hypertrophic RPE cell containing numerous lysosomes, an inclusion (∗) containing lamellar material (thought to be undigested photoreceptor segments), and a paucity of melanosomes. Bar = 5 μm.

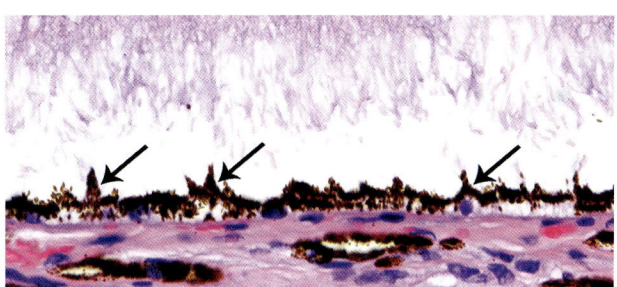

CHAPTER 24, FIGURE 7 Artifactual "tenting" of the retinal pigmented epithelium (RPE) in a dog. Note the distinct angular apices of the RPE cells (arrows), compared to the rounded apices of RPE with genuine hypertrophy (Figure 6, same magnification). Fragments of photoreceptor segments remain attached to tented RPE cells at several points along the retina, which is a common occurrence with artifactual separation. Paraffin-embedded section, H&E, 400×.

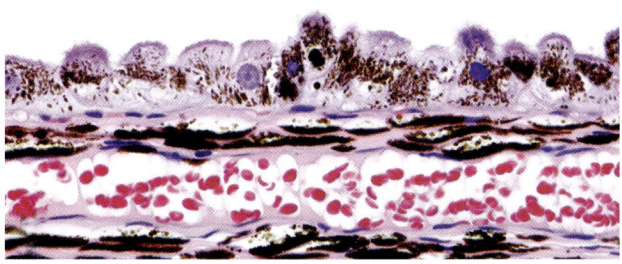

CHAPTER 24, FIGURE 6 Hypertrophy of the retinal pigmented epithelium (RPE) in a dog. Note the swollen ("tombstone") appearance of the cells with apical melanosome dispersion. Apical localization of melanosomes may also occur with artifactual separation. Paraffin-embedded section, H&E, 400×.

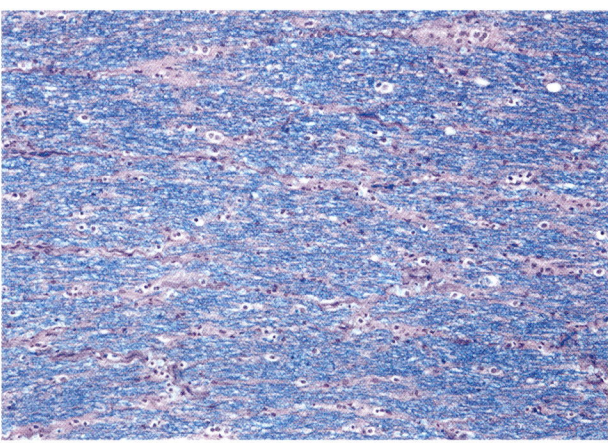

CHAPTER 24, FIGURE 8 Normal myelin distribution in the optic nerve of an adult dog. The bright blue staining for myelin is widespread. Paraffin-embedded section, Luxol fast blue/H&E, 100×.

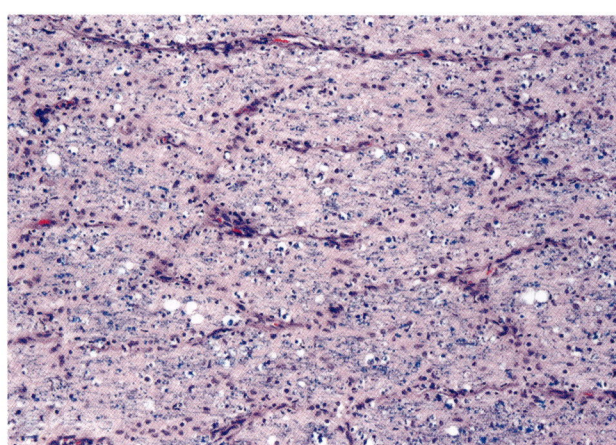

CHAPTER 24, FIGURE 9 Optic nerve hypertrophy with demyelination and gliosis in an adult dog. Note the paucity of blue myelin staining and overall increase in cellularity (due to increased numbers of astrocytes) relative to the structure in control animals (Figure 8). Paraffin-embedded section, Luxol fast blue/H&E, 100×.

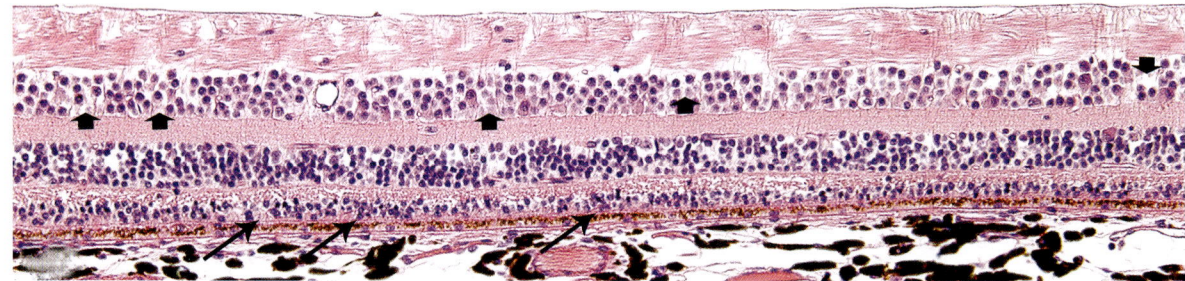

CHAPTER 24, FIGURE 10 Photoreceptor atrophy in the retina of an adult nonhuman primate following intravitreal (IVT) administration of sodium iodate ($NaIO_3$, 1000 μg). Marked atrophy with complete loss of the photoreceptor outer segments is present (thin arrows). Nuclear loss has also occurred in the inner nuclear layer (thick arrows), indicating a reduction in several neuron populations that are involved in the initial processing of visual sensory input. Paraffin-embedded section, H&E, 200×.

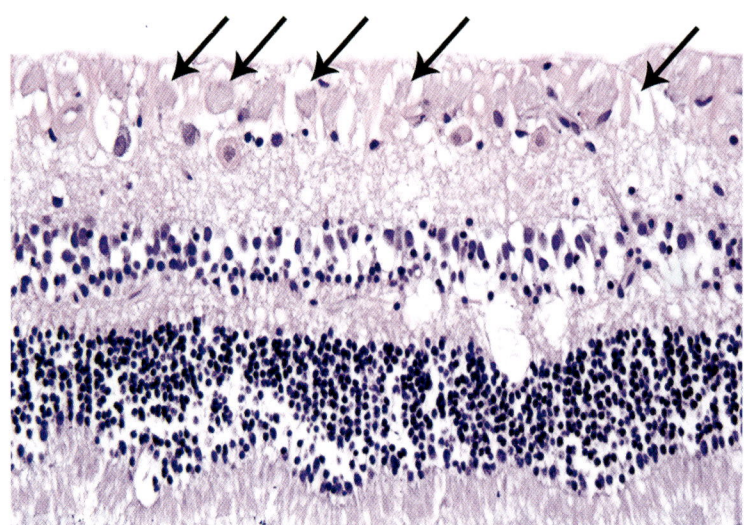

CHAPTER 24, FIGURE 11 Loss of retinal ganglion cells in the dog. Note the almost total lack of "kite-shaped" ganglion cells (neurons) in their normal location in the inner retina (arrows). The few remaining neurons in this location have hypereosinophilic cytoplasm (an early change in degenerating cells). Other retinal layers have been spared. Paraffin-embedded section, H&E, 400×.

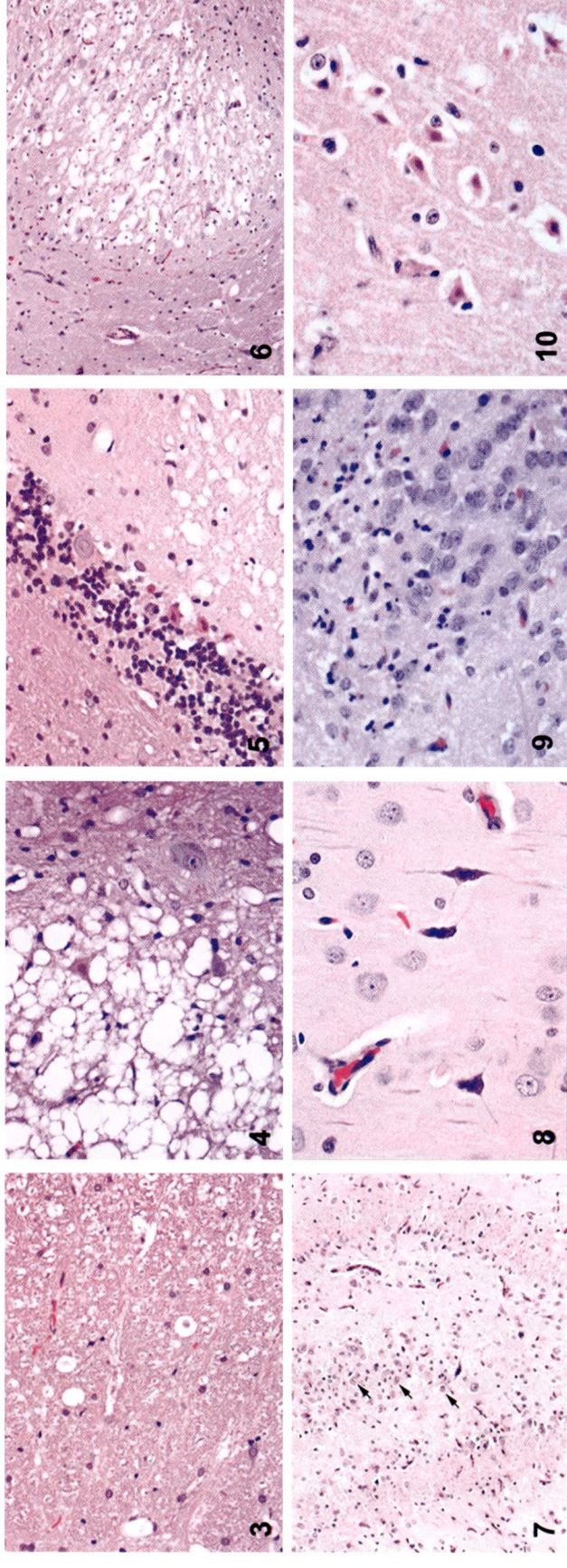

CHAPTER 30, FIGURES 3 to 10

Figure 3   Rat cervical cord, note the swollen axon also referred to as a spheroid, surrounded by a circular space.   H&E, 40X.

Figure 4   Rat cerebellar roof nuclear region. Note the acute eosinophilic necrosis of the neuron top center surrounded by spongiotic response to injury and the neuron with mild central chromatolysis and nuclear eccentricity lower right. H&E, 40X.

Figure 5   Rat cerebellar cortex. Acute eosinophilic necrosis of Purkinje cells and adjacent spongiotic white matter response with pyknotic glia. A more normal Purkinje cell in the adjacent Purkinje cell layer is seen for size comparison. H&E, 40X.

Figure 6   Rat anterior olivary nucleus. Note the round empty spaces and marked reduction in neuronal numbers. This is an example of past neuronal degeneration and resulting neuronal loss or "fall out". H&E, 40X.

Figure 7   Rat hippocampus with a large cluster of gemistocytic astrocytes indicated by arrows. Their cytoplasm is visible because of cell hypertrophy and accumulation of intermediate filaments composed of glial fibrillary acidic protein (GFAP). H&E, 40X.

Figure 8   Rat cerebral cortex. Note the artifactually crenated basophilic neurons in comparison to adjacent normal neurons.   H&E, 40X.

Figure 9   Rat habenular nucleus. Note the many neuronal apoptotic bodies among surviving neurons. H&E, 60X.

Figure 10   Cerebral cortex from a calf with polioencephalomalacia, (PEM) the result of conditioned thiamin deficiency. Note the shrunken eosinophilic necrotic neurons. Close inspection of such neurons reveals nuclear basophilic granules. H&E, 40X.

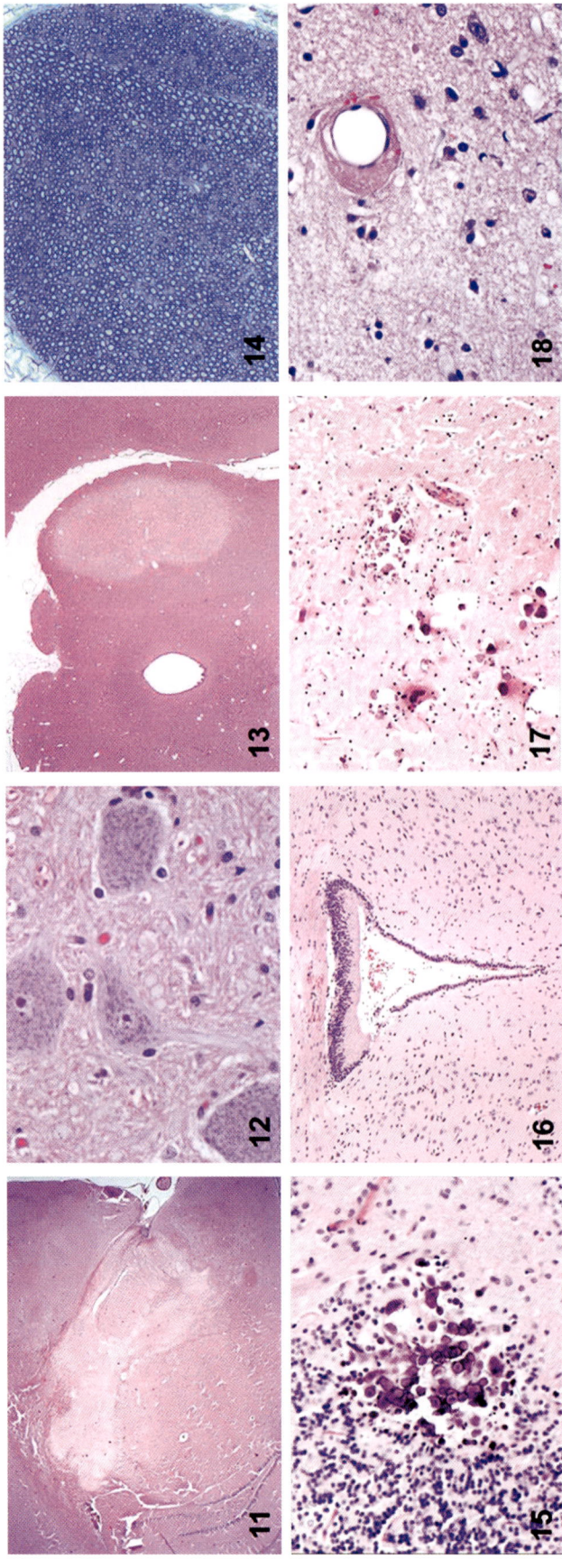

CHAPTER 30, FIGURES 11 to 18

Figure 11 Rat hippocampus, thalamus and optic radiation with a large pale well-defined infarct. The occluded artery is visible at the right center margin. H&E, 2X.

Figure 12 This image depicts a normal canine spinal motor neuron. Neuronal morphologies are diverse throughout the central, peripheral and autonomic nervous system from small cerebellar internal granule cell neurons to large Red nucleus neurons. H&E, 60X.

Figure 13 Rat with a pale region of posterior collicular necrosis, the result of carbonyl sulfide intoxication. Note the artifactually enlarged vascular spaces; the result of perfusion fixation under excessive pressure. H&E, 2X.

Figure 14 Peripheral nerve cross section fixed with glutaraldehyde and plastic embedded. Note the well preserved axonal myelin sheaths devoid of artifactual separation of lamellae allowing for better interpretation of peripheral nerve antemortem changes. Toluidine blue, 20X.

Figure 15 Rat brain old necrosis of internal granule cells with formation of concentric laminated calcospherites. H&E, 40X.

Figure 16 Rat mesencephalon, subcommissural organ. Note the normally elongated and pseudostratified appearance of the ependymal cells in this location. H&E, 20X.

Figure 17 Mouse incidental thalamic mineralized foci. These are commonly bilateral and may focus on the adventitia of vessels or have an apparent random distribution. H&E, 10X.

Figure 18 Posterior colliculus of a mature rat given carbonyl sulfide by inhalation 500 ppm, 5 hours per day for 3 days. Note the prominent perivascular protein-rich edema and fine edematous vacuolation of the adjacent neuropil. There is also necrosis of neurons and pyknotic glial nuclei within this edematous zone. H&E, 60X.

20

MOLECULAR TECHNIQUES IN TOXICOLOGICAL NEUROPATHOLOGY

ARUN R. PANDIRI, STEPHANIE A. LAHOUSSE, AND ROBERT C. SILLS

Cellular and Molecular Pathology Branch, National Toxicology Program, National Institute of Environmental Health Sciences, Research Triangle Park, North Carolina

INTRODUCTION

The modern practice of toxicological neuropathology is evolving from the traditional reliance on morphological assessment of the nervous system to incorporate and even emphasize the correlation of structural, functional, and mechanistic information for any given neural domain. Functional data can be mapped to particular locations using many recent innovative technologies, especially noninvasive imaging. Imaging modalities are especially useful because they allow real-time dynamic analysis of multiple neuroanatomical and neurophysiological parameters in both health and disease, and can facilitate repeated measurements in the same person over time to provide a better understanding of disease progression. Mechanistic data may be obtained from such functional techniques, but an even more useful means of evaluating the pathogenesis of neural diseases is to examine molecular elements (e.g., gene and protein expression) using in situ neurohistologic procedures. Such methods yield qualitative data regarding the potential roles of a given molecule or biochemical pathway in region-specific neural functions during health and disease, and serve as critical diagnostic tools for identifying and following disease progression. In some instances, however, quantitative information is required to permit a more rigorous assessment of risk and resilience in the nervous system. An example for toxicological neuropathologists is the current global effort to define molecular signatures to detect exposures to and the degree of damage induced by neurotoxic agents. In general, quantitative assays fall in modern interdisciplinary fields such as genomics, proteomics, transcriptomics, metabolomics, etc. Currently, the term *genomics* is often employed as a catch-all designation for studies of the genome, transcriptome, and proteome (U.S. Environmental Protection Agency, "Interim Policy on Genomics," http://epa.gov/osa/spc/pdfs/genomics.pdf).

Genomics is the large-scale study of the genomes of organisms. This field is very diverse and is rapidly expanding due to the sequencing of many genomes, including humans as well as several species relevant to toxicologic neuropathology research (e.g., mouse, rat, chicken, rabbit, dog, and rhesus monkey). There has been a rapid emergence of new technologies and new applications for older technologies, enabling the use of the genomic sequence information to solve various complex questions. These high-throughput technologies and applications are very efficient, enabling researchers in diverse disciplines to investigate and solve multifaceted biological problems. Before the advent of genomics, researchers studied the role and function of single genes using low-throughput molecular biology techniques and attempted to determine the physiological and pathological states of organisms and cells. The biology of an organism is complex due to the interplay among many pleotrophic genes and proteins. Thus, the newer high-throughput techniques are very relevant and essential for continued progress in biological research.

The classical molecular biology methods include polymerase chain reaction (PCR) and its numerous variants, such as reverse transcriptase (RT)-PCR, quantitative (Q)-PCR, real-time PCR, nested PCR, methylated PCR,

Fundamental Neuropathology for Pathologists and Toxicologists: Principles and Techniques, First Edition. Edited by Brad Bolon and Mark T. Butt.
© 2011 John Wiley & Sons, Inc. Published 2011 by John Wiley & Sons, Inc.

and allele-specific PCR; gel electrophoresis (agarose, polyacrylamide); macromolecular blots and probes [Southern (DNA), northern (RNA), western (protein), eastern (post-translational modification of proteins)]; expression cloning; serial analysis of gene expression (SAGE); in situ hybridization (ISH) and numerous other techniques. Most of these methods are widely used for routine molecular biology experiments and also as complementary techniques to validate genomic data. Technical details are beyond the scope of this chapter but are widely available in reference books and on the Internet.

Lesions within the nervous system can result from many etiologies, such as developmental, degenerative, microbial, neoplastic, traumatic, vascular, and xenobiotics (chemicals, drugs, toxins). The gross and histological lesions may be characteristic or even pathognomonic in several neuropathological conditions. However, the initial lesions or the earliest tissue reactions in most neuropathological conditions are submicroscopic (ultrastructural) or apparent at only the gene or protein expression level; therefore, the routinely used gross and histological features that comprise such a dominant portion of toxicological neuropathology analysis are absent in the initial stages of many neurological diseases. Accordingly, in order to have insight into the early events in the pathogenesis, it is imperative to study perturbations at the cellular level with nucleic acid and protein resolution. The classical molecular biology methods are useful to study one variable at a time, but the biology of the disease is always complex and multifaceted, especially in the nervous system. To study the pathogenesis of any disease in a meaningful way, it is essential to gain a high-resolution snapshot into the earliest events responsible for its initiation. The focus of this chapter is on genomic technologies such as microarrays and proteomic techniques. In addition, emphasis is placed on the importance of tissue acquisition, sample integrity, and the significance of laser capture microdissection (LCM) for the efficient use of genomic technologies to study complex issues in neuropathology. Finally, brief descriptions of some of the common genomic databases and genomic tools for data analysis are discussed.

FACTORS AFFECTING BRAIN AND NERVE SAMPLE QUALITY

Newer investigative neuropathology studies are multidisciplinary, thus requiring the coordinated expertise of several neuroscience specialists. Everyone involved in a study needs to be aware of the experimental design and the importance of appropriate sample collection. The pathologist is an invaluable collaborator and the very keystone in ensuring that appropriate sample collection and preservation take place.

The newer, more powerful molecular techniques are very sensitive to the quality of the tissues used to extract the relevant starting material for the assay of interest (e.g., protein, DNA, and RNA). This sensitivity requires not only faultless tissue collection methods but also the ability to demonstrate that tissue quality is acceptable. The most important step in tissue collection is achieving the shortest possible postmortem interval (PMI, which is the time between death and sample collection). A brief PMI can reliably be achieved by following a standardized method of autopsy/necropsy, organ collection, tissue dissection, and specimen storage. In some cases, even with prolonged PMI, the RNA and protein quality in neural tissues may be preserved if extreme physiologic disruptions near the time of death are avoided, such as severe pH alterations and agonal states (the term *agonal state* refers to the nature of the terminal events that precede death). The duration and type of agonal state also influences tissue quality. Prolonged postmortem intervals introduce various parameters in neural tissues and cells, such as hypoxia, hypoglycemia, dehydration, and decreased pH, all of which accelerate the degradation of tissue quality.

In all assays involving rodent models or other experimental animals, tissue sample collection can be planned in advance to ensure the shortest PMI. A brief PMI can also be obtained for human or animal neural tissues collected during ante mortem biopsy procedures. In general, human tissue acquired at autopsy or tissue samples collected improperly from experimental rodent studies may not always retain the best sample quality. Next, we describe the sample integrity of DNA, RNA, and protein in tissue specimens that have been collected under less than ideal conditions.

DNA Integrity

The quality of the DNA is fairly well retained compared to RNA and protein under adverse conditions of sample preservation. DNA or tissues stored at −80°C for several years are typically still suitable for genetic analysis. However, DNA is vulnerable to degradation when samples are stored for prolonged periods of time in fixatives such as formalin, due to progressive acidification (via conversion of formaldehyde to formic acid). Unbuffered formalin is more deleterious to DNA than is buffered formalin. The quality of DNA in formalin-fixed, paraffin-embedded (FFPE) tissues is fairly well preserved for long periods of time if the duration of formalin fixation was limited to between 12 and 24 h. However, DNA of the highest quality can be obtained from fresh tissues, followed by samples frozen at −80°C, and then FFPE tissues (if fixation was brief and undertaken in a buffered solution). The QIAamp Micro (Qiagen, Valencia, CA), Illustra GenomiPhi (GE Healthcare, Piscataway, NJ), and TaKaRa Ex taq (Takara Bio, Inc., Madison, WI) protocols yield good-quality DNA from FFPE or frozen archival samples. The efficiency of DNA extraction can be measured using a spectrophotometer, and the 260 : 280 ratio should be greater than 1.9 for good DNA quality without protein contamination. Measurements of absorbance at 230 and 320 nm can be used to measure other impurities within the DNA sample extracted. The quality of DNA can also be assessed using various techniques, such as gel electrophoresis, Southern blots, gene-specific PCR [using housekeeping genes such as glyceraldehyde 3-phosphate dehydrogenase (GAPDH) as an internal control gene], and randomly amplified polymorphic DNA (RAPD)-PCR.[1]

The study of epigenetic events is becoming very important in molecular neuropathology studies as well. For example, it has been shown that there are regional differences in the methylation of gene promoters in the brain and that they may be important in region-specific functional specialization.[2] An age-related increase in DNA methylation in the rat brain has been suggested to alter chromatin conformation.[3] The DNA methylation state is reported to be stable in postmortem brain tissue and is not influenced by pH.[4] In addition, even after a PMI of 30 h and at various tissue pH values, the methylated histone residues and bulk of nucleosomal DNA attached to histones are preserved in human and mouse brain samples that were treated by micrococcal nuclease digestion, but they are not preserved in formaldehyde-fixed and sonicated brain samples.[5]

RNA Integrity

Most large-scale array techniques are based on RNA, and the reliability of the data is directly proportional to the quality of the RNA. The integrity of RNA depends on several variables, such as length and severity of the agonal state (hypoxia, acidosis), duration of PMI, tissue pH, handling, and number of freeze–thaw cycles. Contrary to the commonly accepted belief, there is little correlation between the length of PMI (up to 48 h) and RNA integrity, as long as the agonal state was short.[6] Prompt snap-freezing of tissues in liquid nitrogen and storage at $-80°C$ is the best method to preserve RNA integrity. Direct immersion of 5 mm $\times$ 5 mm $\times$ 5 mm tissue samples in RNAlater RNA stabilization reagent (Ambion, Inc., Austin, TX) can also be used to preserve RNA: for one week (at room temperature), one month (at 4°C), and several years (at either -20 or $-80°C$). RNAlater-treated tissues archived at $-80°C$ can be subjected to 10 freeze–thaw cycles without affecting the quality of subsequently isolated RNA. In addition, immersion fixation in HOPE (HEPES glutamic acid buffer–mediated organic solvent protection effect) fixative (at 0–4°C) (DCS Innovative, Hamburg, Germany) for 14 to 36 h following incubation in acetone (at 0–4°C) and subsequent routine paraffin embedding helps in preserving RNA quality for up to five years at room temperature.[7–9] The efficiency of RNA extraction can be measured using a spectrophotometer, and the 260 : 280 ratio should be greater than 1.8 for good RNA quality without protein contamination. The

agarose gel method for calculation of RNA integrity based on the 28S/18S ratio (where a ratio $\geq$ 2 is better) is subjective and prone to errors. A newer objective method for evaluating RNA quality is called the *RNA integrity number* (RIN).[10] This value is based on an algorithm that includes several electrophoretic RNA measurements derived from an automated, high-throughput, microcapillary electrophoresis method using the Agilent 2100 bioanalyzer (Agilent Technologies, Palo Alto, CA). An RIN value of 10 indicates a sample with perfectly intact RNA, and an RIN of 0 indicates one with completely degraded RNA. In general, RIN values lower than 7 are not adequate for mRNA analysis. In some cases, however, RNA samples with RIN between 6 and 7 may be adequate since not all mRNAs have the same vulnerability to degradation.[11] The RIN values may vary from one region of the brain to another.[11] Furthermore, the integrity of the RNA is not constant over time, so monitoring the RIN of the controls as well as the test RNA samples should be done before every assay to minimize inter- and intraassay variation. Several studies have shown that low pH and having more than two freeze–thaw cycles lower the RIN significantly.[12]

Research on microRNA (miRNA) as regulators of mRNA is crucial in modern neuroscience research into mechanisms of such conditions as neurodegeneration (Alzheimer's disease, Parkinson's disease), schizophrenia, and brain cancer.[6,13] The role of miRNA in toxicant-induced neuropathology will probably become a fertile realm of investigation in the near future. miRNAs have high adenine and uridine content, relatively short half-lives, and limited stability when not complexed with target mRNA.[14] The long-term stability of miRNA has not been fully explored.[15]

Protein Integrity

The integrity of protein in postmortem samples depends largely on the PMI, choice of fixative solution and temperature, and storage time. The postmortem protein integrity may depend on the size and chemical structure of the protein complexes. Aggregated proteins seem to be quite resistant to protein degradation compared to soluble proteins of low molecular weight.[11] This principle is exemplified by the ready degradation of normal soluble α-synuclein with postmortem delay, whereas α-synuclein aggregates (Lewy body fibrils) are fairly resistant to postmortem delay.[11] No definite relationship has been shown between a particular structural or functional family of proteins and the ability to maintain structural integrity in postmortem samples, with the exception of members of the proteasome complex, many proteases, and cathepsins.[16] Proteins that are more resistant to postmortem degradation include subunits of the 19S and 20S proteasome, β-actin, active p38 kinase, synaptophysin, and several enzymes related to energy metabolism such as glycogen 3-phosphate dehydrogenase, malate dehydrogenase, and aldolase A. Vulnerable proteins include peroxiredoxin, ATP synthase, superoxide dismutase 1, synaptic proteins (rab3a, rabphilin, α-synuclein, β-synuclein, syntaxin), kinases (MAP kinase, stress-activated protein kinases, cdk5), trophic factors, and β-tubulin.[11]

The PMI also influences the state of phosphorylation, oxidation, nitration, and function of proteins. The density of the proteins may increase or decrease based on the PMI delay and related factors. For example, the extent of phosphorylation of tau, a microtubule associated protein, is very important for the study of various tauopathies that occur in neurodegenerative conditions such as Alzheimer's disease and Pick's disease. The cardinal histological feature of tauopathies is neurofibrillary tangles in brain neurons, while the biochemical correlate is hyperphosphorylation of tau proteins in sarkosyl-insoluble fractions. The triplet of 68-, 64- and 60-kDa τ isoforms is well preserved in 24-h-old samples stored at 4°C. However, with increased PMI, there is decreased density of these bands (as shown by western blotting).[17] Oxidation and nitration of proteins are particularly prominent processes in many neurodegenerative disorders. The density of some neural proteins is altered (increased or decreased) following lengthy PMI due to oxidation and/or nitration. Hence, it is of paramount importance to evaluate the quality of the proteins within samples obtained during a specific PMI range before further analysis and interpretation are conducted. The abundance and density of the proteins can be measured semiquantitatively by Western blotting, two-dimensional gel electrophoresis (2DGE), and immunohistochemistry. Several antibodies are used to detect oxidation and nitration of lipids, proteins, and carbohydrates. Common lipoxidation markers include malondialdehyde lysine (MDA-L) and hydroxynonenal (HNE); glycoxidation markers include N-(carboxymethyl)lysine (CML) and N-(carboxyethyl)lysine (CEL); and a useful nitration marker is nitrotyrosine (N-Tyr). The enzymatic activity of various proteins is also highly susceptible to postmortem artifacts, so to avoid erroneous conclusions it is vital to compare in parallel the enzymatic activity of samples collected similarly from control and treatment groups.[11]

CONSIDERATIONS IN SAMPLING NERVOUS TISSUE FOR MOLECULAR ANALYSES

Different parts of the nervous system are selectively vulnerable to various insults. Examples from neuropathology literature include Alzheimer's disease (entorhinal cortex, CA1 (cornu ammonis, zone 1) of the hippocampus), mercury (granular cells of the cerebellar cortex), methanol (putamen, retina), carbon monoxide (globus pallidus, pars reticularis of substantia nigra), carbonyl sulfide (inferior colliculus), poliomyelitis (ventral horn of the spinal cord), ischemia/hypoxia/hypoglycemia (CA1; neocortical layers III, V, and

VI; ventral horn), and Cuprizone (corpus callosum, external capsule, caudate/putamen, and dorsal hippocampal commissure).[18–21] Thus, prior knowledge of the target cells and affected neural domains is essential for proper sampling of neural tissues for high-throughput quantitative molecular assays.

All the tissues in a given study should be collected in the same manner, and ideally, similar procedures should be utilized across all equivalent studies to facilitate training of the technicians charged with sample acquisition and to permit the construction of meaningful historical databases regarding molecular expression in specific neural structures. To ensure uniform sample collection across studies, the personnel involved in sample collection should receive comparable and appropriate training in autopsy/necropsy technique and subsequent dissection of the central and peripheral nervous systems to minimize the PMI as well as to avoid introduction of artifacts into the tissue collected.

Neural tissue samples in toxicological neuropathology studies are typically acquired in the following manner. For animal experiments in which molecular pathology analyses will be conducted post hoc based on prior assessment of the histopathology data, samples are taken at necropsy using routine collection methods. The primary requirement is to ensure that the dissection surface and instruments are free of excessive blood, digestive tract contents, and hair (i.e., "clean"), which is accomplished by using clean (preferably sterile) stainless steel dissection tools and clean cutting surfaces. Once taken, small tissue samples for quantitative molecular analysis are processed according to the scientific question to be answered, typically by transfer to a small vial of ice-cold lysis buffer or RNAlater (Ambion, Inc., Austin, TX) (for nucleic acid assays) or by wrapping in aluminum foil (which can be stored for months in resealable plastic bags at −80°C or lower until tissue homogenization and analysis is undertaken). Samples for in situ molecular pathology evaluation are often mounted in cryomolds using OCT compound (Sakura Finetek, Torrance, CA), and flash-frozen by immersion in isopentane cooled in dry ice or liquid nitrogen, and then wrapped in aluminum foil and/or plastic and stored at −80°C. Delays in sample acquisition can be minimized by prelabeling all sample containers prior to initiating the necropsy and dissection. Another means of diminishing the potential for contaminating samples by extraneous material derived from the prosector is to assign a second team member to keep records, thereby reducing the number of extra surfaces that must be touched by the prosector.

For experiments in which molecular pathology analyses are the main focus of the investigation, sample collection depends on the assay that needs to be conducted. Prosectors should arrange their dissection stations in such a manner that all necessary instruments and solutions are available within arm's reach, without the need for reaching over clean or sterile areas. To minimize contamination of the samples, an ideal approach is to employ single-use (i.e., disposable) cutting boards and instruments for each person. The remainder of the sample acquisition and allocation process is comparable to that related in the preceding paragraph. The primary advantage of such a purposeful approach to collecting samples for molecular analysis is that the prosector may immediately remove and subdivide the organ of interest, which minimizes the PMI.

A main component of the experimental design in molecular analysis of the nervous system is to define the means by which the neural structure is to be sampled. In practical terms, this consideration is most important for the brain. A common choice in toxicological neuropathology studies is to hemisect the brain, after which one half is fixed in appropriate fixative (10% neutral buffered formalin, ethanol, Trump's) and then processed into paraffin or plastic for routine histopathological evaluation, while the other half is trimmed into brain-region-defined tissue blocks, mounted in OCT compound, and stored at −80°C. This approach permits identical regions of the brain to be assessed using distinct methods suited to optimal preservation of the tissue structure and molecular integrity. The left and right hemisections may be randomly sorted for fixing or freezing. Alternatively, the study design of a toxicological neuropathology experiment can dictate the specific sampling protocol. Based on the assay to be undertaken, the dissected samples may be frozen and homogenized for extraction of protein, RNA and DNA, or frozen or fixed and then sectioned for in situ demonstration of the relevant elements. It is very important that the fixed or frozen sections of the brain be cut and matched uniformly following the standard skull landmarks, such as bregma, interaural line, lambda, or other appropriate stereotaxic markers.[22] Frozen samples should not be allowed to undergo multiple freeze–thaw cycles, since it is detrimental to RNA and protein quality.

The number of cells required for molecular analysis may be calculated based on published data. For example, about 50 to 200 μg of total RNA is usually required for microarray gene expression studies. On average, one cell contains 10 to 30 pg of total RNA, so the number of cells required to extract 50 to 200 μg ranges from 1.6×10^6 to 2×10^7.[23] The percentage of mRNA within the total RNA varies (up to 3%) with the cell type. Hence, the total amount of RNA to be extracted depends on the size of the cell population or tissue of interest. The low cellular yield of cells that occurs occasionally using microdissection techniques such as LCM may provide amounts of RNA so small that they may not be sufficient for differential gene expression studies. Therefore, RNA samples are usually subjected to a few rounds of linear amplification before enough mRNA is available for hybridization. The most widely used amplification method is the T7 RNA polymerase-based linear antisense RNA (aRNA) amplification protocol (Eberwine protocol), which is preferred due to its high fidelity.[24] Briefly, the mRNA is reverse transcribed to complementary DNA (cDNA) by an oligo dT primer containing a T7 RNA polymerase promoter site. RNase H is then used to digest the mRNA strand in the mRNA–cDNA hybrid. Next, the resulting cDNA is converted to a double-stranded cDNA by DNA polymerase I. Finally, the T7 RNA polymerase transcribes the cDNA into aRNA. The amplified products are usually in the range of 200 to 1000 base pairs (bp) in length. One round of amplification yields about 10^3 fold of the estimated amount of starting mRNA, while two rounds yield about 10^5 fold.[25] The first round of linear amplification involves directional priming from the 3′ end of the mRNA, whereas subsequent rounds of amplification are initiated by random pairing. Thus, antisense RNA (aRNA) amplification is 3′ biased, and complete coverage of the 5′ end is not ensured. This 3′-end bias is counteracted by the 3′-end-biased design of probe sets of cDNA and oligonucleotide arrays.[23] The 3′ bias of the aRNA may also be overcome by the template switching effect at the 5′ end of the mRNA transcript to ensure synthesis of the full-length dsDNA.[25] In most cases, a high level of concordance exists between the amplified and unamplified mRNA, but the sequence variation increases with increasing numbers of amplification cycles and decreased initial mRNA amounts.[25–27] There is an amplification bias after the second round of amplification, which is characterized by the loss of 30% of differentially expressed genes due to long probe–poly (A)-tail distances.[27] Hence, the amplification protocols and microarray probe design are critical in cases where the initial RNA sample is limited. Difficulties encountered during small-sample amplification, potential solutions, and a list of commercially available kits have been reviewed in detail elsewhere.[23]

Tissue Acquisition by Gross Dissection Versus Laser Capture Microdissection

Compared to many tissues, brain has numerous anatomically and functionally distinct domains. Even within the same structures (nuclei), adjacent neurons expressing different neurotransmitters are intermingled with varying numbers of astrocytes, microglia, oligodendrocytes, endothelial cells, and matrix. Thus, to come to meaningful conclusions precise sampling based on the experimental need is of paramount importance. The resolution of sampling (gross dissection vs. microdissection) is ultimately dictated by the research goals.

Gross dissection of anatomically defined brain regions assumes tissue homogeneity at the macroscopic level and may be useful for initial exploratory or discovery studies of differential gene expression or protein studies. In addition, gross dissection is justified in diseases where the pathogenesis involves multiple cell types as well as their microenvironment. The utility of this approach has been demonstrated in the experimental study of carbonyl sulfide–induced lesions within the posterior (inferior) colliculus of the rat brain[21] and the study of a mesial temporal lobe rat model of epilepsy.[28] However, the differential gene expression of grossly dissected tissue homogenates invariably results in averaging the expression levels of different genes within the sample. Data derived from grossly dissected material is reflective of the heterogeneous cellular mixtures and tissue matrix in such large samples. The primary disadvantage of this high-level screen is that it may fail to detect subtle alterations in molecular events that occur in discrete bodies of neural cells. Nevertheless, this sampling method is useful provided that its benefits and limitations are taken into account in the experimental design.

In contrast, LCM is a more focused sampling method where a defined cell population intermixed within heterogeneous cell populations and tissue matrix is differentially extracted for further defined gene expression or protein studies (Figure 1).[29] This technique is particularly well adapted to sampling discrete cell populations within heterogeneous tissue sections, cytological preparations, or living cell cultures, and it is compatible with both frozen and fixed tissues stained by histological stains, immunohistochemistry, in situ hybridization, or other neuropathology methods that can mark the cell phenotype.[30–32] The LCM approach is indispensable to a study of the roles played by individual cell types in the pathogenesis of some diseases. For example, one field of neuropathological research that has benefitted greatly from the advent of LCM is hippocampal biology, due to the heterogeneous pathophysiology of its closely clustered and often intermingled components in CA1 and CA3 of Ammon's horn as well as the dentate gyrus (DG). Pyramidal

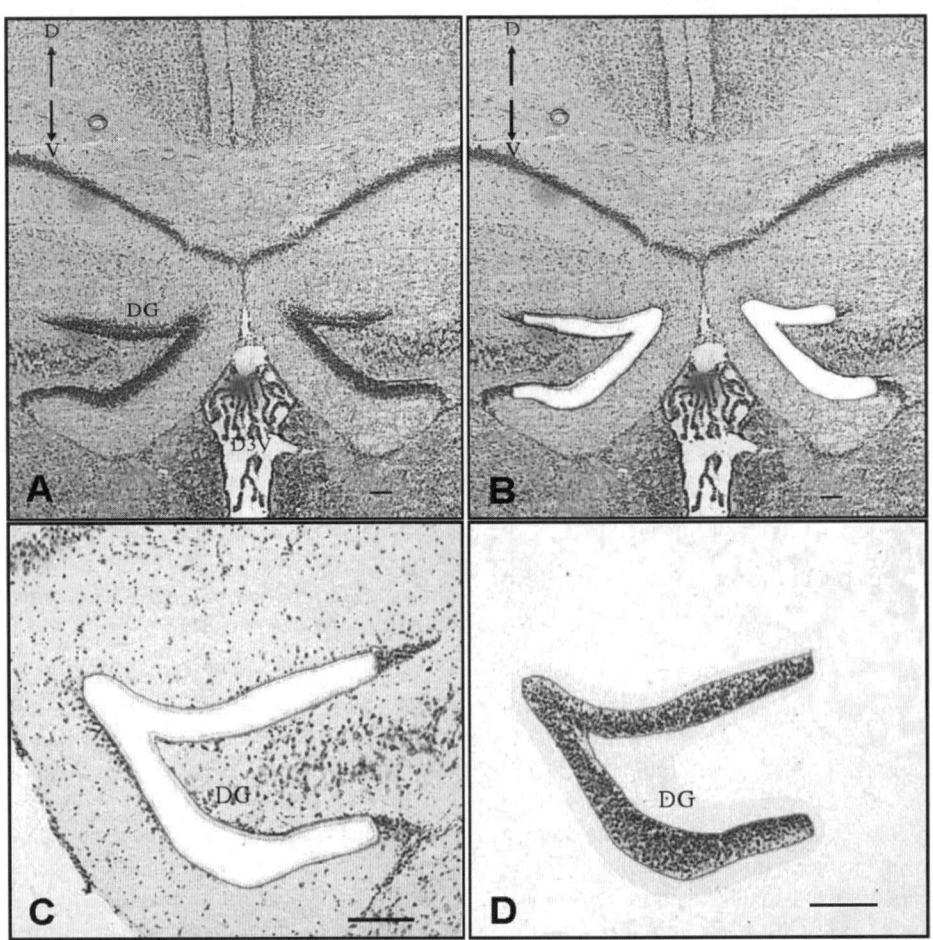

FIGURE 1   Use of laser capture microdissection (LCM) to isolate the hippocampus dentate gyrus from an adult mouse: (A) intact coronal brain section at about Bregma −1.35, stained with cresyl violet; (B, C) the dentate gyrus (DG) has been microdissected with a laser; (D) the dentate gyrus (DG) has been isolated and captured for total RNA extraction. Scale bar = 100 mm. D; dorsal; V; ventral. (From D'Souza et al.[131], with permission. Copyright © 2008 BioMed Central, Ltd.)

neurons of CA1 are particularly sensitive to hypoxia/ischemia- and seizure-induced neurodegeneration, and CA3 pyramidal neurons are extremely vulnerable to seizure- and trauma-related damage, while the granule cells of the DG are remarkably resistant to most insults, including hypoxia/ischemia and seizures.[33–35] Differential gene expression data derived from LCM material is reflective of materials extracted from a specific cell population. The primary disadvantages of this technique relative to gross dissection are the greater costs in terms of instrumentation (for the LCM apparatus and its disposable components), labor (due to the specialized training required for technical staff), and time. These disadvantages are outweighed by the potential for more precise evaluation of molecular events in neuropathological states.

Various categories of LCM are classified based on the type of laser capture system [infrared (IR) or ultraviolet (UV)] and the extraction method (positive or negative). The positive extraction LCM apparatus directs a pulsed IR beam through an optically transparent thermolabile polymer onto the cells of interest within a 5- to 15-μm-thick tissue section to form a polymer–cell composite for microaspiration into a collecting tube (Arcturus PixCell IIe system; MDS Analytical Technologies, Sunnyvale, CA). The negative extraction LCM instrument directs a pulsed UV beam to promote photovolatilization of cells surrounding a selected area, which catapults the desired area by laser pressure into the collecting microfuge tube (PALM Microlaser Technologies, Bernreid, Germany).[36] After LCM harvesting, the live cells retain functional and structural integrity, and therefore can be used for many downstream assays. Commercially available laser-based tissue microdissection systems are listed in Table 1.

Optimal LCM resolution is achieved with sections between 5 to 15 μm in thickness. Sections less than 5 μm thick may not capture the entire cell thickness and thus require the microdissection of more cells. Sections greater than 15 μm thick may not be completely microdissected, which may result in loss of integral cell components left adhered to the slide.[31] The sampling resolution of most LCM machines range from 5 to 20 μm in diameter, so these instruments can effectively sample single cells. Isolation of single cells (especially neurons) is critical since morphologically similar cells may be fundamentally different at the neurochemical and gene expression levels.[37]

Proper upstream handling of LCM samples is as important as the downstream molecular assays. Tissue fixation and staining degrade the quality of the resulting RNA and protein isolates. Ideal samples for LCM are cryosections from snap-frozen tissue blocks, with or without prior tissue fixation. If the sample is to be fixed prior to freezing, molecular integrity will be better preserved using coagulating fixatives such as 70% ethanol or 95% methanol or pure acetone, rather than aldehyde-based cross-linking agents. Traumatic disruption of intermolecular cross-links during extraction and processing results in protein fragments rather than intact proteins and reduced RNA quality. Antigen retrieval through reversal of aldehyde-induced cross-linking of proteins may be achieved using citraconic anhydride and heat.[38] However, for optimal retention of RNA integrity in fixed tissues, a reversible cross-linker such as dithiobis (succinimidyl propionate) (DSP, Lomant's reagent) offers a feasible alternative to formalin as a sample preservative.[39] In addition, another superior alternative to formalin fixation is the PAXgene® tissue system that offers better preservation of nucleic acids for downstream molecular biology studies. If frozen blocks are not available, fixed tissues embedded in paraffin (including FFPE) may be tested. However, the suitability

TABLE 1 Characteristics of Commercial Laser Capture Microdissection Systems

Company	Instrument	Laser/Extraction[a]	Collection
Arcturus Engineering (Mountain View, CA)	PixCell II	IR laser with positive extraction	CapSure cap
Cell Robotics (Albuquerque, NM)	Laser-Scissors Pro300	UV laser with positive extraction	Pick-up sticks
PALM Microlaser Technologies AG (Bernried, Germany)	PALM microbeam	UV laser with negative extraction	Laser pressure catapulting
Leica Microsystems (Bannockburn, IL)	Leica AS LMD	UV laser with negative extraction	Gravity collection
Molecular Machines and Industries AG (Glattbrugg, Switzerland)	CellCut	UV laser with negative extraction	Adhesive layer on collection cap

[a] IR, infrared; UV, ultraviolet; positive extraction, laser directed at cells of interest for extraction; negative extraction, laser photovolatilization of areas surrounding the targeted cells and subsequent extraction of intact cells of interest.

of paraffin sections for molecular analysis following LCM should be tested in advance in each laboratory before new prospective studies are undertaken using this approach.

Preparation of Tissue Sections for In Situ Molecular Analysis

Tissue sections, whether frozen or fixed, are collected on uncharged slides to reduce interference during cell extraction. Such interference can result in incomplete release of the material extracted from the slide. The choice of the slide material (glass/foil membranes) depends on the technology of the LCM instrument. The tissue sections are then dehydrated in graded ethanol (70% > 95% > 100%) and xylene before LCM to prevent hydrostatic forces from interfering with the cell harvest. The dehydrated tissue sections are not cover slipped for LCM since direct contact is needed for harvesting the microdissected cells. Ideally, unstained histologic sections are the best samples for LCM since they do not introduce any artifacts associated with staining and subsequent degradation of RNA and protein. However, without a coverslip, the significant reduction in optical resolution will prevent the recognition of architectural features needed to identify cell populations of interest, especially in the brain. In practice, this difficulty is typically overcome by staining sections to provide greater contrast, using such histochemical techniques as Mayer's or Gill's aqueous hematoxylin, acridine orange, methyl green pyronine, nuclear fast red, methylene blue, Wright–Giemsa, toluidine blue, cresyl violet (Nissl) or HistoGene™ stains. The detrimental effect of routine histological staining methods for LCM may be reduced using higher stain concentrations and shorter incubation periods. Also, addition of RNase inhibitors (2 units/μL) to aqueous staining solutions may reduce some of the RNA-degrading effects of the histological staining procedure. However, the histochemical stains cannot easily discriminate between the various neuronal and glial subtypes with confidence, especially within the brain, so more complicated techniques may be required to adequately discriminate a discrete cell population. The most common procedures used for this purpose are immunohistochemistry (IHC) and in situ hybridization (ISH).[30,32] Accurate histological selection of the target cells for positive extraction, as well as exclusion of undesired cells and stroma for negative extraction, would increase the specificity, precision, and efficiency of the LCM process. The conventional IHC and ISH processes are prolonged and involve multiple incubation steps in aqueous solutions, which will promote activation of endogenous and exogenous RNases and proteases, with subsequent degradation of RNA and proteins, respectively. Addition of a DNase I treatment step is also indispensable for procuring DNA-free RNA. To reduce RNA and protein degradation, IHC and ISH techniques should be performed rapidly, with fewer

steps, and with less corrosive reagents. Variables that can be manipulated include shortening the length of staining and incubation periods, using antibodies with higher affinity and/or at higher concentrations, reducing the number of steps (e.g., by using fluorescently labeled primary antibodies), and employing less harsh techniques of fixation (e.g., precipitating agents such as acetone, methanol, or ethanol instead of formaldehyde or paraformaldehyde) and processing (e.g., low-temperature paraffin embedding).[32] To avoid fixation and staining artifacts and to extract high-quality RNA and protein, "navigated LCM" may be used instead of conventional LCM sampling of stained histologic sections. In navigated LCM, multiple sections from the region of interest are cut and one central section is used to stain and localize the area of interest (e.g., a given brain nucleus). The landmark measurements from the stained section are then used as a guide for microdissection of the neighboring unstained and unfixed sections.[40] Navigated LCM should be used with caution since errors in sampling may occur depending on the abundance or availability of the target cells in the adjacent unstained section and the ability of the histotechnician to orient the tissue sections reliably in the exactly the same spot from slide to slide. Another approach to avoid the degradation artifacts associated with staining and fixation is to use transgenic animals in which the specific neural cell types have been labeled genetically with an appropriate fluorescent marker (e.g., a nuclear variant of green fluorescent protein) followed by fluorescence-guided LCM.[41]

In practice, one set of microdissected cells is used for one downstream application (assessment of DNA, RNA, or protein) due to the requirement of different solubilization schemes, extraction buffers, and denaturing temperatures for each molecular class. Hence, multiple sets of microdissected cells should be collected if analysis of several applications is planned.[31] The number of microdissected cells required for each downstream analysis depends on the yield and efficiency of extraction of the DNA, RNA, and protein. Comprehensive recommendations on cellular yields from microdissected tissue for various downstream analyses have been published.[31]

An alternative method to LCM studies for molecular expression analysis involves neural cell culture–based experiments, with measurements being conducted by flow cytometry and fluorescence-activated cell sorting (FACS) analysis. In this procedure, the cells are immunolabeled and sorted based on the specific light scattering and fluorescent label on each cell. The technique is well suited for suspended cells but does not lend itself to the analysis of cells from intact tissues. Disruption of the intact tissue by enzymatic and/or traumatic processing methods may profoundly alter gene and protein expression patterns and thus may make interpretation of the results with respect to the in vivo situation difficult.[31]

MICROARRAY TECHNOLOGY

Differential gene expression microarray technologies are being used widely to study increasingly complex biological questions and the molecular interrelationships between various genes under physiological as well as pathological conditions. Microarray technology provides a biological snapshot and allows for the simultaneous analysis of numerous parameters, such as gene expression, methylation status, chromatin immunoprecipitation (ChIP)-on-chip (Roche Nimblegen, Madison, WI), miRNA, splice variant analysis, and oligo comparative genomic hybridization (CGH) of thousands of genes in a single RNA/DNA sample.[42] This technology provides a high-throughput platform to generate vast amounts of data requiring the expertise of numerous disciplines, such as molecular biology, bioinformatics, and engineering sciences. A detailed explanation of the technical specifications of this technology is outside the scope of this chapter, so in the current section we introduce currently available chip technologies, experimental design considerations, principles of data analysis, and methods for knowledge extraction to provide biologically relevant information.

The most commonly used gene microarray platforms are based on cDNA (~300 bp) and oligonucleotides (typically, 25 to 60 bp).[43,44] All DNA microarrays work on the principle of base pairing (hybridization) of complementary nucleic acids between the probes (representatives of genes) and labeled targets (cDNA derived from isolated mRNA). Defined sets of probes (cDNA or oligonucleotides) are immobilized on a solid support (e.g., glass slides, silicon chips, nylon membranes), and the labeled targets (cDNA or aRNA) from the experimental samples are hybridized to the probes. The hybridized microarray slides are washed and subsequently scanned with lasers of appropriate wavelength. The intensities of each of the hybridized spots are measured to arrive at the differential gene expression levels.

The arrays may be designed by the user (in-house) or purchased from a commercial provider such as Affymetrix (Santa Clara, CA), Illumina (San Diego, CA), Agilent (Santa Clara, CA), or Roche Nimblegen (Madison, WI). Array platforms based on oligonucleotide probes are more popular than cDNA-based probes because of the ability to screen a greater number of genes as well as the availability of standardized chips from several commercial venders. The prefabricated oligonucleotide arrays are usually employed for genome-wide screening and may be adapted for use in a particular tissue, including neural tissues. They are available for humans as well as several species of relevance to toxicity research and testing, including mouse, rat, zebrafish, monkey, canine, chicken, bovine, and porcine. A cDNA array that has been designed and manufactured in-house is typically cheaper and is generally used when the knowledge of genomic sequence for the species under study is minimal or when the focus is on a single biological pathway or process rather than a general screen of multiple pathways. However, several commercial vendors now provide cheaper and more efficient alternatives to in-house-designed cDNA arrays.

DETECTION METHODS FOR GENE ARRAY TECHNOLOGIES

The method of detection used for gene arrays may be based on one- or two-color detection systems. The one-color technique involves labeling of targets from a single sample with a single fluorophore (Cy3 or phycoerythrin labeled streptavidin) and posthybridization scanning of the array slide at a defined wavelength (Figure 2b). The two-color technique involves labeling the targets from two samples with two different fluorophores (Cy3 and Cy5) and posthybridization scanning at two defined wavelengths (Figure 2a).

The one-color detection system is generally used in oligonucleotide-based arrays. Only one fluorophore [Cy3 (Agilent) or phycoerythrin-labeled streptavidin and biotinylated antibody (Affymetrix)] is used. Comparison of the test and control samples requires two separate arrays labeled with the same fluorophore.

The two-color detection system is usually used for in-house-designed spotted cDNA arrays to control for the variability in probe amount from region to region within the array. This method is also used with some oligonucleotide arrays, such as Agilent and ChIP-chip samples. The fluorescent dyes commonly used in the two-color detection system include the water-soluble cyanine dyes Cy3 and Cy5, which have fluorescence emission wavelengths of 570 nm (green) and 670 nm (red), respectively. The test and control samples may be labeled with Cy3 and Cy5, respectively, and competitively hybridized (in equal proportions) to a single microarray that is later scanned using a confocal scanner at defined wavelengths to visualize the dye fluorescence. The ratio of the relative intensities of test and control samples labeled with respective fluorophores may be used to identify up- and downregulated genes in a single array chip. One must be aware of fluorophore bias due to ozone exposure when interpreting microarray data. Some fluorophores, such as Cy5 and Alexa 647 (a red-fluorescing dye of the Alexa Fluor family that emits at 665 nm), show decreased intensities even under low ozone exposures (5 to 10 ppb for 10 to 30 s), whereas Cy3 and Alexa 555 (which emits at 565 nm) are more resistant to ozone exposure (>100 ppb).

Fluorophore flop (dye swap) designs help in minimizing the dye bias. This adds to the overall cost as well as increases the complexity of the experimental design. Comparison of one- and two-color microarray platforms

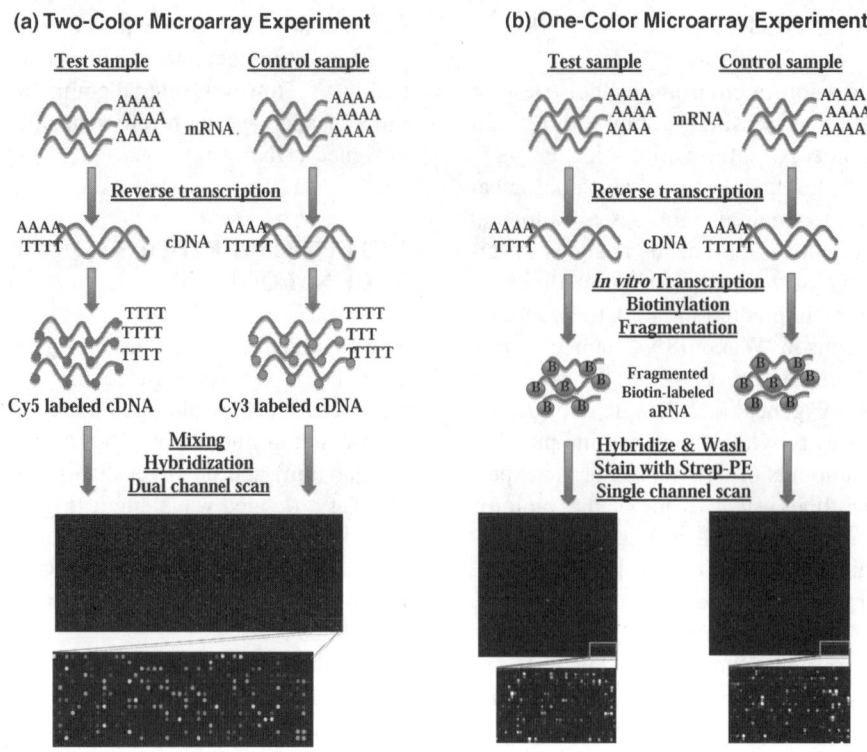

FIGURE 2 Workflow in a typical one- and two-color microarray experiment. In the two-color experiment (a), the mRNA isolated from each of the test and control (or reference) samples is subjected to reverse transcription to generate cDNA targets. The single-stranded cDNA targets from each sample are labeled with fluorophores (Cy3 and Cy5) and mixed in equal proportions. Equal amounts of the labeled cDNA targets are hybridized with the cDNA microarray for a period of about 12 h to attain equilibrium. Ideally, at most probes on the cDNA microarray chip, equal amounts of the labeled cDNA targets should be available for hybridization. The hybridized microarray chip is scanned under a confocal laser microscope and a raw image that is pseudocolored is obtained for data analysis. When equal amounts hybridize, the color is yellow; if Cy5-labeled targets preferentially hybridize, the color is red and if Cy3-labeled targets preferentially hybridize, the color is green; and if none of the labeled targets hybridize, the color is black. The differential fluorescent signal ratios are calculated to derive the differential gene expression within the test and control (or reference) samples from a single microarray chip. Two-color experiments are also done with long oligomer nucleotide probes, such as in Agilent microarray chips. A one-color experiment (b) is performed in a similar manner except that the cDNA [or aRNA (antisense RNA)] from a single test sample is labeled with a single fluorophore and hybridized to a microarray [without a control (or reference) sample as in the two-color technique]. The hybridized microarray is scanned under a confocal laser microscope at an appropriate wavelength (single channel), and the fluorescent values of all the hybridized probes are calculated. This process is repeated with the external control (or reference) sample. The ratio of the fluorescent values of the test and control samples is calculated to derive the differential gene expression data. The Affymetrix Genechip uses a one-color method and includes labeling the aRNA with biotin. The labeled aRNA are fragmented and hybridized to a Genechip. The hybridized Genechip is washed and stained with Streptavidin-Phycoerythrin and subsequently scanned at a defined wavelength. (From Kevin Gerrish, with permission.) (*See insert for color representation of the figure.*)

indicated that both the platforms have a good agreement with high correlation coefficients and high concordance of differential expressed gene lists within each platform, and this variable need not be a primary factor in the experimental design decisions.[45]

cDNA Arrays

These arrays are constructed with picoliter spots of PCR products amplified from cDNA libraries or by plasmid preparation of individual cDNAs and/or expressed sequence tags (ESTs) of cDNAs. These cDNAs are printed on glass slides or nylon membranes in a two-dimensional grid pattern. The mRNA from the test and control samples are converted into cDNA and respectively labeled with fluorescent dyes (Cy3 and Cy5), and subsequently hybridized onto the cDNA-spotted array surface. The hybridized spots are scanned with lasers of appropriate wavelength, and the relative intensities are measured to arrive at the differential gene expression–level data. The cDNA-based arrays, due to the longer length of their probes (~300 bp), may be more specific and less sensitive than oligonucleotide-based arrays, which use much shorter probes (usually between 25 to 60 bp). An advantage of using cDNAs (~300 bp) is that the hybridization is robust, so small mutations or deletions usually do not affect the hybridization process. This property is especially useful in studying polymorphic genes with minor nucleotide sequence mismatches. Another advantage of cDNA arrays is that they include the full representation of the 3'-ends of the genes, which is especially important for detecting the expression of low levels of RNA requiring linear amplification.[46] A third advantage is that they are inexpensive and can be fabricated in-house. However, a disadvantage of using cDNA arrays is that the cost–benefit advantage often disappears when the time and effort for generating the cDNA libraries, array optimization, quality control, and the expensive spotting equipment are taken into account. Another disadvantage is that the hybridization and background signal may not be uniform throughout the slide since the cDNAs vary in length and in their GC (guanine/cytosine) content.

Oligonucleotide Arrays

This technology is the most commonly used microarray platform, with numerous applications. Instead of cDNA, 25- to 60-mer oligonucleotides are used for making the probes. The probes are assembled *in silico* or deposited on beads. Some of the prefabricated commercially available oligo-based microarray platforms include the Affymetrix Genechip (Santa Clara, CA), Illumina Beadchip (San Diego, CA), Agilent (Santa Clara, CA), and Roche Nimblegen (Madison, WI) devices. These platforms are automated and are capable of genome-wide scans as well as other microarray applications, such as methylation arrays, miRNA arrays, and exon-specific arrays.

The Affymetrix Genechip synthesizes probes (25-mer oligonucleotides) *in silico* using photolithographic technology. In this process, the predetermined nucleotides are deposited onto a silica wafer, base by base, using lithographic masks. This technology allows for abundant redundancy for accurate transcript detection, as each transcript is detected by multiple probe sequences. Each probe pair consists of 11 perfectly matched (PM) oligonucleotide sequences and 11 "mismatched" (MM) oligonucleotides sequences (each having a 1-bp mismatch) for detecting

each unique transcript. The intensity readings of the 22 probes are averaged for each transcript according to MM and PM probe sets, and the degree of nonspecific hybridization can be calculated from the mismatch probes.

The Illumina Beadchip synthesizes 50-mer oligonucleotides on 3-μm-diameter beads that are randomly sprayed into wells etched into a wafer substrate, and the identity of each bead in the wells is decoded using an address sequence.[47] Unlike the Affymetrix Genechip, the Illumina Beadchip uses the same oligonucleotide sequence 30 times within each array. This platform is also able to accommodate more samples per array, due to the increased surface area in the bead available for reaction.

The Agilent microarray chip technology relies on the in situ synthesis by phosphoramidite chemistry of 60-mer oligonucleotide probes at or near the array surface using inkjet printing. These longer probes provide better specificity than shorter ones (e.g., the 25-mer reagents used by Affymetrix), due to the larger area available for hybridization. Only one nucleotide probe is used to detect each transcript. Agilent also uses inkjet technology to generate spotted cDNA arrays from PCR products.

The Roche Nimblegen microarray chip employs light-directed synthesis of nucleotide probes by digital micromirror device (DMD) technology. This process uses an array of aluminum mirrors, which can pattern from 786,000 to 4.2 million discrete light pixels, to synthesize a single array having 385,000 to 2.1 million unique probe features. In addition to several ready-made chips, this platform also offers a full-service microarray where the user is required only to submit the genomic DNA.

A comparison of these arrays indicates that each platform has its respective strengths and weaknesses (Table 2).[48] Therefore, the choice of any given platform must be based on the experimental design for either hypothesis-driven research or purely exploratory research.

EXPERIMENTAL DESIGN IN MICROARRAY STUDIES

Microarray experiments are expensive and generate enormous amounts of data. The efficient design of microarray studies is thus very important and should be planned using the intellectual resources of an entire interdisciplinary research team—ideally, one consisting of biologists, toxicologists, pathologists, bioinformaticians, and microarray core personnel. A well-designed microarray study is essential in order to properly test a hypothesis, avoid potential biases, and make optimal use of rare specimens and other research resources. Hence, the objective of the study design is to make the data analysis and interpretation as simple and as definitive as possible within the constraints of the experimental material and logistics.[49]

TABLE 2 Comparison of In-House and Commercially Available Microarray Technologies

Platform	In-House Spotted Array	Affymetrix Genechip Mouse Gene 1.0 Sense Target (ST) Array	Illumina Beadchip Mouse WG-6 v2.0 Expression Beadchip	Agilent Mouse Whole Genome 4X44 Array	Roche Nimblegen Mouse Whole Genome 385K Array
Array format	cDNA or 30–70 mer Oligos	25-mer Oligos	50-mer Oligos + 29-mer address sequence	60-mer Oligos	60-mer Oligos
# of probes/array	10,000–30,000 spot types	750,000 probes	45,281 probes	44,000 probes	385,000 probes
Resolution (# probes per target)	1+ spot per transcript	27 probes/target	One probe/target	One probe/target	9 probes/target
Starting total RNA requirement	5–10 µg	100–300 ng	50–100 ng	50 ng	10 µg
Hybridized product[a]	cDNA or aRNA	cDNA	aRNA	aRNA	cDNA
Amount of hybridized product per array	2–5 µg	2 µg	1.5 µg	0.825 µg of each labeled aRNA	6 µg
Detection method	Protocol dependent	Streptavidin-phycoerythrin	Streptavidin-Cy3	Cy3 and Cy5	Cy3
Analysis	1–4 color (usually 1-2)	1 color	1 color	2 color (or 1 color)	1 color
Samples per Array	1–4	1	6	4	1

Source: Adapted from Elvidge (2006).[48]
[a] aRNA, antisense RNA; cDNA, complementary DNA.

The experimental design of microarray studies should allow basic biological questions to be translated into high-quality and informative data sets suitable for providing a meaningful answer (and additional questions). The primary investigator should have a sophisticated understanding of the disease phenotype and/or potential mechanism(s) and also consult with a good statistician and/or bioinformatician before the study is begun. To begin with, the biological questions that need to be answered should be clear. For example, investigations may evaluate differential gene expression due to differences in a single physiological parameter (e.g., age, gender, genetic background), disease state (disease, tumor subtype, etc.), or manipulation (e.g., genotype or treatment). Some microarray experiments may be exploratory in nature, in which case they are used to produce preliminary data to support subsequent hypothesis generation. The design choice thus depends on the scientific goals of the study. Other aspects of experimental design for microarray studies include the feasibility of the experiment proposed; independent verification of microarray results using other assay methods (such as RT-PCR, SAGE, ISH, IHC, northern blots, and western blots); inclusion of appropriate controls, variation, and replicates (technical and biological); sample collection (cell lines, or fresh or frozen or FFPE tissues); nucleic acid extraction methods; sample size; power; data normalization; and statistical analysis.[49]

Several sources of variation must be accounted for and controlled (if at all possible) when conducting microarray experiments, such as biological variation, technical variation, and measurement error.[50] Biological variation among individuals, and even for a given individual over time, is intrinsic to all animal-based studies and probably reflects fundamental differences in genetics and environment. Technical variations may be introduced during sample extraction, nucleic acid isolation, labeling, and hybridization. Measurement error is associated with evaluation of fluorescent signals. The experimental design and statistical methods used in each microarray experiment should depend on the types of variation that will be encountered (i.e., the experimental design and statistical methods should be based on biological variation if one is interested in the treatment effect on different biological populations and on technical variation if one is interested in variations within treatment groups in a population). Biological and technical variation within a study may be reduced by biological and technical replication since repeated experimentation allows for estimation of effects with greater precision and less variation. Replication is also essential to estimate the variance of log ratios across arrays.[49] Biological replicates usually involve mRNA from different subjects/tissues (pool) and extractions. Technical replicates depend on the equipment and protocols, and should include multiple samples of mRNA from the same subject/tissue (pool) and extraction. Biological replicates improve the biological validity of the experiment, while technical replicates improve the precision of the measurements. To reduce variability in cDNA microarray experiments, some researchers have suggested using at least three technical replicates and at least five biological replicates per sample.[51,52] Therefore, in order to draw biologically meaningful conclusions from a microarray study, biological and technical replication should be incorporated in every study if resources permit.

Researchers confronted with limited mRNA and resources sometimes consider sample pooling. The rationale for this practice is that pooling reduces the cost, although this is true only if the per-subject cost is lower than the cost of the array. Sample pooling is also done to reduce the effects of biological variation since the expression of mRNA from the pool is theoretically the average expression of all the mRNA samples within the pool (biological averaging). However, the danger of losing information exists if subtle but statistically significant gene expression changes were to be masked by pooling (averaging) with other samples that do not have similar genetic changes. This issue is especially important if pooling is under consideration for clinical specimens or samples from genotypically diverse populations. For example, differential gene expression levels in schizophrenic brain samples are moderate and often do not exceed a 2-fold change over baseline levels in tissues from healthy individuals. Pooling such samples will generally result in loss of information due to gene fold averaging.[53] Subtle ($<$2 fold) but unique expression changes of several genes within the hepatic transcriptome have been detected when using individual mRNA samples, but they were not measured when the pooled mRNA samples were used.[54] In addition, the individual sample analysis generated more hits per biological pathway than that of analysis of the pooled sample. However, the individual and pooled sample analyses yielded similar results using cluster analysis. The individual sample approach is valuable when the differential gene expression is subtle and when associating expression data to variable phenotypic responses.[54] The pooled approach is valuable when differential gene expression is robust and fewer than three arrays are used in each biological condition. Pooling of two or three subjects per array is not recommended since the benefits realized are unlikely to be worth the loss of individual-specific information. However, pooling may be advantageous when large numbers of unique biological samples are combined.[55]

After having chosen the appropriate biological and technical replicates, the researcher often encounters the question of statistical power and sample size (i.e., the number of array slides to use). In order to calculate power, the researcher should have (1) some knowledge of the variance among individual measurements, (2) the magnitude of the effect to be determined, (3) the acceptable false-positive rate, and

(4) the desired power of the calculation. In most cases, especially for microarray experiments, information regarding the first two points is not known.[49] The interplay of power and sample size may be addressed by an analysis of the trade-off between power and the false-positive rate. In microarray experiments, the results are usually validated by RT-PCR, western blots, or northern blots. Hence, in situations where the resources are scarce, the researcher may be willing to accept a lower sample size (i.e., number of replicate micro-array analyses), lower power, and a higher false-positive rate, with the intention of subsequent validation by other methods.

The comparisons of greatest interest with respect to the scientific question should determine the study design. In addition to the type of array technology (cDNA vs. oligonucleotide, two-color vs. one-color) used; sometimes, logistics such as limited or unlimited mRNA and reference samples may also determine the study design. The one-color designs are much simpler than the two-color designs. The two-color designs require much planning and more complex analyses. The study designs for microarray analysis may be broadly classified into single- or multiple-factor designs. A single-factor design has one independent variable with different levels or groups, while a multiple-factor design has more than two independent variables with corresponding different levels or groups. Examples of single-factor designs for comparing differential gene expression include reference sample, loop, and all-pairs designs (Figure 3).[49] The reference sample design is the most commonly used arrangement for microarray experiments in several research disciplines, including toxicology and neuropathology. Treatment samples A, B, C, and D may be compared to a common reference sample (R) (Figure 3a); this design is optimal if samples are limited in quantity. The main advantages of the reference sample design are that it enables comparison of various microarray studies conducted at different time points, all comparisons are of equal efficiency, and every new sample is handled in the same way, thereby reducing laboratory error

and increasing the effectiveness of sample handling. However, other researchers argue that the reference sample designs are inefficient since a full half of the measurements are made on a single reference sample, which is not of any inherent interest to the scientific question being evaluated. In addition, the variance is also higher than other designs. The ideal reference sample should be plentiful, genetically homogeneous, and stable over time. The reference sample may be the biological negative control, pooled samples from the study, pooled samples and tissues that light up every spot on the array, or any other biologically relevant sample.[50]

If the sample quantity is not a limiting factor, either a loop or all-pairs arrangement may be a more efficient design for two-color microarray experiments. In the loop design (Figure 3b), all the samples will be compared to one another in a circular fashion. The average variance by this approach is lower, which affords greater efficiency than the reference sample design.[49,56] Each sample is labeled with both red and green fluorophores, thus doubling the number of labeling reactions; thus, using the same number of arrays as the reference sample design, the loop design generates twice as much data. In general, small loops provide good average precision, whereas larger loops with more samples may be inefficient. In addition, the efficiency of the loop design is compromised by loss of just one array. Hybrid designs incorporating both loop and reference samples may be more efficient for some studies.[50] In contrast, in the all-pairs design (Figure 3c) all treatment samples are hybridized to one another. Each sample is labeled twice with Cy3 and twice with Cy5 and hybridized together in a pairwise fashion. This design enables more precise comparisons among different treatment samples, but it may not be suitable for a large number of comparisons, due to the requirement for large amounts of mRNA and the associated increase in cost.

A more complex experimental class consists of the multifactorial designs, which take into account not only the effects due to single factors but also the effects due to several

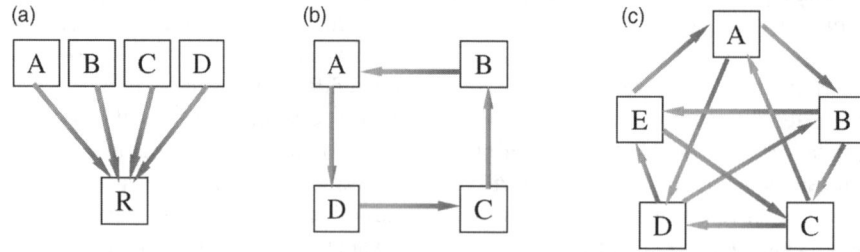

FIGURE 3   The basic types of two-color microarray experimental designs: by convention, the nodes represent the samples and the edges represent the arrays. By convention, the samples labeled with the red fluorophore (Cy5) are placed at the head of the arrow, while the samples labeled with the green fluorophore (Cy3) are placed at the tail. (a) Reference sample design, where the four treatment samples (A, B, C, D) are labeled with one fluorophore and hybridized, respectively, with the common reference sample (R) that is labeled with another fluorophore. (b) Loop design, where each of the treatment samples (A, B, C, D) is labeled once with Cy3 and once with Cy5. In total, there are four successive hybridizations. (c) All-pairs design, where all the treatment samples (A, B, C, D, E) are hybridized to one another, and each sample is labeled twice with Cy3 and twice with Cy5. In total, there are 10 pairwise hybridizations.

factors in combination. A review of multifactorial designs is beyond the scope of this chapter, but technical details have been reviewed.[49]

A simple way to assess the adequacy of the experimental design is to determine the residual degrees of freedom (df). This evaluation is calculated by counting the number of independent units and subtracting from it the number of distinct treatments (i.e., count all the combinations due to multiple treatment factors). For a single-factor experiment with n animals divided into p treatment groups, the df is $n - p$. A good study design should have at least 5 df. If there are no df in a study, the statistical variance will depend on the technical variance but not the biological variance. In such cases, the conclusions may not have biological relevance.[50]

Image Acquisition and Analysis

Once the in vivo or in vitro experiments are completed, high-quality RNA samples are usually submitted to microarray core facilities for analysis. The probes (cDNA or oligonucleotides) and the labeled targets (cDNA or aRNA) are hybridized and scanned to enable measurement of the respective fluorophore (spot) intensities. The fluorophore intensity measurements determine the relative mRNA levels of each test sample compared to that of the control/reference sample (or other sample, based on the experimental design). After the array hybridization phase is completed, image acquisition, analysis, normalization, modeling, and identification of differentially expressed genes is undertaken. The final step is to validate the data with other complementary molecular techniques.

Posthybridization scanning to detect fluorescence intensities of the arrays is performed using a confocal laser scanner. Details of the scanner settings and different strategies to obtain images of optimal quality and data analysis are reviewed elsewhere.[57,58] Briefly, the settings are adjusted so that the brightest pixels are just below the saturation point. The image should have the highest possible resolution, as the raw confocal images represent data sets with the least possible amount of processing. The microarray image format is usually a 16-bit TIFF file that is used to calculate the estimated intensity of each spot. The signal intensities for each spot are extracted in a series of steps: gridding (to determine the approximate localization of the spot), segmentation (to define spot and background identification), and finally, extraction of the foreground (spot) and background intensities.[59] In the case of Affymetrix GeneChip, for example, the raw image with data on the number of pixels in each spot is called a .DAT file, and the intensity calculations on the pixel values of the .DAT file are stored within an accompanying .CEL file. Researchers typically receive the .CEL file from their respective microarray core facilities. Numerous software packages for image analysis of microarray data can be obtained commercially or as freeware.

Data Normalization

Microarray experiments are complex, with the potential to introduce bias at any of the numerous steps. Bias can result from differences in sample and RNA integrity, contamination, linear amplification efficiency and yield, spotting, labeling and washing efficiencies, the choice of fluorescent dye, array spatial effects, hybridization efficiency, background, saturation, and other known or unknown factors. Hence, the main purpose of data normalization is to remove all the systemic biases within and between microarray experiments so that as little noise as possible is introduced into the data, and to arrive at biologically relevant information. Details of microarray data normalization and transformation are beyond the scope of this chapter but have been well reviewed elsewhere.[60–62]

Data Analysis

Before statistical analysis of the microarray data can be undertaken, the data should be assessed to make sure that it satisfies the basic assumptions of the statistical tests to be used in the analysis. For example, parametric analyses require that the data be normally distributed and have uniform variance. It is a good practice to transform the microarray raw data even if a nonparametric approach is used. The most common approach is the logarithmic (base 2) transformation (i.e., a conversion to determine the intensity of probes after background subtraction). This procedure is very useful since variation in the log-transformed intensities and log-transformed ratios of intensities is less dependent on the absolute magnitude; log transformation of the data makes the data more symmetrical and brings skewed data closer to a normal distribution. In addition, fold-change calculations are simplified since the ratios of expressions are converted into differences. The use of nonparametric methods can reduce the reliance on prior assumptions about the data, but may also incur loss of power or be inappropriate when sample sizes are small.[63]

Principal Components Analysis

The principal components analysis (PCA) method is an exploratory multivariate statistical technique used to reduce the dimensionality (i.e., complexity) of large data sets, including microarray data, and is used routinely to evaluate outlier subjects and/or genes[64] (Figure 4b). This procedure is essentially an analysis of covariance between factors; it is often used as a diagnostic step to evaluate the relationship between samples, but it is not a clustering method per se (see

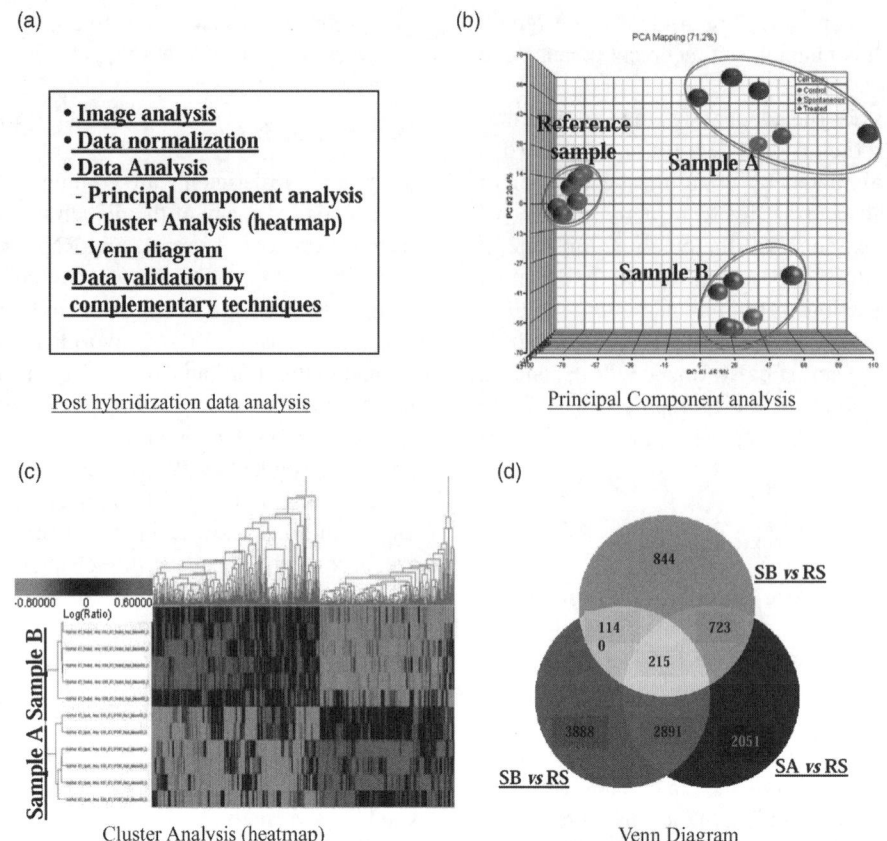

FIGURE 4 Data analysis post microarray experiment. After hybridization of the labeled targets with probes, the microarray chips are scanned with a confocal laser scanner to obtain fluorescent raw images. Subsequent analyses include image analysis, data normalization, and principle component analysis (b), hierarchical cluster analysis (heat maps) (c), and Venn diagrams (d) of selected genes. The principal component analysis helps in visualizing the prominent gene expression patterns and also in identifying the outlier samples. The hierarchical cluster analysis helps in visualizing segregation of the gene expression and comparison of the samples with the reference/control samples. The Venn diagram helps in comparing the gene lists from different samples and aids in visualizing, at a glance, the degree of containment, intersections, and disjunction between gene lists from various samples. SA = sample A, SB = sample B, RS = reference sample. (From Mark Hoenerhoff and Pierre Bushell, with permission.) (*See insert for color representation of the figure.*)

below) because it does not allow user-defined criteria. When PCA is applied to gene data sets, it helps to identify prominent gene expression patterns across subjects. When it is applied to conditions or subjects, it helps to categorize them. However, a key point to keep in mind is to assure that the first few principal components capture a reasonable amount (say, >70%) of the variation in the data, or else the projection of the samples or genes is difficult to interpret. Further technical details for this powerful analytical approach have been described elsewhere.[65]

Cluster Analysis

Cluster analysis can be applied in a data driven manner to find patterns in the data set that are not known a priori.[66] Shared patterns are commonly assumed to be involved in similar biological pathways (i.e., guilt by association). This technique is often used as an exploratory approach since the researcher is often looking for stratification (subclasses) within a filtered data set. Cluster analysis is divided into two categories: supervised learning (which requires a priori knowledge of the samples or the affected genes) and unsupervised learning (with no a priori knowledge of the samples or genes).[67] Supervised clustering requires that the data contain two components: (1) information on gene expression from arrays run on a set of samples and (2) data showing the nature of the sample (normal or diseased tissues, neoplasms, cultured cells, etc.). The goal is to use a mathematical model to predict the characteristics of the sample by the group it clusters with. In the unsupervised clustering, the data are grouped by gene expression values only, and the goal is to find various gene clusters with similar characteristics.[67] Hierarchical clustering is used to produce the conventional "heat maps," in which gene clusters are represented by a multicolored grid (usually, green for reductions and red for increases in expression), where the columns and rows correspond to samples and genes, respectively. Thus, a heat map gives a rapid visual summary of gene clusters using differential expression levels of numerous genes within several samples (Figure 4c).

Microarray Data Validation

In a microarray experiment, the differential gene expression is usually calculated using statistical tests such as a t-test for each of the thousands of genes within the array. A p-value set at 0.05 indicates that there is 5% chance of making a type I error (false positive) on one gene. So when evaluating 10,000 genes, one can expect 500 type I errors (false-positives), i.e., 500 genes are picked up by chance alone. If the t-tests select 1000 genes at a p-value of 0.05, then 500 (50%) out of the 1000 genes may be false-positive (type 1 error). Thus, in microarray analysis, performing statistical tests for tens of thousands of genes creates a multiple-hypothesis testing problem. Type I errors due to multiple testing may be addressed by controlling the family-wise error rate (FWER) by using a Bonferroni correction. This is a very conservative approach that limits the power to identify significant differentially expressed genes. An alternative, less stringent, approach, with greater power adjustment, is the false discovery rate (FDR) control (such as Benjamini and Hochberg method) that estimates the false positives among all the genes identified as differentially expressed, i.e., the FDR is the expected false-positive rate among all significant tests. For example, if the FDR is fixed at 5%, then of 1000 significant calls, 50 of the observations would be expected to be false positives.[68] Also, there are reasons why false positives can occur besides as a consequence of multiple testing. For example, nonspecific hybridization or cross-hybridization of closely related genes to oligonucleotide probes or cDNA probes can result in false positives. Hence, validation of at least a subset of genes from the differential gene expression data by independent complementary quantitative assays is essential in the high-throughput methods such as microarray technology. The expression changes are usually validated by quantitative RT-PCR, northern blot analysis, and/or ISH.

The mRNA expression may not always correlate to the corresponding protein data, due to post-translational modifications. Therefore, the differential gene expression data should be interpreted cautiously, and any hypothesis based on mRNA expression should be validated by one or more protein analyses, such as Western blots, enzyme-linked immunosorbent assays (ELISA), immunohistochemistry, or mass spectrometry. The high-throughput proteomic approaches, such as peptide arrays, tissue arrays, and antibody arrays, can be complementary to the high-throughput genomic technologies. Some of these techniques are reviewed later in the chapter.

Microarray technology must deal with numerous complex variables. Replication of microarray results is not very common, and efforts to do so are hindered further by lack of uniformity between experiments and across platforms. The ability to replicate experiments from published literature is one of the most important aspects of scientific research. This principle is true for microarray studies as well. To standardize microarray data acquisition from various groups and platforms, the Microarray Gene Expression Data (MGED) Society has advocated the use of MIAME (minimum information about a microarray experiment) guidelines for each microarray experiment when presenting and exchanging microarray data.[69] The MIAME guidelines provide standards for experimental design, array design, sample acquisition and processing, hybridization, measurements, and normalization controls; these details are available at the MIAME Web site (http://www.mged.org/Workshops/ MIAME/miame.html).

To address concerns over the reliability and reproducibility of microarray data within and across platforms, the U.S. Food and Drug Administration (FDA) and the U.S. Environmental Protection Agency (EPA) have issued guidelines to facilitate the use of microarray data for drug development and toxicological risk assessment. A consortium led by FDA scientists formed a microarray quality control (MAQC) group to address these concerns. The MAQC group confirmed that microarray data are reproducible following careful experimental design, with appropriate data transformation and analysis. They also demonstrated that fold-change results from microarray data correlate closely with results from other genomic assays, such as quantitative RT-PCR. Levels of variation observed between microarray runs were relatively low and attributable to probe binding to alternatively spliced transcripts or to cross hybridization of probes.[70,71]

Microarray technology is now a mature discipline with new variations and novel applications emerging for both genomic and proteomic analyses. The arrays are packing a higher density of probes with numerous measures to reduce various biases. Improvements in data analysis and data mining are an integral part of the high-throughput microarray technology, and new research methods are enabling the more rapid and reliable assimilation of the enormous amounts of data generated by this technology. With time, the arrays will become cheaper and the methods of data analysis and data mining will make additional advances, which will further boost the popularity and applicability of this approach.

EXAMPLES OF MICROARRAY TECHNOLOGY AS APPLIED TO NEUROPATHOLOGY RESEARCH

Microarray technology is used widely in neurobiological investigations. Basic aspects of neurobiology are a fertile area for microarray-based experimentation. For example, astrocyte biology has been examined using oligonucleotide microarrays, including the molecular diversity of astrocytes as well as their varied functional heterogeneity.[72] Such studies have shown that neocortical astrocytes, but not astrocytes from other regions, have decreased expression of NF-1, a brain tumor suppressor gene, suggesting that this may contribute to regional predilection of gliomas in humans.[73] A growing number of genomics applications have also been directed to neuropathology studies, including the analysis of developmental and neurodegenerative disorders, pathogen infections, neural neoplasia, neuropharmacology, and neurotoxicity.

Microarray technology is an important new tool for elucidating mechanisms of neurological conditions. One obvious application is the study of the pathogenic events that drive neurodegenerative (i.e., structurally based) diseases, such as Alzheimer's disease (AD). This important neurodegenerative disease is characterized by intracellular accumulation of neurofibrillary tangles (NFTs, composed of hyperphosphorylated tau protein) and β-amyloid plaques within the entorhinal cortex and the CA1 region of the hippocampus. Selective sampling of CA1 neurons (with NFTs) from normal and AD human brains followed by microarray analysis has been used to demonstrate that neurons harboring NFTs have significant reductions of several genes implicated in AD neuropathology, such as phosphatases/kinases, cytoskeletal proteins, synaptic proteins, glutamate receptors, and dopamine receptors.[74] Evaluation of differential gene expression in AD patients and normal brains using microarray and brain imaging in combination has revealed downregulation of retromer trafficking molecules VPS35 and VPS26 within brain regions (e.g., entorhinal cortex, dentate gyrus) selectively vulnerable to AD.[75] The retromer complex binds and transports the transmembrane receptor VPS10 from the endosome back to the trans-Golgi network and thus reciprocally regulates the β-amyloid levels. Similarly, microarray can play an important role in evaluating molecular mechanisms of psychiatric (functionally based) diseases such as schizophrenia. cDNA microarray analysis of the prefrontal cortex from matched pairs of schizophrenic and control subjects has revealed a consistent differential downregulation of the *PSYN* functional gene group, *N*-ethylmaleimide sensitive factor, and synapsin II within the schizophrenic patients.[53] These changes are associated with abnormal presynaptic function.

Microarray in combination with other molecular diagnostic techniques is a useful approach to evaluating and classifying microbial infections of the nervous system.

For example, sequence-independent PCR amplification and microarray analysis have been used simultaneously to detect and genotype more than 140 viruses.[76] Rabies is a zoonotic disease of the central nervous system that affects several mammalian species, including humans. Rabies virus (genotype 1) is the type virus of the lyssavirus genus within the Rhabdoviridae family, but six other related lyssavirus genotypes could also cause rabies. Even in countries that have been declared rabies-free, sporadic reports of human rabies arise due to contact with rabid animals, and in some instances these cases may not be due to the rabies virus (genotype 1) but rather, to other, related lyssaviruses. Microarray technology has been used to detect and genotype all seven lyssavirus genotypes simultaneously.[77] Such techniques may prove useful in future efforts to identify and categorize microbial diseases.

Genomic technology has been used to construct databases for many neoplasms, including tumors of the nervous system. For example, the Cancer Genome Atlas (TCGA; http://cancergenome.nih.gov/) has been compiled by a consortium of scientists from multiple U.S. institutions. One study comprehensively characterized the genomic pathways affected in 206 human glioblastoma (the most common primary brain tumor in adult humans) using extensive genome-wide scans from multiple microarray platforms.[78] Genomic heterogeneity is marked within multiple molecular subclasses of these tumors. The TCGA reported that the most frequently altered genes were components of the RTK/RAS/PI(3)K (88%), p53 (87%), and RB (78%) signaling pathways, along with numerous other genes.[78]

Microarray analysis plays a growing role in neuropharmacological experiments. Common molecular pathways associated with drug abuse (e.g., cocaine, cannabis, phencyclidine) have been studied using microarrays on postmortem prefrontal cortex samples. A decrease in the transcription of calmodulin-related genes (e.g., *CALM1, CALM2, CAMK2B*) and an increase in the transcription of genes related to lipid/cholesterol metabolism (*FDFT1, APOL2, SCARB1*), and Golgi and endoplasmic reticulum function (*SEMA3B, GCC1*) were reported.[79] This molecular signature may be related to changes in the synaptic function and plasticity that could affect the decision-making ability of drug abusers.

Microarray technology has also been used extensively in neurotoxicity research. For example, following exposure to the thiocarbamate herbicide/fumigant carbonyl sulfide (COS), differential gene expression changes were detected within the posterior (inferior) colliculus of rats before any apparent histological or immunophenotypical changes occurred.[21] Specific alterations included upregulation of genes associated with DNA damage and G1/S checkpoint regulation, apoptosis, pro-inflammatory mediators, and vascular mediators that were linked to necrosis and inflammatory process within the posterior colliculus. In like manner, microarray studies were also used to study lead-induced encephalopathy. Lead increases the permeability of the blood–brain barrier by acting on astrocytes and thereby negatively affecting the brain microvessel development.[80] Differential gene expression studies on astrocytes exposed to lead in vitro demonstrate upregulation of VEGF, Annexin 5, and other genes involved in angiogenesis.[81] Similarly, in vitro exposure to sublethal concentrations of two distinct genotoxic alkylating agents, methylazoxymethanol (MAM) and nitrogen mustard (HN2), elicit unique gene expression signatures in neonatal cerebellar granule cells.[82] The genes affected by MAM are involved in neuronal differentiation, stress, and immunologic responses, and signal transduction, whereas HN2-induced changes are related to genes involved in protein synthesis and apoptosis.

PROTEOMIC TECHNOLOGIES

Proteomics is the large-scale study of the protein complement of a cell or tissue at a given time point. Several proteomic methods have been developed that make it possible to identify, characterize, and comparatively quantify the expression of hundreds of proteins. The purpose is to obtain a snapshot of the concentration of proteins associated with different cellular states.[83–87] Proteins are the main components of the physiological metabolic pathways that carry out the majority of biological events in the cell. Over 50% of all genes are subject to transcriptional variation by RNA splicing, which accounts for both the production of multiple protein isoforms from a single gene and the specific expression of certain isoforms within various cells and tissues.[88,89]

Proteomics represents the natural extension of functional genomic analysis. Studying the proteome is more complicated than genetic analysis because the proteome of a cell is changing continuously, due to factors such as age and shifts in the spectrum of environmental stimuli.[84] Proteomics enables a more precise understanding of a biological process than does genomics. Gene transcription (i.e., the mRNA profile) only gives a rough estimate of expression and activity for its corresponding protein. However, it is well recognized that the complexity of the human proteome far exceeds that of the genome. When variables such as alternative gene splicing, post-translational modifications, and individual coding variants are taken into account, the number of protein species in humans appears to be one to two orders of magnitude greater than the number of genes available. In addition, proteomic analysis is further complicated by the fact that proteins have no base pairing or other stereotypical chemical signature, so there is no simple means of amplification as there is for nucleic acids.[88–90]

TECHNIQUES FOR ANALYZING PROTEINS

In the past, the analysis of proteins in neuropathology was limited to immunohistochemistry, Western blot analysis, and ELISA. Histochemical stains, immunohistochemical tags, and radiolabels are common methods used to visualize and identify molecular targets. There are inherent limits to the specificity of these methods, in particular to the number of molecular targets that can be monitored simultaneously. In addition, the use of specific tags requires prior knowledge of the target protein, thus limiting their use in molecular discovery. These techniques also depend on the quality of the antibody used to probe for a given molecule, especially its specificity for the protein of interest.[91] Newer technologies, including two-dimensional gel electrophoresis, mass spectrometry, and protein arrays, have improved the efficacy of protein identification and quantification in neural tissues. These novel methods make it possible to examine alterations in the protein profile of cells or tissue in response to disease or toxicant exposure.[83,85,88]

Two-Dimensional Gel Electrophoresis

The two-dimensional gel electrophoresis (2DGE) technique is the most commonly used method for separating the entire complement of proteins in a tissue specimen.[87–89] In 2DGE, mixtures of proteins are separated sequentially by two different biochemical properties. In the first dimension, polypeptides are separated according to their isoelectric points; in the second dimension they are separated by their molecular weight. In principle, this method allows all cellular proteins to be separated on a single large gel. The proteins separated can be detected by various means, most commonly by application of silver or Coomassie blue staining. The bidirectional separation results in a distinct pattern of protein spots for each tissue domain defined. Individual proteins can be excised from the gel, purified, and identified using mass spectrometry methods.[85,89]

The main advantage of 2DGE is that it can often distinguish between different isoforms of a protein. Therefore, it is highly suitable for studying the post-translational modification of proteins. The main limitations of this technique are that it requires a large amount of tissue, does not allow localization of the proteins to specific anatomical structures or cell populations, and is labor intensive. Other disadvantages are that it cannot be applied to proteins or peptides smaller than 10 kDa, and that interpretation can be difficult if a spot contains two or more co-migrating proteins. It also has a limited use for highly hydrophobic proteins.[85,89]

Mass Spectrometry

Mass spectrometry (MS) is an important tool for the detection, identification, and characterization of proteins.

The three primary applications of MS in proteomics are in the measurement of protein expression, to characterize protein–protein interactions, and to identify sites of protein modification. The MS apparatus measures the mass-to-charge ratio of gas-phase ions. Mass spectrometers consist of an ion source that converts the sample molecules into gas-phase ions, a mass analyzer that separates ionized samples on the basis of mass-to-charge ratios, and a detector that records the number of ions at each mass-to-charge value.[83,84,86,87,89,91] Proteomic analysis via MS begins with sample preparation in which proteins are either digested into peptides enzymatically (bottom-up) or analyzed in their intact forms (top-down).[84,87]

Bottom-up analysis, also known as "shotgun" proteomics, is the analysis of enzymatically or chemically produced peptide fragments of proteins. The bottom-up approach is the most popular method when evaluating highly complex samples for large-scale analysis. Proteins are digested without fractionation or separation, and peptides are frequently separated by liquid chromatography followed by tandem MS. Peptide detection is used to infer protein presence, and the resulting peptide masses and sequences are used to identify corresponding proteins. The advantages of this approach are that (1) there is better front-end separation of the peptides compared with proteins, and (2) the sensitivity of this approach is higher than that of the top-down method. The drawbacks are that there is limited protein sequence coverage by the identified peptides. Peptides present at high levels are preferentially identified, while information on peptides present at low levels may not be obtained. Another concern is that the digestion of the proteins results in loss of labile post-translational modifications (PTMs).[84,87,89]

In the top-down strategy, the MS analysis is carried out on whole proteins. The gas-phase ionization of intact protein is followed by their direct fragmentation inside the mass spectrometer without prior digestion, which enables more complete characterization of protein isoforms and PTMs. It is especially useful in characterizing specific combinations of multiple PTMs. This type of analysis is performed primarily on a single protein or a simple protein mixture (attained by preseparation of tissue homogenates using another method, such as 2DGE). The key to this analytical approach is the ability to fragment intact proteins. The preferred fragmentation method is collision-induced dissociation combined with electron capture dissociation or electron transfer dissociation because the combinations yield more complete backbone sequencing and labile PTMs are retained. The major limitations of this method are that large proteins cannot be analyzed readily, and reliable separation of intact proteins is more challenging than separation of peptide mixtures.[84,87,89]

Most mass spectrometers can be categorized into one of two types: electrospray ionization (ESI) and matrix-assisted laser desorption/ionization (MALDI) instruments. The most widely used mass analyzers include ion trap, triple quadrupole, time-of-flight, and Fourier transform ion cyclotron (FTICR) units. These instruments are very different in their mechanism of ion separation, mass accuracy, and resolution, but complementary in their ability to support protein identification.[83–85,87,90] A comparison of the most commonly used mass spectrometers is shown in Table 3.

Electrospray Ionization

Electrospray ionization (ESI) produces ions from solutions of suspended proteins. ESI is driven by high voltage applied between an emitter at the end of the separation pipeline and the inlet of the mass spectrometer. It can be performed in positive or negative ion mode. A sample solution is sprayed as a mist of droplets from a charged capillary tip. As the solvent evaporates, the surface area of the droplets are reduced while the total charge of the proteins in the droplet remains unchanged. The droplet shrinks until the individual ions leave the droplet and enter the analyzer to be detected. The presence of multiple charges reduces the mass-to-charge ratio of each ion compared to that of a singly charged species, which allows mass spectra to be obtained for large molecules.[87,89]

Matrix-Assisted Laser Desorption/Ionization

The matrix-assisted laser desorption/ionization (MALDI) approach is a laser-based soft ionization method, and it is one of the most successful ionization methods for investigating the structure of large molecules.[83,86,87,89,91] The advantage of MALDI is that it avoids the need for time-consuming tissue extraction, purification, and separation steps, which have the potential to produce artifacts via traumatic and chemical degradation of the proteins. MALDI utilizes a matrix composed of a small acidic aromatic molecule that absorbs energy at the wavelength of the irradiating laser, as a platform to dissect molecular structure more efficiently. The dissolved proteins or peptides are mixed with the matrix molecule in solution, small amounts of the mixture are deposited on a surface and allowed to dry, and the sample and matrix co-crystallize as the solvent evaporates. The matrix plays multiple roles in desorption and ionization. Sinapinic acid provides the best signals for high-molecular-weight proteins, while α-cyano-4-hydroxycinnamic acid is more suitable for low-molecular-weight peptides. Matrix application protocols include manual methods such as spraying using an airbrush or thin-layer chromatography sprayer or dipping the tissue sections into matrix-containing solutions. A disadvantage of manual application is poor reproducibility, probably arising from irregular dispersion of the prayed material. Higher reproducibility can be achieved using automated sample preparation devices that fall into two classes: spotting and spraying devices.[86]

TABLE 3   Comparison of Commonly Used Mass Spectrometers[a]

Instrument	Mass Resolution	Mass Accuracy	Sensitivity	Scan Rate	Ion Source	Applications
LTQ	2,000	100 ppm	Femtomole	Fast	ESI	Bottom-up protein identification; PTM identification; high-throughout analysis
Q-q-Q	1,000	100 ppm	Femtomole	Moderate	ESI	Protein quantification
Q-q-LIT	2,000	100 ppm	Femtomole	Fast	ESI	Bottom-up protein identification; protein quantification
TOF	20,000	<5 ppm	Femtomole	Fast	MALDI	Protein identification
TOF-TOF	20,000	<5 ppm	Femtomole	Fast	MALDI	Protein identification
Q-q-TOF	20,000	<5 ppm	Femtomole	Moderate	MALDI; ESI	Bottom-up or top-down protein identification; PTM identification
FTICR	750,000	<2 ppm	Femtomole	Slow	MALDI; ESI	Top-down protein identification; protein quantification; PTM identification
LTQ-Orbitrap	100,000	<5 ppm	Femtomole	Moderate	MALDI; ESI	Top-down protein identification; protein quantification; PTM identification

[a] ESI, electrospray ionization; FTICR, Fourier transform ion cyclotron resonance; LTQ, linear trap quadrupole; MALDI, matrix-assisted laser desorption/ionization; PTM, post-translational modification; Q-q-Q, triple quadrupole; Q-q-LIT, dual quadrupole in linear trap; Q-q-TOF, dual quadrupole time of flight; TOF, time of flight; TOF-TOF, tandem time of flight.

Preparation methods for tissue imaging using MALDI must be performed carefully to maintain the original spatial arrangement of compounds and avoid delocalization and degradation of the sample. Careful handling of the tissue samples includes freezing in liquid nitrogen immediately after procurement to stabilize the native conformation of molecules in the tissue. Samples should not be fixed in formalin because this procedure induces protein cross-linking. Generally, cryosectioning is performed on frozen samples without any external support by a viscous embedding medium. Standard materials such as OCT compound should be avoided because the polymers can smear across the sample surface and compromise the biochemical integrity of the proteins. In some cases, small or fragile samples can be encased in simple media such as gelatin or agarose to facilitate handling. Sections 10 to 20 mm in thickness are optimal for both handling and MS analysis. Gel-based or chromatographic separation is used to reduce the sample complexity prior to MALDI analysis.[86]

The matrix-covered tissue section is subjected to short laser pulses, resulting in molecules in the sample being desorbed. In the desorption process, the molecules become protonated and carry a positive charge. These ions accelerated by a high-voltage grid, traverse a flight tube, and strike a detector at the end of the tube. The time of flight is proportional to the mass.[86] The most intense signals come from the most abundant protein species. The exact protein concentration is not known, so the exact sensitivity of the technology is hard to estimate.[86]

Matrix-Assisted Laser Desorption/Ionization-Imaging Mass Spectrometry

Matrix-assisted laser desorption/ionization-imaging mass spectrometry (MALDI-IMS) is a relatively new method for the simultaneous mapping of hundreds of peptides and proteins present in thin sections, with a lateral resolution of 30 to 50 mm. This technique uses the same sample preparation protocols as does MALDI. However, in MALDI-IMS, proteins are desorbed from discrete spots or pixels upon irradiation of the sample via an ordered projection (array) of the laser across the surface. Each pixel is keyed to a full mass spectrum consisting of signals from protonated species of molecules desorbed from that tissue region. A plot of intensity of any one signal produces a map of the relative amount of that compound over the entire surface imaged.[86,92] The technique allows for the direct correlation of the proteomic data with the neuroanatomic structure in which it is expressed, which provides a better understanding of normal and pathological molecular pathways.

MALDI-IMS software superimposes the MALDI images representing the molecular data over a macroscopic or microscopic optical image of the sample taken before MALDI measurement. Two main approaches are used to correlate histology with MALDI. In the first approach, adjacent tissue sections are used for the analysis, one for MALDI and the next for histological staining. The advantage of this approach is that any staining protocol may be used. The disadvantage is that there is no direct correlation between the data sets, as the MALDI image is derived from a different section from that from which the histological image is derived. The second approach involves histological staining of the tissue section prior to MALDI analysis, which permits direct correlation of histomorphology and MALDI data. Cresyl violet and methylene blue are often used as stains because they do not compromise the overall mass spectra quality but do allow specific structural features of the tissue to be easily recognized and analyzed. Cresyl violet stain is especially important in neuropathology since it is already used to demonstrate neuronal cytoarchitecture, especially Nissl substance (i.e., rough endoplasmic reticulum) and cell nuclei. Routine H&E staining cannot be used for MALDI-IMS because it compromises the quality of mass spectra obtained significantly. Thus, the disadvantages of this second approach are that it subjects the samples to additional handling steps prior to MALDI, and it is not suitable for all staining protocols. The available stains are sufficient to localize cells in neural tissue, but they do not yield the same analytical information as can be obtained from an H&E-stained section. Thus, the second approach often still requires production of a second section, to be processed for H&E or immunohistochemistry, to ensure that no crucial information about the sample has been missed.[86,92]

Surface-Enhanced Laser Desorption/Ionization

Surface-enhanced laser desorption/ionization (SELDI) is a MALDI variant that has come into use in recent years. In SELDI, the protein mixture is spotted on a surface that has been modified to achieve biochemical affinity with the sample. Some proteins bind to modified chemical groups on the surface, whereas others are removed by washing. A matrix is applied to the surface and allowed to crystallize with the sample peptides. Binding to the SELDI surface acts as a separation step to fractionate and enrich subpopulations of proteins from complex mixtures, making them easier to analyze. Common surfaces include a weak-positive ion exchange (CM10), a hydrophobic surface (H50), a metal-binding surface (IMAC30), and a strong anion exchanger (Q10). Surfaces can also be made protein-specific by modification with antibodies, other proteins, or specific DNA sequences. Samples spotted on a SELDI surface are typically analyzed using time-of-flight MS (SELDI-TOF-MS).[85,88]

QUANTITATION OF PROTEINS

A necessary step in proteomics is protein quantitation. Traditionally, proteins are quantified by 2DGE. In the past several years, many MS-based quantitative proteomic methods have also been developed. These methods include the introduction of isotopic tags at specific functional groups on peptides or proteins. The advantage provided by the use of stable isotope-labeled synthetic peptides is that absolute measurement of protein abundance can be achieved.[84,85,87,89,93] For example, isotope-coded affinity tags (ICATs) label the side chains of cysteine residues in two reduced protein samples using the isotopic light or heavy reagent, thereby generating mass signatures that identify sample origin and serve as the basis for accurate quantitation. This procedure allows a comparison of two samples simultaneously. ICATs have several limitations, including the fact that quantification is restricted to cysteine-containing proteins and that ICATs can compare only two samples at a time.[84,85,87,89,93] Another option is isobaric tags for relative and absolute quantitation (iTRAQ), which is based on chemically tagging the N-terminus of peptides generated from protein digests. This technique allows for the identification and quantification of all peptides as well as the comparison of up to eight samples simultaneously. The iTRAQ tags are all the same size, and it is only upon fragmentation that the various mass tags are observed. The tag consists of an amine-specific reactive group, a balancer group, and a reporter mass group. The peptide amino termini and lysine side chains are targeted by the amine-specific reactive group.[84,85,87,89,93]

Protein and Peptide Arrays

Protein arrays are solid-phase ligand-binding assay systems using proteins immobilized on surfaces such as glass, membranes, microtiter wells, mass spectrometer plates, and beads. The assays are rapid and automated, economical, capable of high sensitivity and throughput, and produce a large amount of data. The same instrumentation used for scanning DNA microarrays is applicable to protein arrays. Thus, fluorescence labeling and detection methods are used widely.[94,95]

Protein arrays are a powerful means to detect proteins, quantify protein expression levels, and investigate protein interactions and functions.[94,95] The objective of the protein array is to carry out efficient and sensitive high-throughput protein analysis. Protein arrays make it possible to screen thousands of interactions in parallel, including protein–antibody, protein–protein, protein–ligand, and protein–drug interactions. Most of the proteins analyzed in the array are produced by recombinant methods (i.e., are not subject to alternative splicing), resulting in direct connectivity between protein array results and DNA sequence information.[94,95]

There are three general types of protein arrays: analytical capture arrays, large-scale functional arrays, and lysate arrays.[94,95] Large-scale functional arrays contain full-length proteins or protein domains and are used to study protein interactions. Lysate arrays contain lysate obtained from cells of interest. The array is probed with antibodies to determine the presence and posttranslational modification of proteins. The most common type of protein array is the analytical capture array or antibody array. In this method, each spot in the array contains a different antibody, and the presence and relative amount of protein bound to an antibody in the chip can be assessed specifically for each unique point of the array. The technical aspects of array design include the choice of capture antibodies, the surface for attachment, and the signal generation strategy. Monoclonal antibodies are used in current arrays to allow the generation of reproducible results, using one of two experimental designs. In the direct labeling approach, the proteins to be measured are labeled with a fluorescent tag to allow detection following incubation on an antibody array. In the dual-antibody method, antibody pairs against different antigenic sites on a specific protein are used to capture and detect the target protein; the labeled detection antibody is introduced to capture the protein in the dual-antibody approach, thus obviating the necessity to label the proteins.[96]

A variation on the protein array is the peptide array, which is similar in concept but uses smaller molecules as a substrate. Peptide arrays have an advantage over protein arrays because peptides are more stable on arrays and can be rapidly synthesized as large, defined libraries. Peptides are chemically quite resistant and retain partial aspects of protein function. Peptide arrays have been instrumental in the study of enzyme selectivity and determining preferred substrate motifs for kinases, proteases, and phophatases. Two methods have been employed for fabricating peptide arrays.[97–99] In the first, in situ fabrication (ISF), amino acids are coupled to preprogrammed sites across a two-dimensional surface. The panel of peptide sequences grows on the array layer by layer until the desired length is achieved. The maximum length is 40 residues. The entire array is built and organized on the surface where the screening will take place. Efficient coupling chemistries are required to ensure good purity of the peptides. The advantage of ISF is that library synthesis takes place at the same time as array creation, eliminating extra steps for synthesis and immobilization and reducing the amount of reagents consumed. The main disadvantage of ISF is the limiting rate at which the arrays can be produced.[97–99] The second method is spotting fabrication, in which the peptide libraries are presynthesized and then applied robotically to a defined area on glass slides. Three types of immobilization techniques are used in this approach. In region-specific immobilization and nondirected immobilization, the exact orientation of the peptides is undefined, which results in a mixed presentation of peptides from multiple potential interaction sites on either the arrays or the peptides. In site-specific immobilization, directed and uniform presentation of peptides occurs, with interacting groups on the arrays and peptides clearly defined. The use of covalent attachment strategies is the predominant approach for peptide immobilization. Immobilization may consume biologically relevant side chains or cause steric effects that could abrogate downstream interactions.[97–99]

The major advantages and limitations of the most common proteomic techniques are listed in Table 4. The selection of proteomic technique is usually based on the experimental goals and the complexity of the protein sample being analyzed.

Phosphoprotein Analysis

Protein phosphorylation is a central mechanism for cell regulation and signaling. It is estimated that one-third of all eukaryotic proteins are phosphorylated. A common method for analyzing and identifying phosphoproteins is MS because this technique measures changes in molecular mass. Each phosphorylated amino acid in a peptide or protein can be detected by an 80 Da increase in the amino acid residue mass. Based on MS data, a database search can be conducted to identify phosphopeptide mass values that are increased by 80 Da increments compared to the calculated value for the amino acid sequence. Difficulties arise when modifications are transient, such as when phosphorylation serves a regulatory role. Phosphoprotein enrichment may be used to overcome this difficulty. One common strategy is to combine a separation technique [e.g., immobilized metal affinity chromatography (IMAC), immunoprecipitation, metal-oxide affinity chromatography, and string cation-exchange chromatography] with an affinity-based enrichment method (e.g., use of antibodies against the phosphoamino acids, magnetic materials, and metal ion–phosphopeptide precipitations).[87,89,100] Phosphorylation is labile during both enrichment and MS assessment,[89,101] so considerable care must be taken to ensure that information is not lost when preparing samples for phosphoprotein analysis.

Protein Identification

A number of approaches have been developed to identify proteins based on peptide analysis. The original method was known as peptide mass mapping or mass fingerprinting. This technique compares the experimentally determined MS peak mass values with a database of mass values predicted for peptides generated using theoretical digestion of proteins. Another approach involves collision-induced dissociation (CID) spectra of individual peptides from tandem mass spectrometry (MS/MS), which is the currently accepted standard for protein identification.

TABLE 4 Advantages and Limitations of Common Proteomics Techniques[a]

Technique	Advantages	Limitations
Western blot, IHC	Can distinguish between protein isoforms; can detect and characterize small amounts of protein	Need prior knowledge of target; dependent on antibody quality and affinity
2DGE	Can distinguish between protein isoforms; suitable for studying post-translational modifications	Requires a large amount of tissue; doesn't allow protein localization; labor intensive; spots may contain more than one protein; limited use for hydrophobic proteins
Mass spectrometry	Identify proteins by their mass-to-charge ratios; able to identify unknown proteins using picogram amounts, fast	Identification of proteins in a large number of samples is expensive and time consuming; quantification requires ICAT or iTRAQ
Electrospray ionization	Reliable and rapid method to characterize protein complexes; does not require highly purified samples; can identify covalently modified proteins	Requires a restricted set of buffers; sensitive to adducting salts; sensitive to pH of solvent; sensitive to contaminants
MALDI	Avoids time-consuming protein extraction, purification, or separation; less artifact production	Exact protein concentration is not known; best suited for measuring peptide masses, not intact proteins; highly dependent on quality of sample; highly sensitive to contaminants
MALDI-IMS	Allows localization of proteins in a sample; can map hundreds of proteins and peptides in a single section; does not require molecular tags or chemical modifiers	Subjects samples to additional handling prior to MALDI; limited choice of tissue stains
SELDI	Small sample amount required; sample preparation is fast and straightforward; no pre-separation of sample required; high sensitivity; can resolve low molecular weight proteins	Unable to identify specific proteins that correspond to peaks of interest; higher molecular weight proteins are not well resolved
Protein array	Rapid, automated, economical, capable of high sensitivity; produce large amounts of data; can evaluate thousands of proteins in a single experiment	Stringent demands on production, storage, and experimental conditions due to protein stability, native folding, or activity of immobilized proteins
Peptide array	More stable than protein arrays; can be rapidly synthesized as large, defined libraries; peptides are chemically resistant and retain partial protein function	Peptides frequently need to be tagged to visualize binding

[a] IHC, immunohistochemistry; 2DGE, two-dimensional gel electrophoresis; ICAT, isotope-coded affinity tags; iTRAQ, isobaric tags for relative and absolute quantitation; MALDI-IMS, matrix-assisted laser desorption/ionization–imaging mass spectrometry; SELDI, surface-enhanced laser desorption/ionization.

Proteins are also identified using database searches based on comparisons between the experimentally observed ion fragments and all fragments predicted for all hypothetical peptides of the appropriate molecular mass, based on known protein fragmentation rules. Each peptide match can be linked to a protein match. Although it is possible that a single peptide will identify a protein correctly, identical sequences might be present in closely related proteins. Therefore, matching multiple peptide sequences will provide greater statistical confidence that a protein has been identified correctly.[89]

Several search engines are available for database searching. A number of Web sites allow free access to Web-based database search programs for peptide mass fingerprinting. These include the ExPASy proteomics server (http://ca.expasy.org/), Mascot (http://www.matrixscience.com/search_form_select.html), and Protein Prospector (http://prospector.ucsf. edu/prospector/mshome.htm). Alternatively, commercially available software may be obtained from MS instrument manufacturers such as Sequest (Thermo Fisher Scientific, Waltham, MA) and Spectrum Mill (Agilent Technologies, Santa Clara, CA).[84,86,89]

Validation of Proteomic Data

Because protein identification is based on the probability of a match between a spectrum and a peptide, the proteins identified in proteomics experiments need to be validated. The most widely used methods for protein validation are western blotting, ELISA, and immunohistochemical staining. The major difficulties associated with these approaches is the scarcity of specific antibodies against novel proteins and trouble in detecting changes in proteins and peptides of low abundance.[85]

EXAMPLES OF PROTEOMIC TECHNOLOGY AS APPLIED IN NEUROPATHOLOGY

Proteomics analysis is used increasingly in experimental neurobiology, including assessment of both basic research questions and applied evaluation of neurological diseases. It is a valuable approach to examining protein expression and identifying new protein targets in development and disease, leading to a comprehensive understanding of complex biological processes. For example, the complex signaling pathways that control neural development have been studied by 2DGE and MALDI-TOF-MS to compare protein expression in the hippocampus of rats at different developmental time points (three days, three weeks, and three months after birth). Annexin A3, GTP-binding nuclear protein RAN, phosphatidylethanolamine-binding protein, adenylyl cyclase–associated protein 1, rho-associated protein kinase 1, nucleoside diphosphate kinase A, LIM, and SH3 domain protein 1 were discovered to be regulated developmentally.[102] The differential protein expression was confirmed using western blot analysis.

Proteomic techniques are particularly important in the investigation of neurodegenerative conditions. In Alzheimer's disease (AD), proteomics techniques have been used to identify the components present in senile plaques. High-resolution liquid chromatography coupled with MS/MS has identified 81 proteins in the detergent-insoluble fraction of frontal cortex of AD patients that display significant changes relative to control in brains. The presence of well-established AD-related proteins were identified including β-amyloid, tau, and apolipoprotein E, as well as new proteins such as protease 15, ankyrin B, and 14-3-3.[103] Subsequently, the results were validated using western blot analysis and immunohistochemistry. Similarly, extensive proteomic evaluation of brains from patients with Parkinson's disease revealed 1263 proteins that were differentially expressed in the substantia nigra pars compacta compared to controls.[104] The proteins were identified and quantified by labeling samples with iTRAQ or ICAT and analyzing them using MALDI-TOF-TOF and linear trap quadrupole MS.

As with genomic techniques, proteomic methods are commonly used to evaluate neurovirulent microbes. For example, the pathogenic mechanism by which the rabies virus results in neuronal dysfunction and death has been studied using 2DGE, MALDI-TOF, and peptide mass fingerprinting. The brains of mice infected with wild-type or attenuated rabies virus were examined to profile the neural response to viral infection. The wild-type virus altered the expression of proteins involved in ion homeostasis, reduced intracellular concentrations of sodium and calcium, and downregulated proteins involved in the docking and fusion of synaptic vesicles to the presynaptic membrane.[105] Infection with the attenuated virus resulted in increased expression of proteins involved in apoptosis.[105]

Proteomic technology will be a critical platform for elucidating the pathogenesis of signaling pathways involved in neurocarcinogenesis. A combination of 2DGE and MALDI-TOF has been used to compare protein expression in glioblastoma samples from patients with long-term vs. short-term survival. The results, subsequently confirmed using RT-PCR, western blot analysis, and immunohistochemical staining, suggested that decreased expression of manganese superoxide dismutase in patients with long-term survival represents a candidate marker for predicting glioblastoma prognosis.[106] In an additional study, similar proteomic techniques examined the intratumoral histological heterogeneity of glioblastoma, an increase in which is associated with intractable disease. High expression of ubiquitin carboxyl-terminal esterase L1 was evident in regions with tumor cells having a low histological grade (i.e., more differentiated), while high expression of transthyretin occurred in areas with a high histological grade.[107] The identification of such differentially expressed proteins in histologically heterogeneous areas within glioblastoma may provide potential therapeutic targets or further improve prognostic predictions.

Proteomics technology is also an important platform for neuropharmacological and neurotoxicological research. Analysis of protein expression in the nucleus accumbens of rats with different vulnerabilities to cocaine addiction using 2DGE followed by MALDI-TOF has demonstrated differential expression of five proteins that differ with the individual sensitivity to cocaine addiction (ATP synthase subunit α, fumarate hydratase, transketolase, NADH dehdrogenase flavoprotein 2, and glutathione transferase omega-1).[108] These molecules might be potential biomarkers for differential sensitivity to drug abuse among humans. Postnatal exposure to methylazoxymethanol (MAM), a widely used genotoxic developmental neurotoxicant, has been shown by standard neuropathology methods to induce cerebellar microencephaly. Fluorescence 2-DGE and MALDI-TOF of cerebellar samples taken 24 h or 19 days after exposure on postnatal day 3 identified altered expression of cytoskeletal, cell cycle and differentiation, ion binding, transport, chaperone, and metabolic proteins.[109] The MAM exposure also altered the expression of structural proteins that contribute to the maturation of neurons and glia.

CORRELATION OF GENOMIC AND PROTEOMIC DATA WITH BIOLOGICAL FUNCTIONS AND CONVENTIONAL NEUROPATHOLOGY ANALYSIS

Bioinformatics Databases

Genomic and proteomic experiments generate large amounts of biological data. These large data sets must be further investigated to elucidate the relevant pathways and networks

TABLE 5 Biological Databases for Genomic Data Integration

Database	Example[a]	Website
Genome-level	Ensembl	http://www.ensembl.org/index.html
	Entrez Genome	http://www.ncbi.nlm.nih.gov/sites/entrez?db=genome
	UCSC Genome Browser	http://genome.ucsc.edu
Sequence-level	GenBank	http://www.ncbi.nlm.nih.gov/Genbank
	UniGene	http://www.ncbi.nlm.nih.gov/unigene
	RefSeq	http://www.ncbi.nlm.nih.gov/RefSeq
Annotation	Entrez Gene	http://www.ncbi.nlm.nih.gov/gene
	OMIM	http://www.ncbi.nlm.nih.gov/omim
	GO	http://www.geneontology.org
Microarray	LIMS - dbZach,	http://dbzach.fst.msu.edu
	EDGE	http://mddprod.niddk.nih.gov:7777/EDGE
	Repositories - CEBS,	http://cebs.niehs.nih.gov
	GEO	http://www.ncbi.nlm.nih.gov/geo
Protein-level	UniProt	http://www.uniprot.org/
	RefSeq	http://www.ncbi.nlm.nih.gov/RefSeq
Protein interaction	BIND	http://binddb.org
	MINT	http://mint.bio.uniroma2.it/mint/Welcome.do
	DIP	http://dip.doe-mbi.ucla.edu/dip/Main.cgi
Digital Atlas	AGEA	http://www.brain-map.org/agea
	GENSAT	http://www.ncbi.nlm.nih.gov/projects/gensat
	EMAP	http://genex.hgu.mrc.ac.uk
	Gene Paint	http://genepaint.org
	Gene Atlas	http://geneatlas.org
Integration of genomic databases	Ingenuity	http://www.ingenuity.com
	KEGG	http://www.genome.jp/kegg
	Gene Ontology	http://www.geneontology.org
	EASE	http://david.abcc.ncifcrf.gov/ease/ease1.htm
	NextBio	http://www.nextbio.com

[a] AGEA, anatomic gene expression atlas; BIND, biomolecular interaction network data; CEBS, chemical effects in biological systems; DIP, database of interacting proteins; EDGE, effectual database for gene expression; EMAP, Edinburgh mouse atlas project; GENSAT, gene expression nervous system atlas; GEO, gene expression omnibus; GO, gene ontology; LIMS, local laboratory information management systems; MINT, molecular interaction; OMIM, online Mendelian inheritance in man; RefSeq, reference sequence; UCSC, University of California–Santa Cruz; UniProt, universal protein resource.

and to determine the functional significance of the data. This data mining uses bioinformatics tools to define the genes and proteins identified in the experiments.[110–114] Mining is performed in biology-oriented databases, which are organized in one of six forms (Table 5).[112]

Genome-Level Databases

The first type of database is the genome-level database, which manages and catalogs genome sequence data with respect to the sequence of the full genome. Ensembl (http://www. ensembl.org/index.html) is a joint project between the European Bioinformatics Institute and the Wellcome Trust Sanger Institute that aims to provide a centralized resource for genomes of vertebrates and other eukaryotic species.[112,115]

The Entrez Genome database (http://www.ncbi.nlm.nih. gov/sites/entrez?db=genome) is a searchable collection of complete and incomplete large-scale sequencing, assembly, annotation, and mapping projects for cellular organisms and includes complete chromosomes, organelles, and plasmids as

well as draft genome assemblies.[112] The University of California Santa Cruz Genome Browser (http://genome. ucsc.edu) contains reference sequence and working draft assemblies for a large collection of genomes integrated with a large collection of annotations.[112,116]

Sequence-Level Databases

The second category of bioinformatic databases is sequence-level databases, which are designed to organize genomic information acquired from analysis of expressed sequence tags (ESTs) and cDNA. These databases provide the first level of annotation for microarray data.[112] GenBank (http:// www.ncbi.nlm.nih.gov/Genbank/) is an open-access, annotated collection of all publicly available nucleotide sequences and their protein translations. This database is produced at the National Center for Biotechnology Information (NCBI) as part of the International Nucleotide Sequence Database Collaboration (INSDC). GenBank and its collaborators have received sequences produced in laboratories throughout the world for more than 100,000 organisms.[112,117] UniGene (http://www.ncbi.nlm.nih.gov/unigene) is an NCBI database system for automatically partitioning GenBank sequences into a nonredundant set of gene-oriented clusters, each of which contains sequences that represent a unique gene, information on the tissues in which the gene is expressed, and map location.[112] The Reference Sequence (RefSeq) database (http://www.ncbi.nlm.nih.gov/RefSeq/) is a nonredundant collection of annotated DNA, RNA, and protein sequences from diverse organisms, including plasmids, organelles, viruses, archaea, bacteria, and eukaryotes. Each sequence represents a single naturally occurring molecule from one organism. The goal of RefSeq is to provide a comprehensive, standardized data set that represents sequence information for a species. RefSeq biological sequences are derived from GenBank records but differ in that each sequence is a synthesis of information, not an archived unit of primary research data.[112,118]

Annotation Databases

The third class of bioinformatic database is the annotation database, which provides functional information for genes and their products and also catalogs the structure of genes. This information serves as the initial point for interpretation of microarray data. The major annotation databases are Entrez Gene (described above), which holds the curated and automatically integrated data from NCBI's Reference Sequence (RefSeq) project[112,119] and Online Mendelian Inheritance in Man (OMIM) (http://www.ncbi.nlm.nih.gov/omim/). The OMIM database was created in 1985 as a collaboration between the National Library of Medicine (NLM) and the William H. Welch Medical Library at Johns Hopkins. It is a compendium of human genes and genetic phenotypes. The full-text, referenced overviews in OMIM contain information on all known Mendelian disorders and over 12,000 genes. This database focuses on the relationship between phenotype and genotype.[112,120]

Microarray Databases

The fourth type of bioinformatic database is the microarray database, which is designed to ensure that data obtained from microarray experiments are managed properly. Such databases support the analysis, long-term archiving, and public sharing of information. The Minimum Information About a Microarray Experiment (MIAME) standards provide guidance on the types of supporting information (e.g., the clones, genes, protocols, and sample characteristics) that must be captured and reported in support of a microarray study to ensure that independent investigators can replicate and interpret the data properly. Microarray databases include local laboratory information management systems (LIMSs)

and data repositories. An example of a LIMS is dbZach (http://dbzach.fst.msu.edu), a modular relational database with associated insertion, retrieval, and mining tools to facilitate management, integration, analysis, and sharing of data. The database consists of four core subsystems (clones, genes, sample annotation, protocols); four experimental subsystems (microarray, affymetrix, real-time PCR, toxicology); and three computational subsystems (gene regulation, pathways, orthology) and complies with the MIAME standard.[112,121] The two main data repositories are the Gene Expression Omnibus (GEO) database (http://www.ncbi.nlm.nih.gov/geo/) and the Chemical Effects in Biological Systems (CEBS) database (http://cebs.niehs.nih.gov). GEO is a public repository that archives and distributes high-throughput gene expression data submitted by the scientific community; it holds a billion individual gene expression measurements from over 100 different organisms. The CEBS database is a public archive of data from academic, industrial, and government laboratories derived from multiple gene expression platforms. CEBS contains data on protein expression and protein–protein interaction as well as changes in the levels of low-molecular-weight metabolites aligned by their detailed toxicological context. CEBS integrates study design (treatment parameters and time course), clinical pathology, and histopathology data from all studies and enables discrimination of critical study factors.[112,122]

Protein-Level Databases

The fifth type of bioinformatics database is the protein-level database, which provides annotated lists of proteins linked to given genes of interest. The lists provide information on protein sequences, families, and domain structures. The major protein-level database is the Universal Protein Resource (UniProt) (http://www.uniprot.org/).[112] This database of nonredundant protein sequences has been assembled from the translation of sequences within gene sequence databases. The mission of UniProt is to provide a comprehensive, high-quality, and freely accessible resource of protein sequence and functional information. It is a central repository of protein data created by combining the Swiss-Prot (http://www.ebi.ac.uk/swissprot/), Translated European Molecular Biology Laboratory (TrEMBL) (http://www.genome.jp/dbget-bin/www_bfind?trembl), and Protein Information Resource (PIR) databases (http://pir.georgetown.edu/).[112,123]

Protein Interaction Databases

The final type of bioinformatics database is the protein interaction database, which provides information on the interaction of proteins with other proteins, genes, chemicals, and small molecules. The Biomolecular Interaction Network Data (BIND; http://binddb.org) database is designed to store full descriptions of interactions between any two molecules, molecular complexes, or pathways; it also holds information on chemical reactions, photochemical activation, and conformational changes. The data can be used to study networks of interactions, to map pathways across taxonomic branches, and to generate information for kinetic simulations.[112,124] The Database of Interacting Proteins (DIP; http://dip. doe-mbi.ucla.edu) documents experimentally determined protein–protein interactions and provides an integrated set of tools for browsing and extracting information about protein interaction networks from more than 80 organisms. It is useful for understanding protein function and protein–protein relationships, studying the properties and evolution of networks of interacting proteins, and benchmarking predictions of protein–protein interactions.[112,125] The Molecular Interaction (MINT) database (http://cbm.bio.uniroma2.it/mint/index. html) focuses on experimentally verified protein–protein interactions mined from the scientific literature, emphasizing data on functional interactions. MINT contains descriptive information about interactions and, whenever available, information about kinetic and binding constants and the specific domains participating in the interactions.[112,126]

PROGRAMS FOR INTEGRATING "OMICS" DATABASES

The experimental data obtained from genomic and proteomic experiments must be processed further to determine the biological significance of the data. To make the best use of the biological databases, information from different sources must be integrated and annotated. A number of programs that perform this task for data held in publicly available databases have been developed.

Ingenuity Pathway Analysis

Ingenuity pathway analysis (IPA) (Ingenuity Systems, Inc., Redwood City, CA) (http://www.ingenuity.com/) enables researchers to identify biological mechanisms, pathways, and functions most relevant to their experimental data sets or genes of interest. It is useful for understanding differential gene expression and protein–protein interactions within the context of metabolic and signaling pathways, for understanding how proteins operate and form pathways, and for organizing members of gene families. IPA contains molecular and biological data from *Homo sapiens*, *Mus musculus*, and *Rattus norvegicus*. The molecular biology data are compiled from databases such as OMIM, MGI, and NCBI Gene.[127]

Kyoto Encyclopedia of Genes and Genomes

The Kyoto encyclopedia of genes and genomes (KEGG) (http://www.genome.jp/kegg/) collection of online databases contains a vast array of data associated with pathways, genes, genomes, chemical compounds, and reactions. In KEGG, the genome is a graph of genes that are connected in one dimension, and the pathway is a graph of gene products with more complicated patterns of interconnection. By matching genes in the genome and gene products in the pathway, KEGG can be utilized to predict protein interaction networks and associated cellular functions. The pathways in KEGG are drawn manually and derived from textbooks, literature, and expert knowledge. Genomic information is derived from publicly available resources such as RefSeq.[114,128]

The sixteen databases in the KEGG collection are compiled into three categories: systems information, genomic information, and chemical information. The genomic and chemical information categories represent the molecular building blocks of life, while the systems information category represents functional aspects of the biological systems. The systems information category contains databases such as KEGG Pathway, KEGG BRITE, and KEGG Genes, the genomic information category contains databases such as KEGG SSDB, and the chemical information category contains databases such as KEGG Enzyme. KEGG Pathway contains information regarding generalized protein interaction networks. KEGG Genes catalogs completely sequenced genomes and partial genomes. KEGG SSDB coordinates ortholog and paralog relations for all protein-coding genes in complete genomes. KEGG BRITE is a hierarchical classification of genes, proteins, compounds, reactions, drugs, diseases, cells, and organisms.[114,128]

Gene Ontology

Gene ontology (GO) (http://www.geneontology.org/) is a major informatics initiative that unifies the representation of gene and gene product attributes across all species. The aims of the project are to develop and maintain a vocabulary suitable for describing such attributes, to annotate genes and gene products in biological databases, and to provide a centralized public resource to allow universal access to the information in these entities.[111,129] The GO project has three divisions: Molecular Function, which describes activities at the molecular level, such as catalytic or binding activities: Biological Process, which relates the biological objective in which a gene product participates, such as cell communication, behavior, or development; and Cellular Component, which refers to the location at which the gene product acts.[111,129]

Expression Analysis Systematic Explorer

The Expression Analysis Systematic Explorer [(EASE) EASE Software, Inc., Portland, OR] (http://david.abcc.ncifcrf.gov/ ease/ease1.htm) is an easy-to-use customizable desktop application that facilitates the biological interpretation of gene lists derived from the results of genomic and proteomic experiments. It was developed to automate the process of biological theme determination for differentiated lists of genes, such as up- vs. downregulated genes in a single experiment, and to serve as a customizable gateway to online analysis tools.[130] EASE rapidly converts a gene list into an ordered table of robust themes that summarize the biological result of the experiment.

Genomic and proteomic experiments generate vast amounts of biological data. Bioinformatics databases that hold all the molecular data are powerful tools for correlating genomic and proteomic information with normal and pathological processes in biology. They have been particularly critical instruments in speeding the discovery and characterization of genes and proteins previously unknown to be involved with neurobiological processes.

NextBio

NextBio is a powerful web-based interface (www.nextbio. com) that has a large archive of literature and microarray data sets derived from repositories in public domain. High throughput genomic data from diverse experiments can be imported into NextBio and context-specific queries can be performed on the data in public domain. It can be used to perform meta-analysis of genomic and proteomic data from diverse biological and clinical contexts across 6 species. Also, it may be used to discover new genetic signatures and biomarkers from genomic data (of the researcher as well as the data in public domain) derived from various biological states. However, NextBio's methods of data normalization and algorithms for searching and ranking are proprietary and that should be kept in mind when using this software.

ANATOMICAL CORRELATION OF GENE AND PROTEIN EXPRESSION DATA WITHIN THE BRAIN

The brain is a very complex organ with an astounding diversity at the functional, molecular, and structural levels. Extensive differences in gross, microscopic, and ultrastructural features exist in all three dimensions, across very short distances, within this organ. In addition, these extensive differences were also demonstrated at the gene expression level in the brain.[131] The complexity of the brain is due to large cell populations (10^{11} neurons), the approximately 1000 different cell phenotypes, and the complex synaptic interconnectivity and neural circuitry.[132] Neurons have traditionally been classified based on morphology, neurotransmitter phenotype, electrophysiological properties, and gene expression. However, none of these classifications is complete. For example, attempts to classify forebrain neuronal populations using differential gene expression data revealed

several important findings.[133] First, each of 12 different neuronal types express up to 50% of the genes within the entire forebrain array, indicating that the expression of many genes is common to those of closely related cells. Second, several genes were expressed differentially in each of the neuronal populations profiled, thereby helping to explain the considerable phenotypic divergence that can be encoded by a limited set of genes. Next, greater genetic diversity exists between different GABAergic interneurons than occurs between glutaminergic projection neurons, thus indicating that the degree to which members of a given cell type are related varies substantially, due to many factors. Finally, no single gene or limited gene set can be used to classify neuronal populations satisfactorily. These results underscore the great need to continue mapping and refining our knowledge of the gene expression patterns, with cellular-level resolution, within the brain.

The advent of high-throughput genomic and proteomic technologies has generated vast stores of molecular data related to fundamental neurobiological and neuropathological questions. Neuroscientists are therefore faced with the task of correlating genomic and proteomic data with the complex neuroanatomical structures as well as the dynamic functional phenotypes of the brain. Fortunately, a few online databases (digital atlases) catalog the extensive gene expression data within the context of regional gross and cellular neuroanatomy (i.e., a spatially mapped transcriptome of the brain). These digital atlases provide interactive graphic user interface (GUI) tools, and users can correlate their own data with the background spatial gene expression data from these digital atlases. For example, if a researcher grossly dissects the cerebellum or the hippocampus and derive differential gene expression data from those samples, he/she can correlate his/her data in the context of exact neuroanatomical regions and their corresponding spatially mapped gene expression data cataloged within these digital atlases. Popular digital atlases linking molecular and structural data include the Anatomic Gene Expression Atlas (AGEA; http://www. brain-map.org/agea), and Gene Expression Nervous System Atlas (GENSAT; http://www.ncbi.nlm.nih.gov/projects/gensat/), Edinburgh Mouse Atlas Project (EMAP; http://genex. hgu.mrc.ac.uk), Gene Paint (http://www.genepaint.org) and Gene Atlas (http://geneatlas.org). Most of these atlases are based on high-throughput automated ISH (radioactive or colorimetric) data acquired from thousands of slides (ranging up to a million) labeled to detect thousands of unique probes. Some of these Web sites also provide reference brain atlases which contain region-specific neural diagrams and corresponding coronal and sagittal sections stained with any of numerous neurohistological stains to show specific architectural features (e.g., Nissl stain), cell-type-specific proteins (e.g., myelin, neurofilament protein, parvalbumin), or functional attributes (e.g., acetylcholinesterase, cytochrome oxidase). Some of the atlases also provide images acquired using other platforms, such as electron microscopy and magnetic resonance imaging.

The Allen Brain Institute's AGEA digital atlas maps the brain transcriptome (21,500 genes) of a 56-day-old (young adult) male C57Bl6/6J mouse. The colorimetric ISH-derived expression of each gene is aligned in three dimensions to neuroanatomic structures in the Nissl-based reference atlas and illustrated with the stand-alone, online Brain Explorer application. The voxel mapping resolution of AGEA is 200 µm per side, which may potentially limit the spatial resolution required to distinguish small nuclei. Nevertheless, this important tool resolves details of differential gene expression in several brain regions. The AGEA consists of three components: browser-accessible correlation maps in three dimensions based on average gene expression profiles; a hierarchical, transcriptome-based and navigable spatial ontology of the mouse brain; and a tool to allow users to retrieve gene lists that exhibit localized enrichment and directly access the high-resolution ISH data.[134] Using the AGEA brain transcriptome, about 80% of genes have been shown to be expressed above the background levels, which is a higher extent than would be predicted from microarray analysis of the same genes.[135] Gene expression data mapped to specific neuroanatomical structures in AGEA are able to delineate different classes of neurons and glia as well as to discriminate the laminar layers in the cortex, even though very few genes are expressed uniquely within a single neuroanatomic region.[136]

The GENSAT online digital atlas maps differential gene expression within the mouse central nervous system using radioactive isotopic ISH and transgenic techniques. It is funded by the U.S. National Institute of Neurological Disorders and Stroke (NINDS) and is a collaborative project between St. Jude's Children's Research Hospital [Mouse Brain Gene Expression Map (BGEM)] and Rockefeller University (GENSAT). The BGEM measures gene expression in dark-field images (with reference serial sections stained with cresyl violet) from developing and adult C57Bl6/6J mice using 2400 expert-selected unique probes hybridized to more than 129,000 brain sections. About 65% of these genes are temporally and/or spatially restricted during brain development.[137] The GENSAT takes candidate genes from BGEM and generates transgenic mouse lines that incorporate bacterial artificial chromosome (BAC)-enhanced green fluorescent protein (GFP) reporter and BAC-Cre recombinase elements to permit more efficient gene characterization. The BGEM and GENSAT databases are complementary since users can view gene expression patterns in BGEM and then compare them with GFP fluorescent images in GENSAT.[137] The BAC reporter mice can be requested from the mutant mouse regional centers through the linked Web sites (Table 5).

The major online digital atlases are constantly updated to include more genes, additional ages of mice, and other regions within the central nervous system, as well as to upgrade the tools to provide an improved user interface. The data within the online digital atlases are often complementary, so researchers may substantially speed their investigations by comparing their respective gene lists with multiple digital atlases to gain maximum insight into the neuroanatomical relationships between the expression of various genes and proteins.

These high-throughput genomic technologies have helped enormously in elucidating many complex physiological and pathological processes within the nervous system. In addition, the enormous amounts of molecular data generated from these studies also provide unanticipated supplementary information that forms the basis for new hypotheses that need to be tested and validated by complementary methods as well as in vivo functional studies. With the further evolution of high-throughput genomic technologies and innovative imaging methods, twentieth-first-century neuropathologists and neurotoxicologists can be expected to aid in interdisciplinary efforts to correlate data acquired from routine gross and microscopic neuromorphological methods with the molecular events that drive them.

Acknowledgments

We would like to thank Pierre Bushell, Mark Hoenerhoff, Kevin Gerrish, Keith Shockley, Ken Tomer, and Pat Stockton for their comments on various sections of this chapter.

REFERENCES

1. Siwoski A, et al. An efficient method for the assessment of DNA quality of archival microdissected specimens. *Mod Pathol.* 2002;15:889–892.

2. Ladd-Acosta C, et al. DNA methylation signatures within the human brain. *Am J Hum Genet.* 2007;81:1304–1315.

3. Rath PC, Kanungo MS. Methylation of repetitive DNA sequences in the brain during aging of the rat. *FEBS Lett.* 1989;244:193–198.

4. Ernst C, et al. The effects of pH on DNA methylation state: In vitro and post-mortem brain studies. *J Neurosci Methods.* 2008;174:123–125.

5. Huang HS, et al. Chromatin immunoprecipitation in postmortem brain. *J Neurosci Methods.* 2006;156:284–292.

6. Harrison PJ, et al. The relative importance of premortem acidosis and postmortem interval for human brain gene expression studies: selective mRNA vulnerability and comparison with their encoded proteins. *Neurosci Lett.* 1995;200:151–154.

7. Goldmann T, et al. The HOPE-technique permits Northern blot and microarray analyses in paraffin-embedded tissues. *Pathol Res Pract.* 2004;200:511–515.

8. Olert J, et al. HOPE fixation: a novel fixing method and paraffin-embedding technique for human soft tissues. *Pathol Res Pract.* 2001;197:823–826.

9. Wiedorn KH, et al. HOPE: a new fixing technique enables preservation and extraction of high molecular weight DNA and RNA of >20 kb from paraffin-embedded tissues: HEPES–glutamic acid buffer mediated organic solvent protection effect. *Pathol Res Pract.* 2002;198:735–740.

10. Schroeder A, et al. The RIN: an RNA integrity number for assigning integrity values to RNA measurements. *BMC Mol Biol.* 2006;7:3.

11. Ferrer I, et al. Brain banks: benefits, limitations and cautions concerning the use of post-mortem brain tissue for molecular studies. *Cell Tissue Bank.* 2008;9:181–194.

12. Kingsbury AE, et al. Tissue pH as an indicator of mRNA preservation in human post-mortem brain. *Brain Res Mol Brain Res.* 1995;28:311–318.

13. Hebert SS, De Strooper B. Alterations of the microRNA network cause neurodegenerative disease. *Trends Neurosci.* 2009;32:199–206.

14. Sethi P, Lukiw WJ. Micro-RNA abundance and stability in human brain: specific alterations in Alzheimer's disease temporal lobe neocortex. *Neurosci Lett.* 2009;459:100–104.

15. Kretzschmar H. Brain banking: opportunities, challenges and meaning for the future. *Nat Rev Neurosci.* 2009;10:70–78.

16. Compaine A, et al. Limited proteolytic processing of the mature form of cathepsin D in human and mouse brain: postmortem stability of enzyme structure and activity. *Neurochem Int.* 1995;27:385–396.

17. Ferrer I, et al. Brain protein preservation largely depends on the postmortem storage temperature: implications for study of proteins in human neurologic diseases and management of brain banks: a BrainNet Europe Study. *J Neuropathol Exp Neurol.* 2007;66:35–46.

18. Auer RN, et al. Hypoxia and related conditions. In: Love S, et al., eds. *Greenfield's Neuropathology.* 8th ed. London: Hodder-Arnold; 2008.

19. Nohda K, et al. Selective vulnerability to ischemia in the rat spinal cord: a comparison between ventral and dorsal horn neurons. *Spine (Phila Pa 1976).* 2007;32:1060–1066.

20. Yang HJ, et al. Region-specific susceptibilities to cuprizone-induced lesions in the mouse forebrain: implications for the pathophysiology of schizophrenia. *Brain Res.* 2009;1270: 121–130.

21. Morrison JP, et al. Gene expression studies reveal that DNA damage, vascular perturbation, and inflammation contribute to the pathogenesis of carbonyl sulfide neurotoxicity. *Toxicol Pathol.* 2009;37:502–511.

22. Paxinos G, et al. Bregma, lambda and the interaural midpoint in stereotaxic surgery with rats of different sex, strain and weight. *J Neurosci Methods.* 1985;13:139–143.

23. Nygaard V, Hovig E. Options available for profiling small samples: a review of sample amplification technology when combined with microarray profiling. *Nucleic Acids Res.* 2006;34:996–1014.

24. Van Gelder RN, von Zastrow ME, Yool A, Dement WC, Barchas JD, et al. Amplified RNA synthesized from limited quantities of heterogeneous cDNA. *Proc Natl Acad Sci USA.* 1990;87:1663–1667.

25. Wang E, et al. High-fidelity mRNA amplification for gene profiling. *Nat Biotechnol.* 2000;18:457–459.

26. Baugh LR, et al. Quantitative analysis of mRNA amplification by in vitro transcription. *Nucleic Acids Res.* 2001;29: E29.

27. Boelens MC, et al. Microarray amplification bias: loss of 30% differentially expressed genes due to long probe–poly(A)-tail distances. *BMC Genomics.* 2007;8:277.

28. Sharma AK, et al. Kainic acid-induced F-344 rat model of mesial temporal lobe epilepsy: gene expression and canonical pathways. *Toxicol Pathol.* 2009;37:776–789.

29. Emmert-Buck MR, et al. Laser capture microdissection. *Science.* 1996;274:998–1001.

30. Bernard R, et al. Gene expression profiling of neurochemically defined regions of the human brain by in situ hybridization-guided laser capture microdissection. *J Neurosci Methods.* 2009;178:46–54.

31. Espina V, et al. Laser-capture microdissection. *Nat Protoc.* 2006;1:586–603.

32. Fend F, et al. Immuno-LCM: laser capture microdissection of immunostained frozen sections for mRNA analysis. *Am J Pathol.* 1999;154:61–66.

33. Greene JG, et al. Quantitative transcriptional neuroanatomy of the rat hippocampus: evidence for wide-ranging, pathway-specific heterogeneity among three principal cell layers. *Hippocampus.* 2009;19:253–264.

34. Mathern GW, et al. The clinical–pathogenic mechanisms of hippocampal neuron loss and surgical outcomes in temporal lobe epilepsy. *Brain.* 1995;118(1):105–118.

35. Ordy JM, et al. Selective vulnerability and early progression of hippocampal CA1 pyramidal cell degeneration and GFAP-positive astrocyte reactivity in the rat four-vessel occlusion model of transient global ischemia. *Exp Neurol.* 1993;119: 128–139.

36. Ginsberg GG. The art and science of painting in early gastric cancer: Is there a role for ablation therapy? *Gastrointest Endosc.* 2006;63:55–59.

37. Kamme F, et al. Single-cell microarray analysis in hippocampus CA1: demonstration and validation of cellular heterogeneity. *J Neurosci.* 2003;23:3607–3615.

38. Namimatsu S, et al. Reversing the effects of formalin fixation with citraconic anhydride and heat: a universal antigen retrieval method. *J Histochem Cytochem.* 2005;53:3–11.

39. Xiang CC, et al. Using DSP, a reversible cross-linker, to fix tissue sections for immunostaining, microdissection and expression profiling. *Nucleic Acids Res.* 2004;32:e185.

40. Mouledous L, et al. Navigated laser capture microdissection as an alternative to direct histological staining for proteomic analysis of brain samples. *Proteomics.* 2003;3:610–615.

41. Rossner MJ, et al. Global transcriptome analysis of genetically identified neurons in the adult cortex. *J Neurosci.* 2006;26: 9956–9966.

42. Imbeaud S, Auffray C. "The 39 steps" in gene expression profiling: critical issues and proposed best practices for microarray experiments. *Drug Discov Today.* 2005;10:1175–1182.

43. Chee M, et al. Accessing genetic information with high-density DNA arrays. *Science.* 1996;274:610–614.

44. Schena M, et al. Quantitative monitoring of gene expression patterns with a complementary DNA microarray. *Science.* 1995;270:467–470.

45. Patterson TA, et al. Performance comparison of one-color and two-color platforms within the MicroArray Quality Control (MAQC) project. *Nat Biotechnol.* 2006;24:1140–1150.

46. Altar CA, Vawter MP, Ginsberg SD. Target identification for CNS diseases by transcriptional profiling. *Neuropsychopharmacology.* 2009;34:18–54.

47. Gunderson KL, et al. Decoding randomly ordered DNA arrays. *Genome Res.* 2004;14:870–877.

48. Elvidge G. Microarray expression technology: from start to finish. *Pharmacogenomics.* 2006;7:123–134.

49. Yang YH, Speed T. Design issues for cDNA microarray experiments. *Nat Rev Genet.* 2002;3:579–588.

50. Churchill GA. Fundamentals of experimental design for cDNA microarrays. *Nat Genet.* 2002;32(1):490–495.

51. Lee ML, et al. Importance of replication in microarray gene expression studies: statistical methods and evidence from repetitive cDNA hybridizations. *Proc Natl Acad Sci USA.* 2000;97:9834–9839.

52. Pavlidis P, et al. The effect of replication on gene expression microarray experiments. *Bioinformatics.* 2003;19:1620–1627.

53. Mirnics K, et al. Molecular characterization of schizophrenia viewed by microarray analysis of gene expression in prefrontal cortex. *Neuron.* 2000;28:53–67.

54. Jolly RA, et al. Pooling samples within microarray studies: a comparative analysis of rat liver transcription response to prototypical toxicants. *Physiol Genom.* 2005;22:346–355.

55. Kendziorski C, et al. On the utility of pooling biological samples in microarray experiments. *Proc Natl Acad Sci USA.* 2005;102:4252–4257.

56. Kerr MK, Churchill GA. Experimental design for gene expression microarrays. *Biostatistics.* 2001;2:183–201.

57. Leung YF, Cavalieri D. Fundamentals of cDNA microarray data analysis. *Trends Genet.* 2003;19:649–659.

58. Yang YH, et al. Comparison of methods for image analysis on cDNA microarray data. *J Comput Graph Stat.* 2002;11:108–136.

59. Sievertzon M, et al. Improving reliability and performance of DNA microarrays. *Expert Rev Mol Diagn.* 2006;6:481–492.

60. Li C, Wong WH. Model-based analysis of oligonucleotide arrays: expression index computation and outlier detection. *Proc Natl Acad Sci USA.* 2001;98:31–36.

61. Quackenbush J. Microarray data normalization and transformation. *Nat Genet.* 2002;32(1):496–501.

62. Bolstad BM, et al. A comparison of normalization methods for high density oligonucleotide array data based on variance and bias. *Bioinformatics.* 2003;19:185–193.

63. Pavlidis P. Using ANOVA for gene selection from microarray studies of the nervous system. *Methods.* 2003;31:282–289.

64. Hilsenbeck SG, et al. Statistical analysis of array expression data as applied to the problem of tamoxifen resistance. *J Natl Cancer Inst.* 1999;91:453–459.

65. Ma S, Kosorok MR. Identification of differential gene pathways with principal component analysis. *Bioinformatics.* 2009;25:882–889.

66. Shannon W, et al. Analyzing microarray data using cluster analysis. *Pharmacogenomics.* 2003;4:41–52.

67. Ringner M, et al. Analyzing array data using supervised methods. *Pharmacogenomics.* 2002;3:403–415.

68. Cui X, Churchill GA. Statistical tests for differential expression in cDNA microarray experiments. *Genome Biol.* 2003;4:210.

69. Brazma A, et al. Minimum information about a microarray experiment (MIAME)-toward standards for microarray data. *Nat Genet.* 2001;29:365–371.

70. Shi L, et al. The MicroArray Quality Control (MAQC) project shows inter- and intraplatform reproducibility of gene expression measurements. *Nat Biotechnol.* 2006;24:1151–1161.

71. Canales RD, et al. Evaluation of DNA microarray results with quantitative gene expression platforms. *Nat Biotechnol.* 2006;24:1115–1122.

72. Bachoo RM, et al. Molecular diversity of astrocytes with implications for neurological disorders. *Proc Natl Acad Sci USA.* 2004;101:8384–8389.

73. Yeh TH, et al. Microarray analyses reveal regional astrocyte heterogeneity with implications for neurofibromatosis type 1 (NF1)-regulated glial proliferation. *Glia.* 2009;57:1239–1249.

74. Ginsberg SD, et al. Expression profile of transcripts in Alzheimer's disease tangle-bearing CA1 neurons. *Ann Neurol.* 2000;48:77–87.

75. Small SA, et al. Model-guided microarray implicates the retromer complex in Alzheimer's disease. *Ann Neurol.* 2005;58:909–919.

76. Wang D, et al. Microarray-based detection and genotyping of viral pathogens. *Proc Natl Acad Sci USA.* 2002;99:15687–15692.

77. Gurrala R, et al. Development of a DNA microarray for simultaneous detection and genotyping of lyssaviruses. *Virus Res.* 2009;144:202–208.

78. Chin L, Gray JW. Translating insights from the cancer genome into clinical practice. *Nature.* 2008;452:553–563.

79. Lehrmann E, et al. Transcriptional changes common to human cocaine, cannabis and phencyclidine abuse. *PLoS One.* 2006;1:e114.

80. Laterra J, et al. Formation and differentiation of brain capillaries. *NIDA Res Monogr.* 1992;120:73–86.

81. Bouton CM, Hossain MA, Frelin LP, Laterra J, Pevsner J. Microarray analysis of differential gene expression in lead-exposed astrocytes. *Toxicol Appl Pharmacol.* 2001;176:34–53.

82. Kisby GE, et al. Genotoxicants target distinct molecular networks in neonatal neurons. *Environ Health Perspect.* 2006;114:1703–1712.

83. Bayes A, Grant SG. Neuroproteomics: understanding the molecular organization and complexity of the brain. *Nat Rev Neurosci.* 2009;10:635–646.

84. Han X, et al. Mass spectrometry for proteomics. *Curr Opin Chem Biol.* 2008;12:483–490.

85. Shi M, et al. Biomarker discovery in neurodegenerative diseases: a proteomic approach. *Neurobiol Dis.* 2009;35:157–164.

86. Walch A, et al. MALDI imaging mass spectrometry for direct tissue analysis: a new frontier for molecular histology. *Histochem Cell Biol.* 2008;130:421–434.

87. Yates JR, et al. Proteomics by mass spectrometry: approaches, advances, and applications. *Annu Rev Biomed Eng.* 2009;11: 49–79.

88. Netto GJ, Saad R. Diagnostic molecular pathology: 2. Proteomics and clinical applications of molecular diagnostics in hematopathology. *Proc (Bayl Univ Med Cent).* 2005;18:7–12.

89. Tannu NS, Hemby SE. Methods for proteomics in neuroscience. *Prog Brain Res.* 2006;158:41–82.

90. Domon B, Aebersold R. Mass spectrometry and protein analysis. *Science.* 2006;312:212–217.

91. Johnson MD, et al. Proteomics in diagnostic neuropathology. *J Neuropathol Exp Neurol.* 2006;65:837–845.

92. Cornett DS, et al. MALDI imaging mass spectrometry: molecular snapshots of biochemical systems. *Nat Methods.* 2007;4:828–833.

93. Steen H, Mann M. The ABC's (and XYZ's) of peptide sequencing. *Nat Rev Mol Cell Biol.* 2004;5:699–711.

94. Hall DA, et al. Protein microarray technology. *Mech Ageing Dev.* 2007;128:161–167.

95. Kerschgens J, et al. Protein-binding microarrays: probing disease markers at the interface of proteomics and genomics. *Trends Mol Med.* 2009;15:352–358.

96. Lv LL, Liu BC. High-throughput antibody microarrays for quantitative proteomic analysis. *Expert Rev Proteomics.* 2007;4:505–513.

97. Breitling F, et al. High-density peptide arrays. *Mol Biosyst.* 2009;5:224–234.

98. Tapia VE, et al. 2009 Exploring and profiling protein function with peptide arrays. *Methods Mol Biol.* 2009;570:3–17.

99. Uttamchandani M, Yao SQ. Peptide microarrays: next generation biochips for detection, diagnostics and high-throughput screening. *Curr Pharm Des.* 2008;14: 2428–2438.

100. Zhao Y, Jensen ON. Modification-specific proteomics: strategies for characterization of post-translational modifications using enrichment techniques. *Proteomics.* 2009;9:4632-4641.

101. Witze ES, et al. Mapping protein post-translational modifications with mass spectrometry. *Nat Methods.* 2007;4:798–806.

102. Weitzdorfer R, et al. Changes of hippocampal signaling protein levels during postnatal brain development in the rat. *Hippocampus.* 2008;18:807–813.

103. Gozal YM, et al. Proteomics analysis reveals novel components in the detergent-insoluble subproteome in Alzheimer's disease. *J Proteome Res.* 2009.

104. Jin J, Hulette C, Wang Y, Zhang T, Pan C, et al. Proteomic identification of a stress protein, mortalin/mthsp70/GRP75: relevance to Parkinson disease. *Mol Cell Proteom.* 2006;5: 1193–1204.

105. Dhingra V, et al. Proteomic profiling reveals that rabies virus infection results in differential expression of host proteins involved in ion homeostasis and synaptic physiology in the central nervous system. *J Neurovirol.* 2007;13:107–117.

106. Park CK, et al. Tissue expression of manganese superoxide dismutase is a candidate prognostic marker for glioblastoma. *Oncology.* 2009;77:178–181.

107. Park CK, et al. Multifarious proteomic signatures and regional heterogeneity in glioblastomas. *J Neurooncol.* 2009;94:31–39.

108. del Castillo C, et al. Proteomic analysis of the nucleus accumbens of rats with different vulnerability to cocaine addiction. *Neuropharmacology.* 2009;57:41–48.

109. Kisby GE, et al. Proteomic analysis of the genotoxicant methylazoxymethanol (MAM)-induced changes in the developing cerebellum. *J Proteome Res.* 2006;5: 2656–2665.

110. Baxevanis AD. The importance of biological databases in biological discovery. *Current Protocols in Bioinformatics.* 2009: Unit 1.1.

111. Blake JA, Harris MA. The Gene Ontology (GO) Project: structured vocabularies for molecular biology and their application to genome and expression analysis. In: *Current Protocols in Bioinformatics.* 2008: Unit 7.2.

112. Burgoon LD, Zacharewski TR. Bioinformatics: databasing and gene annotation. *Methods Mol Biol.* 2008;460: 145–157.

113. Ganter B, Giroux CN. Emerging applications of network and pathway analysis in drug discovery and development. *Curr Opin Drug Discov Dev.* 2008;11:86–94.

114. Kanehisa M, et al. The KEGG databases at GenomeNet. *Nucleic Acids Res.* 2002;30:42–46.

115. Hubbard TJ, et al. Ensembl 2009. *Nucleic Acids Res.* 2009;37: D690–D697.

116. Karolchik D, et al. The UCSC Genome Browser Database. *Nucleic Acids Res.* 2003;31:51–54.

117. Benson DA, et al. *GenBank. Nucleic Acids Res.* 2009;37: D26–D31.

118. Pruitt KD, et al. NCBI reference sequences: current status, policy and new initiatives. *Nucleic Acids Res.* 2009;37: D32–D36.

119. Maglott D, et al. Entrez Gene: gene-centered information at NCBI. *Nucleic Acids Res.* 2007;35:D26–D31.

120. Amberger J, et al. McKusick's online Mendelian inheritance in man (OMIM). *Nucleic Acids Res.* 2009;37: D793–D796.

121. Burgoon LD, Zacharewski TR. dbZach toxicogenomic information management system. *Pharmacogenomics.* 2007;8: 287–291.

122. Waters M, et al. CEBS—Chemical Effects in Biological Systems: a public data repository integrating study design and toxicity data with microarray and proteomics data. *Nucleic Acids Res.* 2008;36:D892–D900.

123. Bairoch A, et al. The Universal Protein Resource (UniProt). *Nucleic Acids Res.* 2005;33:D154–D159.

124. Bader GD, et al. BIND: the Biomolecular Interaction Network Database. *Nucleic Acids Res.* 2003;31:248–250.

125. Salwinski L, et al. The Database of Interacting Proteins: 2004 update. *Nucleic Acids Res.* 2004;32: D449–D451.

126. Chatr-aryamontri A, et al. MINT: the Molecular INTeraction database. *Nucleic Acids Res.* 2007;35:D572–D574.

127. Jimenez-Marin A, et al. Biological pathway analysis by ArrayUnlock and ingenuity pathway analysis. *BMC Proc.* 2009;3(4):S6.

128. Aoki KF, Kanehisa M. Using the KEGG database resource. In: *Current Protocols in Bioinformatics.* 2005: Unit 1.12.

129. Harris MA, et al. The Gene Ontology (GO) database and informatics resource. *Nucleic Acids Res.* 2004;32: D258–D261.

130. Hosack DA, et al. Identifying biological themes within lists of genes with EASE. *Genome Biol.* 2003;4:R70.

131. D'Souza CA, et al. Identification of a set of genes showing regionally enriched expression in the mouse brain. *BMC Neurosci.* 2008;9:66.

132. Hatten ME, Heintz N. Large-scale genomic approaches to brain development and circuitry. *Annu Rev Neurosci.* 2005;28:89–108.

133. Sugino K, et al. Molecular taxonomy of major neuronal classes in the adult mouse forebrain. *Nat Neurosci.* 2006;9:99–107.

134. Ng L, et al. An anatomic gene expression atlas of the adult mouse brain. *Nat Neurosci.* 2009;12:356–362.

135. Sandberg R, et al. Regional and strain-specific gene expression mapping in the adult mouse brain. *Proc Natl Acad Sci USA.* 2000;97:11038–11043.

136. Sunkin SM, Hohmann JG. Insights from spatially mapped gene expression in the mouse brain. *Hum Mol Genet.* 2007;6 (Spec No 2):R209–R219.

137. Magdaleno S, et al. BGEM: an in situ hybridization database of gene expression in the embryonic and adult mouse nervous system. *PLoS Biol.* 2006;4:e86.

PART 3

TOXICOLOGICAL NEUROPATHOLOGY: CURRENT PRACTICES

21

EVALUATION OF THE ADULT NERVOUS SYSTEM IN PRECLINICAL STUDIES

Tox Path Specialists, LLC, Hagerstown, Maryland

INTRODUCTION

During the development of every therapeutic designed to address a medical condition and many chemicals, preclinical studies are performed to determine its safety with regard to intended or accidental human exposure. With regard to general toxicity studies, currently the approach usually followed is to examine three or four sections of the brain (full coronal sections for rodents[1]; partial coronal sections for larger animals): two to three transverse sections of spinal cord and a section of sciatic nerve (transverse, longitudinal, or both). The heterogeneity of the brain[2,3] makes such an approach suboptimal even for a screening study. For a study specifically designed to evaluate the nervous system, this approach is woefully inadequate. Even a casual review of any of the fine brain atlases available to pathologists will reveal the difficulty in capturing a representative sampling of the brain in three sections (especially in animals larger than rodents).[4–7]

The nervous system is composed of three main parts: central, peripheral, and autonomic. Every pathologist knows this, and most toxicologists know this. When developing a plan to evaluate the nervous system of an adult animal, an initial decision must be made: The pathologist/study director must decide into which of three categories the evaluation is going to be placed: general/screening examination, general examination with enhanced neuropathology, or comprehensive examination of the nervous system.

The degree of sectioning followed in many general preclinical toxicity studies was stated above. A general examination with enhanced neuropathology typically involves a regular screening evaluation scheme plus more sections of brain and/or nerves, depending on the known or suspected target tissue. The main purpose of this examination is to further characterize morphological effects found in a portion of the nervous system (usually, the brain) during a general screening study. In other words, an enhanced neuropathology investigation is generally implemented to salvage a general screening study with the ultimate goal of satisfying the scientific concerns, as indicated by the results of the initial tissue review. A study with enhanced neuropathology endpoints may also be designed based on prior knowledge of a specific test article or compound or a class of compounds.

A comprehensive nervous system examination is one that is preconceived to maximize the study resources in regard to providing a practical but thorough morphological evaluation of the nervous system. This chapter deals primarily with this third category, but aspects are useful when attempting to augment a general evaluation from any particular study.

For studies that have directed neuropathology endpoints, particular attention must be given to proper dose selection, appropriate sacrifice intervals, adequate sectioning (at least somewhat standardized between animals), evaluation of a sufficient number of animals, methods to decrease artifactual change, examination of appropriate sites (particularly within the brain), and appropriate embedding and staining techniques.[8–15] This chapter is focused primarily on four crucial aspects of the neuropathology assessment: necropsy, trimming and embedding, staining, and evaluation.

In this chapter, *large animal* means an animal rabbit size and larger. *Rodent* refers to mice, rats, and animals up to guinea pig size. *Coronal* means transverse (right to left)

Fundamental Neuropathology for Pathologists and Toxicologists: Principles and Techniques, First Edition. Edited by Brad Bolon and Mark T. Butt.
© 2011 John Wiley & Sons, Inc. Published 2011 by John Wiley & Sons, Inc.

sections of the brain. The term "coronal" as applied to brain sectioning is confusing, but in most atlases and in this chapter, the definition above applies. *Sagittal* refers to sections taken parallel to the long axis (rostral to caudal) of the brain.

NECROPSY

Timing

It is an inescapable truth that some test articles or chemicals are toxic to neurons. Based on a test article's particular characteristics, we are able to predict whether or not the brain/spinal cord/nerves will be exposed by crossing the blood–brain barrier (maintained by tight junctions between endothelial cells and associated astrocytes) or the blood–cerebrospinal fluid (CSF) barrier (maintained by the choroid plexus).[16] Small hydrophobic molecules or those that are actively transported across the endothelial cells (or choroid plexus) may access the central nervous system (CNS) directly. Therapies and chemicals may also enter through more porous areas such as many of the circumventricular organs in the brain.[17] Naturally, any therapy administered directly to the CNS accesses the brain and spinal cord. Direct delivery is a common means of administering larger molecules (proteins, antibodies), anesthetics, and cells to the CNS. Some test articles not expected to enter the brain do either via active transport, deficiencies in the blood–brain barrier, or other mechanisms.

Chemicals, devices, or drugs that cause, directly or indirectly, adverse effects in the nervous system should be evaluated with specialized testing procedures designed to detect neurotoxicity.[14] Any therapy or chemical that has one of the following characteristics should be fully characterized for morphological effects in the nervous system:

- Morphological lesion detected in the nervous system during a general or screening toxicity study
- Known neuroactive test article or chemical
- Similar in structure or function to a known neuroactive test article or chemical
- Molecular weight that indicates likely penetration of the blood–brain barrier
- Lipophilic molecules
- Biologics (including monoclonal antibodies) that target a nervous system structure or target a nervous system disorder or have been shown to penetrate the nervous system tissues or are suspected to penetrate the nervous system tissues
- Any therapy or chemical applied, injected, or infused directly into or adjacent to the central or peripheral nervous system

- Any therapy or chemical that causes clinical signs (seizures, ataxia, disorientation, etc.) attributable to the nervous system
- Any therapy or chemical that causes an alteration in brain weight
- Any therapy or chemical that shows a positive result on ancillary tests, including but not limited to changes in brain weight or abnormalities detected with electrophysiology, motor activity, functional observational battery (FOB), grip strength, or cerebrospinal fluid examination[9]

Many toxicants that enter the brain case will cause peak cell death in 1 to 4 days.[13,15,18] Conventional wisdom may dictate that if a particular test article is toxic to neurons, repeated administration of that test article is likely to continue to cause neuronal death, and therefore examining the brain from animals that have been dosed for 14 or 28 or 90 days (or more) should increase your likelihood of seeing neuronal necrosis. Although this situation can (perhaps rarely) happen, the actual situation seems to differ. When exposed to a neurotoxicant, the susceptible neuronal population(s) may be maximally affected and even die upon initial exposure to the neurotoxicant.[13,15,18] To observe this cell death, it is crucial to look at a time when the necrotic neurons are still visible. Two weeks or two months (or longer) after neuronal death occurs may preclude finding the dead neurons, as they are cleaned up rapidly and efficiently following death.[13] In other words, standard acute 14-, 28-, or 90-day studies may not reveal the full spectrum of neurotoxicity for a given test article, especially if a limited number of sections are evaluated.

At some time during the development of a known or suspected neuroactive test article or chemical, a study designed specifically to investigate neuronal necrosis during this 1- to 4-day time period should be conducted. Animal numbers do not have to be large, and using a single group of negative control animals, a single group of a positive control (something that will reliably produce neuronal necrosis, such as MPTP,[19] kainic acid,[15] MK-801,[18] or trimethyl tin[20]) and one or more test article–dosed groups (provided that the dose is similar to or higher than that to be used in subsequent or previous toxicity studies) may be enough to investigate the potential for neuronal death. For that study, sectioning of the brain should be exhaustive (see the discussion below on sectioning): at least every 1 mm for rodents (and submillimeter sectioning is preferable) and 1–2 mm for large animals. A review of the possible target areas for neurotoxicants[13] and the small sizes of these structures, especially in rodents, demonstrate the necessity for this degree of sectioning to feel confident that no areas were missed.

Other than this special requirement for a study to investigate the day 1 to 4 window for neuronal death, the majority of neuropathology studies will be conducted at time points more common to regulatory-type toxicology. These include 4, 28, 90 day and one-year studies (depending on the particular development plan for the particular test article). Two-year studies are (generally) designed to detect carcinogenicity. Neurotoxic effects (except for neoplasia) may occasionally be discovered in a two-year rodent carcinogenicity study, but this would be unusual.

Gross Examination

If any part of a comprehensive examination of the nervous system must be completed accurately, it is the animal's necropsy. The necropsy will determine if the proper tissues are harvested and if those tissues are suitable for the microscopic examination to follow. Necropsy, like death itself, is permanent. A proper necropsy includes planning, tissue collection, and fixation. All are important.

Planning

Necropsies should be performed accurately but also quickly. Although some academic investigations may involve a few animals, preclinical toxicity studies may require the necropsy of 80 rats or 48 sheep (as examples) over a 4-day period. To do this correctly, necropsy technicians must be thoroughly trained. It is one thing to teach a person, during the course of a necropsy, to efficiently harvest the liver, kidneys, or mammary gland. It is a very different task to expect someone to remove a brain without damaging it, locate the sural nerve, or remove the dorsal spinous processes to expose and/or harvest the spinal cord with attached spinal nerve roots and dorsal root ganglia.

Intravascular perfusion (discussed below) and many of the tissues harvested for a complete neuropathology examination add time to the necropsy. The better trained the technical staff, the more efficient the necropsy. Although this is always true, it is of particular importance for necropsies harvesting the various components of the nervous system.

Adequate preparation is essential. Often, multiple fixatives are used for various components. The necessary labels, vials, cassettes, and containers should be prepared prior to the day of necropsy. The necessary fixatives should be available.

The most important factors to consider when planning a necropsy are the study endpoints. A common reason to preclude intravascular perfusion fixation is the need to use an immunohistochemical stain that will not work properly on fixed tissues or to perform an ex vivo characterization/distribution of a protein (for therapeutic protein-based therapies) or nucleic acid characterization/distribution (for therapies utilizing a viral vector as a delivery method). If a necessary study endpoint precludes perfusion fixation, an alternative method of fixation must be decided upon. If half of the brain is required for various investigations, either the number of animals in each group needs to be increased (to allow for all the necessary investigations) or a scheme for dividing the brain at necropsy is required. (Note: Cutting a fresh, nonperfused brain of any species is a challenge.)

Fixation

The purposes of fixation are to stop the process of autolysis (the breakdown of tissue) and to preserve the tissue in a condition that is as close as possible to the in-life situation, decrease artifacts of handling and fixation (as the process of fixation can cause tissue changes), and render the tissues more suitable for subsequent processing for microscopic examination. There are several references that cover the topic of fixation of the nervous system in a comprehensive manner.[21–24]

Most general toxicity studies utilize immersion fixation. As compared to intravascular perfusion fixation, immersion is faster and cheaper. Artifactual and autolytic changes are generally more pronounced with immersion-fixed tissues, but an experienced pathologist is usually able to "read through" these changes. Immersion fixation also allows for the harvest of fresh (unfixed) tissue and the distribution of various tissues into whatever fixative may be most suitable. For example, in many studies, the testes are immersion-fixed in Bouin's solution, the eyes go into Davidson's solution, and the remainder of the tissues go into 10% neutral buffered formalin. It is easier to control the fumes from the fixation chemicals when using immersion fixation, as the fixative containers can remain closed except for the time during which tissues are placed into them, and the containers can be placed in a fume hood or next to a fume removal system. For many studies, and even for many directed neuropathology investigations, the use of immersion fixation is suitable for doing a complete examination of the nervous system provided that the various tissues are harvested in a time frame that minimizes autolysis and artifactual changes. However, perfusion fixation is typically preferred for optimal preservation of tissues and to decrease artifactual changes.[11,23,24]

Since the great majority of necropsies conducted in general toxicity studies utilize immersion fixation in 10% neutral buffered formalin, perfusion-fixed tissues will generally not be available for retrospective evaluations. Retrospective refers to those studies where an effect was detected or suggested by the standard sections examined, usually in a general toxicity study, and further evaluation of residual wet tissue is attempted. This does not necessarily prevent a thorough evaluation of the nervous system, but the pathologist and toxicologist should be aware that immersion fixation increases the possibility of false positives (artifactual changes being interpreted as real lesions) and false negatives (failure to observe or correctly interpret subtle changes such as cellular vacuolation or degeneration, which may be masked by artifactual changes).[25]

Information on the technique and apparatus required to conduct an intravascular perfusion is available.[22,26] For intravascular perfusion, special systems to remove the fumes that arise during intravascular perfusion procedure and from the perfused carcass (during the necropsy) are absolutely necessary to keep necropsy personnel safe from the detrimental effects of the fixative fumes. Intravascular perfusions (except with saline or other nontoxic liquid) should not be performed without a sufficient fume and liquid removal system in place. As (typically) the person in attendance with the best medical training, it is the pathologist's responsibility to refuse to conduct a perfusion fixation under any conditions that will unnecessarily expose anyone to a toxic level of fumes from the fixatives. Those in attendance should wear monitoring devices (usually, badges) that can be analyzed for exposure levels to the fixative in use. Despite the potential dangers, unless particular endpoints preclude it, intravascular perfusion is the preferred method of fixation in order to decrease the degree of artifactual changes present in the brain (from handling and autolysis) and spinal cord.[9,11,18,23] Vacuolation in the neuropil and shrunken, dark neurons are two notable artifacts generally decreased by intravascular fixation with an aldehyde fixative.[11]

The choice of fixative may depend on the endpoints of the study. The most common fixatives are 10% neutral buffered formalin (which is actually 37% formaldehyde diluted 1 : 10 and thus equivalent to 3.7% formaldehyde), 4% formaldehyde (prepared from paraformaldehyde powder), variable concentrations of glutaraldehyde (up to 4%), or variable mixtures of glutaraldehyde and formaldehyde. Each has advantages and disadvantages.[27]

Because of its small size and reactive aldehyde group, formaldehyde penetrates tissues rapidly and is an excellent fixative.[27] Formaldehyde from stock solutions (usually, commercially available 10% neutral buffered formalin) may be used unless stabilizing chemicals in those solutions will interfere with subsequent testing; then formaldehyde prepared from paraformaldehyde powder (purchased or prepared) is preferred. Glutaraldehyde penetrates tissues more slowly[24] than formaldehyde but improves the cross-linking of proteins and preservation of the lipid components (i.e., myelin) of the nervous system.[27] Glutaraldehyde or a glutaraldehyde–formaldehyde mixture is often preferred when ultrastructural (using the electron microscope) examination will be utilized.[24]

For glutaraldehyde fixation, a prefixation flushing of the vascular system using a solution buffered similarly to the fixative should be used; pre-flushing is recommended but not absolutely necessary if a purely formaldehyde-based fixative is used. Whether using a gravity feed, pressurized perfusion apparatus, or pump, the perfusate should be delivered at a pressure that is at least systolic pressure or preferably slightly greater since once the right side of the heart is opened (to allow fluids to escape the system), the perfusion is going into an essentially open system. Intermittent interruptions in the pressure in the vascular system, as may occur if the perfusion process is interrupted even briefly, may cause premature collapse of small blood vessels and suboptimal fixation. Systolic pressure in the various laboratory animal species varies but is generally between 100 and 120 mmHg, 100 mmHg equating (weight-wise) to a column of water around 120 cm tall. So for a gravity-feed system, a fluid level that is maintained approximately 120 cm above the heart of the animal will provide pressure in the proper range. Osmolality, pH, buffering capacity, and temperature are all important characteristics of the flushing and fixative solutions.[23,24]

Perfusion with ice-cold buffered saline (followed as soon as possible by immersion fixation) may reduce artifactual changes by slowing postmortem autolysis. This technique is sometimes used for studies where a particular endpoint precludes intravascular perfusion fixation. Fixatives of reduced concentration (e.g., 1 or 2% formaldehyde) may also be used if a higher concentration of fixatives will prevent the retrieval of specific antigens.

The perfusion fixation procedure can be conducted in various ways. Some useful references are available on the Internet. In general, regardless of species, the following procedure, which describes whole-body perfusion, is followed (it is also possible to do a largely head only perfusion by canulating the carotid artery on each side). A recent compact disk produced by a joint effort of the National Toxicology Program, National Institute of Environmental Health Sciences, and Charles River Laboratories (June 2010) is available and provides an excellent training aid for the conduct of the intravascular perfusion procedure.

A gravity-feed system requires two containers: one containing the fixative and one containing the preflushing solution. The line coming from each container must contain a valve or stopcock that allows the line to be completely blocked or completely open. These two lines join via a "Y"-type connector. A single tube coming from the Y container terminates in a needle/canula/catheter that will be inserted through the wall of the left ventricle and to the base of the aorta. For rodents, a blunted 20- to 16-gauge needle can be used to enter the heart. A gavage needle also works and can easily be advanced to rest at the base of the aorta, which is the preferred location for the tip of the perfusion needle/canula. For larger animals, a large-gauge needle (12 or more) or a canula (approximately 3 to 4 mm inside diameter) fabricated from steel tubing can be used. Some have achieved success with a Foley-type catheter, although this will require some means of slicing the left ventricle to facilitate entry of the catheter into the proper area. The tube running from the Y connector to the needle or catheter has a valve that can be set to allow for a variable flow rate.

Prior to the first perfusion, the flow rate is set by opening the valve on the single or final tube section and measuring the rate of flow using a watch with a second hand or another timer and a graduated cylinder. The opening on the valve is adjusted to achieve the flow rate desired. For rodents, standard intravenous infusion sets (with drip rates that accurately predict the flow rate) can be used. The flow rate varies according to the size of the animal. Generally, the following flow rates (for the flushing solution and the perfusion solution) will result in a satisfactory perfusion:

- *Mice:* 5 mL/min
- *Rats:* 25 to 50 mL/min
- *Monkeys:* 300 to 500 mL/min
- *Dogs:* 500 mL to 1 L/min
- *Sheep/large animals:* 700 mL to 1.5 L/min

Once the flushing or perfusion rate is set, the valve on the terminal tube is left alone (i.e., it is not further adjusted). The valves on the tubes coming from the flushing or perfusion solution are opened or closed depending on what solution is being administered. At the end of each perfusion it is important to flush the terminal tube with the flushing solution (so residual fixative solution is not the first solution to enter the next animal).

Times vary between the people who do perfusions; each has a preferred technique. One procedure that works well is to infuse the flushing solution for 2 to 5 min, followed by the perfusion solution for 10 min. Times will vary based on each animal. Some prefer to run the flushing solution until various organs become pale, indicating the absence of most of the blood. The liver is a useful reference organ for this technique.

With the animal deeply anesthetized to a plane that absolutely and unequivocally prevents the animal from sensing any pain from even a severe noxious stimulus, the animal is placed in either dorsal recumbency or lateral recumbency, depending on the technique preferred to access the left side of the heart. Dorsal recumbency provides easier access but requires sufficient removal of the sternum and adjacent ribs. Lateral recumbency requires an incision and spreading of the ribs overlying the left ventricle. The correct location for the incision with the lateral approach can be checked by flexing the thoracic limb caudally and dorsally; the location of the elbow over the rib cage is approximately the area that should be incised.

The heart is exposed and any bleeders are controlled; bleeding is seldom an issue if the animal is in lateral recumbency. Some bleeding is sometimes encountered at the thoracic inlet when the animal is perfused in dorsal recumbency.

The canula or needle is inserted into the left ventricle. Usually, a blunted needle or a beveled canula can be inserted through the wall of the left ventricle without any incision of the left ventricle. If a catheter is used, the wall of the ventricle can be partially incised to facilitate entry. The tip of the device is advanced under the leaflets of the atrioventricular valve and to the base of the aorta. In larger animals, this location can be checked manually. In rodents, this location can usually be assessed visually.

The flow of the flushing solution is started prior to entry of the device into the left ventricle. This helps to preclude any sudden drop in pressure during the process. Once the device is at the base of the aorta (or even just in the left ventricle), there is typically a noticeable increase in vascular pressure (visible as a slight expansion of the right side of the heart). The right auricle is then incised sufficiently to allow for the egress of fluid returning to the heart. This steady state of perfusion with escape of returning fluid keeps the pressure relatively steady. The returning fluid, which will at first be a diluted solution of blood and the flushing solution and will finally become nearly pure perfusion fixative, should be collected and disposed of in accordance with local standards. Rodents can easily be perfused on racks over pans that greatly facilitate collection of the perfusate. For larger animals, some other procedure is required, usually either collection of the freely flowing liquid or by direct collection from the right ventricle using catheters that equal or exceed the diameter of the perfusing device.[28]

If an animal is being perfused successfully, there is typically a characteristic, whole-body muscle fasciculation that may begin as an extension of the limbs and tail. This should begin approximately 1 min following the initiation of the fixative portion of the perfusion. If this muscle fasciculation does not begin, the system should be checked. There can be multiple reasons for a nonsuccessful perfusion, including:

- Failure to open the proper valve(s) of the perfusion device
- Failure to place fixative in the proper container
- Penetration of the ventricle/aorta such that the perfusate is not entering the vascular system
- Failure to open the right side of the heart to allow for the escape of returning fluids
- Elevated pressure on the vascular system, such as the abdominal viscera putting excess pressure on the descending aorta
- Positioning of the animal [especially a problem if the animal is perfused in lateral recumbency, as the up limb (and sometimes the down limb) may not perfuse properly]
- Unknown

Regardless of how careful the staff is, for whatever reason, some perfusions just do not work as intended. It is useful to record, on the necropsy sheet, an evaluation of the thoroughness of the perfusion so that the pathologist who will later be reviewing the tissues microscopically has some idea as to the degree of fixation achieved during the perfusion procedure.

The tissues of a perfused animal lose some compliance, making the dissection procedure more difficult. The higher the concentration of glutaraldehyde in the fixative, the firmer the tissue will be and the more extreme the limb extensions or muscle fasciculations will be during the procedure. Perfusion fixation does allow some relaxation of the need to get tissues into an immersion fixative as quickly as possible, as the tissues are already exposed to fixative.

Supplemental Fixation

As noted, for studies where immersion fixation is being used (and for many when intravascular perfusion is being used), 10% neutral buffered formalin is typically the fixative of choice, mostly because of the ease of use of this fixative (it is easily obtained commercially) and its cost. Fewer artifacts will be encountered with methanol-free formaldehyde solutions (such as those prepared from paraformaldehyde powder). The inclusion of glutaraldehyde enhances the preservation of myelin.[27]

Glutaraldehyde immersion (including immersion following perfusion fixation) is preferred for any samples being used for electron microscopic examination.[21,23] Typically, these samples are later further immersion-fixed in osmium tetroxide. Because osmium does not penetrate tissues well, samples should be less than 2 mm thick.

Because nerves are generally relatively thin, they lend themselves well to immersion fixation in glutaraldehyde and later to fixation in osmium.[29,30] The best fixation of nerves is achieved by cutting the nerves into 2- to 4-mm lengths with very sharp razor blades (using a crossing technique in which the blades cross, straddling the nerve and slicing it cleanly as the blades are crossed) and then fixing them in glutaraldehyde (and later osmium). This is especially important for nerves that are surrounded by a prominent fatty sheath, such as the vagosympathetic trunk in the neck region.

Post-fixation in osmium tetroxide is recommended for resin-embedded/toluidine blue–stained peripheral nerve sections and/or ultrastructural examination and is specifically preferred for examination of myelin.[21]

Tissue Collection: Brain

If an animal is not perfusion fixed, the brain should be removed as soon as possible after the animal's death in order to get the brain into the immersion fixative as quickly as possible. Excessive handling of nervous system components will increase artifactual change. For example, removal of the brain from the cranial cavity, even in perfused specimens, can increase the propensity for artifactual cell shrinkage and increased staining of some neurons (called "dark"

neurons).[11,23] So for perfused specimens, it is acceptable to allow the brain to remain in the skull for 24 h following perfusion fixation. (Note: For perfusion-fixed animals, it is the author's preference to remove the skullcap in rodents, then remove the brain 24 h later, and in larger animals, to remove the brain entirely, at necropsy). The practice of immersing the brain in a saline solution, followed at many facilities that lack optimal ventilation in the necropsy room, may result in a surface vacuolar change.[23] For studies where perfusion fixation is not conducted, immediate immersion (after prompt weighing) into a suitable fixative is the preferred method.

In rodents, the brain is removed by skinning the skull and using a small ronjeur to remove the flat portion of the caudal skull down to the foramen magnum. Once this area is removed, the entire caudal portion of the cerebellum and cut surface of the medulla oblongata is exposed. Using the same ronjeur, the bone between the orbits is cut (but not removed). Moving to the back of the skull, the midline is clipped (again with the ronjeur) about 3 mm. This frequently (especially with an experienced technician) completely separates the right and left sides of the skull (along a suture line) up to the level of the orbit. Then the two halves are folded back and removed. The brain, including the olfactory lobes (especially in rodents, rabbits, and dogs), is gently removed rostrally to caudally by elevating the brain with a thin, flat instrument (spatula, scalpel handle, forceps) and cutting any attachments on the floor of the skull (the cranial nerves) and any dura that is attached to indentations in the brain. Once the brain is reflected to the back of the skull, it can be removed completely. With practice, this technique yields a brain that is free of any artifactual cuts.

In larger animals, a bone-cutting saw is required to remove the brain. After skinning the head, the bone is cut approximately 1 cm caudal to the caudal most portions of the orbit. This cut is extended to approximately the edge of the orbit. The edge of this cut is extended along the side of the skull, above the auditory canal, to approximately the 10 and 2 o'clock aspects of the foramen magnum. Once these cuts are made completely through the skull, the cap can be removed. Then the brain is removed in the manner as described above for rodents. Practice is required to perfect this technique.

Tissue Collection: Spinal Cord, Spinal Nerve Roots, and Dorsal Root Ganglia

The spinal cord and nerve roots and the dorsal root ganglia are important tissues in any complete examination of the nervous system. The endothelial cells lining the vascular of the dorsal root ganglia (DRG) have larger fenestrations than similar cells in the brain and spinal cord, so toxins that may not gain entrance into the brain and spinal cord may still exert an effect on the sensory neurons of the DRG.[31]

The spinal nerve roots and ganglia are best removed by first removing the dorsal spinous processes of the spinal column. Several rootlets coming from the cord coalesce to form the roots. If the dura is opened, the rootlets can be removed from the dorsal surface of the cord (the ventral rootlets are not as readily accessible). These are dissected free from the cord and separated from the dura. The ventral root is cut and then comes along with the dorsal root. The roots are followed to the vertebral foramen, and sufficient bone is removed to access the dorsal root ganglion and the combined spinal nerve as it exits the vertebrae.

In a perfusion-fixed animal, the spinal cord and attached spinal nerve roots and ganglia should be adequately fixed. If time permits at necropsy (and it seldom does), the cord with attached spinal nerve roots and ganglia are removed from the carcass from C1 to at least the level of the lumbosacral junction. Since the cord terminates in the lumbar spinal canal in all species, this ensures harvest of the entire spinal cord (with sufficient terminal spinal nerve roots). In the cervical and thoracic segments of the spinal cord, the spinal nerve roots exit the vertebral formamen essentially directly laterally. Because the vertebrae continue to grow or elongate past the time when the spinal cord elongates, the spinal nerve roots arising from the caudal lumbar and sacral and caudal segments run down the spinal canal before leaving, at a variable distance caudally, through the corresponding intervertebral foramen. This means that the lumbar spinal nerve roots are easier to find and remove, but to get the corresponding dorsal root ganglion, the roots must be followed to the appropriate foramen.

If the cord and roots cannot be removed at necropsy, the next-best option is to remove the dorsal spinous processes, which allows access of the fixative to the cord. Although initially seemingly daunting, removal of the dorsal spinous processes, at least in rats, can be accomplished rather quickly (in 5 to 10 min) by an experienced person. Larger animals obviously present a bit more of a challenge. Removal of the dorsal spinous processes is a good practice whether the animal was perfusion-fixed or if immersion fixation alone is being utilized.

If time is short, the cord with attached spinal nerve roots and ganglia and the surrounding vertebrae (with as much excess tissue as possible removed) are removed and placed in the immersion fixative chosen. If only immersion fixation is being used, it is important to provide a means for the fixative to adequately access the spinal cord, spinal nerve roots, and ganglia. This is usually accomplished by cutting the spinal cord or spinal column into multiple pieces and/or removing the dorsal spinal processes. Many laboratories separate the spinal column into cervical, thoracic, and lumbar segments. It is preferred to separate the spinal column between the T_2–T_3 vertebrae (because the brachial plexus is formed by approximately C_5 to T_2) and then again just proximal to the thoracolumbar junction.

The actual spinal cord segments and the required spinal nerves and associated dorsal root ganglia are generally dissected out by the histology laboratory at the time of tissue trimming. This is acceptable provided that the fixation is adequate. For general/screening studies, typically only two or three spinal cord segments (cervical, thoracic, and lumbar) are saved, especially in the larger laboratory species. For studies where the main endpoint is a thorough neuropathology evaluation, the entire spinal cord should be saved.

Tissue Collection: Peripheral Nerves

The peripheral nerves to be harvested are determined by the objectives of the study. For test articles or chemicals with a suspected or likely change in peripheral nerves, nerves from the thoracic and pelvic limbs may be harvested. For most neuropathology studies, the sciatic, tibial, and sural nerves are harvested. For some studies the brachial plexus, radial nerve, and peroneal nerves are included.

The sciatic nerve is harvested in most studies, not just those studies particularly interested in neuropathological endpoints. The sciatic is removed from the lateral thigh between the sciatic notch (upper femur) and the site where the nerve splits into the tibial and peroneal nerves.

The tibial nerve is harvested between the lateral and medial bellies of the gastrocnemius muscle at approximately mid-tibia.

The lateral cutaneous sural nerve splits off the sciatic nerve at a level slightly above where the nerve branches into the tibial and peroneal nerves. The sural nerve can be traced down the leg and, after coming through a fat pad at the caudal aspect of the knee, runs along the lateral surface of the lateral gastrocnemius muscle. It is small, but is fairly easily harvested from all species, including mice.

The peroneal nerve splits from the sciatic and courses slightly cranially; it can be harvested as it branches off the sciatic.

With practice, the brachial plexus can be harvested relatively easily by leaving the thoracic limbs attached to the carcass. After exposure of the spinal cord in that area via removal of the dorsal spinous processes, the spinal nerves are dissected from the cord out through the vertebrae as the scapula and attached muscle is reflected laterally, revealing the brachial plexus.

Tissue Collection: Cranial Nerves

The optic nerve is typically harvested with the eye. The optic nerve in all species is centrally myelinated (the optic vesicle is an outgrowth of the neural tube; the depth of central myelination of the optic disk varies slightly between species).

The trigeminal ganglion (ganglion of cranial nerve 5) is frequently harvested. This ganglion lies just under the dura just caudolateral to the pituitary fossa. It is visible under the dura in all species as a grayish brown structure at the junction of a single nerve (the sensory root of cranial nerve 5) coming from the ganglion (to the brain) and three nerves (although usually only two are visible) entering the ganglion (from the direction of the face).

Tissue Collection: Representative Autonomic Nervous System Components

The autonomic nervous system consists of two portions: the sympathetic nervous system and the parasympathetic nervous system. For the parasympathetic side, preganglionic efferent (*efferent* means leaving the CNS) neurons are in the midbrain (nucleus of cranial nerve 3), the medulla oblongata (nuclei of cranial nerves VII, IX, and X), and in sacral segments of the spinal cord.[32] So these preganglionic neurons may be examined with a complete exam of the brain and spinal cord. Naturally, it takes some time and practice to identify the exact sites in the brain. The postganglionic neurons reside either in four ganglia in the head (these are seldom examined unless there is a specific need to do so) or in the target organs, including the heart, kidney, and gastrointestinal organs. Therefore, these postganglionic neurons can be examined in the regular tissue sections usually processed for those organs. The afferent neurons of the parasympathetic nervous system are in the dorsal root ganglia.

The preganglionic neurons in the sympathetic nervous system reside in the thoracic and lumbar segments (T_1 to L_2)[32] of the spinal cord. The postganglionic neurons lie in the paravertebral ganglia, many of which lie in the sympathetic trunk in the thoracic cavity and in prevertebral ganglia in the thoracic and abdominal cavities. The sympathetic trunk runs parallel to the spinal column. With the lungs removed, the sympathetic trunk appears as a thin nerve beneath the parietal pleura, running in parallel with the vertebral column about at the junction of the ribs and vertebral column. There are numerous ganglia within the sympathetic trunk, although even in the larger species, these are not readily evident. One ganglion that is easy to harvest is the cervicothoracic ganglion, which is the conjoined caudal cervical and cranial thoracic ganglion of the sympathetic trunk. This ganglion is harvested by following the sympathetic trunk cranially through the chest cavity. The ganglion lies approximately at the level of the first rib; it appears as a slightly triangular, flattened, grayish brown ("ganglion color") structure that is an obvious enlargement of the sympathetic trunk.

Another ganglion that is relatively easy to harvest is the superior/cranial mesenteric ganglion. This ganglion, often fused with portions of the celiac ganglion, is located near the left adrenal gland and surrounds the cranial mesenteric artery near its origin from the aorta. When the tissue is harvested, there should be a prominent artery running through the tissue. Sectioning usually reveals the ganglia. The most thoroughly studied sympathetic ganglion, at least in the rat, is probably the cranial superior cervical ganglion. This ganglion can be located by following the sympathetic trunk into the proximal cervical region. The ganglion is deep to the bifurcation of the carotid artery (on each side).

The adrenal medulla is a modified sympathetic ganglion. Postganglionic neurons of the sympathetic system reside within the adrenal medulla. They do not produce axons that innervate distant structures but rather, release noradrenaline and adrenaline into the blood.[33]

The following tissue harvest scheme will allow for an examination of at least portions of the autonomic nervous system:

- *Brain, midbrain, and medulla oblongata:* preganglionic nuclei of the parasympathetic nervous system
- *Sacral segments of the spinal column:* preganglionic neurons of the parasympathetic nervous system
- *Target organs* (heart, kidney, gastrointestinal organs): postganglionic neurons of the parasympathetic nervous system
- *Spinal cord* (T_1 to L_2 segments): preganglionic neurons of the sympathetic nervous system
- *Cervicothoracic ganglion and/or cranial/superior cervical ganglion:* postganglionic neurons of the sympathetic nervous system
- *Sympathetic trunk:* postganglionic neurons of the sympathetic nervous system and axons from those neurons
- *Cranial/superior mesenteric ganglion:* Postganglionic neurons of the sympathetic nervous system
- *Adrenal medulla:* postganglionic neurons of the sympathetic nervous system

Tissue Collection: Eyes

Perfusion fixation even with 10% neutral buffered formalin will generally result in excellent preservation of most eye structures, including the retina. If immersion fixation is used, Bouin's fixative or Davidson's fixative will preserve the retina better than formalin[34] or other readily available fixatives.

Tissue Collection: Muscle

Several portions of muscle, typically sections of the biceps femoris and gastrocnemius, should be harvested. Some special staining of muscle fibers (used for fiber typing) requires the harvest of fresh, frozen sections. If detailed histology and evaluation of muscle tissue will be required, special fixation and harvest procedures will probably be required. These topics are covered elsewhere in the book.

TRIMMING AND EMBEDDING

The manner in which the brain is trimmed and how many levels are to be processed and examined depend on the information that is sought. Because of the complexity of brain structure, you can never evaluate too much brain in any study. But balancing scientific merit with some degree of practicality (in science, the term *practical* translates as "making the best use of available time and money"), there are two main types of investigations: a neuronal necrosis investigation and a standard neuropathology study.

Brain Trimming: Neuronal Necrosis Investigation

A neuronal necrosis study is designed specifically to address the possibility that a given therapy or chemical may cause neuronal necrosis following initial exposure. This investigation is conducted in 1 to 7 days (typically, 2 to 4 days) following exposure to a given potential neurotoxicant. The time frame is dictated by the knowledge that days 2 to 4 tend to be the days of peak neuronal necrosis[13] for many neuronal toxicants—although, of course, this can vary.

A review of the coronal or sagital sections from many of the excellent atlases of the brain[4–7] reveals the remarkable heterogeneity of the brain. Any of the many specific neuronal areas could be affected by a given neuroactive test article. So to be complete, a sectioning scheme that is going to provide an examination of the various neuronal areas is a necessity. Based on a review of various brain atlases, this sectioning scheme should be at intervals of around 0.5 mm (or less) for rodents and not more that 1 to 2 mm for larger animals in order to capture most of the distinct nuclear groups. The sectioning scheme will dictate how the brain is fixed, trimmed, embedded, sectioned, and stained. Due to the limits of trimming for paraffin embedding, for intervals of 1 mm or less, it is generally preferred to use frozen sections to achieve the evaluation intervals desired.

Paraffin sections or frozen sections are the two main ways in which brain tissue is typically assessed. Paraffin sections are usually cut in the 4 to 8 μm range. Such thin sections and the paraffin embedding process provide sections that are excellent for morphologic evaluation. Frozen sections are typically cut at 10 to 60 μm. Because the sections are thicker, the ability to assess the sections morphologically is somewhat compromised compared to paraffin sections. However, some staining procedures require frozen sections, and for specific purposes, they can reveal abundant morphological detail, especially with special staining procedures.

For paraffin embedding, it is possible to trim the brain of rodents into 1-mm slices and larger animals into 2-mm slices without much difficulty. Various brain-trimming matrices exist that allow for a repeatable, standardized sectioning scheme. The matrix consists of a metal or acrylic mold in which the brain rests (usually, top side down). A series of slits in the matrix guide razor blades. Depending on the matrix, the slits are usually 1 to 3 mm apart. The slits may allow for the production of coronal or sagittal sections. The brain-trimming matrix works well, but like most techniques, requires some experience to perfect, as the brain has a tendency to shift during the trimming procedure. This shift may result in the production of wedge-shaped partial sections. The brain-trimming matrix tends to be the best method for producing thin (down to 1 mm thick) sections of a rodent brain.

In the author's laboratory, a trimming guide (Figure 1) is used for most brain-trimming procedures for larger animals. The guide is produced by a machine shop utilizing a block of noncorroding metal. On each side of the block, the central 10-cm area (approximately) is removed to a depth of 1, 2, 3, or 4 mm, leaving 1-cm-wide bars on each side of the central

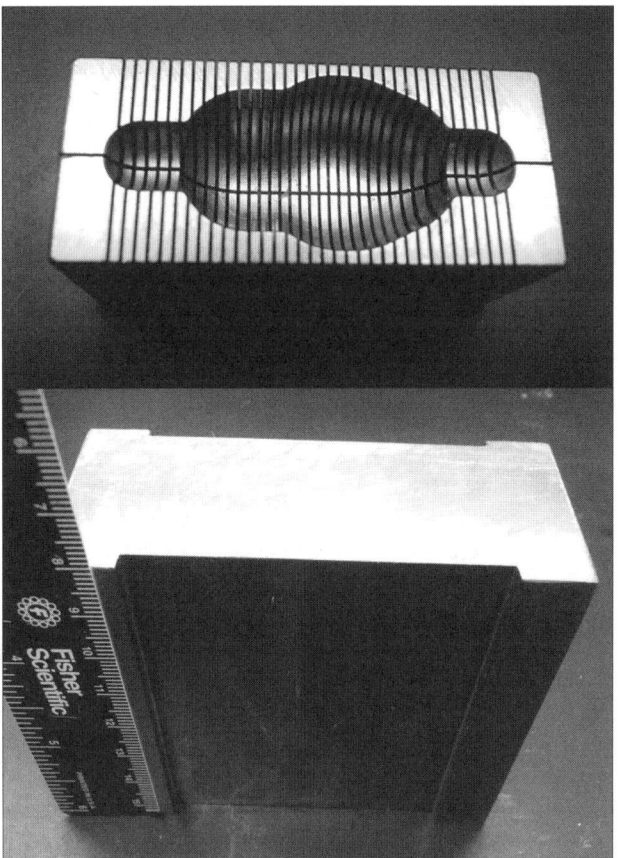

FIGURE 1 Two examples of brain trimming devices. In the top device, the brain is placed top side down. The precut slits are used to guide a razor blade. In this device the slits are 1 mm apart. The bottom device is a solid block of metal, machined to produce a wide, flat recess on two sides. One recess is 2 mm deep; the one closest is 3 mm deep. Once the brain is initially cut (usually at the rostral border of the temporal lobes) into two or more slabs, the flat surfaces can be placed against the recessed surface of the metal device, and the lateral bars used as a guide for a long razor blade. Both devices allow for very precise trimming of the brain.

channel. For trimming, the brain is first cut into two pieces at the rostral edge of the temporal lobes. This allows a symmetrical cut to be made and produces two pieces, each with a flat side. The flat side is then placed against the channel of the block (using the side corresponding to the desired thickness of the slice) and cuts are made until the brain is completely sectioned. Somewhere about the level of the caudal midbrain or pons, the occipital lobes will separate from the rest of the brain. At this level and caudal, the midbrain and brain stem and each occipital lobe must be sectioned separately.

Typically, full coronal sections of rodent brains are embedded in standard-sized cassettes and examined. If the test article is given by a means other than direct delivery to the central nervous system, it is acceptable to look at only one side of the brain in any species. Unless a laboratory has the equipment and expertise to produce full coronal sections of the brain in larger species, it will be necessary to cut full coronal sections into four separate sections (or right/left sides of the brain into two sections). This approach is necessary in animals larger than rabbits since the full coronal sections are too large to fit into a standard-sized cassette. Full coronal sections of any species are strongly preferred if there is direct delivery into the brain via injection, cell implantation, or catheter.

Frozen sections allow for the most flexibility in terms of thorough brain sectioning. Embedding the entire brain or sections of the brain and using a sliding microtome, the entire brain can be cut into 10 to 60 (or thicker) sections. All the sections can be saved into a series of vials or wells and made available for subsequent mounting and staining. Using these techniques, many different markers can be assessed at known intervals. This technique also allows the use of amino cupric silver stain for neuronal necrosis.[12,13] Because frozen sections can be prepared reliably thicker than paraffin sections, they are the most common technique used when stereological investigations (neuronal counts in the brain) are to be performed. The topic of stereology, including performing such investigations on thin paraffin sections, is covered in Chapter 15.

Regardless of the methods used, the intent of the neuronal necrosis study is to perform an exhaustive examination of the brain to fully assess the many neuronal areas of the brain. Frozen sections allow a more complete sectioning scheme to be performed, allow the use of amino cupric silver stain for neuronal necrosis, and provide a useful means of assessing a variety of biomarkers. If paraffin sections are used, and even for frozen sections, the Fluoro-Jade B or C stains[15] also allow the detection of neuronal necrosis (see below).

Brain Trimming: Standard Neuropathology Study

The degree of sectioning recommended for the neuronal necrosis study is the best way to assess brain pathology. But such an extensive sectioning scheme would be considered impractical if repeated for each of the studies typically performed to satisfy scientific and regulatory requirements for the development of a therapy or chemical. For standard neuropathology studies, and to produce additional sections in general toxicity studies where there is a desire to investigate the brain further, the following sectioning scheme is recommended as one possible starting point. Other approaches have been published.[25]

- *Brain, rodent.* Examine a minimum of eight full coronal sections at regular intervals and, in addition, include longitudinal sections of the olfactory lobes. The rodent brain is conveniently sized to enable (generally, old and large rats may defy this slightly) four coronal sections to fit in a standard-sized cassette. The rodent brain can be sectioned using a brain matrix or other trimming aid (or just trimming freehand) and producing a series of 2- to 3-mm sections. It is best to use enough cassettes (three will be sufficient) to include all the brain sections. Alternatively, a scheme that fits into two standard-sized cassettes can be used.[35] Although it may be beneficial to use a standard trimming scheme utilizing landmarks that assist with reproducibility by a trained technical staff, some slight variation in sectioning between animals is not necessarily undesirable. Some variation in sectioning allows, across animals and groups, a wider range of brain structures to be reviewed.

- *Brain, large animals (administration by any means other than direct delivery).* A minimum of 15 sections (which should fit into 14 standard-sized cassettes) will be sufficient to include portions of the forebrain, multiple portions of the cerebrum, caudate putamen area, thalamus (rostral and caudal), midbrain including the substantia nigra, pons, cerebellum, and multiple sections of the medulla oblongata (Figure 2). These sections will include neurons known be especially susceptible to excitotoxicity mechanisms of damage (i.e., Purkinje cells, hippocampus, cerebral cortex)[9] and provide a somewhat comprehensive evaluation of the brain. Sectioning the brain into 3-mm slices (using a matrix or the trimming block) provides sufficient slices to allow a selection of brain sections that will fit into standard-sized cassettes. Unless there is some reason to suspect that it would not be appropriate to do so, evaluation of one side of the brain is acceptable. Note that this approach of 15 sections, primarily from one side of the brain for large animals, should be considered the minimal number reviewed in order to attempt to perform a somewhat comprehensive evaluation of the brain.

- *Brain (direct delivery).* For studies where direct central nervous system delivery is utilized (e.g., intraparenchymal injection, intracerebroventricular, intrathecal administration, etc.), examine at least 8 to 10 full coronal sections (all species) with some emphasis around the area of test article administration (if the brain is

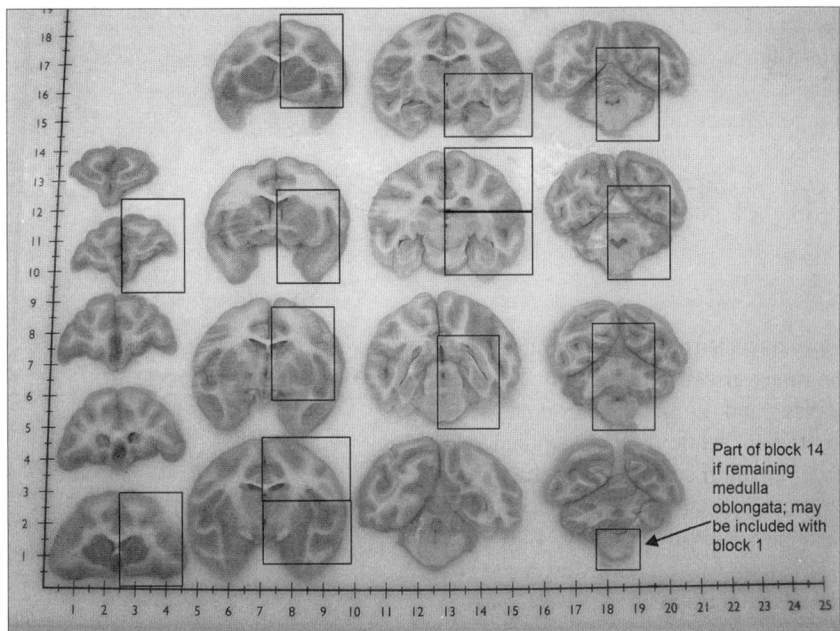

Part of block 14 if remaining medulla oblongata; may be included with block 1

FIGURE 2   Set of transverse sections from a primate brain trimmed using the device shown as the bottom image in Figure 1. Note the uniformity of the sections. The black boxes are drawn to indicate the sections to be processed and examined. Such a diagram can be used by technical staff to produce relatively consistent and homologous sections between animals.

the area of administration). For direct delivery it is important to look at both sides of the brain, allowing comparison of the ipsilateral brain (the side of administration) with the contralateral side. If delivery is to both sides of the brain (bilateral intraparenchymal infusions as an example), the usefulness of looking at both sides of the brain is obvious. Evaluation of full coronal sections of the brain for large animals is best done with the complete section mounted on oversized (usually, 2 × 3 inch) slides, or the sections can be split into however many regular-sized cassettes is necessary.

Spinal Cord

- *Spinal cord, nondirect delivery.* Examine transverse and oblique sections at the level of the cervical and lumbar intumescences (rodents) or transverse and oblique sections at the level of the cervical, thoracic, and lumbar segments for larger animals. Evaluating only transverse sections of spinal cord may not be sufficient to detect subtle changes in the various nerve fiber tracts that comprise the white matter. Sections taken along an oblique dorsal–ventral plane (at a 30° angle or less) provide tissue oriented more closely to the course of the nerve fibers traversing the spinal cord, maintain the overall architecture of the cord, and are generally preferred to longitudinal sections because the overall architecture of the cord is recognizable and all nerve tracts are present in the section.

- *Spinal cord, direct delivery.* For intrathecal or spinal column implant studies, sections should be taken at the levels listed above plus sufficient sections to include at least three levels of the delivery device or implant (if possible). For example, for a study utilizing an intrathecal catheter, three transverse sections at the catheter tip (one just in front of the tip, one at the tip, and one caudal to the tip, including an oblique section between the second and third transverse sections), transverse/oblique sections mid-catheter, and transverse/oblique sections at the level of the catheter insertion (to evaluate possible complications of catheter insertion, including damage to spinal nerve roots) is recommended. In addition, a transverse section through the area of the cauda equina (the spinal nerve roots that fill the spinal canal caudal to the termination of the spinal cord) is useful in evaluating the morphology of the terminal nerve roots.

Spinal Nerve Roots with Dorsal Root Ganglia

- Roots and ganglia from the same levels as harvested for the spinal cord should be examined. Although it is feasible (and sometimes necessary) to look at transverse sections of the spinal nerve roots, it is generally sufficient to look at just longitudinal sections that include the dorsal root with the ganglion and the ventral root. Generally, paraffin embedding is sufficient to achieve excellent morphological detail. If a detailed study of a cross section of the spinal nerves is required, special post-fixation with osmium and staining would be necessary (detailed below). Technically, spinal nerve roots and ganglia are components of the peripheral nervous system. Some guidelines require that peripheral nervous system tissues be embedded in "plastic"[36] if a peripheral neuropathy is known or suspected.

Peripheral Nerves

- The sciatic nerve at two levels (transverse and longitudinal), tibial nerve (transverse and longitudinal), and sural nerve (transverse and longitudinal) should be examined. The longitudinal sections can be embedded in paraffin or a harder medium such as glycol methacrylate, Epon resin, or Spurr's resin. Cross sections should be post-fixed in osmium tetroxide (usually, 1 to 2% osmium for 1 to 2 h) to assist with myelin preservation[21,29,30] (Figure 3). Note that even if the nerve cross section is embedded in paraffin, osmium post-fixation greatly improves preservation of the myelin sheath.

Autonomic Nervous System

- The ganglia are embedded in paraffin and examined. Use of a harder embedding medium is also perfectly acceptable and may even result in sections that provide detail not easily obtained in paraffin sections.
- The sympathetic trunk should be handled as a peripheral nerve (longitudinal section; transverse section post-fixed in osmium)

Eyes with Optic Nerves

- Both eyes are sectioned in a sagittal plane that includes the optic disk and nerve. If needed, a transverse section of the optic nerve should be handled as a transverse section of a peripheral nerve.

Skeletal Muscle (Thigh Musculature ± Gastrocnemius)

- An evaluation of the nervous system should include examination of a cross section of one or more muscles.

STAINING

Neuronal Necrosis Study

The purpose of the neuronal necrosis study (typically performed 2 to 7 days following initial exposure) is to detect necrotic neurons and/or disintegrating processes (dendrites, axons, synaptic terminals) from those neurons. With standard hematoxylin and eosin (H&E) staining, necrotic neurons

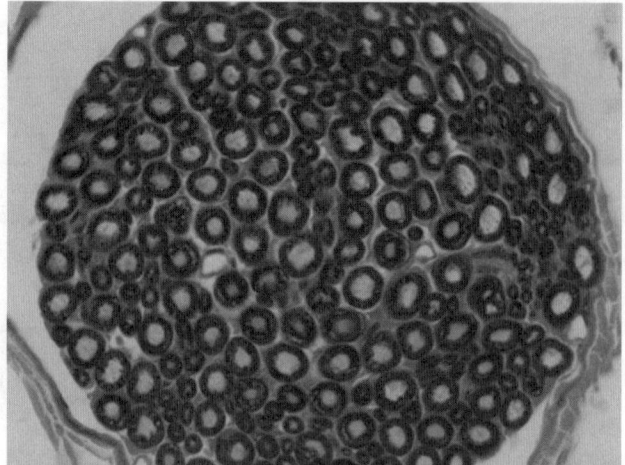

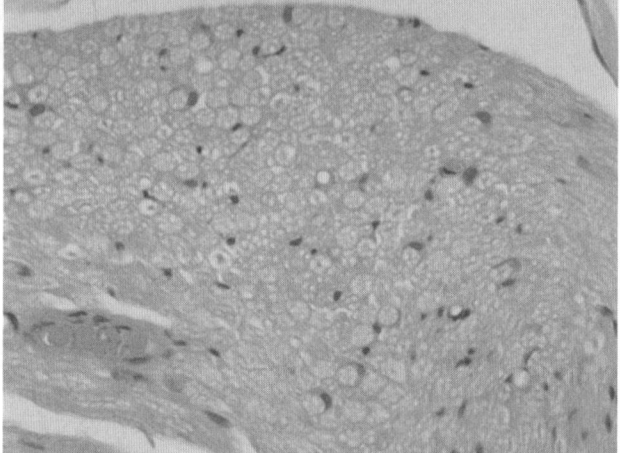

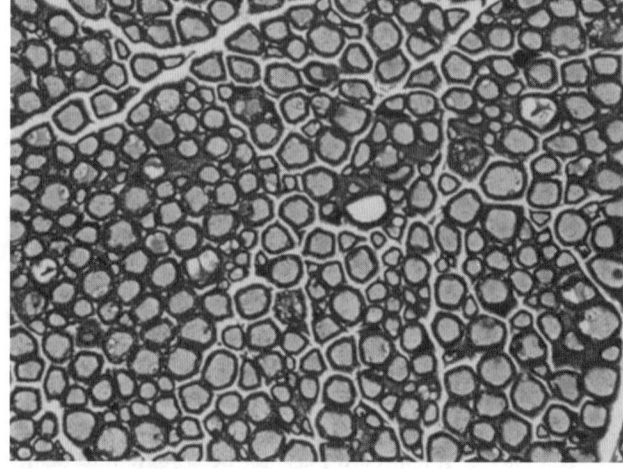

FIGURE 3 The top image is a cross section of a perfusion-fixed (2% paraformaldehyde/2.5% glutaraldehyde) nerve post fixed in 2% osmium for 2 h, embedded in paraffin, sectioned at 5 μm and stained with H&E. The bottom section was similarly fixed, then embedded in Spurr's resin, sectioned at 2 μm and stained with toluidine blue. The center specimen is a formalin immersion–fixed cross section of a nerve embedded in paraffin and stained with H&E but with no osmium post-fixation. The top and bottom specimens provide excellent visualization of the myelin sheaths and demonstrate the usefulness of osmium post-fixation when evaluating nerve cross sections.

appear as shrunken cells with eosinophilic (bright pink) cytoplasm and a condensed nucleus. With extremely careful examination of brain sections, most, and perhaps all, necrotic neurons will be visualized with the H&E stain. Disintegrating axons and dendrites will generally not be visualized readily with this type of staining. For at least some chemically induced lesions, necrotic neurons are autofluorescent even with standard H&E staining.[37]

The use of a specific neuronal necrosis (disintegration) stain is strongly recommended in order to survey the brain for the presence of necrotic neurons and/or disintegrating neuronal processes. The two main staining procedures for the detection of neuronal necrosis are the amino cupric silver stain and the Fluoro-Jade B and C stains.[12,13,15,18] Each has advantages and disadvantages.

The following information is from the perspective of a morphological pathologist examining brain sections for the presence of necrotic neurons and disintegrating neuronal processes. Because pathologist and laboratory expertise and experience vary, so might any individual pathologist's perspective on the best stain to use for a particular circumstance.

Silver disintegration stains (notably, amino cupric silver) offer numerous advantages for the detection of disintegrating and necrotic neurons and their processes. The stain dramatically highlights affected neurons as black staining against a pale to yellow background.[12,13,18] If a group of neurons is affected, it is often possible initially to detect the affected area even before examination with a microscope. Proper staining with a degenerative silver stain requires special fixation (usually, intravascular perfusion with paraformaldehyde and cacodylate-buffered solutions) and the preparation of frozen sections; therefore, this technique is not useful for retrospective evaluations of paraffin-embedded material. The technique has a reputation for being notoriously difficult to master; still, these stains are available. Cupric silver stains may be more effective than Fluoro-Jade at demonstrating persistent, disintegrating portions of axons, dendrites, and synaptic terminals, although both stains do provide for the visualization of these processes.[12,13,15] The exact structure staining with this technique is still unknown but is believed to be due to silver ions complexing with areas of various amino acids that become exposed as proteins degenerate in dead and dying cells.[13]

The disadvantages of the cupric silver stain are that the brains must be perfusion-fixed (for the best results), frozen sections must be used, and the stain is relatively difficult to perform. Still, in my experience, the use of cupric silver stain is the absolute best way to visualize necrotic neurons and, especially, disintegrating neuronal processes. Even when the actual dead cell body has been removed via phagocytosis, the remaining disintegrated elements (especially axons) can be visualized for an extended period of time.

Fluoro-Jade B (or Fluoro-Jade C) is a sensitive stain for neuronal necrosis; this stain has been tested against a variety of known neurotoxicants.[15,38] The advantages of the

Fluoro-Jade B and C procedures are that they are easy to perform, work on paraffin-embedded tissue, do not require special fixation procedures, and can even be performed on immersion-fixed tissues. Fluoro-Jade B may also assist with the identification of degenerating dendrites, axons, and synaptic terminals, although these structures are usually visualized more easily with amino cupric silver stain. Disintegrating axons, dendrites, and synaptic terminals are generally impossible to identify with H&E since the neuropil stains a relatively uniform pink.

The disadvantages of the Fluoro-Jade technique (compared to the amino cupric silver stained sections) are that it requires a fluorescent microscope, reveals less overall morphological detail [although this can be enhanced by the addition of DAPI (diamidino-2-phenylindole), a fluorescent probe that binds DNA and allows visualization of the nucleus when viewed with an ultraviolet filter], and does not generally provide visualization of the disintegrating neuronal processes as effectively as does amino cupric silver.

The exact entity that is staining with Fluoro-Jade B in degenerating neurons is unknown. It is theorized that the process of degeneration, apoptosis, or necrosis produces a strongly basic "degeneration" molecule[38] that attracts the acidic Fluoro-Jade B moiety. Fluorescent stains have a reputation of fading over time especially when exposed to intense light. However, Fluoro-Jade-stained slides are relatively resistant to severe fading under most conditions.[15]

Regardless of the technique used (amino cupric silver stain or a Fluoro-Jade stain), when applied to sections sufficiently close together to provide an exhaustive view of the brain, and when applied to brain sections taken at a time point most likely to contain necrotic or disintegrating neurons and/or neuronal processes, these stains will greatly enhance the pathologist's ability to detect neuronal effects of a neurotoxic compound or therapy. Use of a disintegrative neuronal stain can actually save time in the evaluation of numerous brain sections. Since these stains are sensitive and easily evaluated, brain sections can be surveyed for neuronal damage more quickly than is possible with H&E staining. These stains can also help sort out the question of artifactual change from actual neuronal death.

Other stains may also prove useful for the neuronal necrosis study. Figure 4 provides an example of the usefulness of the neuronal necrosis study when performed at a time point optimal for capturing cell death and glial cell reaction. That figure also shows the futility of using a more traditional stain (in Figure 4, the Nissl stain on a frozen section) and attempting to visualize all the cell changes that may be taking place. Still, for frozen sections, a Nissl stain to identify neurons will be useful for evaluation of the overall morphology of the brain. However, frozen sections will not provide the detail present in a paraffin section.

For paraffin sections, a standard H&E stain will provide excellent morphologic detail and allow any areas of neuronal

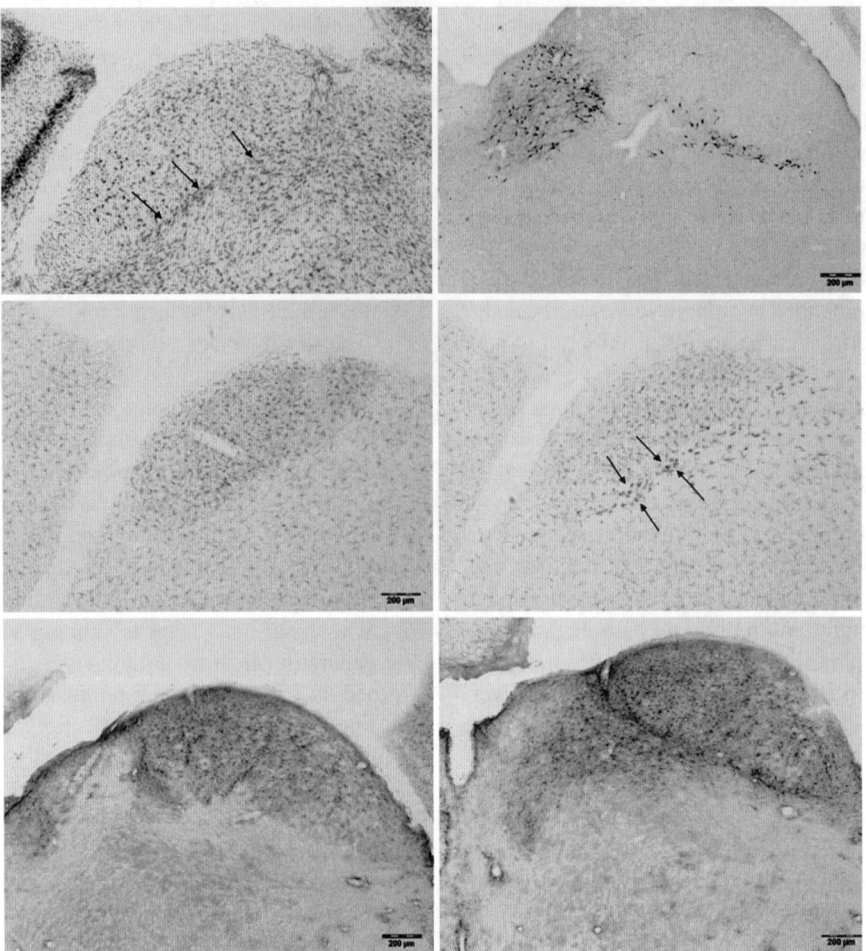

FIGURE 4 Series of images (all taken with a 4× objective with the same camera, microscope, and software) from a mouse treated with MPTP (1-methyl-4-phenyl-1,2,3,6-tetrahydropyridine). MPTP can cause necrosis of dopaminergic neurons of the pars compacta region of the substantia nigra (and the adjacent, medial ventral tegmental region) when administered to some strains of mice and also subhuman primates and humans. These images show the usefulness of various special stains when applied at the correct time to the correct region of the brain. Both of the top two images show the region of the substantia nigra from the same animal. Neuronal disintegration is present in both, but the affected neurons are only visible, at this magnification, in the amino cupric silver stained section on the right where the affected neurons appear as black structures against a pale orange to pink background. The arrows in the left top image indicate the pars compacta region of the substantia nigra. The center images are IBA1-stained (for microglia) sections of the substantia nigra from two animals: a negative control on the left and a MPTP-treated animal on the right. The arrows on the right image show the increased IBA1 staining, indicating a microglial reaction in the pars compacta region. The bottom images are GFAP-stained (for astrocytes) sections of the substantia nigra from two animals: a negative control on the left and a MPTP-treated animal on the right. Note the increased GFAP staining in the right image, staining that forms a linear line in the pars compacta. There is also increased GFAP staining in the adjacent ventral tegmental area (to the left of the pars compacta) of the right lower image. (*See insert for color representation of the figure.*)

necrosis to be viewed in association with the surrounding brain. Necrotic neurons are also visible with H&E stain.

For frozen or paraffin sections, the use of an immunohisto-chemical stain for microglial cells may provide information regarding the presence of clustering or reactive microglia cells. There are numerous specific stains for microglial cells; our laboratory prefers the IBA1 (ionized calcium binding adaptor molecule 1) stain. Microglial reaction, especially in response to an administered biologic that gains entrance into the brain or in response to neuronal necrosis, may take place very quickly (unpublished data). A sensitive method to detect microglial reactions is a very useful tool in detecting even subtle neurotoxicity.

Astrocytes may react somewhat quickly to damage in the central nervous system (Figure 4), although typically less quickly than a microglial reaction (unpublished data). Still, a specific immunohistochemical stain for astrocytes (usually, a stain for glial fibrillary acidic protein) may be useful in the neuronal necrosis study. GFAP (immunohistochemical stain for glial fibrillary acidic protein) is a major constituent of astrocyte filaments. Enhanced expression of GFAP is considered a marker for astrocytic response and therefore a diverse marker for neurotoxicity.[39] Astrocytic reactions may be detectable for months following the initial injury.

Standard Neuropathology Study

Most preclinical toxicology studies are evaluated by embedding the tissues in paraffin. The need to employ special sections/stains not compatible with paraffin embedding (e.g., for stereological investigations) would be one reason to use frozen sections in addition to or instead of paraffin embedding. But many standard neuropathology investigations, especially those done as a part of a larger, general preclinical toxicology study, are performed using paraffin embedding. What follows is one proven approach to evaluating neurotoxicity in a variety of study types. The final approach, built through discussions between the study pathologist, study sponsor, and testing facility, should be matched to the particular study during protocol development. The principal evaluation of the nervous system tissues will generally be performed with H&E-stained slides.

The approach that follows has proven to be successful at detecting neurotoxicity, satisfying internal scientific concerns, and generally addressing regulatory concerns concerning neurotoxicity (although obviously, each study is different). The stains all have a specific purpose, and all may reveal issues not readily apparent with H&E staining. The decision as to whether or not to apply these stains broadly or selectively will be based on the experience of the examining pathologist, the time frame for submission of the pathology report, and the resources (scientific and financial) available.

- *All brain sections* (including recuts of the original sections for retrospective examinations): H&E for general morphology, Fluoro-Jade B for neuronal degeneration (assuming paraffin sections; amino cupric silver or Fluoro-Jade B or C may be used on frozen sections), GFAP as a general marker of neurotoxicity, IBA1 for microgliosis, and a nonselective silver stain (e.g., Bielschowsky's) or neurofilament stain (immunohistochemical stain) to visualize axons, dendrites, and other neurofilament-containing neuronal processes. With current staining techniques, it is possible to do double

or triple labeling of immunohistochemical stains. For example, GFAP, IBA1, and a neurofilament stain can all be applied to the same paraffin section. This use of double and triple labeling is very suitable for associating the various glial reactions and how they may be associated with other areas of damage. Note that the need to perform a neuronal necrosis stain in studies of 28 days or more is somewhat variable if a specific neuronal necrosis study has been performed at a dose sufficient to detect the damage. Of those mentioned, increased GFAP expression may be the most persistent marker of cell injury in the central nervous system[18,40] and therefore in some studies may be the only evidence of a prior injury.

- *All spinal cord sections:* H&E, GFAP, IBA1, and Bielschowsky's silver or neurofilament stain. Neuronal disintegration stains are less important for spinal cord sections since the number of neurons is somewhat limited and the organization of the neurons allows a sensitive evaluation with just the H&E stain. However, there is no disadvantage to using neuronal disintegration on spinal cord sections.

- *All peripheral nerve sections:* H&E and Bielschowsky's silver for longitudinal sections embedded in paraffin, toluidine blue staining on cross sections (post-fixed in osmium) embedded in resin (Epon, Spurr's), or H&E or toluidine blue staining on cross sections (post-fixed in osmium) (Figure 3) provides a very sensitive assessment of the myelin content of the nerve. These sections, if they are of sufficient quality, may also allow a morphometric determination of nerve fiber sizes if there is a suspicion that a nerve lesion may be preferentially affecting a particular size range of nerve fibers.

- *All spinal nerve root and ganglia sections:* H&E and Bielschowsky's silver if embedded in paraffin or glycolmethacrylate, toluidine blue if post-fixed in osmium tetroxide and embedded in resin or comparable media.

Other stains that may augment evaluation of the nervous systems include luxol fast blue or myelin basic protein stains for myelin, lectin stains for microglial cells, Nissl stains for neuronal detail, and a variety of immunohistochemical stains to characterize neuronal populations. All may be useful for a particular study, but a complete discussion is beyond the scope of this chapter. In particular, myelin stains (such as Luxol fast blue) are commonly used to detect neurotoxicity. This stain can be combined with a silver stain to produce sections that specifically stain myelin and the axonal components of nerve fibers.

EVALUATION

The evaluation of nervous system tissue is no different than any other tissue. The pathologist is usually looking for nothing in particular and everything in general. For example, departures from normal structure, evidence of inflammation or other cellular reactions (astrocyte enlargement; microglial clustering), axonal swellings, dilation of myelin sheaths, neuronal or glial cell necrosis, vacuolation, aberrant staining patterns, and dilation of the ventricular system are all fairly common microscopic lesions in the nervous system.

It is useful to present the data from an in-depth examination of the nervous systems (utilizing appropriate sectioning and staining methods) in a manner that conveys to a reviewer the extent of the microscopic evaluation. In other words, just listing "brain" may not alert a reader immediately that extra efforts were conducted to examine the nervous system. Internal standard operating procedures, individual preferences of the study sponsors, or specific needs based on regulatory or scientific concerns may influence how the data are presented. A much more in-depth discussion of the neuropathology report is provided in Chapter 32.

One sample list, which is by no means exhaustive, would be to record diagnoses for the following: brain, meninges; brain, ventricular system; brain, choroid plexus; brain, olfactory bulbs (dogs, rodent); brain, basal forebrain; brain, basal nuclei; brain, amygdala; brain, frontal cortex; brain, piriform cortex; brain, parietal cortex; brain, temporal cortex; brain, occipital cortex; brain, thalamus, anterior; brain, thalamus, posterior; brain, hypothalamus; brain, hippocampus; brain, midbrain; brain, pons; brain, cerebellum; and brain, medulla oblongata. Other references are available to provide assistance in this area.[25]

The method of evaluation will be determined by each individual pathologist. Typically, tissues are examined by animal; that is, all the tissues for one animal are examined prior to beginning the examination of another animal. For some studies, reading a given tissue or tissue level (specifically for the brain and spinal cord) across all animals may be beneficial both scientifically and from a time management aspect. In studies where a device or drug delivery system is present (e.g., a catheter in the intrathecal space in the spinal cord region), it is very important to judge the reaction associated with the device in a very consistent manner. Pathologists are familiar with diagnostic "drift,"[41] where slight shifts in diagnostic terminology and/or grading take place in studies that are evaluated over extended periods of time. To decrease the likelihood of this occurring, and to produce the highest level of consistency, it is recommended to evaluate the area traversed by the device in as concise a period of time as possible. In addition, it is useful to have a predetermined set of diagnoses that are going to be graded (of course, other diagnoses can be added if they are encountered). This requires experience with the particular type of device and/or study.

Each pathologist has his or her own method of tissue evaluation, based on experience, knowledge, and the time available. The method may change from study to study. One approach to the evaluation of the brain is a systematic method applicable to the determination of the no-observed-effect level and/or the no-observed-adverse-effect level and is suitable to most studies:

- Evaluate several control animals first. This helps to establish a baseline for the spontaneous changes that may be present in the population of animals used in the specific study.

- After at least several control animals are evaluated, one or more high-dose animals are examined. This helps to establish the potential test article–related changes.

- Following evaluation of the high-dose animals and possibly the remainder of the control animals, the mid-dose and then the low-dose animals are examined to determine to what dose level the test article–related changes are present.

Within each group, the following order for slide evaluation generally proves effective in the detection of test article–related changes.

- Review the positive control slides for the various special stains. The purpose of a positive control stain is to verify the accuracy of the staining procedure. If the positive controls did not stain properly, restaining is necessary.

  - For the neuronal disintegration stains, a separate positive control is absolutely essential. For example, sections of brain from rats treated with kainic acid will consistently have necrotic neurons in the hippocampus and other locations.[18,42] These sections are generally reliable positive controls for the neuronal disintegration stains amino cupric silver and the Fluoro-Jade stains.

  - For the glial stains (GFAP, IBA1, etc.), and the nonselective silver stains for neurofilamentous structures, pathologists vary as to their preference for a separate positive control, as each brain, or at least the brains of the study control animals, can serve as a suitable positive control. Each laboratory and each pathologist, based on their expertise and scientific experience, will have their own procedure for positive controls for these stains.

- Examine the stains for the detection of neuronal necrosis. These stains (cupric silver; Fluoro-Jade B or C) represent a very specific and sensitive means of detecting neuronal necrosis. If no neuronal necrosis is detected following the proper evaluation of these stains, it is very unlikely that neuronal necrosis will be noted on the H&E-stained sections. This may actually decrease the time required to examine the H&E-stained sections adequately.

- Examine the H&E-stained sections (for those studies where the tissue is paraffin embedded). Regardless of the study or the staining battery, it is likely that most of the information on the morphological changes in the tissues is going to be derived from the H&E-stained sections. The H&E stain is the procedure used for most preclinical toxicity studies, and it is this stain with which most pathologists have the most experience.

- Examine the other special stains (GFAP, IBA1, luxol fast blue, nonselective silver, neurofilaments, etc.).

  - Each special stain generally has one primary specific purpose. Generally, much less time will be spent on any given special staining procedure (as compared to the H&E-stained sections) because each stain has such a narrow focus.

  - Although all stained sections should be examined, specific attention needs to be directed to sections where a lesion was noted previously with neuronal necrosis and/or H&E-stained sections.

  - Gross and microscopic changes noted previously (with the stains evaluated previously) may provide some hints as to possible or probable lesions that may be detectable with the special staining procedures. For examples, areas of neuronal necrosis may be associated with an astrocyte and/or microglial reaction (Figure 4). Areas of nerve fiber degeneration, noted with the H&E staining, may contain areas of swollen axons easily detectable using nonselective silver stain. Areas devoid of myelin can usually be detected with the H&E stain and will usually be easily seen with a Luxol fast blue (or similar) stain.

  - The osmium post-fixed cross sections of the peripheral nerves will allow for very precise interpretation concerning the normality of the myelin content.

REFERENCES

1. Solleveld H, Boorman G. Brain. In: Boorman G, Montgomery S, MacKenzie W, eds. *Pathology of the Fischer Rat Reference and Atlas.* San Diego, CA: Academic Press; 1990:155–177.

2. Bolon B. Comparative and correlative neuroanatomy for the toxicologic pathologist. *Toxicol Pathol.* 2000;28:6–27.

3. Garman R. Evaluation of large-sized brains for neurotoxic endpoints. *Toxicol Pathol.* 2003;31:32–43.

4. Paxinos G, Watson C. *The Rat Brain in Stereotaxic Coordinates,* 4th ed. San Diego, CA: Academic Press; 1998.

5. Singer M. *The Brain of the Dog in Section.* Philadelphia: W. B. Saunders; 1962.

6. Paxions G, Franklin KB. *The Mouse Brain in Stereotaxic Coordinates,* 2nd ed. San Diego, CA: Academic Press; 2001.

7. Saleem KS, Logothetis NK. *A Combined MRI and Histology Atlas of the Rhesus Monkey Brain in Stereotaxic Coordinates.* San Diego, CA: Academic Press; 2007.

8. Broxup B. Neuropathology as a screen for neurotoxicity assessment. *J Am Coll Toxicol.* 1991;10:689–695.

9. Mattsson J, Eisenbrandt D, Albee R. Screening for neurotoxicity: complementarity of functional and morphologic techniques. *Toxicol Pathol.* 1990;18:115–126.

10. Garman RH, Fix AS, Jortner BS, et al. Methods to identify and characterize developmental neurotoxicity for human health risk assessment: II. Neuropathology. *Environ Health Perspect.* 2001;109(1):93–100.

11. Garman R. Artifacts in routinely immersion fixed nervous tissue. *Toxicol Pathol.* 1990;18:149–153.

12. Switzer R. Strategies for assessing neurotoxicity. *Neurosci Biobehav Rev.* 1991;15:89–93.

13. Switzer R. Application of silver degeneration stains for neurotoxicity testing. *Toxicol Pathol.* 2000;28:70–83.

14. Eisenbrandt DL, Allen SL, Berry PH, et al. Evaluation of the neurotoxic potential of chemicals in animals. *Food Chem. Toxicol.* 1994;32(7):655–669.

15. Schmued LC, Hopkins KJ. Fluoro-Jade B: a high affinity fluorescent marker for the localization of neuronal degeneration. *Brain Res.* 2000;874:123–130.

16. Vinters HV, Kleinschmidt-Demasters BK. General pathology of the central nervous system. In: Love S, Louis DN, Ellison DW, eds. *Greenfield's Neuropathology,* 7th ed. London: Hodder Arnold; 2008:1–62.

17. Oldfield BJ, McKinley MJ. Circumventricular organs. In: Paxinos G, ed. *The Rat Nervous System,* 2nd ed. San Diego, CA: Academic Press; 1995:391–403.

18. Fix A, Ross J, Stitzel S, Switzer R. Integrated evaluation of central nervous system lesions: stains for neurons, astrocytes, and microglia reveal the spatial and temporal features of MK-801-induced neuronal necrosis in the rat cerebral cortex. *Toxicol Pathol.* 1996;24:291–304.

19. Jackson-Lewis V, Jakowec M, Burke RE, Przedborski S. Time course and morphology of dopaminergic neuronal death caused by the neurotoxin 1-methyl-4-phenyl-1,2,3,6-tetrahydropyridine. *Neurodegeneration.* 1995;4:257–269.

20. Chang LW. Neuropathology of trimethyl tin: a proposed pathogenetic mechanism. *Fundam Appl Toxicol.* 1986;6: 217–232.

21. Bagnell R, Langaman C, Madden V, Suzuki K. Ultrastructural methods for neurotoxicology and neuropathology. In: Chang L, Slikker W, eds. *Neurotoxicology: Approaches and Methods.* San Diego, CA: Academic Press; 1995:81–98.

22. Cassella J, Hay J, Lawson S. *The Rat Nervous System: An Introduction to Preparatory Techniques.* New York: John Wiley; 1997.

23. Fix A, Garman R. Practical aspects of neuropathology: a technical guide for working with the nervous system. *Toxicol Pathol.* 2000;28:122–131.

24. McDowell EM, Trump BF. Histologic fixatives suitable for diagnostic light and electron microscopy. *Arch Pathol Lab Med.* 1976;100:405–414.

25. Garman R. Evaluation of large-sized brains for neurotoxic endpoints. *Toxicol Pathol.* 2003;31:32–43.

26. Switzer RC. Perfusion Protocol. http://www.neuroscience-associates.com/perf-protocol.htm. Accessed June 14, 2010.

27. Kiernan JA. Formaldehyde, formalin, paraformaldehyde and glutaraldehyde: what they are and what they do. *Microsc Today.* 2000;1:8–12.

28. Santoreneos S, Stoodley MA, Jones NR, Brown CJ. A technique for in vivo vascular perfusion fixation of the sheep central nervous system. *J Neurosci Methods.* 1998;79:195–199.

29. CMiko TL, Gschmeissner. Histological methods for assessing myelin sheaths and axons in human nerve trunks. *Biotech Histochem.* 1994;69(2):68–77.

30. Pender MP. A simple method for high resolution light microscopy of nervous tissue. *J Neurosci Methods.* 1985;15:213–218.

31. Mimenez-Andrade JM, Herrera MB, Ghilardi JR, Vardanyan M, Melemedjian OK, Mantyh PW. Vascularization of the dorsal root ganglia and peripheral nerve of the mouse: implications for chemical-induced peripheral sensory neuropathies. *Mol Pain.* 2008;4:10.

32. Evans H. *Miller's Anatomy of the Dog*, 3rd ed. Philadelphia: W.B. Saunders; 1993:780–787.

33. Silverthorn DU. *Human Physiology: An Integrated Approach*, 4th ed. San Francisco: Benjamin Cummings; 2009:379–386.

34. Latendresse JR, Warbrittion AR, Jonassen H, Creasy DM. Fixation of testes and eyes using a modified Davidson's fluid: comparison with Bouin's fluid and conventional Davidson's fluid. *Toxicol Pathol.* 2002;30(4):524–533.

35. Jordan WH, Hall DG, Young JK, Hyten MJ. Practical rat neuropathology. *Histotechnol.* 2007;30(2):115–120.

36. Organisation for Economic Co-operation and Development. *Neurotoxicity Study in Rodents.* Test 424. Paris: OECD; 1997. Available at: http://titania.sourceoecd.org/vl=564424/cl=52/nw=1/rpsv/ij/oecdjournals/1607310x/v1n4/s24/p1.

37. Sharma AK, Jordan WH, Reams RY, Hall DG, Snyder PW. Temporal profile of clinical signs and histopathologic changes in an F-344 rat model of kainic acid induced mesial temporal lobe epilepsy. *Toxicol Pathol.* 2008;36:932–943.

38. Ye X, Carp R, Schmued L, Scallet A. Fluoro-Jade and silver methods: application to the neuropathology of scrapie, a transmissible spongiform encephalopathy. *Brain Res Brain Res Protoc.* 2001;8:104–112.

39. O'Callaghan J, Sriram K. Glial fibrillary acidic protein and related glial proteins as biomarkers of neurotoxicity. *Expert Opin Drug Saf.* 2005;4:433–442.

40. Buss A, Brook GA, Kakulas B, et al. Gradual loss of myelin and formation of an astrocytic scar during Wallerian degeneration in the human spinal cord. *Brain.* 2004;127:34–44.

41. Crissman JW, Goodman DG, Hildebrandt PK, et al. Best practices guideline: toxicologic histopathology. *Toxicol Pathol.* 2004;32:126–131.

42. Schmued LC, Stowers CC, Scallet AC. Fluoro-Jade results in ultra high resolution and contrast labeling of degenerating neurons. *J Brain Res.* 2005;1035:24–31.

22

PATHOLOGY METHODS IN NONCLINICAL NEUROTOXICITY STUDIES: THE DEVELOPING CENTRAL NERVOUS SYSTEM

WOLFGANG KAUFMANN

Experimental Toxicology and Ecology, BASF SE, Ludwigshafen, Germany

INTRODUCTION

In 1990, the first guideline for testing "developmental neurotoxicity" in an experimental animal model was discussed and published by the U.S. Environmental Protection Agency (EPA) (OPPTS EPA 870.6300). The background for this initiative was primarily the increasing epidemiological information on the impact of certain environmental chemicals and drugs on central nervous system (CNS) development in children. For the first time, this guideline included neurohistopathological examination of the CNS at stages where neurodevelopment is still ongoing. Questions arose immediately regarding whether pathology methods used for the neuropathology assessment of the adult nervous system would be sufficient for examination of the developing nervous system as well. The answer given by the scientific community at that time was "in principle, yes," but there are some important age-specific characteristics that must be recognized if the correct histotechnical procedures are to be performed or the correct neuropathological distinction between normal development or toxicant-induced maldevelopment of evolving neurological structures is to be made. This chapter focuses on currently recommended procedures for performing developmental neuropathology evaluations,[1-3] especially those that address the regulatory requirements of the EPA[4,5] and the Organization for Economic Cooperation and Development (OECD).[6,7] In some cases, the scientific community has not decided on a single best practice. In these instances, several alternative approaches are presented so that the researcher can decide the most appropriate study design based on the most recent scientific discussions, coupled with knowledge of the specific chemical properties of the test article.

DESIGNING THE DEVELOPMENTAL NEUROPATHOLOGY

Selection of the Species

The current developmental neurotoxicity (DNT) guidelines (OPPTS EPA 870.6300 and OECD 426 Technical Guideline) recommend outbred rats as the preferred species (Wistar or Sprague–Dawley strains; not Fischer 344, as developmental timings are different). The distinct advantages in using rats for DNT testing are summarized by Meyer[8]: (1) small size and relatively low purchase and maintenance costs; (2) ease of handling, treatment, and testing; (3) simplicity of breeding in the laboratory setting; (4) short gestation periods; and (5) relatively large litter sizes, thereby providing many offspring for the study. Rats have long been the general species of choice for neurobiology research, so the methods and normal characteristics of macro- and microneuroanatomy, neurophysiology, and behavior are well mapped in this species. The significant similarities between rat and human neural development[9,10] make the rat model a good predictor for the impact of xenobiotic exposure on brain development in humans. The concordance between the behavioral and structural changes in humans and rats exposed to the same

Fundamental Neuropathology for Pathologists and Toxicologists: Principles and Techniques, First Edition. Edited by Brad Bolon and Mark T. Butt.
© 2011 John Wiley & Sons, Inc. Published 2011 by John Wiley & Sons, Inc.

developmental neurotoxicants is, in many cases, impressively high.[11] For example, animal and human data on the neurodevelopmental impacts of ethanol correlate quite well, even with regard to subtle changes; this good parallel response is also shown for other well-known developmental neurotoxicants.[12] However, rats, mice, and hamsters are *altricial* (born at relatively underdeveloped stages with eyes shut), so many neurogenic events occur postnatally in these species which take place during late gestation in primates.[13] Humans, rhesus macaques (the leading laboratory primate species), and guinea pigs are *precocial*, close to independence at birth. These classifications reflect the different neurodevelopmental stages at birth between rats and humans and have to be kept in mind when considering the experimental design of a DNT study. The human day of birth is translated to rat postnatal day (PND) 12 to 13.[14]

Although to the author's knowledge mice have not been used in regulatory DNT studies to this date, they are currently the most important model species for brain research purposes, as they are most amenable to genetic manipulations. For DNT studies, mice may be considered in specific cases only, when the rat model is proven to be less susceptible. For examples DNT studies with test articles with structural similarity to MPTP (1-methyl-4-phenyl-4-tetrahydropyridine) or its neurotoxic metabolite MPP^+ should consider mice as the species of choice, as the dopaminergic neurons (the target site of MPP^+) in rats are reported to be relatively resistant to MPTP-induced neurotoxicity.[15]

Number of Animals

For DNT testing, the neuropathology cohort should have a minimum of 10 animals per sex, group, and time point to provide enough statistical power for neuropathologic and morphometric analysis. In the classical DNT study, a minimum of 20 litters per group with ideally four male and four female offspring per litter are necessary. One pup per litter[6] is chosen to serve as the litter representative for the DNT neuropathology examinations.[2] On a case-by-case basis, the standard neuropathology examination can be augmented with an additional neuropathology cohort, as long as the integrity of the original tests required is not compromised.[6] Depending on the mode of action of the compound under test, satellite animals might need to be specially prepared to enable evaluation by special morphologic techniques in an extended, second-tier neuropathology approach. For example, if a neurotoxicant is known to interfere with the development of certain neurotransmitter systems (e.g., for compounds with a suspected Parkinson's disease–like pattern), examination of sagittal brain sections may be indicated in addition to the routinely used coronal ("transverse") sections to ensure that the appropriate neurotransmitter pathways may be examined in as much detail as possible. When extensive morphometric measurements are needed (e.g., additional stereological assessments of brain area, volume, or number of neurons), additional brain tissue is necessary in order to undertake this special histotechnical approach.[3]

Time Points for Examination

The right time to look, where to look, and what to look for are extremely important considerations when planning the study design for the neuropathology part in DNT studies.[3] The existing guidelines for DNT testing require neuropathology examination of one juvenile age (PND 11 *or* PND 22, *or* at an earlier time point between PND 11 and PND 22), and at a young adult age (PND 60 to 70).

PND 11 (Pups)/PND 22 (Adolescents) The juvenile age evaluation addresses primarily the impact of toxicants on such fundamental neurodevelopmental processes as neuronal and glial cell proliferation, migration, and settling, which are mainly completed by PND 22 at weaning in rats (but still ongoing at PND 11). This time point also addresses chemical disruption of the *brain growth spurt*, the stage during which rapid oligodendrogliogenesis (sometimes termed *myelination gliosis*) leads to a doubling of brain weight between birth (PND 0, Wistar rats: mean values around 500 mg) to PND 11 (Wistar rats, mean values around 1050 mg), and a further gain in brain weight of approximately 25% between PND 11 and PND 22 (Wistar rats, mean values around 1250 mg). As a rule, in DNT studies this first time point of neuropathology examinations is also the time when treatment of dams with the test agent is halted.

The PND 11 time point is still a requirement in the EPA OPPTS 870.6300 Developmental Neurotoxicity Study Guideline.[4] Selection of PND 11 will permit examination of the dynamic structural changes that occur in rapidly growing neural tissues and transient structures (e.g., external germinal matrix layer of the cerebellar cortex). However, examinations at PND 11 will often render recognition of neurotoxicant-induced damage difficult, as the rapid remodeling in rodent neural centers at this time may mask the effects of neurotoxic agents on the rapidly evolving neural structures. The examination of pup brains during the brain growth spurt will generate a greater divergence in brain size and interindividual variability in brain weights, which might obscure the ability to detect subtle dose-dependent differences related to xenobiotic exposure.[1-3] In addition, due to the lack of myelination at PND 11, the brains of pups are much more water-rich and friable in nature, which can cause substantial problems during histological sectioning (e.g., when attempting to produce homologous coronal sections from multiple animals for morphometry). These biological and technical factors have to be balanced carefully when selecting the age at which to perform the DNT neuropathology evaluation.[2]

The PND 22 time point is recommended by the recent OECD 426 Developmental Neurotoxicity Study Guideline.[6] This document allows some flexibility in choosing a time point between PND 11 and PND 22, if scientifically justifiable. However, the PND 22 time point best covers the purpose of the DNT study to expose the developing nervous system to test agents at all critical developmental periods during gestation and early childhood in the rat model; this approach is desirable, as it encompasses more of the neurodevelopmental events that occur in human infants and toddlers. At the time of weaning, the PND 22 brain is structurally mature, so the tissue is well suited to all standard procedures for neuropathology preparation and analysis. It is more firm (as myelination is well advanced, although not complete) and the brain size is less variable. This consistency significantly improves the acquisition of homologous brain sections for quantitative analysis. Thus, at this time the PND 22 time point is favored as the most serviceable early time point for neuropathology evaluation in DNT studies.[2]

PND 60–70 (Young Adults)   The adult age of PND 60–70 remains the time point recommended by regulatory guidelines where residual neurotoxic effects on brain development and maturation well after treatment has ended are to be evaluated.[4,6] At this age, the rat brain shows adultlike functional as well as anatomical maturity. This time point serves to differentiate whether neuropathology changes observed at the earlier (PND 11 or 22) time point were transient or persistent. The fundamental processes of brain maturation during the young adult period—*neuronal differentiation*, *synaptogenesis*, and *myelination*—may be compromised due to treatment-related neurodevelopmental alterations that arose during the fetal, embryonic, and/or postnatal treatment periods (GD 6-PND 10/21). For example, toxicant-induced reductions in synaptogenesis may lead to a decreased brain size, which can be detected initially at PND 60–70. As true neurodevelopmental impacts always result in persistent qualitative and/or quantitative changes in the adult brain, in DNT studies the neuropathology examination on PND 60–70 is therefore the crucial structural examination capable of covering all basic neurodevelopmental processes that occur during the life of the animal.

For the moment, later time points are not a focus for developmental neuropathology examinations in DNT studies. However, exposures in utero have long been thought to be important determinants of certain cancers that occur in children and young adults. It is now generally accepted that the fetus and young child may be more susceptible than the adult to the effects of ionizing radiation, a premise first indicated by epidemiological observations made more than 40 years ago (Oxford survey of childhood cancer) when diagnostic radiographs of pregnant women were related to the subsequent development of leukemia and other cancer in their children.[16,17] Experimental studies in pregnant animals or neonates have shown that mutagen exposures (e.g., ethylnitrosourea, methylazoxymethanol) can promote tumor induction at an early age, including neoplasia in such atypical locations as the nervous system. Nervous system tumors after transplacental or neonatal exposures of chemical carcinogens are described for rats, mice, hamster, opossum, and the rabbit.[16,18] This capacity is unusual because the rate of neural cell division and the number of neural stem cells are relatively low compared to other tissues. More research on these topics may give a clearer indication of the future value of incorporating later time points during DNT testing.

FIXATION PROCEDURE

In principle, an in vivo fixation method (perfusion fixation) is required for proper tissue preservation during the neuropathology part of neurotoxicity studies in the current adult[5] and developmental[4,6] neurotoxicity study guidelines. However, immersion or perfusion fixation may be performed at the early (PND 11 or PND 22) time points during conventional DNT studies as a concession to the difficulty of perfusing small animals.[6]

Perfusion Fixation

Perfusion fixation of adolescent (PND 22) and young adult (PND 60–70) rats is considered the best practice for preserving the neural architecture for the DNT neuropathology evaluation.[2,3] It is the favored procedure for the nervous system (including the eyes and other sense organs) because it provides superior tissue quality and resolution power for qualitative assessment and detection of subtle structural lesions during light-microscopic examination. A whole-body fixation procedure is necessary for the nervous system. It can also be applied to PND 11 pups with good success, when the infusion pressure is carefully controlled and run at the lower range of the systolic blood pressure (approximately 100 to 120 mmHg). This will prevent vascular rupture and artifactual distortion of brain tissue.[2]

Fixative   For a conventional DNT study without an electron microscopic examination, a neutral buffered 4% formaldehyde solution (i.e., 10% formalin) is the fixative of choice. It readily enables a suitable light-microscopic assessment and any further immunohistochemistry that may be requested. However, other fixatives are often preferred for the fixation of neural tissue (e.g., mixtures of 2.5% formaldehyde and 2% glutaraldehyde, according to Karnovsky).[19] Glutaraldehyde provides superior (because irreversible) cross-linking of proteins, which is particularly important in nervous tissue because of its high lipid content and general resistance to fixation.[20] Glutaraldehyde is the fixative of choice for all electron microscopic applications in neural tissue. When using glutaraldehyde as the chief or only fixative, one must remember that this is an inappropriate solution for long-term storage of fixed tissue. Extended holding in glutaraldehyde will lead to significant shrinkage and hardening of tissue samples over time, which will irreversibly impede or prevent any quantitative analysis of brain structures at a later time. For a review of appropriate fixatives for the nervous system, including their advantages and disadvantages, see also the article by Fix and Garman.[20]

Perfusion Procedure Directly before anesthesia, the test animals are given 60 mg/kg sodium heparin i.p. to prevent intravascular coagulation of blood, especially in small vessels. Before perfusion fixation, test animals are deeply anesthetized (e.g., with isoflurane). The whole-body perfusion fixation procedure is initiated with a 1 to 2-min pre-flush (until blood is removed from the vascular tree) with the buffer solution used in preparing the fixative. This step is performed by implanting a blunt cannula (smaller gauge for younger animals, no. 21 to 25; for adults, no. 19 to 21) into the left heart ventricle. The solution is infused in a constant fashion, ideally using a perfusion pump, by strictly maintaining the systolic pressure between 100 and 150 mmHg. When starting infusion, the right heart auricle is opened rapidly (within 5 to 10 s, when the right atrium is just starting to enlarge) to prevent any rise of pressure in the vascular system that might rupture microvessels in the brain. It is extremely important that the entire procedure not be interrupted and that the infusion pressure stay within the optimal pressure range over the entire fixative time of 7 to 10 min. Successful perfusion is recognized by blanching of the internal organs and stiffening (extension) of the distal limbs. This appearance is especially true when using Karnovsky's or glutaraldehyde fixatives; the rigidity of the carcass is dependent on the concentration of the glutaraldehyde solution. Stiffening of the distal limbs is not a reliable signal of a successful perfusion with formaldehyde fixative only.

With respect to performing morphometric measurements in DNT studies, one technical artifact that occurs after perfusion fixation has to be recognized. This change is *dilation of the ventricles*, which has been reported in perfusion-fixed rat brains from both adult neurotoxicity (ANT) and DNT studies.[3] This change may affect the width of periventricular brain areas that are subject to linear measurements performed according to the requirements of the current DNT guidelines, which require a "simple" morphometric measurement. Pathologists should be aware of this technical artifact and should exclude from the quantitative analysis brain areas exhibiting this effect.

Following perfusion, the calvaria over the brain and all dorsal arches of all vertebrae are removed as gently and as soon as possible. The skin and muscles of the hindlimbs are partially dissected to expose peripheral nerves. All of this is done with the greatest care not to touch, press, or stretch the underlying neural tissues. After nervous system dissection, the carcass is immersed in the same fixative, taking care that the exposed neural tissues of floating carcass are in contact with the fixative solution. This procedure increases the fixation time, which is desirable in those instances when the initial perfusion was suboptimal. After 24 to 48 h of post-fixation, the neural tissues are removed from the carcass. Eyes are removed and post-fixed further in formalin. Under no circumstances should a full dissection of the nervous system be attempted directly after the perfusion procedure, as manipulation of incompletely fixed tissues will induce significant pressure artifacts in the delicate neural parenchyma. The chief such artifact at the microscopic level are *dark neurons*, which are frequently misinterpreted as true lesions by inexperienced workers tasked with the neuropathology portion of the DNT study.[21] The occurrence of dark neurons complicates the detection of true alterations in neuronal elements. Therefore, some investigators recommend that the calvaria and vertebral arches not be opened until the carcass has been post-fixed by immersion for several hours in a refrigerator (4°C). This delay can minimize pressure artifacts, especially in superficial tissues,[2] but can result in modestly less preservation of neural tissues. In the neuropathology arm of a conventional DNT study, the immediate removal of the calvaria and vertebral arches to permit continued immersion fixation of neural tissues in situ is the best compromise for achieving good-to-excellent tissue quality where (1) 80 animals have to be perfusion-fixed over a brief period (typically about 4 days, depending on the delivery date of the dams), and (2) some animals will not have perfect perfusion (even by skilled and experienced technicians).

Immersion Fixation

Immersion fixation of the central nervous system is an accepted procedure for the early developmental age of PND 11[4,6] and is occasionally employed at PND 22 (even though perfusion fixation is favored at the later time). The disadvantages of immersion fixation in adult brains are not as relevant in the PND 11 rat brain, because the immature organ is smaller in size and contains much less myelin (i.e., has a relatively larger proportion of proteins available for cross-linking) and significantly more water.

Fixative Neutral buffered 10% formalin is the fixative of choice for immersion fixation in DNT studies. Bouin's fixation may be considered for PND 11 pup brains to harden the neural tissue, which can improve the integrity of sections and homology of levels across animals within the study.[2] However, the marked dehydration elicited by Bouin's solution may cause such substantial brain shrinkage that meaningful quantitative measurements of brain structures cannot be obtained.[2] It should be emphasized that immersion fixation in formalin is not the optimal procedure for preserving the delicate sensory and nonsensory retina. Instead, eyes and optic nerves should be removed and immersed in either Bouin's or modified Davidson's solution, with the latter given preference because it is less toxic but gives equivalent fixation.

Perfusion Procedure At necropsy, the head is separated from the carcass at the atlas vertebra (i.e., C_1), and the calvaria is removed as soon as possible. The brain is then fixed in situ for at least 48 h, after which it is removed with great care to prevent fragmenting of the tissue. Brain removal can be a difficult challenge for technicians, especially if morphometric measurements of brain require that homologous brain sections be obtained. Therefore, we do not recommend taking the PND 11 time point as the only early time point in DNT studies, except when scientific reasons clearly justify it. As a rule, the design of DNT studies permits the inclusion of a third cohort at PND 11 for additional neuropathology examinations. In these cases, the PND 11 brains should be collected, fixed in situ, and preserved as wet tissue. Depending on the neuropathology data acquired at the other time points (PND 22 and PND 60–70), the PND 11 brains may be analyzed post hoc to clarify borderline data.

BRAIN WEIGHT DETERMINATION

Brain weights may serve as good first indicators of any toxicant-induced aberrations on neural growth. The gains in body weight and brain weight during development strictly follow an *allometric relationship*. Accordingly, the relative brain weight (i.e., brain weight as a percentage of total body weight) is a valuable means of rapidly assessing major changes in neurodevelopment. Typical values for relative brain weight in Wistar rats are 5.65% (PND 11), 3.29% (PND 22), and 0.71% (PND 60–70) for males, and 5.55% (PND 11), 3.28% (PND 22), and 1.04% (PND 60–70) for females. Historical controls from one test facility are given in Table 1. A significant decrease in the relative brain weight during a DNT study indicates a direct toxicant-induced impact on brain growth. In contrast, a significant decrease only in absolute brain weight is seen after severe general developmental toxicity leading to a retardation of body weight gain.

If perfusion fixation is performed (as is typical in DNT studies), brain weights can be determined only in post-fixed specimens. The weight of all fixed brains is taken after a consistent period of in situ fixation. Such standardization (with respect to the amount of fixative and the duration of fixation) is required to reduce potential confounding influences on the brain weight values. Given standard conditions, weighing of fixed brains will give reliable information. Weights from fixed and unfixed brains should not be pooled

for analysis, as fixation affects brain weight unpredictably by removing water content from the neuropil and by filling the ventricular cavities of the brain.[2]

Ideally, brain weights are taken at necropsy for unfixed organs. For this reason, a separate neuropathology cohort is recommended in the OECD 426 guideline[6] for both the early (PND 11/22) and late (PND 60–70) time points. As delayed fixation necessitated by brain removal and weighing will alter the tissue quality significantly, those brains used for weighing will not be evaluated by histopathology. The major disadvantage of this arrangement is that qualitative and quantitative light-microscopy assessments in perfusion-fixed animals cannot be directly correlated to the brain weights acquired in fresh tissue. The primary advantage of taking unfixed brain weights is that they exactly reflect the in vivo situation. However, as brain weights are relatively crude indices of neural abnormalities relative to the qualitative and quantitative light-microscopic measurements, the necessity of obtaining unfixed brain weights can often be avoided if post-fixation brain weights are to be acquired instead.

Regional brain weight measurements[22] have been proposed as a less time-consuming alternative to acquiring morphometric measurements from histological sections.[23] However, this approach has not been used frequently in DNT studies, mainly because it leads to significant structural distortion of brain tissue that makes qualitative and quantitative light-microscopic analyses difficult, if not impossible.[2]

TISSUE SELECTION AND NEUROHISTOLOGICAL PROCESSING

In DNT studies, tissue samples from all major regions of the nervous system should be taken for analysis at the light microscopic level.[4,6] The usual DNT tissue list—DRG, dorsal and ventral spinal nerve roots, sciatic nerve (proximal and distal segments), and tibial nerve—is similar to that used in an ANT study. In some protocols, more distal nerve branches may also be included. The sural nerve is a preferred site for nerve biopsy in humans, although in rats, findings at this site are rarely described. A more suitable site for sampling the integrity of distal nerves in rats is the plantar nerve.[2]

Collection of specimens for potential electron microscopic assessment is not guided by regulatory guidelines.[4,6] Special samples for electron microscopy are usually harvested only during the course of purposefully designed mechanistic studies.

Basic Considerations for Tissue Selection

The central nervous system has been subdivided into seven major divisions, the spinal cord and six brain domains: cerebral hemispheres (or telencephalon), diencephalon,

TABLE 1   Absolute and Relative Brain Weight Data for Control Wistar Rats from Selected Developmental Neurotoxicity Studies

Rat Strain	No. of Rats	Year(s)	Male Absolute Weight (g)	Male Relative Weight (%)	Female Absolute Weight (g)	Female Relative Weight (%)
			PND 11			
Wistar (Janvier)	10/10	1999	1.27	5.60	1.25	5.28
Wistar (Charles River)	10/10	2000	1.30	6.00	1.27	6.07
	10/10	2003	1.15	5.34	1.15	5.29
Total	30/30	Mean	1.24	5.65	1.22	5.55
		Max.	1.30	6.00	1.27	6.07
		Min.	1.15	5.34	1.15	5.28
			PND 22			
Wistar (Janvier)	10/10	1999	1.69	3.28	1.62	3.22
Wistar (Charles River)	10/10	2001/2002	1.67	3.45	1.67	3.44
	10/10	2002/2003	1.64	3.17	1.56	3.19
	10/10	2003	1.67	3.47	1.64	3.40
	10/10	2003/2004	1.61	3.13	1.56	3.04
	10/10	2003/2004	1.65	3.17	1.63	3.25
	10/10	2004/2005	1.62	3.27	1.56	3.38
	10/10	2005	1.68	3.38	1.57	3.39
	10/10	2008/2009	1.68	3.27	1.57	3.23
Total	90/90	Mean	1.66	3.29	1.60	3.28
		Max.	1.69	3.47	1.67	3.44
		Min.	1.61	3.13	1.56	3.04
			PND 60–70			
Wistar (Janvier)	10/10	1999	2.06	0.62	1.96	0.94
Wistar (Charles River)	10/10	2000	2.02	0.77	1.91	1.11
	10/10	2001/2002	2.08	0.69	1.92	1.00
	10/10	2002/2003	2.01	0.75	1.90	1.05
	10/10	2003	2.03	0.68	1.80	1.01
	10/10	2003/2004	1.96	0.69	1.88	1.03
	10/10	2003/2004	2.03	0.71	1.83	1.04
	10/10	2004/2005	1.96	0.67	1.86	1.03
	10/10	2005	2.01	0.76	1.86	1.11
	10/10	2008/2009	2.02	0.72	1.86	1.04
Total	100/100	Mean	2.02	0.71	1.88	1.04
		Max.	2.08	0.77	1.96	1.11
		Min.	1.96	0.62	1.80	0.94

midbrain, pons, cerebellum, and medulla.[24,25] In the following paragraphs a short summary of the relevant anatomical and functional characteristics of these major regions is given as a rationale for evaluating the specific brain levels that are typically selected in DNT studies. These overviews are derived primarily from Kandel[24] and Amaral.[25]

Cerebral Hemispheres

The two cerebral hemispheres consist of the cerebral cortex and three deep structures: the basal ganglia (with four nuclei: striatum comprised of the caudate/putamen, globus pallidus, subthalamic nuclei, and substantia nigra), the amygdala, and the hippocampal formation. The basal ganglia participate in regulating voluntary movement, the amygdala coordinates the autonomic and endocrine responses of emotional states, and the hippocampus is involved in aspects of memory storage.

Many areas of the cerebral cortex are involved primarily in processing sensory input or delivering motor commands. The cerebral cortex is divided into four lobes (frontal, parietal, temporal, and occipital), each of which has distinct specialized functions. The frontal cortex is concerned largely with integration and planning (i.e., "thought") as well as the control of movement; the parietal cortex with somatic sensation, with forming a body image, and with relating one's body image with extrapersonal space; the temporal cortex with hearing; and the occipital cortex with vision.[24] The corpus callosum interconnects the two hemispheres and is the largest of the commissures linking similar regions of the left and right sides of the brain.

In DNT studies, coronal sections of the telencephalon taken through the frontal cortex (to include the striatum and corpus callosum), the parietal cortex (which also incorporates the hippocampal formation and diencephalon), and the temporal and occipital cortices (which overlap the midbrain) address the major functional domains mediated by the cerebral hemispheres.

Morphological evidence for neurodevelopmental alterations includes disorganization of cerebrocortical layers and misplaced neuronal cell clusters (mainly in the entorhinal, prefrontal, and cingulate cortices). These changes have been observed in the brains of schizophrenic patients.[26] Methyl mercury is known to produce bilateral cerebral hypoplasia with significant neuronal disorganization in the cerebral cortex in children with the fetal Minamata disease (FMD). Hydrocephalus ex vacuo (brain ventricle dilation) is a further consequence in the hypoplastic brains of FMD.[27] When given on gestational day (GD) 15 in rats, Methylazoxymethanol has been shown to produce significant hypoplasia of the cerebral cortex and progressive dilatation of the lateral ventricles during the brain growth spurt.[3]

Diencephalon   This region rests beneath the cerebral hemispheres and contains two main components. The thalamus serves as the gating and modulating portal for transferring motor and sensory information from the entire central nervous system (except the olfactory system) to the cerebral cortex. Specific thalamic nuclei are linked to specific cerebrocortical domains in a strict topographical arrangement. The hypothalamus regulates autonomical functions (e.g., circadian rhythms, cyclic behaviors), endocrine pathways (e.g., hormonal systems of the pituitary gland), and visceral functions (e.g., growth, eating, drinking, self-defense, reproductive activity). Hormonal impacts on sexual differentiation appear to be mediated by the sexually dimorphic nucleus of the preoptic area, which is sensitive to gonadal steroids. Such gonadal hormones exert permanent (or organizational) effects on the developing CNS and transient functional effects in the adult brain. The organizational effects are conceptually important during neurodevelopment in regulating neural circuitry.[28]

As noted above, the conventional coronal section through the parietal cortex favored in DNT studies will encompass the thalamus and hypothalamus as well. Partial or complete agenesis of the corpus callosum and reduced numbers of pyramidal neurons in the hippocampus have been observed in fetal brains of children with fetal alcohol syndrome (FAS).[29] Hypoplasia of the corpus callosum, hypo- and dysmyelina-

tion of white matter, and reduced pyramidal cells in the hippocampus are also found in the brains of children with FMD.[27] Hypoplasia of the arcuate nucleus in the hypothalamus seems to be a relatively consistent morphological finding in sudden infant death syndrome (SIDS).[30] Hormonal imbalance with respect to endogenous levels of gonadal steroid or high concentrations of xenoestrogens during the sensitive period of sexual differentiation (PND 7–9 in rats) may irreversibly affect sexual behavior in the adult.[31]

Midbrain   The midbrain includes relay centers of the visual (lateral geniculate nucleus) and auditory (medial geniculate nucleus) systems that help control eye movement and coordinate visual and auditory reflexes as well as nuclei involved in motor control (e.g., the substantia nigra provides important input to regulate voluntary movements).

As noted above, the coronal sections through the temporal and occipital lobes that are taken in conventional DNT studies will include the midbrain. However, such sampling will seldom be sufficient to evaluate all the major structures noted above (especially the geniculate nuclei) unless special care is taken to standardize the trimming level and sectioning plan.

Adverse effects on substantia nigra neurons (nigrostriatal pathways) during neurodevelopment are suspected to cause a change (hypersensitivity) in the developing animal's postnatal reaction to certain chemicals (e.g., the fungicide maneb or the herbicide paraquat). These two compounds have been used in two experimental models with C57BL mice, one where dams were treated intraperitoneally (i.p.) from gestational days 10 to 17 with high doses and the other where the offspring were treated i.p. from postnatal days 5 to 19. In both models, quantitative analysis of dopaminergic neurons in the substantia nigra pars compacta has demonstrated a decreased number of neurons in this structure in the brains of the offspring. Such a developmental deficit has been proposed as the stimulus for an earlier onset of Parkinson-like symptoms.[32]

Pons   The pons (pontine nuclei), the rostral part of the brain stem, relays information about movement and sensation from the cerebral cortex to the cerebellum. In DNT studies, standard coronal sections through the brain stem that resides beneath the rostral cerebellum will include the pons.

Cerebellum   The cerebellar cortex is divided into multiple lobes separated by distinct fissures. It is connected to the brain stem by several major fiber tracts (peduncles). Deep nuclei within the white matter just dorsal to the peduncles serve as relay centers for signals from the fiber tracts to specific topographical domains of the cerebellar cortex. The cerebellum helps in maintaining posture, in coordinating head and eye movements, in fine-tuning movements, and in learning motor skills.

In DNT studies, coronal sections through the cerebellum survey this region adequately. The primary consideration in coronal-oriented trimming is that adequate care must be taken if the deep cerebellar nuclei are to be included consistently in sections taken for histopathological analysis. However, the most complete choice for evaluating the cerebellum will also include samples taken in the longitudinal (midaxial or para-sagittal) orientation. Examination in this plane provides a perspective of all the folia in the vermis of the cerebellar cortex, which would be lacking if only a single coronal section were taken.[33]

Development of the cerebellum may be altered after methyl mercury intoxication. Children afflicted with FMD have cerebellar cortices with significantly reduced numbers of granule cells.[27] Focal cerebellar dysplasia, including the heterotopic location of Purkinje cells and granule cells, has been observed in the offspring of pregnant rats treated with methyl mercury throughout gestation and after parturition.[34] Human FAS brains may also show hypoplasia of the cerbellar cortex and the cerebellar vermis, and may have reduced numbers of Purkinje cells.[29] Hypoplasia of the cerebellar cortex has been reported in rats after developmental exposure to MAM, although the extent of the lesions varies depending on the time of exposure.[23]

Medulla Oblongata The medulla is the caudal part of the brain stem and is the direct extension from the cervical spinal cord. It contains neuronal groups responsible for vital autonomical functions such as regulation of digestion, breathing, and the control of heart rate and blood pressure. In DNT studies, coronal sections of this region are used to evaluate the structural elements responsible for these major functions.

Spinal Cord

The spinal cord is subdivided anatomically into cervical, thoracic, lumbar, and sacral regions based on the vertebrae through which their spinal nerves pass. At the microscopic level, all these domains can be separated into the central gray matter and the surrounding white matter. The spinal cord receives sensory information from the somatic sensory receptors in the skin, joints, and muscles of the limbs and trunk and transfers it the sensory relay neurons of the dorsal horn, which transmit information to the brain and other segments of the spinal cord, and the motor neurons of the ventral horn, which are responsible for both voluntary and reflex movement of muscles. The white matter is made up of longitudinal tracts of myelinated axons that form parallel ascending sensory and descending motor tracts.[25] The topographical arrangement of tracts within the white matter is quite specific.[35] Spinal nerves link the spinal cord with the periphery via dorsal sensory and ventral motor roots. Dorsal root ganglia contain clusters of sensory neurons

that convey information to the spinal cord from the skin, the muscles, and the joints.

In DNT studies, the spinal cord is typically sampled using transverse and/or longitudinal sections harvested from the cervical, thoracic, and lumbar spinal cord. If the spinal cord is removed from the vertebral column prior to trimming (which is the usual practice in DNT studies), technicians must remember that the spinal cord segments in mammals occupy a position in the vertebral canal that is somewhat cranial to the location of their corresponding vertebral body. However, the cranial shift is modest in juvenile rats (about one segment) at PND 22[36] and negligible at PND 11.

Ganglia

There are two principal approaches that may be followed in processing the ganglia (e.g., DRG, autonomic ganglia) and their associated nerves. One protocol is to harvest three DRG associated with both the cervical and lumbar swellings. This number works well because three DRG fit readily into one plastic block (the embedding medium required by current EPA guidelines), which helps to reduce the cost (in money and time) of this low-throughout procedure. The alternative is to fix and process the spinal cord and DRG in situ. The major advantage of the latter technique is that the DRG as well as their dorsal and ventral root fibers can be evaluated in their native state. However, the two main disadvantages are that the extended decalcification during processing of the vertebral column can induce subtle artifacts in neural tissues, and that it is often difficult to visualize more than one or two DRG in each sample.[2]

SAMPLING RECEPTOR AND EFFECTOR ORGANS

Neurodevelopmental effects of toxicant exposure may affect sensory organs, so conventional DNT studies assess selected receptor organs as a matter of routine. Typical sites include both eyes (for retina and optic nerve), ideally after perfusion fixation, and one level from the caudal nasal cavity that includes the olfactory neuroepithelium (level III[34]).

Skeletal muscle is taken as an indirect index of nervous system function (by looking at an effector organ controlled by complex neural pathways). Evaluation of the nerve trunks within the muscle sample also serves as a direct means of assessing neural integrity, as the small intramuscular branches of the tibial nerve are highly sensitive to neurotoxic agents.[39] The gastrocnemius is generally selected if no muscular effects are anticipated; this muscle contains predominantly anaerobic *type II* (often termed *white* or *fast twitch*) *fibers* used for short-term force. Other muscles may be included when a neurodevelopmental impact on muscle

TABLE 2 Central Nervous System and CNS-Associated Neural Tissues for Paraffin Embedding at Early (Postnatal Day 11 Pups or PND 22 Adolescents) and Late (60–70 Young Adults) Time Points in Conventional Rodent Developmental Neurotoxicity Studies[a]

Sources of Sample	Test Groups (Dose)			
	0 (Control)	1 (Low)	2 (Mid)	3 (High)
Brain (coronal sections)				
Olfactory bulb	A10	F10	F10	A10
Frontal lobe (with corpus callosum, corpus striatum)	A10	B10	B10	A10
Parietal lobe with diencephalon (with corpus callosum, hippocampus, thalamus, and hypothalamus)	A10	B10	B10	A10
Midbrain with occipital and temporal lobes (with substantia nigra)	A10	F10	F10	A10
Pons	A10	F10	F10	A10
Cerebellum (coronal and longitudinal planes of section)[b]	A10	B10	B10	A10
Medulla oblongata	A10	F10	F10	A10
Spinal cord (longitudinal and transverse sections)[c]				
Cervical level I (C_1–C_3: $\mathbf{C_1}$)	A10	F10	F10	A10
Cervical level II (C_3–C_5: $\mathbf{C_5}$)	A10	F10	F10	A10
Thoracic level (T_5–T_8: $\mathbf{T_8}$)	A10	F10	F10	A10
Lumbar level (L_1-L_4: $\mathbf{L_4}$)	A10	F10	F10	A10
Brain-associated intracranial organs				
Pituitary gland	A10	F10	F10	A10
Trigeminal (Gasserian) ganglia with nerve (cranial nerve V)	A10	F10	F10	A10
Brain-associated receptor organs				
Eyes (for retina) and optical nerve	A10	F10	F10	A10
Olfactory neuroepithelium (nasal cavity, level III)	A10	F10	F10	A10
All gross neural lesions	A2	A2	A2	A2

[a] Methods/scope of examinations: A, paraffin embedding, sectioning, and staining with hematoxylin & eosin; B, paraffin embedding, archived without sectioning and analysis; F, preservation in neutral buffered 4% formaldehyde solution (10% formalin); 10, all perfused animals per group and gender; 2, all affected animals per group and gender.

[b] The cerebellum is separated from the brain stem and divided along the mid-sagittal plane into two halves. A cross section through the mid-cerebellum (the widest region) is made in one half, and a longitudinal (para-sagittal) section is made through the vermis in the other.

[c] Boldface type indicates the level at which the transverse section should be taken.

innervation is suspected, so that sites with different skeletal fiber compositions are investigated more fully. The soleus is often chosen for this purpose, as this muscle harbors more of the aerobic *type I* (also termed *red* or *slow-twitch*) *fibers* needed for sustained activity.

A practical example of tissues addressed in a conventional DNT study and the scope of routine histological processing and light-microscopic evaluation is summarized in Tables 2 and 3.

TRIMMING PROCEDURE

Trimming the brains taken during a DNT study is one of the great challenges to the skill of a histotechnician. The organ has to be sectioned exactly in the coronal orientation at the standard levels required for qualitative and quantitative analysis. Trimming at least six to eight levels is necessary to adequately survey the major brain areas known to be susceptible to many neurotoxic agents. Homologous coronal sections are best obtained by using definitive anatomical structures,[1] especially such external landmarks on the ventral aspects of the brain as the mammilary bodies and the optic chiasm (see Figures 25 and 26). Sample collection for DNT neuropathology studies is typically performed by freehand trimming and is identical for all DNT time points (Figures 25-1 to 25-9, 26-7, and 26-8). Cuts must be obtained in the same coronal (transverse) and vertical planes in all animals, with no oblique angles in any axis, to provide sections suitable for morphometric procedures. Exact placement of cuts is facilitated by plastic or metal brain matrix molds, which stabilize the soft juvenile brains while strictly maintaining the orientation of brain blocks via a series of grooves placed at 1-mm intervals. Brains of different dimensions require molds made to accommodate organs of their size (see Figure 24). A disadvantage of using brain molds for DNT studies is that the softer tissues of early (PND 11 and 22) brains are more easily damaged by pressure when cradled in the brain mold. In addition, the knife grooves in such molds do not always line up exactly with the desired anatomical landmarks in younger animals.[3] In the end, the final selection of sections for neuropathology evaluation is guided by the internal landmarks, representing the major brain areas[2] (see Figure 27-1 to 27-9). Step-

TABLE 3 Peripheral Nervous System and PNS-Associated Tissues for Plastic Embedding (PND 60–70 Young Adults) in Conventional Rodent Developmental Neurotoxicity Studies[a]

	Test Groups (Dose)			
Sources of Sample	0 (Control)	1 (Low)	2 (Mid)	3 (High)
Peripheral nervous system				
Dorsal root ganglion (C_3–C_6), three sites	T10	P10	P10	T10
Dorsal spinal nerve roots (C_3–C_6)	T10	P10	P10	T10
Ventral spinal nerve roots (C_3–C_6)	T10	P10	P10	T10
Dorsal root ganglion (L_1–L_4), three sites	T10	P10	P10	T10
Dorsal spinal nerve roots (L_1–L_4)	T10	P10	P10	T10
Ventral spinal nerve roots (L_1–L_4)	T10	P10	P10	T10
Proximal sciatic nerve (cross and longitudinal sections)	T10	P10	P10	T10
Proximal tibial nerve (at knee, cross and longitudinal sections)	T10	P10	P10	T10
Distal tibial nerve (at lower leg, cross and longitudinal sections)	T10	P10	P10	T10
Plantar nerve	T10	P10	P10	T10
PNS-associated effector organs	T10	P10	P10	T10
Gastrocnemius muscle (longitudinal and cross sections)	A10	F10	F10	A10

[a] Methods/scope of examinations: A, paraffin embedding, sectioning, and staining with hematoxylin and eosin; F, preservation in neutral buffered 4% formaldehyde solution (10% formalin); P, secondary fixation in 5% glutardaldehyde solution and retention of fixed specimens in buffer solution; T, secondary fixation in 5% glutardialdehyde solution, plastic embedding (epoxy resin), semi-thin sectioning (0.5 μm), and staining with azure II–methylene blue–basic fuchsin (AMbF); 10, all perfused animals per group and gender.

sectioning may be needed to reach the desired brain levels and produce sections with sufficient structural homology to support the quantitative analysis.[1,2]

With respect to trimming the cerebellum, a slightly different approach is undertaken[23] in which the cerebellum is removed from the brain stem and split in the mid-sagittal plan. One half is used to acquire mid-sagittal sections of the vermis suitable for morphometric assessment of the folium pyramis (VIII). The other half is cut in the coronal orientation through the flocculus/paraflocculus (the most lateral extension of the cerebellar cortex) for qualitative light-microscopic examination (Figures 26-7 and 8 and 27-7 and 8).

The spinal cord is typically sampled at three or four sites representing the cervical, thoracic, and lumbar divisions. At a minimum, cross and longitudinal sections are taken from the caudal cervical cord (C_4 to C_7, with the cross section acquired at the widest portion of the cervical swelling), the middle thoracic cord, and the lumbar cord (L_1 to L_4, with the cross section taken at the widest portion of the lumbar swelling). The cervical and lumbar swellings are sampled in this fashion to view the greatest number of motor neurons that control motor functions in the forelimbs and hindlimbs, respectively. Some investigators also sample C_1–C_2 at the transition between the spinal cord and the brain stem, as the ascending sensory tracts terminating at this site are known to be exquisitely sensitive to neurotoxic agents.[37]

A variety of special neural elements are also sampled routinely for the neuropathology arm of conventional DNT studies. The paired trigeminal (Gasserian) ganglia (cranial nerve V), which are functionally equivalent to dorsal root ganglia, are harvested from the basal and lateral aspects of the pons (Figures 25-9a + b and 27-9a + b). The pituitary gland is removed and embedded in the coronal orientation to show at least some portion of all three zones (anterior, intermediate, and posterior lobes). Olfactory neuroepithelium (in situ in the nasal cavity, in cross sections taken at level III[38]) and cross sections of the rostral olfactory bulb are sampled. Eyes are oriented to include the retina and optic nerve.

Skeletal muscle sampling should include both cross and longitudinal sections. Preparation of specimens does not require forced muscle extension (i.e., muscle clamps or needle fixation to cork plates) to prevent postmortem contraction as it follows perfusion–fixation and at least 24 to 48 h of immersion post-fixation.

Fiber typing to assess the type I/type II fiber ratio requires flash-frozen muscle samples suitable for enzyme histochemistry, and thus the DNT study design will require an additional cohort of animals that will not be perfused. Type I and type II fibers are detected by using procedures for succinate dehydrogenase (SDH) and adenosine triphosphatase (ATPase), respectively. Experimental denervation during development will alter the normal differentiation of muscle fibers so that the muscle fiber ratio shifts significantly. Other changes include fewer myofibrils per fiber, lower fiber calibers, and reduced or no histochemical differentiation to either type I or type II muscle fibers. An intact innervation is mandatory for normal development and maturation of skeletal muscle fibers, so any neurodevelopmental impact on neural transmission will lead to significant skeletal muscle hypotrophy with compensatory interstitial lipomatosis.[40,41]

Embedding

Trimmed sections of the CNS, receptor organs (olfactory neuroepithelium, eye and optic nerve), and effector organs (skeletal muscle) are embedded in paraffin wax. Different rates of tissue shrinkage may be produced by automated processing schedules during the dehydration and embedding process, due to the differing permeabilities and compositions of these tissues. A histology laboratory engaged in DNT neuropathology work must therefore optimize its processing protocols to provide the best compromise between mass handling of samples and the requirement of these diverse tissues for special care to ensure that morphometric measurements can be performed later. Prolonged storage in ethanol during the paraffin infiltration process may cause artifactual vacuolation in the white matter of the CNS, which may mislead inexperienced pathologists into diagnosing a treatment-related lesion. Therefore, brain and spinal cord tissues from all treatment groups should be processed within the same time frame. In particular, brain samples destined for quantitative analysis should be processed as a single batch.

Special Procedures for the Developing PNS

In DNT studies, the PNS is examined only at the later (PND 60–70) time point. As a rule, this single time point has been shown to be sufficient in a first-tier approach to detect any adverse consequences of developmental neurotoxicity on the peripheral nervous system (PNS) and the effector organs. Analysis of PNS samples in these young adult animals basically follows the same procedures as those used to assess mature adult rats for neurotoxicity.[5,7]

There are two principal approaches that may be followed in processing PNS samples for a DNT neuropathology evaluation. In the first, tissues are harvested and post-fixed by sequential immersions in 5% glutaraldehyde solution for 2 h followed by immersion in 1 to 2% osmium tetroxide for 1 to 2 h (inside a sealed glass box at 4°C). This fixation protocol provides optimal preservation of myelin and also permits a meaningful electron microscopic analysis, should such become necessary. It must be remembered that osmium tetroxide is a highly toxic, volatile compound [gaseous at about 20°C (i.e., room temperature)], so it must be handled only in a fume hood; laboratories equipped with downdraft tables rather than a fume hood should be maintained at 16 to 18°C to prevent volatilization. Fixed samples are then embedded in plastic (Table 3) because it is required in the EPA guidelines[4,5] as a means of obtaining higher-resolution views of neurons and neuronal processes. The resolution of fin structures is enhanced because the thickness of glycol methacrylate sections (1 to 2 μm) is much thinner than that of paraffin sections (4 μm or greater). Resolution of cellular detail can be improved further if plastic embedding is performed with epoxy resin, as even thinner sections (down to 0.5 μm) can be acquired.[1] By the current DNT and ANT guidelines,[4–7] other samples to evaluate the PNS except for trigeminal (Gasserian) ganglia and skeletal muscle should be embedded in plastic.

NEURODEVELOPMENTAL HISTOPATHOLOGY

Qualitative Analysis

The approach to performing histopathology of the developing nervous system is in principle identical to the approach used to evaluate tissues from the more stable adult stage. The same routine stains are employed. Paraffin-embedded tissues are stained with hematoxylin and eosin (H&E), while plastic sections are stained with toluidine blue or AmbF (azure methylene blue–basic fuchsin). Some laboratories may also use cresyl violet as a basic stain for highlighting neurons and their Nissl bodies in paraffin-embedded sections, although there is no definitive scientific reason for changing the routine staining battery to accommodate the occasional DNT study, given the long experience that most pathologists have with interpreting H&E-stained neural tissues. If neuropathology changes are actually observed or suspected based on evaluation of the H&E-stained tissue, an entire battery of special neurohistological methods are available; their utility has been summarized in detail elsewhere.[42–52] These special neuron and myelin stains may be extremely useful in a second-tier assessment as a means of revealing the precise character of neurotoxicological impact.

Should special neurohistological stains be employed, the pathologist must keep in mind that the staining properties of neural tissues may vary substantially depending on the maturational stage of the brain structure being evaluated. This inconsistency presumably reflects the dynamic flux in the molecular and biochemical constituents within a rapidly evolving neural structure. Such variation may not be relevant for the general quality of the basic stains (e.g., H&E), but it may be quite important for procedures addressing special neural components (e.g., myelin) or immunohistochemical epitopes (e.g., astrocyte markers). Therefore, the neurodevelopmental time point is always an important consideration. For example, in rat DNT studies the luxol fast blue/Kluver–Barrera myelin stain, which binds complex lipids not affected by solvent extraction during paraffin processing,[53] exhibits no or very low activity in the white matter of the brain stem or cerebellum of PND 11 pups, a very weak positive reactivity in PND 21 adolescents at weaning, and a clear positive staining for PND 60–70 young adults (Figures 1–6). This age-related divergence in the myelin staining pattern has been used to recognize hypomyelination induced by carbon monoxide,[54] 6-aminonicotinamide,[55] hexachlorophene,[56] or triethyltin.[57] For the detection of *hypomyelination* in the

FIGURES 1 to 3 Coronal sections of control Wistar rat brains showing the parietal cortex, hippocampus, and diencephalon. Myelination (blue stain) is absent at postnatal day (PND) 11 (Figure 1), slight at PND 21 (Figure 2), and clearly evident at PND 62 (Figure 3). Luxol fast blue stain, original magnification 10×. (*See insert for color representation of the figures.*)

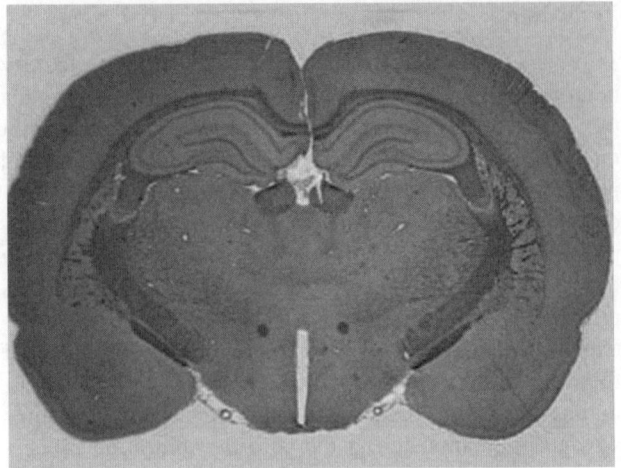

FIGURE 3

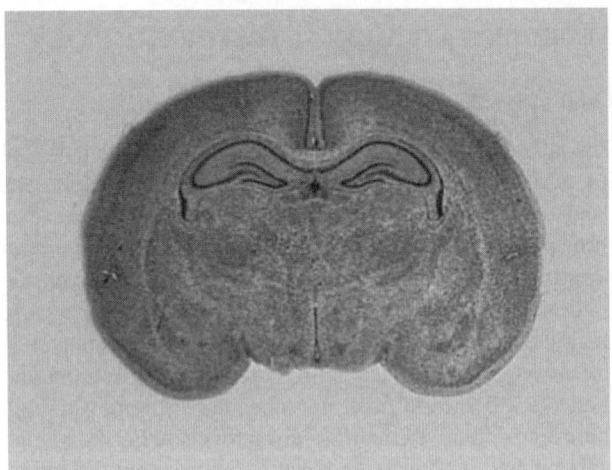

FIGURE 1

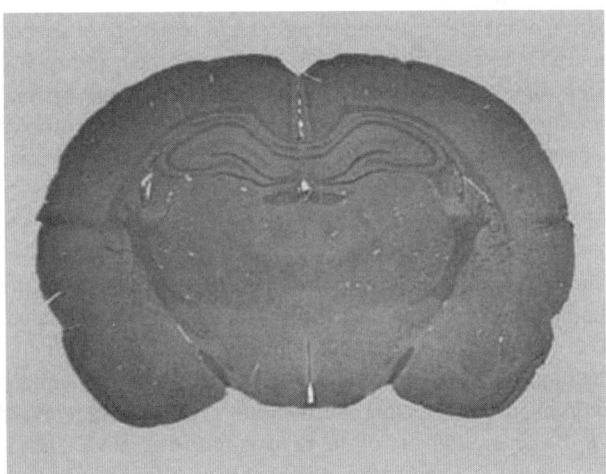

FIGURE 2

neonatal brain, the classic cresyl violet stain is performed to demonstrate myelinating glial cells which have strongly basophilic cytoplasm at a paranuclear unilateral position.[53] Immunohistochemistry with polyclonal antibodies to myelin basic protein (MBP) may also be employed to evaluate hypomyelinating disorders after intrauterine exposure, an approach verified for carbon monoxide.[54]

Different staining intensities in neural tissues may also occur as an artifact of histological processing. To prevent misinterpretation due to procedural artifacts, a strictly

standardized processing protocol must be followed, preferably by including at least one brain level of all animals from one study in the same processing run. In general, control and all treatment groups should also always be processed together in one run. The use of an automated histostainer will reduce the variability of staining as well.[2]

The developing brain does possess a number of transient but normal anatomical structures that must not be confused with toxicant-induced lesions. These features, including the external and ventricular germinal matrix layers, subcortical cell clusters, and higher neuronal densities, are evident in the developing brain but are no longer present in adult brains. In rats, these transient entities are visible at PND 11 but are no longer a consistent morphological feature by PND 21 (Figures 7–9).

Neurohistologic analysis should focus on identifying types of neuropathological alterations and cataloging their severity. Indications of developmental insults to the brain are of particular importance. Such indices, which might include deviations in the normal size and shape of the cerebral hemispheres or the normal foliation pattern in the cerebellum, evidence of ventricular dilation (e.g., hydrocephalus), aberrant cell aggregates (e.g., heterotopiae), functional disturbances (e.g., altered proliferation or excessive apoptotic cell death), and retention of transient structure, should be noted.[3]

Neurotoxicants may induce significant malformations that are visible by mere macroscopic inspection during necropsy (e.g., "periventricular overgrowth"[58]; Figures 10–13). These can often be identified during a conventional developmental toxicity study; therefore, the focus of the neuropathology arm of a DNT study is to discover neurodevelopmental alterations at the microscopic level. A typical outcome evident by qualitative examination is the presence of *heterotopiae* (used synonymously with the term *ectopiae*). Heterotopic cells are defined as misplaced neuronal cell clusters due to a

FIGURES 4 and 5 Coronal section of control Wistar rat brains showing the anterior commissure. Myelination (blue stain) is absent at postnatal day (PND) 11 (Figure 4) but clearly evident at PND 62 (Figure 5). Luxol fast blue stain, original magnification 200×. (*See insert for color representation of the figures.*)

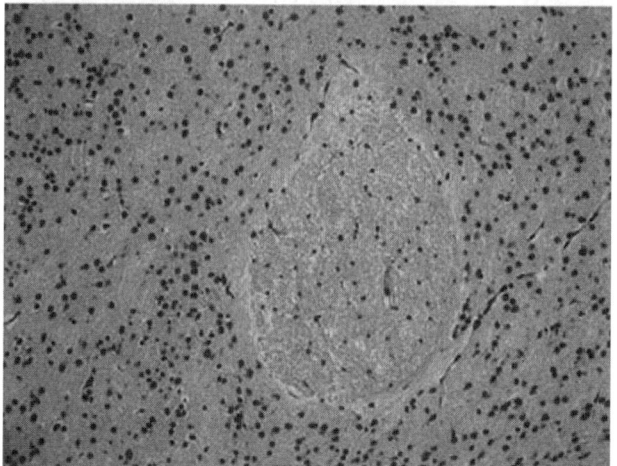

FIGURE 4

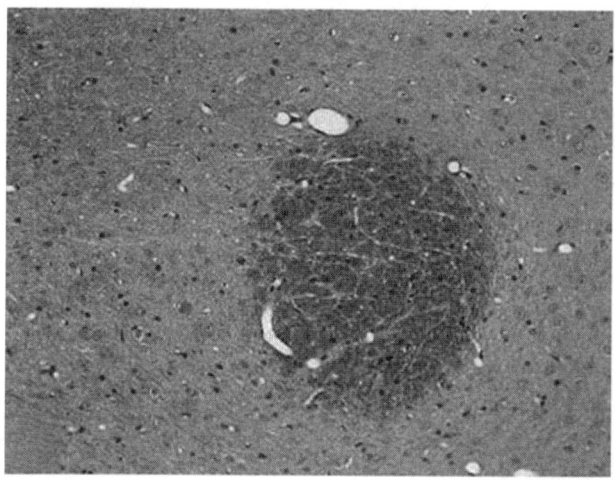

FIGURE 5

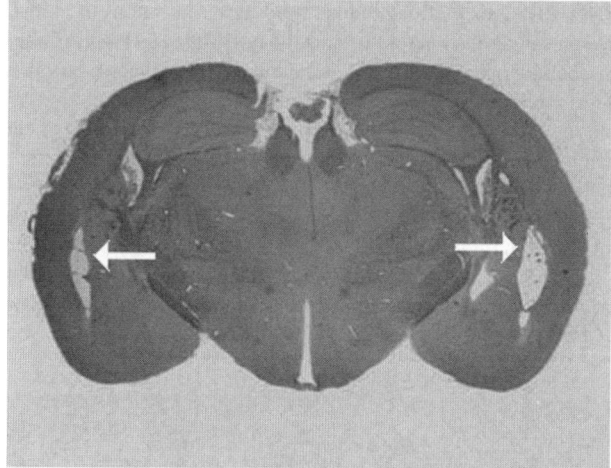

FIGURE 6 Coronal section of a neurotoxicant-exposed Wistar rat brain at postnatal day (PND) 62 showing the parietal cortex, hippocampus, and diencephalon. Relative to an age-matched control animal (Figure 3), the size of the brain is reduced significantly, but the extent and intensity of myelination (blue stain) is not altered. Hypoplasia of the cerebral cortex results in hydrocephalus internus in the adult brains (arrows), the first stage of which is apparent in this section. The neurotoxicant, methylazoxymethanol, was given to the dam on gestational day 15 at 30 mg/kg body weight i.p.. Luxol fast blue stain, original magnification 10×. (*See insert for color representation of the figure.*)

disturbance of early neuronal migration and/or terminal differentiation.[59] Nodular neuronal heterotopiae around the ventricles and within the cerebral cortex have been identified in humans with trisomy 18 and following prenatal exposure to methyl mercury, hypoxia, alcohol, and x-irradiation.[9,60–63] In experimental animals, cytotoxic compounds such as MAM or ethylnitrosourea (ENU)[60,64–66] are known to induce abnormal neuronal aggregations around the ventricles and result in neuronal dislocation into the hippocampus. Prenatal exposure of dams to MAM (25 mg/kg per day i.p.) between GD 11 and 16 has been shown to induce specific patterns of

heterotopic cells in neocortical layer I and the periventricular region, with the exact outcome dependent on the timing of treatment. As the age of exposure was increased, the distribution showed a clear temporospatial gradient in which layer I heterotopiae were located progressively more medially, dorsally, and rostrally, while periventricular heterotopiae moved progressively more rostrally. Interestingly, the nodular heterotopiae in the hippocampus, located in the cornu ammonis 1 (CA1) and CA2 subfields, did not shift with the time of treatment.[56] These results indicate that morphologic screening of the developing brain always needs a standardized topographic approach if one is to detect neurotoxic lesions reliably.[3] Such anatomical disruption is correlated with functional deficits. Nodular heterotopiae in the hippocampus of rats exposed in utero to a single dose of MAM (30 mg/kg) on GD 15 (Figures 14 to 21) are related to hyperexcitability and may contribute to the increased seizure susceptibility observed in treated animals.[3] Neurons within ectopiae have been shown to receive direct excitatory and inhibitory input from adjacent normotopic cortex, to display multiple types of action-potential firing patterns, and to display epileptiform activity.[67] The findings in this rat model may have relevance for humans, as nodular heterotopiae (neocortical layer I ectopiae) are recognized increasingly in epileptic persons.[68] Ectopic neuronal clusters in the neocortex may combine with altered neurotransmitter and receptor properties to elicit the functional outcome of

FIGURES 7 to 9   Longitudinal (midsagittal) sections of control Wistar rat cerebellum. At postnatal day (PND) 11 (Figure 7), the cerebellum-specific external germinal matrix layer (arrows) is quite wide due to the high demand for neuronal production for this brain region. By PND 21 (Figure 8), only remnants of the external germinal matrix layer (arrows) are left, indicating that the end of neurogenesis stage is imminent. No external germinal matrix layer is found at PND 62 (Figure 9, arrows). White matter myelination (blue stain) of the white matter is lacking at PND 11, slight at PND 21, and prominent at PND 62. Luxol fast blue stain, original magnification 100×. (*See insert for color representation of the figures.*)

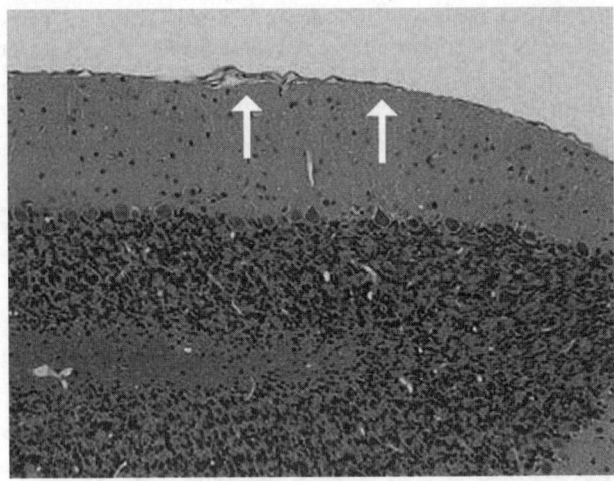

FIGURE 9

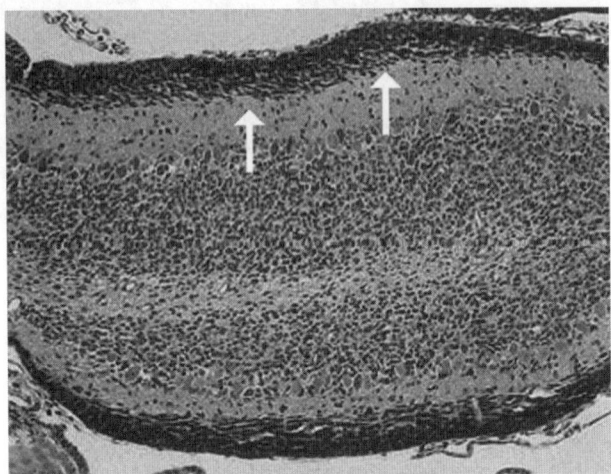

FIGURE 7

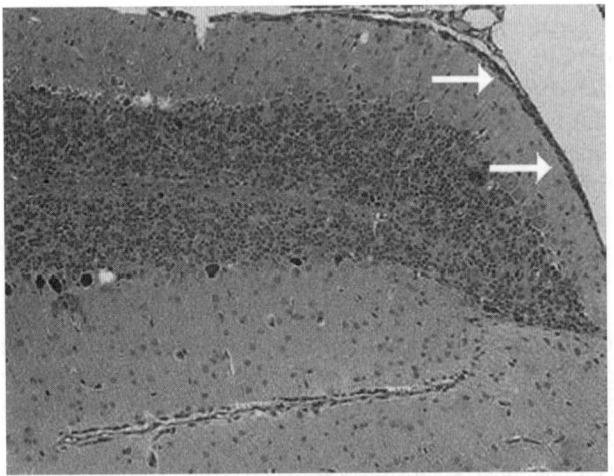

FIGURE 8

hyperexcitability and epileptogenicity (Figure 22). Major ectopiae represent a massive developmental disturbance that is often accompanied by severe cell loss and are as a rule indicative of a severe functional disorder. Small ectopiae may

also occur spontaneously in young animals and infants and are even reported to disappear over time, but major ectopiae are permanent anatomical features of the brain.[9,69]

The neurodevelopmental finding hydrocephalus ex vacuo typically involves the lateral brain ventricle. It is most evident in older (adult) brains. This lesion often occurs due to an imbalance of brain mass (e.g., due to cerebral cortex hypoplasia). However, other etiologies have to be excluded, especially artificial ventriclar dilation due to supraphysiological infusion pressure during perfusion fixation.

Quantitative Analysis

Morphometric analysis of carefully selected brain areas is of great diagnostic value in detecting neurotoxicant-induced effects on neurodevelopment. It has been shown to be a sensitive indicator for xenobiotic-induced neuroanatomical size alterations.[70,71] For regulatory DNT studies, morphometrical analysis is a required component of the standard neurohistological assessment of the brain.[4,6] The requirement for brain morphometry as a first-tier approach in DNT studies is based on fundamental differences between the adult and developing brain in their reaction to neurotoxic events. In the developing brain, an early loss of cells reduces the overall volume of various neuronal populations and possibly the size of the entire organ but does not alter the cell density of the structure; in contrast, neurotoxic exposure in the adult brain leads to a decreased cell density in the target cell population. With the exception of some reduction in cell size (which may be but often is not evident via qualitative light microscopy), no other obvious cytoarchitectural abnormalities may be discerned. Gliosis, the classic neurohistologic marker for neurotoxicant-induced neuronal elimination in the adult CNS, is not a

FIGURES 10 to 13   Coronal sections of near-term [gestational day (GD) 20] rat fetuses showing the parietal cortex, diencephalon, and/or hippocampus. In Figure 10 (gross photograph), a visible mass (arrow) is evident in one lateral ventricle, leading to asymmetry of the left and right hemispheres. The wide space between the skull and outer brain surface is a shrinkage artifact associated with immersion fixation in Bouin's solution. In Figure 11, a periventricular overgrowth (PVO) is apparent as a mass filling one lateral ventricle (arrow). Such PVO is comprised of undifferentiated cells and indicates a disturbance in migration and settling of neural cells. It may be much smaller [compare similar regions of unaffected control brain (Figure 12) and abnormal neurotoxicant-exposed brain (Figure 13)]. The fungicide Benomyl [methyl 1-(butylcarbomoyl)-2-benzimidazolecarbamate] did cause this PVO, when given to the dam by gavage on gestational days 7 to 20 at 62.5 mg/kg body weight. H&E stain (Figures 11 to 13), original magnification 10× (Figures 10 and 11) and 40× (Figures 12 and 13).

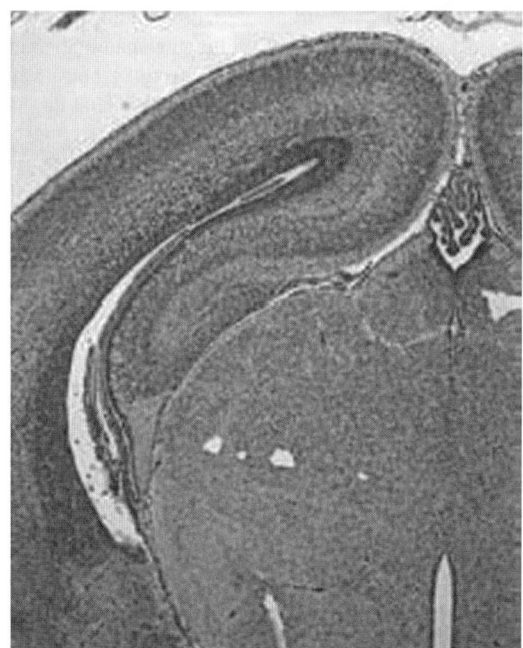

FIGURE 12

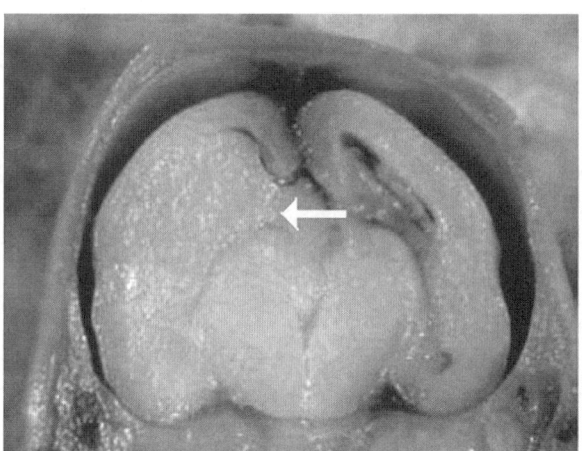

FIGURE 10

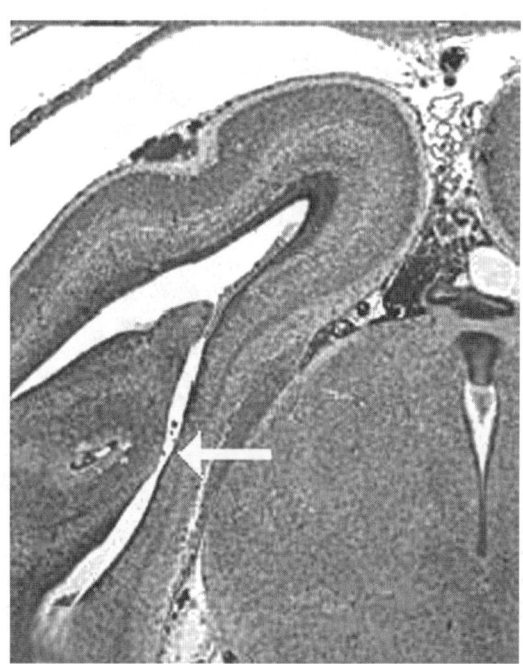

FIGURE 13

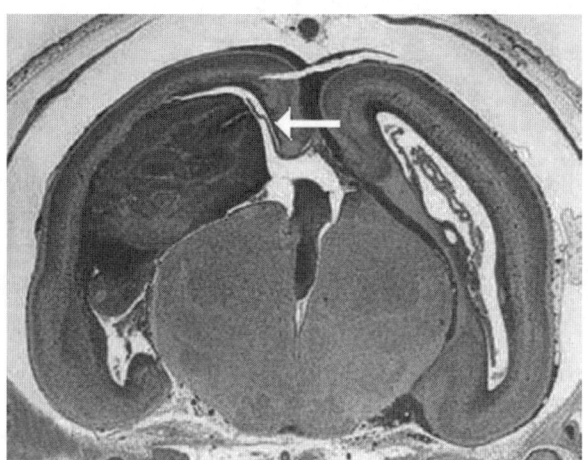

FIGURE 11

regular feature of cell death during the developmental stages characterized by extensive cell proliferation. Reduction in brain size without an obvious histologically recognizable lesion in the normal cytoarchitecture led Rodier and Gramann[70] to suggest the use of "simple morphometry" as a valuable or even essential tool for evaluating cell loss in the developing brain.

FIGURES 14 to 21 Spectrum of neuropathology lesions induced in developing Wistar rats following maternal treatment with methylazoxymethanol (MAM) at 30 mg/kg body weight i. p. on gestational day (GD) 15. Grossly, at postnatal day (PND) 21 the cerebral cortex in control rat brains covers the caudal portion of the olfactory bulbs and all of the mesencephalon (Figure 14), while MAM-related cerebrocortical hypoplasia results in retraction of the cerebral cortex and exposure of the caudal olfactory bulbs and the mesencephalon (Figure 18, at PND 21). In H&E-stained coronal sections, the normal layering of neurons in control rat parietal cortex (Figure 15), hippocampus (Figures 16 and 17) is disrupted by persistent, MAM-related, diffuse reductions in neuronal numbers, altered neuronal layering in the parietal cortex (Figure 19, at PND 11, arrows), and multifocal neuronal heterotopie (ectopiae) in the hippocampus (Figure 20, at PND 11; Figure 21, at PND 62, arrows). Original magnification 40× (Figures 15, 16, 19, and 20) or 200× (Figures 17 and 21).

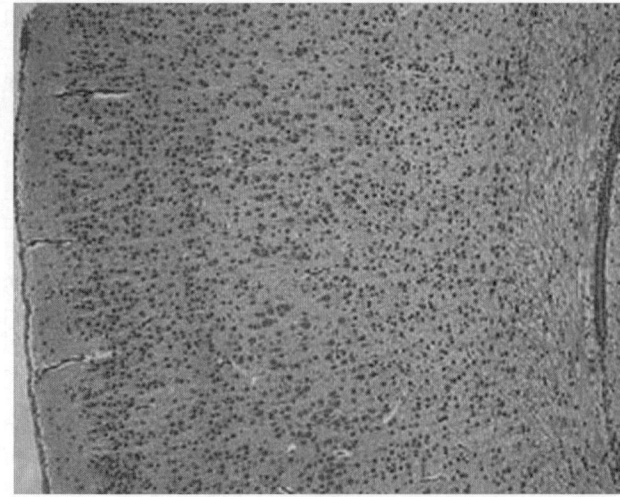

FIGURE 15

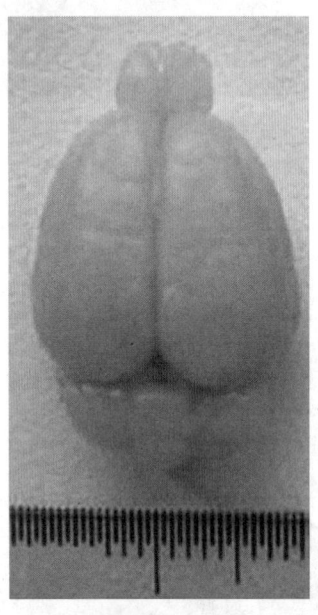

FIGURE 14

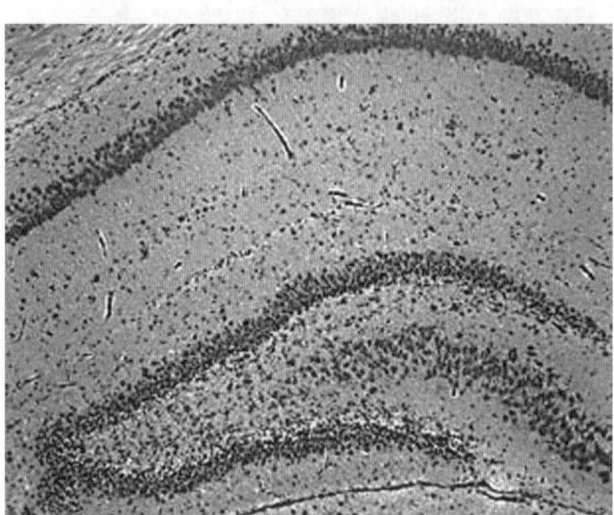

FIGURE 16

LINEAR MEASUREMENTS

In DNT studies, morphometrical analysis seeks to discover deviations in the normal rate of brain growth in different sites. The need for strictly homologous brain sections has to be emphasized, as closer resemblance improves the discriminative power of the linear (two-dimensional) measurements, especially in the hippocampus and cerebral cortex.[71] The best means for ensuring proper section homology is to trim the brain blocks correctly using external anatomical landmarks and then to select the optimal sections based on internal landmarks evident prior to transfer of the floating sections to slides. One additional way to reduce or prevent variations in section size is to mount sections dry without intermediate

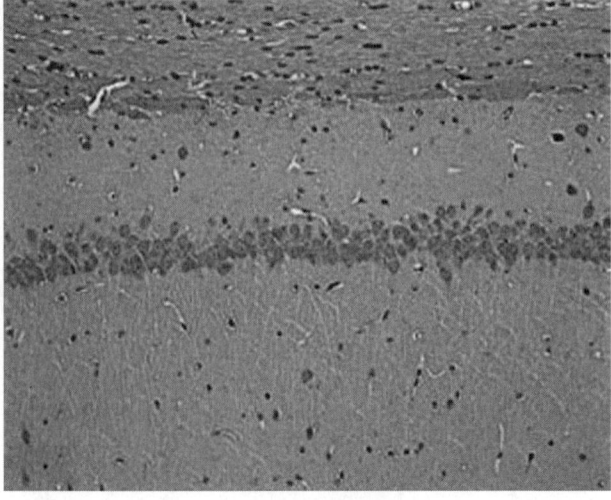

FIGURE 17

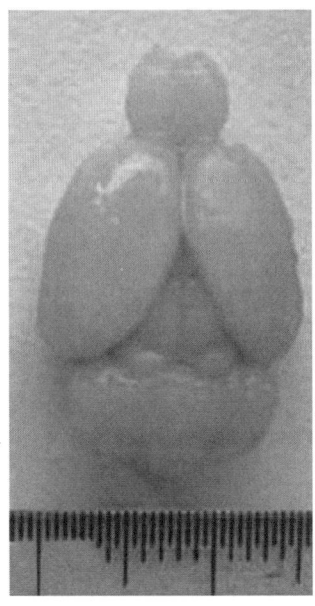

FIGURE 18

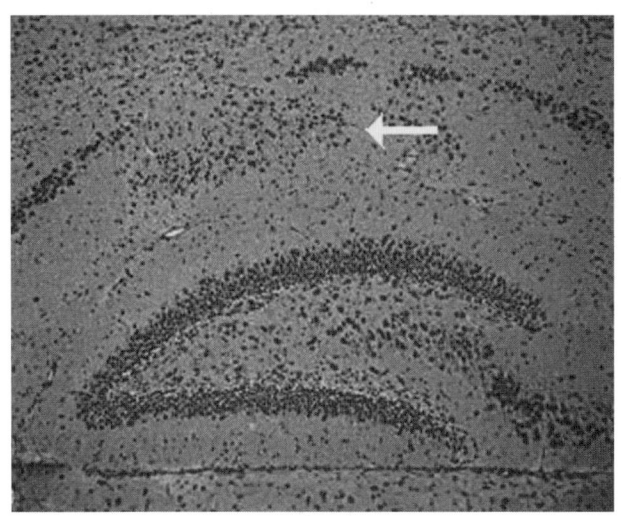

FIGURE 20

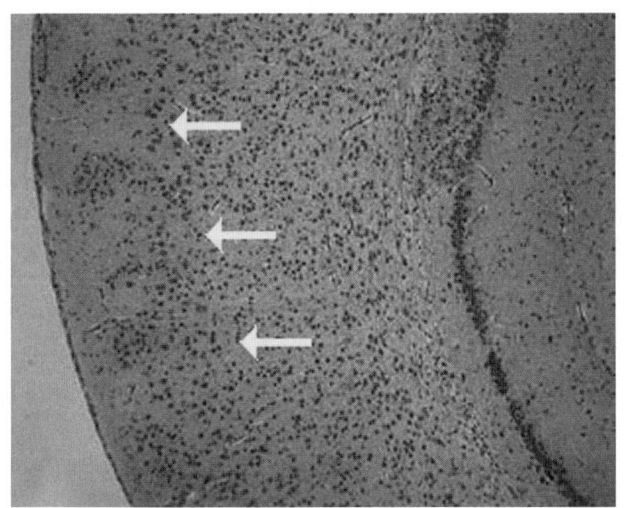

FIGURE 19

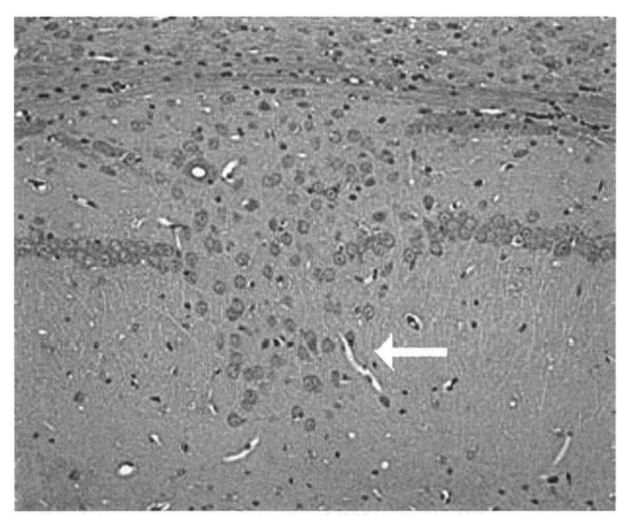

FIGURE 21

floating in a water bath. This procedure indirectly reduces variations in linear measurements and enhances the ability of linear morphometry to detect early developmental neurotoxicity.[71] The current EPA guidelines[4] require, at a minimum, that histological measurements be collected for the neocortex, the hippocampus, and the cerebellum. Linear measurements of the dissected but untrimmed brain (e.g., length and width of the whole brain and/or cerebrum and cerebellum) are also often gathered at PND 11, PND 22, and PND 60–70[2] (see Figure 23).

Linear measurements at the light-microscopic level are made most readily in sections with high contrast between structures (Figures 24 to 27), a feature that is not always optimal in H&E-stained sections. The contrast can be enhanced with either a combined neuronal/myelin stain (e.g., cresyl violet with Luxol fast blue) or by using an image analysis system equipped with a high-definition filter. Measurements in two-dimensional sections are performed separately for the left and right brain (Figures 28 to 31). Bilateral measurements are indicated as the left and right brain hemispheres mediate different functional and emotional properties that may be affected selectively by neurotoxicants. Despite their superficial similarity, the left and right brain are not symmetrical. For example, the thickness of the

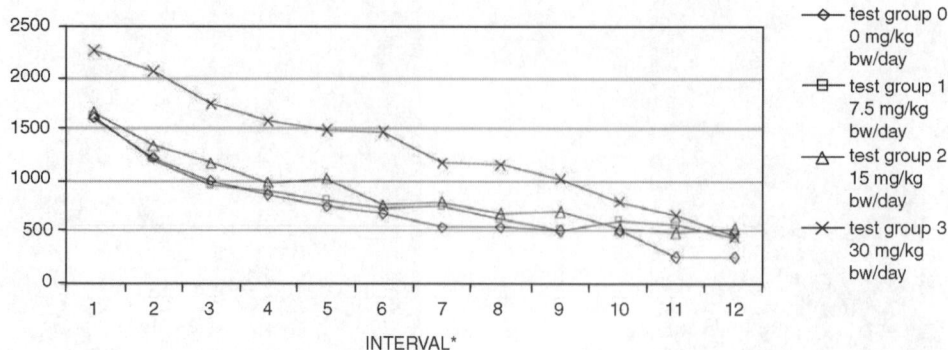

METHYLAZOXYMETHANOL:TOTAL MONEMENT DISTANCE, POSTNATAL DAY 60

*12 interval measurements of the movement distance are taken within one testing hour after each 5 minutes

FIGURE 22 The presence of widespread neuronal heterotopiae is highly correlated with the clinical finding of hyperexcitability found after motor activity measurements of young adult test animals on postnatal (PND) 60 after a single maternal treatment methylazoxymethanol (MAM) on gestational day (GD 15). Hyperexitability and heterotopiae were induced by the high dose [at 30 mg/kg body weight (BW) i.p.] but not at lower doses (15 or 7.5 mg/kg BW).

left cerebral cortex in male but not female rats is greater than that of the right, so the natural asymmetry in the brains of males exceeds that which occurs in females.[28]

Morphometrical measurements in the developing rat brain are taken routinely for forebrain [thickness of the frontal and parietal cortices and the width of the corpus striatum (caudatus/putamen)], corpus callosum (thickness at the midline), hippocampus (gyral width), and cerebellum (cerebellum height at the midline in a coronal section).[2] Alternatively, the width of selected cerebellar folium (e.g., folium pyramis) may be determined in a mid-sagittal (longitudinal) section[3] (Figures 30 and 31). Experienced

laboratories may include other linear measurements, depending on the endpoints desired and the qualitative findings in the study. The major brain areas noted above are included in the morphometrical analysis of conventional DNT studies, as they undergo extensive neural differentiation and synaptogenesis (cerebral cortex, hippocampus, and cerebellum,[44,70] contain high concentrations of major neurotransmitters (dopamine and acetylcholine in corpus striatum[72]), and they exhibit different developmental profiles with respect to neuron proliferation.[73] The corpus callosum is included, as it is the major myelinated tract that carries associative fibers between cerebral hemispheres.

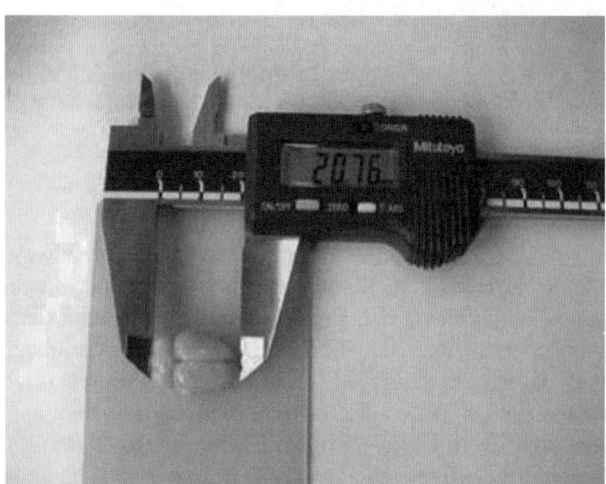

FIGURE 23 Gross quantitative measurements of toxicant-induced developmental neuropathology include simple linear measurements of brain dimensions, such as whole brain length and width acquired with a Vernier caliper.

FIGURE 24 Brain molds for tissue trimming may help to achieve strictly vertically placement of coronal sections. Trimming levels should be selected based on external anatomical landmarks.

FIGURES 25 and 26 Placement of trimming cuts that should be taken as the minimum acceptable tissue set for qualitative and quantitative analysis during a standard developmental neuropathology evaluation. The sections [from a postnatal day (PND) 62 rat] will permit the examination of the major brain areas that should be addressed in a conventional developmental neurotoxicity study at early and later time-points of examination (Table 2): olfactory bulbs (Figure 25-1); rostral forebrain (Figure 25-2, frontal cortex and striatum); middle forebrain (Figure 25-3, parietal and temporal cortices, hippocampus, diencephalon); caudal forebrain (Figure 25-4, occipital cortex and mesencephalon); pons (Figure 25-5); medulla oblongata (Figure 25-6); trigeminal (Gasserian) ganglia (Figure 25-9 a + b); cerebellum, coronal orientation (Figure 26-7); and cerebellum, mid-sagittal orientation (Figure 26-8).

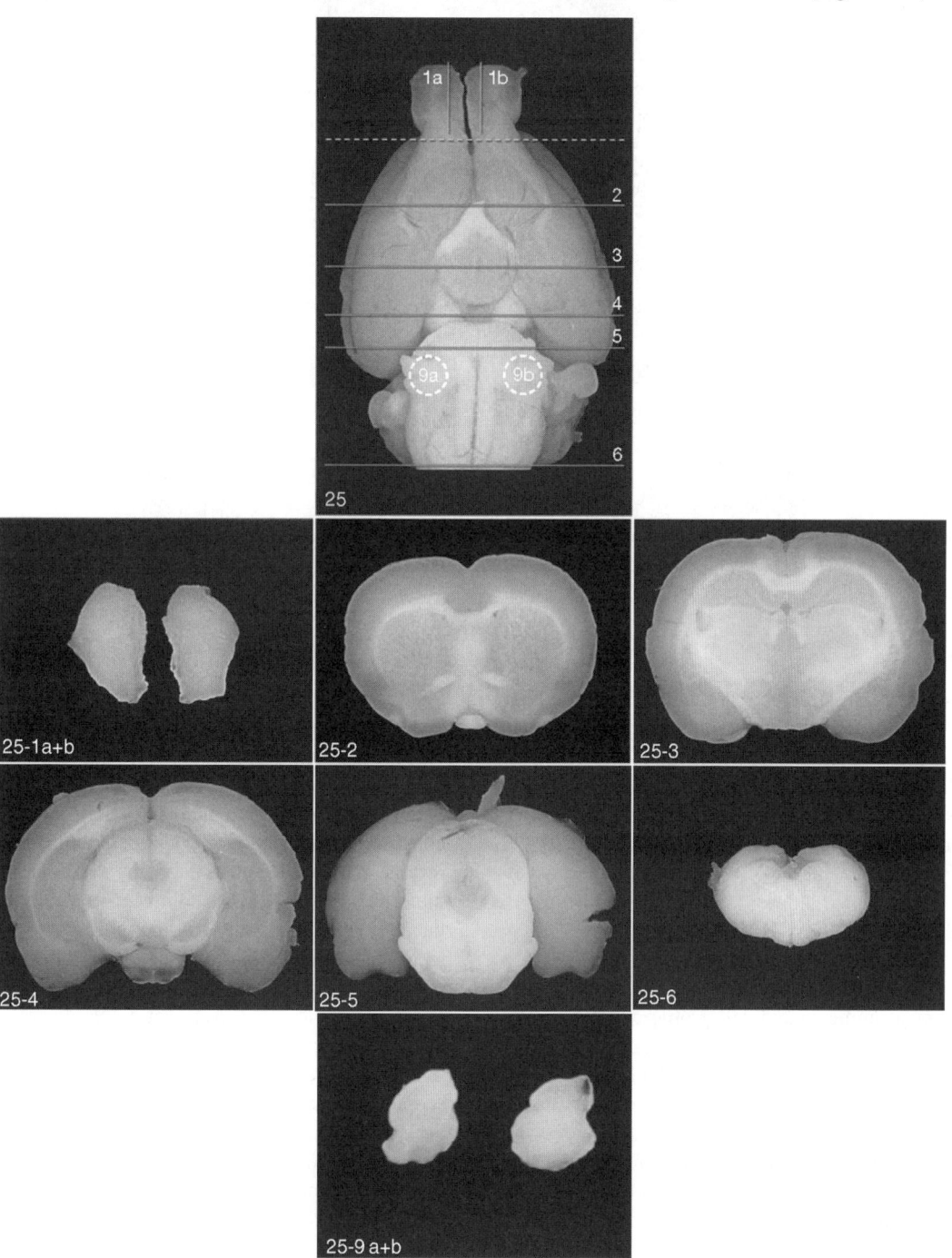

FIGURE 25

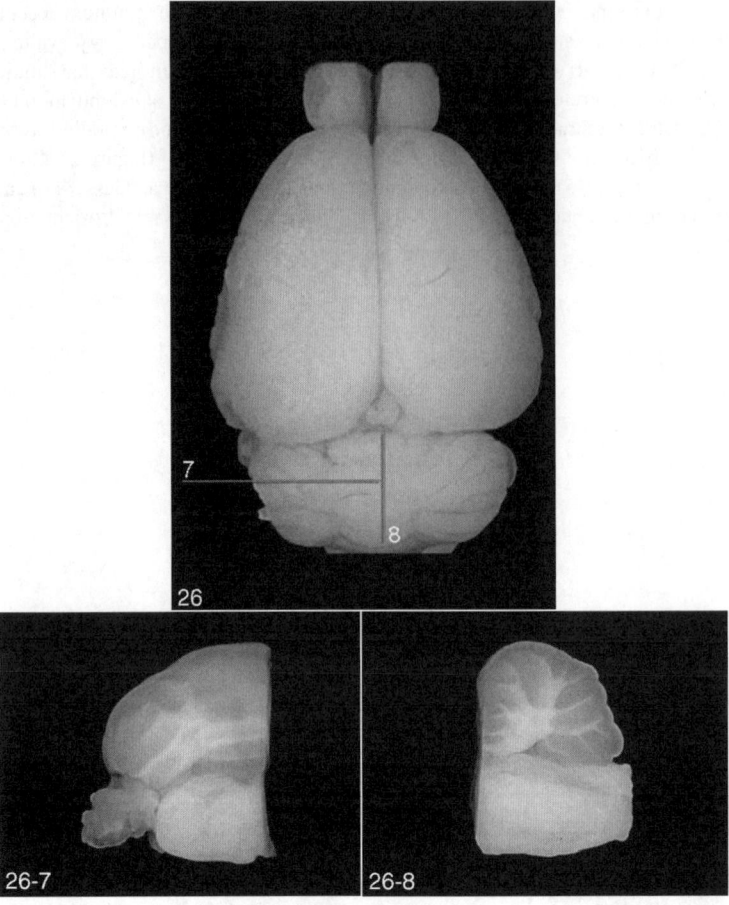

FIGURE 26

Hypoplasia of brain structures, including hypomyelination of white matter tracts or altered synaptogenesis in various brain regions, may be reliably indicated by statistically significant deviations in linear measurements on homologous section levels when compared to the respective control brains (G. Krinke, personal communication).

SECOND-TIER APPROACHES

Advanced morphometric techniques (e.g., stereology), including area measurements and cell profile counts may be performed in a second-tier approach on potential target regions identified by gross pathology, light microscopy, and/or linear morphometry, or in the case where inconclusive or inconsistent linear measurements have provided no definitive conclusion. Stereology may be of special value when altered neuronal cell counts are suspected. However, this approach needs additional tissue samples from a further cohort of test animals in a routine DNT study if an appropiate stereological analysis is to be undertaken.[71] In a developmental neurobehavioral study of ammonium perchlorate, a compound known to induce hypothyroidism, male rats at

PND 10 and PND 20 had slightly but significantly increased linear measurements for a few major brain areas, especially for the corpus callosum, relative to concurrent controls while females experienced a decrease in the same measurements.[74] A dose–response relationship was not present, nor was any consistent neuropathologic lesion evident. Based on these results, the authors postulated the existence of a sexually dimorphic response in the control of programmed cell death, but they concluded that more sophisticated morphometric techniques were required to determine the true neurotoxic potential of ammonium perchlorate.[74] A more sophisticated approach employing a systematic, tiered approach has been described recently using the MAM model.[71] The bottom line is that each tier of the neuropathology assessment (macroscopic measurements of brain size → linear measurements → stereology) improved the understanding of predilection areas for MAM developmental neurotoxicity in the brain. For this study, dams were treated with MAM at 7.5 mg/kg i.p. on GD 13 to 15; this dose does not elicit heterotopiae or ectopic neuronal clusters in the PND 22 brain. Each tier indicated that the major effects of MAM were concentrated in the cerebrum, including a considerable loss of neurons in the hippocampal CA1 pyramidal layer; the cerebellum and,

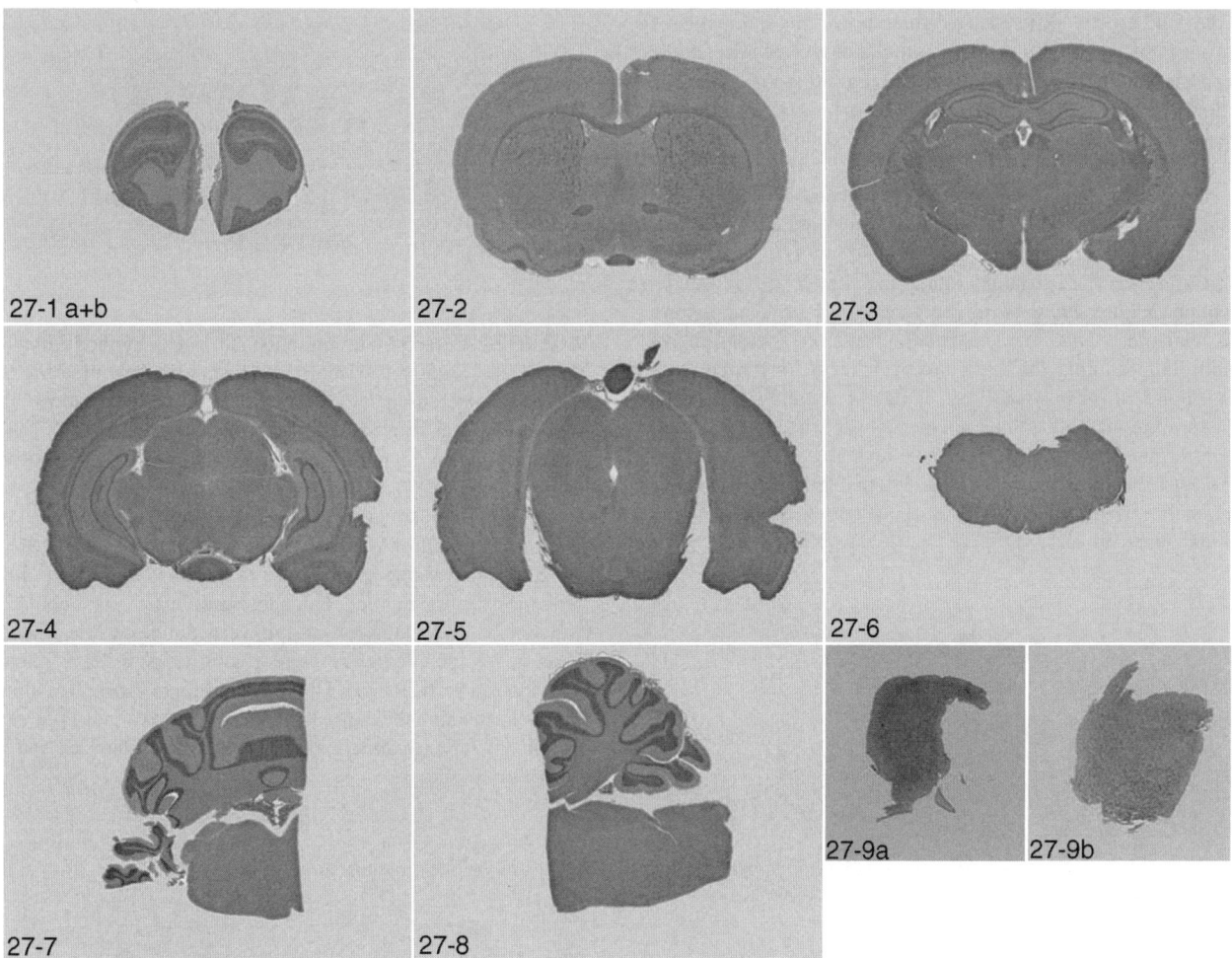

FIGURE 27   Histological appearance of the paraffin-embedded H&E-stained thin sections (Figures 27-1 to 27-9) corresponding to the trimming cuts defined above (Figures 25-1 to 25-6, 25-9 a + b, and 26-8 and 26-9). H&E stain, original magnification 10×.

more specifically, the neurons in the granule cell layer were not changed. This pattern reflects the absence of any substantive impact of treatment on cerebellar growth during the prenatal period.[75] In summary, this tiered approach demonstrated the usefulness and discriminative strength of a tiered quantitative analysis to demonstrate developmental neurotoxicity, in an instance where a qualitative analysis by light microscopy alone would have failed.

An advanced morphometric technique can be combined with another contrast-enhancing procedure to provide even more discriminating power. For example, stereological analysis in conjunction with immunohistochemistry for tyrosine hydroxylase (TH, an enzyme involved in dopamine synthesis) has been applied to test compounds that induce a Parkinson's disease-like pattern of neuropathologic lesions, including MPTP (1-methyl-4-phenyl-4-tetrahydropyridine), paraquat (N,N-dimethyl-4-4'-bipiridinum, which exhibits a striking structural similarity to MPP$^+$, the toxic metabolite of MPTP), and maneb (manganese ethylene

bisdithiocarbamate). These agents affect the dopaminergic neurons of the substantia nigra pars compacta (SNpc). Immunohistochemistry for TH will selectively illuminate dopaminergic neurons in the SNpc (Figures 32 to 34), after which a stereological, quantitative analysis can be used to estimate the mean cross-sectional area and cytoplasmic volume of the TH$^+$ neuronal population. In two DNT studies with paraquat and maneb using male and female C57BL mice, a significant loss of dopaminergic neurons was found in the offspring by stereology following maternal treatment between either GD 10 and 17 or between PND 5 and 19.[32,76] These examples clearly demonstrate that second-tier quantitative neuropathologic methods represent an important approach for the neuropathology part of conventional DNT studies.

In some cases, stereology may be the only means by which an adequate examination of neural domain may be undertaken, and as such this advanced morphological method would move from an "optional" to a "required" element of

FIGURES 28 to 31 Microscopic quantitative measurements of toxicant-induced developmental neuropathology include simple linear measurements of various brain regions on histological sections. The internal structural landmarks used to site the measurements (indicated by dark lines) should be reliably produced for all animals, even if additional step sections must be taken at a later time to achieve them. In the rostral forebrain (Figure 28), typical measurements include thickness of the cerebral cortex (the vertical and oblique lines) and width of the corpus striatum (horizontal lines). In the caudal forebrain (Figure 29, showing the corpus callosum, hippocampus, thalamus, and hypothalamus), common measurements include hippocampal thickness (vertical lines). In the cerebellum [Figure 30, at postnatal day (PND) 11; Figure 31, at PND 62], conventional linear measurements—shown here for cerebellar folium VIII—span the cerebellar cortex from the brain surface to the deep margin of the internal granule cell layer. H&E stain, original magnification x10 (Figures 28 and 29) and 40× (Figures 30 and 31).

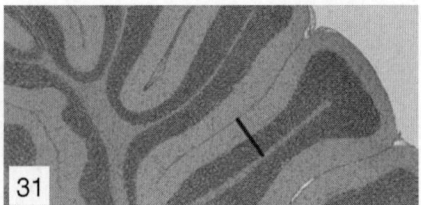

FIGURE 31

FIGURES 32 to 34 Immunohistochemical methods may be applied to qualitatively visualize and quantitatively measure (via second-tier stereological counts of neuronal populations) the impact of toxicants on specific neurotransmitter pathways. Typical surveys incorporate both parasagittal (Figure 32) and coronal (Figures 33 and 34) sections, usually taken at multiple levels. This example illustrates dopaminergic neurons and projections (arrows) in a control Wistar rat brain at postnatal day (PND) 62 using an anti-tyrosine hydroxylase (TH) procedure. Major dopaminergic centers include the substantia nigra pars compacta (Figure 32, horizontal arrow at right vertical line), while principal target zones include the corpus striatum (Figure 32, horizontal arrows at left vertical line). The vertical lines in Figure 32 indicate the levels at which the coronal sections were taken for Figures 33 and 34. TH stain, original magnification 10× (Figures 31 and 32) and 100× (Figure 33).

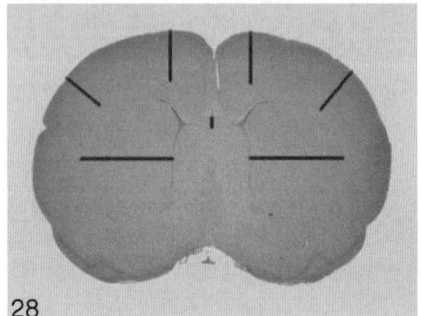

FIGURE 28

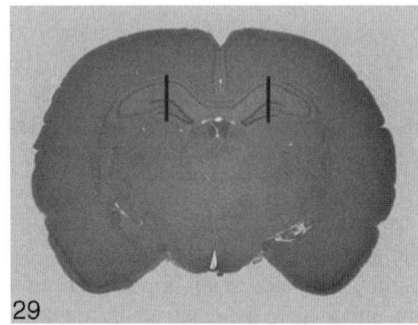

FIGURE 29

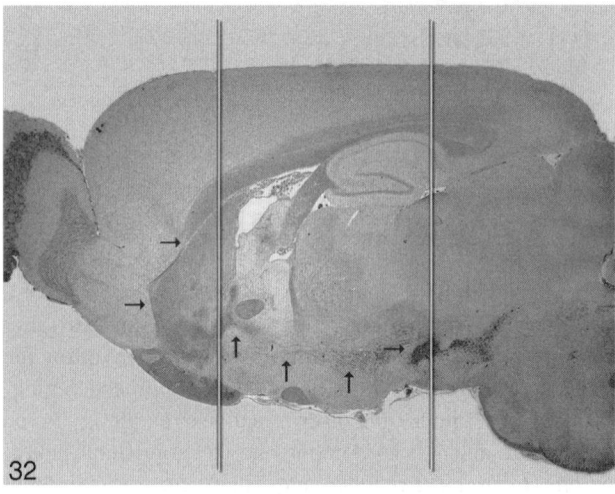

FIGURE 32

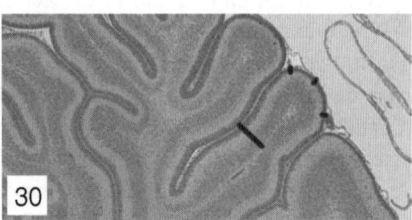

FIGURE 30

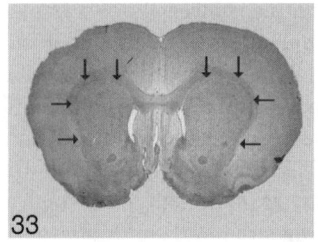

FIGURE 33

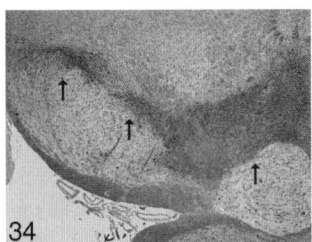

FIGURE 34

the DNT neuropathology evaluation. A primary example of this need is illustrated by the occasional requirement for enumerating the cell populations of various sexually dimorphic nuclei. The most important sexually dimorphic nuclei are located in the preoptic area of the hypothalamus, but other such elements are scattered throughout the brain, including the olfactory bulb and the amygdala of the limbic system.[31] The sexually dimorphic nucleus of the preoptic area is five times larger in male rats than in females. This divergence is explained by a much higher incidence of apoptotic cell death in females between PND 4 and 10; the activity of testosterone prevents this site-specific apoptosis in males.[28] Treatment of female rats with testosterone propionate injections for 9 days immediately after birth results in larger nuclei compared with the size of the same structure in untreated control females. In contrast, castration of male rats immediately after birth significantly decreases the size of the sexually dimorphic nucleus. These differences can be detected reliably only by stereology and not by qualitative light microscopy or simple linear measurements.

REFERENCES

1. Garman RH, et al. Methods to identify and characterize developmental neurotoxicity for human health risk assessment: II. Neuropathology. *Environ Health Perspect*. 2001;109 (Suppl 1):93–100.

2. Bolon B, et al. A "best practices" approach to neuropathologic assessment in developmental neurotoxicity testing—for today. *Toxicol Pathol*. 2006;34:296–313.

3. Kaufmann W, Gröters S. Developmental neuropathology in DNT studies: a sensitive tool for the detection and characterization of developmental neurotoxicants. *Reprod Toxicol*. 2006;22(2):196–213.

4. US, Environmental Protection Agency. *Health Effects Test Guidelines: Developmental Neurotoxicity Study*. OPPTS 870.6300. Washington, DC: US EPA; 1998. Available at: http://www.epa.gov/epahome/research.htm.

5. US Environmental Protection, Agency. *Health Effects Test Guidelines: Neurotoxicity Screening Battery*. OPPTS 870.6200. Washington, DC: US EPA; 1998. Available at: http://www.epa.gov/epahome/research.htm.

6. Organisation for Economic Co-operation and Development. *Guideline for the Testing of Chemicals 426: Developmental Neurotoxicity Study*. Paris: OECD; 2007. Available at: http://www.oecd.org/ehs/test/health.htm. Accessed Apr 17, 2010.

7. Organisation for Economic Co-operation and Development. *Guideline for the Testing of Chemicals 424: Neurotoxicity Study in Rodents*. Paris: OECD; 1997. Available at: http://www.oecd.org/ehs/test/health.htm. Accessed Apr 17, 2010.

8. Meyer JS. Behavioral assessment in developmental neurotoxicology. In: Slikker W, Chang LW, eds. *Handbook of Developmental Neurotoxicology*. San Diego, CA: Academic Press; 1998: 403–426.

9. Bayer SA, et al. Timetables of neurogenesis in the human brain based on experimentally determined patterns in the rat. *Neurotoxicology*. 1993;14:83–144.

10. Wood SL, et al. Species comparison of postnatal CNS development: functional measures. *Birth Defects Res B*. 2003;68: 391–407.

11. Schardein JL. Animal/human concordance. In: Slikker W, Chang LW, eds. *Handbook of Developmental Neurotoxicology*. San Diego, CA: Academic Press; 1998: 687–708.

12. Stanton ME, Spear LP. Workshop on the qualitative and quantitative comparability of human and animal developmental neurotoxicity. Work Group I report: Comparability of measures of developmental neurotoxicity in humans and laboratory animals. *Neurotoxicol Teratol*. 1990;12(3): 261–267.

13. Clancy B, et al. Extrapolating brain development from experimental species to humans. *Neurotoxicology*. 2007;28: 931–937.

14. Romijn HJ, et al. At what age is the developing cortex cerebri of the rat comparable to that of the full-term newborn human baby? *Early Human Dev*. 1991;26:61–67.

15. Bové J, et al. Toxin-induced models of Parkinson's disease. *Neuro Rx*. 2005;2:484–494.

16. Anderson LM, et al. Critical windows of exposure for children's health: cancer in human epidemiological studies and neoplasm in experimental animal model. *Environ Health Perspect*. 2000;108(3):573–594.

17. Bithell JF, Stewart AM. Pre-natal irradiation and childhood malignancy: a review of British data from the Oxford survey. *Br J Cancer*. 1975;31:271–287.

18. Goldman L, et al. Environmental pediatrics and its impact on government health policy. *Pediatrics*. 2004; 113: 1146–1157.

19. Karnovsky MJ. A formaldehyde–glutaraldehyde fixative of high osmolarity for use in electron microscopy. *J Cell Biol*. 1965;2:137A.

20. Fix AS, Garman RH. Practical aspects of neuropathology: a technical guide for working with the nervous system. *Toxicol Pathol*. 2000;28:122–131.

21. Jortner BS. The return of the dark neuron. A histological artifact complicating contemporary neurotoxicologic evaluation. *Neurotoxicology*. 2006;27:628–634.

22. Glowinski J, Iversen LL. Regional studies of catecholamines in the rat brain: I. *J Neurochem*. 1966;13:655–669.

23. Kaufmann W. Developmental neurotoxicity. In: Krinke GJ, ed. *The Laboratory Rat*. San Diego, CA: Academic Press; 2000: 227–250.

24. Kandel ER. The brain and behavior. In: Kandel ER, et al, eds. *Principles of Neural Science*, 4th ed. New York: McGraw-Hill; 2000: 5–18.

25. Amaral DG. The anatomical organization of the central nervous system. In: Kandel ER, et al., eds. *Principles of Neural Science*, 4th ed. New York: McGraw-Hill; 2000: 317–336.

26. Weinberger DR. Schizophrenia as a neurodevelopmental disorder. In: Hirsch SR, Weinberger DR, eds. *Schizophrenia*. Oxford; UK: Blackwell; 1995: 293–323.

27. Chang LW, Guo GL. Fetal minamata disease: congenital methylmercury poisoning. In: Slikker W, Chang LW, eds. *Handbook of Developmental Neurotoxicology*. San Diego, CA: Academic Press; 1998: 507–516.

28. Gorski RA. Sexual differentiation of the nervous system. In: Kandel ER, et al., eds. *Principles of Neural Science*, 4th ed. New York: McGraw-Hill; 2000: 1131–1148.

29. Nulman I, et al. The effect of alcohol on the fetal brain: the central nervous system tragedy. In: Slikker W, Chang LW, eds. *Handbook of Developmental Neurotoxicology*. San Diego, CA: Academic Press; 1998: 567–586.

30. Kinney HC. Abnormalities of the brainstem serotonergic system in the sudden infant death syndrome: a review. *Pediatr Dev Pathol*. 2005;8:507–524.

31. Maeda K, et al. Physiology of reproduction. In: Krinke GJ, ed. *The Laboratory Rat*. San Diego, CA: Academic Press; 2000: 145–176.

32. Cory-Slechta DA, et al. Developmental pesticide models of the Parkinson disease phenotype. *Environ Health Perspect*. 2005; 113:1263–1270.

33. Palay SL, Chan-Palay V. *Cerebellar Cortex–Cytology and Organization* Vol. 3 *The Design of the Cerebellar Cortex*. Berlin: Springer-Verlag; 1974: 5–10.

34. Sakamoto M, et al. Evaluation of changes in methylmercury accumulation in the developing rat brain and its effect: a study with consecutive and moderate exposure throughout gestation and lactation. *Brain Res*. 2002;949:51–59.

35. Bolon B. Comparative and correlative neuroanatomy for the toxicologic pathologist. *Toxicol Pathol*. 2000; 28: 6–27.

36. Mitsumori K, Boorman GA. Spinal cord and peripheral nerves. In: Boorman GA, et al., eds. *Pathology of the Fischer Rat: Reference and Atlas*. San Diego, CA: Academic Press; 1990: 179–191.

37. Schaeppi U, Krinke G. Differential vulnerability of 3 rapidly conducting somatosensory pathways in the dog with vitamin B6 neuropathy. *Agents Actions*. 1985;16:567–579.

38. Kittel B, et al. Revised guides for organ sampling and trimming in rats and mice: 2. *Exp Toxicol Pathol*. 2004;55:413–431.

39. Krinke G, et al. Differential susceptibility of peripheral nerves of the hen to triorthocresyl phosphate and to trauma. *Agents Actions*. 1979;9:227–231.

40. Kaufmann W. Histometrische und enzymhistochemische Untersuchungen zur Pathogenese der Muskelveränderungen nach Neurektomie bei wachsenden Schweinen [Histometric and enzyme histochemical examinations on the pathogenesis of muscle lesions following neurectomy in growing pigs]. *J Vet Med A*. 1985;32:683–698.

41. Kaufmann W. Lichtmikroskopische und ultrastrukturelle Untersuchungen zur Pathogenese der Muskelveränderungen nach Neurektomie bei wachsenden Schweinen [Light microscopic and ultrastructural examinations on the pathogenesis of muscle lesions following neurectomy in growing pigs]. *J Vet Med A*. 1985;32:699–716.

42. Fix AS, et al. Integrated evaluation of central nervous system lesions: stains for neurons, astrocytes, and microglia reveal the spatial and temporal features of MK-801-induced neuronal necrosis in the rat cerebral cortex. *Toxicol Pathol*. 1996;24: 291–304.

43. De Olmos JS, et al. Use of an amino-cupric-silver technique for the detection of early and semiacute neuronal degeneration caused by neurotoxicants, hypoxia, and physical trauma. *Neurotoxicol Teratol*. 1994;16:545–561.

44. Barone S Jr, et al. Vulnerable processes of nervous system development: a review of markers and methods. *Neurotoxicology*. 2000;21:15–36.

45. Krinke GJ, et al. Detecting necrotic neurons with Fluoro-Jade stain. *Exp Toxicol Pathol*. 2001;53:365–372.

46. Olney JW, et al. Environmental agents that have the potential to trigger massive apoptotic neurodegeneration in the developing brain. *Environ Health Perspect*. 2000;108 (Suppl 3):383–388.

47. Olney JW. New insights and new issues in developmental neurotoxicology. *Neurotoxicology*. 2002;23:659–668.

48. Schmued LC, et al. Fluoro-Jade C results in ultra high resolution and contrast labeling of degenerating neurons. *Brain Res*. 2005;1035:24–31.

49. Schmued LC, Hopkins KJ. Fluoro-Jade B: a high affinity fluorescent marker for the localization of neuronal degeneration. *Brain Res*. 2000;874:123–130.

50. Schmued LC, Slikker W Jr., Black-Gold: a simple, high-resolution histochemical label for normal and pathological myelin in brain tissue sections. *Brain Res*. 1999;837:289–297.

51. Streit WJ. An improved staining method for rat microglial cells using the lectin from *Griffonia simplicifolia* (GSA I-B4). *J Histochem Cytochem*. 1990;38:1683–1686.

52. Kaur C, et al. Lectin labelling of amoeboid microglial cells in the brain of postnatal rats. *J Anat*. 1990;173:151–160.

53. Schlote W. Pathologie des Nervensystems II. In: Doerr W, Seifert G, eds. *Spezielle pathologische Anatomie*. Berlin: Springer-Verlag; 1983: 2–172.

54. Carratu MR, et al. Developmental neurotoxicity of carbon monoxide. *Arch Toxicol Suppl*. 1995;17:295–301.

55. Aikawa H, Suzuki K. Neurotoxic effects of 6-aminonicotinamide, mouse. In: Jones TC, et al., eds. *Nervous System, Monographs on Pathology of Laboratory Animals*. Sponsored by the International Life Science Institute. Berlin: Springer-Verlag; 1988: 53–57.

56. Nieminen L, et al. Effect of hexachlorophene on the rat brain during ontogenesis. *Food Cosmet Toxicol*. 1973;11(4): 635–639.

57. Cook L, et al. Tin distribution in adult and neonatal rat brain following exposure to triethyltin. *Toxicol Appl Pharmacol.* 1984;72:75–81.

58. Ellis WG, et al. Relationship of periventricular overgrowth to hydrocephalus in brains of fetal rats exposed to benomyl. *Terato Carcino Mutag.* 1988;8:377–391.

59. Friede RL. Dysplasias of cerebral cortex. In: Friede RL, ed. *Developmental Neuropathology.* Berlin: Springer-Verlag; 1989: 330–346.

60. Collier PA, Ashwell KW. Distribution of neuronal heterotopias following prenatal exposure to methylazoxymethanol. *Neurotoxicol Teratol.* 1993;15:439–444.

61. Norman MG. Bilateral encephaloclastic lesions in a 26 week gestation fetus: effect on neuroblast migration. *Can J Neurol Sci.* 1980;7:191–194.

62. Streissguth AP, et al. Teratogenic effects of alcohol in humans and laboratory animals. *Science.* 1980;209:353–361.

63. Rakic P. Cell migration and neuronal ectopias in the brain. *Birth Defects Orig Artic Ser.* 1975;11:95–129.

64. Spatz M, Laqueur GL. Transplacental chemical induction of microencephaly in two strains of rat: I. *Proc Soc Exp Biol Med.* 1968;129:705–710.

65. Singh SC. Ectopic neurones in the hippocampus of the postnatal rat exposed to methylazoxymethanol during foetal development. *Acta Neuropathol.* 1977;40: 111–116.

66. Hallas BH, Das GP. An aberrant nucleus in the telencephalon following administration of ENU during neuroembryogenesis. *Teratology.* 1979;19:159–164.

67. Gabel LA, LoTurco JJ. Electrophysiological and morphological characterization of neurons within neocortical ectopias. *J Neurophysiol.* 2001;85:495–505.

68. Farrell MA, et al. Neuropathologic findings in cortical resections (including hemispherectomies) performed for the treatment of intractable childhood epilepsy. *Acta Neuropathol (Berl).* 1992;83:246–259.

69. Schmechel DE, Rakic PA. Golgi study of radial glial cells in developing monkey telencephalon: morphogenesis and transformation into astrocytes. *Anat Embryol.* 1979;156: 115–152.

70. Rodier PM, Gramann WJ. Morphologic effects of interference with cell proliferation in the early fetal period. *Neurobehav Toxicol.* 1979;1:129–135.

71. de Groot DM, et al. 2D and 3D assessment of neuropathology in rat brain after prenatal exposure to methylazoxymethanol, a model for developmental neurotoxicity. *Reprod Toxicol.* 2005;20:417–432.

72. Paxinos G. *The Rat Nervous System,* 3rd ed. San Diego, CA: Academic Press; 2004.

73. Altman J, Bayer SA. *Atlas of Prenatal Rat Brain Development.* Boca Raton, FL: CRC Press; 1995.

74. York RG, et al. Refining the effects observed in a developmental neurobehavioral study of ammonium perchlorate administered orally in drinking water to rats: II. Behavioral and neurodevelopmental effects. *Int J Toxicol.* 2005;24: 451–467.

75. Himwich WA. Problems in interpreting neurochemical changes occuring in developing and aging animals. In: Ford DH, ed. *Neurobiological Aspects of Maturation and Aging.* Progress in Brain Research, Vol 40 Amsterdam: Elsevier; 1973: 13.

76. Barlow BK, et al. A fetal risk factor for Parkinson's disease. *Dev Neurosci.* 2004;26:11–23.

23

NEUROPATHOLOGICAL ANALYSIS OF THE PERIPHERAL NERVOUS SYSTEM

GEORG J. KRINKE

Pathology Evaluations, Frenkendorf, Switzerland

ANATOMY OF THE PERIPHERAL NERVOUS SYSTEM

The functional cells—the neurons—connect the central nervous system (CNS) with the "periphery": the organs and tissues that monitor the conditions of the body's internal environment and external surroundings and exert active responses to the signals received. Each neuron is connected to other neurons or to peripheral nerve endings with either sensory, motor, or secretory function. The peripheral nerves associated with the spinal cord are termed *spinal nerves*; those associated with the brain are designated *cranial nerves*. The first cranial nerve, the olfactory nerve, is formed by the axonal processes of sensory cells located in the olfactory neuroepithelium. The second cranial nerve, the optic nerve, is not a peripheral nerve but, instead, is formed by cell types that also occur in the CNS. The remaining cranial nerves have compositions comparable to those of the spinal nerves.

The nerve cell bodies that give origin to peripheral nerves may be located either outside the CNS, as in the sensory neurons in the sensory ganglia, or inside the CNS, as in the motoneurons. The sensory ganglia of spinal nerves are the dorsal root ganglia, which are associated with the dorsal spinal nerve roots containing sensory nerve fibers. The ventral spinal nerve roots contain motor fibers formed by the motoneurons located in the ventral horns of the spinal cord gray matter. The sensory neurons in the dorsal root ganglia have two long processes: one of them extending into the peripheral nerve, the other one passing through the dorsal spinal nerve root into the spinal cord either to ascend in the dorsal spinal columns as primary sensory spinal tracts

(fasciculi gracilis and cuneatus) or to provide input to secondary sensory neurons (such as those forming the spinocerebellar tract).

Owing to this arrangement of sensory and motor neurons, there is no sharp anatomical border between the CNS and the peripheral nervous system (PNS); the cell bodies of the motoneurons and the central processes (axons) of the sensory neurons are located within the CNS. Therefore, many neurological diseases, including the toxic neuropathies, present as a central/peripheral neuropathy (Figure 1).[81]

In addition to somatic sensory and somatic motor neurons, there is an autonomic nervous system composed of the parasympathetic and sympathetic ganglia and nerve fibers, located mainly outside the CNS. The sympathetic peripheral neurons are adrenergic, using noradrenaline (norepinephrine) as the neurotransmitter. The peripheral sympathetic neurons are positioned in ganglia located outside the organs that they innervate (extramural neurons). The parasympathetic neurons are cholinergic, using acetylcholine as the neurotransmitter. The peripheral parasympathetic neurons are located within the organ wall (intramural neurons). Both sympathetic and parasympathetic neurons are under the control of the CNS and receive innervation from "proximal" CNS neurons.[85] Collectively, the autonomic nervous system regulates involuntary functions such as those residing in the cardiovascular system and the viscera; the sympathetic and parasympathetic divisions mediate opposing actions of these organs. The nerve fibers within the autonomic nervous system are both myelinated and unmyelinated, but in the terminal networks the thin, unmyelinated fibers are predominant. A separate division of the autonomic nervous system is

Fundamental Neuropathology for Pathologists and Toxicologists: Principles and Techniques, First Edition. Edited by Brad Bolon and Mark T. Butt.
© 2011 John Wiley & Sons, Inc. Published 2011 by John Wiley & Sons, Inc.

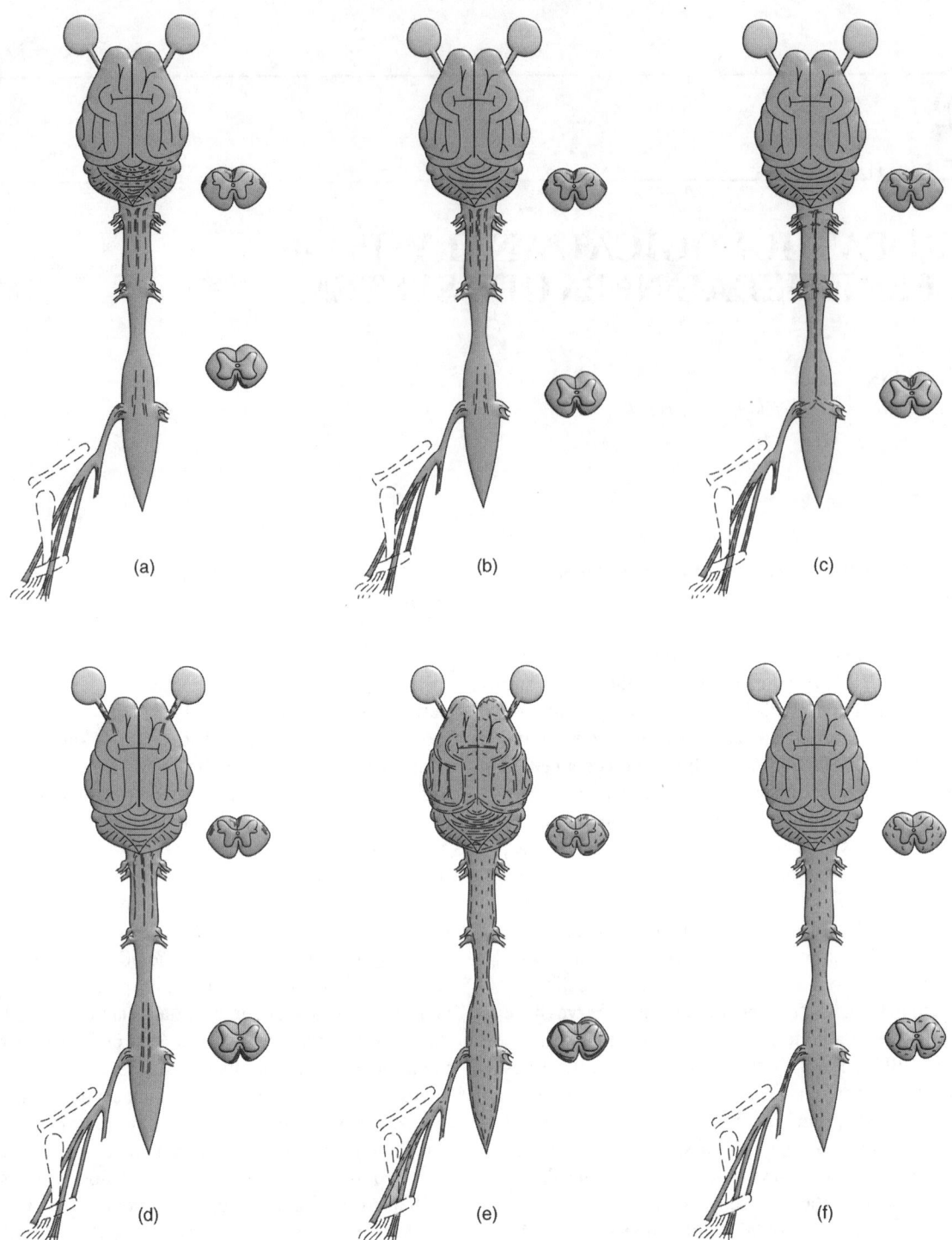

FIGURE 1   Selected distribution patterns of toxic neuropathy. In most cases it is impossible to discriminate "central" from "peripheral" neuropathy. (a) Acrylamide induces changes in the dorsal root ganglia, distal peripheral and spinal tract axons, but also necrosis of the cerebellar Purkinje cells. (b) Organophosphorus-induced delayed neuropathy (OPIDN) induces changes in the distal peripheral and spinal tract axons. (c) Pyridoxine induces changes in the dorsal root ganglia and primary sensory axons in the peripheral nerves and spinal tracts. (d) Clioquinol induces changes in the distal spinal axons and the optic tract. (e) Toxic myelin edema diffusely affects the central and peripheral myelinated areas. (f) Segmental demyelination is most prominent in the spinal nerve roots and may diffusely affect all columns of the spinal cord.

the peptidergic enteric nervous system, consisting of the intrinsic neurons of the gut. This enteric network is capable of local independent action to maintain such automatic processes as peristalsis, but its activities are modulated by parasympathetic and sympathetic nerves. This complex pattern of functions is mediated by multiple (mainly peptidergic) neurotransmitters, including substance P, vasoactive intestinal polypeptide (VIP), opioid peptides, neuropeptide Y, calcitonine gene–related peptide (CGRP), and other peptides.[18] In addition to the peptidergic enteric nervous system, peristalsis is also regulated by the interstitial cells of Cajal, which are stellate mesenchymal pacemaker cells that regulate the basic electrical rhythm.[23]

CELL STRUCTURE AND FUNCTION IN THE PNS

The nerve cell body (perikaryon) contains the genetic information (nuclear DNA) and the organelles (rough endoplasmic reticulum) needed for protein synthesis. Structures within the long nerve cell processes, the axons, are maintained by the metabolic machinery of the nerve cell bodies and receive needed materials by means of axonal transport. Given the large size of axons in the PNS (some axons traverse the total length of the spinal cord or the peripheral nerve), their maintenance is very demanding. It renders the long and large-diameter axons especially highly susceptible to damage.

Slow anterograde (proximodistal) axonal transport (moving at a few millimeters per day) carries large molecules, especially the neurofilament proteins, whereas fast axonal transport (about 400 mm/day) carries smaller molecules, including the neurotransmitters. Retrograde (distoproximal) axonal transport conveys used elements back to the perikaryon for degradation and recycling, and shifts neurotropic substances from the nerve endings toward the nerve cell body. Retrograde axonal transport also contributes to the spread of noxious entities from the peripheral site of inoculation into centrally located neurons, including such infectious agents as herpesvirus, enterovirus, or rabies virus, as well as toxins taken up at the nerve endings.

Neuronal cytoskeleton is formed by the neurofilaments, which are 10-nm intermediate filaments, and microtubules. Initially, when produced in the nerve cell body, the neurofilaments are not phosphorylated, but they become phosphorylated for their export into the axons. The axonal neurofilaments can be subdivided according to their molecular weight into three categories: low, medium, and large, corresponding to 68, 150, and 200 kDa, respectively. Each subset can be demonstrated by using specific antibodies, but for most practical applications in PNS analysis, staining all neurofilaments simultaneously using a triple antibody directed against phosphorylated neurofilaments is suitable. The neurofilament content is a major determinant of axonal size

(diameter). The microtubules are composed of tubulin protein dimers, about 10 to 20 μm long. They have been hypothesized to serve as "rails" for fast axonal transport.

All PNS axons are ensheathed to some extent by Schwann cells. The axons become myelinated (i.e., covered by multiple layers of duplicated Schwann cell membrane) when they are larger than about 1 mm in diameter. Smaller axons are covered by simple layers of Schwann cell membrane and form unmyelinated nerve fibers. Most unmyelinated nerve fibers are about 1.0 to 1.6 mm in diameter. There is a correlation between axon diameter and the thickness of the myelin sheath, so that large-diameter axons have thicker myelin sheaths, and vice versa. In the PNS, myelin segments of myelinated nerve fibers are formed for each axon by a single Schwann cell. This arrangement is different from that found in the CNS, where a single oligodendroglial cell provides myelin sheaths for many (up to about 50) axons (Figure 2). Due to this difference, remyelination and regeneration occur more readily in the PNS. Interestingly, when oligodendrocytes and astrocytes in the spinal cord were destroyed by injection of ethidium bromide, the demyelinated areas became invaded by Schwann cells.[4]

The myelin sheaths are lipoprotein structures containing 70 to 80% lipids, especially cholesterol and cerebrosides (galactolipids). The proteins characteristic for myelin are proteolipid protein (PLP), located at the intraperiod line (minor dense line, representing the contact between the outer leaflets of the cell membrane of myelinating cells); myelin basic protein (MBP), located at the major dense line (contact of inner leaflets of the cell membrane of myelinating cells); and myelin-associated glycoprotein (MAG), located in the periaxonal space. Protein 0 (P0) and peripheral membrane protein 22 (PMP-22) occur only in the PNS.

Unlike the CNS, the PNS lacks astrocytes and a tight blood–nerve barrier, so that blood-borne agents can easily penetrate the nerve tissue. The nerve cell bodies in the peripheral ganglia are surrounded by satellite cells, which are similar to Schwann cells. The PNS fibers are surrounded by a mesh of connective tissue forming the endoneurium

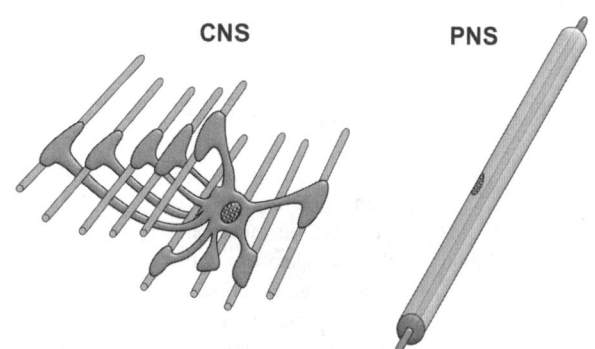

FIGURE 2 Difference in number of myelin segments produced by oligodendroglia cells (CNS) and Schwann cells (PNS).

(between axons), perineurium (separating axonal bundles), and epineurium (surrounding entire nerves). The optic nerve, consistent with its CNS origin, is invested by meninges.

The particular cell elements in the PNS may represent primary disease targets. Primary damage to nerve cell bodies is described as *neuronopathy*, primary damage to axons is designated *axonopathy*, and primary damage to myelinating (Schwann) cells and myelin is called *myelinopathy*. In toxicology, the recognition of primary cell targets is essential for hazard identification and risk assessment. To find out which cell elements are primarily affected in the earliest stage of a PNS lesion, experimental studies are generally designed so that various areas of the nervous system can be repeatedly examined in *spatiotemporal studies*. Such a time-course strategy is required, as primary damage to one PNS element typically causes secondary responses from other PNS cells.[81]

TYPES OF NERVE FIBERS IN THE PERIPHERAL NERVES

The peripheral nerves of mammals are equipped with various types of nerve fibers, depending on their function and targets of innervation.[23] An overview of the major fiber types is presented in Table 1. The conduction velocity of nerve impulses is proportional to the nerve fiber diameter: myelinated fibers conduct faster that unmyelinated, and large-diameter fibers conduct faster than those with smaller diameters. The nerve fibers, particularly motor fibers, substantially decrease their size within the field of their terminal ramification by extensive branching. Therefore, for a quantitative study of nerve composition, the more proximal "parent trunks" are most suitable for assessment, as at this level it is very unlikely that the small-diameter fibers were merely branches of larger

nerves.[9,92] In myelinated nerves there is a linear relationship between the axon diameter and the thickness of the myelin sheath.[76] An exception to this rule is found in the nerve fibers of the dorsal root ganglia that comprise the initial complex (region of axonal bifurcation). In such fibers, the myelin sheath is unusually thin.[80] In a similar manner, the relationship of the number of myelin lamellae to the axon circumference is approximately linear; it can be taken as a measure of the degree of myelination.[76] The length of myelin segments is also proportional to the nerve fiber diameter. In developing nerves, the segment length increases during the period of early myelination and subsequent longitudinal growth of the peripheral nerve.[75] In regenerated peripheral nerves, which usually exhibit a decreased conduction velocity, the thickest axons are thinner than normal, the myelin sheaths are also relatively thin, and the myelin segments are shorter.

TECHNICAL COMMENTS ON PNS SAMPLING

Sampling

The choice of PNS specimens for neuropathology examination depends on the effects expected and the clinical signs observed. Toxic axonopathies generally present themselves in the PNS as lesions to long and large-diameter axons. Accordingly, the sciatic nerve and its distal branches, including the tibial and plantar nerves, as well as the branch to the medial gastrocnemius muscle, are usually evaluated to detect axonopathies. In contrast, toxic myelinopathies affect mainly the long and large myelin segments present in the lumbosacral spinal nerve roots and the proximal sciatic nerve. Selective damage to the peripheral components of the autonomic nervous system typically requires deliberate sampling. The sympathetic system is best evaluated by

TABLE 1 Types of Nerve Fibers in the Mammalian Peripheral Nerves[a]

Fiber Type[b] According to Erlanger and Gasser[21]		Function	Fiber Diameter (μm)	Conduction Velocity (m/s)		Numerical Classification of Sensory Fibers
A	α	Proprioceptive, somatic motor	12–20	70–120	Ia	Muscle spindle, annulospiral ending
					Ib	Golgi tendon organ
A	β	Touch, pressure, motor	2–12	30–70	II	Muscle spindle, flower-spray endings: touch, pressure
A	γ	Motor to muscle spindles	3–6	15–30		
A	δ	Pain, cold, touch	2–5	12–30	III	Pain and cold receptors; some touch receptors
B		Preganglionic autonomic	<3	3–15		
C	Dorsal root	Pain, temperature, some mechanoreception, reflex responses	0.4–1.2	0.5–2	IV	Pain, temperature, and other receptors
C	Sympathetic	Postsynaptic sympathetic	0.3–1.3	0.7–2.3		

[a] *Source:* Ganong.[23]

[b] Type A and B fibers are myelinated; type C fibers are unmyelinated.

assessment of discrete ganglia [e.g., cranial cervical (superior cervical), in the neck] or plexuses (e.g., celiacomesenteric, at the root of the mesentery), although the nerves can be examined in the densely innervated vas deferens. The parasympathetic system is usually explored by examining the vagus nerve (cranial nerve X) with nodose ganglion as well as the intramural ganglia and nerve networks of the digestive tract.

Given the complexity of the PNS, prospective neurotoxicity studies must be designed carefully to ensure that appropriate tissue samples are taken for processing and analysis, or retained for potential future analysis.[7,36] A major decision in this respect is whether or not to collect PNS tissues unilaterally or bilaterally. Toxic and metabolic neuropathies are typically bilateral (including in transverse sections of the spinal cord segments from which they arise), so a case can be made for unilateral acquisition. However, it is the author's recommendation to use bilateral collection with unilateral processing, thereby preserving one side for additional special neuropathology examination (e.g., bilateral comparison) if dose-related effects are observed in the initial analysis.

Sample Orientation

Qualitative screening for PNS neuropathology can be undertaken effectively in paraffin-embedded material. Two adjacent samples should be assessed, one each in longitudinal and transverse orientations. Quantitative morphometry of PNS changes is performed on *transverse* hard-plastic sections. Typical PNS parameters subject to quantitative analysis include axon circularity, axon diameter, myelin sheath thickness, the number of axons/nerve fibers per square unit of area, or the axon/fiber caliber distribution.

Teased-Fiber Specimens

Teased-fiber preparations[48] are helpful for detailed characterization of changes in myelin segments and discrimination between segmental demyelination from Wallerian/secondary axonal degeneration. This preparation is very tedious work, so it is typically undertaken only for experiments (such as toxicity studies) in which all individuals can reliably be expected to develop the same type of lesion, and then only in a few high-dose specimens.

Teased-fiber preparations are unsuitable for quantitative evaluation, although attempts to do this have been made. Even with careful preparation, not all fibers can be isolated successfully, so the quantitative results are certain to be biased to some degree. Preparation of teased fibers in liquid epoxy resin following impregnation with osmic acid is expensive and quite difficult, and is justified only when selected nerve areas are to be examined by electron microscopy.

PRINCIPAL LESION TYPES OBSERVED IN TOXIC NEUROPATHIES

The general rules of peripheral neuropathy are summarized in Table 2.

Wallerian Degeneration

Augustus Voney Waller, a British physiologist who lived between 1816 and 1870, immortalized his name by postulating a law governing the events in a severed peripheral nerve: "The anatomical and physiological integrity of the nerve fibre is preserved only as long as it is connected to its cell of origin. When nerve fibres are cut, the distal portions of the fibres degenerate".[90] The cut nerve fiber has two stumps. The proximal stump, still connected to the nerve cell body, can regenerate (i.e., grow out toward the originally innervated target), whereas the distal stump, detached from the nerve cell body, degenerates (Figure 3). This *Wallerian degeneration* of the distal stump is characterized by fragmentation of fiber segments that starts in the middle part of segments (i.e., in the mid-internodal area, near the nucleus of a Schwann cell) (Figure 4). The fragments are formed at the Schmidt–Lanterman incisures,[42] which are the sites in the peripheral myelin segments where the myelin lamellae detach from each other to allow penetration of Schwann cell cytoplasm in the direction of the axon.[79] Fragmentation of the degenerating nerve fibers is an active process undertaken by Schwann cells that surround their nonviable axons. The result of fiber fragmentation is the formation of ovoid or round *digestion chambers* containing axon and myelin fragments (Figure 5). It has been suggested that phagocytosis of degenerating fibers was performed by Schwann cells, but there is overwhelming evidence that the majority of macrophages in Wallerian degeneration are actually of hematogenous origin.[2,52,74] The signals that draw bloodborne phagocytic cells to sites of PNS degeneration have not been defined. Fragmentation of degenerating nerve fibers at the Schmidt–Lanterman incisures must be discriminated from the artifact known as *beading*, which is produced by constriction at the Schmidt–Lanterman incisures following forceful pulling on the nerve (e.g., during the dissection process).[68,69]

The breakdown and removal of nerve fibers in the distal stump is followed by proliferation of initially anaxonal Schwann cells that form solid columns of cells, often termed *bands of Büngner*, connecting the severed site with the location of distal axon termination. Axons regenerating from the proximal stump grow distally into these newly formed columns of Schwann cells, where they are guided toward their appropriate targets for innervation. Regeneration of axons from the proximal stump is enabled by a transient increase in protein synthesis within the neuronal body. The organelles inside the cell body become rearranged so that the

TABLE 2   Summary of General Rules in Analyzing Peripheral Neuropathies

Kind of Lesion	Changes in Nerve Cell Body	Changes in Axon	Changes in Schwann Cells/Myelin Sheaths
Wallerian degeneration (secondary to nerve transection or crush)	Nerve cell body undergoes "axon reaction" (e.g., central chromatolysis, peripheral dislocation of the nucleus)	Proximal stump of severed nerve fiber regenerates; distal stump degenerates. "Digestion chambers" in degenerating distal stump contain ovoid fragments of myelin and axons. Axons regrow from the proximal stump into the bands of von Büngner.	Breakdown (fragmentation) takes place at Schmidt–Lanterman incisures and starts mid-internodally. Chainlike columns of proliferated Schwann cells form the bands of von Büngner.
"Dying back," distally accentuated axonopathy	Injured nerve cell body fails to maintain the axon. This may be a metabolic condition with or without morphological changes, or may be due to a proximal failure in axonal transport.	Distal portions of long and large-diameter axons are especially susceptible; therefore, the nerves in hindlimbs are predisposed to damage. "Dying back" axonal lesion progresses from distal to more proximal axonal positions. "Distally accentuated lesions" of the peripheral nerves can be produced by "summing up" of multifocal axonal damage. In the spinal cord, the distribution of axonal lesions follows that of anatomical spinal tracts; with distally accentuated lesions the sensory tracts are most affected at the upper cervical level, the motor tracts at the lumbosacral level.	Myelin sheaths devoid of their axons collapse and break down, as myelinating cells are unable to persist without reciprocal support from their axons. This is secondary demyelination resulting from primary axonal damage and loss.
Primary myelinopathy	Intact.	In primary demyelination the axons remain intact and shrink in diameter, but persist and can become remyelinated.	Myelin edema is due to increased presence of water and electrolytes in lipid-rich myelin; it leads to splitting of myelin lamellae; it is diffusely distributed and affects CNS and PNS. Myelin bubbles contain continuous axons and usually macrophages that are engaged in removing myelin debris. Segmental demyelination is most prominent in the proximal portion of peripheral nerves and the nerve roots (especially the motor nerve roots) because the Schwann cells at this level are especially large. Segmental demyelination starts in the paranodal areas of myelin segments and is associated with rapid remyelination of the persisting axons.

TABLE 2 (*Continued*)

Kind of Lesion	Changes in Nerve Cell Body	Changes in Axon	Changes in Schwann Cells/Myelin Sheaths
			Massive demyelination leads to accumulation of cholesterol and formation of "cholesterol granulomas" in the peripheral nerves or their roots.
			In the spinal cord, demyelination can be distributed rather diffusely in all white matter columns.
			Hypertrophic neuropathy is characterized by formation of "onion bulbs" by supernumerary Schwann cells as a result of recurrent demyelination and remyelination, as in response to nerve compression

rough endoplasmic reticulum is displaced to cell periphery (central chromatolysis) and the cell nucleus is translocated into an eccentric position. The cell body is intensely synthesizing proteins needed for renewal of the lost axonal portions. This structural reaction of the nerve cell body to axonal damage is known as the *axon reaction* (Figure 6).[35,86] Fragmentation starts within hours after the injury, and the regeneration advances with the velocity of slow axonal transport, millimeters per day.

The spatiotemporal pattern of distal stump destruction during Wallerian degeneration has been a matter of controversy. Some workers, using teased-fiber preparations, have demonstrated centrifugal (i.e., proximodistal) spread of fragmentation from the point of nerve interruption toward the periphery.[55,56] Other researchers assessed the pattern based on axonal microtubule density and concluded that it was nondirectional, affecting the entire distal stump, due to simultaneous microtubule dissolution along the length of the distal stump.[61]

The onset of Wallerian degeneration is almost twice as rapid: about 26 h after injury in thin fibers (caliber 2 to 4.5 μm) relative to thick fibers (caliber 8.5 to 11 μm), where it takes about 45 h. The distance of nerve transection from the site of nerve origin (where nerve cell bodies are located) influences the time of onset. After distal transection, degeneration appears earlier and is, for an equal amount of time, more severe than the changes that develop after more

FIGURE 3 Demonstration of the Wallerian law: When the nerve fibers are cut, the distal portions of the fibers degenerate.

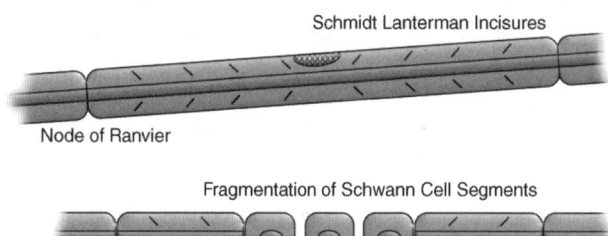

(a)

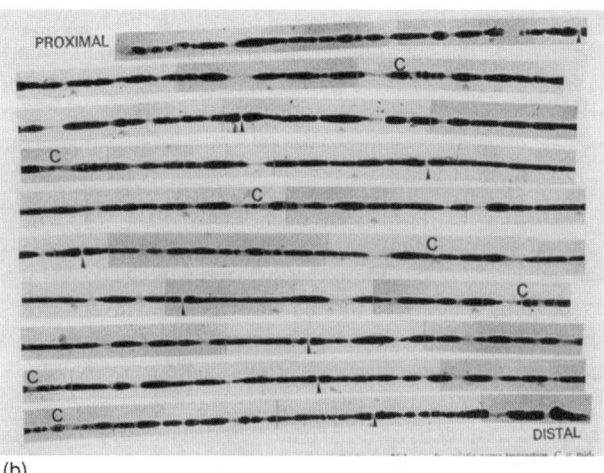

(b)

FIGURE 4 (a) Demonstration that in Wallerian degeneration the fiber fragmentation occurs at the Schmidt–Lanterman incisures, starting in the mid-internodal area of the segments. (b) Wallerian degeneration, rat tibial nerve, 36 h after sciatic nerve transection, formalin fixation, sudan black, teased fiber. C, Mid-internodal areas where the fiber fragmentation starts. (From Krinke et al.,[48] with permission of Sage Publications).

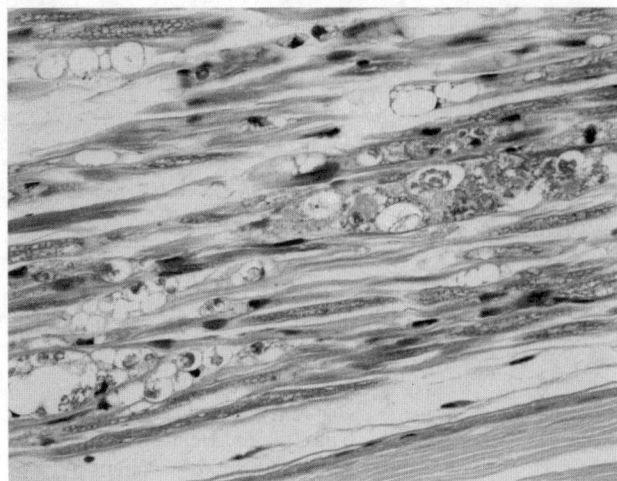

FIGURE 5 Digestion chambers in the dog peripheral nerve undergoing secondary axonal degeneration. Paraffin, hematoxylin and eosin.

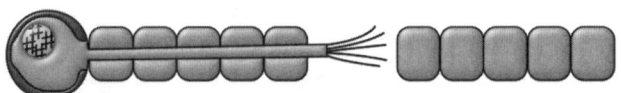

Axonal Reaction   Axonal Sprouts   Bands of Büngner

FIGURE 6 Regeneration after Wallerian degeneration: The nerve cell body undergoes axonal reaction, the regenerating axons sprout from the proximal stump; in the distal stump the proliferated Schwann cells form the *bands of Büngner*, ready to receive and guide the outgrowing axons.

proximal transection. When the peripheral nerve fiber is transected distally, the nerve cell body undergoes the axon reaction and the fiber regenerates from the proximal stump. In contrast, very proximal axon transection near where the axon originates from the neuronal perikaryon (i.e., the axon hillock) severely disrupts neuronal physiology and leads to cell death.[38]

Regeneration of an injured peripheral nerve is simpler following a crush injury, in which the nerve fibers are damaged but nerve continuity is preserved by the connective tissue. Total transection through the nerve with displacement of the stumps or removal of a long nerve portion makes regeneration impossible and brings about atrophy of the nerve fibers in the proximal stump[44] and denervation of the regions supplied by the disconnected distal stump.

Primary Axonopathy Versus Secondary Degeneration of the Peripheral Nerve Fibers

Primary Axonopathy Some axonopathies exhibit distinct morphological changes: for example, the *giant axonal neuropathy* (see axonopathy). These axonopathies could represent primary effects on the axon.

Secondary Axon Degeneration Secondary degeneration of nerve fibers has features similar to changes that occur in the distal stump in Wallerian degeneration. The only difference is that in Wallerian degeneration the nerve fiber is totally severed, whereas in secondary degeneration the ability of nerve cell body to maintain its axon is impaired in some fashion, so that the export of materials needed in the axon and/or nerve terminal is decreased below the critical minimum required for maintenance, resulting in breakdown of the axon. Secondary degeneration begins distally and over time will move proximally if the metabolic disruption is continued. Apparently, normal morphology of the nerve cell body and proximal axon does not exclude functional, biochemical changes of enough significance to induce structural changes in the distal axon. The nerve fibers affected undergo secondary degeneration ascending from the terminal portions toward the nerve cell body, called *dying-back neuropathy*. In primary sensory neurons with one central and one peripheral axon, dying back may occur simultaneously in the two processes, affecting the sensory spinal tracts and the peripheral nerves. In the spinal cord, axonal damage shows the anatomical distribution of selectively affected ascending and descending spinal tracts, most affected in their terminal (cranial for ascending tracts and caudal for descending tracts) portions. Identification of the spinal tracts affected enables discrimination of sensory from motor neuropathies or from combined sensory/motor lesions.

Neuroaxonal Dystrophy *Neuroaxonal dystrophy* is characterized by axonal swellings (*spheroids*) in preterminal portions of axons and in synaptic terminals. By light microscopy, axonal spheroids appear as large, relatively homogeneous eosinophilic bodies. Ultrastructurally, they contain a mixture of neurofilaments, tubulovesicular structures, mitochondria, lysosomes, and membranous bodies. Neuroaxonal dystrophy occurs in various forms as inherited or acquired neurodegenerative disorders in humans and animals. In laboratory animals, neuroaxonal dystrophy is usually encountered as a spontaneous, age-related change, characteristically occurring in the terminal (cranial) portions of the ascending sensory tracts in the fasciculus gracilis and fasciculus cuneatus of the spinal cord.

Primary and Secondary Demyelination

Primary Demyelination Primary demyelination results from damage to myelin sheaths with preservation of their axons. In the PNS, damaged and lost myelin sheaths are replaced rapidly in the process of remyelination. In the peripheral nerves, primary demyelination affects single segments of myelin sheaths; therefore, it is known as *segmental demyelination*. Longer and thicker Schwann cell segments are generally most susceptible, and because such segments occur in spinal nerve roots, segmental demyelination usually

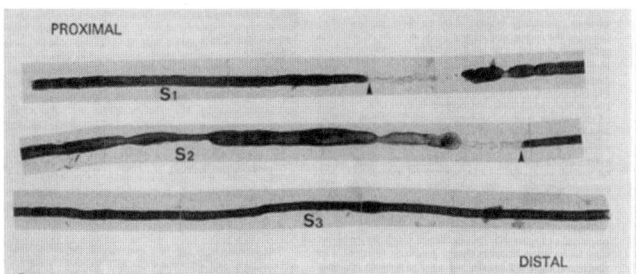

FIGURE 7 Segmental demyelination, rat lumbar ventral nerve root, formalin fixation, sudan black, teased fiber. Initial stage of segmental demyelination manifested by paranodal demyelination at both ends of segment S2. (From Krinke et al.,[48] with permission of Sage Publications.)

is most prominent at the level of the spinal nerve roots, especially the motor roots, which contain large-diameter nerve fibers.[8,22] Segmental demyelination begins in paranodal areas, which are the periphery of the myelin sheath formed by a single Schwann cell. In the process of remyelination the newly formed Schwann cell segments are shorter and thinner than the original segments (Figures 7 and 8).[19,25,26,51] Unlike post-mitotic neurons, in Schwann cells there is a natural turnover with regular loss and replacement of the myelin segments.

Degradation of damaged myelin sheaths is associated with swelling of the affected segments, forming *myelin bubbles* (also known as *myelin bubbling*, *blebbing*, or *ballooning*). In contrast to digestion chambers in degenerating fibers, which contain fragments of myelin and disintegrating axons, the myelin bubbles retain continuous, viable axons even though nearby macrophages are removing myelin debris (Figures 9 and 10). The axons within myelin bubbles are preserved but shrunken, because normal axonal diameter is preserved only within intact myelin sheaths.

Observation of shrunken axons within myelin bubbles sometimes leads to the presumption that *axonal atrophy* was a reason for the bubbling. However, experimental induction of axonal atrophy leads to the formation of *infolded myelin loops*, not myelin bubbling.[44] Advanced massive demyelination releases large amounts of cholesterol, leading to

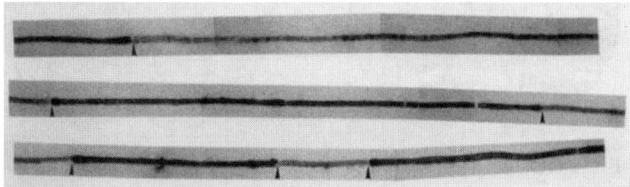

FIGURE 8 Segmental demyelination, rat tibial nerve, fixation in glutaraldehyde, impregnation with Dalton's solution, teased fiber. The demyelinated segments are pale, thin and shorter than the intact segments. (From Krinke et al.,[48] with permission of Sage Publications)

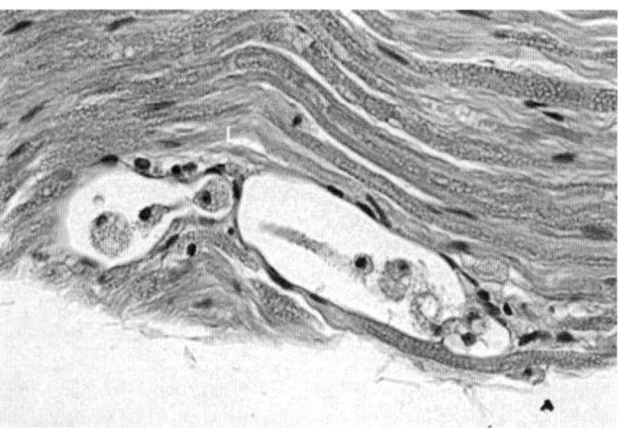

FIGURE 9 Rat sciatic nerve containing a myelin bubble. Within the bubble the axon is still preserved, whereas the macrophages are removing the myelin. Paraffin section, hematoxylin and eosin.

formation of *cholesterol granulomas* within the nerves (Figure 11).

Myelin Edema Myelinated tracts in the CNS (*white matter*), as well as the PNS, can develop myelin edema, in which increased fluid within the myelin results in distension and splitting of myelin lamellae. Myelin edema is reversible to a certain extent, but excessive edema can constrict nerve processes to a degree that produces secondary damage to neuronal elements.

Secondary Demyelination Secondary demyelination comprises the breakdown and removal of myelin sheaths that have lost their axons. It is irreversible because myelin sheaths are unable to persist without their axons. Secondary collapse of myelin sheaths is a characteristic feature of primary axonal damage. In microscopic preparations the

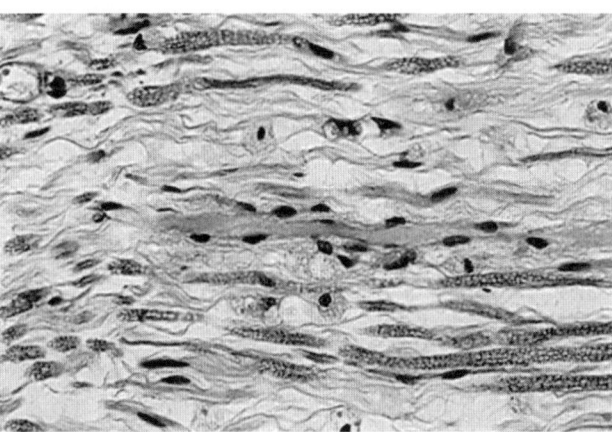

FIGURE 10 Rat sciatic nerve with a demyelinated ("naked") axon. The axon is surrounded by multiple nuclei of Schwann cells, which will provide rapid remyelination. Paraffin section, hematoxylin and eosin.

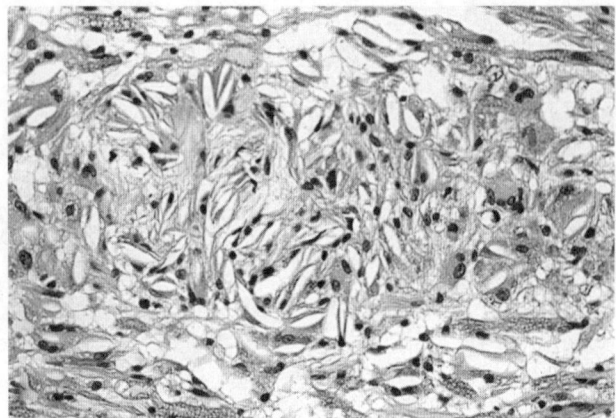

FIGURE 11 Rat sciatic nerve, containing "cholesterol granuloma" resulting from massive demyelination. Paraffin section, hematoxylin and eosin.

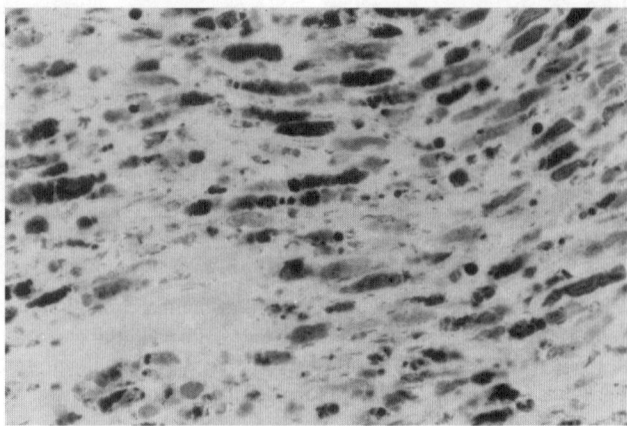

FIGURE 13 Rat sciatic nerve, advanced stage (day 12) of Wallerian degeneration, paraffin, Marchi stain. The breakdown of myelin sheaths in the process of secondary demyelination is demonstrated using the Marchi method — the myelin fragments stain black. However, the primary lesion consisting in axonal damage and loss is not evident from this stain.

damaged axons are absent, indicating that the visible myelin change is merely a reflection of the initial axon injury. This incongruous appearance has misled some workers to describe axonal lesions as *demyelination* (Figure 12). Such mistakes illustrate that effective analysis of the PNS typically requires special procedures to specifically highlight the axonal and myelin portions of the trunks when mechanisms of nerve damage are being investigated.

Breakdown of myelin sheaths in secondary demyelination of the PNS takes about two weeks: until the myelin lipoprotein is degraded to neutral fat and starts to stain positively with a fat stain (e.g., sudan red, oil red O). During the first two weeks of myelin breakdown, the damaged myelin sheaths are positive for Marchi stain,[83] which contains osmium tetroxide

and an oxidizing agent such as potassium chlorate (*Marchi stage* of myelin breakdown) (Figure 13). Compared to axonal lesions affecting the selected spinal tracts, primary demyelination can be distributed rather diffusely in all spinal columns (Figure 14).

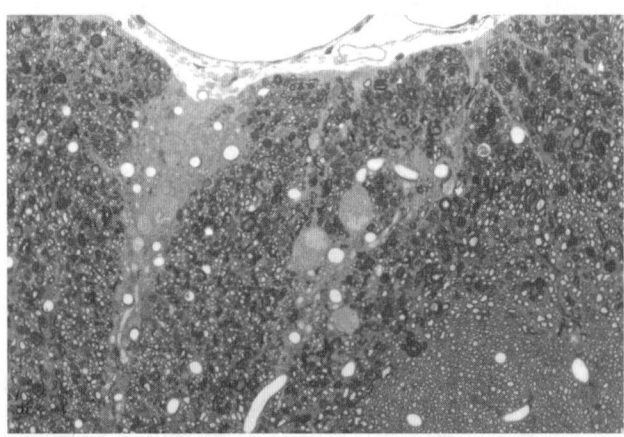

FIGURE 12 Rat cervical spinal cord, dorsal columns, pyridoxine-induced neuronopathy. In the sensory nerve fibers the axons are lost and the myelin sheaths collapsed. The motor fibers (bottom right) are intact. The presence of motor nerve fibers in the rat dorsal spinal columns is an anatomical feature of the rat. Epoxy resin, toluidine blue.

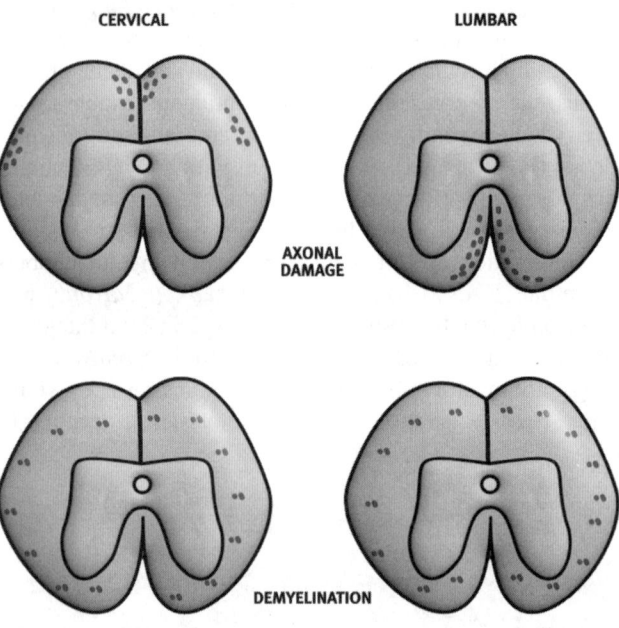

FIGURE 14 Difference in distribution of axonopathy and demyelination in the spinal cord. With axonopathy the ascending sensory spinal tracts are most affected at the terminal cervical level, in dorsal and dorsolateral columns, and the descending motor tracts are most affected at the terminal lumbosacral level, in the ventral columns. In contrast, primary demyelination can distribute diffusely at all spinal cord levels and in all columns.

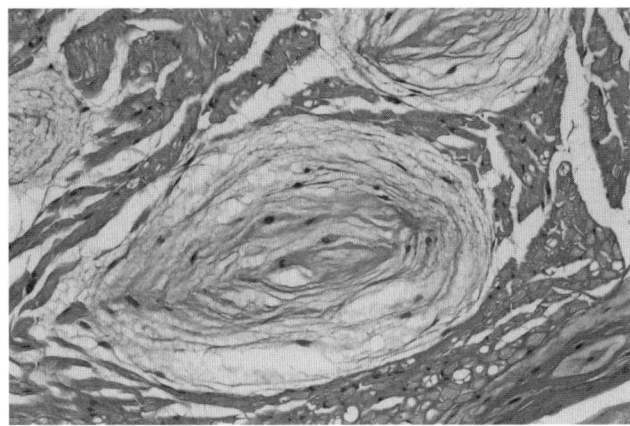

FIGURE 15 *Renaut bodies* in the dog sciatic nerve. Paraffin section, hematoxylin and eosin.

HYPERTROPHIC NEUROPATHY

Hypertrophic neuropathy is a lesion consisting of the local proliferation of Schwann cells that become arranged in concentric layers, called *onion bulbs* (the layered structure resembles the layering of an onion). This change is considered to result from recurrent demyelination and remyelination and can be induced by repeated compression on the peripheral nerve, sufficient to damage the Schwann cells but insufficient to induce axonal damage leading to Wallerian degeneration. Schwann cell–derived onion bulbs need to be discriminated from comparably arranged proliferative lesions with concentrically layered cells (see our description of perineuriomas below).

Renaut Bodies

Renaut bodies are accumulations of mucus-rich connective tissue within peripheral nerves, especially of dogs and horses. They serve a supportive or cushioning function and can be induced experimentally by mechanical stress or repeated nerve compression. They can participate in hypertrophic neuropathy (Figure 15).

PATHOLOGY OF UNMYELINATED PERIPHERAL NERVE FIBERS

Unmyelinated nerve fibers are small-diameter, slow-conducting nerve fibers serving both somatic and autonomic functions. Instead of being invested by multiple layers of Schwann cell membrane (i.e., a myelin sheath), they are clustered in groups encompassed by a Schwann cell, all surrounded by a single continuous basal lamina. The unmyelinated nerve fibers participate in the same diseases as those affecting the myelinated nerve fibers except for demyelination, which occurs only in myelinated fibers.

The general spectrum of lesions that develops in unmyelinated fibers follows that seen in myelinated fibers: fiber damage, loss, cellular proliferation, and finally, regeneration. Due to their small caliber, the unmyelinated fibers typically escape notice in routine light-microscopic preparations and are better examined by immunohistochemistry to reveal their cytoskeleton, neurotransmitters, or related molecules, or by electron microscopy (EM). Small-caliber axons seen by EM may represent normal unmyelinated nerve fibers, regenerating unmyelinated fibers, or regenerating myelinated nerve fibers in which myelin integrity has not been fully restored; it is not easy to recognize the difference, especially in chronic neuropathies. Other neuropathic changes that can be detected by EM include axonal swelling or shrinkage. A finding of Schwann cell columns devoid of axons, called *empty Schwann cell bands*, is regarded as a sign of damage to unmyelinated fibers and is a counterpart to the bands of Büngner that develop in myelinated fibers. *Collagen pockets* are bundles of collagen fibers surrounded by Schwann cell processes. They occur both in normal unmyelinated nerves and pathological specimens. In normal nerves they increase with age, but in neuropathies they may become very abundant. They are considered to mark the original location of degenerated unmyelinated fibers.[67]

Unmyelinated and small-caliber myelinated nerve fibers are the target of small-fiber neuropathies, diseases characterized by defective pain and temperature sensation as well as autonomic dysfunction. This specific group of diseases includes amyloidosis, familial α_1-lipoprotein deficiency (Tangier disease), α-galactosidase deficiency (Fabry's disease), familial dysautonomia, and some forms of hereditary sensory neuropathy. Toxic neuropathies targeting small-caliber fibers include capsaicin (associated with damage to small neurons in dorsal root ganglia)[81] and "chemical sympathectomy" by agents preferentially damaging the adrenergic innervation of the vas deferens (toxic neuronopathy).

NEURONOPATHY

Storage Diseases and Induced Phospholipidosis

Storage diseases (thesaurismoses) result from an abnormal metabolic process, typically an inborn error in metabolism associated with the deficiency of certain enzymes. However, some neuronal storage diseases are acquired due to disease or failure of critical endocrine or metabolizing organs (e.g., liver). Neuronal injury results in the accumulation of material within the cytoplasm of nerve cell bodies. Frequently, neurons (*cytons*) in dorsal root ganglia and the spinal cord are affected (e.g., in gangliosidoses G_{M1} and G_{M2}, Niemann–Pick disease, sphingomyelinic lipidosis, fucosidosis, or several types of generalized glycogenosis). The effects of poisonous plants (e.g., of the genus *Swainsona*) can induce conditions resembling a storage disease.[33]

Drug-induced *phospholipidosis* (originally called *lipidosis*) is the accumulation of phospholipids leading to the accumulation of lamellated (myeloid) bodies in the cytoplasm. This change has been observed in dorsal root ganglia and spinal cord neurons after treatment with tricyclic antidepressants.[57,58] In contrast to storage diseases, there is no primary enzyme failure, but the xenobiotic substrate is not disassembled and accumulates.

Motor Neuron Disease

Motor neuron disease consists of a group of disorders usually affecting the lower motor neurons [located in the ventral (anterior) horns of the spinal cord]. The most important disease in humans is *amyotrophic lateral sclerosis* (ALS). The mechanisms leading to various forms of this disease appear to be heterogeneous. In some cases there is abnormal neurofilament accumulation in the neuronal perikaryon or proximal axonal segment and their inappropriate phosphorylation. Other cases are characterized by neuronal loss, attributed to oxidative stress due to excessive activity of free oxygen radicals or insufficient activity of enzymes (e.g., superoxide dismutase) responsible for inactivating free oxygen radicals.[34] The disease has been reported in various breeds of dogs, horses, cattle, pigs, goats, cats, and mice.

Mutant Models of Neuronopathy

The *muscle deficient* (*mdf*) mouse mutant is an autosomal recessive mutation (chromosome 19) in which homozygotes (*mdf/mdf*) develop progressive motoneuron disease. There is a "posterior waddle" at four to eight weeks of age. Soon thereafter the hindlimbs become paralyzed and the forelimbs become weak. The disease progresses slowly but is fatal; the mean lifespan is reduced to eight months. The neurons within the spinal cord ventral horns and some motor nuclei in the brain stem develop cytoplasmic vacuolar degeneration. Peripheral nerves display axonal degeneration, while their skeletal muscle targets undergo neurogenic atrophy. This mutant was proposed as a model for juvenile motoneuron diseases with chronic evolution.[5]

The *motor neuron degeneration* (*mnd*) mouse is a mutation (chromosome 8) on the C57BL/6 background, leading to an ascending loss of central motor neurons. The lesion begins by deposition of lipopigment in spinal motoneurons (lumbar first, with later cranial progression) and subsequently in cranial motoneurons, but also in other cell types and peripheral tissues. The neurons accumulate inclusion bodies, probably abnormally retained cell breakdown products. Axonal swelling can occur, and the nerve cell bodies can develop axon reaction. This mutant is considered, together with the *wasted* and *wobbler* mutants, to be a spontaneous model for ALS.[62,66]

The *wasted* (*wst*) mouse is an autosomal recessive mutation (chromosome 2) that arose in the HRS/J strain. In homozygous mice, motoneurons in the spinal cord and cranial nerve nuclei show vacuolation and accumulation of phosphorylated 200-kDa neurofilament protein.

The *wobbler* (*wr, wb*) mouse is an autosomal recessive mutant derived from the C57BL/Fa strain. Homozygous mice have degeneration of motor neurons, more prominent in the brain stem and cervical spinal cord than in the lumbar cord. The neurons affected have cytoplasmic vacuolation, chromatolysis, and breakdown of the entire neuron.

In all three mouse models of ALS (*mnd, wb, wst*), primary damage to CNS motoneurons is inevitably associated with substantial secondary damage to their processes in the peripheral nerves.

Toxic Neuronopathy

Artemether Toxic oxidative stress, resembling motor neuron disease, has been induced experimentally with artemether, an active agent of a herbal antimalarial medicine, when administered at high dose levels parenterally to dogs. Necrotic neurons occurred mainly in the brain stem, but the motoneurons in the spinal cord were affected as well.[17]

Pyridoxine, Zoniporide Toxic neuronopathy that selectively affects the primary sensory neurons [dorsal root ganglia, trigeminal (Gasserian) ganglia] can be induced with excessive doses of vitamin B_6 (pyridoxine).[40] Very high dose levels induce cytoplasmic vacuolation (hydropic degeneration) in the perikaryon, which can progress to cell death. The neurons that vanish may be replaced by *nodules of Nageotte*, produced by proliferating satellite cells. Some surviving neurons exhibit the axon reaction and accumulation of neurofilaments in the cytoplasm, presumably because they cannot be exported into the axon. Lower doses may produce distal sensory axonopathy with preservation of nerve cell bodies. This axonopathy has features of unspecific secondary degeneration. The distal change results from impairment of cell metabolism at the level of the nerve cell body. Various species are susceptible, including humans, dogs, and rats. Selective primary sensory neuropathy has also been reported with administration of zoniporide to rats and dogs[70] (Figure 16).

Cytotoxic Lectins Experimental neuronopathy can be induced by ricin, abrin, modeccin, or volkensin. These agents, when injected into the peripheral nerves or their target tissues, are taken up and transported retrogradely into the cell body, where they inactivate ribosomes, inhibit protein synthesis, and bring about cell death.[91]

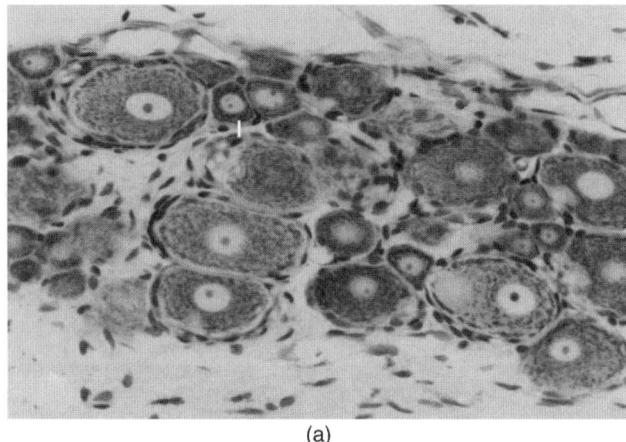

(a)

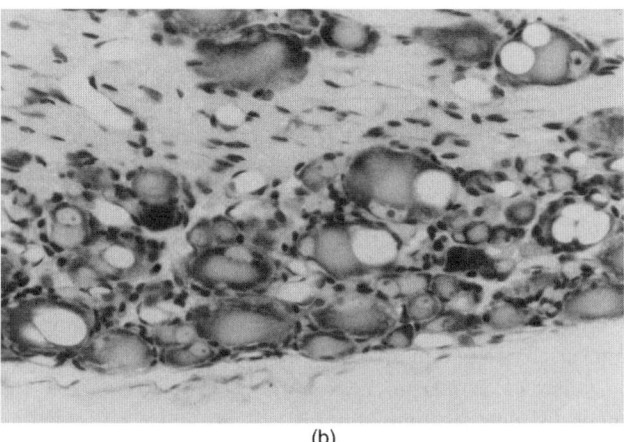

(b)

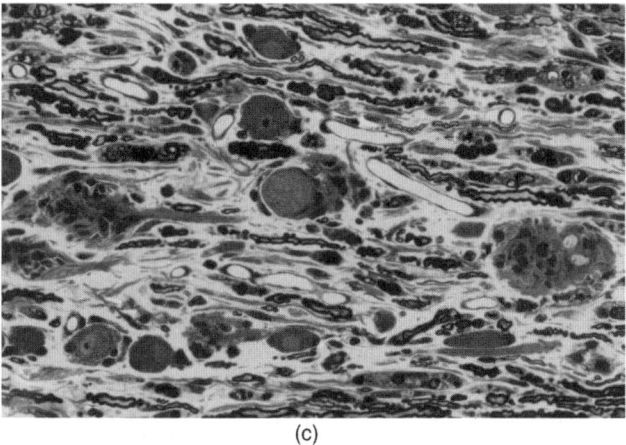

(c)

FIGURE 16 (a) Rat dorsal root ganglion, intact in control animal. Paraffin section, cresyl violet. (b) Rat dorsal root ganglion, pyridoxine-induced neuronopathy characterized by chromatolysis and vacuolation of the nerve cell bodies. Paraffin section, cresyl violet. (c) Rat dorsal root ganglion, advanced stage of pyridoxine neuronopathy manifested by loss of ganglionic cells and their replacement by the *nodules of Nageotte.* Epoxy resin, toluidine blue.

Acrylamide Neurons of the PNS, especially those in the dorsal root ganglia, are probably affected during many peripheral toxic neuropathies, although this finding is rarely reported. For example, in acrylamide neuropathy some researchers have reported changes in DRG neurons prior to the onset of the hallmark axonopathy.[82] The scarcity of toxic lesions reported in the DRG probably reflects the difficulty in collecting these structures.

Ganglionic Blockers *Ganglionic blockers* such as guanethidine, guanacline, and bretylium tossylate were used for some time to treat hypertension. The molecular structure of these agents is similar to that of the sympathetic neurotransmitter noradrenaline. They are taken up by the amine pump into sympathethic neurons, where they substitute for the natural transmitter and decrease sympathethic pathway activity. When administered at high dosage levels, they accumulate in sympathetic neurons (especially in developing adrenergic neurons of newborn rodents), thereby inducing chemical sympathectomy.[1,29,30] Such effects have been reported most frequently in the readily accessible cranial (superior) cervical ganglion. However, high sensitivity (at least in the rat) is also exhibited by the short adrenergic neurons in the pelvic plexus, especially those innervating the vas deferens (ganglion hypogastricum).[24] Damage to the sympathetic innervation in the vas deferens results in stagnation of sperm, reduced fertility, and formation of spermatic granulomas in rats.

INFLAMMATORY LESIONS

The knowledge of inflammatory lesions in neural tissues as possible background pathology confounding the results of chemical safety studies is indispensable for all toxicological pathologists and toxicologists. Enterovirus-induced encephalomyelitis affects motor and sensory neurons in several species. The known forms are Teschen–Talfan disease in pigs, "epidemic tremor" in chickens, Theiler's disease in mice, and poliomyelitis in humans. Ganglionitis occurs as a consequence of acute or chronic viral infection, such as in herpesvirus infections in pigs [pseudorabies (Aujeszky's disease, paralysis bulbaris infectiosa)] and humans, and rabies virus in many species, especially carnivores.[84] Rabies virus also induces widespread polioencephalomyelitis.

AXONOPATHY

Breed-Specific Axonopathies in Dogs

Each member of this group of spontaneous conditions exhibits similar if not identical background and features, regardless of the mechanism. Systematic extensive examination of the nervous system has not been carried out in these syndromes, so that a critical comparative assessment of neural changes is not possible. This group of diseases includes hound ataxy in hunting dogs, including beagles, which is hypothesized to result from an unknown dietary abnormality.[84] Therefore, spontaneous ataxy must be considered as a differential diagnosis for axonal lesions in canine neurotoxicity studies.

Mutant Model of Axonopathy

The *gracile axonal dystrophy* (*gad*) mouse is an autosomal recessive mutant that develops a predominantly sensory neuropathy. The initial lesion is axonal degeneration of central and peripheral axon terminals of the dorsal root ganglion cells. At a later stage, the changes extend to the peripheral motor neurons and the brain stem and also affect the dorsal spinocerebellar tract and the trigeminal tract. This mutant is considered to be a model for human neuropathies characterized by dying back of the axons, especially *neuroaxonal dystrophy* and *Friedreich's ataxia*.[66]

Toxic Axonopathy

Proximal Axonopathy   Administration of β,β'-iminodipropionitrile (IDPN) produces neurofilament-filled axonal swelling in the proximal regions of motor and sensory nerve fibers. Distal to these changes, axonal swelling and secondary degeneration may occur. The effects of IDPN are associated with an impairment of slow axonal transport.[28,87,88]

Distal Axonopathy   Distal axonopathy represents the most frequently reported type of peripheral toxic neuropathy. Its distribution pattern is characterized by distal accentuation so that distal nerve branches are more severely affected than proximal portions of the sciatic nerve. Over the course of time, continued exposure drives progression of the lesion in the proximal direction. Examples of distally accentuated central and peripheral neuropathy are afforded by low toxic doses of pyridoxine, acrylamide, or neuropathic organophosphorus (OP) agents such as tri-*o*-cresyl phosphate (TOCP).[16,47] As noted above, the distal neuropathy induced by pyridoxine or acrylamide probably represents secondary axonal changes resulting from primary effects on the neuronal bodies.[13,43] The exact mechanism of organophosphorus-induced "delayed" neuropathy (OPIDN) is unknown, although a correlation between the neuropathic propensity of various OPs and the capability to strongly inhibit enzyme neuropathy target esterase (NTE) has been demonstrated.

Neurofilamentous axonopathy is a central or peripheral, distally accentuated neuropathy characterized by large ("giant") swellings of distal axons that are filled with neurofilaments (Figure 17). With continued exposure, nerve fiber degeneration develops distal to the axonal swellings. This type of neuropathy has been induced with carbon

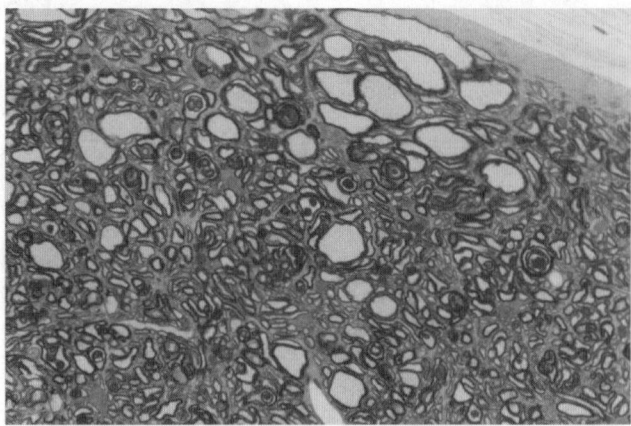

FIGURE 17   "Giant axonal neuropathy" induced with 2,5-hexanedione in the baboon spinocerebellar tract. The swollen axons are unusually large and their myelin sheaths are stretched and thinner. Epoxy resin, toluidine blue.

disulfide or by certain hexacarbons exhibiting the γ-diketone structure.[27] Most of the axonal swellings occur proximal to the nodes of Ranvier and are accompanied by extensive cross-linking of neurofilament protein. The mechanism of fiber degeneration proposed is mechanical obstruction from neurofilament accumulation at the nodes of Ranvier, the main "bottleneck" point where PNS axons are anatomically constricted and intraaxonal blockages will prevent distal passage of axoplasmic materials.[14] The requirement for neurofilament accumulation is not absolute, however, as administration of γ-diketones or acrylamide to crayfish—in which the axons are devoid of neurofilaments—does not prevent axonal degeneration. Based on these results, it has been proposed that axonal atrophy is a specific morphological component of γ-diketone neuropathy; it could be the primary pathogenic event, whereas the characteristic axonal swelling is of unclear significance.[53,54,78]

Axonopathy of Unmyelinated Nerve Fibers   Although the large-diameter myelinated nerve fibers as a rule undergo the earliest damage following toxicant exposure, the small-diameter unmyelinated axons can be susceptible as well, especially in prolonged intoxication. Damage and loss of unmyelinated axons has been demonstrated in feline acrylamide neuropathy[71] and thalidomide neuropathy in humans.[77]

MYELINOPATHY

Leukodystrophies and Hypomyelinogenesis

Leukodystrophies are disorders of myelin synthesis and maintenance. Many of these diseases are inherited disorders of metabolism; others are due to the infection of myelinating cells during gestation. These conditions affect primarily the CNS, including those portions of the spinal cord that transmit signals to and from the PNS. They have been reported in various animal species, notably in several dog breeds. The most common such condition affecting the PNS is *globoid cell leucodystrophy* (*Krabbe's disease*), a spontaneous leukodystrophy resulting from deficiency of the lysosomal enzyme galactosylceramidase. Animal models of this disease have been reported in several dog breeds of dogs and in the Twitcher mutant mouse.[10] In many canine cases, involvement of the peripheral and autonomic nerves is characterized by myelin loss, irregular swelling of axons, and clustering of the globoid macrophages in perivascular groups.

The basic, of a number of diseases leading to insufficient myelination during development is known as *hypomyelinogenesis*. Chemically induced disturbance of peripheral nerve myelination can be induced in weanling rats given tellurium between days 15 to 35. Peripheral nerves rapidly develop accumulation of lipid droplets in Schwann cell cytoplasm and then undergo segmental demyelination. The peripheral nerves of adult rats are more resistant.[65,73] Examples of hereditary hypomyelinating diseases are provided in the following overview of mutant and engineered models of myelinopathy.

Mutant or Engineered Models of Myelinopathy

Mouse The *enervated* (*Enr*) mouse[72] is a transgene-induced insertional mutation (BPFD 36) with disrupted Schwann cell and axon interactions. This abnormality leads to insufficient myelination of regenerating axons during Wallerian degeneration.

The *jimpy* (*jp*, *jp/Y*) mouse is a spontaneous, juvenile, lethal, gender-linked, recessive mutation in which homozygous males lack the proteolipid protein C-terminal domain. At necropsy a striking paucity of myelin is detected in the CNS, although the PNS is normally myelinated. The normal-appearing central axons are insufficiently myelinated, and myelin breakdown and removal by phagocytes take place, especially late in the course of the disease. Oligodendrocytes proliferate but are unable to differentiate and contain inclusions representing dystopic membranes. Jimpy is considered to be a model of human *Pelizaeus–Merzbacher disease*.

The *myelin synthesis deficiency* (*msd*) is a gender-linked, recessive mutant, allelic (an alternative gene form) to the jp locus. The lesions include lack of central myelin, similar to that which occurs in the jp mutants, but also abnormal myelin in peripheral nerves.

The *Niemann–Pick type C* (*NPC*) mouse[31] is a mutation on the BALB/c background that induces changes similar to those of human *Niemann–Pick disease type C*. This sphingolipidosis is characterized by impaired intracellular transport of low density lipoprotein–derived cholesterol, resulting in massive accumulation of unesterified cholesterol in lysosomes of many cell types throughout the body. The NPC mouse shows an accumulation of small myelin figures in paranodal swellings of Schwann cell cytoplasm and hypomyelination of large myelinated nerve fibers in the PNS.

The *protein zero* (*P0*) mutant mouse (chromosome 1) results from an engineered complete loss of P0 function. P0 is an integral membrane glycoprotein unique to Schwann cells and the most abundant protein of peripheral myelin; homozygous P0 mutants are affected, heterozygous animals are normal. In the peripheral nerves, Schwann cells fail to assemble normal myelin. This mutant is considered to be a model for human Charcot–Marie–Tooth disease, type 1b (a peripheral demyelinating neuropathy with a dominant inheritance).

The *quaking* (*qk*) mutant is a recessive autosomal mutation first discovered in a DBA/2 subline. In homozygous mice, there is severe myelin deficiency in all areas of the central nervous system. The density of oligodendroglia cells is increased, and they contain lamellar inclusions, dense bodies, and vacuoles. Myelin in the PNS is thinner than in controls.

Shiverer (*shi*) is an autosomal recessive mutation in the mouse myelin basic protein (Mbp) gene (chromosome 18). The oligodendrocytes fail to assemble compacted myelin, thereby leading to hypomyelination of the CNS, characterized by the absence of the "major dense line." Peripheral myelin is only mildly changed.[84]

Trembler (*Tr*, allele *Tr-J*) is a dominant mutation of the PMP-22 gene (chromosome 11). PMP-22 is a glycosylated membrane protein of peripheral myelin. Schwann cells of Tr mice proliferate but fail to myelinate the axons. Consequently, Tr mutants have about a 10-fold increase in the number of Schwann cells. This mutant is considered to be a model for human Charcot–Marie–Tooth disease, type 1a, Déjérine–Sottas disease, and hereditary neuropathy with liability to pressure palsies.

The *twitcher* (*twi*) mutant arose on the C57BL/6J strain as an autosomal recessive mutant (chromosome 12). The principal defect is a deficiency in galactosylceramidase activity. Homozygous mice develop demyelination associated with astrocytic gliosis and infiltration by macrophages ("globoid cells"). Increased GFAP expression is observed in the CNS as well as the PNS; interestingly, in the PNS this is attributed to Schwann cells and satellite cells.[37] This mutant is considered to be a model for human *globoid cell leukodystrophy* or Krabbe's disease.

Rat Zitter (zi/zi), Tremor (tm/tm), and SER mutants[39] Zitter arose in the Sprague–Dawley strain, tremor in the Kyo–Wistar strain. SER (spontaneous epileptic rat) was produced by hybridization between a zitter homozygous female and a tremor heterozygous male. All three mutants exhibit tremors and, especially in the SER, epileptic seizures. They have CNS hypomyelination and vacuolation, mainly in the brain stem and thalamus, consisting of swollen astrocytic processes and enlargement of the extracellular and periaxonal space. The peripheral nerves are generally unaffected despite changes in the spinal cord in the pathways associated with the PNS.

Taiep is an autosomal recessive mutant of the Sprague–Dawley strain.[59,64] The name is derived from clinical signs and their chronology: T tremor (age about three weeks), A ataxia (4 months), I immobility, E epilepsy (after six months), and P paresis or paralysis (older animals). The animals may live for up to 18 months. They have a defect in CNS myelination which worsens with age and is associated with abnormal accumulation of microtubules in the oligodendrocytes. On biochemical analysis, myelin yield of the brain hemispheres was only 10 to 15% of controls, and that of the spinal cord only 20 to 25% of controls. Myelin-associated glycoprotein (MAG) was more reduced than other myelin constituents. The PNS is unaffected despite changes in the spinal cord in the pathways associated with the PNS.

Toxic Myelin Edema

The lesion is most prominent in the central nervous system but often involves the peripheral myelin as well. At necropsy, the brain of affected animals is pale and enlarged, and its surface is flattened, owing to increased intracranial pressure. Microscopic examination reveals diffuse vacuolation of the white matter without degeneration of myelinating cells. In the brain of clinically recovered animals, no vacuolation is seen and no loss of myelinated fibers is obvious. Electron microscopy shows that myelin vacuoles are formed by splitting the myelin lamellae at the intraperiod line, representing the fused outer layers of the cell membrane of the oligodendrocytes or Schwann cells. The extracellular space in the brain is not widened, and the structure and permeability of blood vessels are not damaged. The lipid composition of myelin is unaltered. The edema fluid consists of water containing high levels of sodium and chloride. In advanced severe lesions, breakdown of myelin and secondary changes in other tissue elements may occur. In the PNS, findings usually exhibit a proximodistal gradient of damage, with the ventral (anterior) nerve roots more involved than their dorsal (posterior) counterparts, and the proximal sciatic nerve less involved than the roots. The optic nerve, an extension of the CNS, is very susceptible to chemically induced myelin edema. Examples of inducing agents that can cause myelin edema include triethyltin (contaminant of Stalinon, an oral antibacterial concoction associated with dozens of deaths), hexachlorophene, bromethalin,[89] isonicotinic acid hydrazide, and salicylanilides. In contrast, the myelin edema induced by the copper chelator cuprizone (biscyclohexanone oxalyldihydrazone) is characterized by early degenerative changes in the oligodendroglia.[3]

Segmental Demyelination

Spontaneous, age-related segmental demyelination, known as *radiculopathy*, occurs in various species but is most prominent and best known in aging rats. It characteristically affects the motor nerve roots in the lumbosacral region. In the rat it develops fully by late in the second year of life, but it can be accelerated by treatment with selected neurotoxicants.[15,41]

Toxicant-induced segmental demyelination is usually a subtle lesion. It is difficult to recognize because the initial clinical signs (typically, weakness) tend to improve; this clinical reversal results from rapid remyelination of the affected peripheral nerves. For the same reason, on light-microscopic examination of routine paraffin sections, the only subtle finding in the remyelinated peripheral nerves might be increased cellularity (i.e., more cell nuclei per square unit of nerve section). In toxicological experiments, segmental demyelination can easily be overlooked and probably is generally underreported.

Immune-Mediated Demyelination

Immune-mediated demyelination of nerves, or *experimental allergic neuritis*, can be induced in rats by immunization with an adjuvant containing nerve root myelin, or more specifically, with purified P_2 protein (which is typically confined to PNS myelin).[32] This model is used as a surrogate of human inflammatory polyneuropathy (Guillain–Barré syndrome). Certain strains of Theiler's murine encephalomyelitis virus (TO) produce persistent infection with chronic CNS but not PNS demyelination. These have been considered as a model of human multiple sclerosis.[46] The white matter disease form of canine distemper may involve spinal cord white matter but typically does not lead to alterations in PNS served by the regions affected.

PROLIFERATIVE AND NEOPLASTIC PNS LESIONS

Peripheral nerve tumors are derived from the cells in the peripheral ganglia (and ganglia-like entities) and the supporting cells in the nerves forming the nerve sheaths. The neural parenchyma in the PNS (axons) do not exhibit neoplastic growth. The tumors derived from the ganglionic cells include benign ganglioneuroma and malignant ganglioneuroblastoma. The counterparts of these lesions in the adrenal medulla (which is a "modified" sympathetic ganglion) are benign and malignant adrenal medullary tumors. Outside the adrenal medulla, the noradrenergic ganglionic cells of sympathetic ganglia can form *paragangliomas*. Nonadrenergic cells of the dispersed chemoreceptor system form *chemodectomas* (nonchromaffin paragangliomas). Such tumors have been induced experimentally in developing transgenic mice that overexpress artemin, a neurotrophic factor supporting sympathethic neuron development.[6]

The most common tumors derived from PNS supporting cells are benign neurofibromas and benign schwannomas, and their malignant counterparts, neurofibrosarcomas and malignant schwannomas. Differential diagnosis of these particular tumor types is complicated by the difficulty of discriminating between neoplastic and reactive cells. Owing to the great regenerative potential of the peripheral nerves, vigorous proliferation of Schwann cells in reaction to nerve damage can occur in connective tissue neoplasms, so that reactive Schwann cells ultimately form a substantial proportion of the tumor tissue. Furthermore, the identity of peripheral nerve sheath cells has been a matter of controversy; neuroectodermal origin has been proposed for the perineural cell, while mesenchymal properties have been attributed to the Schwann cell. The common immunophenotypic expression of certain antigens indicates a close relationship between the Schwann cells, perineural cells, and the primitive fibroblasts in the nerve sheaths. Chemically induced PNS tumors, especially those induced with administration of *N*-ethylnitrosourea (ENU), have features of schwannoma, but following successive transplantation, they become highly anaplastic tumors characterized by a sarcomatous profile.[11]

Most pathologists appear to agree that schwannomas (also called neurinomas and neurilemmomas) are characterized by exclusive proliferation of neoplastic Schwann cells, whereas neurofibromas consist of combined proliferation of Schwann cells, endoneural cells, and perineural cells. In benign well-differentiated tumors, the cellular elements can be discriminated according to their characteristic morphology. The Schwann cells do not produce collagen, are S-100 protein positive, and surrounded by a basal lamina. Other markers for Schwann cells have been suggested, such as antibodies to myelin-specific proteins. However, neoplastic Schwann cells do not produce myelin, so that reliance on such markers can be a source of false-negative diagnoses. In neurofibromas, the cell nuclei are often curved and usually smaller than those observed in schwannomas. Typically, the neoplastic cells in neurofibromas are surrounded by collagen fibers and an alcian-blue-positive myxoid matrix. In malignant anaplastic tumors, the characteristic structural features are missing; many such lesions represent a mixture of Schwann cells, perineural cells, and primitive fibroblasts, so that there is a gray zone of neoplasms that exhibit features of both malignant schwannomas and neurofibrosarcomas. Many pathologists approach the dilemma posed by these complex tumors by lumping them into the single category of peripheral nerve sheath tumors (PNST), with modifiers of benign and malignant, depending on the apparent differentiation of the various elements. For practical purposes, rodent neoplasms having areas of Schwann cell differentiation, especially when supported by a distinct S-100 protein positivity on

immunohistochemistry, can be diagnosed as schwannomas, whereas tumors negative for S-100 that exhibit features characteristic of fibromas or fibrosarcomas should be diagnosed as connective tissue tumors.[12,45,49,50,63]

A schwannoma is an expansive, compressing lesion that can have a capsule derived from the perineurium. Anaxonal Schwann cells are elongated and form nuclear palisades (Antoni A pattern): two adjacent palisades of nuclei and the intervening cytoplasm form a *Verocay body*. Portions of the neoplasm can be sparsely cellular with a clear matrix (Antoni B pattern) (Figure 18). The evidence of S-100 protein positivity and the presence of basement membrane (reaction for laminin, or evident by electron microscopy) support the diagnosis of a schwannoma. A differential diagnosis is the perineurioma, which is a mass comprised of perineural cells that are arranged in concentric rings and negative for S-100

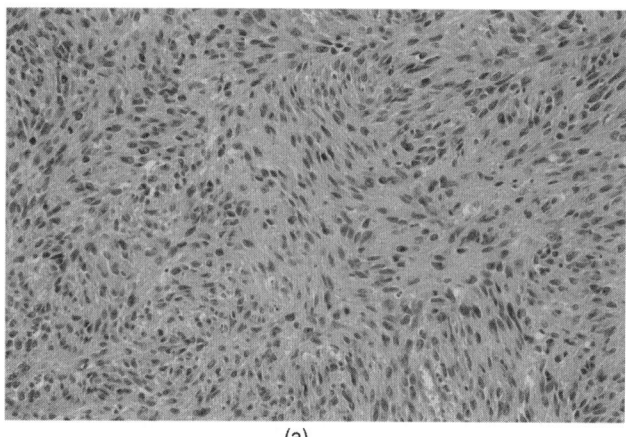

(a)

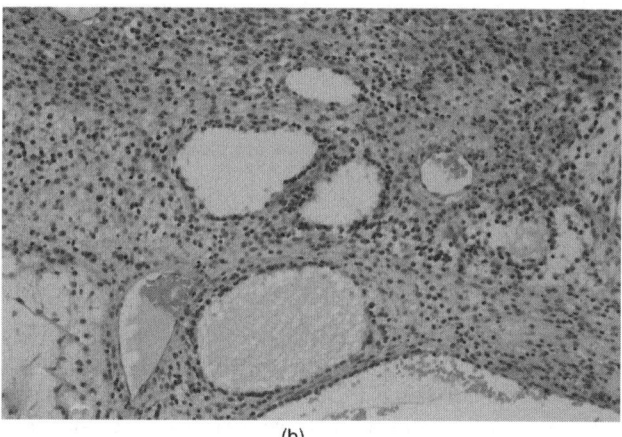

(b)

FIGURE 18 Characteristic cell arrangement in the schwannoma. (a) Antoni A pattern. The Schwann cells are elongated and form nuclear palisades; two adjacent palisades of nuclei and the intervening cytoplasm form a "Verocay body"; area of a spontaneous schwannoma in the rat heart. (b) Antoni B pattern. The neoplasm is sparsely cellular, with a clear matrix, occasional cystic cavities; area of a spontaneous schwannoma from the rat mandibular salivary gland. Paraffin, hematoxylin and eosin.

protein but positive for claudin-1, epithelial membrane antigen (EMA) and glucose transporter-1 (GLUT-1).[60]

The reason for arrangement of neoplastic Schwann cells in either Antoni A or B pattern is unclear. It has been observed that neoplasms initially arranged in the type A pattern can change to preponderantly type B pattern after multiple transplantations, so that the type B pattern could represent a more primitive, anaplastic state. Particular rodent schwannomas can be composed of exclusively one pattern. Characteristic topographic sites for PNS schwannomas in mice include the cranial and spinal nerve roots and the genital organs, such as the uterus or epididymis. In rats, they include the abdominal cavity, subcutaneous tissue and specific forms that occur in the area of the mandibular salivary gland (mostly type B pattern), the external ear (mostly type A pattern), and the orbit. A particular type of schwannoma also arises in the subendocardial layers of rat heart.

Peripheral nerve tumors must be discriminated from focal nerve thickenings at the site of a local PNS injury. These proliferative but nonneoplastic *neuromas* are reactive hyperplastic lesions representing a disordered repair process, often resulting from aberrant growth of regenerating axons outside the normal pathway to the innervation target. The presence of axons in neuromas is helpful in the differential diagnosis of schwannomas, which are mostly anaxonal. Neuromas are the result of trauma (usually, nerve transection with displacement of the cut ends). Toxicants cannot induce neuromas because the action of xenobiotic agents on peripheral nerves can damage axons, but it is not known to directly induce their proliferation (a function that is controlled by the nerve cell bodies).

RECOMMENDED READING

Readers interested in specific details and comparative aspects of PNS pathology are referred to the excellent book edited by P. J. Dyck and P. K. Thomas.[20]

REFERENCES

1. Angeletti PU, et al. Structural and ultrastructural changes in developing sympathetic ganglia induced by guanethidine. *Brain Res.* 1972;43:515–525.

2. Berner A, et al. Origin of macophages in traumatic lesions and Wallerian degeneration in peripheral nerves. *Acta Neuropathol, (Berl).* 1973;25:228–236.

3. Blakemore WF. Demyelination of the superior cerebellar peduncle in the mouse induced by cuprizone. *J Neurol Sci.* 1973;20:63–72.

4. Blakemore WF. Ethidium bromide induced demyelination in the spinal cord of the cat. *Neuropathol Appl Neurobiol.* 1982;8:365–375.

5. Blot S, et al. The mouse mutation muscle deficient (*mdf*) is characterized by a progressive motoneuron disease. *J Neuropathol Exp Neurol.* 1995;54:812–825.

6. Bolon B, et al. The candidate neuroprotective agent artemin induces autonomic neural dysplasia without preventing peripheral nerve dysfunction. *Toxicol Pathol.* 2004;32:275–294.

7. Bolon B, et al. A "best practices" approach to neuropathologic assessment in developmental neurotoxicity testing — for today. *Toxicol Pathol.* 2006;34:296–313.

8. Bouldin TW, et al. Schwann cell vulnerability to demyelination is associated with internodal length in tellurium neuropathy. *J Neuropathol Exp Neurol.* 1988;47:41–47.

9. Boyd IA, Davey MR. *Composition of Peripheral Nerves.* Edinburgh; UK: E&S Livingstone; 1968.

10. Bradl M, Linington C. Animal models of demyelination. *Brain Pathol.* 1996;6:303–311.

11. Cardesa A, et al. Tumors of the peripheral nervous system. In: Turusov V, Mohr U, eds. *Pathology of Tumours in Laboratory Animals,* Vol. 1. Tumours of the Rat. 2nd ed. Lyon, France: IARC; 1990;699–724.

12. Carlton WW, et al. *International Classification of Rodent Tumors, Part I: The Rat.* Mohr U., ed. *Part 2. Soft Tissue and Musculoskeletal System.* IARC Scientific Publication 122. Lyon, France: IARC; 1992;1–62.

13. Cavanagh JB. The pathokinetics of acrylamide intoxication: a reassessment of the problem. *Neuropathol Appl Neurobiol.* 1982;8:315–336.

14. Cavanagh JB. The pattern of recovery of axons in the nervous system of rats following 2,5-hexandiol intoxication: a question of rheology. *Neuropathol Appl Neurobiol.* 1982;8:19–34.

15. Classen W, et al. Functional and morphological characterization of neuropathy induced with 5-lipoxygenase inhibitor CGS 21595. *Exp Toxicol Pathol.* 1994;46:119–125.

16. Classen W, et al. Susceptibility of various areas of the nervous system of hens to TOCP-induced delayed neuropathy. *Neurotoxicology.* 1996;17:597–604.

17. Classen W, et al. Differential effects of orally versus parenterally administered quinghaosu derivative artemether in dogs. *Exp Toxicol Pathol.* 1999;51:507–516.

18. Dockray GJ. Physiology of enteric neuropeptides. In: Johnson LR, ed. *Physiology of the Gastrointestinal Tract.* New York, Raven Press; 1994;169–209.

19. Dyck PJ, et al. Lead neuropathy: 2. Random distribution of segmental demyelination among "old internodes" of myelinated fibers. *J Neuropathol Exp Neurol.* 1977;36:570–575.

20. Dyck PJ, Thomas PK. *Peripheral Neuropathy,* 4th ed. Philadelphia: W.B. Saunders; 2005: 2992.

21. Erlanger J, Gasser HS. *Electrical Signs of Nervous Activity.* Philadelphia: University of Pennsylvania Press; 1937.

22. Friede RL. Variance in relative internode length (*l/d*) in the rat and its presumed significance for the safety factor and neuropathy. *J Neurol Sci.* 1983;60:89–104.

23. Ganong WF. *Review of Medical Physiology,* 22nd ed. New York: Lange Medical Books/McGraw-Hill; 2005;912.

24. Gerkens JF. Guantehidine and guanacline on the rat vas deferens. *Br J Pharmacol.* 1974;52:191–195.

25. Goodrum JF, Bouldin TW. The cell biology of myelin degeneration and regeneration in the peripheral nervous system. *J Neuropathol Exp Neurol.* 1996;55:943–953.

26. Goodrum JF. Role of organotellurium species in tellurium neuropathy. *Neurochem Res.* 1998;23:1313–1319.

27. Graham DG, et al. Pathogenetic studies of hexane and carbon disulfide neurotoxicity. *Crit Rev Toxicol.* 1995;25:91–112.

28. Griffin JW, et al. IDPN neuropathy in the cat: coexistence of proximal and distal axonal swellings. *Neuropathol Appl Neurobiol.* 1982;8:351–364.

29. Heath JW, et al. Degeneration of adrenergic neurons following guanethidine treatment: an ultrastructural study. *Virchows Arch B.* 1972;11:182–197.

30. Heath JW, et al. Axon retraction following guanethidine treatment: studies in sympathetic neurons in vivo. *Z. Zellforsch.* 1973;146:439–451.

31. Higashi Y, et al. Peripheral nerve pathology in Niemann–Pick type C mouse. *Acta Neuropathol (Berl).* 1995;90:158–163.

32. Hughes RAC, et al. Immune responses in experimental allergic neuritis. *J Neurol Neurosurg Psychiatry.* 1981;44:565–569.

33. Huxtable CR, Dorling PR. Mannoside storage and axonal dystrophy in sensory neurones of swainsonine-treated rats: morphogenesis of lesions. *Acta Neuropathol (Berl).* 1985;68:65–73.

34. Jaarsma D, et al. Neuron-specific expresion of mutant superoxide dismutase is sufficient to induce amyotropic lateral sclerosis in transgenic mice. *J Neurosci.* 2008;28:2075–2088.

35. Jones HB, Cavanagh JB. Comparison between the early changes in isoniazid intoxication and the chromatolytic response to nerve ligation in spinal ganglion cells of the rat. *Neuropathol Appl Neurobiol.* 1981;7:489–501.

36. Jortner BS. Mechanisms of toxic injury in the peripheral nervous system: neuropathologic considerations. *Toxicol Pathol.* 2000;28:54–69.

37. Kobayashi S, et al. Expression of glial fibrillary acidic protein in the CNS and PNS of murine globoid cell leukodystrophy, the twitcher. *Am J Pathol.* 1986;125:227–243.

38. Koliatsos VE, Price DL. Axotomy as an experimental model of neuronal injury and cell death. *Brain Pathol.* 1996;6:447–465.

39. Kondo A, et al. CNS pathology in the neurological mutant rats zitter, tremor and zitter-tremor double mutant (spontaneously epileptic rat, SER). *Brain.* 1991;114:979–999.

40. Krinke G, et al. Pyridoxine megavitaminosis produces degeneration of peripheral sensory neurons (sensory neuronopathy) in the dog. *NeuroToxicology.* 1981;2:(1):13–24.

41. Krinke GJ. Spinal radiculoneuropathy in aging rats: demyelination secondary to neuronal dwindling. *Acta Neuropathol (Berl).* 1983;59:63–96.

42. Krinke G, et al. The role of Schmidt–Lanterman incisures in Wallerian degeneration. *Acta Neuropathol (Berl).* 1986;69:168–170.

43. Krinke GJ, Fitzgerald RE. The pattern of pyridoxine-induced lesion: difference between the high and the low toxic level. *Toxicology.* 1988;49:171–178.

44. Krinke G, et al. Adjustment of the myelin sheath to axonal atrophy in the rat spinal root by the formation of infolded myelin loops. *Acta Anat.* 1988;131:182–187.

45. Krinke GJ. Nonneoplastic and neoplastic changes in the peripheral nervous system. In: Mohr U, et al., eds. *Pathobiology of the Aging Mouse,* Vol. 2 Washington, DC: ILSI Press; 1996;83–103.

46. Krinke GJ, Zubriggen A. Spontaneous demyelinating myelopathy in aging laboratory mice. *Exp Toxicol Pathol.* 1997;49:501–503.

47. Krinke GJ, et al. Optimal conduct of the neuropathology evaluation of organophosphorus induced delayed neuropathy in hens. *Exp Toxicol Pathol.* 1997;49:451–458.

48. Krinke GJ, et al. Teased-fiber technique for peripheral myelinated nerves: methodology and interpretation. *Toxicol Pathol.* 2000;28:113–121.

49. Krinke GJ, et al. Morphologic characterization of spontaneous nervous system tumors in mice and rats. *Toxicol Pathol.* 2000;28:178–192.

50. Krinke GJ, et al. Nonneoplastic and neoplastic changes in the peripheral nervous system. In: Mohr U, et al., eds. *Pathobiology of the Aging Dog.* Ames, IA: Iowa State University Press; 2001;10–21.

51. Lampert PW, Schochet SS. Demyelination and remyelination in lead neuropathy. *J Neuropathol Exp Neurol.* 1968;27:527–545.

52. Liu HM, et al. Schwann cell properties: 3. C-fos expression, bFGF production, phagocytosis and proliferation during Wallerian degeneration. *J Neuropathol Exp Neurol.* 1995;54:487–496.

53. LoPachin RM, Lehning EJ. The relevance of axonal swelling and atrophy to γ-diketone neurotoxicity: a forum position paper. *Neurotoxicology.* 1997;18:7–22.

54. LoPachin RM, Lehning EJ. Response to commentaries on forum position paper: the relevance of axonal swelling and atrophy to -diketone neurotoxicity. *Neurotoxicology.* 1997;18:37–40.

55. Lubinska L, Jastreboff P. Early course of Wallerian degeneration in myelinated fibers of the rat phrenic nerve. *Brain Res.* 1977;130:47–63.

56. Lubinska L. Patterns of Wallerian degeneration of myelinated fibers in short and long peripheral stumps and in isolated segments of rat phrenic nerve: interpretation of the role of axoplasmic flow of the trophic factor. *Brain Res.* 1982;233:227–240.

57. Lüllmann-Rauch R. Lipidosis like alterations in dorsal root ganglion cells of rats treated with trycyclic antidepressants. *Naunyn-Schmied Arch Pharm.* 1974;283:219–222.

58. Lüllmann-Rauch R. Lipidosis-like alterations in spinal cord and cerebellar cortex of rats treated with chlorphentermine or tricyclic antidepressants. *Acta Neuropathol (Berl).* 1974;29:237–249.

59. Lunn KF, et al. The temporal progression of the myelination defect in the taiep rat. *J Neurocytol.* 1997;26:267–281.

60. Macarenco RS, et al. Perineuroma: a distinctive and under-recognized peripheral nerve sheath neoplasm. *Arch Pathol Lab Med.* 2007;131:625–636.

61. Malbouisson AMB, et al. The non-directional pattern of axonal changes in Wallerian degeneration: a computer-aided morphometric analysis. *J Anat.* 1984;139:159–174.

62. Messer A. Mutant mouse models of ALS. *Neurobiol Aging.* 1994;15:247–248.

63. Mohr U, eds. *International Classification of Rodent Tumors: The Mouse.* Berlin: Springer-Verlag; 2001;474.

64. Möller JR, et al. Biochemical analysis of myelin proteins in a novel neurological mutant: the taiep rat. *J Neurochem.* 1997;69:773–779.

65. Morell P, Toews AD. Schwann cells as targets for neurotoxicants. *Neurotoxicology.* 1996;17:685–696.

66. Notterpek L, Tolwani RJ. Experimental models of peripheral neuropathies. *Lab Anim Sci.* 1999;49:588–599.

67. Ochoa J. Recognition of unmyelinated fiber disease: morphologic criteria. *Muscle Nerve.* 1978;1:375–387.

68. Ochs S. Beading of myelinated nerve fibers. *Exp Neurol.* 1965;12:84–95.

69. Ochs S, et al. The origin and nature of beading: a reversible transformation of the shape of nerve fibers. *Prog Neurobiol.* 1997;52:391–426.

70. Pettersen JC, et al. Neurotoxic effects of zoniporide: a selective inhibitor of the Na$^+$/H$^+$ exchager isoform 1. *Toxicol Pathol.* 2008;36:608–619.

71. Post EJ. Unmyelinated nerve fibers in feline acrylamide neuropathy. *Acta Neuropathol (Berl).* 1978;42:19–24.

72. Rath EM, et al. Impaired peripheral nerve regeneration in a mutant strain of mice (Enr) with a Schwann cell defect. *J Neurosci.* 1995;15:7226–7237.

73. Reuhl KR, Polunas MA. Tellurium. In: Spencer PS, et al. eds. *Experimental and Clinical Neurotoxicology*, 2nd ed. New York: Oxford University Press; 2000;1140–1143.

74. Scheidt P, Friede RL. Myelin phagocytosis in Wallerian degeneration: properties of millipore diffusion chambers and immunohistochemical identification of cell populations. *Acta Neuropathol (Berl).* 1987;75:77–84.

75. Schlaepfer WW, Myers FK. Relationship of myelin internode elongation and growth in the rat sural nerve. *J Comp Neurol* 1973;147:255–266.

76. Schröder JM. Altered ratio between axon diameter and myelin sheath thickness in regenerated nerve fibers. *Brain Res.* 1972;45:49–65.

77. Schröder JM, Gibbels E. Unmyelinated nerve fibers in senile nerves and in late thalidomide neuropathy: a quantitative electron microscopy study. *Acta Neuropathol (Berl).* 1977;39:271–280.

78. Sickles DW, et al. Toxic axonal degeneration occurs idependent of neurofilament accumulation. *J Neurosci Res.* 1994;39:347–354.

79. Sotnikov OS. Structure of Schmidt–Lanterman incisures. *Arkh Anat, Gistol Embriol.* 1965;43:31. Reprinted in *Fed Proc Fed Am Soc Exp Biol.* 1966;25:204–210.

80. Spencer PS, et al. Axon diameter and myelin thickness—unusual relationship in dorsal root ganglia. *Anat Rec.* 1973;176:225–244.

81. Spencer PS, eds. *Exper Clin Neurotoxicol.* 2nd ed. New York: Oxford University Press; 2000;1310.

82. Sterman AB. Acrylamide induces early morphologic reorganization of the neuronal cell body. *Neurology.* 1982;32:1023–1026.

83. Strich SJ. Notes on the Marchi mathod for staining degenerating myelin in the peripheral and central nervous system. *J Neurol Neurosurg Psychiatry.* 1968;31:110–114.

84. Summers BA, et al. *Veterinary Neuropathology.* St. Louis, MO: Mosby; 1995;527.

85. Sun MK. Central neural organization and control of sympathetic nervous system in mammals. *Prog Neurobiol.* 1995;47:157–233.

86. Torvik A. Central chromatolysis and the axon reaction: a reappraisal. *Neuropathol Appl Neurobiol.* 1976;2:423–432.

87. Tsahala-Katumbay DD, et al. A new murine model of giant proximal axonopathy. *Acta Neuropathol (Berl).* 2005;109:405–410.

88. Tsahala-Katumbay DD, et al. Monocyclic and dicyclic hydrocarbons: structural requirements for proximal giant axonopathy. *Acta Neuropathol (Berl).* 2006;112:317–324.

89. Van Lier RBL, Cherry LD. The toxicity and mechanism of action of bromethalin: a new single feeding rodenticide. *Fundam Appl Toxicol.* 1988;11:664–672.

90. Waller AV. Experiments on the section of glossopharyngeal and hypoglossal nerves of the frog, and observations of the alterations produced thereby in the structure of their primitive fibers. *Phil Trans R Soc Lond.* 1850;140:423–469.

91. Wiley RG, Stirpe F. Neurotoxicity of axonally transported toxic lectins, abrin, modeccin and volkensin in rat peripheral nervous system. *Neuropathol Appl Neurol.* 1987;13:39–53.

92. Zenker W, Hohberg E. Motorische Nervenfaser: Axonquerschnittsfläche von Stammfaser und Endästen. *Z Anat Entwickl-Gesch.* 1973;139:163–172.

24

TOXICOLOGICAL PATHOLOGY OF THE RETINA AND OPTIC NERVE

MEG RAMOS

Drug Safety Evaluation, Allergan, Inc., Irvine, California

CHRISTOPHER M. REILLY

Department of Pathology, Microbiology, and Immunology, School of Veterinary Medicine, University of California–Davis, Davis, California

BRAD BOLON

GEMpath, Inc., Longmont, Colorado

INTRODUCTION

Neuropathological examination of the retina and optic nerve shares many similarities with that of the analytical techniques used for other parts of the nervous system. However, assessment of intraocular neural components entails several additional considerations, including interspecies anatomical differences, distinctive processing methods, and logistical concerns unique to the eye, and to the retina and optic nerve in particular. In this chapter we review these unique features of ocular neuroanatomy, histology, neuropathology, and neurotoxicology as they pertain to the retina and/or optic nerve.

OCULAR ANATOMY

Retina

The eye is an important "peripheral" site for toxicant-induced neural lesions. The ocular components that are most commonly affected by toxicants are the retina and optic nerve (cranial nerve II), both of which are derived from bilateral evaginations of the prosencephalon early during embryogenesis. The tip of the optic stalks ultimately invaginates to form the bilayered optic cup. The inner layer differentiates into the photosensitive retina, while the outer layer becomes the retinal pigmented epithelium (RPE). These two walls fuse later in embryonic development to close the gap between them. The RPE is a syncytium of epithelial cells located between the sensory retina and the vascular choroid. The optic nerve is unique among neural projections located outside the central nervous system (CNS) because it is a direct extension of the CNS. Axons in the optic nerve are supported by astrocytes and myelinated by oligodendroglia, while the nerve itself is surrounded by all three layers of meninges.

The overall anatomy of the retina is similar for all mammalian species used routinely in toxicologic neuropathology research, with a few key differences. The retina is organized into multiple layers, or laminae (Figure 1). The relationship of these layers is best understood conceptually by performing pathological examinations from outside to inside, which conforms to the pathway of the neural impulse generated by incident light. The outer nonsensory retinal layer is the RPE, which consists of a single row of cuboidal cells with many cytoplasmic pigment granules. The RPE melanosomes have a unique elliptical shape compared to the round conformation assumed by their uveal (and systemic) counterparts. This difference can be used to determine the origin of pigmented cells in the retina and/or subretinal

Fundamental Neuropathology for Pathologists and Toxicologists: Principles and Techniques, First Edition. Edited by Brad Bolon and Mark T. Butt.
© 2011 John Wiley & Sons, Inc. Published 2011 by John Wiley & Sons, Inc.

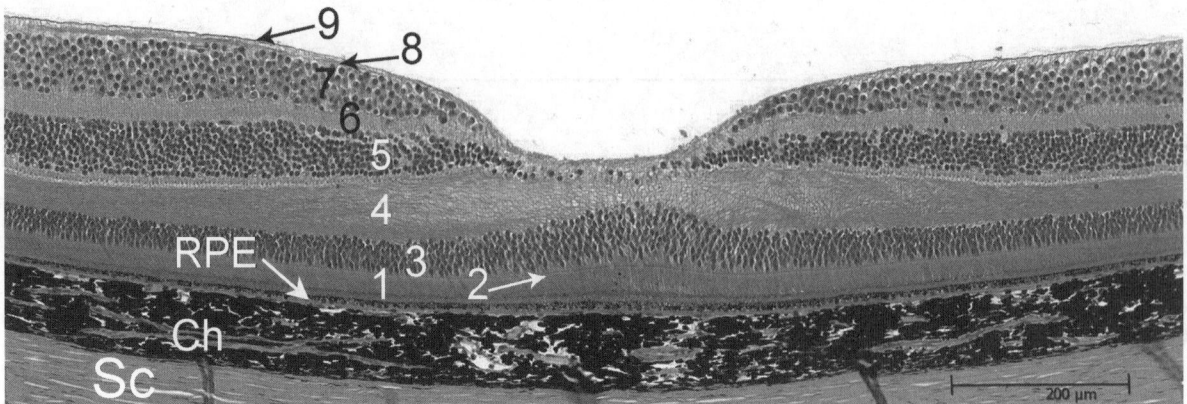

FIGURE 1 Normal retina from an adult nonhuman primate near the fovea centralis (the central depressed region). The photosensitive retina has nine layers: 1, photoreceptors (rods and cones); 2, outer limiting membrane; 3, outer nuclear layer (photoreceptor cell bodies); 4, outer plexiform layer, 5, inner nuclear layer [amacrine, bipolar, and horizontal neurons as well as Müller (glial) cells]; 6, inner plexiform layer; 7, ganglion cell layer; 8, layer of optic nerve fibers; and 9, inner limiting membrane. The retinal pigmented epithelium (RPE) separates the photosensitive inner retina from the vascular choroid (Ch) and fibrous sclera (Sc). Paraffin-embedded section, H&E, 100×. (*See insert for color representation of the figure.*)

space. Some investigators also consider the combination of the RPE basement membrane and the basement membrane of the choriocapillaris (a penta-laminar mesh of highly vascularized connective tissue that abuts the RPE) as another distinct, nonsensory layer of the retina. This structure is referred to as Bruch's membrane in humans.[1,2]

The photosensitive retina is segregated into nine levels (Figure 1), discussed here from outer to inner.[1] The layer of rods and cones, which abuts the RPE, contains the light-sensitive cytoplasmic extensions of photoreceptor cells. These light-sensitive segments are connected to the perikarya of their photoreceptors by narrow cytoplasmic bridges.[2] The outer limiting membrane is formed at the level of these bridges by the processes of specialized glial cells (Müller cells), which surround them.[3] The outer nuclear layer contains the cell bodies of the photoreceptors. These cells typically have a distinctive, cleaved nuclear morphology. Their axons extend into the outer plexiform layer,[4] where they contribute to the middle limiting membrane (a mat of desmosome-like synaptic attachments between photoreceptors) and also synapse with the dendrites of neurons whose cell bodies are located in the inner nuclear layer.[5] These neurons—the amacrine, bipolar, and horizontal cells—collectively perform the initial processing of visual sensory input. The inner nuclear layer also harbors the nuclei and perikarya of the Müller cells. The different cell types in the inner nuclear layer cannot readily be distinguished by routine light microscopy.[6] The inner plexiform layer is comprised chiefly of axons of the neurons in the inner nuclear layer (especially bipolar cells) and dendrites from neurons in the ganglion cell layer.[7] The neurons in this layer are the largest retinal neurons. They are characterized by a diamond-shaped

cell body, round nucleus, prominent nucleoli, and abundant dispersed Nissl substance [except at the origin of the axon (axon hillock), which is often not in the plane of section]. The ganglion cells tend to be concentrated at the central regions of the retina and are particularly numerous near the boundaries of the fovea centralis (or fovea).[8] The layer of optic nerve fibers is composed of the axons of ganglion cells traveling toward the optic disc; the layer thickens near the disc as more and more axons are added to it.[9] The inner limiting membrane is an indistinct basal lamina arising from the basement membrane of Müller cells, which separates the retina from the vitreous humor; this layer is unapparent in some histological preparations because it is pulled away toward the vitreous humor. In addition to defining structural changes according to their effects on retinal layering, the retina may also be considered conceptually to have distinct inner and outer zones, each of which may be affected differentially by various pathological processes [e.g., inner retinal atrophy as the result of excessive intraocular pressure (glaucoma) vs. outer retinal degeneration of the progressive retinal atrophies].

Retinal organization varies somewhat depending on the location within the eye. For example, the cones, suited to discriminating color and detail, are collected in the center, while the rods, adapted for performance in dim light (i.e., low resolution), are more common near the periphery. The cones are particularly concentrated in the fovea (also termed the *fovea centralis* or *area centralis*), a small circular or oval zone in the back of the retina responsible for sharp central vision (Figure 1). To accomplish this increase in acuity, the cones in the foveal pit have a smaller diameter so that they can be more densely packed. Furthermore, the fovea is the region of highest ganglion cell density in the eye; the ganglion cells are arranged in layers, typically four to five cells deep. In most laboratory animal species, the fovea is slightly dorsal and temporal (i.e., lateral) to the optic disc, but rabbits have a horizontal band of increased ganglion cell density (referred to as the *visual streak*) located ventral to the optic nerve. In primates, the fovea is surrounded by the *macula* ("spot"), a larger region of increased ganglion cell density in which the thinner layering (two to three cells deep) of ganglion cells is suboptimal for providing the best acuity. In humans, the number of cones is approximately 20-fold less than the number of rods. Approximately 50% of optic nerve axons carry information from the fovea, while the other 50% carry information from the rest of the retina; thus, the fovea communicates with more than 50% of the visual cortex even though the fovea represents less than 1% of the retina. Photoreceptors are absent over the optic disc, which is the point at which the ganglion cell axons come together and enter the optic nerve.

Optic Nerve

Conceptually, the optic nerve (cranial nerve II) can be considered to consist of two parts: the head and the trunk. The head is the intraocular portion of the optic nerve. It represents the locus where the ganglion cell axons of the nerve fiber layer (retinal layer 8) collect at the optic disc (also called the optic nerve head) and turn toward the brain (Figure 2). The fiber bundle passes through the lamina cribrosa, a sievelike modification of the sclera, to form the trunk at the caudal (posterior) pole of the eye. The trunk is the extraocular connection that links the globe to the optic chiasm.

Significant anatomical variations of these structures exist among species. For example, in most species the lamina cribrosa is evident just behind the optic disc as multiple discontinuous layers of collagen fibers running perpendicular to the optic nerve fibers (Figure 2). However, in rodents and rabbits, the presence of a lamina cribrosa is diminished or absent.[3,4] The optic disc is myelinated in only a few species (e.g., dog, rabbit), while in other laboratory animal species and primates (including humans) the nerve only becomes myelinated once the fibers have exited through the lamina cribrosa to form the optic nerve trunk. The trunk is usually heavily myelinated throughout its length.

NONNEURAL STRUCTURES IN THE EYE

Vasculature

In general, the inner retina is supplied by the retinal or supraretinal vasculature, while the outer retina is fed from the choriocapillaris. The source of blood to the neural components of the eye differs substantially across species. The arrangement of retinal vessels assumes one of several general patterns. The existence of these divergent schemes has practical implications with respect to appropriately interpreting vascular distribution for a given species (i.e., within normal limits vs. lesion) and understanding the potential for direct intravascular delivery of neurotoxicants to the retina.

The most common arrangement is holangiotic, in which the entire retina is supplied with blood, except for the macula. Two vascular patterns have been described in this regard, one for primates (nonhuman and human) and the second for such laboratory animal species as rodents (mice, rats, gerbils) and small carnivores (cats, dogs, ferrets). The primate pattern is to supply both the retina and the optic nerve via a central retinal artery, which is a branch of the internal ophthalmic artery. In other animal species, the optic nerve is still fed by the internal ophthalmic artery, but the retina is supported by the long and short caudal (posterior) ciliary arteries and the external ophthalmic artery, which are all branches of the external maxillary artery.[5] The retina is also supported by diffusion from a separate capillary bed in the choroid layer.

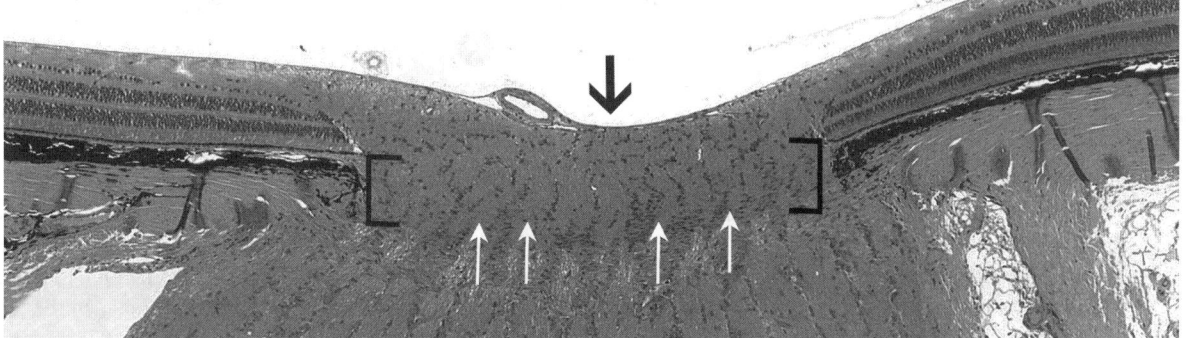

FIGURE 2   Lamina cribrosa of the optic nerve from an adult nonhuman primate. Axons extend from the ganglion cells in the retina to collect at the optic disc (also called the optic nerve head; thick arrow), where they pass through the lamina cribrosa (a sieve-like fibrous component of the sclera; in brackets, with the deep margin delineated by the thin arrows) to form the trunk of the optic nerve (located below the tails of the arrows). Paraffin-embedded section, H&E, 50×.

Other arrangements of retinal vessels are uncommon. For example, the rabbit has a merangiotic pattern in which only a portion of the retina is supplied with blood. Guinea pigs and horses exhibit a paurangiotic array in which a few small vessels supply the retina for only a short distance from the optic nerve. Finally, birds (chickens) and some rodents (chinchillas, guinea pigs) have an anangiotic arrangement in which the retina lacks blood vessels. In the absence of a retinal vascular supply, the retina receives blood-borne materials by diffusion from the choroid layer and, to a minor extent, the vitreous humor. The avian retina may also be supplied by the pectin, a richly vascular and variably ornate extension of the choroid into the vitreous, which is adjacent to the optic disc.

Tapetum

The reflective inner layer of the choroid is referred to as the *tapetum lucidum*, or tapetum (reviewed by Ollivier et al.[6]). The tapteum serves to direct light back through the retina for "second-pass" visualization, especially in species needing to move (e.g., hunt or avoid predation) in low-light environments. The RPE is present, but not pigmented, in tapetal regions to facilitate the passage of light.

A few laboratory animal species (cats, dogs, ferrets, horses, ruminants) have a tapetum; others do not (rodents, rabbits, primates). The structure of the tapetum in histological sections varies by species. Carnivores have cellular tapeta, with a relatively sparse matrix component between cell nuclei. The cellular tapetum is most striking in the cat, due to the bright red, abundant cytoplasm. In contrast, large herbivores have fibrous tapeta, with a more pronounced connective tissue matrix and relatively fewer nuclei.

CLINICAL ASSESSMENT OF THE RETINA AND OPTIC NERVE

The same optical qualities of the eye necessary for vision permit the unique opportunity for noninvasive structural and functional assessment of this neurosensory organ. In many instances, newer imaging modalities come within micrometers of the resolution offered by microscopic evaluation, thus allowing repeated "sampling" for morphological evidence of xenobiotic-induced ocular toxicity throughout the course of a spontaneous disease or a toxicity study. This ability can be an important diagnostic and prognostic aid for morphological changes in the eye, particularly in the retina, and often precedes the onset of functional deficits. Many ocular diseases in humans have been well characterized through imaging modalities that may readily be adapted to animal model applications. The present section provides a brief review of the major techniques used to examine the functional integrity of the retina and optic nerve.

Noninvasive Structural Assessment of the Retina and Optic Nerve

Several noninvasive means are used routinely to acquire anatomical data about the neural components of the eye. The most direct approach is to view structural traits of the retina by peering through the pupil. Innovative new methods can reconstruct the retinal cytoarchitecture using tissue-specific properties. With experience, data acquired using these techniques can be used to predict the likely microscopic appearance of the retina and optic nerve and to select the most efficient and effective approach to performing the histological assessment.

Indirect Ophthalmoscopy   Bilateral indirect ophthalmoscopy is the standard parameter assessed in medical and veterinary clinical practice as a first-tier diagnostic screen. This technique is also an industry standard for assessing visual system integrity when assigning animals to specific treatment groups before toxicity studies are initiated. Indirect ophthalmoscopy enables the practitioner to evaluate the surface morphology of intraocular tissues in the fundus, especially the retina and the vitreous body. A handheld biconvex lens is positioned a few centimeters from the dilated pupil and oriented to focus light onto the retina. An inverted image of the retina is visualized in midair between the examiner, who is wearing a binocular ophthalmoscope, and the biconvex lens.[7] The fundus is examined for such retinal abnormalities as altered contours, discolorations (usually indicating hemorrhage, pigment changes, or scarring), and vascular distribution. The optic disc (i.e., the head of the optic nerve) can also be assessed for abnormalities of color, structure (e.g., cupping, swelling, or masses), and vascularity.

Examination of the fundus is facilitated if the pupils are dilated in advance using a topical mydriatic agent (e.g., 1% tropicamide). Pupil dilation is helpful for evaluating the central retina, but is absolutely essential for adequate assessment of the peripheral retina. Mydriasis is achieved in most species within 5 to 20 min but may be delayed or incomplete in eyes that are highly pigmented, probably due to binding of the drug to melanin within the iris.[7] Dilation may be enhanced in such cases by the additional instillation of 2.5% phenylephrine hydrochloride.

Slit Lamp Biomicroscopy   Slit lamp examinations employ a table-mounted binocular instrument to direct an intense narrow light beam into the fundus. The beam can be rotated about a vertical axis so that the entire expanse of the retina is open to evaluation. Stereoscopic imaging of the fundus is obtained by positioning a handheld or mounted lens over the surface of the cornea, or through the temporary application of a diagnostic contact lens to the cornea, to project a magnified aerial image of the fundus contents.[7] The scatter of the light beam and the resulting shadows from the retina and vitreous humor demonstrate such morphologic irregularities as altered retinal contours (e.g., via detachment) and floating debris ("floaters"), respectively.

Autofluorescence Photography Fundic cameras are designed with an aspherical lens that, when focused to the optics of the subject's eye, matches the plane of focus to the curvature of the fundus using a light beam delivered axially through the filters and optics of the camera. Variable-angle camera systems are capable of imaging between 45° to 60° arcs of the retina, and are used to capture fluorescence intensity, pigment distribution, and the retinal vasculature pattern.

Natural autofluorescence (AF) of the retina is attributed to excitation of intraretinal lipofuscin in RPE cells by short wavelength emissions. The extent of AF increases with age and may be exacerbated in some disease conditions. For example, drusen are punctuate, extracellular AF deposits of lipids, lipoproteins, and complement that occur under the detached RPE, between the RPE and Bruch's membrane (the combined basement membranes of the RPE and choriocapillaris), or within Bruch's membrane during the course of age-related macular degeneration (ARMD).[1,2,8] Drusen are not static, but grow or regress, and substantial deposits may lead to global RPE atrophy. Conversely, areas of hypofluorescence may signify RPE death or loss of photoreceptors. Photographic AF analysis is thus a useful noninvasive tool to monitor retinal structure over time, and can be employed effectively during drug discovery and development as an adjunct method to characterize and monitor the efficacy and toxicity of novel xenobiotics.

Fluorescein Angiography (FA) This method is used clinically to diagnose and monitor the impact of therapeutic interventions on the retinal and choroidal vasculature. It is particularly well suited to follow such conditions as diabetic retinopathy, retinal vein occlusion, neovascularization, and ischemia. Fluorescein sodium will enter the ocular vasculature within seconds after injection into a peripheral vein. The majority of the agent is quickly bound to plasma proteins, but approximately 20% circulates freely and is subject to peak excitation upon exposure to blue light (wavelength range, 465 to 490 nm), with transient photoluminescence also occurring at green-yellow wavelengths of 520 to 530 nm.[9,10] Sequential digital photographs or video captured using a fundic camera will track the dye movement through the vasculature, thereby delineating both structural integrity and normal circulatory function.

Proper interpretation of the angiogram to identify vascular changes reliably requires a thorough understanding of the normal pattern of ocular circulation for the test species. For example, leakage from the retinal capillaries is prevented by tight junctions between the endothelial cells. In contrast, the choroidal capillaries are fenestrated and thus may leak under "normal" physiological conditions. Different patterns of blood vessel distribution within the retina may be categorized using FA.

Optical Coherence Tomography (OCT) This two-dimensional, real-time imaging modality uses tissue-specific properties of light reflectivity to generate tomographic images of retinal architecture.[11,12] The portion of light reflected back from a tissue varies with such regional morphological features as cell density, so logarithmic integration and conversion of data points captured in a series of axial and longitudinal scans can be used to construct a cross-sectional likeness of its structure.[13] Thus, the cell layers in the retina may be distinguished on OCT due to their different densities and light-backscattering properties. Retinal OCT images closely reflect the layer-specific architecture that can be observed in histological cross sections (Figure 3) and can

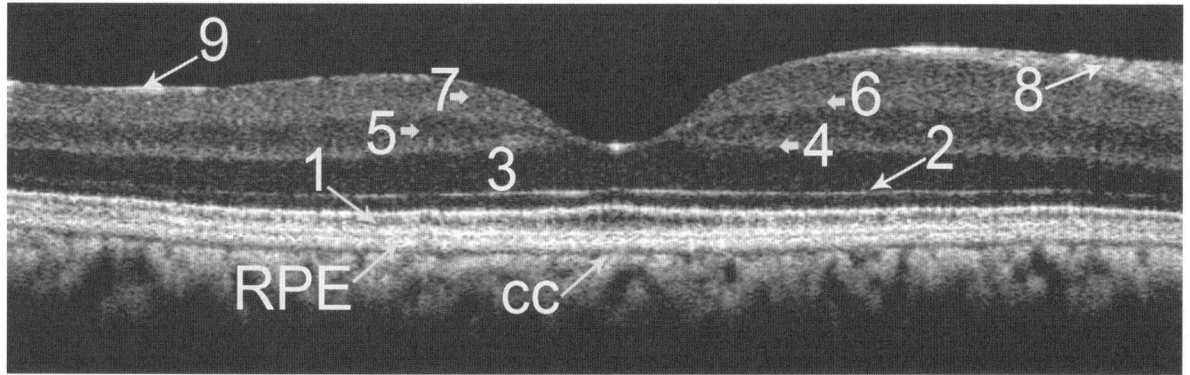

FIGURE 3 Optical coherence tomography (OCT) image of the retina from a normal adult nonhuman primate. Retinal layers are visualized as alternating light and dark bands based on differential light reflectivity from structures with distinct densities. Retinal layers are numbered as in Figure 1. cc, Choriocapillaris; RPE, retinal pigmented epithelium. In-vivo image acquired using a spectral domain OCT (SDOCT) instrument (Bioptigen, Inc., Research Triangle Park, NC). Axial resolution, 5 μm.

be used to quantitatively evaluate anatomic structure, thickness, and structural integrity. Although lacking in cellular detail, an axial resolution of 2 to 3 μm is obtainable with current-generation OCT.[14]

Retinal OCT images can be obtained longitudinally and used both in clinical practice and in the experimental setting to follow disease progression over time without the need for enucleation. The anatomical characteristics of many retinal diseases have been well characterized using OCT, including atrophy (resulting from glaucoma), cysts, degeneration (in ARMD), detachments, edema of the sensory retina and breaks, drusen, and hypertrophy of the RPE.[13,15] Similar findings can be identified in rabbit and primate models of these diseases, and correlation of OCT and histologic findings may be a useful approach for assessing drug efficacy and safety. OCT can guide histologic evaluation of the eye in drug development to ensure relevant interpretation.

Functional Assessment of the Retina and Optic Nerve

Functional alterations in the retina may precede, follow, or even occur without morphological changes. Several noninvasive techniques of assessing visual function used in the clinical setting have been adapted by industrial toxicology laboratories to objectively evaluate the impact of a compound on the visual system. Functional tests may be used to differentiate transient alterations in vision from irreversible toxicant-induced changes. More recently, these techniques have been employed to demonstrate the efficacy of ophthalmic drugs in animal models of human ocular disease.

Electroretinography (ERG)   This technique collects a recorded summation of electrical activity arising in the retina in response to light stimulation. Therefore, the ERG represents the collective functional assessment of the various cell layers of the retina. Several types of ERG may be collected. Full-field ERGs are elicited simultaneously using a Ganzfeld dome to provide uniform illumination of the entire retinal field, with the amplitude proportional to the area of functional retina that is stimulated.[16] The flash ERG (fERG) is the summation of the electrical potential that is produced by the dark-adapted retina in response to a bright flash of light. Multifocal ERG (mfERG) measures the focal photopic (i.e., cone-driven) response of the central retina using a black-and-white hexagonal array as the stimulus. All types of ERG are measured by placing the electrode at the corneal surface and are driven primarily by the responses of the photoreceptors, bipolar cells, and Müller cells.

A typical ERG is generated by radial currents that arise from changes in the extracellular potassium ion (K^+) concentration and occur as two successive potentials: a cornea-negative a-wave followed by a cornea-positive b-wave. The negative deflection (a-wave) is generated by the hyperpolarization of the photoreceptor cell membrane induced by light

activation of rhodopsin, subsequent hydrolysis of cyclic guanosine monophosphate (cGMP), and closure of potassium ion channels in the cell membrane.[17,18] Hyperpolarization results in the diminished release of glutamate, the primary excitatory neurotransmitter of the retina, from the photoreceptor activated by photons. This visual signal is conducted to the inner retina by bipolar cells, which are excitatory interneurons juxtaposed between the photoreceptor and ganglion cells. Bipolar cells either depolarize (ON-center) or hyperpolarize (OFF-center) in response to the diminished release of glutamate following photoreceptor activation. OFF-center bipolar cells contribute to the a-wave when the retina is dark-adapted and then stimulated with light, as with flash ERG.[19,20]

The positive b-wave represents the subsequent activity of the cell processes within the inner plexiform layer, which arise primarily from the ON bipolar cells that depolarize to light, but also from Müller cell activity.[21] In addition, significant contributions from third-order neurons (amacrine, interplexiform, and ganglion cells) have recently been demonstrated.[22] Oscillatory potentials (OP) are high-frequency, low-amplitude potentials superimposed on the ascending front of the b-wave as wavelets subsequent to a brief light flash. These OPs have been attributed to several components of the retina, but recent pharmacological studies suggest that OPs are generated mainly from photoreceptors and molecules within the ON-pathway of the inner retina.[23] An overly simplified interpretation of the two ERG potentials is that a diminished a-wave represents photoreceptor loss; a diminished b-wave represents loss of the photoreceptor and/or neuronal layers; and an intact a-wave with a diminished b-wave represents loss of the neuronal layer.[20,24–27]

An additional element of the ERG, the c-wave, is driven by hyperpolarization of the RPE, with additional contributions from Müller cell activity. However, the c-wave is not analyzed in conventional animal toxicity studies because of technical and interpretative difficulties due to the slower response for generation. If specific functional assessment of the RPE is desirable, intravenous injections of sodium azide (< 1 mg/kg) will produce a marked, transient depolarization of the RPE that is analogous in shape and temporal properties to the ERG c-wave. The "azide response" thus is dependent upon RPE integrity and is thought to measure the potential generated in response to a reduction in extracellular potassium triggered by light absorption in the photoreceptors.[28]

Functional contributions from rod and cone photoreceptors can be differentiated using ERG recordings elicited by qualitatively and quantitatively divergent light stimuli. A dim short-wavelength stimulus (at a frequency of 0.5 Hz) under scotopic (dark-adapted) conditions primarily stimulates rods, while a long-wavelength stimulus induces a mixed response from both rods and cones. A flicker response (30 Hz) under photopic (light) conditions will suppress rod activity. Chromatic stimuli can be employed to stimulate cones selectively.

The fERG represents electrical activity of the entire retina. Since it assumes that all parts of the retina contribute equally to ERG production, focal regions with functional deficits can be missed. Multifocal ERG (mfERG) is a more sophisticated measurement of cone-driven function that is acquired simultaneously over multiple areas within the central retina. The mfERG is collected using a black-and-white hexagonal array as the retinal stimulus, where the luminous intensity of each hexagon is temporally modulated between black and white according to a binary m-sequence.[29] The resulting focal recordings, similar to the a-wave and b-wave obtained in the fERG, are converted to a three-dimensional plot of response amplitude that can then be used to either identify regions within the retina with deficient function or to compare the function of retinal domains having or lacking lesions. Studies in glaucomatous human patients and non-human primate models of ocular hypertension suggest a significant contribution from the retinal ganglion cells to the mfERG. Loss of these cells and their axons correlates with diminished mfERG.[30,31]

The components of the ERG are sensitive to pharmacological manipulations, which can be exploited in the research setting to further assess the function of relay neurons that synapse with the photoreceptors. For example, intravitreal injection of the glutamatergic receptor agonist 2-amino-4-phosphonobutyric acid (APB) or the glutamatergic receptor antagonist *cis*-2,3-piperidine dicarboxylic acid (PDA) transiently blocks synaptic transmission between photoreceptors and depolarizing ON-center or hyperpolarizing OFF-center bipolar cells, respectively.[32,33] The interpretation of the effects provided by these two compounds is that the ERG generated from conditions near the threshold for stimulating cone photoreceptors is dominated by the activity of postsynaptic cells residing in the inner retina.

Altered ERG patterns have been well characterized in many human retinal diseases, including diabetic retinopathy, glaucoma, macular degeneration, and retinitis pigmentosa.[29,34,35] The ERG has also become an invaluable noninvasive technique for the development of ocular disease models in laboratory animals, including the impact of potential ocular toxicants on the function of specific cell types within the retina.[36] However, meaningful interpretation of the ERG in the context of drug discovery and development necessitates strict attention to species differences and standardization of the conduct and interpretation of efficacy and toxicity studies. For example, amplitude of the recordings can vary considerably with placement of the electrode and the precise conditions of illumination. Consideration must be given to the anatomical and functional differences in the visual system between nocturnal and diurnal species, with specific consideration of such factors as the animal's physiological status, state of consciousness, body temperature,

and glucose levels—all of which may influence test results. Age, gender, and time of day for performing the ERG may contribute to variations of response. Some animals (e.g., the rabbit) can be readily restraint-adapted to tolerate this relatively quick, noninvasive procedure. However, sedation or general anesthesia may be required for adequate ERG examination of most laboratory animal species. Ketamine may dampen the amplitude and increase the response time of ERG recordings, while phenobarbital has been shown to increase the amplitude of the ERG.[24] Recent published guidelines and protocols for ERG procedures that are in compliance with good laboratory practice (GLP) standards should be consulted for additional information to guide drug safety development.[20,24,26]

Visual Evoked Potential (VEP)   The ERG affords an assessment of retinal function in response to light but does not provide information regarding the complete visual pathway. Indeed, the ERG may be normal in optic nerve atrophy despite substantial loss of retinal ganglion cells.[16] In contrast, the visual evoked potential (VEP) is a direct measurement of the visual cortex response to retinal stimulation by a light flash or contrast-reversing pattern, and thus provides an objective assessment of visual acuity and contrast sensitivity for the entire visual system.[37,38] Recordings are acquired from electrodes placed subcutaneously in the scalp overlying the occipital cortex. The VEP waves have both positive and negative electrical potentials, and both the response amplitude and latency are used to assess the integrity of the visual pathway.[25] Normal VEP recordings are an indication that the visual pathway from the ganglion cells to the visual cortex is intact. A prolonged VEP is an indicator of myelin defects, such as those resulting from ethambutol toxicity.[27] Abolition of the VEP in one eye results from severing the optic nerve that serves it; interestingly, the formation of the VEP in the contralateral eye is also delayed.

Several VEP variants are available, including pattern VEP (pVEP), flash VEP (fVEP), and sweep VEP (sVEP). These methods differ with respect to the visual stimulus used to launch the VEP. The pVEP is more sensitive in detecting axonal conduction defects that result from optic nerve compression or demyelination, and is considered to have less variability in normal subjects compared to fVEP. The sVEP is a steady-state version of the pVEP, which examines the visual pathway response to a pattern of elements that varies over time (i.e., pattern reversal images of increasing spatial frequency). An advantage of the sVEP over conventional pVEP is that it can be collected using shorter recording times of approximately 10 s.[37] Visual acuity is extrapolated from the sVEP by determining the highest spatial frequency, considered to be the limit of resolution, at which the visual system responds during a sweep from low to high spatial frequency.[38,39]

Contrast sensitivity is determined from the lowest contrast to which the visual system responds when contrast is swept within a fixed spatial frequency.[37] Because of the rapid recording of the threshold for these functions, sVEP is successfully used in the clinical setting to assess visual function in nonverbal patients (young children, patients unable to complete an eye chart examination). Similarly, sVEP is used in drug discovery and development as an indicator of visual acuity in test animals. For example, the sVEP of nonhuman primates is comparable to that of humans,[39] a correlation that is consistent with the many anatomical and functional similarities between them. In nonclinical drug development studies, repeated VEP/sVEP recordings can detect trends outside normal variation that suggest the possibility of ocular toxicity, and can monitor enhanced function that implies the potential for efficacy.[36]

Interspecies differences, patient preparation practices, and the methods of visual stimulation and electrical recording are manifested in the VEP, so protocols for collecting the VEP or sVEP must be standardized for each species if meaningful interpretations are to be achieved. Placement of the scalp electrodes depends on the species to be examined. Recordings from the electrodes are subject to marked variation among individuals and can even differ for a given individual across time. Proper focus of the stimulus onto the retina is essential to avoid artifactual decreases in amplitude and increases in latency; sedation or anesthesia also contributes to diminished amplitudes and augmented latencies in the VEP. Due to the rapidity of the test procedure, some species (e.g., rabbits) that can be trained to accept restraint may be recorded in a fully conscious state.

SPECIMEN ACQUISITION AND PROCESSING FOR OCULAR NEUROPATHOLOGY STUDIES

Proper specimen handling from procurement through sectioning is essential to obtaining samples of sufficient quality to support the neuropathologic examination of the eye. This organ is particularly sensitive to several structural artifacts that can impede or preclude a meaningful analysis; global distortion during sectioning, lens luxation, retinal autolysis, retinal detachment, and retinal intralaminar separation are a few of the most significant. Severe artifactual damage can be produced by even skilled technicians at any of several stages during ocular tissue processing. With experience and attention to detail, however, many of these artifacts can be greatly reduced or eliminated.

The choice of procurement and processing conditions for the eye depends on the nature of the question and the sample. For diagnostic purposes from a single individual, eyes can be removed carefully using standard equipment and then processed routinely by immersion fixation in neutral buffered 10% formalin (NBF) followed by paraffin embedding. This practice allows ocular tissues to be prepared in tandem with other tissues collected during the autopsy or necropsy and is

suitable for immunohistochemical methods that are commonly employed to diagnose particular disease processes. The primary disadvantage is that neural artifacts in formalin-fixed eyes are relatively common (especially retinal detachment and intralaminar separation), and they may prevent the identification and detailed analysis of subtle neuroanatomic changes. In general, neurotoxicological studies are conducted in a prospective fashion (i.e., designed in advance) with many subjects to evaluate novel xenobiotics for neurotoxic potential and possibly to define the progression and/or mechanism of such adverse effects. Identification and characterization of subtle lesions requires exceptional structural preservation, which requires more specialized fixatives and often plastic embedding.

Basic approaches to processing the eye effectively for various neuropathological applications are reviewed in this section. Where necessary, references containing more specialized technical information have been provided.

Procurement

For globes of most laboratory animal species, preservation of the retina and optic nerve is best accomplished by surgical or postmortem enucleation. If done as part of a postmortem procedure, enculeation should be performed as soon as possible after death to minimize autolysis and maintain the best preservation of morphological detail. Subsequently, all extraneous tissues (lids/adnexae, periorbital fat, extraocular muscles, and episcleral connective tissue) are detached completely, with the exception of the optic nerve. In general, these tissues should be removed prior to immersion fixation to facilitate penetration of the fixative and embedding media and thus optimize retinal preservation. The extraocular tissues can also be removed from fixed globes (including those subjected to perfusion fixation during dedicated neurotoxicity experiments) prior to trimming, although cleaning after fixation must be undertaken on a downdraft table or in a fume hood to reduce exposure of technicians to toxic fixative fumes. Regardless of when it is detached, removal of the extraocular tissues both reduces the severity of globe distortion caused by shrinkage of extraocular connective tissues during fixation and eases the acquisition of intact histologic sections.

Enucleation is generally carried out in one of two ways. The transpalpebral approach is used for rapid and reliable removal of the globes of larger animals (cats, dogs, primates, etc.). Briefly, the eyelids are clamped shut (ideally with towel clamps, but other locking instruments, such as hemostats or forceps, may be used). Next, an elliptical incision is made using a scalpel around the entire circumference of the palpebral fissure, dissecting the subcutaneous plane lining the orbit. A combination of blunt and sharp dissection is used to separate the loose orbital connective tissues and transect the extraocular muscles, fascial insertions, lateral and medial canthal ligaments (which are quite robust in relatively large species such as the rabbit, cat, and dog), and the optic nerve as far proximal (i.e., toward the brain) as possible. If the nerve is transected too close to the globe, myelin can be forced into the eye, where it comes to rest in the subretinal space ("myelin artifact").[3] Although gentle traction is required to dissect behind the globe, care should be used not to pull unduly on the optic nerve prior to cutting it, as this traumatic retraction can produce artifactual axonal disruption and/or vacuolation within the nerve parenchyma. A second method of enucleation, a transconjunctival approach, is used in some instances. Incisions are made in the skin at the lateral canthus and subsequently in the conjunctiva just caudal to the limbus. A similar combination of blunt and sharp dissection is used to remove the eye, although most of the subcutaneous tissues remain attached to the wall of the eye socket.

Similar techniques for enucleating the globes of small animals (e.g., rodents) are detailed elsewhere.[3] It is possible to remove rodent eyes using scissors alone to quickly "pop" the globes from their sockets. This procedure may be suitable for rapid, high-throughput screening of the retina for substantial anatomical defects, but it is unacceptable for evaluating the optic nerve (as the sample is small and heavily traumatized by this procedure) and for discerning subtle retinal lesions. In our experience, additional time and money often must be expended for follow-up studies if this high-throughput approach is employed in toxicological neuropathology testing. For very small globes (e.g., young mice), fixation and processing in situ can be utilized for screening purposes. However, processing the whole head may limit the orientations available for histologic analysis of the eye; multiple coronal sections are typically required to evaluate both the retina and the optic nerve. One means of in situ ocular processing that permits better positioning of the globe for sectioning is to hemisect the head in the longitudinal plane, so that each half of the skull can be sectioned in a sagittal plane (i.e., with the globe and its attached optic nerve sectioned along their midaxial planes). A variety of techniques for processing mouse ocular tissues are discussed elsewhere, most of which can be applied to other species.[3] In situ processing also requires extended decalcification with acid solutions that often distort fine histological detail and hamper immunohistochemical applications.

Fixation

Many options exist for suitable fixation of ocular tissue (Table 1), each with its own advantages and disadvantages. The choice of a given alternative generally represents a trade-off between processing speed, the degree to which fine structural details are preserved, and the need to evaluate molecular markers within the specimen.

The most readily available option is routine formalin fixation. The solution of choice is neutral buffered 10% formalin (NBF) rather than acidic formalin. The usual procedure is to immerse the isolated globe and optic nerve in approximately 10 volumes of NBF per volume of tissue. Suitable fixation of the retina and optic nerve is generally achieved within 48 to 96 h; when immunohistochemistry to detect intraocular antigens is anticipated, fixation time should be minimized to 24 to 48 h. Fixation for less than 24 h is insufficient to promote scleral rigidity, which may permit shifting of part, or all, of the ocular wall during sectioning and the induction of substantial intraocular artifacts. Immersion in NBF provides adequate fixation and histological detail for most diagnostic pathology purposes and for experimental pathology screening. However, discrimination of subtle toxicant-induced retinal lesions will usually require the superior tissue preservation provided by specialty ocular fixatives or whole-body perfusion fixation, particularly for publication.

Specialized ocular fixatives such as Bouin's solution or Davidson's solution (Table 1) are often employed in dedicated ocular pathology studies. Both fixatives are commercially available as ready-to-use solutions, but can easily be "homemade" as well. As with NBF, immersion fixation in Bouin's solution or Davidson's solution should employ 10 volumes of fluid for each volume of tissue. A difference with respect to NBF is that tissues immersed in either Bouin's solution or Davidson's solution should be transferred to either 70% ethanol or NBF after 24 h (up to 48 h if simultaneous decalcification is being undertaken with Bouin's) for postfixation and storage. Specimens fixed in Bouin's solution may be rinsed in a series (generally, 8 to 10) of water baths prior to immersion in 70% ethanol, or in sequential 70% ethanol changes, to reduce tissue discoloration.

Both Bouin's solution and Davidson's solution offer several major advantages for processing ocular tissues relative to NBF. Immersion in either of them increases the hardness of the specimens, which provides better support for the delicate neural tissues during trimming and sectioning, and improves fixative penetration, which enhances histological preservation of the retina. An added advantage of Bouin's solution is that it simultaneously fixes and decalcifies (via the action of picric acid) when whole head samples are employed to process the eyes in situ, making it a fixative of choice when globes must be examined within the orbit.

Although these traits offer advantages, the specialty fixatives also have some distinct disadvantages. The most significant problems are confined to Bouin's solution. The most important drawback to using this fixative is that the picric acid component is corrosive to metals and violently explosive when dry, thus leading to substantial difficulties associated with personnel safety as well as fixative storage and disposal. Bouin's solution also imparts a strong and persistent yellow hue to fixed tissues, which greatly reduces the contrast in gross photographs and stains any surface the solution or specimen contacts. Bouin's also delivers less pristine cellular preservation and prevents detection of many molecular entities using special techniques [such as immunohistochemistry (IHC)].

TABLE 1   Composition of Common Ocular Fixatives

Fixative	Constituent	Quantity
10% Formalin (neutral buffered)	Formaldehyde (37 to 40%)	100 mL
	Distilled water	900 mL
	Disodium diphosphate (Na_2HPO_4)	6.5 g
	Monosodium phosphate (NaH_2PO_4)	4.0 g
Bouin's solution	Picric acid (saturated aqueous)	75 mL
	Formaldehyde (37% w/w)	25 mL
	Glacial acetic acid	5 mL
Davidson's solution	Ethanol (95%)	30 mL
	Formaldehyde (37 to 40%)	20 mL
	Glacial acetic acid	10 mL
	Distilled water	30 mL
Smith–Rudt fixative	Glutaraldehyde (25%)	12.2 mL
	Paraformaldehyde (16%)	12.5 mL
	Phosphate buffer (0.2 M, pH 7.2)	100 mL
	Distilled water	125 mL
Telly's (Tellyesniczky–Fekete) fixative	Ethanol (70%)	100 mL
	Formaldehyde (37% w/w)	10 mL
	Glacial acetic acid	5 mL

Davidson's solution also has a few minor shortcomings, such as corneal opacification, lens discoloration, and vacuolation of the optic nerve. However, the degree of contrast in Davidson's-fixed globes is greater than that in Bouin's-fixed samples. Furthermore, Davidson's solution does not contain picric acid, so it does not present the same safety, storage, and disposal issues as those presented by Bouin's solutions. The optic nerve is best examined in cross sections fixed in NBF.

Two other practical disadvantages may preclude the routine utilization of Bouin's solution or Davidson's solution in toxicological neuropathology studies. First, research institutions typically prohibit the mixing of these composite specialty fixatives and NBF in the same container during disposal. Thus, studies that require special disposal methods for unused Bouin's solution, Davidson's solution, and tissues contaminated with them may raise the expense of ocular neuropathology experiments (especially those that use explosive-prone Bouin's solution). More important, routine use of Bouin's solution or Davidson's solution to fix ocular tissue will increase study costs because the specially fixed eyes will have to be processed separately from other routinely fixed study tissues. The latter disadvantage is mitigated in specialty neuropathology studies by the routine use of whole-body perfusion fixation.

Special processing procedures may be required for unique investigational needs. For example, suitable ocular fixation for plastic embedding is often achieved using Smith–Rudt fixative (Table 1).[40] Eyes are placed in Smith–Rudt fixative for 24 h at 4°C, and then either dehydrated on a shaker (1 h each in 25, 50, and 75% ethanol followed by 2 h in 95% ethanol) and immediately embedded in plastic or,

alternatively, transferred to 0.1 M phosphate buffer (pH 7.2) and retained at 4°C until processing can occur. Plastic embedding, especially if it is to support transmission electron microscopy, often employs post-fixation overnight by immersion in 1% osmium tetroxide in 0.1 M phosphate buffer (pH 7.2) at 4°C. If IHC is planned, eyes should instead be fixed in Telly's (Tellyesniczky–Fekete) fixative (Table 1) for 24 h or phosphate-buffered 4% paraformaldehyde (PFA) and then embedded in paraffin. Telly's fixative is a preferred choice because it penetrates rapidly and is compatible with a wide range of IHC protocols.

Detailed methods for whole-body perfusion of neural tissues have been reported elsewhere (see the article by Fix and Garman[41] and Chapter 21 for reviews). No special consideration for ocular fixation relative to preservation of other nervous system components is required when perfusion fixation is employed.

Regardless of fixative type (routine NBF vs. a specialty ocular fixative) or technique (immersion vs. perfusion), some pathologists advocate taking special steps to ensure adequate fixation of the retina. One approach is to inject larger globes (i.e., from dogs, pigs, primates, and rabbits) directly with a small amount (0.1 to 0.2 mL) of the chosen fixative using a fine needle (22 gauge or smaller). An alternative is to make a small (approximately 0.2 to 1 cm, depending on the diameter) nick through the full thickness of the globe, at about the level of the *pars plana* (i.e., the region of the ciliary body just caudal to the lens). However, while these techniques may improve fixative penetration and thus retinal morphology, they can also lead to undesirable anatomical artifacts, most notably retinal detachment.

In some instances it may be desirable to evaluate the optic nerve as a distinct entity rather than as a small portion of the entire globe. An advantage of this approach is that the soft neural tissue is not distorted by any artifacts that arise when trying to cut the hard ocular tissues. Thus, subtle lesions (e.g., loss of neural fibers) will be less likely to be masked by artifactual changes. In this event, the optic nerve is typically immersed in a mixture of 0.8% glutaraldehyde/1.2% para-formaldehyde in 0.1 M phosphate buffer (pH 7.2) overnight at 4°C.

Trimming

The ideal orientation of eye sections for neurohistopathologic analysis in both the diagnostic and experimental settings is to obtain a midaxial ("sagittal") representation through the entire globe so that all major ocular structures, including the largest expanse of retina and the optic nerve, may be seen. This view may be achieved with ease in small globes (e.g., rodents) without any trimming whatsoever. However, larger globes must be trimmed to provide the proper orientation (Figure 4). Such manipulation must be performed with great care to avoid inducing trauma-associated structural artifacts in the retina as the globe is penetrated.

Sharp, fairly rigid blades, such as Thomas Tissue Slicer Blades (Thomas Scientific, Swedesboro, NJ), are ideal for performing the initial cut. The blades are longer and sharper than the standard razor blade, while maintaining sufficient stiffness to section large fibrous globes relatively easily. The globe is placed cornea-side down and is supported on two sides by either the thumb and forefinger or by a pair of forceps (closed only enough to hold the specimen without exerting pressure on the intraocular contents). The blade is positioned on the caudal (posterior) surface about 1 to 3 mm away from the stump of the optic nerve and oriented in the desired plane (Figure 4A). A gentle sawing motion is used to penetrate the sclera and continued until the lens is engaged, at which point the cut is finished with a single firm, guillotine-like downstroke. A final sawing motion is sometimes necessary to sever the cornea completely. The dense fibrous tissues of the sclera and lens dull blades very quickly. Therefore, a fresh blade should be employed for each major cut to avoid producing substantial histological artifacts upon sectioning.

This initial cut will yield two approximately equal halves, or calottes. These hemispheres are generally still too large to process effectively, so the equatorial region must be separated to allow processing reagents to circulate through the specimen without the risk of bubbles becoming entrapped under the intact scleral cap. This second incision, made parallel to the first to isolate the equatorial region with the optic nerve, may be made with the calotte held in either the vertical or horizontal orientation. If the vertical position is

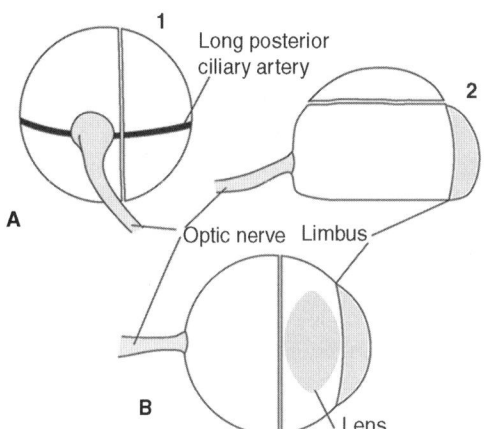

FIGURE 4 Schematic diagram for trimming large animal eyes. (A) Whole eye preparation (parasagittal blocking). (1) The globe is balanced on the corneal surface while held at the equator with thumb and forefinger (or forceps). The initial incision into the globe is made using a Tissue Slicer Blade (Thomas Scientific, Swedesboro, NJ) placed on the caudal (posterior) surface of the sclera 1 to 3 mm away from the optic nerve, depending on the size of the globe. A gentle sawing motion is used to penetrate the sclera and cut through the retina and vitreous humor, while a single firm guillotine-like downstroke is employed to complete the cut once the lens is engaged. (2) To isolate the center of the globe, a second equatorial incision is made parallel to the first, but on the opposite side of the optic nerve, using a flexible, double-edged carbon steel blade (Feather Blades, Ted Pella, Inc., Redding, CA). The detached calotte (half) is approximately 5 to 10 mm in width and must be placed into a deep cassette. (B) Retinal isolation (frontal blocking). For separating the caudal (posterior) pole for specific evaluation of the retina, a cross-sectional incision is made using a Thomas blade (or other very sharp blade) a few millimeters behind the limbus, which places the incision just behind the lens. (The lens is indicated in the diagram, but is not visible while making this incision.) The vitreous humor is then removed, and the retina can be carefully sectioned further as needed (plastic embedding, etc.).

selected, the calotte is held corneal surface down with a cork board placed against the flat (previously cut) surface to reduce the structural artifacts that commonly arise if insufficient support is not provided during a vertical second cut. Alternatively, a horizontal orientation may be selected (i.e., the flat surface of the calotte placed down on the benchtop; Figure 4A), but in many instances this approach will provide insufficient support for the ocular tissues once the blade engages the lens. In the authors' experience, the best instrument for this second cut is a flexible double-edged carbon steel blade (e.g., Feather Blades, Ted Pella, Inc., Redding, CA), which can incise the fibrous sclera with minimal pressure on the retina. A similar equatorial slice should be acquired from the other calotte if both sides of the globe are to be processed. For most globes large enough to prepare in this fashion, the trimmed calottes should be placed in extra-deep cassettes (Tissue-TEK Mega Cassette, Sakura Finetek USA, Inc., Torrance, CA) for processing.

To obtain whole eye sections with minimal anatomical distortion, large globes should be trimmed from the caudal (posterior) to the rostral (anterior) pole. The angle of the cut depends on the question and the species. In animals with a tapetum (e.g., cats, dogs, ferrets, ruminants), the initial cut should be in the vertical plane—perpendicular to the long caudal (posterior) ciliary artery—as the tapetum is located at the dorsal (superior) aspect of the globe. Sampling of the fovea centralis can be accomplished by sampling on the temporal (lateral) side of the optic nerve or by positioning the specimen at a more horizontal angle. For primates and other species without a tapetum, the trimmed calotte should typically be oriented in the horizontal plane so that the fovea will be available for examination. In rabbits, specimens should be positioned both vertically and horizontally (i.e., below the optic disc) so that the visual streak will be sampled across (cross section) and along (longitudinal sections) its axes, respectively.

In some instances, the most appropriate means of evaluating retinal morphology is to examine multiple regions of the retina from within a single large globe; this analysis is often performed by isolating retinal samples from the eye rather than processing the entire globe. For this preparation, an initial circumferential cut along the equator is made with a Thomas blade to open the eye. The cut should be placed at approximately the level of the pars plana (the widest part of the limbus), or just caudal to the position of the lens (Figure 4B), so that the lens will remain attached to the anterior segment and the risk of traumatic retinal detachment will be minimal. The caudal ocular wall (consisting of retina, choroid, and sclera) is restrained at one margin using a pair of serrated forceps; restraint is applied at only one margin to avoid altering the curvature of the retina, which can traumatize the retina. The vitreous humor is then gently removed, and the desired portions of the caudal ocular wall are acquired. A surgical scalpel blade (No. 10, 11, or 20) wielded with a handle often affords the most practical means for making the precise cuts needed to isolate small retinal samples. The margin of the ocular wall that was used to hold the sample should obviously be discarded.

Routine processing of whole globes should naturally incorporate a longitudinal section of the optic nerve trunk in the calotte. If feasible, a cross section of the optic nerve trunk should also be taken, although the portion near the severed end should be avoided, as it will probably include trauma-induced artifacts related to removal of the eye. In some instances, multiple cross sections at different levels along the nerve are harvested to provide more tissue for analysis. The optic chiasm and optic tracts may also be evaluated, either by specific dissection away from the brain or by including them in regional brain samples.

Processing

The choice of processing method will depend to some extent on the nature of the embedding medium to be used and the histology protocols employed in each laboratory. The general protocols used in facilities that prepare tissues from general toxicity studies will probably process eyes using the standard paraffin-embedding protocol, while those facilities that specialize in ocular research often use special processing conditions. For routine studies, both eyes of a single rodent may be processed in a single cassette, while for larger species each eye must be placed in a single cassette.

The following method is one example of an ocular-specific processing protocol used in our laboratories to process small eyes (e.g., rodents). Fixed specimens are placed, one per container, into 20-mL glass vials along with a paper label on which the sample identifying number is written in pencil (i.e., is not soluble in organic solvents). This label is kept in the vial throughout processing and can be used to identify the block during embedding. The vials are filled with dehydrating solutions (given below) and placed on a rocker or rotator at room temperature to facilitate penetration of the fluid into the sample. Dehydration consists of 70% ethanol (2 h) followed by two exchanges of 95% ethanol (for 1 h each). No further processing is required if globes are to be embedded in plastic, whereas one or two exchanges of xylene (1 h) may be employed if the sample is to be placed into paraffin. Typically, however, this laborious technique is not used unless specimens are to be embedded in plastic.

Embedding

The two media used to support ocular tissues during embedding are paraffin wax and plastic [e.g., glycol methacrylate (GMA) or methyl methacrylate (MMA)], both utilized according to standard procedures. Paraffin is suitable for clinical diagnostic applications, including routine screening for retinal and optic nerve lesions in general toxicity studies. Plastic is preferred by many laboratories that specialize in ocular research because it provides superior section quality and morphological preservation. The main advantage of plastic relative to paraffin is the ease of sectioning, which results from the similar firmness of the embedding medium and the hardest ocular tissues (lens and sclera). In contrast, paraffin is relatively soft, so eye sections tend to tear or fold during sectioning, as the microtome blade separates the hard and soft eye tissues and dislodges the harder tissues from the soft wax. An additional advantage of hard plastic resins is that they offer enough stability to permit the cutting of very thin tissue sections (less than 3 μm thick) with minimal folding and tearing. Such thin sections provide finer resolution of subtle cellular and subcellular features. Therefore, we recommend that plastic be used as the embedding medium for special neurotoxicity studies in which neuropathological lesions are anticipated in the retina and/or optic nerve; this medium should always be used if the optic nerve is to be examined in isolation from the globe. The plastic medium (e.g., Technovit 7100 GMA Kit, Energy Beam Sciences, East Granby, CT) should be reconstituted according to the manufacturer's instructions during the second 95% ethanol dehydration wash (see above) for optimal effect.

When embedding, eyes should be placed into the mold so that the globe will be sectioned in a midaxial plane that will include the optic nerve. If necessary, a magnifier should be used for small (e.g., rodent) eyes to ensure that the correct plane is attained. Paraffin will begin to harden quickly, so repositioning of the eye will usually not be required if this medium is used. However, both GMA and MMA are slower to polymerize, so specimens will need to be checked regularly to ensure that the desired orientation is maintained during the plastic polymerization process. Mounting chucks must not be placed onto plastic-mounted specimens too soon during polymerization, as eyes tend to roll in semisoft plastic when pressure is applied.

Sectioning

Histological sectioning of paraffin-embedded globes can be challenging, but with a little experience sections of acceptable quality can be produced regularly. In general, larger globes (e.g., carnivores, primates) should be faced down until the optic nerve is evident before sections are acquired, while small rodent globes may be step-sectioned to sample various portions of the retina and optic nerve. Strips of isolated ocular wall (retina with associated choroid and sclera) are not as challenging to section well, due to the absence of the lens. If globes are to be sectioned in situ, the skull must be positioned carefully in the block to optimize orientation of the retina and optic nerve.

For detailed morphologic assessment of retina, thin sections (2 to 3 μm in thickness, cut with a standard steel microtome blade) are ideal but are technically challenging to achieve with paraffin-embedded tissue. The divergent tissue densities of extraocular tissues and the vitreous humor relative to the lens and sclera are the biggest challenges to effective microtomy of paraffin-embedded eyes; these differences can be reduced by detaching all extraocular tissues from enucleated globes. The ease of sectioning can also be improved by cooling the paraffin block on an ice cube after initial facing, with or without applying 3 to 4 drops of 28% sodium hydroxide (to soften the lens and reduce shattering) to the surface of the cube.[3] Some histologists and pathologists advocate the intentional removal of the lens from the globe prior to embedding because its tendency to fall out of the block during sectioning can substantially disrupt the tissue morphology. For most evaluations of toxicant-induced retinal and optic nerve toxicological pathology, the risk of excessive tissue damage may be less important. However, in

the authors' opinion, the trauma of forcibly dislocating the lens may prevent the discovery of important anatomical damage (i.e., cataracts, lens luxations, lens capsule ruptures, etc.) and/or pathogenetic mechanisms, and thus should not be undertaken if an evaluation of the integrity of the entire globe is required. Adding a drop of Elmer's Glue (Elmer's Products, Inc., Columbus, OH) or poly-L-lysine (from a 0.1% w/v stock solution) to the water bath can reduce the likelihood that the lens will detach from the section during staining.[3]

Plastic sections of eyes can routinely be cut at thicknesses of less than 2 μm. Such fine sections must be acquired using a diamond or glass blade rather than a standard steel blade. In addition, the fine degree of control on sectioning speed required for such thin sections usually is best provided by a motorized microtome.

Staining

As for other tissues, the standard stain for evaluating eye tissues is hematoxylin and eosin (H&E). This stain provides a good view of the general architecture of the retina and optic nerve, as well as the other nonneural components of the globe, for both paraffin- and plastic-embedded tissue. Toluidine blue is often used as an ancillary stain, especially in plastic-embedded sections, to demonstrate more clearly the fine detail of subcellular organelles.

Common neurohistological methods may be used when evaluating the neural elements of the eye in paraffin sections. The integrity of axons and myelin in the optic nerve may be examined using silver stains (e.g., Bodian's) or Luxol fast blue (LFB), respectively. If desired, LFB can be followed by H&E to obtain added information from one section (see Figures 8 and 9).

ELECTRON MICROSCOPY

Principal techniques for transmission electron microscopic (TEM) examination of neural tissues are provided elsewhere in the book, so only a brief overview of their applicability to the retina is given here. Detailed protocols for TEM of ocular tissues are given elsewhere.[3,40]

In general, preparation of retinal tissue for TEM is performed only during studies that have been designed especially for this purpose.[3] Retinal preservation for TEM is best accomplished using a mixture of glutaraldehyde and paraformaldehyde in phosphate buffer (i.e., Karnovsky's solution) to cross-link proteins, followed by osmium tetroxide as a post-fixative to stabilize lipids. A fairly standard mixture is 0.8% glutaraldehyde and 1.2% paraformaldehyde in 0.1 M Sorenson's phosphate buffer (pH 7.2), fixing the entire globe for at least 2 to 3 h at 4°C prior to further trimming. Next, the globe is incised circumferentially just caudal to the lens, as described previously (Figure 4), and cut into 1-mm^3 cubes that incorporate the entire retina and choroid. The tissue is then fixed again in the primary aldehyde-based fixative at 4°C for at least 4 h, followed by two to three washes (15 min each) in phosphate-buffered saline (PBS), pH 7.2. Specimens are post-fixed in 1 to 2% osmium tetroxide (OsO$_4$) in 0.1 M phosphate buffer (pH 7.2) for from 2 h to overnight at 4°C, followed by two to three more PBS washes. Specimens are then stained en bloc in 2% uranyl acetate for 1 h at room temperature, washed twice for 10 min in acetate buffer, and then dehydrated through graded ethanol [10 min each in 60, 80, 95, and 100% at room temperature (RT)]. Finally, the samples are incubated in propylene oxide twice for 10 min at RT before infiltrating for 24 h each (on a rotator or rocker) in a 1:1 mixture of 100% propylene oxide and hard plastic resin (e.g., Epon) followed by 100% plastic resin. The resin is hardened by curing for 24 h at 70°C. A thick section (1 μm in thickness) is acquired and stained with toluidine blue to confirm that the orientation of the ocular sample permits evaluation of all desired structures. Finally, a definitive (80- to 100-nm-thick) thin section is obtained and stained with uranyl acetate for analysis.

An additional method for accentuating myelin damage in optic nerves and tracts is to employ *p*-phenylenediamine (PPD) as a supplemental step during processing.[42,43] The PPD stain is made as a 1% solution in 70% alcohol and then applied to OsO$_4$-post-fixed tissues either *en bloc* before embedding or after embedded (paraffin or plastic) tissues have been sectioned.[42] The PPD chelates unprecipitated osmium (resulting in pallor) everywhere except degenerating neural processes, which preferentially bind osmium (yielding a heightened dark profile over sites of damage).[43]

QUANTITATIVE ANALYSIS

Morphometric and stereologic analyses for quantitative differences in retinal and optic nerve morphology are complex, and thus typically fall outside the scope of general toxicologic neuropathology investigations. Basic concepts for quantitative analysis of neural tissues are presented elsewhere. Specialized fixation and sectioning techniques for quantitative analysis of retinal neuron numbers[44,45] and optic nerve axons[45] are detailed elsewhere.

Briefly, the key to effective quantitative analysis is to ensure that equivalent specimens are taken for all individuals. In the retina, samples should be acquired at the fovea centralis (or visual streak in rabbits, i.e., where the ganglion cell layer is thickest) and, if warranted, at other sites located at a uniform distance from this focus. Optic nerve specimens should be taken at a standard distance from the sclera. All specimens should be handled in an identical fashion during tissue acquisition and processing to minimize the impact of artifacts on the analysis.

OCULAR PROCESSING RECOMMENDATIONS FOR ROUTINE TOXICOLOGICAL NEUROPATHOLOGY STUDIES

Many options exist for processing eye specimens for ocular neuropathology assessment. Obviously, the method of choice will depend on the nature of the question being asked and the experimental design. In our experience, prospective neurotoxicity studies in laboratory animal species should generally employ the following practices to provide the optimal retinal and optic nerve specimens for neuropathology analysis.

Eyes may be removed by transpalpebral dissection or left in situ for fixation. Tissue preservation should be accomplished either by immersion in a special ocular fixative (Bouin's solution or Davidson's solution) or by whole-body perfusion fixation using an aldehyde fixative (e.g., NBF). For screening studies, eyes and the attached optic nerves should be fixed for at least 48 h after removal from the skull and detachment of extraneous fascia, after which they should be trimmed (if necessary) to isolate the calottes and processed into paraffin. A single H&E-stained section is then obtained for the initial retinal and optic nerve analysis. Follow-up studies to further characterize potential retinal or optic nerve lesions observed during a screening study should limit the length of fixation to 24 h (to permit the use of immunohistochemical or other special histology methods), use a soft plastic embedding medium, and include, at minimum, serial sections stained with H&E and toluidine blue.

PRINCIPLES OF OCULAR TOXICOLOGICAL NEUROPATHOLOGY

In this section we illustrate fundamental principles of ocular neuropathology and ocular neurotoxicity, using anatomical subdivisions to promote clarity. The histological appearance of typical retinal and optic nerve lesions will be highlighted, and, where possible, the induction of such lesions will be linked to specific ocular toxicants.

Fundamental Principles of Ocular Pathology and Toxicology

Toxicant Delivery to Ocular Neural Tissues Ocular toxicity may result from either local or systemic delivery of toxicants. As with other organ systems, the impact of toxicants on ocular neural tissues is based on agent-specific pharmacokinetic and pharmacodynamic parameters.

Topical absorption of ocular neurotoxicants may occur through the cornea, the conjunctiva, the lacrimal system, or the sclera. Nonionized lipophilic molecules penetrate these structures more readily. Their entry across the cornea may be further enhanced by the addition of preservatives or surfactants that disrupt the surface epithelium. The bioavailability to the internal ocular tissues, however, is estimated at 1 to 10% of the topical dose.[46,47] Exposure of the retina to topically administered compounds is further limited by aqueous humor flow; by physical barriers imposed by the ciliary body, iris, and lens; and by enzymatic systems for metabolism distributed in the various ocular tissues.

Substantial systemic exposure to xenobiotics can occur through either the uveal or retinal circulation. The eye has a rich vascular supply relative to its small mass, and both the retina and the choroid (a portion of the uvea) are amply fed by large arterial branches. However, vascular delivery of xenobiotics to the retina is limited by the blood–ocular barrier, an obstacle resembling the blood–brain barrier of the CNS. The blood–ocular barrier consists of the tight junctional complexes (zonula occludens) between epithelial cells of the RPE and between endothelial cells of the retinal capillaries. Passive transfer across the blood–ocular barrier is generally limited to lipid-soluble substances of low molecular weight ($< 500\,kDa$),[48] although molecules of higher weight may enter readily if the integrity of retinal vessels has been compromised by the effects of vasoactive compounds or by existing vascular disease (e.g., perivascular inflammation). On the other hand, the endothelial lining of the choroidal capillaries is fenestrated, thereby allowing larger molecules to penetrate the choroid with ease. Agents accumulating within the choroid can reach the retina by crossing Bruch's membrane. In principle, ocular exposure may also occur by transfer of xenobiotics from the brain into the eye along the optic nerve.

Intravitreal injections offer greater opportunity for experimental evaluations of retinal toxicity, although in clinical practice this route of exposure is considered relatively safe. The aqueous fraction of the vitreous humor offers no barrier to exposure of the ocular neural tissues. Therefore, the main advantages that recommend this route for toxicologic neuropathology studies in the eye is that both the retina and optic nerve head are directly exposed to the potential toxicant, and the exact dose of the agent can be determined with great precision.

Physiological Attributes Affecting Toxicant Availability in Intraocular Tissues

Toxicity to the sensory retina is dependent on multiple factors. The exposure level, the frequency of exposure, the concentration of the toxicant (or its toxic metabolite) that reaches the target site, the extent of metabolism, and the rate of clearance all affect the degree to which a xenobiotic may induce ocular toxicity. The balance between metabolic pathways for bioactivation and detoxification represents one of the most critical factors that influence the ability of neurotoxicants to affect the eye. Cytochromes P450 and other phase I and phase II enzymes (i.e., alkaline phosphatase, monoamine oxidase, aryl hydrocarbon hydroxylase, etc.) are found in the corneal epithelium, conjunctiva, ciliary epithelium, RPE, and retina.[48,49] Although biotransformation may protect the eye through detoxification, these metabolic reactions also generate free radicals that can induce lipid peroxidation in cell membranes. The sensitive tissues of the retina contain high levels of long-chain polyunsaturated fatty acids that are particularly vulnerable to peroxidative damage given continual exposure to light and the high oxygen demand of the intraocular tissues.[49,50] Such factors contribute to natural degeneration of the retina, the progressive increase in cell death (such as that which occurs during ARMD), and ultimately, the loss of vision.[1,8,51] Short of cell death, lipid peroxidation in retinal cells leads to leaky membranes, loss of cellular homeostasis, and altered Na^+ and K^+ gradients, all of which are essential components of phototransduction that are necessary for normal visual function. Several antioxidant mechanisms, including glutathione, glutathione-coupled enzymes, superoxide dismutase, certain vitamins (C and E), and endogenous scavengers of free radicals, are also present in ocular tissues. However, the effectiveness of these antioxidant pathways is subject to depletion with age and is affected by concurrent exposure to multiple xenobiotics.[50,52] The simultaneous presence of two or more drugs in the eye may alter the normal metabolism and clearance of either or both compounds, thereby prolonging exposure to toxic metabolites; it may also result in reduced efficacy through steric or functional interference.

A unique factor that contributes to the sensitivity of ocular neural tissues is the photoactivation of retinal chromophores. The chemical reactions that underlie this process generate free radicals, promote lipid peroxidation, and cross-link proteins, especially in the outer segments of the photoreceptors.[53] Endogenous chromophores found in the retina include visual pigments (rhodopsins), heme protein, flavoproteins, and (within the RPE) melanin and lipofuscin. Some xenobiotics (e.g., allopurinol, amiodarone, phenothiazines) are toxic via their photosensitizing activity and its impact on the membranes of intraocular target cells.[48,53,54]

Many of the enzymatic mechanisms within the eye responsible for bioactivation and detoxification are known to be subject to genetic influences. Slower drug metabolism may result from mutations in the P450 system. Genetic polymorphisms may contribute to individual susceptibility to the toxic effects of certain drugs and account for variability in therapeutic response.[48] For example, enhanced susceptibility to retinal degeneration has been linked to induction of aryl hydrocarbon hydroxylase activity in the RPE in susceptible individuals.[49]

Exaggerated Pharmacology as a Mechanism of Ocular Neurotoxicity

Agonists or antagonists of signal cascades, enzyme inhibitors or activators, or molecules that induce allosteric modifications can produce toxicity if the impact at their active site is excessive. The multilaminar structure, the multitude of neurotransmitters active within the various layers, and the complexity of the modulators and mechanisms involved in processing and transmitting information from the photoreceptors to the optic nerve render the retina exquisitely susceptible to altered electrophysiological function, whether or not structural integrity is compromised. Such pharmacological effects may present as temporary visual disturbances rather than more lasting retinal toxicity, but permanent functional alterations in the retina can be induced at the molecular level without eliciting any corresponding structural lesions. The outer segments of the photoreceptors are frequent targets where exaggerated pharmacology can cause functional visual disturbance, probably by inhibition of Na^+/K^+-ATPase.[27] Abnormalities in dark adaptation, color vision, and visual acuity as well as scotomas (i.e., an area within the visual field of diminished vision) are manifestations of exaggerated pharmacology that are de facto indices of retinal toxicity. Such deficits can occur in the absence of anatomical evidence for photoreceptor degeneration, as has been shown with cardiac glycosides.[5] Transient visual disturbances also occur with systemic administration of sildenafil citrate, an inhibitor of phosphodiesterase-5 activity used to treat erectile dysfunction, due to the off-target inhibition of phosphodiesterase-6 within photoreceptors.[55] Permanent visual deficits in the absence of structural changes may occur after treatment with the antiepileptic drug vigabatrin, which irreversibly binds γ-aminobutyric acid (GABA) transaminase, thereby preventing retinal metabolism of this inhibitory neurotransmitter.[27]

Contribution of the Retinal Pigmented Epithelium to Retinal Neurotoxicity

Although not part of the sensory retina, the adjacent RPE plays several critical roles in sustaining retinal integrity and thus preserving visual function (reviewed by Mecklenburg and Schraermeyer[56]). Major functions of the RPE in preventing primary photoreceptor degeneration include maintenance and phagocytosis of senescent rod outer segments (ROS), upkeep of the blood–ocular barrier and extracellular ion gradients within the retina, amelioration of photooxidative damage, and the regulation of retinol metabolism.[27,57–59] Some RPE constituents (e.g., melanin) also quench deflected light, thereby enhancing visual acuity, and can bind some xenobiotics, thus reducing exposure to retinal target cells. These RPE roles are all vulnerable to both degenerative aging changes and toxic insult. Thus, loss of the RPE may result in secondary degeneration of retinal photoreceptors (especially the outer segments).

Maintenance of photoreceptor integrity is a primary responsibility of the RPE. The apical villi of RPE surround the ROS and scavenge the debris that accumulates during the diurnal cycle of ROS shedding. Phagosomes within RPE cells are transported to the basal compartment, where they fuse with lysosomes. The digestion by-products are then reused by the RPE cell in its own metabolism or released for recycling by the photoreceptor cells. The reprocessing capacity for the shed ROS diminishes with both age and oxidative stress, so the by-products gradually accumulate as lipofuscin, a pigmented complex of protein, lipids, carbohydrates, metals, and vitamin A aldehyde. Retention of lipofuscin eventually leads to a decline in RPE function while (as noted above) simultaneously acting as a hub for the formation of light-generated free radicals.[8,57,60]

Other primary tasks of RPE cells are to uphold the structures responsible for (1) maintaining the ionic gradients that permit the perception of light stimuli and (2) providing the blood–ocular barrier that protects the tissue that creates the ion gradient and is capable of perceiving light. The polarized distribution of proteins (especially ionic pumps and channels) within cell membranes controls ionic homeostasis, while the tight junctions between RPE cells provide the anatomical basis for the barrier needed to eliminate uncontrolled intercellular transport. Transepithelial movement of molecules and water between the sensory retina and the choriocapillaris is controlled largely through RPE receptors and membrane proteins. Normal adhesion of the retina to the RPE is in part regulated by Na^+ and K^+ pumps located in the apical portion of the RPE cells. These pumps control the flow of Na^+ and K^+ across the RPE cell membranes; in addition to producing electrical potentials, this flux actively adjusts ion concentrations and thus passively regulates water retention in the subretinal space (i.e., a potential cavity between the retina and RPE that can serve as a site for fluid accumulation).[27,58] Tight junctions in the RPE are pharmacologically sensitive. Toxicant-induced alterations in zonula occludens integrity (e.g., from sodium iodate exposure) can alter the permeability of the blood–ocular barrier in the absence of other retinal changes, thereby leading to retinal edema and eventually to the secondary degeneration of photoreceptors and/or other retinal cells.[61,62]

Another RPE role is to control retinol metabolism. Retinol is a major constituent of the molecular machinery in photoreceptor cells that is required for light perception. The RPE cells contain receptors and transport mechanisms for the uptake and storage of retinol from the blood and for the exchange of vitamin A analogs between the photoreceptors and RPE. Recycling of vitamin A analogs to the active 11-*cis*-retinaldehyde form is crucial to the visual cycle; this task is performed in the RPE.[63] Compounds that affect vitamin A transport or metabolism may result in functional alterations causing reduced contrast sensitivity and diminished dark adaptation.[27] Isotretinoin, an inhibitor of retinol dehydrogenase activity in the RPE,[64] is one such compound.

Melanin granules within the RPE fill several important roles that are essential to proper retinal function. Melanin binds and sequesters oxidizing agents such as metal ions (e.g., aluminum, lead) and polycyclic aromatic compounds.[60] Melanin also enhances retinal image processing and visual acuity by absorbing extraneous light and reducing scatter. Melanin can alter the pharmacological profile of xenobiotics within the eye, depending on the binding affinity of the compound for melanin. Melanin may enhance toxicity by acting as a sink, prolonging exposure of the retina through sustained release. The pathogenesis of chloroquine-induced ocular toxicity is at least partially due to the high-binding affinity of chloroquine for melanin, allowing for accumulation within the eye,[65] while melanin-bound phenothiazines serve as a focus for phototoxic retinopathy.[66] Conversely, melanin may reduce toxicity by tightly sequestering compounds until they can be metabolized and/or removed. All of these activities can generate free radicals and oxidative stress in melanin-containing cells, which may result in diminished melanin content, aberrant melanin function, or RPE cell loss.[54] Phenothiazines bound to RPE melanin cause pigment clumping and have also been associated with RPE proliferation.[65]

CATEGORIES OF LESIONS IN OCULAR TOXICOLOGICAL NEUROPATHOLOGY

Toxicological neuropathology in ocular tissues follows several main principles. First, lesions often occur in multiple regions of the retina simultaneously following toxicant exposure. For example, a primary lesion in one retinal lamina (e.g., ganglion cell necrosis in layer 7) may induce secondary lesions at other retinal sites (e.g., attenuation of the nerve fibers in layer 8 and/or axonal loss in the optic nerve trunk). A second major principle is that the timing of the lesion and its progression is critical. Secondary lesions may progress so rapidly that they cannot be distinguished from a primary toxic effect. For example, RPE hypertrophy associated with retinal detachment may be a primary consequence of toxicant-associated RPE damage or a secondary sequel to primary retinal detachment. A third basic concept is that toxicant-induced lesions affecting ocular neural tissues will eventually progress to full-thickness retinal atrophy with disorganization and formation of glial or chorioretinal scars, and may result in atrophy of the optic nerve as well. Therefore, efforts to identify and characterize the primary lesion and its mechanism and then differentiate them from secondary sequelae will require one or more carefully designed time-course studies in laboratory animals. A final fundamental principle is that ocular neurotoxicants generally affect both globes to an equal extent following systemic exposure. Unilateral exposure by topical or intravitreal administration usually results in toxicity only to the treated eye, because systemic absorption of agents applied in these fashions is negligible.[67]

Toxicant-Induced Lesions of the Retina

Detachment Retinal detachment (Figure 5) is a common processing artifact, but this lesion is induced by a handful of retinal toxicants. Some toxicants may induce retinal detachment in many species, such as benzoic acid.[68] Other toxicants may elicit the lesion in only a single species, such as ethylene glycol in cats[69] and hydroxypyridine-thione[70] and oxygen[71] in dogs. Three microscopic features have been used to differentiate between artifactual loosening of the retina and genuine detachments; real lesions are usually characterized by photoreceptor atrophy, subretinal deposits, and RPE hypertrophy. Photoreceptors begin to regress within 12 to 24 h of retinal detachment, starting as attenuation of the outer segments and eventually proceeding to complete structural dissolution beyond the outer limiting membrane (Figure 5). The rate of photoreceptor degeneration is dependent on the extent to which nutrients can reach the metabolically active neural cells in the retina; rabbits, which possess a fairly limited extrachoroidal retinal blood supply, exhibit degeneration quite rapidly after detachment.[72] Loss of the photoreceptor segments is at least partially reversible for many hours after detachment, because the atrophied photoreceptor cell bodies of the outer nuclear layer can rebuild the segments if the retina and RPE can regain contact. However, photoreceptor regeneration after longer periods of detachment (three or more days) may be incomplete for weeks.[73] The presence of a subretinal deposit represents definitive evidence that a detached retina is a true lesion, because matter cannot gain access to this potential space unless an active pathologic process has occurred. Such

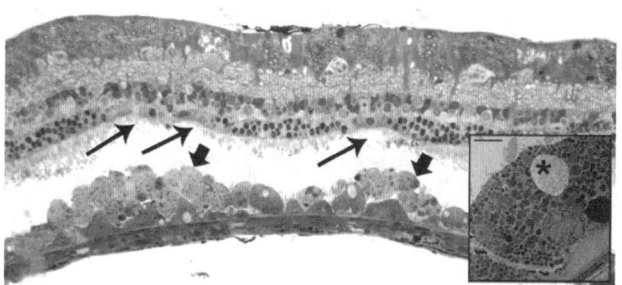

FIGURE 5 Spontaneous degeneration of the retinal pigmented epithelium (RPE) with retinal detachment in a rabbit. Degenerated and detached RPE are present in the subretinal space (thick arrows), corresponding to regions of photoreceptor loss in the segments and outer nuclear layer (thin arrows). The remaining RPE are hypertrophied and lack cellular detail. Enlarged vacuoles/inclusions (whose contents were lost during processing) are occasionally observed in the RPE. Plastic-embedded section, Richardson's stain, 200×. *Inset*: Transmission electron micrograph of a hypertrophic RPE cell containing numerous lysosomes, an inclusion (∗) containing lamellar material (thought to be undigested photoreceptor segments), and a paucity of melanosomes. Bar = 5 μm. (*See insert for color representation of the figure.*)

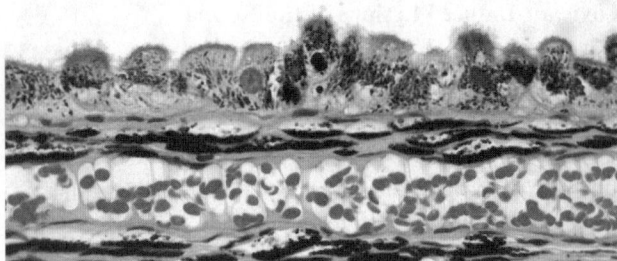

FIGURE 6 Hypertrophy of the retinal pigmented epithelium (RPE) in a dog. Note the swollen ("tombstone") appearance of the cells with apical melanosome dispersion. Apical localization of melanosomes may also occur with artifactual separation. Paraffin-embedded section, H&E, 400×. (*See insert for color representation of the figure.*)

subretinal accumulations may be eosinophilic, cell-free material (i.e., protein-rich fluid); exudates comprised of one or more leukocyte types; hemorrhage; degenerate RPE (Figure 5); or a combination of them all. Material may be presumed to have entered this space antemortem (rather than becoming impacted there during trimming or sectioning) if it fills the cavity and/or can be seen within adjacent tissues. Hypertrophy of the RPE that occurs within 12 h of retinal detachment is evident as swelling and rounding (i.e., "tombstone" conformation) of the cells (Figure 6). This change is distinguished from RPE "tenting" artifact (Figure 7) by the retention of photoreceptor outer segments attached to the angular apices of RPE cells in the latter instance. "Tenting" results from traumatic shearing of the retina during tissue trimming.

Lesions in Photoreceptors: Outer Segments and/or Nuclei

Alterations in photoreceptor numbers is a common finding associated with direct neurotoxicity to the retina. Routine methods (e.g., analysis of NBF-fixed, paraffin-embedded,

FIGURE 7 Artifactual "tenting" of the retinal pigmented epithelium (RPE) in a dog. Note the distinct angular apices of the RPE cells (arrows), compared to the rounded apices of RPE with genuine hypertrophy (Figure 6, same magnification). Fragments of photoreceptor segments remain attached to tented RPE cells at several points along the retina, which is a common occurrence with artifactual separation. Paraffin-embedded section, H&E, 400×. (*See insert for color representation of the figure.*)

FIGURE 8 Normal myelin distribution in the optic nerve of an adult dog. The bright blue staining for myelin is widespread. Paraffin-embedded section, Luxol fast blue/H&E, 100×. (*See insert for color representation of the figure.*)

H&E-stained tissue sections; Figures 8 and 9) reveal only overt changes, such as thinning (i.e., atrophy) of the photoreceptor outer segments in retinal lamina 1 and fewer photoreceptor nuclei (i.e., cell degeneration) in the outer nuclear layer (lamina 3) (Figures 5 and 10). More subtle lesions can be discerned with special processing methods (e.g., fixation in Davidson's solution followed by plastic embedding, or preparation for TEM analysis), including such abnormalities as membranous inclusions, retained outer segments, and vacuoles in photoreceptor cells.

Many toxicants injure the photoreceptors (Table 2). Overt ocular neurotoxicity to this target cell population occurs in

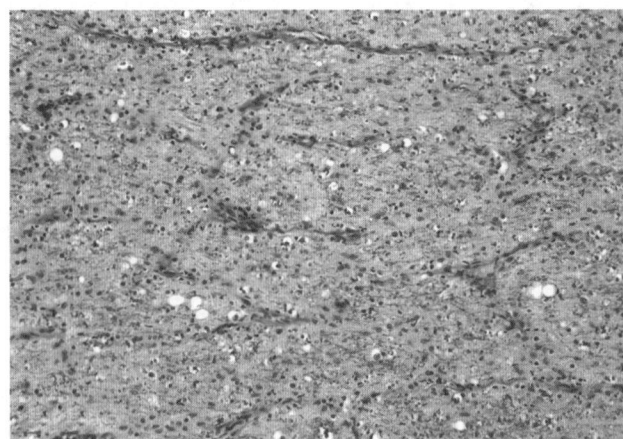

FIGURE 9 Optic nerve hypertrophy with demyelination and gliosis in an adult dog. Note the paucity of blue myelin staining and overall increase in cellularity (due to increased numbers of astrocytes) relative to the structure in control animals (Figure 8). Paraffin-embedded section, Luxol fast blue/H&E, 100×. (*See insert for color representation of the figure.*)

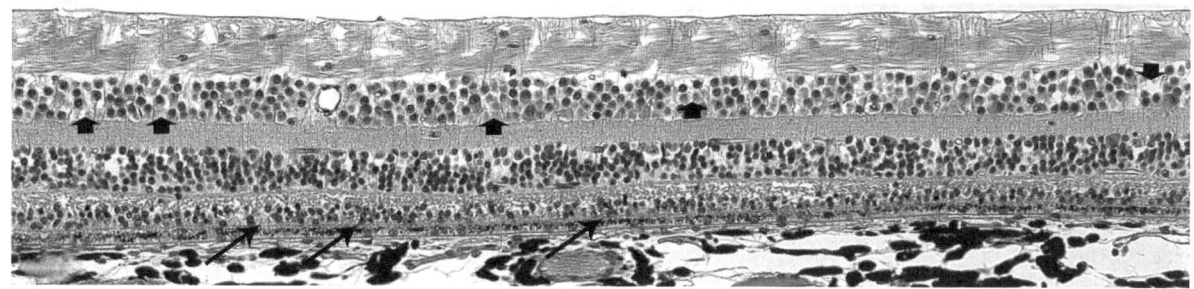

FIGURE 10 Photoreceptor atrophy in the retina of an adult nonhuman primate following intravitreal (IVT) administration of sodium iodate (NaIO₃, 1000 μg). Marked atrophy with complete loss of the photoreceptor outer segments is present (thin arrows). Nuclear loss has also occurred in the inner nuclear layer (thick arrows), indicating a reduction in several neuron populations that are involved in the initial processing of visual sensory input. Paraffin-embedded section, H&E, 200×. (*See insert for color representation of the figure.*)

sheep fed bracken fern (*Pteridium aquilinum*), where blindness is associated with fragmentation and degeneration of the photoreceptor outer segments.[5,74] Iodoacetate is employed experimentally to induce a rodent model of retinitis pigmentosa, causing structural damage to the photoreceptor membrane resulting from inhibition of glycolysis.[65] The same toxicants also generally damage other retinal cell types that are closely associated with photoreceptor activity, including the RPE and the neurons of the ganglion cell layer (lamina 7) and their processes in the nerve fiber layer (lamina 8). For example, 1,4-bis(4-aminophenoxy)-2-phenylbenzene (2-phenyl-APB-144) disrupts photoreceptor outer segments as well as the RPE.[75] Parenteral treatment with enrofloxacin, a fluoroquinolone antibiotic, has been reported to cause degeneration of the photoreceptors and the outer nuclear layer in cats.[76]

Lesions of Ganglion Cells and the Inner Retina Lesions in retinal ganglion cells (RGCs, the main neuronal population in lamina 7) are closely linked to the other main layers of the inner retina [inner nuclear layer (lamina 5), nerve fiber layer

(lamina 8)]. The RGCs are responsible for visual acuity and are the target of several known retinal toxicants (Table 3). A few toxicants preferentially target the inner nuclear layer, such as the cholinergic agonist ethylcholine mustard aziridinium ion (AF64A).[77] In particular, neurotransmitter-like agents have been suggested to exert greater adverse effects on bipolar, amacrine, and horizontal cells in this layer, rather than impacting RGCs.[5]

As in other neural tissues, acute neurotoxicant-induced changes observed in RGCs at the light microscopic level include cell swelling, central chromatolysis, and necrosis. Dying cells are characterized by cytoplasmic hypereosinophilia and nuclear fragmentation (pyknosis) or dissolution (karyolysis) (Figure 11). Electron microscopic evidence of RGC damage is also similar to that seen in other neural tissues and consists of cell and organelle swelling, chromatin clumping, and disrupted mitochondria. Drugs that cause generalized lipidosis may also induce the formation of lipid-filled lysosomal inclusions in many retinal cells. Chloroquine toxicity is one example, in which membranous inclusions within RGCs are thought to result from altered

TABLE 2 Selected Toxicants that Induce Lesions in Photoreceptors and the Outer Retina

Agent	Species Affected	Histopathological Lesions of the Outer Retina	References
Ammeline	Mouse	Photoreceptor degeneration (secondary to RPE[a] damage)	68
Closantel	Sheep	Acute photoreceptor necrosis	97
Colchicine	Rat	Photoreceptor degeneration	104
Ethylenamine 4,4'-diaminodiphenylmethane	Cat	Rod and cone atrophy	105
Piperidylchlorophenothiazine	Cat, human	Disruption of rod outer segments	68
Pteridium aquilinum (bracken fern)	Sheep	Degeneration of photoreceptor outer segments	74
Stypandra imbricata (blind grass)	Rat, sheep	Photoreceptor degeneration	106,107
Vincristine	Rat	Degeneration of photoreceptor inner segments	108

[a] RPE, retinal pigmented epithelium.

TABLE 3   Selected Toxicants That Induce Lesions in the Retinal Ganglion Cells and the Inner Retina

Agent	Species Affected	Histopathological Lesions of the Inner Retina	References
Amphotericin B	Rabbit	Ganglion cell loss	109
Arsanilic acid	Human	Ganglion cell degeneration	68
Carbon disulfide	Human, rabbit, rat	Ganglion cell degeneration	110,68
Chloramphenicol	Human	Ganglion cell loss	111
L-Cysteine	Rat	Ganglion and amacrine cell degeneration	112,113
Doxorubricin	Rat	Neurofilament accumulation in ganglion cells (early), ganglion cell loss (eventual)	114
Glutamate	Rat (neonate)	Ganglion cell degeneration	115
Kainic acid	Rat	Amacrine cell mitosis and pyknosis, edema of the inner plexiform layer	116
Lidocaine	Rat	Ganglion cell necrosis	117
Locoweed (*Astragalus mollissimus*)	Cattle, sheep	Ganglion and bipolar cell vacuolation	118
Quinine	Dog	Ganglion cell degeneration	5
Vincristine	Primate	Ganglion cell atrophy	119

protein synthesis and lipid peroxidation.[78] This change is most frequently associated with the systemic administration of amphiphilic cationic compounds, including antimalarial, antidepressant, anorectic, and hypocholesterolemic drugs.[59]

Exaggerated pharmacological activity rather than direct cytotoxicity may also result in the indirect loss of target neurons within the retina. For example, the loss of RGCs, a hallmark of many ophthalmic diseases, is attributed to glutamate receptor-mediated hyperactivity that induces an influx of extracellular calcium (Ca^{2+}) and the intracellular accumulation of sodium (Na^+) ions, eventually leading to widespread RGC death with atrophy of the optic nerve. Excitatory toxicity of retinal neurons can be experimentally induced with agonists that act as glutamate analogs,[79,80] and has been suggested to be one of the mechanisms responsible for ethambutol-associated ocular toxicity.[81] The glutamate homolog DL-α-aminoadipate (DL-AAA) preferentially causes necrosis of Müller cells rather than neurons,[65,82,83] although neurons of the inner nuclear layer are also affected over time. This time course suggests that primary neurotoxic activation of the Müller cells (i.e., retinal glia) can contribute to secondary loss of neurons in the inner retina.[84] An additional mechanism of DL-AAA retinal toxicity is blood vessel proliferation, leading some investigators to employ this agent as a possible animal model of ocular neovascularization (Figure 12).

Lesions of the Intraocular Vasculature In general, systemic or localized vascular lesions will induce retinal pathology resulting from hypoxia. The most common presentations are neuronal atrophy or necrosis. The extent of the retinal neuropathology depends on the severity (partial or total) and length of the hypoxic insult. Vascular-based necrotic lesions tend to have a less selective, though often focal/segmental, distribution involving multiple retinal layers or retinal cell types. The predominant portion of the retina affected (inner vs. outer) depends largely on the vessels affected (retinal vs. choriocapillaris). Focal retinal necrosis is characterized by loss of cell detail, cytoplasmic hypereosinophilia, and nuclear pyknosis.

Several agents cause primary histologic lesions in the retinal blood vessels (Table 4). These changes typically manifest as vessel narrowing, which in severe instances can lead to occlusion. However, dilation or engorgement can also occur. Ergotamine causes constriction of retinal arterioles in humans, and barbiturates can also cause retinal vasoconstriction.[5] Oxygen toxicity may produce vasoconstriction in humans and rats.[5,85] Vasodilation and/or engorgement have

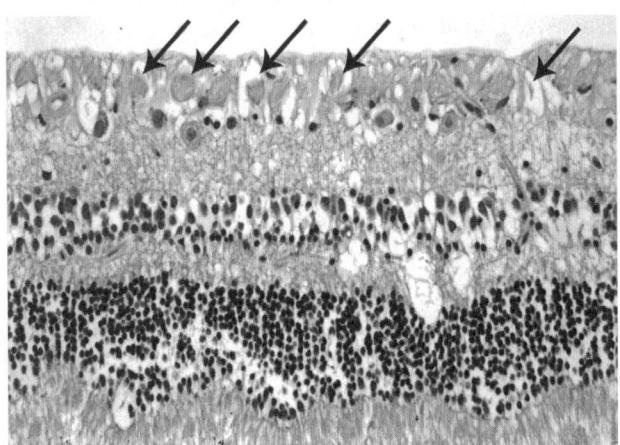

FIGURE 11 Loss of retinal ganglion cells in the dog. Note the almost total lack of "kite-shaped" ganglion cells (neurons) in their normal location in the inner retina (arrows). The few remaining neurons in this location have hypereosinophilic cytoplasm (an early change in degenerating cells). Other retinal layers have been spared. Paraffin-embedded section, H&E, 400×. (*See insert for color representation of the figure.*)

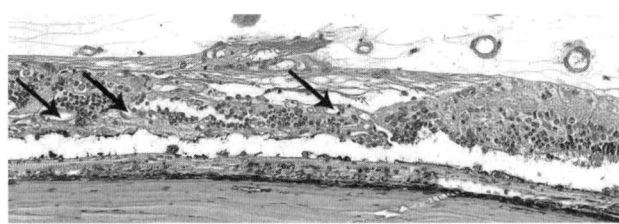

FIGURE 12 Neovascularization of the retina in a Dutch Belt rabbit following intravitreal (IVT) administration of 100 μl of 0.025 M DL-α-aminoadipate (DL-AAA). Profiles of new blood vessels are present at the inner retina surface and within the retina (arrows) in association with necrosis and loss of the laminar organization. Paraffin-embedded section, H&E, 200×.

been reported following exposure to amyl nitrate, carbon dioxide, and methanol.[5] Vascular proliferative lesions can be induced with intravitreal (IVT) delivery of angiogenic factors, such as basic fibroblast growth factor (bFGF) and vascular endothelial growth factor (VEGF) in primates[86] and rabbits.[87] Intraocular injection of VEGF also induces dilation and leakage of retinal vessels and, in rabbits, neovascularization;[88,89] repeated IVT injection of VEGF is now a recognized model of ARMD in drug development programs (Figure 13). Proliferation of retinal vessels may also be an eventual consequence of prolonged exposure to other retinal toxicants such as naphthalene.[90] An unusual finding in the retinal vasculature of humans associated with canthaxanthin exposure (a naturally occurring carotenoid pigment) is deposition of red or yellow birefringent crystals in the macula, probably due to interactions with lipids located in the retinal vessel walls; this damage may lead to neovascularization.[91] Perimacular deposits of refractile crystals related to axonal degeneration have been reported in humans given tamoxifen.[65,92]

Retinal hemorrhage associated with lesions of the vascular wall is variable depending on the nature and extent of the vascular injury. Hemorrhage is a straightforward diagnosis as it consists of extracellular red blood cells within the retinal parenchyma, posterior vitreous humor, and/or subretinal space. Care must be taken when interpreting the source of hemorrhage near the retina, as erythrocytes may be present in the vitreous or subretinal space from extraretinal sources as

well (choroid, ciliary body). With chronicity, hemosiderin-laden macrophages may accumulate. These phagocytic cells contain irregularly shaped, golden brown pigment granules that are distinctly different from the relatively uniform, oblong black melanin granules of the RPE. Obvious toxic causes of retinal hemorrhage include warfarin and related anticoagulants.[5] Agents causing hyperviscosity, such as dextrans, have also been associated with retinal hemorrhage in primates.[93]

Blood vessel proliferation alone can be a vision-threatening lesion, even in the absence of primary toxic damage to the retinal neurons and retinal vessel walls. Some molecules are capable of inciting retinal neovascularization as a primary change, such as bFGF and VEGF (Table 4). Indeed, several VEGF inhibitors have been approved to treat the exudative ("wet") form of ARMD. As stated earlier, reliable interpretation of vascularity in the absence of definite lesions in the vessel walls requires familiarity with the normal vascularization schema of the species being studied.

End-Stage Retinal Atrophy Due to the interconnectedness of retinal neurons, most toxic insults to the eye will eventually lead to end-stage (full thickness) retinal atrophy. This lesion is typified by marked reduction in retinal thickness, loss of layer stratification, disorganization of the remaining cells, and a variable degree of reactive glial proliferation (i.e., a "glial scar"; Figures 14 to 16). This end-stage lesion can develop and progress rapidly in some instances, so assessment of earlier time points may be required to determine the initial site of retinal injury when characterizing animal models of progressive retinal disease.

Toxicant-Induced Lesions of the Retinal Pigment Epithelium

The high metabolic activity of the RPE renders it exquisitely sensitive to several ocular toxicants, including fluoride, lead, naphthalene, and methanol (reviewed by Mecklenburg and Schraermyer[56]; Table 5). Typical primary responses by injured RPE cells include hypertrophy (swelling), loss of cell polarity, depigmentation, necrosis, exfoliation into the subretinal space, and membrane deposits (Figure 5). These changes may occur individually or together. Many of these

TABLE 4 Selected Toxicants That Induce Lesions in the Retinal Blood Supply

Agent	Species Affected	Histopathological Lesions of Ocular Vessels	References
2-Butoxyethanol	Rat	Retinal hemorrhage	120
Gentamicin	Pig, primate	Capillary narrowing, hemorrhagic necrosis, leukocyte plugging, thrombosis	121
3,3-Iminidopropionitrile	Rat	Hemorrhage	122
Oxygen (hyperbaric)	Rat	Hemorrhage, neovascularization	123
Prostaglandins	Rabbit	Hemorrhage, leakage, occlusion	124

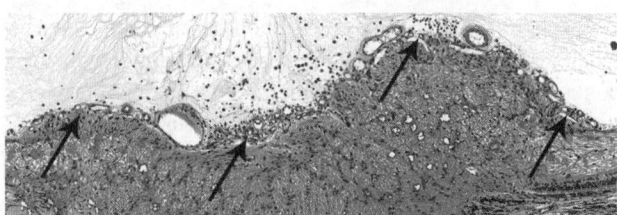

FIGURE 13   Neovascularization and inflammation of the optic nerve head in an adult pigmented rabbit associated with biweekly intravitreal (IVT) administration of 0.5 μg of recombinant human vascular endothelial growth factor (rhVEGF)-165. Profiles of new blood vessels (arrows) are present on the surface of the optic disc. The vitreous humor contains a dispersed, mixed leukocytic infiltrate. Paraffin-embedded section, H&E, 100×.

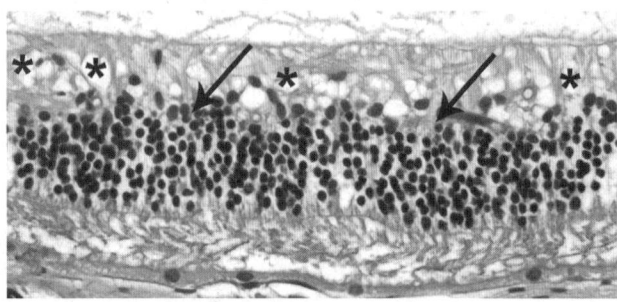

FIGURE 14   Advanced atrophy of the inner retina in a dog. Note the thinning of the nerve fiber layer (top), absence of retinal ganglion cells (∗), and fusion of the inner and outer nuclear layers (arrows) relative to the normal retina (Figure 1) and a "limited" lesion consisting only of ganglion cell loss (Figure 11). The outer retina (photoreceptor segments) is relatively spared. Paraffin-embedded section, H&E, 400×.

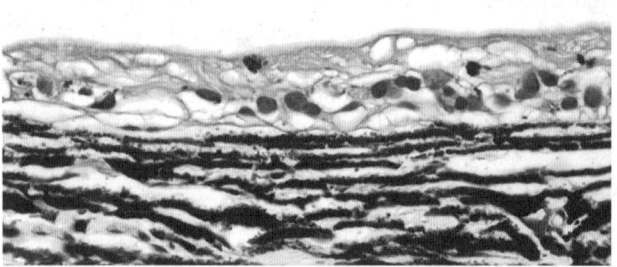

FIGURE 15   End-stage retinal atrophy in a dog. The entire retina is markedly thin, and no identifiable layers are evident (compared to Figure 14, at the same magnification). Scattered pigment [representing melanosomes from the retinal pigmented epithelium (RPE)] is apparent within the remaining retina. Paraffin-embedded section, H&E, 400×.

changes have several potential causes, often discernible at the ultrastructural level. For example, RPE hypertrophy may result from organelle hypertrophy, accretion of lysosomes or phagosomes, or the accumulation of materials such as crystals, lipofuscin, or other inclusions (Figure 5). Toxicant-induced RPE necrosis is characterized by cell swelling, cytoplasmic hypereosinophilia, and nuclear degeneration; RPE melanosomes may also be released into the subretinal space.

The RPE can also develop secondary changes in response to primary retinal damage (e.g., photoreceptor degeneration, retinal detachment). For example, 1,4-bis(4-aminophenoxy)-2-phenylbenzene (2-phenyl-APB-144) has been shown by light microscopy to disrupt photoreceptor outer segments and elicit RPE necrosis within 1 to 2 days.[75] Early ultrastructural changes in RPE cells include mitochondrial swelling, vesiculation of the smooth endoplasmic reticulum, and increased cytoplasmic density, followed by vacuoles and electron-dense granules in photoreceptor outer segments. Over time, RPE necrosis becomes more widespread, photoreceptor disruption intensifies, and abnormal elements such as lysosomal inclusion bodies, myelin bodies, and disintegrating outer segment lamellae begin to collect in RPE cells (Figure 5).

Toxicant-Induced Lesions of the Tapetum Lucidum

Although not part of the retina per se, tapetal lesions have been reported to occur with several retinal toxicants, often in conjunction with other retinal lesions (detachment, edema, etc.). The two most commonly reported lesions are edema and necrosis. These lesions have been described in dogs following exposure to hydroxypyridinethione,[70] pyridine-thiol,[94] and zinc pyridinethione.[95] The latter two agents have been implicated in tapetal atrophy in cats as well.[95]

Toxicant-Induced Lesions of the Optic Nerve

The optic disc or nerve may be the primary target of ocular toxicity (Table 6), and insults to these sites may result in visual disturbances, inflammation, and/or atrophy of the optic nerve. Ethambutol-induced ocular toxicity presents mainly as a primary retrobulbar neuritis, with dose-dependent central visual deficits.[65] Methanol-associated visual toxicity in primates, but not rodents, is attributed to metabolic acidosis from the formation of the toxic metabolite formic acid, a mitochondrial poison, leading to loss of RGC and eventual optic nerve atrophy secondary to axonal loss and demyelination.[65,68,96] Atrophy, demyelination, and fibrosis of the optic nerve secondary to necrosis of the photoreceptors have been reported following exposure to closantel, a halogenated salicylanide.[97] Hexachlorophene causes axonal degeneration and secondary demyelination of the optic nerve.[68] In humans, optic neuritis from drug-related toxicity

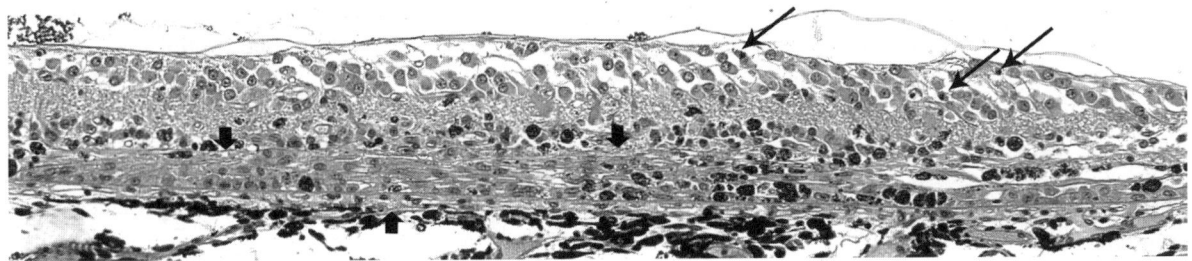

FIGURE 16 Retinal scarring and atrophy in a nonhuman primate following intravitreal (IVT) administration of sodium iodate (NaIO$_3$, 1000 µg). The outer layers of the retina and retinal pigmented epithelium (RPE) have been replaced by a linear array of cells (differentiated RPE cells) that form a plaque (thick arrows). The adjacent ganglion cell layer is sparsely populated and contains numerous apoptotic cells (thin arrows). Paraffin-embedded section, H&E, 200×.

frequently involves ganglion cell fibers within the papillo-macular bundle while sparing the peripheral fibers, suggesting that some xenobiotics have a selective, site-specific affinity for some component of centrally located RGC axons.[5,65]

Axonal Degeneration   Acute injury to optic nerve axons presents histologically with structural changes similar to those found in other CNS white matter tracts following neurotoxicant exposure. Typical lesions include axonal or

TABLE 5   Selected Toxicants That Induce Lesions of the Retinal Pigmented Epithelium

Agent	Species Affected	Histopathological RPE Lesions	References
Aluminum chloride	Rat	Thinning	125
2'3'-Dideoxyinosine	Human	Altered thickness (atrophy or hypertrophy), hypopigmentation, loss of choriocapillaris	126
Lead	Rabbit	Cell swelling, lipofuscin accumulation	127
Methanol	Human, nonhuman primate, rat	Vacuolation	102
Methoxyflurane	Human	Deposition of oxalate crystals	128
N-Methyl-N-nitrosourea	Hamster	Degeneration	129
Phenothiazines	Cat, human	Cell swelling, accumulation of lipofuscin and melanolysosomes; eventual atrophy	5,130
Quinolones	Cat, rabbit	Myeloid bodies (membranous phospholipid inclusions)	76,131
Sodium iodate	Mouse	Degeneration and necrosis	61
Tricyclic antidepressants	Rat	Lipidosis	132

TABLE 6   Selected Toxicants That Induce Lesions in the Optic Nerve

Agent	Species Affected	Histopathological Optic Nerve Lesions	References
Amiodarone	Human	Papilledema	100,101
Closantel	Sheep	Papilledema (optic disc); degeneration leading to fibrosis, secondary to compressive neruopathy (optic nerve)	97
Ethambutol	Primate, rat	Optic neuritis (especially at the chiasm)	133,134
Formaldehyde	Rabbit	Vacuolation	135
Hexachlorophene	Rat	Axonal degeneration, demyelinization	5
Methanol	Human, rats	Mitochondrial swelling (by electron microscopy)	102,136
Oxaliplatin	Human	Papilledema	103
Plasmocid	Cat, dog, human, rabbit	Optic neuritis leading eventually to atrophy	68
Stypandra imbricata (blind grass)	Goat, rat, sheep	Axonal degeneration and swelling, myelin vacuolation	106,107
Vincristine	Primate	Optic nerve atrophy, demyelinization, gliosis	119

axon sheath swelling, which is followed in time by secondary demyelination (Figure 9) and infiltration by macrophages (gitter cells). These changes have been reported following intravitreal tumor necrosis factor-alpha (TNFα) treatment in rats.[98]

Inflammation Infiltration with leukocytes is not the main mode by which most agents act, but inflammation is a consequence of exposure to several ocular toxicants. Features of optic neuritis include an influx of inflammatory cells (generally a mixed population), possibly with appreciable edema. Agents reported to cause confirmed optic neuritis include closantel in dogs[99] and plasmocid in humans, cats, dogs, and rabbits.[68] Some reports of toxicant-induced optic neuritis are based not on histological features but rather on clinical evidence [e.g., decreased visual acuity, decreased color perception, and field deficits (e.g., from scotomas, zones of impaired acuity surrounded by regions with less impaired or normal perception)], particularly in humans, with visual improvement following drug discontinuation.[66]

Atrophy Chronic injury to the optic nerve leads to regression of the tract as a final common response to many types of more acute injuries, including primary axonal injury, optic neuritis, or hypoxia/ischemia. Time-course studies are necessary to define the earliest lesions and to ascertain the underlying cause. The two most common manifestations of this end-stage response are axonal dropout and gliosis (Figure 9), and they typically occur in tandem. The hallmark microscopic feature of optic nerve atrophy is a relative increase in the prominence of the pial trabeculae (the thin bands of collagen, arising from the pia mater, that invest the optic nerve) due to the extensive loss of axons. Species differences in the structure of the lamina cribrosa are important when interpreting the extent of atrophy near the site where the optic nerve passes through the sclera. Both a relative and absolute increase in glial cells accompanies chronic atrophy; the former is due to the extensive loss of myelin and axonal substance, while the latter results from reactive proliferation of glial cells. Because optic nerve atrophy is a chronic response, it is important to determine the progenitor lesions in toxicological studies. For example, agents that cause RGC atrophy or necrosis will eventually lead to optic nerve atrophy. The reverse process (nerve atrophy leading to RGC lesions) is also possible, but retrograde neuronal effects typically take much longer to develop.

Papilledema Lesions of the retina or extraocular optic nerve can affect the optic nerve head. One unique lesion of the optic nerve head is papilledema, the collection of fluid within the head. In this lesion, the nuclear layers of the retina adjacent to the optic nerve head will be pushed peripherally. This finding may be the result of local edema or a manifestation of generalized CNS edema. Ocular toxicants associated with this change include amiodarone,[100,101] closantel,[97] methanol,[96,102] and oxaliplatin[103] (Table 6).

Acknowledgments

The authors thank Oded Foreman (The Jackson Laboratory, West, Sacramento, CA) for technical assistance with ocular embedding techniques and John Doval (University of California–Davis) for preparing the schematic diagrams for ocular trimming.

REFERENCES

1. Guymer R, Bird AC. Bruch's membrane, drusen, and age-related macular degeneration. In: Marmor MF, Wolfensberger TJ, eds. *The Retinal Pigment Epithelium*. New York: Oxford University Press; 1998:693–705.

2. Guymer R, et al. Changes in Bruch's membrane and related structures with age. *Prog Retin Eye Res*. 1999;18:59–90.

3. Smith RS, et al, eds. *Systematic Evaluation of the Mouse Eye: Anatomy, Pathology, and Biomethods*. Research Methods for Mutant Mice Series. Boca Raton, FL: CRC Press; 2002.

4. May CA, Lütjen-Drecoll E. Morphology of the murine optic nerve. *Invest Ophthalmol Vis Sci*. 2002;43:2206–2212.

5. Millichamp NJ. Toxicity in specific ocular tissues. In: Chiou GCY, ed. *Ophthalmic Toxicology*. Ann Arbor, MI: Taylor & Francis; 1999: 43–87.

6. Ollivier FJ, et al. Comparative morphology of the tapetum lucidum (among selected species). *Vet Ophthalmol*. 2004;7:11–22.

7. Friberg TR. Examination of the retina: ophthalmoscopy and fundus biomicroscopy. In: Albert DM, Miller JW, eds. *Albert & Jakobiec's Principles and Practice of Ophthalmology*. Philadelphia: W.B. Saunders; 2008: 1677–1688.

8. Bressler SB, et al. Age-related macular degeneration: drusen and geographic atrophy. In: Albert DM, Miller JW, eds. *Albert & Jakobiec's Principles and Practice of Ophthalmology*. Philadelphia: W.B. Saunders; 2008: 1901–1916.

9. Bennett TJ, et al. Principles of fluorescein angiography. In: Albert DM, Miller JW, eds. *Albert & Jakobiec's Principles and Practice of Ophthalmology*. Philadelphia: W.B. Saunders; 2008: 1689–1704.

10. Novotny HR, Alvis DL. A method of photographing fluorescence in circulating blood in the human retina. *Circulation*. 1961;24:82–86.

11. Hassenstein A, Meyer CH. Clinical use and research applications of Heidelberg retinal angiography and spectral-domain optical coherence tomography: a review. *Clin Exp Ophthalmol*. 2009;37:130–143.

12. Sakata LM, et al. Optical coherence tomography of the retina and optic nerve: a review. *Clin Exp Ophthalmol*. 2009;37:90–99.

13. Altaweel MM, Johnson DL. 2008. Optical coherence tomography. In: Albert DM, Miller JW, eds. *Albert & Jakobiec's*

Principles and Practice of Ophthalmology. Philadelphia: W.B. Saunders; 2008:1725–1740.

14. Anger EM, et al. Ultrahigh resolution optical coherence tomography of the monkey fovea: identification of retinal sublayers by correlation with semithin histology sections. *Exp Eye Res.* 2004;78:1117–1125.

15. Jaffe GJ, Caprioli J. Optical coherence tomography to detect and manage retinal disease and glaucoma. *Am J Ophthalmol.* 2004;137:156–169.

16. Ogden TE. Clinical electrophysiology. In: Ryan SJ, ed. *Retina.* Philadelphia: Mosby; 2006: 351–372.

17. Breton ME, et al. Analysis of ERG a-wave amplification and kinetics in terms of the G-protein cascade of phototransduction. *Invest Ophthalmol Vis Sci.* 1994;35:295–309.

18. Hood DC, Birch DG. The a-wave of the human electroretinogram and rod receptor function. *Invest Ophthalmol Vis Sci.* 1990;31:2070–2081.

19. Bush RA, Sieving PA. A proximal retinal component in the primate photopic ERG a-wave. *Invest Ophthalmol Vis Sci.* 1994;35:635–645.

20. Rosolen SG, et al. Retinal electrophysiology for toxicology studies: applications and limits of ERG in animals and ex vivo recordings. *Exp Toxicol Pathol.* 2008;60:17–32.

21. Miller RF, Dowling JE. Intracellular responses of the Müller (glial) cells of mudpuppy retina: their relation to b-wave of the electroretinogram. *J Neurophysiol.* 1970;33:323–341.

22. Dong CJ, Hare WA. Contribution to the kinetics and amplitude of the electroretinogram b-wave by third-order retinal neurons in the rabbit retina. *Vis Res.* 2000;40:579–589.

23. Dong CJ, et al. Origins of the electroretinogram oscillatory potentials in the rabbit retina. *Vis Neurosci.* 2004;21: 533–543.

24. Bee WH. Standardized electroretinography in primates: a non-invasive preclinical tool for predicting ocular side effects in humans. *Curr Opin Drug Discov Dev.* 2001;4:81–91.

25. Perlman I. Testing retinal toxicity of drugs in animal models using electrophysiological and morphological techniques. *Doc Ophthalmol.* 2009;118:3–28.

26. Rosolen SG, et al. Recommendations for a toxicological screening ERG procedure in laboratory animals. *Doc Ophthalmol.* 2005;110:57–66.

27. Zrenner E. The role of electrophysiology and psychophysics in ocular toxicology. In: Fraunfelder FT, et al., eds. *Clinical Ocular Toxicology.* Philadelphia: W.B. Saunders; 2008: 21–38.

28. Ohtaka K, et al. Protective effect of hepatocyte growth factor against degeneration of the retinal pigment epithelium and photoreceptor in sodium iodate–injected rats. *Curr Eye Res.* 2006;31:347–355.

29. Feigl B, et al. Objective functional assessment of age-related maculopathy: a special application for the multifocal electroretinogram. *Clin Exp Optom.* 2005;88:304–312.

30. Hare W, et al. Electrophysiological and histological measures of retinal injury in chronic ocular hypertensive monkeys. *Eur J Ophthalmol.* 1999;9(Suppl 1):S30–S33.

31. Hare WA, et al. Characterization of retinal injury using ERG measures obtained with both conventional and multifocal methods in chronic ocular hypertensive primates. *Invest Ophthalmol Vis Sci.* 2001;42:127–136.

32. Hare WA, Ton H. Effects of APB, PDA, and TTX on ERG responses recorded using both multifocal and conventional methods in monkey: effects of APB, PDA, and TTX on monkey ERG responses. *Doc Ophthalmol.* 2002;10: 189–222.

33. Knapp AG, Schiller PH. The contribution of on-bipolar cells to the electroretinogram of rabbits and monkeys: a study using 2-amino-4-phosphonobutyrate (APB). *Vis Res.* 1984;24:1841–1846.

34. Hare WA, et al. Efficacy and safety of memantine treatment for reduction of changes associated with experimental glaucoma in monkey: II. Structural measures. *Invest Ophthalmol Vis Sci.* 2004;45(8):2640–2651.

35. Rangaswamy NV, et al. Photopic ERGs in patients with optic neuropathies: comparison with primate ERGs after pharmacologic blockade of inner retina. *Invest Ophthalmol Vis Sci.* 2004;45(10):3827–3837.

36. Brigell M, et al. An overview of drug development with special emphasis on the role of visual electrophysiological testing. *Doc Ophthalmol.* 2005;110:3–13.

37. Almoqbel F, et al. The technique, validity and clinical use of the sweep VEP. *Ophthalmic Physiol Opt.* 2008;28:393–403.

38. Glickman RD, et al. Noninvasive techniques for assessing the effect of environmental stressors on visual function. *Neurosci Biobehav Rev.* 1991;15:173–178.

39. Ver Hoeve JN, et al. VEP and PERG acuity in anesthetized young adult rhesus monkeys. *Vis Neurosci.* 1999;16:607–617.

40. Smith RS, Rudt LA. Ultrastructural studies of the blood-aqueous barrier: 2. The barrier to horseradish peroxidase in primates. *Am J Ophthalmol.* 1973;76:937–947.

41. Fix AS, Garman RH. Practical aspects of neuropathology: a technical guide for working with the nervous system. *Toxicol Pathol.* 2000;28:122–131.

42. Ledingham JM, Simpson FO. Intensification of osmium staining by *p*-phenylenediamine: paraffin and epon embedding; lipid granules in renal medulla. *Biotech Histochem.* 1970;45:255–260.

43. Sadun AA, et al. Paraphenylenediamine: a new method for tracing human visual pathways. *J Neuropathol Exp Neurol.* 1983;42:200–208.

44. Danias J, et al. Cytoarchitecture of the retinal ganglion cells in the rat. *Invest Ophthalmol Vis Sci.* 2002;43:587–594.

45. Fileta JB, et al. Efficient estimation of retinal ganglion cell number: a stereological approach. *J Neurosci Methods.* 2008;170:1–8.

46. Davies NM. Biopharmaceutical considerations in topical ocular drug delivery. *Clin Exp Pharmacol Physiol.* 2000;27:558–562.

47. Fraunfelder FT. Ocular drug delivery and toxicology. In: Fraunfelder FT, et al, eds. *Clinical Ocular Toxicology.* Philadelphia: W.B. Saunders; 2008: 9–14.

48. Koneru PB, et al. Oculotoxicities of systemically administered drugs. *J Ocul Pharmacol.* 1986;2:385–404.

49. Song ZH, Schroeder A. Molecular basis of ophthalmic toxicology. In: Chiou GCY, ed. *Ophthalmic Toxicology.* Ann Arbor, MI: Taylor & Francis; 1999: 27–41.

50. Winkler BS, et al. Oxidative damage and age-related macular degeneration. *Mol Vis.* 1999;5:32.

51. Eldred GE. Lipofuscin and other lysosomal storage deposits in the retinal pigment epithelium. In: Marmor MF, Wolfensberger TJ, eds. *The Retinal Pigment Epithelium.* New York: Oxford University Press; 1998: 651–668.

52. Komeima K, et al. Antioxidants slow photoreceptor cell death in mouse models of retinitis pigmentosa. *J Cell Physiol.* 2007;213:809–815.

53. Glickman RD. Phototoxicity to the retina: mechanisms of damage. *Int J Toxicol.* 2002;21:473–490.

54. Dayhaw-Barker P. Retinal pigment epithelium melanin and ocular toxicity. *Int J Toxicol.* 2002;21:451–454.

55. Cordell WH, et al. Retinal effects of 6 months of daily use of tadalafil or sildenafil. *Arch Ophthalmol.* 2009;127:367–373.

56. Mecklenburg L, Schraermeyer U. An overview on the toxic morphological changes in the retinal pigment epithelium after systemic compound administration. *Toxicol Pathol.* 2007;35:252–267.

57. Besharse JC, Defoe DM. Role of the retinal pigment epithelium in photoreceptor membrane turnover. In: Marmor MF, Wolfensberger TJ, eds. *The Retinal Pigment Epithelium.* New York: Oxford University Press; 1998: 152–174.

58. Hughs BA, et al. Transport mechanisms in the retinal pigment epithelium. In: Marmor MF, Wolfensberger TJ, eds. *The Retinal Pigment Epithelium.* New York: Oxford University Press; 1998:103–134.

59. Wolfensberger TJ. Toxicology of the retinal pigment epithelium. In: Marmor MF, Wolfensberger TJ, eds. *The Retinal Pigment Epithelium.* New York: Oxford University Press; 1998:621–650.

60. Boulton M. Melanin and the retinal pigment epithelium. In: Marmor MF, Wolfensberger TJ, eds. *The Retinal Pigment Epithelium.* New York: Oxford University Press; 1998: 68–85.

61. Franco LM, et al. Decreased visual function after patchy loss of retinal pigment epithelium induced by low-dose sodium iodate. *Invest Ophthalmol Vis Sci.* 2009;50:4004–4010.

62. Nishimura T, et al. Effects of sodium iodate on experimental subretinal neovascularization in the primate. *Ophthalmologica.* 1990;200:28–38.

63. Chader GJ, et al. Retinoids and the retinal pigment epithelium. In: Marmor MF, Wolfensberger TJ, eds. *The Retinal Pigment Epithelium.* New York: Oxford University Press; 1998: 135–151.

64. Campochiaro PA. Seeing the light: new insights into the molecular pathogenesis of retinal diseases. *J Cell Physiol.* 2007;213:348–354.

65. Potts AM. Toxic responses of the eye. In: Klaassen CD, ed. *Casarett & Doull's Toxicology: The Basic Science of Poisons.* New York: McGraw-Hill; 1996:583–616.

66. Li J, et al. Drug-induced ocular disorders. *Drug Saf.* 2008;31:127–141.

67. Ziemssen F, Zierhut M. Principles of therapy. In: Fraunfelder FT, et al., eds. *Clinical Ocular Toxicology.* Philadelphia: W.B. Saunders; 2008: 1–7.

68. Grant WM, Schuman JS. *Toxicology of the Eye: Effects on the Eyes and Visual System from Chemicals, Drugs, Metals and Minerals, Plants, Toxins and Venoms; also Systemic Side Effects from Eye Medications.* 4th ed. Springfield, IL: Charles C Thomas; 1993.

69. Barclay SM, Riis RC. Retinal detachment and reattachment associated with ethylene glycol intoxication in a cat. *J Am Anim Hosp Assoc.* 1979;15:719–724.

70. Delahunt CS, et al. The cause of blindness in dogs given hydroxypyridinethione. *Toxicol Appl Pharmacol.* 1962;4: 286–291.

71. Beehler CC, et al. Retinal detachment in adult dogs resulting from oxygen toxicity. *Arch Ophthalmol.* 1982;71:665–670.

72. Kubay OV, et al. Retinal detachment neuropathology and potential strategies for neuroprotection. *Surv Ophthalmol.* 2005;50:463–475.

73. Lewis GP, et al. Experimental retinal reattachment: a new perspective. *Mol Neurobiol.* 2003;28:159–175.

74. Barnett KC, Watson WA. Bright blindness in sheep: a primary retinopathy due to feeding bracken. *Res Vet Sci.* 1970;11: 289–290.

75. Lee KP, Valentine R. Pathogenesis and reversibility of retinopathy induced by 1,4-bis(4-aminophenoxy)-2-phenylbenzene (2-phenyl-APB-144) in pigmented rats. *Arch Toxicol.* 1991;65:292–303.

76. Gellatt KN, et al. Enrofloxacin-associated retinal damage in cats. *Vet Ophthalmol.* 2001;4:99–106 [Erratum: *Vet Ophthalmol.* 2001;4:231].

77. Gómez-Ramos P, et al. Neuronal and microvascular alterations induced by the cholinergic toxin AF64A in the rat retina. *Brain Res.* 1990;520:151–158.

78. Pasadhika S, Fishman GA. Effects of chronic exposure to hydroxychloroquine or chloroquine on inner retinal structures. *Eye (Lond).* 2010;340–346.

79. Hare WA, Wheeler L. Experimental glutamatergic excitotoxicity in rabbit retinal ganglion cells: block by memantine. *Invest Ophthalmol Vis Sci.* 2009;50:2940–2948.

80. Sucher NJ, et al. Molecular basis of glutamate toxicity in retinal ganglion cells. *Vis Res.* 1997;37:3483–3493.

81. Vistamehr S, et al. Ethambutol neuroretinopathy. *Semin Ophthalmol.* 2007;22:141–146.

82. Kato S, et al. DL-α-aminoadipate is a toxin to Müller cells. *Prog Retin Eye Res.* 1996;15:435–456.

83. Welinder E, et al. Effects of intravitreally injected DL-α-aminoadipic acid on the c-wave of the D.C.-recorded electroretinogram in albino rabbits. *Invest Ophthalmol Vis Sci.* 1982;23:240–245.

84. Lebrun-Julien F, et al. Excitotoxic death of retinal neurons in vivo occurs via a non-cell-autonomous mechanism. *J Neurosci.* 2009;29:5536–5545.

85. Torbati D, et al. Experimental retinopathy by hyperbaric oxygenation. *Undersea Hyperb Med.* 1995;22:31–39.

86. Cui JZ, et al. Natural history of choroidal neovascularization induced by vascular endothelial growth factor in the primate. *Graefe's Arch Clin Exp Ophthalmol.* 2000;238: 326–333.

87. Wong CG, et al. Intravitreal VEGF and bFGF produce florid retinal neovascularization and hemorrhage in the rabbit. *Curr Eye Res.* 2001;22:140–147.

88. Campochiaro PA, Hackett SF. Ocular neovascularization: a valuable model system. *Oncogene.* 2003;22:6537–6548.

89. Ozaki H, et al. Intravitreal sustained release of VEGF causes retinal neovascularization in rabbits and breakdown of the blood–retinal barrier in rabbits and primates. *Exp Eye Res.* 1997;64:505–517.

90. Orzalesi N, et al. Subretinal neovascularization after naphthalene damage to the rabbit retina. *Invest Ophthalmol Vis Sci.* 1994;35:696–705.

91. Sujak A. Interactions between canthaxanthin and lipid membranes: possible mechanisms of canthaxanthin toxicity. *Cell Mol Biol Lett.* 2009;14:395–410.

92. Bartlett JD. Ophthalmic toxicity by systemic drugs. In: Chiou GCY, ed. *Ophthalmic Toxicology.* Ann Arbor, MI: Taylor & Francis; 1999: 225–283.

93. Mausolf FA, Mensher JH. Experimental hyperviscosity retinopathy: preliminary report. *Ann Ophthalmol.* 1973;5: 205–209.

94. Synder FH, et al. Safety evaluation of zinc 2-pyridinethiol 1-oxide in a shampoo formulation. *Toxicol Appl Pharmacol.* 1965;7:425–437.

95. Cloyd GG, et al. Ocular toxicology studies with zinc pyridinethione. *Toxicol Appl Pharmacol.* 1978;45:771–782.

96. Eells JT. Methanol-induced visual toxicity in the rat. *J Pharmacol Exp Ther.* 1991;257:56–63.

97. van der Lugt JJ, Venter I. Myelin vacuolation, optic neuropathy and retinal degeneration after closantel overdosage in sheep and in a goat. *J Comp Pathol.* 2007;136:87–95.

98. Kitaoka Y, et al. TNF-α-induced optic nerve degeneration and nuclear factor-κB p65. *Invest Ophthalmol Vis Sci.* 2006;47:1448–1457.

99. McEntee K, et al. Closantel intoxication in a dog. *Vet Hum Toxicol.* 1995;34:234–236.

100. Gittinger JWJ, Asdourian GK. Papillopathy caused by amiodarone. *Arch Ophthalmol.* 1987;105:249–251.

101. Shinder R, et al. Regression of bilateral optic disc edema after discontinuation of amiodarone. *J Neuroophthalmol.* 2006;26:192–194.

102. Eells JT, et al. Development and characterization of a rodent model of methanol-induced retinal and optic nerve toxicity. *Neurotoxicology.* 2000;21:321–330.

103. O'Dea D, et al. Ocular changes with oxaliplatin. *Clin J Oncol Nurs.* 2006;10:227–229.

104. Chiou GCY. Ophthalmic toxicity by local agents. In: Chiou GCY, ed. *Ophthalmic Toxicology.* Ann Arbor, MI: Taylor & Francis; 1999: 285–353.

105. Schilling B, et al. Retinal changes in cats in poisoning by peroral or percutaneous administered chemicals. *Verh Dtsch Ges Pathol.* 1966;50:429–435.

106. Main DC, et al. (1981). *Stypandra imbricata* ("blindgrass") toxicosis in goats and sheep: clinical and pathologic findings in 4 field cases. *Aust Vet J.* 1981;57:132–135.

107. Huxtable CR, et al. Myelin oedema, optic neuropathy and retinopathy in experimental *Stypandra imbricata* toxicosis. *Neuropathol Appl Neurobiol.* 1980;6:221–232.

108. Hansson HA. Retinal changes induced by treatment with vincristine and vinblastine. *Doc Ophthamol.* 1972;31: 65–88.

109. Cannon JP, et al. Comparative toxicity and concentrations of intravitreal amphotericin B formulations in a rabbit model. *Invest Ophthalmol Vis Sci.* 2003;44:2112–2117.

110. Ide T. Histopathological studies on retina, optic nerve and arachnoidal membrane of mouse exposed to carbon disulfide poisoning. *Acta Soc Let Ophthalmol Jpn.* 1958;62A: 85–108.

111. Cogan DG, et al. Optic neuropathy, chloramphenicol, and infantile genetic agranulocytosis. *Invest Ophthalmol.* 1973;12:534–537.

112. Karlsen RL, Pedersen OO. A morphological study of the acute toxicity of L-cysteine on the retina of young rats. *Exp Eye Res.* 1982;34:65–69.

113. Pedersen OO, Karlsen RL. The toxic effect of L-cysteine on the rat retina: a morphological and biochemical study. *Invest Ophthalmol Vis Sci.* 1980;19:886–892.

114. Parhad IM, et al. Doxorubicin intoxication: neurofilamentous axonal changes with subacute neuronal death. *J Neuropathol Exp Neurol.* 1984;43:188–200.

115. Lucas DR, Newhouse JP. The toxic effect of sodium L-glutamate on the inner layers of the retina. *AMA Arch Ophthalmol.* 1957;58:193–201.

116. Lessell S, et al. Kainic acid induces mitoses in mature retinal neurones in rats. *Exp Eye Res.* 1980;30:731–738.

117. Grosskreutz CL, et al. Lidocaine toxicity to rat retinal ganglion cells. *Curr Eye Res.* 1999;18:363–367.

118. Van Kampen KR, James LF. Sequential development of the lesions in locoweed poisoning. *Clin Toxicol.* 1972;5: 575–580.

119. Green WR. Retinal and optic nerve atrophy induced by intravitreous vincristine in the primate. *Trans Am Ophthalmol Soc.* 1975;73:389–416.

120. Nyska A, et al. Ocular thrombosis and retinal degeneration induced in female F344 rats by 2-butoxyethanol. *Hum Exp Toxicol.* 1999;18:577–582.

121. Tanu T, et al. Gentamicin and other antiobiotic toxicity. *Ophthalmol Clin North Am.* 2001;14:611–624.

122. Barone SJ, et al. Effects of 3,3′-iminodipropionitrile on the peripheral structures of the rat visual system. *Neurotoxicology.* 1995;16:451–467.

123. Benedetto R, Calogero G. Oxygen-induced retinopathy in newborn rats: effects of normobaric and hyperbaric oxygen supplementation. *Pediatrics.* 1988;82:193–198.

124. Peyman GA, et al. Effects of intravitreal prostaglandins on retinal vasculature. *Ann Ophthalmol.* 1975;7:279–288.

125. Lu ZY, et al. Aluminum chloride induces retinal changes in the rat. *Toxicol Sci.* 2002;66:253–260.

126. Whitcup SM, et al. A clinicopathologic report of the retinal lesions associated with didanosine. *Arch Ophthalmol.* 1994;112:1594–1598.

127. Brown DV. Reaction of the rabbit retinal pigment epithelium to systemic lead poisoning. *Trans Am Ophthalmol Soc.* 1974;72:404–447.

128. Albert DM, et al. Flecked retina secondary to oxalate crystals from methoxyflurane anesthesia: clinical and experimental studies. *Trans Sect Ophthalmol Am Acad Ophthalmol Otolaryngol.* 1975;79:OP817–OP826.

129. Herrold KM. Pigmentary degeneration of the retina induced by *N*-methyl-*N*-nitrosourea: an experimental study in Syrian hamsters. *Arch Ophthalmol.* 1967;78:650–653.

130. Miller FS, et al. Clinical-ultrastructural study of thioridazine retinopathy. *Ophthalmology.* 1982;89:1478–1488.

131. Rampal S, et al. Ofloxacin-associated retinopathy in rabbits: role of oxidative stress. *Hum Exp Toxicol.* 2008;27:409–415.

132. Lüllmann-Rauch R. Retinal lipidosis in albino rats treated with chlorphentermine and with tricyclic antidepressants. *Acta Neuropathol.* 1976;35:55–67.

133. Schmidt IG, Schmidt LH. Studies of the neurotoxicity of ethambutol and its racemate for the rhesus monkey. *J Neuropathol Neurol.* 1966;25:40–67.

134. Lessell S. Histopathology of experimental ethambutol intoxication. *Invest Ophthalmol Vis Sci.* 1976;15:765–769.

135. Hayasaka Y, et al. Ocular changes after intravitreal injection of methanol, formaldehyde, or formate in rabbits. *Pharmacol Toxicol.* 2001;89:74–78.

136. Sharpe JA, et al. Methanol optic neuropathy: a histopathological study. *Neurology.* 1982;32:1093–1100.

25

TOXICOLOGICAL NEUROPATHOLOGY OF THE EAR

ANDREW FORGE AND RUTH TAYLOR
Centre for Auditory Research, UCL Ear Institute, London, United Kingdom

BRAD BOLON
GEMpath, Inc., Longmont, Colorado

INTRODUCTION

The inner ear contains the sensory end organs of hearing, in the cochlea, and balance, in the vestibular apparatus. These organs detect and convert mechanical stimuli—sound vibrations in the cochlea, motion and head position changes in the vestibular apparatus—into neural signals. Several chemicals and drugs can induce pathology in the sensory units or neural pathways associated with these end organs. Such neurotoxic events are referred to as *ototoxicity*. The methods required to assess ototoxic neuropathology deserve special attention here because many neuroscientists, including neuropathologists, have had little experience in the evaluation of inner ear pathology. In this chapter we emphasize the proper performance of conventional processing and analytical approaches that are used most often by pathologists engaged in nonclinical and clinical drug development. More detailed descriptions of functional and specialized microscopic procedures for evaluating ear tissues are available in a recent review.[1]

ANATOMY AND PHYSIOLOGY OF THE INNER EAR

The inner ear consists of a series of membranous canals enclosed in interconnected bony channels (Figure 1). In humans and nonhuman primates, the inner ear resides within the temporal bone at the base of the skull. In other mammals, the inner ear and middle ear are both confined inside the auditory bulla, a thin-walled protuberance that is not fused with the skull base and therefore can be isolated with relative ease (Figure 2).

The inner ear can be divided into multiple compartments. The hearing organ, or cochlea, consists of a triangular membranous canal subdivided into three parallel canals (scala media, scala tympani, and scale vestibuli; Figure 3) arranged in a spiral around a central axis (the modiolus) that contains the neural and vascular supplies.[2] The number of turns in the spiral from base to apex varies among species. The organ of balance, or vestibular apparatus, consists of three semicircular membranous canals arising from paired sacs, the saccule and utricle (Figure 1). The scala media of the cochlea and the membranous canals of the vestibular apparatus form a continuous endolymphatic compartment that is filled with endolymph, a potassium (K^+)-rich fluid. The endolymphatic compartment terminates in the endolymphatic sac via the endolymphatic duct (Figure 1). The scala tympani and scala vestibuli (which are connected at the apical end of the cochlear spiral) comprise a single perilymphatic compartment that is filled with perilymph, a sodium (Na^+)-rich fluid. The perilymphatic compartment is linked to the subarachnoid space via the cochlear aqueduct (Figure 1).

At the cochlear base, two small openings penetrate the bony wall separating the middle ear from the inner ear. One opening, the oval window, is bracketed by the stapes footplate in the middle ear side and the broad end of the scala vestibuli on the inner ear side. The other opening, the round window, is a membrane-covered orifice over the broad end of the scala tympani; in the cynomolgus monkey (*Macaca fascicularis*) the round window membrane contains blood

Fundamental Neuropathology for Pathologists and Toxicologists: Principles and Techniques, First Edition. Edited by Brad Bolon and Mark T. Butt.
© 2011 John Wiley & Sons, Inc. Published 2011 by John Wiley & Sons, Inc.

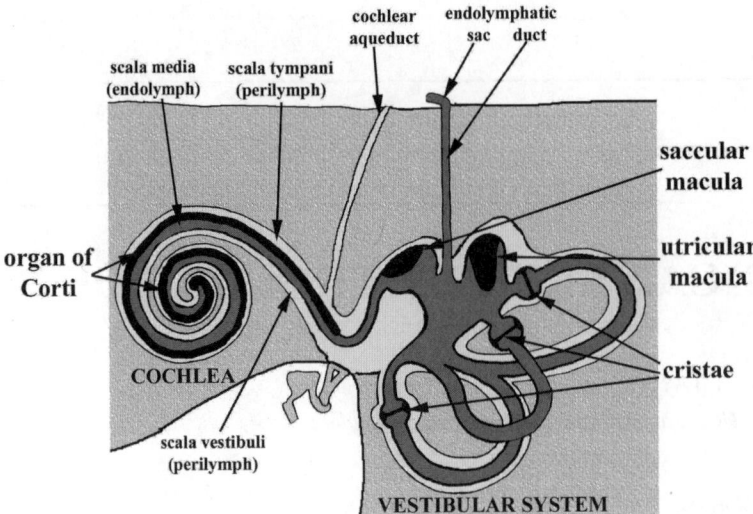

FIGURE 1 Diagrammatic representation of the human inner ear. A series of membranous, connected canals are partitioned into the cochlea (the spiralling organ of hearing) and the vestibular apparatus (the organ of balance, with three semicircular canals arising from the saccule and utricle). Sound waves in the environment are transmitted to the cochlea by the ossicles of the middle ear, which bridge the gap between the tympanic membrane (not shown) and the oval window. The vibrations of the stapes in the oval window cause fluid waves to propagate within the sodium (Na^+)-rich perilymph, first in the scala vestibuli and subsequently in the scala tympani; the scala vestibuli and scala tympani are connected at the "helicotrema," a small hole at the apical end of the cochlear spiral. These two perilymph-filled canals are linked to the subarachnoid space near the brainstem via the cochlear aqueduct. The scala media of the cochlea and the canals of the vestibular apparatus form a continuous compartment that contains potassium (K^+)-rich endolymph.

and lymphatic vessels, leukocytes, and glands which are thought to provide the middle ear with a specific immunoprotective mechanism for disposal of foreign and noxious substances before they reach the inner ear.[3] Movement of the stapes by sound waves causes propagation of perilymph waves in the scala vestibuli which are then transmitted up the cochlear spiral, transferred into the scala tympani, and finally, dampened by displacement of the elastic membrane covering the round window.

At the microscopic level, the inner ear is composed chiefly of epithelium, neurons, and connective tissue. The epithelial subpopulations are specialized to detect waves in the endolymph (sensation), to transport ions (to support sensory cell function), or to line membranous canals. The connective tissue elements help maintain the microenvironment for the sensory epithelia. The roles of these cellular elements are comparable for both balance and hearing functions.

Although both the cochlea and the vestibular apparatus have sensory cells with equivalent structures serving comparable sensory tasks, functional changes in one system can occur without a corresponding alteration in the other. For example, significant vestibular loss without concurrent hearing loss has been described in Ce/J, MRL/MpJ, and SJL/J mice but not in other mouse strains.[4] Functional and structural changes in both sense organs are cumulative over time.[4,5]

Cochlea

The scala media is partitioned from the scala vestibuli by an impermeable, cellular (Reissner's) membrane, and from the scala tympani by a permeable acellular basilar membrane upon which rests the sensory organ (of Corti; Figure 3). The Na^+-rich perilymph inside the scala tympani readily penetrates the basilar membrane to bathe the cell bodies and basal axon projections of the sensory epithelium. Intercellular tight junctions at the apices of the sensory cells segregate the K^+-rich endolymph in the scala media from mixing with the perilymph, thereby maintaining an ionic gradient across the organ of Corti which allows the production of a K^+ current through the sensory organ cells upon detection of a sound. The lateral wall of the scala media is formed by the stria vascularis, an ion-transporting epithelium that generates the endocochlear potential (EP) by actively recycling K^+ from the perilymph.[6] The stria vascularis rests upon the spiral ligament, a population of fibrocytes that participates in K^+ recycling as well as cochlear homeostasis.[7]

The functional component of the organ of Corti (Figure 3) is the sensory epithelium, and specifically the "hair" cells (HCs). The elongate HCs are arranged in a mosaic in parallel rows, with each HC separated from its nearest neighbors by intervening support cells and large basal extracellular spaces

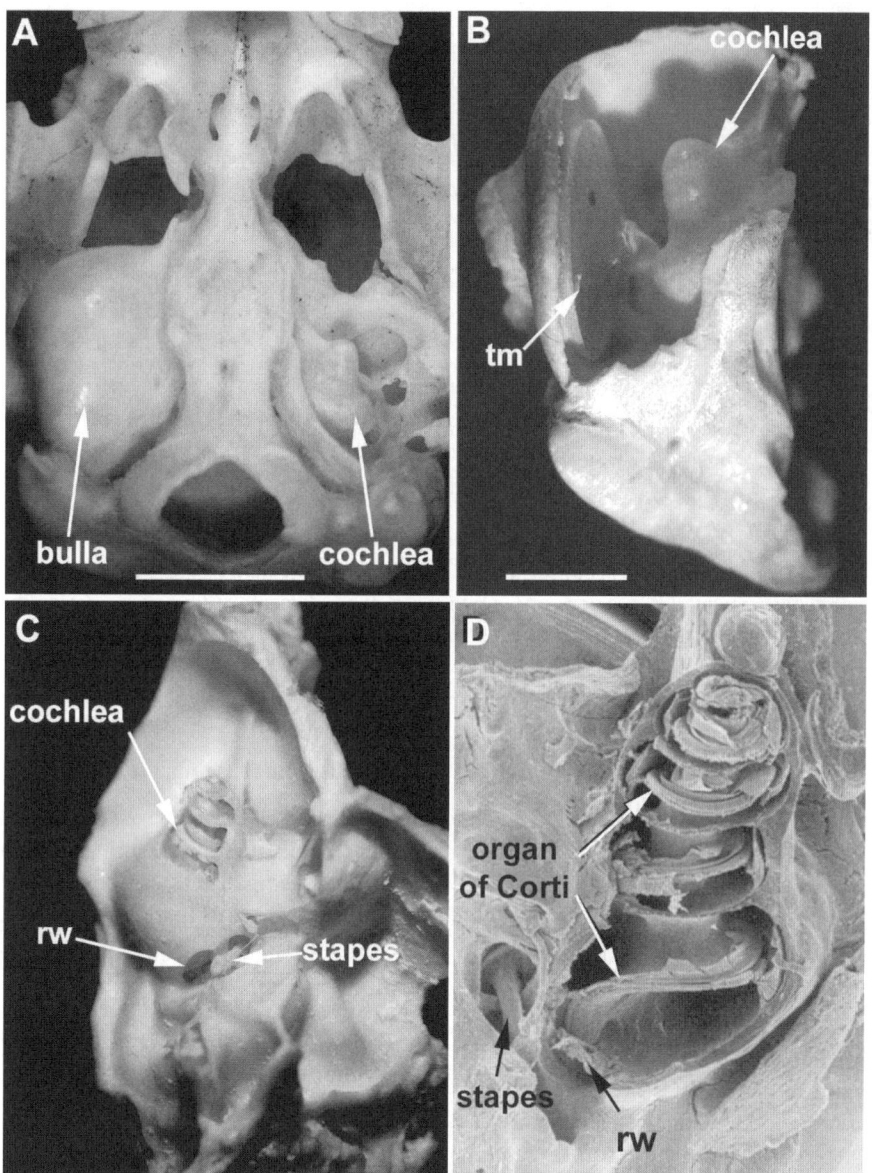

FIGURE 2   Gross structure of the guinea pig bulla and cochlea. (A) Base of the skull (observed from below). The bulla of the right ear (on the left) is intact, while that of the left ear (on the right) has been opened to expose the cochlea. (B) An isolated bulla has been opened to reveal the cone-shaped tympanic membrane (tm) and the cochlea (covered in a bony wall). (C) Isolated, opened bulla with the cochlea partially dissected to expose the apical coils of the cochlear spiral. The round window (rw) and stapes (in position in the oval window) at the base of the cochlea are indicated. (D) Scanning electron micrograph of the guinea pig cochlea, with the bony wall opened to expose the interior. The spiralling organ of Corti is indicated, and the tissues that form the lateral wall of the scala media can be seen. The round window (rw) and the stapes (in position in the oval window) are indicated. Scale bars = 0.5 mm.

filled with perilymph. The typical arrangement is a single row of inner hair cells (IHCs) and from three to five rows (depending on the species) of outer hair cells (OHCs). The actual number of HCs varies by species. The CBA/Ca mouse strain (in which "normal" hearing acuity is maintained for most of its lifespan[8]) averages 725 IHCs and 2300 OHC,[9] the

guinea pig has about 2000 IHCs and 7000 OHCs,[10] and humans have approximately 3000 IHCs and 9000 OHCs.[11] The two HC types in the cochlea have different roles. The IHC is the primary receptor cell, each HC responding to a stimulus of a particular frequency. Each IHC is directly innervated by myelinated dendrites from several bipolar

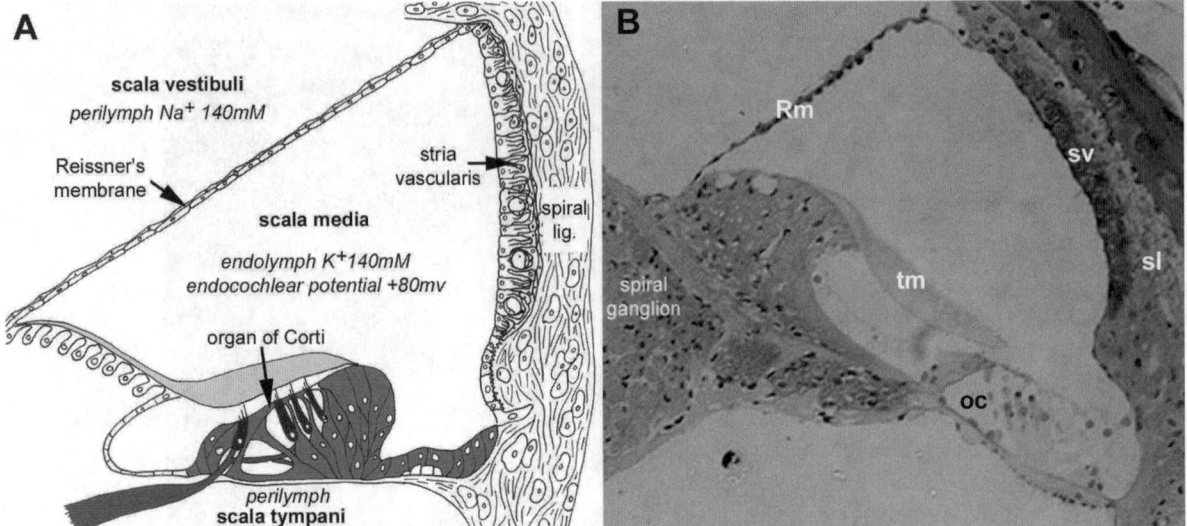

FIGURE 3   Microscopic structure of a single cross section of the cochlea. (A) Diagrammatic representation showing the positions of the major structural elements of the inner ear. (B) Photomicrograph of a single turn of the mouse cochlea. The location of the cell bodies of the auditory neurons that form the spiral ganglion is labeled. oc, Organ of Corti; Rm, Reissner's membrane; sl, spiral ligament; sv, stria vascularis; tm, tectorial membrane.

neurons of the spiral ganglion. In contrast, OHCs signal via unmyelinated processes to actively modulate the sound-induced movements of the basilar membrane, thus amplifying the signal that reaches the IHCs and increasing sensitivity.[12,13] Death of OHCs in particular can be triggered by failures to maintain the physiological environment within the cochlea. These proteins, which are involved in maintaining the physiological environment of the cochlea, are potential targets for xenobiotics.

An acellular, fibrous tectorial membrane covers the organ of Corti. The apices of the HCs are accented by projecting bundles of "stereocilia" (actin-packed microvilli) that increase in height in one direction across the cell surface. The tips of the longest stereocilia are embedded in the underside of the tectorial membrane. Deflections of the hair bundle toward and away from the longest row of stereocilia alter the permeability of nonselective cation channels,[14] thereby stimulating the HC. The tips of the stereocilia are linked in various fashions[15–17] to better coordinate the genesis, maintenance, and function of the stereocilia bundles.

Several site-specific variations in cochlear dimensions dictate the mechanical properties of the system and thus direct the sites at which sounds of a particular frequency are heard. High-frequency (high-pitched) sounds cause maximum fluid displacement, and thus maximal stimulation of the HCs at the cochlear base, while low frequencies are detected at the apical end.[18] Thus, differential damage along the length of the organ of Corti will be reflected in differential loss of frequency perception.

The Vestibular Apparatus

The sensory epithelia consist of arrays of HCs in the maculae (spots) inside the utricle and saccule as well as the cristae within the ampullae (swellings) located at one end of each semicircular canal (Figure 1). As in the organ of Corti, the HCs are stimulated by deflection of their stereocilia, although the stimuli detected in this apparatus relate to the static (via the maculae) or moving (in the cristae) position of the head in relation to the direction of the force of gravity. Motion detection by the cristae depends on fluid inertia in the endolymph; displacement of the endolymph is delayed relative to motions of the head, thus permitting a transient fluid surge to deflect the hair bundles in the cristae. Unlike the organ of Corti, the vestibular sensory neuroepithelia are compact tissues with no large extracellular spaces, and the basement membrane (equivalent to the basilar membrane of the organ of Corti) rests directly on connective tissue. However, similar to the organ of Corti, the thin, basal intercellular spaces within the sensory epithelia are continuous with the perilymphatic compartment, while the apical surface is bathed in endolymph. The maculae of the utricle and saccule are covered by an acellular fibrous membrane upon which sit the otoconia (calcium-containing crystals), while the cristae of the semicircular canals are overlaid by a gelatinous material (the cupula).

Two types of HCs reside in the vestibular apparatus, type 1 HCs resembling cochlear IHCs and type 2 HCs similar to cochlear OHCs. However, both vestibular HC types have a single true cilium, the kinocilium, located behind the row of longest stereocilia. Together, the position of the kinocilium and the longest stereocilia define the morphological and functional polarity of the hair bundle of each HC. This cellular orientation in mirrored in a larger scale by the structural and functional polarity in the vestibular sensory organs. In the vestibular maculae an abrupt change in bundle polarity along a centrally located "striola" separates two regions of HCs, each of which has hair bundles oriented at 90° to those in the other region. Furthermore, the two HC types are differentially distributed in the maculae, with type 1 HCs near the striola and type 2 HCs concentrated near the periphery. In the cristae, the hair bundles of all HCs are oriented in the same direction across the entire epithelium, but the type 1 HCs predominate in the center, whereas type 2 HCs are more prevalent in the borders.

Vestibular endolymph is maintained by the cells of the dark cell regions located around the utricular macula and at the base of the saddle-shaped cristae. The dark cell monolayer is equivalent to the marginal cells of the cochlear stria vascularis and serves to actively transport K^+ ions from the perilymph bathing the cell base into the endolymph. However, despite the ionic gradient, the vestibular dark cells do not generate an endovestibular potential equivalent to that observed in the cochlea.

ACCESS OF OTOTOXICANTS TO THE INNER EAR

The inner ear is protected from direct exposure to exogenous ototoxicants by the surrounding bone. Toxic agents can reach the inner ear via the blood, endolymph, or perilymph. Most ototoxic agents reach the inner ear via the circulation. The existence of a *blood–perilymph barrier* is indicated by the differing compositions of perilymph in the scala tympani and scala vestibuli relative to the compositions of blood plasma and cerebrospinal fluid (CSF).[19] This barrier is capable of restricting the entrance of blood-borne materials, as shown by the requirement for facilitated diffusion of glucose into this compartment.[19,20] Little is known about the ability of the blood–perilymph barrier with respect to the passage of potential ototoxicants in the blood, but any agents that can enter the perilymph have free access to the basolateral membranes of HCs and their associated nerve fibers.

Entry to endolymph in the inner ear is even more restricted.[19] The principal boundaries between the endolymph and perilymph are formed by selectively permeable or impermeable membranes separating the cochlear scala as well as the apical tight junctions between epithelial cells. Any toxicant that succeeded in crossing these boundaries would have access to the hair-bearing apices of the HCs.

Access to the perilymph can occur from the middle ear by passive transfer across the semipermeable round window membrane.[21] Initial entry by this route would deliver the toxicant (usually, bacterial toxins or antibacterial drugs) to the vestibule near the cochlear base, after which the molecule would slowly diffuse to the remaining perilymphatic compartment. Potential ototoxicants (e.g., bacterial toxins) in the CSF may move into the perilymph through the cochlear aqueduct in some instances. However, this aqueduct tends to close with age, so this route is not likely to offer a major potential for exposure.

METHODS FOR STUDYING THE INNER EAR

Routine Gross and Microscopic Analytical Procedures

In most mammals the auditory bulla can readily be isolated from the base of the skull and opened to expose the cochlea (Figure 2). The inner ear may be prepared for perfusion fixation by removing (1) the elastic membrane covering the round window, (2) the stapes closing the oval window, and (3) the bony tip of the cochlear apex. The inner ear is then flushed gently with at least 0.5 mL of fixative (typically, 4% paraformaldehyde for light microscopy or 2.5% glutaraldehyde for ultrastructural studies) using a 28-gauge needle, followed by immersion in fresh fixative for at least 2 h at room temperature. Adequate preservation can be attained by whole-body perfusion followed by isolation of the bulla, opening of the cochlea, and post-fixation by immersion. In our experience, tissue trimming and morphologic integrity is seldom, if ever, acceptable if the inner ear evaluation is attempted following immersion fixation of the intact skull without penetration of the auditory bulla and cochlea. As with other tissues, the finest structures of the inner ear (e.g., stereocilia cross-links) are especially vulnerable to degradation if fixation is not optimal.[17]

The most straightforward means of assessing the neuroepithelia of the inner ear is to examine the surface features in whole-mount preparations. Following fixation, the bone encasing the inner ear can be removed in small segments and the neuroepithelia dissected away. Gentle decalcification (e.g., serial immersion in sodium formate or sodium EDTA) after fixation may assist dissection. The length of decalcification will depend on both the decalcifying agent (formate acts rapidly, EDTA gradually) and the species, and should be established by experimentation in each laboratory. In general, the decalcification needed for suitable inner ear preparations is a couple of days in mouse, about a week in guinea pigs and rats, and several weeks in humans. The organ of Corti can be removed in segments, usually beginning at the apex and moving toward the base. The sensory tissue in both the cochlea and vestibular apparatus can then be examined by either phase-contrast or differential interference contrast (DIC) microscopy, or by fluorescence microscopy following staining with conjugated phalloidin (which interacts with actin to intensely label stereocilia and the cell junctions in the reticular lamina). Alternatively, scanning electron microscopy (SEM) can be employed to examine the cochlear or vestibular sensory epithelia.[22] These techniques are typically employed to "map" the position of each HC along the entire length of the organ of Corti (a *cytocochleogram*[23]) or across the vestibular maculae and cristae[24] so that the extent and location of damaged or missing HCs can be defined. The cytocochleogram can be related to in-life physiological data from the same cochlea. An advantage of SEM for HC evaluation is that it allows assessment of early, relatively minor HC damage which may be invisible by light microscopy but may still have major functional implications.[25–27]

An alternative approach can be obtained by conventional histopathological examination of the inner ear. After appropriate fixation and decalcification (as described above), the entire inner ear is embedded in paraffin or plastic, and sections are acquired parallel to the modiolus (i.e., in "longitudinal" orientation) to present multiple loops of the cochlear spiral simultaneously. The advantages of this technique are that the cochlear tissues are retained in their original relationship, an adequate evaluation can be made of functionally distinct cochlear domains (e.g., apex, middle, and base) in one or two sections, and all toxicant-susceptible tissues (e.g., sensory epithelia, stria vascularis, spiral ganglia and their projections) can be assessed at once for lesions that might account for hearing impairment: damage to or loss of sensory cells or neurons, swelling or atrophy of the stria vascularis, loss of spiral ligament fibrocytes, and swelling or collapse of Reissner's membrane. "Cytocochleograms" can be acquired from such cochlear preparations by serial reconstruction of the cochlear spiral. The serial sectioning required for the latter procedure is much more laborious than surface examination of whole mounts but can give a more complete picture of overall inner ear pathology.[28] The standard stain used in such conventional inner ear preparations is hematoxylin and eosin (H&E). Special stains [e.g., periodic acid-Schiff (PAS) for membranous structures, antineurofilament protein for neuronal process, anti-caspase 3 for cells undergoing apoptosis, extravasation of intravenously administered fluorescein-labeled albumen for capillary integrity in the stria vascularis] may be used as required to delineate specific features of inner ear anatomy relevant to potential target tissues for ototoxic agents.

Novel Methods to Examine the Direct Effects of Ototoxic Agents

Studies of ototoxic agents typically involve evaluation of inner ear structure and/or function after systemic application of the agent, a situation that mimics the clinical setting with regard to the distribution of molecules into cochlear and vestibular tissues. Such a design has disadvantages when investigating cellular or molecular mechanisms, especially the difficulty in separating the initial action of the agent from the ensuing sequelae. Therefore, mechanistic investigations in the inner ear usually employ direct exposure of tissues or isolated cells, either in situ or in vitro, by perfusion with ototoxic substances.[29]

The molecular and biochemical characteristics of all HC types are broadly similar, and the vulnerability of HCs to particular damaging agents is conserved across vertebrate classes. Thus, explant cultures of the inner ears from chickens have been used to explore the molecular basis of HC susceptibility to ototoxic agents.[30–32] The HCs in the neuromasts of the fish lateral line are susceptible to aminoglycoside toxicity in much the same way as are mammalian inner ear HCs,[33] thereby allowing the fish lateral line to serve as a potential model for identifying agents that kill HCs.[34–36] The lateral line organs of larval zebrafish are easily visualized by light microscopy, and certain fluorescent dyes [e.g., 2-(4-(dimethylamino)styryl)-*N*-ethylpyridinium iodide (DAS-PEI)] are taken preferentially into viable neuromast HCs.[37] This ability provides a simple means for accurate and rapid counting of HC numbers and for scoring the extent of HC death in response to toxic agents.

EFFECTS AND ACTIONS OF OTOTOXIC DRUGS

A diverse range of drugs and a few industrial chemicals have been shown to be ototoxic (Table 1). The adverse effects of ototoxicants vary from relatively minor, reversible functional deficits to marked, permanent structural impairment resulting from substantial functional damage. Agents that damage the inner ear may act on the sensory neuroepithelia (almost always the OHCs; Figure 4), the ion-transporting cells (Figure 5), or both. Hearing loss is the most common effect

TABLE 1   Ototoxicants Known to Induce Structural Lesions in the Inner Ear[a]

Classification	Agents
Aminoglycoside antibiotics	Amikacin, dibekacin, dihydrostreptomycin, framycetin, gentamicin, kanamycin, neomycin, netilmicin, ribostamycin, sisomicin, streptomycin, tobramycin
Macrolide antibiotics	Erythromycin, azithromycin, clarithromycin
Other antibiotics	Viomycin
Antineoplastic agents	***cis*-Platinum, carboplatin**
Anti-inflammatory agents	**Salicylate, aspirin**
Antimalarial agents	**Quinine**
Loop diuretics	Bumetanide, ethacrynic acid, furosemide, piretanide
Iron chelators	Desferrioxamine
Industrial chemicals	Styrene, toluene, trichloroethylene, trimethyl tin, xylene

[a] Ototoxicants discussed in detail in the chapter are indicated in bold face type.

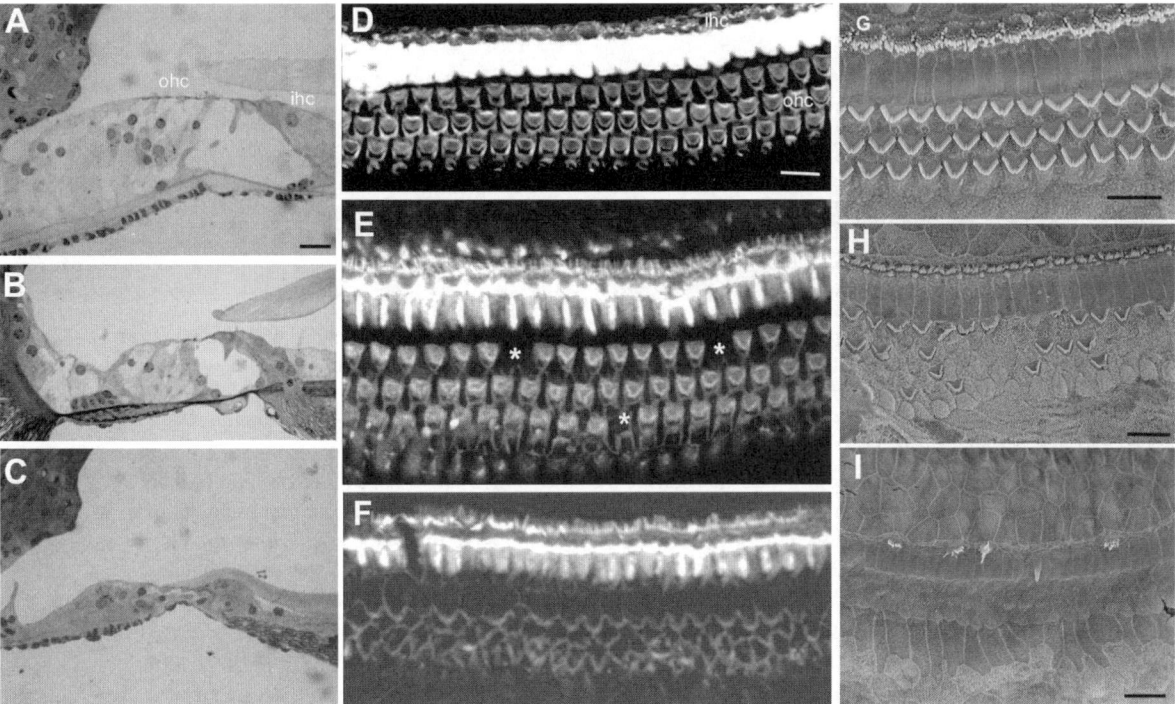

FIGURE 4 Toxicants that damage the organ of Corti typically induce hair cell loss. (A–C) Sections of plastic-embedded mouse cochleae stained with toludine blue. (A) Normal tissue. There are three rows of outer hair cells (ohc) and a single row of inner hair cells (ihc), all with their nuclei located in the upper third of the tissue. The nuclei of the various supporting cells are found at lower levels. Scale bar [also refers to (B) and (C)] = 10 μm. (B) The ohc have been lost. Only supporting cell nuclei are evident in the region where ohc would be present in the normal tissue (A). Expansion of the bodies of supporting cells fills some of the extracellular spaces normally present in the sensory epithelium. (C) The organ of Corti has flattened with loss of both ohc and ihc. (D, E) Phalloidin-labeled whole mounts of the mouse organ of Corti, imaged by confocal microscopy. Projection stack. (D) Normal organ of Corti, mid-basal coil. Phalloidin labels actin in the stereocilia of the hair bundles in each hair cell, actin in the junctional regions between adjacent hair cells, and the actin bundles in the heads of the pillar cells that separate the row of inner hair cells (ihc) from the three rows of outer hair cells (ohc). The characteristic W-shape of the hair bundles on each ohc and the more linear arrangement of the hair bundles on the ihc are evident. Scale bar [also refers to (E) and (F)] = 10 μm. (E) Loss of a few ohc in the apical coil following aminoglycoside damage. Asterisks (∗) indicate locations of lost ohc, where expansion of the heads of adjacent supporting cells has repaired the lesion site at the apical surface. Note the greater width of the organ of Corti in the apical coil compared with the width of the basal coil [shown in (D)]. (F) Complete loss of ohc in the basal coil following aminoglycoside exposure. The heads of supporting cells expand to close the sites of the lesions from where the ohc are lost. (G–I) Scanning electron microscopy of the apical surface of the mouse organ of Corti. (G) Normal organ of Corti. The characteristic morphology of the hair bundles on the inner hair cells (upper row with open U-shaped stereociliary bundles and outer hair cells (three lower rows with W-or V-shaped stereociliary bundles) is clearly evident. Each hair cell is separated from its immediate neighbors by the heads of the intervening supporting cells. (H) Scattered loss of outer hair cells that develops soon after aminoglycoside exposure. Some hair cells have been fixed while undergoing acute degeneration, indicated (at arrow) by fusion of stereocilia in the bundle or extrusion of the apical end of the hair cell into the endolymphatic space. The sites from which hair cells have been lost are now filled by the expanded heads of the supporting cells that close the lesion sites. (I) Loss of all outer hair cells and most inner hair cells at a later stage in the progression of aminoglycoside-induced damage. The sites of the missing ihc, like those of the missing ohc, are closed by the expanded heads of the supporting cells that had surrounded the hair cells. Scale bars = 10 μm.

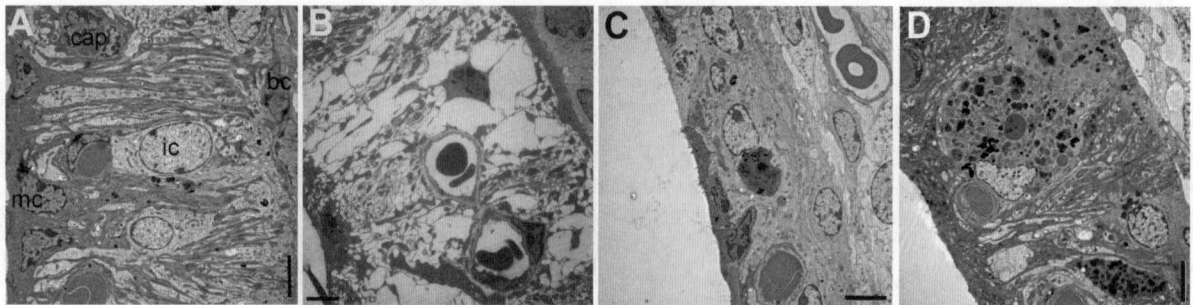

FIGURE 5 The stria vascularis and some of its possible pathological changes. (A) Normal stria vascularis of the mouse (apical surface located to the left). bc, Basal cells, which separate the stria from the underlying spiral ligament; ic, intermediate cells; mc, marginal cells, which line the endolymphatic space. (B) Swelling and intercellular edema in the guinea pig stria vascularis following a single systemic injection of a loop diuretic (ethacrynic acid). The thickness of the tissue from the endolymphatic surface of the marginal cells to the basal limit of the basal cells adjacent to the spiral ligament is greater than normal. (Note the higher magnification of this panel in comparison with the others.) (C) Atrophy of the stria vascularis following aminoglycoside exposure. The arrows denote the boundary between the residual stria vascularis and the spiral ligament. A blood vessel containing two erythrocytes is present in the upper right quadrant of the image. (D) Large inclusions (within intermediate cells in particular) in the stria vascularis of an aged mouse (27 months old) that was not exposed to any damaging stimulus (noise, ototoxicant, etc.). The inclusions contain numerous pigment granules (as intermediate cells are melanocytes). Scale bars = 5 μm.

of ototoxicants. In mammals, functional deficits resulting from HC loss are permanent because the organ of Corti does not regenerate, unlike the sensory epithelia in the inner ear of birds and other nonmammalian vertebrates.[38–41] Many of these agents may also affect the vestibular system, but few agents have been identified that are exclusively vestibulotoxic. The mammalian vestibular apparatus can regenerate HCs to some extent.[42–44] In general, supporting cells that surround HC are resistant to ototoxic insults. Instead, the death of an HC leads to expansion of the supporting cells as a reparative response.[45–48]

The pattern of HC damage induced by many otic insults is quite distinct and reproducible. In general, OHCs die first, beginning at the cochlear base and spreading to the apex, while loss of IHC is delayed. This pattern is induced by the ototoxic aminoglycoside antibiotics[49] and *cis*-platinum (*cis*-dichlorodiammine platinum II or *cis*-DDP[50,51]), but also by aging,[11] some genetic mutations,[52–55] and excessive noise.[56] However, some ototoxic agents can induce a different pattern of HC damage. For example, toluene and trichloroethylene (TCE) damage OHCs first in the middle portion of the cochlea. Carboplatin is a notable exception because its effects, although similar to those of *cis*-DPP in many species, preferentially injure IHCs in chinchillas.[57] The nature of this species- and site-specific vulnerability is not known.

The occurrence and extent of ototoxicity depend on many factors, such as the health status of the exposed individual and the potential for synergistic interactions among drugs. For example, stresses such as malnutrition and infection may increase inner ear sensitivity.[58] Certain genetic factors

may predispose the ear to ototoxicant-induced damage; an example is the A1555G mutation in a mitochondrial gene that results in enhanced sensitivity to aminoglycosides.[59,60] Simultaneous treatment with aminoglycosides and loop diuretics induces rapid, extensive HC loss[49] and profound deafness. Many ototoxic agents (e.g., aminoglycosides, polypeptide antibiotics, anti-neoplastics) are also nephrotoxic, which may reduce drug clearance and maintain high serum levels for extended periods. The multiplicity of factors indicates that it is often difficult to predict the effect of administering a potentially ototoxic drug. Furthermore, the level of exposure required to elicit permanent damage is variable. Aminoglycosides, *cis*-DPP, toluene, and TCE are usually detrimental only after repeated treatment, whereas trimethyltin (TMT) is ototoxic after one administration.

CLASSES OF OTOTOXIC AGENTS

Ototoxic chemicals may be categorized into the three broad groups based on their selective actions on different tissues within the inner ear. The following discussion is limited to those agents that elicit structural changes which may be detected by morphological means.

Agents That Selectively and Reversibly Affect Sensory Epithelia

Salicylate and *quinine* cause temporary threshold shift (TTS) across most of the detectable frequency range, and also tinnitus (the detection of a sound in the absence of a sound stimulus). These agents cause TTS across all frequencies, indicating that the entire cochlear spiral is affected at the same time. Salicylate and quinine act by entering the perilymph and disrupting the function of OHC.[61,62] Only quinine has been shown to induce ultrastructural changes in the OHC, and then only in vitro at extremely high concentrations.[63,64] Electrophysiological evidence suggests that the primary site of quinine toxicity might be on IHC synaptic transmission rather than OHCs, but morphological evidence to support this hypothesis has not been reported.

Agents That Selectively Affect Ion-Transporting Epithelia

Some ototoxicants primarily alter endolymph production from the stria vascularis in the cochlea and/or the dark cells in the vestibular apparatus. Such agents, exemplified by the loop diuretics (e.g., bumetanide, ethacrynic acid, furosemide, and piretanide) and macrolide antibiotics (e.g., erythromycin), generally elicit acute, reversible disruption after a single high-dose intravenous administration.[65,66] Repeated diuretic treatment does not appear to cause permanent damage except in rare instances.[67]

Histopathological lesions in the inner ear of patients who died undergoing diuretic therapy[68,69] and in experimental animals[70–72] exhibit profound edema of the stria vascularis. Similar swelling has been described in inner tissues from human patients who died during macrolide antibiotic treatment.[73] The impact of the antibiotics appears to be less pronounced than that of the diuretics.

The rapid onset of tissue swelling in the stria vascularis suggests that diuretics gain direct access to their site of action via the strial vasculature. The edema suggests that the major mechanism is inhibition of ion transport; the resulting accumulation of ions in the extracellular spaces would be confined by the tight junctions sealing the basal cell and marginal cell layers of the stria, resulting in osmotic uptake of fluid. Both cochlear marginal cells[74] and vestibular dark cells[75] are susceptible to the transient diuretic-induced disruption of $Na^+/K^+/Cl^-$ co-transporters in their basolateral membranes. A similar target on the basolateral membrane of marginal cells is Na^+/K^+-ATPase, which occurs at high concentrations.[76] A second mechanism proposed for the ototoxic effects of diuretics is disturbed oxidative metabolism, as the stria vascularis is one of the most metabolically active tissues in the body.[19] An exact biochemical pathway by which diuretics might affect oxygen delivery and/or utilization has not been identified. Similarly, mechanisms that might explain the ototoxicity of the macrolide antibiotics have not been defined.

Agents Inducing Permanent Damage to Sensory and Ion-Transporting Epithelia

Many ototoxicants cause permanent hearing impairment and/or balance dysfunction by killing sensory HCs. The major agents for which this capacity has been demonstrated are the aminoglycoside antibiotics, cis-platinum, and TMT. In most instances, ototoxicity develops only after repeated systemic administration. However, TMT and on occasion some aminoglycosides can produce permanent deficits after a single exposure.

The *aminoglycoside antibiotics* are the most important group of ototoxic agents in the clinical setting[58] as these molecules are selectively toxic to HCs in all vertebrate classes, including the developing HCs in the fetus[77] as well as nematocyst cells in marine and aquatic invertebrates.[78] All aminoglycosides are potentially both cochleotoxic and vestibulotoxic, but the different variants exhibit divergent preferences in their toxic potential and target organ. Electrophysiological and in vitro assessment[79,80] have shown that neomycin is the most toxic; gentamicin, kanamycin, and tobramycin less so; and amikacin and netilmicin least toxic. In the human ear, gentamicin and streptomycin are more vestibulotoxic, whereas amikacin and neomycin are primarily cochleotoxic. The mechanisms for such site predilections are not known but are not related to tissue-specific drug concentration.[58] Strain- and species-specific variations in vulnerability can be used to investigate potential mechanisms of aminoglycoside cochleotoxicity and vestibulotoxicity.[81] However, pronounced species differences in susceptibility to the different aminoglycosides[82] can foil predictions of ototoxic potential in new molecules of this class. For example, dihydrostreptomycin is markedly toxic to the human cochlea but is relatively innocuous to the cochlea in all animal models, including macaques, except for the patas monkey (*Erythrocebus patas*);[83] streptomycin produces hearing loss in patas monkeys when given at a dose that is not cochleotoxic to humans.[84] Interestingly, the marked sensitivity of the patas cochlea to aminoglycoside toxicity appears to be specific for streptomycin, as the lesions induced in the patas by other aminoglycosides are similar to those produced in other species.[85]

The effects of aminoglycosides usually appear only after prolonged parenteral treatment. The severity of the impairment progresses over time, even after treatment has ceased, probably as a consequence of the very slow turnover of the agent within the inner ear.[86,87] Early HC damage is concentrated in the basal cochlea,[88] but HC damage spreads toward the cochlear apex over time and/or with higher doses. In a given cochlear region, the OHCs are more susceptible than IHCs.[49] Loss of HCs leads to supporting cell expansion[45–48] that finally replaces the entire organ of Corti. With the loss of the IHCs, their associated spiral ganglion neurons also die,[84,89] probably because IHC-derived neurotrophic factors are no longer available.[90,91] In contrast, in the vestibular apparatus aminoglycoside-induced HC loss is seen initially in the central regions of the sensory epithelia (i.e., at the crests of the saddle-shaped cristae and across the central "striola" of the maculae), after which it spreads peripherally.[92] The HCs of the vestibular organs exhibit differential sensitivity by both cell type and location. Type I HC are more vulnerable than type II cells;[93] loss of HC is greatest in the cristae, somewhat reduced in the utricle, and least in the saccule.[89] The reason for this differential vulnerability is not known. Damage to the vestibular HCs in mammals can be reversed to a limited extent[43] by HC regeneration[38,40] and/or transdifferentiation[41] from the adjacent supporting cells.

Several mechanisms appear to underlie the ototoxic potential of aminoglycosides. First, these agents are specifically taken into cochlear HCs,[94,95] apparently through the transduction channels[96] concentrated at the apical poles of the sensory cells.[14,26] Mutant mice with HCs that lack functional transduction machinery cannot imbibe aminoglycosides and thus do not develop HC damage.[97] The ability of these molecules to enter via the transduction channels probably explains the selective sensitivity of the OHCs in general, and those of the basal cochlea in particular. The probability that the transduction channels of HCs are in the "open" state is approximately 50% for the basal OHCs, progressively less for the OHCs nearer the apex,[98] and only about 5% for IHC. However, differential absorption is not the sole mechanism, as both OHCs and IHCs in cultured murine organs of Corti take up gentamicin, even though only OHCs are susceptible.[97] A second proposed mechanism is free-radical damage to HCs following aminoglycoside exposure,[99] possibly mediated by the formation of iron–aminoglycoside complexes.[58] Administration of agents that deplete[100] or augment[24,101–103] free radical–scavenging systems can enhance or attenuate ototoxicant-induced HC damage, respectively. A third potential mechanism is the irreversible interaction of aminoglycosides with phosphatidylinositol 4,5-bisphosphate (PIP$_2$),[79] a cell membrane component located on the cytosolic side of the plasmalemma. Tissues susceptible to aminoglycoside toxicity, including the inner ear and kidney, have a more active metabolism of phosphoinositides than do more resistant tissues. Possible means by which an aminoglycoside–PIP$_2$ combination might act include conformational alterations in the membrane[104] and generation of free radicals (as the preferred electron donor arachidonic acid is one building block of PIP$_2$).[105] Regardless of the mechanism(s), loss of HCs in both the cochlea[24,45,106] and vestibular apparatus[48] appears to proceed by apoptosis. Caspase inhibitors can prevent HC death in cultured utricular maculae exposed to gentamicin,[24] and may have wider value in the clinical setting as well.[107,108]

Although HCs are the primary site of aminoglycoside action, the stria vascularis may also be affected in the initial stages of the cochleotoxic response. Strial alterations including apoptosis of marginal cells have been identified in conjunction with the earliest organ of Corti lesions in animals given aminoglycosides.[109] Similarly, the volume of the stria vascularis in human patients is decreased within two weeks of aminoglycoside treatment,[110] probably as a consequence of marginal cell atrophy.[111] These cells accumulate gentamicin as rapidly as do HCs.[112] A possible mechanism for the strial lesions is that the level of PIP$_2$ in the stria is even higher than that of the organ of Corti.[79] It is not yet clear whether the strial changes are a secondary response to sensory epithelial damage and altered endolymphatic homeostasis or a primary insult to an independent cell population. The endocochlear potential can be maintained for some time by the shrunken stria, suggesting that the tissue has either significant functional redundancy or a reserve.

cis-Platinum (*cis*-DDP) is an ototoxic and nephrotoxic anticancer agent[113] that causes hearing impairment in 75% or more of human patients. This ototoxicant induces dose-dependent, progressive OHC loss in the cochlea[114] beginning at the cochlear base,[50,115] as well as HC loss in the vestibular apparatus.[116,117] As with aminoglycosides, the IHCs are relatively resistant.[51] In addition, *cis*-DDP appears to induce direct toxicity on the neural elements in the cochlea,[118] including a progressive rise in myelin detachment from spiral ganglion neuronal processes (i.e., the cells that innervate the unaffected IHCs) that parallels the loss of cochlear OHCs. The site specificity may be altered by manipulating the dosing regimen.[113] Exposure to *cis*-DDP results in atrophy of the stria vascularis in association with a reduced endocochlear potential.[119,120] The metal accumulates in the strial marginal cells as well as some fibrocytes of the spiral ligament,[118] and it can initiate cochlear-wide marginal cell apoptosis[121,122] even when damage to the sensory epithelium remains confined to the cochlear base. Apoptosis also occurs in spiral ligament fibrocytes cultured in the presence of *cis*-DDP,[123] but to date this lesion has not been confirmed in vivo. Thus, the organ of Corti and the stria vascularis appear to represent separate targets for the cochleotoxic actions of *cis*-DDP.

The mechanisms of *cis*-DDP ototoxicity are not fully understood. A known mechanism of toxicity for both the organ of Corti and the stria vascularis is that this drug leads to free radical generation. The pathways involved appear to include inhibition of enzymes or depletion of factors essential for free radical disposal.[113,124–126] Agents that preserve free radical scavenging capacity *cis*-DDP ototoxicity both in vitro[127] and in vivo.[128] A second possibility is that *cis*-DDP blocks the apical transduction channels in HCs,[129] but evidence to support this premise are not yet conclusive.

Trimethyl tin (TMT) is an industrial chemical with known neurotoxic potential. Interestingly, TMT is ototoxic at levels far below those that are neurotoxic.[130,131] The cochleotoxic lesions induced by TMT in guinea pigs[132] and rats[133] resemble those produced by aminoglycosides and *cis*-DDP, namely a loss of OHCs in the basal cochlea that progresses to involve the more apical cochlea over time. However, unlike the drugs, this organic metal elicits profound functional and structural changes after a single parenteral injection. The consequence is a precipitous rise in TMT within the perilymph, followed by chronologically distinct actions at two different targets on the perilymphatic side of the organ of Corti. The first effect is an acute, reversible decrease in afferent neurotransmission at the IHCs along the entire cochlear spiral. This functional change is followed quickly by region-specific OHC death beginning at the cochlear base. Exposure to TMT can also cause lesions in the stria vascularis, including reduced tissue thickness, increased capillary diameter, but reduced vascular density.[134] These strial changes are most prominent in the middle and apical cochlear turns, whereas OHC loss is greatest in the basal coil, suggesting that the effects in the stria and organ of Corti are independent. The cochleotoxic mechanisms of TMT are not well characterized. Hypotheses applicable to the organ of Corti are altered OHC length, leading to aberrant micromechanical responsiveness[130,131,134] and disruption of ionic gradients.[135] Altered intracellular calcium storage within spiral ganglion neurons (the relay center connecting the cochlear IHCs with the brain stem) may also be involved.[135] Uncoupling of oxidative phosphorylation in the metabolically active stria vascularis has been proposed as a toxic mechanism for this tissue.[134]

Organic solvent exposures at high concentrations are typically known for their neurotoxic effects, but solvent-induced ototoxic damage has been recognized in chemical industry workers and solvent abusers.[136] Solvents confirmed to be ototoxic in rats include allylbenzene, ethylbenzene, α-methylstyrene, *trans*-β-methylstyrene, *n*-propylbenzene, styrene, toluene, trichloroethylene (TCE), and *p*-xylene.[137–139] Some species, including guinea pigs[140,141] or chinchillas,[142] are quite resistant to ototoxic damage. Therefore, the rat is considered to be a more appropriate model for assessing ototoxic risk to humans.[143]

The ototoxic effects of solvents are usually ascribed to the cochlea, but vestibular dysfunction has also been reported.[144] Concomitant exposure to solvents and high noise levels may exacerbate the hearing impairment induced by either condition alone.[145] All solvents yield a similar pattern of ototoxicity[137–139,143] resulting from selective OHC elimination.[146,147] However, unlike most ototoxicants, the middle region of the cochlea seems to be more susceptible than the base,[147,148] and supporting cells can also sustain damage.[147] Neurons in the spiral ganglion may also be a target of TCE.[138]

The mechanism of solvent-induced ototoxicity probably reflects tissue deposition of the chemicals because these agents are essentially insoluble in the inner ear fluids. Solvents are thought to reach the inner ear via the vessels in the stria vascularis and then pass to the organ of Corti within the tissues.[143,149] The initial damage in the organ of Corti appears to be to the supporting cells that surround the HCs.[150] Dysfunction of the supporting cells prevents their removal of K^+ from the sensory epithelium, thereby leading to excessive K^+ accumulation near OHCs. Solvents may also alter membrane integrity of supporting cells and HCs, leading to cell swelling and death of the organ of Corti. Cell death in the inner ear following solvent exposure has been attributed to both apoptosis[150] and necrosis.[143] The impact (if any) of solvents on the stria vascularis is not known.

Morphological Effects of Other Putative Ototoxiants

Other than the agents listed above, incontrovertible evidence for the ototoxicity of the other agents listed in Table 1 is sparse. Viomycin elicits vestibulotoxic damage following chronic administration[151,152] as a result of HC death.[153] Repeated high-dose administration of the iron-chelating agent desferrioxamine (DFX) induces HC loss in the avian basal cochlea[154] and appears to cause high-frequency hearing loss in about 20 to 40% of human patients.[155–157] Direct DFX-induced ototoxicity has not been confirmed in other clinical studies,[158,159] nor in experimental studies with a mammalian model (chinchilla).[160] A suitable explanation for this discrepancy has not been achieved.

OTOTOXIC INTERACTIONS

Human epidemiological and animal experimental studies have demonstrated that simultaneous exposure to both an ototoxic agent and extreme noise amplify the hearing impairment relative to either condition alone. This synergistic effect has been shown for both aminoglycosides[161] and organic solvents.[136] This effect may be evident at levels of noise and doses of ototoxicant that would not be expected to cause damage if experienced in the absence of the other factor. In a like manner, coexposure to multiple ototoxic agents at nontoxic doses (e.g., an aminoglycoside or *cis*-platinum given with a loop diuretic[49,70,162]) may produce significant damage to the organ of Corti (HC loss) and the stria vascularis (marginal cell death). The mechanism of toxicity appears to be heightened uptake of the co-toxicant[112] resulting from a diuretic-induced increase in co-toxicant distribution into endolymph.[163] Interestingly, animals raised in relatively quiet rooms take up less gentamicin then do animals raised in a facility having "normal" levels of noise,[164] suggesting that background noise may prime transduction channels to assume an "open" state suitable for toxicant entry. An elevated noise level also facilitates generation of free radicals in the inner ear.[165,166] However, no systematic studies have been performed to assess potential mechanisms by which concomitant exposures to ototoxic stimuli exert their combinatorial effects.

SUMMARY

The inner ear is a histologically complex structure in which many different cell types serve a range of functions necessary for the effective detection and transduction of mechanosensory data. All these cell types are potential targets for ototoxic agents. Structural integrity of the various cell and tissues types can be evaluated effectively only in properly prepared samples, typically those in which the cochlea has been isolated and perfusion-fixed prior to embedding. The main structural lesion associated with impaired hearing (cochleotoxicity) or balance (vestibulotoxicity) is loss of hair cells in the sensory epithelia. However, atrophy of the cochlear stria vascularis as a consequence of marginal cell death is an important secondary lesion induced by some ototoxic agents (e.g., aminoglycosides and *cis*-platinum) and certain environmental (e.g., excessive noise) and physiological (e.g., aging) factors. The sensory epithelia and stria vascularis are independent targets for some agents (e.g., *cis*-platinum), while the sequence in which damage develops at these two sites as well as the relationship (independent or linked) is unclear for other molecules. Damage to the stria vascularis and/or the spiral ligament[56,167] is likely to affect cochlear homeostasis, which will affect the function and perhaps the very survival of sensory cells. This interdependence emphasizes the importance of evaluating the entire cochlea rather than only the sensory epithelium to determine the ototoxic potential of any molecule.

REFERENCES

1. Forge A, Taylor R, Harpur ES. Ototoxicity. In: Ballantyne B, Marrs T, Syversen T, eds. *General and Applied Toxicology*. Hoboken, NJ: Wiley-Blackwell; 2009.

2. Raphael Y, Altschuler RA. Structure and innervation of the cochlea. *Brain Res Bull*. 2003;60:397–422.

3. Engmér C, Laurell G, Bagger-Sjöbäck D, Rask-Andersen H. Immunodefense of the round window. *Laryngoscope*. 2008;118:1057–1062.

4. Jones SM, Jones TA, Johnson KR, Yu H, Erway LC, Zheng QY. A comparison of vestibular and auditory phenotypes in inbred mouse strains. *Brain Res*. 2006;1091:40–46.

5. Spicer SS, Schulte BA. Spiral ligament pathology in quiet-aged gerbils. *Hear Res*. 2002;172:172–185.

6. Wangemann P. K$^+$ cycling and the endocochlear potential. *Hear Res*. 2002;165:1–9.

7. Spicer SS, Schulte BA. The fine structure of spiral ligament cells relates to ion return to the stria and varies with place-frequency. *Hear Res*. 1996;100:80–100.

8. Zheng QY, Johnson KR, Erway LC. Assessment of hearing in 80 inbred strains of mice by ABR threshold analyses. *Hear Res*. 1999;130:94–107.

9. Ding D, McFadden SL, Salvi R. Cochlear hair cell densities and inner-ear staining techniques. In: Willott JF, ed. *Handbook of Mouse Auditory Research:From Behavior to Molecular Biology*. Boca Raton, FL: CRC Press; 2001: 189–204.

10. Thorne PR, Gavin JB. The accuracy of hair cell counts in determining distance and position along the organ of Corti. *J Acoust Soc Am*. 1984;76:440–442.

11. Wright A, Davis A, Bredberg G, Ulehlova L, Spencer H. Hair cell distributions in the normal human cochlea. *Acta Otolaryngol Suppl*. 1987;444:1–48.

12. Ashmore JF, Kolston PJ. Hair cell based amplification in the cochlea. *Curr Opin Neurobiol*. 1994;4:503–508.

13. Dallos P, Fakler B. Prestin, a new type of motor protein. *Nat Rev Mol Cell Biol*. 2002;3:104–111.

14. Hudspeth AJ. How the ear's works work. *Nature*. 1989;341:397–404.

15. Goodyear RJ, Marcotti W, Kros CJ, Richardson GP. Development and properties of stereociliary link types in hair cells of the mouse cochlea. *J Comp Neurol*. 2005;485:75–85.

16. Pickles JO, Comis SD, Osborne MP. Cross-links between stereocilia in the guinea pig organ of Corti, and their possible relation to sensory transduction. *Hear Res*. 1984;15:103–112.

17. Rhys Evans PH, Comis SD, Osborne MP, Pickles JO, Jeffries DJ. Cross-links between stereocilia in the human organ of Corti. *J Laryngol Otol*. 1985;99:11–19.

18. Pickles JO. *An Introduction to the Physiology of Hearing*, 2nd ed. London: Academic Press; 1988.

19. Wangemann PSJ. Homeostatic mechanisms in the cochlea. In: Dallos P, Popper AN, Fay RR, eds. *The Cochlea*. New York: Springer-Verlag; 1996: 130–185.

20. Ito M, Spicer SS, Schulte BA. Immunohistochemical localization of brain type glucose transporter in mammalian inner ears:comparison of developmental and adult stages. *Hear Res*. 1993;71:230–238.

21. Salt AN, Plontke SK. Local inner-ear drug delivery and pharmacokinetics. *Drug Discov Today*. 2005;10: 1299–1306.

22. Davies S, Forge A. Preparation of the mammalian organ of Corti for scanning electron microscopy. *J Microsc*. 1987;147:89–101.

23. Viberg A, Canlon B. The guide to plotting a cochleogram. *Hear Res*. 2004;197:1–10.

24. Forge A, Li L. Apoptotic death of hair cells in mammalian vestibular sensory epithelia. *Hear Res*. 2000;139:97–115.

25. Assad JA, Shepherd GM, Corey DP. Tip-link integrity and mechanical transduction in vertebrate hair cells. *Neuron*. 1991;7:985–994.

26. Bryant J, Forge A, Richardson GP. The differentiation of hair cells. In: Kelley MW, Wu DK, Popper AN, Fay RR, eds. *Development of the Inner Ear*. New York: Springer-Verlag; 2005.

27. Osborne MP, Comis SD. High resolution scanning electron microscopy of stereocilia in the cochlea of normal, postmortem, and drug-treated guinea pigs. *J Electron Microsc Tech*. 1990;15:245–260.

28. Liberman MC. Quantitative assessment of inner ear pathology following ototoxic drugs or acoustic trauma. *Toxicol Pathol.* 1990;18:138–148.

29. Nuttall AL. Perfusion of aminoglycosides in perilymph. In: Lerner SA, Matz GJ, Hawkins JE, eds. *Aminoglycoside Ototoxicity.* Boston: Little, Brown; 1981: 51–61.

30. Matsui JI, Gale JE, Warchol ME. Critical signaling events during the aminoglycoside-induced death of sensory hair cells in vitro. *J Neurobiol.* 2004;61:250–266.

31. Matsui JI, Ogilvie JM, Warchol ME. Inhibition of caspases prevents ototoxic and ongoing hair cell death. *J Neurosci.* 2002;22:1218–1227.

32. Mangiardi DA, McLaughlin-Williamson K, May KE, Messana EP, Mountain DC, Cotanche DA. Progression of hair cell ejection and molecular markers of apoptosis in the avian cochlea following gentamicin treatment. *J Comp Neurol.* 2004;475:1–18.

33. Williams JA, Holder N. Cell turnover in neuromasts of zebrafish larvae. *Hear Res.* 2000;143:171–181.

34. Ton C, Parng C. The use of zebrafish for assessing ototoxic and otoprotective agents. *Hear Res.* 2005;208:79–88.

35. Ou HC, Raible DW, Rubel EW. Cisplatin-induced hair cell loss in zebrafish *Danio rerio* lateral line. *Hear Res.* 2007;233:46–53.

36. Chiu LL, Cunningham LL, Raible DW, Rubel EW, Ou HC. Using the zebrafish lateral line to screen for ototoxicity. *J Assoc Res Otolaryngol.* 2008;9:178–190.

37. Balak KJ, Corwin JT, Jones JE. Regenerated hair cells can originate from supporting cell progeny: evidence from phototoxicity and laser ablation experiments in the lateral line system. *J Neurosci.* 1990;10:2502–2512.

38. Cotanche DA. Structural recovery from sound and aminoglycoside damage in the avian cochlea. *Audiol Neurootol.* 1999;4:271–285.

39. Staecker H, Van De Water TR. Factors controlling hair-cell regeneration/repair in the inner ear. *Curr Opin Neurobiol.* 1998;8:480–487.

40. Stone JS, Oesterle EC, Rubel EW. Recent insights into regeneration of auditory and vestibular hair cells. *Curr Opin Neurol.* 1998;11:17–24.

41. Taylor RR, Forge A. Hair cell regeneration in sensory epithelia from the inner ear of a urodele amphibian. *J Comp Neurol.* 2005;484:105–120.

42. Forge A, Li L, Corwin JT, Nevill G. Ultrastructural evidence for hair cell regeneration in the mammalian inner ear. *Science.* 1993;259:1616–1619.

43. Forge A, Li L, Nevill G. Hair cell recovery in the vestibular sensory epithelia of mature guinea pigs. *J Comp Neurol.* 1998;397:69–88.

44. Kopke RD, Jackson RL, Li G, et al. Growth factor treatment enhances vestibular hair cell renewal and results in improved vestibular function. *Proc Natl Acad Sci USA.* 2001;98: 5886–5891.

45. Forge A. Outer hair cell loss and supporting cell expansion following chronic gentamicin treatment. *Hear Res.* 1985;19:171–182.

46. Raphael Y, Altschuler RA. Scar formation after drug-induced cochlear insult. *Hear Res.* 1991;51:173–183.

47. Meiteles LZ, Raphael Y. Scar formation in the vestibular sensory epithelium after aminoglycoside toxicity. *Hear Res.* 1994;79:26–38.

48. Li L, Nevill G, Forge A. Two modes of hair cell loss from the vestibular sensory epithelia of the guinea pig inner ear. *J Comp Neurol.* 1995;355:405–417.

49. Taylor RR, Nevill G, Forge A. Rapid hair cell loss: a mouse model for cochlear lesions. *J Assoc Res Otolaryngol.* 2008;9:44–64.

50. Laurell G, Bagger-Sjöbäck D. Degeneration of the organ of Corti following intravenous administration of cisplatin. *Acta Otolaryngol.* 1991;111:891–898.

51. Kaltenbach JA, Church MW, Blakley BW, McCaslin DL, Burgio DL. Comparison of five agents in protecting the cochlea against the ototoxic effects of cisplatin in the hamster. *Otolaryngol Head Neck Surg.* 1997;117:493–500.

52. Steel KP, Kros CJ. A genetic approach to understanding auditory function. *Nat Genet.* 2001;27:143–149.

53. Boettger T, Hübner CA, Maier H, Rust MB, Beck FX, Jentsch TJ. Deafness and renal tubular acidosis in mice lacking the K–Cl co-transporter Kcc4. *Nature.* 2002;416:874–878.

54. Cohen-Salmon M, Ott T, Michel V, et al. Targeted ablation of connexin 26 in the inner ear epithelial gap junction network causes hearing impairment and cell death. *Curr Biol.* 2002;12:1106–1111.

55. Rozengurt N, Lopez I, Chiu CS, Kofuji P, Lester HA, Neusch C. Time course of inner ear degeneration and deafness in mice lacking the Kir4.1 potassium channel subunit. *Hear Res.* 2003;177:71–80.

56. Wang Y, Hirose K, Liberman M. Dynamics of noise-induced cellular injury and repair in the mouse cochlea. *J Assoc Res Otolaryngol.* 2002;3:248–268.

57. Hofstetter P, Ding D, Powers N, Salvi RJ. Quantitative relationship of carboplatin dose to magnitude of inner and outer hair cell loss and the reduction in distortion product otoacoustic emission amplitude in chinchillas. *Hear Res.* 1997;112:199–215.

58. Forge A, Schacht J. Aminoglycoside antibiotics. *Audiol Neurootol.* 2000;5:3–22.

59. Prezant TR, Agapian JV, Bohlman MC, et al. Mitochondrial ribosomal RNA mutation associated with both antibiotic-induced and non-syndromic deafness. *Nat Genet.* 1993;4:289–294.

60. Usami S, Abe S, Tono T, Komune S, Kimberling WJ, Shinkawa H. Isepamicin sulfate-induced sensorineural hearing loss in patients with the 1555 A → G mitochondrial mutation. *J Otorhinolaryngol Relat Spec.* 1998;60:164–169.

61. Shehata WE, Brownell WE. Effects of salicylate on shape, electromotility and membrane characteristics of isolated outer hair cells from guinea pig cochlea. *Acta Otolaryngol.* 1991;111:707–718.

62. Tunstall MJ, Gale JE, Ashmore JF. Action of salicylate on membrane capacitance of outer hair cells from the guinea-pig cochlea. *J Physiol.* 1995;485:739–752.

63. Karlsson KK, Flock A. Quinine causes isolated outer hair cells to change length. *Neurosci Lett.* 1990;116:101–105.

64. Karlsson KK, Flock B, Flock A. Ultrastructural changes in the outer hair cells of the guinea pig cochlea after exposure to quinine. *Acta Otolaryngol.* 1991;111:500–505.

65. Rybak LP, Whitworth C, Scott V. Comparative acute ototoxicity of loop diuretic compounds. *Eur Arch Otorhinolaryngol.* 1991;248:353–357.

66. Liu J, Marcus DC, Kobayashi T. Inhibitory effect of erythromycin on ion transport by stria vascularis and vestibular dark cells. *Acta Otolaryngol.* 1996;116:572–575.

67. Rybak LP. Ototoxicity of ethacrynic acid a persistent clinical problem. *J Laryngol Otol.* 1988;102:518–520.

68. Arnold W, Nadol JB, Weidauer H. Temporal bone histopathology in human ototoxicity due to loop diuretics. *Scand Audiol.* 1981;14(Suppl):201–213.

69. Matz GJ. The ototoxic effects of ethacrynic acid in man and animals. *Laryngoscope.* 1976;86:1065–1086.

70. Brummett RE. Effects of antibiotic–diuretic interactions in the guinea pig model of ototoxicity. *Rev Infect Dis.* 1981;3 (Suppl):216–223.

71. Brummett RE, Fox KE. Studies of aminoglycoside ototoxicity in animal models. In: Whelton A, Neu HC, eds. *The Aminoglycosides Microbiology, Clinical Use and Toxicology.* New York: Marcel Dekker; 1982: 419–451.

72. Pike DA, Bosher SK. The time course of the strial changes produced by intravenous furosemide. *Hear Res.* 1980;3:79–89.

73. McGhan LJ, Merchant SN. Erythromycin ototoxicity. *Otol Neurotol.* 2003;24:701–702.

74. Wangemann P, Liu J, Marcus D. Ion transport mechanisms responsible for K^+ secretion and the transepithelial voltage across marginal cells of stria vascularis in vitro. *Hear Res.* 1995;84:19–29.

75. Marcus DC, Marcus NY. Transepithelial electrical responses to Cl^- of nonsensory region of gerbil utricle. *Biochim Biophys Acta.* 1989;987:56–62.

76. Tawackoli W, Chen G-D, Fechter LD. Disruption of cochlear potentials by chemical asphyxiants: cyanide and carbon monoxide. *Neurotoxicol Teratol.* 2001;23:157–165.

77. Rasmussen F. The ototoxic effect of streptomycin and dihydrostreptomycin on the foetus. *Scand J Resp Dis.* 1969;50:61–67.

78. Watson GM, Mire P, Hudson RR. Hair bundles of sea anemones as a model system for vertebrate hair bundles. *Hear Res.* 1997;107:53–66.

79. Schacht J. Molecular mechanisms of drug-induced hearing loss. *Hear Res.* 1986;22:297–304.

80. Kotecha B, Richardson GP. Ototoxicity in vitro: effects of neomycin, gentamicin, dihydrostreptomycin, amikacin, spectinomycin, neamine, spermine and poly-L-lysine. *Hear Res.* 1994;73:173–184.

81. Wu WJ, Sha SH, McLaren JD, Kawamoto K, Raphael Y, Schacht J. Aminoglycoside ototoxicity in adult CBA, C57BL and BALB mice and the Sprague–Dawley rat. *Hear Res.* 2001;158:165–178.

82. Harpur ES. Ototoxicity: morphological and functional correlates between experimental and clinical studies. In: Ballantyne J, ed. *Perspectives in Basic and Applied Toxicology.* London: Wright; 1987: 42–69.

83. Hawkins JE Jr, Stebbins WC, Johnsson L-G, Moody DB, Muraski A. The patas monkey as a model for dihydrostreptomycin ototoxicity. *Acta Otolaryngol.* 1977;83:123–129.

84. Hawkins JE Jr, Johnsson L-G. Histopathology of cochlear and vestibular ototxicity in laboratory animals. In: Lerner SA, Matz GJ, Hawkins JE Jr, eds. *Aminoglycoside Ototoxicity.* Boston: Little, Brown; 1981:175–195.

85. Stebbins WC, Moody DB, Hawkins JE Jr, Johnsson L-G, Norat MA. The species-specific nature of the ototoxicity of dihydrostreptomycin in the patas monkey. *Neurotoxicology.* 1987;8:33–44.

86. Tran Ba Huy P, Meulemans A, Wassef M, Manuel C, Sterkers O, Amiel C. Gentamicin persistence in rat endolymph and perilymph after a two-day constant infusion. *Antimicrob Agents Chemother.* 1983;23:344–346.

87. Tran Ba Huy P, Bernard P, Schacht J. Kinetics of gentamicin uptake and release in the rat: comparison of inner ear tissues and fluids with other organs. *J Clin Invest.* 1986;77: 1492–1500.

88. Fausti SA, Rappaport BZ, Schechter MA, Frey RH, Ward TT, Brummett RE. Detection of aminoglycoside ototoxicity by high-frequency auditory evaluation: selected case studies. *Am J Otolaryngol.* 1984;5:177–182.

89. Wersäll J. Structural damage to the organ of Corti and the vestibular epithelia caused by aminoglycoside antibiotics in the guinea pig. In: Lerner SA, Matz GJ, Hawkins JE Jr, eds. *Aminoglycoside Ototoxicity.* Boston: Little, Brown; 1981: 197–214.

90. Ylikoski J, Pirvola U, Moshnyakov M, Palgi J, Arumäe U, Saarma M. Expression patterns of neurotrophin and their receptor mRNAs in the rat inner ear. *Hear Res.* 1993;65:69–78.

91. Ernfors P, Van de Water TR, Loring J, Jaenisch R. Complementary roles of BDNF and NT-3 in vestibular and auditory development. *Neuron.* 1995;14:1153–1164.

92. Lindeman HH. Regional differences in sensitivity of the vestibular sensory epithelia to ototoxic antibiotics. *Acta Otolaryngol.* 1969;67:177–189.

93. Lyford-Pike S, Vogelheim C, Chu E, Della Santina CC, Carey JP. Gentamicin is pimarily localized in vestibular type I hair cells after intratympanic administration. *J Assoc Res Otolaryngol.* 2007;8:497–508.

94. Hiel H, Bennani H, Erre JP, Aurousseau C, Aran JM. Kinetics of gentamicin in cochlear hair cells after chronic treatment. *Acta Otolaryngol.* 1992;112:272–277.

95. Hiel H, Erre JP, Aurousseau C, Bouali R, Dulon D, Aran JM. Gentamicin uptake by cochlear hair cells precedes hearing impairment during chronic treatment. *Audiology.* 1993;32:78–87.

96. Marcotti W, van Netten SM, Kros CJ. The aminoglycoside antibiotic dihydrostreptomycin rapidly enters mouse outer hair cells through the mechano-electrical transducer channels. *J Physiol.* 2005;567:505–521.

97. Richardson GP, Forge A, Kros CJ, Fleming J, Brown SD, Steel KP. Myosin VIIA is required for aminoglycoside accumulation in cochlear hair cells. *J Neurosci.* 1997;17:9506–9519.

98. Russell IJ, Kössl M. Sensory transduction and frequency selectivity in the basal turn of the guinea-pig cochlea. *Philos Trans R Soc Lond B.* 1992;336:317–324.

99. Sha S-H, Taylor R, Forge A, Schacht J. Differential vulnerability of basal and apical hair cells is based on intrinsic susceptibility to free radicals. *Hear Res.* 2001;155:1–8.

100. Hoffman DW, Jones-King KL, Whitworth CA, Rybak LP. Potentiation of ototoxicity by glutathione depletion. *Ann Otol Rhinol Laryngol.* 1988;97:36–41.

101. Song B-B, Schacht J. Variable efficacy of radical scavengers and iron chelators to attenuate gentamicin ototoxicity in guinea pig in vivo. *Hear Res.* 1996;94:87–93.

102. Garetz SL, Altschuler RA, Schacht J. Attenuation of gentamicin ototoxicity by glutathione in the guinea pig in vivo. *Hear Res.* 1994;77:81–87.

103. Sha S-H, Zajic G, Epstein CJ, Schacht J. Overexpression of copper/zinc-superoxide dismutase protects from kanamycin-induced hearing loss. *Audiol Neurootol.* 2001;6:117–123.

104. Forge A, Zajic G, Davies S, Weiner N, Schacht J. Gentamicin alters membrane structure as shown by freeze-fracture of liposomes. *Hear Res.* 1989;37:129–139.

105. Priuska EM, ClarkBaldwin K, Pecoraro VL, Schacht J. NMR studies of iron–gentamicin complexes and the implications for aminoglycoside toxicity. *Inorg Chim Acta.* 1998;73:85–91.

106. Jiang H, Sha S-H, Forge A, Schacht J. Caspase-independent pathways of hair cell death induced by kanamycin in vivo. *Cell Death Differ.* 2006;13:20–30.

107. Rybak LP, Whitworth CA. Ototoxicity: therapeutic opportunities. *Drug Discov Today.* 2005;10:1313–1321.

108. Lynch ED, Kil J. Compounds for the prevention and treatment of noise-induced hearing loss. *Drug Discov Today.* 2005;10:1291–1298.

109. Forge A, Fradis M. Structural abnormalities in the stria vascularis following chronic gentamicin treatment. *Hear Res.* 1985;20:233–244.

110. Kusunoki T, Cureoglu S, Schachern PA, et al. Effects of aminoglycoside administration on cochlear elements in human temporal bones. *Auris Nasus Larynx.* 2004;31:383–388.

111. Forge A, Wright A, Davies SJ. Analysis of structural changes in the stria vascularis following chronic gentamicin treatment. *Hear Res.* 1987;31:253–266.

112. Dai CF, Steyger PS. A systemic gentamicin pathway across the stria vascularis. *Hear Res.* 2008;235:114–124.

113. Rybak LP, Whitworth CA, Mukherjea D, Ramkumar V. Mechanisms of cisplatin-induced ototoxicity and prevention. *Hear Res.* 2007;226:157–167.

114. Hoeve LJ, Mertens zur Borg IR, Rodenburg M, Brocaar MP, Groen BG. Correlations between *cis*-platinum dosage and toxicity in a guinea pig model. *Arch Otorhinolaryngol.* 1988;245:98–102.

115. Laurell G, Bagger-Sjöbäck D. Dose-dependent inner ear changes after i.v. administration of cisplatin. *J Otolaryngol.* 1991;20:158–167.

116. Wright CG, Schaefer SD. Inner ear histopathology in patients treated with *cis*-platinum. *Laryngoscope.* 1982;92:1408–1413.

117. Cunningham LL. The adult mouse utricle as an in vitro preparation for studies of ototoxic-drug-induced sensory hair cell death. *Brain Res.* 2006;1091:277–281.

118. van Ruijven MW, de Groot JC, Klis SF, Smoorenburg GF. The cochlear targets of cisplatin:an electrophysiological and morphological time-sequence study. *Hear Res.* 2005;205:241–248.

119. Laurell G, Engström B. The ototoxic effect of cisplatin on guinea pigs in relation to dosage. *Hear Res.* 1989;38:27–33.

120. Klis SF, O'Leary SJ, Hamers FP, De Groot JC, Smoorenburg GF. Reversible cisplatin ototoxicity in the albino guinea pig. *Neuroreport.* 2000;11:623–626.

121. Alam SA, Ikeda K, Oshima T, Suzuki M, Kawase T, Kikuchi T, et al. Cisplatin-induced apoptotic cell death in Mongolian gerbil cochlea. *Hear Res.* 2000;141:28–38.

122. Watanabe K, Inai S, Jinnouchi K, Baba S, Yagi T. Expression of caspase-activated deoxyribonuclease CAD and caspase 3 CPP32 in the cochlea of cisplatin CDDP-treated guinea pigs. *Auris Nasus Larynx.* 2003;30:219–225.

123. Liang F, Schulte BA, Qu C, Hu W, Shen Z. Inhibition of the calcium- and voltage-dependent big conductance potassium channel ameliorates cisplatin-induced apoptosis in spiral ligament fibrocytes of the cochlea. *Neuroscience.* 2005;35:263–271.

124. Ravi R, Somani SM, Rybak LP. Mechanism of cisplatin ototoxicity: antioxidant system. *Pharmacol Toxicol.* 1995;76:386–394.

125. Lee JE, Nakagawa T, Kita T, et al. Mechanisms of apoptosis induced by cisplatin in marginal cells in mouse stria vascularis. *J Otorhinolaryngol Relat Spec.* 2004;66:111–118.

126. Mukherjea D, Whitworth CA, Nandish S, Dunaway GA, Rybak LP, Ramkumar V. Expression of the kidney injury molecule 1 in the rat cochlea and induction by cisplatin. *Neuroscience.* 2006;139:733–740.

127. Kopke RD, Liu W, Gabaizadeh R, et al. Use of organotypic cultures of Corti's organ to study protective effects of antioxidant molecules on cisplatin-induced damage of auditory hair cells. *Am J Otol.* 1997;18:559–571.

128. Campbell KC, Meech RP, Rybak LP, Hughes LF. D-Methionine protects against cisplatin damage to the stria vascularis. *Hear Res.* 1999;138:13–28.

129. Kimitsuki T, Nakagawa T, Hisashi K, Komune S, Komiyama S. Cisplatin blocks mechano-electric transducer current in chick cochlear hair cells. *Hear Res.* 1993;71:64–68.

130. Fechter LD, Young JS, Nuttall AL. Trimethyltin ototoxicity: evidence for a cochlear site of injury. *Hear Res.* 1986;23:275–282.

131. Fechter LD, Clerici WJ, Yao L, Hoeffding V. Rapid disruption of cochlear function and structure by trimethyltin in the guinea pig. *Hear Res.* 1992;58:166–174.

132. Clerici WJ, Ross B, Fechter LD. Acute ototoxicity of triakyltins in the guinea pig. *Toxicol Appl Pharmacol.* 1991;109:547–556.

133. Hoeffding V, Fechter LD. Trimethyltin disrupts auditory function and cochlear morphology in pigmented rats. *Neurotoxicol Teratol.* 1991;13:135–145.

134. Fechter LD, Carlisle L. Auditory dysfunction and cochlear vascular injury following trimethyltin exposure in the guinea pig. *Toxicol Appl Pharmacol.* 1990;105:133–143.

135. Liu Y, Fechter LD. Comparison of the effects of trimethyltin on the intracellular calcium levels in spiral ganglion cells and outer hair cells. *Acta Otolaryngol.* 1996;116:417–421.

136. Morata TC, Dunn DE, Sieber WK. Occupational exposure to noise and ototoxic organic solvents. *Arch Environ Health.* 1994;49:359–365.

137. Crofton KM, Lassiter TL, Rebert CS. Solvent-induced ototoxicity in rats: an atypical selection mid-frequency hearing deficit. *Hear Res.* 1994;80:25–30.

138. Fechter LD, Liu Y, Herr DW, Crofton KM. Trichloroethylene ototoxicity: evidence for a cochlear origin. *Toxicol Sci.* 1998;42:28–35.

139. Gagnaire F, Langlais C. Relative ototoxicity of 21 aromatic solvents. *Arch Toxicol.* 2005.

140. Fechter LD. Effects of acute styrene and simultaneous noise exposure on auditory function in the guinea pig. *Neurotoxicol Teratol.* 1993;15:151–155.

141. Lataye R, Campo P, Pouyatos B, Cossec B, Blachere V, Morel G. Solvent ototoxicity in the rat and guinea pig. *Neurotoxicol Teratol.* 2003;25:39–50.

142. Davis RR, Murphy WJ, Snawder JE, Striley CA, Henderson D, Khan A, et al. Susceptibility to the ototoxic properties of toluene is species specific. *Hear Res.* 2002;166:24–32.

143. Campo P, Maguin K. Solvent-induced hearing loss: mechanisms and prevention strategy. *Int J Occup Med Environ Health.* 2007;20:265–270.

144. Morata TC, Nylen P, Johnson AC, Dunn DE. Auditory and vestibular functions after single or combined exposure to toluene. *Arch Toxicol.* 1995;69:431–443.

145. Morata TC. Assessing occupational hearing loss:beyond noise exposures. *Scand Audiol.* 1988;27:111–116.

146. Pryor GT, Dickinson J, Feeney E, Rebert CS. Hearing loss in rats first exposed to toluene as weanlings or as young adults. *Neurobehav Toxicol Teratol.* 1984;6:111–119.

147. Campo P, Latayer R, Cossesc B, Placidi V. Toluene-induced hearing loss: a mid-frequency location of the cochlear lesions. *Neurotoxicol Teratol.* 1997;19:129–140.

148. Johnston AC, Canlon B. Toluene exposure affects the functional activity of the outer hair cells. *Hear Res.* 1994;72: 189–196.

149. Campo P, Lataye R, Loquet G, Bonnet P. Styrene-induced hearing loss: a membrane insult. *Hear Res.* 2001;154:170–180.

150. Chen GD, Chi LH, Kostyniak PJ, Henderson D. Styrene induced alterations in biomarkers of exposure and effects in the cochlea:mechanisms of hearing loss. *Toxicol Sci.* 2007;98:167–177.

151. Daly JF, Cohen NL. Viomycin ototoxicity in man: a cupulometric study. *Ann Otol Rhinol Laryngol.* 1965;74:521–534.

152. Nakayama M, Miura H, Kamei T. Investigation of vestibular damage by antituberculous drugs. *Acta Otolaryngol Suppl.* 1991;481:481–485.

153. Kanda T, Igarashi M. Ultrastructural changes in vestibular sensory end organs after viomycin sulfate intoxication. *Acta Otolaryngol.* 1969;68:474–488.

154. Ryals B, Westbrook E, Schacht J. Morphological evidence of ototoxicity of the iron chelator deferoxamine. *Hear Res.* 1997;112:44–48.

155. Olivieri N, Buncic J, Chew E, et al. Visual and auditory neurotoxicity in patients receiving subcutaneous deferoxamine infusions. *N Engl J Med.* 1986;314:869–873.

156. Chiodo AA, Alberti PW, Sher GD, Francombe WH, Tyler B. Desferrioxamine ototoxicity in an adult transfusion-dependent population. *J Otolaryngol.* 1997;26:116–122.

157. Karimi M, Asadi-Pooya A, Khademi B, Asadi-Pooya K, Yarmohammadi H. Evaluation of the incidence of sensorineural hearing loss in beta-thalassemia major patients under regular chelation therapy with desferrioxamine. *Acta Haematol.* 2002;108:79–83.

158. Masala W, Meloni F, Gallisai D, et al. Can deferoxamine be considered an ototoxic drug? *Scand Audiol Suppl.* 1988;30:237–238.

159. Cohen A, Martin M, Mizanin J, Konkle DF, Schwartz E. Vision and hearing during deferoxamine therapy. *J Pediatr.* 1990;117:326–330.

160. Shirane M, Harrison RV. A study of the ototoxicity of deferoxamine in chinchilla. *J Otolaryngol.* 1987;16:334–339.

161. Brummett RE, Fox KE. Quantitative relationships of the interaction between sound and kanamycin. *Arch Otolaryngol HNS.* 1992;118:498–500.

162. Brummett RE. Ototoxicity resulting from the combined administration of potent diuretics and other agents. *Scand Audiol Suppl.* 1981;14:215–224.

163. Tran Ba Huy P, Manuel C, Meulemans A, Sterkers O, Wassef M, Amiel C. Ethacrynic acid facilitates gentamicin entry into endolymph of the rat. *Hear Res.* 1983;11:191–202.

164. Hayashida T, Hiel H, Dulon D, Erre J-P, Guilhaume A, Aran J-M. Dynamic changes following combined treatment with gentamicin and ethacrynic acid with and without noise. *Acta Otolaryngol.* 1989;108:404–413.

165. Yamashita D, Jiang HY, Schacht J, Miller JM. Delayed production of free radicals following noise exposure. *Brain Res.* 2004;1019:201–209.

166. Henderson D, McFadden SL, Liu CC, Hight N, Zheng XY. The role of antioxidants in protection from impulse noise. *Ann NY Acad Sci.* 1999;884:368–380.

167. Hequembourg S, Liberman MC. Spiral ligament pathology: a major aspect of age-related cochlear degeneration in C57BL/6 mice. *J Assoc Res Otolaryngol.* 2001;2:118–129.

26

NEUROPATHOLOGY OF THE OLFACTORY SYSTEM

MARY BETH GENTER

Department of Environmental Health, University of Cincinnati, Cincinnati, Ohio

BRAD BOLON

GEMpath, Inc., Longmont, Colorado

INTRODUCTION

Olfaction in mammals is mediated by two distinct intranasal organs.[1] The olfactory mucosa (OM) is the main olfactory neuroepithelium directing the conscious perception of volatile odorants.[2,3] The vomeronasal organ (VNO) manages a series of behavioral and neuroendocrine outcomes in response to the unconscious detection of volatile pheromones.[3,4] Accordingly, the OM has broad responsibility for an organism's interaction with its environment, while the VNO is engaged in the narrower task of regulating scent-based signaling among individuals of a given species. These two sensory epithelia have primary projections to unique brain domains: the OM to the main olfactory bulb (MOB), and the VNO to the accessory olfactory bulb (AOB). The biology of these paired olfactory systems (emphasizing the OM and olfactory brain) in humans and animals has been described in detail elsewhere.[2]

Xenobiotic exposure has been shown to elicit neurotoxicity in both the OM[5–7] and to a lesser extent the VNO.[8–11] However, in most instances only the OM is evaluated when assessing the susceptibility of the olfactory system to toxicants. This choice is based on three factors: the vestigial nature of the VNO in humans, with associated transfer of human pheromone sensation to the OM;[8,12] our incomplete understanding of functional/structural correlates for the VNO across species; and the availability of simple antemortem olfactory function tests only for sensory responses mediated by the OM. Accordingly, in the current chapter we present a basic review of the salient anatomical,

biochemical, and functional characteristics of the OM and MOB that must be considered when undertaking a neuropathology evaluation of a putative olfactory neurotoxicant, along with the most useful methods for performing a morphological evaluation of the OM.

ANATOMY AND FUNCTION OF THE OLFACTORY MUCOSA AND OLFACTORY TRACT

The OM is a multilayered neuroepithelium in the nose that exhibits three unique features relative to other sensory organs of the nervous system. First, the primary sensory neurons in the OM interact directly with the external environment, as the receptors on these cells are separated from the air in the nasal cavity by only a thin layer of mucus which covers their dendritic projections resting on the mucosal surface. Second, olfactory nerves are the only sensory modality that synapses within the central nervous system without an intervening thalamic relay to filter the neural input that ultimately reaches the cerebral cortex. Finally, the OM possesses a population of stem cells that divide continuously and then differentiate into new sensory neurons throughout life. The continual turnover allows sensory neurons in the OM to regenerate after a toxic insult[13] or damage to the olfactory brain.[14]

Macroarchitecture of the Olfactory Mucosa

The OM is located in the caudodorsal regions of the nasal cavity. As expected, the best animal model for the gross

Fundamental Neuropathology for Pathologists and Toxicologists: Principles and Techniques, First Edition. Edited by Brad Bolon and Mark T. Butt.
© 2011 John Wiley & Sons, Inc. Published 2011 by John Wiley & Sons, Inc.

structure of the olfactory region in humans is the nonhuman primate,[15] because the OM is a small domain located in recesses high in the olfactory cleft. In contrast, in rodents and dogs the OM covers most of the ethmoid turbinates and lines the major dorsal passageways (i.e., dorsal meatus) by which inspired air streams through the nasal cavity. Olfactory neuroepithelium is also found in some portions of the frontal sinus in the dog.[16] The surface area of the rat OM increases continuously with age,[17] but in other species the functional tissue is altered (e.g., mice[18]) or replaced by progressive encroachment of respiratory mucosa (e.g., humans) with age (discussed below).

A key difference among animal species is the highly variable fraction of the total nasal surface that is represented by OM (Figure 1).[19] The canine OM is approximately 20 times larger than that in humans.[20–22] This wide divergence is in keeping with the relative importance of the olfactory sense across species.[23] For example, the human nasal cavity has a very small percentage of the epithelial surface devoted to olfaction, perhaps as little as 3%.[24] In contrast, the nasal surface devoted to olfaction in mice, rats, and hamsters is approximately 50 to 60%,[19,25,26] whereas that in rabbits is about 80% (calculated from published data[27]).

Microarchitecture of the Olfactory Mucosa

The OM is histologically similar across animals species.[19] The OM has a very precise multilayered structure, which renders it appreciably thicker than the adjacent respiratory epithelium. It is separated from the underlying tissue by a basement membrane. The normal organization of the OM is shown in Figure 2. The major cell types include bipolar olfactory receptor neurons (ORNs), sustentacular cells (akin to glial support cells in function), and basal cells (the stem cell population).[3,28]

Each ORN has a thick apical dendrite and a thin basal axon that is enveloped by glial cell processes at the base of the epithelium (i.e., just before it exits the mucosa).[28] In mature ORNs, the bulbous, ciliated, protruding tip of the dendrite, termed the *olfactory vesicle*, reaches the OM surface, whereas the nascent dendrite of immature (basally located) ORNs does not. The topographical organization of ORNs within the OM is reported to differ among species. In humans, immature and mature ORNs are interspersed throughout the OM,[29] whereas in rodents immature and mature ORNs are localized to discrete epithelial layers in the basal two-thirds of the OM (Figure 2).

The sustentacular cells are columnar elements that span the entire thickness of the olfactory epithelium. Their superficial width exceeds that of their basal portion, due to the presence apically of the nucleus and numerous organelles. Their narrow basal stalks branch to form foot processes that spread between the basal cells and follow the basement membrane.[28]

The basal (stem) cells reside adjacent to the basement membrane of the OM. There are at least two structurally distinct populations, horizontal and globose basal cells.[30] The functional differences between these two cell phenotypes (if any) have yet to be characterized. Together, these two elements continuously produce new neurons (and to a lesser extent glia) to replace senescent cells, even in the absence of a toxic insult. Sensory neuron regeneration is supported in part by olfactory ensheathing cells (OEC), the specialized glia which support the axonal extension that accompanies ORN regeneration by producing growth-promoting adhesion and extracellular matrix molecules.[31,32] Production of neurons is also dependent on the normal function of the olfactory brain, as lesions of the main olfactory bulb (MOB) prevent regeneration of the OM. The proposed explanation for this phenomenon is that a target-derived molecule produced in the MOB guides regenerating ORN axons to their proper synaptic fields within the MOB.[33]

Other specialized cells have also been described in the OM. One recently explored component is the microvillar cell (MVC), which has been demonstrated in rodents, canines, and primates.[34] To date, the exact morphological and molecular characteristics of these cells have not been defined. The MVC have been proposed to coordinate the waves of apoptosis and regeneration in ORNs, possibly via a chemosensory ability coupled with their stimulus-induced capacity to release neuropeptide Y.[35]

The basement membrane of the OM rests on a lamina propria comprised of tubuloalveolar gland acini (Bowman's glands), unmyelinated axonal bundles (derived from ORNs), and blood vessels, all of which are surrounded by a loose connective tissue matrix. The Bowman's glands are lined by thin acinar cells that produce a serous secretion and also contain many of the metabolic enzymes of the olfactory mucosa. The straight ducts of these glands pass through the basement membrane and across the OM to connect with the nasal cavity lumen. The olfactory ensheathing cells (OEC) that encompass the ORN axons are particularly interesting, as they express traits of both astrocytes and Schwann cells.[31] The basic OEC phenotype boasts broad functional and structural plasticity, depending on the constellation of microenvironmental stimuli.[32]

Functional Specialization of the Olfactory Mucosa

The localization of the major known molecules characteristic of the OM in adult, intact rodents is given in Table 1. Recent research efforts have identified cell type–specific immunohistochemical markers for many elements of the OM. Despite interspecies similarities in structure and function, the molecular signatures in the OM may exhibit substantial differences among rodents, carnivores, and primates.[19]

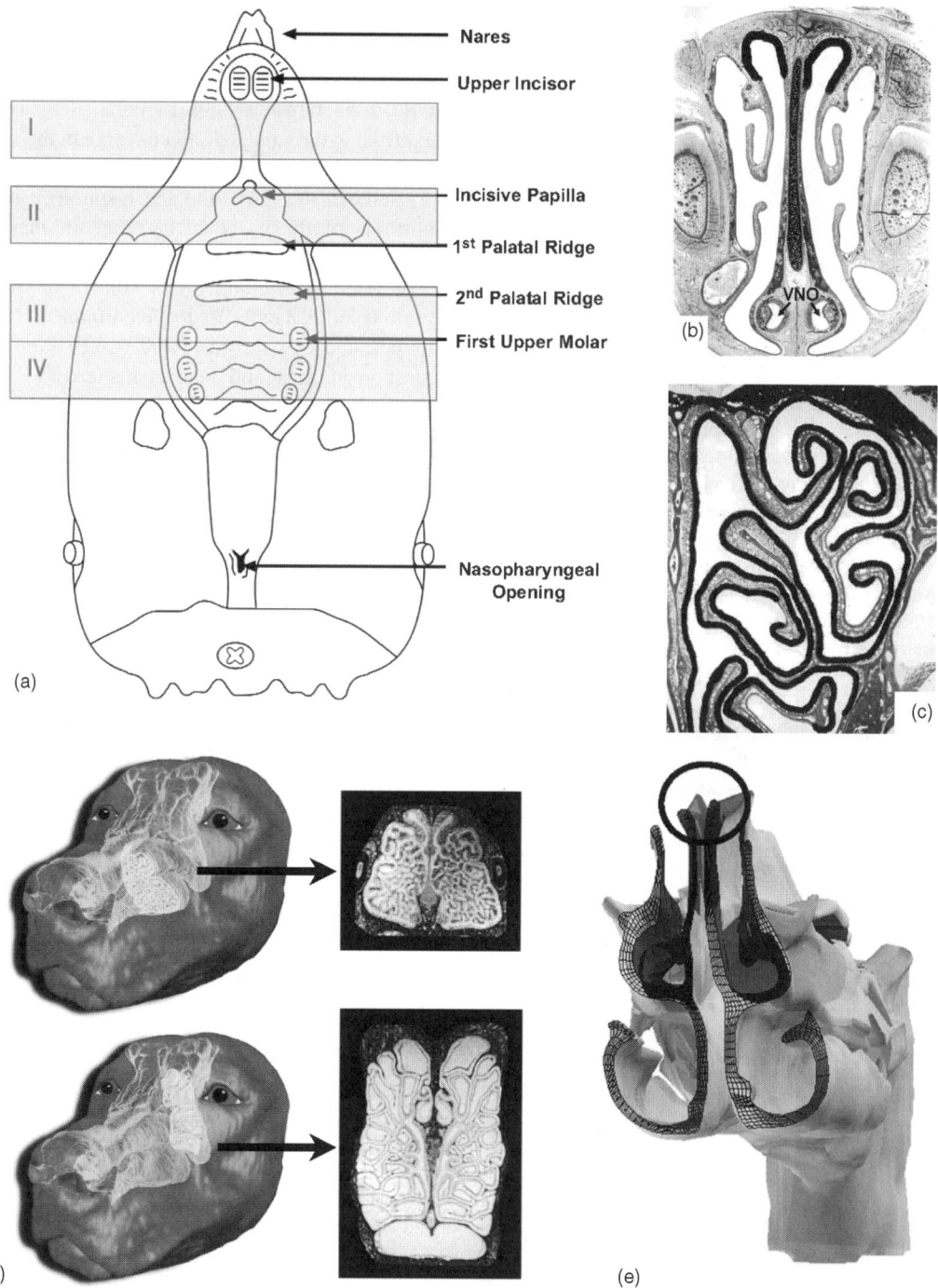

FIGURE 1 Comparative distribution of olfactory mucosa in the nasal cavity of the rat, dog, and human. (a) Schematic diagram showing landmarks on the ventral (intra-oral) surface of the palate used to trim decalcified rodent skulls to acquire standard nasal cavity cross sections. (Adapted from Young,[57] by permission of Elsevier.) (b) Olfactory mucosa in a level I[57] section of rat nasal cavity is limited to the dorsal medial meatus (indicated by the heavy black line). Stain: H&E. (c) Olfactory mucosa (heavy black line) covers a much larger fraction of the ethmoid turbinates in the caudal nasal cavity in this level III[57] section of the rat nose. Stain: H&E. (d) In the dog (*Canis familiaris*), the branched maxilloturbinates in the rostral nose (top panels) are covered with respiratory epithelium, while the turbinates in the caudal nose (bottom panels) are covered with olfactory mucosa. (Upper right panel is adapted from Craven et al.[20] with permission of John Wiley & Sons, Inc. Copyright © 2007 Wiley-Liss, Inc. The remaining portions of this figure are courtesy of Brent A. Craven, Pennsylvania State University.) (e) Three-dimensional reconstruction of a human nasal cavity showing the extremely limited distribution of olfactory mucosa in the most superior recesses of the nasal cavity (circle). (Courtesy of Julia Kimbell, University of North Carolina–Chapel Hill, and Jeffry Schroeter, the Hamner Institutes for Health Sciences.)

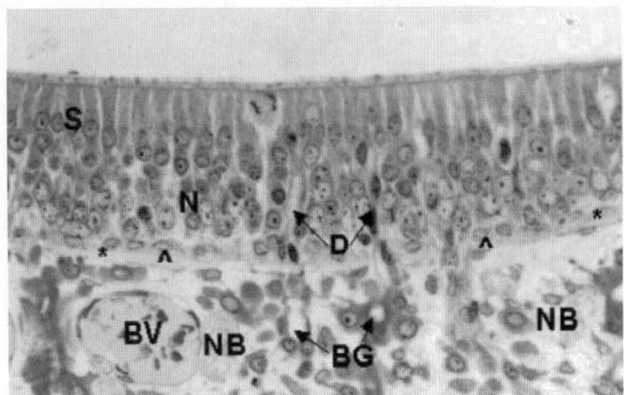

FIGURE 2 Photomicrograph of normal olfactory mucosa in the adult rat. Key features in the neuroepithelium proper include a superficial layer of sustentacular (supporting) cells (S), a middle layer of mature olfactory receptor neurons (N), and a one-cell-thick deep layer of basal (stem) cells (∧), and straight ducts of Bowman's glands (D) which empty into the airway lumen. The epithelial basement membrane (∗) runs horizontally beneath the mucosa. Subepithelial structures include nerve bundles comprised of olfactory receptor neuron axons (NB), Bowman's glands (BG), and blood vessels (BV). Stain: H&E.

The best-known molecular flag for a distinct OM cell type is *olfactory marker protein* (OMP), an antigen expressed in mature ORNs and their axons in many vertebrate species.[36] This protein was the first olfactory-specific protein to be isolated and cloned. The function of OMP is not fully understood, but studies in mice lacking OMP indicate that one task is to modify the perception of odorant quality.[37] Another important ORN attribute is defined by each cell's choice of odorant receptor gene (OR). The OR superfamily is

the largest in the human genome.[38] During differentiation, the OR complement undergoes random selection, so that each mature ORN expresses only one of approximately 1000 ORs, derived from a single parental allele; axons from ORNs expressing the same OR all converge in the same glomerulus within the olfactory bulb.[39]

Cells in the OM (and the respiratory mucosa in other portions of the nasal cavity) contain many enzymes for metabolizing xenobiotics, probably as an evolutionary adaptation to ongoing direct contact of the airway lining with volatile agents in the environment.[40,41] As in other tissues, the cytochrome P450 (CYP) family is by far the most prominent and best characterized of these enzyme systems. The CYPs are heme-containing proteins, generally with a molecular weight of 50 to 60 kDa, which are localized to the endoplasmic reticulum. They catalyze oxidation reactions by transferring reducing equivalents from the flavoprotein NADPH–cytochrome P450 oxidoreductase (POR) onto any xenobiotic that enters the cell, whether delivered in the inhaled air or via the blood. The various CYP families and subfamilies have some degree of substrate specificity. For example, the CYP1A subfamily has high metabolic activity toward many polycyclic aromatic hydrocarbons. Multiple subfamilies of CYPs have been found in the OM of at least one vertebrate species (Table 2; reviewed by Ding and Dahl[42]). The OM of rats, dogs, and nonhuman primates has approximately four-fold more CYP activity than that of the respiratory mucosa.[43] Substantial differences in the olfactory CYP content exist among species.[43] Immunohistochemical studies consistently localize CYPs to nonneuronal elements in the OM, most often within sustentacular cells and/or in both the acinar and ductal cells of Bowman's glands.

TABLE 1   Antigens and Their Distribution in Intact Olfactory Mucosa of Adult Rodents

Antigen (Designation)	Localization	References
1F4	Sustentacular cell microvilli	123
Cytochromes P450[a]	Bowman's glands and ducts; sustentacular cells	47,124–126
Carnosine	Olfactory neurons	127
Cytokeratins	Horizontal basal cells	30
GAP43 (neuron-specific phosphoprotein)	Immature olfactory neurons	128
GBC-1 (unknown antigen in globose basal cells)	Globose basal cells	129
Nestin	Foot processes of sustentacular cells	130
Neural cell adhesion molecules (NCAM)	Globose basal cells and olfactory neurons	131
Neuron-specific enolase (NSE)	Olfactory neurons	132
Neuron-specific tubulin (NST)	Immature and mature olfactory neurons	133
Protein gene product 9.5 (PGP9.5; UCHL1)	Olfactory sensory neurons	132
Olfactory 1 (Olf1 transcription factor)	Mature and immature olfactory neurons	134
Olfactory marker protein (OMP)	Mature olfactory neurons and their axons	135
Sus1–Sus3 (unknown antigens in sustentacular cells)	Sustentacular cells; Bowman's glands	136
Vimentin (intermediate filament)	Mature olfactory neurons and their axons	137
Vomeromodulin	Lateral nasal gland; mucus of vomeronasal organ	121

[a] Representative examples.

TABLE 2   Olfactory Mucosal Cytochromes P450 (CYP) and Their Substrates

Family/Subfamily	Representative Substrate[a]	References
CYP1A1	Lidocaine	139
CYP2A	Coumarin, nicotine	140,141
CYP2B	Chloracetanilide herbicides[b]	142
CYP2C	S-Mephenytoin	143,144
CYP2E1	Chloroform	145
CYP2F (mouse isoform 2)	Naphthalene	47
CYP2G1[c]	Coumarin	141
CYP2J	Arachidonic acid, vitamin D	146
CYP3A	Cocaine	147
CYP4A	Fatty acids	148
CYP4B	4-Ipomeanol, 2-aminoanthracene	142

Source: Modified from Ding and Dahl[42] and Hodgson and Goldstein.[138]
[a] Not intended to be an exhaustive list.
[b] Examples include alachlor, acetochlor, butachlor, and metolachlor.
[c] Rodents only; both human alleles have a loss-of-function mutation.[149]

Multiple non-CYP metabolizing enzymes are also well represented in the OM.[40,41] Inhaled aldehydes (e.g., acetaldehyde, formaldehyde) and esters (e.g., ethyl acrylate, polypropylene glycol monomethylether acetate) are metabolized by various aldehyde dehydrogenases, carboxylesterase, or naphthylbutyrate esterase in the OM.[44,45] The sustentacular cells, basal cells, and Bowman's glands are richly supplied with multiple glutathione-S-transferase (GST) variants.[46] High expression of paraoxonase-1 (PON1), responsible for both normal lipid homeostasis and detoxification of organophosphorus compounds, has been found in the OM using GeneChip analysis;[47] this activity was subsequently localized by in situ hybridization primarily within Bowman's glands (M. B. Genter et al., unpublished observation). Rhodanese, the enzyme responsible for cyanide detoxification, has been localized to supporting cells in the rat OM.[48]

Architecture and Function of the Olfactory Brain

The primary central projections from the OM are olfactory nerve fascicles extending through the cribriform plate to the main olfactory bulb (MOB). The secondary projections from the MOB extend via the olfactory tract (OT) directly to several regions within the basal telencephalon that comprise the olfactory cortex (OC). As with the OM itself, the sizes of the MOB, OT, and OC in a given species are proportional to the importance of olfaction as a modality for exploring the external environment.[23] Accordingly, the MOB and OT of rodents and dogs ("macrosmatic" species) are prominent landmarks on the rostral and ventral faces of the brain, respectively, where they comprise approximately 15% (in midsized dogs) to 35% (in rodents) of the entire brain

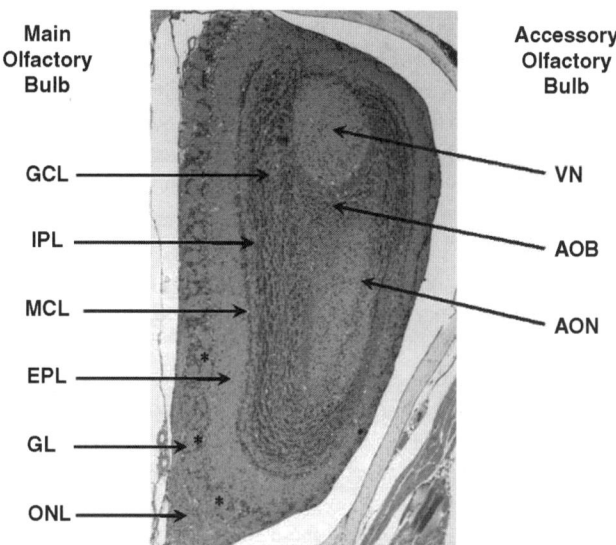

FIGURE 3   Photomicrograph of the olfactory bulb in the adult rat (low magnification cross section) demonstrating the laminar organization. Unmyelinated axons from the olfactory mucosa pass through the cribriform plate to the main olfactory bulb (MOB), where they cross the outer olfactory nerve layer (ONL) to synapse on the dendrites of mitral cells and tufted cells in the glomerular layer (GL). The individual glomeruli (*) are circular areas with pale cores of dense neuropil surrounded by rims of dark nuclei. Deep to the GL, the external plexiform layer (EPL) is a pale, wide, relatively acellular region that contains the dendrites of mitral, tufted, and granule cells and the cell bodies of tufted cells. The mitral cell layer (MCL) is a thick band (two or three cells wide) deep to the EPL that harbors the cell bodies of mitral cells—the principal output cells of the MOB—and some granule cells. The pale internal plexiform layer (IPL) resides beneath the MCL and holds the axons of the mitral and tufted cells as well as dendrites of the granular cells. The granule cell layer (GCL) is the innermost lamina and consists of numerous granule cells arranged in parallel rows. The anterior (rostral) olfactory nucleus (AON) is a small domain of olfactory cortex that extends forward into the caudal portion of the MOB. Unmyelinated axons from the vomeronasal organ pass along the vomeronasal nerve (VN) to the accessory olfactory bulb (AOB), which resides along the caudodorsal aspect of the MOB. (Anatomical features classified after Kruger et al.[183]) Stain: H&E.

mass. In contrast, in humans and nonhuman primates ("microsmatic" species) the MOB and OT form a very thin band of tissue confined to a narrow olfactory sulcus on the ventral brain surface.

The MOB exhibits a characteristic laminar organization (Figure 3).[3] The outer olfactory nerve layer (ONL) contains unmyelinated axons from the ORNs and their supporting glia (which are derived from both the central and peripheral nervous systems). The axons pass through the ONL to synapse on the dendrites of mitral cells and tufted cells within glomeruli, which are ovoid structures (comprised of several distinctive neuron types and also specialized astrocytes) where primary sensory neurons (ORNs) synapse with relay neurons. Immediately deep to the glomerular layer is the external plexiform layer (EPL), which consists chiefly of dense neuropil formed by dendrites of mitral, tufted, and granule cells and the cell bodies of tufted cells. The next level in, the mitral cell layer (MCL), contains the soma of mitral cells—the principal output cells of the MOB—and some granule cells. The internal plexiform layer (IPL) resides beneath the MCL and harbors the axons of the mitral and tufted cells as well as dendrites of the granular cells. The innermost neuronal lamina, the granule cell layer (GCL), contains the bulk of the MOB granule cells. The MOB sends outputs to the OC mainly via axons of the mitral cells and more deeply located tufted cells. The MOB output is transmitted in the lateral olfactory tract (LOT) to the ipsilateral OC. The primary OC is a composite of regions, including (from rostral to caudal), but not limited to, the anterior olfactory nucleus (AON); olfactory tubercle; the piriform, periamygdaloid, and entorhinal cortices; and the nucleus of the lateral olfactory tract (NLOT).

The organization of the olfactory brain is defined by the need to integrate a complex combination of sensory inputs.[3] This integration occurs at both the cellular and organ levels. The glomerular layer of the MOB is the first locus of integration within the olfactory brain, as each glomerulus receives signals from several thousand ORNs. In the rat, for example, each glomerulus is supplied by approximately 25 mitral cells, but each mitral cell extends to only one glomerulus. Each MOB domain is innervated by ORNs from widely dispersed regions of the OM, but ORNs sharing the same odorant sensitivity (i.e., neurons that express the same OR) project to the same glomerulus.[3] Incoming signals are further integrated by widespread reciprocal synapses between the dendrites of mitral and tufted cells, neurites from other MOB interneuron populations, and granule cells. The function of mitral and tufted cells is to excite the granule cells, while the action of granule cells is to inhibit the other neuronal types. The granule cells are also coupled by gap junctions to provide better synchronization. The organ-level substrates of functional integration are classic fiber tracts which provide reciprocal feedback. For example, the OC is linked to the MOB by projections from the AON, entorhinal cortex, and NLOT. In addition, collateral projections in the olfactory brain course widely within the ipsilateral MOB and are also common (as both ipsilateral and contralateral connections) within and between the various structures that comprise the primary OC.

PREPARATION OF THE OLFACTORY MUCOSA FOR NEUROPATHOLOGY EXAMINATION

Consistent processing procedures—fixation, decalcification, trimming, sectioning, and staining—should be employed when preparing the OM and the olfactory brain for evaluation. Such standardization is essential if comparisons across studies, and especially among different species, are to provide useful information.

In this section we describe conventional histological methods for harvesting and preparing OM specimens. Processing of the olfactory brain has been considered as a component of preparing the adult central nervous system for assessment.

Fixation

Multiple fixation protocols have been use to preserve the OM. Regardless of the fixative, the most important means of achieving optimal structural protection is to flush the nasal cavity with fixative in a retrograde fashion (i.e., from back to front) via the nasopharynx. This step ensures that the fixative penetrates the olfactory recesses of the nasal passages bilaterally before the delicate OM can degrade. The nasopharyngeal opening (a narrow tube or slit near the caudal border of the palate) is exposed by separating the head from the body and then detaching the mandible. A beveled hypodermic needle attached to a fixative-filled syringe is inserted no more than 2 to 3 mm into the opening; the needle diameter should approximate that of the nasopharynx and thus should be 25 to 27 gauge for use in adult mice and 20 to 23 gauge for use in adult rats. Fixative is then flushed slowly into the nasal cavity. Fixation has been initiated properly if the solution exits the nares during the flush procedure. Our laboratory flushes the nasal passages with 1 to 2 mL of fixative for adult mice and 5 to 10 mL for adult rats. After the fixative flush, the entire skull is immersed in the same fixative for an additional 12 to 48 h. The skin may be removed from the dorsal surface of the skull to facilitate fixative penetration. We have found that this two-stage fixation protocol (flush followed by immersion) provides superior preservation of OM structures relative to that obtained by mere immersion fixation of the intact skull or even transcardial (whole body) perfusion with fixative.

Many different agents can be employed successfully to preserve the OM. Common choices include neutral buffered 10% formalin (NBF: 3.7% formaldehyde), buffered 4% paraformaldehyde, and Bouin's, Carnoy's, Fowler's, Helly's, and Zenker's fixatives.[49–51] A comprehensive head-to-head comparison of four alternatives—Bouin's, Fowler's, NBF, and Zenker's—demonstrated that OM sections maintain better microscopic detail with substantially fewer artifacts during routine histopathological evaluation if fixation had been performed in Bouin's, followed by Zenker's, NBF, and finally Fowler's.[51] An added advantage of Bouin's solution is that decalcification of bone (both the nasal turbinates supporting the OM and the surrounding skull) begins in conjunction with tissue fixation. However, major disadvantages associated with Bouin's or Zenker's—corrosiveness, toxicity (Zenker's contains mercuric chloride), potential instability (Bouin's can explode under certain storage conditions), and tendency to degrade fragile molecular epitopes—precludes their routine use as OM preservatives. The preferred choices for nasal fixation in our laboratory are NBF for routine microscopic examination and buffered 4% paraformaldehyde for studies that include in situ molecular pathology methods.

Decalcification

The OM is surrounded by bony tissues, and the delicate nasal turbinates contain bone. Therefore, optimal OM sections can be obtained only when the specimen has been softened by removal of calcium in the bony structures prior to trimming and thin sectioning. The two primary decalcification solutions are acids or calcium (Ca^{2+}) chelators. Acidic solutions allow more rapid Ca^{2+} removal, but typically preclude in situ molecular analysis of all but the most robust protein epitopes and messenger RNAs. A head-to-head comparison of hydrochloric acid–based solution and a formic acid–resin combination did not reveal differences in OM architecture or ease of sectioning, but the hydrochloric acid provided faster decalcification.[51] Use of chelators such as ethylenediaminetetraacetic acid (EDTA) to remove Ca^{2+} is highly recommended when evaluating in situ functional (e.g., enzyme histochemistry) and molecular endpoints (e.g., immunohistochemistry, comparative genomic hybridization, in situ hybridization) because chelation is less harsh on many antigens and mRNAs.[52,53] For example, antibodies to the metal transporters ZIP8 and ZIP14 react very weakly on sections from archived blocks that have been decalcified using 10% formic acid but bind robustly on EDTA-decalcified tissues (M.B. Genter, unpublished observation). In general, EDTA is purchased as a disodium or tetrasodium salt and dissolved in water as either a 10% solution (corresponding to 0.24 M for tetrasodium EDTA) or 0.5 M solution (i.e., essentially a saturated solution). The EDTA-based decalcifying solutions should be adjusted to pH 7.4. The preferred option for nasal decalcification in our research laboratory is a 0.5 M EDTA solution. However, for routine screening studies to detect any olfactory neurotoxic potential, an acid solution will probably afford an acceptable compromise between the need for structural preservation of the OM and the short time lines deemed desirable for product development.

Regardless of the solution choice (acidic or chelating), specimens are decalcified by total immersion only. The volume of decalcifying solution should be 5 to 10 times that of the specimen. No retrograde flushing of the nasal cavity is required, as the solution is either changed repeatedly until decalcification is complete or is supplemented with a Ca^{2+}-chelating cation exchange resin (e.g., Rexyn 101 [H] beads, available commercially from multiple vendors). Utilization of EDTA is more expensive on a per-experiment basis because the solution must be changed daily (over a time course of days for mouse skulls, and two to three weeks for rat heads); in contrast, acid solutions may be used repeatedly, and replaced at longer intervals. Agitation (e.g., by placing specimen cups containing samples submersed in EDTA on a shaker, or by putting a magnetic stir bar into the acid solution and placing the vat on a stir plate) can decrease the time needed for decalcification. The success of the decalcification is gauged by using a blade to cut through an incisor tooth.

In larger animal species and humans, carefully selected OM specimens may require minimal or no decalcification. This is the case when the OM has been dissected away from the underlying bone of the nasal turbinates and walls. The most common scenarios for procuring such specimens are to be found in human medical practice (e.g., surgical biopsies) or research using human cadavers.[54–56]

Tissue Trimming

For routine toxicity studies, the decalcified rodent nasal cavity is typically sampled at three to five different levels to reveal the OM, respiratory mucosa (located rostrally), and if desired, the VNO.[51,57] A far more extensive sectioning protocol has been developed for mapping toxicant-induced lesions in nasal epithelia.[58]

In rodents, trimming is typically undertaken using external anatomical landmarks on the surface of the palate[57] and is normally performed in transverse orientation (cross section) so that bilateral symmetry of olfactory mucosal lesions may be evaluated. Tissue blocks are isolated by freehand trimming using razor blades. The OM in rodents is readily sampled by acquiring cross sections bracketed by the molar teeth. Tissue blocks from unfixed, undecalcified specimens may be isolated using an low-speed electric rotary saw equipped with a diamond blade; the standard anatomical landmarks are employed to guide the saw cuts.[59] These blocks may then be fixed and decalcified for abbreviated periods, thereby preserving the function and/or structure of particularly labile nucleic acids and proteins (especially enzymes).

Trimming the OM of larger vertebrates will require careful planning. In humans the OM is approached by opening the nasal passages and harvesting the tissue lining the upper nasal septum and superior (dorsal) nasal turbinates. The same method may be utilized in large animals (e.g., dogs, nonhuman primates). However, the most elegant means of evaluating the OM, and especially the symmetry of lesions, is to use complete transverse sections through the region. Again, these blocks are isolated using external landmarks on the pallet.[60,61] Such sections can be obtained in decalcified specimens using thin knives or long razor blades. Similar sections may be harvested from undecalcified samples using a low-speed rotary saw.[61]

Embedding Olfactory Mucosa

The two common media used to support OM for sectioning are paraffin wax and plastic [typically glycol methacrylate (GMA)]. Both are suitable for routine studies, but preservation of subtle cellular and subcellular structures has been reported to be better in 3-μm-thick GMA sections than in standard 5-μm-thick paraffin sections.[51,61,62] Fixed but undecalcified samples may be sectioned if they are supported in plastic.[61] We routinely employ paraffin in our laboratory as it is less expensive and the components are less toxic than plastic. Our standard section thickness is 5 μm.

Embedding unfixed, undecalcified nasal tissue in 1% carboxymethylcellulose followed by freezing in isopentane cooled in liquid nitrogen has been used to produce frozen sections without inducing process-related reductions in molecular structure and function.[62] These preparations offered excellent preservation of enzyme activity but provided very poor morphological detail.

Staining Olfactory Mucosa

The usual stain for examining the OM (and the VNO) is hematoxylin and eosin (H&E). The good contrast afforded by this method provides suitable delineation of the various cell types in the OM, and the familiarity that general pathologists and neuropathologists have with the appearance of sections stained in this manner makes H&E the first choice for routine screening studies as well as many mechanistic studies. Other enzyme histochemical, immunohistochemical, or in situ molecular methods can be performed on serial sections as necessary to showcase specific cell types (e.g., immunohistochemistry for OMP to detect mature ORNs) or distinct biological processes [e.g., cell proliferation (see below)].

SPECIAL PROCEDURES FOR NEUROPATHOLOGY EVALUATION OF THE OLFACTORY MUCOSA

Cell Proliferation

Regeneration of the OM can be readily assessed using standard methods. The most common procedures detect incorporation of exogenous [³H]thymidine[63] or 5-bromo-2′-deoxyuridine (BrdU, a thymidine analog)[64] into dividing (S-phase) cells or reveal the endogenous marker proliferating cell nuclear antigen (PCNA). Labeling of S-phase cells is most evident in the basal cell layer of the OM (Figure 4).

At present, the preferred method for evaluating OM cell proliferation is anti-BrdU immunohistochemistry. The typical protocol involves administration of a BrdU "pulse" (50 to 100 mg/kg) by intraperitoneal (i.p.) injection 2 h prior to sample acquisition. Proper placement of the i.p. injection is essential, as inadvertent placement in the intestine or subcutis will prevent adequate BrdU distribution into the

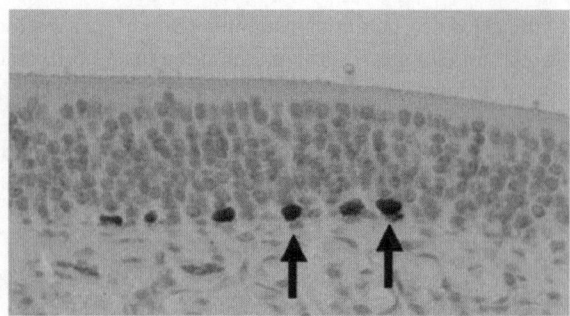

FIGURE 4 Photomicrograph of proliferating basal (stem) cells in the olfactory mucosa of an adult rat. Labeled cells (dark nuclei near arrows) have incorporated 5-bromo-2′-deoxyuridine (BrdU, a thymidine analog) while replicating DNA during the S phase. Details of the BrdU immunohistochemistry protocol are presented by Genter et al.[64]

OM following a 2-h pulse; correct parenteral delivery is confirmed by collecting a small intestine segment from each animal so that it can be processed and stained simultaneously with the OM. Alternatively, BrdU may be infused continuously via subcutaneously implanted osmotic minipumps.[65] In our laboratory, we prepare nasal tissues for cell proliferation studies using the standard conditions for general toxicity studies: fixation in NBF, decalcification in either formic acid (up to 10%) or EDTA, embedding in paraffin, and sectioning at 5 μm. We employ commercially available BrdU and anti-BrdU antibodies at dilutions recommended by the manufacturers. Because many anti-BrdU antibodies are directed specifically against BrdU in single-stranded DNA, anti-BrdU protocols typically include a DNA-denaturing step such as incubation in 1 N hydrochloric acid at room temperature for 1 h or microwaving in weak acid (usually citric) solutions; the manufacturer's protocol will recommend the method best suited for their antibody. Because denaturing protocols are quite harsh, it is critical to mount OM sections on slides that have been optimized for tissue adhesion. After immunostaining, sections are generally counterstained (often with hematoxylin) and cover-slipped using a mounting medium that is compatible with the chromagen selected to visualize anti-BrdU antibody binding.

Autoradiographic assessment of [³H]thymidine incorporation in the OM preceded the use of BrdU for cell proliferation studies in the olfactory epithelium. As with BrdU, [³H]thymidine is usually administered as an i.p. or intravenous (i.v.) pulse, after which labeled nasal tissues are processed by routine methods. Slides bearing sections are then dipped into a suitable photographic emulsion and exposed for prolonged periods (typically, 10 to 12 weeks) at subzero temperatures (−25°C) in the dark.[49,63] The threshold for the number of silver grains that must overlie an olfactory epithelial nucleus for it to be counted as positive is determined by examining the labeling in a highly proliferating tissue (e.g., the basal cell layer of the oral mucosa or small intestinal crypts). At the present time, this isotopic method should be avoided for all but the most specialized studies due to the slow and laborious nature of such experiments, the expense of radionuclides, the need for substantially greater degrees of personnel training and oversight, and the generation of extensive radioactive liquid (fixative, decalcification, and washing solutions) and solid (paraffin shavings) waste during processing.

In contrast to the S-phase-specific labeling yielded by BrdU and [³H]thymidine, the endogenous nuclear protein PCNA[66] can be used to discriminate among multiple phases of the cell cycle in the OM.[67] Cell proliferation analysis must be undertaken with a clear knowledge of the PCNA expression pattern within the OM cells. A potential advantage of the PCNA method is that it can be used for the retrospective analysis of archived tissues from individuals who were not treated with BrdU or [³H]thymidine.

Cell proliferation data from OM is typically calculated in one of two manners. The first way is to measure the *labeling index*, which is the percentage of labeled nuclei in the entire section. This value is readily attained but can be misleading if the amount of OM is not uniform among the various animals. An alternative measure is to calculate the *unit length labeling index* (ULLI[17,68]), where the number of proliferating cells in the OM is divided by the length of the underlying basement membrane so that the resulting data are expressed as the number of labeled cells per millimeter. The ULLI effectively compensates for differences in the conformation of nasal sections (i.e., the quantity of OM available for evaluation) among subjects. Again, the credibility of cell proliferation data for a given animal is checked qualitatively by evaluation of the LI of the small intestinal crypts in the same individual. A low ULLI in the OM can be assigned correctly to either the existing olfactory physiological status (if the intestinal LI is high) or an injection error (if the intestinal LI is low) by means of this comparison.

Enzyme Histochemistry

Multiple protocols have been devised to explore the metabolic potential of the OM. Enzyme histochemistry is a particularly useful tool in this regard, as it combines the specificity of anatomic localization with a demonstration of functional capacity. This technique can be used to evaluate tissue sections or tissue whole mounts, and it can determine the cell type–specific distribution of a protein within a tissue. Histochemistry relies on a number of factors, including minimal tissue fixation, application of the necessary cofactors for enzyme activity, and the availability of a chromagen that can be used under the conditions of the enzymatic reaction.

As noted above, many biotransformation enzymes have been identified in the OM. Prominent among them are many

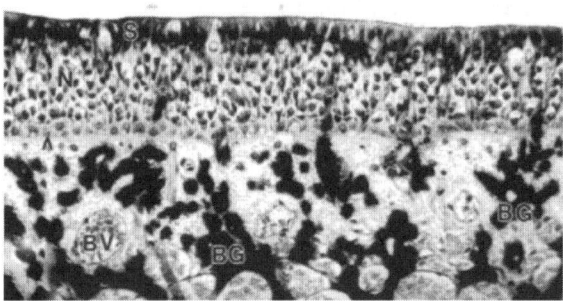

FIGURE 5 The olfactory mucosa has an extensive capability to metabolize toxicants delivered in the air or blood. This photomicrograph demonstrates the distribution of naphthyl butyrate esterase (NBE, indicated by the intense black reaction product) activity in the olfactory mucosa. Labeled structures include the apical cytoplasm of sustentacular cells (S), straight ducts of Bowman's glands within the neuroepithelium (∗) and acinar cells of Bowman's glands (BG). Note that NBE activity is absent in the olfactory receptor neurons (N) and basal (stem) cells (∧). Details of the NBE enzyme histochemistry protocol are presented by Randall et al.[62] BV, blood vessel.

cytochromes P450, P450 reductase, aldehyde dehydrogenases, β-napthylbutyrate esterase, glutathione-*S*-transferases, and epoxide hydrolase. Most enzymes are found in several or all cell types within the OM except for the ORNs.[44,46] For example, naphthylbutyrate esterase activity is found in sustentacular cells and Bowman's glands (Figure 5). In contrast, carbonic anhydrase has been demonstrated histochemically in ORNs and their axonal processes in glutaraldehyde-fixed cryostat sections.[69]

Transgene distribution in the OM may be monitored histochemically by coupling expression of the engineered gene to a reporter element, such as bacterial β-galactosidase (lacZ). However, the OM (and the MOB) contains endogenous β-galactosidase activity (M. B. Genter, unpublished observation), as indicated by the ability to hydrolyze 5-bromo-4-chloro-3-indolyl-D-galactopyranoside (X-gal) into galactose (which is colorless) and 4-chloro-3-bromo-indigo (which forms an intense blue precipitate). Therefore, tissue processing to reveal lacZ localization in the OM and brain requires special processing conditions. In our laboratory, the protocol for demonstrating lacZ in the olfactory system of engineered rodents is to flush the nasal cavity with ice-cold 2% paraformaldehyde [in 150 mM NaCl, 15 mM Na phosphate, pH 7.3 (PBS)], followed by immersion fixation in the same fixative solution for 1 h at 4°C. Nasal cavities (either intact or opened using a longitudinal split along the midline) are then rinsed in PBS and immersed overnight at 30°C in PBS containing 1 mg/mL of X-gal, 5 mM potassium ferricyanide, 5 mM potassium ferrocyanide, and 2 mM magnesium chloride.[70] Further modifications to the procedure may also be undertaken to achieve even greater specific inhibition of endogenous galactosidases but not lacZ.[71]

Olfactory Epithelial Biopsies

In animals, olfactory tissues are generally harvested at necropsy. In contrast, OM samples from humans can be acquired by clinicians as a means of disease diagnosis in patients and by olfactory researchers evaluating the biology of sensory pathways in volunteer research subjects. For this purpose, OM biopsies have been shown to contain ORNs that respond to odorant stimuli[72] as well as olfactory progenitor cells,[56] and they have been used in mechanistic studies to develop and test hypotheses regarding mechanisms of neurological disease.[29,73] The biopsy procedure is made possible by the facts that ORNs project directly into the nasal cavity, and are therefore accessible for biopsy using minimally invasive surgical techniques, and can regenerate after injury (if advanced neuropathology is not present in the OM). Follow-up studies of patients who have undergone OM biopsies indicate no subsequent compromise of their olfactory function.[29,56,74]

Procedures for OM biopsy have been described in detail by several groups.[29,54,74] In general, a rigid endoscope is inserted into one naris and directed into the caudal nasal cavity, after which a local anesthetic is applied and a cutting or punch forceps is introduced to remove tissue from the appropriate site. The routine biopsy of olfactory tissue in humans is hampered by the irregular and patchy distribution of OM in this species. The extension of respiratory epithelial foci (i.e., metaplasia) into areas of the nasal cavity normally occupied by the OM (e.g., upper nasal septum; superior nasal turbinate) seems to be characteristic of the human OM and may increase as a function of age.[55] Once acquired, OM biopsies can be processed by routine histological practices and subjected to the standard array of immunohistochemical procedures.[73,75–77]

NEUROPATHOLOGY OF THE OLFACTORY MUCOSA AND OLFACTORY TRACT: BASIC PRINCIPLES

It is intuitive that inhalation exposure to many agents may cause toxicity to the OM, either reversible or irreversible, by direct contact (Table 3). Damage may be caused by the parent agent (e.g., propylene oxide[65]) or following local bioactivation by enzyme systems within sustentacular cells or Bowman's gland epithelium (e.g., naphthalene[78]). For highly water-soluble or reactive airborne toxicants, nasal airflow patterns play a major role in determining lesion distribution.[79] The usual pattern of toxicant-induced OM lesions following exposure to inhaled agents results in disruption of the nasal mucosa in the dorsal (superior) meatus and over part (in human and nonhuman primates) or most (for rodents) of the nasal turbinates and adjacent recesses in the most caudal regions of the nasal cavity.[6] What is less intuitive is

TABLE 3   Inhaled Compounds That Induce Olfactory Mucosal Damage (D) or Neoplasia (N) in Rodents

Compound	Endpoint	References
Acetaldehyde	D, N	150
Acrolein	D	151
Acrylic acid	D	152
Allyl glycidyl ether	D, N	153
1,3-Butadiene	D	154
Bis(chloromethyl)ether	C	155
n-Butyl propionate	D	156
Chlorine	D	157
Chloroform	D	158
Chloropicrin	D	159
Chloroprene	D	160
1,2-Dibromo-3-chloropropane	D, N	161,162
1,2-Dibromoethane	D	162
Dimethylamine	D	163
Dimethyl sulfate	D	164
Ethyl acrylate	D	165
Furfuraldehyde	D	166
1,6-Hexamethylene diisocyanate	D	167
Hexanethylphosphoramide	D	168
Isoprene	D	169
3-Methylfuran	D	170
N-Methylformiminomethyl ester	D	171
Methyl bromide	D	172
4,4-Methylenediamine	D	173
Methyl isocyanate	D	174
Naphthalene	D, N	78
Nickel subsulfide	D	101
Nickel sulfate	D	175
Propargyl alcohol	D	176
Propylene	D	65
Propylene oxide	D	177
Styrene	D	178
Vinyl chloride monomer	D	179

that compounds administered by noninhalation routes (i.e., systemically administered substances) can also induce degeneration and even promote neoplasia of the OM (Table 4). Olfactory neurotoxicity following systemic exposure is facilitated by the high density of blood vessels in the lamina propria of the OM (194 vessels/mm^2 in the rat, vs. 66 vessels/mm^2 beneath the respiratory epithelium[80]). Compounds that cause nasal lesions following systemic administration are typically bioactivated in the liver or in situ in Bowman's glands. Thus, the nasal epithelia, particularly the OM, must not be overlooked as a potential target tissue following systemic toxicant delivery.

The vigorous regenerative capacity of the OM means that neurotoxicity at this site, even if profound, is likely to be temporary. If a xenobiotic exposure induces cell death in the OM, progressive recovery that eventually leads to the full reconstitution of both histological structure and sensory

TABLE 4 Systemically Administered Agents Associated with Olfactory Mucosal Damage in Rodents

Compound	Reference
Acetaminophen	180
Alachlor	180
Antimicrotubule chemotherapeutic drugs[a]	181
Benomyl	180
Benzophenone	182
Benzyl acetate	97
Bromobenzene	180
Butanal oxime	97
Caffeine-derived N-nitroso compounds	180
Carbimazole	180
Chloroform	182
Chlorthiamid	180
Coumarin	180
p-Cresidine	180
Cyclohexanone oxime	97
2,3-Dibromopropanol	182
2,6-Dichlorobenzamide	180
2,6-Dichlorobenzonitrile (dichlobenil)	180
2,6-Dichlorothiobenzamide	180
Diethyldithiocarbamate	180
Dihydropyridines	180
2,6-Dimethylaniline	182
Dimethylvinyl chloride	182
Dipropylene glycol	97
Disulfiram	180
1,4-Dithiane	180
Hexamethylphosphoramide	182
N-Bis(2-hydroxypropyl)nitrosamine	180
N-Hydroxy-3,3-iminodipropionitrile	180
3,3-Iminodipropionitrile	180
Mercuric chloride	97
Methacrylonitrile	97
Methimazole	180
Methylethyl ketoxamine	97
3-Methylindole	180
Methylsulfonyl-2,6-dichlorobenzene	180
Monochloroacetic acid	97
Naphthalene	182
N-Nitrosodiethylamine	180
o-Nitrotoluene sulfone	97
4-(N-methyl-N-nitrosamino)-1-(3-pyridyl)-1-butanone	180
Pentachloroanisole	97
Pentachlorophenol	97
2-Pentenenitrile	180
Phenacetin	180
Piclamist	182
Polychlorinated biphenyls	97
Procarbazine	182
RP 73401	180
trans-Cinnamaldehyde	97

Source: Modified from Baker and Genter.[180]

[a] Includes vincristine sulfate, vinblastine sulfate, vindesine sulfate, and paclitaxel.

function is the norm. Functional restoration generally precedes histological recovery.[13,81] Significant morphologic changes in the OM in the absence of functional deficits may be induced by some toxicants.[82] Structural regeneration of the OM can occur in the face of continued toxicant exposure.[13] Coexposure to inhaled toxicants may protect some aspects of normally vulnerable olfactory mucosa.[83]

The proximity of the OM surface to the rostral brain coupled with the direct exposure of ORNs to materials in the nasal cavity affords a straightforward route for xenobiotics in the nasal passage (either soluble or particulate) to access the brain.[84–90] A study in which cadmium was instilled unilaterally into the nasal cavity of rats has shown that the toxicant collects in the ipsilateral OM and MOB but does not reach the contralateral structures or pass to the olfactory cortex.[91] Potential mechanisms by which transfer occurs have only recently been addressed at the molecular level. Many transporters have been identified, and these have been implicated in both the uptake and exclusion of compounds from the brain. For example, divalent metal transporter-1 (DMT1), which serves primarily to facilitate iron uptake in the gastrointestinal tract, is expressed in the sustentacular cells of the rodent OM.[92] The Belgrade rat, which has a point mutation in the *Dmt1* gene that markedly reduces transporter function, transfers much lower quantities of the neurotoxic metal manganese (Mn) from the nasal cavity into the brain relative to rats with wild-type *Dmt1* alleles.[92] Uptake of Mn from the nasal cavity is dramatically reduced by transection of the olfactory nerves, indicating that the ORN axons act as conduits for transferring materials from the nasal cavity into the brain. We have shown that ZIP8 and ZIP14, two transport proteins that primarily transfer divalent metal ions, are expressed by olfactory nerves themselves, so it appears that both ORNs and supporting cells can transmit neurotoxic metals into the brain[90,93] (also, M. B. Genter, unpublished observation).

The cells of the OM have tight junctions near their apical surfaces, which are thought to contribute to the blood–brain barrier in the nasal cavity.[90] These structures prevent access of most large molecules and particulate materials across the OM unless they enter via a transcellular route (i.e., through cells).[89] Passage of materials such as small proteins from the nasal cavity can be enhanced by enzyme (protease) inhibitors and agents that disrupt these tight junctions.[94] Some small molecules and tiny particulates in the nasal lumen can also gain access to the brain via a "paracellular" mechanism, whereby substances are passed across the mucosa in the spaces between OM cells.[90]

Neuropathology Evaluation of the Olfactory Mucosa

Microscopic assessment of the OM is typically performed in adult animals (usually rodents) during general toxicity studies or on biopsy specimens acquired from humans. The best regions to evaluate the OM in the rodent are the linings of the dorsal meatus and the ethmoid turbinates, which should in normal cases be lined by continuous fields of OM. If present in the rostral septum, the VNO should also be evaluated. Standardized nomenclature[95] should be employed to permit more ready extrapolation of findings among studies and across species.

The usual practice in conducting the neuropathological examination of the OM is first to examine the tissue in a single section using a common histological stain such as H&E. This initial review is a qualitative screen for obvious structural changes in the sensory neuroepithelium and the structures (e.g., olfactory nerve bundles, Bowman's glands) in the lamina propria. A serial section may be cut and processed concurrently if a particular additional endpoint was prospectively implemented in the study design (e.g., antemortem BrdU administration to permit quantification of proliferating cell numbers). However, additional sections are usually procured only if an abnormal finding is detected in the H&E-stained screening section.

Common Lesions in the Olfactory Mucosa of Rodents

Several common nonproliferative and proliferative lesions of OM have been observed in rodents following inhalation or systemic exposure to olfactory toxicants.[5,7,96,97]

Nonproliferative Lesions The most common finding is acute *degeneration* or *necrosis* (Figure 6) of the OM. This change is often the earliest morphological change that develops following exposure to an airborne toxicant. It is characterized at low magnification by disruption of the pseudostratified arrangement and/or a "moth-eaten" (vacuolated, due to cell depletion) appearance of the OM. At higher magnification, sensory and sustentacular cells are observed to be dying, fragmenting, or absent.[81,83] Many olfactory neurotoxicants affect both ORNs and sustentacular cells,[13] but in some cases the primary effect may be the loss of the sustentacular support cells[82] or Bowman's glands[98] rather than primary deficits in the ORN population. In more extensive lesions, the OM may be extensively folded or entirely sloughed except for a few residual basal cells lining the basement membrane. If the ORNs do represent a major target cell population, their axon bundles in the lamina propria are usually attenuated or absent as well, and the Bowman's glands often exhibit changes.[81,83]

A common sequel to OM degeneration is *regeneration*. Recovery of the denuded neuroepithelium is dependent on survival of the basal (stem) cell layers. Regenerating OM is characterized by florid, often disorganized proliferation of highly basophilic basal cells. The normal laminar structure (Figure 2) is restored after some weeks

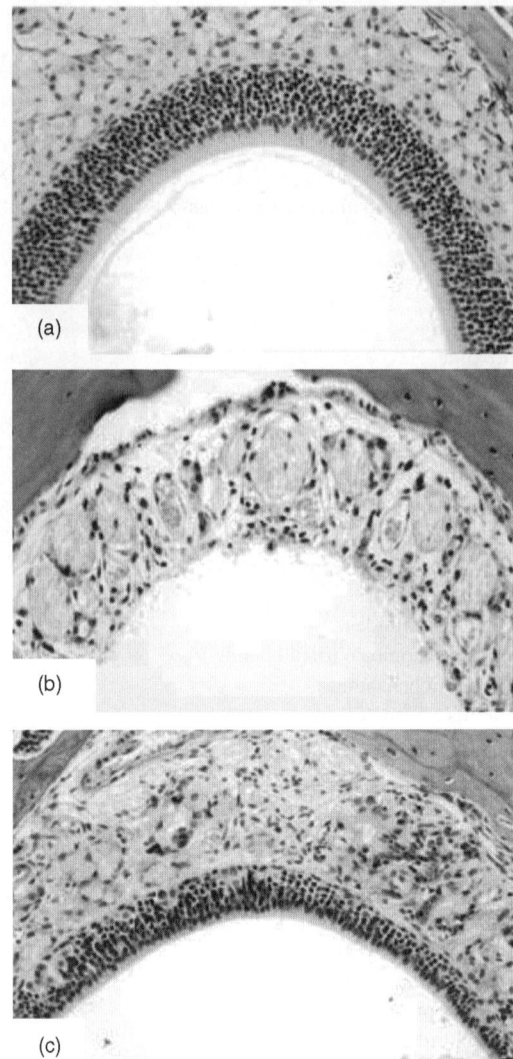

(a)

(b)

(c)

FIGURE 6 Olfactory mucosal degeneration and regeneration are the most frequent olfactory neuropathology findings following neurotoxicant exposure (in this case naphthalene, 100 mg/kg i.p., given once to a C57BL/6J mouse). (a) Olfactory mucosa from a vehicle-treated control, with normal appearance of olfactory epithelium lining the dorsal medial meatus of the nasal cavity. (b) Severe acute degeneration of the olfactory mucosa 24 h after treatment has resulted in complete sloughing of the neuroepithelial layer. (c) Early regeneration (14 days after treatment) is characterized by a thin neuroepithelial layer of uneven thickness, indicating partial recovery. The neuroepithelium generally recovers completely by 30 days after treatment. Stain: H&E.

in younger animals. If the basement membrane is breached during the acute necrotic episode, the OM will not regenerate to any great degree. The void left by the missing OM will usually be filled in one of two fashions: bridging the turbinates by columns or sheets of fibrous connective tissue (scarring),[99,100] and/or extension of respiratory epithelium (metaplasia) to cover the surfaces that are normally covered by OM.[83,96] Alternatively, postdegenerative atrophy characterized by the absence of ORNs and sustentacular cells may develop in one or multiple foci;[6,82,101] this change may be unilateral.

Basal cell hyperplasia along the OM basement membrane may be evident following extended inhalation exposure to toxicants.[6] The proliferating basal cells form a distinct layer (typically, two to four cells thick). The overlying OM may consist of neuroepithelial cells exhibiting some degree of degeneration and/or regeneration, or it may consist chiefly of respiratory-type epithelium instead.

Several other nonproliferative alterations have been reported to affect the OM at low incidence, including inflammation (acute or chronic), mineral deposition (generally in the lamina propria), and thrombi.[96] An occasional finding in olfactory epithelial cells is the widespread occurrence of intracytoplasmic eosinophilic protein droplets.[96] These membrane-bound inclusions appear to develop in the rough endoplasmic reticulum.[102] This change sometimes seems to increase in conjunction with the dose of an inhaled toxicant (e.g., pyridine[103]). However, eosinophilic droplets are found regularly in control rodents and tend to increase in severity and incidence with age.[102]

Proliferative Lesions Toxicant-induced proliferative lesions occasionally originate in the OM. Neoplasms of the OM proper—commonly termed *olfactory esthesioblastomas* or *esthesioneuroblastomas*—have highly variable morphological features which often require special histopathological methods to confirm the cell of origin. These masses have been attributed to expansion of ORNs, sustentacular cells, and even Bowman's gland epithelium.[97] Toxicant-induced OM tumors may arise at multiple sites, and over time they can aggressively invade adjacent tissues (including through the cribriform plate, thereby entering the MOB). The OM represents a primary target tissue of some carcinogens. For example, repeated subcutaneous administration of a tobacco-specific carcinogen, either N'-nitrosonornicotine or 4-(N-methyl-N-nitrosamino)-1-(3-pyridyl)-1-butanone, induces OM tumors in more than 75% of male and female F344 rats.[104] Simultaneous exposure to multiple agents can enhance the OM carcinogenic effect of a proven nasal toxicant, as has been shown for the tobacco-specific carcinogen N-nitrosopyrrolidine (given i.p.) and ethanol (delivered p.o.).[105]

Not all neoplasms affecting the olfactory region are derived from neuroepithelial elements. Sustained exposure to alachlor induces both benign and malignant masses in the olfactory region of adult rats which actually arise from regions of metaplastic respiratory epithelium.[64] The OM can also be affected by the invasion of nonolfactory neoplasms which extend through the lumen from the rostral nasal cavity into the caudal nasal passages.

Common Lesions in the Olfactory Mucosa of Humans

Abnormalities of the OM have been described in both normosmic and dysosmic patients.[106] Anosmia in patients with a history of head trauma has been associated with ultrastructural disorganization in the OM, consistent with the inability of regenerating axons from primary ORNs in the OM to reach their usual site of synaptic contact with mitral cells in the glomeruli of the MOB.[107]

Interest in the use of olfactory biopsies for diagnosis of neurodegenerative diseases has expanded rapidly since the publication of data showing unique neuropathological changes in the morphology, distribution, and immunoreactivity of neuronal structures in OM biopsies of patients with Alzheimer's disease (AD) and idiopathic Parkinson's disease (IPD).[77,108] Similar lesions have been detected in OM biopsies of patients suffering from mood disorders and other neurodegenerative diseases[73,75,76,109–111] as well as neurodevelopmental disorders.[112,113] However, comparable changes have also been observed in OM biopsies from unaffected elderly patients.[108] Therefore, at present the morphology of ORNs in OM biopsies from elderly patients with dementia is not sufficiently specific to confirm a diagnosis of incipient AD or IPD.[114] Nevertheless, a currently popular hypothesis that at least some neurodegenerative diseases are the result of chronic neurotoxicity[115] will sustain OM biopsies as a practical (i.e., relatively noninvasive) means of obtaining injured neural tissue rapidly and repeatedly throughout the course of disease progression, thereby providing researchers with a reasonable avenue for investigating earlier diagnostic techniques, disease mechanisms, and the efficacy of proposed interventions.

COMPARATIVE NEUROPATHOLOGY OF THE OLFACTORY MUCOSA AND OLFACTORY BRAIN

Investigations comparing responsiveness of the OM and the olfactory brain across species typically have been reserved for mice and rats. This situation arises from such practical issues as the scarcity of olfactory neurotoxicity studies on nonrodent large animal species and the ethical constraints against performing prospective evaluations of olfactory neurotoxicity in humans. Despite the similar microscopic structure[19] and function across species, one cannot always extrapolate OM changes in one species to predict the response of the OM in another, even among closely related species.[11,15,19,60,98,108]

For the time being, risk assessments of neurotoxic potential to the human olfactory mucosa will continue to be undertaken in the conventional manner: using rodent data to set relatively arbitrary factors to adjust for interspecies extrapolations, differences in the duration and time of exposure, and so on. Imaging modalities and image reconstruction are emerging as regular means for assessing olfactory mucosal toxicant-induced damage.[116,117] We anticipate that assessing the risk of toxicant-induced neuropathology in the olfactory brain will be performed in conjunction with a more global assessment of risk to the central nervous system, at least until such time that definitive evidence of toxicants as the inciting factor in one or more neurodegenerative diseases accelerates the race to perfect the diagnostic uses of olfactory epithelial biopsies.

TOXICOLOGICAL NEUROPATHOLOGY OF THE VOMERONASAL ORGAN

The VNO is a small, recessed, tubular, mucus-filled structure in the rostroventral wall of the nasal septum (Figure 7).[3,4,118] The VNO functions in macrosmatic species (e.g., rodents) as a separate and parallel sensory modality. The sensory neurons detect certain classes of odors such as pheromones. Rather than the broad-spectrum sensitivity of the ORNs, the vomeronasal receptor neurons (VRNs) are tuned narrowly to respond to a single sensory signal.[3] In addition, they respond to much lower odorant concentrations

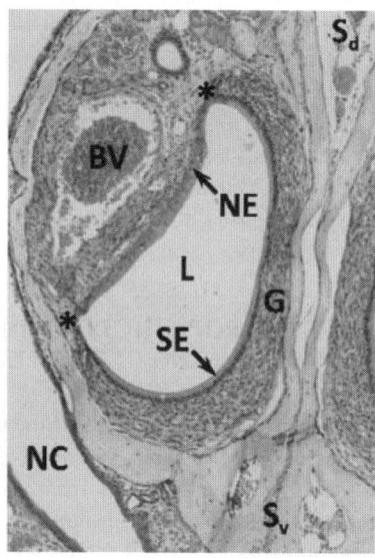

FIGURE 7   Vomeronasal organ (VNO) in a level I nasal section (acquired as shown in Figure 1a) from a normal adult C57BL/6J mouse. The VNO is a recessed, mucus-filled tube with a single opening located bilaterally in the rostroventral nasal septum. It is lined by two epithelial types, a sensory neuroepithelium (SE) containing bipolar vomeronasal receptor neurons with surface microvilli, and a ciliated nonsensory epithelium (NE). The boundaries between these epithelia are delineated by asterisks (∗). As with the olfactory mucosa, the SE is associated with supporting cells and mucous glands (G). Odorants enter the VNO in nasal secretions that are drawn into the lumen (L) by constriction of regional blood vessels (BV). Stain: H&E. NC, nasal cavity; S_d, dorsal and S_v, ventral nasal septum. (From Young.[57])

than do the ORNs.[3] The VNO is dependent on constriction of regional blood vessels to expand the lumen, thereby drawing odorant-containing nasal secretions into the tube where they can contact the VRNs.

In rodents, the medial (concave) face of the VNO is lined by a multilayered neuroepithelium comprised of three main cell types: VRNs, supporting cells, and basal cells. The supporting cells are distributed superficially, while the basal cells are localized near the basement membrane. The VRNs arise during embryonic development from the olfactory placode along with the ORN precursors that populate the OM. A thin duct that opens onto the floor of the nasal cavity inside the nostril is the only route of access for stimulus chemicals.[24]

The axons of the VRNs extend to the accessory olfactory bulb (AOB), which is located at the caudodorsal end of the MOB (Figure 3). The AOB Is smaller than the MOB but exhibits a similar layered cytoarchitectural organization, except that the AOB has fewer and smaller glomeruli.[3] The AOB projects centrally to brain regions that do not overlap with those supplied by the MOB. Some specific targets for AOB-derived circuits include the amygdala, the hippocampus, and the nucleus of the accessory olfactory tract, which in turn are connected widely to various hypothalamic nuclei that control neuroendocrine responses.[3]

Olfactory neurotoxicants can disrupt VMN structure, often while also affecting the morphology of the OM.[8,10,11] The typical toxicant-induced changes observed in the VMN, neuroepithelial degeneration or single-cell necrosis, are equivalent to those observed in the OM. The basal cell population of the VMN proliferates in response to toxicant-induced damage,[10] indicating that this sensory neuroepithelium is also capable of regeneration. Another response to toxicant exposure is to alter the expression of various molecules [e.g., cytokeratin,[119] vomeromodulin[120] (a pheromone-binding protein[121,122])]. To our knowledge, such molecular adaptations are not specific to any given class of toxicants.

REFERENCES

1. Rouquier S, Giorgi D. Olfactory receptor gene repertoires in mammals. *Mutat Res.* 2007;16:95–102.

2. Doty RL, ed., *Handbook of Olfaction and Gustation*, 2nd ed. New York: Marcel Dekker; 2003.

3. Shipley MT, Ennis M, Puche AC. Olfactory system. In: Paxinos G, ed. *The Rat Nervous System*. New York: Elsevier; 2004: 923–964.

4. Brennan PA, Keverne EB. The vomeronasal organ. In: Doty RL, ed. *Handbook of Olfaction and Gustation*. New York: Marcel Dekker; 2003: 967–979.

5. Brown HR, Monticello TM, Maronpot RR, Randall HW, Hotchkiss JR, Morgan KT. Proliferative and neoplastic lesions in the rodent nasal cavity. *Toxicol Pathol.* 1991;19:358–372.

6. Hardisty JF, Garman RH, Harkema JR, Lomax LG, Morgan KT. Histopathology of nasal olfactory mucosa from selected inhalation toxicity studies conducted with volatile chemicals. *Toxicol Pathol.* 1999;27:618–627.

7. Monticello TM, Morgan KT, Uraih L. Nonneoplastic nasal lesions in rats and mice. *Environ Health Perspect.* 1990;85.

8. Gaafar H, Tantawy A, Hamza M, Shaaban M. The effect of ammonia on olfactory epithelium and vomeronasal organ neuroepithelium of rabbits: a histological and histochemical study. *J Otorhinolaryngol Relat Spec.* 1998;60:88–91.

9. Barron S, Riley EP. The effects of prenatal alcohol exposure on behavioral and neuroanatomical components of olfaction. *Neurotoxicol Teratol.* 1992;14:291–297.

10. Suzuki Y, Takeda M, Obara N, Suzuki N. Colchicine-induced cell death and proliferation in the olfactory epithelium and vomeronasal organ of the mouse. *Anat Embryol (Berl).* 1998;198:43–51.

11. Kai K, Sahto H, Yoshida M, Suzuki T, Shikanai Y, Kajimura T, et al. Species and sex differences in susceptibility to olfactory lesions among the mouse, rat and monkey following an intravenous injection of vincristine sulphate. *Toxicol Pathol.* 2006;34:223–231.

12. Witt M, Woźniak W. Structure and function of the vomeronasal organ. *Adv Otorhinolaryngol.* 2006;63:70–83.

13. Hurtt ME, Thomas DA, Working PK, Monticello TM, Morgan KT. Degeneration and regeneration of the olfactory epithelium following inhalation exposure to methyl bromide: pathology, cell kinetics, and olfactory function. *Toxicol Appl Pharmacol.* 1998;94:311–328.

14. Weiler E, Farbman AI. Mitral cell loss following lateral olfactory tract transection increases proliferation density in rat olfactory epithelium. *Eur J Neurosci.* 1999;11:3265–3275.

15. DeSesso JM. The relevance to humans of animal models for inhalation studies of cancer in the nose and upper airways. *Qual Assur Good Pract Reg Law.* 1993;2:213–221.

16. Skinner AP, Pachnicke S, Lakatos A, Franklin RJ, Jeffery ND. Nasal and frontal sinus mucosa of the adult dog contain numerous olfactory sensory neurons and ensheathing glia. *Res Vet Sci.* 2005;78:9–15.

17. Weiler E, Farbman AI. Proliferation in the rat olfactory epithelium: age-dependent changes. *J Neurosci.* 1997;17:3610–3622.

18. Genter MB, Goss KH, Groden J. Strain-specific of alachlor on murine olfactory mucosal responses. *Toxicol Pathol.* 2004;32:719–725.

19. Harkema JR, Comparative pathology of the nasal mucosa in laboratory animals exposed to inhaled irritants. *Environ Health Perspect.* 1990;85:231–238.

20. Craven BA, Neuberger T, Paterson EG, Webb AG, Josephson EM, Morrison EE, et al. Reconstruction and morphometric analysis of the nasal airway of the dog (*Canis familiaris*) and implications regarding olfactory airflow. *Anat Rec.* 2007;290:1325–1340.

21. Issel-Tarver L, Rine J. The evolution of mammalian olfactory receptor genes. *Genetics.* 1997;145:185–195.

22. Thorne C. Feeding behaviour of domestic dogs and the role of experience. In: Serpell J, ed. *The Domestic Dog: Its Evolution, Behaviour and Interactions with People.* Cambridge, UK: Cambridge University Press; 1995: 104–114.

23. Smith TD, Bhatnagar KP. Microsmatic primates: reconsidering how and when size matters. *Anat Rec B.* 2004;279:24–31.

24. Sorokin SP. The respiratory system. In: Weiss L, ed. *Cell and Tissue Biology: A Textbook of Histology.* Baltimore: Urban & Schwarzenberg; 1988: 751–814.

25. Clancy AN, Schoenfeld TA, Forbes WB, Macrides F. The spatial organization of the peripheral olfactory system of the hamster: II. Receptor surfaces and odorant passageways within the nasal cavity. *Brain Res Bull.* 1994;34:211–241.

26. Gross EA, Swenberg JA, Fields S, Popp JA. Comparative morphometry of the nasal cavity in rats and mice. *J Anat.* 1982;135:83–88.

27. Mulvaney BD, Heist HE. Mapping of rabbit olfactory cells. *J Anat.* 1970;107:19–30.

28. Nomura T, Takahashi S, Ushiki T. Cytoarchitecture of the normal rat olfactory epithelium: light and scanning electron microscopic studies. *Arch Histol Cytol.* 2004;67: 159–170.

29. Hahn CG, Han LY, Rawson NE, Mirza N, Borgmann-Winter K, Lenox RH, et al. In vivo and in vitro neurogenesis in human olfactory epithelium. *J Comp Neurol.* 2005;483: 154–163.

30. Schwartz Levey M, Chikaraishi DM, Kauer JS. Characterization of potential precursor populations in the mouse olfactory epithelium using immunocytochemistry and autoradiography. *J Neurosci.* 1991;11:3556–3564.

31. Barnett SC, Riddell JS. Olfactory ensheathing cells (OECs) and the treatment of CNS injury: advantages and possible caveats. *J Anat.* 2004;204:57–67.

32. Vincent AJ, West AK, Chuah MI. Morphological and functional plasticity of olfactory ensheathing cells. *J Neurocytol.* 2005;34:65–80.

33. Margolis FL, Verhaagen J, Biffo S, Huang FL, Grillo M. Regulation of gene expression in the olfactory neuroepithelium: a neurogenetic matrix. *Prog Brain Res.* 1991;89:97–122.

34. Miller ML, Andringa A, Evans JE, Hastings L. Microvillar cells of the olfactory epithelium: morphology and regeneration following exposure to toxic compounds. *Brain Res.* 1995;669:1–9.

35. Montani G, Tonelli S, Elsaesser R, Paysan J, Tirindelli R. Neuropeptide Y in the olfactory microvillar cells. *Eur J Neurosci.* 2006;24:20–24.

36. Keller A, Margolis FL. Immunological studies of the rat olfactory marker protein. *J Neurochem.* 1975;24:1101–1106.

37. Youngentob SL, Margolis FL, Youngentob LM. OMP gene deletion results in an alteration in odorant quality perception. *Behav Neurosci.* 2001;115:626–631.

38. Olender T, Lancet D, Nebert DW. Update on the olfactory receptor (OR) gene superfamily. *Hum Genomics.* 2008;3: 87–97.

39. Imai T, Sakano H. Odorant receptor gene choice and axonal projection in the mouse olfactory system. *Results Probl Cell Differ.* 2009;47:57–75.

40. Reed CJ. Drug metabolism in the nasal cavity: relevance to toxicology. *Drug Metab Rev.* 1993;25:173–205.

41. Bond JA, Harkema JR, Russell VI. Regional distribution of xenobiotic metabolizing enzymes in respiratory airways of dogs. *Drug Metab Dispos.* 1988;16:116–124.

42. Ding X, Dahl AR. Olfactory mucosa: composition, enzymatic localization, and metabolism. In: Doty RL, ed. *Handbook of Olfaction and Gustation.* New York: Marcel Dekker; 2003: 51–73.

43. Dahl AR. Possible consequences of cytochrome P-450-dependent monooxygenases in nasal tissues. In: Barrow CS, ed. *Toxicology of the Nasal Passages.* New York: Hemisphere; 1986: 263–273.

44. Bogdanffy MS. Biotransformation enzymes in the rodent nasal mucosa: the value of a histochemical approach. *Environ Health Perspect.* 1990;85:177–186.

45. Keller DA, Heck HD, Randall HW, Morgan KT. Histochemical localization of formaldehyde dehydrogenase in the rat. *Toxicol Appl Pharmacol.* 1990;106:311–326.

46. Banger KK, Foster JR, Lock EA, Reed CJ. Immunohistochemical localisation of six glutathione S-transferases within the nasal cavity of the rat. *Arch Toxicol.* 1994;69:91–98.

47. Genter MB, Van Veldhoven PP, Jegga AG, Sakthivel B, Kong S, Stanley K, et al. Microarray-based discovery of highly expressed olfactory mucosal genes: potential roles in the various functions of the olfactory system. *Physiol Genom.* 2003;16:67–81.

48. Lewis JL, Rhoades CE, Bice DE, Harkema JR, Hotchkiss JA, Sylvester DM, et al. Interspecies comparison of cellular localization of the cyanide metabolizing enzyme rhodanese within olfactory mucosa. *Anat Rec.* 1992;232:620–627.

49. Chang JC, Gross EA, Swenberg JA, Barrow CS. Nasal cavity deposition, histopathology, and cell proliferation after single or repeated formaldehyde exposures in B6C3F1 mice and F-344 rats. *Toxicol Appl Pharmacol.* 1983;68:161–176.

50. St. Clair MB, Gross EA, Morgan KT. Pathology and cell proliferation induced by intra-nasal instillation of aldehydes in the rat: comparison of glutaraldehyde and formaldehyde. *Toxicol Pathol.* 1990;18.

51. Uraih LC, Maronpot RR. Normal histology of the nasal cavity and application of special techniques. *Environ Health Perspect.* 1990;85:187–208.

52. Alers JC, Krijtenburg PJ, Vissers KJ, van Dekken H. Effect of bone decalcification procedures on DNA in situ hybridization and comparative genomic hybridization: EDTA is highly preferable to a routinely used acid decalcifier. *J Histochem Cytochem.* 1999;47:703–710.

53. Arber JM, Weiss LM, Chang KL, Battifora H, Arber DA. The effect of decalcification on in situ hybridization. *Mod Pathol.* 1997;10:1009–1014.

54. Lovell MA, Jafek BW, Moran DT, Rowley JC. Biopsy of human olfactory mucosa: an instrument and a technique. *Arch Otolaryngol.* 1982;108:247–279.

55. Paik SI, Lehman MN, Seiden AM, Duncan HJ, Smith DV. Human olfactory biopsy: the influence of age and receptor distribution. *Arch Otolaryngol Head Neck Surg.* 1992;118:731–738.

56. Winstead W, Marshall CT, Lu CL, Klueber KM, Roisen FJ. Endoscopic biopsy of human olfactory epithelium as a source of progenitor cells. *Am J Rhinol.* 2005;19:83–90.

57. Young JT. Histopathologic examination of the rat nasal cavity. *Fundam Appl Toxicol.* 1981;1:309–312.

58. Méry S, Gross EA, Joyner DR, Godo N, Morgan KT. Nasal diagram: a tool for recording the distribution of nasal lesions in rats and mice. *Toxicol Pathol.* 1994;22:353–372.

59. Marit GB, Cooper MK, Latendresse JR. Processing and immunohistochemical staining of rat nasal sections. *J Histotechnol.* 1995;18:111–114.

60. Walsh K, Courtney CL. Nasal toxicity of CI-959, a novel anti-inflammatory drug, in Wistar rats and Beagle dogs. *Toxicol Pathol.* 1998;26:717–723.

61. Randall HW, Monticello TM, Morgan KT. Large area sectioning for morphologic studies of nonhuman primate nasal cavities. *Stain Technol.* 1988;63:355–362.

62. Randall HW, Bogdanffy MS, Morgan KT. Enzyme histochemistry of the rat nasal mucosa embedded in cold glycol methacrylate. *Am J Anat.* 1987;179:10–17.

63. Monticello TM, Morgan KT, Everitt JI, Popp JA. Effects of formaldehyde gas on the respiratory tract of rhesus monkeys: pathology and cell proliferation. *Am J Pathol.* 1989;134:515–527.

64. Genter MB, Burman DM, Dingeldein MW, Clough I, Bolon B. Evolution of alachlor-induced nasal neoplasms in the Long–Evans rat. *Toxicol Pathol.* 2000;28:770–781.

65. Pottenger LH, Malley LA, Bogdanffy MS, Donner EM, Upton PB, Li Y, et al. Evaluation of effects from repeated inhalation exposure of F344 rats to high concentrations of propylene. *Toxicol Sci.* 2007;97:336–347.

66. Bravo R, Frank R, Blundell PA, Macdonald-Bravo H. Cyclin/PCNA is the auxiliary protein of DNA polymerase-delta. *Nature.* 1987;326:515–517.

67. Ohta Y, Ichimura K. Proliferation markers, proliferating cell nuclear antigen, Ki67, 5-bromo-2′-deoxyuridine, and cyclin D1 in mouse olfactory epithelium. *Ann Otol Rhinol Laryngol.* 2000;109:1046–1048.

68. Monticello TM, Morgan KT, Hurtt ME. Unit length as the denominator for quantitation of cell proliferation in nasal epithelia. *Toxicol Pathol.* 1990;18:24–31.

69. Brown D, Garcia-Segura LM, Orci L. Carbonic anhydrase is present in olfactory receptor cells. *Histochemistry.* 1984;80:307–309.

70. Sanes JR, Rubenstein JL, Nicolas JF. Use of a recombinant retrovirus to study post-implantation cell lineage in mouse embryos. *EMBO J.* 1986;5:3133–3142.

71. Bolon B. Whole mount enzyme histochemistry as a rapid screen at necropsy for expression of β-galactosidase (LacZ)-bearing transgenes: considerations for separating specific LacZ activity from nonspecific (endogenous) galactosidase activity. *Toxicol Pathol.* 2008;36:265–276.

72. Restrepo D, Okada Y, Teeter JH, Lowry LD, Cowart B, Brand JG. Human olfactory neurons respond to odor stimuli with an increase in cytoplasmic Ca^{2+}. *Biophys J.* 1993;64: 1961–1966.

73. Trojanowski JQ, Newman PD, Hill WD, Lee VM. Human olfactory epithelium in normal aging, Alzheimer's disease, and other neurodegenerative disorders. *J Comp Neurol.* 1991;310:365–376.

74. Lanza DC, Deems DA, Doty RL, Moran D, Crawford D, Rowley JC, et al. The effect of human olfactory biopsy on olfaction: a preliminary report. *Laryngoscope.* 1994;104: 837–840.

75. Perry G, Castellani RJ, Smith MA, Harris PL, Kubat Z, Ghanbari K, et al. Oxidative damage in the olfactory system in Alzheimer's disease. *Acta Neuropathol.* 2003; 106:552–556.

76. Tabaton M, Cammarata S, Mancardi G, Cordone G, Perry G, Loeb C. Abnormal tau-reactive filaments in olfactory mucosa in biopsy specimens of patients with probable Alzheimer's disease. *Neurology.* 1991;41:391–394.

77. Talamo BR, Rudel R, Kosik KS, Lee VM, Neff S, Adelman L, et al. Pathological changes in olfactory neurons in patients with Alzheimer's disease. *Nature.* 1989;337: 736–739.

78. Long PH, Herbert RA, Peckham JC, Grumbein SL, Shackelford CC, Abdo K. Morphology of nasal lesions in F344/N rats following chronic inhalation exposure to naphthalene vapors. *Toxicol Pathol.* 2003;31:655–664.

79. Morgan KT, Monticello TM. Airflow, gas deposition, and lesion distribution in the nasal passages. *Environ Health Perspect.* 1990;85:209–218.

80. Yuasa T. Stereographic demonstration of the nasal cavity of the rat with reference to the density of blood vessels. *Acta Anat Nippon.* 1991;66:191–200.

81. Genter MB, Llorens J, O'Callaghan JP, Peele DB, Morgan KT, Crofton KM. Olfactory toxicity of β,β'-iminodipropionitrile in the rat. *J Pharmacol Exp Ther.* 1992;263:1432–1439.

82. Evans JE, Miller ML, Andringa A, Hastings L. Behavioral, histological, and neurochemical effects of nickel(II) on the rat olfactory system. *Toxicol Appl Pharmacol.* 1995; 130:209–220.

83. Bolon B, Bonnefoi MS, Roberts KC, Marshall MW, Morgan KT. Toxic interactions in the rat nose: pollutants from soiled bedding and methyl bromide. *Toxicol Pathol.* 1991; 19:571–579.

84. Fernández-Urrusuno R, Calvo P, Remuñán-López C, Vila-Jato JL, Alonso MJ. Enhancement of nasal absorption of insulin using chitosan nanoparticles. *Pharm Res.* 1999;16: 1576–1581.

85. Jeannet PY, Roulet E, Maeder-Ingvar M, Gehri M, Jutzi A, Deonna T. Home and hospital treatment of acute seizures in children with nasal midazolam. *Eur J Paediatr Neurol.* 1999;3:73–77.

86. Schneider NG, Olmstead R, Mody FV, Doan K, Franzon M, Jarvik ME, et al. Efficacy of a nicotine nasal spray in smoking cessation: a placebo-controlled, double-blind trial. *Addiction.* 1995;90:1671–1682.

87. Yajima T, Juni K, Saneyoshi M, Hasegawa T, Kawaguchi T. Direct transport of 2',3'-didehydro-3'-deoxythymidine (D4T) and its ester derivatives to the cerebrospinal fluid via the nasal mucous membrane in rats. *Biol Pharm Bull.* 1998;21: 272–277.

88. Charlton ST, Whetstone J, Fayinka ST, Read KD, Illum L, Davis SS. Evaluation of direct transport pathways of glycine receptor antagonists and an angiotensin antagonist from the nasal cavity to the central nervous system in the rat model. *Pharm Res.* 2008;25:1531–1543.

89. Mistry A, Glud SZ, Kjems J, Randel J, Howard KA, Stolnik S, et al. Effect of physicochemical properties on intranasal nanoparticle transit into murine olfactory epithelium. *J Drug Target.* 2009;17:543–552.

90. Genter MB, Kendig EL, Knutson MD. Uptake of materials from the nasal cavity into the blood and brain: Are we finally beginning to understand these processes at the molecular level? *Ann NY Acad Sci.* 2009;1170:623–628.

91. Hastings L, Evans JE. Olfactory primary neurons as a route of entry for toxic agents into the CNS. *Neurotoxicology.* 1991;12:707–714.

92. Thompson K, Molina RM, Donaghey T, Schwob JE, Brain JD, Wessling-Resnick M. Olfactory uptake of manganese requires DMT1 and is enhanced by anemia. *FASEB J.* 2007;21: 223–230.

93. Kinoshita Y, Shiga H, Washiyama K, Ogawa D, Amano R, Ito M, et al. Thallium transport and the evaluation of olfactory nerve connectivity between the nasal cavity and olfactory bulb. *Chem Senses.* 2008;33:73–78.

94. Sayani AP, Chien YW. Systemic delivery of peptides and proteins across absorptive mucosae. *Crit Rev Ther Drug Carrier Syst.* 1996;13:85–184.

95. Renne RA, Brix AE, Harkema JR, Herbert RA, Kittel B, Lewis D, et al. Proliferative and nonproliferative lesions of the rat and mouse respiratory tract. *Toxicol Pathol.* 2009;37:5S–73S.

96. Nagano K, Katagiri T, Aiso S, Senoh H, Sakura Y, Takeuchi T. Spontaneous lesions of nasal cavity in aging F344 rats and BDF1 mice. *Exp Toxicol Pathol.* 1997;49:97–104.

97. Sells DM, Brix AE, Nyska A, Jokinen MP, Orzech DP, Walker NJ. Respiratory tract lesions in noninhalation studies. *Toxicol Pathol.* 2007;35:170–177.

98. Jensen RK, Sleight SD. Toxic effects of *N*-nitrosodiethylamine on nasal tissues of Sprague–Dawley rats and golden Syrian hamsters. *Fundam Appl Toxicol.* 1987;8:217–229.

99. Bahrami F, Bergman U, Brittebo EB, Brandt I. Persistent olfactory mucosal metaplasia and increased olfactory bulb glial fibrillary acidic protein levels following a single dose of methylsulfonyl-dichlorobenzene in mice: comparison of the 2,5- and 2,6-dichloronated isomers. *Toxicol Appl Pharmacol.* 2000;162:49–59.

100. Peele DB, Allison SD, Bolon B, Prah JD, Jensen KF, Morgan KT. Functional deficits produced by 3-methylindole-induced olfactory mucosal damage revealed by a simple olfactory learning task. *Toxicol Appl Pharmacol.* 1991;107:191–202.

101. Benson JM, Carpenter RL, Hahn FF, Haley PJ, Hanson RL, Hobbs CH, et al. Comparative inhalation toxicity of nickel subsulfide to F344/N rats and B6C3F1 mice exposed for 12 days. *Fundam Appl Toxicol.* 1987;9:251–265.

102. Popp JA, Morgan KT, Everitt J, Jiang XZ, Martin JT. Morphologic changes in the upper respiratory tract of rodents exposed to toxicants by inhalation. In: Romig ADJ, Chambers WF, eds. *Microbeam Analysis.* San Francisco: San Francisco Press; 1986: 581–582.

103. Nikula KJ, Novak RF, Chang IY, Dahl AR, Kracko DA, Zangar RC, et al. Induction of nasal carboxylesterase in F344 rats following inhalation exposure to pyridine. *Drug Metab Dispos.* 1995;23:529–535.

104. Hecht SS, Chen CB, Ohmori T, Hoffmann D. Comparative carcinogenicity in F344 rats of the tobacco-specific nitrosamines, *N'*-nitrosonornicotine and 4-(*N*-methyl-*N*-nitrosamino)-1-(3-pyridyl)-1-butanone. *Cancer Res.* 1980;40:298–302.

105. McCoy GD, Hecht SS, Katayama S, Wynder EL. Differential effect of chronic ethanol consumption on the carcinogenicity of *N*-nitrosopyrrolidine and *N'*-nitrosonornicotine in male Syrian golden hamsters. *Cancer Res.* 1981;41:2849–2854.

106. Holbrook EH, Leopold DA, Schwob JE. Abnormalities of axon growth in human olfactory mucosa. *Laryngoscope.* 2005;115:2144–2154.

107. Moran DT, Jafek BW, Rowley JC, Eller PM. Electron microscopy of olfactory epithelia in two patients with anosmia. *Arch Otolaryngol.* 1985;111:122–126.

108. Talamo BR, Feng WH, Perez-Cruet M, Adelman L, Kosik K, Lee MY, et al. Pathologic changes in olfactory neurons in Alzheimer's disease. *Ann NY Acad Sci.* 1991;640:1–7.

109. Hahn CG, Gomez G, Restrepo D, Friedman E, Josiassen R, Pribitkin EA, et al. Aberrant intracellular calcium signaling in olfactory neurons from patients with bipolar disorder. *Am J Psychiatry.* 2005;162:616–618.

110. Tabaton M, Monaco S, Cordone MP, Colucci M, Giaccone G, Tagliavini F, et al. Prion deposition in olfactory biopsy of sporadic Creutzfeldt-Jakob disease. *Ann Neurol.* 2004;55:294–296.

111. Zanusso G, Ferrari S, Cardone F, Zampieri P, Gelati M, Fiorini M, et al. Detection of pathologic prion protein in the olfactory epithelium in sporadic Creutzfeldt-Jakob disease. *N Eng J Med.* 2003;348:711–719.

112. Johnston MV, Blue ME, Naidu S. Rett syndrome and neuronal development. *J Child Neurol.* 2005;20:759–763.

113. Ronnett GV, Leopold D, Cai X, Hoffbuhr KC, Moses L, Hoffman EP, et al. Olfactory biopsies demonstrate a defect in neuronal development in Rett's syndrome. *Ann Neurol.* 2003;54:206–218.

114. Hawkes C. Olfaction in neurodegenerative disorder. *Adv Otorhinolaryngol.* 2006;63:133–151.

115. Doty RL. The olfactory vector hypothesis of neurodegenerative disease: Is it viable? *Ann Neurol.* 2008;63:7–15.

116. Carey SA, Minard KR, Trease LL, Wagner JG, Garcia GJ, Ballinger CA, et al. Three-dimensional mapping of ozone-induced injury in the nasal airways of monkeys using magnetic resonance imaging and morphometric techniques. *Toxicol Pathol.* 2007;35:27–40.

117. Robinson DA, Foster JR, Nash JA, Reed CJ. Three-dimensional mapping of the lesions induced by β,β'-iminodiproprionitrile, methyl iodide and methyl methacrylate in the rat nasal cavity. *Toxicol Pathol.* 2003;31:340–347.

118. Wysocki CJ. Neurobehavioral evidence for the involvement of the vomeronasal system in mammalian reproduction. *Neurosci Biobehav Rev.* 1979;3:301–341.

119. Schlage WK, Bülles H, Friedrichs D, Kuhn M, Teredesai A, Terpstra PM. Cytokeratin expression patterns in the rat respiratory tract as markers of epithelial differentiation in inhalation toxicology: II. Changes in cytokeratin expression patterns following 8-day exposure to room-aged cigarette sidestream smoke. *Toxicol Pathol.* 1998;26:344–360 (erratum: *Toxicol Pathol.* 1998; 326:585).

120. Genter MB, Warner BM, Medvedovic M, Sartor MA. Comparison of rat olfactory responses to carcinogenic and non-carcinogenic chloracetanilides. *Food Chem Toxicol.* 2009;47:1051–1057.

121. Khew-Goodall Y, Grillo M, Getchell ML, Danho W, Getchell TV, Margolis FL. Vomeromodulin, a putative pheromone transporter: cloning, characterization, and cellular localization of a novel glycoprotein of lateral nasal gland. *FASEB J.* 1991;5:2976–2982.

122. Krishna NS, Getchell ML, Getchell TV. Expression of the putative pheromone and odorant transporter vomeromodulin mRNA and protein in nasal chemosensory mucosae. *J Neurosci Res.* 1994;39:243–259.

123. Pixley SK, Farbman AI, Menco BP. Monoclonal antibody marker for olfactory sustentacular cell microvilli. *Anat Rec.* 1997;248:307–321.

124. Foster JR, Elcombe CR, Boobis AR, Davies DS, Sesardic D, McQuade J, et al. Immunocytochemical localization of cytochrome P-450 in hepatic and extra-hepatic tissues of the rat with a monoclonal antibody against cytochrome P-450 C. *Biochem Pharmacol.* 1986;35:4543–4554.

125. Genter MB, Yost GS, Rettie AE. Localization of CYP4B1 in the rat nasal cavity and analysis of CYPs as secreted proteins. *J Biochem Mol Toxicol.* 2006;20:139–141.

126. Piras E, Franzén A, Fernández EL, Bergström U, Raffalli-Mathieu F, Lang M, et al. Cell-specific expression of CYP2A5 in the mouse respiratory tract: effects of olfactory toxicants. *J Histochem Cytochem.* 2003;51:1545–1555.

127. Margolis FL. Carnosine in the primary olfactory pathway. *Science.* 1974;184:909–911.

128. Verhaagen J, Oestreicher AB, Gispen WH, Margolis FL. The expression of the growth associated protein B50/GAP43 in the olfactory system of neonatal and adult rats. *J Neurosci.* 1989;9:683–691.

129. Goldstein BJ, Schwob JE. Analysis of the globose basal cell compartment in rat olfactory epithelium using GBC-1, a new

monoclonal antibody against globose basal cells. *J Neurosci.* 1996;16:4005–4016.

130. Doyle KL, Khan M, Cunningham AM. Expression of the intermediate filament protein nestin by sustentacular cells in mature olfactory neuroepithelium. *J Comp Neurol.* 2001;437:186–195.

131. Miragall F, Kadom G, Husmann M, Schachner M. Expression of cell adhesion molecules in the olfactory system of the adult mouse: presence of the embryonic form of N-CAM. *Dev Biol.* 1988;129:516–531.

132. Weiler E, Benali A. Olfactory epithelia differentially express neuronal markers. *J Neurocytol.* 2005;34:217–240.

133. Roskams AJ, Cai X, Ronnett GV. Expression of neuron-specific bIII tubulin during olfactory neurogenesis in the embryonic and adult rat. *Neuroscience.* 1998;83:191–200.

134. Dubois L, Vincent A. The COE—Collier/Olf1/EBF—transcription factors: structural conservation and diversity of developmental functions. *Mech Dev.* 2001;108:3–12.

135. Hartman BK, Margolis FL. Immunofluorescence localization of the olfactory marker protein. *Brain Res.* 1975;96:176–180.

136. Hempstead JL, Morgan JI. A panel of monoclonal antibodies to the rat olfactory epithelium. *J Neurosci.* 1985;5:438–449.

137. Schwob JE, Farber NB, Gottlieb DI. Neurons of the olfactory epithelium in adult rats contain vimentin. *J Neurosci.* 1986;6:208–217.

138. Hodgson E, Goldstein JA. Metabolism of toxicants: phase I reactions and pharmacogenetics. In: Hodgson E, Smart RC, eds. *Introduction to Biochemical Toxicology.* New York: Wiley-Interscience; 2001: 67–113.

139. Deshpande VS, Genter MB, Jung C, Desai PB. Characterization of lidocaine metabolism by rat nasal microsomes: implications for nasal drug delivery. *Eur J Drug Metab Pharmacokinet.* 1999;24:177–182.

140. Su T, Bao Z, Zhang QY, Smith TJ, Hong JY, Ding X. Human cytochrome P450 CYP2A13: predominant expression in the respiratory tract and its high efficiency metabolic activation of a tobacco-specific carcinogen, 4-(methylnitrosamino)-1-(3-pyridyl)-1-butanone. *Cancer Res.* 2000;60:5074–5079.

141. Zhuo X, Gu J, Zhang QY, Spink DC, Kaminsky LS, Ding X. Biotransformation of coumarin by rodent and human cytochromes P-450: metabolic basis of tissue-selective toxicity in olfactory mucosa of rats and mice. *J Pharmacol Exp Ther.* 1999;288:463–471.

142. Coleman S, Linderman R, Hodgson E, Rose RL. Comparative metabolism of chloroacetamide herbicides and selected metabolites in human and rat liver microsomes. *Environ Health Perspect.* 2000;108:1151–1157.

143. Genter MB, Apparaju S, Desai PB. Induction of olfactory mucosal and liver metabolism of lidocaine by 2,3,7,8-tetrachlorodibenzo-*p*-dioxin. *J Biochem Mol Toxicol.* 2002;16: 128–134.

144. Yokose T, Doy M, Taniguchi T, Shimada T, Kakiki M, Horie T, et al. Immunohistochemical study of cytochrome P450 2C and 3A in human non-neoplastic tissues. *Virchow's Arch.* 1999;434:401–411.

145. Méry S, Larson JL, Butterworth BE, Wolf DC, Harden R, Morgan KT. Nasal toxicity of chloroform in male F-344 rats and female B6C3F1 mice following a 1-week inhalation exposure. *Toxicol Appl Pharmacol.* 1994;125:214–227.

146. Aiba I, Yamasaki T, Shinki T, Izumi S, Yamamoto K, Yamada S, et al. Characterization of rat and human CYP2J enzymes as vitamin D 25-hydroxylases. *Steroids.* 2006;71:849–856.

147. Pellinen PH, Stenbäck F, Niemitz M, Alhava E, Pelkonen O, et al. Cocaine *N*-demethylation and the metabolism-related hepatotoxicity can be prevented by cytochrome P450 3A inhibitors. *Eur J Pharmacol.* 1994;270:35–43.

148. Ding X, Kaminsky LS. Human extrahepatic cytochromes P450: function in xenobiotic metabolism and tissue-selective chemical toxicity in the respiratory and gastrointestinal tracts. *Annu Rev Pharmacol Toxicol.* 2003;43:149–173.

149. Sheng J, Guo J, Hua Z, Caggana M, Ding X. Characterization of human CYP2G genes: widespread loss-of-function mutations and genetic polymorphism. *Pharmacogenetics.* 2000;10:667–678.

150. Woutersen RA, Appelman LM, Van Garderen-Hoetmer A, Feron VJ. Inhalation toxicity of acetaldehyde in rats: III. Carcinogenicity study. *Toxicology.* 1986;41:213–231.

151. Dorman DC, Struve MF, Wong BA, Marshall MW, Gross EA, Willson GA. Respiratory tract responses in male rats following subchronic acrolein inhalation. *Inhal Toxicol.* 2008;20:205–216.

152. Miller RR, Ayres JA, Jersey GC, McKenna MJ. Inhalation toxicity of acrylic acid. *Fundam Appl Toxicol.* 1981;1: 271–277.

153. Renne RA, Brown HR, Jokinen MP. Morphology of nasal lesions induced in Osborne–Mendel rats and B6C3F1 mice by chronic inhalation of allyl glycidyl ether. *Toxicol Pathol.* 1992;20:416–425.

154. Melnick RL, Huff JE, Haseman JK, McConnell EE. Chronic toxicity results and ongoing studies of 1,3-butadiene by the National Toxicology Program. *Ann NY Acad Sci.* 1988;534:648–662.

155. Kuschner M, Laskin S, Drew RT, Cappiello V, Nelson N. Inhalation carcinogenicity of alpha halo ethers: III. Lifetime and limited period inhalation studies with bis(chloromethyl) ether at 0.1 ppm. *Arch Environ Health.* 1975;30:73–77.

156. Banton MI, Tyler TR, Ulrich CE, Nemec MD, Garman RH. Subchronic and developmental toxicity studies of *n*-butyl propionate vapor in rats. *J Toxicol Environ Health A.* 2000;61:79–105.

157. Jiang XZ, Buckley LA, Morgan KT. Pathology of toxic responses to the RD50 concentration of chlorine gas in the nasal passages of rats and mice. *Toxicol Appl Pharmacol.* 1983;71:225–236.

158. Larson JL, Wolf DC, Morgan KT, Méry S, Butterworth BE. The toxicity of 1-week exposures to inhaled chloroform in female B6C3F1 mice and male F-344 rats. *Fundam Appl Toxicol.* 1994;22:431–446.

159. Buckley LA, Jiang XZ, James RA, Morgan KT, Barrow CS. Respiratory tract lesions induced by sensory irritants at the

RD50 concentration. *Toxicol Appl Pharmacol.* 1984;74: 417–429.

160. Melnick RL, Elwell MR, Roycroft JH, Chou BJ, Ragan HA, Miller RA. Toxicity of inhaled chloroprene (2-chloro-1,3-butadiene) in F344 rats and B6C3F(1) mice. *Toxicology.* 1996;108:79–91.

161. Reznik G, Reznik-Schüller H, Ward JM, Stinson SF. Morphology of nasal-cavity tumours in rats after chronic inhalation of 1,2-dibromo-3-chloropropane. *Br J Cancer.* 1980;42:772–781.

162. Reznik G, Stinson SF, Ward JM. Respiratory pathology in rats and mice after inhalation of 1,2-dibromo-3-chloropropane or 1,2-dibromoethane for 13 weeks. *Arch Toxicol.* 1980;46:233–240.

163. Buckley LA, Morgan KT, Swenberg JA, James RA, Hamm TEJ, Barrow CS. The toxicity of dimethylamine in F-344 rats and B6C3F1 mice following a 1-year inhalation exposure. *Fundam Appl Toxicol.* 1985;5:341–352.

164. Mathison BH, Frame SR, Bogdanffy MS. DNA methylation, cell proliferation, and histopathology in rats following repeated inhalation exposure to dimethyl sulfate. *Inhal Toxicol.* 2004;16:581–592.

165. Frederick CB, Udinsky JR, Finch L, Buckley LA, Morgan KT, et al. The regional hydrolysis of ethyl acrylate to acrylic acid in the rat nasal cavity. *Toxicol Lett.* 1994;70:49–56.

166. Feron VJ, Kruysse A, Dreefvan der Meulen HC. Repeated exposure to furfural vapour: 13-week study in Syrian golden hamsters. *Zentralbl Bakteriol B.* 1979;168:442–451.

167. Foureman GL, Greenberg MM, Sangha GK, Stuart BP, Shiotsuka RN, Thyssen JH. Evaluation of nasal tract lesions in derivation of the inhalation reference concentration for hexamethylene diisocyanate. *Inhal Toxicol.* 1994;6 (Suppl):341–355.

168. Harman AE, Voigt JM, Frame SR, Bogdanffy MS. Mitogenic responses of rat nasal epithelium to hexamethylphosphoramide inhalation exposure. *Mutat Res.* 1997;380:155–165.

169. Melnick RL, Roycroft JH, Chou BJ, Ragan HA, Miller RA. Inhalation toxicology of isoprene in F344 rats and B6C3F1 mice following 2 week exposures. *Environ Health Perspect.* 1990;86:93–98.

170. Haschek WM, Morse CC, Boyd MR, Hakkinen PJ, Witschi HP. Pathology of acute inhalation exposure to 3-methylfuran in the rat and hamster. *Exp Mol Pathol.* 1983;39:342–354.

171. Rehn B, Breipohl W, Naguro T, Schmidt U. Effect of *N*-methyl-formimino-methylester on the vomeronasal neuroepithelium of mice. *Cell Tissue Res.* 1982;225:465–468.

172. Hurtt ME, Morgan KT, Working PK. Histopathology of acute toxic responses in selected tissues from rats exposed by inhalation to methyl bromide. *Fundam Appl Toxicol.* 1987;9:352–365.

173. Reuzel PGJ, Arts JHE, Lomax LG, Kuijpers MHM, Kuper CF, Gembardt C, et al. Chronic inhalation toxicity and carcinogenicity study of respirable polymeric methylene diphenyl diisocyanate (polymeric MDI) aerosol in rats. *Fundam Appl Toxicol.* 1994;22:195–210.

174. Boorman GA, Uraih LC, Gupta BN, Bucher JR. Two-hour methyl isocyanate inhalation and 90-day recovery study in B6C3F1 mice. *Environ Health Perspect.* 1987;72:63–69.

175. Program NT. *NTP Toxicology and Carcinogenesis Studies of Nickel Sulfate Hexahydrate (CAS No. 10101-97-0) in F344 Rats and B6C3F1 Mice (Inhalation Studies).* National Toxicology Program Technical Report Series. Research Triangle Park, UK: National Institute of Environmental Health Sciences, 1996: 1–380.

176. Program NT. *Toxicology and Carcinogenesis Studies of Propargyl Alcohol (CAS No. 107-19-7) in F344/N Rats and B6C3F1 Mice (Inhalation Studies).* National Toxicology Program Technical Report Series. Research Triangle Park, UK: National Institute of Environmental Health Sciences, 2008;1–172.

177. Eldridge SR, Bogdanffy MS, Jokinen MP, Andrews LS. Effects of propylene oxide on nasal epithelial cell proliferation in F344 rats. *Fundam Appl Toxicol.* 1995;27:25–32.

178. Cruzan G, Cushman JR, Andrews LS, Granville GC, Miller RR, Hardy CJ, et al. Subchronic inhalation studies of styrene in CD rats and CD-1 mice. *Fundam Appl Toxicol.* 1997;35:152–165.

179. Gaskell BA. Nonneoplastic changes in the olfactory epithelium: experimental studies. *Environ Health Perspect.* 1990;85:275–289.

180. Baker H, Genter MB. The olfactory system and the nasal mucosa as portals of entry of biruses, drugs, and other exogenous agents into the brain. In: Doty RL, ed. *Handbook of Olfaction and Gustation.* New York: Marcel Dekker; 2003: 549–573.

181. Kai K, Satoh H, Kajimura T, Kato M, Uchida K, Yamaguchi R, et al. Olfactory epithelial lesions induced by various cancer chemotherapeutic agents in mice. *Toxicol Pathol.* 2004;32:701–709.

182. Jeffrey AM, Iatropoulos MJ, Williams GM. Nasal cytotoxic and carcinogenic activities of systemically distributed organic chemicals. *Toxicol Pathol.* 2006;34:827–852.

183. Kruger L, Saporta S, Swanson LW. *Photographic Atlas of the Rat Brain.* Cambridge, UK: Cambridge University Press; 1995.

PART 4

APPLIED TOXICOLOGICAL NEUROPATHOLOGY

27

SPINAL DELIVERY AND ASSESSMENT OF DRUG SAFETY

Tony L. Yaksh

Department of Anesthesiology, University of California–San Diego, La Jolla, California

SPINAL DRUG DELIVERY

Development of the hypodermic needle and syringe[1] along with the purification of cocaine[2] led to the demonstration that delivery of cocaine adjacent to a nerve would yield a powerful *conduction block*.[3] The demonstration that these blocks could prevent nerves from transmitting information arising from a local region led to the initial efforts by James Leonard Corning to produce such a block of spinal function. Corning predicted that the cocaine "... being transported by the blood to the substance of the cord would give rise to anesthesia of the sensory ... and the motor tracts." Corning is credited with showing such a spinal block by delivering the cocaine into the extradural space of dogs by inserting a needle between the vertebrae. In 1761, Cotugno had described the water in the space surrounding the brain and the spinal cord and suggested that it appeared as exudates of the smallest arteries.[4] In 1825, Magendie emphasized that this fluid circulated around the brain and the spinal column.[5] The routine approach to this fluid-filled spinal intrathecal space evolved first from the work of Quincke (1891), who developed a standardized technique of lumbar puncture for removing cerebrospinal fluid (CSF) to relieve pathology associated with increased intracranial pressure. He employed a relatively sharp, beveled, hollow needle, still referred to as the Quincke needle.[6] Using this approach, Augustus Bier (1899) reported the reversible, therapeutically useful anesthetic effects of lumbar intrathecal cocaine in patients. This technique of *medullary narcosis* with intrathecal cocaine spread rapidly. By the first decade, virtually everything we know about the functional properties of intrathecal anesthetic action were well appreciated, including the incidence of headaches,[7] the role of needle size,[8] hypotension,[9] paralysis,[10] and neurologic sequela, including the cauda equina syndrome and paraplegia.[11] An early effort was made to permit on-going delivery of the spinal drug. The use of malleable needles permitted leaving the needle in place during surgery. A Mayo Clinic surgeon, J. G. Love, employed a chronically placed intrathecal catheter for continuous subarachnoid drainage for meningitis,[12] and Edward Touhy produced a continuous spinal with a local catheter.[13] The development of implantable pumps and their connection to such chronic catheters completed the loop by permitting long-term drug delivery into the spinal space.[14] While the dominant use of the neuraxial drug delivery approach has been to produce anesthesia, our evolving insights into the physiology and pharmacology of spinal systems has led to a ready expansion of the drugs and targets for such delivery. In 1975 we demonstrated that opiates with an action limited to the spinal cord can produce a potent analgesia,[15] a finding that reflected the powerful role of dorsal horn opiate receptors in regulating pain processing. This preclinical finding was subsequently shown in humans after bolus delivery.[16] Shortly thereafter, chronic infusion of morphine with implantable pumps was shown to produce a maintained analgesia in chronic pain patients.[14] The demonstration that baclofen would produce motor weakness in rats[17] was translated into chronic delivery of this agent for the management of spasticity.[18] Other interests in the spinal delivery route included the use of anticancer agents for meningeocarcinomatosis and antibacterial agents for meningitis. All of these approaches took advantage of the fact that local delivery to the region where the drug was targeted to act permitted significant control of spinal function with reduced exposure to either the periphery

Fundamental Neuropathology for Pathologists and Toxicologists: Principles and Techniques, First Edition. Edited by Brad Bolon and Mark T. Butt.
© 2011 John Wiley & Sons, Inc. Published 2011 by John Wiley & Sons, Inc.

or to the brain. Moreover, it permitted the use of agents that had minimal central bioavailability after systemic delivery, such as large peptides and proteins.

FACTORS AFFECTING THE ACTIONS OF INTRATHECAL AGENTS

The profile of spinal drug delivery varies in several important ways from the profile of systemic delivery of the same agent. These differences reflect the properties of the intrathecal space and the targets upon which the agent acts.

Drug Injectate or Infusate Concentrations

The examination of spinal drug formulations often reveals drug concentrations that are extraordinarily high. Local anesthetics may have concentrations of 7.5 mg/mL (bupivacaine) to 50 mg/mL (lidocaine). Baclofen is used in a concentration of 2 mg/mL or higher, and morphine is used in concentrations of 25 to 50 mg/mL. The high concentrations employed reflect two unique aspects of spinal drug delivery. The first aspect is that drug effect is considered proportional to dose with systemic drug delivery. Normally, the dose can be manipulated by increasing the volume of drug delivered in a fixed formulation. The spinal intrathecal space, however, represents a limited volume, ranging from 50 µL in the rat to 1 to 5 mL in the dog to 30 to 50 mL in the human. Drug volumes must be limited accordingly to maintain a sustainable rostrocaudal gradient between the brain and the lumbar spinal cord. Higher drug volumes contribute to a nonspinal drug effect; thus, bolus intrathecal deliveries are on the order of 10 to 20 µL in rats, 100 to 200 µL in dog, and 200 to 500 µL in humans. Changing drug doses typically involves changing not the volume of injection, but the concentration of the injectate. Second, it should be emphasized that the site of drug action is not in the intrathecal space, but typically lies at the sites within the parenchyma that are distributed rostrally and caudally over several segments of the spinal cord. Thus, opiate acting on opiate receptors and baclofen acting on γ-aminobutyric acid (GABA) B receptors must make their way through the cord parenchyma to those receptors on cell bodies and synapses where the drugs are known to act. These targets may be deep in the parenchyma, approximately 50 µm in rats, 100 to 150 µm in dogs, and 300 to 500 µm in humans. Moreover, the drug must diffuse from the site of delivery across the spinal levels at which the pathology is mediated. Thus, to treat pain from the foot and leg, the drug must diffuse over spinal segments ranging from L_5 to L_1, a length of 5 cm or more in humans. Even agents such as local anesthetics exert their effects by diffusing into the root, where it acts upon sodium channels. A drug gradient to drive diffusion is required to obtain such rostrocaudal and intraparenchymal movement. The combination of these properties has led to the

use of elevated injectate concentrations. In the case of continuous infusion, parameters are also driven by device parameters such as pump reservoir volumes. In an effort to increase the interval between refills, the practitioner may reduce the infusion rate and increase the concentration correspondingly, to maintain a given range of doses. This principle implies that local rate of infusion or volume plays no role; this is an erroneous assumption, as noted below. Accordingly, drug dosing for spinal delivery typically employs surprisingly high drug concentrations; higher concentrations are the rule rather than the exception.

Blood–Brain Barrier

It is evident that unlike a systemically delivered drug, there are few barriers limiting the exposure of the spinal tissue to the locally delivered intrathecal drug. The pia represents a minimal diffusion barrier to intrathecal agents.[19] Thus, local spinal tissue and meninges are exposed directly to the high-injectate concentrations.

CSF Circulation

Although Magendie recognized the likelihood of CSF pressure and flow, the principal finding in the last 10 years has been the appreciation that lumbar CSF is a surprisingly static medium. Flow studies using MRI have repeatedly emphasized that net flow is minor along the spinal axis and heterogeneous at best, with areas of essentially little or no flow. Measurement of pressure waves emphasizes that the spinal fluid volumes are compressed periodically by the cardiac cycle, with increasing intracranial blood volumes from above, and locally through filling of the local venous plexi, as with increased thoracic or abdominal compression of the large veins that alternately restrain and drain the spinal epidural venous plexus. These forces lead to a modest degree of mixing of the local space.[20,21] Conceptually, the intrathecal space may be modeled as a tidal backwater as opposed to a river.

Drug Clearance

Specific studies on drug dilution after intrathecal delivery have reliably confirmed that small volumes or low rates of infusion lead to a highly restricted redistribution of the injectate.[22] CSF sampling from sites adjacent to the lumbar injection site have suggested that there is an acute local dilution of the injectate into local CSF volume and then a slow redistribution after bolus delivery. Drugs that are very lipid soluble (e.g., fentanyl, lidocaine) are cleared at short intervals by movement into the parenchyma, but more readily though the meninges, and show a short second-phase half-life. In contrast, larger molecules appear to be restrained within the CSF and often display a bulk dilution in the local CSF that may give a several-hour half-life.[23–26] Together the local exposure of the tissue to elevated concentrations of injectate may persist for an extended period of time depending on the drug properties.

Structures Exposed to an Intrathecal Drug

After spinal delivery, the drug exists in the local environment formed by the roots and the afferent cell bodies in the dorsal root ganglia, the parenchyma, and the meninges. The meninges are of particular interest as the arachnoid is a highly cellular and modestly vascularized tissue[27] with a full complement of inflammatory elements such as mast cells and macrophages.[28,29]

PRECLINICAL ASSESSMENT OF THE TOXICOLOGY OF INTRATHECAL AGENTS

An important consideration in the clinical implementation of a novel agent for the spinal delivery route is the direct effect of the spinal agent on the local tissue.[30] Even from the earliest days of intrathecal injection, the potential issues of drug safety were appreciated. Barker (1904) reported on CSF pleocytosis after a local anesthetic, and Wossidlo (1908) reported on changes in Nissl substance in dog dorsal root ganglion (DRG) after 5% procaine administration. As reviewed above, the intrathecal route and space have several properties that inherently raise safety concerns. Notably, the high concentrations that are used routinely, the minimal redistribution leading to accumulation of material in limited areas around the injection site, and the extended exposure of that region to the drug, exacerbated by continued/continuous drug delivery. It is now apparent that several types of toxicological events can be identified which are specific to the intrathecal actions of the drug. Three examples of intrathecal toxicity are outlined below to show how the preclinical models predict the human condition and what insights can be derived in developing strategies to diminish the potential of toxicity when a novel agent is moved into humans.

Local Anesthetics

Clinical Observations   Local anesthetics have been used for more than 100 years. As noted above, from the earliest days, incidences of neurological symptoms have been reported, but in general, these have been considered minimal and often reflect a misadventure with the injection. Since the early 1990s, symptoms characterized by perineal sensory loss, lower-limb weakness, as well as bladder and bowel dysfunction, were reported after continuous spinal anesthesia and described as cauda equina syndrome and is described as damage to sensory roots at S2 to S4.[31–34] A variant on that syndrome is a transient sensory dysesthesia which involves injury limited to L5 and is referred to as *transient radicular*

irritation (TRI).[35] The problems were observed in patients receiving several anesthetics and appeared to be associated with local concentrations. It is important to note that the incidence of radiculopathies was evidently augmented when the local anesthetics were delivered through microbore intrathecal catheters. Given the poor redistribution that has been demonstrated from these catheters, the likelihood that there was a tendency of the local anesthetic to remain locally and coat the adjacent nerve roots has been an important aspect of considering the role of concentration in this lesion.[36]

Preclinical Studies   The observation of a persistent effect following the clearance of the molecule was taken as an evident sign of a direct effect upon neural integrity. In a series of preclinical studies using rats chronically prepared with lumbar intrathecal catheters, it was shown that bolus delivery of several anesthetics, including lidocaine, bupivacaine, and tetracaine, produced a concentration incidence of pathological signs and an accompanying hindlimb motor disability. Histological examination revealed white matter axonal degeneration, gray matter vacuolization, and infiltration of macrophages and degeneration of Schwann cell sheaths.[37,38] An important question relates to whether this was a target (sodium channel block)-related effect. This possibility was excluded by failing to see persistent blocks with tetrodotoxin.[39] As glucose is often added to the local anesthetic, studies considering the effects of glucose were undertaken and were negative.[40] Comparable results have been observed in rabbits with tetracaine, lidocaine, bupivacaine, and ropivacaine.[41–43] Bolus intrathecal studies with bupivacaine in dogs with 72 h survival reported concentration-dependent incidences of nerve injury with hyperbaric bupivacaine solutions,[44] while a longer intrathecal infusion of bupivacaine reported no morphological changes.[45]

Mechanism   The mechanisms of the concentration and drug-dependent nerve injury include increased release of glutamate, which has a neurotoxic effect mediated by α-amino-3-hydroxyl-5-methyl-4-isoxazole-propionate (AMPA) receptors.[46] Although this explains the parenchymal toxicity, it less readily explains demyelination. Direct effects have been reported on DRG cells, leading to neurotoxic concentrations of free calcium.[47] A third possibility is these agents may form micelles at the high concentrations employed, and the micelles may yield a detergent-like disruptive effect on the lipid membranes.[48]

Opiates

Clinical Observations   Following the initial characterization of the analgesic actions of intrathecal opiates in animals,[15] spinal delivery of morphine was widely initiated in humans for the management of acute pain. Implementation of chronic indwelling intrathecal catheters with implantable pumps permitted the routine continuous delivery of spinal opiate analgesics.[14] Implementation of bolus and chronic intrathecal infusion of opiates for pain management grew rapidly.[49] Aside from the well-appreciated issues of urinary retention, pruritus, and respiratory depression (resulting from supraspinal drug redistribution), evidence of toxicity was not observed and not sought. Extensive preclinical work has been undertaken, but with few exceptions, no attempts were made to define the histopathological effects or the role of chronic delivery on elevated drug concentrations. Several studies that undertook basic spinal histopathology revealed no effect of repeated intrathecal bolus deliveries of several opiates in rats, cats, and dogs.[50,51] Beginning in 1991, case reports described patients receiving chronic morphine infusion who presented with motor/sensory dysfunction secondary to a local compressive mass.[52–58] Retrospective reviews suggested an incidence of approximately 0.1% in a population of approximately 13,000 intrathecal pump patients, based on the presence of neurological signs.[59] However, in seven patients, one displayed neurological symptoms. Of these seven patients, MRI revealed that the symptomatic patient and two nonsymptomatic patients displayed a granuloma. Without exception, these patients received high concentrations of morphine (25 to 50 mg/mL) delivered at low rates (10 to 20 μL/h). Histology revealed no infectious process where the mass was resected, but did identify the presence of macrophages, neutrophils, and monocytes, all with a necrotic center and was referred to as a *granuloma*.[55,60] The time of onset required for mass formation or its reversibility is not certain given that intrathecal therapy involves progressive increases in dose over an extended period of time and the potential disconnect between the presence of a granuloma and the appearance of neurological signs. However, even with those caveats, four of the seven patients who had received infusions for less than six months had neurological symptoms.[59]

Preclinical Studies In the intrathecally catheterized canine model, morphine sulfate produced a local mass when infused for more than 28 days (see Figures 1 to 3). The local mass consisted of multifocal accumulations of neutrophils, monocytes, macrophages, and plasma cells at the catheter tip that produced a local spinal cord compression. An important observation that we made was that the cellular mass arose from the local vascular collaterals in the dura/arachnoid layer and not from the spinal parenchyma.[61] Aside from spinal cord compression, there were no other changes in spinal morphology, indicating that even high concentrations of morphine had no direct effect on axons or cell bodies. Comparable effects were observed in sheep.[62] Using serial MRIs, an identified mass was observed at the catheter tip as early as 3 to 7 days, with a maximal mass occurring by 14 days.[63] Termination of morphine administration but continuing vehicle administration typically resulted in a progressive reduction in mass size over the remaining 14 to 28 days. The

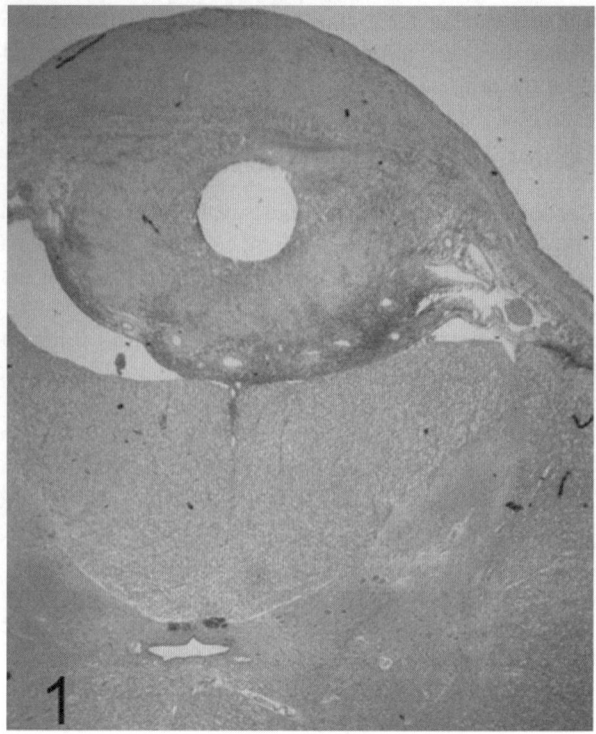

FIGURE 1 Histopathology (H&E) of the lumbar spinal cord of a dog showing an intrathecal mass with a catheter orifice in the center. The mass is compressing the dorsal portion of the cord.

toxicology of the opiate initiated mass had several defining properties: (1) control experiments and other drug studies showed that the granuloma was not secondary to the catheter or the infusion, and substitution of saline vehicle for the morphine dose resulted in a reduction in granuloma volume; (2) cultures and stains were negative for infection; (3) all solutions had osmolarity of about 300 milliosmols and pH in the range 6.5 to 7 for these agents and for the vehicle; (4) the incidence of granuloma formation in the dog was heavily dependent on concentration, > 12 mg/mL, with no granulomas observed when the same dose was infused at a lower concentration and higher rate[63]; (5) infusion of 12 mg/mL per day resulted in a granuloma, whereas bolus delivery for 28 days of 10 mg/mL had no effect, suggesting a role for persistent exposure[51]; (6) granulomas were produced by equianalgesic intrathecal infusion doses of morphine,-hydromorphone, L/D-methadone (racemic mixture), and the μ-opioid peptide DAMGO, but not fentanyl.[64] This discrepancy argues against a simple μ-opioid effect.

Mechanism The essential question is what leads to the migration of inflammatory cells from the meninges. Several possibilities have been suggested:

1. NO (nitric oxide) release. Mu-opioids acting through μ-opioid receptors can initiate release of nitric oxide in

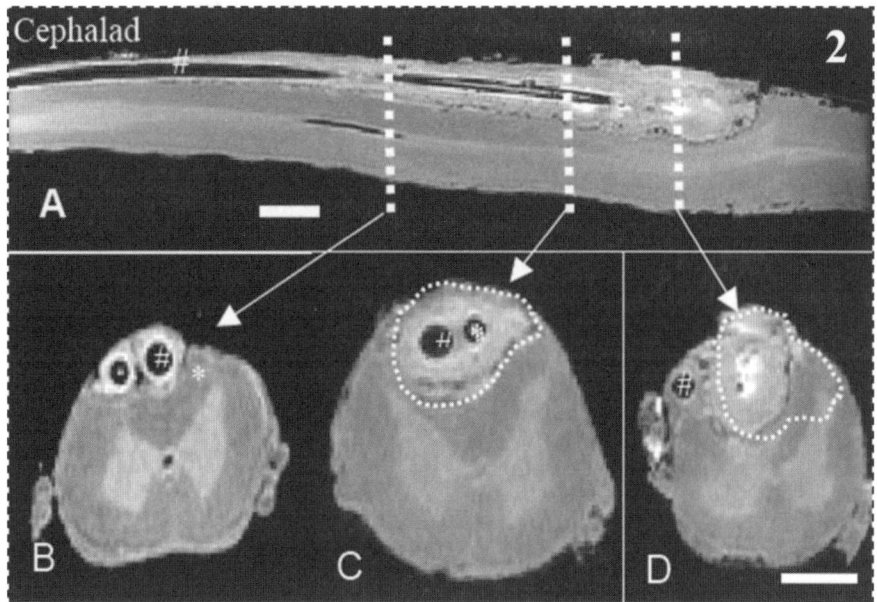

FIGURE 2 Post-sacrifice MRI of the lumbar spinal cord taken from a dog with a PE (polyethylene) 50 sampling catheter (#) and a PE10 infusion catheter (∗) that had received a 28-day infusion of intrathecal morphine sulfate [12.5 mg/mL/(40 μL/h)].

human endothelial cells.[65] These observations lead to a working hypothesis that morphine at elevated concentrations and persistent exposure may activate nitric oxide synthase in meningeal vasculature and initiate a cascade that serves to increase local capillary permeability to these activated cells. The pharmacology of a μ-opioid effect reported, however, would suggest that the ordering of "granuloma-producing" activity should be fentanyl > morphine, which it is not.

2. Mast cells. There are large numbers of mast cells that are in residence in brain and spinal meninges.[66,67] Degranulated mast cells release numerous compounds, including proteolytic enzymes (tryptase and chymase), vasodilators (histamine and serotonin), compounds, that increase vascular and blood–brain barrier permeability and act as chemoattractants (TNF-α).[68] Tryptase is present in large amounts in all mast cells and has been shown to play a crucial role in increasing local vascular permeability.[69] In skin, opiates produce skin mast cell degranulation and mediator release,

but the ordering of activity is not representative of a μ-opioid receptor: morphine = hydromorphone > fentanyl.[70–72] This ordering of activity for mast cell degranulation resembles what has been found for granuloma induction.

NMDA Antagonists

Clinical Observations Studies on the spinal pharmacology of pain processing have pointed strongly to the role of n-methyl-d-aspartate (NMDA) ionophores in regulating the facilitated pain signal.[73] As a result, there has been a long interest in developing NMDA antagonists for human intrathecal use. This desire is enhanced as the supraspinal effects of the NMDA antagonists give the systemic effects of these drugs a narrow therapeutic ratio.[74] In humans, the therapeutic activity of several NMDA antagonists, including ketamine and CPP [3-(2-carboxypiperazin-4-yl)propyl-1-phosphonic

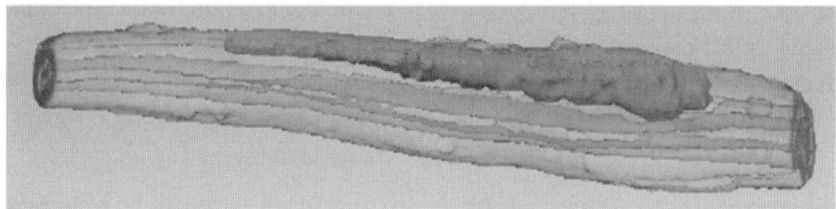

FIGURE 3 Reconstruction of a granuloma. (Courtesy of Miriam Scadeng at the Keck MRI Center, University of California–San Diego.)

acid] after epidural or intrathecal delivery has been reported.[75–78] In terms of safety, case reports have indicated postmortem histological observations. While concerning, such observations are made controversial by the fact that these patients had typically received other agents, chemotherapy and radiation, and/or suffered extensive metastatic disease, all of which may have contributed to the spinal changes observed.[77,79,80]

Preclinical Studies A number of preclinical studies have been reported with several NMDA blockers, and a complicated picture has emerged. Several studies with single- and multiple-bolus doses of ketamine in rabbits[81,82] and pigs[83] reveal little pathology. However, in other rabbit studies, notable pathology was found, which, as in humans, included necrotizing subpial lesions.[84] Intrathecal CPP was studied in rats with multiple-bolus doses and no untoward histological findings were noted.[85] In sheep, chronic intrathecal infusion of memantine, dextrorphan, and dextromethorphan was found to produce a comparable spinal pathology demonstrating parenchymal necrosis.[86] In more recent work, a variety of agents with significant NMDA blocking activity, including AP5 [(2R)-amino-5-phosphonopentanoate], amitriptyline, ketamine, MK801 [Dizocilpinej (+)-5-methyl-10,11-dihydro-5H-dibenzo[a,d]cyclohepten-5,10-imine maleate], memantine, and S-methadone, were studied in dogs with chronic infusion administration. These animals showed a concentration-dependent hindlimb paresis by days 3 to 5, which progressed over the 28-day infusion interval. Compared to the other agents examined, AP5 showed a decidedly less prominent effect. A gradient of increasing pathology from cervical to lumbar segments was noted ranging from local demyelination to necrotizing lesions of spinal parenchyma proximal to the catheter tip.

Mechanism The mechanisms of the observed pathology are not known. It is possible that this is a target-mediated pathology. NMDA receptors are expressed in dorsal horn neurons and nonneuronal cells. Systemic NMDA antagonists, including MK801, AP7, AP5, ketamine, memantine, dextromethorphan, and CPP, produce vacuolization in the central nervous system, notably in the retrosplenial cortex, accompanied by a reduction in mitochondria and endoplasmic reticulum.[87] More recently, transient blockade of NMDA receptors in the newborn rat activates cell death.[88] Whether this occurs in adult rat spinal cord is not known.

An alternative hypothesis is that outlined previously for local anesthetics, which is the formation of micelles. Amitriptyline has local anesthetic actions after spinal delivery, and it has been demonstrated that at concentrations of 0.3% or greater, amitriptyline produces irreversible neural impairment.[89] Amitriptyline undergoes aggregation at approximately 5 mg/mL. In the dog studies, this range results in significant intrathecal toxicity. The high degree of localiza-

tion of the toxicity strongly suggests a phenomenon dependent on a steep concentration gradient around the catheter tip. It is interesting that methadone, but not morphine, has been reported to form micelles,[90,91] a finding consistent with the associated pathology (parenchymal necrosis vs. granuloma).

SOME ORGANIZING PRINCIPLES FOR ROBUST PRECLINICAL EVALUATIONS OF SPINAL DRUG SAFETY

We now consider several challenges that arise in determining whether a drug formulation developed preclinically poses unacceptable risks when given by an intrathecal route of delivery. All of the studies described to date have emphasized the importance of concentration and time course to which the spinal tissues are exposed to the drug. These issues have been discussed elsewhere.[92,93]

Test Article and Formulation

1. A drug cannot be presumed to be safe for spinal delivery because it is used systemically. Concentrations at the target sites after spinal injection will far exceed those achieved by systemic delivery. Nevertheless, the availability of a safety profile for a test article novel to the spinal cord is a virtue, as it provides information on drug manufacture [of GMP (good manufacturing practice) grade], stability of the formulation, pyrogenicity, teratogenicity, and the safety of the drug when it is cleared into the systemic circulation after spinal delivery (note that perineural and spinal local anesthetic doses are routinely limited by their systemic dose because of potential systemic cardiac toxicity).

2. The drug formulation examined must be identical to what is intended to be used in humans. In general, excipients such as metabisulfite and ascorbate (antioxidants), and glycine (buffering and for lyophilization to form a cake) are examples of agents that are generally considered as safe. Novel excipients, on the other hand, present additional safety-related questions. Deviations of the formulation from normal CSF parameters (i.e., osmolarity, pH, etc.) are not forbidden, but the more variation from normal, the greater the possibility of an untoward effect related to the formulation. Note, for example, that a water-based formulation with a pH of 3.5 would raise concerns.

Route of Delivery

The epidural route can be employed to deliver drugs preferentially to the neuraxis if the agent can pass the meninges. The barrier presented by the meninges is reflected in the typically higher doses of drugs that are used by the epidural as opposed to the intrathecal route, and by the fact that some spinally active agents may be essentially inactive by that route (see, e.g., the intrathecal vs. epidural activity of a large peptide).[94] Safety evaluations must therefore mimic the clinical route to be targeted. The original work with clonidine was targeted for epidural infusion, and the test model used that route accordingly.[95]

Delivery Mode

Bolus intrathecal delivery results in an acute high concentration that briefly reflects the concentration in the injectate. This injectate undergoes rapid dilution in the local CSF volume and subsequent progressive clearance. Small molecules may be cleared rapidly whereas larger molecules may persist in the CSF with a time course that parallels the dilution of the drug in the intrathecal fluid. Continuous infusion yields a steady state proportional to the infusate concentration and rate that is highest at the catheter tip and declines rostrally and caudally to the catheter tip.[22] Accordingly, it is evident that the drug exposure of the local tissue to the drug will be dependent on the delivery paradigm. This observation was reviewed above in that the daily continuous infusion of a dose of morphine that yields a granuloma in dogs will not produce a granuloma when given as a once-daily bolus.

Animal Model

Chronic intrathecal and epidural models have been reported in small animals (typically, rat or rabbit) and in large animals (typically, dogs, beagles, hounds, sheep, and to a lesser extent, primates). Although percutaneous needle punctures can be undertaken in all of these species, the use of a catheter is generally preferred, as (1) it permits a secure route of delivery which can, upon necropsy or by imaging, assure that the drug delivery has been made into the appropriate space; (2) the injections can be made in the absence of any anesthesia or sedation that might obscure peri-injection phenomena; and (3) multiple injections and continuous infusions can be undertaken routinely. The models are, however, characterized by a number of limitations, including differences in their robustness. Rodents have been readily used for intrathecal bolus and infusion studies (with subcutaneous osmotic pumps) with intervals up to 4 weeks. Although the problem is minimized by the use of very small polyurethane and polyethylene catheters, the longer the infusion, the greater the signal that is produced by the spinal catheter independent of the drug. Larger animals have been prepared routinely using surgical exposure and direct catheter placement. Although percutaneous catheter placement is potentially feasible, the spinal space requires smaller needles that can accommodate an appropriate catheter. Smaller catheters that can be used with small needles become harder to maintain, due to fracture over longer study intervals. Continuous delivery requires a pump. External pumps can be fitted in a vest in dogs and sheep. The need to use an implantable pump is a requirement for larger models, such as dogs and sheep. An overview of the properties of models using different species is provided in Table 1. See additional comments on models elsewhere.[96]

Epidural catheters may be placed, but they pose special problems because of the degree to which they will evoke a local reaction that may prevent free injection into the epidural space. In rats and dogs, reports of complete localization of the catheter have been reported at intervals of 7 to 10 days.[97,98] In humans, significant fibrosis has also been reported, precluding their use for extended intervals of drug delivery.[99]

The U.S. Food and Drug Administration requires the use of a small and large species for drug safety evaluation. The telling issue, independent of which species is chosen, is whether the model can indeed predict toxicity when such toxicity is expected. Thus, for local anesthetics, as reviewed above, demyelination has been reported in all species, including rats, rabbits, and dogs. For opiates, dogs and sheep have been highly predictive with presenting granulomas, as

TABLE 1 Summary of Preclinical Spinal Injection Models and Comparison of CSF Parameters

Species	Route[a]	Acute Bolus	Implanted Catheter	Chronic Infusion (Pump)	CSF Formation Rate (mL/h)	Estimated Total CSF Volume (mL)	Intrathecal Injection Bolus (mL)
Mouse	IT	Yes	No	No	0.038	0.04	0.005
Rat	EP/IT	Yes	Yes	Yes	0.25	0.25	0.01
Dog	EP/IT	Yes	Yes	Yes	4.0	23	0.5
Sheep	EP/IT	Yes	Yes	Yes	9.6	60	1.0
Pig	EP/IT	Yes	Yes	Yes	7.8	50	1.0
Primate	EP/IT	Yes	Yes	Yes	2.4	15	0.5
Human	EP/IT	Yes	Yes	Yes	21.6	130	2.0

[a] IT; intrathecal; EP; epidural.

they have for NMDA antagonist pathology. Among the larger species, with regard to local drug toxicity, there is no evidence that primates are inherently superior to any of the quadruped species. In general, it is the author's opinion that when a single species is to be used for predictive safety data, a larger species is preferred over rodents given the issues of scaling of the target structures in the spinal canal.

Exposure Factor

As reviewed above, aside from the identity of the model test system, there is no variable that appears more relevant to the safety assessment than the drug concentration to which the local tissue is exposed and the time course of that exposure. At one extreme, the higher the concentration and the longer the exposure, relative to what will be employed in the human being, the more robust the safety assessment. For any given route (epidural/intrathecal), factors governing the exposure are (1) drug concentration, (2) volume or rate of delivery, (3) duration of exposure, and (4) the animal model. The choices to be made in the selection of the several parameters are predicated in part on how the drug can be formulated and how it is intended to be used therapeutically. Maximum concentration may be determined by solubility and stability. If the test article is to be provided in a solution, then arguably the worst-case scenario is delivery of the undiluted material. A lyophilized formulation (such as the opiate remifentanil) may raise concerns as to what might be the maximum concentration that the physician might deliver. The volume or rate of delivery will define the dose delivered once the concentration is chosen. It should be stressed that unlike a parenteral route of delivery where volume is of little concern, the safety study is limited to spinal delivery volumes that can be tolerated which mimic the local action sought in the human. Thus, in the beagle, intrathecal bolus injection volumes of 1 mL or greater will clearly move dye marker into the cisterna, and infusion rates of greater than 8 mL/day (320 µL/h) are tolerated but will elevate cisternal CSF pressures. Calibration studies may be required for different species and models where higher rates or volumes are sought.

A difficult question relates to the duration of drug exposure that is adequate for assessing potential toxicity. Again, the selection of exposure may in part depend on the intended therapeutic use. Drugs for single-bolus use, as with postoperative pain, may reasonably be limited to several repeated boluses delivered at intervals that reflect the drug half-life. The safety evaluation of the single-dose use with liposomal morphine formulation (DepoDur)[92,100] or the two-dose epidural treatment with etanercept[101] employed such a paradigm.

More complex is the question of long-term delivery, as in pumps, where therapy may potentially persist for years. The unknown is the time dependency of the development of any anticipated pathology. For granulomas, it is evident that in the

therapeutic range of morphine doses, there is a high incidence of granulomas, and they appear at intervals of 28 days along with neurological signs (secondary to cord compression). Studies using MRI imaging reveal morphine-evoked masses to be evident as early as 10 days (see above). Based on the presumed role of accumulating cells from the meninges when there is a continuous opiate exposure, it is reasonable to conclude that the problem is inherent to the spinal actions of morphine and, as such, poses a clinical risk that requires watchful observation but is not a reason to preclude drug implementation. These findings were made in the face of information gathered with 28-day studies (cf.[61,62]). Local anesthetics have been shown to have demyelinating effects at intervals of several days after bolus delivery.[37] Accordingly, for these targets, failure of an agent to produce pathology within 28 days would represent a safety profile that greatly exceeds what is currently available. Previous safety work with several candidate agents, notably the calcium channel blocker (Prialt)[102,103] and the α2∂ binding agent gabapentin (Richard Rauch, personal communication, 2008), have demonstrated acceptable clinical safety which correlates with preclinical studies in dogs with 28-day exposures. This is an instance where the ability of the animal model to tolerate, without pathology signs, higher concentrations or doses than are anticipated for use in the humans may be considered as supporting a more limited time of exposure.[104]

A final comment on the exposure factor relates to use of the preclinical model. Previous work has suggested that dose for dose; the animal model presents a smaller intrathecal space than does the human. If one assumes that with the bolus delivery of a given dose (volume × concentration), the local peak concentrations shortly after injection will be inversely proportional to the relative volume of the spinal space into which the drug is being injected. Based on the simple assumption that the local diffusion area is a function of the circumference of the lumbar spinal cord and that the dog and human circumferences are approximately 35 and 16 mm, respectively, this suggests that for any given injection or infusion dose, acute dilution of the injectate will be 2.2 times greater in humans than in dogs.[105] Thus, for any given dose, the dog spinal cord proximal to the catheter tip will experience approximately a 2.2-fold greater exposure for any concentration or dose than will the human.

Study Groups

At the very minimum, a study consists of a vehicle and drug treatment group. A single-dose group may be justified in the face of the delivery of a formulation that represents the maximum concentration (as when the agent is provided as a fixed solution). Should toxicological effects be observed at that formulation, lower-dose groups would have to be tested for further development of the agent to demonstrate "no (pathological) effect dose." Additional groups might be required should it be determined that post-treatment recovery intervals are required. In the case of an acute treatment, there is the possibility that deleterious effects leading to apoptosis would be seen at short 2 to 3-day intervals after treatment, while longer-term effects such as glial activation and scarring may not appear for a more extended period. This would entail additional groups. The issues relating to the choice of recovery times is discussed elsewhere in the book.

SUMMARY

It is important to appreciate that the principal aim is to determine the possibility that a given spinal drug treatment can be given in humans with a minimal risk of misadventure. The degree to which the model and paradigm exceed the drug–tissue exposure anticipated in humans, the more robust the assessment and the more likely it is that we will later discover a deleterious action. Conversely, as we approach in humans the limiting parameters that were used to define safety, the less certain we can be that the hypothesis of safety will hold. For that reason, increases in concentration, altered formulations (such as addition of products to render the solution hyperbaric), or changes in the delivery paradigm (bolus vs. infusion, changes in volumes or rates) must all be considered in the context of whether they are going to increase the tissue–drug exposure. The approval of high concentrations of morphine for intrathecal delivery is an important example of how increasing the concentrations of a widely used agent can result in an altered outcome (granuloma).[106] In this case it is worthwhile to note that in a paraphrase of Paracelsus (1493–1541), "... All things [drugs] are poison.... Solely the dose determines that a thing is not a poison." In summary, an important consideration in the intrathecal delivery of novel agents relates to their safety. Here, safety explicitly considers not only the potential effect on function (e.g., motor, sensory, autonomical) but changes in spinal morphology. Assertions of safety should be based on the use of robust assessments that involve validated animal models, concentrations equaling or exceeding those in humans, and preferably multiple or continuous deliveries through routes used in the intended human therapy. Although not specifically addressed in this review, there is an increasing utilization of the spinal route of delivery in nepnates and juveniles. This warrants the study of spinal delivery in suitable animal models to evaluate the safety of these delivery methods in younger patients.[107,108] In the course of our wish to advance drug therapy, we should not forget the dictum *Primare non nocere*.

Acknowledgments

The work described by Yaksh and colleagues on opiates was supported by DA 15353 (TY) and Medtronic Corporation. The work on adenosine, neostigmine, ketorolac, and several NMDA antagonists was completed with National Institutes of Health (NIH) grant GM51245 (TY) and partly in collaboration with Jim Eisenach at Wake Forrest with NIH GM48085 (JE).

REFERENCES

1. Howard-Jones N. A critical study of the origins and early development of hypodermic medication. *J Hist Med Allied Sci.* 1947;2:201–249.
2. Niemann N. *Über eine neue organische Base in den Cocablättern.* Göttingen, Germany; 1860.
3. Olch PD. William S. Halsted and local anesthesia: contributions and complications. *Anesthesiology.* 1975;42:479–486.
4. Pearce JM. Cotugno and cerebrospinal fluid. *J Neurol Neurosurg Psychiatry.* 2004;75:1299.
5. Hajdu SI. A note from history: discovery of the cerebrospinal fluid. *Ann Clin Lab Sci.* 2003;33:334–336.
6. Quincke H. Die Lumbalpunction des Hydrocephalus. *Wiener Klini Wochenschrit.* 1891;28:929–965.
7. Bier A. Versuche uber cocainisirunge Ruckenmarkes. *Dtsch Z Chir.* 1899;51:361–369.
8. Babcock WW. The technique of spinal anesthesia. *NY J Med.* 1914;50:637–702.
9. Tuffier T, Hallion F. Effects circulatoires des injections soud-aracho de coca dans la regionlombaire. *CR Hebd Seances Mem Soc Biol.* 1990;52:897–899.
10. Morton AW. The subarachnoid injection of cocaine for operations on all parts of the body. *Am Med.* 1901;2:176–179.
11. Lusk WC. The anatomy of spinal puncture with some considerations on technic and paralytic sequels. *Ann Surg.* 1911;54:449–484.
12. Love JG. Continuous subarachnoid drainage for menigitis by means of a ureteral catheter. *J Am Med Assoc.* 1935;104: 1595–1597.
13. Tuohy EB. Continuous spinal anesthesia: a new method utilizing a ureteral catheter. *Surg Clin N Am.* 1945;25: 834–840.
14. Onofrio BM, et al. Continuous low-dose intrathecal morphine administration in the treatment of chronic pain of malignant origin. *Mayo Clin Proc.* 1981;56:516–520.
15. Yaksh TL, Rudy TA. Analgesia mediated by a direct spinal action of narcotics. *Science.* 1976;192:1357–1358.
16. Wang JK, et al. Pain relief by intrathecally applied morphine in man. *Anesthesiology.* 1979;50:149–151.
17. Wilson PR, Yaksh TL. Baclofen is antinociceptive in the spinal intrathecal space of animals. *Eur J Pharmacol.* 1978;51:323–330.
18. Campbell SK, et al. The effects of intrathecally administered baclofen on function in patients with spasticity. *Phys Ther.* 1995;75:352–362.
19. Bernards CM, Hill HF. Morphine and alfentanil permeability through the spinal dura, arachnoid, and pia mater of dogs and monkeys. *Anesthesiology.* 1990;73:1214–1219.
20. Schroth G, Klose U. Cerebrospinal fluid flow: I. Physiology of cardiac-related pulsation. *Neuroradiology.* 1992;35:1–9.

21. Schroth G, Klose U. Cerebrospinal fluid flow: II. Physiology of respiration-related pulsations. *Neuroradiology*. 1992;35:10–15.

22. Bernards CM. Cerebrospinal fluid and spinal cord distribution of baclofen and bupivacaine during slow intrathecal infusion in pigs. *Anesthesiology*. 2006;105:169–178.

23. Ummenhofer WC, et al. Comparative spinal distribution and clearance kinetics of intrathecally administered morphine, fentanyl, alfentanil, and sufentanil. *Anesthesiology*. 2000; 92:739–753.

24. Bernards CM. Understanding the physiology and pharmacology of epidural and intrathecal opioids. *Best Pract Res Clin Anaesthesiol*. 2002;16:489–505.

25. Yaksh TL, et al. Kinetic and safety studies on intrathecally infused recombinant-methionyl human brain-derived neurotrophic factor in dogs. *Fundam Appl Toxicol*. 1997;38: 89–100.

26. Kern SE, et al. The pharmacokinetics of the conopeptide contulakin-G (CGX-1160) after intrathecal administration: an analysis of data from studies in beagles. *Anesth Analg*. 2007;104:1514–1520.

27. Mack J, et al. Anatomy and development of the meninges: implications for subdural collections and CSF circulation. *Pediatr Radiol*. 2009;39:200–210.

28. Levy D. Migraine pain, meningeal inflammation, and mast cells. *Curr Pain Headache Rep*. 2009;13:237–240.

29. Yamate J, et al. Macrophage populations and expressions of regulatory proinflammatory factors in the rat meninx under lipopolysaccharide treatment in vivo and in vitro. *Histol Histopathol*. 2009;24:13–24.

30. Eisenach JC, et al. New epidural drugs: primum non nocere. *Anesth Analg*. 1998;87:1211–1212.

31. Rigler ML, et al. Cauda equina syndrome after continuous spinal anesthesia. *Anesth Analg*. 1991;72:275–281.

32. Schell RM, et al. Persistent sacral nerve root deficits after continuous spinal anaesthesia. *Can J Anaesth*. 1991;38:908–911.

33. Schneider MC, et al. Transient neurologic toxicity after subarachnoid anesthesia with hyperbaric 5% lidocaine. *Anesth Analg*. 1994;79:610.

34. Snyder R, et al. More cases of possible neurologic toxicity associated with single subarachnoid injections of 5% hyperbaric lidocaine. *Anesth Analg*. 1994;78:411.

35. Arai T, Hoka S. Neurotoxicity of intrathecal local anesthetics. *J Anesth*. 2007;21:540–541.

36. Ross BK, et al. Local anesthetic distribution in a spinal model: a possible mechanism of neurologic injury after continuous spinal anesthesia. *Reg Anesth*. 1992;17:69–77.

37. Drasner K, et al. Persistent sacral sensory deficit induced by intrathecal local anesthetic infusion in the rat. *Anesthesiology*. 1994;80:847–852.

38. Sakura S, et al. The comparative neurotoxicity of intrathecal lidocaine and bupivacaine in rats. *Anesth Analg*. 2005;101:541–547.

39. Sakura S, et al. Local anesthetic neurotoxicity does not result from blockade of voltage-gated sodium channels. *Anesth Analg*. 1995;81:338–346.

40. Hashimoto K, et al. Comparative toxicity of glucose and lidocaine administered intrathecally in the rat. *Reg Anesth Pain Med*. 1998;23:444–450.

41. Ready LB, et al. Neurotoxicity of intrathecal local anesthetics in rabbits. *Anesthesiology*. 1985;63:364–370.

42. Yamashita A, et al. A comparison of the neurotoxic effects on the spinal cord of tetracaine, lidocaine, bupivacaine, and ropivacaine administered intrathecally in rabbits. *Anesth Analg*. 2003;97:512–519.

43. Malinovsky JM, et al. Intrathecal ropivacaine in rabbits: pharmacodynamic and neurotoxicologic study. *Anesthesiology*. 2002;97:429–435.

44. Ganem EM, et al. Neurotoxicity of subarachnoid hyperbaric bupivacaine in dogs. *Reg Anesth*. 1996;21:234–238.

45. Kroin JS, et al. The effect of chronic subarachnoid bupivacaine infusion in dogs. *Anesthesiology*. 1987;66:737–742.

46. Koizumi Y, et al. The effects of an AMPA receptor antagonist on the neurotoxicity of tetracaine intrathecally administered in rabbits. *Anesth Analg*. 2006;102:930–936.

47. Gold MS, et al. Lidocaine toxicity in primary afferent neurons from the rat. *J Pharmacol Exp Ther*. 1998;285:413–421.

48. Kitagawa N, et al. Possible mechanism of irreversible nerve injury caused by local anesthetics: detergent properties of local anesthetics and membrane disruption. *Anesthesiology*. 2004;100:962–967.

49. Yaksh TL. Spinal opiate analgesia: characteristics and principles of action. *Pain*. 1981;11:293–346.

50. Yaksh TL, et al. Studies of the pharmacology and pathology of intrathecally administered 4-anilinopiperidine analogues and morphine in the rat and cat. *Anesthesiology*. 1986;64: 54–66.

51. Sabbe MB, et al. Spinal delivery of sufentanil, alfentanil, and morphine in dogs: physiologic and toxicologic investigations. *Anesthesiology*. 1994;81:899–920.

52. Schuchard M. Neurologic sequelae of intraspinal drug delivery systems. *Neuromodulation*. 1998;1:137–148.

53. North RB, et al. Spinal cord compression complicating subarachnoid infusion of morphine: case report and laboratory experience. *Neurosurgery*. 1991;29:778–784.

54. Langsam A. Spinal cord compression by catheter granulomas in high-dose intrathecal morphine therapy: case report. *Neurosurgery*. 1999;44:689–691.

55. Cabbell KL, et al. Spinal cord compression by catheter granulomas in high-dose intrathecal morphine therapy: case report. *Neurosurgery*. 1998;42:1176–1180 (discussion: 1180–1171).

56. Blount JP, et al. Intrathecal granuloma complicating chronic spinal infusion of morphine: *report of three cases. J Neurosurg*. 1996;84:272–276.

57. Bejjani GK, et al. Intrathecal granuloma after implantation of a morphine pump: case report and review of the literature. *Surg Neurol*. 1997;48:288–291.

58. Aldrete JA, et al. Paraplegia in a patient with an intrathecal catheter and a spinal cord stimulator. *Anesthesiology*. 1994;81:1542–1545 (discussion: 1527A–1528A).

59. Coffey RJ, Burchiel K. Inflammatory mass lesions associated with intrathecal drug infusion catheters: report and observations on 41 patients. *Neurosurgery.* 2002;50:78–86 (discussion: 86–77).

60. Langman MJ, et al. Risks of bleeding peptic ulcer associated with individual non-steroidal anti-inflammatory drugs. *Lancet.* 1994;343:1075–1078.

61. Yaksh TL, et al. Chronically infused intrathecal morphine in dogs. *Anesthesiology.* 2003;99:174–187.

62. Gradert TL, et al. Safety of chronic intrathecal morphine infusion in a sheep model. *Anesthesiology.* 2003;99:188–198.

63. Allen JW, et al. Time course and role of morphine dose and concentration in intrathecal granuloma formation in dogs: a combined magnetic resonance imaging and histopathology investigation. *Anesthesiology.* 2006;105:581–589.

64. Allen JW, et al. Opiate pharmacology of intrathecal granulomas. *Anesthesiology.* 2006;105:590–598.

65. Stefano GB. Autoimmunovascular regulation: morphine and anandamide and ancondamide stimulated nitric oxide release. *J Neuroimmunol.* 1998;83:70–76.

66. Theoharides TC, et al. The role of mast cells in migraine pathophysiology. *Brain Res Brain Res Rev.* 2005;49:65–76.

67. Artico M, Cavallotti C. Catecholaminergic and acetylcholine esterase containing nerves of cranial and spinal dura mater in humans and rodents. *Microsc Res Tech.* 2001;53:212–220.

68. Bradding P, Holgate ST. Immunopathology and human mast cell cytokines. *Crit Rev Oncol Hematol.* 1990;31:119–133.

69. He S, Walls AF. Human mast cell tryptase: a stimulus of microvascular leakage and mast cell activation. *Eur J Pharmacol.* 1997;328:89–97.

70. Hermens JM, et al. Comparison of histamine release in human skin mast cells induced by morphine, fentanyl, and oxymorphone. *Anesthesiology.* 1985;62:124–129.

71. Feldberg W, Paton WD. Release of histamine from skin and muscle in the cat by opium alkaloids and other histamine liberators. *J Physiol.* 1951;114:490–509.

72. Blunk JA, et al. Opioid-induced mast cell activation and vascular responses is not mediated by mu-opioid receptors: an in vivo microdialysis study in human skin. *Anesth Analg.* 2004;98:364–370.

73. Yaksh TL, et al. The spinal biology in humans and animals of pain states generated by persistent small afferent input. *Proc Natl Acad Sci USA.* 1999;96:7680–7686.

74. Chen HS, Lipton SA. The chemical biology of clinically tolerated NMDA receptor antagonists. *J Neurochem.* 2006;97:1611–1626.

75. Hocking G, Cousins MJ. Ketamine in chronic pain management: an evidence-based review. *Anesth Analg.* 2003;97:1730–1739.

76. Subramaniam K, et al. Ketamine as adjuvant analgesic to opioids: a quantitative and qualitative systematic review. *Anesth Analg.* 2004;99:482–495.

77. Vranken JH, et al. Treatment of neuropathic cancer pain with continuous intrathecal administration of S + -ketamine. *Acta Anaesthesiol Scand.* 2004;48:249–252.

78. Kristensen JD, et al. The NMDA-receptor antagonist CPP abolishes neurogenic 'wind-up pain' after intrathecal administration in humans. *Pain.* 1992;51:249–253.

79. Stotz M, et al. Histological findings after long-term infusion of intrathecal ketamine for chronic pain: a case report. *J Pain Symptom Manage.* 1999;18:223–228.

80. Karpinski N, et al. Subpial vacuolar myelopathy after intrathecal ketamine: report of a case. *Pain.* 1997;73:103–105.

81. Malinovsky JM, et al. Is ketamine or its preservative responsible for neurotoxicity in the rabbit? *Anesthesiology.* 1993;78;109–115.

82. Borgbjerg FM, et al. Histopathology after repeated intrathecal injections of preservative-free ketamine in the rabbit: a light and electron microscopic examination. *Anesth Analg.* 1994;79:105–111.

83. Errando CL, et al. Subarachnoid ketamine in swine—pathological findings after repeated doses: acute toxicity study. *Reg Anesth Pain Med.* 1999;24:146–152.

84. Vranken JH, et al. Severe toxic damage to the rabbit spinal cord after intrathecal administration of preservative-free S(+)-ketamine. *Anesthesiology.* 2006;105:813–818.

85. Kristensen JD, et al. Laser-Doppler evaluation of spinal cord blood flow after intrathecal administration of an N-methyl-D-aspartate antagonist in rats. *Anesth Analg.* 1994;78:925–931.

86. Hassenbusch SJ. Preclinical toxicity study of intrathecal administration of the pain relievers dextrorphan, dextromethorphan, and memantine in the sheep model. *Neuromodulation.* 1999;2:230–240.

87. Olney JW, et al. Pathological changes induced in cerebrocortical neurons by phencyclidine and related drugs. *Science.* 1989;244:1360–1362.

88. Ikonomidou C, et al. Blockade of NMDA receptors and apoptotic neurodegeneration in the developing brain. *Science.* 1999;283:70–74.

89. Kitagawa N, et al. A proposed mechanism for amitriptyline neurotoxicity based on its detergent nature. *Toxicol Appl Pharmacol.* 2006;217:100–106.

90. Attwood D, Tolley JA. Self-association of analgesics in aqueous solution: association models for codeine, oxycodone, ethylmorphine and pethidine. *J Pharm Pharmacol.* 1980;32:761–765.

91. Attwood D, Tolley JA. Self-association of analgesics in aqueous solution: micellar properties of dextropropoxyphene hydrochloride and methadone hydrochloride. *J Pharm Pharmacol.* 1980;32:533–536.

92. Yaksh TL, et al. Preclinical safety evaluation for spinal drugs. In: Yaksh T,ed. *Spinal Drug Delivery.* Amsterdam: Elsevier Science; 1999:417–437.

93. Yaksh TL, Allen JW. Preclinical insights into the implementation of intrathecal midazolam: a cautionary tale. *Anesth Analg.* 2004;98:1509–1511.

94. Allen JW, et al. An assessment of the antinociceptive efficacy of intrathecal and epidural contulakin-G in rats and dogs. *Anesth Analg.* 2007;104:1505–1513.

95. Yaksh TL, et al. Pharmacology and toxicology of chronically infused epidural clonidine. HCl in dogs. *Fundam Appl Toxicol.* 1994;23:319–335.

96. Yaksh TL, Malkmus SA. Animal models of intrathecal and epidural drug delivery. In: Yaksh TL, ed. *Spinal Drug Delivery.* Amsterdam: Elsevier Science; 1999:317–344.

97. Durant PA, Yaksh TL. Epidural injections of bupivacaine, morphine, fentanyl, lofentanil, and DADL in chronically implanted rats: a pharmacologic and pathologic study. *Anesthesiology.* 1986;64:43–53.

98. Lebeaux MI. Experimental epidural anaesthesia in the dog with lignocaine and bupivacaine. *Br J Anaesth.* 1973;45:549–555.

99. Rodan BA, et al. Fibrous mass complicating epidural morphine infusion. *Neurosurgery.* 1985;16:68–70.

100. Yaksh TL, et al. Safety assessment of encapsulated morphine delivered epidurally in a sustained-release multivesicular liposome preparation in dogs. *Drug Deliv.* 2000;7:27–36.

101. Cohen SP, et al. Randomized, double-blind, placebo-controlled, dose-response, and preclinical safety study of transforaminal epidural etanercept for the treatment of sciatica. *Anesthesiology.* 2009;110:1116–1126.

102. Wermeling DP. Ziconotide, an intrathecally administered N-type calcium channel antagonist for the treatment of chronic pain. *Pharmacotherapy.* 2005;25:1084–1094.

103. Yaksh TL. Ziconotide: a viewpoint by Tony L. *Yaksh. CNS Drugs.* 2006;20:340–341.

104. Eisenach JC, Yaksh TL. Safety in numbers: How do we study toxicity of spinal analgesics? *Anesthesiology.* 2002;97:1047–1049.

105. Sabbe MB, et al. Toxicology of baclofen continuously infused into the spinal intrathecal space of the dog. *Neurotoxicology.* 1993;14:397–410.

106. Yaksh TL, et al. Inflammatory masses associated with intrathecal drug infusion: a review of preclinical evidence and human data. *Pain Med.* 2002;3:300–312.

107. Walker SM, Westin BD, Deumens R, Grafe M, Yaksh TL. Effects of intrathecal ketamine in the neonatal rat: evaluation of apoptosis and long-term functional outcome. *Anesthesiology.* 2010;113:147–159.

108. Westin BD, Walker SM, Deumens R, Grafe M, Yaksh TL. Validation of a preclinical spinal safety model: effects of intrathecal morphine in the neonatal rat. *Anesthesiology.* 2010;113:183–199.

28

DIAGNOSTIC NEUROPATHOLOGY

Ana Alcaraz

College of Veterinary Medicine, Western University of Health Sciences, Pomona, California

Mark T. Butt

Tox Path Specialists, LLC, Hagerstown, Maryland

INTRODUCTION

The main objective of this chapter is to introduce common changes that are encountered during gross and histopathologic study of the nervous system. Although much of this book deals with toxicological neuropathology, this chapter is more oriented toward those changes encountered with spontaneous disease in domestic and companion animals. Most of the terminology and most of the lesions are applicable to the common laboratory species, however. Also, whereas animals in preclinical toxicity studies emphasizing neuropathology are often terminated during a perfusion fixation procedure, most tissues from diagnostic cases are preserved using immersion fixation into 10% neutral buffered formalin; so the images included in this chapter are primarily of tissues immersion fixed in formalin.

The presence of clinical neurologic signs require that a complete diagnostic evaluation, including gross and microscopic components, be performed on the entire brain, spinal cord, and representative nerves.[1,20,29] Only a thorough and detailed morphological and topographical description of the findings will permit in accurate comparison of the pathological findings with specific nervous system clinical signs.[3]

GROSS EXAMINATION AND TISSUE COLLECTION

It is crucial to conduct a thorough gross examination and a proficient collection of tissues in order to achieve an adequate and relevant microscopic evaluation.

At necropsy, the central nervous system tissue should be removed completely from the cranial cavity and the vertebral canal. After the surface of the extracted tissue is examined, complete specimens are preserved in 10% formalin before any other processing.[29] The brain and spinal cord must be preserved for a minimum of 3 days for small animals and 7 days for large animals in a formalin/tissue ratio of 10 : 1. Two or three cuts should be made in the cerebrum of large animal brains from the dorsal surface into the lateral ventricles to facilitate adequate immersion fixation. The trimming of unpreserved brain is difficult because of the soft consistency of these tissues. Prefixation trimming produces uneven transverse sections which lead to artifacts that may impede accurate microscopic analysis.

For fixation, the spinal cord should be removed from the vertebral canal and sectioned into its three regions (cervical, thoracic, and lumbo/sacro/caudal) and labeled appropriately. In large animals, is then sectioned further the thoracic segment into cranial and caudal components. These spinal cord segments are laid in a flat container or suspended in a tall container if possible. For small animals, both the brain and spinal cord segments may be placed in a single container; for large animals, separate containers are typically used for brain and spinal cord. Once the specimens are preserved, transverse sections of the brain, approximately 0.5 cm thick in small animals and 1 cm (or less) thick in large animals, are cut beginning rostrally to facilitate the precise morphological and topographical description of the lesion and lesions (Figure 1). The sections are laid out for examination in a

---

Fundamental Neuropathology for Pathologists and Toxicologists: Principles and Techniques, First Edition. Edited by Brad Bolon and Mark T. Butt.
© 2011 John Wiley & Sons, Inc. Published 2011 by John Wiley & Sons, Inc.

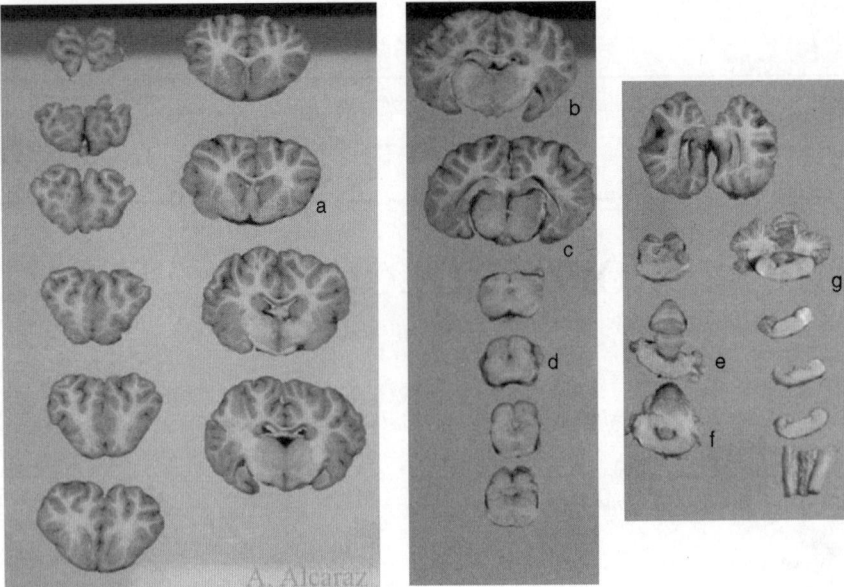

FIGURE 1   Transverse sections of the brain in a rostral-to-caudal sequence. Consistently displaying the sections with the caudal surface up maintains right–left orientation.

rostral-to-caudal sequence with the caudal surface facing up in order to maintain left and right orientation. A thorough gross examination of the transverse sections is recommended with particular attention to distribution and location of any changes observed in symmetry, color, appearance, and consistency.[3,6,13]

When assessing the spinal cord, the location of the lesion is important in developing a set of differential diagnoses.[9,29] Specific locations include the extradural space, the intradural extramedullary space, and the intramedullary region (otherwise known as the parenchyma).

For routine sectioning of the spinal cord, incise the dura on the dorsal median plane to remove the dura (or skip this

FIGURE 2   Spinal cord samples need to include a transverse and longitudinal sections for a thorough evaluation. H&E and Luxol fast blue stains.

step to include the dura with the sections of cord). In small animals make transverse sections about every 0.5 to 1 cm and 1 to 2 cm in large animals. When removing a sample for histological examination, make one transverse section and two adjacent longitudinal sections, one in a dorsal plane and one in a sagittal plane (Figure 2). Label the specific spinal cord segments. When examining a spinal cord lesion it is important to refer to the anatomical location, such as ventral, dorsal, or lateral funiculi, when describing the white matter, and dorsal and ventral horns when describing the gray matter.[3]

To sample nerves, the nerve is placed on a tongue depressor or similar flat surface prior to immersion in the preservative to prevent artifacts such as curling. Both transverse and longitudinal sections of the nerve should be included in the microscopic examination.

In cases that are suspected to involve neuropathology, a diagnostic protocol should include at least seven brain sections plus any other lesions found during gross examination. A sample of a suggested sectioning scheme is provided in Table 1.

BASIC CELLULAR CHANGES

Neurons

A *neuron* is composed of a cell body (soma or perikaryon) and the cell processes (dendrites and axon). One or all parts can be affected by pathological changes.[5] The neuron should not be confused with a similar term, *neuropil,* which is the interwoven cytoplasmic processes of neurons (i.e., dendrites

TABLE 1   Suggested Sectioning Scheme

An example of those sections required for a basic evaluation of the major divisions of the brain is provided in Figure 1. These sections (and preferably more) should be examined in all cases of undiagnosed neurological disease. At a minimum, anatomical areas determined during a neurological examination to be responsible for clinical signs should also be examined. The anatomical areas are provided for reference. Not all areas may belong to the specific major brain subdivision.

Telencephalon

One section rostral to the optic chiasm to show cerebral cortex, corona radiata, internal capsule, corpus callosum, basal nuclei, and rostral lateral ventricle. See Figure 1a.

Diencephalon

One section between the optic chiasm and mamillary bodies to show thalamus, hypothalamus, third and lateral ventricles, fornix, corpus callosum, amygdala, internal capsule, corona radiata, and cerebral cortex. See Figure 1b.

One section just caudal to the mamillary bodies to show the geniculate bodies and pretectal area of the thalamus, junction of third ventricle and mesencephalic aqueduct, hippocampus and lateral ventricle, internal capsule, corona radiata, and cerebral cortex. See Figure 1c.

Mesencephalon

One section through the rostral colliculi to show rostral colliculi, mesencphalic aqueduct, crus cerebri, and substantia nigra. Include a section of the adjacent occipital lobe. See Figure 1d.

Metencephalon

One section through the center of the transverse fibers of the pons to show these fibers, cerebellar peduncles. cranial nerve V, and the rostral portion of the cerebellum. See Figure 1e.

Myelencephalon

One section at the level of the trapezoid body and cochlear nuclei to show the rostral medulla with the dorsal nucleus of the trapezoid body, confluence of the cerebellar peduncles, and the cerebellar medulla with cerebellar nuclei and folia with cerebellar cortex. See Figure 1f.

One section at the level of the obex to show the hypoglossal nuclei, olivary nuclei, and vestibular nuclei. See Figure 1g.

and axons) and glial cells forming a meshwork in the gray matter of the central nervous system.

Chromatolysis is a condition in which the cytoplasm of the neuronal body has lost the prominent granular endoplasmic reticulum known as the Nissl substance. Dispersion or loss of the Nissl substance is classified as central or peripheral chromatolysis according to the location relative to the neuronal nucleus. The area affected is characterized by weakly stained, pale, homogeneous, eosinophilic cytoplasm.[9,14]

In neurons with central chromatolysis, the large round nucleus is displaced by the affected cytoplasm from the center to the periphery of the cell (Figure 3).[4] This is sometimes referred to as *margination* of the nucleus. Central chromatolysis occurs in many different disease states and is usually associated with injury of the peripherally projecting

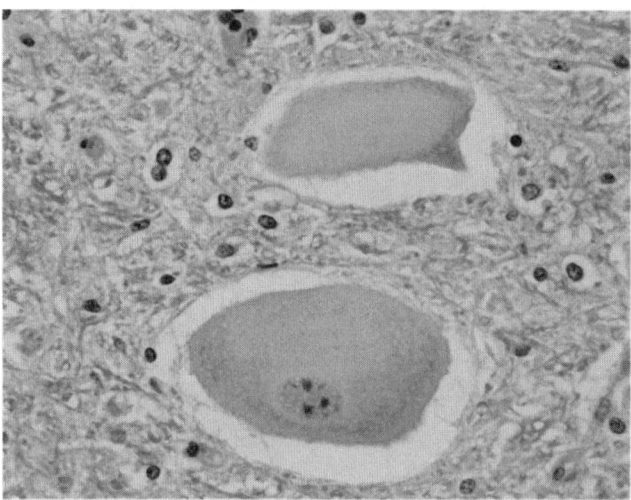

FIGURE 3   Spinal cord of a horse. Neurons with central chromatolysis. Notice the loss of Nissl substance and the presence of pale, homogeneous, smooth cytoplasm and a displaced, marginated nucleus. H&E, 60×.

axon, especially if the injury happens close to the neuronal soma.[4] At the ultrastructural level, the central pallor is often identified as abundant smooth endoplasmic reticulum, Golgi membranes, and neurofilaments, instead of the orderly parallel arrays of rough endoplasmic reticulum with a large number of free ribosomes.[6,29]

Ischemia/hypoxic damage is one of the most common forms of acute injury to neurons.[29] The underlying pathophysiological mechanisms may include impaired blood flow, reduced oxygen tension in circulating blood, or toxic factors that produce some sort of energy deficiency or even excitotoxicity. Hypoglycemia and thiamin deficiency will induce similar morphological neuronal changes.[9,11,19]

Ischemic neurons are characterized by shrunken bright red cytoplasm, loss of Nissl substance, and small, triangular pyknotic nuclei (Figure 4). Handling may produce artifactual changes in neurons, including shrunken basophilic cytoplasm and a condensed nucleus.[20,29,31] Artifactual changes are described in Chapter 13.

Other changes found in hypoxic damage are increased numbers of glial cells, prominence of the vasculature, and occasional neuronophagia. The blood vessels in areas of hypoxia may actually appear to have increased branching with very prominent and hypertrophied endothelial cells (Figure 5).[14]

As the name implies, *neuronophagia* occurs when neurons are consumed. This happens when a nerve cell undergoes necrosis and phagocytic cells respond to remove the cell body and the proximal processes. The responding cells can be neutrophils, microglial cells, or macrophages. The microglial phagocytes surround not only the cell body but also adjacent processes. Occasionally, groups of

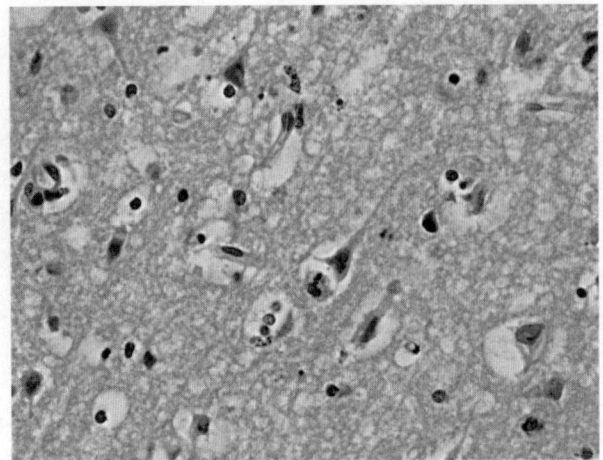

FIGURE 4   Ischemic/necrotic neurons showing shrunken triangular pyknotic nuclei and scant eosinophilic cytoplasm. H&E, 60×.

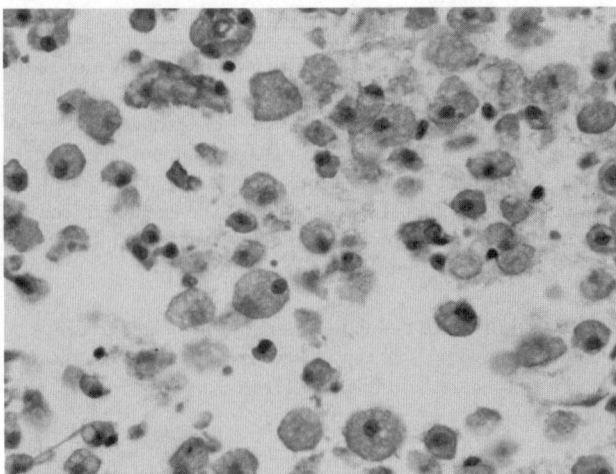

FIGURE 6   Close-up view of the Gitter cells in a necrotic area. The cells have abundant granular and vacuolated cytoplasm with an eccentric round dark nucleus. H&E, 60×.

microglial cells can be seen, indicating the original position of a dead neuron that can no longer be identified. This change should be distinguished from *satellitosis*.[9,14]

Which is the reaction of an oligodendrocyte (or other glial cell) around a neuronal cell body with or without degenerative changes. In satellitosis, the neuronal body is intact, and increased numbers of glial cells surround the neuron. It is generally accepted that satellitosis may progress to neuronophagia.[14]

Glial Cells (Astrocytes, Oligodendrocytes, Microglia)

Gliosis is an abnormal accumulation of any of the three glial cell types. Glial cells include astrocytes (associated with neuronal metabolism), oligodendrocytes (responsible for myelin production), and microglial cells (with phagocytic function).[15,24] More specifically, gliosis refers to the focal

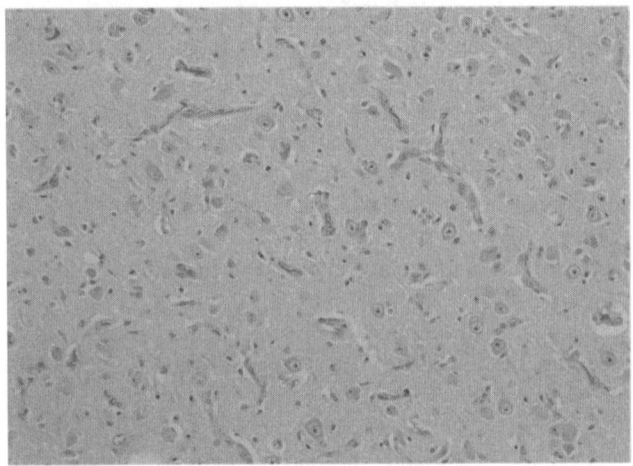

FIGURE 5   Polioencephalomalacia in a calf. Prominent blood vessels showing branching. 40×. (Image by A. Miller.)

proliferation of any type of glial cell as a response to damage to the central nervous system (CNS).

Microglial cells, with phagocytic functions, may contribute to gliosis. Microglia phagocytize damaged tissue after ischemic events or infarcts.[20,24] These cells, called *Gitter cells*, are characterized by abundant granular to vacuolated cytoplasm with a displaced small round nucleus (Figure 6).[9,14,30]

Astrocytes are commonly referred to as the interstitial cells of the central nervous system. However, it is now recognized that they have a complex functional role in neuronal activity. Astrocyte functions are numerous and variable, from guiding neuronal migration during brain development, to having a structural role in the parenchyma, to participating in the synaptic activity of neurons.[7,21,23]

Astrocytes have a round-to-oval nucleus with clear homogeneous chromatin and (usually) indistinct cytoplasm. Two types of astrocytes are generally recognized: type 1, or protoplasmic, found mainly in gray matter; and type 2, or fibrous, found in white matter or around blood vessels. The distinction between protoplasmic and fibrous astrocytes is imperceptible at the level of light microscopy, their location being the main difference that will aid in their classification.[21,29]

Astrocytosis

Astrocytosis is an increase in the number and size of astrocytes.[14,23] Astrocyte changes are commonly observed in nearly every type of central nervous system injury and include hyperplasia, hypertrophy, and/or degenerate alterations.[14] Astrocytes (phenotypically) respond to injury by becoming larger, acquiring distinct cytoplasm, and/or becoming binucleate.[23] Enlarged, reactive astrocytes are also known as gemistocytes or gemistocytic astrocytes (from the

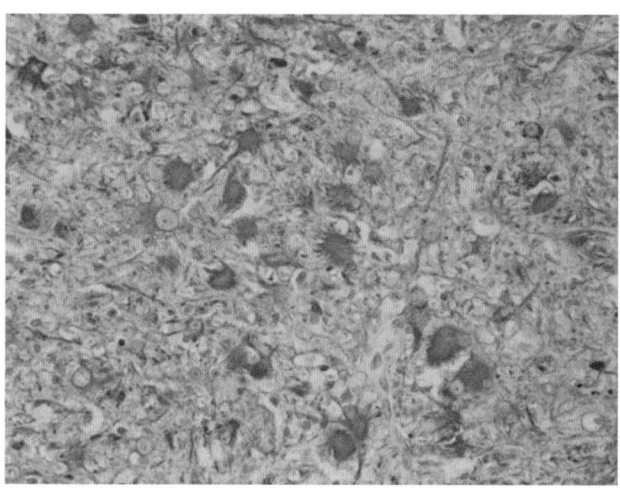

FIGURE 7 Reactive astrocytes stain strongly with an immuno-histochemical stain for GFAP. 40×. (Image by A. Miller.)

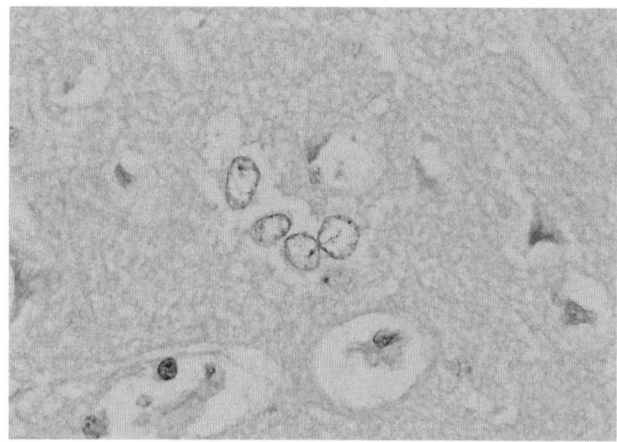

FIGURE 8 Horse with hepatic insufficiency. Cluster of Alzheimer type II astrocytes with prominent vesicular nuclei and no visible cytoplasm. These cells are commonly found in cases of hepatic encephalopathy. H&E, 60×.

Greek *gemistos,* meaning laden or full). They have a pale, homogeneous, eosinophilic, abundant well-delineated cytoplasm with an eccentric, clear, oval nucleus. An immunohistochemical stain for glial fibrillary acidic protein (GFAP; an intermediate filament mostly unique to astrocytes) is used to detect reactive astrocytes[21,25] and also stains nonreactive astrocytes (Figure 7). When there is slow destruction of the basic architecture of the tissue, proliferation of astrocytic processes forms a pale amphophilic meshwork known as the astrocytic scar.[26]

Alzheimer Astrocytes

Alzheimer type II astrocytes are typically binucleated and their nuclei are moderately enlarged and may be lobulated with dispersed, lacy chromatin; they often occur in clusters (Figure 8). The cell body or karyon is not seen with routine hematoxylin and eosin (H&E) stain, but is enhanced with silver stains such as Bielschowsky stain.[29] Alzheimer type II astrocytes can be found in cases of hepatic encephalopathy[10,29] and some other metabolic abnormalities. *Alzheimer type I astrocytes* are enlarged or karyomegalic, with large bizarre lobulated nuclei. This particular cell phenotype is very rarely encountered in veterinary medicine.[20]

Nerves

A *nerve* is a collection of neuronal processes (axons with or without myelin) outside the central nervous system. These axons are associated with Schwann cells [the myelinating cells of the peripheral nervous system (PNS)], which provide varying amounts of myelin around the axons. The Schwann cells are surrounded by endoneurial connective tissue.[5,6] The spinal nerve roots and most of the cranial nerves are peripheral nerves.

Wallerian degeneration includes the morphological changes found in nerves distal to the site of axonal injury.[8,17] It extends from the injury site to the termination of the axon and consists of fragmentation to complete loss of an axon. This is followed by degradation of the myelin sheath, which is secondary demyelination due to loss of axonal integrity. Wallerian degeneration also occurs in the CNS; however, the process of myelin loss is slower to occur and (generally) slower to repair. A myelin ellipsoid (or digestive chamber) is a focal distention of the myelin sheath found in an area of Wallerian degeneration that encloses a fragmented axon. In the PNS, the nerve affected eventually becomes hypercellular with secondary proliferation of Schwann cells and endoneurium, forming what is known as the *von Büngner band.* Schwann cells have little or no phagocytic activity, and macrophages typically infiltrate the area of damage to remove debris.[8]

Spheroids are focal axonal swellings containing degenerate organelles. A focal axonal swelling is characterized by an enlarged, pale, ovoid, eosinophilic, homogeneous structure (Figure 9). Spheroids represent an "injury" to an axon which might progress to axonal fragmentation.[8,9]

Primary demyelinating changes are observed when there is destruction of the myelin sheath without axonal damage. The axon can survive deprived from myelin, but the opposite is not the case. The primary spontaneous demyelinating diseases in veterinary medicine are mostly viral (e.g., canine distemper, caprine arthritis encephalitis virus), and they usually show concurrent axonal injury in variable degrees of severity.[28] All of the diseases affecting white matter produce some degree of demyelination. The histological changes associated with primary

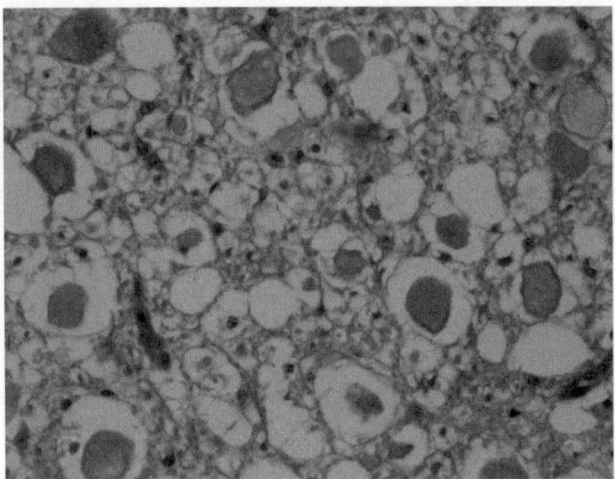

FIGURE 9 Spheroids. Focal axonal swelling. Transverse section of spinal cord. H&E, 60×.

demyelination are myelin vacuolation, which produces loss of stain affinity and sponginess in the white matter, astrogliosis, or an increase in the number of macrophages that contain myelin debris.[9,29]

Pigments

Lipofuscin is a brown granular pigment found in the cytoplasm of the neuronal cell body. It accumulates on one side of the cytoplasm and may impart a polarized appearance to cells. The presence of lipofuscin in neurons has no diagnostic value, except that it is commonly found in older animals (Figure 10).[22,29] Ultrastructurally, lipofuscin is a dense granular and lamellated material associated with lipid droplets.[6]

Calcium or iron deposits are found in *mineralized neurons*, also known as *ferruginated* or *encrusted neurons*. The presence of these mineral deposits is associated with

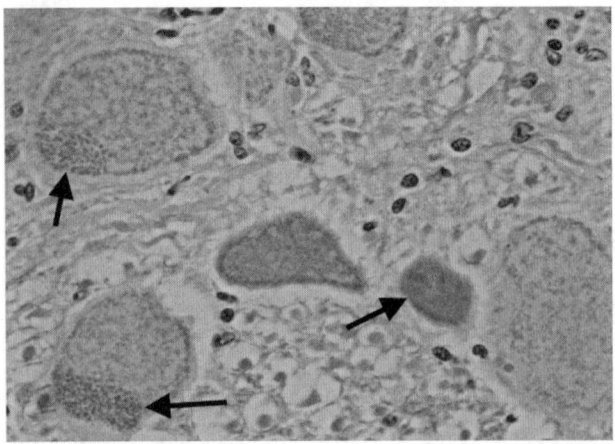

FIGURE 10 Neuron. Fine peripheral brown granular cytoplasmic pigment interpreted as lipofuscin. H&E, 60×.

neurons undergoing degeneration or necrotic neurons; the change is commonly seen around aging, hemorrhagic areas. Mineralized neurons have granular and basophilic cytoplasmic deposits and are positive to VanKossa stain for calcium or Prussian blue for iron.[29]

Marginal siderosis is a brownish discoloration of nervous tissue due to chronic subarachnoid hemorrhage. Hemosiderin accumulation gives a yellow-orange discoloration to the meninges and sometimes affects the adjacent nervous tissue. Histologically, there is superficial gliosis and a moderate number of macrophages with intracellular pigment granules that are positive for Prussian blue.[12,16]

Neuromelanin pigment occurs in some hypothalamic neurons and the substantia nigra of some species, including humans.[29] This is a normal structure and typically of no diagnostic value.

BASIC PARENCHYMAL CHANGES

Spongiosis, also called *status spongoisis*, is defined as a microcystic degeneration of nervous tissue seen at the level of light microscopy. The loosely arranged ("spongy") parenchyma shows vacuoles described as variably sized, empty spaces expanding the neuroparenchyma. The loosening of the parenchyma is due to vacuolation of neuronal processes, cytoplasmic swelling of astrocytes (astrocytic swelling), or the vesiculation of myelin sheaths. Demyelination can be seen as vacuolation or loss of density in the white matter characterized by diminished tinctorial affinity with specific myelin stains such as toluidine blue or Luxol fast blue. These stains are particularly useful in identifying and evaluating myelin since the stain detects the proteolipids in the myelin sheath.[9,14,29]

Rarefaction, characterized by a decrease in parenchymal density, may result from mild edema where there is loss of tissue density but not volume.

Malacia refers to the softening of the CNS parenchyma due to necrosis. The term should be reserved for gross descriptions of pathological findings, as it only means softening of the tissue. Histologically, necrosis in the nervous tissue is always described as liquefactive necrosis, due to the loss of cellular detail and softening of the parenchyma with formation of cavities due to loss of tissue.[9,14] In some cases, the cavities are replaced by engorged, foamy macrophages called Gitter cells, described previously.

Edema of the CNS is variable in appearance. It is often characterized grossly by the swelling of the cerebral parenchyma, and it can affect both gray and white matter. However, there may be no gross evidence of edema. CNS edema is a nonspecific change that can occur in inflammatory, degenerative, neoplastic, or traumatic CNS lesions. The areas affected are normal in texture but can be drastically thickened and distended. Because the parenchyma of the CNS does not have lymphatic vessels, the edema fluid accumulates intracellularly as well as in the interstitial spaces.[2,9,29]

Edema can be classified as either cytotoxic or vasogenic. *Cytotoxic edema* is the result of a direct injury to glial cells characterized microscopically by astrocytic swelling and spongiotic changes in the white matter if oligodendrocytes and/or myelin are affected. *Vasogenic edema* occurs when there is damage to the blood–brain barrier, causing fluid and proteins to escape, affecting primarily the white matter, since the gray matter neuropil meshwork is more resistant to fluid accumulation than the interstitium of the white matter. The histological description of vasogenic edema includes a loosening of the white matter due to the accumulation of interstitial fluid.[14,29]

Inflammation of the CNS may include the presence of perivascular cuffs, gliosis, neuronal satellitosis, and/or neuronophagia. Accumulations of any inflammatory cells, mainly lymphocytes, in the perivascular or Virchow–Robin space are referred to as perivascular cuffs. The Virchow–Robin space is continuous with the meningeal space, so infiltrates in the meninges may be observed to variable depths in the Virchow–Robin space. In neuropathology, perivascular nonsuppurative encephalitis refers to the inflammatory lymphocytic reaction that is found in any viral disease affecting the CNS.[20,29]

Special care needs to be taken not to overinterpret artifacts when conducting a neuropathological examination. Postmortem autolysis due to poor tissue preservation is a common artifact that needs to be identified as such. Nervous tissue that is properly collected, well preserved, and optimally sectioned and stained allows for clear differentiation between the many cell types found in the neuropil. If there are no obvious differences in nuclear size and chromatin features between astrocytes, other glial cells and neurons, a high-quality microscopic examination may be feasible.[2]

LYSOSOMAL STORAGE DISEASES

Enzymatic disorders associated with certain genetic diseases can lead to an accumulation of abnormal intraneuronal material. Accumulation and storage of metabolites within the neurons cause malfunctions. Affected tissues show enlarged swollen neurons with abundant cytoplasm that vary in appearance and staining affinity, depending on the material abnormally stored in the neuron (Figure 11). GM-2 gangliosidosis is one of the more common lysosomal storage diseases in animals, with lesions found in neurons throughout the CNS, spinal ganglia, and retina. The neuronal cell bodies affected are enlarged with abundant, pale, smooth, eosinophilic cytoplasm. The spherical-to-oval centrally located single nucleus is displaced to the periphery, with increased

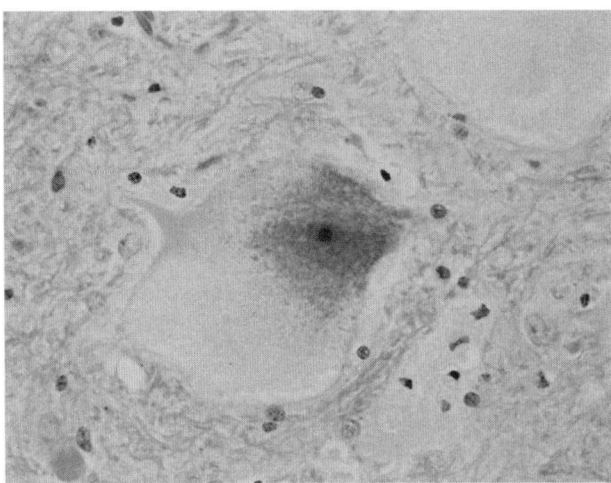

FIGURE 11 Neuron: Distended, granular cytoplasm enlarged by lysosomes filled with abnormal accumulations of a proteinaceous material. H&E, 60×.

cytoplasmic density.[27] Ultrastructurally, the cytoplasm is filled with membrane-bound vacuoles of membranous material organized in concentric layers.[14]

VIRAL DISEASES

Viral diseases are diagnosed by the presence of viral inclusion bodies in neural tissue, particularly when the distribution of the inflammatory reaction is taken into consideration. It is also relevant to properly describe the location of the inclusion bodies and to identify the cells affected.

Rabies is an RNA virus that produces intracytoplasmic inclusion bodies in concurrence with polioencephalitis (the term *polio* refers to gray matter) or polioencephalomyelitis. The inclusions, also called *Negri bodies*, are found predominately in neurons present in areas with nonsuppurative perivascular inflammation, which is characteristic of most viral disease encephalitides (Figure 12).

Herpesviruses have been identified in cases of encephalitis in many species. Herpes is a DNA virus that produces intranuclear inclusion bodies, displacing the nuclear chromatin, and surrounded by a clear halo (Figure 13). It should be noted that not all encephalopathic herpesviruses produce intranuclear inclusions, especially in the horse, in which there are no inclusion bodies noted in CNS cases of equine herpesvirus-1 infection.

Adenovirus encephalitis is associated with severe vasculitis and intranuclear inclusion bodies found in the nucleus of endothelial cells (Figure 14). The neurotropic or enteric strains of mouse hepatitis virus (a coronavirus) have the ability to produce syncytial cells (merged cells that appear as single, multinucleated cells) in the brain or in epithelial cells at the tip of the intestinal villi.

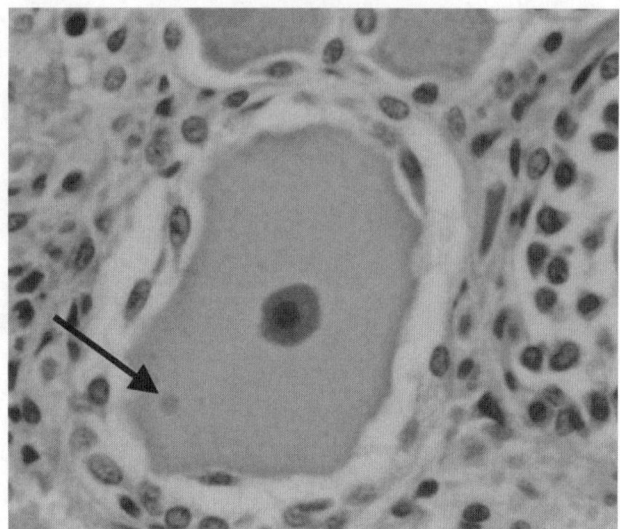

FIGURE 12   Negri bodies. Intracytoplasmic viral inclusion bodies found in neurons infected with rabies virus. H&E, 60×.

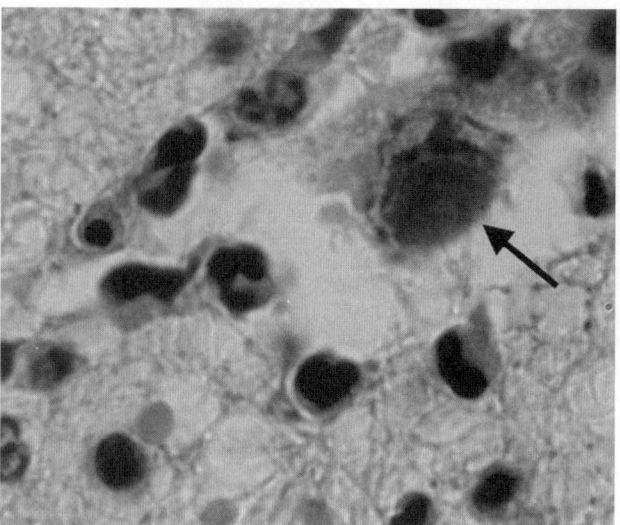

FIGURE 14   Bovine, adenoviral encephalitis; Vasculitis and intranuclear viral inclusion body (arrow) in an endothelial cell. H&E, 100×.

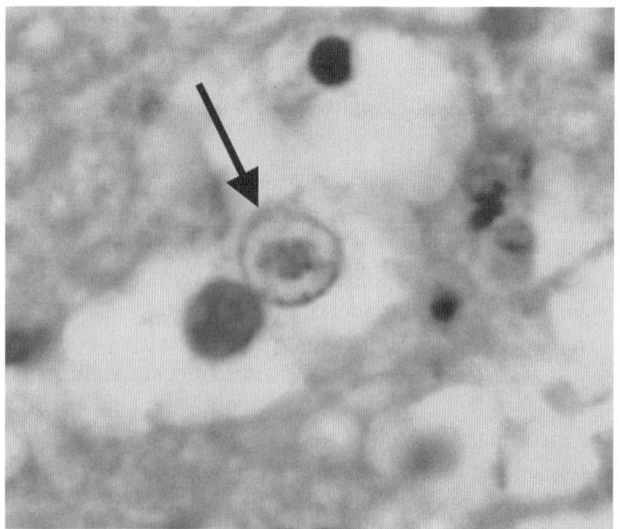

FIGURE 13   Intranuclear viral inclusion body (arrow). Rabbit, herpesvirus encephalitis. H&E, 100×. (Image by D. Russell.)

In rodents, intracytoplasmic, hypereosinophilic inclusion bodies in the thalamic neurons are of no significance or diagnostic consequence. These inclusions are associated with age and have no viral etiology.

NEOPLASTIC DISEASES[14,18,29]

Because the brain and spinal cord are encased in rigid, protective structures, neoplastic diseases have no place to go, resulting in the replacement and/or compression of preexisting soft tissues. As a result, even benign neoplasms in the central nervous system can be catastrophic.

Meningioma is a neoplasm of meningothelial arachnoidal cells. These neoplasms are usually discrete, with granular-to-smooth, occasionally gritty surfaces. There are nine histological variants: meningothelial, fibroblastic, transitional, psamommatous, angiomatous, papillary, granular, myxoid, and anaplastic. The first eight have similar biological behavior, being discrete benign tumors with a (usually) fair-to-good prognosis. The last variant, anaplastic meningioma, is the exception and is typically highly invasive and malignant. These tumors are commonly seen in cats and dogs. In the spinal cord, they are found in the intradural extramedullary space.

Astrocytoma is a common neoplasm originating from astrocytes. Astrocytomas can be composed of any of the astrocytic cells, including fibrillar, protoplasmic, or gemistocytic, with fibrillar astrocytomas being the most common type. Grossly astrocytic tumors are yellow or grayish white, poorly demarcated, vaguely infiltrative, and commonly associated with hemorrhage (especially in cases of high-grade astrocytomas). Microscopically, astrocytomas are loosely arranged neoplasms of low-to-moderate cellularity (Figure 15). Two types of astrocytomas are described: low grade, characterized by well-differentiated astrocytes; and high grade or anaplastic, some of which are classified as glioblastoma multiforme. The low-grade astrocytomas are microscopically characterized as unencapsulated, ill-defined masses composed of low-to-moderate cellularity with loosely arranged, relatively well-differentiated astrocytes with little cellular atypia. These tumors are occasionally infiltrative and may have other glial cells, such as

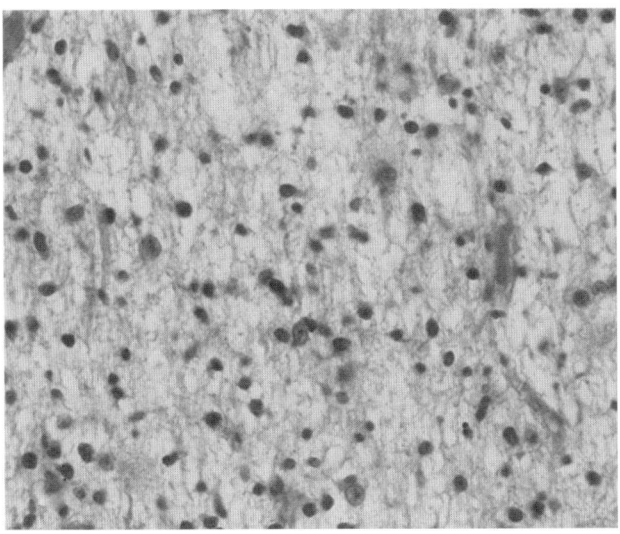

FIGURE 15   Astrocytoma. H&E, 60×.

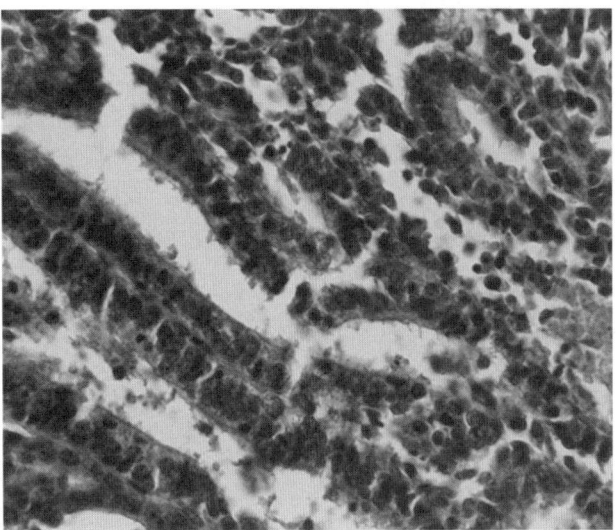

FIGURE 17   Ependymoma. H&E, 60×.

oligodendrocytes, scattered throughout. The presence of mitotic figures, necrosis, and vascular proliferation varies from minimal in low-grade variants to moderate in high-grade astrocytomas.

Oligodendrogliomas are neoplasms composed of oligodendrocytes in various stages of differentiation. Grossly oligodendroglial tumors are usually soft pinkish red to grayish pink gelatinous masses. Microscopically, oligodendrogliomas are moderately to highly cellular. The neoplastic cells are homogeneous and uniform with a distinct, darkly stained nucleus, lightly stained cytoplasm, and clear cytoplasmic membrane, giving it the appearance of a fried egg or a honeycomb (Figure 16).

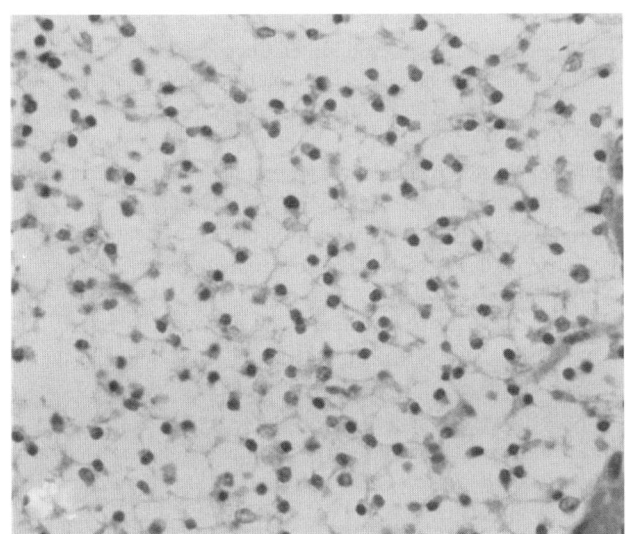

FIGURE 16   Oligodendroglioma. H&E, 60×.

Ependymomas are neoplasms composed of ependymal cells. Typically, these are slow-growing tumors and usually benign. Grossly, ependymal tumors are soft, gray, relatively discrete masses that arise anywhere in the ventricular system. Histologically, there is a highly cellular, rather uniform neoplastic cell population. The cells are arranged in sheets, pseudorosettes, or rosettes. The neoplastic cells are round to oval, with a hyperchromatic nucleus and scant eosinophilic cytoplasm (Figure 17).

Choroid plexus tumors are neoplastic growths composed of choroid plexus epithelium. These tumors arise from the epithelial cells of the choroid plexus and extend into the ventricular system. They are well-defined granular masses and vary from gray to white. There are two variants: choroid plexus papilloma and choroid plexus carcinoma. The degree of cellular anaplasia and the presence of metastasis are the only criteria used to differentiate between the two variants. Histologically, there are elongated exophytic proliferations of fine fibrovascular stroma covered by cuboidal-to-columnar epithelium. The cells have a usually round (but variable shaped) nucleus. Anaplastic features include nuclear atypia, increased mitotic index, and focal areas of necrosis.

Lymphoma is a tumor of lymphoid tissue. This type of neoplasm can be primary or metastatic to the CNS. Regardless of its origin, lymphoma is a malignant neoplasm. Malignant lymphoma or lymphosarcoma are redundant terms: lymphoma is the proper diagnostic terminology. CNS lymphoma is characterized by grossly swollen and discolored areas, usually associated with the white matter or the meninges. Microscopically, lymphomas have a characteristic perivascular orientation that can be confused with nonsuppurative perivascular cuffing. The latter are confined to the perivascular space, while lymphoma cells extend and

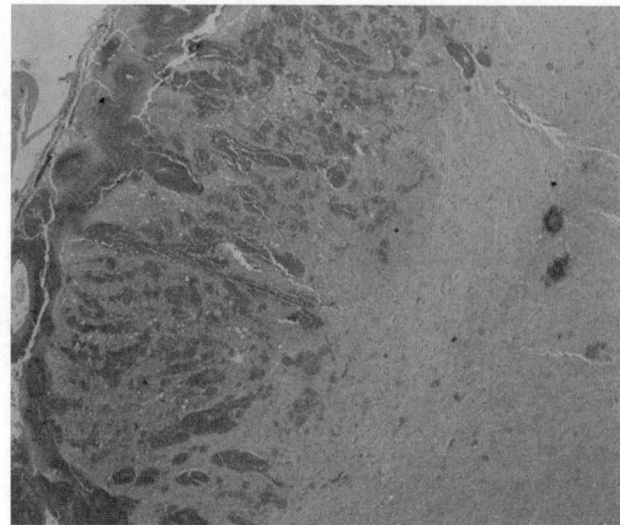

FIGURE 18 Lymphoma. Ventral part of the brain with meningeal and perivascular neoplastic lymphocytes. H&E 2×.

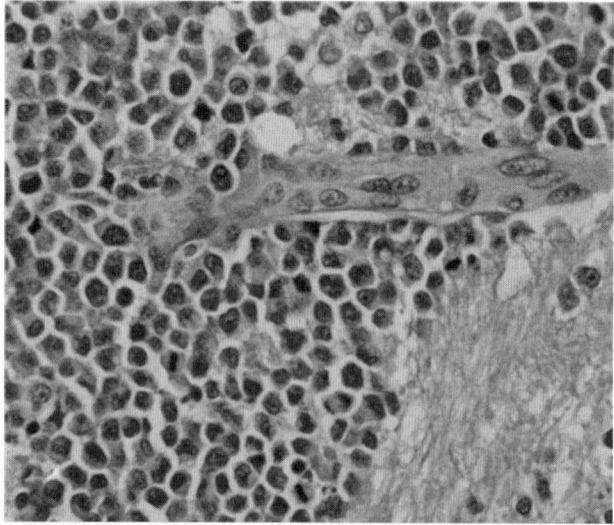

FIGURE 19 Neoplastic lymphocytes extending to and invading the adjacent perivascular stroma. H&E, 60×.

invade the perivascular nervous tissue (Figures 18 and 19). Macroscopically, lymphoma can resemble adipose tissue, especially in the extradural tissue in the spinal cord region. Of all the neoplastic diseases, lymphoma may have the most varied clinical, gross, and microscopic presentation.

REFERENCES

1. Brayton C, Justice M, Montgomery CA. Evaluating mutant mice: anatomic pathology. *Vet Pathol.* 2001;38:1–19.

2. deLahunta A.Cornell University, College of Veterinary Medicine. Personal communication.

3. deLahunta A, Glass E. *Veterinary Neuroanatomy and Clinical Neurology*, 3rd ed. Philadelphia: W.B. Saunders; 2009.

4. Divers TJ, Mohammed HO, Cummings JF, Valentine BA, De Lahunta A, Jackson CA, Summers BA. Equine motor neuron disease: findings in 28 horses and proposal of a pathophysiological mechanism for the disease. *Equine Vet J.* 1994;26(5):409–415.

5. Evans HE. *Miller's Anatomy of the Dog*, 3rd ed. Philadelphia: W.B. Saunders; 1993.

6. Fawcet DW. *The Cell.* Philadelphia: W.B. Saunders; 1981.

7. Girvin AM, Gordon KB, Welsh CJ, Clipstone NA, Miller SD. Differential abilities of central nervous system resident endothelial cells and astrocytes to serve as inducible antigen-presenting cells. *Blood.* 2002;99(10):3692–3701.

8. Glass JD. Wallerian degeneration as a window to peripheral neuropathy. *J Neurol Sci.* 2004;15:220(1–2):123–124.

9. Gray F, Poirier J, De Girolami U. *Escourolle and Poirier's Manual of Basic Neuropathology.* Burlington, MA: Elsevier; 2003.

10. Hasel KM, Summers BA, deLahunta A. Encephalopathy with idiopathic hyperamonemia and Alzheimer type II astrocytes in equidae. *Equine Vet J.* 1999;31(6):478–482.

11. Himsworth CG. Polioencephalomalcia in a llama. *Can Vet J.* 2008;49(6):598–600.

12. Huxtable CR, de Lahunta A, Summers BA, Divers T. Marginal siderosis and degenerative myelopathy: a manifestation of chronic subarachnoid hemorrhage in a horse with a myxopapillary ependymoma. *Vet Pathol.* 2000;37(5):483–500.

13. Jackson CA, de Lahunta A, Dykes NL, Divers TJ. Neurological manifestation of cholesterinic granulomas in three horses. *Vet Rec.* 1994;135(10):228–230.

14. Jubb KVF, Huxtable CR. The nervous system. In: Jubb KVF, Kennedy PC, Palmer N, eds. *Pathology of Domestic Animals*, 4th ed. 1993:1(3).

15. Kim SU, de Vellis J. Microglia in health and disease. *J Neurosci Res.* 2005;81(3):302–313.

16. Koeppen AH, Michael SC, Li D, Chen Z, Cusack MJ, Gibson WM, Petrocine SV, Qian J.The pathology of superficial siderosis of the central nervous system. *Acta Neuropathol.* 2008;116(4):371–382.

17. Koeppen AH. Wallerian degeneration: history and clinical significance. *J Neurological Sci.* 2004;220:115–117.

18. Koestner A, Bilzer T, Fatzer R, Schulman FY, Summers BA, Van Winkle TJ. *Histological Classification of the Tumors of the Nervous System of Domestic Animals.* Washington, DC: American Registry of Pathology; 1999.

19. Kul O, Karahan S, Basaln M, Kabakci N. Polioencephalomalacia in cattle: a consequence of prolonged feeding of barley malt sprouts. *J Vet Med.* 2006;53:123–128.

20. Love S, Louis DN, Ellison DW. In: Love S, et al., eds. *Greenfield's Neuropathology.* 8th ed. New York: Oxford University Press; 2008.

21. Ludwin SK. Reaction of oligodendrocytes and astrocytes to trauma and implantation: a combined autoradiographic

and immunohistochemical study. *Lab Invest*. 1985;52(1): 20–30.

22. Márquez M, Serafin A, Fernández-Bellon H, et al. Neuropathologic findings in an aged albino gorilla. *Vet Pathol*. 2008;45: 531–537.

23. Montgomery DL. Astrocytes: form, functions, and roles in disease. *Vet Pathol*. 1994;31(2):145–167.

24. Napoli I, Neumann H. Microglia clearance function in health and disease. *Neuroscience 6*. 2009;158(3):1030–1038.

25. Pekny M, Pekna M. Astrocyte intermediate filaments in CNS pathologies and regeneration. *J Pathol*. 2004;204:428–437.

26. Rolls A, Shechter R, Schwarts M. The bright side of the glial scar in CNS repair. *Nat Rev. Neurosci*. 2009;10:235–241.

27. Singer HS, Cork LC. Canine GM2 gangliosidosis: morphological and biochemical analysis. *Vet Pathol*. 1989;26(2):114–120.

28. Summers BA, Appel MJ. Demyelination in canine distemper encephalomyelitis: an ultrastructural analysis. *J Neurocytol*. 1987;16(6):871–881.

29. Summers BA, Cummings JF, De Lahunta A. *Veterinary Neuropathology*. Orlando, FL: Mosby; 1994.

30. Williams KJ, Summers BA, De Lahunta A. Cerebrospinal cuterebriasis in cats and its association with feline ischemic encephalopathy. *Vet Pathol*. 1998;35(5):330–343.

31. Jortner BS. The return of the black neuron. A histological artifact complicating contemporary neurotoxicologic evaluation. *Neurotoxicology*. 2006;27:628–634.

29

TOXICOLOGICAL NEUROPATHOLOGY IN MEDICAL PRACTICE

THOMAS J. MONTINE
Department of Pathology, University of Washington, Seattle, Washington

DOUGLAS C. ANTHONY
University of Missouri School of Medicine, Columbia, Missouri

INTRODUCTION

The toxic effects of compounds on the nervous system have a significance that reaches beyond the identification of exogenous neurotoxicants (xenobiotics) or endogenous neurotoxins. Neurotoxicology has helped define mechanisms of action and patterns of response of the nervous system to specific quantifiable injuries resulting from exposure to these agents. Although considered separate fields of study, neurotoxicology, neurodegeneration, stroke, trauma, and metabolic diseases of the nervous system inform each other about normal metabolic pathways, mechanisms of neuronal dysfunction and death, and response to injury in the nervous system. Several neurotoxicants first identified in humans have subsequently come into use as models of human neurodegenerative disease. Perhaps the most striking example is 1-methyl-4-phenyl-1,2,3,6-tetrahydropyridine (MPTP), identified initially as a contaminant of an illicit drug causing Parkinsonism,[1] which provided an unparalleled model of the naturally occurring disease in a very short time frame. Many other well-known compounds initially identified as neurotoxicants have become fundamental tools used by neuroscientists, including tetrodotoxin, curare, and kainic acid.[2] Conversely, progress in other areas of neuroscience continually advances the understanding of the mechanisms of neurotoxicants.

In medical practice, diagnostic neuropathological assessment of toxic exposures poses a challenge that is quite different from experimental neurotoxicology. The fundamental neuropathological diagnostic evaluation begins with an assessment of neuropathological findings and whether they are outside the normal range of variation. Abnormal findings are identified in different areas of the brain in order to determine whether there is a pattern of injury. It is from this pattern of injury (or the sum of diagnostic abnormalities) that a potential etiology is implicated, and when a toxic etiology is within the differential diagnosis, documentation of exposure is sought. For example, selective loss of pyramidal cells in Sommer's sector in the hippocampus and Purkinje cells in the cerebellum are common in ischemic injury. The finding that both areas are involved implies a hypoxic or ischemic injury, or a toxic exposure that mimics ischemic injury. In this way, a specific type of injury may be implicated. To evaluate whether a particular pattern of injury indicates toxicant exposure, it is necessary first to understand the cellular changes that commonly accompany neurotoxicant exposures.

COMMON NEUROPATHOLOGICAL CHANGES CAUSED BY NEUROTOXICANTS

Many individual pathological changes in the brain following neurotoxicant exposure are not specific and in isolation could represent several different etiologies. The implication of a specific neurotoxicant comes from the sum of findings and the clinical toxicological evidence.[3,4]

Fundamental Neuropathology for Pathologists and Toxicologists: Principles and Techniques, First Edition. Edited by Brad Bolon and Mark T. Butt.
© 2011 John Wiley & Sons, Inc. Published 2011 by John Wiley & Sons, Inc.

Edema is a striking macroscopic finding and occurs in some acute toxic encephalopathies. The brain may be heavy and swollen, even after fixation, with broadened cortical gyri and obliterated sulci. In severe cases, transtentorial and cerebellar tonsilar herniations may occur. Cerebral edema secondary to vascular damage, as in lead encephalopathy, or direct damage to central nervous system (CNS) myelin, as in triethyltin encephalopathy, is confined largely to white matter. In contrast, edema resulting from diffuse cytotoxic damage, as in thallium intoxication, affects both gray and white matter and has been observed to impart a "moth-eaten" appearance to cerebral cortex.[5] The histological manifestations of cerebral edema may be slight: myelin pallor, mild gliosis, and perineuronal vacuoles may be all that is observed by standard histochemical stains (Figure 1).

Cytological changes within individual neurons are uncommon in toxic exposures, in contrast to naturally occurring disease. Notable exceptions are MPTP exposure, which can lead to formation of Lewy body–like intraneuronal inclusions that mimic those observed in Parkinson's disease (Figure 2), and the inclusions that have been reported following lead intoxication.[6,7]

Neuronal degeneration may be either acute, or remote if following exposure that occurred in the distant past. In acute phases, there are necrotic neurons in various stages of degeneration. When remote, histological findings of neuronal degeneration may be recognized only by the absence of normal neuron populations. Neuronal loss may be diffuse and involve many areas of the brain, such as the widely distributed effects of thallium intoxication.[5] In contrast, a number of neurotoxicants are associated with neuronal loss that is focused within specific structures, such as the predilection of methyl mercury to damage neurons within the calcarine cortex and cerebellar cortex.[3,8] Whether neuronal loss is focal or generalized, secondary degeneration of the fiber pathways that originate from these neurons occurs.

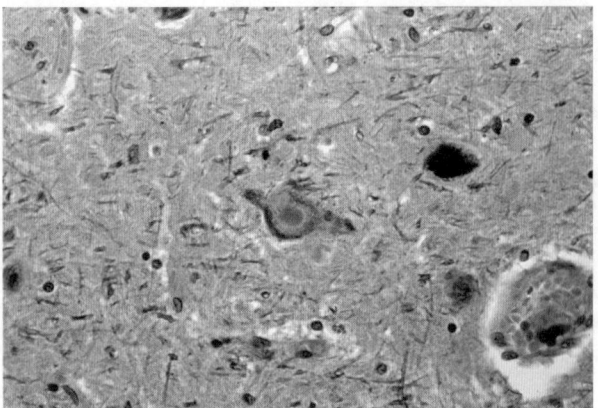

FIGURE 2 Lewy body in the substantia nigra. These inclusions characteristically appear as a smooth, eosinophilic, spherical body with surrounding pallor located in the cytoplasm of a pigmented neuron. Substantia nigra, H&E–LFB.

Gliosis is the most common glial response to injury, and includes changes both in astrocyte structure and in the number of astrocytes. Changes in astrocyte structure most often include cytoplasmic hypertrophy with development of a prominent eosinophilic cell body; these hypertrophic astrocytes are often called *gemistocytes* (Figure 3). Proliferation of astrocytes is evident on routine histological preparations by the increased number of astrocytic nuclei and a fibrillar background that is denser than normal neuropil. Reactive astrocytes are strongly immunoreactive for glial fibrillary acidic protein (GFAP), and in many instances, GFAP immunoreactivity is most prominent around blood vessels (Figure 4). When subtle, gliosis can be highlighted with immunohistochemistry for GFAP, with the network of glial

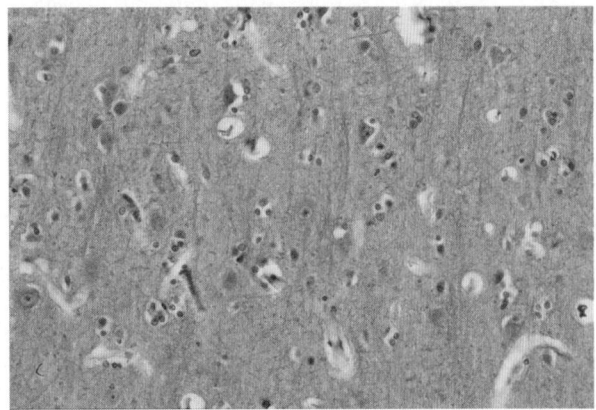

FIGURE 1 Cerebral edema (following acute arsenic toxicity) may be subtle histologically. The neuropil is pale, some myelinated axons are splayed, and clear spaces are evident in perineuronal and perivascular spaces. Cerebral cortex, H&E–Luxol fast blue (LFB).

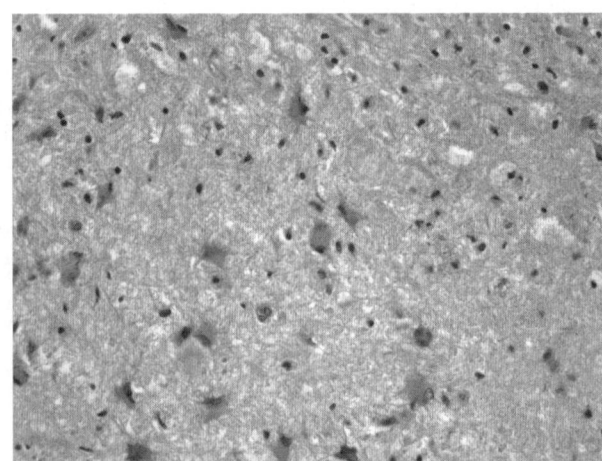

FIGURE 3 Reactive astrocytes are enlarged and have eosinophilic cytoplasm and short radiating processes. The term *gemistocyte* is sometimes used when there is abundant cytoplasm, as seen in several cells in this focus of reactive gliosis. Binucleation (large cell in center) is common. White matter, H&E–LFB.

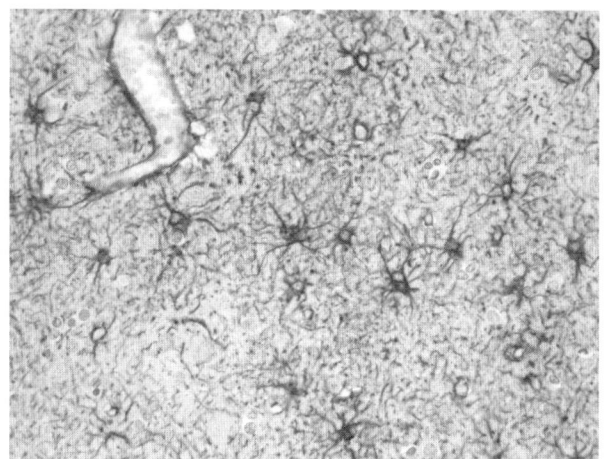

FIGURE 4 Immunohistochemistry (IHC) to reveal reactive astrocytes expressing glial fibrillary acidic protein (GFAP) is used frequently in diagnostic neuropathology. Reactive astrocytes appear as evenly spaced, star-shaped cells with long processes that radiate toward blood vessels. Dark perivascular staining is due to numerous astrocyte processes in this perivascular location. White matter, anti-GFAP immunohistochemistry (GF IHC).

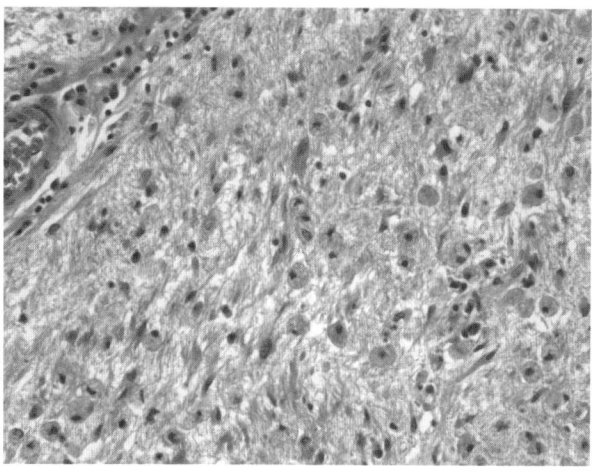

FIGURE 6 Sites of CNS injury may be highlighted by the presence of phagocytic monocyte-derived cells, sometimes referred to as *gitter cells*. Typical features include abundant, pale, foamy cytoplasm with small, eccentrically located nuclei and well-defined cytoplasmic edges. White matter, H&E–LFB.

processes staining intensely. With time, individual astrocytes may lose the gemistocytic appearance and have less abundant cytoplasm. Another glial response to injury is the formation of Alzheimer type II astrocytes, which are associated with encephalopathy from hyperammonemic states (Figure 5). The characteristic features of these cells are clearing of the nuclear center and a peripheral rim of chromatin at the nuclear boundary.

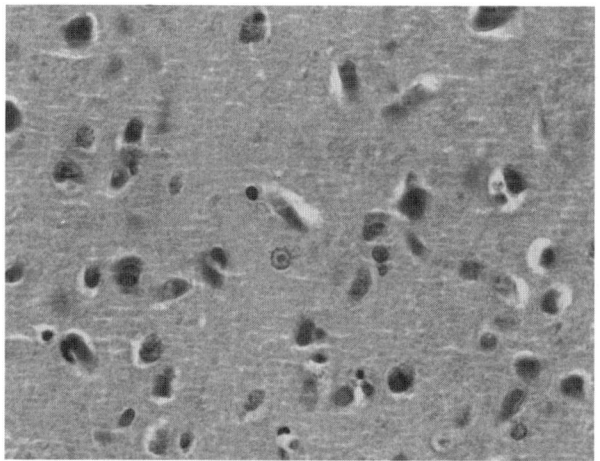

FIGURE 5 Alzheimer type II astrocytes are common in the encephalopathy associated with hepatic failure and may be recognized by their widely varying shapes with clearing of the nuclear center and a peripheral rim of chromatin at the nuclear boundary (round cell at center). Periodic acid-Schiff (PAS) stain, which demonstrates glycogen, highlights the central clear zone of the nucleus. Cerebral cortex, periodic acid-Schiff (PAS).

Microglia are inconspicuous elements of the normal CNS, but can assume prominence following tissue damage (Figure 6). Widespread proliferation of microglia has been reported following neurotoxicant exposure.[9] Following neurotoxicant exposure, myelin pallor is more commonly due to a combination of edema and gliosis. However, examples of demyelination exist, such as glue-sniffer encephalopathy.

Axonal degeneration is a common finding in neurotoxicity, and is especially well documented in toxicant-induced peripheral neuropathies. In the CNS, axonal degeneration accompanies any loss of neurons, and this can be documented with axonal stains such as Glees axonal stain, immunohistochemistry for neurofilament proteins, or protein gene product (PGP) 9.5.[10,11] Because axonal degeneration necessarily accompanies all neuronal loss, use of the term is often intended to imply axonal degeneration in the absence of neuronal loss, or "dying-back" axonal loss, with the neuronal cell body remaining viable. In the peripheral nervous system (PNS), an ideal way to view axons is teased fiber preparations (Figure 7), which allow observation of axons along several internodes with linear collections of phagocytic cells that are digesting myelin and axonal debris, indicating axonal degeneration (Figure 8). Some neurotoxicant-induced diseases are characterized by giant axonal swellings, often in proximity to axonal degeneration. These large eosinophilic collections within axons are composed mostly of massive accumulations of neurofilaments.[12] Similar neurofilament-filled axonal swellings are also seen in a rare familial neuropathy termed *giant axonal neuropathy*.[13] This is distinct from the morphologically similar axonal spheroids containing tubulovesicular material that have been observed

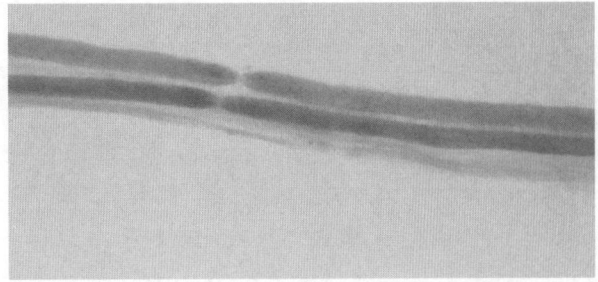

FIGURE 7 Normal isolated nerve fibers typically show regular axonal outlines with gaps at regular intervals representing nodes of Ranvier. The two parallel fibers in this image have nodes located in close proximity. Peripheral nerve biopsy, teased-fiber preparation.

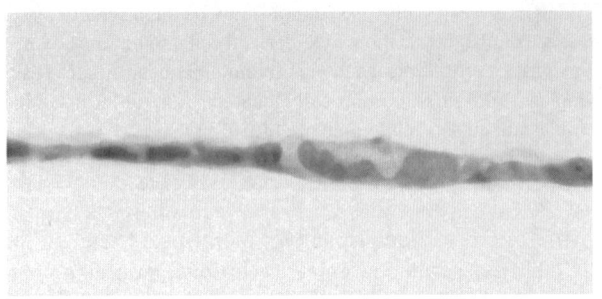

FIGURE 8 Axonal degeneration in an isolated nerve fiber is characterized by loss of the regular axonal contour with a chain of intermingled osmophilic (debris-laden) and clear vacuoles. Peripheral nerve biopsy, teased-fiber preparation.

in characteristic locations in older persons as well as in patients with neuroaxonal dystrophies.[14]

Demyelination is another common CNS response to injury. It may occur within discrete areas of white matter; more often it represents a diffuse loss of myelin within the CNS. Demyelination may be accompanied by intramyelinic edema, with vacuole formation within myelin layers. This type of myelin injury follows hexachlorophene or triethyltin intoxication[15] and ultimately is best visualized by myelin stains. Within the PNS, segmental demyelination may be demonstrated on tease fibers as a segment of the axon with reduced myelination; this is a common pathological lesion in the PNS following toxicant exposure. Abnormally thin myelin sheaths, shortening of internodal distances, and variation of myelin thickness among internodal segments of the same axon are observed during the remyelination that follows demyelination. Recurrent episodes of demyelination with remyelination may lead to Schwann cell hyperplasia.

BASIC PATTERNS AND MECHANISMS OF INJURY

The mechanisms of injury due to neurotoxicants take on several distinct forms. Although presented as separate topics,

these processes are not distinct and commonly conspire in producing damage to the nervous system. However, there are patterns of injury that implicate particular toxic exposures. Experimental studies have suggested that there are sets of cells that are particularly vulnerable to specific injuries, presumably due to their distinct molecular profiles and/or region-specific divergence in metabolic and physiological traits. When that type of injury occurs, a specific cellular response can be anticipated, sometimes referred to as the "topography" of injury associated with that type of molecular event. At one time, the term *pathoclisis*[3] was used to convey the idea that the unit of cells that responded to a particular injury was a specific substrate and that it was this "unit" of cells that was injured. Current understanding is that some cells have greater vulnerability to particular types of molecular stress, and the topography of pathological changes occurs in a predictable pattern within the subset of cells most susceptible to a specific stressor.[4,16]

ISCHEMIC OR HYPOXIC PATTERNS

The CNS is especially vulnerable to inadequate perfusion and oxygen delivery because of its high energy requirements and unique metabolic needs.[17] It is therefore not surprising that several toxicants that target aerobic respiration have their primary functional manifestation in the CNS.[16,17] For global ischemia-related injuries, there is a tendency for hippocampal CA1 region pyramidal neurons to be affected severely, and with cardiac arrest of limited duration in humans, it is this set of cells that shows the greatest injury (Figures 9 and 10). Next in vulnerability to ischemia are Purkinje cells of cerebellar cortex, followed by layers III and V of cerebral cortex.[16,17] The relative vulnerabilities of these neuronal populations to ischemic events are evident in nontoxic situations as well, and finding this pattern of injury strongly implicates an ischemic mechanism of action. However, it is important to realize that the manifestations are protean and can be much more subtle than classic cardiovascular arrest. When ischemia/hypoxia is generalized, the pattern of relative ischemic vulnerability is accentuated, and when localized the pattern is only within the region affected.

One example of a commonly abused xenobiotic that increases the risk of localized ischemic injury (stroke) is cocaine, which has been shown to cause cerebral perfusion defects and cerebral atrophy, among other changes. Most likely to be related to cocaine-induced vasospasm, autopsies of relatively young persons who used cocaine can show changes of typical localized ischemic injury, including infarcts and hemorrhages.[18,19]

Carbon monoxide (CO), through its action of displacing oxygen from hemoglobin, creates a relatively pure generalized hypoxic/ischemic injury to the brain.[20,21] The brain often is swollen, and blood maintains a cherry-red

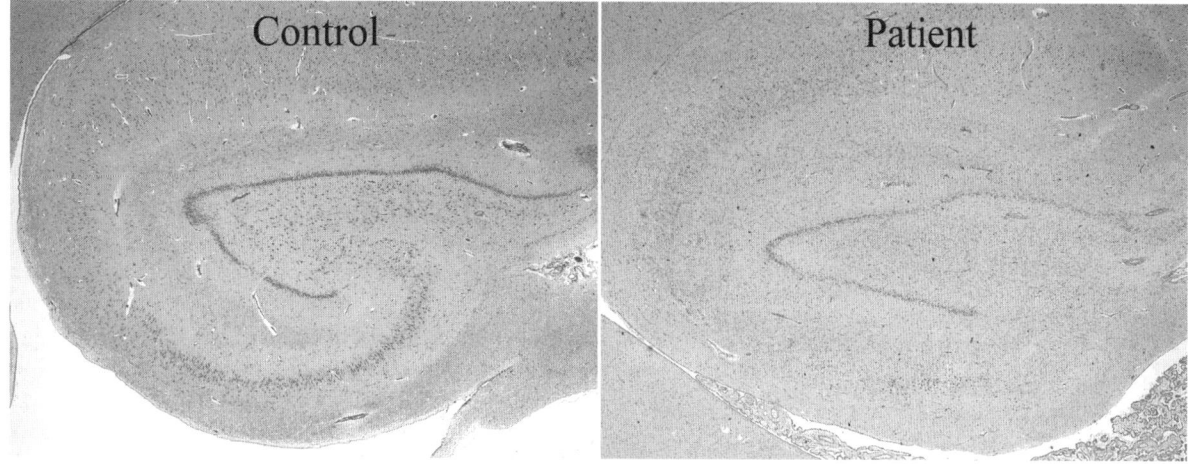

FIGURE 9 Extensive neuronal loss within the CA1 region of the hippocampus in an affected patient (right) compared to the same region of an unaffected person (left). The normal dense band of pyramidal neurons is affected severely. This pattern implies a hypoxic/ischemic etiology due to the selective vulnerability of CA1 pyramidal neurons to this type of injury. Hippocampus, cresyl violet.

appearance even after death. Neuronal necrosis in the hippocampus, Purkinje cells of cerebellum, and layers III and V of the neocortex are often involved.[16,17] In addition to the ischemic topography of neuronal injury, CO may also produce localized areas of infarction, including the globus pallidus of the basal ganglia (Figure 11).[20] However, these are infrequent in CO exposure and are also seen infrequently in patients dying of drowning or strangulation, suggesting that they are associated with hypoxic (rather than ischemic) events. CO exposure may also result in a delayed onset white matter necrosis.[16]

Cyanide exposure, which inhibits mitochondrial cytochrome oxidase, thereby blocking aerobic oxidation, can yield changes of global ischemic damage. The brain may be swollen with cerebral edema, and there is often injury in the topography of ischemic injury, including bilateral necrosis of the globus pallidus and striatum, laminar necrosis of cerebral cortex, and loss of cerebellar Purkinje cells.[16,22]

MITOCHONDRIAL DYSFUNCTION

Neuronal mitochondrial dysfunction leading to neuronal damage and death has been established clearly by toxicants, perhaps the best studied being MPTP. Interruption of electron transport, impairment of the proton gradient, or inhibition of complex V all result in reduced ATP production, release of pro-apoptotic factors such as cytochrome *c*, and cell death.

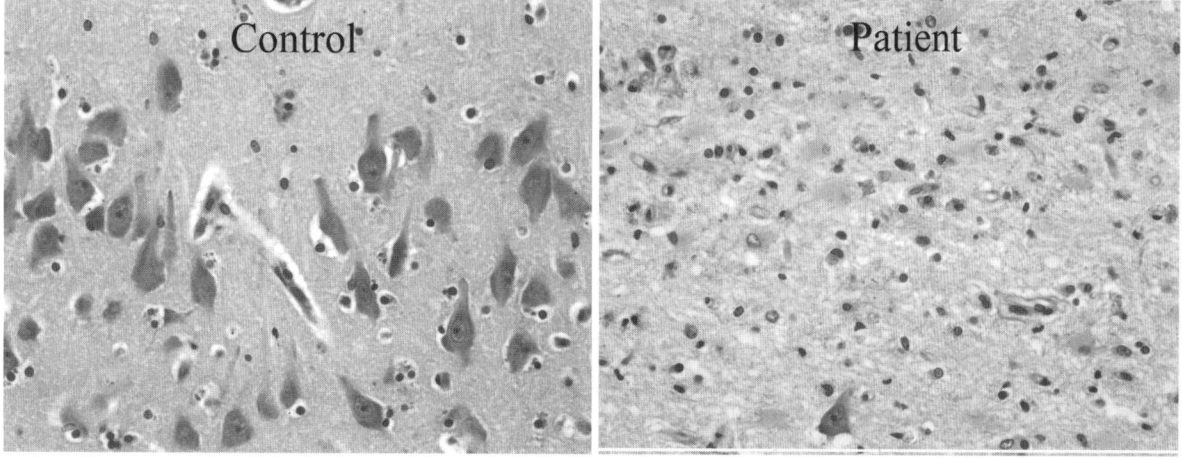

FIGURE 10 Neuronal loss within the CA1 domain of the hippocampus (same case as Figure 9, but at higher magnification). In contrast to the normal population of pyramidal cells (left), the few pyramidal neurons remaining in the patient (right) have been replaced by reactive astrocytes. Hippocampus CA1, cresyl violet.

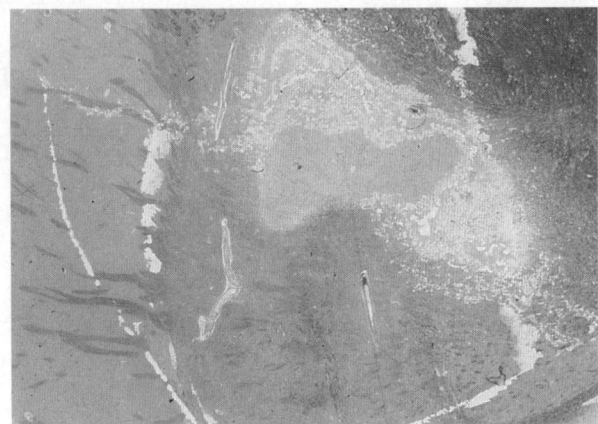

FIGURE 11   Discrete region of necrosis in the basal ganglia may be seen in carbon monoxide (CO) poisoning, as in this fatal case. The zone of necrosis is visible at the boundary of the globus pallidus with the internal capsule (above). Both internal and external segments of the globus pallidus are affected, but the putamen (left) is relatively spared. Basal ganglia, H&E–LFB.

One mechanism of neuron death specific to the CNS following mitochondrial dysfunction is called *indirect excitotoxicity*.[23]

The history of MPTP-induced Parkinsonism in young adults who inadvertently injected themselves with this compound has been well described.[1] MPTP is a protoxicant that after crossing the blood–brain barrier is metabolized by glial monoamine oxidase (MAO)-B to a pyridinium intermediate that undergoes further two-electron oxidation to yield the toxic metabolite 1-methyl-4-phenylpyridinium (MPP$^+$). MPP$^+$ is then selectively transported into nigral neurons via the mesencephalic dopamine transporter (DAT). Once inside these neurons, MPP$^+$ is thought to act primarily as a mitochondrial toxin by inhibiting complex I activity in the mitochondrial electron transport chain, thereby reducing ATP production and increasing the generation of reactive oxygen species (ROS).[24,25] Indeed, MPTP-induced dopaminergic neurodegeneration can be diminished by free-radical scavengers, by inhibitors of the inducible form of nitric oxide synthase (iNOS), and by excitatory amino acid receptor (EAAR) antagonists. In addition, transgenic mice lacking some elements of antioxidant defenses are significantly more vulnerable to MPTP-induced dopaminergic neurodegeneration. So far, the search for xenobiotics that may act similarly to MPTP and could be potential environmental toxicants that promote Parkinson's disease has not yielded clear candidates.

Hydrogen sulfide (H$_2$S), another mitochondrial poison, is a colorless gas at room temperature with a distinctive unpleasant odor reminiscent of rotten eggs. It is highly toxic and acts as a mitochondrial poison by inhibiting oxidative phosphorylation.[26] Humans acutely exposed to H$_2$S develop fatigue, vertigo, and lose consciousness, and although it has broad spectrum toxicity, the effects on the nervous system are prominent. Interestingly, an early effect is on the olfactory system, with loss of the sense of smell so that subjects lose the ability to detect their own exposure.

GENERALIZED NEURONAL INJURY PATTERN

A number of neurotoxic compounds have a generalized pattern of involvement of the CNS, creating an overall loss of neurons diffusely throughout the CNS. Following acute high-level exposures, diffuse neuronal injury often leads to cerebral edema, with swelling of the brain and herniation. Chronic low-level exposures with a generalized pattern of neuronal injury often cause changes in level of consciousness or cognitive abilities, with widespread involvement of neurons, associated with microcephaly in children or cerebral atrophy in adults.

Lead exposure has broad neurotoxic effects, creating a generalized pattern of neuronal injury.[16,27] Acute exposures (most common in children) lead to a rapid-onset encephalopathy. The brain shows edema, sometimes with herniation, necrosis of scattered neurons, and there are often microscopic foci of hemorrhage. The principal injury appears to be the impact of lead toxicity on endothelial cells. Adults are much less likely to develop acute encephalopathy and, instead, tend to develop a peripheral neuropathy with chronic lead exposure.

Arsenic exposure is similarly associated with diffuse neuronal injury.[3,16] Arsenic reacts with sulfhydryl groups and has toxicity beyond the nervous system. However, in the CNS, acute high-level exposure is associated with an acute-onset encephalopathy. There is often cerebral edema accompanied by punctate foci of hemorrhage. As with lead, chronic exposure to arsenic is more often associated with peripheral neuropathy.[28]

LOCALIZED NEURONAL INJURY PATTERN

There are additional distinct topographies of neurotoxicity beyond the hypoxic–ischemic pattern. The injury may affect neurons of a certain size, neurons in particular locations or systems, or subsets of cells within systems. It is insufficient to know that a neurotoxicant has caused "neuronal loss," but rather, information about which locations and which subsets of neurons are affected is required. It is the topography of injury that provides the most useful information in determining whether a particular neurotoxicant is the cause of the injury.

Neurological injury is commonly associated with excessive ethanol consumption. There continues to be debate over what proportion of the neurological injury is due to direct toxic effects of ethanol, and what proportion is due to

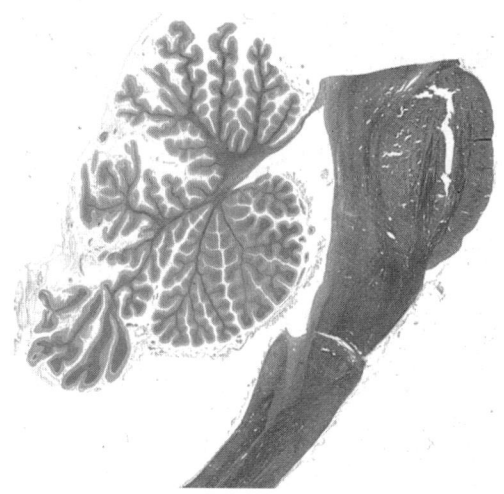

FIGURE 12 Involvement of the anterior vermis (above) relative to the posterior vermis (below) is evident in this case of ethanol-related cerebellar atrophy by increased spacing between both primary and secondary folia branches. Cerebellar atrophy involving predominantly anterior vermis is common with chronic ethanol exposure. Cerebellum and brain stem, H&E–LFB.

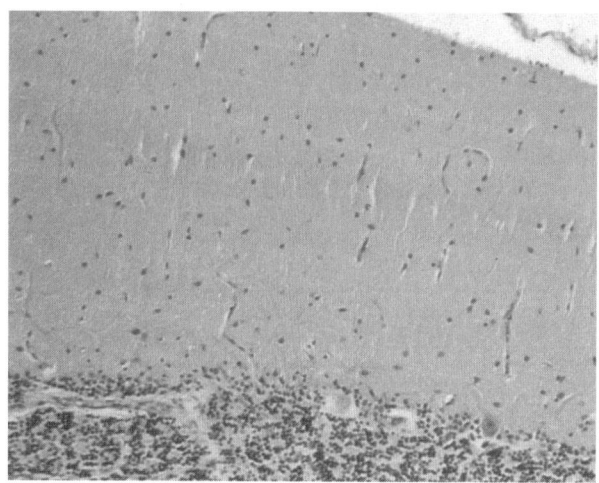

FIGURE 13 Chronic ethanol-induced neurotoxicity in the cerebellum occurs as a loss of Purkinje cells with proliferation of Bergmann astrocytes (cells with small, round nuclei at the interface between the granular cell layer and molecular layer) (left side). A single residual Purkinje cell is seen on the right. This pattern of Purkinje cell loss in cerebellar atrophy is typical of anterior vermis lesions induced by ethanol. Cerebellar cortex, H&E.

the nutritional deficiencies that often accompany chronic alcoholism. For example, it is clear that hemorrhagic necrosis within mammillary bodies is due to thiamine deficiency rather than to direct toxicity of ethanol. Nonetheless, the association between localized neuronal injuries and ethanol exposure is sufficient to warrant classifying them together.

The neuropathological findings associated with chronic ethanol use include a generalized decrease in white matter and atrophy of the anterior vermis of the cerebellum that manifests as increased space between folia (Figure 12).[16,29] Decreased cerebral weight (atrophy) has also been known for some time, but only more recently have quantitative approaches shown that the principal volume loss is within white matter.[4,30] Microscopically, localized cerebellar atrophy is associated with diminished number of Purkinje cells and proliferation of Bergmann astrocytes (Figure 13).

In contrast, cerebellar atrophy associated with mercury toxicity is characterized by loss of cerebellar cortical granule cells[3,8,31] (Figures 14 and 15) with relative sparing of Purkinje cells. Clinically, mercury toxicity often is characterized by delayed onset of tremor, confusion, and a cerebellar syndrome characterized by ataxia. In addition, there may be involvement of the entire cerebral cortex, leading to diffuse cerebral atrophy with accentuation in the calcarine cortex.

Methanol neurotoxicity has a predilection for the visual system. The toxicity of methanol requires metabolic activation, with generation of toxic metabolites, including formates and formic acid.[32] Symptoms typically include visual disturbances and may cause blindness. There is edema of the optic disk with involvement of the optic nerve, which may have areas of necrosis. There may also be foci of necrosis in the basal ganglia, most characteristically including the putamen.[33]

Manganese neurotoxicity has a predilection for involvement of the extrapyramidal motor system.[34] Symptoms of chronic manganese toxicity include Parkinsonism and dystonia. Neuronal loss most often involves the multiple nuclei in the basal ganglia as well as cerebral and cerebellar cortex.

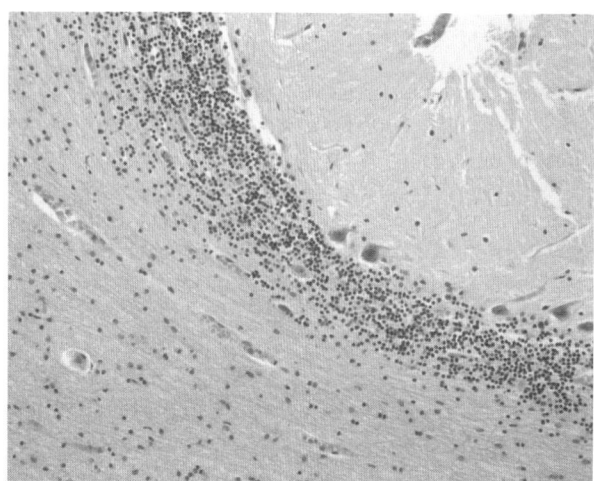

FIGURE 14 Loss of granular cells in the cerebellum (induced by mercury exposure) is characterized by marked depletion of cells in the granule cell layer in the presence of all or most Purkinje cells. Cerebellar cortex, H&E.

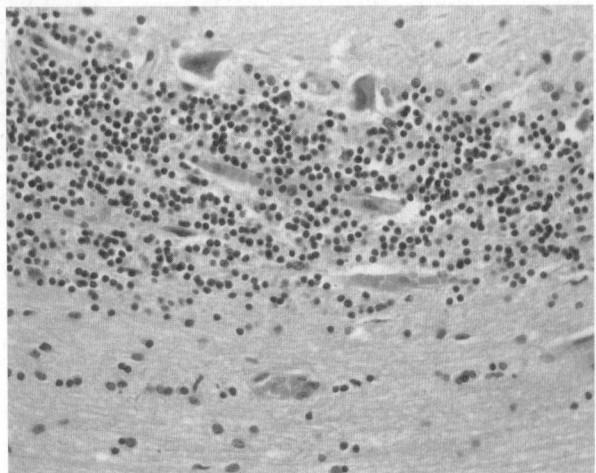

FIGURE 15 Cerebellar granule cell loss in mercury neurotoxicity, observed at higher magnification, Note the residual population of Purkinje cells. Cerebellar cortex, H&E.

Extrapyramidal involvement is also characteristic of Parkinson's disease and related neurological disorders. Catechols such as dopamine and norepinephrine are vulnerable to oxidation, rearrangement, and condensation under physiological conditions. Within synaptic vesicles, high concentrations of antioxidants and metal ion chelators stabilize catechols. However, conditions in extracellular fluid and cytosol are more favorable for chemical transformation of the catechol nucleus. Several catechol products have been proposed to participate in dopaminergic neurodegeneration; these include dopamine quinine, 6-hydroxydopamine, and isoquinolines.

TRANSMISSION-RELATED AND EXCITOTOXIC PATTERN OF NEURONAL INJURY

Excitatory amino acids (EAAs), primarily L-glutamate, and their cell surface ligand-activated ion channels, such as α-amino-3-hydroxyl-5-methyl-4-isoxazole-propionate (AMPA) and N-methyl-D-aspartic acid (NMDA) receptors (named for the selective agonists AMPA and NMDA that activate them), participate in a wide array of neurological functions. The glutamatergic presynaptic terminal releases glutamate upon depolarization that then binds to postsynaptic glutamate receptors such as the AMPA receptor to initiate influx of Na^+ into the postsynaptic element. Glutamate also binds to the NMDA receptor, but unless the postsynaptic membrane is sufficiently depolarized, the voltage-dependent influx of cations such as Ca^{2+} into the postsynaptic neuron is blocked by extracellular Mg^{2+}. Glutamate is removed from the synapse by a number of transporters on neurons and astrocytes, where it can enter the glutamine cycle ultimately to replenish neurotransmitter pools. Excessive

EAA receptor stimulation sets in motion a cascade of events that can contribute to neuronal injury and death through a process called *excitotoxicity*.[35,36]

The key initiator in direct excitotoxicity is increased synaptic concentration of EAAs either by release from the presynaptic terminal, decreased reuptake, or exposure to an excitatory neurotoxicant [such as seafood contaminated with domoic acid, produced in algal blooms, or ingestion of β-oxalyl-L-α,β-diaminopropionic acid (ODAP), the principle of certain legumes responsible for lathyrism]. Extensive activation of postsynaptic glutamate receptors sufficiently depolarizes the postsynaptic membrane to relieve the Mg^{2+} block of the NMDA receptor. Rapid injury and neuronal lysis can follow this depolarization-dependent increase in Ca^{2+} ion influx. Delayed toxicity can develop even after EAA has been removed secondary to activation of Ca^{2+}-dependent enzymes, mitochondrial damage, increased generation of free radicals, and transcription of pro-apoptotic genes.

In contrast to direct excitotoxicity, the key event in indirect excitotoxicity is impaired mitochondrial function with reduced ATP generation.[23] One consequence is reduced activity of Na^+, K^+-ATPase with partial depolarization of the postsynaptic membrane, which when combined with normal levels of glutamate release is sufficient to relieve the Mg^{2+} block of NMDA-mediated ion conductance. This again leads to increased intraneuronal Ca^{2+} with consequences similar to those described for direct excitotoxicity.

Several EAAR agonists are now widely used as tools in neuroscience. The cyclic glutamate analog kainate was initially isolated from seaweed in Japan as the active component of an herbal treatment for ascariasis. Kainate is extremely potent as an excitotoxin, being 100-fold more toxic than glutamate, and is selective at a molecular level for the kainate receptor.[37]

Proof of the deleterious effects of EAAR agonists in people comes from toxic exposures. Perhaps the most striking example is the domoic acid intoxications that occurred in the Maritime Provinces of Canada in late 1987.[38] A total of 107 patients were identified who suffered an acute illness that most commonly presented as gastrointestinal disturbance, severe headache, and short-term memory loss within 24 to 48 h after ingesting mussels. A subset of the more severely afflicted patients were subsequently shown to have chronic memory deficits, motor neuropathy, and decreased medial temporal lobe glucose metabolism by positron emission tomography (PET).[39] Neuropathological evaluation of patients who died within four months of intoxication disclosed neuronal loss with reactive gliosis that was most prominent in the hippocampus and amygdala, but also affected regions of the thalamus and cerebral cortex. The responsible agent was identified as domoic acid, a potent structural analog of L-glutamate that had been concentrated in cultivated mussels.

Botulism is caused by ingestion of food contaminated by *Clostridium botulinum*. Botulinum toxin is a protease[40] that acts presynaptically, preventing the fusion protein synapto-brevin from catalyzing the anchoring of neurotransmitter vesicles to the presynaptic cell membrane, thereby blocking release of acetylcholine from PNS axon terminals at neuro-muscular junctions. Toxicity is characterized by sudden onset of weakness,[41] often initially involving cranial nerves with weakness of facial and extraocular muscles. Diplopia (double vision), facial weakness, and difficulty talking are common initial symptoms, but when the muscles of respiration are affected, death may ensue.

IMMUNE OR INFLAMMATORY PATTERN OF NEURONAL INJURY

Activation of the innate immune response in the brain, primarily mediated by microglia, is a feature shared by several neurodegenerative diseases. Moreover, microglial-mediated phagocytosis of potentially neurotoxic protein aggregates (see below) might be an important means of neuroprotection. Indeed, whereas some aspects of innate immune activation are an appropriate response to the stressors presented by neurodegenerative diseases, other facets of glial activation, especially when protracted, may contribute to neuronal damage. This balance between beneficial and deleterious actions of immune activation is well appreciated by students of pathology in other organs and is currently an area of very active investigation among neuroscientists, with the goal of promoting neurotrophic or neuroprotective actions while suppressing paracrine damage to neurons. Key elements in paracrine neurotoxicity from activated glia appear to include IL-1β, TNF-α, and prostaglandin E$_2$, among others, with activation of cyclooxygenase 2 (Cox2), iNOS, and NADPH oxidase that interface with excitotoxicity and free-radical stress as mechanisms in this process.

FREE-RADICAL STRESS PATTERN OF NEURONAL INJURY

Free radicals have an unpaired electron in their outer orbital and are normal components of second messenger signaling pathways; however, excess or uncontrolled free-radical production is detrimental to cells either by direct damage to macromolecules or liberation of toxic by-products. A number of sites for free-radical generation exist in the nervous system. A major source appears to be oxidative phosphorylation in mitochondria. Other sources of free radicals may become significant during pathological states. These include excitotoxicity, myeloperoxidase, NADPH oxidase, MAO, autooxidation of catechols such as dopamine, and amyloid β-peptides.

Research in biological systems has focused on oxygen-, nitrogen-, and carbon-centered free radicals, with oxygen-centered free radicals being the most extensively studied. During the sequential four-electron reduction of molecular oxygen to water, two free radicals, superoxide anion ($O_2^{\bullet-}$) and hydroxyl radical ($^{\bullet}OH$), are produced along with hydrogen peroxide (H_2O_2). Since H_2O_2 is not a radical, these three molecules are collectively referred to as reactive oxygen species (ROS). Both $O_2^{\bullet-}$ and H_2O_2 are signaling molecules under normal physiological conditions. In contrast, hydroxyl radical is the most reactive and is capable of oxidizing lipids, carbohydrates, proteins, and nucleic acids at a rate limited by its own diffusion. Hydroxyl radical is formed from $O_2^{\bullet-}$ and H_2O_2 by the Haber–Weiss reaction or from H_2O_2 by the Fenton reaction. Several metal ions may participate in the Fenton reaction, including Fe(II), Cu(I), Mn(II), Cr(V), and Ni(II).

Nitric oxide ($^{\bullet}NO$), a free-radical product of NOS-catalyzed oxidation of L-arginine to citrulline, initiates a cascade of reactive nitrogen species (RNS). NO has several physiological functions, including vasodilator and neuromodulator. NO reacts with O_2 to produce peroxynitrite ($ONOO^-$), a potent oxidant that can modify cellular macromolecules. In turn, reaction of ONOO with ubiquitous CO_2 forms nitrosoperoxy carbonate, which cleaves spontaneously to form a nitrating agent, nitrogen dioxide radical, and an oxidant, carbonate anion radical. Finally, ONOO may decompose to nitrous oxide and $^{\bullet}OH$ upon protonation.

Carbon-centered free radicals are generated during the metabolism of some toxicants, such as carbon tetrachloride, and also are produced on fatty acyl chains during *lipid peroxidation*, a term that describes free radical–mediated damage to polyunsaturated fatty acids. Lipid peroxidation differs from other forms of free-radical damage to macromolecules because it is a self-propagating process that generates neurotoxic by-products. Lipid peroxidation begins with abstraction of a hydrogen atom from a fatty acyl chain (LH), typically a polyunsaturated fatty acid (PUFA), because the carbon–hydrogen bond is made more acidic by the adjacent carbon–carbon double bond. This leaves a lipid radical (L·) that rearranges to a conjugated diene that can accept molecular oxygen to form a peroxyl radical (LOO$^{\bullet}$), which in turn propagates the reactions by abstracting a hydrogen atom to generate a lipid hydroperoxide (LOOH). Lipid hydroperoxides are reactive molecules that can participate in many reactions, including the Fenton reaction, to generate fragmentation products of many biomolecules. Thus, there are two deleterious outcomes to lipid peroxidation: structural damage to membranes and generation of bioactive secondary products. Membrane damage derives from the generation of fragmented fatty acyl chains, lipid–lipid cross-links, and lipid–protein cross-links. In addition, lipid peroxyl radicals can undergo endocyclization to produce novel fatty acid esters that may disrupt membranes. Fragmentation of lipid hydroperoxides, in addition to producing abnormal fatty acid esters, liberates a number of diffusible products, some of which are potent electrophiles; the most abundant are chemically reactive aldehydes such as acrolein[42] and 4-hydroxy-2-nonenal.[43]

AXONOPATHY PATTERN OF NEURONAL INJURY

Distal sensorimotor polyneuropathy is probably the most common clinical manifestation of neurotoxicant exposure in humans. A variety of toxicants, including *n*-hexane, methyl *n*-butylketone (2-hexanone), carbon disulfide (CS_2), acrylamide, and organophosphorus esters, result in degeneration of the distal portion of the longest, largest myelinated axons in the PNS and CNS, an observation encapsulated in the diagnostic term *central-peripheral distal axonopathy.*

Two of these toxicants, *n*-hexane and CS_2, are especially interesting because despite their very different chemical structures, chronic exposure to either can produce unusual pathological nerve changes.[12] The characteristic lesion produced by both *n*-hexane and CS_2 is axonal swellings (Figure 16). In the longitudinal plane, these swellings are multifocal fusiform enlargements on the proximal side of nodes of Ranvier (Figure 17) in distal but preterminal nerve segments. The swellings consist of massive accumulations of disorganized 10-nm-diameter neurofilaments, decreased number of microtubules, thin myelin, and segregation of axoplasmic organelles and cytoskeletal components. Distal to swellings, axons may become shrunken and then degenerate (Figure 18). With continued exposure, more proximal swellings occur with subsequent axonal degeneration.

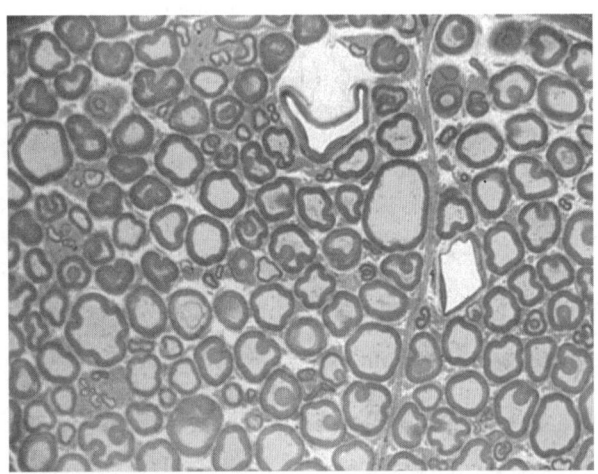

FIGURE 16 Axonal swelling in a neurofilamentous axonopathy induced by 2,5-hexanedione (2,5-HD). The distended contours of many axons are filled with pale, homogeneous axoplasm, while myelin thickness appears reduced relative to the enlarged axonal caliber. By electron microscopy, swollen axons are filled with 10-nm-diameter neurofilaments. Peripheral nerve biopsy, toluidine blue.

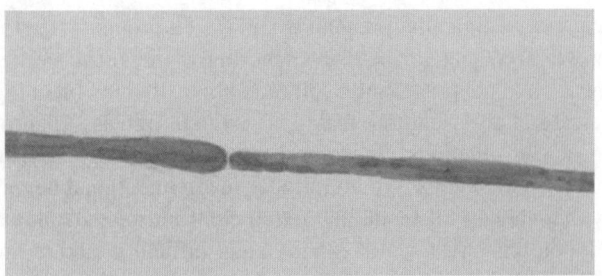

FIGURE 17 Axonal swellings in neurofilamentous neuropathies occur on the proximal side of nodes of Ranvier (on the left in this image), giving the two sides of the node an asymmetrical caliber. Peripheral nerve biopsy, teased-fiber preparation.

Investigations of *n*-hexane and CS_2 have determined that the key to their shared clinical and pathological profiles appears to be the ability of each compound to generate protein-bound electrophilic species that can covalently cross-link proteins.

Hexane neuropathy, presenting as a distal sensorimotor peripheral neuropathy with swollen axons, was first identified in workers who used *n*-hexane as a solvent. A similar syndrome with comparable distal sensorimotor lesions was subsequently observed in workers exposed to 2-hexanone (methyl *n*-butyl ketone) who developed a similar distal sensorimotor peripheral neuropathy with swollen axons. The common toxic metabolite of both solvents (2,5-hexanedione; 2,5-HD) clarified the relationship between the toxicity of these two solvents, with the neurotoxicity of both compounds deriving from 2,5-HD, which reacts with free amino groups to form an aromatic ring (pyrrole) with a subsequent cross-linking step. Both reactions appear to be key determinants of neurotoxicity, based on structure–activity relationships.

CS_2 has also been used as a solvent, and its neurotoxicity shows a similar distal sensorimotor axonal neuropathy with formation of enlarged neurofilament-filled axonal swellings. Covalent binding of CS_2 occurs by an initial formation of a dithiocarbamate that is capable of further reaction and protein cross-linking.[44]

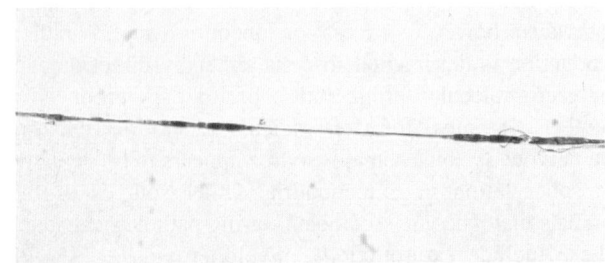

FIGURE 18 Axonal degeneration in neurofilamentous neuropathies occurs distal to axonal swellings, with disintegration of the axon into a chain of osmophilic ovoids. Peripheral nerve biopsy, teased-fiber preparation.

In addition to acute cholinergic toxicity, organophosphates (OPs) have been shown to produce delayed neurotoxicity. This syndrome was first identified in patients exposed to pesticides or who consumed a Prohibition Era medicinal product ("Ginger Jake") adulterated with tri-*o*-cresyl phosphate (TOCP). Coined *organophosphate-induced delayed neurotoxicity* (OPIDN), the potency of a series of related OP congeners identified a structure–activity relationship between delayed toxicity and inhibition of an esterase activity, subsequently termed *neurotoxic esterase*. The pathological effect is a relatively pure axonal degeneration involving both peripheral and central long-tract axons, such as those of the spinal cord.[45]

Vincristine neurotoxicity is related predominantly to use of this agent as a chemotherapeutic agent in the treatment of certain cancers. Development of this peripheral polyneuropathy is a dose-limiting side effect of vincristine[46] and develops in a substantial proportion of patients. The major pathological change is an axonal neuropathy without massive accumulation of axoplasmic contents, with axonal degeneration affecting the longest myelinated axons in the PNS. Axonal regeneration and remyelination can occur once vincristine exposure is halted.

SUMMARY

Patterns of injury commonly identified in human neuropathological assessment suggest mechanisms of injury to the nervous system and provide diagnostic clues that can be useful in identifying neurotoxic exposures and characterizing the nature and prognosis of neurotoxic conditions. Processes especially vulnerable to neurotoxicants highlight those aspects of the nervous system that distinguish it from other organs: high-energy need with unique metabolic demands, excitatory neurotransmission, relative immune sanctuary, high content of polyunsaturated fatty acids as ready-made sources of toxic metabolites, and cellular transport across long cellular processes. In medical practice, the pattern of injury is usually the initial clue to etiology. Toxic agents are usually uncovered clinically either due to known exposure with subsequent development of specific clinical syndromes, or due to identification of a particular pattern of injury that implicates a specific neurotoxicant, with subsequent documentation of exposure. Continued discovery of neurotoxicants and their mechanisms of action will further illuminate processes critical to nervous system physiology and pathophysiology.

REFERENCES

1. DiMonte DA, Langston JW. MPTP and analogs. In: Spencer PS, Schaumburg HH, eds. *Experimental and Clinical Neurotoxicology*. 2nd ed. New York: Oxford University Press; 2000: 812–818.

2. Kandell ER, et al. *Principles of Neural Science*. 3rd ed. East Norwalk, CT: Appleton & Lange; 1991.

3. Ropper AH, Brown RH. Disorders of the nervous system due to drugs, toxins, and other chemical agents. In: *Adams and Victor's Principles of Neurology*. 8th ed. New York: McGraw-Hill; 2005:1016–1045.

4. Harris J, et al. Nutritional deficiencies, metabolic disorders and toxins affecting the nervous system. In: Love S, et al, eds. *Greenfield's Neuropathology*. 8th ed. London: Hodder Arnold; 2008:675–732.

5. Davis LE, et al. Acute thallium poisoning: toxicological and morphological studies of the nervous system. *Ann Neurol.* 1981;10:38–44.

6. Toscano CD, Guilarte TR. Lead neurotoxicity: from exposure to molecular effects. *Brain Res.* 2005;49:529–554.

7. Silberberg EK. (1992). Mechanisms of lead neurotoxicity, or looking beyond the lamppost. *FASEB J.* 1992;6:3201–3206.

8. Clarkson TW, Magos L. The toxicology of mercury and its chemical compounds. *Crit Rev Toxicol.* 2006;36:609–662.

9. Gudi V, et al Regional differences between grey and white matter in cuprizone induced demyelination. *Brain Res.* 2009;1283:127–138.

10. Trapp BD, et al. Axonal transaction in the lesions of multiple sclerosis. *N Eng J Med.* 1998;338:278–285.

11. Chiang M-C, et al. Cutaneous innervation in chronic inflammatory demyelinating polyneuropathy. *Neurology.* 2002;59: 1094–1098.

12. Graham DG, et al. Pathogenetic studies of hexane and carbon disulfide neurotoxicity. *Crit Rev Toxicol.* 1995;25:91–112.

13. Bomont P, et al. (2000) The gene encoding gigaxonin, a new member of the cytoskeletal BTB/kelch repeat family, is mutated in giant axonal neuropathy. *Nat Genet.* 2000;26:370–374.

14. Malik I, et al. Disrupted membrane homeostasis and accumulation of ubiquitinated proteins in a mouse model of infantile neuroaxonal dystrophy caused by PLA2G6 mutations. *Am J Pathol.* 2008;172:406–416.

15. Spencer PS. Biological principles of chemical neurotoxicity. In: Spencer PS, Schaumburg HH, eds. *Experimental and Clinical Neurotoxicology*. 2nd ed. New York: Oxford University Press; 2000:3–54.

16. Schochet SS Jr, Gray F. Acquired metabolic disorders. In: Gray F, et al., eds. *Escourelle and Poirier Manual of Basic Neuropathology*. 4th ed. Philadelphia: Butterworth Heinemann; 2004:197–217.

17. Auer RN, et al. Hypoxia and related conditions. In: Love S, et al, eds. *Greenfield's Neuropathology*. 8th ed. London: Hodder Arnold; 2008:63–119.

18. Wang AM, et al. (1990). Cocaine- and methamphetamine-induced acute cerebral vasospasm: an angiographic study in rabbits. *Am J Neuroradiol.* 1990;11:1141–1146.

19. Conway JE, Tamargo RJ. Cocaine use is an independent risk factor for cerebral vasospasm after aneurysmal subarachnoid hemorrhage. *Stroke.* 2001;32:2338–2343.

20. Ropper AH, Brown RH. The acquired metabolic disorders of the nervous system. In: *Adams and Victor's Principles of Neurology*. 8th ed. New York: McGraw-Hill; 2005:959–982.

21. Prockop LD, Chichkova RI. Carbon monoxide intoxication: an updated review. *J Neurol Sci*. 2007;262:122–130.

22. Salkowski AA, Penney DG. Cyanide poisoning in animals and humans: a review. *Vet Hum Toxicol*. 1994;36:455–466.

23. Jacquard C, et al. Brain mitochondrial defects amplify intracellular [Ca2 +] rise and neurodegeneration but not Ca^{2+} entry during NMDA receptor activation. *FASEB J*. 2006;20: 1021–1023.

24. Miller RL, et al. Oxidative and inflammatory pathways in Parkinson's disease. *Neurochem Res*. 2009;34:55–65.

25. Yokoyama H, et al. Targeting reactive oxygen species, reactive nitrogen species, and inflammation in MPTP neurotoxicity and Parkinson's disease. *Neurol Sci*. 2008;29:293–301.

26. Reiffenstein RJ, et al. Toxicology of hydrogen sulfide. *Annu Rev Pharmacol Toxicol*. 1992;1992:109–134.

27. White LD, et al. New and evolving concepts in the neurotoxicology of lead. *Toxicol Appl Pharmacol*. 2008;225:1–27.

28. Vallat J-M, et al. Peripheral nerve diseases. In: Gray F, ed. *Escourelle and Poirier Manual of Basic Neuropathology*. 4th ed. Philadelphia: Butterworth Heinemann; 2004:315–343.

29. Ropper AH, Brown RH. Diseases of the nervous system due to nutritional deficiency. In: *Adams and Victor's Principles of Neurology*. 8th ed. New York: McGraw-Hill; 2005: 983–1003.

30. De la Monte SM. Disproportionate atrophy of cerebral white matter in chronic alcoholics. *Arch Neurol*. 1988;45:990–992.

31. Chang LW, Verity MA. Mercury neurotoxicity: effects and mechanisms. In: Chang LW, Dyer RS, eds. *Handbook of Neurotoxicology*. New York: Marcel Dekker; 1995:31–59.

32. Medinsky MA, Dorman DC. Recent developments in methanol toxicity. *Toxicol Lett*. 1995;82–83:707–711.

33. Feany MB, et al. Two cases with necrosis and hemorrhage in the putamen and white matter. *Brain Pathol*. 2000;11:121–122.

34. Milatovic D, et al. Oxidative damage and neurodegeneration in manganese-induced neurotoxicity. *Toxicol Appl Pharmacol*. 2009;240:219–225.

35. Milanese M, et al. Glutamate release from astrocytic gliosomes under physiological and pathological conditions. *Int Rev Neurobiol*. 2009;85:295–318.

36. Forder JP, Tymianski M. Postsynaptic mechanisms of excitotoxicity: involvement of postsynaptic density proteins, radicals, and oxidant molecules. *Neuroscience*. 2009;158:293–300.

37. Wang Q, et al. Kainic acid-mediated excitotoxicity as a model for neurodegeneration. *Mol Neurobiol*. 2005;31:3–16.

38. Pulido OM. Domoic acid toxicologic pathology: a review. *Mar Drugs*. 2008;6:180–219.

39. Gjedde A, Evans AC. PET studies of domoic acid poisoning in humans: excitotoxic destruction of brain glutamatergic pathways, revealed in measurements of glucose metabolism by positron emission tomography. *Can Dis Week Rep*. 1990;16 (Suppl 1E):105–109.

40. Brunger AT, et al. Highly specific interactions between botulinum neurotoxins and synaptic vesicle proteins. *Cell Mol Life Sci*. 2008;65:2296–2306.

41. Domingo RM, et al. Infant botulism: two recent cases and literature review. *J Child Neurol*. 2008;23:1336–1346.

42. Roy J, et al. Acrolein induces a cellular stress response and triggers mitochondrial apoptosis in A549 cells. *Chem-Biol Interact*. 2009;181;154–167.

43. Forman HJ, et al. The chemistry of cell signaling by reactive oxygen and nitrogen species and 4-hydroxynonenal. *Arch Biochem Biophysics*. 2008;477:183–195.

44. Valentine WM, et al. Covalent cross-linking of erythrocyte spectrin by carbon disulfide in vivo. *Toxicol Appl Pharmacol*. 1993;121:71–77.

45. Kamanyire R, Karalliedde L. Organophosphate toxicity and occupational exposure. *Occup Med (Oxford)*. 2004;54:69–75.

46. Tarlaci S. Vincristine-induced fatal neuropathy in non-Hodgkin's lymphoma. *Neurotoxicology*. 2008;29:748–749.

30

TOXICOLOGICAL NEUROPATHOLOGY IN VETERINARY PRACTICE

PETER B. LITTLE

Pathology Associates, Charles River Laboratories, Durham, North Carolina; Professor Emeritus, Department of Pathobiology, Ontario Veterinary College, University of Guelph, Guelph, Ontario, Canada

DAVID C. DORMAN

College of Veterinary Medicine, North Carolina State University, Raleigh, North Carolina

INTRODUCTION

In this chapter we provide an approach to neurotoxicological investigation. We use several classical neurotoxicants and the lesions that they produce in animals as examples. A basic principle in the deduction of any neurological problem is finding answers to two key questions: "Where is the lesion?" and "What is the lesion?" A well-rounded exposure to the clinical aspects and diagnostic process is a firm foundation for any neuropathologist or neurotoxicologist.

Clinical training in neurology helps guide the identification of the location and definition of the neuropathologic process. The student of neurotoxicological pathology therefore needs to have a strong working knowledge of the nervous system and an excellent neuroanatomical foundation of the species being investigated. Pathology, unlike that taught in some institutions, is not a game of "hide and seek" where the mentor provides a tissue without any hint of species or clinical background in what often appears to be an attempt to humble and embarrass students and expose their current lack of knowledge. Instead, learning pathology must be done in the context of the real world, which is based on an accurate revelation of what is known clinically and what can be determined from the gross and ultimately the microscopic pathological findings.

ANTE MORTEM NEUROLOGICAL EVALUATIONS

It is important for the neuropathologist to consider carefully the clinical signs and the results of specialized neurological evaluations performed on an animal. Neurophysiological evaluations [e.g., electroencephalography (EEG), electromyography, evoked potentials] may occasionally provide ancillary clinical evidence consistent with neurotoxicity. Experimental studies of neurotoxicity conducted in rodents often incorporate behavioral and functional tests into the study design. Alterations in behavior are a sensitive indicator of neurotoxicity since behavior reflects the coordinated function of a large portion of the neural network. Behavioral endpoints are generally non-invasive and can be used to assess subjects repeatedly during the course of an experiment. Endpoints commonly used in neurotoxicity bioassays include acoustic startle, motor activity, and tests of learning and memory. In addition, the functional observational battery (FOB) is used widely in neurotoxicology.[1] Behavioral tests often lack specificity for the nervous system. Their results must be interpreted within the context of other tests of neurotoxicity and potential confounders such as systemic toxicity with indirect neurological effects (e.g., hepatic encephalopathy). Functional effects on motor activity may result from damage to the central nervous system (CNS) or the peripheral nervous system (PNS).

Fundamental Neuropathology for Pathologists and Toxicologists: Principles and Techniques, First Edition. Edited by Brad Bolon and Mark T. Butt.
© 2011 John Wiley & Sons, Inc. Published 2011 by John Wiley & Sons, Inc.

Neurotoxic agents that decrease motor activity include many pesticides and metals. Some neurotoxicants (e.g., toluene, xylene) produce transient increases in activity, presumably by stimulating neurotransmitter release. Others (e.g., trimethyl tin) produce persistent increases in motor activity by destroying specific regions of the brain (e.g., hippocampus).

The normal acoustic startle response depends on a complex neural circuit (cochlea, cranial nerve VIII, cochlear nucleus, caudal colliculus, the reticular nucleus in midbrain, the reticulospinal tract, motor neurons in the spinal cord ventral horn, and motor nerves to the limbs). Consequently, when a toxicant alters the reflex, the only conclusion is that the agent has altered some aspect(s) of this sensory-motor reflex pathway. However, toxic effects in several brain regions distant from the primary startle pathway may also alter the startle response. Damage to the olfactory bulbs, frontal cortex, brachium of the caudal colliculus, mesencephalic reticular formation, septal area, periaqueductal gray matter, and the median raphe nuclei all increase startle amplitude. Conversely, lesions of the locus coeruleus depress startle activity.

Some neurochemical endpoints have also proven particularly useful in understanding neurotoxic mechanisms of action. Neurotoxicants can block reuptake of neurotransmitters and/or their precursors, overstimulate receptors, block transmitter release, or inhibit transmitter synthesis or metabolism. Neuroactive agents may increase or decrease neurotransmitter levels, but such changes are not necessarily indicative of a neurotoxic effect unless they induce neurophysiological, neuropathologic, or neurobehavioral alterations. In addition, a common reaction to CNS damage is astrocytic hypertrophy and enhanced expression of glial fibrillary acidic protein (GFAP), the major intermediate filament protein of astrocytes. Assays quantifying GFAP levels are a sensitive, simple approach for assessing CNS damage. Increases in GFAP may occur in the absence of histopathological changes. Therefore, in adult rodents, increases in GFAP above control levels may be indicative of neurotoxicity. However, increased GFAP expression can also result from events other than neurotoxic injury (e.g., increased corticosteroid levels, normal aging). Several neurotoxicants known to increase levels of GFAP expression include the chemicals 3-acetylpyridine and 1-methyl-4-phenyl-1,2,3,6-tetrahydropyridine (MPTP) and the metals cadmium, methyl mercury, and trimethyl tin.

Moreover, in diagnostic neuropathology the pathologist should also consider the history of exposure; a determination of whether the exposure dose is sufficient to induce toxicity; and whether compatible clinical signs emerged. Although tempting, a positive diagnosis should not be based on any single factor. There are no pathognomonic signs of neurotoxicity.

MACROSCOPIC EXAMINATION OF THE BRAIN AND NERVOUS SYSTEM

At the core of neuropathology are an appropriate examination of the nervous system and interpretation of the changes. In the *veterinary clinical* setting, a major issue in the neuropathological examination is the postmortem interval and the associated artifacts due to delayed fixation. In human medical practice, formalin fixation of the whole brain at the time of necropsy followed by detailed examination of the brain by a neuropathologist significantly increases the detection rate of brain pathology at autopsy.[2] Veterinarians should anticipate a similar outcome in veterinary species.

In experimental studies, neuropathologists often use perfusion fixation techniques to obtain better preservation of the nervous system. Each animal is deeply anesthetized, the thoracic cavity is opened, and the fixative solutions are slowly introduced into the left cardiac ventricle (in small animals such as rodents) or the base of the aorta (in larger animals) using a blunt needle. Typical perfusion protocols begin with infusion of a flushing solution (e.g., heparinized isotonic saline) followed by an appropriate fixative such as neutral buffered 10% formalin. Systemic drainage is provided through puncture of the right auricle of the heart. At the end of the perfusion procedure, the brain is often removed and further preserved by immersion in fresh fixative. Perfusion is generally superior to immersion fixation for preserving the CNS because aqueous fixatives cannot penetrate the lipid-rich brain tissue at a rapid rate. In addition, perfusion leads to fewer artifacts from postmortem cellular hypoxia such as dark shrunken neurons and swollen glia. Rodents are generally perfused under physiological pressure, which for the rat vascular system is approximately 85 mmHg. In an industrial or experimental setting, the main issue with respect to nervous system preservation is thus not so much the adequacy of fixation as a numbers game involving cost and time.

HARVESTING THE NERVOUS SYSTEM

Brain

Whether or not fixed in situ by perfusion, the brain should be removed quickly and carefully, reducing the degree of handling and compression to avoid the induction of artifactual basophilic neuronal change and vacuolation. In addition, fresh and fixed neural tissues should not be allowed to dry during removal and trimming.

Spinal Cord

Anyone who has personally removed the spinal cord in its entirety from a horse knows the meaning of hard work. Unfortunately, horses are commonly affected by a host of neural conditions that require removal of the cord. The dissected vertebral column is often left in the cooler until the next day, when time and assistance are available. This delay is not a good idea.

The same is true of experimental neurotoxicity studies in which the time required for spinal cord removal often makes it one of the last postmortem activities. Needless to say, there are always priorities in formulating a dissection plan, so good preparation and adequate assistance with spinal cord removal are essential, particularly if lesions may be expected. With whole-body perfusion, the removal of the spinal cord, dorsal root ganglia, and spinal nerves may receive less priority since the neural tissues are fixing as the necropsy process proceeds. Nevertheless, prompt removal and additional immersion fixation (with the dura mater incised longitudinally first) will further improve the quality of spinal cord preservation.

Ganglia and Nerves

This portion of the neuropathological examination often lacks priority in veterinary neuropathological evaluations unless the clinical signs clearly establish these sites as prime candidates for lesions. A good working knowledge of the nerves and their principal distributions and functions is important. Nerves should be removed carefully, avoiding the unnecessary stretching that can distort the structure and arrangement of neural fibers. Nerves from unfixed carcasses should be allowed to adhere by their own collagenous stickiness to an identified cardboard card before being placed in fixative. This gentle handling will avoid the contraction and "wavy" appearance that commonly affects longitudinal nerve sections and permits effective evaluation of longer pieces of a particular nerve fiber. With perfusion-fixed material it will be necessary to use stainless steel pins to attach the nerve to the card, as prior fixation destroys the stickiness. Routine removal of dorsal root ganglia with their attached spinal nerve roots is a common practice for dedicated neuropathological studies; trained personnel familiar with this process are essential to attaining good-quality specimens. The time involved in harvesting intact ganglia/nerve samples is considerable. To this day, an evaluation of the autonomic nervous system is seriously neglected in almost all neurotoxicity studies, and when performed at all, such an examination is rarely systematic. Evaluation of the vagus nerve (cranial nerve X) and dedicated in situ examination of the autonomic ganglia and plexuses in the intestine and urinary bladder are a bare minimum in any comprehensive toxicologic neuropathological examination.

TRIMMING NEURAL TISSUES

The organization of the nervous system is intricate, with still unresolved functional and structural complexity. The neuropathological evaluation should be thorough, yet focused particularly on the neuroanatomical structures that are correlated with the clinical findings of neurological dysfunction, and any known neurotoxic effects of the toxicant.

As a general rule, examination of more brain sections increases the likelihood of identifying a toxicant-induced lesion. Although it may be desirable to serially section the entire brain, the time and cost required to process even one brain in this manner makes this procedure impractical. In routine adult rodent commercial toxicity studies, it has been common practice to examine three coronal levels of the brain: (a) a rostral section that includes frontal cortex and the basal ganglia; (b) a mid-diencephalic section through the infundibulum to examine the parietal cortex, hippocampus, and principal thalamic and hypothalamic nuclei; and (c) a mid-metencephalic section to demonstrate the cerebellum, deep cerebellar nuclei, and cochlear nuclei. Clearly, this limited sampling strategy is inadequate.

Instead, seven or more coronal brain sections should be examined. Sections preferred by the authors are depicted in Figure 1. In addition to evaluating the brain, the trigeminal (cranial nerve V) ganglion and nerve, the sciatic and tibial nerves, and at least three spinal cord cross sections (cervical 1, midthoracic, and midlumbar levels) should also be examined. Sections are placed rostral face down in the cassette.

TISSUE EMBEDDING AND THE ASSESSMENT OF NEUROPATHOLOGICAL LESIONS

The use of a traditional hematoxylin and eosin (H&E) stain remains the principal first step in neuropathological examination. Several neuropathology texts will be useful for interpretation of brain lesions in animals and humans.[3–5] The use of more specialized staining and immunohistochemical techniques can assist in the identification of toxicant-induced neural lesions.

The U.S. Environmental Protection Agency (EPA) provides important guidance for the conduct of neuropathological assessment of tissues from studies designed for the assessment of potential neurotoxicity in rats.[6,7] Instructions for histopathological evaluation of the nervous system are relatively specific and state that:

> Tissues should be prepared for histological analysis using in situ perfusion and paraffin and/or plastic embedding procedures. Paraffin embedding is acceptable for tissues from the central nervous system. Plastic embedding of tissue samples from the central nervous system is encouraged, when feasible. Plastic embedding is required for tissue samples from the peripheral nervous system. Subject to professional judgment and the type of neuropathological alterations observed, it is recommended that additional methods such as Bodian's and Bielschowsky's silver methods, and/or glial fibrillary acidic protein (GFAP) immunohistochemistry be used in conjunc-

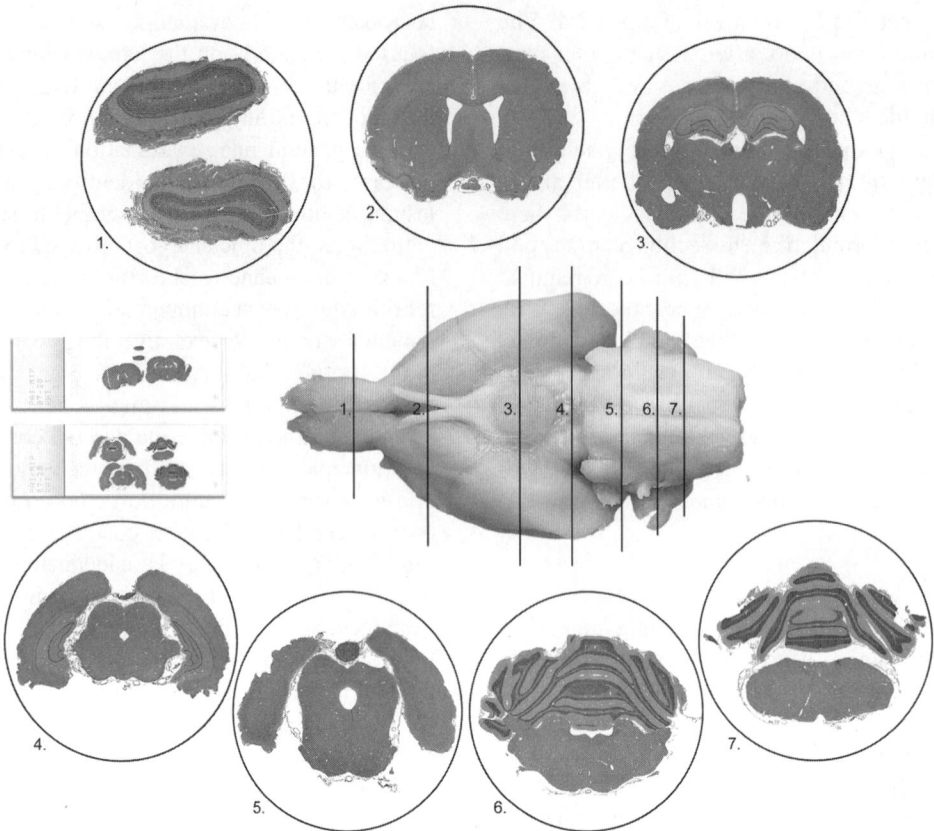

FIGURE 1   Scheme recommended for trimming the rodent brain to produce the seven coronal (transverse) sections that the authors believe are required for a minimally acceptable neuropathology examination. The appearance of the resulting histological sections are shown in the labeled circles; the numbers by each circle correspond to the numbered lines where the trimming cuts are made. The main topographic markers on the brain surface that are used to orient each trimming cut (to see the particular studies noted in boldface type) are: level 1, midpoint of the olfactory bulbs (either surface); level 2, 2 mm rostral (anterior) to the optic chiasm (ventral surface) to reveal the septal nucleus; level 3, the midpoint of the infundibulum (ventral surface) to demonstrate the cerebral cortex, hippocampus, and diencephalon; level 4, midpoint of the rostral colliculi (dorsal surface); level 5, midpoint of the caudal colliculi (dorsal surface); level 6, midpoint of the cerebellum (dorsal surface) and the underlying pons associated with the vestibulocochlear nerves (cranial nerve VIII); and level 7, 2 mm rostral to the termination of the cerebellum to show the medulla oblongata. Stain: H&E.

tion with more standard stains to determine the lowest dose level in which neuropathological alterations are observed.

It is important to note that clinical signs of neurological dysfunction or changes in brain neurochemistry can occur in the absence of a recognizable neuropathological response. A lack of structural changes is seen following strychnine exposure. The strychnine alkaloid blocks postsynaptic spinal cord and motor neuron receptors of the inhibitory neurotransmitter glycine, resulting in enhanced activity of excitatory synapses leading to motor disturbance, increased muscle tone, hyperactivity of sensory, and visual and acoustic perception, with higher doses resulting in tonic convulsions and, finally, death through respiratory or spinal paralysis or by cardiac arrest. Animals that die from acute strychnine poisoning often have no CNS lesions, although brain macrophage infiltration and loss of interneurons and motor neurons can occur following repeated strychnine exposure.[8]

COMMON ARTIFACTS

There are many common artifacts in histological preparations of nervous tissue. Some artifacts are induced by rough handling (such as the maceration of fresh brain and spinal cord tissue during removal), especially by an inexperienced prosector. The primary artifact induced in brain by such rough manipulation is the focally extensive, superficial *cortical neuronal basophilia,* or "dark neuron artifact" (see Figure 8, Panel 1; Figure 12, Panel 2). This artifact is the bane of neuropathologists since it is so common in immersion-fixed material. This change occurs when fixation is delayed and tissues dry out during the postmortem collection and trimming procedures. The change is commonly present superficially on brain surfaces prone to being pressed or pulled during handling but may also be distributed at random throughout the brain core and within other neural tissues. The greatest problem is that its presence may cloud the pathologist's detection of real neuronal degeneration (see below). Artifactual neuronal basophilia has some helpful identifying characteristics. The nucleus, while basophilic, retains a relatively normal size and configuration of chromatin, while the perikaryon is variably shrunken and basophilic. In contrast, authentic cases of acute to subacute neuronal necrosis associated with neuronal basophilia are characterized by neurons in which the nucleus is shrunken and distorted and frequently contains prominent basophilic granules. Of importance in differentiating between neuronal degeneration and dark neuron artifact is close examination of adjacent tissue for the presence of other criteria of antemortem neuronal injury. These features include localized spongiform change in the neuropil adjacent to the affected tissue, capillary endothelial hypertrophy, perivascular edematous protein deposits, and the presence of activated astrocytes, microglia or mononuclear inflammatory cells. The use of perfusion fixation avoids much of the basophilic neuronal artifact and is therefore highly recommended for dedicated neuropathological evaluations performed during the course of neurotoxicity experiments in animals.

Distortion of spinal cord gray and white matter by overforceful handling may impart a dysplastic appearance to the tissue. This occurs because spinal cord tissue within the dura mater can easily be compressed when fresh, and when it is subsequently fixed within the unopened dura artifacts are created by tissue shrinkage during the fixation process. This distortion and displacement makes the artificial separation of the tissue less discernible on microscopic examination.

Fragments of bone commonly become embedded in CNS tissue, especially if mechanical saws are used to extract unfixed organs. These small fragments are evident in histological sections as fine, jagged, basophilic spicules that are often lodged deep in the brain. Whole-body perfusion fixation usually is not complicated by these types of artifacts since the tissue is firm and much less vulnerable to mechanical trauma by the time the brain and spinal cord are removed.

It is common for fixed neural tissues to be conveyed to other fluids or buffers and then held for some time pending further preparatory steps. This transfer can create vacuolar artifacts, especially if the secondary solution contains significant amounts of alcohol. The superficial regions that are most exposed to the alcohol, especially myelinated tracts, are often most affected. When conducting studies on fetal or immature brain, however, special alcoholic fixatives are generally necessary to impart firmness to the tissue that is unattainable using aldehyde fixatives. Hyperperfusion using primarily mechanical pumps during fixation commonly leads to abnormal distension of the capillaries and arterioles. This

should be avoided since it produces abnormally large vascular spaces in the tissue that can distract and may impair the pathologist's evaluation.

Eosinophilic neuronal change is a clear presentation of common acute neuronal injury that is seldom confused with dark neuronal artifact by an experienced toxicological neuropathologist (see Figures 4, 5, and 10, Panel 1). The neuronal perikaryon is shrunken, hypereosinophilic, and has an angular profile, while the nucleus is also shrunken but contains coarse basophilic fragments. One problem inherent with describing this change is the variety of descriptive terms that have been used to describe it by pathologists. Acute eosinophilic neuronal necrosis is a common finding in brain infarction such as occurs in acute vascular occlusion—*stroke* in humans. As a result, the eosinophilic neuronal change in human neuropathology texts is usually referred to as *ischemic neuronal necrosis*. For those people practicing in commercial neuropathology laboratories where the neuronal effects of many metabolic poisons are encountered, eosinophilic neuron change is more an indicator of arrested neuronal respiratory mechanisms such as cytochrome oxidase activity and the entire related respiratory chain of complex biochemical events. In the industrial setting, the more general term *acute metabolic arrest* for such eosinophilic change is preferred unless a more obvious cause, such as thrombosis, can be determined. Many pathologists use the term *red and dead* to describe such a change, but this usage should be limited to casual conversation.

Spongiform change is another change associated with potential confusion for neuropathologists. This finding consists of many small vacuolated spaces in the brain's background substance (neuropil), which is constituted of astroglial processes, vasculature, neurite, and myelinated nerve fibers within which the neuronal population is supported. Vacuolation of the neuropil may be a postmortem artifact common in immersion-fixed neural tissue. It is particularly common in lamina 1 of the cerebral cortex, in the white matter of the cerebellar folia, and in the region of the cerebellar roof nuclei, and usually occurs in a nonsymmetric fashion. Authentic foci of spongiform lesions signifying axonal injury, demyelination, and edema are usually marked by some accompanying criteria of injury, such as symmetrical localization to specific tracts, swollen axons within the vacuoles, and the presence of macrophages, if only a few, in some of the spaces (see Figures 4 and 5, Panel 1). Other indicators of neural injury may include capillary endothelial hypertrophy or the presence of activated microglia and/or mononuclear inflammatory cells.

DEVELOPMENT OF NEURAL LESIONS

An understanding of the chronology of neural tissue reactions is important when aging a lesion. Acquisition of this information can be obtained from the literature when the

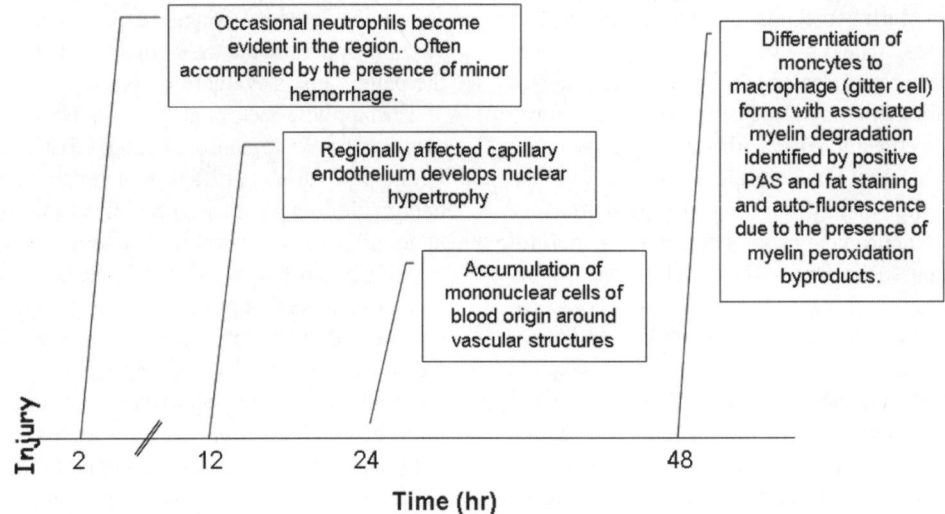

FIGURE 2   Chronology of events seen in the CNS following an acute infarct or trauma.

critical time course of lesions is included in the report; unfortunately, this information is often omitted in manuscripts.

The time course of tissue response in brain is similar among animal species. Figure 2 provides a chronology of events seen in the CNS following an acute infarction or CNS trauma. Microglial cells, the resident immunologically active elements of the nervous system, can be recognized by their elongated "banana"-shaped nuclei in normal H&E-stained tissue. These cells respond quickly to injury and migrate through the neuropil, with the nucleus taking on a more elongated structure over time. These elements are then often referred to as an activated microglia or "rod" cells, and they serve as valuable markers of prior neural damage of even a subtle nature. They participate with blood-derived monocytes in forming macrophages at the injured site. Recognizable microglial cells first appear 12 to 24 h after injury and may be increased both at the lesion site and more peripherally.

The reaction of the astrocyte population is another important consideration in the genesis of neural lesions. Immediately after injury in their vicinity, these cells begin to generate increased amounts of glial fibrillary acidic protein (GFAP) in their cytoplasm. These cytoplasmic intermediate filaments can be visualized at the earliest stages of lesion development by silver stains and also by immunohistochemical techniques. This astrocytic change constitutes an easily appreciable criterion of neural injury of even minimal magnitude. The astrocyes with GFAP accumulation change their appearance, becoming characterized by an eccentric nucleus surrounded by lightly eosinophilic cytoplasm that trails off into the surrounding neuropil as multiple processes in routine H&E staining (see Figure 7, Panel 1). These cells are referred to as gemistocytic astrocytes after the German term *gemasten*, meaning "fattened." This metamorphosis can easily be appreciated in H&E-stained sections by 16 to 18 days after the onset of many types of neural injury. The number and appearance of these reactive astrocytes is dependent on the degree and type of injury. If tissue necrosis and loss has occurred in an area of approximating 1 to 2 mm in diameter, the astrocytic population undergoes hyperplasia to fill in the defect and leaves an asymmetric (anisomorphic) distorted region of astrocytosis. Larger sites of necrosis are often retained as cystic spaces around which an array of reactive astrocytes is present. Minor nonnecrotic foci of neural injury are characterized by scattered reactive astrocytes with a similar (isomorphic) pattern that does not distort the architecture of the neural tissue.

Nerve fiber pathology presents in several forms associated with traumatic injury, the disturbed neural transport arising primarily from toxic agents and inherited metabolic disturbances. In traumatic injury of the spinal cord and peripheral nerves, nerve fibers and their myelin wrappings undergo a process referred to as *Wallerian degeneration* (first described by Waller while performing nerve-crushing experiments). This degenerative process begins with swelling (see Figure 3, Panel 1) followed by fragmentation of the axon distal to the site of injury followed in the PNS by expansion of adjacent Schwann cells and an influx of circulating monocytes which act together to form "digestion chambers" and undertake phagocytosis of disintegrating axon fragments and degenerating myelin. Frequently, as the process advances, a clear space remains in longitudinal sections of the damaged nerve and an identifiable degenerate macrophage with pyknotic nucleus may be observed within the space. The fragmented axon may best be visualized by silver stains; the degrading myelin can be visualized with lipid stains such as sudan III or oil red O. Nerve fiber injury in the brain and spinal cord are commonly seen as spherical to oval-shaped eosinophilic bodies in the neuropil; fortuitously placed longitudinal sections may show these bodies to be focal enlargements of the nerve fiber connected at one or both ends to apparently normal nerve fiber. These "spheroids" do need to be differentiated from tangential cross sections of degenerate neurons, and the presence of such spheroids in a defined nucleus of neurons should heighten the suspicion that a neuronal lesion is present. Axonal dystrophic processes are also characterized by variably sized eosinophilic spheroids and occur in aged or nutritionally deprived animals and in those with inherited metabolic disorders in which axonal transport is disturbed. Examples include vitamin E deficiency and inherited giant axonopathy.[9]

Meningeal changes reflect processes occurring in the brain and spinal cord. The subtlest change is characterized by hypertrophy of the meningeal cell nuclei, particularly the pia mater and arachnoid membranes. Careful examination of the meninges may uncover the presence of a few scattered macrophages or plasma cells, and this should always initiate a closer examination of underlying brain sections for subtle degenerative processes.

The ependymal lining of the ventricles is only modestly affected by injurious processes in the brain. However, it is important to recognize the variation in structure that occurs normally in specific CNS regions. The normal ependyma is cuboidal to low columnar in most locations except the regions of the circumventricular organs and particularly the subcomissural organ, where they are high columnar (see Figure 16, Panel 2). In the infundibular region of the third ventricle the ependyma is also varied in appearance, blending dorsally from the regular ependymal morphology to narrow columnar to pseudostratified cells more ventrally.

Neuronal morphology in response to toxicity varies over time. The earliest change is typically swelling of the nucleus and perikaryon. The nuclear and cytoplasmic membranes progressively fade, and the neuron may progress quickly from swelling to lysis, referred to as *ghost forms*. This may eventually leave a cell-sized (20 to 30 μm) round-to-oval space in the neuropil with little to no cellular response nearby (see Figure 6, Panel 1). The loss of neurons is sometimes called neuronal "fallout." Another common change is acute necrosis. This finding was mentioned above as eosinophilic neuronal change or metabolic arrest necrosis, in which the perikaryon is eosinophilic, shrunken, and angular. Necrotic neurons injured by viral attack usually incite an early phagocytic response. In contrast, toxicant-induced necrosis often leaves dead eosinophilic cells in situ for weeks or months without inciting a phagocytic response. The prolonged presence of necrotic neurons may initiate a process of mineralization called *ferrugination* in the cell remains. This change has a granular basophilic appearance on H&E-stained sections, with the outline of the neuron and some processes still recognizable. This process of mineralization may also proceed to replacement of dead neurons by round concentrically laminated densely basophilic structures called calcospherites (see Figures 15 and 17, Panel 2). The presence of less severe neuronal injury is characterized by central *chromatolysis*, in which the cytoplasmic Nissl granules, consisting ultrastructurally of aggregated ribosomes, are dispersed progressively

from the region of the cell body around the nucleus. During this process the nucleus and filamentous cytoskeletal organelles that suspend the nucleus in the center of the cell are disturbed, and the nucleus is displaced more peripherally. If the toxicity is not lethal, these changes are reversible; the ribosomes reaggregate into Nissl substance beginning around the nucleus, so that recovering neurons may be detected with peripheral chromatolysis. Neurons also undergo apoptosis in which the nucleus quickly degenerates into distinct punctate basophilic fragments (apoptotic bodies) while the cytoplasm shrinks to an almost invisible coating over these fragments (see Figure 9, Panel 1). Such forms may be difficult to detect, and they often disappear rapidly with little evidence of an inflammatory cellular response.

These varied degenerative changes commonly occur together, depending on the severity and time course of the insult, the vulnerability of the individual neuronal populations, and the proximity of the neuron to the insult. The response to these forms of neuronal degeneration is also variable and consists of fluid accumulation in the neuropil (evident as regional spongiosis and/or perivascular protein pooling); nuclear hypertrophy of endothelial cells; early increases of monocytes in the perivascular spaces and activated microglia; and later, amplification of the astrocytic population in the same region.

Depending on the severity and type of neurotoxicity, necrosis in the CNS may affect either individual neurons or the entire tissue, including astroglia and oligodendroglial elements. The vascular components usually survive and respond with endothelial nuclear hypertophy and angiogenesis. This type of lesion, in which all the regional tissue is necrotic, is referred to as *malacia*, from the Latin meaning "soft." The use of this term is often confused and should be restricted to descriptions of total necrosis of an entire brain region (see Figures 11 and 13, Panel 2). The classic example of this process is infarction, but it also occurs when large portions of the brain are vulnerable to a particular toxicant or metabolic interruption. Other good examples are the cortical and deep brain necrosis seen in acute thiamine deficiency[5] and the cortical and caudal collicular necrosis observed in carbonyl sulfide toxicity.[10] The term *malacia* as a microscopic descriptor derives its origin from gross assessments of brain texture affected by large regions of necrosis where the tissue is unusually soft and friable. Where the gliovascular structure of the brain is relatively unaffected, the brain retains most of its normal turgor and relative resilience to handling.

REGIONAL AND TISSUE SPECIFICITY

One unique characteristic of the brain, spinal cord, ganglia, and nerves is the particular vulnerability of their subanatomic sites to toxic insult. This vulnerability often depends on an individual cell's properties, metabolic profile, and location in

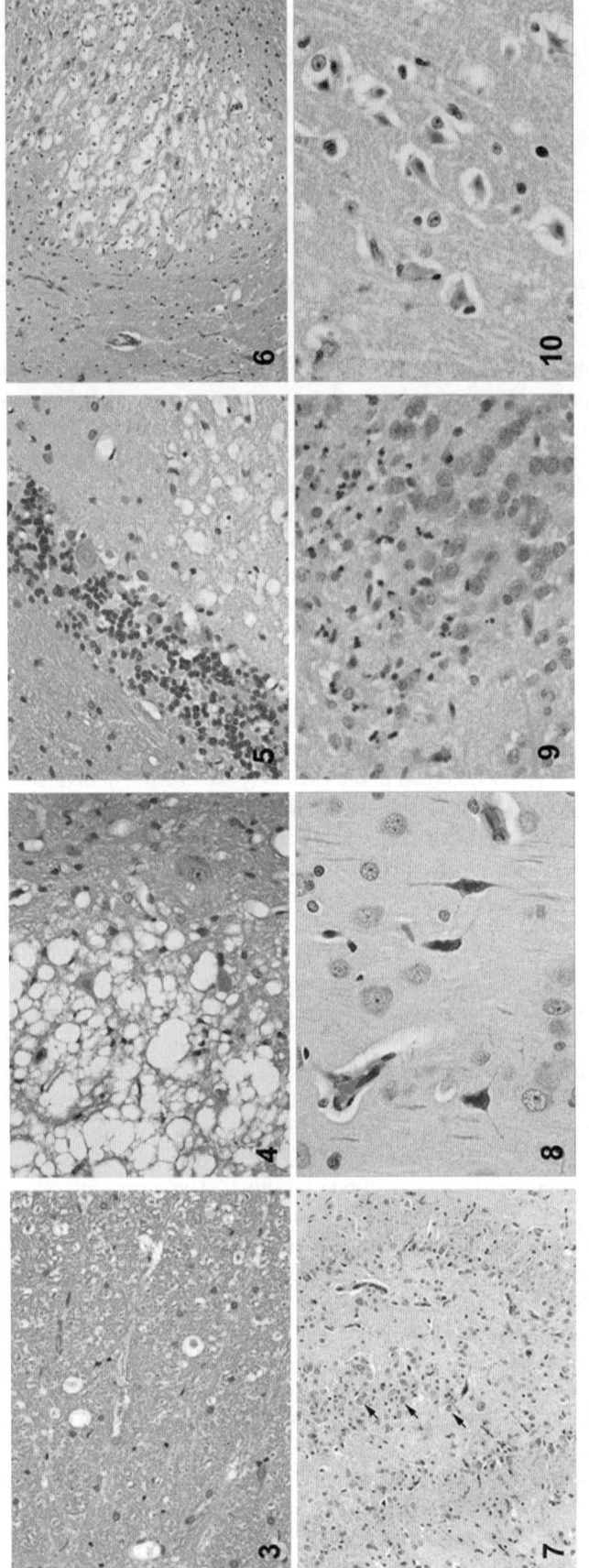

PANEL 1, FIGURES 3 to 10 (*See insert for color representation of the figures.*)

Figure 3 Rat cervical cord, note the swollen axon also referred to as a spheroid, surrounded by a circular space. H&E, 40X.

Figure 4 Rat cerebellar roof nuclear region. Note the acute eosinophilic necrosis of the neuron top center surrounded by spongiotic response to injury and the neuron with mild central chromatolysis and nuclear eccentricity lower right. H&E, 40X.

Figure 5 Rat cereballar cortex. Acute eosinophilic necrosis of Purkinje cells and adjacent spongiotic white matter response with pyknotic glia. A more normal Purkinje cell in the adjacent Purkinje cell layer is seen for size comparison. H&E, 40X.

Figure 6 Rat anterior olivary nucleus. Note the round empty spaces and marked reduction in neuronal numbers. This is an example of past neuronal degeneration and resulting neuronal loss or "fall out". H&E, 40X.

Figure 7 Rat hippocampus with a large cluster of gemistocytic astrocytes indicated by arrows. Their cytoplasm is visible because of cell hypertrophy and accumulation of intermediate filaments composed of glial fibrillary acidic protein (GFAP). H&E, 40X.

Figure 8 Rat cerebral cortex. Note the artifactually crenated basophilic neurons in comparison to adjacent normal neurons. H&E, 40X.

Figure 9 Rat habenular nucleus. Note the many neuronal apoptotic bodies among surviving neurons. H&E, 60X.

Figure 10 Cerebral cortex from a calf with polioencephalomalacia, (PEM) the result of conditioned thiamin deficiency. Note the shrunken eosinophilic necrotic neurons. Close inspection of such neurons reveals nuclear basophilic granules. H&E, 40X.

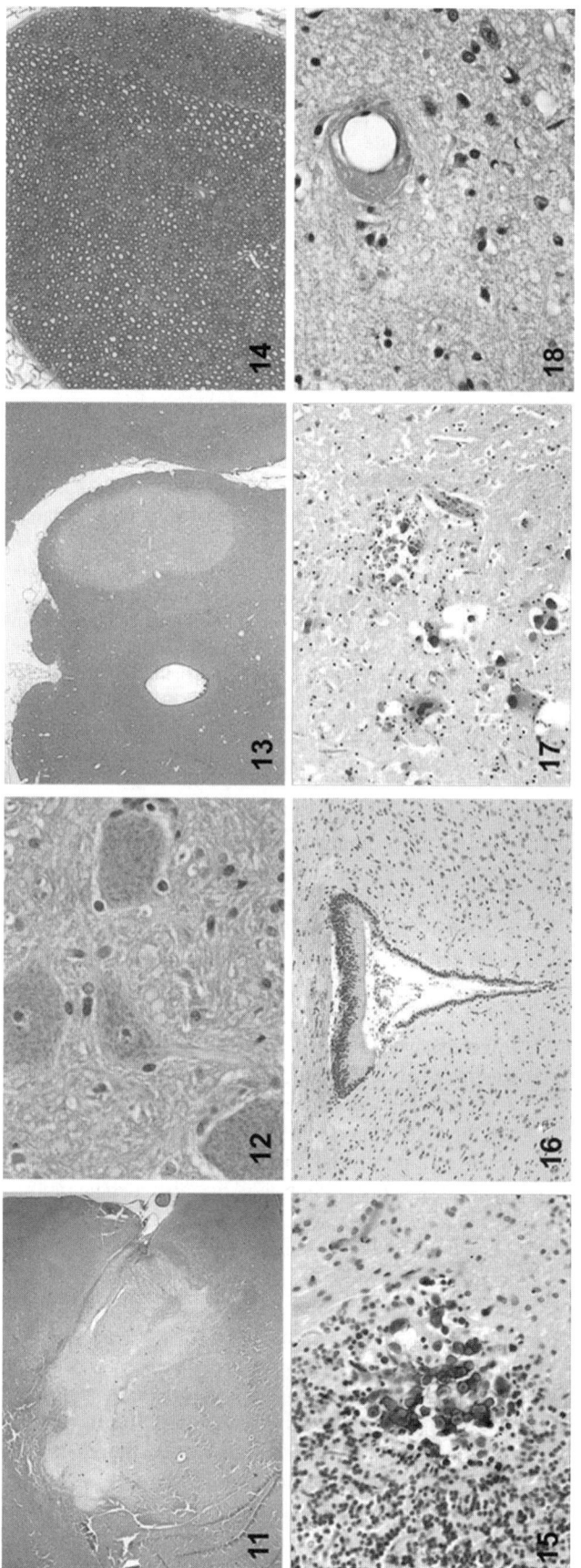

PANEL 2, FIGURES 11 to 18 *(See insert for color representation of the figures.)*

Figure 11 Rat hippocampus, thalamus and optic radiation with a large pale well-defined infarct. The occluded artery is visible at the right center margin. H&E, 2X.

Figure 12 This image depicts a normal canine spinal motor neuron. Neuronal morphologies are diverse throughout the central, peripheral and autonomic nervous system from small cerebellar internal granule cell neurons to large Red nucleus neurons. H&E, 60X.

Figure 13 Rat with a pale region of posterior collicular necrosis, the result of carbonyl sulfide intoxication. Note the artifactually enlarged vascular spaces; the result of perfusion fixation under excessive pressure. H&E, 2X.

Figure 14 Peripheral nerve cross section fixed with glutaraldehyde and plastic embedded. Note the well preserved axonal myelin sheaths devoid of artifactual separation of lamellae allowing for better interpretation of peripheral nerve antemortem changes. Toluidine blue, 20X.

Figure 15 Rat brain old necrosis of internal granule cells with formation of concentric laminated calcospherites. H&E, 40X.

Figure 16 Rat mesencephalon, subcommissural organ. Note the normally elongated and pseudostratified appearance of the ependymal cells in this location. H&E, 20X.

Figure 17 Mouse incidental thalamic mineralized foci. These are commonly bilateral and may focus on the adventitia of vessels or have an apparent random distribution. H&E, 10X.

Figure 18 Posterior colliculus of a mature rat given carbonyl sulfide by inhalation 500 ppm, 5 hours per day for 3 days. Note the prominent perivascular protein-rich edema and fine edematous vacuolation of the adjacent neuropil. There is also necrosis of neurons and pyknotic glial nuclei within this edematous zone. H&E, 60X.

the context of vascular supply. For example, in the case of peripheral nerves, the most distal part is often particularly vulnerable to toxic agents (e.g., organophosphates) and the progressive process of ascending degeneration referred to as "dying back" neuropathy. All of these variables account for the relatively unique distribution of lesions with each type of toxic agent. As neuropathologists, we depend on these patterns of lesion type and distribution to assist in the diagnosis of the probable nature of the toxic agent involved. In the commercial setting, it is important to establish and document the exact pattern of lesion distribution to understand the clinical effect of the toxicant and to predict the resulting hazards to the organism of their occurrence.

Neurotoxicants That Primarily Cause Metabolic Injury to Neurons

The nervous system has a very high metabolic rate that is almost exclusively dependent on aerobic glucose metabolism. To support this high metabolic demand, the brain receives approximately 15% of the total cardiac output and accounts for 20% of the oxygen consumption for the entire body, despite accounting for only 1.5 to 2% of the total body weight. Thus, the nervous system is extremely sensitive to neurotoxicants that disrupt mitochondrial function and energy metabolism or blood pressure/vascular perfusion. The brain's high metabolic rate, elevated concentrations of polyunsaturated fatty acids, and low-to-moderate levels of antioxidant enzymes also predispose it to oxidative damage.

The brain's high metabolic demand makes it exquisitely sensitive to agents that interfere with the metabolism of living cells (Table 1). Especially sensitive regions include the visual, olfactory, and auditory centers and to a lesser degree other regions that control motor function. As a result, lesions induced by metabolic poisons are commonly relatively restricted to the frontoparietal, occipital cortex, and the caudal colliculi. The basal ganglia motor regions (consisting of the caudate nucleus, putamen, and globus pallidus) and the cerebellum are also frequent sites for lesions associated with disruption of energy metabolism. The hippocampus also has a unique neuronal vulnerability to arrest of energy metabolism, with the CA1 and CA3 regions most vulnerable and the CA2 domain relatively resistant. This area-specific susceptibility results from molecular divergence of the affected neuronal populations; both CA1 and CA2 are closely linked to memory functions, but the CA1 (and CA3) neurons have a biochemical profile different from that of CA2. This pattern of hippocampal lesion distribution is a good indication that the toxic insult has affected some aspect of neuronal energy production.

Neurotoxicants That Primarily Injure Axons

Axonopathies are diseases in which the primary site of toxicity is the axon. Conceptually, most toxic axonopathies

TABLE 1   Selected Toxicants That Produce a "Metabolic" Neuronopathy

Chemical	Effects
Carbon monoxide	Selective, severe structural damage in the globus pallidus, and widespread neuronal necrosis in the hippocampus, cerebellar cortex, and other vulnerable areas
Hydrogen sulfide	Olfactory neuron necrosis
Manganese	Extrapyramidal movement disorder characterized by loss of dopaminergic neurons in the nigrostriatal pathway
Methyl mercury	Neuronal degeneration of granule neurons of layer IV in the visual cortex and the granular cell layer of the cerebellum, as well as sensory neurons in the dorsal root ganglia
1-Methyl-4-phenyl-1,2,3,6-tetrahydropyridine	Apoptotic loss of midbrain dopaminergic neurons
Trimethyl tin	Limbic–cerebellar syndrome characterized by neuronal degeneration of the granular neurons of the fascia dentate and the pyramidal cells of the hippocampal Ammon's horn

represent a chemical transection of the axon, with points distal to the lesion undergoing degeneration. Disruption of axonal transport appears to be the toxic mechanism for most axonotoxic chemicals. Damage to the axon will result in secondary myelin degeneration, but the neuron cell body will remain intact. Longer axons (e.g., ascending sensory axons and descending motor axons in the spinal cord and PNS) are affected first. If the insult is transient, PNS axons can regenerate, but affected CNS axons cannot. Sensory effects often occur first in the hands and feet of intoxicated workers, and in severe cases, the limb muscles become chronically weak and atrophied. These signs accompany selective peripheral and distal central nerve fiber degeneration in the medulla oblongata and cerebellum.

A number of chemicals associated with axonal injury include acrylamide; β,β′ iminodipropionitrile (IDPN); carbon disulfide; n-hexane; and some dithiocarbamate and organophosphorus compounds. Acute exposure to some organophosphate and carbamate insecticides can result in significant depression of brain acetylcholinesterase (AChE) activity. This toxic syndrome often results in profound cholinergic effects—salivation, lacrimation, urination, defecation, diarrhea, muscle tremors, seizures, and pulmonary edema—even though histological changes are typically lacking. Some organophosphate insecticides are associated with the development of a distal axonopathy [organophosphate-induced delayed neuropathy (OPIDN)]. OPIDN can occur in many species, including humans, cats, chickens, and water buffalo. Clinical manifestations of OPIDN include progressive irreversible ataxia that develops weeks to months after exposure. Lesions are found in peripheral nerves and the spinal cord. OPIDN is induced only by organophosphorus compounds that inhibit neuropathic target esterase (NTE). Organophosphorus compounds are tested for their potential to cause OPIDN in adult chickens before they are registered for use as insecticides. In addition to a distal axonopathy, IDPN also causes degeneration of the olfactory mucosa in rodents following systemic exposure. Acrylamide affects sensory fibers, particularly the medium and large myelinated fibers that supply Pacinian corpuscles (touch mechanoreceptors for deep pressure and high-frequency vibration) and primary muscle spindle afferents, respectively. Pacinian corpuscles are affected early. Both humans and macaques develop reduced sensitivity to touch following extended acrylamide exposure.

Axonopathies are generally assessed in screening bioassays using thin plastic sections of peripheral nerve that have been stained with toluidine blue (see Figure 14, Panel 2). However, more specific localization, characterization, and quantification of axonal lesions are best accomplished using teased nerve fiber preparations. This specialized technique, although tedious, allows visualization of individual longitudinal segments of axons and their myelin sheaths.

Neurotoxicants That Primarily Cause Injury to Oligodendroglia

Oligodendroglial function is to concentrically wrap the axons and dendrites in the CNS with their complex proteolipid membranes. This myelination process permits more rapid saltatory conduction of nerve impulses in these insulated fibers. Oligodendroglial injury may lead to temporary accumulation of fluid within the oligodendroglial cytoplasm, leading to disruption of the tight wrapping in myelin. The result is the presence of spongy spaces in the CNS neuropil, which may be confused with artifact in less than optimally preserved tissue. Investigation of damage to the oligodendroglia requires electron microscopic inspection of the myelin wrapping and differentiation from artifactual separation. This requires extreme attention to immediate postmortem collection, correct sampling, glutaraldehyde fixation, storage, osmification, and tissue preparation, which is best accomplished at the time of necropsy. Returning to formalin or perfusion-fixed tissue for preparation of EM specimens is unlikely to lead to satisfactory results since artifactual separation and disruption of the myelin sheath in CNS or peripheral nerves is a serious problem in such material. At the light-microscopic level, lethal toxic injury to the oligodendroglial cell is characterized by nuclear pyknosis and/or apoptotic cell death. Such change is frequently unappreciated

since the nucleus of the oligodendroglial cell is relatively dense in its normal state. The consequence of lethal oligodendroglial injury is demyelination, which may be diffuse, patchy, or (more commonly) localized symmetrically in specific vulnerable tracts. These patterns can be best defined with luxol fast blue stain or an immunohistochemical procedure to detect myelin basic protein (MBP). In the diagnostic neuropathology setting, active demyelination by macrophages as part of viral or immune processes is often an issue in evaluating potential neurotoxic lesions, but in the commercial toxicology setting the process of demyelination is seldom the result of a primary self-destructive process. As demyelination proceeds by either toxic or immune process, inflammation will lead to the presence of microglia, macrophage activity, and astroglial responses in the affected region, thus confounding identification of the demyelination's cause. In commercial studies, necropsy of a subgroup of animals at timed intervals will best identify the initiation and course of the demyelinating process. Lipid breakdown products may be identified in macrophages using a periodic acid-Schiff (PAS) or lipid stain, while an anti-GFAP method will define the extent and likely duration of astroglial response.

Examples of toxic oligodendroglial injury include bromethalin, triethyltin, and hexachlorophene. These agents are all uncouplers of mitochondrial oxidative phosphorylation. The resulting inhibition of energy production reduces the ability of myelin to segregate fluid from the interlamellar space, resulting in intramyelinic edema. The associated lesion is partially reversible.

Neurotoxicants That Impair Vascular Integrity

Vessels, particularly those that supply the nervous system, have three major functions: to deliver oxygen and essential nutrients, to clear waste products, and to exclude harmful blood-borne substances. Protection from some potential toxicants is afforded by the blood–brain barrier (BBB) in the CNS and a similar blood–neural barrier present in the PNS. The anatomical basis of the BBB is thought to reside in specialized microvascular endothelial cells, which decrease the capillary wall permeability due to their tight intercellular junctions, lack of fenestrae, and low endocytotic activity. Astrocytes, whose foot processes encircle the abluminal surfaces of capillaries, are also critical for the maintenance, functional regulation, and repair of the BBB. Transport of xenobiotics into the brain is influenced by the metabolic functions of the BBB. For example, d-glucose transporters in the BBB actively facilitate transport of glucose to the CNS; comparable carriers also exist for certain amino acids. P-glycoproteins, expressed at the luminal surface of the capillary endothelial cells, impart additional protection by limiting the CNS uptake of some xenobiotics. Certain neural regions (e.g., dorsal root and autonomic ganglia, the circumventricular organs) lack a functional BBB and may therefore be exposed to higher levels of blood-borne neurotoxicants.

A number of agents impair the BBB to some degree by increasing vascular permeability (see Figure 18, Panel 2). Although the vasculature is the primary target for some chemicals, other CNS cells may suffer additional neurotoxic effects once the BBB has been crossed. Acute lead and cadmium encephalopathy in humans is associated with breakdown of the BBB. Brain capillaries may be dilated, narrowed, necrotic, or thrombosed, and endothelial cells often swell. The consequent extravasation of fluid results in widespread cerebral and cerebellar edema. Accompanying these vascular changes are neuronal necrosis (cerebrocortical and Purkinje cells) with secondary reactive gliosis and astrocytic scar formation. Neuronal lesions may be caused directly by lead rather than by defective vascular function since neuronal necrosis without vascular injury has been observed in acute experimental lead toxicity.[11] In the PNS, lead neuropathy in humans and experimental animals is manifested by Wallerian axonal degeneration and segmental demyelination affecting primarily motor nerves.

VETERINARY DIETARY NEUROTOXICANTS OF NOTE

A variety of toxins and other toxicants are associated with neuropathology in animals. A detailed description of these agents is beyond the scope of this chapter. However, we have included several common agents to highlight the range of lesions seen in animals with these syndromes.

Phalaris Grass Poisoning

Neuronal lipofuscinosis has been described in ruminants consuming *Phalaris aquatica* for extended periods. The most striking gross lesions include diffuse brown to olive green to bluish discoloration of the cerebral and cerebellar gray matter, thalamus, brainstem, and medulla. Histological examination reveals scattered neurons with pigment either loosely dispersed or in a distinct eccentric clump near the nucleus. These pigment granules were strongly positive for lipofuscin (with Schmorls' stain) but negative for hemosiderin (by Prussian blue stain). Isolated glial cells in the neuropil between affected neurons exhibit nuclear pyknosis. Phalaris poisoning can present as an acute ("sudden death") syndrome, most often characterized by unexpected development of neurological signs in the absence of CNS microscopic changes and a chronic syndrome (Phalaris staggers) featuring the gradual development of neurological signs and the presence of characteristic lesions in the CNS. Clinical signs are thought to be due to 3-tryptamine alkaloids, which are structurally similar to the excitatory neurotransmitter serotonin.

Locoweed Poisoning

Locoweed poisoning (*locoism*) is a chronic disease that develops in livestock grazing for several weeks on certain *Astragalus* and *Oxytropis* spp. that contain swainsonine, an endophyte-produced indolizidine alkaloid. Swainsonine inhibits lysosomal α-mannosidase and Golgi mannosidase II, resulting in cellular vacuolation and degeneration due to increased lysosomal retention of partially processed oligosaccharides and glycoproteins. Reported histological lesions include widespread neuronal perikaryon swelling, cytoplasmic vacuolation, and axonal dystrophy (present in the cerebrum, basal ganglia, hippocampus, thalamus, midbrain, cerebellum, and autonomic ganglia) as well as vacuolation of the kidney, liver, and various endocrine cells. Locoweed poisoning is relatively common in cattle, horses, sheep, and goats.

Yellow Star Thistle Poisoning

Yellow star thistle (*Centaurea solstitialis* L.) is a common contaminant of rangeland and cultivated fields in the western United States (especially California, Oregon, and Idaho). A neurological disorder of horses known as *chewing disease* or *yellow star thistle poisoning* was linked experimentally to the ingestion of large amounts of this plant. Horses feeding on this thistle develop immobility of the facial musculature, idle chewing and tongue flicking, impaired eating and drinking, and eventually, hypokinesia and a lack of reactivity that persists until death. Microscopic examination of the brain reveals bilateral necrosis of the rostral globus pallidus and the zona reticulata of the substantia nigra. Therefore, the disease has also been named *nigro-pallidal encephalomalacia*. Similar lesions are associated with ingestion of Russian knapweed (*Centaurea repens*). Recent studies of repin, a sesquiterpene lactone extracted from both *C. solstitialis* and *C. repens*, in rodents and cultured PC12 neurons (derived from a rat adrenal pheochromocytoma) suggests that the toxin interferes with striatal dopamine release and may be involved in the pathogenesis of the nigropallidal lesions.[12]

Equine Leukoencephalomalacia

This syndrome occurs when horses eat corn that is contaminated with the fungus *Fusarium moniliforme*. The exact pathogenesis of this syndrome remains unknown, but it is related to the presence of the fungal toxin fumonsin B_1.[13] Early clinical signs may include anorexia, lethargy, and CNS depression. Neurological abnormalities including weakness of the face and pharyngeal muscles, ataxia, conscious proprioception deficits, facial desensitization, and a tendency to lean to one side are associated with lesions in the sensorimotor cortex. Horses with the neurological syndrome have liquefactive necrosis of the subcortical white matter of one or both cerebral hemispheres. These necrotic areas can vary in size from pinpoint to greater than 4 cm in diameter and will be yellow-orange with a creamy mucoid consistency. Histologically, a center of necrosis with no recognizable structure is often observed. The transition between normal and necrotic tissue will often show hemorrhage, edema, congested blood vessels, and neuronophagia.

Polioencephalomalacia

Polioencephalomalacia (PEM) is an important acute neurological disease of cattle, sheep, goats, deer, and camelids characterized by cortical blindness, medial dorsal strabismus, recumbency, and seizures. In veterinary medicine, PEM is commonly associated with conditioned thiamine deficiency associated with a high-carbohydrate diet, which promotes the overgrowth of a digestive tract microflora that produces abundant thiaminases. Similar brain lesions in cattle may be induced by toxic levels of hydrogen sulfide in feed and water supplies, primarily in the western United States.[14] The classic lesion of PEM is laminar cortical necrosis, which is often quite subtle, especially early in the disease. Acutely affected animals may have brain swelling with gyral flattening and coning of the cerebellum due to herniation into the foramen magnum. Slight yellowish discoloration of the necrotic cortical tissue may be present. The brains of acutely affected animals may also have yellow-green autofluorescent (at 365 nm) bands of affected cerebral cortex evident on meningeal and cut surfaces of the brain. With survival over several weeks, the necrotic cerebrocortical regions become cavitated. The most prominent acute histological lesion is necrosis of parietal and occipital cortical neurons and caudal collicular necrosis (see Figure 13, Panel 2). Cortical spongiosis adjacent to necrotic cortex is present, especially in the early phases of the acute form.

Sodium Toxicosis or Water Deprivation

Salt toxicity, or more appropriately *water deprivation/sodium ion toxicosis*, occurs when excessive quantities of sodium chloride are ingested and the intake of potable water is limited. Pigs, cattle, and poultry are most commonly affected. Early clinical signs seen in pigs include increased thirst, pruritus, blindness, deafness, and constipation. Affected pigs may wander aimlessly, bump into objects, circle, or pivot around a single limb. After 1 to 5 days of limited water intake, intermittent seizures occur in which the pig sits on its haunches, jerks its head backward and upward, and finally, falls on its side in clonic–tonic seizures and opisthotonos; dying pigs may lie on their sides, paddling in a coma, and expire within a few hours or days. During the first few days, swine develop eosinopenia, eosinophilic cuffs around small blood vessels in the cerebral cortex and adjacent meninges, and cerebral edema or necrosis. After 3 to 4 days, eosinophilic cuffs are usually no longer present. Cattle do not develop eosinophilic perivascular cuffs.

Chastek Paralysis

This condition in mink, foxes, and cats results from thiamine deficiency caused by excessive feeding of certain raw fish that contain the enzyme thiaminase type II. Affected animals gradually become anorectic, lose weight, and die after terminal convulsions and paralysis. Lesions are bilaterally symmetrical and include vascular dilatation and endothelial cell hypertrophy, hemorrhage, and encephalomalacia commonly in the oculomotor nuclei, dorsal vagal nucleus, and vestibular and caudal collicular nuclei.[15]

PYRIDOXINE NEUROPATHY

Human and veterinary (canine) patients have developed neuropathy affecting large fibers with severe loss of proprioceptive function following high-dose vitamin B_6 (pyridoxine) exposure.[16] In people, ingestion of 6 g/day for 12 to 40 months resulted in progressive sensory neuropathy, diminished proprioceptive function and loss of vibration sense, and paraesthesias and hyperesthesias. Pyridoxine overexposure resulted in a severe loss of large and small peripheral sensory (dorsal root, trigeminal) fibers with secondary degeneration of axons of affected cells. Cell body changes include cytoplasmic vacuolization, increased dense bodies, neurofilament aggregates, and chromatolysis. Myelinated fibers are also affected with changes being most pronounced distally. Observed changes are consistent with secondary axonal injury.

REFERENCES

1. Moser VC. Application of a neurobehavioral screening battery. *J Am Coll Toxicol.* 1992;10:661–669.

2. Katelaris A, et al. Brains at necropsy: To fix or not fix? *J Clin Pathol.* 1994;47:718–720.

3. Davis RL, Robertson DM. *Textbook of Neuropathology*, 2nd ed. Baltimore: Williams & Wilkins; 1991.

4. Hume Adams J, Duchen LW. *Greenfield's Neuropathology*, 5th ed. New York: Oxford University Press; 1992.

5. Summers BA, et al. *Veterinary Neuropathology*. St Louis, MO: Mosby; 1994.

6. US Environmental Protection Agency. *Health Effects Test Guidelines Neurotoxicity Screening Battery.* OPPTS 870.6200. Washington, DC: US EPA; 1998. Available at: http://www.epa.gov/oppts/pubs/frs/publications/Test_Guidelines/series870.htm. Accessed May 10, 2010. (Listed under "Group E—Neurotoxicity Test Guidelines.")

7. US Environmental Protection Agency. *Health Effects Test Guidelines Developmental Neurotoxicity Study.* OPPTS 870.6300. Washington, DC: US EPA; 1998. Available at: http://www.epa.gov/oppts/pubs/frs/publications/Test_Guidelines/series870.htm. Accessed May 10, 2010. (Listed under "Group E—Neurotoxicity Test Guidelines.")

8. Rodríguez-Ithurralde D, et al. Motor neurone acetylcholinesterase release precedes neurotoxicity caused by systemic administration of excitatory amino acids and strychnine. *J Neurol Sci.* 1998;160 (Suppl 1):S80–S86.

9. Duncan ID, Griffiths IR. Canine giant axonal neuropathy, some aspects of its clinical pathological and comparative features. *Small Anim Pract.* 1981;22:491–496.

10. Morgan DL, et al. Neurotoxicity of carbonyl sulfide in F344 rats following inhalation exposure for up to 12 weeks. *Toxicol Appl Pharmacol.* 2004;200(2):131–145.

11. White LD, et al. New and evolving concepts in the neurotoxicology of lead. *Toxicol Appl Pharmacol.* 2007; 225:1–27.

12. Robles M, et al. Repin-induced neurotoxicity in rodents. *Exp Neurol.* 1998;152:129–136.

13. Wilson TM, et al. Fumonsin B1 levels associated with an epizootic of equine leukoencephalomalacia. *J Vet Diagn Invest.* 1990;2:213–216.

14. Loneragan GH, et al. Association of excess sulfur intake and an increase in hydrogen sulfide concentrations in the ruminal gas cap of recently weaned beef calves with polioencephalomalacia. *J Am Vet Med Assoc.* 1998;213: 1599–1604.

15. Evans CA, et al. The pathology of Chastek paralysis in foxes: a counterpart of Wernicke's hemorrhagic polioencephalitis of man. *Am J Pathol.* 1942;18:79–91.

16. Jortner BS. Mechanisms of toxic injury in the peripheral nervous system: neuropathologic considerations. *Toxicol Pathol.* 2000;28:54–69.

31

REGULATORY CONSIDERATIONS IN TOXICOLOGICAL NEUROPATHOLOGY

KARL F. JENSEN

Neurotoxicology Division, National Health and Environmental Effects Research Laboratory, Office of Research and Development, U.S. Environmental Protection Agency, Research Triangle Park, North Carolina

KATHLEEN C. RAFFAELE

National Center for Environmental Assessment, Office of Research and Development, U.S. Environmental Protection Agency, Washington, DC

Disclaimer: This document was reviewed in accordance with the U.S. Environmental Protection Agency (EPA) policy and has been approved for publication. Mention of trade names or commercial products does not constitute endorsement or recommendation for use. The views expressed in this chapter are those of the authors and do not necessarily reflect the views or policies of the EPA.

INTRODUCTION

Neuropathology is an essential aspect of neurotoxicity assessment and is included in several U.S. Environmental Protection Agency (EPA) and Organization for Economic Cooperation and Development (OECD) testing guidelines for health effects of chemicals. A variety of considerations are important to anyone who relies on a pathology report to determine whether or not effects are treatment related, adverse, and dose dependent. These considerations include the number of animals employed, appropriate sampling and preparation of tissues, control groups, the qualitative and quantitative assessments, and the content of the report.

Neuropathology is a cornerstone of neurotoxicity assessment and is included in three different sets of EPA and OECD guidelines for testing chemicals for health effects: delayed neurotoxicity of organophosphates EPA 870.6100 and OECD 418 and 419,[1–3] adult rodent neurotoxicity EPA 870.6200 and OECD 424,[4,5] and developmental neurotoxicity EPA 870.6300 and OECD 426.[6,7] Each type of study is designed to provide dose–response data for use in risk assessment. The criteria developed by the EPA are designed to help ensure that the data can be used to determine whether an effect is, or is not, treatment related. The most notable discussions of such issues in the past several years have focused on developmental neurotoxicity testing. With few exceptions, however, the issues raised regarding neuropathologic assessments for developmental neurotoxicity studies are also applicable to other types of neurotoxicity studies. These include the number of animals, tissue sampling, tissue preparation, control groups, qualitative examination, quantitative examination, and the content of the report.

Although many of these issues may appear obvious to the experienced toxicologic neuropathologist, they are included here because of the frequency with which they have been encountered as issues in determining the acceptability of neuropathology reports submitted to the EPA. Procedures that enhance the accuracy and consistency of histopathology in general, and as such are applicable to neuropathology, have been discussed by Crissman et al.[8] (and is detailed in Chapter 4).

Fundamental Neuropathology for Pathologists and Toxicologists: Principles and Techniques, First Edition. Edited by Brad Bolon and Mark T. Butt.
© 2011 John Wiley & Sons, Inc. Published 2011 by John Wiley & Sons, Inc.

NUMBER OF ANIMALS

For delayed neurotoxicity (acute and 28-day), sufficient animals (hens) are to be employed to allow six animals per dose to survive the observation period for histopathology. For adult rodent neurotoxicity studies, five animals per gender per dose are to survive until termination of treatment and observation for histopathology. Current recommendations for developmental neurotoxicity studies include evaluation of 10 animals per gender per dose per survival time. An important challenge is that to achieve the specified number of animals for examination, consideration needs to be given to how many animals will survive through treatment and observation. If insufficient animals survive at the high dose, examination of intermediate-dose animals is recommended.

TISSUE SAMPLING

For a study to be acceptable, tissue sampling should include adequate representation of the nervous system. Lists of regions that need to be examined for such adequate representation have been developed.[9–11] If information regarding the mode of action for a chemical suggests that a particular region of the nervous system may be more sensitive, care should be taken to include relevant tissues in the examination. Sampling needs to ensure that any neuropathological alterations observed qualitatively can be subjected to quantitative analysis. In developmental neurotoxicity studies, tissue sampling is to be adequate to support morphometric analysis. Perhaps the most challenging aspect of requirements for morphometric analysis is assuring the homology of sections at a given level from all animals.

TISSUE PREPARATION

A number of artifacts can arise during tissue preparation that may preclude successful qualitative and quantitative assessments. With regard to qualitative assessments, a number of processing-related artifacts can be mistaken for treatment-related effects.[12–14] (see also Chapter 13.) The recognition that factors such as time in fixative and processing schedules can differentially influence the extent of tissue shrinkage has led to the strong recommendation that all tissue samples from all dose groups be embedded within the same time frame. In addition, a balanced design is necessary, where all dose groups are included in each batch of tissue samples to be embedded. Histological processing of tissue samples from young animals in developmental neurotoxicity studies can be particularly challenging and care is required to avoid the loss of tissues during extraction, processing, sectioning, and staining. The loss of tissue samples during preparation can result in inadequate number of sections for a particular region of the nervous system and thereby compromise the acceptability of a study.

CONTROL GROUPS

Control groups important to neuropathological assessments include positive control groups as well as concurrent negative (vehicle) controls and historical negative controls.

Positive Control Groups

The primary purpose of positive control studies is to demonstrate the proficiency of the laboratory in detecting treatment-related and dose-dependent alteration in the structure of the adult and developing nervous system. For delayed neurotoxicity studies in hens, positive control chemicals are often included in each study design. For other types of neurotoxicity studies, positive control studies need not be run concurrently with the test chemical, but should be run within a similar time frame. The results of positive control studies should be submitted along with the studies of test substances. The absence of appropriate positive control studies has been identified as a frequent deficiency in developmental neurotoxicity studies,[15] and suggestions for addressing such shortcomings to demonstrate proficiency have been discussed.[16]

Concurrent Negative (Vehicle) Control Group and Historical Negative Control Groups

While the failure to include a concurrent negative (vehicle) control group is seldom a problem in the studies submitted, problems do arise when differences between the concurrent vehicle control group and treatment groups are small with respect to overall variability, particularly with respect to morphometric assessments. In such cases, range and variability of historical controls is frequently invoked by study authors to dismiss the biological significance of differences between the concurrent vehicle control and treatment groups. There is no clear-cut guidance for best practices for such comparisons; however, concurrent controls are generally considered the most appropriate comparison group for treated animals within a given study. In such cases, evaluation of intermediate-dose groups can provide additional information regarding dose–response relationships for any differences between control and treated animals, and can also provide more information regarding measurement variability. This is especially useful given the relatively small sample size for neuropathology in some studies (i.e., as few as five animals in adult neurotoxicity screening studies). A sound biological basis for dismissing differences between treated and concurrent control animals is not likely to be found until the basis is such that differences can be accounted for by variables other than treatment.

QUALITATIVE EXAMINATION: DETECTION OF TREATMENT-RELATED EFFECTS

The identification of treatment-related alterations in the nervous system is the core of the neuropathologic assessment. For reliable and consistent detection, three factors are essential: (1) artifact is excluded,[12–14] (2) the frequency of spontaneous or 'background' lesions[17,18] in both control and treated animals is documented, and (3) alterations are clearly identified by the appropriate nomenclature (see chapter 13) and their specific location. If no effects are observed in the high-dose group, no further analysis is necessary. If effects are observed in the high-dose group, it is necessary to evaluate dose dependence by examining the medium- and low-dose groups for the occurrence of similar effects. One reoccurring problem area is when an effect observed in the high-dose group is considered by the pathologist not to be biologically significant. In such cases, a clear and compelling basis for such a decision needs to be stated explicitly in the report. This is a critical aspect of the report since such a decision determines whether additional dose groups are examined. Any controversy arising regarding this decision can result in a request by the EPA for further examination of additional dose groups. In the absence of a compelling rationale for determination that a given finding is not treatment related, it may be prudent to proceed to an evaluation of lower-dose groups. Inclusion of these data in the study report will preclude a need for generation of additional data at a later time, which often results in artifacts and delays in completion of the study review.

DOSE DEPENDENCE OF TREATMENT-RELATED EFFECTS

When a treatment-related effect is observed in the high-dose group, it is necessary that the intermediate dose groups be examined to assess the severity and incidence of the treatment-related lesion to determine its dose dependence (and to identify a dose without adverse effects). In qualitative examinations from adult and developmental neurotoxicity studies, severity of lesions are typically subjectively graded ($+1$, $+2$, etc.), while morphometric assessments in developmental studies can employ methods such as linear or areal measurements or even stereological assessments.[9,10] The use of coded ("blinded") and randomized slides for evaluation of the dose dependence of treatment-related effects may be required for select studies. Nonparametric statistics are appropriate for incidence and graded scores based on qualitative assessments, whereas parametric statistics are most appropriate for measurements that are comprised of continuous data, provided that data are normal and homogeneous.[9] It is important to note, however, that statistical significance is not always identical with biological significance and that either one can occur in the absence of the other.

Content of the Report

Key elements of the pathology report and responsibilities of the pathologist when pathology results are incorporated into integrated toxicology reports have been reviewed.[19] In addition to a detailed presentation of the data (both individual and summary), a comprehensive description of the methods and procedures is critical. For example, the absence of an adequate description of how measurements are made frequently creates a problem with understanding what measurements were performed. As noted above, another critical component of the report is the basis on which the pathologist has determined that any effect is, or is not, treatment-related. In either case, it is essential that the pathologist provide a clear and compelling basis for his or her decision.

REFERENCES

1. US Environmental Protection Agency. (EPA). *Acute and 28-Day Delayed Neurotoxicity of Organophosphorus Substances.* 870.6100. Washington, DC: US EPA; 1998. Available at: http://www.regulations.gov/search/Regs/contentStreamer?objectId=09000064809bc92d&disposition=attachment&contentType=pdf.

2. Organisation for Economic Co-operation and Development. *Delayed Neurotoxicity of Organophosphorus Substances Following Acute Exposure.* Test 418. Paris: OECD; 1995. Available at: http://puck.sourceoecd.org/vl=3201244/cl=23/nw=1/rpsv/ij/oecdjournals/1607310x/v1n4/s18/p1.

3. Organisation for Economic Co-operation and Development. *Delayed Neurotoxicity of Organophosphorus Substances: 28-Day Repeated Dose Study.* Test 419. Paris: OECD; 1995. Available at: http://puck.sourceoecd.org/vl=642618/cl=32/nw=1/rpsv/ij/oecdjournals/1607310x/v1n4/s19/p1.

4. US Environmental Protection Agency. *Neurotoxicity Screening Battery.* 870.6200. Washington, DC: US EPA; 1998. Available at: http://www.regulations.gov/search/Regs/contentStreamer?objectId=09000064809bc92e&disposition=attachment&contentType=pdf.

5. Organisation for Economic Co-operation and Development. *Neurotoxicity Study in Rodents.* Test 424. Paris: OECD; 1997. Available at: http://titania.sourceoecd.org/vl=564424/cl=52/nw=1/rpsv/ij/oecdjournals/1607310x/v1n4/s24/p1.

6. US Environmental Protection Agency. *Health Effects Test Guidelines: Developmental Neurotoxicity Study.* OPPTS 870.6300. Washington, DC: US EPA; 1998. Available at: http://www.regulations.gov/search/Regs/contentStreamer?objectId=09000064809bc92f&disposition=attachment&contentType=pdf.

7. Organisation for Economic Co-operation and Development. *Developmental Neurotoxicity Study*. Test 426. Paris: OECD; 2007. Available at: http://puck.sourceoecd.org/vl=848909/cl=12/nw=1/rpsv/ij/oecdjournals/1607310x/v1n4/s26/p1.

8. Crissman JW, et al. Best practices guideline: toxicologic histopathology. *Toxicol Pathol*. 2004;32:126–131.

9. Bolon B, et al. A "best practices" approach to neuropathologic assessment in developmental neurotoxicity testing—for today. *Toxicol Pathol*. 2006;34:296–313.

10. Garman RH, et al. Methods to identify and characterize developmental neurotoxicity for human health risk assessment: II. Neuropathology. *Environ Health Perspect*. 2001;109 (Suppl 1):93–100.

11. Kaufmann W, Groters S. Developmental neuropathology in DNT-studies: a sensitive tool for the detection and characterization of developmental neurotoxicants. *Reprod Toxicol*. 2006;22:196–213.

12. Fix AS, Garman RH. Practical aspects of neuropathology: a technical guide for working with the nervous system. *Toxicol Pathol*. 2000;28:122–131.

13. Garman RH. Artifacts in routinely immersion fixed nervous tissue. *Toxicol Pathol*. 1990;18:149–153.

14. Garman RH. The return of the dark neuron: a histological artifact complicating contemporary neurotoxicologic evaluation. *Neurotoxicology*. 2006;27:11–26.

15. Crofton KM, et al. A qualitative retrospective analysis of positive control data in developmental neurotoxicity studies. *Neurotoxicol Teratol*. 2004;26:345–352.

16. Crofton KM, et al. Undertaking positive control studies as part of developmental neurotoxicity testing: a report from the ILSI Research Foundation/Risk Science Institute expert working group on neurodevelopmental endpoints. *Neurotoxicol Teratol*. 2008;30:266–287.

17. McMartin DN, et al. Non-proliferative lesions of the nervous system in rats. *Guides Toxicol Pathol*. 1997;NS-1:3–18.

18. Eisenbrandt DL, et al. Spontaneous lesions in subchronic neurotoxicity testing of rats. *Toxicol Pathol*. 1990;18:154–164.

19. Morton D, et al. Best practices for reporting pathology interpretations within GLP toxicology studies. *Toxicol Pathol*. 2006;34:806–809.

NOTE ADDED IN PROOF

An up-to-date list of worldwide regulatory guidelines for preclinical toxicologic neuropathology research has been recently published: Bolon B, et al. Compilation of international regulatory guidance documents for neuropathology assessment during nonclinical toxicity studies. *Toxicol Pathol*. 2011;39:92–96.

32

THE NEUROPATHOLOGY REPORT AND THE NEUROPATHOLOGY PEER REVIEW REPORT

MARK T. BUTT

Tox Path Specialists, LLC, Hagerstown, Maryland

INTRODUCTION

There are two primary types of neuropathology reports. The diagnostic report typically reports the gross and microscopic examination of one or more related specimens from a single animal for the purpose of determining the cause of some clinical condition. The diagnostic report is primarily descriptive. A diagnostic report describing a brain biopsy or nerve biopsy is this type of report. The main objective of the diagnostic report is to describe the morphological changes in sufficient detail to support the diagnosis and to allow a clinician to correlate the clinical findings with the structural (and associated functional) changes in order to generate a diagnostic plan and/or prognosis. For terminal cases, the diagnostic report seeks to explain the cause of death or moribundity.

The second type of neuropathology report, the type described in this chapter, is the pathology study report. The pathology study report, hereafter referred to as the pathology report, is concerned with reporting and interpreting gross and microscopic changes noted in various populations of animals. Typically, those populations are differentiated by the various experimental conditions, such as different devices, different therapies, or different dosages of the same therapy. Each study eventually culminates in a series of pathology reports: typically, an initial draft report due within a matter of weeks after the final necropsy, followed by a final report due within weeks to months to years (depending on the particular sponsor of the study).

What makes or constitutes an acceptable pathology report has been covered in previous publications.[1-4] The basic requirements for an acceptable pathology report also apply to a report detailing neuropathology findings (for this chapter, neuropathology studies are defined as investigations conducted to determine the potential neurotoxicity of a neuroactive device, therapy, or chemical).

The quality and usefulness of a report, like the quality of any type of writing, be it a novel, newspaper column, or comic book, can only be judged by the reader. Effective writing, like beauty, is in the eye of the beholder. The writer is often the worst judge of the quality of the writing. Each type of writing requires its own "ideal reader," a moniker from Stephen King in his excellent book on writing titled *On Writing*.[5] The ideal reader decides whether the report suits its intended purpose. The writer nearly always thinks what has been written is excellent; the reader is not always so sure.

Every writer needs to have on his or her shelf at least two references: Stephen King's *On Writing: A Memoir of the Craft*[5] and Strunk and White's *The Elements of Style*.[6] Familiarity with these books provides an understanding about what separates excellent from mediocre writing. Like a good neuropathology report, these books are concise, informative, and satisfying.

In this chapter we delve into what makes a good report, then describe ways of achieving that goal. How best to prepare a report is a bit subjective, so it is inevitable that some of the author's experience and specific likes and dislikes concerning the best way to write and present information crept into this chapter. Some of the information in this chapter will agree with previously published wisdom; some will not. You decide. For this chapter, you are the ideal reader.

Fundamental Neuropathology for Pathologists and Toxicologists: Principles and Techniques, First Edition. Edited by Brad Bolon and Mark T. Butt.
© 2011 John Wiley & Sons, Inc. Published 2011 by John Wiley & Sons, Inc.

WHAT MAKES A REPORT A GOOD REPORT?

Most pathologists are very competent at the microscope: They get the diagnoses right more often than not. Things get missed; that is true in all aspects of medicine. But generally, study pathologists with sufficient experience to be working independently see and record the lesions. There is always difficulty in differentiating artifactual or spontaneous changes from meaningful test article effects; deciding whether or not a change is related to the test article, and then whether or not the change is adverse. Recognition and interpretation of the changes are factors of the pathologist's experience, other information available from the study, published literature, knowledge of the compound, and possibly, knowledge based on historical data. Recording the actual lesions typically gets done correctly. It is the conveyance (the report) of the observations that sometimes goes awry.

So what makes a good report? Table 1 provides a list of what differentiates a "good" from a "not so good" neuropathology report.

STRUCTURE OF THE NEUROPATHOLOGY REPORT

If the report achieves the parameters set forth in Table 1, the actual structure of the report becomes less and less important.

Meeting the needs of the intended reader is paramount. Those needs are usually available locally from the study objectives, study sponsor, or study director. Published guidelines as to what would be preferred by regulatory personnel may also be available.[3] The pathology report is a medical report; the information contained therein, and its ability to be understood by readers, is more important than the presentation. It is convenient to the reader if this useful information is provided in an organized fashion, but there is no real best way to do this. What follows is one approach.

Title Page

The title page should contain all specific study identifiers/numbers, the test site (where was the pathology evaluation performed), the testing facility, the study sponsor, and the report date. Dating every version of a report assists with version control and helps reconstruct the historical development of the report, allowing everyone a means of making sure that they are working on the correct version.

Table of Contents

The table of contents can be generated manually or produced through the magic of word-processing programs. But some manner of a table of contents will allow readers to get to

TABLE 1   What Makes a Good Neuropathology Report

Good report	Not So Good Report
Uses terminology that readers (pathologists, study directors, regulatory personnel, other readers) understand	Uses terminology that only a pathologist could possibly understand
Is concise	Is unnecessarily wordy and/or lengthy
Provides sufficient detail on the neuroanatomic sites that were reviewed for the reader to judge the completeness of the review	Does not provide sufficient detail on the completeness of the examination
Concentrates on the test article- or chemical-related lesions	Too much of the report is devoted to lesions that are not test article- or chemical-related
Is divided into sections that convey the nature of the morphological evaluation	Is disorganized without specific sections detailing the gross and microscopic evaluations
Specifically lists or states each test article- or chemical-related gross and microscopic change, using terminology that is easily traced to the data tables	Fails to state specifically what the test article- or chemical-related lesions were
Provides the reader with all the information required to meet the objectives of the study	Fails to address the study objectives and generates questions from the study director and/or regulatory personnel
Provides a summary that can easily be read, understood, and restated by a reader to a colleague	Fails to include a summary or has a summary that is difficult to understand or cannot be restated easily to a nonpathologist colleague
Provides a summary that is easy to find since many readers will only be looking at the summary	Fails to include a summary or has a summary hidden and difficult to find
Makes conclusions specifically supported by information provided in the results and discussion section(s)	Makes conclusions that are not supported in the results section
Is a stand-alone report signed by the study pathologist or, if within a comprehensive report, is absolutely clear regarding what was contributed by the study pathologist	Is part of a comprehensive report that does not specifically indicate what portion was contributed by the study pathologist

where they want to be (in case they want to look at something other than the summary). If the pathology report will be submitted as an electronic document, the table of contents can be made to link to various portions of the report.

Summary Page

Many readers never get past the summary page. Put the summary page as close to the front as possible. On that summary page, include a short paragraph on study design, then summarize the test article–related changes and the groups or dosages that contained or produced those changes. If there are many test article- or chemical-related changes, specifically include those used (or likely to be used) to define the *no-observed-adverse-effect level* (NOAEL) by either the pathologist or the study director. Remember: The reader should be able to read the summary page and then, in a few minutes or less, convey to a colleague the important aspects of the pathology evaluation. Summaries often contain too much detailed information; they become regurgitations of the results section. The summary should be a summary; it should provide a synopsis of the study design, followed by the conclusions of the study. If the reader is interested in the data that supported the conclusions in the summary, he or she can consult the results section and the data tables.

Compliance Statement

Compliance statements tend to grow insidiously over time and can reach a formidable size. Many pathology reports are ultimately going to be submitted to support a regulatory submission for a particular drug, chemical or device. The compliance statement states specifically what regulatory guidelines were satisfied during conduct of the pathology portion of the study. It is the pathologist, not the quality assurance unit, who is ultimately responsible for the regulatory compliance of the pathology portion of the study. There are often issues that need to be mentioned in a compliance study, especially if there is some peculiarity regarding an aspect of the pathology evaluation. For example, it may be necessary to clarify what the raw data were for the pathology evaluation. It is generally accepted that the raw data for a pathology report are the slides, the final signed report, and any other specimens, such as digital images, used to make specific, original, distinct observations such as measurements.[7–9] If images are included in the report for the purposes of illustration only (i.e., the observations recorded were made on gross tissues and/or the slides), it is useful to clarify that all original observations were made on glass slides and that any images in the report were included purely for descriptive purposes and thus are similar to the text in the report (providing sufficient documentation for the images exists). If morphometric investigations were performed on a nonvalidated system or with software

typically not validated (ImageJ, the public domain image-processing software is a prime example), the compliance statement is an ideal place to explain this.

Some reports include the signature line for the pathologist (the signature that will determine when the report has been finalized) on the same page as the compliance statement, so the signature can serve a double role: certifying the compliance of the pathology evaluation and ultimately signifying that the report has been finalized. Typically, draft reports are not signed by the pathologist. If only the final report is signed, everyone knows which report is final. Some studies have aspects that are out of the control of the pathologist, so consider a statement immediately preceding the signature line that reads similar to the following:

> The undersigned pathologist, by signing this document, certifies that the information and data included in this report accurately reflect the morphologic findings of this study, and the conclusions are based on the best information available to the pathologist at the time of the signature.

Information changes over time. Things we believe today may be proven to be false tomorrow. Your signature signifies that the report was accurate when you signed it.

Responsible Personnel

List the responsible personnel (those people most important for the conduct of the pathology portion of the study, typically the pathologist and perhaps the histology laboratory director) including all pertinent contact information so that it is easy for a reviewer or reader to contact the proper person with any questions (although a really good report leaves no questions unanswered).

Introduction

The body of the report should include those sections everyone is familiar with: Introduction, Methods, Results, Discussion (if needed), and Conclusions. The introduction should restate, in a form that is pertinent to pathology or is lifted verbatim out of the study protocol, the study objectives. If the study objectives pertinent to the pathology report are buried in Protocol Amendment 14 (for example) or wherever, it may be helpful (to the reader) to include the location of those objectives. Whatever other information the pathologist feels is useful to the reader prior to reviewing the details of the report can also be placed in the introduction.

Methods

The methods sections should be complete enough to allow a reader who reads only the pathology report to understand the study design. Particularly if the pathology report is a stand-alone report, the study design as it pertained to pathology should be repeated (from the study protocol) in the pathology report.[3] Be selective: Include what the reader needs to know, but it is reasonable to assume that the reader will also have access to other study information, including the protocol. If information is lifted from the study protocol, state that up front in the methods section. The methods section must state explicitly what work was performed by the pathologist or pathology laboratory and what portions of the study were conducted elsewhere.

Any methods specific to pathology must be included, including various staining methodologies. Provide an explanation for anything that would not be evident to a nonpathologist scientist. Although it may be obvious to a pathologist why a special stain was used, the reader might not be as well informed. The inclusion of a statement such as "Silver stains were added to provide an improved visualization of neurofilament rich structures [axons, dendrites, etc.]" will assist the reader in understanding the use and need for special staining procedures. "Fluoro-Jade B staining sensitively and specifically stains necrotic neurons" quickly explains why a fluorescent dye–based staining procedure was utilized.

The grading scheme used during the microscopic examination needs to be defined.[1] This is important so that the reader has some idea what *minimal, mild, moderate,* and *severe* actually mean or at least how they relate to each other. Obviously, lesion grades vary depending on the particular change and the particular study. But there should be something that assists the reader in determining how the various grades actually equate to severity. If told that someone is severely ill, a mental picture of a very sick person is developed. Similarly, a morphological change that is graded as "severe" should equate to a change that has a meaningful and pronounced effect on the tissue.

A specific explanation for severity grades for some test article- or device-related effects may be necessary. A few lymphocytes in the meninges of an animal with an implanted intracerebroventricular catheter have essentially no clinical significance to that animal. A few lymphocytes in the meninges of an animal with viral encephalitis may be highly significant. Defining severity grades can be a never-ending process, but provide an explanation of the basic grading scheme, then augment that with more specifics if deemed necessary for report interpretation. The key is to provide sufficient explanation of the grading schemes to allow the reader to understand the significance of the changes without unduly burdening the pathologist with having to explain the reason that each grade was assigned to each specific lesion.

Results

All early deaths, moribund sacrifices, and unscheduled sacrifices must be presented in the results discussion. Death is the ultimate adverse effect, so address every animal that did not survive until its scheduled termination date specifically and thoroughly. The important results are those that address the objectives of the study. Generally, this means specifically and unequivocally (if possible) identifying the morphological changes associated with the test article, chemical or therapy and the various dose groups affected. Gross and microscopic changes, and the correlations between the two, need to be included. Although the report should flow and be cohesive, bulleted or numbered lists often suffice to specifically identify the test article–related effects.

If a bulleted list of lesions is too complicated, a summary table of test article–related changes can serve as a very efficient means of conveying the overall effect of a test article and quickly demonstrate what dose groups were affected. These tables can be generated from your data collection program or be produced manually. But if there are more than three or four test article–related lesions, consider a table to simplify or summarize things for the reader. Do not overburden the reader with having to guess which lesions in the data sections are test article related; there should be no doubt as to what the pathologist considers related to treatment.

How to record the microscopic data represents a major differentiator between a general pathology report and a neuropathology report. Although the results for any pathology report should convey the extent of the examination, this is especially critical for a neuropathology evaluation.

It is usually not enough just to list "Brain" in the data tables. The results should indicate the extent of the examination, but not be so exhaustive in terms of anatomical regions to decrease the possibility of recognizing a test article effect that may affect different areas.

Various approaches to listing the regions examined can be used. One approach comes from various guidelines that specifically list regions of the nervous system to be examined. As an example, OECD guideline 424[10] specifically lists the brain regions, spinal cord levels, and other regions to be examined in a rodent neurotoxicity study.

Another approach is to generate a list of structures that are of importance to the specific study. Although this may not be ideal for standardizing the data tables, it can be a sound approach for addressing individual concerns. For example, a test article that might target the dopaminergic system, specifically listing the substantia nigra and caudate/putamen (striatum) as areas in the data tables may be beneficial to the study director, sponsor, and any reviewers. If a drug may target the cholinergic areas of the brain, it may be informative to examine or record data for specific areas, including the nucleus basalis of Meynert (not a large area, but a frequent target of test articles designed to promote the function of the acetylcholine system), caudate putamen area and the diagonal band. For an evaluation of NMDA (*N*-methyl-D-aspartic acid) antagonists, inclusion of the cingulate gyrus as a tissue in the data tables is important because of the well-documented occurrence of necrotic neurons that occurs in that area in rats following the administration of certain antagonists of the NMDA receptor.[11]

Another approach, having the advantage of being somewhat tied to a defined trimming scheme, is provided in Robert Garman's article on the evaluation of large-sized brains.[12] This sectioning scheme lists the following general areas: neocortex, archicortex/paleocortex/transitional cortex, basal ganglia, limbic system, thalamus/hypothalamus, midbrain, cerebellum, pons, and medulla oblongata.

Depending again on the nature of the test article, an even more extensive list may be required. An example is: brain, meninges; brain, olfactory bulbs; brain, frontal cortex; brain, basal nuclei (listing the specific nuclei when necessary); brain, cerebral cortex (cingulate) (for studies involving NMDA antagonists); brain, cerebral cortex (parietal); brain, cerebral cortex (temporal); brain, cerebral cortex (occipital); brain, thalamus; brain, hypothalamus; brain, midbrain; brain, substantia nigra; brain, pons; brain, medulla oblongata; brain, cerebellum. The advantage of listing the areas this way is that most database programs will still allow all the sites to be searched (by selecting "brain" as the filtering parameter) for the purposes of reviewing historical control data. Whatever list you choose, it is important to make sure that the microscopic evaluation remains primarily an examination and not a quality control exercise to make sure that all the separate areas listed in your data tables are actually present on the slides.

Pick a reasonably complete list of structures that covers most studies, then augment that with other sites when necessary. Base the basic list on your particular trimming scheme or standard operating procedures to keep the quality control to a minimum. For more information on brain trimming, see Chapter 21.

Discussion

The discussion section is the portion of the report used to preempt subsequent questions. Anything that might logically be asked following a review of the pathology data and pathology report needs to be addressed in the discussion section.

Discussion is not always necessary in a report. Often, the results are concise and easily gleaned from the various data tables. If there is a situation or lesion that requires further explanation, put those details in the discussion. While the portions of the report devoted to non-test article-related change should be kept to a minimal, some spontaneous or incidental changes might require further mention. This includes those lesions that were not interpreted to be related to the device or treatment but where group differences did exist.[3] Neoplasms often stir interest so they may need to be addressed specifically, especially in studies less than one year in duration.

Conclusions

Succinctly stated, the conclusions sections should contain your conclusions. It should be brief and exactly to the point. There is no need to repeat the study design in the conclusions; save that for the summary section.

Next to the summary, the conclusions section is the most reviewed portion of the report (although a proper summary should make a review of the conclusions section superfluous). Don't invent anything new here. The conclusions must represent your results. It is acceptable to repeat things in this section. If you stated in the results section that the test article caused necrosis of neurons in the cerebral cortex, an acceptable conclusion is: "Administration of the test article at a particular dose caused necrosis of neurons in the cerebral cortex." There is no need to be clever.

In addition to concluding what the test article–related changes were, there should be some indication of the significance (i.e., adversity) of the tissue changes. The FDA publication commonly known as the "Redbook"[13] provides some guidance: "Alterations that significantly compromise an organism's ability to function appropriately in its environment are considered *adverse*. **Neurotoxicity** refers to any adverse effects of exposure to chemical, biological, or physical agents on the structure or functional integrity of the developing or adult nervous system" (italics and boldface in the original). Another reference is provided by Lewis et al.[14] In that publication, a stepwise approach to the interpretation of findings is presented, with discriminating factors provided to attempt to assist the pathologist (or any scientist) in classifying a treatment effect as either adverse or nonadverse.

The pathologist is the medical professional. The pathologist, in consultation with the other scientists involved and the pertinent scientific reports (organ weights, clinical observations, clinical chemistry, hematology, etc.), should be the one to determine the significance of the morphological changes. Expecting a person without medical training to determine the significance of a meningeal infiltrate (for example) is not practical and is not wise.

If you will be reporting a NOAEL, based strictly on the morphological findings (the study director always sets the final NOAEL), state what it is as your last sentence in the conclusions and summary. The issue of whether or not to mention the NOAEL in the pathology report is controversial. Some study directors want it; some don't. Since it is the study director's responsibility to ultimately make the call concerning the NOAEL, then the entire study report is the place for that information. But a statement in the pathology report similar to: "Based solely on the gross and microscopic tissue changes, the NOAEL was . . ." is not only reasonable but appropriate. Recognition and interpretation (including degree of significance) of the morphologic changes are the reasons the pathologist is involved in this process.

THE NEUROPATHOLOGY PEER REVIEW REPORT

The Society of Toxicologic Pathologists has published several guides regarding pathology peer review,[15,16] including an upcoming "Best Practices" position paper that at the time of this chapter preparation was still in draft form.[17]

Most pathologists (if they have the opportunity) seek and receive frequent limited informal peer reviews of individual slides or related slides concerning a particular morphological change. Documentation of these informal reviews is unnecessary; the study pathologist is merely seeking additional information to make an informed opinion.[17] These minireviews are akin to a physician in a hospital stopping another physician and asking them what they think of this or that. Requiring these informal peer reviews to be documented would constitute an unnecessary burden on pathologists in particular and on drug development in general.

But if a formal peer review is being performed (a formal peer review is one that is documented as having taken place, or will take place, in a protocol amendment or other study documentation), for whatever reason, a proper peer review report/statement is as necessary as a proper pathology report (see Table 2).

The purpose of the neuropathology peer review report is to provide an independent appraisal of the study pathologist's findings in order to verify the original findings and conclusions, to assess the dose-group relationship of the test article–related findings, and to provide an additional screening of a selected group of tissues and animals in order to see if something was missed or misinterpreted.[15] In short, the peer review pathologist provides a second opinion.

All materials available to the study pathologist should be available to the peer review pathologist. Otherwise, the conclusions of the study pathologist cannot be assessed in the proper context.

Although the actual conduct of a peer review is beyond the scope of this chapter, a brief mention of some basics of peer review is necessary since the generation of an accurate and concise report depends on these items. It is important to realize that, typically, the study pathologist has put much more time, energy, and thought into the initial pathology examination than the peer reviewer is going to invest in providing a second opinion. (Obviously, there are exceptions.) So, unless there is a good reason to do so, the findings of the study pathologist should not be changed. (There are, of course, many good reasons to change a study pathologist's finding, and none of those good reasons should ever be ignored.)

The peer reviewer should concern himself or herself with the big picture: What are the test article- or chemical-related changes, what dose levels are affected, and what if anything was missed, misinterpreted, or misstated in terms of significance to the animal and to the study? The terminology of the study pathologist should be kept if at all possible. Severity grades are not worth arguing over unless they have some significance to the conclusions of the study or the study pathologist has not been consistent in applying his or her own stated scheme for assigning a grade. It is best if the peer review is conducted in a cooperative vs. an adversarial atmosphere. The goal of all parties should be to produce the most accurate diagnoses and conclusions possible. There is no room for ego or unjustified dogma.

TABLE 2   What Makes a Good Neuropathology Peer Review Report

Good Peer Review Report	Not So Good Peer Review Report
Produces a report that is clear to colleagues and regulatory personnel concerning exactly what was done and by whom	Fails to produce a report, or produces a report leaving colleagues and regulatory personnel questioning what was examined
Is a stand-alone report	Is incorporated as part of the study pathologist's report
Specifically states all the animals and slides that were examined	Fails to state exactly what animals and slides were examined
Specifically states what versions (and the date of the version) of the pathology report and data tables were examined	States that reports and/or data tables were examined, but does not provide the version and/or date of what was examined
Specifically states what conclusions by the study pathologist were agreed to or disagreed with; if the peer reviewer was in complete agreement with the study pathologist and/or if all differences are reconciled prior to the signing of the peer review report, the report should state that there was complete agreement	Leaves colleagues or regulatory personnel room to question if the peer review pathologist and the study pathologist were actually in agreement
Has a signature date that is after the date of all materials that were examined; If the peer review statement or report is limited to a statement something like, "The reviewing pathologist and study pathologist were in complete agreement in terms of the diagnoses and conclusions," the peer review statement/report should be signed after the study pathologist signs the final pathology report and after the peer reviewer has reviewed that final signed report	States that the peer reviewer is in complete agreement with all findings and conclusions in the final report but is signed prior to the final pathology report being signed

Structure of the Peer Review Report

The peer review report should contain the following:

- The objectives of the review.
- The responsible personnel.
- The materials, animals, or specimens that were reviewed (including the reports and the versions and dates of the reports). This list should be specific down to the exact animals, exact slides, and exact data reports reviewed.
- What study conclusions (by the study pathologist) the peer reviewer agrees or disagrees with.

Any unresolved differences between the peer reviewer and the study pathologist should be noted and explained. If there were no differences, or if all differences are reconciled prior to the signing of the peer review report, the report can, as a conclusion, state that there was complete agreement between the study pathologist and the peer review pathologist.

One question that arises is the issue of the generation of additional data during the conduct of the peer review. If the review includes (as it should) a review of the study pathologist's findings, any notes made by the peer reviewer relevant to possible diagnostic differences will be discussed with the study pathologist. Therefore, the notes made by the peer review pathologist are similar to any notes made by the study pathologist during the original evaluation. These notes do not constitute raw data but, rather, developing opinions[8,9] eventually culminating in the final diagnoses provided in the final, signed report (which is raw data). In other words, it is generally not necessary to generate separate data tables during a peer review unless there is a specific reason to do so.

The content of the peer review report depends on several factors. If the peer review report is signed after the signature date of the final pathology report and the reviewer does review that final report, it is acceptable to limit the peer review report to a statement signifying the level of agreement between the study pathologist and the reviewing pathologist. Obviously, any differences need to be stated specifically. If the peer review report is signed prior to the signing of the final path report, then the exact conclusions agreed with (or disagreed with) and the last version of the data/path report reviewed must be specifically recorded.

Typically, the study pathologist and the peer review pathologist can come to an agreement. But if not, a failure to arrive at a total consensus is not the terrible circumstance that some study directors and sponsors want to make it. Although a consensus is the ideal goal, some differences can exist and should be reported. The sponsor or testing facility can support or refute the differences with other documentation using additional expert review, literature, or historical information.

REFERENCES

1. Crissman JW, Goodman DG, Hildebrandt PK, et al. Best practices guideline: toxicologic histopathology. *Toxicol Pathol.* 2004;32:126–131.

2. Morton D, Kemp R, Francke-Carroll S, et al. Best practices for reporting pathology interpretations within GLP toxicology studies. *Toxicol Pathol.* 2006;34:806–809.

3. Dua PN, Jackson BA Review of pathology data for regulatory purposes. *Toxicol Pathol.* 1988;16(4):443–450.

4. Society of Toxicologic Pathologists. Society of Toxicologic Pathologists' position paper on audit trails on microscopic pathology data. *Toxicol Pathol.* 1987;15:377.

5. King S. *On Writing: A Memoir of the Craft.* New York: Pocket Books; 2000.

6. Strunk W, White EB. *The Elements of Style*, 4th ed. New York: Longman; 2000.

7. Tuomari DL, Kemp RK, Sellers R, et al. Society of toxicologic pathology position paper on pathology image data: compliance with 21 CFR parts 58 and 11. *Toxicol Pathol.* 2007;35: 450–455.

8. Preamble to the Good Laboratory Practice Regulations. *U.S. Code of Federal Regulations.* 1987;52(172):33768–33782.

9. Geoly FJ, Kerlin RL. Pathology raw data in nonclinical laboratory studies for the pharmecuetical industry: the pathologists' view. *Qual Assur J.* 2004;8:161–166.

10. Organisation for Economic Co-operation and Development. *Neurotoxicity Study in Rodents.* Test 424. Paris: OECD; 1997. Available at: http://titania.sourceoecd.org/vl=564424/cl=52/nw=1/rpsv/ij/oecdjournals/1607310x/v1n4/s24/p1.

11. Fix AS, Ross JF, Stitzel SR, Switzer RC. Integrated evaluation of central nervous system lesions: stains for neruons, astrocytes, and microglia reveal the spatial and temporal features of MK-801-induced neuronal necrosis in the rat cerebral cortex. *Toxicol Pathol.* 1996;24: 291–304.

12. Garman RH. Evaluation of large-sized brains for neurotoxic endpoints. *Toxicol Pathol.* 2003;31(1):32–43.

13. U.S. Food and Drug Administration. Toxicological principles for the safety assessment of food ingredients. Section IV. C. 10. Neurotoxicity studies. *Redbook 2000.* Washington, DC: US FDA; July 2000.

14. Lewis RW, Billington R, Debryune E, et al. Recognition of adverse and nonadverse effects in toxicity studies. *Toxicol Pathol.* 2002;30(1):66–74.

15. Society of Toxicologic Pathologists. *Peer review in toxicologic pathology: some recommendations. Toxicol Pathol.* 1991;19 (3):290–292.

16. Society of Toxicologic Pathologists. Documentation of pathology peer review. *Toxicol Pathol.* 1997;25:655.

17. Society of Toxicologic Pathologists. Pathology peer review: best practice recommendations. Toxicol Pathol. 2010:1–17. Available at: http://www.toxpath.org/positions.asp.

33

PREPARATION OF PERSONNEL FOR PERFORMING NEUROPATHOLOGICAL ASSESSMENT

Jamie K. Young and Mary Jeanne Kallman

Covance Laboratories, Inc., Greenfield, Indiana

INTRODUCTION

When conducting studies with neuropathological endpoints, optimal results are usually dependent on the training and skills of numerous support personnel. The neuropathological assessment will be compromised if the resulting histology sections do not include the specified neuroanatomic sites or have artifacts that limit the ability of the pathologist to detect lesions. Correlating neuropathological lesions with clinical signs requires the detection, and subsequent thorough and objective characterization, of clinical observations, especially those that relate to the nervous system. This is often accomplished by collecting functional observational data. The following sections are targeted toward laboratories that routinely conduct preclinical toxicology studies and have designated technical support staff for live phase observations, necropsy, and histology. The species used as an example is the rat, but similar strategies can be adopted for other species.

FUNCTIONAL OBSERVATIONAL DATA

Functional observational data, when combined with neuropathological evaluation, provide a perspective on the functional consequences of brain changes. Assessment of neurological and behavioral capabilities is more focused than general observations that accompany a traditional toxicology study. The typical toxicology study includes daily cage-side observation of the animals in a study. Although this approach can provide adequate information about mortality and serious physiological changes, it is rarely sufficient for capturing

the functional ability of the animal. To assess functional capability, the animal must move or behave to permit identification of deficits. If an animal is asleep in the back of the cage, it is unlikely that neurological and behavioral deficits will be detected. The most direct approach to obtain meaningful observations is to remove the animal from the home cage and place it in a different environment to carry out the observation. Handling the animal produces arousal of the subject and the opportunity to detect physical abnormalities that may interfere with normal function. An animal with an injured limb clearly will not be able to ambulate normally. When placed in the center of the new environment, the animal will move. The activity that occurs will provide the opportunity to document locomotion and movement, potential stereotypy, response to objects in the environment, and general activity level. Observations can be collected by live observation of the animal and documentation of the behavioral signs observed, or the animal can be filmed and the film can then be coded for behavioral signs after the animal is returned to the home environment.

In a recent laboratory study, animals exposed to a neurotoxic compound displayed significant neuronal degeneration in the cerebellum. On the 30-day general toxicology study, no aberrant motor signs were observed. This seemed very unusual given the degree of neuronal damage observed in the cerebellum and previous studies that reported disruption of locomotion with cerebellar damage.[1] Before concluding that the damage observed had no functional consequence, a second study was conducted that included repetitive observation of the animals outside the home cage. The subsequent study that was focused on collecting

Fundamental Neuropathology for Pathologists and Toxicologists: Principles and Techniques, First Edition. Edited by Brad Bolon and Mark T. Butt.
© 2011 John Wiley & Sons, Inc. Published 2011 by John Wiley & Sons, Inc.

observational data outside the home cage indicated that the animals displayed a relatively severe disruption of motor functioning, including hindlimb splay, a hunched waddling posture, awkward gait, and tremor. Clearly, the context for collecting the observations on these animals was critical for detecting the motor changes that resulted from exposure to this neurotoxic compound. This observation also illustrates the importance of collecting functional observations in conjunction with the pathological assessment earlier in investigations.

Two basic approaches for conducting functional observations have been used. The first is an unstructured approach in which the observer watches the animal for some fixed period of time and records any abnormalities observed. Unstructured observations are no better than the training level of the observer since one must understand normal behavior in great detail and be able to record all observed deviations from normal. Because of the difficulty in conducting unstructured observations, most laboratories conduct structured observations based on a variety of detailed checklists to document and observe specific functional characteristics. The best known of these structured approaches include the Irwin Battery[2] and Moser's FOB (Functional Observational Battery).[3,4] Both of these approaches for collecting observations includes a home cage section, an animal manipulation section, and an open environment section where specific behaviors or physiological parameters are noted as present or absent or the degree of the abnormality is recorded. Both of these procedures require considerable evaluation time per animal. Because of the time required, especially with Irwin's FOB, most typically a modified or shortened version of the Irwin is used. Good observational techniques require the ability to clearly document observations that are descriptive of the behaviors displayed rather than observations that are judgmental or that contain clinical conclusions in the documentation. Signs and observations are best integrated by one with significant scientific training in the integration of functional capabilities and the relationship of functional loss with specific neurological syndromes.

To provide good-quality observation of functional capabilities, a number of critical variables should be maintained. Observations should always be conducted in a regimented manner. The environment for the observation should be quiet, free of noisy interruptions, well lighted, and temperature controlled. The environment for making observations should be the same for all animals, and the duration of the observation should be for a fixed length of time. Using a plastic rodent cage with bedding material in the bottom makes an adequate environment for observations. Sometimes rats have been placed on an elevated platform in the center of a cage or arena for observation. Since rats have a natural tendency to move to the outside of the arena, they will typically step down from the center platform providing the opportunity to document the motor capacity of the rat in the step-down

situation. Other additions or designs can provide specific behavioral challenges for making observations and can be designed for many different technical needs.

The most critical component of observational evaluations is the person performing the observations. The observer must be superb at describing normal behavior, and the observer must understand the abnormal possibilities that might occur. The only way to grasp the basic knowledge of normal functioning is to spend time making observations, handling rodents, and training with animals treated with many different positive control compounds. Observers should be trained to conduct FOBs by conducting functional observations of rodents treated with multiple doses of known behaviorally active compounds that affect behavior in different ways. Some central nervous system compounds that could be included for training might include amphetamine, morphine, chlorpromazine, and ketamine. Training observers to use multiple doses of the active compounds is important, as it permits the detection of various degrees of the behavioral changes from minimal to moderate to severe. It also provides some index of the individual threshold for detecting effects. Continued training with these compounds can permit consistency among staff and a clear understanding of their sensitivity level for detecting effects. Probably the most precise approach for this training is to film the treated animals and make observations from the film. This approach limits the number of animals that will be used for training purposes and provides an opportunity to see precisely where new technicians need extra training to be consistent with other observers.

There are two types of observational variability that should be evaluated in a laboratory: interrater variability or the consistency between observers, and intrarater variability or how consistent an individual observer is for the same observations. Filming of drug- and vehicle-treated rats offers the best opportunity to document both types of variability. Competency evaluations to determine successful training are critical for maintaining a high-quality observational staff. Abbreviated retraining or competency reevaluation may be warranted when long lapses have occurred in technical observational opportunities.

How do you evaluate observational data? The most direct approach is to tabulate the number of rats displaying a particular characteristic or the frequency of a particular behavior for each animal. Frequency data may be sufficient for making correlations between functional changes and neuropathological data. The addition of functional data provides a broader set of information for understanding the importance of brain changes that might occur. Functional data also provide information about premonitory events that could be followed in the clinical situation. These premonitory events may even precede neuropathological changes. Temporal information will provide clarification of the progression of functional changes and the potential for reversibility of chemical-induced alterations.

NECROPSY PERSONNEL TRAINING

Although most experienced necropsy personnel are cognizant of the need to handle tissue gently, they are often not aware of what actually causes dark neuron and other nervous-tissue handling artifacts. They may not be aware that seemingly gentle pressure with gloved fingers on an unfixed brain can result in significant neuronal artifacts or that hyperflexing the neck to make it easier to sever the spinal cord can result in dark neuron artifacts in the hind brain. One teaching tool that can reinforce this concept is to have the prosectors purposely handle some practice brains traumatically and some brains correctly, and then make histology slides or photomicrographs to share and compare. The visual picture of a linear array of dark neurons tracking down from a superficial cut deep into the underlying parenchyma can impress on the prosectors the need to be careful. This can be reinforced further by challenging the prosectors to assume the role of the pathologist and attempt to pick out rare necrotic neurons from a photomicrograph that includes many dark neurons.

If a laboratory does not routinely collect tissues for detailed neuropathological assessment, it is advisable to review the laboratory's current collection methods to determine their suitability for the neuropathology study proposed. This may include reviewing the standard operating procedures (SOPs) and/or observing the prosectors while they remove brain, spinal cord, or peripheral nerves as part of other standard studies. Microscopically evaluating appropriate tissues from control animals from other studies may enable a pathologist to determine whether the tissue quality using the current methodology will be acceptable for the planned neuropathological assessment. If tissue quality is questionable, the pathologist or training staff can review appropriate collection techniques with the prosectors, observe the prosectors collecting tissues from a small number of stock animals, and submit those tissues for histopathology with identification that would allow the slides to be traced back to individual prosectors. In this way, the laboratory can hopefully identify at least one prosector with good technique who can then train his or her colleagues; or, by characterizing the sites and nature of the artifacts, the pathologist may be able to identify probable causes and coach the prosectors on how to improve their technique.

If a laboratory does not have SOPs or training manuals, generating such documents can be a valuable tool for developing consistent collection techniques if recurrent study requests are anticipated. Table 1 includes an example from an unpublished necropsy laboratory manual of how to remove a rat brain for preclinical toxicology studies. To ensure that the needs of specific studies are met and to improve consistency, limit the number of study-related decisions that necropsy personnel are required to make by providing clear directives in study protocols or similar documents. For example, it may be appropriate to specify

TABLE 1   Example from an Unpublished Preclinical Toxicology Laboratory Necropsy Manual Describing How to Remove a Rat Brain

- Euthanize and exsanguinate the animal.
- From the ventral side, sever the spinal cord near the atlantooccipital junction with the spine and head in a neutral position.
- Remove the skin and eyes from the cranium.
- Using appropriate tools, remove the skull cap, taking care to avoid damage to the brain.
- Carefully remove the meninges.
- Sever the olfactory bulbs from the brain (assuming that olfactory bulbs are not being collected).
- Working rostral to caudal, gently remove the brain from the cranial vault by cutting the cranial nerves and gently pulling in a dorsal direction.
- Fix at least 48 h in 10% neutral buffered formalin prior to trimming.

the order of nervous-tissue removal; or additional detail may be provided for how to fix peripheral nerves (e.g., dissect and mount on paper or other support prior to fixation). If orientation is important (i.e., identifying distal and proximal aspects of a peripheral nerve), the protocol may specify that the proximal end of the nerve be marked with tissue ink. In-house visual training aids such as photos, diagrams, and/or videotapes are very helpful, especially if the techniques are not utilized on a regular basis or there is frequent turnover of necropsy personnel. Some neuropathological assessments may require prosectors to be able to dissect out specific brain regions, ganglia, or nerves. Many people who excel at necropsy are visual learners, and asking them to learn neuroanatomy from a reference book or didactic style lecture may not be productive. Using the hands-on approach of working with practice tissue and laboratory-specific SOPs and training aids is typically a more productive method of training. Do not clutter up training aids with unnecessary detail, but concentrate on what the prosectors need to know to accomplish the task. Prosectors may appreciate having their own copies of the training aid on which they can make additional notes. Larger laboratories may have a structured training program that includes proficiency testing before a prosector is allowed to collect specimens for a given type of study. If a technique has not been used for many months or there has been staff turnover, doing a proficiency recheck prior to the initiation of the study is advisable. Table 2 includes a checklist of things to review should prosection techniques not be satisfactory. Simple things such as switching from a sharp to a blunt forcep, transferring a small rodent brain by cradling with a slightly opened pair of blunt tipped curved scissors rather than grasping with fingers or forceps, avoiding use of a shaker table, or keeping the speed of the shaker table very low may all reduce handling and prefixation artifacts.

If perfusion fixation is used, additional concerns of appropriate anesthesia and ventilation may need to be addressed. This may include interactions with the veterinary staff,

TABLE 2   Specifics to Address If Prosection Techniques Are Not Satisfactory

- Tools and workstation
- Euthanasia and exsanguination
- Order of tissue removal
- Duration of time from euthanasia to tissue fixation
- Handling of tissues during removal, weighing, and transfer to fixative
- Appropriate packaging of tissues (e.g., nerves mounted on paper prior to fixation, appropriately sized tissue cassettes to avoid compression)
- Appropriate fixative, fixative volume and temperature, and/or use of a shaker table

Institutional Animal Care and Use Committee (IACUC), and/or environmental health and safety personnel. One very common misunderstanding with perfusion fixation is the belief that the brain is fully fixed immediately after perfusion. Perfusion technicians may need to be reminded that precautions still need to be taken to prevent dark neuron artifact after perfusion of fixative. Aggressive dissection of the head immediately following perfusion, such as removal of the mandibles from young rodents, can put enough pressure on the encased brain to create dark neuron artifacts in the temporal regions. Depending on the fixative and nature of the study, a postperfusion fixation time is usually required prior to trimming. This is typically carried out by immersing the perfused tissue in fixative for a specified length of time. A common mistake is for a prosector to open the cranial vault before immersing the tissue in fixative, due to the mistaken belief that getting more fixative in contact with the brain is beneficial, but doing so will result in numerous dark neurons.

HISTOLOGY PERSONNEL TRAINING

Most histology personnel are experienced in handling nervous system tissues, but basic procedures occasionally need to be reviewed or specified in protocols, such as fixation times prior to trimming, thickness of trimmed tissues, tissue processor settings, and microtoming and mounting techniques. If the laboratory does not already have brain trimming matrices, these may need to be ordered and personnel familiarized with their use. Collecting eight or more coronal brain sections from an average-sized rat brain results in relatively thin sections that require extra care to properly process, embed, and microtome, and may require shorter tissue-processing times. If orientation is critical, the top faces of trimmed brain or spinal cord slices may be marked with tissue ink after placement in tissue cassette to facilitate orientation during embedding. Rodent brain sections may be embedded two to four coronal sections per standard tissue block to decrease the number of slides generated, but sections

must be embedded evenly to allow for microtoming with minimal loss of tissue during microtoming. If the pathologist is concerned about the amount of tissue that is being lost during microtoming, a practice brain can be trimmed with opposing faces of a couple of sections being embedded face down. Bregma levels can be determined from the resulting slides and the distance between the microtomed faces determined, since Bregma levels are recorded as millimeters from the Bregma point. The resulting difference between opposing section faces divided by 2 is the approximate amount of tissue being lost from each section during microtoming. This can vary considerably between laboratories and histology technicians. If an unacceptable amount of tissue is being lost, it may be necessary to embed fewer tissue sections per block. It may also be necessary to limit the number of trimmers or microtomists for a given study to produce more uniform specimens. Protocols for most special stains are readily available in manuals or the literature and were not addressed in this chapter.

Additional training is usually required if tissues are to be trimmed for evaluation of specific regions of interest. Although it is worth sharing various brain atlases and other references with histology technicians, development of simplified and focused in-house guides, with an emphasis on photos and diagrams, is advised. Do not clutter up such aids with unnecessary detail, but concentrate on what the histology technicians need to know to accomplish the tasks. Providing each trimmer with their own copy of the trimming guide on which they can make notes is also useful. Assuming that such studies are the minority of the workload, it is advisable to train only a limited number of designated "specialists" instead of attempting to train all the personnel. Proficiency is best gained through experience, and limiting the number of personnel working on a given aspect of a study will increase consistency of a given study. If the laboratory staff are expected to do quality control for identification of key neuroanatomical sites on resultant slides, having images of the representative sections available for comparison is useful. Images of stained histology slides can be captured easily using a flatbed document scanner or as digital photomicrographs, saved as electronic files, and then annotated and printed (Figure 1).

In summary, a high-quality neuropathological assessment is dependent on the expertise of the numerous support staff. Since there are no readily available manuals or training courses available for learning such skills, developing in-house training programs is usually necessary. Assuming that such studies will continue to be conducted on an ongoing basis, taking the time to produce meaningful SOPs, manuals, and training aids will facilitate skill transfer and consistency. Providing timely, meaningful feedback to personnel in regard to specimen quality can be used to reward high-quality work, address concerns, and maintain consistent standards of quality.

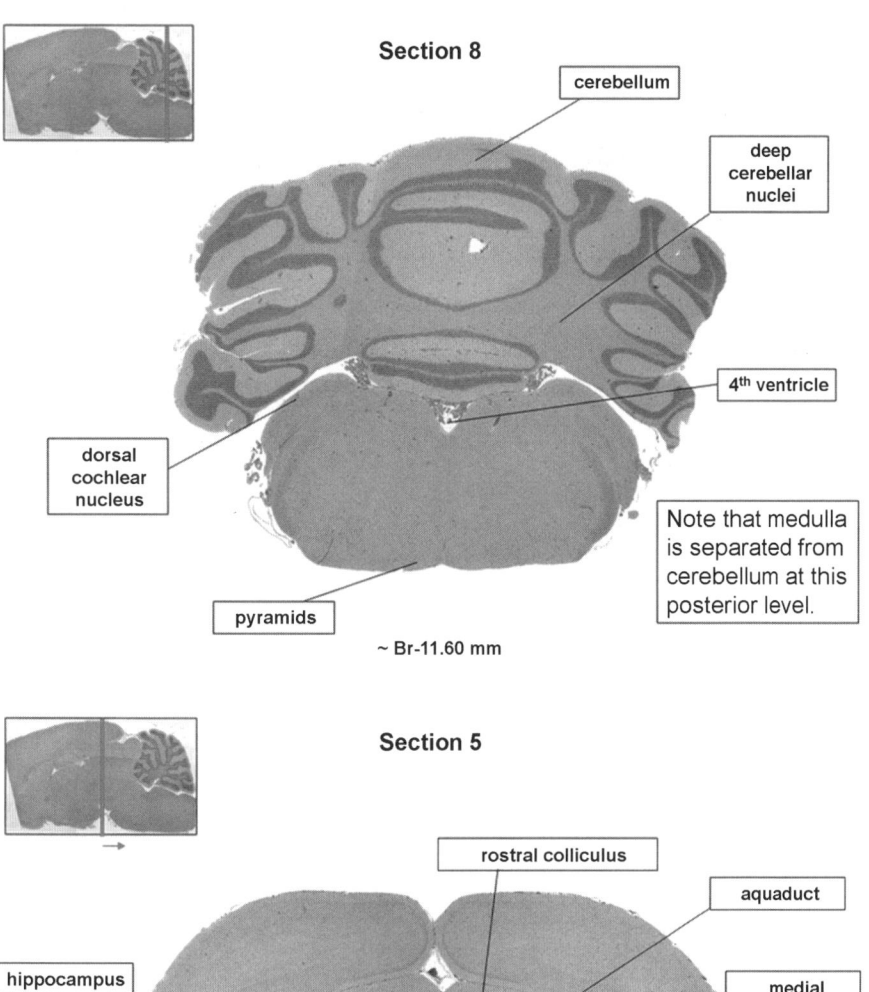

FIGURE 1 Example of a scanned H&E-stained microscopic slide of a coronal rat brain section with annotations for use in training or as a quality control reference.

REFERENCES

1. Ito M. Historical review of the significance of the cerebellum and the role of purkinje cells in motor learning. In: *The Cerebellum: Recent Developments in Cerebellar Research*. New York: Academy of Sciences; 2002: 273–288.

2. Irwin S. Comprehensive observational assessment: 1a. A systematic, quantitative procedure for assessing the behavioral and physiologic state of the mouse. *Psychopharmacologia*. 1968;13:222–257.

3. Moser V. Applications of a neurobehavioral screening battery. *J Am Coll Toxicol*. 1991;10, 661–669.

4. Moser VC, et al. Profiles of chemical effects using a neurobehavioral screening battery. *Toxicologist*. 1992;12:349.

34

REGULATORY GUIDE TO THE HISTOPATHOLOGICAL ASSESSMENT OF NEUROTOXICITY STUDIES

HOPE SALVO

SAIC-Frederick, Inc., Frederick, Maryland

MARK T. BUTT

Tox Path Specialists, LLC, Hagerstown, Maryland

INTRODUCTION

Regulatory agencies worldwide have developed test methods, with the help of experts in the field, to study the effect of chemicals in animal models (see Chapter 31). The test methods become adopted guidelines that help to fulfill requirements for specific laws. Although these guidelines are considered non-binding, they do represent an industry standard and are expected by the agencies involved. The U.S. Environmental Protection Agency (EPA), the U.S. Food and Drug Administration (FDA), and the Organisation for Economic Co-operation and Development (OECD) are organizations that provide current guidelines on neurotoxicity testing. Guidelines prepared by the Japanese Ministry of Health (JMH) are not referenced specifically. Regulatory agencies continually assess current guidelines and propose additional ones or update current ones as needed.

Regulatory guidelines are accessible electronically through the Internet or by contacting the agencies directly. Thousands of guidelines on a variety of topics are provided by these agencies. Accessing specific guidelines from the different regulatory forums can be cumbersome. This section provides "A Regulatory Guide to the Histopathological Assessment of Neurotoxicity Studies" (hereafter referred to as the "Guide") as a concise reference to the most prominent neurotoxicity guidelines currently available. The purpose of this Guide is to provide key test methods for pathologists and toxicologists studying the chemical effects on nervous tissue.

The Guide takes guidelines from the EPA and FDA and extracts pertinent information about the histopathological test methods while consolidating them into a condensed, organized manner. Because harmonization has largely occurred with the EPA and OECD guidelines, only references of additional information from the histopathological recommendations of OECD have been incorporated in the EPA section of this guide and are denoted by an asterisk.

The guidelines for general, developmental, and delayed neurotoxicity test methods are included in this Guide. The Guide was initially produced as part of a graduate thesis. In its initial version, information from the various regulatory guidelines concerning clinical observations was included since the correlation of structural changes to clinical abnormalities is one of the most important duties of the pathologist. However, in the interest of space, guideline information was deleted from this chapter. Pathologists are referred to individual study protocols, and to in-life portions of the guidelines, for information on the conduct of the in-life behavioral testing.

RESOURCES AND METHODS FOR DEVELOPMENT OF A REGULATORY GUIDE FOR NEUROTOXICITY TESTING

Currently, the most relevant regulatory guidelines for neurotoxicity studies are provided by the EPA, FDA, and OECD.

Fundamental Neuropathology for Pathologists and Toxicologists: Principles and Techniques, First Edition. Edited by Brad Bolon and Mark T. Butt.
© 2011 John Wiley & Sons, Inc. Published 2011 by John Wiley & Sons, Inc.

The topics chosen for this Guide provide a tool to preserve and process tissues and produce stained slides of specific nervous tissue to correlate morphology to observations for unbiased microscopic evaluation. The topics include universally accepted methods with appropriate references. Other test methods available for neurotoxicity involve the biochemical and behavioral evaluation of animals and are not included in this Guide.

To provide consistency in representing the information in the guidelines, the test methods for general, delayed, and developmental neurotoxicity are each organized under four sections of evaluation methodologies: necropsy and tissue collection, tissue processing, histopathological examination, and evaluation of the data. Each of these methodologies is further categorized by relevant subtopics. Table 1 provides the evaluation methodologies and the subtopics that are included in the Guide.

Methods for perfusion, gross observations, tissue sampling and trimming, and proper specimen storage are included in the Guide. The purpose of fixation of tissue is to preserve the tissue to as close to the living state as possible and to eliminate artifactual changes. Tissues tend to shrink, distort, and lose resolution if they are not preserved, collected, handled, and processed appropriately. It is important that technicians and pathologists are trained in the identification, harvest, and fixation of specific regions and tissues of the nervous system. The guidelines presented in the "Necropsy and Tissue Collection" section of the Guide provide test methods developed from experts in the field with references to specific topics as needed.

Properly processing tissue allows the embedding medium to infiltrate the tissues and prepare the tissues for microtomy. This process includes stepwise dehydration and clearing of the trimmed tissues followed by paraffin or plastic embedding. The tissues are then sectioned and stained using specific techniques included in the Guide in the "Tissue Processing"

sections. Information on special staining techniques is provided in the references cited with the guidelines.

Suggested qualitative and subjective diagnosis procedures using microscopy have been extracted from the guidelines and included in the "Histopathological Evaluation by Microscopy" section of the Guide. Trained, experienced pathologists provide the diagnosis of the tissues in order to properly distinguish between artifacts and actual morphological changes. A qualitative microscopic evaluation identifies regions with pathological alterations and allows the assignment of a severity grade to that change. Subjective diagnoses emphasize the need for consistency and reliability when evaluating the raw data. It is important to sample all major areas of the brain, spinal cord, and peripheral nervous tissue.

Neuropathological evaluation should be complemented by other neurotoxicity studies such as behavioral and neurophysiological studies which may be relevant to the study; this information is presented in the "Evaluation of the Data" section of the Guide. The test methods provided in this section emphasize an evaluation based on gross necropsy findings and microscopic pathology observations that include relationships between the animal's exposure to the chemical, test article, or food additive and the frequency and severity of the lesions observed. Evaluation of dose–response results for various groups, if they exist, should be provided, and a description of the statistical methods *must* be presented if a statistical analysis is performed.

Test methods for general, delayed, and developmental neurotoxicity studies are represented in the Guide because they are the guidelines for neurotoxicity that specifically apply to histopathological techniques. A general neurotoxicity study is conducted if a chemical substance has been found to be a potential neurotoxin and requires further characterization. These methods represent the broadest area of testing and frequently the most often used.

Delayed neurotoxicity may be a manifestation of exposure to organic solvents and other substances. The guidelines define organophosphorus-induced delayed neurotoxicity (OPIDN) as a neurological syndrome generating signs of weak limbs and upper motor neuron spasiticity.[1] Chronic exposure to organophosphorus substances leads to changes in peripheral nerve function that last from months to years. The hen is used as the animal model because the effects of organophosphates are reliably reproducible in avian species. The neurochemical indications of toxicity include inhibition of neurotoxic esterase (NTE). The methods for evaluating this unique neurologic insult are included in the Guide.

Developmental neurotoxicity can be defined as any effect of a toxicant on the developing nervous system before or after birth that interferes with normal nervous system structure or function. The complexity of managing a study for developmental neurotoxicity comes from the multiple endpoints of toxicity that must be gathered during different stages of gestational and postnatal development. The guidelines

TABLE 1   Evaluation Methodologies and Subtopics

Evaluation Methodologies	Subtopics
Necropsy and tissue collection	General information
	Perfusion technique
	Gross observations
	Tissue sampling and specimen storage
Tissue processing	Dehydration, clearing, and embedding
	Microtomy and staining techniques
Histopathological evaluation by microscopy	Qualitative diagnosis
	Subjective diagnosis
	Electron microscopy
Evaluation of the data	Final analysis

provide methods for evaluating all the stages required and are included in the Guide.

The guidelines chosen are germane to producing slides from fixed tissues. The test article has already been administered, clinical observations have been made, animals sacrificed, and the tissues must now be prepared for microscopic evaluation with correlations made to macroscopic observations. The Guide is an adequate representation of the views of regulatory forums at the forefront of producing reliable data for neurotoxicity. It assists with the sometimes cumbersome process of searching for the necessary test methods developed by the EPA, FDA, and OECD that are appropriate for the histopathological evaluation of neurotoxicity studies. As more data are gathered about neurotoxic substances, guidelines are published to help scientists study them. This process allows regulatory agencies to oversee thousands of new and existing chemicals. Table 2 lists a collection of international regulatory guidelines that contain histopathological test methods for neurotoxicity studies. While this Guide is designed as a quick reference of specific

TABLE 2 Regulatory Guidelines for the Assessment of Neurotoxicity

EPA—General Neurotoxicity Test Methods
1. U.S. EPA OPPTS 870.6200: Neurotoxicity Screening Battery
2. U.S. EPA 40 CFR 799.9620: TSCA Neurotoxicity Screening Battery
3. U.S. EPA 40 CFR 798.6400: Neuropathology
4. U.S. EPA 40 CFR 79.66: Neuropathology Assessment
EPA—Delayed Neurotoxicity Test Methods
1. U.S. EPA OPPTS 870.6100: Acute and 28-Day Delayed Neurotoxicity of Organophosphorus Substances
2. U.S. EPA 40 CFR 798.6560: Subchronic Delayed Neurotoxicity of Organophosphorus Substances
EPA—Developmental Neurotoxicity Test Methods
1. U.S. EPA OPPTS 870.8600: Developmental Neurotoxicity Screen
2. U.S. EPA OPPTS 870.6300: Developmental Neurotoxicity Study
3. U.S. EPA 40 CFR 795.250: Developmental Neurotoxicity Screen
4. U.S. EPA 40 CFR 799.9630: Developmental Neurotoxicity
FDA—General and Developmental Neurotoxicity Test Methods
1. U.S. FDA, CFSAN, Redbook 2000 V.C.10: Neurotoxicity Studies
2. U.S. FDA, CFSAN, Redbook 2000 IV.C.9.a: Guidelines for Reproduction Studies
OECD—General and Developmental Neurotoxicity Test Methods
1. OECD Guideline 424: Neurotoxicity Study in Rodents
2. OECD Guideline 426: Developmental Neurotoxicity Study
OECD—Delayed Neurotoxicity Test Methods
1. OECD Guideline 418: Delayed Neurotoxicity of Organophosphorus Substances Following Acute Exposure
2. OECD Guideline 419: Delayed Neurotoxicity of Organophosphorus Substances: 28-Day Repeated Dose Study

test methods from EPA, FDA, and OECD, the official guidelines from these agencies should be consulted as well.

REGULATORY PERSPECTIVE FOR NEUROTOXICITY STUDIES

Government agencies have the authority under specific laws to develop and enforce regulations. Regulations instruct users about what is required under the law for a specific topic. Guidelines are available to suggest test methods for producing data for an agency to fulfill a regulation enforced by a specific law. Guidelines are therefore nonbinding scientific methods for conducting specific studies. They are created by experts in the field from public group meetings, scientific reviews, and advisory panels. They are considered industry standards and often expected to be a part of a toxicity testing strategy. They are often organized by testing endpoints such as neurotoxicity. These test standards can be modified as needed for an individual test substance.

The EPA and FDA propose a regulation or guideline, research it, list it in the *Federal Register* for public comment, revises it, and issues a final rule. After finalization, the regulation or guideline is published in the *Code of Federal Regulations* (CFR), which is the official record of all rules published in the *Federal Register* by various agencies.[2] Title 40 of the CFR is reserved for most of the environmental regulations. The CFR is revised yearly and updated every three months, and Title 40 is revised in July each year. Methods are now in place for scientists to keep current on the test methods for neurotoxicity by accessing the Internet. This is an efficient way to access the most current guidelines.

The Federal Insecticide, Fungicide and Rodenticide Act (FIFRA), promulgated in 1947, provides the EPA with legislation to enforce the registration of pesticides through the Office of Pesticide Programs (OPP) based on health effect studies. Regulations indicate what types of studies are required for a specific type of chemical pesticide. The OPP presents regulations and guidelines to the Scientific Advisory Panel (SAP) for expert advice on pesticides and how they affect human health and the environment.[3]

The Toxic Substances Control Act (TSCA), promulgated in 1976, provides the EPA with legislation to enforce regulations for testing existing and new industrial chemicals, other than pesticides, through the Office of Prevention, Pesticides and Toxic Substances (OPPTS). Neurotoxicity guidelines are found in 40 CFR 795-799 and are published by the Office of Pollution, Prevention, and Toxics (OPPT) to collect data required under the TSCA for a specific chemical.

The Federal Food, Drug and Cosmetic Act (FFDCA) of 1938 provides the FDA with legislation to enforce the safety of a food additives through the Center for Food Safety and Applied Nutrition (CFSAN) based on toxicity testing in animals. The toxicity tests determine the dose of a chemical that produces an adverse effect and the no observable effect level (NOEL).

In 1982, the FDA published *Redbook 1* or *Toxicological Principles for the Safety Assessment of Direct Food Additives and Color Additives Used in Food* to provide protocols for subchronic, chronic, reproduction, and teratology toxicity testing in animals. The FDA revised its *Redbook* to assess the neurotoxic potential of food additives. *Redbook II* was drafted in 1993 and in November 2003 was revised to include neurotoxicity as an endpoint in short-term, subchronic, and one-year toxicity studies in rodent and nonrodents. It is now entitled *Toxicological Principles for the Safety Assessment of Food Ingredients.*[4] *Redbook 2000* chapters are now available on the FDA Web site.

The *Redbook 2000* chapters now substitute for, or supplement, guidance available in the 1982 *Redbook I* and in the 1993 *Draft Redbook II,* which can be obtained from the Office of Food Additive Safety. As additional chapters of *Redbook 2000* are completed, they will become available electronically. *Draft Redbook II* also included general guidelines for assessing effects on male reproductive function and optional neurotoxicity screens. Since *Draft Redbook II* was released in 1993, additional refinements have been made in the procedures for the multigenerational reproduction study and for the assessment of effects on male reproduction. The references to developmental neurotoxicity from the Draft *Redbook II* have been included in this Guide.

The OECD is an internationally accepted compliance forum for toxicity testing. It is an organization of 30 member countries whose purpose is to respond to various regulatory compliance issues. The OECD promotes international harmonization through its various programs. Countries exchange information, harmonize data requirements for registration, and standardize testing practices in an attempt to reduce the risks associated with pesticide use and minimize resources used for testing. The OECD Guidelines for the Testing of Chemicals are a collection of relevant testing methods to evaluate the toxicity of new and existing chemicals through the Environmental Health and Safety Program. These guidelines are used by government, industry, and independent laboratories to assess the safety of chemical products and chemical preparations, including pesticides and industrial chemicals.[5] The guidelines cover tests for the physical and chemical properties of chemicals, human health effects, environmental effects, and degradation and accumulation of chemicals in the environment. OECD Section 4 Health Effect Guidelines for the Testing of Chemical 418, 419, 424, and 426, are applicable to general and developmental neurotoxicity and provide additional test methods for the histopathological evaluation of neurotoxicity studies. These guidelines can be freely accessed via SourceOECD (http://www.sourceoecd.org), where all updated and new tests are made available online.

In 1985, the EPA established a set of TSCA Test Guidelines in 40 CFR Parts 795 to 798. These guidelines were established as standardized protocols for laboratory testing to evaluate the environmental hazards of a chemical. Standardized guidelines are necessary for the establishment of enforceable test standards in test rules promulgated under Section 4 of TSCA. To reduce the text of the CFR, the EPA deleted those guidelines, which had not been cited in any test rules.

Harmonization of test guidelines between the EPA and the OECD began in the early 1990s to reduce inconsistencies. The EPA guidelines developed by the OPP, OPPT, and OECD guidelines are currently harmonized to make the OPPTS Series of EPA Test Guidelines. EPA scientists develop or modify guidelines for specific endpoints, which are then reviewed by experts and become public drafts. Seven of the 11 Health Effects Test Guidelines that are codified in Subpart H of 40 CFR Part 799 originated from this harmonization process.

Producing a single set of guidelines for use by the OPP under the legislation of the FIFRA and the FFDCA and the OPPT under the legislation of the TSCA has been difficult because of requirements for testing specified by each of these entities. Pesticide test protocols under the FIFRA tend to be more flexible to allow for individual circumstances and promote an interactive process between the EPA and the registrants. Industrial chemical protocols under Section 4 of TSCA require the EPA to have very specific requirements for testing. Therefore, TSCA-specific guidelines found in 40 CFR 799 are nearly identical to the OPPTS harmonized Health Affects Test Guidelines with minor changes in wording to promote enforceability. The 40 CFR 795-798 guidelines from 1985 have not been considered in the harmonization. Future TSCA Section 4 test rules will only cross-reference 40 CFR 799, which were codified in 1997 to distinguish them from the preharmonized 40 CFR 795-798 series. The preharmonized guidelines will be retained until their use has expired.

There are 10 series of harmonized guidelines. The OPPTS Series 870 represents test guidelines for health effects and for neurotoxicity studies. Table 3 provides a chart of the neurotoxicity guidelines and their harmonized versions. The EPA OPPT, the EPA OPP, and the OECD guidelines are all harmonized into EPA's OPPTS guidelines.

An important aspect to this entire process of determining neurotoxicity of a specific chemical is to produce reliable data. The use of good laboratory practices (GLPs) ensures that safety studies are performed in a manner to ensure data integrity. The EPA, FDA, and OECD each have their own GLPs. Although separated by the organization that wrote them, they represent the same ideals. It is critical that personnel be well trained, that adequate facilities and equipment are available, and that all data are properly documented and archived. The guidelines incorporated in this Guide intend for the GLPs to be applied. With the GLPs behind

TABLE 3 Neurotoxicity Guideline Harmonization Chart

Name	EPA OPPT	EPA OPP	OECD	EPA OPPTS
Acute/28-day	798.6450	81–7	418	870.6100
delayed	798.6540	82–5	419	
	798.6560	82–6		
neurotoxicity	798.6050	81–8	424	870.6200
screening	798.6200	82–7		
battery	798.6400	83–1		
Developmental	None	83–6	426	870.6300
neurotoxicity				
study				

each neurotoxicity study conducted, information can be gathered into a reliable database for further characterization of new and existing chemicals.

DEFINITIONS

The following scientific definitions were extracted from the guidelines.

Adverse effect: any treatment-related alteration from baseline that diminishes an organism's ability to survive, reproduce, or adapt to the environment.

Delayed neurotoxicity: a syndrome associated with prolonged delayed onset of ataxia, distal axonopathies in spinal cord and peripheral nerves, and inhibition and aging of neuropathy target esterase in neural tissue.

Developmental neurotoxicity studies: designed to develop data, including dose–response characterizations, on the potential functional and morphological hazards to the nervous system, which may arise in the offspring from exposure of the mother during pregnancy and lactation.

Developmental toxicity: the property of a chemical that causes *in utero* death, structural or functional abnormalities, or growth retardation during the period of development.

Dosage: general term comprising of dose, its frequency, and the duration of dosing.

Dose: the amount of test substance administered. Dose is expressed as weight of test substance (g, mg) per unit weight of a test animal (e.g., mg/kg).

Motor activity: any movement of the experimental animal.

Nerve fiber teasing: separation under a dissecting microscope of individual immersion-fixed or perfusion-fixed peripheral nerve fibers.

Neuropathy target esterase (NTE): a membrane-bound protein that hydrolyzes phenyl valerate. The inhibition and "aging" of the phosphorylated NTE [i.e., the covalent binding of the organophosphate (OP) to the enzyme] is highly correlated with the initiation of organophosphorus-induced delayed neurotoxicity (OPIDN). Not all OPs that inhibit NTE cause OPIDN, but all OPs that cause OPIDN inhibit NTE.

Neurotoxicant: any chemical, biological, or physical agent having the potential to cause neurotoxicity.

Neurotoxicity (NTX) or neurotoxic effect: any adverse effect or change in the structure or function of the nervous system related to exposure to a chemical, biological, or physical substance.

No-observed-effect level (NOEL): the highest dose level where no adverse treatment-related findings are observed.

Organophosphorus-induced delayed neurotoxicity (OPIDN): neurological syndrome in which limb weakness and upper motor neuron spasitcity are the predominant clinical signs; distal axonopathy of peripheral nerve and spinal cord are the correlative pathological signs, and inhibition and aging of the neurotoxic esterase in neural tissues are the correlative biochemical effects. Clinical signs and pathology first appear between one and two weeks following exposures that typically inhibit and subsequently age neurotoxic esterase.

Organophosphorus substances: include uncharged organophosphorus esters, thioesters, or anhydrides of organophosphoric, organophosphoric, or organophosphoramidic acids or of related phosphorothioic, phosphorothioic, or phosphorthioamidic acids, or other substances that may cause the neurotoxicity sometimes seen in this class.

Subchronic delayed neurotoxicity: prolonged, delayed-onset locomoter ataxia resulting from repeated daily administration of the test substance.

Test guidelines: a standardized set of test procedures or protocols organized by health effect or other testing endpoint to produce data, which are accurate, reliable, reproducible, and necessary for the regulatory programs under the Toxic Substances Control Act (TSCA).

Toxic effect: an adverse change in the structure or function of an experimental animal because of exposure to a chemical substance.

U.S. ENVIRONMENTAL PROTECTION AGENCY GENERAL NEUROTOXICOLOGY SCREENING

Necropsy and Tissue Collection

General Information

U.S. EPA 40 CFR 798.6400: Neuropathology

- The goal of the techniques outlined for sacrifice of animals and preparation of tissues is to preserve the tissue's morphology to simulate the living state of the cell.

Perfusion Technique

U.S. EPA OPPTS 870.6200: Neurotoxicity Screening Battery

U.S. EPA 40 CFR 798.6400: Neuropathology

U.S. EPA 40 CFR 799.9620: TSCA Neurotoxicity Screening Battery

U.S. EPA 40 CFR 79.66: Neuropathology Assessment

* OECD Guideline 424: Neurotoxicity Study in Rodents

- Animals should be perfused in situ by a generally recognized technique. For fixation suitable for light or electronic microscopy, saline solution followed by buffered 2.5% glutaraldehyde or buffered 4.0% paraformaldehyde is recommended. Although some minor modifications or variations in procedures are used in different laboratories, a detailed and standard procedure for vascular perfusion may be found in the literature.[6–9]

* The OECD guideline recommends collecting tissues from at least five animals per gender per group.

U.S. EPA 40 CFR 79.66: Neuropathology Assessment

- For studies designed to develop data in animals on morphological changes in the nervous system associated with repeated inhalation exposures to motor vehicle emissions, the lungs should be instilled with fixative via the trachea during the fixation process in order to preserve the lungs and achieve whole-body fixation.

Gross Observations

U.S. EPA OPPTS 870.6200: Neurotoxicity Screening Battery

U.S. EPA 40 CFR 799.9620: TSCA Neurotoxicity Screening Battery

- Any gross abnormalities should be noted during necropsy and the processing of tissues.

Tissue Sampling and Specimen Storage

U.S. EPA 40 CFR 798.6400: Neuropathology

U.S. EPA 40 CFR 79.66: Neuropathology Assessment

* OECD Guideline 424: Neurotoxicity Study in Rodents

- After perfusion, the bony structure (cranium and vertebral column) should be exposed. Animals should then be stored in fixative-filled bags at 4°C for 8 to 12 h. The cranium and vertebral column should be removed carefully by trained technicians without physical damage of the brain and cord. Detailed dissection procedures may be found in the literature.[6,7] After removal, simple measurement of the size (length and width) and weight of the whole brain should be

made. Any abnormal coloration or discoloration of the brain and cord should be noted and recorded. Any observable gross changes should be recorded.

- Unless a given test rule specifies otherwise, cross sections of the following areas should be examined:
  - The forebrain, the center of the cerebrum, (*the OECD guideline suggests including a section through the hippocampus), the midbrain, the cerebellum and pons, and the medulla oblongata.
  - * The OECD guideline suggests examining the eye with the optic nerve and retina.
  - The spinal cord at cervical and lumbar swelling (C_3 to C_6 and L_1 to L_4).
  - *The OECD guideline suggests including both cross or transverse and longitudinal sections of the spinal cord and peripheral nerve sections.
  - Gasserian (trigeminal) ganglia, dorsal root ganglia (C_3 to C_6, L_1 to L_4), dorsal and ventral root fibers (C_3 to C_6, L_1 to L_4), proximal sciatic nerve (midthigh and sciatic notch), sural nerve (at knee), and tibial nerve (at knee).
  - Other sites and tissue elements (e.g., gastrocnemius muscle) should be examined if deemed necessary.
- Tissue samples from both the central and peripheral nervous system should be further immersion fixed and stored in appropriate fixative for future examination, such as:
  - 10% buffered formalin for light microscopy
  - 2.5% buffered glutaraldehyde or 4.0% buffered paraformaldehyde for electron microscopy
- The volume of fixative vs. the volume of tissues in a specimen jar should be no less than 25 : 1.

U.S. EPA OPPTS 870.6200: Neurotoxicity Screening Battery

U.S. EPA 40 CFR 799.9620: TSCA Neurotoxicity Screening Battery

- Tissue samples taken should adequately represent all major regions of the nervous system.
- Tissue samples should be postfixed and processed according to standardized published histological protocols.[6,10–14]

Tissue Processing

Dehydration, Clearing, and Embedding

U.S. EPA 40 CFR 798.6400: Neuropathology

U.S. EPA 40 CFR 79.66: Neuropathology Assessment

- All stored tissues should be washed with buffer for at least 2 h after sampling and storage, prior to further tissue processing. Tissue specimens stored in 10% buffered formalin may be used for histopathological examination. All tissues must be immersion fixed in fixative for at least 48 h prior to further tissue processing.

- All tissue specimens are washed for at least 1 h with water or buffer, prior to dehydration. (A longer washing time is needed if the specimens have been stored in fixative for a prolonged period.)

- Dehydration can be performed with increasing concentration of graded ethanol up to absolute alcohol.

- After dehydration, tissue specimens are cleared with xylene and embedded in paraffin or paraplast.

- Multiple tissue specimens (e.g., brain, cord, ganglia) may be embedded together in one single block for sectioning.

- All tissue blocks should be labeled showing at least the experiment number, animal number, and specimens embedded.

U.S. EPA OPPTS 870.6200: Neurotoxicity Screening Battery

U.S. EPA 40 CFR 799.9620: TSCA Neurotoxicity Screening Battery

- Paraffin embedding is acceptable for tissues of the central nervous system. Plastic embedding of tissue samples from the central nervous system is encouraged when feasible.

- Plastic embedding is required for tissue samples from the peripheral nervous system.

- Tissue blocks and slides should be identified appropriately when stored.

Microtomy and Staining Techniques

U.S. EPA 40 CFR 798.6400: Neuropathology

U.S. EPA 40 CFR 79.66: Neuropathology Assessment

U.S. EPA OPPTS 870.6200: Neurotoxicity Screening Battery

U.S. EPA 40 CFR 799.9620: TSCA Neurotoxicity Screening Battery

- Tissue sections, 5 to 6 μm in thickness, are prepared from the tissue blocks and mounted on standard glass slides.

- It is recommended that several additional sections be made from each block at this time for possible future needs for special staining. All tissue blocks and slides are filed and stored in properly labeled files or boxes.

- Subject to professional judgment and the type of neuropathological alterations observed, it is recommended that additional methods, such as Bodian's or

Bielschowsky's silver methods and/or Glial fibrillary acidic protein (GFAP) immunohistochemistry, be used in conjunction with more standard stains to determine the lowest dosage level at which neuropathological alterations are observed. When new or existing data provide evidence of structural alterations it is recommended that the GFAP immunoassay also be considered.[15]

- A general staining procedure is performed on all tissue specimens in the highest treatment group. Hematoxylin and eosin (H&E) is used for this purpose or a compatible stain. Comparable stains may be used according to standard published protocols.[10,12,13]

- The staining is differentiated properly to achieve bluish nuclei with a pinkish background.

- Based on the results of the general staining with H&E, selected sites and cellular components are further evaluated by the use of specific techniques such as:

  - Neuronal body (e.g., Einarson's gallocyanin)
  - Axon (e.g., Bodian)
  - Myelin sheath (e.g., Kluver's luxol fast blue)
  - Neurofibrils (e.g., Bielschowsky's)

- Peripheral nerve fiber teasing should be used in addition. Detailed staining methodology is available in standard histotechnological manuals.[13,16,17] The nerve fiber teasing technique is discussed in the literature.[6]

- A section of normal tissue is included in each staining to assure that adequate staining has occurred. Any changes should be noted and representative photographs should be taken.

- If a lesion(s) is observed, the special techniques are repeated in the next lower treatment group until no further lesion is detectable.

- If the anatomical locus of expected neuropathology is well defined, plastic (epoxy) embedded sections stained with toluidine blue may be used for small tissue samples. This technique obviates the need for special stains for cellular components.[6]

Histopathological Examination by Microscopy

Qualitative Diagnosis

U.S. EPA OPPTS 870.6200: Neurotoxicity Screening Battery

U.S. EPA 40 CFR 799.9620: TSCA Neurotoxicity Screening Battery

- Representative histological sections from the tissue samples should be examined microscopically by an appropriately trained pathologist for evidence of neuropathological alterations.

- The nervous system should be examined thoroughly for evidence of any treatment-related neuropathological alterations. Particular attention should be paid to regions known to be sensitive to neurotoxic insult or those regions likely to be affected based on the results of functional tests.

- Such treatment-related neuropathological alterations should be clearly distinguished from artifacts resulting from influences other than exposure to the test substance. Guidance for both regions to be examined and the types of neuropathological alterations that typically result from toxicant exposure may be found in the literature.[14]

- A stepwise examination of tissue is recommended.
    1. The high-dose group is compared to the control group.
    2. If no neuropathological alterations are observed in the high dose, subsequent analysis is not required.
    3. If neuropathological alterations are observed in the high dose, samples from the intermediate- and low-dose groups are examined sequentially.

U.S. EPA 40 CFR 79.66: Neuropathology Assessment

- All stained microscopic slides should be examined with a standard research microscope. Examples of cellular alterations (e.g., neuronal vacuolation, degeneration, necrosis) and tissue changes (e.g., gliosis, leukocytic infiltration, cystic formation) should be recorded and photographed.

Subjective Diagnosis

U.S. EPA OPPTS 870.6200: Neurotoxicity Screening Battery

U.S. EPA 40 CFR 799.9620: TSCA Neurotoxicity Screening Battery

- If any evidence of neuropathological alterations is found in the qualitative examination, a subjective diagnosis should be performed for evaluating dose–response relationships.

- All regions of the brain exhibiting any evidence of neuropathological changes should be included in a subjective diagnosis.

- Sections of each region from all dosage groups should be coded as to treatment and examined in randomized order. The frequency of each type and the severity of each lesion should be recorded.

- After all sections from all dosage groups including all regions have been rated, the code should be

broken and statistical analyses performed to evaluate dose–response relationships. For each type of dose-related lesion observed, examples of different ranges of severity should be described.

- Photomicrographs of typical examples of treatment-related regions are recommended to augment the descriptions. The examples should serve to illustrate a rating scale, such as 1+, 2+, and 3+ for the degree of severity, ranging from very slight to very extensive.

Electron Microscopy

U.S. EPA 40 CFR 798.6400: Neuropathology

U.S. EPA 40 CFR 79.66: Neuropathology Assessment

- Based on the results of light-microscopic evaluation, specific tissue sites that reveal a lesion(s) are further evaluated by electron microscopy in the highest treatment group that does not reveal any light-microscopic lesion. If a lesion is observed, the next lower treatment group is evaluated until no significant lesion is found.[8]

- Since the size of the tissue samples that can be examined is very small, at least three or four tissue blocks from each sampling site must be examined. Tissue sections must be examined with a transmission electron microscope.

- Three main categories of structural changes must be considered:
    - *Neuronal body.* The shape and position of the nucleus and nucleolus and any change in the chromatin patterns are noted. Within the neuronal cytoplasm, cytoplasmic organelles such as mitochondria, lysosomes, neurotubules, neurofilaments, microfilaments, endoplasmic reticulum and polyribosomes (Nissl substance), Golgi complex, and secretory granules are examined.
    - *Neuronal processes.* The structural integrity or an alteration of dendrites, axons (myelinated and unmyelinated), myelin sheaths, and synapses is noted.
    - *Supporting cells.* Attention must also be paid to the number and structural integrity of the neuroglial elements (oligodendrocytes, astrocytes, and microglia) of the central nervous system, and the Schwann cells, satellite cells, and capsule cells of the peripheral nervous system. Any change in the endothelial cells and ependymal lining cells is noted whenever possible. The nature, severity, and frequency of each type of lesion in each specimen must be recorded. Representative lesions must be photographed and labeled appropriately.

Evaluation of Data

U.S. EPA 40 CFR 79.66: Neuropathology Assessment

- An evaluation of the data based on gross necropsy findings and microscopic pathology observations should be made and supplied. The evaluation should include the relationship, if any, between the animal's exposure to the test atmosphere and the frequency and severity of any lesions observed.

- The evaluation of dose–response, if it exist, for various groups should be given and a description of statistical method must be presented. The evaluation of neuropathology data should include, where applicable, an assessment in conjunction with any other neurotoxicity, electrophysiological, behavioral, or neurochemical studies, which may be relevant to this study.

- For each animal, data must be submitted showing its identification (animal number, treatment, dose, duration), neurological signs, location(s), nature of, frequency, and severity of lesion(s). A commonly used scale such as 1+, 2+, 3+, and 4+ for degree of severity ranging from very slight to extensive may be used. Any diagnoses derived from neurological signs and lesions, including naturally occurring diseases or conditions, should also be recorded.

- Data of counts and incidence of lesions by test group should be tabulated to show:
  - The number of animals in each group
  - The number of animals displaying specific neurological signs
  - The number of animals in which any lesion was found
  - The number of animals affected by each different type of lesion
  - The average grade of each type of lesion
  - The frequency of each different type and/or location of lesion

U.S. EPA OPPTS 870.6200: Neurotoxicity Screening Battery

U.S. EPA 40 CFR 799.9620: TSCA Neurotoxicity Screening Battery

- The findings from the screening battery should be evaluated in the context of preceding and/or concurrent toxicity studies and any correlated functional and histopathological findings.

- The evaluation should include the relationship between the doses of the test substance and the presence or absence, incidence and severity, of any neurotoxic effects.

- The evaluation should include appropriate statistical analyses: for example, parametric tests for continuous data and nonparametric tests for the remainder. The choice of analyses should consider tests appropriate to the experimental design, including repeated measures. There may be many acceptable ways to analyze the data.

- The following information should be arranged by test group dosage level in tabular form.
  - For each animal:
  - The animal's identification number
  - Body weight
  - Score on each sign at each observation time
  - Time and cause of death (if appropriate)
  - Total session activity counts
  - Intrasession subtotals for each day measured
  - For each group:
  - Number of animals at the start of the test
  - Number of animals showing each observation score at each observation time
  - The mean and standard deviation for each continuous endpoint at each observation time
  - Results of statistical analyses for each measure, where appropriate

- All neuropathological observations should be recorded and arranged by test groups.

U.S. ENVIRONMENTAL PROTECTION AGENCY DELAYED NEUROTOXICITY SCREENING

Necropsy and Tissue Collection

General Information

U.S. EPA OPPTS 870.6100: Acute and 28-Day Delayed Neurotoxicity of Organophosphorus Substances

* OECD guideline 418: Delayed Neurotoxicity of Organophosphorus Substances Following Acute Exposure

* OECD guideline 419: Delayed Neurotoxicity of Organophosphorus Substances: 28-Day Repeated Dose Study

- Three hens from each group are sacrificed at 48 h after the last dose for neuropathy target esterase (NTE) assay. Depending on the duration of acute signs as an indication of the disposition of the test material, the time for sacrifice for NTE and AchE assessment may be chosen at a different time to optimize the detection of effects.

- * The OECD guidelines suggests that three hens from each group (treatment and control) may be sacrificed and assayed for NTE activity after 24 h and then three hens at 48 h after the last dose.

Perfusion Technique

U.S. EPA OPPTS 870.6100: Acute and 28-Day Delayed Neurotoxicity of Organophosphorus Substances

U.S. EPA 40 CFR 798.6560: Subchronic Delayed Neurotoxicity of Organophosphorus Substances

- Tissues from all animals should be fixed in situ by whole-body perfusion, with a fixative appropriate for the embedding media.

Gross Observations

U.S. EPA OPPTS 870.6100: Acute and 28-Day Delayed Neurotoxicity of Organophosphorus Substances

- Gross necropsies are recommended for all survivors and are to include observation of the appearance of the brain and spinal cord.
- All animals should be prepared for microscopic examination.

Tissue Sampling and Specimen Storage

U.S. EPA OPPTS 870.6100: Acute and 28-Day Delayed Neurotoxicity of Organophosphorus Substances

U.S. EPA 40 CFR 798.6560: Subchronic Delayed Neurotoxicity of Organophosphorus Substances

* OECD guideline 418: Delayed Neurotoxicity of Organophosphorus Substances Following Acute Exposure

* OECD guideline 419: Delayed Neurotoxicity of Organophosphorus Substances: 28-Day Repeated Dose Study

- Sections should include:
  - Medulla oblongata
  - Spinal cord
  - Peripheral nerves
- The spinal cord sections should be taken from the:
  - Rostral cervical (upper cervical bulb)
  - The midthoracic
  - The lumbosacral regions
- Sections should be taken from the proximal region of the tibial nerve and its branches.
- A section of the sciatic nerve is requested under U.S. EPA 798.6560.
- * The OECD guidelines suggests further that the sections taken should include the cerebellum (mid-longitudinal level) and that sections of the distal region of the tibial nerve and the branches to the gastrocnemial muscle be taken.

Tissue Processing

Microtomy and Staining Techniques

U.S. EPA OPPTS 870.6100: Acute and 28-Day Delayed Neurotoxicity of Organophosphorus Substances

U.S. EPA 40 CFR 798.6560: Subchronic Delayed Neurotoxicity of Organophosphorus Substances

- Sections should be stained with appropriate myelin and axon-specific stains.

Histopathological Examination by Microscopy

Qualitative Diagnosis

U.S. EPA OPPTS 870.6100: Acute and 28-Day Delayed Neurotoxicity of Organophosphorus Substances

U.S. EPA 40 CFR 798.6560: Subchronic Delayed Neurotoxicity of Organophosphorus Substances

- For 28-day studies a stepwise examination of tissue is recommended:
  - The high-dose group is compared to the control group
  - If no neuropathological alterations are observed in the high dose, subsequent analysis is not required.
  - If neuropathological alterations are observed in the high dose, samples from the intermediate- and low-dose groups are examined sequentially.

Evaluation of Data

U.S. EPA OPPTS 870.6100: Acute and 28-Day Delayed Neurotoxicity of Organophosphorus Substances

- The findings of these delayed neurotoxicity studies should be evaluated in terms of the incidence and severity of:
  - Behavioral effect
  - Neurochemical effect
  - Histopathological effects.
- Evaluation of any other effects observed in the treated and control groups, as well as any information known or available to the authors, such as published studies, are to be included.
- For a variety of the results seen, further studies may be necessary to characterize these effects.

U.S. EPA 40 CFR 798.6560: Subchronic Delayed Neurotoxicity of Organophosphorus Substances

- The findings of a subchronic delayed neurotoxicity study should be evaluated in conjunction with the findings of preceding studies and considered in terms of the incidence and severity of observed neurotoxic effects and any other observed effects and histopathological findings in the treated and control groups.

- A properly conducted subchronic test should provide a satisfactory estimation of a no-effect level based on lack of clinical signs and histopathological changes.

U.S. ENVIRONMENTAL PROTECTION AGENCY DEVELOPMENTAL NEUROTOXICITY SCREENING

Necropsy and Tissue Collection

General Information

U.S. EPA OPPTS 870.6300: Developmental Neurotoxicity Study

U.S. EPA OPPTS 870.8600: Developmental Neurotoxicity Screen

U.S. EPA 40 CFR 799.9630: Developmental Neurotoxicity

U.S. EPA 40 CFR 795.250: Developmental Neurotoxicity

* OECD Guideline 426: Developmental Neurotoxicity Study

- Neuropathological evaluation should be conducted on animals on postnatal day (PND) 11 and at the termination of the study.

- The OECD guideline suggests conducting evaluations on animals at PND 22 or between PND 11 and PND 22 and at study termnination.

- At 11 days of age, one male or female pup should be removed from each litter such that equal numbers of male and female offspring are removed from all litters combined. Of these, six male and six female pups will be sacrificed for neuropathological analysis. The pups will be killed by exposure to carbon dioxide, and immediately thereafter the brains should be removed, weighed, and immersion-fixed in an appropriate aldehyde fixative. The remaining animals will be sacrificed in a similar manner and immediately thereafter their brains removed and weighed.

- At the termination of the study, one male or one female from each litter will be killed by exposure to carbon dioxide, and immediately thereafter the brain should be removed and weighed. In addition, six animals sex per dose group (one male or female per litter) should be sacrificed at the termination of the study for neuropathological evaluation.

- Neuropathological analysis of animals sacrificed at the termination of the study should be performed in accordance with the OPPTS 870.6200.

- The OECD guideline suggests that dams be euthanized after the offspring have weaned.

U.S. EPA OPPTS 870.8600: Developmental Neurotoxicity Screen

U.S. EPA 40 CFR 795.250: Developmental Neurotoxicity Screen

- One male and one female per litter are sacrificed at weaning and the remainder following the last behavioral measures.

- Neuropathology and brain weight determinations are made on animals sacrificed at weaning and after the last behavioral measures.

Perfusion Technique

U.S. EPA OPPTS 870.8600: Developmental Neurotoxicity Screen

U.S. EPA 40 CFR 795.250: Developmental Neurotoxicity Screen

* OECD Guideline 426: Developmental Neurotoxicity Study

- Animals should be perfused in situ by a generally recognized technique.

- * The OECD guideline suggests that animals killed on or before PND 22 can be fixed via immersion or perfusion, and those (adults) killed at termination should be fixed by perfusion.

U.S. EPA OPPTS 870.6300: Developmental Neurotoxicity Study

U.S. EPA 40 CFR 799.9630: Developmental Neurotoxicity

- For fixation of tissue samples for postnatal day 11 animals, immediately following removal, the brain should be weighed and immersion fixed in an appropriate aldehyde fixative.

- The brains should be post fixed according to standardized published histological protocols found in the literature.[11,12,18,19]

Gross Observations

U.S. EPA OPPTS 870.8600: Developmental Neurotoxicity Screen

U.S. EPA 40 CFR 795.250: Developmental Neurotoxicity Screen

- After perfusion, the brain and spinal cord should be removed, the brains weighed, and gross abnormalities noted.

- At least 10 animals that are not sacrificed for histopathology should be used to determine brain weight. The animals should be decapitated and the brains carefully removed, blotted, chilled, and weighed.

The following dissection should be performed on an ice-cooled glass plate:

- The rhombencephalon is separated by a transverse section from the rest of the brain and dissected into the cerebellum and the medulla oblongata/pons.
- A transverse section is made at the level of the optic chiasm which delimits the anterior part of the hypothalamus and passes through the anterior commissure.
- The cortex is peeled from the posterior section and added to the anterior section. This divides the brain into four sections: the telencephalon, the diencephalon/midbrain, the medulla oblongata/pons, and the cerebellum.
- Sections should be weighed as soon as possible after dissection to avoid drying.
- Detailed methodology is available in the literature.[20]

Tissue Sampling and Specimen Storage

U.S. EPA OPPTS 870.6300: Developmental Neurotoxicity Study

U.S. EPA 40 CFR 799.9630: Developmental Neurotoxicity

- Tissue blocks and slides should be identified appropriately when stored.

U.S. EPA OPPTS 870.8600: Developmental Neurotoxicity Screen

U.S. EPA 40 CFR 795.250: Developmental Neurotoxicity Screen

* OECD Guideline 426: Developmental Neurotoxicity Study

- For the cross-section areas recommended for developmental neurotoxicity studies, refer to OPPTS 870.6200.
- Cross sections of the following areas should be examined:
  - Forebrain
  - Center of the cerebrum and midbrain
  - Cerebellum and pons
  - Medulla oblongata
  - Spinal cord at the cervical and lumbar swelling
  - Gasserian ganglia
  - Dorsal root ganglia
  - Dorsal and ventral root fibers
  - Proximal sciatic nerve (midthigh and sciatic notch)
  - Sural nerve (at the knee)
  - Tibial nerve (at the knee)

- * The OECD guideline suggests also sampling olfactory bulbs, hippocampus, basal ganglia, thalamus, and the tibial nerve calf muscle branches.
- Tissue samples from both the central and peripheral portions of the nervous system should be further immersion-fixed and stored in appropriate fixative for further examination.
- * The OECD guideline emphasizes that the spinal cord and peripheral nerves be sampled in cross or transverse and longitudinal sections.

Tissue Processing

Dehydration, Clearing, and Embedding

U.S. EPA OPPTS 870.6300: Developmental Neurotoxicity Study

U.S. EPA 40 CFR 799.9630: Developmental Neurotoxicity

* OECD Guideline 426: Developmental Neurotoxicity Study

- Paraffin embedding is acceptable, but plastic embedding is preferred and recommended.
- The OECD guideline suggests that brain tissue for morphometric analysis be embedded from all dose levels at the same time to avoid shrinkage that may produce artifacts.

U.S. EPA OPPTS 870.8600: Developmental Neurotoxicity Screen

U.S. EPA 40 CFR 795.250: Developmental Neurotoxicity Screen

- After dehydration, tissue specimens are cleared with xylene and embedded in paraffin or paraplast except for the sural nerve, which should be embedded in plastic. A method for plastic embedding is described in the literature.[19]

Microtomy and Staining Techniques

U.S. EPA OPPTS 870.8600: Developmental Neurotoxicity Screen

U.S. EPA 40 CFR 795.250: Developmental Neurotoxicity Screen

- A general staining procedure should be performed on all tissue specimens in the highest treatment group. Hematoxylin and eosin (H&E) stain should be used for this purpose. The staining should be differentiated properly to achieve bluish nuclei with a pinkish background.
- Based on the results of the general staining, selected sites and cellular components should be further

evaluated by use of specific techniques. If H&E screening does not provide such information, a battery of stains should be used to assess the following components in all appropriate required samples:

- Neuronal body (e.g., Einarson's gallocyanin)
- Axon (e.g., Kluver's luxol fast blue)
- Neurofibrils (e.g., Bielschowsky's)
- In addition, nerve fiber teasing should be used.
- A section of normal tissues should be included in each staining to assure that adequate staining has occurred.
- Any changes should be noted and representative photographs should be taken.
- If lesions are observed, the special techniques should be repeated in the next lower treatment group until no further lesions are detectible.
- If the anatomical locus of expected neuropathology is well defined, epoxy-embedded sections stained with toluidine blue may be used for small tissue samples. This technique obviates the need for special stains.

U.S. EPA OPPTS 870.6300: Developmental Neurotoxicity Study

U.S. EPA 40 CFR 799.9630: Developmental Neurotoxicity

- Histological sections should be stained for hematoxylin and eosin, or a similar stain, according to standard published protocols.[10,16,18]
- For animals sacrificed at the termination of the study, the regions to be examined and the types of alterations that should be assessed are identified in OPPTS 870.6200 or 799.9620.
- Subject to professional judgment and the type of neuropathological alterations observed, it is recommended that additional methods such as Bodian's or Bielschowsky's silver methods and/or immunohistochemistry for glial fibrillary acid protein be used in conjunction with more standard stains to determine the lowest dosage level at which neuropathological alterations are observed.

Histopathological Examination by Microscopy

Qualitative Diagnosis

U.S. EPA OPPTS 870.6300: Developmental Neurotoxicity Study

U.S. EPA 40 CFR 799.9630: Developmental Neurotoxicity

- The purposes of the qualitative examination are:

- To identify regions within the nervous system exhibiting evidence of neuropathological alterations.
- To identify types of neuropathological alterations resulting from exposure to the test substance.
- To determine the range of severity of the neuropathological alterations.

- Representative histological sections from the tissue samples should be examined microscopically by an appropriately trained pathologist.
- A stepwise examination of tissue is recommended.
  - Sections from the high-dose group are compared with those of the control group.
  - If no evidence of neuropathological alterations is found in animals of the high-dose group, no further analysis is required.
  - If evidence of neuropathological alterations is found in the high-dose group, animals from the intermediate- and low-dose group are examined.
- Adequate samples should be taken from all major brain regions and examined for any evidence of treatment-related neuropathological alterations, such as:
  - Olfactory bulbs
  - Cerebral cortex
  - Hippocampus
  - Basal ganglia
  - Thalamus
  - Hypothalamus
  - Midbrain (tectum, tegmentum, and cerebral peduncles)
  - Brain stem and cerebellum
- Types of alterations:[21,22]
  - Typical types of cellular alterations include neuronal vacuolation, degeneration, and necrosis.
  - Typical tissue changes include astrocytic proliferation, leukocytic infiltration, and cystic formation.
- Emphasis should be paid to structural changes indicative of developmental insult, including but not restricted to:
  - Gross changes in the size or shape of brain regions such as alterations in the size of the cerebral hemispheres or the normal pattern of foliation of the cerebellum.
  - The death of neuronal precursors, abnormal proliferation, or abnormal migration, as indicated by pyknotic cells or ectopic neurons, or gross alterations in regions with active proliferative and migratory zones, and alterations in transient

developmental structures (e.g., the external germinal zone of the cerebellum).[23]

- Abnormal differentiation, although more apparent with special stains, may also be indicated by shrunken and malformed cell bodies.
- Evidence of hydrocephalus, in particular enlargement of the ventricles, stenosis of the cerebral aqueduct, and general thinning of the cerebral hemispheres.
- Guidance for neuropathological examination for indications of developmental insult to the brain can be found in OPPTS 870.6200.

Subjective Diagnosis

U.S. EPA OPPTS 870.6300: Developmental Neurotoxicity Study

U.S. EPA 40 CFR 799.9630: Developmental Neurotoxicity

- If any evidence of neuropathological alterations is found in the qualitative examination, a subjective diagnosis will be performed for the purpose of evaluating dose–response relationships.
- All regions of the brain exhibiting any evidence of neuropathological changes should be included in this analysis.
- Sections of each region from all dose groups will be coded as to treatment and examined in randomized order. The frequency of each type and the severity of each lesion will be recorded.
- After all sections from all dose groups including all regions have been rated, the code will be broken and statistical analyses performed to evaluate dose–response relationships.
- For each type of dose-related lesion observed, examples of different ranges of severity should be described. The examples will serve to illustrate a rating scale, such as $1+$, $2+$, and $3+$ for the degree of severity ranging from very slight to very extensive.
- Since the disruption of developmental processes is sometimes more clearly reflected in the rate or extent of growth of particular brain regions, some form of morphometric analysis should be performed on postnatal day 11 and at the termination of the study to assess the structural development of the brain. At a minimum, this would consist of a reliable estimate of the thickness of major layers at representative locations within the neocortex, hippocampus, and cerebellum.[24]
- * OECD Guideline 426 suggests a quantitative evaluation including the use of morphometry and stereology to detect treatment-related effects.

Evaluation of Data

U.S. EPA OPPTS 870.6300: Developmental Neurotoxicity Study

U.S. EPA 40 CFR 799.9630: Developmental Neurotoxicity

- A description of the test system and test methods should be provided in the final evaluation of the data to include:
  - A detailed description of the procedures used to standardize observations and procedures as well as operational definitions for scoring observations.
  - Positive control data from the laboratory performing the test that demonstrates the sensitivity of the procedures being used. These data do not have to be from studies using prenatal exposures. However, the laboratory must demonstrate competence in evaluation effects in neonatal animals perinatally exposed to chemicals and establish test norms for the appropriate age group.
  - Procedures for calibrating and ensuring the equivalence of devices and the balancing of treatment groups in testing procedures.
  - A short justification explaining any decisions involving professional judgment.
- The results must be arranged by each treatment and control group as follows:
  - In tabular form, data for each animal must be provided showing:
    - Its identification number and the litter from which it came
    - Its body weight and score on each developmental landmark at each observation time
    - Total session activity counts and intrasession subtotals on each day measured
    - Auditory startle response amplitude per session and intrasession amplitudes on each day measured
    - Appropriate data for each repeated trial (or session) showing acquisition and retention scores on the tests of learning and memory on each day measured
    - Time and cause of death (if appropriate); any neurological signs observed; a list of structures examined as well as the locations, nature, frequency, and extent of lesions; and brain weights
  - The following data should also be provided, as appropriate:
    - Inclusion of photomicrographs demonstrating typical examples of the type and extent of the

neuropathological alterations observed is recommended.

- Any diagnoses derived from neurological signs and lesions, including naturally occurring diseases or conditions, should also be recorded.

- Summary data for each treatment and control group must include:
  - The number of animals at the start of the test
  - The body weight of the dams during gestation and lactation
  - Litter size and mean weight at birth
  - The number of animals showing each abnormal sign at each observation time
  - The percentage of animals showing each abnormal sign at each observation time
  - The mean and standard deviation for each continuous endpoint at each observation time (these will include body weight, motor activity counts, auditory startle responses, performance in learning and memory tests, regional brain weights and whole brain weights, both absolute and relative)
  - The number of animals in which any lesion was found
  - The number of animals affected by each different type of lesion, and the location, frequency, and average grade of each type of lesion for each animal
  - The values of morphometric measurements made for each animal listed by the treatment group

- An evaluation of test results must be made to include the relationship between the doses of the test substance and the presence or absence, incidence, and extent of any neurotoxic effect.

- The evaluation should include appropriate statistical analyses.

- The choice of analyses should consider tests appropriate to the experimental design and needed adjustments for multiple comparisons.

- The evaluation should include the relationship, if any, between observed neuropathological and behavioral alterations.

U.S. FOOD AND DRUG ADMINISTRATION GENERAL NEUROTOXICOLOGY SCREENING

Necropsy and Tissue Collection

Perfusion Technique

U.S. FDA, CFSAN, Redbook 2000 V.C.10: Neurotoxicity Studies

- For screening, either immersion fixation or in situ perfusion of tissues is acceptable.
- For a previously screened chemical that undergoes special neurotoxicity testing and to enhance detection of subtle neuropathological findings, tissues should be perfusion-fixed in situ.

Gross Observations

- A specific histopathological examination should be made of tissue samples representative of all major areas and elements of the brain, spinal cord, and peripheral nervous system.
- Emphasis should be placed more on the carefulness of the histopathological examination of the neuronal tissue and the documentation of the findings rather than on the numbers of sections used, provided that all major areas and elements of the nervous system are included.
- The initial examination may be carried out on tissues from the control and highest dose group. Positive findings would then be followed by examination of tissues from the other dosage groups.

Tissue Processing

Microtomy and Staining Techniques

U.S. FDA, CFSAN, Redbook 2000 IV.C.10: Neurotoxicity Studies

- To enhance the detection of subtle neuropathological findings, tissues should have a detailed histopathological examination (more thorough than the histopathology examination performed during screening) carried out involving the use of special stains to highlight relevant neural structures.

Histopathological Examination by Microscopy

Qualitative Diagnosis

U.S. FDA, CFSAN, Redbook 2000 V.C.10: Neurotoxicity Studies

- Tissues from the control and the highest-dose groups are examined first. Positive findings are followed by examination of tissues from the other dosage groups. The concept of age appropriateness is considered in the morphological evaluation of immature nervous systems.[25]
- For special neurotoxicity testing where a chemical has been identified as producing neurotoxicity, tissues should be perfusion fixed in situ and a detailed histopathological examination (more thorough than

the histopathological examination performed during screening) should be carried out involving special stains to highlight relevant neural structures.[26]

Evaluation of Data

U.S. FDA, CFSAN, Redbook 2000 V.C.10: Neurotoxicity Studies

- Experimental data should be accurately recorded, documented, and reported to the FDA.
- Summary tables of all positive effects should be presented, and all data collected (positive and negative) should be submitted to the FDA to enable review personnel the opportunity of examining the actual study results.
- As appropriate, data should be analyzed using suitable and acceptable statistical procedures and this information, together with any other pertinent toxicity data, should be incorporated into an integrated assessment of the potential for the test chemical to adversely affect the structural or functional integrity of the nervous system.
- Based on this assessment, an explicit statement should be made as to whether or not the test chemical represents a potential neurotoxic hazard that may require special neurotoxicity testing.
- Study protocols for additional neurotoxicity testing should be developed using valid state-of-the-art methodology.
- Throughout the process of protocol design and testing in the assessment of neurotoxic potential, the opportunity for consultation with FDA is available and encouraged.
- A critical element used in defining a chemical's neurotoxic hazard is the no-observed-adverse-effect level (NOAEL), typically using the most relevant and sensitive endpoint identified in previous testing.
- To enable a more quantitative determination of the NOAEL, ample data should be obtained to thoroughly characterize the dose–response and dose–time relationships in repeated exposure studies (e.g., intermittent and continuous exposure regimes), typically using the most relevant and sensitive endpoint.
- At the stage of special neurotoxicity testing, efforts to develop additional relevant information for a more comprehensive assessment of neurotoxic hazard are certainly encouraged. For example, information regarding the occurrence of treatment-related neurochemical changes, the pharmacokinetic properties of the test compound, or the factors that may modulate the sensitivity of the organism to the test compound

could contribute to a better understanding of the neurobiological processes underlying the chemically induced neurotoxicity.

- This mechanistic type of information would enable a more reliable interpretation of the available animal data for predicting neurotoxic risk in humans.

U.S. FOOD AND DRUG ADMINISTRATION DEVELOPMENTAL NEUROTOXICITY SCREENING

General Information

U.S. FDA, CFSAN, Redbook 2000 IV.C.9.a: Guidelines for Reproduction Studies

- Nearly all the information in this guideline pertinent to pathology has to do with the reproductive tissues and therefore is beyond the scope of this chapter. For reference, the reader is referred to this guideline, but in terms of the collection, processing, and evaluation of the nervous system tissues, the conduct of a developmental neurotoxicity screen that follows the EPA/OECD guidelines and an article by Bolon et. al.[27] should be considered as the best and most comprehensive approach.

REFERENCES

1. Organisation for Economic Co-operation and Development. Chairman's Report of the Meeting of the Ad Hoc Working Group of Experts on Systemic Short-term and (Delayed) Neurotoxicity, Paris, Feb 1992.
2. US Environmental Protection Agency. Laws, Regulations, Guidance and Dockets. Available at: http://www.epa.gov/lawsregs/index.html. Accessed Apr 26, 2009.
3. US Environmental Protection Agency. Scientific Advisory Panel. Available at: http://www.epa.gov/scipoly/sap/index.htm. Accessed Apr 26, 2009.
4. US Food and Drug Administration. Center for Food Safety and Applied Nutrition. 2001. Toxicological Principles for the Safety Assessment of Food Ingredients. *Redbook 2000.* Assessment http://www.cfsan.fda.gov/~redbook/red-toca.html. Accessed Apr 26, 2009.
5. Organisation for Economic Co-operation and Development. About the OECD. Assessment http://www.oecd.org. Accessed Apr 26, 2009.
6. Spencer PS, Schaumberg HH. *Experimental and Clinical Neurotoxicology.* Baltimore: Williams & Wilkins; 1980: Chap. 50.
7. Palay SL, Chan Palay V. *Cerebellar Cortex: Cytology and Organization.* New York: Springer-Verlag; 1974.
8. Hayat MA.Vol. 1. Biological applications. In: *Principles and Techniques of Electron Microscopy.* New York: Van Nostrand Reinhold; 1970.

9. Zeman W, Innes JRM. *Craigie's Neuroanatomy of the Rat.* New York: Academic press; 1963.

10. Bennet HS, et al. Science and art in the preparing tissues embedded in plastic for light microscopy, with special reference to glycol methacrylate, glass knives and simple stains. *Stain Technol.* 1976;51:71–97.

11. Di Sant Agnese PA, De Mesy Jensen K. Dibasic staining of large epoxy sections and application to surgical pathology. *Am J Clini Pathol.* 1984;81:25–29.

12. Pender MP. A simple method for high-resolution light microscopy of nervous tissue. *J Neurosci Methods.* 1985;15:213–218.

13. Armed Forces Institute of Pathology. *Manual of Histologic Staining Methods.* New York: McGraw-Hill; 1968.

14. World Health Organization. *Principles and Methods for the Assessment of Neurotoxicity Associated with Exposure to Chemicals.* Environmental Health Criteria Document 60. New York: WHO Press; 1986.

15. O'Callaghan JP. Quantification of glial fibrillary acidic protein: comparison of slot-immunobinding assays with a novel sandwich ELISA. *Neurotoxicol Teratol.* 1991;13:275–281.

16. Ralis HM, Beesley RA, Ralis ZA. *Techniques in Neurohistology.* London: Butterworth; 1973.

17. Chang LW. *Color Atlas and Manual for Applied Histochemistry.* Springfield, IL: Charles C Thomas; 1979.

18. Luna LG. *Manual of Histologic Staining Methods of the Armed Forces Institute of Pathology*, 3rd ed. New York: McGraw-Hill; 1968:1–31.

19. Spencer PS, Bischoff MC, Schaumburg HH. Neuropathological methods for the detection of neurotoxic disease. In: *Experimental and Clinical Neurotoxicology.* Baltimore: Williams & Wilkins; 1980:743–757.

20. Glowinski J, Iversen LL. Regional studies of catecholamines in the rat brain: I. *J Neurochem.* 1966;13:655–669.

21. Friede RL. *Developmental Neuropathology.* New York: Springer-Verlag; 1975:1–23, 297–313, 326–351.

22. Suzuki K. Special vulnerabilities of the developing nervous system to toxic substances. In: *Experimental and Clinical Neurotoxicology.* Baltimore: Williams & Wilkins; 1980: 48–61.

23. Miale IL, Sidman RL. An autoradiographic analysis of histogenesis in the mouse cerebellum. *Exp Neurol.* 1961;4: 277–296.

24. Rodier PM, Gramann WJ. Morphologic effects of interference with cell proliferation in the early fetal period. *Neurobehav Toxicol.* 1979;1:129–135.

25. O'Donoghue JL. Screening for neurotoxicity using a neurologically based examination and neuropathology. *J. Am. Coll Toxicol.* 1989;8:97–115.

26. Slikker W Jr, Chang LW. *Handbook of Developmental Neurotoxicology.* San Diego, CA: Academic Press; 1998.

27. Bolon B, Garman R, Jensen K, Krinke G, Stuart B. A "best practices" approach to neuropathologic assessment in developmental neurotoxicity testing—for today. *Toxicol Pathol* 2006;34:296–313.

NOTE ADDED IN PROOF

An up-to-date list of worldwide regulatory guidelines for preclinical toxicologic neuropathology research has been recently published: Bolon B, et al. Compilation of international regulatory guidance documents for neuropathology assessment during nonclinical toxicity studies. *Toxicol Pathol.* 2011; 39:92–96.

35

TOXICOLOGICAL NEUROPATHOLOGY: THE NEXT TWO DECADES

BRAD BOLON

GEMpath, Inc., Longmont, Colorado

MARK T. BUTT

Tox Path Specialists, LLC, Hagerstown, Maryland

A proverb from Shakespeare's *The Tempest* holds that "What's past is prologue." The truth of this adage is obvious in a world steeped in evolution and the imperative of constant change. Nevertheless, if history is indeed the forerunner of tomorrow, we still have to ask ourselves another proverbial question: Where do we go from here?

Most chapters of this book reviewed the classic themes, fundamental knowledge, and conventional methods and tools that comprise the toxicologic neuropathologist's stock-in-trade. However, a few chapters went beyond the familiar to introduce some relatively new realms of inquiry in this field: noninvasive imaging systems (Chapter 18) and molecular genomics of the nervous system (Chapter 20). These two fields, while accelerating in importance, do not exhaust the avalanche of changes that will transform the practice of toxicologic neuropathology during the next two decades. In the remainder of this chapter we briefly address four of the more vital and far-reaching trends that we predict will drive the toxicological neuropathology profession during this time.

First, the number and complexity of regulations related to neurotoxicity testing will expand substantially. For good or bad, the guidelines provided by the various regulatory agencies greatly affect research, so their importance can only be characterized as being paramount. In general, those institutions developing new drugs and devices are going to at least meet (and hopefully some will exceed, as needed) the scientific rigors provided by the regulatory guidelines.

The reason for this increased regulatory oversight is readily understood. The adverse impact of neurotoxic exposures on human endeavors is not going away.[1] Contact with neurotoxicants occurs in both our personal[2] and professional[3,4] lives. Increased exposure to neurotoxic agents is increasingly associated with the mounting incidence (the rate at which new cases arise during a given period) and prevalence (the proportion of a population that is affected at a given point in time) of addictions,[5] congenital neurological defects,[6–9] chronic neurodegenerative diseases,[10] and neural tumors.[11] Such events are individually devastating, but when entire communities are affected, the outcome is tragic.[8]

The key to an effective interaction between toxicological neuropathologists and regulators will be for both sides to negotiate meaningful changes to improve the existing guidelines. The goal for generating new regulations should be to craft a rational set of common tests that will effectively identify and characterize novel neurotoxic hazards; this objective is obvious, and it continues building on the current body of knowledge. More important, regulatory activity in the next two decades should work vigorously to remove the present confusion[12] regarding study design and data collection that afflicts both regulators and the regulated. A single set of detailed but flexible guidelines for general toxicity studies and dedicated neurotoxicity studies should be negotiated and then implemented around the world, or at least within a given region.

Fundamental Neuropathology for Pathologists and Toxicologists: Principles and Techniques, First Edition. Edited by Brad Bolon and Mark T. Butt.
© 2011 John Wiley & Sons, Inc. Published 2011 by John Wiley & Sons, Inc.

A second broad trend is that toxicological neuropathologists of the future will require much greater theoretical understanding of the nature of the normal and abnormal reactions of nervous system cells. Standard subjects taught in a medical setting (i.e., neuroanatomy, neuroembryology, neurophysiology, neuropathology, neuropharmacology, neurotoxicology) will remain vital to success. However, practitioners of this art will also require in-depth, advanced expertise of many other fundamental biomedical concepts (i.e., functional neuroanatomy, functional neurochemistry, neural biomarkers, neurobehavioral analysis) as well as a greater familiarity with conventional but underutilized tools (e.g., alternative microscopy, such as confocal microscopy and stereology, digital imaging, noninvasive imaging, and telepathology). Neuropathologists of tomorrow will have to be even more specialized than their counterparts of today.

Having now introduced alternative methods, the third area of future growth is expected to be the increased use of more available and practical quantitative techniques. Hopefully, the days of selecting a single "representative section" on which to perform morphometrics are gone. More and more journals are refusing, rightfully so, this type of biased, largely meaningless data. Faster, more efficient and more available methods of quantifying objects in a three-dimensional space (the type of space in which pathologists are always working) require stereological methods. As an example, the concepts of stereology as applied to the quantification of neurotoxic endpoints are introduced in Chapter 15. Although stereology as a workable set of techniques has been available for awhile, automation of some of the processes, the ability to more easily and accurately perform quantitative investigations on paraffin sections (as representatives of the three-dimensional structures of the brain), and the use of computers not only to rapidly and efficiently (in terms of personnel and therefore expenditures) evaluate but also collect data in an automated way has tremendous possibilities. Quantification on a tissue and even cellular level will improve and be more available. Imagine being able to readily determine phenotypic changes in a neuronal population by being able to rapidly and efficiently count receptor density.

As more specific immunohistochemical (fluorescent and nonfluorescent) or nucleic acid based markers are developed, as hardware such as confocal microscopes, digitizing tissue scanners, and laser dissection systems is improved (and becomes more cost-effective), and as improved software increases the speed and accuracy of data collection and interpretation, the routine quantification of neuronal subpopulations, glial reactions, protein concentrations, and receptor density will improve. This will lead to the ability to easily detect what currently are (generally) unrecognized markers of neurotoxicity. In the past and present, far too many neuropathology investigations have depended almost exclusively on the semiquantitative grading of morphologic changes in nervous system tissues. Although this system (when applied by a qualified pathologist) has proved its value over at least several decades, developing methodologies that more easily and more accurately detect cellular changes currently not able to be recognized with available light microscopic techniques will understandably enhance the detection of neurotoxicity. The crude morphologic examinations of today may not only be missing safety issues, they may be overlooking truly amazing breakthroughs in the treatment of neurologic dysfunction simply because we are not able to recognize the effect.

Although we do not predict the replacement of morphological pathologists by supercomputers that read slides, we do see the need for pathologists to embrace and utilize improving technologies in order to quantify a broader range of neurotoxicity endpoints.

Finally, the next two decades will see much greater emphasis on discerning disease mechanisms responsible for neurotoxic conditions. Practitioners of toxicological neuropathology will have to be comfortable dealing with transformational medicine (which employs individual-specific functional, molecular, and structural data acquired in animals) to refine health hazard identification and characterization.

In summary, toxicological neuropathologists will no longer be highly trained but focused scientists dealing only with the acquisition, analysis, interpretation, and communication of traditional data sets. Instead, tomorrow's toxicological neuropathologist will broaden and deepen their skills to take on new roles as "super-specialists."

REFERENCES

1. Claudio L. An analysis of the U. S. Environmental Protection Agency neurotoxicity testing guidelines. *Regul Toxicol Pharmacol*. 1992;16:202–212.

2. Bearer CF. Developmental neurotoxicity: illustration of principles. *Pediatr Clin N Am*. 2001;48:1199–1213.

3. Gobba F. Occupational exposure to chemicals and sensory organs: a neglected research field. *Neurotoxicology*. 2003;24:675–691.

4. Connelly JM, Malkin MG. Environmental risk factors for brain tumors. *Curr Neurol Neurosci Rep*. 2007;7:208–214.

5. Jones DC, Miller GW. The effects of environmental neurotoxicants on the dopaminergic system: a possible role in drug addiction. *Biochem Pharmacol*. 2008;76:569–581.

6. Castoldi AF, Johansson C, Onishchenko N, et al. Human developmental neurotoxicity of methylmercury: impact of variables and risk modifiers. *Regul Toxicol Pharmacol*. 2008;51:201–214.

7. Costa LG, Aschner M, Vitalone A, Syversen T, Soldin OP. Developmental neuropathology of environmental agents. *Annu Rev Pharmacol Toxicol.* 2004;44:87–110.

8. Ekino S, Susa M, Ninomiya T, Imamura K, Kitamura T. Minamata disease revisited: an update on the acute and chronic manifestations of methyl mercury poisoning. *J Neurol Sci.* 2007;262:131–144.

9. Jurewicz J, Hanke W. Prenatal and childhood exposure to pesticides and neurobehavioral development: review of epidemiological studies. *Int J Occup Med Environ Health.* 2008;21:121–132.

10. Reuhl KR. Delayed expression of neurotoxicity: the problem of silent damage. *Neurotoxicology.* 1991;12: 341–346.

11. Bassil KL, Vakil C, Sanborn M, Cole DC, Kaur JS, Kerr KJ. Cancer health effects of pesticides: systematic review. *Can Fam Physician.* 2007;53:1704–1711.

12. Bolon B, Bradley A, Butt M, Jensen K, Krinke G. Compilation of international regulatory guidance documents for neuropathology assessment during nonclinical toxicity studies. *Toxicol Pathol.* 2011;39:92–96.

APPENDIXES

READY REFERENCES OF NEUROBIOLOGY KNOWLEDGE

BRAD BOLON

GEMpath, Inc., Longmont, Colorado

The contents of this book represent a unique source of knowledge and skills required by neuropathologists and neurotoxicologists for the successful practice of toxicologic neuropathology. These appendixes have been compiled to supplement the mainstream details offered in the chapters, and specifically to bring together in one location many neuroscience facts and figures that are buried in the literature or hidden in relatively obscure Web pages. The tables include information on:

Appendix 1: neural cell markers [listed by both the cell populations in which they are expressed (part A) and in alphabetical order (part B)]

Appendix 2: text-based neuroanatomical atlases

Appendix 3: text-based basic neuroscience references

Appendix 4: Web-based neuroscience references (including online atlases)

Appendix 5: references for peripheral target tissues (ear, eye, skeletal muscle) that might be affected in neurotoxicity studies

Appendix 6: comparative and correlative neurobiological parameters for the brain [at both the gross (part A) and microscopic (part B) levels, spinal cord (part C) and their blood supplies (part D)]

Appendix 7: differences in the relative proportions among brain and spinal cord regions in humans and rats

Appendix 8: chemical composition of neural tissues and fluids

The sites chosen and their myriad links will serve as a useful starting point for researchers looking to improve their understanding and proficiency in toxicologic neuropathology. Obviously, though, new data in this field arise almost daily (especially on the Internet), so these tables represent only a first taste to whet the appetite for further exploration in this arena.

Fundamental Neuropathology for Pathologists and Toxicologists: Principles and Techniques, First Edition. Edited by Brad Bolon and Mark T. Butt.
© 2011 John Wiley & Sons, Inc. Published 2011 by John Wiley & Sons, Inc.

APPENDIX 1   NEURAL CELL MARKERS OF POTENTIAL UTILITY FOR TOXICOLOGICAL NEUROPATHOLOGY APPLICATIONS

A. Cell Type–Specific Listing of Markers

Cell Type	Marker Abbreviation	Marker Name	Comments
Neurons			
Mature	CaBP (CALB)	Calbindin	
	ChAT	Choline acetyltransferase	Cholinergic neurons (especially in the basal ganglia and spinal cord intermediate and ventral horns)
	CgA	Chromogranin A	Selected neuroendocrine populations only
	DBH	Dopamine-β-hydroxylase	Autonomic neurons
	GAD1	Glutamic acid decarboxylase 1	GABAergic neurons
	GAP43	Growth-associated protein 43 (AKA: neuromodulin)	Axons (especially growth cones) and cell bodies of regenerating axons and neurons
	LINGO 1	Leucine-rich repeat- and Ig domain-containing Nogo receptor-interacting protein 1	Neurons (especially in cerebral cortex)
	MAP2	Microtubule-associated protein 2	Cell bodies and dendrites
	NFP	Neurofilament proteins	Especially axons of large cells
	NSE	Neuron-specific enolase	γ subunit is fairly specific for neurons (and neuroendocrine cells)
	PRPH	Peripherin	Autonomic and sensory PNS neurons
	PGP 9.5	Protein gene product 9.5	Neurons and neuronal processes (CNS and PNS)
	SYN	Synapsin I	
	SYP	Synaptophysin	
	τ	Tau	Axons of mature CNS neurons
	TH	Tyrosine hydroxylase	Catecholaminergic neurons
	TUJ1	Neuron-specific class III β-tubulin	Neurons (CNS and PNS)
	VIM	Vimentin	Adult neurons that are dying (e.g., in neurodegenerative diseases or after trauma) or that can regenerate (e.g., olfactory)
Precursors	DCX	Doublecortin	Newly formed precursors
	FZD4 (CD344)	Frizzled 4	Stem cells
	GAP43	Growth-associated protein 43 (AKA neuromodulin)	Axons (especially growth cones) and bodies of neuron progenitors
	INA	α-Internexin	Also called Neurofilament protein 66 kDa (NF66)
	NCAM	Neural cell adhesion molecule	
	NES	Nestin	Stem cells
	NOTCH	Notch	Stem cells
	VIM	Vimentin	Stem cells
Glia			
Astrocytes	AQP4	Aquaporin 4	Foot processes
	GFAP	Glial fibrillary acidic protein	Astrocytes
	GS	Glutamine synthetase	Foot processes
	SYT-IV	Synaptotagmin IV	Astrocytes
	VIM	Vimentin	

A. Cell Type–Specific Listing of Markers (*Continued*)

Cell Type	Marker Abbreviation	Marker Name	Comments
Guiding glia	BLBP	Brain lipid binding protein	Radial glia
	GLAST	Glutamate aspartate transporter	Bergmann and radial glia
	HES1	Hairy and enhancer of split-1	Radial glial cells (presumptive)
	RC1	Radial cell 1	Radial glia and their processes
	SOX9	Sry (sex-determining region Y)-type high mobility group box 9	Bergmann glia
Microglia	CD11b	Cluster of differentiation 11b	
	CD68	Cluster of differentiation 68	Activated cells
	Coro1A	Coronin 1A	
	GS-1-B4	*Griffonia simplicifolia* isolectin B4	
	HLA-DR	Human leukocyte antigen, D-related	
	Iba1	Ionized calcium-binding adaptor molecule 1	Activated cells
	MSR1 (CD204)	Macrophage scavenger receptor type 1	
	RCA-1	*Ricinus communis* agglutinin-1	
Oligodendrocytes	CAII	Carbonic anhydrase II	
	CNPase	2′,3′-Cyclic nucleotide 3′-phosphohydrolase	
	GALC	Galactosylceramidase	
	GLAST	Glutamate aspartate transporter	
	GS-1-B4	*Griffonia simplicifolia* isolectin B4	Immature cells
	MAG	Myelin-associated glycoprotein	
	MBP	Myelin basic protein	
	MOG	Myelin oligodendrocyte glycoprotein	Mature cells
	O1	Oligodendrocyte marker 1	Late oligodendrocyte precursors and oligodendrocytes
	O4	Oligodendrocyte marker 4	Early and late oligodendrocyte precursors
	Olig	Oligodendrocyte-specific transcription factor (isoforms 1, 2, and 3)	Oligodendrocyte precursors, neoplastic oligodendrocytes
	OMgp	Oligodendrocyte myelin glycoprotein	Oligodendrocytes
	PLP	Proteolipid protein (AKA lipophilin)	
Perineural cells	EMA	Epithelial membrane antigen	Perineurial cells
	GLUT-1	Glucose transporter 1	
Schwann cells	CNPase	2′,3′-Cyclic nucleotide 3′-phosphohydrolase	
	GALC	Galactosylceramidase	
	MBP	Myelin basic protein	
	MPZ (P0)	Myelin protein 0	
	PMP22	Peripheral myelin protein 22	Schwann cells
	PLP	Proteolipid protein (AKA lipophilin)	
	S100β	100% Soluble (in ammonium sulfate at neutral pH)	
Other elements			
Choroid plexus	CAII	Carbonic anhydrase II	
	CK	Cytokeratin	
	TTR	Transthyretin	
Endothelium	CD31	Cluster of differentiation 31	
	RCA-1	*Ricinus communis* agglutinin-1	
	VWF	Von Willebrand's factor	

(*continued*)

A. Cell Type–Specific Listing of Markers (*Continued*)

Cell Type	Marker Abbreviation	Marker Name	Comments
Ependyma	GFAP	Glial fibrillary acidic protein	Immature cells
	GLAST	Glutamate aspartate transporter	Immature cells
	S100β	100% Soluble (in ammonium sulfate at neutral pH)	
	VIM	Vimentin	
Extracellular matrix	LAM	Laminin	
Meninges	VIM	Vimentin	Fibroblasts especially
Neuroendocrine cells	CgA	Chromogranin A	Neurons in adrenal medulla or autonomic paraganglia
	NSE	Neuron-specific enolase	γ (gamma) subunit is relatively specific for neuroendocrine cells (and various neuron populations)
	PRPH	Peripherin	Neoplastic cells
	PGP 9.5	Protein gene product 9.5	
	SYP	Synaptophysin	Normal and neoplastic cells

B. Alphabetical Listing of Neural Cell Markers

Marker Abbreviation	Marker Name	Cell Specificity	Function
AQP4	Aquaporin 4	Astrocyte foot processes	Water channel protein that controls water movement across the blood–brain barrier
BLBP	Brain lipid-binding protein	Radial glia (that support neuronal migration during development of the cerebral cortex)	Cytoplasmic protein engaged in fatty acid transport and metabolism
CAII	Carbonic anhydrase II	Choroid plexus epithelium (CPE), oligodendrocytes (ODC)	CPE = transport of bicarbonate ions, sodium ions, and water from blood to the CSF; ODC = supports myelin compaction by removing ions and water
CaBP (CALB)	Calbindin	Neurons	Calcium-binding protein that regulates cytosolic calcium levels and controls calcium-dependent signaling
CD11b	Cluster of differentiation 11b	Microglia	Cell surface marker participating in cell adhesion interactions
CD68	Cluster of differentiation 68	Activated microglia	Lysosomal glycoprotein of unknown function that is expressed during inflammation
ChAT	Choline acetyltransferase	Cholinergic neurons (especially in the basal ganglia and spinal cord intermediate and ventral horns)	Enzyme responsible for producing the excitatory neurotransmitter acetylcholine (ACh)
CgA	Chromogranin A	Neuroendocrine cells, selected neurons	Acidic glycoprotein that helps to assemble and stabilize secretory granules, and to guide various hormones into them
CK	Cytokeratin	Choroid plexus epithelium, ependymal cells	Intermediate filament
CNPase	2′,3′-Cyclic nucleotide 3′-phosphohydrolase	Oligodendrocytes, Schwann cells	Membrane protein regulating the production and maintenance of myelin

B. Alphabetical Listing of Neural Cell Markers (*Continued*)

Marker Abbreviation	Marker Name	Cell Specificity	Function
Coro1A	Coronin 1A	Microglia	Actin-binding cytoskeletal protein of unknown function (perhaps linking the cytoskeleton to the plasma membrane and extracellular matrix)
DBH	Dopamine-β-hydroxylase	Autonomic neurons	Enzyme to convert the excitatory neurotransmitter dopamine (DA) into the excitatory neurotransmitter norepinephrine (NE)
DCX	Doublecortin	Immature neurons	Microtubule-associated protein which, due to nearly exclusive presence in newly formed neurons, can be used as an endogenous marker of neuronogenesis
EMA	Epithelial membrane antigen	Perineural cells	Transmembrane glycoprotein (alternate designation: MUC-1)
FZD4 (CD344)	Frizzled 4	Neuronal stem cells	Transmembrane G-protein-coupled receptor that regulates tissue and cell polarization and proliferation
GAD1	Glutamic acid decarboxylase 1	GABAergic neurons	Enzyme catalyzing formation of the inhibitory neurotransmitter γ-aminobutyric acid (GABA)
GALC	Galactosylceramidase	Oligodendrocytes, Schwann cells	Lysosomal enzyme that degrades galactosylceramide (a principal myelin constituent) during myelin production
GAP43	Growth-associated protein 43 (AKA neuromodulin)	Axons (especially growth cones) and cell bodies of neuronal progenitors and neurons engaged in axonal generation or repair	Transmembrane protein expressed in the motile growth cone during development and regeneration; phosphorylated after long-term potentiation during learning
GFAP	Glial fibrillary acidic protein	Astrocytes, ependymal cells	Intermediate filament protein
GLAST	Glutamate aspartate transporter	Astrocytes, Bergmann and radial glia, immature ependymal cells	Transmembrane protein that transfers glutamate into cells against a concentration gradient
GLUT-1	Glucose transporter 1	Brain endothelial cells, perineurial cells	Transmembrane glycoprotein involved in glucose transfer into cells
GS	Glutamine synthetase	Astrocyte foot processes	Enzyme engaged in regulating concentrations of the excitatory neurotransmitter glutamate and the nitrogen donor molecule glutamine
GS-1	*Griffonia simplicifolia* agglutinin	Microglia, immature astrocytes and immature oligodendrocytes	Agglutinin with affinity for terminal α-D-galactosyl residues; also called *Bandeiraea simplicifolia*
HES1	Hairy and enhancer of split-1	Radial glial cells (presumptive)	Transcription factor that represses NOTCH-mediated gene signaling during early glial differentiation
HLA-DR	Human leukocyte antigen, D-related	Microglia	Transmembrane protein of the major histocompatibility complex (MHC), class II, involved in antigen presentation

(*continued*)

B. Alphabetical Listing of Neural Cell Markers (*Continued*)

Marker Abbreviation	Marker Name	Cell Specificity	Function
Iba1	Ionized calcium-binding adaptor molecule 1	Activated microglia	Calcium-binding protein expressed during chronic inflammation
INA	α-Internexin	Neuronal precursors	Intermediate filament expressed during early neuronal development
LAM	Laminin	Extracellular matrix	Modulates cell conformation, differentiation, and motion; promotes neurite regeneration
LINGO 1	Leucine-rich repeat- and Ig domain-containing Nogo receptor-interacting protein 1	Neurons (especially in cerebral cortex)	Transmembrane protein that regulates neurite outgrowth and inhibits myelination
MAG	Myelin-associated glycoprotein	Oligodendrocytes	Ligand for the Nogo66 receptor (NgR), binding of which inhibits axon regeneration in the CNS
MAP2	Microtubule-associated protein 2	Dendrites and neuron cell bodies	Cytoskeletal protein that stabilizes interactions between microtubules and intermediate filaments
MBP	Myelin basic protein	Oligodendrocytes, Schwann cells	Membrane protein participating in the formation and stabilization of myelin
MOG	Myelin oligodendrocyte glycoprotein	Mature oligodendrocytes	Membrane protein of unknown function
MPZ (P0)	Myelin protein 0	Schwann cells	Adhesion molecule that supports compaction of PNS myelin
MSR1 (CD204)	Macrophage scavenger receptor type 1	Microglia	Receptors mediating endocytosis of modified low-density lipoproteins
NCAM	Neural cell adhesion molecule	Neurons	Membrane protein that guides cell migration during development
NES	Nestin	Stem cells, neural crest cells	Intermediate filament-associated protein
NF66	Neurofilament protein 66 kDa	See entry above for "α-internexin"	
NFP	Neurofilament proteins	Neurons (especially axons of large cells)	Intermediate filament protein (three isoforms: light = 68 kDa, medium = 160 kDa, heavy = 200 kDa)
NMDAR-1	*N*-Methyl-D-aspartate receptor, subunit 1	Neurons that bind the excitatory neurotransmitter glutamate	Receptor for glycine (a glutamate co-agonist); regulates neuronal development and plasticity
NMDAR-2	*N*-Methyl-D-aspartate receptor, subunit 2	Neurons that bind the excitatory neurotransmitter glutamate	Receptor with affinity for the selective agonist NMDA; regulates neuron development and plasticity
NSE	Neuron-specific enolase	γ subunit is relatively specific for neurons and neuroendocrine cells	Glycolytic enzyme that participates in energy metabolism
NOTCH	Notch	Neuronal stem cells	Transmembrane protein that helps regulate neural development and stem cell survival and proliferation
O1	Oligodendrocyte marker 1	Late oligodendrocyte precursors and oligodendrocytes	Surface glycolipid of unknown function
O4	Oligodendrocyte marker 4	Early and late oligodendrocyte precursors	Surface glycolipid of unknown function

B. Alphabetical Listing of Neural Cell Markers (*Continued*)

Marker Abbreviation	Marker Name	Cell Specificity	Function
Olig	Oligodendrocyte-specific transcription factor (isoforms 1, 2, and 3)	Oligodendrocyte precursors, neoplastic oligodendrocytes	Differentiation and maturation of oligodendrocyte precursors in the motor neuron progenitor (pMN) domain of the spinal cord
OMgp	Oligodendrocyte myelin glycoprotein	Oligodendrocytes	Cell surface protein that inhibits axonal outgrowth and promotes growth cone collapse
PRPH	Peripherin	Autonomic and sensory neurons of PNS, neoplastic cells originating from the diffuse neuroendocrine system	Intermediate filament protein
PGP 9.5	Protein gene product 9.5	Neurons and neuronal processes (CNS and PNS), neuroendocrine cells	Soluble cytoplasmic protein of the ubiquitin C-terminal hydroxylase (UCH) family
PMP22	Peripheral myelin protein 22	Schwann cells	Membrane protein that regulates the development and maintenance of PNS myelin; may also control maturation, proliferation, shape, and survival of Schwann cells
PLP	Proteolipid protein (AKA lipophilin)	Oligodendrocytes, Schwann cells	Transmembrane protein that aids in compaction and stabilization of myelin
RC1	Radial cell 1	Radial glia and their processes	Protein that acts during neural development to guide the migration of neural precursors
RCA-1	*Ricinus communis* agglutinin-1	Microglia, endothelium	Binds oligosaccharides ending in galactose and sometimes *N*-acetylgalactosamine
S100β	100% Soluble (in ammonium sulfate at neutral pH)	Ependymal cells, glia (especially Schwann cells)	Calcium-binding protein that regulates cell differentiation and cell-cycle progression
SOX9	Sry (sex-determining region Y)-type high mobility group box 9	Bergmann glia	Transcription factor that regulates neural crest formation and neuron migration
SYN	Synapsin I	Neurons	Phosphoprotein in membrane of presynaptic vesicles involved in neurotransmitter release
SYP	Synaptophysin	Neurons, normal and neoplastic neuroendocrine cells	Membrane glycoprotein of presynaptic vesicles in CNS neurons and also the granules of neuroendocrine cells
SYT-IV	Synaptotagmin IV	Astrocytes	Calcium-binding membrane protein that control glial glutamine release
τ	Tau	Axons of mature CNS neurons	Tubulin-binding protein promoting microtubule assembly and stability; tau-rich inclusions develop in neurodegenerative diseases
TH	Tyrosine hydroxylase	Catecholaminergic neurons	Enzyme catalyzing the conversion of the amino acid L-tyrosine to dihydroxyphenylalanine (DOPA), which is a precursor for dopamine

(*continued*)

B. Alphabetical Listing of Neural Cell Markers (*Continued*)

Marker Abbreviation	Marker Name	Cell Specificity	Function
TTR	Transthyretin	Choroid plexus epithelium	Protein regulating transfer of the thyroid hormone thyroxine (T4) into the cerebrospinal fluid (CSF)
TUJ1	Neuron-specific class III β-tubulin	Neurons (CNS and PNS)	Cytoskeletal protein that promotes microtubule stability and plays a role in axonal transport; does not cross-react with glial β-tubulin
VIM	Vimentin	Astrocytes, ependymal epithelium, fibroblasts, meningeal cells, neural stem cells, adult neurons that are dying (e.g., in neurodegenerative diseases or after trauma) or that can regenerate (e.g., olfactory)	Intermediate filament protein of mesenchymal-derived cells and immature neuron populations

APPENDIX 2   TEXT-BASED NEUROANATOMIC ATLASES FOR TOXICOLOGICAL NEUROPATHOLOGY STUDIES

Species	Age	Features	References[a]
Fish			
Zebrafish	Adult		289
Bird			
Chicken	Adult	Stereotaxic	275
	Adult		290
	Multiple ages *in ovo*	Stereotaxic	152,219
	Multiple ages *in ovo*		20,236
Rodent			
Guinea pig	Adult	Stereotaxic	161
Hamster	Adult	Stereotaxic	183
Mouse	Adult	Stereotaxic	86,87,203,245
	Adult		241,274,285
	Multiple gestational ages		125,204,231,265,285
Rat	Adult	Stereotaxic	76,150,199,208–213,263,293
	Adult	Neurochemical distribution	214
	Multiple gestational and early postnatal ages		5,6,12,202
Rabbit	Adult	Stereotaxic	76,97,182,239,272
Carnivore			
Cat	Adult	Stereotaxic	76,248,276
Dog	Adult		2,243
	Adult	Stereotaxic	65,156,197
Pig	Adult	Stereotaxic	74
	Adult		280
Primate			
Baboon	Adult	Stereotaxic	53,222
Cebus monkey	Adult	Stereotaxic	69,165
Chimpanzee	Adult	Stereotaxic	57
Cynomolgus monkey	Adult		288

APPENDIX 2   Text-Based Neuroanatomic Atlases for Toxicological Neuropathology Studies (*Continued*)

Species	Age	Features	References[a]
Primate (*Continued*)			
Human	Adult	Stereotaxic	59,60,75,162–164,230
	Adult		32,80,133,162,167,269,287
	Multiple gestational and early postnatal ages		14–18,190,191,257
Marmoset	Adult	Stereotaxic	198,258
Rhesus monkey	Adult	Stereotaxic	205,206,249
	Adult		227
Squirrel monkey	Adult	Stereotaxic	70,95

This table excludes Web-based neuroanatomic atlases, which are included in Appendix 4.
[a] Assistance in identifying references for this table was provided by Drs. Alys Bradley, Robert Garman, and Anna Oevermann.

APPENDIX 3   TEXT-BASED NEUROBIOLOGY REFERENCES FOR TOXICOLOGICAL NEUROPATHOLOGISTS

Scientific Discipline	References[a]
Neuroanatomy, adult	
Comparative	26,34,49,50,56,123,186,259
Human	32,43,75,81,101,107,108,167,188,207,234,269,284
Laboratory animal	
Cat	62,96
Chicken	96,121,130
Dog	62,71,96
Guinea pig	217
Hamster	217
Mouse	142,149,217
Primate, nonhuman	7,168
Rabbit	194,217
Rat	40,100,200,201,217,292
Neuroanatomy, developmental	
Comparative	55,171,187,221,236
Human	45,64,226,228
Laboratory animal	
Primate, nonhuman	47,102
Rodent (mouse)	125,126,225,265,266
Neurobiology	19,34,42,104,123,124,128,254
Neurochemistry	23-25,48,242,270
Neurohistology	
Methods	
Fixation	27,37,79
Collecting samples	27,30,37,79,91,94,148
Processing tissues	27,36,46,94,148
Staining specimens	27,36,46,58,148,218,235,278
Structural analysis	
Histology	1,262
Markers	109,116
Ultrastructure	216
Morphometry and stereology	31,54,286

(*continued*)

APPENDIX 3   **Text-Based Neurobiology References for Toxicological Neuropathologists** (*Continued*)

Scientific Discipline	References[a]
Neuropathology	
Anatomic pathology	
Comparative	262
Human	1,99,107,139,159,169,192,237
Laboratory animal	
Chicken	44,145
Dog	179,262
Mouse	118,147,166,181,279
Rat	28,118,147,174,180,251,263
Clinical pathology	
Cerebrospinal fluid	13,41,77,114,122,135,223,267,277
Interpretation	
Artifacts	79,90,120,172,173,283
Basic principles	89,99,115,120,146
Central nervous system	1,262
Developing neural tissue	88,154
Peripheral nervous system	1,67,119,132,148,189,233,262
Neurophysiology	51,101,103,124
Neuroscience, clinical	
Medical (human)	4,78,104,188,254,291
Veterinary (multiple species)	29,158,193
Neurotoxicology	
Pharmacology	33,48,185,256
Toxicology	10,22,38,39,61,160,244,252,253,268,273

[a] Assistance in identifying references for this table was provided by Drs. Alys Bradley, Kathleen Funk, Robert Garman, and Georg Krinke.

APPENDIX 4   WEB-BASED NEUROBIOLOGY REFERENCES FOR TOXICOLOGICAL NEUROPATHOLOGISTS

Scientific Discipline	References[a]	Scientific Discipline	References[a]
Neuroanatomy, adult		Neurohistology	
Comparative	84,134,175,271,282	Methods	
Human	21,66,110,112,138,144	Collecting samples	285
Laboratory animal		Processing tissues	136,285
Cat	21,177	Staining specimens	136
Chicken	151,177	Structural analysis	
Dog	82,177	Histology	3,83,110,136
Mouse	9,63,112,113,175,177	Ultrastructure	136
Primate, nonhuman	21,66,72,112,176,177	Neuropathology	
Rabbit	282	Anatomic pathology	
Rat	112,175,177	Comparative	111,131
Neuroanatomy, developmental		Human	3,117,137
Comparative	271	Clinical pathology	
Human	73,246,250,255,260	Cerebrospinal fluid	3,238
Laboratory animal		Neurophysiology	8
Rodent (mouse)	52,113,175,224,232,260		

[a] Dr. Anna Oevermann provided assistance identifying references (~10) for this table.

APPENDIX 5   TOXICOLOGICAL NEUROPATHOLOGY REFERENCES FOR NERVOUS SYSTEM TARGETS

Scientific Discipline	References
Ear	
Biology	92,105,127,143
Pathology	105,184
Eye	
Biology	85,93,155,247
Handling and processing	98
Pathology	68,93
Skeletal muscle	
Biology	3,220
Handling and processing	3,106
Pathology	3,35,106,170,215

APPENDIX 6   COMPARATIVE AND CORRELATIVE NEUROBIOLOGICAL PARAMETERS FOR THE BRAIN

A. Brain, Macrolevel Measurements

Parameter	Structure	Species	Value	References
Weight (grams)	Brain (whole)	Human, adult	1300–1400	42
		Human, neonate	350–400	42
		Cat	30	42
		Chimpanzee	420	42
		Dog (beagle)	72	42
		Goldfish	0.01	42
		Guinea pig	4	42
		Hamster	1.4	42
		Monkey (baboon)	137	42
		Monkey (rhesus)	90–97	42
		Monkey (squirrel)	22	42
		Mouse (20 to 30 g adult)	0.4–0.5	285
		Rabbit	10–13	42
		Rat (400g adult)	2	42
	Hypothalamus	Human (60 kg)	4	42
	Cerebellum	Human (60 kg)	142	261
		Cat (3.5 kg)	5.3	261
		Dog (3.5 kg)	6.0	261
		Guinea pig (485 g)	0.9	261
		Monkey, macaque (6 kg)	7.8	261
		Mouse (28 g)	0.09	261
		Rabbit (1.8 kg)	1.9	261
Weight (as % of body weight)	Brain	Human	2	42
Weight loss per year (%) after age 30	Brain	Human, adult	0.25	157
Volume	Brain	Human, adult male	1274 (range, 1053 to 1499)	153
		Human, adult female	1131 (range, 975 to 1398)	153
Volume (as % of total intracranial contents)	Brain	Human (1700 mL = 100%, i.e., entire intracranial volume)	1400 mL (80%)	42
	Blood	Human (1700 mL = 100%)	150 mL (10%)	42
	Cerebrospinal fluid	Human (1700 mL = 100%)	150 mL (10%)	42
	Brain	Rat	253 mm^3	141

(*continued*)

A. Brain, Macrolevel Measurements (*Continued*)

Parameter	Structure	Species	Value	References
Ratio of gray matter volume to white matter volume	Cerebral hemispheres	Human, 20 years old	1.3	178
		Human, 50 years old	1.1	178
		Human, 100 years old	1.5	178
Area, surface (cm^2)	Cerebral cortex	Human, adult	2500	186
	Cerebellar cortex	Human, adult	50,000	240
	Cerebral cortex	Cat	83	186
		Rat	6	186
Thickness (mm)	Cerebral cortex	Human, adult	1.5–4.5	42

B. Brain, Microlevel Measurements

Parameter	Structure	Species	Value	References
Numbers (Average)				
Neurons	Brain (whole)	Human, adult	1×10^{10} to 1.3×10^{10}	42,229
	Cerebral cortex	Human, adult	1×10^{10} to 2.5×10^{10}	140,195,196,240
	Neocortex	Human, adult male	2.28×10^{10}	195,196
	Neocortex	Human, adult female	1.93×10^{10}	195,196
	Cerebral cortex, auditory	Human, adult	1×10^8	42
	Cerebral cortex, visual	Human, adult	5.4×10^8	42
	Difference between hemispheres	Human, adult	1.86×10^9 more cells on the left	195,196
	Cerebral cortex	Rat	2.1×10^7	141
	Cerebellum	Human, adult	Over 50% of all brain neurons	157
	Cerebellum Purkinje cells	Human, adult	1.5×10^7 to 2.6×10^7	42
	Geniculate body, lateral (visual center)	Human, adult	5.7×10^5	42
	Geniculate body, medial (auditory center)	Human, adult	5.7×10^5	42
	Nucleus of cranial nerve IV (trochlear)	Human, adult	2000–3500	42
	Nucleus of cranial nerve VII (facial)	Human, adult	7000	42
	Nucleus of cranial nerve XII (hypoglossal)	Human, adult	4500–7500	42
	Olfactory receptor cells	Human, adult	4×10^7	42
	Olfactory receptor cells	Dog	1×10^9	42
	Olfactory receptor cells	Rabbit	1×10^8	42
Axons	Corpus callosum	Human, adult	1.1×10^6	42
	Pyramidal tract (above decussation)	Human, adult	2.5×10^8	42
	Cranial nerve II (optic)	Human, adult	1.2×10^6	42
	Cranial nerve II (optic)	Cat	1.19×10^5	42
	Cranial nerve II (optic)	Rat	7.48×10^4	42
	Cranial nerve III (oculomotor)	Human, adult	1.2×10^6	42
	Cranial nerve IV (trochlear)	Human, adult	2000–3500	42
	Cranial nerve V (trigeminal, motor root)	Human, adult	8100	42
	Cranial nerve V (trigeminal, sensory root)	Human, adult	1.4×10^5	42
	Cranial nerve VII (facial)	Human, adult	9000–10000	42

B. Brain, Microlevel Measurements (*Continued*)

Parameter	Structure	Species	Value	References
Glia	Brain (whole)	Human, adult	10–50 times the number of neurons	11,42
	Neocortex	Human, young adult	3.9×10^{10}	195,196
	Neocortex	Human, older adult	3.6×10^{10}	195,196
Synapses	Neocortex	Human, adult	6.0×10^{13} to 2.4×10^{14}	140,195,196,229,240
	Per "typical" neuron	Human, adult	1×10^3 to 1×10^4	42
	Per Purkinje cell	Human, adult	2×10^5	42
Cell turnover	Neocortical neurons	Human, adult	8.5×10^4/day (3.1×10^7/year)	195,196
Diameter				
Neurons	Granule cell	Human, adult	$4\,\mu m$	42
	Motor neuron (spinal cord)	Human, adult	$100\,\mu m$	42
	Neuronal nucleus	Human, adult	3 to $18\,\mu m$	42
	Synaptic vesicles (nm)	Human, adult	50 (small), to 70 to 200 (large)	42
Cytoskeleton	Microfilament		$5\,nm$	42
	Microtubule		$20–25\,nm$	42
	Neurofilament		$7–10\,nm$	42
Length/thickness/width				
Neurons	Plasma membrane width		$5\,nm$	42
Axons	Internodal distance		$150–1500\,\mu m$	42
Synapse	Synaptic cleft width		$10–20\,nm$	42
Transport rate (mm/day)				
Axons	Slow (actin, tubulin)		0.2–0.4	42
	Intermediate (mitochondrial proteins)		15–50	42
	Fast (glycolipids, peptides)		200–400	42

C. Spinal Cord, All Measurements

Parameter	Structure	Species	Value	References
Length (cm)	Spinal cord	Human, adult female	43	42
		Human, adult male	45	42
		Cat	34	42
		Rabbit	18	42
	Vertebral column	Human, adult female	61	42
		Human, adult male	71	42
Weight (gram)	Spinal cord	Human, adult	35	42
		Rabbit	4	42
		Rat (400 g)	0.7	42
Cross-sectional area (mm^2)	Spinal cord (C$_2$)	Human, adult	110	281
	Spinal cord (C$_4$)	Human, adult	122	281
	Spinal cord (C$_5$)	Human, adult	78	281
	Spinal cord (C$_7$)	Human, adult	85	281
Maximal circumference (mm)	Spinal cord, cervical intumescence	Human, adult	38	42
	Spinal cord, lumbar intumescence	Human, adult	35	42
Number of neurons (average)	Spinal cord	Human, adult	1×10^9	42

D. Blood Supply

Parameter	Structure	Species	Value	References
Blood delivery				
Vascular supply	Brain	Human, adult	~ 645 km	153
Blood flow				
From heart (%)	Brain	Human, adult	15–20	124
Throughput (mL/cm^3 per min)	Brain, whole	Human, adult	750–1000	42
Throughput (mL/100 g per min)	Brain, whole	Human, adult	54	42
	Brain, whole	Human, child	105	42
	Cerebral cortex, whole	Human, adult	55–60	42
	Cerebral cortex (gray matter)	Human, adult	75	42
	Cerebral cortex (white matter)	Human, adult	45	42
Arterial flow (mL/min)	Basilar artery	Human, adult	100–200	124
	Carotid artery	Human, adult	350	124
Oxygen (O_2) utilization				
Resting O_2 used (%)	Brain	Human, adult	20	42
O_2 consumption (cm^3/min)	Brain, whole	Human, adult	46	42
O_2 consumption (%)	Brain, gray matter	Human, adult	94	157
	Brain, white matter	Human, adult	6	157

APPENDIX 7   COMPARISON OF RELATIVE PROPORTIONS IN THE BRAIN AND SPINAL CORD REGIONS OF HUMANS AND RATS

	Human	Rat	References		Human	Rat	References
Regional Proportion (%) by Volume (Relative to Total CNS Volume)				*Number of Spinal Cord Segments*			
Cerebral cortex	77	31	264	Cervical	8	8	42
Diencephalon	4	7	264	Thoracic	12	13	42
Mesencphalon	4	6	264	Lumbar	5	6	42
Cerebellum	10	10	264	Sacral	5	4	42
Brain stem	2	7	264	Coccygeal	1	3	42
Spinal cord	2	35	264				
Regional Proportion (%) by Volume (Relative to Total Cortical Volume)							
Frontal cortex	41		129				
Parietal cortex	19		129				
Temporal cortex	22		129				
Occipital cortex	18		129				

APPENDIX 8   CHEMICAL COMPOSITION OF CENTRAL NERVOUS SYSTEM TISSUES AND FLUIDS

Brain Composition

	Content (%)	Reference
Water	77–78	42
Lipid	10–12	42
Protein	8	42
Carbohydrate	1	42
Soluble organic compounds	2	42
Inorganic salts	1	42
Myelin lipids	70–80	42
Myelin proteins	20–30	42

Fluid Composition

	Cerebrospinal Fluid	Serum	References
Osmolarity (mOsm/L)	295	295	77
Water (%)	99	93	77
Protein (mg/dL)	35	7000	77
Glucose (mg/dL)	60	90	77
Na^+ (meq/L)	138	138	77
K^+ (meq/L)	2.8	4.5	77
Ca^{2+} (meq/L)	2.1	4.8	77
Mg^{2+} (meq/L)	0.3	1.7	77
Cl^- (meq/L)	119	102	77
pH	7.3	7.4	77,124

Physiology

		References
Volume	125–150 mL	42
Turnover	3–4 exchanges/day	124
Half-life	3 h	42
Rate of production	0.35 mL/min (500 mL/day)	124
Specific gravity	1.007	42
Intracranial pressure	150–180 mm H_2O	42
Red blood cells	0–5 mm^3	42
White blood cells	0–3 mm^3	42

REFERENCES

1. Adams JH, Duchen LW. *Greenfield's Neuropathology*, 5th ed. Oxford; UK: Oxford University Press; 1992.

2. Adrianov OS, Mering TA. *Atlas of the Canine Brain*. Arlington, MA: NPP Books; 2010.

3. Agamanolis DP. Neuropathology: An Illustrated Interactive Course for Medical Students and Residents. Available at: http://neuropathology.neoucom.edu. Accessed Sept 21, 2009.

4. Alloway KD, Pritchard TC. *Medical Neuroscience*, 2nd ed. Raleigh, NC: Hayes Barton Press; 2007.

5. Altman J, Bayer SA. *Atlas of Prenatal Rat Brain Development*. Boca Raton, FL: CRC Press; 1995.

6. Alvarez-Bolado G, Swanson LW. *Developmental Brain Maps: Structure of the Embryonic Rat Brain*. New York: Elsevier; 1996.

7. Ankel-Simons F. *Primate Anatomy: An Introduction*, 2nd ed. San Diego, CA: Academic Press; 2000.

8. Anonymous. Clinical Neurophysiology on the Internet. Available at: http://www.neurophys.com/contents.shtml. Accessed Sept 21, 2009.

9. Anonymous. Mouse Atlas Project 2.0. Laboratory of Neuro Imaging (LONI), University of California–Los Angeles. Available at: http://map.loni.ucla.edu. Accessed Sept 21, 2009.

10. Anthony DC, Montine TJ, Valentine WM, Graham DG. Toxic responses of the nervous system. In: Klaassen CD, ed. *Casarett & Doull's Toxicology: The Basic Science of Poisons*, 6th ed. New York: McGraw Hill; 2002: 535–564.

11. Aschner M, Sonnewald U, Tan KH. Astrocyte modulation of neurotoxic injury. *Brain Pathol*. 2002;12:475–481.

12. Ashwell KWS, Paxinos G. *Atlas of the Developing Rat Nervous System*, 3rd ed. San Diego, CA: Academic Press; 2008.

13. Wood JH, ed. *Neurobiology of Cerebrospinal Fluid*. New York: Plenum Press; 1983: 205–231.

14. Bayer SA, Altman J. *The Human Brain During the Early First Trimester*. Vol 5 Boca Raton, FL: CRC Press; 2008.

15. Bayer SA, Altman J. *The Human Brain During the Late First Trimester*. Vol 4 Boca Raton, FL: CRC Press; 2006.

16. Bayer SA, Altman J. *The Human Brain During the Second Trimester*, Vol 3. Boca Raton, FL: CRC Press; 2005.

17. Bayer SA, Altman J. *The Human Brain During the Third Trimester*, Vol 2 Boca Raton, FL: CRC Press; 2004.

18. Bayer SA, Altman J. *The Spinal Cord from Gestational Week 4 to the 4th Postnatal Month*. Vol 1 Boca Raton, FL: CRC Press; 2002.

19. Bear MF, Connors BW, Paradiso MA. *Neuroscience: Exploring the Brain*, 3rd ed. Philadelphia: Lippincott Williams & Wilkins; 2006.

20. Bellairs R, Osmond M. *Atlas of Chick Development*, 2nd ed. San Diego, CA: Academic Press; 2005.

21. Brain Biodiversity Bank Atlases at Michigan State University. Available at: https://www.msu.edu/user/brains/brains/index. html. Accessed Sept 21, 2009.

22. Blain PG, Harris JB. *Medical Neurotoxicology: Occupational and Environmental Causes of Neurological Dysfunction*. New York: Oxford University Press; 1999.

23. Bloom FE, Björklund A, Hökfelt T. *The Primate Nervous System, Part I*. Vol 13 Amsterdam: Elsevier; 1997.

24. Bloom FE, Björklund A, Hökfelt T. *The Primate Nervous System, Part II*. Vol 14 Amsterdam: Elsevier; 1998.

25. Bloom FE, Björklund A, Hökfelt T. *The Primate Nervous System, Part III*, Vol 15 Amsterdam: Elsevier; 1999.

26. Bolon B. Comparative and correlative neuroanatomy for the toxicologic pathologist. *Toxicol Pathol*. 2000;28:6–27.

27. Bolon B, Butt M. *Fundamentals of Toxicologic Neuropathology for Pathologists and Toxicologists*. Hoboken, NJ: Wiley; 2011.

28. Boorman GA, Eustis SL, Elwell MR, Montgomery CAJ, MacKenzie WF. *Pathology of the Fischer Rat: Reference and Atlas*. San Diego, CA: Academic Press; 1990.

29. Braund KG. *Clinical Syndromes in Veterinary Neurology*. Baltimore: Williams & Wilkins; 1986.

30. Braund KG. Nerve and muscle biopsy techniques. *Prog Vet Neurol*. 1991;2:35–56.

31. Broxup BR, Yipchuck G, McMillan I, Losos GJ. Quantitative techniques in neuropathology. *Toxicol Pathol*. 1990;18: 105–114.

32. Bruni JE, Montemurro D. *Human Neuroanatomy: A Text, Brain Atlas and Laboratory Dissection Guide*. New York: Oxford University Press; 2009.

33. Brunton L, Lazo J, Parker K. *Goodman & Gilman's The Pharmacological Basis of Therapeutics*. 11th ed. New York: McGraw-Hill; 2005.

34. Butler AB, Hodos W. *Comparative Vertebrate Neuroanatomy: Evolution and Adaptation*, 2nd ed. Hoboken, NJ: Wiley-Liss; 2005.

35. Carpenter S, Karpati G. *Pathology of Skeletal Muscle*, 2nd ed. New York: Oxford University Press; 2001.

36. Carson FL, Hladik C. *Histotechnology: A Self-Instructional Text*, 3rd ed. Chicago: American Society for Clinical Pathology Press; 2009.

37. Cassella J, Hay J, Lawson S. *The Rat Nervous System: An Introduction to Preparatory Techniques*. New York: Wiley; 1997.

38. Chang LW. *Principles of Neurotoxicology*. New York: Marcel Dekker; 1994.

39. Chang LW, Slikker W Jr., *Neurotoxicology: Approaches and Methods*. San Diego, CA: Academic Press; 1995.

40. Chiasson RB. *Laboratory Anatomy of the White Rat*, 5th ed. Boston: WCB/McGraw-Hill; 1994.

41. Chrisman CL. Cerebrospinal fluid analysis. *Vet Clin North Am Small Anim Pract*. 1992;22:781–810.

42. Chudler EH. Brain Facts and Figures. Available at: http://www.vin.com/WebLink.plx?URL=http://faculty.washington.edu/chudler/facts.html. Accessed Sept 21, 2009.

43. Clark DL, Boutros NN, Mendez M. *The Brain and Behavior: An Introduction to Behavioral Neuroanatomy*, 2nd ed. New York: Cambridge University Press; 2005.

44. Classen W, Gretener P, Rauch M, Weber E, Krinke GJ. Susceptibility of various areas of the nervous system of hens to TOCP-induced delayed neuropathy. *Neurotoxicology*. 1996;17:597–604.

45. Cochard LR. *Netter's Atlas of Human Embryology*. Teterboro, NJ: Icon Learning Systems; 2002.

46. Coling D, Kachar B. Theory and application of fluorescence microscopy. In: Gerfen CR, Holmes A, Sibley D, Skolnick P, Wray S, eds. *Current Protocols in Neuroscience*. Hoboken, NJ: Wiley; 1997: 2.1.1–2.1.11.

47. Compilation. *Embryology of the Rhesus Monkey (Macaca mulatta)*. Vol 1. Washington, DC: Carnegie Institution of Washington; 1941.

48. Cooper JR, Bloom FE, Roth RH. *The Biochemical Basis of Neuropharmacology*. 8th ed. New York: Oxford University Press; 2003.

49. Crosby EC, Humphrey T, Lauer EW. *Correlative Anatomy of the Nervous System*. New York: Macmillan; 1962.

50. Crosby EC, Schnitzlein HN. *Comparative Correlative Neuroanatomy of the Vertebrate Telencephalon*. New York: Macmillan; 1982.

51. Daube JR, Rubin DI. *Clinical Neurophysiology*. 3rd ed. New York: Oxford University Press; 2009.

52. Davidson D. Edinburgh Mouse Atlas Project (emap). Available at: http://emouseatlas.org. Accessed Sept 21, 2009.

53. Davis R, Huffman RD. *A Stereotaxic Atlas of the Brain of the Baboon (Papio papio)*. Austin, TX: University of Texas Press; 1968.

54. de Groot DMG, Hartgring S, van de Horst L, et al. 2D and 3D assessment of neuropathology in rat brain after prenatal exposure to methylazoxymethanol, a model for developmental neurotoxicty. *Reprod Toxicol*. 2005;20:417–432.

55. de Lahunta A, Glass E. Development of the nervous system: malformation. In: de Lahunta A, Glass E, eds. *Veterinary Neuroanatomy and Clinical Neurology*, 3rd ed. Philadelphia: Saunders; 2009:23–53.

56. de Lahunta A, Glass E. *Veterinary Neuroanatomy and Clinical Neurology*, 3rd ed. Philadelphia: Saunders; 2009.

57. de Lucchi MR, Dennis BJ, Adey WR. *A Stereotaxic Atlas of the Chimpanzee Brain (Pan satyrus)*. Berkeley, CA: University of California Press; 1965.

58. de Olmos IS, Beltramino CA, de Olmos de Lorenzo S. Use of an amino-cupric-silver technique for the detection of early and semiacute neuronal degeneration caused by neurotoxicants, hypoxia and physical trauma. *Neurotoxicol Teratol*. 1994;16:545–561.

59. DeArmond SJ, Fusco MM, Dewey MM. *A Photographic Atlas: Structure of the Human Brain*, 2nd ed. New York: Oxford University Press; 1976.

60. DeArmond SJ, Fusco MM, Dewey MM. *Structure of the Human Brain. A Photographic Atlas*. Oxford, UK: Oxford University Press; 1974.

61. Dobbs MR. *Clinical Neurotoxicology: Syndromes, Substances, Environments*. Philadelphia: W.B. Saunders; 2009.

62. Done SH, Goody PC, Evans SA, Stickland NC. *Color Atlas of Veterinary Anatomy*, Vol 3. *The Dog and Cat*, 2nd ed. New York: Mosby; 2009.

63. Dorr AE, Lerch JP, Spring S, Kabani N, Henkelman RM. Neuroanatomy Atlas of the C57BL/6j Mouse. Toronto Centre for Phenogenomics. Available at: http://www.mouseimaging. ca/research/C57Bl6j_mouse_atlas.html. Accessed Sept 21, 2009.

64. Drews U. *Color Atlas of Embryology*. New York: Thieme Medical Publishers; 1995.

65. Dua-Sharma S, Sharma S, Jacobs HL. *The Canine Brain in Stereotaxic Coordinates*. Cambridge, MA: MIT Press; 1970.

66. Dubin M. Anatomy: Primate (NeuraLinks). Available at: http://spot.colorado.edu/~dubin/bookmarks/b/060.html. Accessed Sept 21, 2009.

67. Dyck PJ, Thomas PK. *Peripheral Neuropathy*, 4th ed. Philadelphia: W.B. Saunders; 2005.

68. Eagle RCJ. *Eye Pathology: An Atlas and Basic Text*. Philadelphia: W.B. Saunders; 1999.

69. Eidelberg E, Saldias CA. A stereotaxic atlas for Cebus monkeys. *J Comp Neurol*. 1960;115:102–123.

70. Emmers R, Akert K. *A Stereotaxic Atlas of the Brain of the Squirrel Monkey (Saimiri sciureus)*. Madison, WI: University of Wisconsin Press; 1963.

71. Evans HE. *Miller's Anatomy of the Dog*, 3rd ed. Philadelphia: W.B. Saunders; 1993.

72. Faculty. BrainInfo. National Primate Research Center, University of Washington–Seattle. Available at: http://braininfo. rprc.washington.edu/Default.aspx. Accessed Sept 21, 2009.

73. Faculty. Embryology. Swiss Virtual Campus. Available at: http://www.embryology.ch/indexen.html. Accessed Sept 21, 2009.

74. Felix. B, Leger ME, Albe-Fessard D, et al. Stereotaxic atlas of the pig brain. *Brain Res Bull*. 1999;49:1–137.

75. Felten DL, Józefowicz RF. *Netter's Atlas of Human Neuroscience*. Teterboro, NJ: Icon Learning Systems; 2003.

76. Fifkova E, Marsala J. Stereotaxic atlases for the cat, rabbit and rat. In: Bures J, Petrán M, Zachar J, eds. *Electrophysiological Methods in Biological Research*. New York: Academic Press; 1967:653–731.

77. Fishman RA. *Cerebrospinal Fluid in Diseases of the Nervous System*, 2nd ed. Philadelphia: W.B. Saunders; 1992.

78. FitzGerald MJT, Gruener G, Mtui E. *Clinical Neuroanatomy and Neuroscience*, 5th ed. Philadelphia: W.B. Saunders; 2007.

79. Fix AS, Garman RH. Practical aspects of neuropathology: a technical guide for working with the nervous system. *Toxicol Pathol*. 2000;28:122–131.

80. Fix JD. *Atlas of the Human Brain and Spinal Cord*, 2nd ed. Sudbury, MA: Jones & Bartlett; 2008.

81. Fix JD. *High-Yield Neuroanatomy*, 4th ed. Philadelphia: Lippincott Williams & Wilkins; 2008.

82. Fletcher TF. Canine Brain Transections. University of Minnesota College of Veterinary Medicine. Available at: http://vanat.cvm.umn.edu/brainsect. Accessed Jan 1, 2010.

83. Fletcher TJ. Atlas of Veterinary NeuroHistology. University of Minnesota College of Veterinary Medicine. Available at: http://vanat.cvm.umn.edu/NeuroLectPDFs/Neurohistology LectI.pdf. Accessed Sept 21, 2009.

84. Fletcher TJ. Veterinary Neuroanatomy: Brain Gross Anatomy. University of Minnesota College of Veterinary Medicine.

Available at: http://vanat.cvm.umn.edu/grossbrain. Accessed Sept 21, 2009.

85. Forrester JV, Dick AD, McMenamin PG, Roberts F. *The Eye: Basic Sciences in Practice*. Philadelphia: W.B. Saunders; 2008.

86. Franklin KBJ, Paxinos G. *The Mouse Brain in Stereotaxic Coordinates*, 3rd ed. San Diego, CA: Academic Press; 2007.

87. Franklin KBJ, Paxinos G. *The Mouse Brain in Stereotaxic Coordinates*. San Diego, CA: Academic Press; 1997.

88. Friede RL. *Developmental Neuropathology*, 2nd ed. New York: Springer-Verlag; 1989.

89. Fuller GN, Goodman JC. *Practical Review of Neuropathology*. Philadelphia: Lippincott Williams & Wilkins; 2001.

90. Garman RH. Artifacts in routinely immersion-fixed nervous tissue. *Toxicol Pathol*. 1990;18:149–153.

91. Garman RH. Evaluation of large-sized brains for neurotoxic endpoints. *Toxicol Pathol*. 2003;31:32–43.

92. Geisler CD. *From Sound to Synapse: Physiology of the Mammalian Ear*. Oxford, UK: Oxford University Press; 1998.

93. Gelatt KN. *Veterinary Ophthalmology*. Hoboken, NJ: Wiley-Blackwell; 2007.

94. Gerfen CR. Basic neuroanatomical methods. In: Gerfen CR, Holmes A, Sibley D, Skolnick P, Wray S, eds. *Current Protocols in Neuroscience*. Hoboken, NJ: Wiley; 1997: 1.1.1–1.1.11.

95. Gergen JA, MacLean PD. *A Stereotaxic Atlas of the Squirrel Monkey's Brain* (*Saimiri sciureus*). Public Health Service Publication 933. Bethesda, MD: National Institutes of Health; 1962.

96. Getty R. *Sisson and Grossman's The Anatomy of the Domestic Animals*, 5th ed. Philadelphia: W.B. Saunders; 1975.

97. Girgis M. *A New Stereotaxic Atlas of the Rabbit Brain*. St. Louis, MO: W.H. Green; 1981.

98. Grahn BH, Peiffer RL. Fundamentals of veterinary ophthalmic pathology. In: Gelatt KN, ed. *Veterinary Ophthalmology*. Hoboken, NJ: Wiley-Blackwell; 2007:355–438.

99. Gray F, De Girolami U, Poirier J. *Escourolle and Poirier's Manual of Basic Neuropathology*. 4th ed. New York: Elsevier; 2003.

100. Greene EC. *Anatomy of the Rat*. Braintree, MA: Braintree Scientific; 1935.

101. Greenstein B, Greenstein A. *Color Atlas of Neuroscience: Neuroanatomy and Neurophysiology*. New York: Theime; 1999.

102. Gribnau AAM. *Morphogenesis of the Brain in Staged Rhesus Monkey Embryos*. New York: Springer-Verlag; 1984.

103. Guyton AC, Hall JE. *Textbook of Medical Physiology*. 11th ed. Philadelphia: W.B. Saunders; 2005.

104. Haines DE. *Fundamental Neuroscience for Basic and Clinical Applications*, 3rd ed. New York: Churchill Livingstone; 2006.

105. Hamid M, Sismanis A. *Medical Otology and Neurotology: A Clinical Guide to Auditory and Vestibular Disorders*. New York: Thieme; 2006.

106. Heffner RRJ. *Muscle Pathology*. New York: Churchill Livingstone; 1984.

107. Hegedüs K. Neuroanatomy/Neuropathology on the Internet. Available at: http://www.neuropat.dote.hu. Accessed Jan 1, 2010.

108. Heimer L. *Dissection of the Human Brain*. Sunderland, MA: Sinauer Associates; 2007.

109. Hilbig H, Bidmon HJ, Oppermann OT, Remmerbach T. Influence of post-mortem delay and storage temperature on the immunohistochemical detection of antigens in the CNS of mice. *Exp Toxicol Pathol*. 2004;56:159–171.

110. Hulette C. NeuroAnatomy Web Resources. Available at: http://pathology.mc.duke.edu/neuropath/nawr/nawr_index.html. Accessed Sept 21, 2009.

111. Huxtable CR. Atlas of Veterinary Neuropathology. Available at: http://web.vet.cornell.edu/public/oed/neuropathology/index.asp. Accessed Sept 21, 2009.

112. Laboratory of Neuro Imaging. LONI Atlases. University of California–Los Angeles. Available at: http://www.loni.ucla.edu/Atlases/index.jsp. Accessed Jan 1, 2010.

113. Allen Institute. Allen Institute for Brain Science Atlas Portal. Available at: http://www.brain-map.org. Accessed Sept 21, 2009.

114. Irani DN. *Cerebrospinal Fluid in Clinical Practice*. Philadelphia: W.B. Saunders; 2008.

115. Itabashi HH, Andrews JM, Tomiyasu U, Erlich SS, Sathyavagiswaran L. *Forensic Neuropathology: A Practical Review of the Fundamentals*. San Diego, CA: Academic Press; 2007.

116. Johannessen JN. *Markers of Neuronal Injury and Degeneration*. New York: New York Academy of Sciences; 1993.

117. Johnson KA, Becker JA. The Whole Brain Atlas. Harvard Medical School. Available at: http://www.vin.com/WebLink.plx?URL=http://www.med.harvard.edu/AANLIB/home.html. Accessed Sept 21, 2009.

118. Jones TC, Mohr U, Hunt RD. *ILSI Monographs on the Pathology of Laboratory Animals*, Vol 6 *Nervous System*. New York: Springer-Verlag; 1988.

119. Jortner BS. Mechanisms of toxic injury in the peripheral nervous system: neuropathologic considerations. *Toxicol Pathol*. 2000;28:54–69.

120. Jortner BS. The return of the dark neuron: a histological artifact complicating contemporary neurotoxicologic evaluation. *Neurotoxicology*. 2006;27:628–634.

121. Jortner BS. Selected aspects of the anatomy and response to injury of the chicken (*Gallus domesticus*) nervous system. *Neurotoxicology*, 1982;3:299–310.

122. Joseph JT. *Diagnostic Neuropathology Smears*. Philadelphia: Lippincott Williams & Wilkins; 2006.

123. Kaas JH. *Evolutionary Neuroscience*. San Diego, CA: Academic Press; 2009.

124. Kandel ER, Schwartz JH, Jessell TM. *Principles of Neural Science*, 4th ed. New York: McGraw-Hill; 2000.

125. Kaufman MH. *The Atlas of Mouse Development*, 2nd ed. San Diego, CA: Academic Press; 1992.

126. Kaufman MH, Bard JBL. *The Anatomical Basis of Mouse Development*. San Diego, CA: Academic Press; 1999.

127. Kelley MW, Wu DK, Popper AN, Fay RR. *Development of the Inner Ear.* New York: Springer-Verlag; 2005.

128. Kempermann G. *Adult Neurogenesis.* New York: Oxford University Press; 2006.

129. Kennedy DN, Lange N, Makris N, Bates J, Meyer J, Caviness VSJ. Gyri of the human neocortex: an MRI-based analysis of volume and variance. *Cereb Cortex.* 1998;8:372–384.

130. King AS, McLelland J. *Outlines of Avian Anatomy.* Baltimore: Williams & Wilkins; 1975.

131. King JM. Dr. John M.King's Necropsy Show and Tell. Cornell University College of Veterinary Medicine. Available at: http://www.vin.com/WebLink.plx?URL=http://w3.vet.cornell.edu/nst. Accessed Sept 21, 2009.

132. King R. *Atlas of Peripheral Nerve Pathology.* London: Arnold; 1999.

133. Kingsley RE, Kingsley RD. *Interactive Atlas of the Human Brain.* Totowa, NJ: Humana Press; 2008.

134. Kinser PA. Brains on the Web. Haverford College. Available at: http://serendip.brynmawr.edu/bb/kinser/Home1.html. Accessed Sept 21, 2009.

135. Kjeldsberg CR, Knight JA. *Body Fluids: Laboratory Examination of Amniotic, Cerebrospinal, Seminal, Serous & Synovial Fluids,* 3rd ed. Chicago: American Society of Clinical Pathologists; 1993.

136. Klatt EC. Anatomy-Histology Tutorials. University of Utah Eccles Health Science Library. Available at: http://library.med.utah.edu/WebPath/HISTHTML/HISTO.html. Accessed Sept 21, 2009.

137. Klatt EC. CNS Pathology Index. University of Utah Eccles Health Science Library. Available at: http://www.vin.com/WebLink.plx?URL=http://library.med.utah.edu/WebPath/CNSHTML/CNSIDX.html. Accessed Sept 21, 2009.

138. Klatt EC. Neuroanatomy Tutorial. University of Utah Eccles Health Science Library. Available at: http://library.med.utah.edu/WebPath/HISTHTML/NEURANAT/NEURANCA.html. Accessed Sept 21, 2009.

139. Kleihues P, Burger PC, Scheithauer BW. *Histological Typing of Tumours of the Central Nervous System: World Health Organization (WHO) International Histological Classification of Tumours,* 2nd ed. Berlin: Springer-Verlag; 1996.

140. Koch C. *Biophysics of Computation. Information Processing in Single Neurons.* New York: Oxford University Press; 1999.

141. Korbo L, Pakkenberg B, Ladefoged O, Gundersen HJ, Arlien-Søborg P, Pakkenberg H. An efficient method for estimating the total number of neurons in rat brain cortex. *J Neurosci Methods.* 1990;31:93–100.

142. Kovacs W, Denk H. *Der Hirnstamm der Maus [The Brainstem of the Mouse].* Vienna: Springer-Verlag; 1968.

143. Krause WJ (2006). Ear. University of Missouri–Columbia. Available at: http://video.google.com/videoplay?docid=-1703418087947315988#. Accessed Jan 1, 2010.

144. Krebs C. Neuroanatomy at UBC. University of British Columbia. Available at: http://www.neuroanatomy.ca. Accessed Sept 21, 2009.

145. Krinke G, Ullmann L, Sachsse K, Hess R. Differential susceptibility of peripheral nerves of the hen to triorthocresyl phosphate and to trauma. *Agents Actions.* 1979;9:227–231.

146. Krinke GJ, Classen W, Vidotto N, Suter E, Würmlin CH. Detecting necrotic neurons with fluoro-jade stain. *Exp Toxicl Pathol.* 2001;53:365–372.

147. Krinke GJ, Kaufmann W, Mahrous AT, Schaetti P. Morphologic characterization of spontaneous nervous system tumors in mice and rats. *Toxicol Pathol.* 2000;28:178–192.

148. Krinke GJ, Vidotto N, Weber E. Teased-fiber technique for peripheral myelinated nerves: methodology and interpretation. *Toxicol Pathol.* 2000;28:113–121.

149. Krueger G. Mapping of the mouse brain for screening procedures with the light microscope. *Lab Anim Sci.* 1971;21:91–105.

150. Kruger L, Saporta S, Swanson LW. *Photographic Atlas of the Rat Brain.* Cambridge, MA: Cambridge University Press; 1995.

151. Kuenzel W. Chicken Brain Atlas. AvianBrain.org. A Resource for Brain Researchers (Administrator: Jarvis ED, Duke University Durham, NC). Available at: http://www.avianbrain.org/nomen/Chicken_Atlas.html. Accessed Jan 1, 2010.

152. Kuenzel WJ, Masson M. *A Stereotaxic Atlas of the Brain of the Chick.* Baltimore: Johns Hopkins University Press; 1988.

153. Langtree L, Langtree I. Human Brain Facts and Answers. Available at: http://www.disabled-world.com/artman/publish/brain-facts.shtml. Accessed Oct 1, 2009.

154. Lemke G. *Developmental Neurobiology.* San Diego, CA: Academic Press; 2009.

155. Fischbarg J, *The Biology of the Eye*, Vol. 10. Amsterdam: Elsevier, 2006.

156. Lim RKS, Liu C-N, Moffitt RL. *A Stereotaxic Atlas of the Dog's Brain.* Springfield, IL: Charles C Thomas; 1960.

157. Looi S. Fun Facts About the Brain. Brain Health & Puzzles. Available at: http://www.brainhealthandpuzzles.com/fun_facts_about_the_brain.html. Accessed Oct 1, 2009.

158. Lorenz MD, Kornegay JN. *Handbook of Veterinary Neurology,* 4th ed. Philadelphia: W.B. Saunders; 2004.

159. Love S, Louis DN, Ellison DW. *Greenfield's Neuropathology,* 8th ed. London: Hodder Arnold; 2008.

160. Lowndes HE, Reuhl KR. *Nervous System and Behavioral Toxicology.* New York: Elsevier; 1997.

161. Luparello TJ. *Stereotaxic Atlas of the Forebrain of the Guinea Pig.* Baltimore: Williams & Wilkins; 1967.

162. Mai JK, Assheuer J. *Atlas of the Human Brain.* San Diego, CA: Academic Press; 1998.

163. Mai JK, Paxinos G, Assheuer J. *Atlas of the Human Brain,* 2nd ed. San Diego, CA: Academic Press; 2004.

164. Mai JK, Paxinos G, Voss T. *Atlas of the Human Brain,* 3rd ed. San Diego, CA: Academic Press; 2007.

165. Manocha SL, Shantha TR, Bourne GH. *A Stereotaxic Atlas of the Brain of the Cebus Monkey (Cebus apella).* Oxford, UK: Oxford University Press; 1968.

166. Maronpot RR. *Pathology of the Mouse: Reference and Atlas.* Vienna, IL: Cache River Press; 1999.

167. Martin JH. *Neuroanatomy: Text and Atlas*. New York: McGraw-Hill; 2003.

168. Martin RF, Bowden DM. *Primate Brain Maps: Structure of the Macaque Brain*. Amsterdam: Elsevier; 2000.

169. Mazzioatta JC, Toga AW, Frackowiak RSJ. *Brain Mapping: The Systems*. San Diego, CA: Academic Press; 2000.

170. McDonald W. Muscle Pathology. Available at: http://missing link.ucsf.edu/lm/ids_104_musclenerve_path/student_musclenerve/musclepath.html. Accessed Sept 21, 2009.

171. McGeady TA, Quinn PJ, FitzPatrick ES, Ryan MT. *Veterinary Embryology*. Oxford, UK: Blackwell; 2006.

172. McInnes E. Artefacts in histopathology. *Comp Clin Pathol.* 2005;13:100–108.

173. McLarrin GM. Vacuoles in the fiber tracts of rat CNS tissue. *J Histotechnol.* 1982;5:171.

174. McMartin DN, O'Donoghue JL, Morrissey R, Fix AS. *Guides for Toxicologic Pathology*. Washington, DC: Society of Toxicologic Pathology; 1997.

175. Mercer EH.The Whole Mouse Catalog. Available at: http://wmc.rodentia.com. Accessed Sept 21, 2009.

176. Mikula S. Primate Brain Atlas (MRI). BrainMeta.com. Available at: http://brainmeta.com/mri_primate. Accessed Jan 1, 2010.

177. Mikula S, Trotts I, Stone JS, Jones EG. Brain Maps. Available at: http://brainmaps.org. Accessed Sept 21, 2009.

178. Miller AK, Alston RL, Corsellis JA. Variation with age in the volumes of grey and white matter in the cerebral hemispheres of man: measurements with an image analyser. *Neuropathol Appl Neurobiol.* 1980;6:119–132.

179. Mohr U, Carlton WW, Dungworth DL, Benjamin SA, Capen CC, Hahn FF. *Pathobiology of the Aging Dog*. Washington, DC: ILSI Press; 2001.

180. Mohr U, Dungworth DL, Capen CC. *Pathobiology of the Aging Rat*. Washington, DC: ILSI Press; 1992.

181. Mohr U, Dungworth DL, Capen CC, Carlton WW, Sundberg JP, Ward JM. *Pathobiology of the Aging Mouse*, vol. 2 Washington, DC: ILSI Press; 1996.

182. Monnier M, Gangloff H. *Atlas of Stereotaxic Brain Research on the Conscious Rabbit*. Amsterdam: Elsevier; 1961.

183. Morin LP, Wood RI. *A Stereotaxic Atlas of the Golden Hamster Brain*. San Diego, CA: Academic Press; 2001.

184. Nadol JB. *Schucknect's Pathology of the Ear*, 2nd ed. Philadelphia: BC Decker; 2009.

185. Nestler EJ, Hyman SE, Malenka RC. *Molecular Neuropharmacology: A Foundation for Clinical Neuroscience*, 2nd ed. New York: McGraw-Hill; 2008.

186. Niewenhuys R. *The Central Nervous System of Vertebrates*. New York: Springer-Verlag; 1998.

187. Noden DM, de LaHunta A. *The Embryology of Domestic Animals: Developmental Mechanisms and Malformations*. Baltimore: Williams & Wilkins; 1985.

188. Nolte J. *The Human Brain: An Introduction to Its Functional Anatomy*. St. Louis, MO: Mosby; 2002.

189. Notterpek L, Tolwani RJ. Experimental models of peripheral neuropathies. *Lab Anim Sci.* 1999;49:588–599.

190. O'Rahilly R, Müller F. *The Embryonic Human Brain: An Atlas of Developmental Stages*. New York: Wiley-Liss; 1994.

191. O'Rahilly RR, Müller F. *The Embryonic Human Brain: An Atlas Of Developmental Stages*, 3rd ed. Hoboken NJ: Wiley-Liss; 2006.

192. Oh SJ. *Color Atlas of Nerve Biopsy Pathology*. Boca Raton, FL: CRC Press; 2002.

193. Oliver JE, Hoerlein BF, Mayhew IG. *Veterinary Neurology*. Philadelphia: W.B. Saunders; 1987.

194. Osofsky A, LeCouteur RA, Vernau KM. Functional neuro-anatomy of the domestic rabbit (*Oryctolagus cuniculus*). *Vet Clin North Am Exot Anim Pract.* 2007;10:713–730.

195. Pakkenberg B, Gundersen HJG. Neocortical neuron number in humans: effect of sex and age. *J Comp Neurol.* 1997;384:312–320.

196. Pakkenberg B, Pelvig D, Marner L, et al. Aging and the human neocortex. *Exp Gerontol.* 2003;38:95–99.

197. Palazzi X. *The Beagle Dog Brain in Stereotaxic Coordinates*. New York: Springer-Verlag (in press).

198. Palazzi X, Bordier N. *The Marmoset Brain in Stereotaxic Coordinates*. New York: Springer-Verlag; 2008.

199. Palkovits M, Brownstein MJ. *Maps and Guide to Micro-dissection of the Rat Brain*. New York, Elsevier; 1988.

200. Paxinos G. *The Rat Nervous System*, 2nd ed. San Diego, CA: Academic Press; 1995.

201. Paxinos G. *The Rat Nervous System*, 3rd ed. San Diego, CA: Academic Press; 2004.

202. Paxinos G, Ashwell KWS, Törk I. *Atlas of the Developing Rat Nervous System*, 2nd ed. San Diego, CA: Academic Press; 1994.

203. Paxinos G, Franklin KBJ. *The Mouse Brain in Stereotaxic Coordinates*, 2nd ed. New York: Elsevier; 2001.

204. Paxinos G, Halliday G, Watson C, Koutcherov Y, Wang H. *Atlas of the Developing Mouse Brain at E17.5, P0, and P6*. San Diego, CA: Academic Press; 2007.

205. Paxinos G, Huang X-F, Petrides M, Toga AW. *The Rhesus Monkey Brain in Stereotaxic Coordinates*, 2nd ed. San Diego, CA: Academic Press; 2008.

206. Paxinos G, Huang X-F, Toga AW. *The Rhesus Monkey Brain in Stereotaxic Coordinates*, San Diego, CA: Academic Press; 2000.

207. Paxinos G, Mai JK. *The Human Nervous System*, 2nd ed. San Diego, CA: Academic Press; 2004.

208. Paxinos G, Watson C. *The Rat Brain in Stereotaxic Coordinates*, 2nd ed. San Diego, CA: Academic Press; 1986.

209. Paxinos G, Watson C. *The Rat Brain in Stereotaxic Coordinates*. Sydney, Australia: Academic Press; 1982.

210. Paxinos G, Watson C. *The Rat Brain in Stereotaxic Coordinates*, 3rd ed. San Diego, CA: Academic Press; 1997.

211. Paxinos G, Watson C. *The Rat Brain in Stereotaxic Coordinates*, 6th ed. San Diego, CA: Academic Press; 2007.

212. Paxinos G, Watson C. *The Rat Brain in Stereotaxic Coordinates*, 5th ed. San Diego, CA: Academic Press; 2004.

213. Paxinos G, Watson C. *The Rat Brain in Stereotaxic Coordinates*, 4th ed. San Diego, CA: Academic Press; 1998.

214. Paxinos G, Watson C, Carrive P, Kirkcaldie M, Ashwell K. *Chemoarchitectonic Atlas of the Rat Brain*, 2nd ed. San Diego, CA: Academic Press; 2008.

215. Pestronk A. Neuromuscular Disease Center. Washington University. Available at: http://neuromuscular.wustl.edu. Accessed Sept 21, 2009.

216. Peters A, Palay SL, Webster HD. *The Fine Structure of the Nervous System: Neurons and Their Supporting Cells*. 3rd ed. New York: Oxford University Press; 1991.

217. Popesko P, Rajtova V, Horak J. *Colour Atlas of Anatomy of Small Laboratory Animals*, Vol 2. London: W.B. Saunders; 2003.

218. Prophet EB, Mills R, Arrington JB, Sobin LH. *Armed Forces Institute of Pathology Laboratory Methods in Histotechnology*. Washington, DC: American Registry of Pathology; 1992.

219. Puelles L, Martinez de-la-Torre M, Paxinos G, Watson C, Martinez S. *The Chick Brain in Stereotaxic Coordinates*. San Diego, CA: Academic Press; 2007.

220. Punkt K. *Fibre Types in Skeletal Muscles*. Berlin: Springer-Verlag; 2002.

221. Rao MS, Jacobson M. *Developmental Neurobiology*, 4th ed. New York: Kluwer Academic/Plenum Publishers; 2005.

222. Riche D, Christolomme A, Bert J, Naquet R. *Atlas Stéréotaxique du Cerveau de Babourin (Papio papio)*. Paris: Centre National de la Recherche Scientifique; 1968.

223. Rosenberg GA. *Brain Fluids and Metabolism*. New York: Oxford Unversity Press; 1990.

224. Ruffins S, Jacobs R.Caltech MRI Atlas of Mouse Development. California Institute of Technology. Available at: http://mouseatlas.caltech.edu/index_content.html. Accessed Sept 21, 2009.

225. Rugh R. *The Mouse. Its Reproduction and Development*. Oxford, UK: Oxford University Press; 1990.

226. Sadler TW. *Langman's Medical Embryology*. 11th ed. Philadelphia: Lippincott Williams & Wilkins; 2009.

227. Saleem KS, Logothetis NK. *A Combined MRI and Histology Atlas of the Rhesus Monkey Brain*. San Diego, CA: Academic Press; 2006.

228. Sanes DH, Reh TA, Harris WA. *Development of the Nervous System*, 2nd ed. San Diego, CA: Academic Press; 2006.

229. Saver JL. Time is brain—quantified. *Stroke*. 2006;37:263–266.

230. Schaltenbrand G, Wahren W. *Atlas for Stereotaxy of the Human Brain*. Stuttgart, Germany: Georg Thieme; 1977.

231. Schambra U. *Prenatal Mouse Brain Atlas*. New York: Springer-Verlag; 2008.

232. Schambra UB.Electronic Prenatal Mouse Brain Atlas. Available at: http://www.epmba.org. Accessed Sept 21, 2009.

233. Schaumburg HH, Berger AR, Thomas PK. *Disorders of Peripheral Nerves*, 2nd ed. New York: Oxford University Press; 1992.

234. Schmahmann JD, Pandya DN. *Fiber Pathways of the Brain*. New York: Oxford University Press; 2006.

235. Schmued LC, Hopkins KJ. Fluoro-Jade: novel fluorochromes for detecting toxicant-induced neuronal degeneration. *Toxicol Pathol*. 2000;28:91–99.

236. Schoenwolf GC. *Atlas of Descriptive Embryology*, 7th ed. San Francisco: Benjamin Cummings.

237. Schröder JM. *Pathology of Peripheral Nerves: An Atlas of Structural and Molecular Pathological Changes*. Berlin: Springer-Verlag; 2001.

238. Seehusen DA, Reeves MM, Fomin DA. Cerebrospinal fluid analysis. *Am Fam Physician*. 2003;68:1103–1108.

239. Shek JW, Wen GY, Wisniewski HM. *Atlas of the Rabbit Brain and Spinal Cord*. Basel, Switzerland: S Karger, 1985.

240. Shepherd GM. *The Synaptic Organization of the Brain*. New York: Oxford University Press; 1998.

241. Sidman RL, Angevine JBJ, Taber-Pierce E. *Atlas of the Mouse Brain and Spinal Cord*. Cambridge, MA: Harvard University Press; 1971.

242. Siegel GJ, Albers RW, Brady S, Price D. *Basic Neurochemistry: Molecular, Cellular, and Medical Aspects*, 7th ed. San Diego, CA: Academic Press; 2006.

243. Singer M. *The Brain of the Dog in Section*. Philadelphia: W.B. Saunders; 1962.

244. Slikker WJ, Chang LW. *Handbook of Developmental Neurotoxicology*. San Diego, CA: Academic Press; 1998.

245. Slotnick BM, Leonard CM. *A Stereotaxic Atlas of the Albino Mouse Forebrain*. Rockville, MD: U.S. Department of Health, Education, and Welfare; 1975.

246. Smith BR.The Multi-Dimensional Human Embryo. Available at: http://embryo.soad.umich.edu. Accessed Sept 21, 2009.

247. Snell RS, Lemp MA. *Clinical Anatomy of the Eye*. Malden, MA: Blackwell Science; 1998.

248. Snider RS, Lee JC. *A Stereotaxic Atlas of the Cat Brain*. Chicago: University of Chicago Press; 1961.

249. Snider RS, Lee JC. *A Stereotaxic Atlas of the Monkey Brain (Macaca mulatta)*. Chicago: University of Chicago Press; 1961.

250. Sodicoff M. Embryology of the CNS. Temple University School of Medicine. Available at: http://isc.temple.edu/neuroanatomy/lab/embryo_new. Accessed Sept 21, 2009.

251. Solleveld HA, Gorgacz EJ, Koestner A. Central nervous system neoplasms in the rat. In: *Guides for Toxicologic Pathology*. Washington, DC: Society of Toxicologic Pathology; 1991.

252. Spencer PS, Schaumburg HH. *Experimental and Clinical Neurotoxicology*. Baltimore: Williams & Wilkins; 1980.

253. Spencer PS, Schaumburg HH, Ludolph AC. *Experimental and Clinical Neurotoxicology*, 2nd ed. New York: Oxford University Press; 2000.

254. Squire LR, Berg D, Bloom F, du Lac S. *Fundamental Neuroscience*, 3rd ed. San Diego, CA: Academic Press; 2008.

255. Staff. Atlas of Human Embryology. Chronolab A. G. Available at: http://www.embryo.chronolab.com. Accessed Sept 21, 2009.

256. Stahl SM. *Stahl's Essential Psychopharmacology: Neuroscientific Basis and Practical Applications*, 3rd ed. New York: Cambridge University Press; 2000.

257. Steding G. *The Anatomy of the Human Embryo: A Scanning Electron-Microscopic Atlas*. Basel, Switzerland: S. Karger; 2009.

258. Stephan H, Baron G, Schwerdtfeger WK. *The Brain of the Common Marmoset (Callithrix jacchus): A Stereotaxic Atlas*. Berlin: Springer-Verlag; 1980.

259. Stoffel MH. *Funktionelle Neuroanatomie für die Tiermedizin* [Functional Neuroanatomy in Veterinary Medicine], Stuttgart, Germany: Enke Verlag in MVS Medizinverlage; 2011.

260. Sulik KK, Bream PRJ. Embryo Images: Normal and Abnormal Mammalian Development. University of North Carolina–Chapel Hill. Available at: http://www.med.unc.edu/embryo_images. Accessed Sept 21, 2009.

261. Sultan F, Braitenberg V. Shapes and sizes of different mammalian cerebella: a study in quantitative comparative neuroanatomy. *J Hirnforsch*. 1993;34:79–92.

262. Summers BA, Cummings JF, de Lahunta A. *Veterinary Neuropathology*. St. Louis, MO: Mosby; 1995.

263. Swanson LW. *Brain Maps: Structure of the Rat Brain*, 3rd ed. San Diego, CA: Academic Press; 2003.

264. Swanson LW. Mapping the human brain: past, present, and future. *Trends Neurosci*. 1995;18:471–474.

265. Theiler K. *The House Mouse: Atlas of Embryonic Development*. New York: Springer-Verlag; 1989.

266. Theiler K. *The House Mouse: Development and Normal Stages from Fertilization to 4 Weeks of Age*. Berlin: Springer-Verlag; 1972.

267. Thompson EJ. *The CSF Proteins: A Biochemical Approach*. New York: Elsevier Science; 1988.

268. Tilson HA, Mitchell CL. *Neurotoxicity*. Target Organ Toxicology Series. (Hayes AW, Thomas JA, Gardner DE, eds.). New York: Raven Press; 1992.

269. Toga AW, Mazzioatta JC. *Brain Mapping: The Systems*. San Diego, CA: Academic Press; 2000.

270. Tohyama M, Takatsuji K. *Atlas of Neuroactive Substances and Their Receptors in the Rat*. New York: Oxford University Press; 1993.

271. Trujillo KA. Neuroanatomy and Neural Development on the Internet. California State University–San Marcos. http://courses.csusm.edu/psyc391kt/linksanat.html. Accessed Sept 21, 2009.

272. Urban I, Philippe R. *Stereotaxic Atlas of the New Zealand Rabbit's Brain*. Springfield, IL: Charles C Thomas; 1972.

273. Valciukas J. *Foundations of Environmental and Occupational Neurotoxicology*. Piscataway, NJ: Transaction Publishers; 2002.

274. Valverde F. *Golgi Atlas of the Postnatal Mouse Brain*. New York: Springer-Verlag; 2004.

275. van Tienhoven A, Juhasz LP. The chicken telencephalon, diencephalon and mesencephalon in stereotaxic coordinates. *J Comp Neurol*. 1962;118:185–197.

276. Verhaart WJC. *A Stereotaxic Atlas of the Brain of the Cat*. Assen, The Netherlands: Van Gorcum; 1964.

277. Vernau W, Vernau K, Bailey CS. Cerebrospinal fluid. In: Kaneko JJ, Harvey JW, Bruss ML, eds. *Clinical Biochemistry of Domestic Animals*, 6th ed. San Diego, CA: Academic Press; 2008:769–819.

278. Volpicelli-Daley LA, Levey A. Immunohistochemical localization of proteins in the nervous system. In: Gerfen CR, Holmes A, Sibley D, Skolnick P, Wray S, eds. *Current Protocols in Neuroscience*. Hoboken, NJ: Wiley; 2003: 1.2.1–1.2.17.

279. Ward JM, Mahler JF, Maronpot RR, Sundberg JP. *Pathology of Genetically Engineered Mice*. Ames, IA: Iowa State University Press; 2000.

280. Watanabe H, Andersen F, Simonsen CS, Evans SM, Gjedde A, Cumming P. MR-based statistical atlas of the Göttingen minipig brain. *Neuroimage*. 2001;14:1089–1096.

281. Watson C, Paxinos G, Kayalioglu G. *The Spinal Cord*. Amsterdam: Elsevier; 2009.

282. Welker W, Johnson JI, Noe A. Comparative Mammalian Brain Collections. Available at: http://www.brainmuseum.org. Accessed Sept 21, 2009.

283. Wells GAH, Wells M. Neuropil vacuolation in brain: a reproducible histological processing artefact. *J Comp Pathol*. 1989;101:355–362.

284. Williams PL, Warwick R. *Gray's Anatomy*, 36th ed. Philadelphia: WB Saunders; 1980.

285. Williams RW, Rosen GD. Mouse Brain Library. Available at: http://www.mbl.org. Accessed Jan 10, 2010.

286. Williams RW, von Bartheld CS, Rosen GD. Counting cells in sectioned material: a suite of techniques, tools, and tips. In: Gerfen CR, Holmes A, Sibley D, Skolnick P, Wray S, eds. *Current Protocols in Neuroscience*. Hoboken, NJ: Wiley; 2003:1.11.11–11.11.29.

287. Woolsey TA, Hanaway J, Gado MH. *The Brain Atlas: A Visual Guide to the Human Central Nervous System*. Hoboken, NJ: Wiley-Liss; 2007.

288. Wu J, Bowden DM, Dubach MF, Robertson JE. *Primate Brain Maps: Structure of the Macaque Brain*, New York: Elsevier; 2000.

289. Wulliman MF, Rupp B, Reichert H. *Neuroanatomy of the Zebrafish Brain: A Topological Atlas*. Basel, Switzerland: Birkhäuser Verlag; 1996.

290. Yoshikawa T. *Atlas of the Brains of Domestic Animals*. University Park, PA: Pennsylvania State University Press; 1968.

291. Young PA, Young PH, Tolbert DL. *Basic Clinical Neuroscience*, 2nd ed. Baltimore: Lippincott Williams & Wilkins; 2008.

292. Zeman W, Innes JRM. *Craigie's Neuroanatomy of the Rat*. New York: Academic Press; 1963.

293. Zilles K. *The Cortex of the Rat: A Stereotaxic Atlas*. New York: Springer-Verlag; 1985.

INDEX

Aβ plaques, 132. *See also* Amyloid-beta (Aβ) neuropathology; β-amyloid plaques

Aberrant myelination, 78

Abnormal protein accumulation, in dog brain regions, 132

Absolute size distribution, 225

Accelerated aging studies, central problem with, 130

Accessory olfactory bulb (AOB), 429, 442

Accessory olfactory formation (AOF), 188

Accuracy (ACC), 119, 120

Acetylcholinesterase (AChE) activity, depression of, 496

3-Acetylpyridine (3-AP), 107–108

Acidophilic neurons, 9, 192

Acrylamide, 109, 377

Activity measures, 106

Acute cell death, 152–154

Acute dosing, 152

Acute eosinophilic neuron degeneration, 192

Acute evaluations, 154

Acute metabolic arrest, 491

Acute neurotoxins, 153

Acute response, cell death due to, 153

Adenovirus encephalitis, 469

Adult nervous system, evaluation of, 321–338

Adult neural arrangement, pace of, 69–70

Adverse behavioral changes, 109–110

Afferent neurons, 15

Affymetrix Genechip, 295, 299

Age. *See also* Aging
 oxidative damage and, 131–132
 in predicting Aβ accumulation, 132

Aged dogs, clustering by cognitive ability, 131

Age-related cognitive decline, 128

Age-related macular degeneration (ARMD), 389, 390

Age-related neurological diseases, 81

Agilent microarray chip technology, 295

Aging. *See also* Behavioral aging; Human aging
 Alzheimer's disease and, 125–127
 animal, 125

genetic models of, 129–130
 impact on brain structure/function, 125–138
 successful versus pathologic, 128

Alar plates, 68

Albuminocytologic dissociation, 279

Alcaraz, Ana, xi, 463

Alcohol, cell destruction by, 143, 145

Allergic neuritis, experimental, 380

Allometric relationship, 343

Alphaviruses, 265

Altricial animals, 340

Altricial species, 70

Aluminum, link to Alzheimer's disease, 132–133

Alzheimer, Alois, 4

Alzheimer astrocytes
 type I, 467
 type II, 467, 477

Alzheimer's disease (AD), 125
 animal models of, 127–132
 common form of, 126
 environmental neurotoxicants and, 132–133
 hallmark features of, 126–127
 human aging and, 125–127
 neuron loss and, 131

Alzheimer's disease mouse models, imaging, 259

Amino CuAg method, 186–187. *See also* Cupric silver entries

Aminoglycoside antibiotics, 421–422

Aminoglycosides, ototoxic potential of, 422

Amitriptyline, 456

D-Amphetamine, neurotoxicity of, 145

Amphetamine exposure, in rats, 10

Amyloid-beta (Aβ) neuropathology, 126–127, 132. *See also* Aβ plaques; β-amyloid plaques

Amyotrophic lateral sclerosis (ALS), 376

Analytical procedures, for studying the inner ear, 417–418

Anaplastic tumors, malignant, 381

Anatomical changes, related to developmental age, 73

Anatomical organization, in the developing nervous system, 63–64

Fundamental Neuropathology for Pathologists and Toxicologists: Principles and Techniques, First Edition. Edited by Brad Bolon and Mark T. Butt.
© 2011 John Wiley & Sons, Inc. Published 2011 by John Wiley & Sons, Inc.